Lecture Notes in Computer Science 4519

Commenced Publication in 1973
Founding and Former Series Editors:
Gerhard Goos, Juris Hartmanis, and Jan van Leeuwen

Lecture Notes in Computer Science 4510

Commenced Publication in 1973
Founding and Former Series Editors:
Gerhard Goos, Juris Hartmanis, and Jan van Leeuwen

Editorial Board

David Hutchison
Lancaster University, UK
Takeo Kanade
Carnegie Mellon University, Pittsburgh, PA, USA
Josef Kittler
University of Surrey, Guildford, UK
Jon M. Kleinberg
Cornell University, Ithaca, NY, USA
Friedemann Mattern
ETH Zurich, Switzerland
John C. Mitchell
Stanford University, CA, USA
Moni Naor
Weizmann Institute of Science, Rehovot, Israel
Oscar Nierstrasz
University of Bern, Switzerland
C. Pandu Rangan
Indian Institute of Technology, Madras, India
Bernhard Steffen
University of Dortmund, Germany
Madhu Sudan
Massachusetts Institute of Technology, MA, USA
Demetri Terzopoulos
University of California, Los Angeles, CA, USA
Doug Tygar
University of California, Berkeley, CA, USA
Moshe Y. Vardi
Rice University, Houston, TX, USA
Gerhard Weikum
Max-Planck Institute of Computer Science, Saarbruecken, Germany

Enrico Franconi   Michael Kifer
Wolfgang May (Eds.)

The Semantic Web: Research and Applications

4th European Semantic Web Conference, ESWC 2007
Innsbruck, Austria, June 3-7, 2007
Proceedings

Springer

Volume Editors

Enrico Franconi
Free University of Bozen–Bolzano
Faculty of Computer Science
Piazza Domenicani 3, 39100 Bozen-Bolzano (BZ), Italy
E-mail: franconi@inf.unibz.it

Michael Kifer
State University of New York at Stony Brook
Department of Computer Science
Stony Brook, New York, NY 11794-4400, USA
E-mail: kifer@cs.sunysb.edu

Wolfgang May
Georg-August-Universität Göttingen
Institut für Informatik
Lotzestrasse 16-18, 37083 Göttingen, Germany
E-mail: may@informatik.uni-goettingen.de

Library of Congress Control Number: 2007927308

CR Subject Classification (1998): H.4, H.3, C.2, H.5, I.2, K.4, D.2

LNCS Sublibrary: SL 3 – Information Systems and Application, incl. Internet/Web and HCI

ISSN 0302-9743
ISBN-10 3-540-72666-7 Springer Berlin Heidelberg New York
ISBN-13 978-3-540-72666-1 Springer Berlin Heidelberg New York

Springer is a part of Springer Science+Business Media

springer.com

Printed in Germany

Typesetting: Camera-ready by author, data conversion by Scientific Publishing Services, Chennai, India
Printed on acid-free paper SPIN: 12066971 06/3180 5 4 3 2 1 0

Preface

The papers in this volume represent the technical program of the 4th European Semantic Web Conference, ESWC 2007, that took place June 3–7, 2007 in Innsbruck, the capital of the Tyrol region of Austria.

The ESWC series of conferences is an annual, international forum for dissemination and discussion of the latest research and applications of Semantic Web technologies. It has become a major meeting ground for researchers and practitioners in the field. ESWC is part of the European Semantic Systems Initiative (ESSI), a cluster of major European research projects aiming to improve world-wide research and standardization in the area of the Semantic Web. The ESWC 2007 topics of interest included: ontology management, ontology alignment, ontology learning and metadata generation, multimedia and Semantic Web, semantic annotation of data, Semantic Web trust, privacy, security and intellectual property rights, Semantic Web rules and query languages, logics for the Semantic Web, reasoning on the Semantic Web, behavior in the Semantic Web, searching, querying, visualizing, navigating and browsing the Semantic Web, personalization and user modelling, user interfaces and Semantic Web, Semantic Grid and middleware, Semantic Web Services, Semantic Web-based knowledge management, Semantic Web for e-business, e-culture, e-government, e-health, e-learning, e-science, database technologies for the Semantic Web, data semantics and Web semantics, semantic interoperability, semantic workflows, and Semantic Web mining.

The ESWC 2007 call for papers attracted 278 submissions of research papers, a 54% growth with respect to the previous year. Amongst these, the Program Committee selected 46 papers to be presented at the conference. The quality of the competing papers was high, and we decided to nominate two papers for the Best Paper Award:

- Minimal Deductive Systems for RDF (by Sergio Muñoz, Jorge Pérez and Claudio Gutierrez)
- Empowering Software Maintainers with Semantic Web Technologies (by René Witte, Yonggang Zhang and Jürgen Rilling).

Additionally, 10 submissions were accepted as system descriptions and 37 as posters.

Besides the presentation of the 46 technical articles in 13 sessions, one evening session with reception was devoted to the presentation and demonstration of systems and posters.

Four keynote addresses were given by distinguished scientists: Ron Brachman (VP of Worldwide Research Operations at Yahoo!, Santa Clara CA, USA), Stefano Ceri (Technical University of Milan, Italy), Georg Gottlob (Oxford University, UK), and Ning Zhong (Maebashi Institute of Technology, Japan). As with previous ESWC conferences, metadata describing the conference were published, and during the conference developers had an opportunity to showcase their tools using these and other semantic data.

The conference also included a program of seven tutorials (selected out of ten submissions) and eight associated workshops (selected out of 13 submissions). In addition, a PhD symposium took place immediately after the conference, bringing together doctoral students within the Semantic Web community to showcase their work in a major European forum and to obtain valuable feedback from leading scientists in the field. Furthermore, the OWL-ED 2007 and DL 2007 workshops and the RR 2007 conference were co-located.

The success of this year's conference was due to the hard, voluntary work of many people. The Chairpersons who selected the tutorials, workshops, demonstrations and the PhD symposium as well as the local Organization Committee and several other central tasks are listed on the next page. Last but not least, we would like to thank the authors of all papers that were submitted to ESWC 2007, the members of the Program Committee, and the additional experts who helped with the reviewing process for contributing and ensuring the high scientific quality of ESWC 2007.

ESWC 2007 was sponsored by ESSI (European Semantic Systems Initiative, a group of European Projects known as: Knowledge Web, SUPER and Tripcom), that collectively work together to strengthen European research and industry through world-wide standardization), STI2 (Semantic Technology Institutes International), the EU Project X-Media, CEFRIEL (ICT Center of Excellence for Research, Innovation, Education and Industrial Labs Partnership, Milan, Italy), CTIC Foundation (Center for the Development of Information and Communication Technologies in Asturias), the local host DERI Innsbruck, the BIT (Bolzano – Innsbruck – Trento) Joint School for Information Technology, and the companies Asemantics, Empolis, Ontoprise, Ontotext, and Hanival (who also contributed IT services for the local organization of ESWC 2007).

We thank Springer for professional support and guidance during the preparation of these proceedings. We would also like to thank the developers of the EasyChair conference management system (`http://www.easychair.org/`). EasyChair assisted us in the whole process of collecting and reviewing papers, in interacting with authors and Program Committee members, and also in assembling this volume.

March 2007

Enrico Franconi
Michael Kifer
Wolfgang May

Conference Organization

General Chair

Enrico Franconi (Free University of Bozen-Bolzano, Italy)

Program Chairs

Michael Kifer (State Univ. of New York at Stony Brook, USA)
Wolfgang May (Universität Göttingen, Germany)

Workshops Chair

Diana Maynard (University of Sheffield, UK)

Tutorial Chair

Jörg Diederich (Forschungszentrum L3S, Hannover, Germany)

PhD Symposium Chair

Elena Simperl (Freie Universität Berlin, Germany)

Demo Chair

Andy Seaborne (HP Labs, Bristol, UK)

Semantic Technologies Coordinator

Sean Bechhofer (University of Manchester, UK)

Publicity Chair

Stijn Heymans (DERI and Universität Innsbruck, Austria)

Sponsor Chair

Axel Polleres (Universidad Rey Juan Carlos, Madrid, Spain)

Local Organization

Ilona Zaremba (DERI and Universität Innsbruck, Austria)

Conference Administrator

Melanie Plattner (DERI and Universität Innsbruck, Austria)
Christen Ensor (DERI Galway, Ireland)

Treasurer

Birgit Leiter (DERI and Universität Innsbruck, Austria)

Webmaster

Damian Dadswell (The Open University, UK)

Program Committee

Karl Aberer
José Júlio Alferes
Jürgen Angele
Grigoris Antoniou
Alessandro Artale
Franz Baader
Chitta Baral
Cristina Baroglio
Catriel Beeri
Sonia Bergamaschi
Abraham Bernstein
Leopoldo Bertossi
Harold Boley
Piero Bonatti
Alex Borgida
Jeen Broekstra
Jos de Bruijn
François Bry
Andrea Cali
Silvana Castano
Isabel Cruz
Bernardo Cuenca Grau
John Davies
Stefan Decker
Jörg Diederich
Włodek Drabent
Thomas Eiter
Jérôme Euzenat
Norbert Fuchs
Fabien Gandon
Aldo Gangemi
Fausto Giunchiglia
Carole Goble
Asunción Gómez-Pérez
Guido Governatori
Marko Grobelnik
Nicola Guarino
Volker Haarslev
Manfred Hauswirth
Jeff Heflin
Nicola Henze
Martin Hepp
Stijn Heymans
Pascal Hitzler
Ralph Hodgson
Ian Horrocks
Herman ter Horst
Andreas Hotho
Carlos Hurtado
Mustafa Jarrar
Subbarao Kambhampati
Atanas Kiryakov
Rüdiger Klein
Matthias Klusch
Mieczyslaw Kokar
Manolis Koubarakis
Rubén Lara
Ora Lassila
Georg Lausen
Alain Léger
Nicola Leone
Francesca Alessandra Lisi
Alexander Löser
Bertram Ludäscher
Jan Małuszyński
Massimo Marchiori
David Martin
Ralf Möller
Boris Motik
Saikat Mukherjee
John Mylopoulos
Natasha Noy
Daniel Olmedilla
Jeff Z. Pan
Bijan Parsia
Terry Payne
Sofia Pinto
Axel Polleres
Chris Preist
I.V. Ramakrishnan
Riccardo Rosati
Marie-Christine Rousset

Marta Sabou
Kai-Uwe Sattler
Ulrike Sattler
Sebastian Schaffert
Stefan Schlobach
Luciano Serafini
Nigel Shadbolt
Elena Simperl
Munindar P. Singh
Michael Sintek
Derek Sleeman
Umberto Straccia
York Sure
Vojtěch Svátek
Terrance Swift
Hideaki Takeda
Valentina Tamma
Sergio Tessaris
Bernhard Thalheim
Paolo Traverso
Raphaël Troncy
Gerd Wagner
Michael Wooldridge
Guizhen Yang

Additional Reviewers

Pinar Alper
Alia Abdelmoty
Harith Alani
Ahmed Alasoud
Ricardo Amador
Stanislaw Ambroszkiewicz
Alia Amin
Mathieu d'Aquin
Rudi Araújo
Uwe Aßmann
Robert Baumgartner
Pieter Bellekens
Massimo Benerecetti
Domenico Beneventano
Raffaella Bernardi
Ian Blacoe
Sebastian Blohm
Joel Booth
Yevgen Borordin
Shawn Bowers
Adriana Budura
Tobias Bürger
Johannes Busse
Francesco Calimeri
Amit Chopra
Stijn Christiaens
Oscar Corcho
Philippe Cudré-Mauroux
Vasilios Darlagiannis
Hasan Davulcu
Juri Luca De Coi
Xi Deng
Alistair Duke
Ludger van Elst
Michael Erdmann
Sofia Espinosa
Nicola Fanizzi
Cristina Feier
Alfio Ferrara
Gunar Fiedler
Michael Fink
Giorgos Flouris
David Fowler
Rosella Gennari
Michael Gertz
Nick Gibbins
Adrian Giurca
Rigel Gjomemo
Birte Glimm
Antoon Goderis
Gunnar Grimnes
Tudor Groza
Andrea Gualtieri
Francesco Guerra
Yuanbo Guo
Parisa Haghani
Guillermo Hess
Michiel Hildebrand
Aidan Hogan
Thomas Hornung

Bo Hu
Zhisheng Huang
Gearoid Hynes
Giovambattista Ianni
Luigi Iannone
Antoine Isaac
Kaarel Kaljurand
Alissa Kaplunova
Martin Kavalec
Atila Kaya
Yevgeny Kazakov
Marijke Keet
Mick Kerrigan
Christoph Kiefer
Malte Kiesel
Nick Kings
Roman Korf
Ludwig Krippahl
Markus Krötzsch
Tobias Kuhn
Samir Kumar
Martin Labsky
Joey Lam
Freddy Lécué
Jens Lehmann
Jos Lehmann
Katja Lehmann
Domenico Lembo
Sergey Lukichev
Yue Ma
Jalal Mahmud
Marco Manna
Francisco Martin-Recuerda
Davide Martinenghi
Yutaka Matsuo
Michele Melchiori
Sylvia Melzer
Thomas Meyer
Paolo Missier
Shamima Mithun
Malgorzata Mochol
Stefano Montanelli
Matthew Moran
Igor Mozetic
Ullas Nambiar
Jan Nemrava
Davy van Nieuwenborgh
Malvina Nissim
René Noack
Jean-Pierre Norguet
Hans Jürgen Ohlbach
Ikki Ohmukai
Mirko Orsini
Magdalena Ortiz
Jacco van Ossenbruggen
Ignazio Palmisano
Zhengxiang Pan
Charles Penwill
Jorge Pérez
Laura Po
Antonella Poggi
Valentina Presutti
Abir Qasem
Guilin Qi
Domenico Redavid
Gerald Reif
Quentin Reul
Francesco Ricca
Sebastian Rudolph
Massimo Ruffolo
Marco Ruzzi
Antonio Sala
Brahmananda Sapkota
Luigi Sauro
Peyman Sazedj
Roman Schindlauer
Florian Schmedding
Peggy Schmidt
Roman Schmidt
James Scicluna
Arash Shaban-Nejad
Rob Shearer
Mantas Simkus
Kai Simon
Gleb Skobeltsyn
Kees van der Sluijs
Pavel Smrž
Lucia Specia
Ruud Stegers
Markus Stocker
Heiner Stuckenschmidt
William Sunna

David Thau
Edward Thomas
Ian Thurlow
Hans Tompits
Alessandra Toninelli
Dmitry Tsarkov
Anni-Yasmin Turhan
Miroslav Vacura
Maurizio Vincini
Max Völkel
Johanna Völker
Denny Vrandečić
Le-Hung Vu
Richard Waldinger
Hui Wan
Kewen Wang
Xia Wang
David Warren
Paul Warren
Martin Weber
Fang Wei
Moritz Weiten
Garrett Wolf
Huiyong Xiao
Chang Zhao
Antoine Zimmermann
Daniel Zinn

Sponsors:

Table of Contents

Invited Talks

Best Papers

Semantic Web Services

Ontology Learning, Inference and Mapping I

Case Studies

Social Semantic Web

Ontologies: Requirements and Analysis

Personalization I

Foundations of the Semantic Web

Natural Languages and Ontologies

Applications

Querying and Web Data Models

Ontology Learning, Inference and Mapping II

Emerging Sciences of the Internet: Some New Opportunities (Extended Abstract)

Ron Brachman

Yahoo! Research, New York, NY 10011, USA
rjb@yahoo-inc.com

Semantic Web technologies have started to make a difference in enterprise settings and have begun to creep into use in limited parts of the World Wide Web. As is common in overview articles, it is easy to imagine scenarios in which the Semantic Web could provide important infrastructure for activities across the broader Internet. Many of these seem to be focused on improvements to what is essentially a search function (e.g., "list the prices of flat screen HDTVs larger than 40 inches with 1080p resolution at shops in the nearest town that are open until 8pm on Tuesday evenings" [http://en.wikipedia.org/wiki/Semantic_Web]), and such capabilities will surely be of use to future Internet users. However, if one looks closely at the research agendas of some of the largest Internet companies, it is not clear that the staples of SW thinking will intersect the most important paths of the major broad-spectrum service providers. Some of the emerging trends in the research labs of key industry players indicate that SW goals generally taken for granted may be less central than envisioned and that the biggest opportunities may come from some less obvious directions. Given the level of investment and the global reach of big players like Yahoo! and Google, it would pay us to look more closely at some of their fundamental investigations.

While not all companies see the future in the same way, there are some trends and priorities that are instructive to contemplate. While Web search will continue to play a large role, and services composed from piece-parts offered by multiple vendors will be important, some relatively novel ideas may come to dominate. In one view, there will emerge a set of new sciences that are fundamental to future generations of Internet businesses. By understanding the imperatives of those new areas of thought, we may be able to get a better assessment of the true ultimate impact of SW technologies on the broader Internet. At the very least, framing future Semantic Web directions and examples in terms more aligned with some of these ideas may encourage more attention from some of the major Internet companies, which to date have shown only lukewarm interest.

Search will continue to play a significant role for users of the Internet. While from a user's perspective the most essential thing is the ability to find a website or document or element of data relevant to a task at hand, it is critical to the service provider that, in addition to producing relevant results, it be able to expose the user to advertisements well-suited to his or her immediate and longer-term needs and interests. While not restricted to search settings, high-quality matching of ads is critical to the continued success and growth of the major

E. Franconi, M. Kifer, and W. May (Eds.): ESWC 2007, LNCS 4519, pp. 1–3, 2007.

Internet search and content providers. In the long run, a better understanding of users in general and customers in specific is essential to providing a better experience and providing better opportunities through advertising. This is an area where a huge amount of investment is being made. Does it provide any interesting opportunities for SW technology?

One interesting element here is that with the scale of the Web, companies are increasing their use of machine learning technologies to improve the performance of their search engines and other computing systems, since it is simply impossible to process information at the scale and speed of the Web manually. Machine learning technology, based ultimately on richer knowledge representation technologies, will play an extremely prominent role in the infrastructure that makes the Web successful in the future. Will this development provide interesting opportunities for the Semantic Web community? Can it avoid the relative lack of mutual interest that has plagued the KR and ML communities in the recent past?

Another emerging element in what we might call a new Science of Search is a social one. Products like Yahoo! Answers add the human element back into the process of finding relevant information. A similar human element pervades tagging-based services like Flickr. It's not clear what the success of large-scale popular web services based on ad hoc human tagging says about the future of the Semantic Web - are the two incompatible? Will ontologies matter or do folksonomies rule? Others have begun to address the substantial differences between the social Web world and the Semantic Web world. While sometimes portrayed as diametrically opposed, the sides may benefit from each other if we look more deeply. My intuition is that there is room for synergy, and it would behoove us to investigate.

The social element is evident in many areas other than Search. Community-oriented services are already extremely popular (witness the scale of MySpace and FaceBook, and even older services like Yahoo! Groups). On the face of it, this direction would seem not to hold much promise for SW thinking. But if one looks at some of the underlying infrastructure needed to make new community systems succeed, one sees opportunities for robust search and information extraction and organization technologies, among others.

Another exciting and somewhat unexpected set of developments in new Internet science involves collaboration between Economics and Computer Science. Driven in part by underlying elements of advertising auctions that account for extraordinary revenue streams at Google, Yahoo!, and other companies, technical approaches to constructing novel market mechanisms, deploying them, and understanding their practical ramifications are becoming critical. Prediction markets and other novel mechanisms based on large-scale interactions among masses of people are creating never before seen opportunities because of the Internet, and substantial resources are going into the understanding of fundamental principles governing such interactions. This whole line of thinking is not yet prominent in the worldview of the Web community, but it is likely to grow in importance, and it may provide some novel opportunities for SW research.

Finally, as has been pointed out by many others, text-based documents are not the only media available on the Web, and there has clearly been astronomical growth in user-generated non-textual content over the last few years. Podcasting, Flickr, YouTube, Yahoo! Video, and other media-sharing opportunities have shown that there is extraordinary latent interest in the creation and spread of audio, image, video and other complex content. While not yet receiving as much attention, the study and understanding of how users experience media, both individually and in social settings, is likely to lead to an explosion of novel technologies and perhaps even devices for consuming and sharing media. How these new scientific directions will relate to the SW world is anyone's guess, but it makes sense to take a close look to see what opportunities they present.

It is clear that some very large players on the Internet are making substantial - in some cases huge - investments in novel forms of search and information navigation, social and community systems, economic mechanisms for auctions and advertising, and media experiences. With the scale of public interest and the level of economic investment we are now seeing, it is probable that these directions will have a dramatic impact on the evolution of the Internet. While many of the arenas traditionally targeted by the Semantic Web community will continue to matter in the next generation of the Web, it is conceivable that the really big interest, including the majority of investment and opportunity for broad influence on the world, will go elsewhere. It would be nice if the Semantic Web were not left behind. A quick look indicates that there could be some interesting opportunities, and it would appear to be very much worth the community's while to stop to consider the emerging sciences and their implications for future generations of the Internet.

Design Abstractions for Innovative Web Applications: The Case of the SOA Augmented with Semantics

Stefano Ceri[1], Marco Brambilla[1], and Emanuele Della Valle[2]

[1] Dipartimento di Elettronica e Informazione, Politecnico di Milano
Piazza L. Da Vinci, 32. I20133 Milano, Italy
{ceri,mbrambil}@elet.polimi.it
[2] CEFRIEL, via Fucini 2, 20133 Milano, Italy
dellavalle@cefriel.it

Abstract. This work presents a retrospective analysis of how we have addressed new challenges in Web technologies and applications. WebML, which was first defined about 10 years ago, has been an incubator for research concerning abstractions, methods, tools, and technologies, acting as a glue within a group of people spread among universities, technology transfer centres, and a spin-off. In this paper, we first illustrate the common approach to innovation, and then show such approach at work in two contexts. One of them, dealing with "Service-Oriented Architectures" (SOA), has reached a mature state; the other one, "Semantic Web Services" (SWS), is at its infancy, but promises to deliver very interesting results in the forthcoming years.

1 Introduction and Motivation

Data-intensive Web applications, i.e. applications whose main purpose is to give access to well-organized content, represented the first industrial application of the Web, and still constitute the most important Web application in terms of volumes and commercial value. All companies have an institutional site showing their business and describing their offers, and many companies address their customers either electively or exclusively via the Web. Therefore, these applications have been most covered by methods and tools, which have been available for a long time.

Among them, Web Modelling Language (WebML) [1] was defined, about 8 years ago, as a conceptual model for data-intensive Web applications. Early deployment technologies were very unstable and immature; as a reaction, WebML was thought as a high level, implementation-independent conceptual model, and the associated design support environment, called WebRatio [9], has always been platform-independent, so as to adapt to frequent technological changes.

WebML can be considered, in MDA terms, as a Domain Specific Language in the area of Web application development. It is based upon orthogonal separation of concerns: content, interface logics, and presentation logics are defined as separate components. The main innovation in WebML comes from the interface logics (patented in 2003) that enables the computation of Web pages made up of logical components (units) interconnected by logical links (i.e., not only the units but also the

E. Franconi, M. Kifer, and W. May (Eds.): ESWC 2007, LNCS 4519, pp. 4–15, 2007.

links have a formal semantics); the computation is associated with powerful defaults so as to associate to simple diagrams all the required semantics for a full deployment, through code generators. WebML anticipated the concepts and methods formally proposed by the MDA framework, introducing the idea of model transformation and code generation.

While the Web has gone through waves of innovation, new application sectors have developed, and revolutionary concepts – such as enabling the interaction of software artefacts rather than only humans – are opening up. While the foundations of the WebML model and method are still the same, the pragmatics of its interpretation and use has dramatically changed through the last years. Several new challenges have been addressed within the WebML context, including:

- Web services and service-oriented architectures [3];
- Integration with business processes [4];
- Personalization and adaptation;
- Context awareness and mobility [5];
- Rich client-side applications;
- Embedded Web applications;
- Semantic Web and Semantic Web Services [6,7].

A retrospective consideration of our work shows that, in all the above situations, we have addressed every new challenge by using a common approach, which indeed has become evident to us during the course of time, and now is well understood and consolidated. For every new research directions, we had to address four different kinds of extensions, respectively addressing the development process, the content model, the hypertext meta-model, and the tool framework.

- *Extensions of the development process* capture the new steps of the design that are needed to address the new functionalities, providing as well the methodological guidelines and best practices for helping designers.
- *Extensions of the content model* capture state information associated with providing the new functionalities, in the format of standard model, e.g. a collection of entities and relationship that is common to all applications; this standard model is intertwined with the application model, so as to enable a unified use of all available content[1].
- *Extension of the hypertext meta-model* capture the new abstractions that are required for addressing the new functionalities within the design of WebML specifications, through new kinds of units and links which constitute a functionality-specific "library", which adds to the "previous" ones;
- *Extensions of the tool framework* introduce new tools in order to extend those modelling capability falling outside of standard WebRatio components (content, interface logics, presentation logics), or to empower users with new interfaces and wizards to express the semantics of new units and links in terms of existing ones, or to provided direct execution support for new units and links (e.g. sending an email).

[1] Note that any WebML application includes the standard entities User and Group.

This paper demonstrates how this four-step development occurred in the case of Service Oriented Architectures and how we are naturally extending that approach to deal with Semantic Web Services. The treatment of each extension is necessarily concise and visual, for more details we refer readers to published papers and reports.

2 Support of Service-Oriented Architectures

The specification of a Web application according to WebML [2] consists of a set of orthogonal models: the application *data model* (i.e., an extended Entity-Relationship model), one or more *hypertext models* (i.e., different site views for different types of users), expressing the navigation paths and the page composition of the Web application; and the *presentation model*, describing the visual aspects of the pages. A hypertext site view is a graph of pages; pages are composed of units, which are used for publishing atomic pieces of information, and operations, for modifying data or performing arbitrary business actions (e.g., sending e-mails). Units and operations are connected by links, to allow navigation, passing of parameters between the components, and computation of the hypertext. The need for incorporating external logic was felt relatively early, and the initial solution consisted of "custom units" which allow modelling user-defined computations.

The first WebML extension discussed in this paper is towards the Service Oriented Architectures. The requirement addressed in this case is to provide adequate design tools for Web Services and Service-oriented applications. The outcomes of our work included:

- The extension to the development process and the definition of some methodological guidelines for SOA design;
- Two standard models for representing the services and the business processes to be performed;
- New design primitives (namely, WebML units and links) for covering Web service specification and invocation, together with primitives for enforcing business process constraints;
- The support of the specified solutions through a process modeller, a translator of processes into sketches of hypertexts, and an XML2XML mapping tool.

2.1 Process Extensions

The original design process, explained in chapter 6 of [2], included the classic phases of requirement analysis, data design, hypertext design, and presentation design, followed by architecture design and implementation. The 4-step procedure, going from requirements to data to hypertext to presentation, is iterated multiple times through the use of WebRatio, which can be considered as a rapid prototyping environment; and indeed a lot of the advantage of using the approach comes exactly from the ability to generate a prototype whenever required by the need of interaction with stakeholders.

The extension of the original design process to SOA requires adding a phase for modeling the business process and separating application from service design, as shown in Fig. 1. For each addition, new guidelines and best practices were defined.

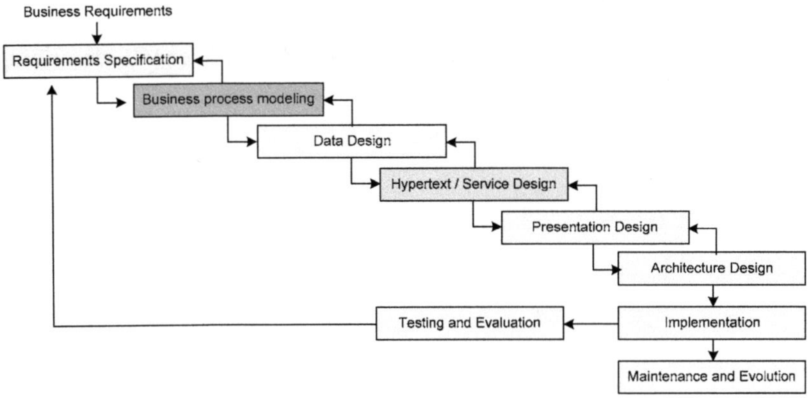

Fig. 1. Development process extensions for SOA

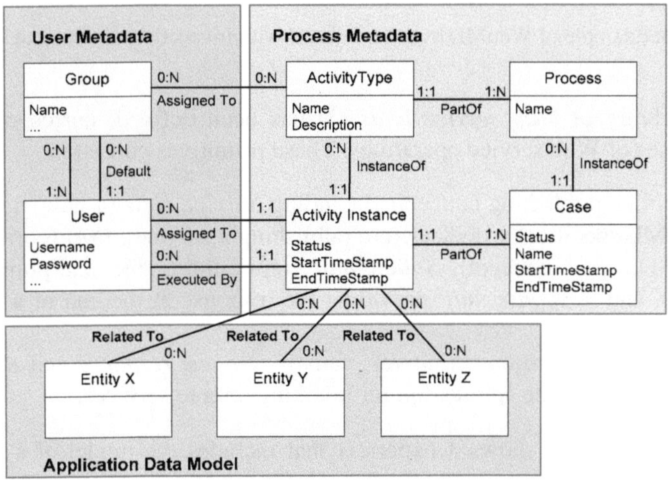

Fig. 2. Standard model for the specification of the business process status

2.2 Content Model Extensions

The standard model for supporting SOA deals with two aspects: a description of Web services and the specification of the workflow state. The first standard model represents Web services according to WSDL, and is omitted here (see [3]); the second standard model represents the information about the implemented business process, shown in Fig. 2 (see [4] for details). In the model, entity Process represents processes and is associated with entity ActivityType, representing the kinds of activities that can be executed in a process. Entity Case denotes an instance of a process and is related to its Process (relationship InstanceOf) and to its activities (via relationship PartOf); entity ActivityInstance denotes the actual occurrences of activities within cases.

2.3 Hypertext Meta-model Extensions

Two groups of new design primitives have been added to WebML, describing Web services and workflow-based applications.

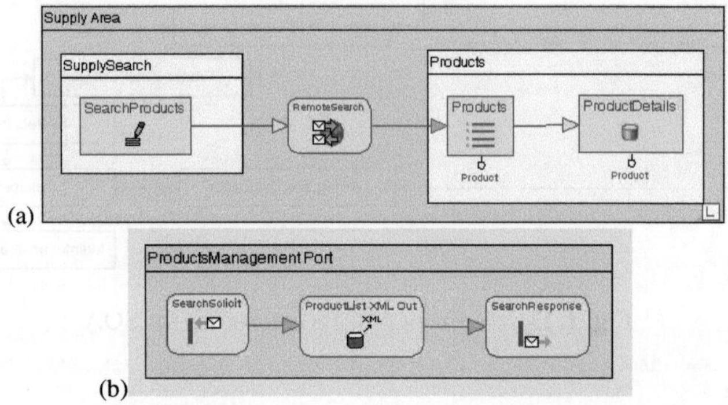

Fig. 3. Example of WebML hypertext model with invocation of a remote service

A new library of *Web service units* [3] has been defined, corresponding to the WSDL classes of Web service operations. These primitives consist in:

- Web service publishing concepts, including *Service view* (a new view supported in WebML specifically dedicated to publishing a service), *Port* (corresponding to the WSDL port concept), *Solicit unit* (representing the end-point of a Web service), and *Response unit* (providing the response at the end of a Web service implementation);
- Web service invocation primitives, namely *Request-response* and *Request* units, to be used within the application for invoking remote services.

For instance, Fig. 3 shows a hypertext that includes the model of a Web service call and of the called Web service. In *Supply Area* of Fig. 3a, the user can browse the *SupplySearch* page, in which the *SearchProducts* entry unit permits the input of search criteria. From this information, a request message is composed and sent to the *RemoteSearch* operation of a Web service. The user then waits for the response message, containing a list of products satisfying the search criteria. From this list, a set of instances of *Product* are created, and displayed to the user by means of the *Products* index unit in the *Products* page; the user may continue browsing, e.g., by choosing one of the displayed products and looking at its details. Fig. 3b represents the model of the *RemoteSearch* service invoked by the previously described hypertext. The interaction starts with the solicit *SearchSolicit* unit, which denotes the reception of the message. Upon the arrival of the message, an XML-out operation extracts from the local data source the list of desired products and formats the resulting XML document. The *SearchResponse* unit produces the response message for the invoker[2].

To cover the development of B2B Web applications implementing business processes, new primitives have been defined for specification of activity boundaries (namely *Activity areas* within the hypertext) and business process-dependent navigation (namely *workflow links*). Fig. 4 shows some of these primitives: site areas marked as "Activity Areas" (A); special incoming links for checking the correctness of the status and starting an activity (i.e., Start and Resume links); special outgoing links for closing an activity (Complete and Suspend links).

Distributed processes and *SOA* can be obtained by combining the workflow primitives with Web services primitives [4].

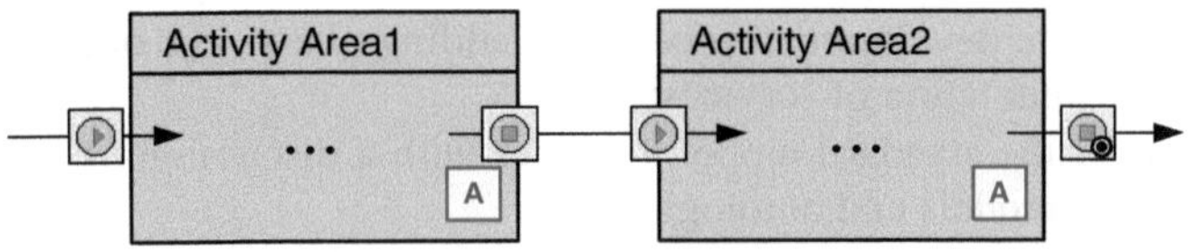

Fig. 4. Two activity areas and corresponding Start and End links

2.4 Tool Framework Extensions

For supporting the design of the new classes of applications, some facilities have been prototyped and are currently being ported to commercial versions of WebRatio:

- A workflow modeling editor that allows to specify business processes according to the BPMN notation.
- A set of model transformations that translate a business process model into a skeleton of WebML hypertext model.
- A visual editor for XML2XML mapping for helping the design of XML transformations to better support messages exchange between Web services.

3 Support of Semantic Web Services

Traditionally, the service requestor and service provider are designed together and then tightly bound together when an application is created. The emerging field of Semantic Web Services (SWS) [10] provides paradigms for semantically enriching the existing syntactic descriptions of Web services; then, the service requestor can search, either at design or at run time, among a variety of Web-enabled service providers, by choosing the service that best fits the requestor's requirement. Such a flexible binding of requestor and providers allows for dynamic and evolving applications to be created utilizing automatic resource discovery, selection, mediation and invocation.

---

2 Service ports are an example of software component that is modelled by using WebML and yet has no interaction with users (hence, no "presentation logics"), and shows that the original motivation of the model has shifted to adapt to new requirements. Even more radical shifts will be needed to deal with semantic web services, as illustrated in the sequel.

Our purpose in approaching the SWS is obviously not to design new methods for performing the above tasks: a community of researchers is working on them. Instead, we aim at extending WebML and WebRatio so as to generate, on top of conventional models (of: processes, data, services, and interfaces), a large portion of the semantic descriptions required by the SWS in a semi-automatic manner; this possibility descends from the fact that WebML is a very rich model, with a lot of embedded semantics - to the point that code can be completely generated from the model with no user intervention. In the same way, some SWS annotations can be automatically generated, conveying a large fraction of the semantics that is typically carried by manual SWS annotations.

In the rest of the section we highlight the following extensions to WebML to cope with SWS[3] requirements:

- Extension of the development process by adding phases for ontology import and for semantic annotation of services;
- Extensions of the standard model, together with a discussion of the relationships between meta models and ontologies;
- Definition of the new primitives in order to manipulate semantic contents;
- Implementation of new tools integrating Semantic Web Service editors and execution environment.

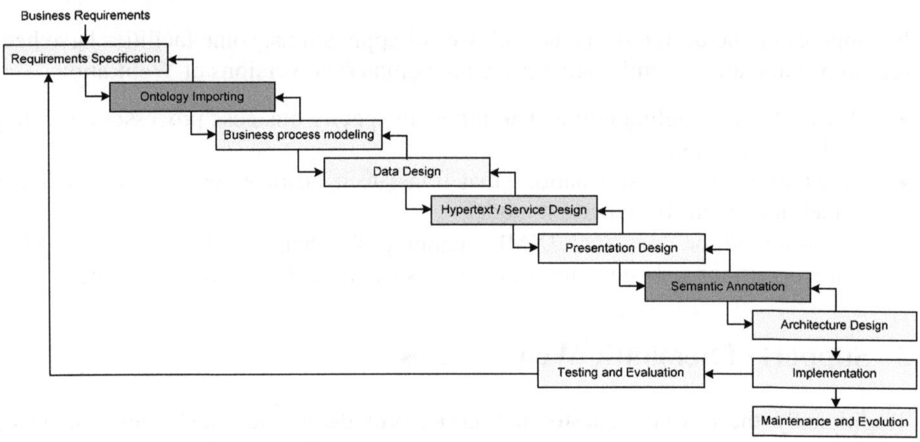

Fig. 5. Development process extensions for SWS

3.1 Process Extensions

To address the new SWS requirements, we extended the process defined for SOA in Fig. 1 with two additional tasks, shown in Fig. 5:

- **Ontology Importing,** for importing existing domain ontologies that may be exploited for describing the domain of the Web application under development.

[3] In our approach we considered WSMO, but being, WSMO the most comprehensive approach to SWS, our experience can be easily extended to OWL-S and WSLD-S approaches.

The imported ontologies should be integrated, at the model level, with the application-specific E-R model, so as to offer an integrated view to the designer.

- **Semantic Annotation,** for specifying (either manually or automatically) how the hypertext pages or services will be annotated using existing ontological knowledge.

3.2 Content Model Extensions

The management of content in Semantic Web applications, thus also in SWS applications, needs to address two main concerns: *(i)* the possibility of importing and integrating existing third-party ontologies and *(ii)* the capability of combining relational data sources with ontological contents.

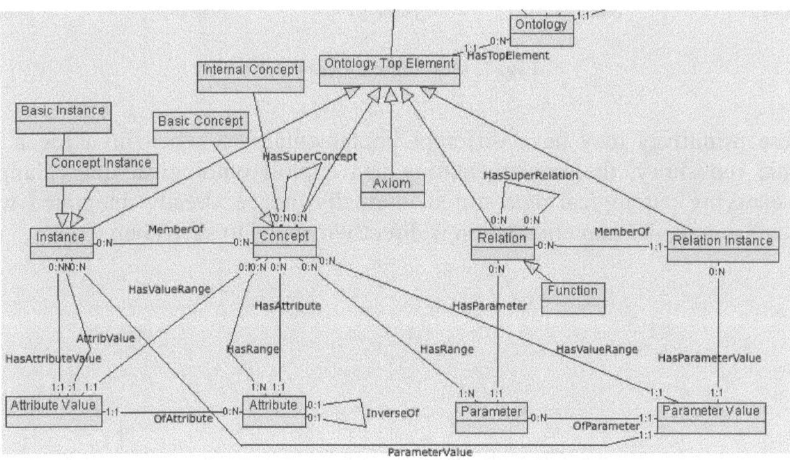

Fig. 6. Standard model of the WSMO ontology structure

We address these two issues by defining a E-R standard model representing ontological concepts, thus allowing to associate in a seamless way semantic content to conventional content defined for the application; Fig. 6 shows a piece of E-R model representing WSMO ontological language. Imported ontological data can be either copied into an application-specific implementation of the E-R model (typically a relational database) or queried directly on a remote ontology. Different implementations of ontology query primitives must be developed in the two cases (see Section 3.3).

3.3 Hypertext Meta-model Extensions

The basic WebML primitives for data retrieval have been used up to now for querying implementations of E-R models, but their generality makes them perfectly fitting in the role of query and navigation of ontologies. The additional expressive power of ontological languages, however, requires some extensions. We have therefore introduced a new set of primitives (inspired by SPARQL and RDF-S syntax) to describe advanced queries over ontological data. These units (see Fig. 7) allow queries on classes, instances, properties, and values; checking the existence of specific concepts; and verifying whether a relationship holds between two resources. Other units import content from an ontology or return the RDF description of a given portion of the standard ontological model for exporting. Operations such as lifting and lowering have renamed specific XML2XML mappings used in the context of SOAs.

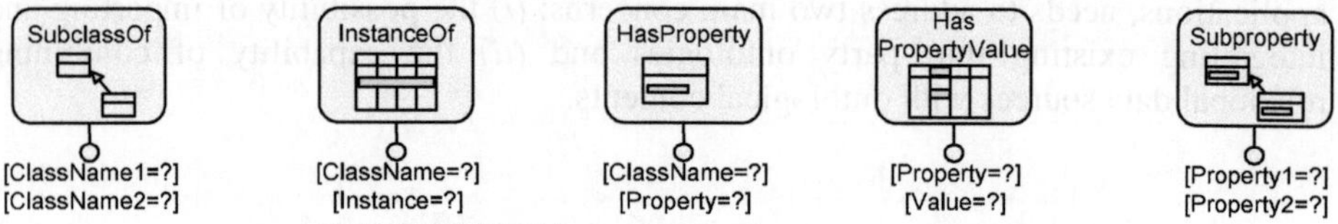

Fig. 7. Ontological query units

These primitives may have different implementations: when invoking a remote semantic repository, the implementation can exploit ontological query languages; when querying ontological data stored internally, hence already integrated within a relational source, the implementation is directly mapped to such source.

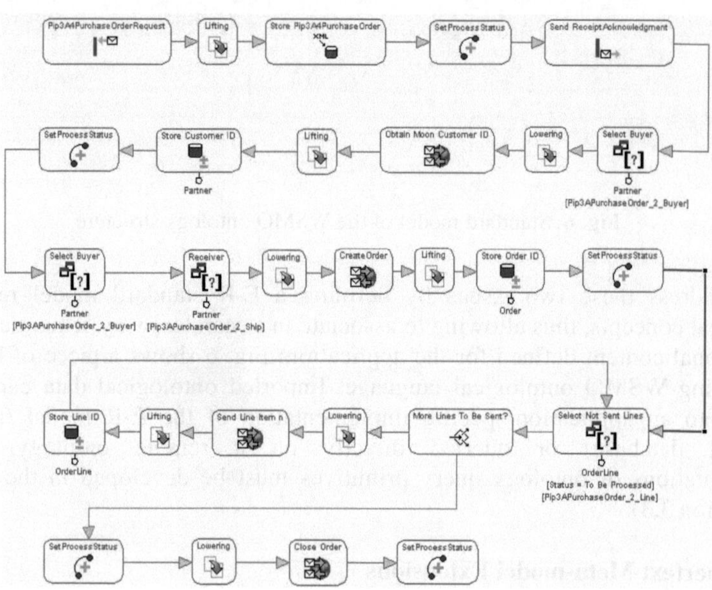

Fig. 8. WebML model of a mediator

These units, together with the standard WebML primitives and the solutions introduced for the SOA, allow specifying completely new kinds of applications with respect to the ones for which WebML was originally conceived. For instance, Fig. shows the WebML model of a WSMO mediator [7] in the context of a B2B purchase interaction for the SWS Challenge 2006 [19]. The logics of the mediator is that of receiving a single purchase order request, containing multiple lines bundled together, and then dispatch each order line to a service which exposes multiple ports, including one for accepting the general information about new orders and one accepting each line separately. We do not expect that the mediator specification can be appreciated in detail, but the reader should notice that the specification is fully graphic, that it embodies a complex workflow, and in particular it incorporates several request-responses for the SWS orchestration. Clearly, no user interaction is involved.

3.4 Tool Framework Extensions

The framework has been extended by providing automatic generators of WSMO-compliant descriptions (goals, choreographies, capabilities, and mediators) from the models already available in WebML, i.e., business processes, content models, and application logics models. Annotations that are automatically generated cannot express the full semantics of SWS services and applications[4], but they give initial descriptions, that can later be integrated manually. In particular, in the contest of the SWS Challenge 2006, we used WSMT [13] as ontology and SWS editor. As a result of the annotation process, applications and services can be deployed on a SWS runtime environment which provides generic services (e.g., service discovery engines, goal matchers, mediators). Again, in the SWS Challenge we have used the Glue discovery engine [18] as reasoner specialized for Web service discovery.

4 Related Work

Our approach has several elements which are common to a number of research centres and companies working towards improving Web Engineering methods and tools; here we list only a few of them. Traditional Web design methodologies (like OOHDM [15]) and new approaches (like Hera [16]) are now focusing on Semantic Web applications. MIDAS is a framework based on MDA for Semantic Web applications [14]. Research efforts are converging on the proposal of combining Semantic Web Services (SWS) and Business Process Management (BPM) to create one consolidated technology, called Semantic Business Process Management (SBPM) [17].

Our approach largely benefits from the WSMO [10] initiative, which provides a comprehensive framework for handling Semantic Web Services; specifically, we benefit from the WSMO conceptual model [10], the WSML language [11], the WSMX execution environment [12], and WSMT design environment [13].

[4] For instance, the process description yields to deriving a specific orchestration of the services, but in a full SWS specification we need to define choreographies, i.e., rules that indicate all the legal sequences of SWS invocations. Such rules must be derived by extending the initial annotations.

5 Conclusions

The "WebML approach" has acted as a framework for continuous innovation and exploration of new research directions. This is made possible by a unique combination of features:

- Availability of well-defined conceptual models;
- Extensibility of the model thanks to a plug-in based structure;
- Formally defined development process for Web applications;
- Availability of a CASE tool for fast prototyping of application and easy integration of new features and components;
- Strong link between the research (mostly performed in university) and the application development (performed within a spin-off);
- Interactions with real world requirements, enabled by interaction with customers.
- Participation to the international research community, through experience and people exchange and several EU-funded projects.

This mix of ingredients has allowed us to follow our peculiar pathway to innovation.

Acknowledgements

We wish to thank all the people who work in the WebML framework: the WebML group at Politecnico di Milano (and particularly Piero Fraternali), the CEFRIEL Semantic Web Activities group, and the WebRatio staff (and particularly Aldo Bongio). Work on the SOA was performed together with Piero Fraternali, Ioana Manolescu, and Sara Comai; work on SWS was performed together with Federico Facca, Christina Tziviskou, Dario Cerizza, Irene Celino and Andrea Turati.

References – WebML

[1] S. Ceri, P. Fraternali, A. Bongio. Web Modeling Language (WebML): a modeling language for designing Web sites. WWW9 / Computer Networks 33, 2000.

[2] S. Ceri, P. Fraternali, A. Bongio, M. Brambilla, S. Comai, M. Matera. Designing Data-Intensive Web Applications. Morgan Kaufmann, 2002.

[3] I. Manolescu, M. Brambilla, S. Ceri, S. Comai, P. Fraternali. Model-Driven Design and Deployment of Service-Enabled Web Applications. ACM TOIT, 5:3, 2005.

[4] M. Brambilla, S. Ceri, P. Fraternali, I. Manolescu. Process Modeling in Web Applications. ACM TOSEM, 15:4, 2006.

[5] S. Ceri, F. Daniel, M. Matera, F. Facca. Model-driven Development of Context-Aware Web Applications, ACM TOIT, 7:1, 2007.

[6] M. Brambilla, I. Celino, S. Ceri, D. Cerizza, E. Della Valle, F. Facca. A Software Engineering Approach to Design and Development of Semantic Web Service Applications. International Semantic Web Conference (ISWC2006), Athens, USA, November 2006, Springer LNCS 4273.

[7] M. Brambilla, S. Ceri, D. Cerizza, E. Della Valle, F. Facca, P. Fraternali, C. Tziviskou. Coping with Requirements Changes: SWS-challenge phase II. SWS Challenge 2006, Phase II, Budva, Montenegro, June 2006.
[8] WebML: http://www.webml.org.
[9] WebRatio: http://www.webratio.com/.

References – Related Work

[10] Fensel, D., Lausen, H., Polleres, A., de Bruijn, J., Stollberg, M., Roman, D., Domingue, J.: Enabling Semantic Web Services – The Web Service Modeling Ontology. Springer (2006)
[11] de Bruijn, J., Lausen, H., Polleres, A., Fensel, D.: The web service modeling language: An overview. In: Proceedings of the 3rd European SemanticWeb Conference (ESWC2006), Budva, Montenegro, Springer-Verlag (2006)
[12] Haller, A. , Cimpian, E., Mocan, A., Oren, E., Bussler, C.: WSMX - A Semantic Service-Oriented Architecture. In Proceedings of the 2005 IEEE International Conference on Web Services (ICWS 2005), Orlando, FL, USA, 321–328, 2005.
[13] Kerrigan, M.: The WSML Editor Plug-in to the Web Services Modeling Toolkit. In Proceedings of 2nd WSMO Implementation Workshop (WIW2005), Innsbruck, Austria, 2005.
[14] Acuña, C. J., Marcos, E.: Modeling semantic web services: a case study. In Proceedings of the 6th International Conference on Web Engineering (ICWE 2006), Palo Alto, California, USA, 32-39.
[15] Schwabe, D., Rossi, G. The Object-Oriented Hypermedia Design Model. In Communications of the ACM, 38 (8), 45-46.
[16] Vdovjak, R., Frasincar, F., Houben, G. J., Barna, P.: Engineering semantic web information systems in Hera. Journal of Web Engineering, Rinton Press, 2(1-2), 3 -26, 2003.
[17] Hepp, M., Leymann, F., Domingue, J., Wahler, A., Fensel, D.: Semantic Business Process Management: A Vision Towards Using Semantic Web Services for Business Process Management. In Proceedings of the IEEE ICEBE 2005, October 18-20, Beijing, China, 535-540.
[18] Della Valle, E., Cerizza, D.: The mediators centric approach to automatic webservice discovery of Glue. First Intl. Workshop on Mediation in Semantic Web Services (MEDIATE 2005), Amsterdam, The Netherlands, December 2005.
[19] Semantic Web Service Challenge 2006: http://www.sws-challenge.org/.

The Lixto Systems Applications in Business Intelligence and Semantic Web

Robert Baumgartner[1], Oliver Frölich[1], and Georg Gottlob[2]

[1] DBAITU Wien Favoritenstr. 9
1040 Vienna Austria
{froelich,baumgart}@dbai.tuwien.ac.at
[2] Oxford University Computing Laboratory
Wolfson Building
Parks Road Oxford, OX1 3QD United Kingdom
georg.gottlob@comlab.ox.ac.uk

Abstract. This paper shows how technologies for Web data extraction, syndication and integration allow for new applications and services in the Business Intelligence and the Semantic Web domain. First, we demonstrate how knowledge about market developments and competitor activities on the market can be extracted dynamically and automatically from semi-structured information sources on the Web. Then, we show how the data can be integrated in Business Intelligence Systems and how data can be classified, re-assigned and transformed with the aid of Semantic Web ontological domain knowledge. Existing Semantic Web and Business Intelligence applications and scenarios using our technology illustrate the whole process.

1 Introduction

1.1 Motivation

Data available on the Web is a crucial asset in the enterprise world today, such as for making decisions on product prices, collecting opinions and getting an overview in fast-changing markets. To make use of Web data, methodologies and software components for harvesting structured facts from the Web are needed. Semantic Web Ontologies can be populated with Web data and sophisticated rule systems can help to leverage data analysis in Business Intelligence scenarios to a new level. In this paper we address the advantages of Web data extraction for Business Intelligence scenarios. Additionally, we consider how Semantic Web technologies can provide helpful means in such a setting.

1.2 Competitive Intelligence and Business Intelligence

Today, the time available for making operative decisions in a business environment is decreasing: decisions must be made within days or even hours. Just two decades, similar decisions still took weeks or months [Ti95]. Therefore, business management is interested both in increasing the internal data retrieval speed and in broadening the

E. Franconi, M. Kifer, and W. May (Eds.): ESWC 2007, LNCS 4519, pp. 16–26, 2007.

external data sources considered to improve information quality. Fortunately, new technologies like the internet and Business Intelligence systems are available to supply this data. Based on the described competitive pressure, a systematic observation of competitor activities becomes a critical success factor for business to early identify chances in the market, anticipate competitor activities, recognize new and potential competitors, and to validate and enhance own strategic goals, processes and products.

This process of collecting and analyzing information about competitors on the market is called "competitive intelligence" (CI) or "competitive analysis" [SCIP04]. Nowadays, much information about competitors can be retrieved legally from public domain information sources, such as Web sites, press releases or public data bases [Ka98]. The Lixto Suite software provides tools to access, extract, transform, and deliver information from various semi-structured sources like Web pages to various customer systems.

CI can be seen as a part of "Business Intelligence" (BI). The term BI is often used as a method box for collecting, representing and analyzing enterprise data to support the decision-making process within a company's management. More generally, BI can be understood as a process providing better insight in a company and its chains of actions.

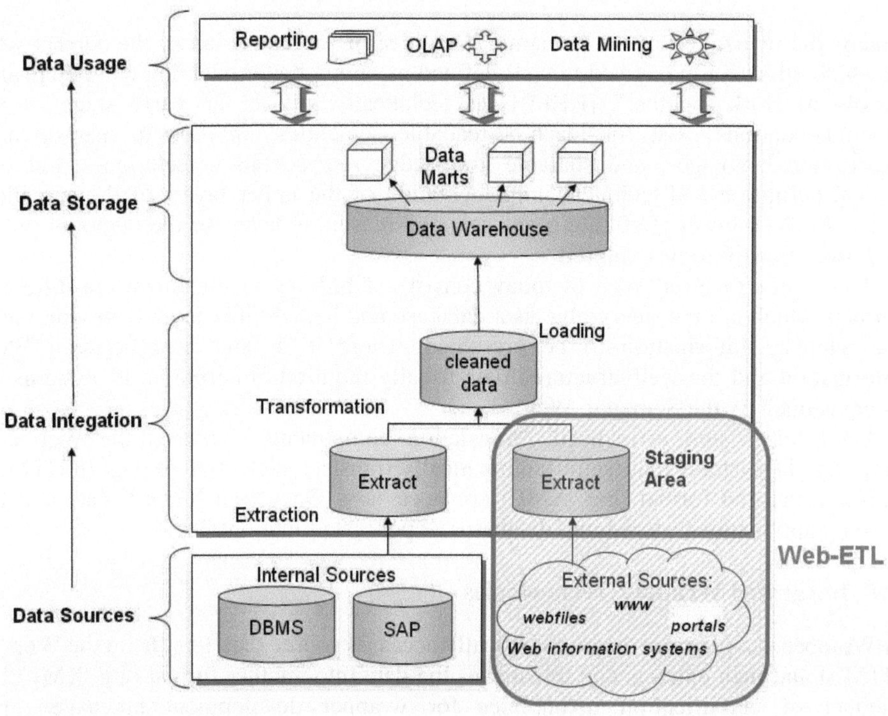

Fig. 1. The Business Intelligence reference process [BFG05]

Technically, the Business Intelligence process covers three main process steps: *data integration*, *data storage* and *data usage* (see fig.1). The most important step concerning our research is *Data integration,* which covers methods to extract data from internal or external data sources. Traditionally, the data is derived for example from database systems in a so-called ETL process (extract, transform, load). We propose using Integrated Wrapper Technologies [Fr06] for Web data extraction and integration in a process we call Web-ETL [BFG05]. This step will be more closely described in the further course of this paper and may also contain data transformations like data "cleaning" and data normalization. A load process with a scheduler regularly uploads (e.g. daily, weekly, or monthly) the processed data into the final data base storage of the BI system, the data warehouse. This *data storage* holds the relevant data for decision makers in a dedicated, homogeneous database. An important characteristic of the data warehouse is the physical storage of data in a single, centralized data pool. It also covers the subject-oriented clustering of data organized by business processes, such as sales, production, or finance. With the information being well-organized in the data warehouse, *Data usage* now can support decision making with predefined reporting for occasional users, ad-hoc data analysis for knowledge workers, or data mining for data analysts.

1.3 Semantic Web

Today, the realization of the Semantic Web idea of "an extension of the current web in which information is given in well-defined meaning, better enabling computers and people to work together" [BHL01] is technically still in an early stage: W3C recommendations exist for machine-readable semantics, appropriate markup and description languages, and sharable knowledge representation techniques, but the logical definition and technical implementation of the upper layers of the so-called Semantic Web tower [Be02], e.g. the rule and reasoning layer, or the layers of proof and trust, are still to be explored.

The semi-structured Web of today consists of billions of documents in different formats which are not query-able as a database and heavily mix layout, structure and the intended information to be presented. There is a huge gap between Web information and the well-structured data usually required in corporate IT systems or as envisioned by the Semantic Web.

Until this vision is realized, "translation components" between the Web and corporate IT systems that (semi-)automatically translate Web content (e.g. in HTML) into a structured format (e.g. XML) are necessary. Once transformed, data can be used by applications, stored into databases or populated into ontologies.

1.4 Integrated Wrapper Technologies

A Wrapper is a program that automatically accesses source data (e.g. from the Web in HTML) and then extracts and transforms the data into another format (e.g. XML). A number of classification taxonomies for wrapper development languages and environments have been introduced in various survey papers. A to our knowledge complete overview of the different approaches and systems is given in [Fr06].

Integrated Wrapper Technology (IWT) systems combine the capabilities of wrapping components with Information Integration (II) components [Fr06]. The latter generally transform the extracted data and integrate it in other (business) IT systems. IWT systems are suitable for implementing advanced information systems for the Semantic Web: Partially, they can bridge the gap between the Web existing today and the today not yet existing Semantic Web that might be used as the largest database on earth where data exists in machine-readable formats and can be integrated easily in other IT systems. IWT systems can extract data from semi-structured Web pages, transform it to a semantically useful structure, and integrate it e.g. with a Web ETL-process into a Business Intelligence system. A solution proposition to this problem will be illustrated in the next chapter.

2 The Lixto Solution

The *Lixto Suite* software is an IWT system which provides tools to access, extract, transform, and deliver information from various semi-structured sources like Web pages to many different customer systems. The Lixto software is based on Java technology and standards such as XML schema, XSLT, SOAP and J2EE. Technically, the main distinguishing criteria to many other approaches are that Lixto embeds the Mozilla browser and is based on Eclipse. This allows Lixto to be always at the cutting edge of Web browser technology and access and extract data from all Web pages even using the newest techniques like Web 2.0, e.g. Ajax. Internally, the software uses the logic-based data extraction language *Elog* [GK02].

The Lixto Suite is comprised of three products: The *Visual Developer* for Wrapper generation, the *Metasearch* for real-time Wrapper queries and the *Transformation Server* as runtime environment for scheduled Wrappers queries and as Information Integration component. In this chapter, we successively describe the process steps for creating and delivering structured data from semi-structured sources.

2.1 Wrapper Generation with Visual Developer

Wrappers generated with Visual Developer extract and translate all relevant information from HTML Web pages to a structured XML format. The underlying extraction language Elog is derived from datalog and is especially designed for wrapper generation. The Elog language operates on Web objects, which are (lists of) HTML elements, and strings. Elog rules can be specified visually in a graphical user interface by a few mouse clicks without knowing the Elog language. Thus, no special programming knowledge is needed, and wrappers can be generated by non-technical personnel, typically by employees with the relevant business expertise for the project, e.g. from a company's marketing department.

Creating a wrapper with the Visual Developer comprises two steps: First, the data model creation. In this phase an XML-schema based model in which extracted data is inserted either is imported (e.g. in case of news extraction typically RSS) or generated from scratch. In the second phase navigation and extraction steps are configured. Such extraction rules are semi-automatically and visually defined by a wrapper designer in an iterative process. Extracted data can subsequently populate an ontology with instance data. The whole wrapper generation process starts by generating a deep Web navigation sequence (such as navigation through forms) and subsequently highlighting the relevant information with two mouse clicks in the integrated standard Mozilla browser window. The software then marks the data in a different colour. Conditions can be defined, allowing the program to identify the desired data even if the structure of the Web page slightly changes. Fig. 2 shows an example where information is extracted from different trip booking websites like *expedia.com* and *opodo.com*.

For a wrapper, an internet page is an *HTML tree structure*. A wrapper does *not* extract just the text from a specified HTML tree node, but uses "intelligent" conditions, so-called *logical patterns*. For the wrapper of fig. 2, such conditions could be *„the relevant area should contain the Dollar-symbol("$") in each line"* or *"some specified company's names should occur"* (these names are stored in a system database). For the *logical pattern* comprised of the conditions, the software searches for the best match within the HTML tree using heuristic methods. So a very high robustness to changes within Web pages over time can be achieved for the wrapper agents.

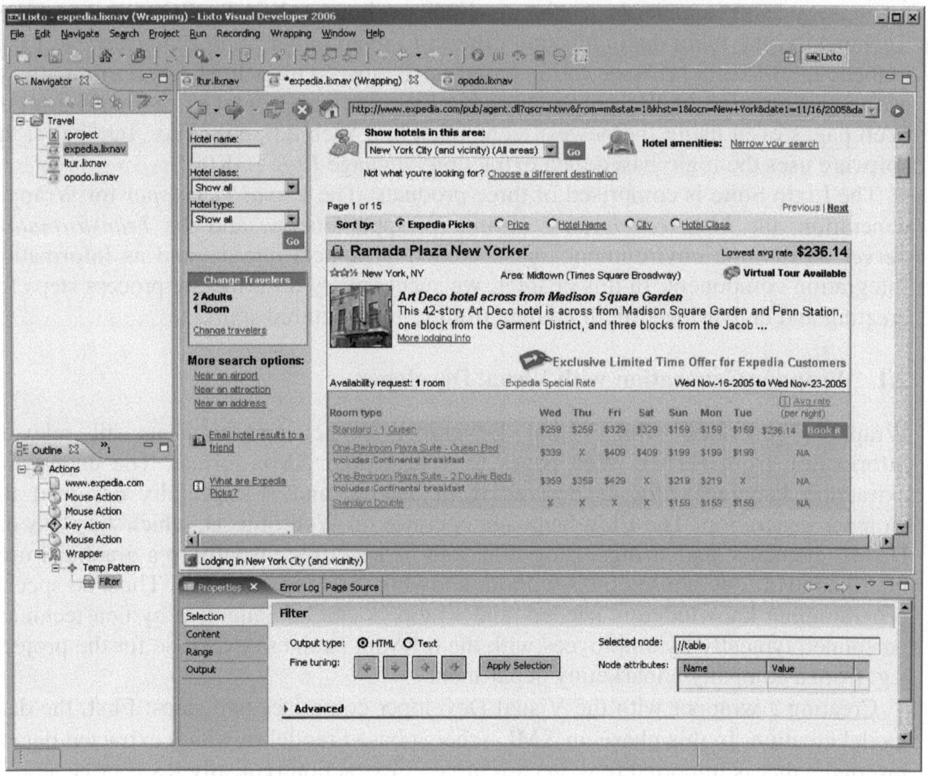

Fig. 2. Visual Developer User Interface

In addition, navigation to further documents during the wrapping process is possible: For example, starting from a Google result list overview page, it is possible to extract defined information from all result sub-pages and, by clicking on the "next"-button on each result list overview page (this action has to be defined only once) to extract all relevant information from all overview pages and their sub pages. The Visual Developer is capable of action-based recording and replaying, i.e. recording actions of a user based on mouse clicks and key actions. This is highly advantageous compared to a plain request-based macro, as it allows the system to deal with dynamically changing HTML pages.

Visual Developer wrapper agents also support automatic logon to password-protected pages, filling in form pages and processing the extraction from corresponding result pages (i.e. for Web interfaces of data bases), dynamic handling of session IDs, and automatic handling of cookies and SSL. Detailed information on further wrapping capabilities can be found in [BFG01].

2.2 The Transformation Server

In a second step, XML data generated by wrappers is processed in the *Lixto Transformation Server* [GH01]. A wrapper in the run-time environment of the Transformation Server retrieves the Web data automatically, with no developer interaction, based on events. Events can be for example a defined schedule, such as Web data retrieval *every hour* or *every 5 minutes*. The usual setting for the creation of *services* based on Web wrappers is that information is obtained from multiple wrapped sources and has to be integrated. Often source sites have to be monitored for changes, and changed content has to be automatically extracted and processed.

The Lixto Transformation Server allows for Information Integration by combining, transforming and re-formatting data from different wrappers. Results of this process can be delivered in various formats and channels to other IT systems, e.g. Business Intelligence systems such as SAP Business Information Warehouse or Microsoft Analysis Server. Transformation Server can interactively communicate with these systems using various interfaces, such as special database formats, XML messaging, and Web services.

The whole process of modelling the workflow and dataflow is done in the graphical user interface of the Lixto Transformation Server. Graphical objects symbolize components, such as an *integrator* for the aggregation of data or a *deliverer* for the transmission of information to other software applications. By drawing connecting arrows between these objects, the flow of data and the workflow are graphically defined (see also fig. 3). A more detailed description of the components is given in [BFG05].

3 Application Business Cases

The following business cases show and illustrate the capabilities of the IWT software *Lixto Suite* in real-world scenarios. The first scenario concentrates more on a BI solution, whereas the second scenario describes an application for the Semantic Web.

3.1 CI Solution for Pirelli

Pirelli is one of the world market leaders in tire production, but also active in other sectors such as cables (energy and telecommunication cables). With headquarters in Milan/Italy, the company runs 21 factories all over the world and has more than thirty-five thousand employees.[1] On account of the growing amount and relevance of Web sites selling tires on the internet (both B2B and B2C), Pirelli analyzed the possibilities of monitoring retail and wholesale tire prices from competitors for their major markets. This external data should be automatically uploaded to their existing BI solution. After an extensive market research concerning available tools for Web data extraction and transformation, Pirelli selected the Lixto software because of its high scalability for back office use, its high robustness concerning data extraction quality and its straightforward administration.

The Lixto Software was integrated in the Pirelli BI infrastructure within a timeframe of two months. Tire pricing information of more than 50 brands and many dozens of tires selling Web sites are now constantly monitored with Lixto (Pirelli prices and competitor prices). A simplified screenshot of the Pirelli service shows the data flow as it is defined in the Lixto Transformation Server (see fig. 3).

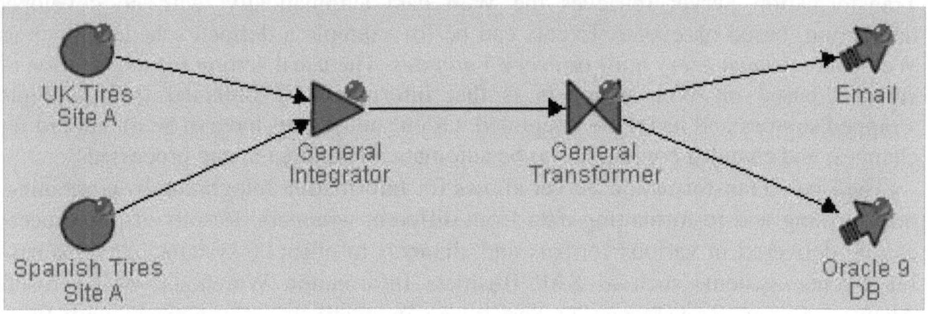

Fig. 3. Modelling the data flow in the Transformation Server [BFG05]

The data is normalized in Transformation Server and then delivered to an Oracle 9 database. From here, the Pirelli BI solution fetches the data and generates i.e. reports in PDF format and HTML format. These reports are automatically distributed to the Pirelli intranet for marketing and sales departments (see fig. 4).

The success of the project can be measured by the more than 1.000 self-registered Pirelli users receiving the Lixto PDF reports regularly by email. Since its introduction, the Lixto reports are in the top 5 list of all most accessed files from the Pirelli intranet. A more detailed description of the Pirelli BI integration project can be found in [BFG05].

3.2 The Personal Publication Reader

The Personal Publication Reader is an advanced personalized information systems using Semantic Web technologies [BHM05] that has been created by the Learning

[1] See http://uk.biz.yahoo.com/p/p/peci.mi.html and [Pi03].

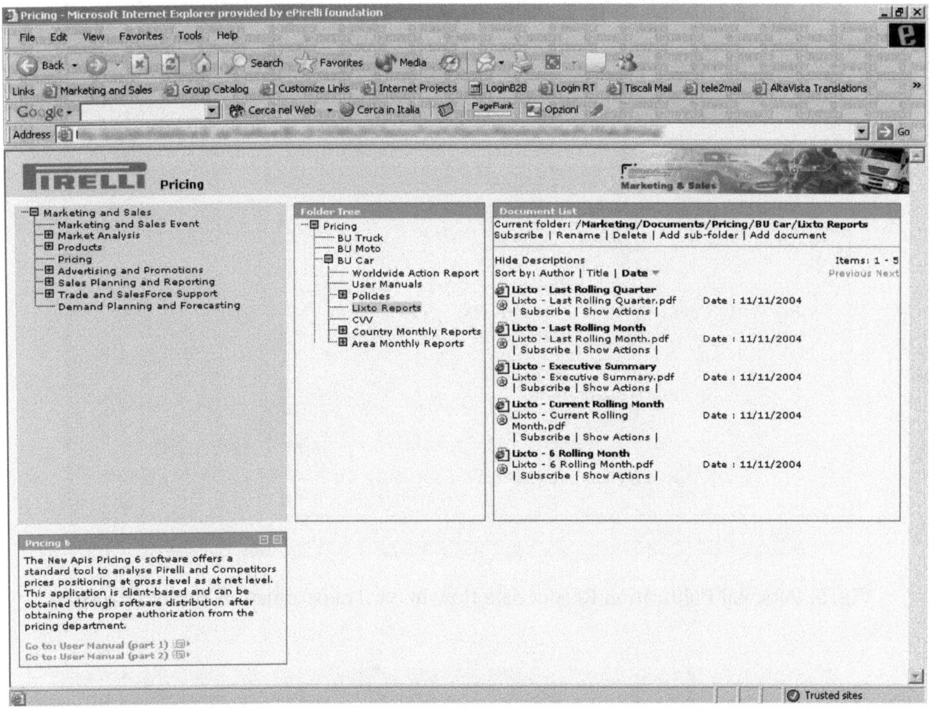

Fig. 4. Lixto Reports within the Pirelli intranet [BFG05]

Lab Hannover. It provides a syndicated and personalized view on non-uniform, distributed Web data. The Personal Publication Reader has been designed and implemented in the context of the Network of Excellence "REWERSE – Reasoning on the Web" and syndicates and personalizes information about the REWERSE project structure, people, objectives, and information about research papers in the context of the project.

The Personal Reader Framework was used for the design and implementation of personalized Web Content Readers. Such readers have been realized for e-Learning and for publication browsing. Web services allow for the development of flexible information systems by encapsulating specific functionality and communication capabilities with other Web services. The aim of the Personal Reader Framework was to develop a toolset for designing, implementing and maintaining Personal Web Content Readers. These Personal Readers allow the user to browse information (the "Reader" part), and to access personal recommendations and contextual information on the currently viewed Web resource (the "Personal" part). The framework features a distributed open architecture designed to be easily extensible. It uses standards such as XML and RDF, and technologies like JSP and XML-based-RPC. The communication between all of these components is syntactically based on RDF descriptions, providing a high level of flexibility for the whole framework.

Content syndication and personalization is achieved by reasoning about ontological knowledge and extracted Web data. The Lixto Software is used for the implementation

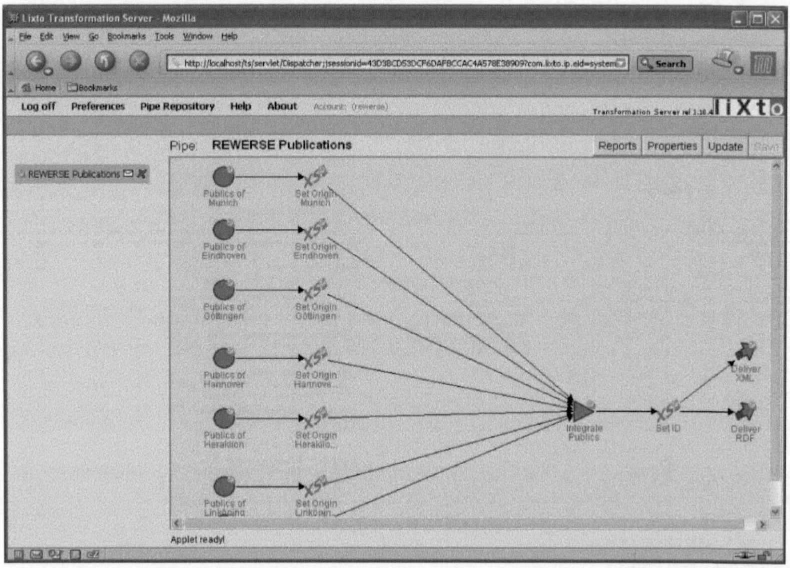

Fig. 5. Personal Publication Reader data flow in the Transformation Server [BHM05]

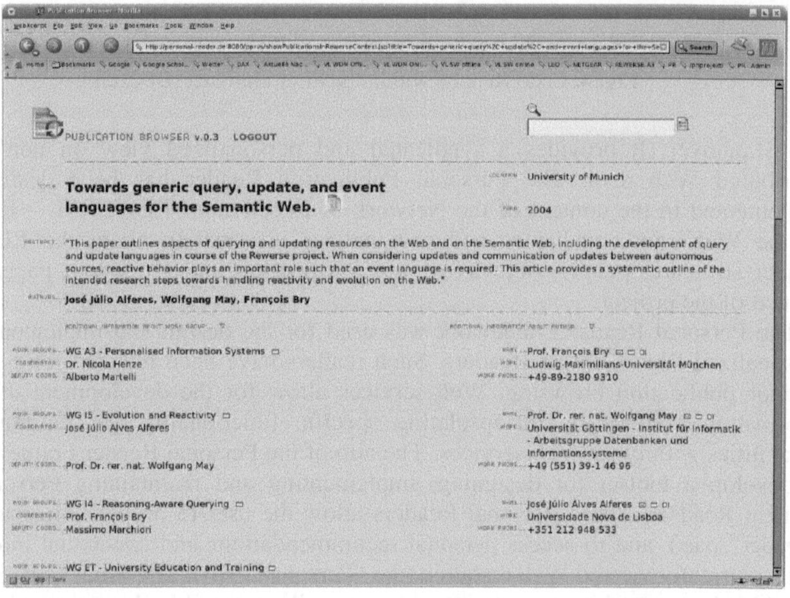

Fig. 6. Screenshot of the Personal Publication Reader [BHM05]

of the information provision part. The process of gathering Web data for the Personal Publication Reader using Lixto is described in the following.

In a first step, wrappers are created for the websites of the REWERSE project participant organizations. Then, these wrappers are loaded to a new service within the Lixto Transformation Server. Here the XML data derived from the various wrappers has to be combined, cleaned, and syndicated into the Framework's ontology. The output format of the Transformation Server is an RDF representation. This process is scheduled regularly and hands over the data to the Personal Publication Reader. Fig. 5 shows the data flow from the wrapper to the RDF output as it is defined in the Transformation Server.

In addition to the extracted information on research papers described above, a "REWERSE-Ontology" has been built in OWL extending the Sementic Web Research Community Ontology (SWRC) [SWRC01]. Here information about research members of the REWERSE project is kept.

Finally, fig. 6 shows the syndicated view on publications in REWERSE together with the corresponding links, and information about the authors of the publications like homepage, phone number, etc. The Personal Publication Reader is available online via the URL *www.personal-reader.de.*

Acknowledgements

This research has been partially supported by REWERSE - Reasoning on the Web (rewerse.net), Network of Excellence, 6th European Framework Program.

The authors would like to thank Giacomo del Felice from Pirelli Pneumatici S.p.A. for his continuous and reliable project support.

References

[BFG01] Baumgartner, R.; Flesca, S.; Gottlob, G.: Visual web information extraction with Lixto. In: Proc. of VLDB, 2001, pp. 119–128.

[BFG05] Baumgartner, R.; Frölich, O.; Gottlob, G.; Harz, P.; Herzog, M.; Lehmann, P.: Web Data Extraction for Business Intelligence: the Lixto Approach, in: Datenbanksysteme in Business, Technologie und Web (BTW), LNI, Series of the Gesellschaft für Informatik, P-65 (2005), pp. 48–65.

[BHL01] Berners-Lee, T.; Hendler, J.; Lassila, O.: The semantic web. Scientific American, May 2001.

[Be02] Berners-Lee, T.: The semantic web - mit/lcs seminar, 2002.. http://www.w3c.org/2002/Talks/09-lcs-sweb-tbl/.

[BHM05] Baumgartner, R.; Henze, N.; Herzog, M.: The Personal Publication Reader: Illustrating Web Data Extraction, Personalization and Reasoning for the Semantic Web, in: European Semantic Web Conference ESWC 2005, LNCS 3532, Springer, Berlin Heidelberg, 2005, pp. 515–530.

[Fr06] Frölich, O.: Erstellung und Optimierung von Geschäftsprozessen durch Integrierte Wrapper-Technologien mit Anwendungsbeispielen aus den Branchen Mobile Services, Competitive Intelligence und dem Verlagswesen. Dissertation, DBAI, TU Vienna, Vienna, 2006.

[GH01] Gottlob, G.; Herzog, M.: Infopipes: A Flexible Framework for M-Commerce Applications, in: Proc. of TES workshop at VLDB, 2001, pp. 175–186.

[GK02] Gottlob, G.; Koch, C.: Monadic datalog and the expressive power of languages for Web Information Extraction, in: Proc. of PODS, 2002, pp. 17–28. Full version: Journal of the ACM 51(1), 2004, pp. 74 – 113.

[Ka98] Kahaner, L.: Competitive Intelligence: How to Gather, Analyse Information to Move your Business to the Top. Touchstone, New York, 1998.

[Pi03] Pirelli & C. SpA: Annual Report 2003. http://www.pirelli.com//investor_relation/bilanciocompl2003.pdf, accessed on 2004-09-28, p. 7.

[SCIP04] Society of Competitive Intelligence Professionals (SCIP): What is CI? http://www.scip.org/ci/index.asp, accessed on 2004-09-28.

[SWRC01] SWRC - Semantic Web Research Community Ontology, 2001. http://ontobroker.semanticweb.org/ontos/swrc.html.

[Ti95] Tiemeyer, E.; Zsifkovitis, H.E.: Information als Führungsmittel: Executive Information Systems. Konzeption, Technologie, Produkte, Einführung; 1st edition; Munich, 1995, p. 95.

Ways to Develop Human-Level Web Intelligence: A Brain Informatics Perspective

Ning Zhong

Department of Life Science and Informatics
Maebashi Institute of Technology, Japan &
The International WIC Institute/BJUT, China
zhong@maebashi-it.ac.jp

Abstract. In this paper, we briefly investigate several ways to develop human-level Web intelligence (WI) from a brain informatics (BI) perspective. BI can be regarded as brain sciences in WI centric IT age and emphasizes on a systematic approach for investigating human information processing mechanism. The recently designed instrumentation (fMRI etc.) and advanced IT are causing an impending revolution in both WI and BI, making it possible for us to understand intelligence in depth and develop human-level Web intelligence.

1 Introduction

The concept of Web intelligence (WI) was first introduced in our papers and books [13, 24, 27, 29, 31]. Broadly speaking, WI is a new direction for scientific research and development that explores the fundamental roles as well as practical impacts of artificial intelligence (AI)[1] and advanced information technology (IT) on the next generation of Web-empowered systems, services, and environments. The WI technologies revolutionize the way in which information is gathered, stored, processed, presented, shared, and used by virtualization, globalization, standardization, personalization, and portals.

As more detailed blueprints and issues of WI are being evolved and specified [13, 29, 31, 36], it has been recognized that one of the fundamental goals of WI research is to understand intelligence in depth and develop wisdom Web based intelligent systems that integrate all the human-level capabilities such as real-time response, robustness, autonomous interaction with their environment, communication with natural language, commonsense reasoning, planning, problem solving, decision making, learning, discovery and creativity.

Turing gave the first scientific discussion of human-level machine intelligence [23]. Newell and Simon made a start on programming computers for general intelligence and investigated human problem solving in a behavior based approach [16]. McCarthy argued that reaching human-level AI requires programs that deal with the commonsense informative situation, in which the phenomena

[1] Here the term of AI includes classical AI, computational intelligence, and soft computing etc.

E. Franconi, M. Kifer, and W. May (Eds.): ESWC 2007, LNCS 4519, pp. 27–36, 2007.

to be taken into account in achieving a goal are not fixed in advance [15]. Laird and Lent proposed using interactive computer games that are the killer application for human-level AI research, because they can provide the environments for research on the right kinds of problem that lead to the type of incremental and integrative research needed to achieve human-level AI [10].

The new generation of WI research and development needs to understand multiple natures of intelligence in depth, by studying integrately the three intelligence related research areas: machine intelligence, human intelligence, and social intelligence, as shown in Figure 1, towards developing truly human-level Web intelligence. Machine intelligence (also called AI) has been mainly studied as computer based technologies for the development of intelligent knowledge based systems; Human intelligence (also called brain sciences) studies the nature of intelligence towards our understanding of intelligence; Social intelligence needs a combination of machine intelligence and human intelligence for establishing social networks that contain communities of people, organizations, or other social entities [29]. Furthermore, the Web can be regarded as a social network in which the Web connects a set of people (or organizations or other social entities). People are connected by a set of social relationships, such as friendship, co-working or information exchange with common interests. In other words, it is a Web-supported social network or called virtual community. In this sense, the study of WI is of social network intelligence (social intelligence for short).

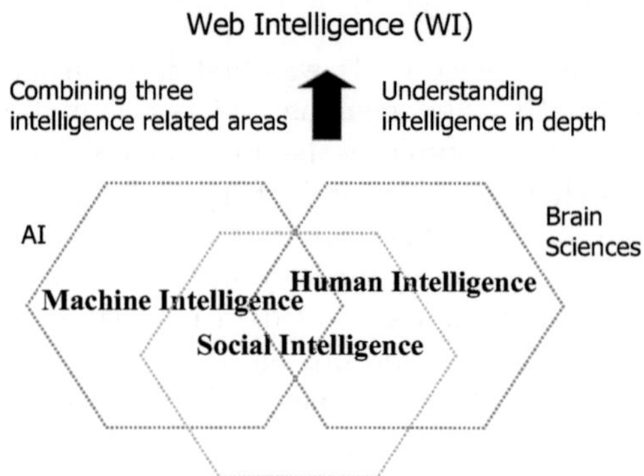

Fig. 1. The relationship between WI and other three intelligence related research areas

In the rest of the paper, we briefly investigate three ways to develop human-level Web intelligence from a brain informatics (BI) perspective. BI can be regarded as brain sciences in WI centric IT age [34, 35]. Although brain sciences have been studied from different disciplines such as cognitive science and neuroscience, BI represents a potentially revolutionary shift in the way that research is undertaken. BI is proposing to study human brain from the viewpoint of informatics (i.e. human brain is an information processing system) and use informatics (i.e. WI centric information technology) to support brain science study, in particular, WI provides urgent research needs.

2 Web Based Problem Solving with Human Level Capabilities

A more concrete issue of WI is the development and application of a Web-based problem-solving system for portal-centralized, adaptable Web services [8,13,22,29,31].

Problem-solving is one of main capabilities of human intelligence and has been studied in both cognitive science and AI [16], where it is addressed in conjunction with reasoning centric cognitive functions such as attention, control, memory, language, reasoning, learning, and so on, using a logic based symbolic and/or connectionist approach. Although logic based problem-solving is "perfect", mathematical systems with no real time and memory constraints, Web-based problem-solving systems need real-time and dealing with global, multiple, huge, distributed information sources.

The core of such a system rests on the Problem Solver Markup Language (PSML) and PSML-based distributed Web inference engines for network reasoning, in which the following essential support functions should be provided:

- The expressive power and functional support in PSML for complex adaptive, distributed problem solving;
- Performing automatic reasoning on the Web by incorporating globally distributed contents and meta-knowledge, automatically collected and transformed from the semantic Web and social networks, with locally operational knowledge-data bases;
- Representing and organizing multiple, large-scale knowledge-data sources for distributed network reasoning;
- Combining multiple reasoning methods in PSML representation and distributed inference engines, efficiently and effectively;
- Modeling user behavior and representing/managing it as a personalized model dynamically;
- Including an emotional factor in developing the Web based reasoning and problem solving system.

A possible way as an immediate step to implement certain distributed reasoning capabilities of the future PSML is to make use of an existing logic language coupled with agent technologies. We have demonstrated one possible implementation of such capabilities. In particular, our proposed implementation, called β-PSML, is based on the combination of OWL with Horn clauses, and able to couple global semantic Web/social networks with local information sources for solving problems in a large-scale distributed Web environment [21,22].

Furthermore, in order to develop a Web based problem-solving system with human level capabilities, we need to better understand how human being does complex adaptive, distributed problem solving and reasoning, as well as how intelligence evolves for individuals and societies, over time and place [20,26,35]. In other words, ignoring what goes on in human brain and focusing instead on behavior has been a large impediment to understand how human being does complex adaptive, distributed problem solving and reasoning.

In the light of BI, we need to investigate specifically the following issues:

– What are the existing thinking/reasoning models in AI, cognitive science, and neuroscience?
– How to design fMRI/EEG experiments and analyze such fMRI/EEG data to understand the principle of human reasoning and problem solving in depth?
– How to build the cognitive model to understand and predict user profile and behavior?
– How to implement human-level reasoning and problem solving on the Web based portals that can serve users wisely?

As a result, the relationships between classical problem solving and reasoning and biologically plausible problem solving and reasoning need to be defined and/or elaborated.

3 Reasoning Centric Thinking Oriented Studies in Human Information Processing System

Human intelligence related research studies the nature of intelligence towards our understanding of intelligence. The capabilities of human intelligence can be broadly divided into two main aspects: perception and thinking. So far, the main disciplines with respect to human intelligence are cognitive science that mainly focuses on studying mind and behavior based cognitive models of intelligence, as well as neuroscience that mainly focuses on studying brain and biological models of intelligence. In cognitive neuroscience, although many advanced results with respect to "perception oriented" study have been obtained, only a few of preliminary, separated studies with respect to "thinking oriented" and/or a more whole information process have been reported [6]. Figure 2 gives a global picture on reasoning centric thinking oriented functions and their relationships in human information processing system.

Our purpose is to understand activities of human information processing system by investigations in the following two levels:

– investigating the spatiotemporal features and flow of human information processing system, based on functional relationships between activated areas of human brain for each given task;
– investigating neural structures and neurobiological processes related to the activated areas [19].

More specifically, at the current stage, we want to understand:

– how a peculiar part (one or more areas) of the brain operates in a specific time;
– how the operated part changes along with time;
– how the activated areas work cooperatively to implement a whole information processing;

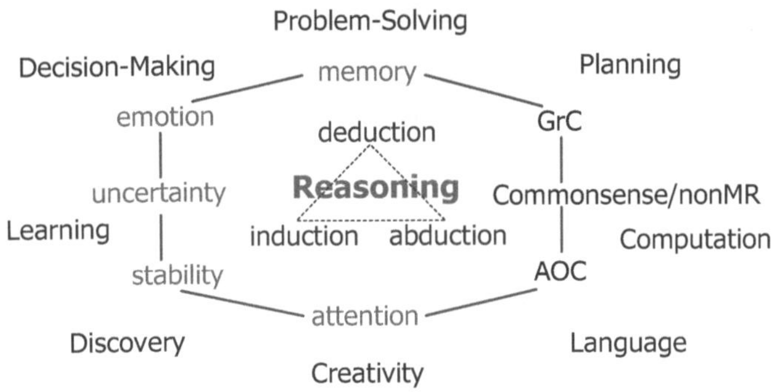

Fig. 2. Reasoning centric thinking oriented functions and their relationships (GrC: Granular Computing [25]; AOC: Autonomy Oriented Computing [14]; nonMR: non-monotonous reasoning)

- how the activated areas are linked, indexed, navigated functionally, and what are individual differences in performance;
- how a cognitive process is supported by neurobiological processes.

We need to study experimental cognitive neuroscience, data mining, intelligent agents, data and knowledge grids, the semantic Web and wisdom Web in a unified way [1, 2, 3, 4, 11, 12, 28, 34]. We have been developing a full process from designing fMRI/EEG experiments based on WI needs for discovering new cognitive WI models. Such a full process means a systematic approach for measuring, collecting, modeling, transforming, managing, and mining multiple human brain data obtained from various cognitive experiments by using fMRI and EEG [33, 34].

As a step in this direction, we observe that fMRI brain imaging data and EEG brain wave data extracted from human information processing mechanism are *peculiar* ones with respect to a specific state or the related part of a stimulus. Based on this point of view, we propose a way of *peculiarity oriented mining (POM)* for knowledge discovery in multiple human brain data, without using conventional imaging processing to fMRI brain images and frequency analysis to EEG brain waves [17, 32, 33, 34]. The proposed approach provides a new way for automatic analysis and understanding of fMRI brain images and EEG brain waves to replace human-expert centric visualization. The mining process is a multi-step one, in which various psychological experiments, physiological measurements, data cleaning, modeling, transforming, managing, and mining techniques are cooperatively employed to investigate human information processing mechanism.

Figure 3 gives the global picture of an example about how to investigate the spatiotemporal features and flow of human information processing system. In the cognitive process from perception to reasoning, data are collected in several event-related time points, and transformed into various forms in which POM

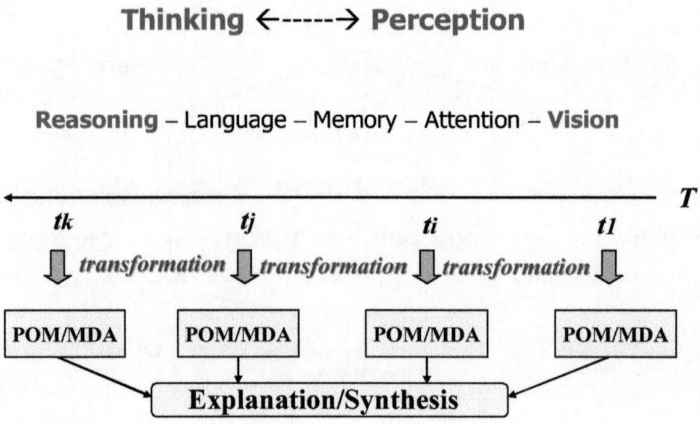

Fig. 3. Investigating the spatiotemporal features and flow of human information processing system

centric multi-aspect data analysis (MDA) can be carried out efficiently and effectively. Furthermore, the results of separate analysis can be explained and combined into a whole flow.

4 A Data-Brain Model and Its Construction

The Data-Brain is a brain database with all of data related to all major aspects and capabilities of human information processing mechanism for systematic investigation and understanding of human intelligence. The Data-Brain provides a holistic view at a long-term, global field of vision to understand the principle, models and mechanisms of human information processing system [9, 34, 35].

The key questions are how to obtain such data by systematic fMRI/EEG experiments, how to manage such huge multimedia data for systematic investigation and understanding of human intelligence, as well as how to analyse such data from multi-aspect and multi-level for discovering new cognitive models. A new conceptual model is needed to represent complex relationships among multiple human brain data sources, which are obtained by systematic fMRI/EEG experiments. Furthermore, the following supporting capabilities are requested to build such a Data Brain:

- It is a grid-based, simulation and analysis oriented, dynamic, spatial and multimedia database;
- It deals with multiple data sources, multiple data forms, multiple levels of data granulation;
- It provides multiple views and organizations;
- It includes various methods for data analysis, simulation, visualization, as well as corresponding knowledge and models.

At first, agents for data collecting, storing and retrieving are deployed on the Grid platform, like Globus, as a standard Grid service. OGSA-DAI is used to build database access applications [5,37]. The aim of OGSA-DAI is to provide the middleware glue to interface existing databases, other data resources and tools to each other in a common way based on the Open Grid Services Architecture (OGSA). This middleware is based on the GGF-defined OGSI specification and layered on top of the Globus toolkit 3 OGSI implementation (GT3 Core).

Multiple data sources are collected by various cognitive fMRI/EEG experiments, modeling and transformation, and they are recorded to the corresponding databases through the Grid service on the distributed sites. Furthermore, the data-flow is a collection of descriptions for the dynamic relationship among multiple data sources on the data-grid. In the current study, data sources from cognitive fMRI/EEG experiments, to be collected on the data-grid, include:

- human multi-perception mechanism for studying the relevance between auditory and visual information processing;
- human deductive/inductive reasoning mechanism for understanding the principle of human reasoning and problem solving in depth;
- human computation mechanism as an example of human problem solving system;
- human decision-making mechanism from developing Web based decision-making support system with an emotional factor;
- human learning mechanism for acquiring personalized student models in an interactive learning process dynamically and naturally.

In order to build a Data Brain, a systematic methodology of cognitive experimental design needs to be developed, so that multiple human brain data sources obtained by fMRI/EEG experiments are interrelated and can be utilized for multi-purpose, not only a specific one. Event-related experimental designs have become an important methodology in EEG/fMRI research to evaluate the high level characteristics of human information processing in the central nervous system [18]. There are, at present, two main methods called event-related potential (ERP) and event-related fMRI for event-related experimental designs. ERP is a tiny signal embedded in the ongoing EEG. By averaging the traces, investigators can extract this signal, which reflects neural activity that is specifically related cognitive events [7]. ERPs are best suited for addressing questions about the time course of cognition rather than elucidating the brain structures that produce the electrical events. ERPs also provide physiological indices of when a person decides to response, or when an error is detected. On the other hand, event-related fMRI follows the same logic as used in ERP/EEG studies and provides the spatial resolution. Thus, event-related fMRI will further allow fMRI and EEG to be combined in paradigms that are identical across methods. By using such techniques, it is now becoming possible to study the precise spatiotemporal orchestration of neuronal activity associated with perceptual and cognitive events [18], as well as systematic collection of human brain data for building a Data Brain.

5 Conclusion

BI emphasizes on a systematic approach for investigating human information processing mechanism, including measuring, collecting, modeling, transforming, managing, and mining multiple human brain data obtained from various cognitive experiments by using fMRI and EEG. The proposed methodology attempts to change the perspective of cognitive/brain scientists from a single type of experimental data analysis towards a holistic view at a long-term, global field of vision to understand the principle, models and mechanisms of human information processing. New generations of WI research and development need to understand multi-nature of intelligence in depth. The recently designed instrumentation (fMRI etc.) and advanced IT are causing an impending revolution in both WI and BI, making it possible for us to understand intelligence in depth and develop human-level Web intelligence.

Acknowledgments

I would like to express my gratitude to Enrico Franconi, Michael Kifer, Wolfgang May, Dieter Fensel and other organizers of ESWC 2007 for the kind invitation and the excellent organization. I am grateful to all my research collaborators, assistants, and students who have, over the years, together contributed to the development of Web Intelligence (WI) and Brain Informatics (BI). Special thanks to Jiming Liu, Yiyu Yao, Jinglong Wu, Benjamin Wah, and Shengfu Lu for our joint projects and discussions. I am very grateful to people who have joined or supported the WI and BI communities, members of the WIC advisory board, WIC technical committee, and WIC research centres, as well as keynote/invited speakers of IEEE/WIC/ACM WI-IAT conferences. This work is partially supported by the grant-in-aid for scientific research (No. 18300053) from the Japanese Ministry of Education, Culture, Sports, Science and Technology.

References

1. T. Berners-Lee, J. Hendler, O. Lassila, "The Semantic Web", *Scientific American*, 284, 34-43 (2001).
2. M. Cannataro and D. Talia, "The Knowledge Grid", *Communications of the ACM*, 46 (2003) 89-93.
3. D. Fensel, *Ontologies: A Silver Bullet for Knowledge Management and Electronic Commerce*, Springer (2001).
4. D. Fensel, "Unifying Reasoning and Search to Web Scale", *IEEE Internet Computing*, 11(2) (2007) 94-96.
5. I. Foster and C. Kesselman (eds.) *The Grid: Blueprint for a New Computing Infrastructure*, Morgan Kaufmann (1999).
6. M.S. Gazzaniga (ed.) *The Cognitive Neurosciences III*, The MIT Press (2004).
7. T.C. Handy, *Event-Related Potentials, A Methods Handbook*, The MIT Press (2004).

8. J. Hu and N. Zhong, "Organizing Multiple Data Sources for Developing Intelligent e-Business Portals", *Data Mining and Knowledge Discovery*, Vol. 12, Nos. 2-3, Springer (2006) 127-150.
9. S.H. Koslow and S. Subramaniam (eds.) *Databasing the Brain: From Data to Knowledge*, Wiley (2005).
10. J.E. Laird and M. van Lent, "Human-Level AI's Killer Application Interactive Computer Games", *AI Magazine* (Summer 2001) 15-25.
11. Y. Li and N. Zhong, "Mining Ontology for Automatically Acquiring Web User Information Needs", *IEEE Transactions on Knowledge and Data Engineering*, Vol. 18, No. 4 (2006) 554-568.
12. J. Liu, N. Zhong, Y.Y. Yao, and Z.W. Ras, "The Wisdom Web: New Challenges for Web Intelligence (WI)", *Journal of Intelligent Information Systems*, 20(1) Kluwer (2003) 5-9.
13. J. Liu, "Web Intelligence (WI): What Makes Wisdom Web?", *Proc. Eighteenth International Joint Conference on Artificial Intelligence (IJCAI'03)* (2003) 1596-1601.
14. J. Liu, X. Jin, and K.C. Tsui, *Autonomy Oriented Computing: From Problem Solving to Complex Systems Modeling*, Springer (2005).
15. J. McCarthy, "Roads to Human Level AI?", Keynote Talk at Beijing University of Technology, Beijing, China (September 2004).
16. A. Newell and H.A. Simon, *Human Problem Solving*, Prentice-Hall (1972).
17. M. Ohshima, N. Zhong, Y.Y. Yao, and C. Liu, "Relational Peculiarity Oriented Mining", *Data Mining and Knowledge Discovery*, Springer (in press).
18. B.R. Rosen, R.L. Buckner, and A.M. Dale, "Event-related functional MRI: Past, Present, and Future", *Proceedings of National Academy of Sciences, USA*, Vol. 95, Issue 3 (1998) 773-780.
19. R.G. Shulman and D.L. Rothman, "Interpreting Functional Imaging Studies in Terms of Neurotransmitter Cycling", *Proceedings of National Academy of Sciences, USA*, Vol. 95, Issue 20 (1998) 11993-11998. [
20. R.J. Sternberg, J. Lautrey, and T.I. Lubart, *Models of Intelligence*, American Psychological Association (2003).
21. Y. Su, L. Zheng, N. Zhong, C. Liu, and J. Liu, "Distributed Reasoning Based on Problem Solver Markup Language (PSML): A Demonstration through Extended OWL", *Proc. 2005 IEEE International Conference on e-Technology, e-Commerce and e-Service (EEE'05)*, IEEE Press (2005) 208-213.
22. Y. Su, J. Liu, N. Zhong, L. Zheng, and C. Liu, "A Method of Distributed Problem Solving on the Web", *Proc. 2005 IEEE/WIC/ACM International Conference on Web Intelligence (WI'05)*, IEEE Press (2005) 42-45.
23. A. Turing, "Computing Machinery and Intelligence", Mind LIX(236) (1950) 433-460.
24. Y.Y. Yao, N. Zhong, J. Liu, and S. Ohsuga, "Web Intelligence (WI): Research Challenges and Trends in the New Information Age", N. Zhong, Y.Y. Yao, J. Liu, S. Ohsuga (eds.) *Web Intelligence: Research and Development*, LNAI 2198, Springer (2001) 1-17.
25. Y.Y. Yao and N. Zhong, Granular Computing Using Information Tables. In T.Y. Lin, Y.Y. Yao, L.A. Zadeh (eds.) *Data Mining, Rough Sets and Granular Computing*, Physica-Verlag (2002) 102-124.
26. L.A. Zadeh, "Precisiated Natural Language (PNL)", *AI Magazine*, 25(3) (Fall 2004) 74-91.

27. N. Zhong, J. Liu, Y.Y. Yao, and S. Ohsuga, "Web Intelligence (WI)", *Proc. 24th IEEE Computer Society International Computer Software and Applications Conference (COMPSAC 2000)*, IEEE Press (2000) 469-470.
28. N. Zhong, C. Liu, and S. Ohsuga, "Dynamically Organizing KDD Process", *International Journal of Pattern Recognition and Artificial Intelligence*, Vol. 15, No. 3, World Scientific (2001) 451-473.
29. N. Zhong, J. Liu, and Y.Y. Yao, "In Search of the Wisdom Web", *IEEE Computer*, 35(11) (2002) 27-31.
30. N. Zhong, "Representation and Construction of Ontologies for Web Intelligence", *International Journal of Foundations of Computer Science*, World Scientific, Vol. 13, No. 4 (2002) 555-570.
31. N. Zhong, J. Liu, and Y.Y. Yao (eds.) *Web Intelligence*, Springer, 2003.
32. N. Zhong, Y.Y. Yao, and M. Ohshima, "Peculiarity Oriented Multi-Database Mining", *IEEE Transaction on Knowlegde and Data Engineering*, Vol. 15, No. 4 (2003) 952-960.
33. N. Zhong, J.L. Wu, A. Nakamaru, M. Ohshima, and H. Mizuhara, "Peculiarity Oriented fMRI Brain Data Analysis for Studying Human Multi-Perception Mechanism", *Cognitive Systems Research*, 5(3), Elsevier (2004) 241-256.
34. N. Zhong, J. Hu, S. Motomura, J.L. Wu, and C. Liu, "Building a Data Mining Grid for Multiple Human Brain Data Analysis", *Computational Intelligence*, 21(2), Blackwell Publishing (2005) 177-196.
35. N. Zhong, "Impending Brain Informatics (BI) Research from Web Intelligence (WI) Perspective", International Journal of Information Technology and Decision Making, World Scientific, Vol. 5, No. 4 (2006) 713-727.
36. N. Zhong, J. Liu, and Y.Y. Yao, "Envisioning Intelligent Information Technologies (iIT) from the Stand-Point of Web Intelligence (WI)", *Communications of the ACM*, 50(3) (2007) 89-94.
37. The OGSA-DAI project: *http://www.ogsadai.org.uk/*.

Empowering Software Maintainers with Semantic Web Technologies

René Witte[1], Yonggang Zhang[2], and Jürgen Rilling[2]

[1] Institute for Progam Structures and
Data Organisation (IPD), Faculty of Informatics
University of Karlsruhe, Germany
witte@ipd.uka.de

[2] Department of Computer Science
and Software Engineering
Concordia University, Montreal, Canada
{rilling,yongg_zh}@cse.concordia.ca

Abstract. Software maintainers routinely have to deal with a multitude of artifacts, like source code or documents, which often end up disconnected, due to their different representations and the size and complexity of legacy systems. One of the main challenges in software maintenance is to establish and maintain the semantic connections among all the different artifacts. In this paper, we show how Semantic Web technologies can deliver a unified representation to explore, query and reason about a multitude of software artifacts. A novel feature is the automatic integration of two important types of software maintenance artifacts, source code and documents, by populating their corresponding sub-ontologies through code analysis and text mining. We demonstrate how the resulting "Software Semantic Web" can support typical maintenance tasks through ontology queries and Description Logic reasoning, such as security analysis, architectural evolution, and traceability recovery between code and documents.

Keywords: Software Maintenance, Ontology Population, Text Mining.

1 Introduction and Motivation

As software ages, the task of maintaining it becomes more complex and more expensive. Software maintenance, often also referred to as software evolution, constitutes a majority of the total cost occurring during the life span of a software system [15, 16]. Software maintenance is a multi-dimensional problem space that creates an ongoing challenge for both the research community and tool developers [8,14]. These maintenance challenges are caused by the different representations and interrelationships that exist among software artifacts and knowledge resources [17,18]. From a maintainer's perspective, exploring [11] and linking these artifacts and knowledge resources becomes a key challenge [1]. What is needed is a unified representation that allows a maintainer to explore, query and reason about these artifacts, while performing their maintenance tasks [13].

E. Franconi, M. Kifer, and W. May (Eds.): ESWC 2007, LNCS 4519, pp. 37–52, 2007.

In this research, we introduce a novel formal ontological representation that integrates two of the major software artifacts, source code and software documentation, thereby reducing the conceptual gap between these artifacts. Discovered concepts and concept instances from both source code and documents are used to explore and establish the links between these artifacts, providing maintainers with support during typical software maintenance tasks [2].

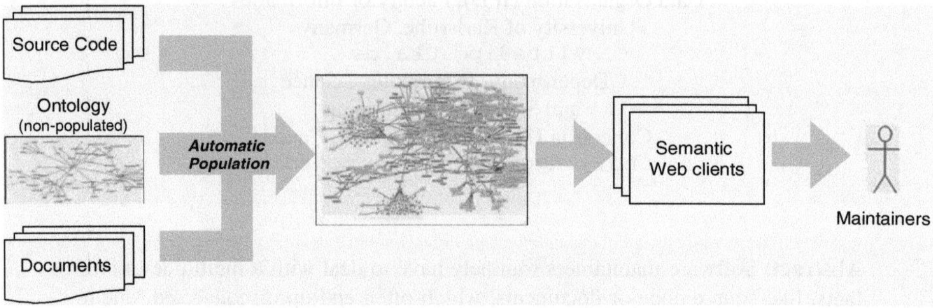

Fig. 1. Ontology-Based Software Maintenance Overview

A general overview of our approach is shown in Fig. 1. In a first step, the existing ontology is automatically populated from both the source code and documentation artifacts. In a second step, the resulting knowledge base is explored, queried and reasoned upon by utilizing various Semantic Web-enabled clients.

Our research is significant for several reasons: *(1)* The fact that we provide a novel approach to unify different software artifacts using a Semantic Web approach. *(2)* We developed fully automatic ontology population that allows us to take advantage of the large body of existing software artifacts, namely software documents and source code and the knowledge they contain. *(3)* We present concrete application examples, illustrating how our ontological representation can benefit software developers during typical maintenance tasks.

In Section 2, we discuss both challenges and requirements for a Semantic Web approach to software maintenance. Section 3 provides a general overview of our system. The design for our software and source code ontologies is discussed in Section 4. In Section 5, we describe in detail the fully automatic population of the ontologies, followed by concrete examples illustrating how our approach can benefit software engineers during typical maintenance tasks in Section 6.

2 Semantic Web and Software Maintenance

In a complex application domain like software maintenance, knowledge needs to be continually integrated from different sources (like source code repositories, documentation, test case results), different levels of scope (from single variables to complete system architectures), and across different relations (static, dynamic, etc.) [7, 9]. No single system is currently capable of supporting a complete domain like software engineering by itself. This makes it necessary to develop focused applications that can deal with individual aspects in a reliable manner, while still being able to integrate their results into a common knowledge base. Ontologies offer this capability: a large body of work exists that deals with ontology alignment and the development of upper level ontologies, while Description Logic (DL) reasoners can check the internal consistency of a knowledge base, ensuring at least some level of semantic integrity. However, before we can design Semantic Web support for software maintenance, we have to analyze the requirements particular to that domain.

2.1 Software Maintenance Challenges

With the ever increasing number of computers and their support for business processes, an estimated 250 billion lines of source code were being maintained in 2000, with that number rapidly increasing [16]. The relative cost of maintaining and managing the evolution of this large software base represents now more than 90% of the total cost [15] associated with a software product. One of the major challenges for the maintainers while performing a maintenance task is the need to comprehend a multitude of often disconnected artifacts created originally as part of the software development process [9]. These artifacts include, among others, source code and software documents (e.g., requirements, design documentation). From a maintainer's perspective, it becomes essential to establish and maintain the semantic connections among these artifacts.

In what follows, we introduce three typical use cases, which we will later revisit to illustrate the applicability of our approach in supporting software maintainers during these typical maintenance tasks.

Use case #1: *Identify security concerns in source code.* As discussed in [11], source code searching and browsing are two of the most common activities during the maintenance of existing software. With applications that become exposed to volatile environments with increased security risks (e.g., distributed environments, web-centric applications), identifying these security flaws in existing software systems becomes one of the major activities in the software maintenance phase.

Use case #2: *Concept location and traceability across different software artifacts.* From a maintainer's perspective, software documentation contains valuable information of both functional and non-functional requirements, as well as information related to the application domain. This knowledge often is difficult or impossible to extract only from source code [12]. It is a well known fact that even in organizations and projects with mature software development processes, software artifacts created as part of these processes end up to be disconnected from each other [1]. As a result, maintainers have to spend a large amount of time on synthesizing and integrating information from various information sources in order to re-establish the traceability links among these artifacts.

Use case #3: *Architectural recovery and restructuring.* With their increasing size and complexity, maintaining the overall structure of software systems becomes an emerging challenge of software maintenance. Maintainers need to comprehend the overall structure of a software system by identifying major components and their properties, as well as linking identified components with their lower-level implementation [14].

2.2 Identified Requirements

Based on the stated use cases, we can now derive requirements for Semantic Web support of software maintenance.

As a prerequisite, a sufficiently large part of the domain must be modeled in form of an ontology, including the structure and semantics of source code and documents to a level of detail that allows relevant queries and reasoning on the properties of existing artifacts (e.g., for security analysis).

Software maintenance intrinsically needs to deal with a large number of artifacts from legacy systems. It is not feasible to manually create instance information for existing source code or documents due to the large number of concept instances that exist in these artifacts. Thus, automatic ontology population methods must be provided for extracting semantic information from those artifacts.

The semantic information must be accessible through a software maintainer's desktop. Knowledge obtained through querying and reasoning should be integrated with existing development tools (e.g., *Eclipse*).

Finally, the acceptance of Semantic Web technologies by software maintainers is directly dependent on delivering added benefits, specifically improving on typical tasks, such as the ones described by the use cases above.

3 System Architecture and Implementation

In order to utilize the structural and semantic information in various software artifacts, we have developed an ontology-based program comprehension environment, which can automatically extract concept instances and their relations from source code and documents (Fig. 2).

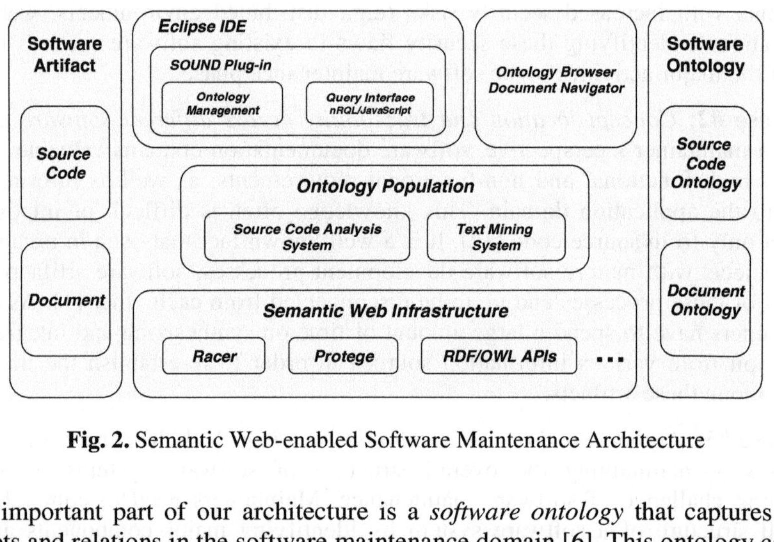

Fig. 2. Semantic Web-enabled Software Maintenance Architecture

An important part of our architecture is a *software ontology* that captures major concepts and relations in the software maintenance domain [6]. This ontology consists of two sub-ontologies: a *source code* and *document* ontology, which represent information extracted from source code and documents, respectively. The ontologies are modeled in OWL-DL[1] and were created using the Protégé-OWL extension of Protégé,[2] a free ontology editor.

Racer [5], an ontology inference engine, is adopted to provide reasoning services. The Racer system is a highly optimized DL system that supports reasoning about instances, which is particularly useful for the software maintance domain, where a large amount of instances needs to be handled efficiently.

Automatic ontology population is handled by two subsystems: The source code analysis, which is based on the JDT Java parser[3] provided by Eclipse[4]; and the document analysis, which is a text mining system based on the GATE (*General Architecture for Text Engineering)* framework [3].

The query interface of our system is a plug-in that provides OWL integration for Eclipse, a widely used software development platform. The expressive query language nRQL provided by Racer can be used to query and reason over the populated ontology. Additionally, we integrated a scripting language, which provides a set of built-in functions and classes using the JavaScript interpreter Rhino[5]. This language simplifies querying the ontology for software engineers not familiar with DL-based formalisms.

4 Ontology Design for Software Maintenance

Software artifacts, such as source code or documentation, typically contain knowledge that is rich in both structural and semantic information. Providing a uniform ontological representation for various software artifacts enables us to utilize semantic information conveyed by these artifacts and to establish their traceability links at the semantic level. In what follows, we discuss design issues for both the documentation and source code ontology used in our approach.

4.1 Source Code Ontology

The source code ontology has been designed to formally specify major concepts of object-oriented programming languages. In our implementation, this ontology is further extended with additional concepts and properties needed for some specific languages (in our case, Java). Examples for classes in the source code ontology are Package, Class, Method, or Variable. Our source code ontology is described in more detail in [20].

Within this sub-ontology, various ObjectProperties are defined to characterize the relationships among concepts. For example, two instances of SourceObject may have a definedIn relation indicating one is defined in the other; or an instance of method may read an instance of Field indicating the method may read the field in the body of the method.

---

[1] OWL Web Ontology Language Guide, W3C, http://www.w3.org/TR/owl-guide/

[2] Protégé ontology editor, http://protege.stanford.edu/

[3] Eclipse Java Development Tools (JDT), http://www.eclipse.org/jdt/

[4] Eclipse, http://www.eclipse.org

[5] Rhino JavaScript interpreter, http://www.mozilla.org/rhino/

Concepts in the source code ontology typically have a direct mapping to source code entities and can therefore be automatically populated through source code analysis (see Section 5.1).

4.2 Documentation Ontology

The documentation ontology consists of a large body of concepts that are expected to be discovered in software documents. These concepts are based on various programming domains, including programming languages, algorithms, data structures, and design decisions such as design patterns and software architectures.

Additionally, the software documentation sub-ontology has been specifically designed for automatic population through a text mining system by adapting the ontology design requirements outlined in [19] for the software engineering domain. Specifically, we included:

A *Text Model* to represent the structure of documents, e.g., classes for *sentences, paragraphs,* and *text positions,* as well as NLP-related concepts that are discovered during the analysis process, like *noun phrases* (NPs) and *coreference chains.* These are required for anchoring detected entities (populated instances) in their originating documents.

Lexical Information facilitating the detection of entities in documents, like the names of common design patterns, programming language-specific keywords, or architectural styles; and lexical normalization rules for entity normalization.

Relations between the classes, which include the ones modeled in the source code ontology. These allow us to automatically restrict NLP-detected relations to semantically valid ones (e.g., a relation like `<variable>` *implements* `<interface>`, which can result from parsing a grammatically ambiguous sentence, can be filtered out since it is not supported by the ontology).

Finally, *Source Code Entities* that have been automatically populated through source code analysis (cf. Section 5.1) can also be utilized for detecting corresponding entities in documents, as we describe in more detail in Sections 5.2.

5 Automatic Ontology Population

One of the major challenges for software maintainers is the large amount of information that has to be explored and analyzed as part of typical maintenance activities. Therefore, support for automatic ontology population is essential for the successful adoption of Semantic Web technology in software maintenance. In this section, we describe in detail the automatic population of our ontologies from existing artifacts: source code (Section 5.1) and documents (Section 5.2).

5.1 Populating the Source Code Ontology

The source code ontology population subsystem is based on JDT, which is a Java parser provided by Eclipse. JDT reads the source code and performs common tokenization and syntax analysis to produce an *Abstract Syntax Tree* (AST). Our population subsystem traverses the AST created by the JDT compiler to identify concept instances and their relations, which are then passed to an OWL generator for ontology population (Figure 3).

As an example, consider a single line of Java source code: *public int sort(){*, which declares a method called ***sort***. A simplified AST corresponding to this line of source code is shown in Fig. 3. We traverse this tree by first visiting the root node *Method Declaration*. At this step, the system understands that a *Method* instance shall be created. Next, the *Name Node* is visited to create the instance of the *Method* class, in this case ***sort***. Then the *Modifier Node* and *Type Node* are also visited, in order to establish the relations with the identified instance. As a result, two relations, ***sort*** *hasModifier* ***public*** and ***sort*** *hasType* ***int,*** are detected.

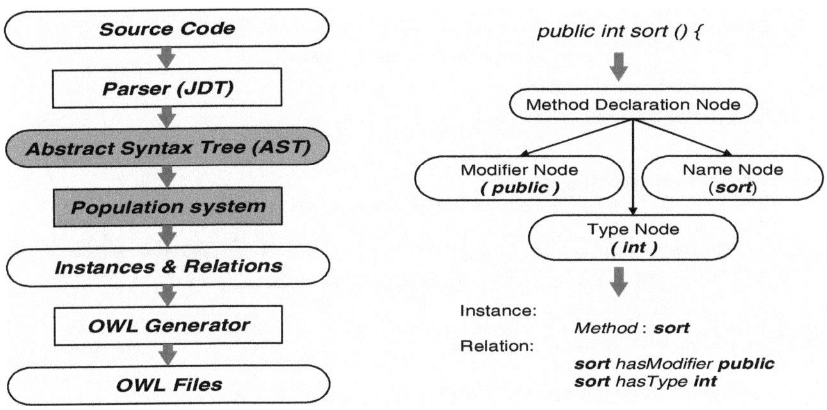

Fig. 3. Populating the source code ontology

The numbers of instances and relations identified by our system depend on the complexity of the ontology and the size of the source code to be analyzed. At the current stage of our research, the source code ontology contains 38 concepts (classes) and 41 types of relations (ObjectProperties). We have performed several case studies on different open source systems to evaluate the size of the populated ontology. Table 1 summaries the results of our case studies, with the size of the software system being measured by lines of code (LOC) and the process time reflecting both AST traversal and ontology population.

Table 1. Source code Ontology size for different open source projects

	LOC	Proc. Time	Instances	Relations	Inst./LOC	Rel./LOC
java.util	24k	13.62s	10140	47009	0.42	1.96
InfoGlue[6]	40k	27.61s	15942	77417	0.40	1.94
Debrief[7]	140k	67.12s	52406	244403	0.37	1.75
uDig[8]	177k	82.26s	69627	284692	0.39	1.61

[6] Infoglue, http://www.infoglue.org
[7] Debrief, http://www.debrief.info
[8] uDig, http://udig.refractions.net

5.2 Populating the Documentation Ontology

We developed a custom text mining system to extract knowledge from software documents and populate the corresponding sub-ontology. The processing pipeline and its connection with the software documentation sub-ontology is shown in Fig. 4. Note that, in addition to the software documentation ontology, the text mining system can also import the instantiated source code ontology corresponding to the document(s) under analysis.

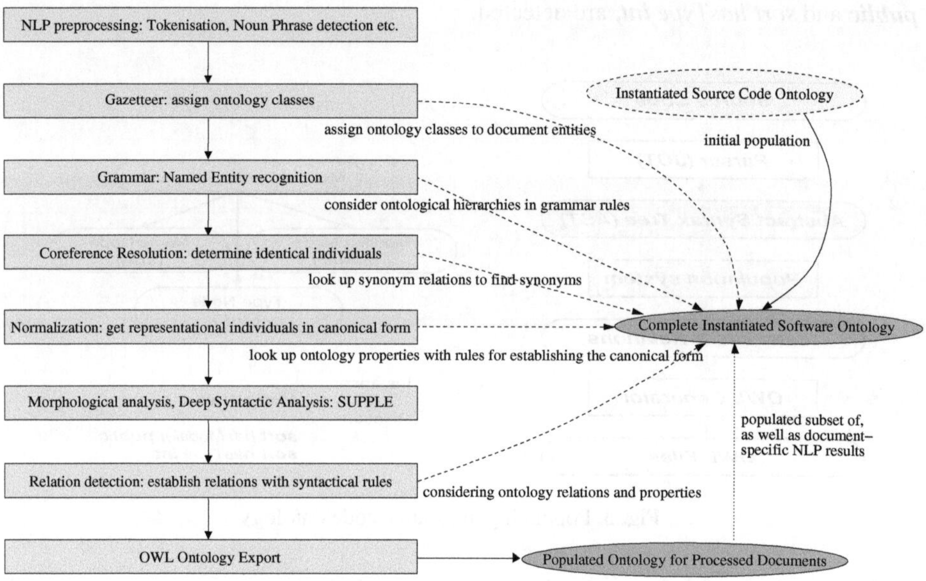

Fig. 4. Workflow of the Ontology-Driven Text Mining Subsystem

The system first performs a number of standard preprocessing steps, such as tokenisation, sentence splitting, part-of-speech tagging and noun phrase chunking.[9] Then, named entities (NEs) modeled in the software ontology are detected in a two-step process: Firstly, an *OntoGazetteer* is used to annotate tokens with the corresponding class or classes in the software ontology (e.g., the word "architecture" would be labeled with the *architecture* class in the ontology). Complex named entities are then detected in the second step using a cascade of finite-state transducers implementing custom grammar rules written in the JAPE language, which is part of GATE. These rules refer back to the annotations generated by the OntoGazetteer, and also evaluate the ontology. For example, in a comparison like `class=="Keyword"`, the ontological hierarchy is taken into account so that a `JavaKeyword` also matches, since a Java keyword *is-a* keyword in the ontology. This significantly reduces the overhead for grammar development and testing.

[9] For more details, please refer to the GATE documentation: http://gate.ac.uk/documentation/

The next major steps are the *normalization* of the detected entities and the resolution of *co-references*. Normalization computes a canonical name for each detected entity, which is important for automatic ontology population. In natural language texts, an entity like a method is typically referred to with a phrase like *"the myTestMethod provides..."*. Here, only the entity *myTestMethod* should become an instance of the *Method* class in the ontology. This is automatically achieved through lexical normalization rules, which are stored in the software ontology as well, together with their respective classes. Moreover, throughout a document a single entity is usually referred to with different textual descriptors, including pronominal references (like *"this method")*. In order to find these references and export only a single instance into the ontology that references all these occurrances, we perform an additional co-reference resolution step to detect both nominal and pronomial coreferences.

The next step is the detection of *relations* between the identified entities in order to compute predicate-argument structures, like *implements(class, interface)*. Here, we combine two different and largely complementary approaches: A deep syntactic analysis using the SUPPLE bottom-up parser and a number of pre-defined JAPE grammar rules, which are again stored in the ontology together with the relation concepts.

Finally, the text mining results are exported by populating the software documentation sub-ontology using a custom GATE component, the *OwlExporter*. The exported, populated ontology also contains document-specific information; for example, for each class instance the sentence it was found in is recorded. Figures 5 and 6 show excerpts of ontologies populated by our text mining system.

6 Application of Semantic Web-Enabled Software Maintenance

In what follows, we describe concrete application scenarios that correspond to the three use cases introduced earlier in Section 2.1.

6.1 Source Code Security Analysis

Existing techniques on detecting and correcting software security vulnerabilities at the source code level include human code reviews, testing, and static analysis. In the following example, we illustrate how our Semantic Web-based approach can facilitate security experts or programmers in identifying potential vulnerabilities caused by unexpected object accessibility.

In this scenario, a maintainer may consider allowing public and non-final fields in Java source code a security risk that may cause the value of the field being modified outside of the class where it was defined. In order to detect this, he can search the ontology through a query[10] that retrieves all `Field` instances that have a `PublicModifier` but no `FinalModifier`:

```
var SecurityConcern1 = new Query();                      // define a new query
SecurityConcern1.declare("F", "MP", "MF" );              // declare three query variables
SecurityConcern1.restrict("F", "Field");                 // variable F must be a Field instance
SecurityConcern1.restrict("MP", "PublicModifier");       // variable MP must be a PublicModifier instance
```

[10] In this and the following examples, we present ontology queries using our JavaScript-based query interface discussed in Section 3.

```
SecurityConcern1.restrict("MF", "FinalModifier");          // variable MF must be a FinalModifier instance
SecurityConcern1.restrict("F", "hasModifier", "MP");       // F and MP have a hasModifier relation
SecurityConcern1.no_relation("F", "hasModifier", "MF");    // F and MF have NO hasModifier relation
SecurityConcern1.retrieve("F");                            // this query only retrieve F
var result = ontology.query(SecurityConcern1);             // perform the query
```

In order to extend the query for more specific tasks, such as: *Retrieve all public data of Java package "user.pkg1" that may potentially be (read or write) accessed by a package "user.pkg2"*, the previous query can be further refined by adding:

```
SecurityConcern1.restrict("F", "definedIn", "user.pkg1");  // F must be definedIn user.pkg1
SecurityConcern1.restrict("M", "Method");                  // variable M must be a Method instance
SecurityConcern1.restrict("M", "definedIn", "user.pkg2");  // M must be definedIn user.pkg2
SecurityConcern1.restrict("M", "access", "F");             // M and F have an access relation
```

It should be noted that fields or methods in Java are defined in classes, and classes are defined in packages. The ontology reasoner will automatically determine the transitive relation definedIn between the concepts Field/Method and Package. In addition, read and write relations between method and field are modeled in our ontology by the readField and writeField ObjectProperties, which are a subPropertyOf access.

Many security flaws are preventable through security enforcement. Common vulnerabilities such as buffer overflows, accessing un-initialized variables, or leaving temporary files in the disk could be avoided by programmers with strong awareness of security concerns. In order to deliver more secure software, many development teams have guidelines for coding practice to enforce security. In our approach, we support maintainers and security experts during enforcement or validation, by checking whether these programming guidelines are followed. For example, to prevent access to un-initialized variables, a general guideline could be: *all fields must be initialized in the constructors.* The following query retrieves all classes that did not follow this specific constructor initialization guideline:

```
var SecurityConcern2 = new Query();                        // define a new query
SecurityConcern2.declare("F", "I", "C");                   // declare three query variables
SecurityConcern2.restrict("F", "Field");                   // variable F must be a Field instance
SecurityConcern2.restrict("I", "Constructor");             // variable I must be a Constructor instance
SecurityConcern2.restrict("C", "Class");                   // variable C must be a Class instance
SecurityConcern2.restrict("F", "definedIn", "C");          // F must be definedIn C
SecurityConcern2.restrict("I", "definedIn", "C");          // I must be also definedIn C
SecurityConcern2.no_relation("I", "writeField", "F");      // I and F have NO writeField relation
SecurityConcern2.retrieve("C", "I");                       // this query only retrieve C and I
var result = ontology.query(SecurityConcern2);             // perform the query
```

These two examples illustrate the power of our Semantic Web-enabled software maintenance approach: Complex queries can be performed on the populated ontology to identify specific patterns in the source code. Such types of queries utilize both the structural (e.g., definedIn) and semantic (e.g., writeField) knowledge of programming languages, which is typically ignored by traditional search tools based on string-matching, such as *grep*[11].

6.2 Establishing Traceability Links Between Source Code and Documentation

After instantiating both the source code and documentation sub-ontologies from their respective artifacts, it is now possible to automatically cross-link instances between

---

[11] Grep tool, http://www.gnu.org/software/grep/

these sub-ontologies. This allows maintainers to establish traceability links among the sub-ontologies through queries and reasoning, in order to find, for example, documentation corresponding to a source code entity, or to detect inconsistencies between information contained in natural language texts vs. the actual code.

For example, our source code analysis tool may identify c_1 and c_2 as classes; and this information can be used by the text mining system to identify named entities – c'_1 and c'_2 – and their associated information in the documents (Fig. 5). As a result, source code entities c_1 and c_2 can now be linked to their occurrences in the documents (c'_1 and c'_2). After source code and documentation ontology are linked, users can perform ontological queries on either documents or source code regarding properties of c_1 or c_2. For example, in order to retrieve document passages that describe both c_1 and c_2 or to retrieve design pattern descriptions referring to a pattern that contains the class currently being analyzed by a maintainer. Furthermore, it is also possible to identify inconsistencies – the documentation might list a method as belonging to a different class than it is actually implemented, for example – which are detected through the linking process and registered for further review by the user.

We performed an initial evaluation on a large open source Geographic Information System (GIS), uDig[12], which is implemented as a set of plug-ins on top of the Eclipse platform. The uDig documents used in the study consist of a set of JavaDoc files and a requirement analysis document.[13]

Links between the uDig implementation and its documentation are recovered by first performing source code analysis to populate the source code ontology. The resulted ontology contains instances of Class, Method, Field, etc., and their relations, such as inheritance and invocation. Our text mining system takes these identified class names, method names, and field names as an additional resource to populate the documentation ontology (cf. Fig. 4). Through this text mining process, a large number of Java language concept instances are discovered in the documents, as well as design-level concept instances such as design patterns or architecture styles. Ontology linking rules are then applied to link the populated documentation and source code ontologies.

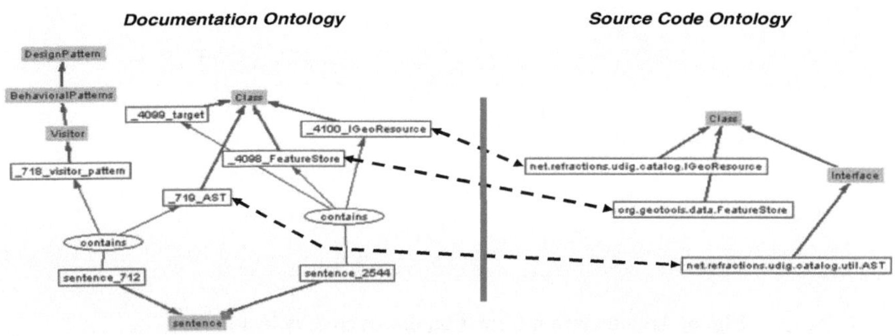

Fig. 5. Linked Source Code and Documentation Ontology

[12] uDIG open source GIS, http://udig.refractions.net/confluence/display/UDIG/Home
[13] uDig documentation, http://udig.refractions.net/docs/

A partial view of a linked ontology is shown in Figure 5; the corresponding sentences are:

Sentence_2544: *"For example if the class FeatureStore is the target class and the object that is clicked on is a IGeoResource that can resolve to a FeatureStore then a FeatureStore instance is passed to the operation, not the IGeoResource".*

Sentence_712: *"Use the visitor pattern to traverse the AST"*

Figure 5 shows that our text mining system was able to discover that *sentence_2544* contains both class instances *_4098_FeatureStore* and *_4100_IGeoResource*. Both of these classes can be linked to the instances in source code ontology, *org.geotools.data.FeatureStore* and *net.refractions.udig.catalog.IGeoResource*, respectively. Additionally, in *sentence_712*, a class instance (*_719_AST*) and a design pattern instance (*_718_visitor_pattern*) are also identified. Instance *_719_AST* is linked in a similar manner to the *net.refractions.udig.catalog.util.AST* interface in the source code ontology. Therefore, the recovery of traceability links between source code and documentation is feasible and implicit relations in the linked ontologies can be inferred.

6.3 Architectural Analysis

The populated ontology can also assist maintainers in performing more challenging tasks, such as analyzing the overall structure of a software system, i.e., architectural analysis. In this case study, we analyzed the architecture of the open source web site content management system, *InfoGlue*[14]. The first step of an architectural analysis is wtypically to identify potential architectural styles [7] and candidate components in the system. By browsing the documentation ontology populated through text mining, we observe that a large number of instances of concept `Layer` are discovered. This information provides us with significant clues that the InfoGlue system might be implemented using a typical *Layered Architecture* [7]. Additionally, the text mining discovered that the application layer contains a set of action classes, as shown in Fig. 6. This information provides important references for our further analysis of the documents and source code.

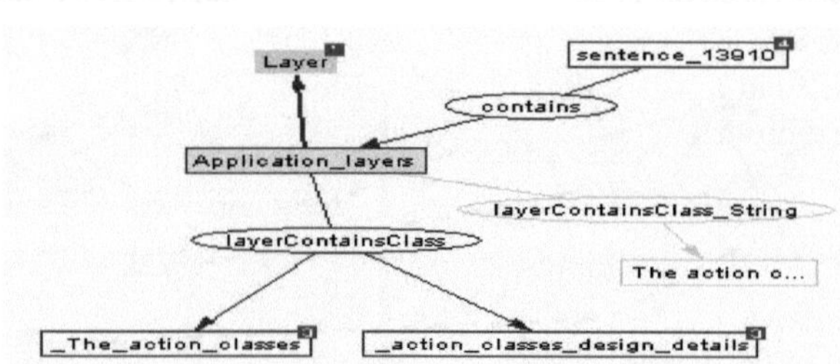

Fig. 6. Architecture information discovered by text mining

We later determined that the action classes refer to classes that implement *webwork.action.Action* interface. Before conducting the analysis, we hypothesized

[14] InfoGlue Open Source Content Management Platform, http://www.infoglue.org/

that the InfoGlue system implements a common layered architecture, in which each layer only communicates with its upper or lower layer. In order to validate our hypothesis, we performed a number of queries on the populated source code ontology to retrieve method calls between layers.

The script first retrieves all layer instances in the ontology, and then iteratively queries method call relations between layers. A similar query is performed to retrieve the number of methods being called.

```
var layers = ontology.retrieve_instance("Layer");

for(var i = 0; i < layers.size(); i++){
  var layer1 = layers.get("Layer", i);
  for(var j = 0; j < layers.size(); j++){
    var layer2 = layers.get("Layer", j);
    if(layer1.equals(layer2)) continue;
    var query = new Query();
    query.declare("M1", "M2");
    query.restrict("M1", "Method");
    query.restrict("M2", "Method");
    query.restrict("M1", "definedIn", layer1);
    query.restrict("M2", "definedIn", layer2);
    query.restrict("M1", "call", "M2");
    query.retrieve("M1", "M2");
    var result = ontology.query(query);
    out.println(layer1 + " calls " + layer2 + " " + result.size() + " times.");
  }
}
```

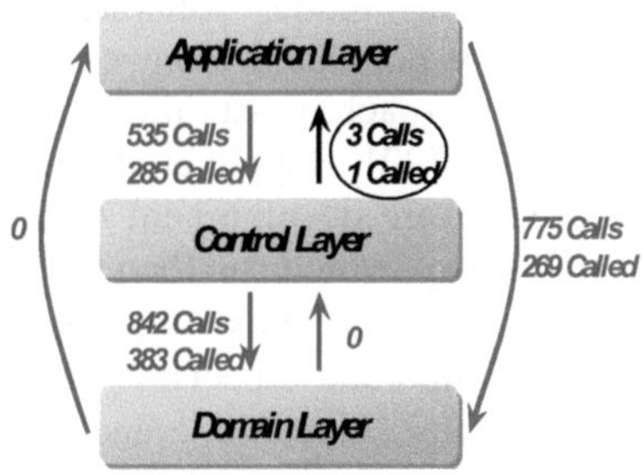

Fig. 7. Example script to detect method calls between layers (left) and results obtained from executing the query on the populated ontology (right)

Fig. 7 summarizes the results of these two queries, by showing both the number of method calls and the number of methods being called. From the analysis of the result one can refute the original hypothesis about the implementation of the common layered architecture. This is due to the fact that one can observe in the InfolGlue system a significant amount of communications from the application layer to domain layer – skipping the control layer. This information is valuable for software maintainers, because it indicates that any changes made in the domain layer may also directly affect the application layer, a situation which one would not expect based on the architectural description found in the InfoGlue system documentation.

In addition, we observed that there is no communication from the domain layer to the control and application layer, i.e., the domain layer can be substituted by other components matching the same interface. This observation also reveals an important property of the domain layer in the InfoGlue system – the domain layer is a self-contained component that can be reused by other applications. Our observation is also supported by the architecture document itself, which clearly states that "*the domain business logic should reside in the domain objects themselves making them self contained and reusable*".

Moreover, by analyzing these results, one would expect that a lower layer should not communicate with its upper layer. The three method calls from the control layer to the application layer can therefore be considered as either implementation defects or as the result of a special design intention not documented. Our further inspection showed that the method being called is a utility method that is used to format HTML content. We consider this to be an implementation defect since the method can be re-implemented in the control layer to maintain the integrity of a common layered architecture style.

7 Related Work and Discussions

Existing research on applying Semantic Web techniques in software maintenance mainly focuses on providing ontological representation for particular software artifacts or supporting specific maintenance task [10]. In [21], Ankolekar et al. provide an ontology to model software, developers, and bugs. This ontology is semi-automatically populated from existing artifacts, such as software interface, emails, etc. Their approach assists the communication between software developers for bug resolution. In [22], Happle et al. present an approach addressing the component reuse issue of software development by storing descriptions of components in a Semantic Web repository, which can then be queried for existing components.

Comparing with the existing approaches, like the LaSSIE system [4], our work differs in two important aspects: *(1)* the automatic population from existing software artifacts, especially source code and its documentation, which are both very different in structure and semantics; and *(2)* the application of queries on the populated ontologies, including DL reasoning, to enhance concrete tasks performed by software maintainers. The first aspect is an important prerequisite to bring a large amount of existing data into the "Software Semantic Web". The inclusion of semantically different and complementary artifacts, in the form of machine-readable code and natural language, provides for real improvement in software maintenance, enabling for the first time an automatic connection between code and its documentation. The second aspect shows the power of DL-based reasoning when applied to the software domain, significantly enhancing the power of conventional software development tools.

8 Conclusions and Future Work

In this paper, we presented a novel approach that provides formal ontological representations of the software domain for both source code and document artifacts. The ontologies capture structural and semantic information conveyed in these artifacts, and therefore allow us to link, query and reason across different software artifacts on a semantic level.

In this research, we address important issues for both the Semantic Web and the software maintenance communities. For the Semantic Web community, we illustrate how the use of the semantic technologies can be extended to the software maintenance domain. Furthermore, we demonstrate how the large body of existing knowledge found in source code and software documentation can be made available through automatic ontology population on the Semantic Web.

From a software maintenance perspective, we illustrate through three concrete use cases how the Semantic Web and its underlying technologies can benefit and support maintainers during typical maintenance tasks.

In future versions, more work is needed on enhancing existing software development tools with Semantic Web capabilities, some of which is addressed in the *Semantic Desktop* community. Many of the ideas presented here obviously also apply to other areas in software engineering besides maintenance; we have also been investigating ontology-enabled software comprehension processes [13], which will complement and further enhance the utility of our "Software Semantic Web" approach.

References

1. G. Antoniol, G. Canfora, G. Casazza, and A. De Lucia, "Information Retrieval Models for Recovering Traceability Links between Code and Documentation". In Proc. of IEEE International Conference on Software Maintenance, San Jose, CA, 2000.
2. R. Brooks, "Towards a Theory of the Comprehension of Computer Programs". International Journal of Man-Machine Studies, pp. 543-554, 1963.
3. H. Cunningham, D. Maynard, K. Bontcheva, V. Tablan. "GATE: A Framework and Graphical Development Environment for Robust NLP Tools and Applications." In Proc. of the 40th Anniversary Meeting of the ACL. Philadelphia, July 2002.
4. P. Devanbu, R.J. Brachman, P.G. Selfridge, and B.W. Ballard, "LaSSIE - a Knowledge-based Software Information System", Comm. of the ACM, 34(5), pp. 36–49, 1991.
5. V. Haarslev and R. Möller, "RACER System Description", In Proc. of International Joint Conference on Automated Reasoning, Siena, Italy, 2002.
6. P. N. Johnson-Laird, "Mental Models: Towards a Cognitive Science of Language, Inference and Consciousness". Harvard University, Cambridge, MI, 1983.
7. A. V. Mayhauser, A. M. Vans, "Program Comprehension during Software Maintenance and Evolution". IEEE Computer, 28(8), pp. 44-55, August, 1995.
8. IEEE Standard for Software Maintenance, IEEE 1219-1998.
9. D. Jin and J. Cordy. "Ontology-Based Software Analysis and Reengineering Tool Integration: The OASIS Service-Sharing Methodology". In Proc. of the 21st IEEE International Conference on Software Maintenance, Budapest, Hungary, 2005.
10. H.-J. Happel, S. Seedorf, "Applications of Ontologies in Software Engineering", In Proc. of International Workshop on Semantic Web Enabled Software Engineering, 2006.
11. T.C. Lethbridge and A. Nicholas, "Architecture of a Source Code Exploration Tool: A Software Engineering Case Study", Department of Computer Science, University of Ottawa, Technical Report, TR-97-07, 1997.
12. M. Lindvall and K. Sandahl, "How well do experienced software developers predict software change?" Journal of Systems and Software, 43(1), pp. 19-27, 1998.
13. W. Meng, J. Rilling, Y. Zhang, R. Witte, P. Charland, "An Ontological Software Comprehension Process Model", In Proc. of the 3rd International Workshop on Metamodels, Schemas, Grammars, and Ontologies for Reverse Engineering, Italy, 2006.
14. C. Riva, "Reverse Architecting: An Industrial Experience Report", In Proc. of the 7th IEEE Working Conference on Reverse Engineering, Australia, 2000.
15. R. Seacord, D. Plakosh, and G. Lewis, "Modernizing Legacy Systems: Software Technologies, Engineering Processes, and Business Practices", Addison-Wesley, 2003.
16. I. Sommerville, "Software Engineering (6th Edition)", Addison-Wesley, 2003.
17. M.A. Storey, S.E. Sim, and K. Wong, "A Collaborative Demonstration of Reverse Engineering tools", ACM SIGAPP Applied Computing Review, Vol. 10(1), pp18-25, 2002.
18. C. Welty, "Augmenting Abstract Syntax Trees for Program Understanding", In Proc. of International Conference on Automated Software Engineering, 1997.

19. R. Witte, T. Kappler, C. Baker, "Ontology Design for Biomedical Text Mining", Chapter 13 in Semantic Web: Revolutionizing Knowledge Discovery in the Life Sciences, Springer Verlag, 2006.
20. Y. Zhang, R. Witte, J. Rilling, V. Haarslev, "An Ontology-based Approach for Traceability Recovery", In Proc. of International Workshop on Metamodels, Schemas, Grammars, and Ontologies for Reverse Engineering, Genoa, Italy, 2006.
21. A. Ankolekar, K. Sycara, J. Herbsleb, R. Kraut, C. Welty, "Supporting Online Problem-solving Communities With the Semantic Web", In Proc. of the 15th International Conference on World Wide Web, 2006.
22. H.-J. Happel, A. Korthaus, S. Seedorf, P. Tomczyk, "KontoR: An Ontology-enabled Approach to Software Reuse", In Proc. of the 18th International Conference on Software Engineering and Knowledge Engineering, 2006.

Minimal Deductive Systems for RDF

Sergio Muñoz[1], Jorge Pérez[2,3], and Claudio Gutierrez[4]

[1] Universidad Católica de la Santísima Concepción, Chile
[2] Pontificia Universidad Católica de Chile
[3] Universidad de Talca, Chile
[4] Universidad de Chile

Abstract. This paper presents a minimalist program for RDF, by showing how one can do without several predicates and keywords of the RDF Schema vocabulary, obtaining a simpler language which preserves the original semantics. This approach is beneficial in at least two directions: (a) To have a simple abstract fragment of RDFS easy to formalize and to reason about, which captures the essence of RDFS; (b) To obtain algorithmic properties of deduction and optimizations that are relevant for particular fragments. Among our results are: the identification of a simple fragment of RDFS; the proof that it encompasses the main features of RDFS; a formal semantics and a deductive system for it; sound and complete deductive systems for their sub-fragments; and an $\mathcal{O}(n \log n)$ complexity bound for ground entailment in this fragment.

1 Introduction

The Resource Description Framework (RDF) is the W3C standard for representing information in the Web [17]. The motivation behind the development of RDF by the W3C was, as Tim Berners-Lee pointed out for the Semantic Web, to have a common and minimal language to enable to map large quantities of existing data onto it so that the data can be analyzed in ways never dreamed of by its creators [2]. If one would like to bring to reality this vision, the processing of RDF data at big scale must be viable. The very future of RDF data deployment and their use will depend critically on the complexity of processing it.

Efficient processing of any kind of data relies on a compromise between two parameters, namely, the size of the data and the expressiveness of the language describing it. As we already pointed out, in the RDF case the size of the data to be processed will be enormous, as examples like Wordnet [12], FOAF [3] and Gene Ontology [19] show. Hence, a program to make RDF processing scalable has to consider necessarily the issue of the expressiveness of RDF. Due to the well known fact that the complexity of entailment using RDF data in its full expressiveness is an untractable problem [7,8,4], such a program amounts essentially to look for fragments of RDF with good behavior w.r.t. complexity of processing. This is the broad goal of the present paper.

The full specification of RDF (that is, including RDFS vocabulary) and their fragments has not yet been studied in detail. Its description is given in [16] and its

E. Franconi, M. Kifer, and W. May (Eds.): ESWC 2007, LNCS 4519, pp. 53–67, 2007.

semantics is defined in [15]. The first observation that arises when dealing with RDFS vocabulary is the difficulty to work with it. An example of this fact is that even the rules of deduction presented in the official RDF Semantics specification are not complete [10,8]. A second empirical observation is that several parts of the RDFS vocabulary have been depreciated, and practice shows that there are others that are hardly used or not being used at all. This makes it very hard for developers to build and optimize sound implementations and algorithms, and for theoreticians to work on this specification.

In order to illustrate the above issues, let us consider two well known RDFS specifications: *WordNet* [12] and *Friend of a Friend* (FOAF) [3]. Both schemas use only a proper subset of the RDFS vocabulary. FOAF schema has no blank nodes. Additionally, there is a point about the real need of explicitly declaring classes via `rdfs:Class`: In both specifications the triples where `rdfs:Class` occurs are redundant (i.e. can be deduced from the rest of the data). Something similar happens with terms defined as properties (`rdf:Property`). Why use all the weight of the full RDFS specification in these cases? Another example where these type of issues will arise, is the SPARQL query language specification [11], which currently does not support RDFS entailment. There is wide agreement that more expressive vocabularies must be treated orthogonally to the rest of the SPARQL features. In practice, each query will use just a small fragment of the RDFS vocabulary. For reasoning and optimization purposes, it would be useful to have a sound and complete theory of each such fragment which preserves the semantics of RDFS.

Among the most important directions of a program to develop solutions to the above mentioned problems are:

- To identify a fragment which encompasses the essential features of RDF, which preserves the original semantics, be easy to formalize and can serve to prove results about its properties.
- To study in detail the semantics of different fragments of RDF, and give sound and complete deductive system for each of them.
- To study the complexity of entailment for the vocabulary in general and in these fragments in particular, and to develop algorithms for testing entailment.

As for the first point, in this paper we identify a fragment of RDFS that covers the crucial vocabulary of RDFS, prove that it preserves the original RDF semantics, and avoids vocabulary and axiomatic information that only serves to reason about the structure of the language itself and not about the data it describes. We lift this structural information into the semantics of the language, hiding them from developers and users.

Regarding the second point, we study thoroughly all fragments of the core fragment showing that they retain the original RDFS semantics. We then study the lattice of the theories induced by these fragments, developing minimal sound and complete proof systems for them. We also calculate what are the minimal sub-theories that should be considered when reasoning with restricted vocabulary.

Finally, regarding the point of complexity of entailment, there are two main aspects of RDF to consider: the built-in vocabulary and the notion of blank nodes. For the complexity of entailment considering blank nodes, good (polynomial) cases can be derived from well known databases and constraint–satisfaction results [4,9,5]. These cases consider special forms of interaction between blank nodes that are very common in practice. On this regard, we prove that there is a notion of normalized proof for RDFS entailment which permits to treat the issue of blank nodes entailment in a way orthogonal to the treatment of RDFS vocabulary. Using this notion, results for blank nodes can be composed modularly with particular results for ground RDFS fragments, that is, not considering blank nodes semantics.

For the the ground case, from a database point of view, even current known bounds seems totally impractical. For example, the naive approach would use closure, and estimates for the size of the closure are high: we show that in the fragment presented, it is quadratic. Nevertheless, this bound is still impractical from a database point of view. On these lines, we prove that entailment can be done in time $\mathcal{O}(n \log n)$ in the worst case, where n is the size of the source data.

The paper is organized as follows. Section 2 presents standard RDF and its semantics and discusses the vocabulary design to conclude with a proposal of core fragment, called ρdf. Section 3 studies the ρdf fragment. Section 4 presents the lattice of minimal fragments of ρdf and their deductive systems. Section 5 studies complexity of entailment in the ρdf fragment. Finally, Section 6 presents the conclusion.

2 RDF Semantics

Assume there are pairwise disjoint infinite sets $\mathbf{U}$ (RDF URI references), $\mathbf{B}$ (Blank nodes), and $\mathbf{L}$ (Literals). Through the paper we assume $\mathbf{U}$, $\mathbf{B}$, and $\mathbf{L}$ fixed, and for simplicity we will denote unions of these sets simply concatenating their names. A tuple $(s,p,o) \in \mathbf{UBL} \times \mathbf{U} \times \mathbf{UBL}$ is called an *RDF triple*. In this tuple, s is the *subject*, p the *predicate*, and o the *object*. Note that –following recent developments [6,11]– we are omitting the old restriction stating that literals cannot be in subject position.

Definition 1. *An* RDF graph *(or simply a graph) is a set of RDF triples. A subgraph is a subset of a graph. The* universe *of a graph G, denoted by* universe(G) *is the set of elements in* **UBL** *that occur in the triples of G. The vocabulary of G, denoted by* voc(G) *is the set* universe(G) $\cap$ **UL**. *A graph is* ground *if it has no blank nodes. In general we will use uppercase letters $N, X, Y, \ldots$ to denote blank nodes.*

In what follows we will need some technical notions. A map is a function $\mu : \mathbf{UBL} \to \mathbf{UBL}$ preserving URIs and literals, i.e., $\mu(u) = u$ for all $u \in \mathbf{UL}$. Given a graph G, we define $\mu(G)$ as the set of all $(\mu(s), \mu(p), \mu(o))$ such that $(s,p,o) \in G$. We will overload the meaning of map and speak of a map μ *from* G_1 *to* G_2, and write $\mu : G_1 \to G_2$, if the map μ is such that $\mu(G_1)$ is a subgraph of G_2.

2.1 Interpretations

The normative semantics for RDF graphs given in [15], and the mathematical formalization in [10] follows standard classical treatment in logic with the notions of model, interpretation, entailment, and so on. In those works the RDFS theory is built incrementally from Simple, to RDF, to RDFS interpretations (or structures) and models for graphs. We present here a single notion of interpretation which summarizes Simple, RDF, and RDFS interpretations in one step, and which will be used later to define the semantics of our fragment.

Definition 2. *An interpretation over a vocabulary V is a tuple*

$$\mathcal{I} = (Res, Prop, Class, Ext, CExt, Lit, Int)$$

such that: (1) Res is a nonempty set of resources, *called the* domain *or* universe *of $\mathcal{I}$; (2) $Prop$ is a set of property names (not necessarily disjoint from Res); (3) $Class \subseteq Res$ is a distinguished subset of Res identifying if a resource denotes a class of resources; (4) $Ext : Prop \rightarrow 2^{Res \times Res}$, a mapping that assigns an* extension *to each property name; (5) $CExt : Class \rightarrow 2^{Res}$ a mapping that assigns a set of resources to every resource denoting a class; (6) $Lit \subseteq Res$ the set of literal values, Lit contains all plain literals in $\mathbf{L} \cap V$; (7) $Int : \mathbf{UL} \cap V \rightarrow Res \cup Prop$, the* interpretation mapping, *a mapping that assigns a resource or a property name to each element of $\mathbf{UL}$ in V, and such that Int is the identity for plain literals and assigns an element in Res to elements in $\mathbf{L}$.*

In [15,10] the notion entailment is defined using the idea of *satisfaction* of a graph under certain interpretation. Intuitively a ground triple (s, p, o) in an RDF graph G will be true under the interpretation $\mathcal{I}$ if p is *interpreted* as a property name, s and o are *interpreted* as resources, and the interpretation of the pair (s, o) belongs to the extension of the property assigned to p.

In RDF, blank nodes work as existential variables. Intuitively the triple (X, p, o) with $X \in \mathbf{B}$ would be true under $\mathcal{I}$ if there exists a resource s such that (s, p, o) is true under $\mathcal{I}$. When interpreting blank nodes, an arbitrary resource can be chosen, taking into account that the same blank node must always be interpreted as the same resource. To formally deal with blank nodes, extensions of the interpretation map Int are used in the following way. Let $A : \mathbf{B} \rightarrow Res$ be a function from blank nodes to resources; we denote Int_A the extension of Int to domain $\mathbf{B}$ defined by $Int_A(X) = A(X)$ when $X \in \mathbf{B}$. The function A captures the idea of existentiality.

The formal definition of model and entailment for RDFS in [15,10] relies on a set of *semantics restrictions* imposed to interpretations in order to model the vocabulary, and the *a priori* satisfaction of a set of *axiomatic triples.* We refer the reader to Appendix A for a complete formal definition of the semantics of RDFS using the notion of interpretation defined here.

2.2 RDFS Vocabulary

The RDF specification includes a set of reserved words, the RDFS vocabulary (RDF Schema [16]) designed to describe relationships between resources as well as to describe properties like attributes of resources (traditional attribute-value pairs). Table 1 (Appendix A) shows the full RDFS vocabulary as it appears in [15], and (in brackets) the shortcuts that we will use in this paper. This vocabulary has a special interpretation (see Definition 6 in Appendix A).

Roughly speaking, this vocabulary can be divided conceptually in the following groups:

(a) a set of properties rdfs:subPropertyOf [`sp`], rdfs:subClassOf [`sc`], rdfs:domain [`dom`], rdfs:range [`range`] and rdf:type [`type`].
(b) a set of classes, rdfs: Resource, rdfs:Class, rdf:Property, rdf:XMLLiteral, rdfs:Literal, rdfs:Datatype.
(c) Other functionalities, like a system of classes and properties to describe lists: rdfs:Container, rdfs:ContainerMembershipProperty, rdfs:member, rdf:List, rdf:Alt, rdf:Bag, rdf:Seq, rdf:first, rdf:rest, rdf:nil, rdf:_1, rdf:_2, ..., and a systems for doing reification: a class rdf:Statement together with properties rdf:subject, rdf:predicate, rdf:object.
(d) Utility vocabulary, like rdfs:seeAlso, rdfs:isDefinedBy, rdfs:comment, rdf:value, rdfs:label.

The groups in (b), (c) and (d) have a very light semantics, essentially describing its internal function in the ontological design of the system of classes of RDFS. Their semantics is defined by "axiomatic triples" [15]which are relationships among these reserved words. Note that all axiomatic triples are "structural", in the sense that do not refer to external data, but talk about themselves. Much of this semantics correspond to what in standard languages is captured via typing. From a theoretical and practical point of view it is inconvenient to expose it to users of the language because it makes the language more difficult to understand and use, and for the criteria of simplicity in the design of the language.

On the contrary, the group (a) is formed by predicates whose intended meaning is non-trivial and is designed to relate individual pieces of data external to the vocabulary of the language. Their semantics is defined by rules which involve variables (to be instantiated by real data). For example, rdfs:subClassOf[`sc`] is a binary property reflexive and transitive; when combined with rdf:type[`type`] specify that the type of an individual (a class) can be lifted to that of a superclass. This group (a) forms the core of the RDF language developers use, as practice is showing.

For all the above considerations, it is that group (a) forms a natural fragment of RDFS to be studied in depth. Section 3 is devoted to study this fragment, and our results will show that there are theoretical reasons that support the convenience of this choice.

3 The ρdf Fragment of RDFS

Define ρdf (read rho-df, the ρ from *restricted* rdf) to be the following subset of the RDFS vocabulary:

$$\rho\text{df} = \{\texttt{sp}, \texttt{sc}, \texttt{type}, \texttt{dom}, \texttt{range}\}.$$

Definition 3. *Let G be a graph over* ρdf. *An interpretation $\mathcal{I}$ is a model of G under* ρdf, *denoted $\mathcal{I} \models_{\rho\mathrm{df}} G$, iff $\mathcal{I}$ is an interpretation over* $\rho\mathrm{df} \cup \mathrm{universe}(G)$ *that satisfies the following conditions:*

1. *Simple:*
   (a) *there exists a function $A : \mathbf{B} \to Res$ such that for each $(s, p, o) \in G$, $Int(p) \in Prop$ and $(Int_A(s), Int_A(o)) \in Ext(Int(p))$, where Int_A is the extension of Int using A.*
2. *Subproperty:*
   (a) *$Ext(Int(\mathtt{sp}))$ is transitive and reflexive over $Prop$*
   (b) *if $(x, y) \in Ext(Int(\mathtt{sp}))$ then $x, y \in Prop$ and $Ext(x) \subseteq Ext(y)$*
3. *Subclass:*
   (a) *$Ext(Int(\mathtt{sc}))$ is transitive and reflexive over $Class$*
   (b) *if $(x, y) \in Ext(Int(\mathtt{sc}))$ then $x, y \in Class$ and $CExt(x) \subseteq CExt(y)$*
4. *Typing I:*
   (a) *$x \in CExt(y) \Leftrightarrow (x, y) \in Ext(Int(\mathtt{type}))$*
   (b) *if $(x, y) \in Ext(Int(\mathtt{dom}))$ and $(u, v) \in Ext(x)$ then $u \in CExt(y)$*
   (c) *if $(x, y) \in Ext(Int(\mathtt{range}))$ and $(u, v) \in Ext(x)$ then $v \in CExt(y)$*
5. *Typing II:*
   (a) *For each $\mathtt{e} \in \rho\mathrm{df}$, $Int(\mathtt{e}) \in Prop$.*
   (b) *if $(x, y) \in Ext(Int(\mathtt{dom}))$ then $x \in Prop$ and $y \in Class$.*
   (c) *if $(x, y) \in Ext(Int(\mathtt{range}))$ then $x \in Prop$ and $y \in Class$.*
   (d) *if $(x, y) \in Ext(Int(\mathtt{type}))$ then $y \in Class$.*

We define G entails H under ρdf, *denoted $G \models_{\rho\mathrm{df}} H$, iff every model under* ρdf *of G is also a model under* ρdf *of H.*

Note that in ρdf–models we do not impose the *a priori* satisfaction of any axiomatic triple. Indeed, ρdf–models does not satisfy any of the RDF/S axiomatic triples in [15,10], because all of them mention RDFS vocabulary outside ρdf. This is also the reason for the inclusion of conditions 5 in ρdf models that capture the semantics restrictions imposed syntactically by the RDF/S axiomatic triples (`dom`, `dom`, `prop`), (`dom`, `range`, `class`), (`range`, `dom`, `prop`), (`range`, `range`, `class`), and (`type`, `range`, `class`), and the fact that every element in ρdf must be interpreted as a property.

The next theorem shows that this definition retains the original semantics for the ρdf vocabulary:

Theorem 1. *Let $\models$ be the RDFS entailment defined in [15,10], and let G and H be RDF graphs that do not mention RDFS vocabulary outside* ρdf. *Then*

$$G \models H \text{ iff } G \models_{\rho\mathrm{df}} H.$$

The issue of reflexivity. There are still some details to be refined in the theory of ρdf. Note that, although in ρdf–models we do not impose the *a priori* satisfaction of any triple, there are triples that are entailed by all graphs, for example the triples $(\mathtt{sp}, \mathtt{sp}, \mathtt{sp})$, $(\mathtt{sc}, \mathtt{sp}, \mathtt{sc})$, $(\mathtt{type}, \mathtt{sp}, \mathtt{type})$, $(\mathtt{dom}, \mathtt{sp}, \mathtt{dom})$, and $(\mathtt{range}, \mathtt{sp}, \mathtt{range})$. These triples are true under every ρdf model due to the fact that sp must be interpreted as a reflexive relation. Also, because blank nodes work as existential variables, the triples above with the subject or the object replaced by any blank node, are also true in every ρdf–model. The good news is that these are the only triples in the ρdf fragment that are satisfied by every model:

Proposition 1. *Let t be an RDF triple such that $\models_{\rho\text{df}} t$. Then, either $t \in \{(\mathtt{sp}, \mathtt{sp}, \mathtt{sp}), (\mathtt{sc}, \mathtt{sp}, \mathtt{sc}), (\mathtt{type}, \mathtt{sp}, \mathtt{type}), (\mathtt{dom}, \mathtt{sp}, \mathtt{dom}), (\mathtt{range}, \mathtt{sp}, \mathtt{range})\}$, or t is obtained from these triples replacing the subject or object by a blank node.*

This is part of a more general phenomena, namely the presence of reflexivity for sp and sc. We will show that reflexivity for sp and sc is orthogonal with the rest of the semantics.

Definition 4 (Semantics without reflexivity of sp and sc). *An interpretation $\mathcal{I}$ is a* reflexive–relaxed *model under* ρdf *of a graph G, written $\mathcal{I} \models^{\text{nrx}}_{\rho\text{df}} G$, iff $\mathcal{I}$ is a* ρdf *model that does not necessarily satisfy the restrictions stating that $Ext(Int(\mathtt{sp}))$ and $Ext(Int(\mathtt{sc}))$ are reflexive relations over $Prop$ and $Class$ respectively.*

Theorem 2. *Let G and H be* ρdf *graphs. Assume that H does not contain triples of the form $(x, \mathtt{sp}, x)$ nor $(x, \mathtt{sc}, x)$ for $x, y \in \mathbf{UL}$, nor triples of the form $(X, \mathtt{sp}, Y)$ nor $(X, \mathtt{sc}, Y)$ for $X \in \mathbf{B}$ or $Y \in \mathbf{B}$. Then,*

$$G \models_{\rho\text{df}} H \text{ iff } G \models^{\text{nrx}}_{\rho\text{df}} H.$$

Essentially the above theorem states that the only use of reflexive restrictions in RDFS models is the entailment of triples of the form $(x, \mathtt{sp}, x)$, $(x, \mathtt{sc}, x)$, or their existential versions replacing the subject or object by blank nodes. Another property of $\models^{\text{nrx}}_{\rho\text{df}}$ is that it does not entail axiomatic triples:

Corollary 1. *There is no triple t such that $\models^{\text{nrx}}_{\rho\text{df}} t$.*

3.1 Deductive System for ρdf Vocabulary

In what follows, we present a sound and complete deductive system for the fragment of RDF presented in the previous section. The system is arranged in groups of rules that captures the semantic conditions of models. In every rule, $\mathcal{A}, \mathcal{B}, \mathcal{C}, \mathcal{X}$, and $\mathcal{Y}$ are meta-variables representing elements in **UBL**.

1. *Simple:*

(a) $\frac{G}{G'}$ *for a map* $\mu : G' \to G$ (b) $\frac{G}{G'}$ *for* $G' \subseteq G$

2. *Subproperty:*

(a) $\frac{(\mathcal{A},\texttt{sp},\mathcal{B})\ (\mathcal{B},\texttt{sp},\mathcal{C})}{(\mathcal{A},\texttt{sp},\mathcal{C})}$ (b) $\frac{(\mathcal{A},\texttt{sp},\mathcal{B})\ (\mathcal{X},\mathcal{A},\mathcal{Y})}{(\mathcal{X},\mathcal{B},\mathcal{Y})}$

3. *Subclass:*

(a) $\frac{(\mathcal{A},\texttt{sc},\mathcal{B})\ (\mathcal{B},\texttt{sc},\mathcal{C})}{(\mathcal{A},\texttt{sc},\mathcal{C})}$ (b) $\frac{(\mathcal{A},\texttt{sc},\mathcal{B})\ (\mathcal{X},\texttt{type},\mathcal{A})}{(\mathcal{X},\texttt{type},\mathcal{B})}$

4. *Typing:*

(a) $\frac{(\mathcal{A},\texttt{dom},\mathcal{B})\ (\mathcal{X},\mathcal{A},\mathcal{Y})}{(\mathcal{X},\texttt{type},\mathcal{B})}$ (b) $\frac{(\mathcal{A},\texttt{range},\mathcal{B})\ (\mathcal{X},\mathcal{A},\mathcal{Y})}{(\mathcal{Y},\texttt{type},\mathcal{B})}$

5. *Implicit Typing:*

(a) $\frac{(\mathcal{A},\texttt{dom},\mathcal{B})\ (\mathcal{C},\texttt{sp},\mathcal{A})\ (\mathcal{X},\mathcal{C},\mathcal{Y})}{(\mathcal{X},\texttt{type},\mathcal{B})}$ (b) $\frac{(\mathcal{A},\texttt{range},\mathcal{B})\ (\mathcal{C},\texttt{sp},\mathcal{A})\ (\mathcal{X},\mathcal{C},\mathcal{Y})}{(\mathcal{Y},\texttt{type},\mathcal{B})}$

6. *Subproperty Reflexivity:*

(a) $\frac{(\mathcal{X},\mathcal{A},\mathcal{Y})}{(\mathcal{A},\texttt{sp},\mathcal{A})}$ (c) $\frac{}{(p,\texttt{sp},p)}$ *for* $p \in \rho\text{df}$

(b) $\frac{(\mathcal{A},\texttt{sp},\mathcal{B})}{(\mathcal{A},\texttt{sp},\mathcal{A})\ (\mathcal{B},\texttt{sp},\mathcal{B})}$ (d) $\frac{(\mathcal{A},p,\mathcal{X})}{(\mathcal{A},\texttt{sp},\mathcal{A})}$ *for* $p \in \{\texttt{dom}, \texttt{range}\}$

7. *Subclass Reflexivity:*

(a) $\frac{(\mathcal{A},\texttt{sc},\mathcal{B})}{(\mathcal{A},\texttt{sc},\mathcal{A})\ (\mathcal{B},\texttt{sc},\mathcal{B})}$ (b) $\frac{(\mathcal{X},p,\mathcal{A})}{(\mathcal{A},\texttt{sc},\mathcal{A})}$ *for* $p \in \{\texttt{dom}, \texttt{range}, \texttt{type}\}$

Note 1 (On rules (5a) and (5b)). As noted in [10,8], the set of rules presented in [15] is not complete for RDFS entailment. The problem is produced when a blank node X is implicitly used as standing for a property in triples like $(a, \texttt{sp}, X)$, $(X, \texttt{dom}, b)$, or $(X, \texttt{range}, c)$. Here we solve the problem following the elegant solution proposed by Marin [10] adding just two new rules of implicit typing (rules 5 above).

An instantiation of a rule is a uniform replacement of the metavariables occurring in the triples of the rule by elements of **UBL**, such that all the triples obtained after the replacement are well formed RDF triples.

Definition 5 (Proof). *Let G and H be graphs. Define $G \vdash_{\rho\text{df}} H$ iff there exists a sequence of graphs $P_1, P_2, \ldots, P_k$, with $P_1 = G$ and $P_k = H$, and for each j $(2 \leq j \leq k)$ one of the following cases hold:*

- *there exists a map $\mu : P_j \to P_{j-1}$ (rule (1a)),*
- *$P_j \subseteq P_{j-1}$ (rule (1b)),*

– *there is an instantiation* $\frac{R}{R'}$ *of one of the rules (2)–(7), such that* $R \subseteq P_{j-1}$ *and* $P_j = P_{j-1} \cup R'$.

The sequence of rules used at each step (plus its instantiation or map), is called a proof *of H from G.*

Theorem 3 (Soundness and completeness). *The proof system* $\vdash_{\rho\mathrm{df}}$ *is sound and complete for* $\models_{\rho\mathrm{df}}$*, that is, given graphs G and H we have*

$$G \vdash_{\rho\mathrm{df}} H \;\; \textit{iff} \;\; G \models_{\rho\mathrm{df}} H.$$

Corollary 2. *Define the proof system* $\vdash^{\mathrm{nrx}}_{\rho\mathrm{df}}$ *as* $\vdash_{\rho\mathrm{df}}$ *by droping rules of reflexivity (rules (6) and (7)). Then for graphs G and H,*

$$G \vdash^{\mathrm{nrx}}_{\rho\mathrm{df}} H \;\; \textit{iff} \;\; G \models^{\mathrm{nrx}}_{\rho\mathrm{df}} H.$$

4 Deductive Systems for Minimal Fragments of ρdf

We will assume in the rest of the paper that the user does not redefine or enrich the semantics of the ρdf-vocabulary. In syntactical terms this means that there is no triple where this vocabulary occurs in subject or object positions. This assumption is light and can be found on almost all published RDF specifications.

To begin with, the following theorem shows that for several purposes blank nodes can be treated in an orthogonal form to ρdf vocabulary.

Theorem 4 (Normal form for proofs). *Assume* $G \vdash_{\rho\mathrm{df}} H$*. Then there is a proof of H from G where the rule (1) is used at most once and at the end.*

Consider the lattice of fragments of ρdf in Figure 1. Given one of the fragments X, by an X-graph we will understand a graph that mention ρdf vocabulary only from X. Similarly, an X-rule is one rule (2-7) that mention ρdf vocabulary only from X.

Theorem 5. *Let X be one of the fragments of* ρdf *in Figure 1, and let G and H be X-graphs. Assume that* $G \vdash_{\rho\mathrm{df}} H$*, then there exists a proof of H from G which only uses X-rules and rule (1).*

The above result is based in the observation that in a proof of H from G we can avoid the following fact: a sequence of graphs $P_i, P_{i+1}, \ldots, P_{i+j}$ produced in the proof may present vocabulary outside X, but with P_i and P_{i+j} X-graphs. This fact may impose new rules obtained from the rules of $\vdash_{\rho\mathrm{df}}$ by a concatenation that result in a sound derivation between X-graphs. It can be shown that the only rules obtained in this way coincide actually with X-rules. A second point is that triples with vocabulary outside X, produced by the application of non X-rules are not needed and can be left out of the proof of H from G.

Theorem 5 implies that X-rules are sound and complete for $\models_{\rho\mathrm{df}}$ in fragment X. As a direct consequence we also obtain that X-rules without considering reflexivity rules, are sound and complete for $\models^{\mathrm{nrx}}_{\rho\mathrm{df}}$ in fragment X.

In what follows $G|_V$ means the subgraph induced by vocabulary V, i.e. those triples having subject, or predicate, or object in V.

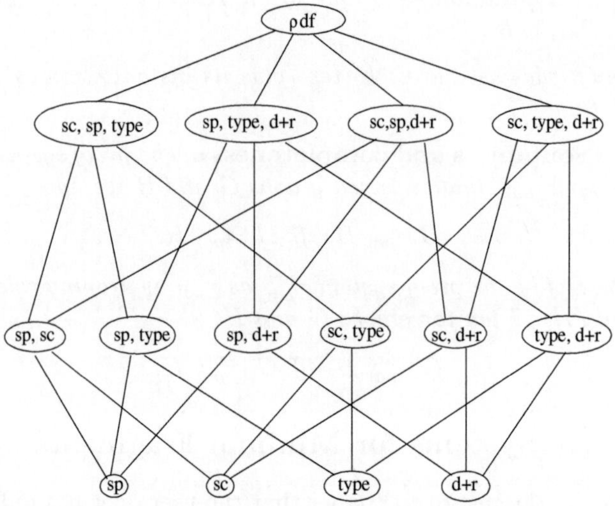

Fig. 1. The lattice of fragments of ρdf

Interpolation Lemmas for RDF. Interpolation lemmas refer to lemmas expressing the role of vocabularies in deduction. They follow from the previous results in this section.

Lemma 1. *Let G and H be graphs. If $(a, b, c) \in G$ and a, b, c do not appear in* voc(H) *nor in* ρdf*, then $G \models_{\rho\text{df}} H$ iff $G - \{(a, b, c)\} \models_{\rho\text{df}} H$.*

Lemma 2. *Let a, b, c be ground terms with b not belonging to* ρdf*. Then: $G \models_{\rho\text{df}} (a, b, c)$ iff $G|_{\{\texttt{sp},a,b,c\}} \models_{\rho\text{df}} (a, b, c)$.*

Lemma 3. *Let $a, b \in$* **UBL***, then*

1. $G \models_{\rho\text{df}} (a, \texttt{dom}, b)$ iff $G|_{\texttt{dom}} \models_{\rho\text{df}} (a, \texttt{dom}, b)$.
2. $G \models_{\rho\text{df}} (a, \texttt{range}, b)$ iff $G|_{\texttt{range}} \models_{\rho\text{df}} (a, \texttt{range}, b)$.

Moreover, if a, b are ground, $\models_{\rho\text{df}}$ reduces to membership in G.

Note 2. Although $(a, \texttt{dom}, b)$ refers to a property a and a class b, inferring a dom statement in the RDFS system does not depend on statements about classes or properties. For example, from the previous lemma follows the non-intuitive fact that $\{(c_1, \texttt{sc}, c_2), (c_2, \texttt{sc}, c_1), (a, \texttt{dom}, c_1)\}$ does not entail $(a, \texttt{dom}, c_2)$.

Lemma 4. *Let $a \neq b$, then*

1. $G \models_{\rho\text{df}} (a, \texttt{sc}, b)$ iff $G|_{\texttt{sc}} \models_{\rho\text{df}} (a, \texttt{sc}, b)$.
2. $G \models_{\rho\text{df}} (a, \texttt{sp}, b)$ iff $G|_{\texttt{sp}} \models_{\rho\text{df}} (a, \texttt{sp}, b)$.

It turns out that type is the most entangled keyword in the vocabulary and deducing $G \models_{\rho\text{df}} (a, \texttt{type}, b)$ can involve all of G (except those triples mentioned in Lemma 1).

5 The Complexity of ρdf Ground Entailment

Let us introduce some notation. For a graph G and a predicate p, define G_p as the subgraph of G consisting of the triples of the form (x, p, y) of G, and define $G_\emptyset$ as the subgraph consisting of triples without ρdf vocabulary. Let $G(\mathtt{sp})$ be the directed graph whose vertices are all the elements v which appear as subject or objects in the triples of G, and in which (u, v) is an edge if and only if $(u, \mathtt{sp}, v) \in G$. Similar definition for $G(\mathtt{sc})$.

The naive approach to test the entailment $G \models H$ in the ground case would be to consider the *closure* of G and check if H is included in it. Recall that for ground G, the closure is the graph obtained by adding to G all ground triples that are derivable from G. The following result shows that this procedure would take time proportional to $|H| \cdot |G|^2$ in the worst case, which is too expensive from a database point of view.

Theorem 6. *The size of the closure of G is $\mathcal{O}(|G|^2)$, and this bound is tight.*

For the upper bound, the result follows by an analysis of the rules. The most important point is the propagation –when applicable– of the triples of the form (x, a, y) through the transitive closure of the $G(\mathtt{sp})$ graph by the usage of rule 2(b): it can be shown that this gives at most $|G_\emptyset| \times |G_{\mathtt{sp}}|$ triples. For triples having a fixed predicate in ρdf the quadratic bound is trivial. For the tightness, consider the graph $\{(a_1, \mathtt{sp}, a_2), \ldots, (a_n, \mathtt{sp}, a_{n+1})\} \cup \{(x_1, a_1, y_n), \ldots, (x_n, a_n, y_n)\}$. The number of triples of the closure of this graph is $2n + 1 + \sum_{k=1}^{n} k$ that is quadratic in n.

The following algorithm presents a much better procedure to check ground entailment in this fragment.

Algorithm (Ground Entailment)
Input: G, triple (a, p, b)

1. `IF` $p \in \{\mathtt{dom}, \mathtt{range}\}$ `THEN` check if $(a, p, b) \in G$.
2. `IF` $p = \mathtt{sp}$, $a \neq b$, `THEN` check if there is a path from a to b in $G(\mathtt{sp})$.
3. `IF` $p = \mathtt{sc}$, $a \neq b$, `THEN` check if there is a path from a to b in $G(\mathtt{sc})$.
4. `IF` $p \in \{\mathtt{sp}, \mathtt{sc}\}$ and $a = b$, `THEN` check if $(a, p, a) \in G$ else check all patterns of triples in the upper part of rules 6 (for `sp`) and rule 7 (for `sc`).
5. `IF` $p \notin$ ρdf `THEN` check $(a, p, b) \in G_\emptyset$, if it is not
   `LET` $G(\mathtt{sp})^*$ be the graph $G(\mathtt{sp})$ with the following marks:
   For each $(a, v, b) \in G_\emptyset$, if $v \in G(\mathtt{sp})$ then mark it green.
   `IN` Check in $G(\mathtt{sp})^*$ if there is a path from a vertex marked green to p
6. `IF` $p = \mathtt{type}$ `THEN`
   `LET` $G(\mathtt{sp})'$ be the graph $G(\mathtt{sp})$ with the following marks:
   - For each triple $(u, \mathtt{dom}, v) \in G_{\mathtt{dom}}$, if $u \in G(\mathtt{sp})$ mark the vertex u with $d(v)$.
   - For each triple $(a, e, y) \in G_\emptyset$, if $e \in G(\mathtt{sp})$, mark the vertex e with a.

LET $G(\texttt{sc})'$ be the graph $G(\texttt{sc})$ with the following marks:
- For vertex u marked $d(v)$ reachable from a vertex marked a in $G(\texttt{sp})'$, if $v \in G(\texttt{sc})$ mark it blue.
- For each $(a, \texttt{type}, w) \in G$, if $w \in G(\texttt{sc})$ mark it blue.

IN Check in $G(\texttt{sc})'$ if there is a path from a blue node to b.
Repeat this point for range instead of dom.

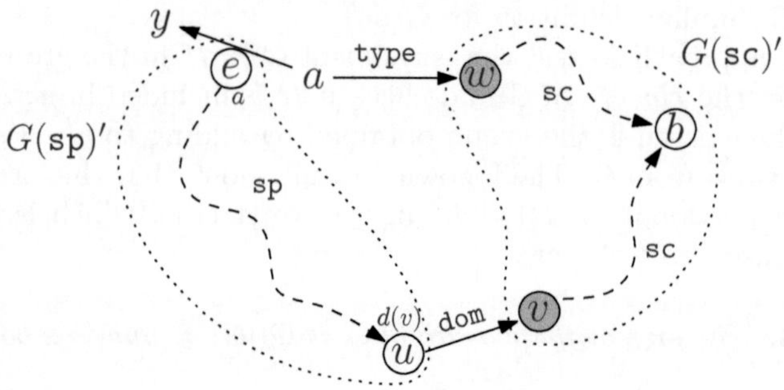

Fig. 2. Point 6 of the *Ground Entailment Algorithm*

Theorem 7. *Let (a, b, c) be a ground triple. The algorithm above can be used to test the entailment $G \models_{\rho\text{df}} (a, b, c)$ in time $\mathcal{O}(|G| \log |G|)$.*

Correctness and completeness of the algorithm follows from an inspection of the rules. The algorithm uses the rules in a bottom-up fashion. There are some subtleties in points 5 and 6. Point 5 follows from Lemma 2 and rule 2(a). The construction of $G(\texttt{sp})^*$ can be done in $|G| \log |G|$ steps: order $G_\emptyset$ and then while traversing $G(\texttt{sp})$ do binary search on $G_\emptyset$. For point 6 (see Figure 2) the crucial observation is that in $G(\texttt{sp})'$, if there is a path from a vertex marked a to a vertex u marked $d(v)$, then $G \models (a, u, y)$ for some y, and hence $G \models (a, \texttt{type}, v)$ using rule 4(a). Note that this checking takes time at most linear in $|G|$. From here, it is easy to see that the checking in $G(\texttt{sc})'$ will do the job.

Corollary 3. *Let H be a ground graph. Deciding if $G \models_{\rho\text{df}} H$ can be done in time $\mathcal{O}(|H| \cdot |G| \log |G|)$.*

The following result shows that the above algorithm cannot be essentially improved, in the sense that, any other algorithm for testing the ground entailment $G \models_{\rho\text{df}} H$ will take time proportional to $|H| \cdot |G| \log |G|$ in the worst case.

Theorem 8. *The problem of testing $G \models_{\rho\text{df}} t$ takes time $\Omega(|G| \log |G|)$.*

The bound is obtained by coding the problem of determining whether two sets are disjoint, which is a well known problem that needs $\Omega(n \log n)$ comparisons in the worst case [1]. The codification is as follows: Given the sets $A = \{a_1, \ldots, a_n\}$ and $B = \{b_1, \ldots, b_n\}$, construct the RDF graph $G = \{(a_{i-1}, \texttt{sp}, a_i)\}_{2 \leq i \leq n} \cup \{(x, b_j, y)\}_{1 \leq j \leq n}$. Then, we have that $G \models (x, a_n, y)$ iff $A \cap B \neq \emptyset$.

6 Conclusions

We presented a streamlined fragment of RDFS which includes all the vocabulary that is relevant for describing data, avoiding vocabulary and semantics that theoretically corresponds to the definition of the structure of the language. We concentrated in studying the semantics, entailment, minimal proof systems, and algorithmic properties of this relevant fragment of RDFS. Our results show a viable proposal to lower the complexity of RDF data processing by using fragments of RDFS.

In this paper we have concentrated primarily on the ground dimension of RDF. Future work includes the refinement of our current results about the interplay between blank nodes semantics and the ground part. We are also working in the applications of our results to practical cases, as well as developing best practices for logical design of RDF specification based on the previous considerations.

Acknowledgments. Pérez was supported by Dirección de Investigación – Universidad de Talca, Gutierrez by Proyecto Enlace DI 2006, ENL 06/13, Universidad de Chile, and the three authors by Millennium Nucleus Center for Web Research, P04-067-F, Mideplan, Chile.

References

1. M. Ben-Or. *Lower bounds for algebraic computation trees. Proc. 15th Annual Symposium on Theory of Computing*, pp 80-86, 1983.
2. T. Berners-Lee. *Principles of Design. Personal Notes*, `http://www.w3.org/DesignIssues/Principles.html`.
3. Dan Brickley, Libby Miller. *FOAF Vocabulary Specification.* July 2005. `http://xmlns.com/foaf/0.1/`
4. J. de Bruijn, E. Franconi, S. Tessaris. *Logical Reconstruction of normative RDF.* In *OWLED 2005*, Galway, Ireland, November 2005
5. Victor Dalmau, P. G. Kolaitis, M. Vardi. *Constraint Satisfaction, Bounded Treewidth, and Finite–Variable Logics Proc. 8th Int. Conf. on Principles and Practice of Constraint Programming*, September, 2002.
6. Jeremy J. Carroll, Christian Bizer, Pat Hayes, Patrick Stickler, *Named graphs*, Journal of Web Semantics vol. 3, 2005, pp. 247 - 267
7. C. Gutierrez, C. Hurtado, A. O. Mendelzon, *Foundations of Semantic Web Databases*, Proceedings ACM Symposium on Principles of Database Systems (PODS), Paris, France, June 2004, pp. 95 - 106.
8. H. ter Horst. *C*ompleteness, decidability and complexity of entailment for RDF Schema and a semantic extension involving the OWL vocabulary. Journal of Web Semantics, vol. 3, 2005.
9. Jean–Francois Baget, *RDF Entailment as a Graph Homomorphism*, In *ISWC 2005.*
10. Draltan Marin, *A Formalization of RDF (Applications de la Logique á la sémantique du web)*, École Polytechnique – Universidad de Chile, 2004. Technical Report Dept. Computer Science, Universidad de Chile, TR/DCC-2006-8. `http://www.dcc.uchile.cl/cgutierr/ftp/draltan.pdf`
11. E. Prud'hommeaux, A. Seaborne. *SPARQL Query Language for RDF.* W3C Working Draft, October 2006. `http://www.w3.org/TR/rdf-sparql-query/`.

12. *RDF/OWL Representation of WordNet.* Edit. Mark van Assem, Aldo Gangemi, Guus Schreiber. Working Draft, April 2006. `http://www.w3.org/2001/sw/BestPractices/WNET/wn-conversion`.
13. *Resource Description Framework (RDF) Model and Syntax Specification*, Edit. O. Lassila, R. Swick, Working draft, W3C, 1998.
14. *RDF/XML Syntax Specification (Revised) W3C Recommendation 10 February 2004*, Edit. D. Beckett
15. *RDF Semantics, W3C Recommendation 10 February 2004* Edit. P. Hayes
16. *RDF Vocabulary Description Language 1.0: RDF Schema, W3C Recommendation 10 February 2004*, Edit. D. Brickley, R.V. Guha.
17. *RDF Concepts and Abstract Syntax, W3C Recommendation 10 February 2004*, Edit. G. Klyne, J. J. Carroll.
18. *RDF Primer, W3C Recommendation 10 February 2004*, Edit. F. Manola, E. Miller,
19. *Gene Ontology.* `http://www.geneontology.org/`

A Appendix: RDFS Semantics

To easy the job of the reader, we reproduce here the definitions and axioms of the normative semantics of RDF [15] consisting of a model theory and axiomatic triples. The set rdfsV stands for the RDFS vocabulary.

Definition 6 (cf. [15,10]). *The interpretation $\mathcal{I}$ is an* RDFS model *for an RDF graph G, denoted by $\mathcal{I} \models G$, iff $\mathcal{I}$ is an iterpretation over vocabulary* rdfsV $\cup$ universe(G) *that satisfies the RDF/S axiomatic triples [15,10] and the following semantic conditions:*

1. *Simple:*
   (a) *there exists a function $A : \mathbf{B} \to Res$ such that for each $(s,p,o) \in G$, $Int(p) \in Prop$ and $(Int_A(s), Int_A(o)) \in Ext(Int(p))$, where Int_A is the extension of Int using A.*
2. *RDF:*
   (a) $x \in Prop \Leftrightarrow (x, Int(\texttt{prop})) \in Ext(Int(\texttt{type}))$
   (b) *If $l \in$ universe(G) is a typed XML literal with lexical form w, then $Int(l)$ is the XML literal value of w, $Int(l) \in Lit$, and $(Int(l), Int(\texttt{xmlLit})) \in Ext(Int(\texttt{type}))$.*
3. *RDFS Classes:*
   (a) $x \in Res \Leftrightarrow x \in CExt(Int(\texttt{res}))$
   (b) $x \in Class \Leftrightarrow x \in CExt(Int(\texttt{class}))$
   (c) $x \in Lit \Leftrightarrow x \in CExt(Int(\texttt{literal}))$
4. *RDFS Subproperty:*
   (a) *$Ext(Int(\texttt{sp}))$ is transitive and reflexive over Prop*
   (b) *if $(x,y) \in Ext(Int(\texttt{sp}))$ then $x, y \in Prop$ and $Ext(x) \subseteq Ext(y)$*
5. *RDFS Subclass:*
   (a) *$Ext(Int(\texttt{sc}))$ is transitive and reflexive over Class*
   (b) *if $(x,y) \in Ext(Int(\texttt{sc}))$ then $x, y \in Class$ and $CExt(x) \subseteq CExt(y)$*
6. *RDFS Typing:*
   (a) $x \in CExt(y) \Leftrightarrow (x,y) \in Ext(Int(\texttt{type}))$
   (b) *if $(x,y) \in Ext(Int(\texttt{dom}))$ and $(u,v) \in Ext(x)$ then $u \in CExt(y)$*
   (c) *if $(x,y) \in Ext(Int(\texttt{range}))$ and $(u,v) \in Ext(x)$ then $v \in CExt(y)$*

7. *RDFS Additionals:*
   (a) *if* $x \in Class$ *then* $(x, Int(\texttt{res})) \in Ext(Int(\texttt{sc}))$.
   (b) *if* $x \in CExt(Int(\texttt{datatype}))$ *then* $(x, Int(\texttt{literal})) \in Ext(Int(\texttt{sc}))$
   (c) *if* $x \in CExt(Int(\texttt{contMP}))$ *then* $(x, Int(\texttt{member})) \in Ext(Int(\texttt{sp}))$

Now, given two graphs G and H we say that G RDFS entails *H and write* $G \models H$, *iff every RDFS model of G is also an RDFS model of H.*

Table 1. RDF/S vocabulary [15,10] with shortcuts in brackets. The first column shows built-in classes, second and third show built-in properties.

rdfs:Resource [res]	rdf:type [type]	rdfs:isDefinedBy [isDefined]
rdf:Property [prop]	rdfs:domain [dom]	rdfs:comment [comment]
rdfs:Class [class]	rdfs:range [range]	rdfs:label [label]
rdfs:Literal [literal]	rdfs:subClassOf [sc]	rdf:value [value]
rdfs:Datatype [datatype]	rdfs:subPropertyOf [sp]	rdf:nil [nil]
rdf:XMLLiteral [xmlLit]	rdf:subject [subj]	rdf:_1 [_1]
rdfs:Container [cont]	rdf:predicate [pred]	rdf:_2 [_2]
rdf:Statement [stat]	rdf:object [obj]	...
rdf:List [list]	rdfs:member [member]	rdf:_i [_i]
rdf:Alt [alt]	rdf:first [first]	...
rdf:Bag [bag]	rdf:rest [rest]	
rdf:Seq [seq]	rdfs:seeAlso [seeAlso]	
rdfs:ContainerMembershipProperty [contMP]		

Web Service Contracting: Specification and Reasoning with SCIFF

Marco Alberti[1], Federico Chesani[2], Marco Gavanelli[1], Evelina Lamma[1], Paola Mello[2], Marco Montali[2], and Paolo Torroni[2]

[1] ENDIF, University of Ferrara
Via Saragat 1, 44100 Ferrara, Italy
Name.Surname@unife.it
[2] DEIS, University of Bologna
V.le Risorgimento 2, 40136 Bologna, Italy
Name.Surname@unibo.it

Abstract. The semantic web vision will facilitate automation of many tasks, including the location and dynamic reconfiguration of web services. In this article, we are concerned with a specific stage of web service location, called, by some authors, *contracting*. We address contracting both at the operational level and at the semantic level. We present a framework encompassing communication and reasoning, in which web services exchange and evaluate goals and policies. Policies represent behavioural interfaces. The reasoning procedure at the core of the framework is based on the abductive logic programming SCIFF proof-procedure. We describe the framework, show by examples how to formalise policies in the declarative language of SCIFF, and give the framework a model-theoretic and a sound proof-theoretic semantics.

1 Introduction

The Service Oriented Computing (SOC) paradigm, and its practical implementation based on Web Services, are rapidly emerging as standard architectures for distributed application development. Different service providers, heterogenous in terms of hardware and software settings, can easily inter-operate by means of communication standards and well-known network protocols. In such a scenario, the use of off-the-shelf solutions and dynamic reconfiguration becomes attractive, both at design level, as well as at execution (run-time) level. However, dynamic reconfiguration of services is possible only if supported by a suitable technology. The Semantic Web vision, based on the idea that adding machine-understandable semantic information to Web resources will facilitate automation of many tasks [18,20], including the location of Web Services, is a promising way to address this issue.

Drawing from [18], we consider a process of searching the right service to match a request as consisting of two stages. During the first one, called *discovery*, the requester states only the things that are desired, thus, using an ontology or other KR formalisms, it seeks for all the services that can potentially satisfy

E. Franconi, M. Kifer, and W. May (Eds.): ESWC 2007, LNCS 4519, pp. 68–83, 2007.

a request of such a kind. During the second stage, called *contracting*, the requester provides in input specific information for an already requested service. The purpose of this second stage is to verify that the input provided will lead to a desired state that satisfies the requester's goal.

Many relevant papers are written about web service discovery; some of them use Logic Programming (LP) techniques. They mostly focus on the discovery stage. In this article, we are concerned with the contracting stage, which we address both at the operational level, and at the semantic level. We present a framework, called SCIFF Reasoning Engine (SRE) encompassing reasoning and communication in a networked architecture inspired to the model of discovery engines. We discuss the problem of communicating policies between web services, and of determining whether the policies of a provider and a requester are compatible with each other. We use a mixture of Abductive Logic Programming (ALP, [17]), and Constraint Logic Programming (CLP, [16]). ALP is used to construct sets of input and output messages, along with assumed data, while CLP is used to maintain a partial temporal order among them. We chose to adopt a hypothetical reasoning framework such as ALP because reasoning is made before execution. A component such as SRE which reasons on possible developments of the interaction among web services has to come up with hypotheses about which messages are to be expected in the future. In fact, a similar approach, also based on hypothetical reasoning, though not on LP, has been followed by others, notably [18].

In this work we assume that a previous discovery stage has already produced multiple candidate services. The intended user of SRE will typically be a web service, although for the sake of presentation we will name the users *alice* and *eShop*. The user query is given in terms of goals and policies. Policies describe the observable behaviour of users, i.e., their *behavioural interface*, in terms of relationships among externally observable events regarding the service. We formalise web services' policies in a declarative language derived from the $\mathcal{S}$CIFF language, born in the context of the EU SOCS project [1]. $\mathcal{S}$CIFF was conceived to specify and verify social-level agent interaction. In this work, a simplified version of the $\mathcal{S}$CIFF language is adopted for defining policies, by means of *social integrity constraints*: a sort of reactive rules used to generate and reason about possible evolutions of a given interaction instance. Distinguishing features of SRE are its declarative and operational approaches combined together into a working framework. Users specify their goals as well as their own policies (related to the goals) by means of rules; in their turn, service providers publish their service descriptions, together with their policies about the usage of the services, again by means of rules. The use of ALP reconciles forward and backward reasoning in a unified reasoning engine: two aspects that frequently, in the context of web services, are treated separately from each other. Moreover, while ALP and CLP have been used to address planning problems before, in this work we want to show how a mixture of known, but little-used techniques can be used to solve a real-world problem with a real implementation.

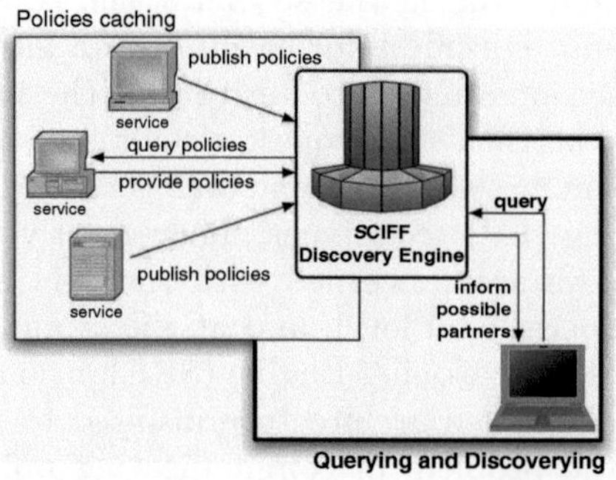

Fig. 1. The architecture of SRE

In the next section, we show the SRE architecture and introduce informally a walk through scenario. In Sect. 3 we explain the notation used to write policies in SRE and in Sect. 4 we present its underlying logic. Sect. 5 develops the scenario in more detail and shows the reasoning in SRE by example, and Sect. 6 concludes by discussing related approaches and future work.

2 Architecture

The SRE architecture, shown in Figure 1, is a client-server architecture augmented with a "smart" discovery engine (which incorporates the SRE itself). We assume that SRE has information available about web services, either gathered in a previous discovery phase from the Internet (in the style of web spiders), or because explicitly published to it by web services. So we can safely assume that the data collected has already been filtered and, if providers refer to different ontologies, equivalences between concepts have already been established.

At the logical level, the retrieved information consists of triplets in the form $\langle s, ws, (\mathcal{KB}_{ws}, \mathcal{P}_{ws})\rangle$, where s identifies a service, ws is the name of a web service that provides s, and $(\mathcal{KB}_{ws}, \mathcal{P}_{ws})$ are the knowledge base and policies that ws associates to s. In particular, for a given provider ws providing s, a set of policies (rules) describes ws's behaviour with respect to s, and a knowledge base, in the form of a logic program, contains information that ws wants to disclose to potential customers, together with its policies. A sample policy could state that the service delivers goods only to certain countries, or zones. The list of such zones could be made available through the knowledge base.

SRE reasons based on a client's query (also called *goal*, in the LP sense) which it matches to a service. Such a query will contain the name of the service that the client (c) needs, a (possibly empty) set of policies $\mathcal{P}_c$ and a (possibly empty) knowledge base $\mathcal{KB}_c$. The goal is an expression consisting of a conjunction of elements, which can represent, for example, events and constraints, like partial orders among events. The output of SRE is a number of triplets $\langle ws, \mathcal{E}, \Delta\rangle$, each one containing the name of a web application which provides the service, plus some additional information: $\mathcal{E}$, which encodes a possible future interaction, i.e., a partially ordered sequence of events, occurring between ws and c and regarding s, and a set Δ containing a number of additional validity conditions for $\mathcal{E}$. For example, ws could be the name of a service that provides a *device*, $\mathcal{E}$ could be "first ws shows evidence of membership to $\mathcal{B}etter\ \mathcal{B}usiness\ \mathcal{B}ureau$ ($\mathcal{BBB}$), then c pays by credit card", and Δ could be "delivery in Europe". These concepts are better instantiated in the following scenario.

2.1 The *alice* and *eShop* Scenario

The scenario we use in this paper is inspired from [11,2]. *eShop* is a web service that sells devices, while *alice* is another web service, which wants to get a *device*. *alice* and *eShop* describe their behaviour concerning sales/payment/... of items through policies (rules), which they publish using some Rules Interchange Format. These two actors find each other via SRE: in particular, *alice* submits a query to the discovery engine, by specifying her policies and the service she is looking for (e.g., getting *device*). Once suitable services (e.g., *eShop*) have been found, SRE, by checking the satisfiability of *alice*'s goal and the compatibility of the rules describing *alice*'s and *eShop*'s behaviour, provides back to *alice* the list of web services that could satisfy her specific need. SRE also defines the conditions that must be fulfilled by each web service, in order to reach the goal.

Let *eShop*'s policies regarding *device* be as follows:

(shop1) if a customer wants to get an item, then, (*i*) if the customer can be accepted, *eShop* will request him/her to pay using an acceptable method, otherwise (*ii*) *eShop* will inform the customer of a failure;

(shop2) if an acceptable customer paid the item, using an acceptable method, then *eShop* will deliver the item;

(shop3) if a customer requests a certificate about *eShop*'s membership to the $\mathcal{BBB}$, then the shop will send it.

eShop publishes a knowledge base $\mathcal{KB}_{eShop}$, which specifies that a customer is *accepted* if it is resident in some zone; it also specifies the accepted payment methods. SRE retrieves information about *eShop* in the triplet: $\langle sell(device)$, $eShop$, $(\mathcal{KB}_{eShop}, \mathcal{P}_{eShop})\rangle$, indicating that *eShop* offers service *sell*(*device*), with a set $\mathcal{P}_{eShop}$ of policies defined as $\mathcal{P}_{eShop} \equiv \{(\mathsf{shop1}), (\mathsf{shop2}), (\mathsf{shop3})\}$ and a knowledge base $\mathcal{KB}_{eShop}$. We consider three different scenaria for *alice*:

Scenario 1. *alice*'s goal is to obtain *device*. Her policies are as follows:

(alice1) if a shop requires that *alice* pays by credit card, *alice* expects that the shop provides a certificate to guarantee that it is a $\mathcal{BBB}$ member;

(alice2) if a shop requires that *alice* pays by credit card, and it has proven its membership to the $\mathcal{BBB}$, then *alice* will pay by credit card;

(alice3) if a shop requires *alice* to pay with any other method than credit card, then *alice* will pay without any further request;

Besides, *alice* is based in Europe. However, for privacy reason, *alice* does not make this information public. $\mathcal{KB}_{alice}$ is an an empty knowledge base.

Scenario 2. Policies are the same as above. However, *alice* will not agree to pay cash, as she specifies in her query to SRE. Moreover, $\mathcal{KB}_{alice}$ is not empty, but instead it contains information about her place of residence and age;

Scenario 3. *alice* has no policies to express in relation to the query she submits to SRE. We can imagine here that *alice* is a human user, and she queries SRE, using a suitable interface, simply because she wishes to know what her options are regarding the purchase of *device*. Later, *alice* may evaluate SRE's answer and possibly re-submit a refined query.

3 Notation

In SRE, policies describe a web service's observable behaviour in terms of *events* (e.g., messages). SRE considers two types of events: those that one can directly control (e.g., if we consider the policies of a web service ws, a message generated by ws itself) and those that one cannot (e.g., messages that ws receives, or does not want to receive). Atoms denoted by **H** denote "controllable" events, those denoted by **E** and **EN** denote "passive" events, also named *expectations*. Since SRE reasons about possible future courses of events, both controllable events and expectations represent *hypotheses* on possible events. We restrict ourselves to the case of events being messages exchanged between the two parties in play. The notation is:

- $\mathbf{H}(ws, ws', M, T)$ denotes a message with sender ws, recipient ws', and content M, which ws expects to be sending to ws' at a time T;
- $\mathbf{E}(ws', ws, M, T)$ denotes a message with sender ws', recipient ws, and content M, which ws expects ws' to be sending at a time T;
- $\mathbf{EN}(ws', ws, M, T)$ denotes a message with sender ws', recipient ws, and content M, which ws expects ws' *not* to be sending at a time T;

Web service specifications in SRE are relations among expected events, expressed by an abductive logic program. This is in general a triplet $\langle \mathcal{KB}, \mathcal{A}, \mathcal{IC} \rangle$, where $\mathcal{KB}$ is a logic program, $\mathcal{A}$ (sometimes left implicit) is a set of literals named *abducibles*, and $\mathcal{IC}$ is a set of *integrity constraints*. Intuitively, in ALP the role of $\mathcal{KB}$ is to define predicates, the role of $\mathcal{A}$ is to fill-in the parts of $\mathcal{KB}$ which are unknown, and the role of $\mathcal{IC}$ is to control the ways elements of $\mathcal{A}$ are hypothesised, or "abduced." Reasoning in ALP is usually goal-directed. It starts from a "goal" $\mathcal{G}$, i.e., an expression which we want to obtain as a logical consequence of the abductive logic program, and it amounts to finding a set of abduced hypotheses Δ built from atoms in $\mathcal{A}$ such that $\mathcal{KB} \cup \Delta \models \mathcal{G}$ and $\mathcal{KB} \cup \Delta \models \mathcal{IC}$. Symbol $\models$ represents logical entailment, which can be associated with one among several semantics. In literature one finds different readings of abduction in LP. Δ can be considered as an "answer" to a query or goal $\mathcal{G}$. In other contexts, one can interpret $\mathcal{G}$ as an observation and Δ as its explanation. This is for example the reading of an abductive anwer in abductive reasoning-based diagnosis. In the domain of web services, we will use ALP as a reasoning paradigm that combines backward, goal-oriented reasoning with forward, reactive reasoning [19]: two aspects that frequently, in the context of web services, are treated separately from each other.

Definition 1 (Web Service Behavioural Interface Specification). *Given a web service ws, its* web service behavioural interface specification $\mathcal{S}_{ws}$ *is an abductive logic program, represented by the triplet* $\mathcal{S}_{ws} \equiv \langle \mathcal{KB}_{ws}, \mathcal{A}, \mathcal{IC}_{ws} \rangle$, *where* $\mathcal{KB}_{ws}$ *is ws's Knowledge Base,* $\mathcal{A}$ *is the set of* abducible literals, *and* $\mathcal{IC}_{ws}$ *is ws's set of Integrity Constraints (ICs).*

$\mathcal{KB}_{ws}$, which corresponds to $\mathcal{KB}_{ws}$ of Sect. 2, is a set of clauses which declaratively specifies pieces of knowledge of the web service. Note that the body of $\mathcal{KB}_{ws}$'s clauses may contain **E**/**EN** expectations about the behaviour of the web services. $\mathcal{A}$ is the set of *abducible literals*. It includes all possible **E**/**EN** expectations, **H** events, and predicates left undefined by $\mathcal{KB}_{ws}$. It is the set of all possible unknowns. Note that $\mathcal{E}_{ws}$ and Δ of Sect. 2 are subsets of $\mathcal{A}$. In the following sometimes we leave $\mathcal{A}$ implicit, as we did in Sect. 2. $\mathcal{IC}_{ws}$, which corresponds to $\mathcal{P}_{ws}$ of Sect. 2, contains *ws*'s policies. In particular, each IC in $\mathcal{IC}_{ws}$ is a rule in the form *Body* → *Head*. Intuitively, the *Body* of an IC is a conjunction of events, literals and CLP constraints; the *Head* is either a disjunction of conjunctions of events, literals and CLP constraints, or *false*. The operational behaviour of ICs is similar to that of forward rules: whenever the body becomes true, the head is also made true. The syntax of $\mathcal{KB}_{ws}$ and $\mathcal{IC}_{ws}$ is given in Equations (1) and (2), respectively, where *Constr* indicates a CLP constraint [16].

$$
\begin{aligned}
\mathcal{KB}_{ws} &::= [\ Clause\ ]^{\star} \\
Clause &::= Atom \leftarrow Cond \\
Cond &::= ExtLiteral\ [\ \wedge ExtLiteral\ ]^{\star} \\
ExtLiteral &::= [\neg]Atom \mid true \mid Expect \mid Constr \\
Expect &::= \mathbf{E}(Atom, Atom, Atom, Atom) \mid \\
&\quad\ \mathbf{EN}(Atom, Atom, Atom, Atom)
\end{aligned}
\tag{1}
$$

$$
\begin{aligned}
\mathcal{IC}_{ws} &::= [\ IC\ ]^{\star} \\
IC &::= Body \rightarrow Head \\
Body &::= (Event \mid Expect)\ [\wedge BodyLit]^{\star} \\
BodyLit &::= Event \mid Expect \mid Atom \mid Constr \\
Head &::= Disjunct\ [\ \vee Disjunct\ ]^{\star} \mid false \\
Disjunct &::= (Expect \mid Event \mid Constr) \\
&\quad\ [\ \wedge (Expect \mid Event \mid Constr)]^{\star} \\
Expect &::= \mathbf{E}(Atom, Atom, Atom, Atom) \mid \\
&\quad\ \mathbf{EN}(Atom, Atom, Atom, Atom) \\
Event &::= \mathbf{H}(Atom, Atom, Atom, Atom)
\end{aligned}
\tag{2}
$$

Let us see how we can implement the walk through scenario in SRE. Note that, following the LP notation, variables (in *italics*) start with upper-case letters. $T_r, T_a, \ldots$ indicate the (expected) time of events.

The first IC in $\mathcal{IC}_{eShop}$, corresponding to (shop1), is the following:

$$\begin{aligned} &\mathbf{H}(Customer, eShop, request(Item), T_r)\\ \rightarrow &accepted\_customer(Customer)\\ &\wedge accepted\_payment(How)\\ &\wedge \mathbf{H}(eShop, Customer, ask(pay(Item, How)), T_a)\\ &\wedge \mathbf{E}(Customer, eShop, pay(Item, How), T_p)\\ &\wedge T_p > T_a \wedge T_a > T_r\\ \vee &rejected\_customer(Customer)\\ &\wedge \mathbf{H}(eShop, Customer, inform(fail), T_i) \wedge T_i > T_r. \end{aligned} \quad \text{(shop1)}$$

All accepted payment modalities are listed in *eShop*'s knowledge base, $\mathcal{KB}_{eShop}$, shown in (kb) below. In our example, *Customer* may pay either by credit card or cash. The concepts of "accepted" and "rejected" customer are defined in the KB_{eShop} too: a *Customer* is accepted if the *Zone* she resides in is a valid destination for *eShop*; *Customer* is *rejected* otherwise. Both payment modalities and accepted destinations are listed as facts. In this example, *eShop* can only send items to Europe. The next element of *eShop*'s policies (shop2) states that if an accepted *Customer* pays for an *Item* using one of the supported payment modalities, then *eShop* will deliver the *Item* to *Customer*:

$$\begin{aligned} &\mathbf{H}(Customer, eShop, pay(Item, How), T_p)\\ &\wedge accepted\_customer(Customer)\\ &\wedge accepted\_payment(How)\\ \rightarrow &\mathbf{H}(eShop, Customer, deliver(Item), T_d)\\ &\wedge T_d > T_p. \end{aligned} \quad \text{(shop2)}$$

Finally, (shop3) states that if a *Customer* asks it to provide a guarantee (i.e., a certificate about its membership to $\mathcal{BBB}$), *eShop* will send such a guarantee:

$$\begin{aligned} &\mathbf{H}(Customer, eShop, request\_guar(\mathcal{BBB}), T_{rg})\\ \rightarrow &\mathbf{H}(eShop, Customer, give\_guar(\mathcal{BBB}), T_g)\\ &\wedge T_g > T_{rg}. \end{aligned} \quad \text{(shop3)}$$

$$\begin{aligned} accepted\_customer(Customer) \leftarrow &resident\_in(Customer, Zone)\\ &\wedge accepted\_dest(Zone).\\ rejected\_customer(Customer) \leftarrow &resident\_in(Customer, Zone)\\ &\wedge not(accepted\_dest(Zone)).\\ accepted\_payment(cc).&\\ accepted\_payment(cash).&\\ accepted\_dest(europe).& \end{aligned} \quad \text{(kb)}$$

When a (generic) *Shop* asks *alice* to pay an *Item* with credit card, then *alice* will request the *Shop* to provide a guarantee, and she will expect to receive it:

$$\begin{aligned} &\mathbf{H}(Shop, alice, ask(pay(Item, cc)), T_a) \\ \rightarrow &\mathbf{H}(alice, Shop, req\_guar(\mathcal{BBB}), T_{rg}) \\ &\wedge \mathbf{E}(Shop, alice, give\_guar(\mathcal{BBB}), T_g) \\ &\wedge T_g > T_{rg} \wedge T_{rg} > T_a. \end{aligned} \qquad \text{(alice1)}$$

If *Shop* provides a guarantee, *alice* will pay for the requested *Item*:

$$\begin{aligned} &\mathbf{H}(Shop, alice, ask(pay(Item, cc)), T_a) \\ &\wedge \mathbf{H}(Shop, alice, give\_guar(\mathcal{BBB}), T_g) \\ \rightarrow &\mathbf{H}(alice, Shop, pay(Item, cc), T_p) \\ &\wedge T_p > T_A \wedge T_p > T_g. \end{aligned} \qquad \text{(alice2)}$$

When the *Shop* asks to use a payment modality other than credit card, *alice* satisfies *eShop*'s request:

$$\begin{aligned} &\mathbf{H}(Shop, alice, ask(pay(Item, How)), T_a) \\ &\wedge How \neq cc \\ \rightarrow &\mathbf{H}(alice, Shop, pay(Item, How), T_p) \wedge T_p > T_A. \end{aligned} \qquad \text{(alice3)}$$

4 Declarative Semantics and Reasoning

In SRE, a client c specifies a goal $\mathcal{G}$, related to a requested service. $\mathcal{G}$ will often be an expectation, but in general it can be any goal, defined as a conjunction of expectations, CLP constraints, and any other literals. c also publishes a (possibly empty) knowledge base $\mathcal{KB}_c$, and a (possibly empty) set of policies $\mathcal{IC}_c$. The idea is to obtain, through abductive reasoning made by SRE, a set of expectations $\mathcal{E}$ and validity conditions Δ about a possible course of events that, together with $\mathcal{KB}_c$ and $\mathcal{KB}_{ws}$, satisfies $\mathcal{IC}_c \cup \mathcal{IC}_{ws}$ and $\mathcal{G}$. Note that we do not assume that all of ws's knowledge base is available to SRE, as it need not be entirely a part of ws's public specifications. $\mathcal{KB}_{ws}$ can even be the empty set. However, in general, ICs can involve predicates defined in the KB: such as "delivery in Europe." If the behavioural interface provided by ws involves predicates that have not been made public through $\mathcal{KB}_{ws}$, SRE makes assumptions about such unknown predicates, and considers unknowns that are neither **H** nor **E**/**EN** expectations as literals that can be abduced. These are kept then in the set Δ, of a returned triplet $\langle ws, \mathcal{E}, \Delta \rangle$ (see Sect. 2), and can be regarded as conditions which must be met to insure the validity of $\mathcal{E}$ as a possible set of expectations achieving a goal.

4.1 Declarative Semantics

We define declaratively the set of abductive answers $\langle ws, \mathcal{E}, \Delta \rangle$ representing possible ways c and ws can interact to achieve $\mathcal{G}$ (we assume that $\mathcal{KB}_c$ and $\mathcal{KB}_{ws}$ are consistent) via the two following equations:

$$\mathcal{KB}_c \cup \mathcal{KB}_{ws} \cup \mathcal{E} \cup \Delta \models \mathcal{G} \qquad (3)$$

$$\mathcal{KB}_c \cup \mathcal{KB}_{ws} \cup \mathcal{E} \cup \Delta \models \mathcal{IC}_c \cup \mathcal{IC}_{ws}. \qquad (4)$$

where $\mathcal{E}$ is a conjunction of **H** and **E**, **EN** atoms, Δ is a conjunction of abducible literals, and the notion of entailment is grounded on the *possible models* semantics defined for abductive disjunctive logic programs [23]. In the possible models semantics, a disjunctive program generates several (non-disjunctive) *split programs*, obtained by separating the disjuncts in the head of rules. Given a disjunctive logic program P, a *split program* is defined as a (ground) logic program obtained from P by replacing every (ground) rule

$$r : L_1 \vee \cdots \vee L_l \leftarrow \Gamma$$

from P with the rules in a non-empty subset of $Split_r$, where

$$Split_r = \{L_i \leftarrow \Gamma \mid i = 1, \ldots, l\}.$$

By definition, P has in general multiple split programs. A *possible model* for a disjunctive logic program P is then defined as an answer set of a split program of P.

Note that in [23] the possible models semantics was also applied to provide a model theoretic semantics for Abductive Extended Disjunctive Logic Programs (AEDP), which is our case. Semantics is given to AEDP in terms of *possible belief sets*. Given an AEDP $\Pi = \langle P, \mathcal{A} \rangle$, where P is a disjunctive logic program and $\mathcal{A}$ is the set of abducible literals, a possible belief set S of Π is a possible model of the disjunctive program $P \cup E$, where P is extended with a set E of abducible literals ($E \subseteq \mathcal{A}$).

Definition 2 (Answer to a goal $\mathcal{G}$). *An* answer E to a (ground) goal $\mathcal{G}$ *is a set E of abducible literals constituting the abductive portion of a possible belief set S (i.e., $E = S \cap \mathcal{A}$) that entails $\mathcal{G}$.*

We rely upon possible belief set semantics, but we adopt a new notion for minimality with respect to abducible literals. In [23], a possible belief set S is $\mathcal{A}$-minimal if there is no possible belief set T such that $T \cap \mathcal{A} \subset S \cap \mathcal{A}$. We restate, then, the notion of $\mathcal{A}$-minimality as follows:

Definition 3 ($\mathcal{A}$-minimal possible belief set). *A possible belief set S is $\mathcal{A}$-*minimal *iff there is no possible belief set T* for the same split program *such that $T \cap \mathcal{A} \subset S \cap \mathcal{A}$.*

More intuitively, the notion of minimality with respect to hypotheses that we introduce is checked by considering the *same* split program, and by checking whether with a smaller set of abducible literals the same consequences can be made true, but in the same split program. Finally, we provide a model-theoretic notion of explanation to an observation, in terms of answer to a goal, as follows.

Definition 4 ($\mathcal{A}$-minimal answer to a goal). *E is an $\mathcal{A}$-*minimal answer to a goal *$\mathcal{G}$ iff $E = S \cap \mathcal{A}$ for some possible $\mathcal{A}$-minimal belief set S that entails $\mathcal{G}$.*

Definition 5 (Possible Interaction about $\mathcal{G}$). *A* possible interaction about a goal *$\mathcal{G}$ between a client c and a web service ws is an $\mathcal{A}$-minimal set $\mathcal{E} \cup \Delta$ such that Equations 3 and 4 hold.*

Among possible interactions, we identify those which are *coherent*:[1]

Definition 6 (Coherent Possible Interaction about $\mathcal{G}$). *A possible interaction $\mathcal{E} \cup \Delta$ about a goal $\mathcal{G}$ is* coherent *iff:*

$$\mathcal{E} \models \mathbf{E}(X, Y, Action, T), \mathbf{EN}(X, Y, Action, T) \rightarrow false \quad (5)$$

Possible interactions about a goal $\mathcal{G}$ generally contain (minimal) sets of events and expectations about messages raised either by c and ws. Moreover, further abducible literals in Δ represent assumptions about unknown predicates (for c and ws).

SRE selects among coherent possible interactions only those where the course of events expected by c about ws's messages is *fulfilled* by ws's messages, and vice-versa, i.e., the course of events expected by ws about c's messages is *fulfilled* by c's messages:

Definition 7 (Possible Interaction Achieving $\mathcal{G}$). *Given a client c, a web service ws, and a goal $\mathcal{G}$, a* possible interaction achieving $\mathcal{G}$ *is a coherent possible interaction $\mathcal{E} \cup \Delta$ satisfying the following equations:*

$$\mathcal{E} \models \mathbf{E}(X, Y, Action, T) \rightarrow \mathbf{H}(X, Y, Action, T) \quad (6)$$

$$\mathcal{E} \models \mathbf{EN}(X, Y, Action, T), \mathbf{H}(X, Y, Action, T) \rightarrow false \quad (7)$$

In practice, Definition 7 requires that any positive expectation raised by c or ws on the behaviour of the other party is fulfilled by an event hypothetically performed by the other party (Equation 6), and that any negative expectation raised by c or ws on the behaviour of the other party does not match any event hypothetically performed by the other party (Equation 7).

4.2 Operational Semantics

The operational semantics of SRE is a modification of the $\mathcal{S}$CIFF proof-procedure [8], that combines forward, reactive reasoning with backward, goal-oriented reasoning, and was originally developed to check compliance of the agent behaviour to interaction protocols in multi-agent systems. Like the IFF proof procedure [13], which inspired it, $\mathcal{S}$CIFF is a rewriting system, defined in terms of *transitions* which turn a state of the computation into another. Since some of the transitions open choice points, a computation can be represented as a tree, whose root node is the initial state and whose leaves can be either the special node *fail*, or a termination node (i.e., a node to which no transition is applicable), that is considered as a *success* node (and, in the original $\mathcal{S}$CIFF setting, represents a response of compliance of the agent behaviour to the interaction protocols)

$\mathcal{S}$CIFF is sound and complete, under reasonable assumptions [8]; it has been implemented in SICStus Prolog and Constraint Handling Rules [12] and integrated in the SOCS-SI software component, in order to interface it to several

[1] This notion is introduced because of **EN** expectations in the SRE framework, and therefore the necessity of stating explicitly the incompatibility between **E** and **EN**.

multi-agent systems [7], and with web services via a RuleML encoding of ICs. The $\mathcal{S}$CIFF version that acts as the core reasoning engine in SRE is designed to reason, off-line, about the web service behaviour: a successful leaf node represents an interaction which achieves the desired goal while respecting the specified policies. SRE is a conservative modification of the SCIFF proof-procedure, in which the happened events are abducibles. The proofs of soundness and completeness can be trivially extended to such a case.

5 The *alice* and *eShop* Scenario Revisited

We provide here a sketched demonstration of the operational behaviour of the SRE engine, by showing how the answers to *alice*'s query are generated. Let us suppose that *alice* sends a query to SRE containing policies (alice1), (alice2) and (alice3), an empty knowledge base and the following goal $\mathcal{G}$:

$$\begin{aligned} \mathcal{G} \equiv\ & \mathbf{H}(alice, Shop, request(device), T_r) \\ & \wedge \mathbf{E}(Shop, alice, deliver(device), T_d) \wedge T_d > T_r . \end{aligned} \tag{goal1}$$

which states that *alice* will send a request to some *Shop* in order to obtain *device* and she expects that *Shop* will deliver it. SRE starts from *alice*'s goal:

$$\begin{aligned} \mathcal{E}_0 = \{ & \mathbf{H}(alice, eShop, request(device), T_r), \\ & \mathbf{E}(eShop, alice, deliver(device), T_d), T_d > T_r \ \} \\ \Delta_0 = \emptyset & \end{aligned}$$

According to (shop1), *eShop* may react to this expectation in different ways, depending on whether *alice* is an accepted customer or not. SRE tries initially to resolve predicate $accepted\_customer(alice)$. By unfolding it, SRE finds atom $resident\_in(alice, Zone)$, which is not known to SRE and, therefore, is abduced. Afterwards, based on $\mathcal{KB}_{eShop}$, the *eShop* public knowledge base, SRE grounds $Zone$ to *europe*: the only destination accepted by *eShop*. As a consequence, hypothesising that *alice* is resident in *europe*, *eShop* would ask *alice* to pay with one of the accepted modalities, and it would expect to receive the payment in response. Moreover, *eShop* specifies in $\mathcal{KB}_{eShop}$ that credit card is an accepted payment modality.

$$\begin{aligned} \mathcal{E}_1 = \{ & \mathbf{H}(alice, eShop, request(device), T_r), \\ & \mathbf{H}(eShop, alice, ask(pay(device, cc), T_a), \\ & \mathbf{E}(alice, eShop, pay(device, cc), T_p), \\ & \mathbf{E}(eShop, alice, deliver(device), T_d), \\ & T_p > T_a, T_a > T_r, T_d > T_r \ \} \\ \Delta_1 = \{ & resident\_in(alice, europe)\} \end{aligned}$$

alice has specified that, in order to perform credit card payments, she requests a guarantee from the shop (alice2); *eShop* volunteers to provide such a document, by (shop3), and *alice*'s expectation about the guarantee is then satisfied (SRE hypothesises that the document is indeed sent):

$$\begin{aligned}\mathcal{E}_2 = \{&\mathbf{H}(alice, eShop, request(device), T_r),\\ &\mathbf{H}(eShop, alice, ask(pay(device, cc), T_a),\\ &\mathbf{H}(alice, eShop, req\_guar(\mathcal{BBB}), T_{rg}),\\ &\mathbf{H}(eShop, alice, give\_guar(\mathcal{BBB}), T_g),\\ &\mathbf{E}(alice, eShop, pay(device, cc), T_p),\\ &\mathbf{E}(eShop, alice, deliver(device), T_d),\\ &T_g > T_{rg}, T_{rg} > T_a, T_p > T_a, T_a > T_r, T_d > T_r\ \}\\ \Delta_2 = \{&resident\_in(alice, europe)\}\end{aligned}$$

Upon receipt of the guarantee, *alice* would proceed with the payment (alice2), and *eShop* would deliver the *device* (shop3). Therefore, the following, (possibly) fruitful, interaction is found by SRE:

$$\begin{aligned}\mathcal{E}_F = \{&\mathbf{H}(alice, eShop, request(device), T_r),\\ &\mathbf{H}(eShop, alice, ask(pay(device, cc), T_a),\\ &\mathbf{H}(alice, eShop, req\_guar(\mathcal{BBB}), T_{rg}),\\ &\mathbf{H}(eShop, alice, give\_guar(\mathcal{BBB}), T_g),\\ &\mathbf{H}(alice, eShop, pay(device, cc), T_p),\\ &\mathbf{H}(eShop, alice, deliver(device), T_d),\\ &T_d > T_p, T_p > T_g, T_g > T_{rg}, T_{rg} > T_a, T_a > T_r\ \}\\ \Delta_F = \{&resident\_in(alice, europe)\}\end{aligned}$$

SRE provides in output also a simpler possible interaction, where instead of selecting "credit card" as payment method, "cash" is now preferred. Policy (alice3) tells us that, in such a case, *alice* would proceed straightforward with the payment, and SRE is indeed able to propose a second fruitful interaction as answer to *alice*'s initial query.

In order to compute these two possibly fruitful interactions, $resident\_in(alice, europe)$ has been abduced. This means that if such interactions are really possible or not, it depends on whether *alice* resides in *europe*, and in fact it may well turn out that such interaction is not possible at all. SRE looks also for other solutions where this hypothesis is not assumed, but all other interactions do not satisfy *alice*'s goal, and hence they are discarded.

5.1 Refined Query

In the second scenario, *alice* submits a different goal, and the $\mathcal{KB}$ below:

$$\begin{aligned}\mathcal{G} \equiv\ &\mathbf{H}(alice, Shop, request(device), T_r)\\ &\wedge \mathbf{E}(Shop, alice, deliver(device), T_d) \wedge T_d > T_r\\ &\wedge \mathbf{EN}(alice, Shop, pay(device, cash), T_p).\end{aligned} \qquad \text{(goal2)}$$

$$resident\_in(alice, europe).\ \ age(alice, 24). \qquad \text{(kb2)}$$

This time, *alice* explicitly prohibits to pay cash (this is expressed using the **EN** notation). Thanks to the piece of knowledge (kb2) that *alice* provides through her

$\mathcal{KB}$, SRE knows that *alice* does indeed resides in the EU, hence this information does not need to be abduced anymore, but it is simply verified. SRE finds a solution which is similar to the one above (Scenario 1). However, since this time the set Δ is empty, this interaction will surely lead to success, provided that both *alice* and *eShop* behave coherently with respect to their own policies.

5.2 Unconstrained Query

As we have pointed out, *alice* is able to query SRE without specifying any policy. In this case, *alice* only wishes to obtain a list of services that are able to accommodate her goal. In such a situation, *alice* only sends the following general policy:

$$\begin{aligned} &E(alice, Shop, DoSomething, T)\\ &\rightarrow H(alice, Shop, DoSomething, T) \end{aligned} \tag{r1}$$

which specifies that *alice* will perform every action that she is expected to do. If *alice* queries SRE by using (r1) and (goal1), the response of SRE will be:

$$\begin{aligned} \mathcal{E}_F = \{&\mathbf{H}(alice, eShop, request(device), T_r),\\ &\mathbf{H}(eShop, alice, ask(pay(device, How), T_a),\\ &\mathbf{H}(alice, eShop, pay(device, How), T_p),\\ &\mathbf{H}(eShop, alice, deliver(device), T_d),\\ &T_d > T_p, T_p > T_a, T_a > T_r, How :: [cc, cash]\ \}\\ \Delta_F = \{&resident\_in(alice, europe)\} \end{aligned}$$

6 Discussion

We described a reasoning engine, SRE, which considers the policies of two web services and a goal of one of them. SRE tries to match such policies and find possible ways the two web services could interact, and eventually achieve the goal. The output of SRE is a sequence of events, which could be messages to be exchanged between the web services and lead to a state in which the goal is achieved. This can be regarded as a possible plan of action. SRE is based on a mixture of ALP and CLP. ALP is used to construct sets of input and output messages, along with assumed data, while CLP is used to maintain a partial temporal order among the plans. In this work we did not address the issue of efficiency of the reasoning process of SRE, but we are aware that this may be a drawback, as it is the case with many expressive logics proposed for the Semantic Web. We intend to evaluate SRE, both as it concerns its complexity and its efficiency, through an empirical analysis based on case studies.

Another aspect we did not look into in detail is the problem of reasoning about equivalences of concepts or ontologies, as much related work instead does. Also our notions of *action*, such as it could be the delivery of goods, are pretty much left at the abstract level. Our proposal could be regarded as a functionality complementary to many proposals, which could further improve the discovery process. To cite some, [3,22] propose ontology languages to define web services. In [4], besides proposing a general language for representing semantic web service specification using logic, a discovery scenario is depicted and an architectural solution is proposed (we draw inspiration for our scenario from the Discovery Engine example). A notion of "mediator" is introduced to overcome differences between different ontologies, and then a reasoning process is performed over the user inputs and the hypothetical effects caused by the service execution.

Our work makes explicit reference to [18], in which the authors present a framework for automated web service discovery which uses the Web Service Modeling Ontology (WSMO) as the conceptual model for describing web services, requester goals and related aspects. Whereas [18] tackles both (mainly) discovery and contracting stage, in our work we are only concerned with the contracting stage. In [18] the authors use F-logic and transaction logic as the underlying formalisms, we rely on extended logic programming. In both the approaches, however, hypothetical reasoning is used for service contracting. Compare to work by Kifer et al. [18,4], in which only the client's goal is considered, in our framework the client can specify its policies. In this way, the client could be considered a web service as well. Therefore, we hope to be able to smoothly extend SRE to dealing with the problem of inter-operability. A proposal in this direction is presented in [6].

The outcome of the reasoning process performed by SRE, which we called a possible plan, could in fact be regarded as a sort of "contract agreement" between the client and the discovered service, provided that each party is tightly bounded to its previously published policies/knowledge bases. For example, the dynamic agreement about contracts (e-contracting) is addressed in SweetDeal [15,10], where Situated Courteous Logic (SCL) is adopted for reasoning about rules that define business provisions policies. The formalism used supports contradicting rules (by imposing a prioritisation and mutual exclusion between rules), different ontologies, and effectors as procedures with side effects. However, SweetDeal is more focussed on establishing the characteristics of a business deal, while our aim is to address the problem of evaluating the feasibility of an interaction. To this end, we perform hypothetical reasoning on the possible actions and consequences; moreover, we hypothesise which condition must hold, in order to inter-operate. This technique in literature is also known as "constructive" abduction.

Other authors propose to rules to reason about established contracts: in [14], for example, Defeasible Deontic Logic of Violation is used to monitor the execution of a previously agreed contract. We have addressed this issue in a companion paper [9], where integrity constraints have been exploited and conciliated with the deontic concepts. Among other work in the area of policy specifications and matching we find PeerTrust [21,5]. Similarly to our work and to SCL, PeerTrust builds upon an LP foundation to represent policy rules and iterative trust declaratively. In PeerTrust, trust is established gradually by disclosing credentials and requests for credentials by using a process of trust negation. An important difference is in the language used in PeerTrust for specifying policies, which can be considered as orthogonal to the one described in this paper. While PeerTrust represents policies and credentials as guarded distributed logic programs, and the trust negotiation process consists of evaluating an initial LP query over a physically distributed logic program, in this work we use ALP, integrity constraints and CLP constraints for expressing policies, perform a local proof and we use abductive reasoning to formulate hypotheses about unknown external behaviour. Moreover, while in our current approach reasoning is done in a single step using $\mathcal{S}$CIFF, an iterative version could be introduced in order to support trust negotiation.

Acknowledgments

We wish to thank the anonymous reviewers for their comments, constructive criticisms and valuable suggestions. This work has been partially supported by the National FIRB project TOCAI.it and by the PRIN projects Nos. 2005-011293 and 2005-015491.

References

1. `http://lia.deis.unibo.it/research/socs/`.
2. `http://www.w3.org/TR/rif-ucr/`.
3. `http://www.daml.org/services/owl-s/`.
4. `http://www.w3.org/Submission/SWSF-SWSL/`.
5. `http://www.l3s.de/peertrust/`.
6. M. Alberti, F. Chesani, M. Gavanelli, E. Lamma, P. Mello, and M. Montali. An abductive framework for a-priori verification of web services. In *Proc. PPDP*, pp. 39–50. ACM Press, 2006.
7. M. Alberti, F. Chesani, M. Gavanelli, E. Lamma, P. Mello, and P. Torroni. Compliance verification of agent interaction: a logic-based tool. *Applied Artificial Intelligence*, **20**(2-4):133–157, 2006.
8. M. Alberti, F. Chesani, M. Gavanelli, E. Lamma, P. Mello, and P. Torroni. Verifiable agent interaction in Abductive Logic Programming: the SCIFF framework. *ACM Transactions on Computational Logic*, **8**, 2007.
9. M. Alberti, M. Gavanelli, E. Lamma, P. Mello, G. Sartor, and P. Torroni. Mapping deontic operators to abductive expectations. *Computational and Mathematical Organization Theory*, **12**(2-3):205–225, 2006.
10. S. Bhansali and N. Grosof. Extending the sweetdeal approach for e-procurement using sweetrules and RuleML. In *Proc. RuleML*, *LNAI* **3791**:113–129, 2005.
11. F. Bry and M. Eckert. Twelve theses on reactive rules for the web. In *Proc. Workshop on Reactivity on the Web*, Munich, Germany, March 2006.
12. T. Frühwirth. Theory and practice of constraint handling rules. *Journal of Logic Programming*, **37**(1-3):95–138, 1998.
13. T. H. Fung and R. A. Kowalski. The IFF proof procedure for abductive logic programming. *Journal of Logic Programming*, **33**(2):151–165, 1997.
14. G. Governatori and D. P. Hoang. A semantic web based architecture for e-contracts in defeasible logic. In *Proc. RuleML*, *LNAI* **3791**:145–159, 2005.
15. B. N. Grosof and T. C. Poon. SweetDeal: representing agent contracts with exceptions using XML rules, ontologies, and process descriptions. In *Proc. 12th WWW*, pp. 340–349. ACM Press, 2003.

16. J. Jaffar and M. Maher. Constraint logic programming: a survey. *Journal of Logic Programming*, **19-20**:503–582, 1994.
17. A. C. Kakas, R. A. Kowalski, and F. Toni. Abductive Logic Programming. *Journal of Logic and Computation*, **2**(6):719–770, 1993.
18. M. Kifer, R. Lara, A. Polleres, C. Zhao, U. Keller, H. Lausen, and D. Fensel. A logical framework for web service discovery. In *Semantic Web Services: Preparing to Meet the World of Business Applications. CEUR Workshop Proc.* **119**, 2004.
19. R. A. Kowalski and F. Sadri. From logic programming towards multi-agent systems. *Annals of Mathematics and Artificial Intelligence*, **25**(3/4):391–419, 1999.
20. S. A. McIlraith, T. C. Son, and H. Zeng. Semantic web services. *IEEE Intelligent Systems*, **16**(2):46–53, 2001.
21. W. Nejdl, D. Olmedilla, and M. Winslett. PeerTrust: Automated trust negotiation for peers on the semantic web. In *Proc. Secure Data Management (SMD 2004)*, *LNAI* **3178**:118–132. Springer-Verlag, 2004.
22. D. Roman, U. Keller, H. Lausen, J. de Bruijn, R. Lara, M. Stollberg, A. Polleres, C. Feier, C. Bussler, and D. Fensel. Web service modeling ontology. *Applied Ontology*, **1**(1):77 – 106, 2005.
23. C. Sakama and K. Inoue. Abductive logic programming and disjunctive logic programming: their relationship and transferability. *JLP*, **44**(1-3):75–100, 2000.

Dynamic Service Discovery Through Meta-interactions with Service Providers*

Tomas Vitvar, Maciej Zaremba, and Matthew Moran

Digital Enterprise Research Institute (DERI),
National University of Ireland, Galway
{tomas.vitvar,maciej.zaremba,matthew.moran}@deri.org

Abstract. Dynamic discovery based on semantic description of services is an essential aspect of the Semantic Web services integration process. Since not all data required for service discovery can always be included in service descriptions, some data needs to be obtained during run-time. In this paper we define a model for service interface allowing required data to be fetched from the service provider during discovery process. We also provide a specification of such interface for WSMO and demonstrate the model on a case scenario from the SWS Challenge implemented using WSMX – a middleware platform built specifically to enact semantic service oriented architectures.

1 Introduction

The Web has a volatile nature where there can only be a limited guarantee of being able to access any specific service at a given time. This leads to a strong motivation for discoverying and binding to services at run-time (*late binding*). Existing XML-based WSDL descriptions of data, messages, or interfaces are insufficient or provide limited expressivity for machines to understand. Service discovery operating on semantic descriptions offer the potential of flexible matching that is more adaptive to changes over services' lifetime. In general, discovery matches definitions of user requests (goals) with those of offered services. Different levels of match are possible e.g. subsumption match, plug-in match, exact match etc.[13,4]. Semantic discovery works on the abstract definitions of services and goals (containing no instance data). This needs to be further elaborated to achieve more accurate results. For example, a request to "buy a Harry Potter book" involves first searching for descriptions of services that sell books, but which then determining if the service sells Harry Potter books and if those books are in stock. Taking Amazon as an example, it is clearly unfeasible to include data for the entire catalogue and its availability directly in the service description. Such information has a dynamic character and therefore should only be fetched from the service at discovery-time.

* This work is supported by the Science Foundation Ireland Grant No. SFI/02/CE1/I131, and the EU projects Knowledge Web (FP6-507482), DIP (FP6-507483) and SUPER (FP6-026850).

E. Franconi, M. Kifer, and W. May (Eds.): ESWC 2007, LNCS 4519, pp. 84–98, 2007.

For this purpose, we propose a general mechanism enabling the definition of an interface on the service to allow the *fetching* of required data from the service during the late binding phase (e.g during service discovery, contracting/negotiation, selection etc.). These tasks are performed in a semantic service environment in a (semi) automated fashion by means of the "intelligence" of intermediary (middleware) services. We define a model for the service interface which provides a mechanism to fetch data from the service provider during the discovery process. Choosing the Web Service Modeling Ontology (WSMO) as our conceptual model, we define an extension for this interface and demonstrate this work through a case scenario of the SWS Challenge[1] implemented using WSMX – a middleware platform built specifically to enact semantic service oriented architectures.

In section 2 we introduce the underlying specifications for our work, namely WSMO, WSML and WSMX providing a conceptual framework, ontology language and execution environment for Semantic Web services. In section 3 we define a model for a service interface and algorithm to fetch data for service discovery and further show how this model can be specified using WSMO service model. In section 4 we illustrate the model on the case scenario implemented in the WSMX environment and describe the evaluation for the implementation. In section 5 we describe related work and in section 6 we conclude the paper and indicate our future work.

2 Semantic Web Services and WSMO

A general aim of Semantic Web Services is to define a semantic mark-up for Web services providing the higher expressivity then traditional XML-based descriptions. One of the initiatives in the area is the Web Service Modeling Ontology (WSMO)[11]. WSMO provides a conceptual model describing all relevant aspects of Web services in order to facilitate the automation of service discovery, composition and invocation. The description of WSMO elements is represented using the Web Service Modeling Language (WSML)[11] – a family of ontology languages – which consists of a number of variants based on different logical formalisms and different levels of logical expressiveness. WSMO also defines the conceptual model for WSMX[9], a Semantic Web Services execution environment. Thus, WSMO, WSML and WSMX form a coherent framework for modeling, describing and executing Semantic Web Services. The WSMO top-level conceptual model consists of *Ontologies*, *Web Services*, *Goals*, and *Mediators*.

- **Ontologies** provide the formal definition of the information model for all aspects of WSMO. Two key distinguishing features of ontologies are, the principle of a shared conceptualization and, a formal semantics (defined by WSML in this case). A shared conceptualization is one means of enabling information interoperability across independent Goal and Web service descriptions.

[1] http://sws-challenge.org

- **Web Services** are defined by the functional capability they offer and one or more interfaces that enable a client of the service to access that capability. The Capability is modeled using preconditions and assumptions, to define the state of the information space and the world outside that space before execution, and postconditions and effects, defining those states after execution. Interfaces are divided into *choreography* and *orchestration*. The choreography defines how to interact with the service while the orchestration defines the decomposition of its capability in terms of other services.
- **Goals** provide the description of objectives a service requester (user) wants to achieve. WSMO goals are described in terms of desired information as well as "state of the world" which must result from the execution of a given service. The WSMO goal is characterized by a requested capability and a requested interface.
- **Mediators** describe elements that aim to overcome structural, semantic or conceptual mismatches that appear between different components within a WSMO environment. Although WSMO Mediators are essential for addressing the requirement of loosely coupled and heterogeneous services, they are out of the scope of our work at this point.

In this paper, we will elaborate on WSMO service definition and in particular on its service interface. The service interface defines choreography as a formal description of a communication pattern the service offers. Two types of such choreography interfaces are defined: (1) *execution choreography* used during the execution phase where the functionality of the service is consumed by a service requester and (2) *meta-choreography* used during late binding to get additional information necessary for communication with the service. Since the WSMO model is open, such definitions may be extended to be used for the particular tasks of the late binding phase. We focus on the definition of one such extension for use for the service discovery phase.

3 Data Fetching for Discovery

A Web service capability can be described in terms of results potentially delivered by the service. A goal describes its capability as the information the user wants to achieve as a result of service provision. We denote the description of the Web service and the goal as $\mathcal{W}$ and $\mathcal{G}$ respectively. For the $\mathcal{W}$ and $\mathcal{G}$ we also introduce the data of these descriptions which we denote $\mathcal{D}_\mathcal{W}$ and $\mathcal{D}_\mathcal{G}$ respectively and which is provided *directly* as part of their respective descriptions.

For purposes of our work and based on grounds of [4,13], we define the following three basic steps when the matching of a goal $\mathcal{G}$ and a Web service $\mathcal{W}$ needs to be performed: (1) *Abstract-level match*, (2) *Instance-level Match*, (3) *Data Fetching*. Abstract-level Match operates on abstract descriptions of $\mathcal{G}$ and $\mathcal{W}$ without their data being taken into account. The matching is defined by the following set-theoretic relationships between objects of $\mathcal{G}$ and $\mathcal{W}$: (1) exact match, (2) subsumption match, (3) plug-in match, (4) intersection match and

(5) disjointness. If the goal and the Web service match (relationships 1-4), it is further checked if the service can provide a concrete service by consulting the data of the goal and the service (Instance-level Match). If all data is not available for step 2, the data needs to be obtained from the service (Data Fetching). Later in this section we further formalize these steps and define the algorithm. For step 1 and step 2, we define a matching function as follows:

$$s \leftarrow matching(\mathcal{G}, \mathcal{W}, \mathcal{B}_{gw}), \tag{1}$$

where $\mathcal{G}$ and $\mathcal{W}$ is a goal and a service description respectively and $\mathcal{B}_{gw}$ is a common knowledge base for the goal and the service. The knowledge base contains data which must be *directly* (through descriptions $\mathcal{G}$ and $\mathcal{W}$) or *indirectly* (through data fetching) available so that the matching function can be evaluated. The result s of this function can be: (1) *match* when the match was found at both abstract and instance levels (in this case all required data in $\mathcal{B}_{gw}$ is available), (2) *nomatch* when the match was not found either at abstract level or at instance level (in this case all required data in $\mathcal{B}_{gw}$ is available), or (3) *nodata* when some required data in $\mathcal{B}_{gw}$ is not available and thus the matching function cannot be evaluated.

We further assume that all data for the goal is directly available in the description $\mathcal{G}$. The data fetching step is then performed for the service when the matching function cannot be evaluated (the result of this function is *nodata*). We then define the knowledge base as:

$$\mathcal{B}_{gw} = \mathcal{D}_{\mathcal{G}} \cup \mathcal{D}_{\mathcal{W}} \cup \{y_1, y_2, ..., y_m\}, \tag{2}$$

where $\{y_i\}$ is all additional data that needs to be fetched from the service in order to evaluate the matching function.

Based on the representation of service interface using abstract state machines [12], we define the *data fetch interface* for service $\mathcal{W}$ as

$$\mathcal{I}_{\mathcal{W}} = (in(\mathcal{W}), out(\mathcal{W}), L), \tag{3}$$

where $in(\mathcal{W})$ and $out(\mathcal{W})$ denotes input and output vocabularies which correspond to input and output data of the interface $in(\mathcal{I}_{\mathcal{W}})$ and $out(\mathcal{I}_{\mathcal{W}})$ respectively, and L is a set of transition rules. The matching function can be then evaluated if

$$\forall y_i \in out(\mathcal{I}_{\mathcal{W}}) : \exists x \in \mathcal{B}_{gw} \wedge x \in in(\mathcal{I}_{\mathcal{W}}) \qquad i = 1, 2, \ldots, m. \tag{4}$$

According to 4, data $\{y_i\}$ can be fetched from the service through the data fetch interface if input data $in(\mathcal{I}_{\mathcal{W}})$ is either initially available in the knowledge base $\mathcal{B}_{gw}$ (data directly available from the goal or web service descriptions) or the input data becomes available during the processing of the interface. For a rule $r \in L$ we denote $r.ant$ and $r.con$ as the rule *antecedent* and the rule *consequent* respectively. The antecedent $r.ant$ defines an expression which if holds during run-time in the *memory*[2], the memory is modified according to the definition

[2] We use the term memory to denote a processing memory through which states of an abstract state machine are maintained during its processing.

of an action in *r.con* (i.e. specified data is *added*, *updated* or *removed* from the memory) (see the algorithm in section 3.1). Please note that each concept of the vocabulary $in(\mathcal{W})$ and $out(\mathcal{W})$ has defined *grounding* to respective message in WSDL. Through this grounding definition it is possible to invoke WSDL operation when instance data of the concept is to be added or updated in the memory (and thus the data is fetched from the service). This definition of the grounding is described in [6].

3.1 Algorithm

In algorithm 1, the matching function is integrated with data fetching. The algorithm operates on inputs, produces outputs and uses internal structures as follows:

Input

- Repository $Q = \{W_1, W_2, ..., W_n\}$, where $W \in Q$ is the web service description. For each web service W we denote D_W as data of the web service and I_W as data fetching interface of the web service with rule base L. This interface is defined according to definition 3 and its description is optional for the web service. In addition, for each rule $r \in L$ we denote action of the rule consequence as *r.con.action* and its corresponding data as *r.con.data*.
- Goal description G for which we denote D_G as data of the goal.

Output

- List $E = \{W_1, W_2, ..., W_m\}$, where $W_i \in Q$ and W_i matches G (the result of the matching function for W_i and G is *match*).

Uses

- Processing memory M containing data fetched during processing of the rules of the data fetching interface.
- Knowledge base B_{gw} which contains data for processing of the matching function.
- Boolean variable $modified$ indicating whether the knowledge base has been modified or not during the processing.

The algorithm performs the matching of the goal with each Web service in the repository using the matching function (line 7). If the matching cannot be evaluated (the result is *nodata*), the algorithm tries to fetch data from the service by processing the service's data fetch interface. Whenever the new data is available from the service, the algorithm updates the knowledge base and process the matching again. This cycle ends when no data can be fetched from the interface or matching can be evaluated (the result is *match* or *nomatch*). In case the matching is evaluated as *match*, the web service is added to the list of matched services and the cycle is performed for the next service in the repository.

Algorithm 1. Data Fetching for Discovery

```
E ← ∅
for all W in Q do
  B_gw ← D_G ∪ D_W
  M ← B_gw
  repeat
    modified ← false
    s ← matching(G, W, B_gw)
    if s = nodata and exists(I_w) then
      while get r from L: holds(r.ant, M) and not modified do
        if r.con.action = add then
          add(r.con.data, M)
          add(r.con.data, B_gw)
          modified ← true
        end if
        if r.con.action = remove then
          remove(r.con.data, M)
        end if
        if r.con.action = update then
          update(r.con.data, M)
          update(r.con.data, B_gw)
          modified ← true
        end if
      end while
    end if
  until s ≠ nodata or not modified
  if s = match then
    E ← E ∪ W
  end if
end for
```

During the processing of the interface, the algorithm allows to hook in a matching function which is called whenever the new data is available from the service. The algorithm uses independent memory (memory M) from the knowledge base (B_{gw}) for processing of the data fetching interface. This allows that already-obtained data cannot be removed from the knowledge base while, at the same time, correct processing of the interface is ensured. The memory M is used not only for data but also for control of interface processing (in general, the content of the memory does not need to always reflect the content of the knowledge base). According to the particular interface definition, the data can be fetched step-wise allowing minimizing of the interactions with the service during discovery. This also follows the strong decoupling principle when services are described semantically and independently from users' requests. For example, during service creation phase a service provider (creator) does not know which particular data will be required for particular data fetching (in general, matching with a goal could require some or all defined data which depends on the definition of the request). The interface defined using rules allows to get only data which is needed for the matching (for example in some cases only price is needed, in some cases price and a location of selling company could be needed if offered by the service).

3.2 WSMO Service Interface for Data Fetching

As stated in [11], Web Service interface defines *choreography* and *orchestration* allowing the modeling of external and internal behavior of the service respectively. In this respect, the interface for data fetching follows the WSMO service interface describing a meta-choreography which allows additional data to be obtained from the service for the discovery process. WSMO service will thus have additional interface defined (WSMO service allows multiple interface definitions). This interface will however only use the choreography describing a meta-choreography for obtaining additional data for the discovery process. We do not specify *orchestration* for this interface as the logic of how data fetch is performed by the service (how data is obtained out of aggregation of other services) is not of interest for discovery and we do not use it in our algorithm. In order to distinguish between the interface used for data fetching and the interface used for execution (defining how actual service is consumed by the service requester), we use non-functional property. For purposes of our work we further use non-functional property *interfacePurpose* with values *execution* and *discovery*. Another possibilities for distinguishing both interfaces would be to define data-fetch interface as specialization of WSMO service interface. The decision on whichever approach will be used will be done in the context of the WSMO WG.

4 Implementation and Evaluation

The model introduced in this paper has been implemented and evaluated through the SWS Challenge discovery scenario. The scenario introduces five different service providers offering various purchasing and shipment options. They provide different availability and pricing for their services with constraints on package destination, weight, dimension and shipment date. Service requesters with different requirements search for the best offers with packages of different weight and shape. Our model for data fetching for discovery fits well into this scenario since not all information can be provided in service descriptions meaning they must be dynamically obtained at discovery-time. In this section we base examples on the Mueller service whose price information is not available in the service description and needs to be fetched during the service discovery via data fetching interface. In section 4.3 we further describe the evaluation of our implementation from the broader context of the SWS Challenge requirements.

4.1 Scenario and Assumptions

In the scenario depicted in figure 1, a user accesses the e-marketplace called Moon where a number of companies such as Mueller or Racer have registered their services. The Moon runs a (1) Web portal through which it provides services to users and (2) the WSMX middleware system through which it facilitates the integration process between users and service providers.

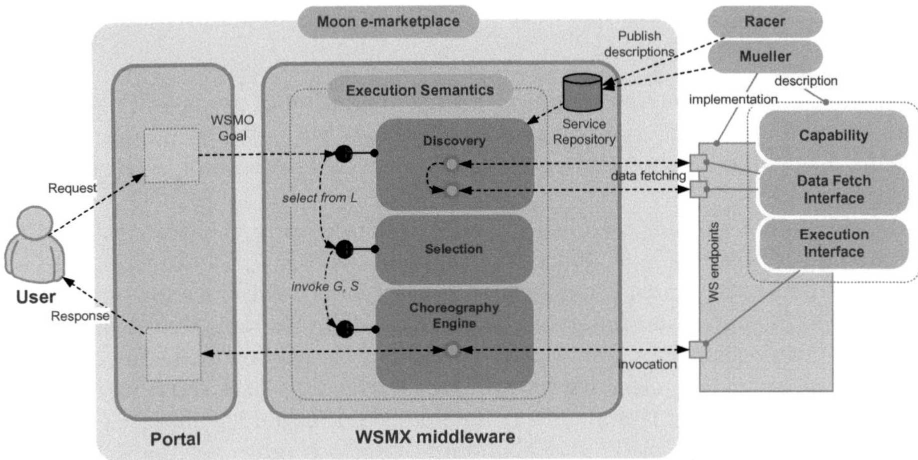

Fig. 1. Architecture for the Scenario

For this scenario and the aims of this paper we make the following assumptions.

- Service providers and Moon use the WSMO formalism for Web service description. We assume that service requesters maintain their own adapters to their back-end systems while at the same time providing lifting/lowering functionality between their existing technology (e.g. WSDL, XML Schema) and WSML.
- All service providers adopt a common ontology maintained by the Moon e-marketplace. Handling data interoperability issues, where service providers and Moon use different ontologies, is out of the scope of this paper.
- During execution, interactions between the user and the Moon are simplified to a single request-response exchange. Either the user submits a goal (our scenario) or a pre-selected service for invocation. Meta-interactions between users and the middleware system are not of our interest in this work.

In our scenario, a user defines her requests through the Web portal's user interface. The Web portal generates a WSMO goal corresponding to the request, sends it to WSMX, receives the response and presents the result back to the user. The execution process, run in WSMX after the receipt of the goal, includes discovery (with data-fetching from services), selection of the best service and invocation. Although the whole process of this scenario is implemented, the contribution of this paper lies in the integration of data fetching with discovery. Other parts, i.e. selection and invocation are not described in detail here.

4.2 Modeling Ontologies, Goals and Services

In order to implement the scenario, we first need to create semantic models for ontologies, goals and services. We describe these models in the following subsections. We present examples of ontologies, services and goals definitions in WSML using the following prefixes to denote their respective namespaces: *mo* – common ontology, *gl* – goal ontology.

Ontologies describe concepts used for the definition of goals and services. In our scenario we use a common scenario ontology with additional ontologies to define specific axioms or concepts used by the descriptions of services and/or goals.

The common ontology defines shared concepts used in the description of the goal and services, such as *Address*, *ShipmentOrderReq*, *Package*, etc. In addition, we use the common ontology to specify named relations for services and goals. Specific ontologies for goals and services declare axioms that define the relations to represent their conditions. An analogy for this approach are interfaces in programming labguages like Java. The interface declares some functionality but does not say how this should be implemented. Using this approach, we define a set of relations in the common ontology which represent the axioms that a service may need to define. Listing 1.1 shows the simple definition for the *isShipped* relation from the common ontology and its implementation in the Mueller's ontology.

```
/* isShipped relation in the common ontology */ relation
isShipped(ofType mo#ShipmentOrderReq)

/* implementation of the isShipped relation in the Mueller's ontology */
axiom isShippedDef
    definedBy
      ?shipmentOrderReq[mo#to hasValue ?to, mo#package hasValue ?package] memberOf mo#
            ShipmentOrderReq and
      ?to[mo#city hasValue ?city] and
      isShippedContinents(?city, mo#Europe, mo#Asia, mo#Africa) and
      ((?package[mo#weightKg hasValue ?weightKg] memberOf mo#Package) and (?weightKg<50))
    implies
      mo#isShipped(?shipmentOrderReq).
```

Listing 1.1. *isShipped* relation

The relation *isShipped* is true if the service provider can ship products according to the shipment order request (*ShipmentOrderReq*). In the second part of the listing 1.1, *isShipped* is defined such that the destination city for the shipment must be in Europe, Asia or Africa and the weight of the package is less then 50kg. This forms part of the Mueller service description.

Services. We focus on the description of the data-fetching interface of the Mueller service showing how and which data can be fetched during discovery.

```
1  interface WSMullerDataFetchInterface
2    nfp
3      _"interfacePurpose" hasValue "discovery"
4      ...
5    endnfp
6
7    choreography WSMullerDataFetchChoreography
8      ...
9      transitionRules WSMullerDataFetchTransitionRules
10         /* Rule 1: Request for product quote */
11         forall {?purchaseQuoteReq} with (
12           ?purchaseRequest memberOf mo#PurchaseQuoteReq
13         ) do
14           add(_# memberOf mo#PurchaseQuoteResp)
15         endForall
16
17         /* Rule 2: Request for shipment quote */
18         forall {?shimpmentQuoteReq} with (
19           ?purchaseQuoteResp[mo#package hasValue ?package] memberOf mo#
                  PurchaseQuoteResp and
20           ?shipmentQuoteReq[mo#to hasValue ?to] memberOf mo#ShipmentOuoteReq and
21           mo#isAvailable(?purchaseQuoteResp) and mo#isShipped(?to, ?package)
22         ) do
23           add(_# memberOf mo#ShipmentQuoteResp)
24         endForall
```

Listing 1.2. Mueller data fetching interface

In listing 1.2, the first rule (line 6) describes how to get the price and the product availability information (the quote request data is part of the goal description). The second rule (line 13) describes how to get a quote for shipment. This rule will be only used if requested product is available (determined through relation *isAvailable*) and Mueller can ship to specified address (determined through relation *isShipped*). Here, shipment address (*to* variable) is taken from the shipment quote request and packaging information (*package* variable) is taken from purchase order response. According to this definition, the first rule is used independently (and could be the only rule used where the user does not request shipment) while for the second rule, the first rule need to be executed first (the rule can be executed if the product is available and shippable which is determined through results of the first rule). Concepts *PurchaseQuoteReq*, *ShipmentOuoteReq* and *PurchaseQuoteResp*, *ShipmentQuoteResp* are defined as input and output vocabularies respectively (including grounding mechanism) (for brevity, this is not shown in the listing).

Goals. The goal for the scenario describes the user's aim to buy certain products and ship them to a specific location. In addition, the goal specifies a preference that price be used for selection of the best service where multiple matching services are discovered. The goal as in listing 1.3 is defined for our scenario with respect to the implementation of the matching function from section 3 (we discuss this implementation in section 4.3). The goal defines the capability postcondition specifying to get a quote for the product while at the same time the product must be available and shippable to location specified by the shipment order request.

```
Goal GoalPurchaseShip
  nfp
    _"preference" hasValue "?price"
      ...
  endnfp
  ...
  capability GoalPurchaseShipCapability
    postcondition
      definedBy
        ( ?x[mo#price hasValue ?price] memberOf mo#PriceQuoteResp and
          mo#isAvailable(go#purchaseOrderReq) and
          mo#isShipped(go#shipmentOrderReq) ).
  ...
```

Listing 1.3. User Goal in WSMO

4.3 Implementation

The scenario is implemented as follows: when the goal is generated out of the request specified by the user, it is sent to the WSMX system. The WSMX starts a new operational thread (execution semantics) which first invokes the discovery component which in turn returns a list of services matching the goal. This list is passed to the selection component to select the service that best fits the user request. Control passes to the choreography engine which uses the choreography descriptions of the goal and service respectively, to drive the message exchange with the discovered service. This section describes the implementation of the algorithm from section 3 within the discovery component of WSMX. The details about other parts of the execution process can be found in our previous work in [2].

Section 3 describes three steps for discovery. A prototype for the *abstract-level* matching is under development in the WSMO working group. The implementation, described here, focuses on the steps of *instance-level* matching and *data-fetching.* A match between the goal and Web services is determined on the knowledge base created out of their descriptions, including instance data (both available from the descriptions and fetched). The goal capability defines a query (listing 1.3) which is used to query the knowledge base.

According to the algorithm 1 in section 3, the knowledge base B_{gw} is created for every goal and web service from the repository as shown in figure 2. Initially, the knowledge base imports all concepts from the common ontology and data from both goal and web service descriptions. In order to evaluate the matching function, we simply query the knowledge base using the goal postcondition. If the result of the evaluation is $true$, we add the web service to the list E of web services to the position determined by the preference. If the result of the evaluation is $false$, we first try to fetch new data by processing the fetching interface. If new data is available we evaluate the matching function again. If new data is available, the matching function is evaluated again. Otherwise, the cycle ends and the next service from the repository is processed. We briefly discuss this implementation in the next section 4.4.

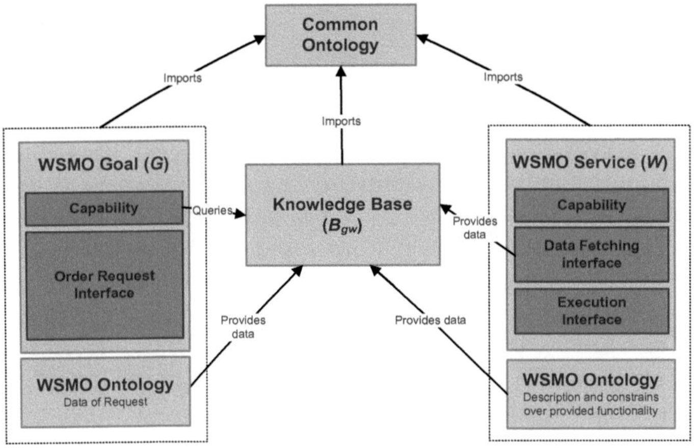

Fig. 2. Knowledge Base B_{gw}

4.4 Evaluation

Our implementation has been evaluated according to the methodology defined by the SWS Challenge. The SWS Challenge is an initiative led by a Semantic Web Services community providing a standard set of increasingly difficult problems, based on industrial specifications and requirements. Entrants to the SWS Challenge are peer-evaluated to determine if semantically-enabled integration approaches reduce costs of establishing and maintaining the integration between independent systems. In each SWS challenge workshop, the entrants first address introduced initial scenario of particular problem (e.g. mediation, discovery) in a testing environment prepared by the SWS Challenge organizers. The organizers then introduce some changes to back-end systems of the testing environment when the adaptivity of solutions is evaluated – solutions should handle introduced changes by modification of declarative descriptions rather than code changes. This evaluation is done by a methodology, developed by the SWS Challenge organizers and participants, which identifies following so called *success levels*. *Success level 0* indicates a minimal satisfiability level, where messages between middleware and backend systems are properly exchanged in the initial scenario. *Success level 1* is assigned when changes introduced in the scenario require code modifications and recompilation. *Success level 2* indicates that introduced changes did not entail any code modifications but only declarative parts had to be modified. *Success level 3* is assigned when changes did not require either modifications to code or the declarative parts, and the system was able to automatically adapt to the new conditions. Our implementation was evaluated to successfully address the scenario based on the location, weight, dimensional weight and price requirements, scoring success level 2. The implementation proved to be generic where only modifications of the WSMO Goals were necessary in order to correctly handle introduced changes. Discovery based on the location was successfully resolved using the common *isShipped* relation (see listing 1.1). Additional criteria imposed on the service such as weight and price have also been evaluated to level 2. No changes in WSMX code or in the descriptions of the services were required – only the Goal requests had to be changed. With respect to the fully-fledged discovery, there are however some limitations. It does not distinguish between the result *nodata* and *nomatch* (as defined in the algorithm) while it treats both results as *nodata*. This means that the whole fetching interface needs to be always processed until new data can be fetched or unless the match is found. This is a forced limitation of our implementation while at the same time it is a temporary solution for our environment before the fully-fledged discovery component will be available. The algorithm presented here however allows to use various implementations of matching functions which adhere to its defined interface. As a consequence, our solution currently offers a limited scalability. It might generate a significant network overhead in large-scale discovery scenarios when detailed interactions with every potential web service needs to be performed. We plan to address the optimality of our algorithm with respect to scalability in our future work. Our current solution also does not directly address security. It is important to ensure that information retrieved from service provider can be accessed after authorization and that data is fetched in a secure way. Such security aspects should be however implemented between the e-marketplace and service providers transparently to data fetching.

5 Related Work

There is no directly comparable work in the extension of the interface description in Web services to allow the fetching of additional data to aid discovery at run-time. However, there are two topics that are closely related. The first is service discovery based on semantic matchmaking which is the research area in which this paper is set and the second is service contracting and negotiation. Research into Goal-based discovery for WSMO and WSMX takes a step-wise approach with both theoretical and implementation concerns addressed at each stage. Three strategies have been investigated in this manner. The first is keyword-based discovery [4], which uses an algorithm that matches keywords, identified from the Goal description, with words occurring in various locations within the Web service description. The second strategy is for a lightweight Semantic Web Services discovery component for the WSMX platform and is described in [1]. This approach models a service in terms of the objects it can deliver. The term object, in this sense, means something of value the service delivers in its domain of interest. A third strategy is based around the use of quality-of-service attributes as described in [3] and implemented by the authors as a WSMX component. Upper level ontologies describing various domains for quality-of-service attributes are provided and non-functional properties are introduced to the service descriptions whose meanings are defined in these QoS ontologies. The approach in this paper is compatible with each of the matching strategies as it extends the matching power by requesting data from the service that is not directly available in its description. In [7], contracting is identified as an activity that may take place between the requester and provider once the initial discovery has identified candidate services. The discovery mechanisms in OWL-S rely on subsumption reasoning to match a service profile of service requesters with candidate service profiles published by service providers ad described in [10]. As with the WSMO efforts, they acknowledge that a *negotiation* phase may be necessary after discovery to allow requesters and providers agree on quality of service issues. Automated negotiation of service provision is related to the topic of this paper as, for negotiation to take place, it must be possible to determine during discovery exactly the terms that are being offered by the service which may be open to negotiation. A substantial body of work is devoted to the development of negotiation systems ranging from the application of intelligent agents for eCommerce in [8] through negotiation using Bayesian Learning [14] to using Web services and BPEL for automated negotiations [5].

6 Conclusion and Future Work

Service discovery which operates on abstract descriptions of services needs to be further elaborated in order to return results of concrete services satisfying concrete goals. For this purpose, instance data needs to be used. Since all data can not be included in service descriptions (usually for practical reasons) it needs to be fetched from the service provider at discovery-time. In this paper we presented an approach to model the service interface allowing such data to be fetched from the service provider. We use the abstract state machine formalism to model the interface allowing scalable interactions with a service provider for specific discovery sessions. This approach allows the use of only the rules and data required, by the service requester at discovery-time (and thus limit data transmission or other costs) while at the same time it is possible to adapt the interface for various purposes of the late binding phase, i.e. discovery, selection, contracting/negotiation, etc. We also showed how, by extending WSMO service interface, the WSMO service choreography definition can be used to implement this interface. In a case scenario, we described the necessary semantic models and presented the algorithm (including creation of the knowledge base, processing the interface, and querying the knowledge base).

In our future work we plan to address the optimality for data fetching to decide on preferences for those interactions which might lead to results without processing all data fetching interface. In addition, we want to extend the data fetching interface to support other parts of the late binding phase. For example, negotiation building on data fetching might use interactions with specific meaning, such as for bidding etc. Layering of specific late-binding interfaces on the top of data fetching allows a modular approach to the definition of such interactions. We also plan to improve the implementation of the matching function for fully-fledged service discovery. In addition, we plan to incorporate run-time data mediation aspects into the discovery process where service requester and service providers use different ontologies.

References

1. Andreas Friesen and Stephan Grimm. DIP WP4 Service Usage, D4.8 Discovery Specification, available at http://dip.semanticweb.org/documents/D4.8Final.pdf. Technical report, 2005.
2. Thomas Haselwanter, Paavo Kotinurmi, Matthew Moran, Tomáš Vitvar, and Maciej Zaremba. Wsmx: A semantic service oriented middleware for b2b integration. In *ICSOC*, pages 477–483, 2006.
3. Manfred Hauswirth, Fabio Porto, and Le-Hung Vu. P2P and QoS-enabled service discovery specification available at http:/dip.semanticweb.org/documents/D4.17-Revised.pdf. Technical report, 2006.
4. Uwe Keller, Ruben Lara, Holger Lausen, Axel Polleres, Livia Predoiu, and Ioan Toma. WSMO D10.2 Sematic Web Service Discovery available at http://www.wsmo.org/TR/d10/v0.2/d10.pdf. Technical report, 2005.
5. J.B. Kim, A.Segev, A.Patankar, and M.G.Cho. Web services and bpel4ws for dynamic ebusiness negotiation processes,. In *International Conference on Web Services*, Las Vegas, Nevada, USA, 2003.
6. Jacek Kopecký, Dumitru Roman, Matthew Moran, and Dieter Fensel. Semantic web services grounding. In *AICT/ICIW*, page 127, 2006.
7. Ruben Lara and Daniel Olmedilla. Discovery and Contracting of Semantic Web Services. Technical report, 2005.
8. L.C. Lee. Progressive multi-agent negotiation. In *International Conference on Multi-Agent Systems*. MIT Press, 1995.
9. Adrian Mocan, Matthew Moran, Emilia Cimpian, and Michal Zaremba. Filling the gap - extending service oriented architectures with semantics. In *ICEBE*, pages 594–601. IEEE Computer Society, 2006.
10. M. Paolucci, T. Kawamura, T. Payne, and K. Sycara. Semantic matching of web services capabilities. In *1st International Semantic Web Conference (ISWC)*, page 333347, 2002.
11. Dumitru Roman, Uwe Keller, Holger Lausen, Jos de Bruijn, Rubn Lara, Michael Stollberg, Axel Polleres, Cristina Feier, Christoph Bussler, and Dieter Fensel. Web Service Modeling Ontology. *Applied Ontologies*, 1(1):77 – 106, 2005.
12. Dumitru Roman, James Scicluna, Dieter Fensel, Axel Polleres, and Jos de Bruijn. D14v0.3. Ontology-based Choreography of WSMO Services, available from http://www.wsmo.org/TR/d14/v0.4/. Technical report, 2006.
13. A. Moormann Zaremski and J. M. Wing. Specification matching of software components. *ACM Transactions on Software Engineering and Methodology*, 6(4):333–369, 1997.
14. D. Zeng and K. Sycara. Bayesian learning in negotiation. In *Working Notes for the AAAI Symposium on Adaptation, Co-evolution and Learning in Multiagent Systems*, pages 99 – 104, 1996.

Two-Phase Web Service Discovery Based on Rich Functional Descriptions

Michael Stollberg, Uwe Keller, Holger Lausen, and Stijn Heymans

Digital Enterprise Research Institute Innsbruck (DERI Austria),
Institute for Computer Science, University of Innsbruck,
Technikerstrasse 21a, A-6020 Innsbruck, Austria
firstname.lastname@deri.org

Abstract. Discovery is a central reasoning task in service-oriented architectures, concerned with detecting Web services that are usable for solving a given request. This paper presents two extensions in continuation of previous works towards goal-based Web service discovery with sophisticated semantic matchmaking. At first, we distinguish goal templates as generic objective descriptions and goal instances that denote concrete requests as an instantiation of a goal template. Secondly, we formally describe requested and provided functionalities on the level of state transitions that denote executions of Web services, respectively solutions for goals. Upon this, we specify a two-phase discovery procedure along with semantic matchmaking techniques that allow to accurately determine the usability of a Web service. The techniques are defined in the Abstract State Space model that supports several languages for describing Web services.

1 Introduction

Discovery is concerned with detecting usable Web services for solving a given request. This is the first central reasoning task in the context of Semantic Web services, followed by contracting and behavioral conformance tests [17]. Several research works present discovery techniques by semantic matchmaking of requested and provided functionalities, e.g. [16,13,2,7,11]. However, due to deficiencies in the expressiveness and the formal semantics of functional descriptions most existing approaches lack in the achievable quality of the matchmaking results for Web service discovery.

In this respect, we present the advancements towards a goal-based approach for semantically enabled Web service discovery with sophisticated matchmaking. Initially presented in [9], the requester and the provider perspective are separated by formally describing client objectives as goals; a Web service is understood to provide access to several services by its invocation with concrete input values. We extend this approach by differentiating two notions of goals. A *goal template* is a generic objective description that is defined at design time, and a *goal instance* denotes a concrete client request that is created at runtime by instantiating a goal template with concrete input values. Apart from better supporting goal formulation by clients, this allows to realize an efficient two-phase Web service discovery. Usable Web services for goal templates are determined at design time and kept in the system. At runtime, the discovery for goal

E. Franconi, M. Kifer, and W. May (Eds.): ESWC 2007, LNCS 4519, pp. 99–113, 2007.

instances only needs to investigate those Web services that are usable for the corresponding goal template, so that the number of matchmaking operations necessary at runtime can be reduced. This paper specifies the semantic matchmaking techniques for this framework.

In order to properly describe provided and requested functionalities, we consider a state-based model of the world. Therein, a particular execution of a Web service denotes a sequence of state transitions; such a sequence is also a solution for a goal if the client objective is solved in the end-state. The functionality provided by a Web service is a set of all its possible executions, and a goal template as well as a goal instance describes a set of possible solutions. We formally describe possible executions and solutions with respect to the start- and end-states in *Abstract State Spaces*, a language independent model that defines precise formal semantics for such functional descriptions [10].

On top of this, we specify semantic matchmaking techniques that allow to precisely determine the usability of a Web service for solving a goal. In particular, we (1) revise the definition of previously identified matching degrees and use these to differentiate the usability of a Web service on the goal template level, (2) present a novel approach for semantic matchmaking on the goal instance level, and (3) finally integrate the matchmaking techniques for the goal template and the goal instance level. We specify the techniques in a first-order logic framework and illustrate the definitions by a running example throughout the paper: a goal specifies the objective of finding the best restaurant in a city, and a Web service provides a search facility for the best French restaurant in a city. As we shall discuss, this Web service is only usable for specific goal instances – namely those that specify a city wherein the best restaurant in French.

The paper is structured as follows. Section 2 introduces the concepts of our two-phase discovery approach, and Section 3 defines the formal functional descriptions for Web services and goals. Section 4 specifies the integrated semantic matchmaking techniques for Web service discovery, and Section 5 demonstrates this in the running example. Section 6 discusses related work and positions our approach therein. Finally, Section 7 concludes the paper. A detailed report on this work is provided in [20].

2 Concepts and Approach

The specification of semantic matchmaking techniques for Web service discovery is strongly dependent on the underlying conception and the formal description of Web services and goals. This section introduces the relevant concepts and then outlines the two-phase Web service discovery by discussing the meaning of a match.

2.1 Web Services and Goals

In accordance to the common understanding, we consider a Web service as a computational facility that is invocable over the Internet via an interface [1]. As an abstraction that is sufficient for our purpose, we define a Web service as a pair $W = (IF, \iota)$ such that $IF = (i_1, \ldots, i_n)$ is a finite set of names that denotes all inputs required for invoking W, and ι is the implementation of W that is executed when W is invoked.

In the Abstract State Space model (ASS, [10]), a particular execution of W denotes a finite sequence of state transitions $\tau = (s_0, \ldots, s_m)$, i.e. a change of the world from

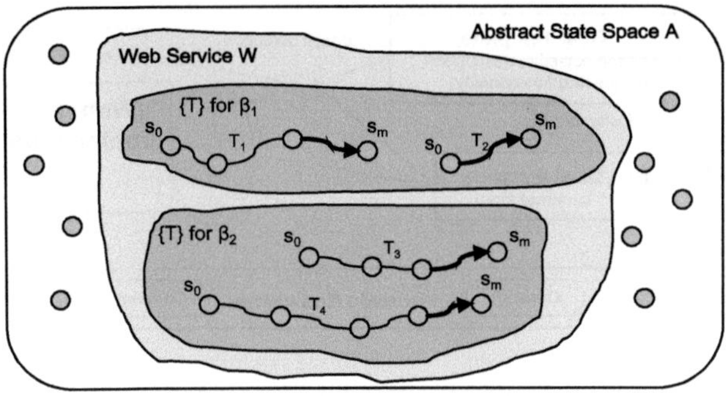

Fig. 1. Web Service, Executions, Input Bindings

a start state s_0 to an end state s_m. Such a τ is triggered by invoking W with concrete input values; we refer to this as an input binding $\beta : \{i_1, \ldots, i_n\} \to \mathcal{U}$, i.e. a total function that assigns objects of some universe $\mathcal{U}$ to the IF-names. In dependence of the start state, there can be different executions of W for the same input binding. Relevant for the context of discovery, we understand the *overall functionality* provided by W as the set of all its possible executions, denoted by $\{\tau\}_W$. As illustrated in Figure 1, this can be further differentiated into the distinct sets of possible executions of W for each valid input binding, such that $\{\tau\}_W = \bigcup \{\tau\}_{W(\beta)}$ with $W(\beta)$ denoting the set of possible executions of W when invoked with a particular input binding β.[1]

Goals in our approach are formally described client objectives. In accordance to related AI research (e.g. [3,15]), we understand a goal as the formal description of the desire of the client to get from the current state of the world into a state wherein the objective is satisfied. This abstracts from technical details irrelevant to the client objective. As promoted by the WSMO framework [12], the overall aim is to enable problem-oriented Web service usage: the client merely specifies the objective to be achieved as a goal, and the system detects and executes suitable Web services for solving this.

We have refined the initial WSMO goal model based on experiences in realizing respective technologies [21]. The extension relevant in the context of discovery is the differentiation of *goal templates* as generic, reusable objective descriptions, and *goal instances* that denote concrete client requests as instantiations of a goal template. Inspired by related system implementations such as IRS [4] and SWF [22], this allows to support goal formulation by client via graphical user interfaces. Instead of requiring the client to specify potentially complex logical formulae for goal formulation, merely pre-defined templates are instantiated with concrete inputs. Figure 2 illustrates this.

While we shall specify their formal description in the next section, a goal template $\mathcal{G}$ defines generic constraints on the initial state and the desired final state to be achieved. In our restaurant search example, the goal template $\mathcal{G}$ defines that the best restaurant

[1] We consider the functionalities provided by Web services to satisfy two properties: (1) *deterministic*, i.e. all outputs and effects of an execution are completely dependent on the provided inputs and the start state; non-deterministic functionalities violate the composability of Web services [17]; (2) *non-adaptive*, meaning that in contrast to intelligent software agents a Web service does not itself change the provided functionality [6].

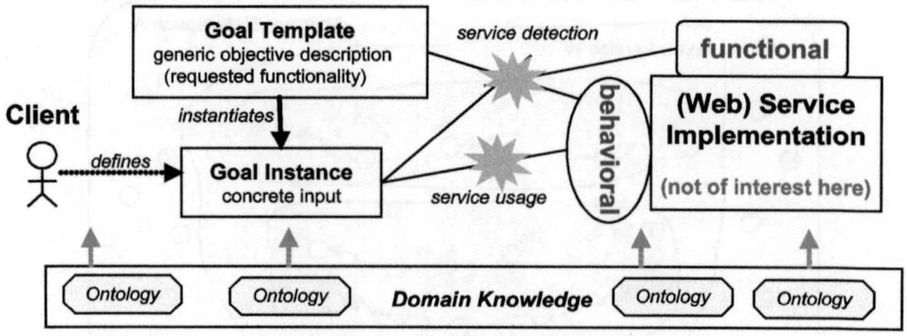

Fig. 2. Goal Templates, Goal Instances, and Web Services

shall be found in a city that is provided as an input by the client. Its meaning in the ASS model is that $\mathcal{G}$ specifies a set of sequences of state transitions $\{\tau\}_{\mathcal{G}}$ as its possible solutions. For each $\tau = (s_0, \ldots, s_m) \in \{\tau\}_{\mathcal{G}}$, the start-state s_0 satisfies the constraints on the initial state, and the end-state s_m satisfies the constraints on the desired state of the world. At runtime, a client creates a goal instance $GI(\mathcal{G})$ by defining concrete values for the inputs specified in $\mathcal{G}$. In the example, this is the concrete city in which the best restaurant shall be found. We refer to this as an input binding β for $\mathcal{G}$; this also constitutes the input binding for invoking a Web service to solve $GI(\mathcal{G})$ as discussed above. Because of this instantiation, the possible solutions for $GI(\mathcal{G})$ are a subset of those for $\mathcal{G}$, so that $\{\tau\}_{GI(\mathcal{G})} \subset \{\tau\}_{\mathcal{G}}$.

2.2 The Meaning of a Match for Web Service Discovery

We now turn towards Web service discovery. With respect to the conception of Web services and goals explained above, the aim is to find a Web service that can provide a τ that is a solution for the goal. Hence, we define the meaning of a match as follows.

Definition 1. *Let W be a Web service, $\mathcal{G}$ a goal template, and $GI(\mathcal{G})$ a goal instance that instantiates $\mathcal{G}$ with an input binding β. Let $\tau = (s_0, \ldots, s_m)$ be a sequence of states in an Abstract State Space $\mathcal{A}$. We define the following sets:*

$$
\begin{array}{ll}
\{\tau\}_{\mathcal{G}} & := \textit{possible solutions for } \mathcal{G} \\
\{\tau\}_{W} & := \textit{possible executions of } W \\
\{\tau\}_{GI(\mathcal{G})} \subset \{\tau\}_{\mathcal{G}} & := \textit{possible solutions for } GI(\mathcal{G}) \textit{ that defines } \beta \\
\{\tau\}_{W(\beta)} \subset \{\tau\}_{W} & := \textit{possible executions of } W \textit{ when invoked with } \beta
\end{array}
$$

We define the usability of a Web service for solving a goal *as:*

$$
\begin{array}{ll}
\textit{(i) } match(\mathcal{G}, W) & : \exists \tau.\ \tau \in (\{\tau\}_{\mathcal{G}} \cap \{\tau\}_{W}) \\
\textit{(ii) } match(GI(\mathcal{G}), W) & : \exists \tau.\ \tau \in (\{\tau\}_{GI(\mathcal{G})} \cap \{\tau\}_{W(\beta)})
\end{array}
$$

This defines the basic matching conditions for Web Service discovery. Clause (i) states that a Web service W is usable for solving a goal template $\mathcal{G}$ if there exists at least one execution of W that is a possible solution for $\mathcal{G}$. Clause (ii) defines that W is usable for solving a goal instance $GI(\mathcal{G})$ if there is at least one execution of W that is also a solution for $GI(\mathcal{G})$ when W is invoked with the inputs defined in $GI(\mathcal{G})$.

Because of $\{\tau\}_{GI(\mathcal{G})} \subset \{\tau\}_{\mathcal{G}}$ it holds that a Web service that is usable for solving a goal instance is also usable for the corresponding goal template. If W can provide a $\tau \in \{\tau\}_{GI(\mathcal{G})}$, then this τ also is also an element of $\{\tau\}_{\mathcal{G}}$. Formally, we can express this as $match(GI(\mathcal{G}), W) \Rightarrow match(\mathcal{G}, W)$. As the logical complement, it also holds that $\neg match(\mathcal{G}, W) \Rightarrow \neg match(GI(\mathcal{G}), W)$, i.e. that a Web service that is not usable for a goal template is also not usable for any of its goal instances.

This constitutes the foundation of our two-phase discovery. Usable Web services for goal templates $\mathcal{G}$ can be determined at design, i.e. when a new goal template is defined. Web service discovery for concrete goal instances $GI(\mathcal{G})$ is performed at runtime. Because of $\neg match(\mathcal{G}, W) \Rightarrow \neg match(GI(\mathcal{G}), W)$, this merely needs to consider the set of Web services that are usable for the corresponding goal template $\mathcal{G}$. While the achievable efficiency increase is discussed elsewhere [22], this paper specifies the semantic matchmaking techniques for evaluating the matching conditions on the basis of formal descriptions. Without such techniques, we would need to perform test runs of W in order to determine its usability for solving a goal.

3 Formal Functional Descriptions

The following defines functional descriptions for Web services and goals that serve as the basis for semantic matchmaking techniques for Web service discovery. To properly describe requested and provided functionalities on the level of state transitions, we apply functional descriptions as defined in the ASS model mentioned above. This section specifies their structure and formal meaning in a first-order logic framework, and illustrates the definitions in our running example.

3.1 Definition and Semantics

The ASS model describes functionalities in terms of preconditions and effects along with explicitly defining in- and outputs. Focussing on the formal meaning of functional descriptions, they are defined independent of the language used for specifying preconditions and effects. The following recalls the definitions, referring to [10] for details.

An Abstract State Space $\mathcal{A}$ is defined over a signature Σ and some domain knowledge Ω. A functional description is described as a 5-tuple $(\Sigma, \Omega, IF, \phi^{pre}, \phi^{eff})$. The signature Σ differentiates *static symbols* Σ_S that are not changed, *dynamic symbols* Σ_D that are changed by execution of a Web service, and Σ_D^{pre} that denote the interpretation of a dynamic symbol in the start state. Preconditions ϕ^{pre} and effects ϕ^{eff} are defined as statements in a logic $\mathcal{L}(\Sigma)$. $IF = (i_1, \ldots, i_n)$ is a set of variables that denote all required inputs. To explicitly specify the deterministic dependency between the start- and end-states with respect to input values, they can occur as the only free variables in ϕ^{pre} and ϕ^{eff}. An input binding $\beta : \{i_1, \ldots, i_n\} \rightarrow \mathcal{U}_{\mathcal{A}}$ is a total function that assigns objects of the universe of $\mathcal{A}$ to each IF-variable. Finally, the symbol *out* denotes the computational outputs that are constrained by ϕ^{eff}.

The meaning of a functional description is defined with respect to the start- and the end-state of a sequence of state transitions. Formally, a $\tau = (s_0, \ldots, s_m)$ in $\mathcal{A}$ is considered to satisfy the described functionality if and only if it holds that if $s_0 \models_{\mathcal{L}(\Sigma)} \phi^{pre}$

then $s_m \models_{\mathcal{L}(\Sigma)} \phi^{eff}$. Here, $s \models_{\mathcal{L}(\Sigma)} \phi$ expresses that the formula ϕ is satisfied by the universe $\mathcal{U}_\mathcal{A}$ in a state s under the logic $\mathcal{L}(\Sigma)$. We refer to this as *implication semantics*: if the precondition is satisfied in s_0, then s_m will satisfy the effect; otherwise, we can not make any statement about the behavior of the described functionality. Because the *IF*-variables occur as free variables in both the precondition ϕ^{pre} and the effect ϕ^{eff}, the end-state s_m is completely dependent on the start-state s_0. This reflects the deterministic nature of functionalities provided by Web services.

While functional descriptions in the ASS model are defined independent of the specification language for preconditions and effects, we use classical first-order logic (FOL, [19]) for illustration throughout this work. In order to ease the handling of functional descriptions, we describe them as a first-order logic structure that maintains the formal semantics as defined in the ASS model.

Definition 2. *A functional description is a 4-tuple $\mathcal{D} = (\Sigma, \Omega, IF, \phi^{\mathcal{D}})$ such that:*

(i) Σ is a signature consisting of Σ_S (static symbols), Σ_D (dynamic symbols), and Σ_D^{pre} (pre-variants of dynamic symbols)

(ii) $\Omega \subseteq \mathcal{L}(\Sigma)$ defines consistent domain knowledge

(iii) IF is a set of variables $i_1, \ldots, i_n$ that denote all required input values; an input binding $\beta : \{i_1, \ldots, i_n\} \rightarrow \mathcal{U}_\mathcal{A}$ is a total function that assigns objects of the universe of $\mathcal{A}$ to each IF-variable

(iv) $\phi^{\mathcal{D}}$ is a FOL formula of the form $[\phi^{pre}]_{\Sigma_D^{pre} \rightarrow \Sigma_D} \Rightarrow \phi^{eff}$ such that
- *ϕ^{pre} is the precondition with IF as the only free variables*
- *ϕ^{eff} is the effect with IF as the only free variables and the outputs are denoted by the predicate* `out`
- *$[\phi]_{\Sigma_D^{pre} \rightarrow \Sigma_D}$ is the formula ϕ' derived from ϕ by replacing every dynamic symbol $\alpha \in \Sigma_D$ by its corresponding pre-variant $\alpha_{pre} \in \Sigma_D^{pre}$.*

Essentially, $\phi^{\mathcal{D}}$ defines a logical implication between the precondition and the effect formulae. The rewriting function for the precondition handles dynamic symbols. For example, consider a functionality for a bank account withdrawal with $\phi^{pre} : account(a) \land balance(a) \geq x$, $\phi^{eff} : account(a) \land balance(a) = balance_{pre}(a) - x$, and $\Sigma_D = balance(a)$. We obtain $\phi^{\mathcal{D}} = (account(a) \land balance_{pre}(a) \geq x) \Rightarrow (account(a) \land balance(a) = balance_{pre}(a) - x)$, so that the relationship between the start- and end-state is specified explicitly. The following specifies the meaning of such a functional description that formally describes the overall functionality provided by a Web service.

Definition 3. *Let W be a Web service with $\{\tau\}_W$ as the set of its possible executions in an Abstract State Space $\mathcal{A}$. Let $\mathcal{D} = (\Sigma, \Omega, IF, \phi^{\mathcal{D}})$ be a functional description. Let $\Omega_\mathcal{A} = \Omega \cup [\Omega]_{\Sigma_D^{pre} \rightarrow \Sigma_D}$ be the domain knowledge extended with $\alpha_{pre} \in \Sigma_D^{pre}$. W provides the functionality described by $\mathcal{D}$, denoted by $W \models_\mathcal{A} \mathcal{D}$, if and only if:*

(i) every Σ-interpretation I with $I \models \Omega_\mathcal{A}$ and $I, \beta \models \phi^{\mathcal{D}}$ under every input binding $\beta : IF \rightarrow \mathcal{U}_\mathcal{A}$ represents a $\tau \in \{\tau\}_W$, and

(ii) every $\tau \in \{\tau\}_W$ is represented by a Σ-interpretation I with $I, \beta \models \phi^{\mathcal{D}}$ and $I \models \Omega_\mathcal{A}$ under every input binding $\beta : IF \rightarrow \mathcal{U}_\mathcal{A}$.

This defines that a Web service W provides the functionality described by $\mathcal{D}$ if and only if every Σ-interpretation I, β that is a model of $\phi^{\mathcal{D}}$ describes a $\tau = (s_0, \ldots, s_m)$

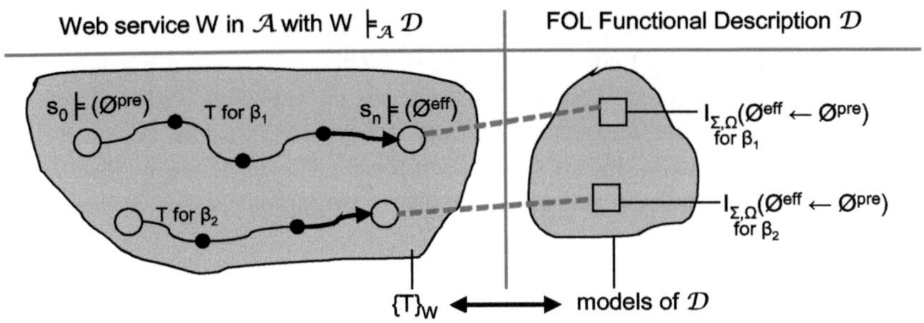

Fig. 3. Illustration of $W \models_{\mathcal{A}} \mathcal{D}$

$\in \{\tau\}_W$. Such a Σ-interpretation describes the objects that exists in the end-state s_m if W is executed for a particular input binding β in a specific start state s_0. For the implication semantics from clause (iv) in Definition 2, it holds that $I, \beta \models \phi^{\mathcal{D}}$ if $I, \beta \models \phi^{pre}$ and $I, \beta \models \phi^{eff}$; if $I \not\models \phi^{pre}$, we can not make any statement about the end-state of a τ. Hence, if a $\tau \in \{\tau\}_W$ can be described by a Σ-interpretation I with $I, \beta \models \phi^{\mathcal{D}}$, then it satisfies the described functionality; if there is a $\tau \in \{\tau\}_W$ that cannot be described by such a Σ-interpretation, then W does not provide the described functionality. Figure 3 illustrates this, while we refer to [20] for the formal explanation of this definition and its relationship to the ASS model.

The meaning of a functional description $\mathcal{D}_{\mathcal{G}}$ of a goal template $\mathcal{G}$ is analogous. Here, $\{\tau\}_{\mathcal{G}}$ is the set of sequences of state transitions that are solutions for $\mathcal{G}$ such that every $\tau \in \{\tau\}_{\mathcal{G}}$ corresponds to a Σ-interpretation that is a model of $\mathcal{D}_{\mathcal{G}}$. To precisely evaluate the usability of a Web service, in some cases we need to consider the concrete value assignments for the IF-variables. These are provided by the creation of a goal instance $GI(\mathcal{G})$ that defines an input binding β for the IF-variables in $\mathcal{D}_{\mathcal{G}}$ of the corresponding goal template $\mathcal{G}$. Subsequently, this β constitutes the inputs for invoking a Web service in order to solve $GI(\mathcal{G})$. We shall discuss this in more detail in the context of discovery on the goal instance level (Section 4.2).

3.2 Illustration in Running Example

In order to illustrate the above definitions, Table 1 shows the formal functional descriptions of the goal template $\mathcal{G}$ and the Web service W in our restaurant search example.

The goal describes the objective of finding the best restaurant in a city. The specific city is an input required for instantiation. Hence, $\mathcal{D}_{\mathcal{G}}$ specifies one IF-variable that is constrained in the precondition ϕ^{pre} to be a *city*. The effect ϕ^{eff} describes the desired state of the world to be given if and only if the received output is a restaurant in the city such that there does not exists any better restaurant in the city. Analogously, $\mathcal{D}_W$ describes the functionality provided by the Web service W. The mere difference occurs in the effect: the output of W is a French restaurant in the city that is provided as input such that there does not exist any better French restaurant in the city.

We use classical first-order logic (FOL, [19]) as the specification language. The signature Σ for both $\mathcal{D}_{\mathcal{G}}$ and $\mathcal{D}_W$ defines the respective symbols. Here, $?\langle name \rangle$ denotes a variable. The domain knowledge Ω is defined in the *best restaurant ontology*. This contains axioms specifying that the predicate $better(\cdot,\cdot)$ denotes a partial order, that any restaurant has exactly one type and that the restaurant types $italian$ and $french$ are distinct from each other, and that restaurants are located in cities. We omit the complete ontology specification due to space limitations. The table shows the functional descriptions with precondition and effects and the corresponding $\phi^{\mathcal{D}}$ in accordance to Definition 2.

Table 1. Functional Descriptions $\mathcal{D}_{\mathcal{G}}, \mathcal{D}_{W}$ in Running Example

Goal "find best restaurant in a city"	**Web Service** "provide best French restaurant in a city"
Ω: best restaurant ontology IF: {?x} ϕ^{pre}: $city(?x)$ ϕ^{eff}: $\forall ?y.\ out(?y) \Leftrightarrow ($ $restaurant(?y)$ $\wedge\ in(?y,?x)$ $\wedge\ \neg\exists ?z.(restaurant(?z)$ $\wedge\ in(?z,?x)$ $\wedge\ better(?z,?y)\ )\ ).$	Ω: best restaurant ontology IF: {?x} ϕ^{pre}: $city(?x)$ ϕ^{eff}: $\forall ?y.\ out(?y) \Leftrightarrow ($ $restaurant(?y)$ $\wedge\ in(?y,?x)\ \wedge type(?y, french)$ $\wedge\ \neg\exists ?z.(restaurant(?z)$ $\wedge\ in(?z,?x)\ \wedge type(?z, french)$ $\wedge\ better(?z,?y)\ )\ ).$
$\phi^{\mathcal{D}_{\mathcal{G}}}$: $city(?x) \Rightarrow ($ $\forall ?y.\ out(?y) \Leftrightarrow ($ $restaurant(?y)$ $\wedge\ in(?y,?x)$ $\wedge\ \neg\exists ?z.(restaurant(?z)$ $\wedge\ in(?z,?x)$ $\wedge\ better(?z,?y)\ )\ )\ ).$	$\phi^{\mathcal{D}_{W}}$: $city(?x) \Rightarrow ($ $\forall ?y.\ out(?y) \Leftrightarrow ($ $restaurant(?y)$ $\wedge\ in(?y,?x)\ \wedge type(?y, french)$ $\wedge\ \neg\exists ?z.(restaurant(?z)$ $\wedge\ in(?z,?x)\ \wedge type(?z, french)$ $\wedge\ better(?z,?y)\ )\ )\ ).$

4 Semantic Matchmaking for Web Service Discovery

On the basis of the formal descriptions we now specify the semantic matchmaking techniques for the two-phased Web service discovery introduced in Section 2.2. The aim is to provide semantic means that allow to precisely determine the usability of a Web service with respect to the matching conditions on the goal template and the goal instance level from Definition 1. We therefore define matchmaking on functional descriptions and input bindings as specified above. These provide sufficiently rich descriptions of possible Web service executions and possible solution for goals. The following first specifies semantic matchmaking on the goal template level, then on the goal instance level, and finally integrates the techniques for both levels. We shall demonstrate the techniques in our running example in Section 5.

4.1 Goal Template Level

We express the usability of a Web service W for solving a goal template $\mathcal{G}$ in terms of matching degrees. Adopting the concept and denotation of the degrees from several previous works on Web service discovery (e.g. [16,13,8]), we define them over the functional descriptions of goals and Web services as defined in Section 3.1.

The distinct degrees denote specific relationships between the possible executions $\{\tau\}_W$ of W and possible solutions $\{\tau\}_\mathcal{G}$ for $\mathcal{G}$. Four degrees – *exact, plugin, subsume, intersect* – denote different situations wherein the matching condition in clause (i) of Definition 1 is satisfied; the *disjoint* degree denotes that this is not given. In our two-phase discovery, these matching degrees serve as a pre-filter for determining the usability of a Web service W for solving a goal instance $GI(\mathcal{G})$ that instantiates the goal template $\mathcal{G}$. We shall discuss this in more detail in Section 4.3.

We define the criteria for each degree over $\mathcal{D}_\mathcal{G}$ and $\mathcal{D}_W$ from Definition 2, along with an explicit quantification of input bindings β. As the condition for the *exact* degree, $\Omega_\mathcal{A} \models \forall\beta.\ \phi^{\mathcal{D}_\mathcal{G}} \Leftrightarrow \phi^{\mathcal{D}_W}$ defines that every possible execution of W is a solution for $\mathcal{G}$ and vice versa. We assume that all functional descriptions $\mathcal{D}$ are consistent, i.e. that there exists a Σ-interpretation I under a β that is a model of $\phi^\mathcal{D}$. Representing a refinement of the matching degree definitions from [8], we therewith obtain a precise means for differentiating the usability of a Web service on the goal template level. Table 2 provides a concise compilation of the matchmaking degree definitions.

Table 2. Definition of Matching Degrees for $\mathcal{D}_\mathcal{G}, \mathcal{D}_W$

Denotation $\mathcal{D}_\mathcal{G} = (\Sigma, \Omega, IF, \phi^{\mathcal{D}_\mathcal{G}})$ $\mathcal{D}_W = (\Sigma, \Omega, IF, \phi^{\mathcal{D}_W})$	**Definition** $\beta : IF \rightarrow \mathcal{U}_\mathcal{A}$ $\phi^\mathcal{D} = [\phi^{pre}]_{\Sigma_D^{pre} \rightarrow \Sigma_D} \Rightarrow \phi^{eff}$ $\Omega_\mathcal{A} = \Omega \cup [\Omega]_{\Sigma_D^{pre} \rightarrow \Sigma_D}$	**Meaning** for $\{\tau\}_\mathcal{G}, \{\tau\}_W$ with $W \models_\mathcal{A} \mathcal{D}_W$
exact($\mathcal{D}_\mathcal{G}, \mathcal{D}_W$)	$\Omega_\mathcal{A} \models \forall\beta.\ \phi^{\mathcal{D}_\mathcal{G}} \Leftrightarrow \phi^{\mathcal{D}_W}$	if and only if $\tau \in \{\tau\}_\mathcal{G}$ then $\tau \in \{\tau\}_W$
plugin($\mathcal{D}_\mathcal{G}, \mathcal{D}_W$)	$\Omega_\mathcal{A} \models \forall\beta.\ \phi^{\mathcal{D}_\mathcal{G}} \Rightarrow \phi^{\mathcal{D}_W}$	if $\tau \in \{\tau\}_\mathcal{G}$ then $\tau \in \{\tau\}_W$
subsume($\mathcal{D}_\mathcal{G}, \mathcal{D}_W$)	$\Omega_\mathcal{A} \models \forall\beta.\ \phi^{\mathcal{D}_\mathcal{G}} \Leftarrow \phi^{\mathcal{D}_W}$	if $\tau \in \{\tau\}_W$ then $\tau \in \{\tau\}_\mathcal{G}$
intersect($\mathcal{D}_\mathcal{G}, \mathcal{D}_W$)	$\Omega_\mathcal{A} \models \exists\beta.\ \phi^{\mathcal{D}_\mathcal{G}} \wedge \phi^{\mathcal{D}_W}$	there is a τ such that $\tau \in \{\tau\}_\mathcal{G}$ and $\tau \in \{\tau\}_W$
disjoint($\mathcal{D}_\mathcal{G}, \mathcal{D}_W$)	$\Omega_\mathcal{A} \models \neg\exists\beta.\ \phi^{\mathcal{D}_\mathcal{G}} \wedge \phi^{\mathcal{D}_W}$	there is no τ such that $\tau \in \{\tau\}_\mathcal{G}$ and $\tau \in \{\tau\}_W$

4.2 Goal Instance Level

A goal instance $GI(\mathcal{G})$ is created by defining an input binding β for the IF-variables in the functional description $\mathcal{D}_\mathcal{G}$ of the corresponding goal template $\mathcal{G}$. Recalling from Definition 1, a match on the goal instance level is given if there exists a $\tau = (s_0, \ldots, s_m)$ in $\mathcal{A}$ that is a solution for $GI(\mathcal{G})$ and can be provided by a Web service W when it is invoked with the concrete input values defined in $GI(\mathcal{G})$. The following specifies a general technique for determining this on the basis of the available descriptions, independent of the matching degree between $\mathcal{D}_\mathcal{G}$ and $\mathcal{D}_W$.

Formally, an input binding $\beta : \{i_1, \ldots, i_n\} \rightarrow \mathcal{U}_\mathcal{A}$ is a total function that defines a variable assignment over the universe $\mathcal{U}_\mathcal{A}$ for the input variables IF defined in a functional description $\mathcal{D}$ (*cf.* Definition 2). We therewith obtain an assignment of concrete values v for all inputs required in $\mathcal{D}$, i.e. $\beta = \{i_1|v_1, \ldots, i_n|v_n\}$. Given such a β, we can instantiate $\mathcal{D}$ by substituting all IF-variables that occur as free variables in ϕ^{pre} and ϕ^{eff} by the concrete values defined in β. We obtain $[\mathcal{D}]_\beta$ as the functional description that is instantiated for the context of β; this can be evaluated because it does no longer contain any free variables. By instantiating the functional descriptions $\mathcal{D}_\mathcal{G}$ of the corresponding goal template $\mathcal{G}$ and $\mathcal{D}_W$ of the Web service W with the input binding β defined in $GI(\mathcal{G})$, we obtain $[\mathcal{D}_\mathcal{G}]_\beta$ as the functionality requested by $GI(\mathcal{G})$ and $[\mathcal{D}_W]_\beta$ as the functionality that can be provided by W when it is invoked with β.

For W to be usable for solving $GI(\mathcal{G})$, there must be a τ such that $\tau \in \{\tau\}_{GI(\mathcal{G})}$ and $\tau \in \{\tau\}_{W(\beta)}$ (*cf.* clause (ii) from Definition 1). To determine this on the basis of the given descriptions, it must hold that – with respect to the domain knowledge – there exists a Σ-interpretation I that is a common model for $\phi^{\mathcal{D}_\mathcal{G}}$ and $\phi^{\mathcal{D}_W}$ when both functional descriptions are instantiated with the input binding β defined in $GI(\mathcal{G})$. Formally, this means that the union of the formulae $\Omega_\mathcal{A} \cup \{[\phi^{\mathcal{D}_\mathcal{G}}]_\beta, [\phi^{\mathcal{D}_W}]_\beta\}$ must be satisfiable, i.e. that there exists a Σ-interpretation that is a model for the extended domain knowledge $\Omega_\mathcal{A}$ and for the instantiated goal description $[\phi^{\mathcal{D}_\mathcal{G}}]_\beta$ and for the instantiated Web service description $[\phi^{\mathcal{D}_W}]_\beta$. In accordance to Definition 3, this I represents a τ that is a solution for $GI(\mathcal{G})$ and can be provided by W if it is invoked with β.

Definition 4. *Let* $\mathcal{D}_\mathcal{G} = (\Sigma, \Omega, IF_\mathcal{G}, \phi^{\mathcal{D}_\mathcal{G}})$ *be a functional description of a goal template* $\mathcal{G}$. *Let* $GI(\mathcal{G})$ *be a goal instance that instantiates* $\mathcal{G}$ *with the input binding* $\beta : IF_\mathcal{G} \rightarrow \mathcal{U}_\mathcal{A}$. *Let* $\mathcal{D}_W = (\Sigma, \Omega, IF_W, \phi^{\mathcal{D}_W})$ *be a functional description, and let* $W = (IF, \iota)$ *be a Web service with* $W \models_\mathcal{A} \mathcal{D}_W$.
$match(GI(\mathcal{G}), W)$ *is given if there exists a* Σ*-interpretation* I *such that:*

$$I \models \Omega_\mathcal{A} \quad and \quad I \models [\phi^{\mathcal{D}_\mathcal{G}}]_\beta \quad and \quad I \models [\phi^{\mathcal{D}_W}]_\beta.$$

Another requirement for W to be usable for solving $GI(\mathcal{G})$ is that the β defined in $GI(\mathcal{G})$ provides concrete values for all inputs that are required to invoke W. This is given if there is a bijection $\pi : IF_{\mathcal{D}_\mathcal{G}} \rightarrow IF_{\mathcal{D}_W}$ such that for every input variable in $\mathcal{D}_W$ there is a corresponding input variable in $\mathcal{D}_\mathcal{G}$, and each $i \in IF_{\mathcal{D}_\mathcal{G}}$ is assigned with the concrete value from β. Subsequently, if there is a second bijection $\pi_2 : IF_{\mathcal{D}_W} \rightarrow IF_W$ such that for each input name required by W there is a corresponding input variable in $\mathcal{D}_W$, then there is a concrete value assignment for each input required by W.[2]

4.3 Integration of Matchmaking Techniques

We complete this section with combing the semantic matchmaking techniques for the goal template and the goal instance level in order to attain an integrated matchmaking

[2] We are aware of that this is requirement is not trivial to realize in practice, as it requires a semantic mapping between the input variables of functional descriptions and the Web service. Moreover, this may require mediation between incompatible ontologies used by the requester and provider [5]. However, to invoke a Web service there must be concrete values for all required inputs – the two bijections denote the basic requirement therefore. [20] discusses ways to weaken the requirements for the necessary compatibility, e.g. by creating existentially quantified ontology instances for input values that are not explicitly defined by the client.

framework for our two-phase Web service discovery. We therefore extend matchmaking degrees from Table 2 with the matchmaking condition for the goal instance level. Due to their definition, we can simplify the matching condition from Definition 4 for the distinct matchmaking degrees as follows.

Theorem 1. *Let $\mathcal{D}_\mathcal{G}$ describe the requested functionality in a goal template $\mathcal{G}$. Let $GI(\mathcal{G})$ be a goal instance of $\mathcal{G}$ that defines an input binding β. Let W be a Web service, and let $\mathcal{D}_W$ be a functional description such that $W \models_\mathcal{A} \mathcal{D}_W$.*
W is usable for solving $GI(\mathcal{G})$ if and only if:

(i) `exact` $(\mathcal{D}_G, \mathcal{D}_W)$ *or*
(ii) `plugin` $(\mathcal{D}_G, \mathcal{D}_W)$ *or*
(iii) `subsume` $(\mathcal{D}_G, \mathcal{D}_W)$ *and* $\bigwedge \Omega_\mathcal{A} \wedge [\phi^{\mathcal{D}_W}]_\beta$ *is satisfiable, or*
(iv) `intersect` $(\mathcal{D}_G, \mathcal{D}_W)$ *and* $\bigwedge \Omega_\mathcal{A} \wedge [\phi^{\mathcal{D}_\mathcal{G}}]_\beta \wedge [\phi^{\mathcal{D}_W}]_\beta$ *is satisfiable.*

This specifies the minimal matchmaking conditions for determining the usability of a Web service for solving a concrete client request that is described by a goal instance. Under both the *exact* and the *plugin* degree, W can be used for solving any goal instance $GI(\mathcal{G})$ because $\{\tau\}_{GI(\mathcal{G})} \subset \{\tau\}_\mathcal{G} \subseteq \{\tau\}_W$ and $\tau \in \{\tau\}_{GI(\mathcal{G})} \Leftrightarrow \tau \in \{\tau\}_{W(\beta)}$. Under the *subsume* degree it holds that $\{\tau\}_\mathcal{G} \supseteq \{\tau\}_W$, i.e. every execution of W can solve $\mathcal{G}$ but there can be solutions of $\mathcal{G}$ that cannot be provided by W. Hence, W is only usable for solving $GI(\mathcal{G})$ if the input binding β defined in $GI(\mathcal{G})$ allows to invoke W. This is given if there is a Σ-interpretation that is a model for $[\phi^{\mathcal{D}_W}]_\beta$ and the conjunction of the axioms in $\Omega_\mathcal{A}$. Under *intersect* as the weakest degree, the complete matchmaking condition for the goal instance level must hold because there can be solutions for $\mathcal{G}$ that can not be provided by W and vice versa. The *disjoint* degree denotes that W is not usable for solving the goal template and thus neither for any of its instantiations. We refer to [20] for the formal proof of this theorem.

5 Evaluation

In order to demonstrate the precision for Web service discovery that is achievable with the presented matchmaking techniques, this section discusses them for our restaurant search example. We have implemented and verified the matchmaking techniques in VAMPIRE [18], a resolution-based theorem prover for classical first-order logic with equality that allows to realize matchmaking exactly as we have specified above. Due to space limitations, we here content ourselves with condensed explanations on the matchmaking techniques for the goal and the Web service as introduced in Section 3.2. A more detailed documentation as well as further examples for discovery under other matchmaking degrees is provided in [20].[3]

The following discusses the matchmaking techniques for the goal of finding the best restaurant in a city and a Web service that provides the best French restaurant in a city

[3] The VAMPIRE implementation along with installation instructions and the proof obligations for the best restaurant search example are available at: `http://members.deri.at/ michaels/software/best-restaurant-example.zip`

(*cf.* functional descriptions in Table 1). This is an example for the *intersect* degree and hence requires the full range of the extended matchmaking for the goal instance level.

For illustration, it is sufficient to consider city A wherein the best restaurant is French and city B wherein the best restaurant is not French. We define two input bindings, $\beta_1 = \{?x|A\}$ and $\beta_2 = \{?x|B\}$, and examine the solutions for $\mathcal{G}$ and the executions of W for each. Table 3 provides a concise overview of the information relevant for our discussion. The first part shows the description of the three best restaurants in A and B as background ontologies $\Omega_1, \Omega_2 \subseteq \Omega$. The second part shows the goal instances, i.e. when $\mathcal{D}_\mathcal{G}$ is instantiated with the concrete values defined in the distinct β as explained in Section 4.2. Analogously, the third part shows the only possible instantiations for W. Finally, the fourth part identifies common Σ-interpretations that serve as a witness for a semantic match between the goal instances and the described Web Services.

We can observe that for the input binding β_1, there is a Σ-interpretation I_1 that is consistent with the background ontology Ω and satisfies both the instantiation of the goal template $[\phi^{\mathcal{D}_\mathcal{G}}]_{\beta_1}$ as well as the instantiation of the Web service $[\phi^{\mathcal{D}_W}]_{\beta_1}$. The

Table 3. Relevant Information for Matchmaking Illustration

City A: $\Omega_1 \subseteq \Omega$	**City B:** $\Omega_2 \subseteq \Omega$
$\Omega_1 = \{city(A)$ $restaurant(r1A)$ $in(r1A, A), type(r1A, french)$ $restaurant(r2A)$ $in(r2A, A), type(r2A, italian)$ $restaurant(r3A)$ $in(r3A, A), type(r3A, french)$ $better(r1A, r2A)$ $better(r2A, r3A)\}$	$\Omega_2 = \{city(B)$ $restaurant(r1B)$ $in(r1B, B), type(r1B, italian)$ $restaurant(r2B)$ $in(r2B, B), type(r2B, french)$ $restaurant(r3B)$ $in(r3B, B), type(r3B, french)$ $better(r1B, r2B)$ $better(r2B, r3B)\}$
$[\phi^{\mathcal{D}_\mathcal{G}}]_{\beta_1}$ **with** $\beta_1 = \{x\|A\}$	$[\phi^{\mathcal{D}_\mathcal{G}}]_{\beta_2}$ **with** $\beta_2 = \{x\|B\}$
$city(A) \Rightarrow ($ $\forall ?y.(out(?y) \Leftrightarrow ($ $restaurant(?y) \wedge in(?y, A)$ $\wedge \neg\exists ?z.(restaurant(?z)$ $\wedge in(?z, A)$ $\wedge better(?z, ?y))))$	$city(B) \Rightarrow ($ $\forall ?y.(out(?y) \Leftrightarrow ($ $restaurant(?y) \wedge in(?y, B)$ $\wedge \neg\exists ?z.(restaurant(?z)$ $\wedge in(?z, B)$ $\wedge better(?z, ?y))))$
$[\phi^{\mathcal{D}_W}]_{\beta_1}$ **with** $\beta_1 = \{x\|A\}$	$[\phi^{\mathcal{D}_W}]_{\beta_2}$ **with** $\beta_2 = \{x\|B\}$
$city(A) \Rightarrow ($ $\forall ?y.(out(?y) \Leftrightarrow ($ $restaurant(?y)$ $\wedge in(?y, A) \wedge type(?y, french)$ $\wedge \neg\exists ?z.(restaurant(?z)$ $\wedge in(?z, A) \wedge type(?z, french)$ $\wedge better(?z, ?y))))$	$city(B) \Rightarrow ($ $\forall ?y.(out(?y) \Leftrightarrow ($ $restaurant(?y)$ $\wedge in(?y, B) \wedge type(?y, french)$ $\wedge \neg\exists ?z.(restaurant(?z)$ $\wedge in(?z, B) \wedge type(?z, french)$ $\wedge better(?z, ?y))))$
I_1 **with** $I_1 \models \Omega \cup \{[\phi^{\mathcal{D}_G}]_{\beta_1}, [\phi^{\mathcal{D}_W}]_{\beta_1}\}$	I_2 **with** $I_2 \models \Omega \cup \{[\phi^{\mathcal{D}_G}]_{\beta_2}, [\phi^{\mathcal{D}_W}]_{\beta_2}\}$
$\Omega_1 \cup \Omega_2 \cup \{out(r1A),$ $better(r1A, r3A), better(r1B, r3B)\}$	No such I_2 can exist!

witnessing execution τ corresponds to the pair (I_1, β_1). Hence, the condition for the *intersect* match is satisfied (*cf.* Table 2). Furthermore, we observe that for the input binding β_2 there can not exist such a common interpretation. Hence, neither the condition for the *subsumes* nor for the *plugin* is satisfied; thus also not the one for the *exact* degree. Assume that there would be such a common interpretation I_2, i.e. a Σ-interpretation that satisfies Ω, $[\phi^{\mathcal{D}_G}]_{\beta_2}$ and $[\phi^{\mathcal{D}_W}]_{\beta_2}$. From the second column of Table 3 we can conclude that any object $?y$ that is the best restaurant in city B is a french restaurant. However, this is not consistent with the background ontology Ω as described above, since then restaurant $r1B$ must be at the same time an italian as well as a french restaurant.

Because of the *intersect* degree on the goal template level, clause (iv) of Theorem 1 must hold for W to be usable for solving a goal instance $GI(\mathcal{G})$ that instantiates $\mathcal{G}$. This requires that there must be a Σ-interpretation that is (a) consistent with the background ontology Ω and (b) a common model for $[\phi^{\mathcal{D}_\mathcal{G}}]_\beta$ and $[\phi^{\mathcal{D}_W}]_\beta$ (*cf.* Definition 4). Let us consider $GI(\mathcal{G})_1$ as the goal instance that instantiates $\mathcal{G}$ with β_1, and $GI(\mathcal{G})_2$ as the goal instance that defines β_2. Analyzing the possible solutions and executions in Table 3 reveals the intuitively expected discovery results: the Σ-interpretation I_1 serves as a witness for a $\tau \in \{\tau\}_{GI(\mathcal{G})_1}$ and $\tau \in \{\tau\}_{W_{\beta_1}}$. Hence, W is usable for solving $GI(\mathcal{G})_1$. On the other hand, as discussed above, there can not exist such a witness for $GI(\mathcal{G})_2$; therefore W can not be used to solve $GI(\mathcal{G})_2$.

6 Related Work

Due to its relevance for service-oriented architectures, Web service discovery is subject to several research efforts. We here discuss directly related works with respect to the quality of matchmaking techniques and the modelling client objectives, referring to more exhaustive overviews, e.g. in [9,10,20].

As early works, [16] presents matchmaking of in- and outputs in OWL-S, and [13] defines matchmaking of requested and provided results in a DL framework. Both define the matching degrees in terms of concept subsumption, and work on OWL-S service advertisements and requests described by inputs, outputs, preconditions, and effects [14]. Although using OWL as an expressive specification language, this description neither explicates the dependency pre- and post execution descriptions nor defines formal semantics for functional descriptions. Hence, the matchmaking algorithms merely allow to detect ontological relationships between corresponding description elements – but not to determine whether the invocation of a Web service in a particular state of the world will satisfy a client request. We can observe the same deficiencies in [2].

In WSMO, provided and requested capabilities are described by preconditions, assumptions, postconditions, and effects, along with *shared variables* to define dependencies between the formulae [12]. However, no formal semantics are defined for these complex functional descriptions – which hampers the specification of accurate matchmaking mechanisms. Our functional descriptions overcome this by explicitly describing dependency of preconditions and effects and defining precise formal semantics. [7] presents a recent approach with a similar focus. Functionalities are described by inputs, outputs, and the relationship between them; a match is given if the requester can provide the input required by the Web service, and the Web service then can provide outputs that satisfy the ones requested. However, this approach is restricted to stateless Web services and hence only covers a subset of the functionalities supported by our approach.

WSMO is the only framework that promotes a goal-based approach for Semantic Web services; most other approaches model client requests as queries for specific Web service descriptions. The differentiation of goal templates and goal instances is a refinement of the WSMO goal model based on experiences in technology realization [21]. A similar two-phased discovery approach is presented in [11]. However, therein goals are described by the desired final state only; the input binding for invoking the discovered Web service is created at runtime. In contrast, we describe the requested functionality in goal templates by preconditions and effects. The reason is that in service-oriented architectures usually the current state of the world is not explicated or is not accessible to the interaction partners. Moreover, defining input bindings on the level of goal instances allows to minimize the client-system interaction as it just needs to be done once.

7 Conclusions

This paper has presented the integrated semantic matchmaking for a two-phased Web service discovery that distinguishes goal templates and goal instances. Continuing previous work, we have defined matchmaking techniques that work on sufficiently rich functional descriptions and can precisely determine the usability of a Web service.

To formally describe client requests on the problem layer, we distinguish goal templates as generic objective descriptions and goal instances that denote a concrete client request as the instantiation of a goal template. We use functional descriptions that precisely describe the start- and end-states of possible executions of Web services as well as of possible solutions for goals. A match is given if a Web service can provide an execution that is a solution for the goal. We have specified semantic matchmaking techniques to evaluate this. On the goal template level, we define matching degrees that differentiate the relationship between possible executions of a Web service and solutions. For a goal instance, a Web service is usable if its execution triggered by the invocation with the concrete inputs is a solution for the instantiated goal description. We therefore have presented a novel matchmaking technique and formally integrated this with the matching degrees on the goal template level. Finally, we have demonstrated that the matchmaking techniques allow to precisely determine the usability of a Web service for solving a concrete client request that is described as a goal instance.

The presented techniques denote the formal foundations for semantic matchmaking in this two-phased discovery approach. We plan to extend this with techniques for efficient management of discovery results, and to continue the integration into frameworks and system implementations for Semantic Web services.

Acknowledgments. This material is based upon works supported by the EU under the **DIP** project (FP6 - 507483) and by the Austrian Federal Ministry for Transport, Innovation, and Technology under the project **RW**^{2} (FFG 809250). The authors like to thank Martin Hepp and Rubén Lara for constructive discussions on the presented work.

References

1. G. Alonso, F. Casati, H. Kuno, and V. Machiraju. *Web Services: Concepts, Architectures and Applications*. Data-Centric Systems and Applications. Springer, Berlin, Heidelberg, 2004.
2. B. Benatallah, M.-S. Hacid, A. Leger, C. Rey, and F. Toumani. On Automating Web Services Discovery. *VLDB Journal*, 14(1):84–96, 2005.
3. M. E. Bratman. *Intention, Plans and Practical Reason*. Harvard University Press, Cambridge, MA (USA), 1987.
4. L. Cabral, J. Domingue, S. Galizia, A. Gugliotta, B. Norton, V. Tanasescu, and C. Pedrinaci. IRS-III – A Broker for Semantic Web Services based Applications. In *Proc. of the 5th International Semantic Web Conference (ISWC 2006), Athens(GA), USA*, 2006.
5. E. Cimpian, A. Mocan, and M. Stollberg. Mediation Enabled SemanticWeb Services Usage. In *Proc. of the 1st Asian Semantic Web Conference (ASWC 2006), Beijing, China*, 2006.
6. I. Dickinson and M. Wooldridge. Agents are not (just) Web Services: Considering BDI Agents and Web Services. In *Proc. of the 2005 Workshop on Service-Oriented Computing and Agent-Based Engineering (SOCABE'2005), Utrecht, The Netherlands*, 2005.
7. D. Hull, E. Zolin, A. Bovykin, I. Horrocks, U. Sattler, and R. Stevens. Deciding Semantic Matching of Stateless Services. In *Proc. of the 21st National Conference on Artificial Intelligence (AAAI'2006)*, 2006.
8. U. Keller, R. Lara, H. Lausen, and D. Fensel. Semantic Web Service Discovery in the WSMO Framework. In J. Cardoses, editor, *Semantic Web: Theory, Tools and Applications*. Idea Publishing Group, 2006.
9. U. Keller, R. Lara, H. Lausen, A. Polleres, and D. Fensel. Automatic Location of Services. In *Proc. of the 2nd European Semantic Web Conference (ESWC 2005), Crete, Greece*, 2005.
10. U. Keller, H. Lausen, and M. Stollberg. On the Semantics of Funtional Descriptions of Web Services. In *Proc. of the 3rd European Semantic Web Conference (ESWC 2006), Montenegro*, 2006.
11. R. Lara. Two-phased Web Service Discovery. In *Proc. of AI-Driven Technologies for Services-Oriented Computing Workshop at AAAI-06, Boston, USA*, 2006.
12. H. Lausen, A. Polleres, and D. Roman (eds.). Web Service Modeling Ontology (WSMO). W3C Member Submission 3 June, 2005.
13. L. Li and I. Horrocks. A Software Framework for Matchmaking based on Semantic Web Technology. In *Proc. of the 12th World Wide Web Conference, Budapest, Hungary*, 2003.
14. D. Martin. OWL-S: Semantic Markup for Web Services. W3C Member Submission 22 November, 2004. online: http://www.w3.org/Submission/OWL-S/.
15. A. Newell. *Unified Theories of Cognition*. Harvard University Press, Cambridge, MA (USA), 1990.
16. M. Paolucci, T. Kawamura, T. Payne, and K. Sycara. Semantic Matching of Web Services Capabilities. In *Proc. of the First International Semantic Web Conference, Springer*, 2002.
17. C. Preist. A Conceptual Architecture for Semantic Web Services. In *Proc. of the 3rd International Semantic Web Conference (ISWC 2004), Hiroshima, Japan*, 2004.
18. A. Riazanov and A. Voronkov. The Design and Implementation of VAMPIRE. *AI Communications*, 15(2):91–110, 2002. Special Issue on CASC.
19. R. M. Smullyan. *First Order Logic*. Springer, 1968.
20. M. Stollberg and U. Keller. Semantic Web Service Discovery. Technical report, DERI, 2006.
21. M. Stollberg and B. Norton. A Refined Goal Model for Semantic Web Services. Proc. of the 2nd International Conference on Internet and Web Applications and Services (ICIW 2007), Mauritius, 2007.
22. M. Stollberg, D. Roman, I. Toma, U. Keller, R. Herzog, P. Zugmann, and D. Fensel. Semantic Web Fred – Automated Goal Resolution on the Semantic Web. In *Proc. of the 38th Hawaii International Conference on System Science (HICSS-38)*, 2005.

A Reasoning Framework for Rule-Based WSML

Stephan Grimm[1], Uwe Keller[2], Holger Lausen[2], and Gábor Nagypál[1]

[1] FZI Research Center for Information Technologies at the University of Karlsruhe, Germany
{stephan.grimm,gabor.nagypal}@fzi.de
[2] Digital Enterprise Research Institute (DERI), University of Innsbruck, Austria
{uwe.keller,holger.lausen}@deri.org

Abstract. WSML is an ontology language specifically tailored to annotate Web Services, and part of its semantics adheres to the rule-based knowledge representation paradigm of logic programming. We present a framework to support reasoning with rule-based WSML language variants based on existing Datalog inference engines. Therein, the WSML reasoning tasks of knowledge base satisfiability and instance retrieval are implemented through a language mapping to Datalog rules and Datalog querying. Part of the WSML semantics is realized by a fixed set of rules that form meta-level axioms. Furthermore, the framework exhibits some debugging functionality that allows for identifying violated constraints and for pointing out involved instances and problem types. Its highly modular architecture facilitates easy extensibility towards other language variants and additional features. The available implementation of the framework provides the first reasoners for the WSML language.

1 Motivation

In the Semantic Web, recently Web Services are annotated by semantic descriptions of their functionality in order to facilitate tasks like automated discovery or composition of services. Such semantic annotation is formulated using ontology languages with logical formalisms underlying them. The matching of semantic annotation for discovery or the checking of type compatibility for composition requires reasoning support for these languages. A relatively new ontology language specifically tailored for the description of Web Services is WSML (Web Service Modeling Language) [6], which comes in variants that follow the rule-based knowledge representation paradigm of logic programming [14]. WSML adds features of conceptual modelling and datatypes, known from frame-base knowledge representation, on top of logic programming rules.

We present a framework for reasoning with rule-based WSML variants that builds on existing infrastructure for inferencing in rule-based formalisms. The framework bases on a semantics-preserving syntactic transformation of WSML ontologies to Datalog programs, as described in the WSML specification [8]. The WSML reasoning tasks of checking knowledge base satisfiability and of instance retrieval can then be performed by means of Datalog querying applied on a transformed ontology. Thus, the framework directly builds on top of existing Datalog inference engines. Besides these standard reasoning tasks, the framework provides debugging features that support an ontology engineer in the task of ontology development: the engineer is pointed out to violated

E. Franconi, M. Kifer, and W. May (Eds.): ESWC 2007, LNCS 4519, pp. 114–128, 2007.

constraints together with some details on the ontological entities that cause the violation. Such a feature helps to improve the error reporting in situations of erroneous modelling. Instead of directly mapping WSML entities, i.e. concepts, instances, attributes, to Datalog predicates and constants, we use special meta-level predicates and axioms which form a vocabulary on reified entities for reproducing the WSML language constructs in Datalog. This way of using Datalog as an underlying formalism facilitates the metamodelling features of WSML. The framework is implemented and can be readily used to reason about ontologies formulated in rule-based WSML. As such, it is the first implementation of a reasoning tool for this language. In contrast to most of the available rule engines and Datalog implementations, this reasoning framework supports the combination of typical rule-style representation with frame-style conceptual modelling, as offered by WSML.

The WSML reasoning framework is jointly developed within, and funded by the European project DIP (IST-FP6-507483) and the Austrian projects SEnSE (FFG 810807) and **RW**^{2} (FFG 809250) .

2 The WSML Language

The Web Service Modeling Language (WSML) [6] is a language for the specification of various aspects of Semantic Web Services (SWS), such as what functionality is provided by a SWS or how to interact with the SWS. It provides a formal language for the Web Service Modeling Ontology (WSMO[1]) [15] and is based on well-known logic-based knowledge representation (KR) formalisms, namely Description Logics [2] and Logic Programming [14]. In fact, WSML is a family of representation languages that comes in several variants with different expressiveness. Besides various SWS-specific language constructs, such as “goal”, “interface”, “choreography” or “capability”, WSML particularly provides means to formulate the domain ontologies in terms of which SWSs are semantically annotated. Since here we are interested in reasoning with such semantic annotation with respect to the underlying ontology formalism, we focus on the ontology-related part of WSML. Furthermore, we use the human-readable syntax of the language in our presentation, while WSML also specifies XML and RDF serialisations to be compatible with existing web standards.

2.1 Language Constructs

WSML makes a clear distinction between the modeling of different conceptual elements on the one hand and the specification of complex axiomatic information on the other. To this end, the WSML syntax is split into two parts: the *conceptual syntax*, and *logical expression syntax*, while elements from both can be combined in a WSML document. We illustrate the interplay of conceptual modelling with logical expressions in WSML by means of an example given in Listing 1.1, which specifies an ontology in the domain of telecommunications taken from a project case study. For a complete account of all WSML syntax elements, we refer to [8].

[1] `http://www.wsmo.org`

Listing 1.1. WSML Example Ontology

```
concept Product
    hasProvider inverseOf(Provider#provides) impliesType Provider
concept ITBundle subConceptOf Product
    hasNetwork ofType (0 1) NetworkConnection
    hasOnlineService ofType (0 1) OnlineService
    hasProvider impliesType TelecomProvider
concept NetworkConnection subConceptOf BundlePart
    providesBandwidth ofType (0 1) _integer
concept DialupConnection subConceptOf NetworkConnection
concept DSLConnection subConceptOf NetworkConnection
axiom DialupConnection_DSLConnection_Disjoint definedBy
  !- ?x memberOf DialupConnection and ?x memberOf DSLConnection.
concept OnlineService subConceptOf BundlePart concept
SharePriceFeed subConceptOf OnlineService axiom
SharePriceFeed_requires_bandwidth definedBy
  !- ?b memberOf ITBundle and ?b[hasOnlineService hasValue ?o]
    and ?o memberOf SharePriceFeed and
    ?b[hasNetwork hasValue ?n] and
    ?n[providesBandwidth hasValue ?x] and ?x < 512.
concept BroadbandBundle subConceptOf ITBundle
    hasNetwork ofType (1 1) DSLConnection
axiom BroadbandBundle_sufficient_condition definedBy
    ?b memberOf BroadbandBundle :- ?b memberOf ITBundle
    and ?b[hasNetwork hasValue ?n] and ?n memberOf DSLConnection.
instance BritishTelekom memberOf TelecomProvider.
instance UbiqBankShareInfo memberOf SharePriceFeed.
instance MyBundle memberOf ITBundle
    hasNetwork hasValue ArcorDSL
    hasOnlineService hasValue UbiqBankShareInfo
    hasProvider BritishTelekom.
instance MSNDialup memberOf DialupConnection
    providesBandwidth hasValue 10.
instance ArcorDSL memberOf DSLConnection
    providesBandwidth hasValue 1024.
```

Conceptual Modelling. The WSML conceptual syntax for ontologies essentially allows for the modeling of concepts, instances and relations.

In ontologies, concepts form the basic elements for describing the terminology of the domain of discourse by means of classes of objects. In the telecommunications domain, a concept like NetworkConnection stands for the class of all network connections. Concepts can be put in a subsumption hierarchy by means of the **subConceptOf**-construct. For example, NetworkConnection is a subconcept of BundlePart, meaning that any network connection is part of some IT product bundle, and has itself the subconcepts DialupConnection and DSLConnection, as can be seen from Listing 1.1.

Attributes, i.e. binary relations, are used to relate concepts in a customary way, while they can point to other concepts or datatypes. In our example, NetworkConnection has a datatype attribute providesBandWidth, whereas concept ITBundle has attributes like hasNetwork or hasOnlineService that point to concepts for the single parts which make up the bundle. Attribute definitions can either be *constraining* (using **ofType**) or *inferring* (using **impliesType**)[2]. Constraining attribute definitions define a type constraint on the values for an attribute, similar to integrity constraints in databases; inferring attribute definitions allow that the type of the values for the attribute is inferred from the attribute definition,

[2] The distinction between inferring and constraining attribute definitions is explained in more detail in [7, Section 2].

similar to range restrictions on properties in RDFS [3] and OWL [9]. Furthermore, an attribute can be marked as **transitive**, **symmetric**, or **reflexive**, and can be constrained by a minimum and a maximum cardinality (using (n_{min} n_{max})), as can be seen from Litsing 1.1. Similar constructs are available to define n-ary relations in ontologies.

Instances represent concrete objects a the domain, such as MSNDialup as a particular dial-up connection in the telecommunications domain. By means of the **memberOf**-construct, instances are associated with concepts, and using **hasValue** they are linked to other instances or data values, as can also be seen in Listing 1.1. Notice, that WSML supports metamodelling and allows an entity to be both a concept and an instance.

Logical Expressions. By means of the **axiom**-construct, arbitrarily complex logical expressions can be included in a WSML ontology, interfering with the conceptual definitions. In our example, the axiom named BroadbandBundle_sufficient_condition specifies that any IT bundle that has a DSL network connection is concluded to be a broadband bundle.

The general logical expression syntax for WSML has a first-order logic style, in the sense that it has constants, function symbols, variables, predicates and the usual logical connectives. Additionally, WSML provides extensions based on F-Logic [12] as well as logic programming rules and database-style integrity constraints.

Besides standard first-order atoms, WSML provides so-called *molecules*, inspired by F-Logic, that can be used to capture information about concepts, instances, attributes and attribute values. A molecule of the form I **memberOf** C denotes the membership of an instance I in a concept C, while a molecule C_1 **subConceptOf** C_2 denotes the subconcept relationship between concepts C_1 and C_2. Further molecules have the form $I[A$ **hasValue** $V]$ to denote attribute values of objects, $C[A$ **ofType** $T]$ to denote a type-constraining attribute signature, or $C[A$ **impliesType** $T]$ to denote an inferring attribute signature. Some of these molecule forms appear in Listing 1.1, e.g. in axiom BroadbandBundle_sufficient_condition.

WSML has the usual first-order connectives: the unary (classical) negation operator **neg**, and the binary operators for conjunction **and**, disjunction **or**, right implication **implies**, left implication **impliedBy**, and bi-implication **equivalent**. Variables, preceeded by the ?-symbol may be universally quantified using **forall** or existentially quantified using **exists**. Apart from first-Order constructs, WSML supports logic programming rules of the form $H\ :-B$ with the typical restrictions on the head and body expressions H and B (see [8]), allowing the symbol **naf** for negation-as-failure on atoms in B. A constraint is a special kind of rule with an empty head expression. While the aforementioned axiom is expressed by a rule, the axiom named DialupConnection_DSLConnection_Disjoint comes in form of a constraint, stating that no instance is allowed to be member of both the concepts DialupConnection and DSLConnection at the same time.

Language Variants. WSML comes in different variants that map to semantically different target formalisms. Therefore, each variant also defines some restrictions on the use of syntactical constructs: ***WSML-Core*** allows only first-order formulae which conform to DLP [11] as the least common denominator of the description logics and logic programming paradigms, by which its semantics is defined. It allows for most of conceptual modelling but is rather restricted in the use of logical expressions. ***WSML-DL*** allows first-order formulae which can be translated to the description logic $\mathcal{SHIQ}(\mathbf{D})$, that defines its semantics. Thus, WSML-DL is very similar to OWL [9]. ***WSML-Flight***

extends WSML-Core by allowing variables in place of instance, concept and attribute identifiers and by allowing relations of arbitrary arity. In fact, any such formula is allowed in the head of a WSML-Flight rule. The body of a WSML-Flight rule allows conjunction, disjunction and default negation. WSML-Flight is based on the well-founded semantics [10] and additionally allows meta-modeling. ***WSML-Rule*** extends WSML-Flight by function symbols and unsafe rules, i.e. variables occurring in the head or in a negative body literal but not in a positive body literal. ***WSML-Full*** does not restrict the use of syntax and allows the full expressiveness of all other WSML variants under a first-order umbrella with nonmonotonic extensions.

In the following, we refer to the WSML-Core, WSML-Flight and WSML-Rule variants jointly as *rule-based WSML* and focus on reasoning in these variants.

2.2 Reasoning in Rule-Based WSML

Various reasoning tasks, such as consistency checking or entailment of implicit knowledge, are considered useful in Semantic Web and SWS applications. Here, we sketch the typical reasoning tasks for rule-based formalisms, and thus for rule-based WSML.

Let O denote a rule-based WSML ontology and $\pi_{c-free}(O)$ denote the constraint-free projection of O, i.e. the ontology which is obtained from O by removing all constraining description elements, such as attribute type constraints, cardinality constraints, integrity constraints etc. (1) **Consistency checking** means to verify whether O is satisfiable, i.e. if $\pi_{c-free}(O)$ has a model in which no constraint in O is violated. (2) **Ground Entailment** means, given some variable-free formula ϕ_g, to check if ϕ_g is satisfied in well-founded model of $\pi_{c-free}(O)$ in which no constraint in O is violated. We denote this by $O \models \phi_g$. (3) **Instance Retrieval** means, given an ontology O and some formula $Q(\vec{x})$ with free variables $\vec{x} = (x_1, \ldots, x_n)$, to find all suitable terms $\vec{t} = (t_1, \ldots, t_n)$ constructed from symbols in O only, such that $O \models Q(\vec{t})$.

3 Mapping WSML to Datalog

The semantics of rule-based WSML is defined via a mapping to Datalog [5,1] with (in)equality, default negation and integrity constraints, as described in [8]. In the following, we refer to this language simply as Datalog. To make use of existing rule engines, the reasoning framework performs various syntactical transformations to convert an original ontology in WSML syntax into a semantically equivalent Datalog program. WSML reasoning tasks are then realized by means of Datalog querying via calls to an underlying Datalog inference engine fed with the rules contained in this program.

3.1 Ontology Transformations

The transformation of a WSML ontology to Datalog rules forms a pipeline of single transformation steps that are subsequently applied, starting from the original ontology.

Axiomatization. In a first step, the transformation τ_{axioms} is applied as a mapping $\mathcal{O} \rightarrow 2^{\mathcal{LE}}$ from the set of all valid rule-based WSML ontologies to the powerset of all logical expressions that conform to rule-based WSML. In this transformation step, all conceptual syntax elements, such as concept and attribute definitions or cardinality and

Table 1. Examples for axiomatizing conceptual ontology modeling elements

Expression α in conceptual syntax	*Resulting logical expression(s): $\tau_{axioms}(\alpha)$*
concept C_1 **subConceptOf** C_2	C_1 **subConceptOf** C_2.
concept C A **ofType** $(0, 1)$ T	C[A **ofType** T]. !- ?x **memberOf** C **and** ?x[A **hasValue** ?y, A **hasValue** ?z] **and** ?y != ?z.
concept C A_1 **inverseOf** A_2 **impliesType** T	C[A **impliesType** T]. ?x **memberOf** C **and** ?v **memberOf** T **implies** (?x[A_1 **hasValue** ?v] **equivalent** ?v[A_2 **hasValue** ?x]).
relation R_1/n **subRelationOf** R_2	$R_1(\vec{x})$ **implies** $R_2(\vec{x})$. where $\vec{x} = (x_1,...,x_n)$
instance I **memberOf** C A **hasValue** V	I **memberOf** C. I[A **hasValue** V].

Table 2. Normalization of WSML logical expressions

original expression	*normalized expression*	*original expression*	*simplified rule(s)*
$\tau_n(\{E_1,\ldots,E_n\})$	$\{\tau_n(E_1),\ldots,\tau_n(E_n)\}$	$\tau_{dlog}(\{E_1,\ldots,E_n\})$	$\{\tau_{dlog}(E_1),\ldots,\tau_{dlog}(E_n)\}$
$\tau_n(E_x$ **and** $E_y.)$	$\tau_n(E_x)$ **and** $\tau_n(E_y)$	$\tau_{dlog}($!- $B.)$	$\square$:- $\tau_{dlog}(B)$
$\tau_n(E_x$ **or** $E_y.)$	$\tau_n(E_x)$ **or** $\tau_n(E_y)$	$\tau_{dlog}(H.)$	$\tau_{dlog}(H)$.
$\tau_n(E_x$ **and** $(E_y$ **or** $E_z).)$	$\tau_n(\tau_n(E_x)$ **and** $\tau_n(E_y)$ **or**	$\tau_{dlog}(H$:- $B.)$	$\tau_{dlog}(H)$:- $\tau_{dlog}(B)$
	$\tau_n(E_x)$ **and** $\tau_n(E_z).)$	$\tau_{dlog}(E_x$ **and** $E_y.)$	$\tau_{dlog}(E_x) \wedge \tau_{dlog}(E_y)$
$\tau_n((E_x$ **or** $E_y)$ **and** $E_z).)$	$\tau_n(\tau_n(E_x)$ **and** $\tau_n(E_z)$ **or**	$\tau_{dlog}($**naf** $E.)$	$\sim \tau_{dlog}(E)$
	$\tau_n(E_y)$ **and** $\tau_n(E_z).)$	$\tau_{dlog}(C_x$ **subConceptOf** $C_y.)$	$p_{sco}(C_x, C_y)$
$\tau_n($ **naf** $(E_x$ **and** $E_y).)$	**naf** $\tau_n(E_x)$ **or** **naf** $\tau_n(E_y)$.	$\tau_{dlog}(I$ **memberOf** $C.)$	$p_{mo}(I, C)$
$\tau_n($ **naf** $(E_x$ **or** $E_y).)$	**naf** $\tau_n(E_x)$ **and** **naf** $\tau_n(E_y)$.	$\tau_{dlog}(I[a$ **hasValue** $V].)$	$p_{hval}(I, a, V)$
$\tau_n($ **naf** $($ **naf** $E_x).)$	$\tau_n(E_x)$	$\tau_{dlog}(C[a$ **impliesType** $T].)$	$p_{itype}(C, a, T)$
$\tau_n(E_x$ **implies** $E_y.)$	$\tau_n(E_y)$:- $\tau_n(E_x)$.	$\tau_{dlog}(C[a$ **ofType** $T].)$	$p_{otype}(C, a, T)$
$\tau_n(E_x$ **impliedBy** $E_y.)$	$\tau_n(E_x)$:- $\tau_n(E_y)$.	$\tau_{dlog}(r(X_1,\ldots,X_n).)$	$r(X_1,\ldots,X_n)$
$\tau_n(X[Y_1,\ldots,Y_n].)$	$X[Y_1]$ **and** ... **and** $X[Y_n]$.	$\tau_{dlog}(X = Y.)$	$X = Y$
		$\tau_{dlog}(X$!= $Y.)$	$X \neq Y$

original expr.	*simplified rule(s)*	*original expression*	*simplified rule(s)*
$\tau_{lt}(\{E_1,\ldots,E_n\})$	$\{\tau_{lt}(E_1),\ldots,\tau_{lt}(E_n)\}$	$\tau_{lt}(H_1$ **and** ... **and** H_n :- $B.)$	$\tau_{lt}(H_1$:- $B.)$, ..., $\tau_{lt}(H_n$:- $B.)$
$\tau_{lt}(H_1$:- H_2 :- $B.)$	$\tau_{lt}(H_1$:- H_2 **and** $B.)$	$\tau_{lt}(H$:- B_1 **or** , ..., **or** $B_n.)$	$\tau_{lt}(H$:- $B_1.)$, ..., $\tau_{lt}(H$:- $B_n.)$

type constraints, are converted into appropriate axioms specified by logical expressions. Table 1 shows the details of some of the conversions performed by τ_{axioms}, based on [8]. The WSML conceptual syntax constructs on the left-hand side are converted to the respective WSML logical expressions on the right-hand side. The meta variables C, C_i range over identifiers of WSML concepts, R_i, A_i over identifiers of WSML relations and attributes, T over identifiers of WSML concepts or datatypes and V over identifiers of WSML instances or data values.

Normalization. The transformation τ_n is applied as a mapping $2^{\mathcal{LE}} \rightarrow 2^{\mathcal{LE}}$ to normalize WSML logical expressions. This normalization step reduces the complexity of formulae according to [8, Section 8.2], to bring expressions closer to the simple syntactic form of literals in Datalog rules. The reduction includes conversion to negation and disjunctive normal forms as well as decomposition of complex WSML molecules. The left part of Table 2 shows how the various logical expressions are normalized in detail. The meta variables E_i range over logical expressions in rule-based WSML, while X, Y_i range over parts of WSML molecules. After τ_n has been applied, the resulting expressions have the form of logic programming rules with no deep nesting of logical connectives.

Lloyd-Topor Transformation. The transformation τ_{lt} is applied as a mapping $2^{\mathcal{LE}} \rightarrow 2^{\mathcal{LE}}$ to flatten the complex WSML logical expressions, producing simple rules according to the Lloyd-Topor transformations [13], as shown in the lower part of Table 2. Again, the meta variables E_i, H_i, B_i range over WSML logical expressions, while H_i and B_i match the form of valid rule head and body expressions, respectively, according to [8]. After this step, the resulting WSML expressions have the form of proper Datalog rules with a single head and conjunctive (possibly negated) body literals.

Datalog Rule Generation. In a final step, the transformation τ_{dlog} is applied as a mapping $2^{\mathcal{LE}} \rightarrow \mathcal{P}$ from WSML logical expressions to the set of all Datalog programs, yielding generic Datalog rules that represent the content of the original WSML ontology. Rule-style language constructs, such as rules, facts, constraints, conjunction and (default) negation, are mapped to the respective Datalog elements. All remaining WSML-specific language constructs, such as **subConceptOf** or **ofType**, are replaced by special meta-level predicates for which the semantics of the respective language construct is encoded in meta-level axioms as described in Section 3.2. The right-hand part of Table 2 shows the mapping from WSML logical expressions to Datalog including the meta-level predicates p_{sco}, p_{mo}, p_{hval}, p_{itype} and p_{otype} that represent their respective WSML language constructs as can be seen from the mapping. The meta variables E, H, B range over WSML logical expressions with a general, a head or a body form, while C, I, a denote WSML concepts, instances and attributes. Variables T can either assume a concept or a datatype, and V stands for either an instance or a data value, accordingly.

The resulting Datalog rules are of the form $H \;\; :- \; B_1 \wedge \ldots \wedge B_n$, where H and B_i are literals for the head and the body of the rule, respectively. Body literals can be negated in the sense of negation-as-failure, which is denoted by $\sim B_i$. As usual, rules with an empty body represent facts, and rules with an empty head represent constraints. The latter is denoted by the head being the empty clause symbol $\Box$.

Ultimately, we define the basic[3] transformation τ for converting a rule-based WSML ontology into a Datalog program based on the single transformation steps introduced before by $\tau = \tau_{dlog} \circ \tau_{lt} \circ \tau_n \circ \tau_{axioms}$. As a mapping $\tau : \mathcal{O} \rightarrow \mathcal{P}$, this composition of the single steps is applied to a WSML ontology $O \in \mathcal{O}$ to yield a semantically equivalent Datalog program $\tau(O) = P \in \mathcal{P}$ when interpreted with respect to the meta-level axioms discussed next.

3.2 WSML Semantics Through Meta-level Axioms

The mapping from WSML to Datalog in the reasoning framework works such that each WSML-identifiable entity, i.e. concept, instance, attribute etc., is mapped to an instance (or logical constant) in Datalog, as depicted in Figure 1. There, the concepts C_1, C_2, C_3 as well as the instances I_1, I_2 and the attribute a are mapped to constants such as I_{C_1}, I_{I_1} or I_a in Datalog, representing the original WSML entities on the instance level.

Accordingly, the various special-purpose relations that hold between WSML entities, such as **subConceptOf**, **memberOf** or **hasValue**, are mapped to Datalog predicates that form a meta-level vocabulary for the WSML language constructs. These are the meta-level predicates that appear in Table 2 for τ_{dlog}, and which are applied to the Datalog constants

[3] Later on, the transformation pipeline is further extended to support datatypes and debugging.

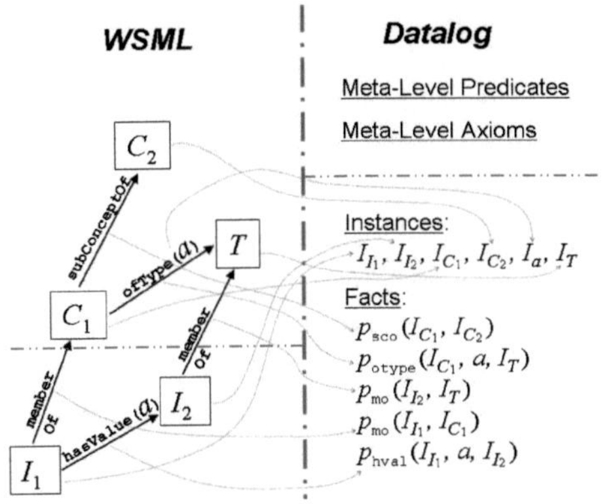

Fig. 1. Usage of meta-level predicates

Meta-Level Axioms

$$
\begin{aligned}
&(1)\ p_{\mathsf{sco}}(C_1, C_3)\ :-\ p_{\mathsf{sco}}(C_1, C_2) \wedge p_{\mathsf{sco}}(C_2, C_3)\\
&(2)\ p_{\mathsf{mo}}(I, C_2)\ :-\ p_{\mathsf{mo}}(I, C_1) \wedge p_{\mathsf{sco}}(C_1, C_2)\\
&(3)\ p_{\mathsf{mo}}(V, C_2)\ :-\ p_{\mathsf{itype}}(C_1, a, C_2) \wedge p_{\mathsf{mo}}(I, C_1) \wedge p_{\mathsf{hval}}(I, a, V)\\
&(4)\ \Box\ :-\ p_{\mathsf{otype}}(C_1, a, C_2) \wedge p_{\mathsf{mo}}(I, C_1) \wedge p_{\mathsf{hval}}(I, a, V) \wedge \sim p_{\mathsf{mo}}(V, C_2)
\end{aligned}
$$

Fig. 2. WSML semantics in Datalog

that represent the WSML entities. The facts listed in Figure 1 illustrate the use of the meta-level predicates. For example, the predicate p_{mo} takes a Datalog constant that represents a WSML instance and one that represents a WSML concept, to state that the instance is in the extension of this concept.

In contrast to a direct mapping from WSML to Datalog with concepts, attributes and instances mapping to unary predicates, binary predicates and constants, respectively, this indirect mapping allows for the WSML metamodelling facilities. Metamodelling allows an entity to be a concept and an instance at the same time. By representing a WSML entity as a Datalog constant, it could, for example, fill both the first as well as the second argument of e.g. the predicate p_{mo}.

A fixed set P_{meta} of Datalog rules, shown in Figure 2, forms the meta-level axioms which assure that the original WSML semantics is properly maintained. Axiom (1) realizes transitivity for the WSML **subConceptOf** construct, while axiom (2) ensures that an instance of a subconcept is also an instance of its superconcepts. Axiom (3) realizes the semantics for the **implisType** construct for attribute ranges: any attribute value is concluded to be in the extension of the range type declared for the attribute. Finally, axiom (4) realizes the semantics of the **ofType** construct by a constraint that is violated whenever an attribute value cannot be concluded to be in the extension of the declared range type.

3.3 WSML Reasoning by Datalog Queries

To perform reasoning over the original WSML ontology O with an underlying Datalog inference engine, a Datalog program $P_O = P_{meta} \cup \tau(O)$ is built up that consists of the meta-level axioms together with the transformed ontology. The different WSML reasoning tasks are then realized by performing Datalog queries on P_O. Posing a query $Q(\vec{x})$ to a Datalog program $P \in \mathcal{P}$ is denoted by $(P,\ ?-Q(\vec{x}))$ and yields the set of all tuples $\vec{t}$ that instantiate the vector $\vec{x}$ of variables in the query such that $Q(\vec{t})$ is satisfied in the well-founded model of P. If $Q(\vec{x})$ contains no variables, in fact a boolean query Q is posed that instead evaluates either to $\{Q\}$ if Q is satisfied in the well-founded model of P or $\emptyset$ otherwise.

Ontology Consistency – The task of checking a WMSL ontology for consistency is done by querying for the empty clause, as expressed by the following equivalence: O is satisfiable $\Leftrightarrow$ $(P_O,\ ?-\ \Box) = \emptyset$. If the resulting set is empty then the empty clause could not be derived from the program and the original ontology is satisfiable, otherwise it is not.

Entailment – The reasoning task of ground entailment by a WSML ontology is done by using queries that contain no variables, as expressed in the following equivalence: $O \models \phi_g \Leftrightarrow (P_O,\ ?-\ \tau'(\phi_g))) \neq \emptyset$. The WSML ground fact $\phi_g \in \mathcal{LE}$ is transformed to Datalog with a transformation $\tau' = \tau_{dlog} \circ \tau_{lt} \circ \tau_n$, similar to the one that is applied to the ontology, and is evaluated together with the Datalog program P_O. If the resulting set is non-empty then ϕ_g is entailed by the original ontology, otherwise it is not.

Retrieval – Similarly, instance retrieval can be performed by posing a WSML query $Q(\vec{x})$ with free variables $\vec{x}$ to the Datalog program P_O, which yields the following set: $\{\vec{o}\,|\,O \models Q(\vec{o})\} = (P_O,\ ?-\ \tau'(Q(\vec{x})))$. The query $Q(\vec{x})$ is transformed to Datalog by τ' and evaluated together with the program P_O. The resulting set contains all object tuples $\vec{o}$ for which an instantiation of the query expression is entailed by the original ontology, while the objects in $\vec{o}$ can be identifiable WSML entities or data values. For example, the query $Q(\text{?x})$ = ?x **memberOf** BroadbandBundle posed to the ontology in Listing 1.1 yields the set $\{(\text{MyBundle})\}$ that contains one unary tuple with the instance MyBundle, which can be inferred to be a broadband bundle due to its high network bandwidth.

3.4 Realising Datatype Reasoning

Although most of the generic Datalog rules are understood by practically any Datalog implementation, realizing datatype reasoning has some intricate challenges. The main challenge is related to Axiom (4) in Figure 2, which checks attribute type constraints. The crucial part of the axiom is the literal

$$\sim p_{\mathsf{mo}}(V, C_2)$$

because for datatype values no explicit membership facts are included in the ontology that could instantiate this literal. Consider, for example, the instance MSNDialup from the WSML ontology in Section 2 – there is no fact $p_{\mathsf{mo}}(10, \_\mathtt{integer})$ for the value of the providesBandwidth attribute. Whenever a value is defined for an attribute constrained by **ofType**, Axiom (4) would cause a constraint violation.

To solve this problem, p_{mo} facts should be generated for all datatype constants that appear as values of attributes having **ofType** constraints in the ontology. I.e., for each such constant in the ontology, axioms of the following form should appear,

$$p_{\mathsf{mo}}(V, D)\ :-\ typeOf(V, D_T)$$

where D denotes the WSML datatype, D_T denotes a datatype supported by the underlying Datalog implementation, which is compatible with the WSML datatype, and *typeOf* denotes a built-in predicate implemented by the Datalog tool, which checks whether a constant value belongs to the specified datatype.

These additional meta-level axioms result in a new set of Datalog rules, denoted by P_{data}, which are no longer in generic Datalog but use tool-specific built-in predicates of the underlying inference engine. The program P_O is extended by these rules as follows.

$$P_O = P_{meta} \cup P_{data} \cup \tau(O)$$

In addition to datatypes, WSML also supports some predefined datatype predicates, such as numeric comparison (see [8] for a full list). The definition of the axiom SharePriceFeed_requires_bandwidth from the WSML ontology in Section 2, for example, uses a shortcut of the WSML **numericLessThan** predicate (denoted by $<$). For translation of these special predicates to the corresponding tool-specific built-in predicates supported by the underlying Datalog reasoner, we introduce a new tool-specific transformation step τ_{dpred} as a mapping $\mathcal{P} \rightarrow \mathcal{P}$. This affects the transformation pipeline τ as follows.

$$\tau = \tau_{dpred} \circ \tau_{dlog} \circ \tau_{lt} \circ \tau_n \circ \tau_{axioms}$$

In summary, the underlying Datalog implementation must fulfill the following requirements to support WSML datatype reasoning: (i) It should provide built-in datatypes that correspond to WSML datatypes. (ii) It should provide a predicate (or predicates) for checking whether a datatype covers a constant and (iii) It should provide built-in predicates that correspond to datatype-related predefined predicates in WSML.

4 Debugging Support

During the process of ontology development, an ontology engineer can easily construct an erroneous model containing contradictory information. In order to produce consistent ontologies, inconsistencies should be reported to engineers with some details about the ontological elements that cause the inconsistency.

In rule-based WSML, the source for erroneous modelling are always constraints, together with a violating situation of concrete instances related via attributes. The plain Datalog mechanisms employed in the reasoning framework according to Section 3 only allow for checking whether some constraint is violated, i.e. whether the empty clause is derived from P_O indicating that the original ontology O contains errors – more detailed information about the problem is not reported. Experience shows that it is a very hard task to identify and correct errors in the ontology without such background information.

In our framework, we support debugging features that provide information about the ontology entities which are involved in a constraint violation. We achieve this by replacing constraints with appropriate rules that derive debugging-relevant information.

4.1 Identifying Constraint Violations

In case of an inconsistent ontology due to a constraint violation, two things are of interest to the ontology engineer: a) the type of constraint that is violated and b) the entities, i.e. concepts, attributes, instances, etc., that are involved in the violation.

To give an example, consider the WSML ontology in Section 2. There, the attribute hasOnlineService of the concept ITBundle is constrained to instances of type OnlineService. Suppose we replace the current value of the attribute hasOnlineService for the instance MyBundle by the instance MSNDialup. Then, this constraint would be violated because MSNDialup is not an instance of the concept OnlineService. For an ontology engineer who needs to repair this erroneous modelling, it is important to know the entities that cause the violation, which in this case are the attribute hasOnlineService together with the range concept OnlineService and the non-conforming instance MSNDialup.

For the various types of constraint violations, the information needed by the ontology engineer to track down the problem successfully is different from case to case.

Attribute Type Violation – An attribute type constraint of the form $C[a$ **ofType** $T]$ is violated whenever an instance of the concept C has value V for the attribute a, and it cannot be inferred that V belongs to the type T. Here, T can be either a concept or a datatype, while V is then an instance or a data value, accordingly. In such a situation, an ontology engineer is particularly interested in the instance I, in the attribute value V that caused the constraint violation, together with the attribute a and the expected type T which the value V failed to adhere to.

Minimum Cardinality Violation – A minimum cardinality constraint of the form **concept** $C\ a\ (n\ *)$, is violated whenever the number of distinguished values of the attribute a for some instance I of the concept C is less than the specified cardinality n. In such a situation, an ontology engineer is particularly interested in the instance I that failed to have a sufficient number of attribute values, together with the actual attribute a. (Information about how many values were missing can be learned by separate querying).

Maximum Cardinality Violation – A maximum cardinality constraint of the form **concept** $C\ a\ (\mathbf{0}\ n)$, is violated whenever the number of distinguished values of the attribute a for some instance I of the concept C exceeds the specified cardinality n. Again, here an ontology engineer is particularly interested in the instance I for which the number of attribute values was exceeded, together with the actual attribute a.

User-Defined Constraint Violation – Not only built-in WSML constraints, but also user-defined constraints, contained in an axiom definition of the form **axiom** Ax_{ID} **definedBy !-** B, can be violated. In this case, the information which helps an ontology engineer to repair an erroneous situation is dependent on the arbitrarily complex body B and cannot be determined in advance. However, a generic framework can at least identify the violated constraint by reporting the identifier Ax_{ID} of the axiom.

To give an example, consider again the ontology from Section 2. Replacing the network connection ArcorDSL of MyBundle by the slower MSNDialup one results in a violation of the user-defined constraint specified by the axiom named SharePriceFeed_requires_bandwidth. This constraint requires a certain bandwidth for connections in bundles with share price feed online services, which is not met by MSNDialup, and thus the ontology engineer is reported the axiom name that identifies the violated constraint.

4.2 Debugging by Meta-level Reasoning

In our framework, we realize the debugging features for reporting constraint violations by replacing constraints with a special kind of rules. Instead of deriving the empty clause, as constraints do, these rules derive information about occurrences of constraint

Table 3. Replacing constraints by rules

Constraint	*Rule*
$\tau_{debug}(\{E_1,\ldots,E_n\})$	$\{\tau_{debug}(E_1),\ldots,\tau_{debug}(E_n)\}$
$\tau_{debug}(\,!\!-B_{mincard}.)$	$p_{\mathsf{v\_mincard}}(a,I) :- B_{mincard}.$
$\tau_{debug}(\,!\!-B_{maxcard}.)$	$p_{\mathsf{v\_maxcard}}(a,I) :- B_{maxcard}.$
$\tau_{debug}(\,!\!-B_{user}.)$	$p_{\mathsf{v\_user}}(Ax_{ID}) :- B_{user}.$
$\tau_{debug}(C[a\ \textbf{ofType}\ T].)$	$p_{\mathsf{v\_otype}}(a,T,I,V) :-$ $C[a\ \textbf{ofType}\ T]$ **and** I **memberOf** C **and** $I[a\ \textbf{hasValue}\ V]$ **and naf** V **memberOf** T.

violations by instantiating debugging-specific meta-level predicates with the entities involved in a violation. In this way, information about constraint violations can be queried for by means of Datalog inferencing.

The replacement of constraints for debugging is included in the transformation

$$\tau = \tau_{dpred} \circ \tau_{dlog} \circ \tau_{lt} \circ \tau_n \circ \tau_{debug} \circ \tau_{axioms}$$

where the additional transformation step τ_{debug} is applied after the WSML conceptual syntax has been resolved, replacing constraints on the level of WSML logical expressions. Table 3 shows the detailed replacements performed by τ_{debug} for the different kinds of constraints.

Minimal cardinality constraints (with bodies $B_{mincard}$) and maximal cardinality constraints (with bodies $B_{maxcard}$) are transformed to rules by keeping their respective bodies and adding a head that instantiates one of the predicates $p_{\mathsf{v\_mincard}}$ and $p_{\mathsf{v\_maxcard}}$ to indicate the respective cardinality violation. The variables for the involved attribute a and instance I are the ones that occur in the respective constraint body B.

Similarly, a user-defined constraint is turned into a rule by keeping the predefined body B_{user} and including a head that instantiates the predicate $p_{\mathsf{v\_user}}$ to indicate a user-defined violation. The only argument for the predicate $p_{\mathsf{v\_user}}$ is the identifier Ax_{ID} of the axiom, by which the constraint has been named.

Constraints on attribute types are handled differently because these constraints are not expanded during the transformation τ_{axioms}; they are rather represented by WSML **ofType**-molecules for which the semantics is encoded in the meta-level axioms P_{meta}. In order to avoid the modification of P_{meta} in the reasoning framework, such molecules are expanded by τ_{debug}, as shown in Table 3.[4]

To maintain the semantics of the replaced constraints, an additional set of meta-level axioms $P_{debug} \subset \mathcal{P}$ is included for reasoning. The rules in P_{debug} have the form $\Box\ :- p_{\mathsf{v}}$ and derive the empty clause for any type and occurrence of a constraint violation.

Including the debugging features, the Datalog program for reasoning about the original ontology then turns to

$$P_O = P_{meta} \cup P_{data} \cup P_{debug} \cup \tau(O) \quad .$$

[4] After this expansion of **ofType** molecules, the respective axiom (4) in P_{meta} for realising the semantics of attribute type constraints does not apply anymore.

Occurrences of constraint violations can be recognized by querying P_O for instantiations of the various debugging-specific meta-level predicates $p_{\mathsf{v\_otype}}$, $p_{\mathsf{v\_mincard}}$, $p_{\mathsf{v\_maxcard}}$ and $p_{\mathsf{v\_user}}$. For example, the set

$$(P_O,\ ?-\ p_{\mathsf{v\_otype}}(a, T, I, V))$$

contains tuples for all occurrences of attribute type violations in P_O, identifying the respective attribute a, expected type T, involved instance I and violating value V for each violation. This set is empty no attribute types are violated.

5 Reasoning Framework Overview

The design goals of our framework are modularity for the transformation steps and flexibility with respect to the underlying inference engine. The high modularity allows to reuse transformation functionality across different WSML variants and reduces the effort for accomplishing other reasoning tasks. By realizing WSML on top of a generic Datalog layer, we have also reduced the effort of integrating other reasoners to a minimum The presented framework has been fully implemented in Java and can be downloaded and tested online[5].

Architecture and Internal Layering. Figure 3 shows the internal architecture of the framework as well as the data flow during a prototypical usage scenario. The outer box outlines a WSML reasoner component that allows a user to register WSML ontologies and to pose queries on them. The inner box illustrates the transformation pipeline introduced in Section 3 and shows its subsequent steps in a layering scheme.

Registered ontologies go through all the transformation steps, whereas user queries are injected at a later stage, skipping the non-applicable axiomatization and constraint replacement steps. Here, the internal layering scheme allows for an easy reorganization and reuse of the transformation steps on demand, assuring high flexibility and modularity. A good example for this is the constraint replacement transformation τ_{debug}: if included in the pipeline, it produces the rules that activate the debugging features according to Section 4; if excluded, the constraints remain in the resulting Datalog program and are mapped to native constraints of the underlying reasoning engine.

The core component of the framework is an exchangeable Datalog inference engine wrapped by a reasoner facade which embeds it in the framework infrastructure. This facade mediates between the generic Datalog program produced in the transformations and the external engine's tool-specific Datalog implementation and built-in predicates.

Interface and Integration with Existing Technology. Our framework is based on the WSMO4J [6] project, which provides an API for the programmatic handling of WSML documents. WSMO4J performs the task of parsing and validating WSML ontologies and provides the source object model for our translations. For a reasoner to be connected to the Framework, a small adapter class needs to be written, that translates generic Datalog elements to their equivalent constructs within the internal representation layer of

[5] `http://dev1.deri.at/wsml2reasoner`

[6] `http://wsmo4j.sourceforge.net`

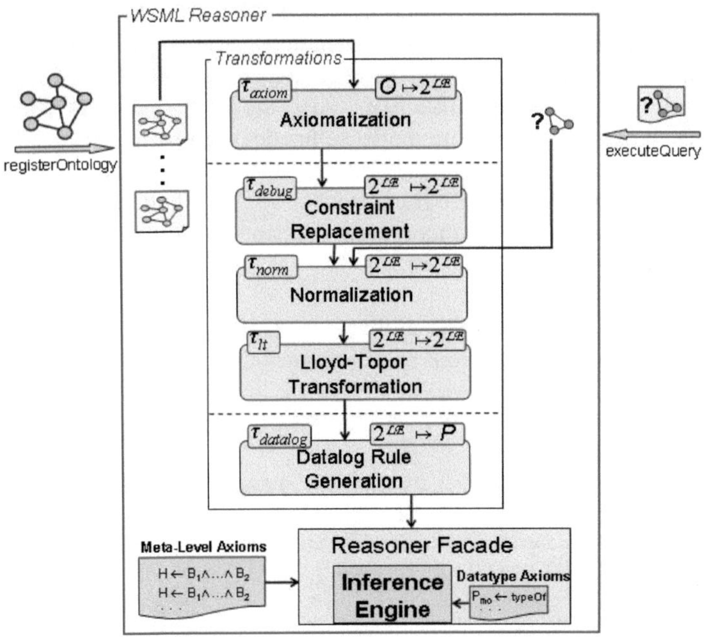

Fig. 3. Internal framework architecture

the underlying reasoner. Our framework currently comes with facades for two built-in reasoners: KAON2[7] and MINS[8]. The initial development was done with the KAON2 inference engine that, with respect to the challenges for datatype reasoning, provides a very flexible type system that allows for user-defined datatypes, together with predicates on these datatypes, including type checking predicates. However, KAON2 cannot be used for reasoning in WSML-Rule as it does not support function symbols and unsafe rules. The second reasoner, MINS, can be used for the WSML-Rule variant but has limited support for datatype reasoning. (For determining the WSML variant of an ontology, one can use the validation facilities built into WSMO4J).

6 Conclusion and Outlook

We have presented a framework for reasoning in rule-based WSML that builds on a mapping to Datalog and on querying a generic Datalog layer. The single well-defined transformation steps can be reused across various adaptations for different scenarios in a highly modular way. We have incorporated debugging features by replacing native constraints with rules to derive debugging-relevant information that can be queried by an ontology engineer. We have implemented our framework with two existing reasoner tools, namely KAON2 and MINS, as alternative implementations of the generic Datalog layer, by which we provide the first available reasoning system for the WSML language.

[7] http://kaon2.semanticweb.org

[8] http://dev1.deri.at/mins

While the current framework focuses on WSML-Core, -Flight and -Rule, efforts are ongoing to extend the transformations to disjunctive Datalog and description logics. The KAON2 system natively supports disjunctive Datalog and DL reasoning, the latter even extended by WSML-Flight-like rules. Also the DLV system [4] (implementing disjunctive Datalog under the stable model semantics) can be used to realise a similar reasoning. Furthermore, we plan to integrate the KRHyper system [16], which allows reasoning with disjunctive logic programs with stratified default negation. Transformations to DL additionally allow to incorporate description logic system APIs to support efficient reasoning with WSML-DL.

Acknowledgements. We would like to thank our colleagues that wrote [8], especially Jos de Bruijn, for fruitful discussions and the contribution of mapping definitions.

References

1. S. Abiteboul, R. Hull, and V. Vianu. *Foundations of Databases*. Addison-Wesley, 1995.
2. F. Baader, D. Calvanese, D. L. McGuinness, D. Nardi, and P. F. Patel-Schneider, editors. *The Description Logic Handbook*. Cambridge University Press, 2003.
3. D. Brickley and R. V. Guha. RDF Vocabulary Description Language 1.0: RDF Schema. Recommendation 10 February 2004, W3C, 2004.
4. S. Citrigno, T. Eiter, W. Faber, G. Gottlob, C. Koch, N. Leone, C. Mateis, G. Pfeifer, and F. Scarcello. The DLV System: Model Generator and Advanced Frontends. In *Workshop LP*, 1997.
5. M. Dahr. *Deductive Databases: Theory and Applications*. International Thomson Publishing, December 1996.
6. J. de Bruijn, H. Lausen, A. Polleres, and D. Fensel. The Web Service Modeling Language WSML: An Overview. In *Proc. of the 3rd Euro. Semantic Web Conference (ESWC)*, 2006.
7. J. de Bruijn, A. Polleres, R. Lara, and D. Fensel. OWL DL vs. OWL Flight: Conceptual Modeling and Reasoning on the Semantic Web. In *Proceedings of the 14th International World Wide Web Conference (WWW2005)*, Chiba, Japan, 2005. ACM.
8. J. de Bruin. The Web Service Modeling Language (WSML) Specification. Tech. Report, Digital Enterprise Research Institute (DERI), Feb. 2005. http://www.wsmo.org/TR/d16/.
9. M. Dean and G. Schreiber, editors. *OWL Web Ontology Language Reference*. 2004. W3C Recommendation 10 February 2004.
10. A. V. Gelder, K. Ross, and J. S. Schlipf. The Well-Founded Semantics for General Logic Programs. *Journal of the ACM*, 38(3):620–650, 1991.
11. B. Grosof, I. Horrocks, R. Volz, and S. Decker. Description Logic Programs: Combining Logic Programs with Description Logics. In *Proceedings of WWW-2003*, 2003.
12. M. Kifer, G. Lausen, and J. Wu. Logical Foundations of Object-Oriented and Frame-Based Languages. *JACM*, 42(4):741–843, 1995.
13. J. Lloyd and R. Topor. Making Prolog More Expressive. *Journal of Logic Programming*, 3:225–240, 1984.
14. J. W. Lloyd. *Foundations of Logic Programming; (2nd extended ed.)*. Springer, New York, NY, USA, 1987. ISBN 3-540-18199-7.
15. D. Roman, U. Keller, H. Lausen, R. L. Jos de Bruijn, M. Stollberg, A. Polleres, C. Feier, C. Bussler, and D. Fensel. Web Service Modeling Ontology. *Applied Ontology*, 1(1):77–106, 2005.
16. C. Wernhard. System Description: KRHyper. Technical report, Fachberichte Informatik 14–2003, Universitat Koblenz-Landau, Institut fur Informatik., 2003.

GenTax: A Generic Methodology for Deriving OWL and RDF-S Ontologies from Hierarchical Classifications, Thesauri, and Inconsistent Taxonomies

Martin Hepp and Jos de Bruijn

Digital Enterprise Research Institute (DERI), University of Innsbruck
mhepp@computer.org, jos.debruijn@deri.org

Abstract. Hierarchical classifications, thesauri, and informal taxonomies are likely the most valuable input for creating, at reasonable cost, non-toy ontologies in many domains. They contain, readily available, a wealth of category definitions plus a hierarchy, and they reflect some degree of community consensus. However, their transformation into useful ontologies is not as straightforward as it appears. In this paper, we show that (1) it often depends on the context of usage whether an informal hierarchical categorization schema is a classification, a thesaurus, or a taxonomy, and (2) present a novel methodology for automatically deriving consistent RDF-S and OWL ontologies from such schemas. Finally, we (3) demonstrate the usefulness of this approach by transforming the two e-business categorization standards eCl@ss and UNSPSC into ontologies that overcome the limitations of earlier prototypes. Our approach allows for the script-based creation of meaningful ontology classes for a particular context while preserving the original hierarchy, even if the latter is not a real subsumption hierarchy in this particular context. Human intervention in the transformation is limited to checking some conceptual properties and identifying frequent anomalies, and the only input required is an informal categorization plus a notion of the target context. In particular, the approach does not require instance data, as ontology learning approaches would usually do.

Keywords: Ontology engineering, ontology learning, OWL, RDF-S, reuse, taxonomies, thesauri, classifications, UNSPSC, eCl@ss, e-business.

1 Introduction

Hierarchical classification standards, thesauri, and such taxonomies that were not initially designed to be used as ontologies exist in many domains. They are likely the most promising sources for the creation of domain ontologies at reasonable costs, because they reflect some degree of community consensus and contain, readily available, a wealth of category definitions plus a hierarchy. For instance, UNSPSC [1], a standard categorization for products and services and often referred to as a products and services ontology for e-business, contains 20,789 categories (in version 7,0901) and the similar but more expressive industrial standard eCl@ss [2] defines 25,658 categories plus 5,525 precisely specified object and datatype properties (in version 5.1de). The products classification in eBay, as an additional example, includes

E. Franconi, M. Kifer, and W. May (Eds.): ESWC 2007, LNCS 4519, pp. 129–144, 2007.

more than 2,000 categories for computer and networking equipment alone. For a quantitative analysis of the content and domain coverage of such standards in the products and services domain, see [3].

While it is tempting to write simple scripts that mechanically create ontology classes for the categories in the source standard and `rdfs:subclassOf` relations for the edges constituting the hierarchy, as has been done by [4] and [5], this straightforward approach often yields ontologies that are of limited practical use, since it implies a particular interpretation of the categories so that the original hierarchical order is a valid subsumption hierarchy. If, for example, "ice" is a subcategory of "beverages" in the original hierarchy, this naïve transformation forces us to read the category "beverages" as something like "beverages and related stuff from a purchasing manager's perspective", because only then holds that all instances of the former class are also instances of the latter. While this choice is a valid transformation, it does often not yield the most useful ontologies, as has been shown in [6] and [7]. In particular, the ontology classes would not be sufficiently narrow to describe actual products or services instances in an unambiguous way.

The main cause for this problem is that, due to the informal nature of the original schemas, the meanings of (1) the categories, (2) the hierarchical relations between them, and (3) the task of assigning an instance to a category are usually blurry, and the meanings of the three components are not clearly separated from each other. This means that such informal specifications entail multiple possible ontologies. For example, we can interpret the categories in a way so that the original hierarchy forms a consistent subsumption hierarchy and can be represented using `rdfs:subClassOf`, or we can interpret the categories in another way but must then use another transitive, binary relation of the kind "A is a subcategory of B in some context" in order to capture the hierarchy [6, 7].

Since most categorizations were not created under rigorous knowledge engineering methodologies, they often suffer from additional conceptual anomalies, e.g. local names or a varying semantics of the hierarchy relation by depth of branching. Such anomalies are sometimes found in only relatively small parts of the categorization schema. They may thus not become apparent by a quick view on a part of the specification.

Besides these difficulties in understanding the original semantics and selecting a useful interpretation for a given application, we are additionally constrained by the expressiveness of popular ontology formalisms. OWL DL, for example, does not allow the definition of transitive relations between ontology classes, which may force us to invent suitable ontology modeling patterns as workarounds.

Finally, it is highly desirable that the generation of derived ontologies is automated as much as possible, because of the high number of categories.

1.1 Classification, Thesaurus, and Taxonomy

The terms thesaurus and taxonomy are well established in ontology research. Basically, a *thesaurus* is a collection of concepts that are augmented by three types of relations: "broader term" (BT) and "narrower term" (NT), which may be read as a hierarchical order, and "related term" (RT), which is used to capture conceptual proximity [cf. e.g. 8]. An important characteristic of the NT/BT relation is that it is semantically less specific than a *subclassOf* relation used in ontology engineering for building a subsumption hierarchy, since an instance that fits one subcategory of a thesaurus needs not to be an instance of a the respective parent category. For example, "ice cubes" may be a narrower term to "beverages", but instances of the category "ice cube" are not instances of the category "beverage" if we read the categories literally.

A *taxonomy* is different from a thesaurus in that it contains a subsumption hierarchy in the form of transitive *subclassOf* relations, i.e. each instance of a class can be assumed to be also an instance of all parent categories. It should be noted that hierarchical classifications are sometimes imprecisely referred to as taxonomies even though they do not include a real subsumption hierarchy.

Classifications are sets of concepts and have been used for ages as a means of grouping entities by similarity. It is important to stress that the initial purpose of classification was not to capture the essence of things, i.e. modeling a part of the world, as in ontology engineering, but aggregating entities for some arbitrary purpose. Also, a hierarchical order is a frequent but not a mandatory property of classifications. Sometimes, classifications are assumed to be limited to hierarchical classifications, which are rooted trees where the semantics of the edges may vary widely depending on the purpose and context of usage [cf. 9].

In this paper, we will subsume all three types, i.e. taxonomies, thesauri, and hierarchical classifications under the term *hierarchical categorization schema*, which all have in common that they include a set of categories and some form of a hierarchical order. There are two main reasons for this unified view on the three variants. First, it may depend of the context of usage whether a given collection is a taxonomy, a thesaurus, or just a hierarchical classification. Second, we want to provide an approach that can be directly applied to all three types, thus allowing us to reuse the wealth of any such schemas for building domain ontologies, which are urgently needed for making the Semantic Web a reality.

1.2 Our Contribution

In this paper, we (1) show that it often depends on the context of usage whether an informal, hierarchical categorization schema is a classification, a thesaurus, or a taxonomy, (2) develop a novel methodology for mechanically deriving consistent, lightweight ontologies for a particular context from hierarchical classifications, thesauri, or taxonomies, even if they contain typical conceptual anomalies, (3) present suitable modeling patterns for RDF-S and OWL-DLP that require no reasoning support beyond `rdfs:subClassOf`, which allows for the use of the resulting ontologies with lightweight, scalable reasoners and repositories like OWLIM [10], and (4) demonstrate the usefulness of our approach by transforming the e-business categorization standards eCl@ss [2] and UNSPSC [1] into fully-fledged ontologies.

We also propose to use deductive statistics for the diagnosis of common anomalies and for selecting modeling options when handling large categorization schemas. This allows us to quantify the likelihood that the resulting ontology is consistent without the need to evaluate the complete schema manually.

The structure of the paper is as follows: In section 2, we present a unified model for hierarchical classifications schemas, thesauri, and taxonomies, which takes into account the role of contexts. In section 3, we present our methodology for deriving ontologies from hierarchical categorization schemas. In section 4, we show how our approach can be successfully applied to the representation of eCl@ss and UNSPSC in RDF-S and OWL. In section 5, we discuss our findings and compare them to related works.

2 A Uniform Model of Classifications, Thesauri, and Taxonomies

In this section, we will present a unified formal model that fits any kind of hierarchical categorization schema, be it a domain classification, a thesaurus, or a taxonomy.

2.1 Overview

When taking the categories found in a hierarchical categorization schema as the basis for the creation of an ontology, we face two fundamental problems: First, the meaning of the categories may vary by context. With context we mean in here a domain of usage over which a category label is interpreted. Second, unless there is a formal definition of the semantics of the arcs constituting the hierarchy, the meaning of the category concepts is not determined independently of the meaning of that hierarchy relationship, i.e. both are tangled. For example, a category labeled "TV Set" can, depending on the context of usage, mean very different things, e.g. (1) any entity that is an actual TV set, (2) all TV sets and somewhat related items, (3) all invoices and cost statements that are related to TV sets, or (4) anything that can in any context be regarded as related to TV sets.

In the original fields of usage, this blurriness constitutes no serious problems, since one usually never expresses that an entity is *an instance* of a particular category, but rather *assigns entities to categories in well-defined contexts*. Thus, incompatible meanings of the categories do usually not become apparent. Since a relation like `rdf:type` is never used, it is no problem that in catalog data exchange contexts, actual TV set makes and models are assigned to the UNSPSC category "TV Set", while for spend analysis, invoices reflecting TV set and TV cabling purchases are put into the same category.

One could easily be tempted to trace back these problems to a lack of understanding of the original context and assume that there was one correct interpretation of the semantics of the categories. However, this is not the case, since we can observe that the very same categorization schemas are used in very different contexts with varying interpretations. When we want to build useful ontologies, however, we need to be clear about the semantics of the resulting ontology classes, i.e. what it means to be an instance of this very class.

Two examples might further illustrate this fundamental problem: The hierarchies of both UNSPSC and eCl@ss were created on the basis of practical aspects of procurement, treating those commodities that "somehow" belong to a specific category, as descendents of this closest category. This makes "ice" a subcategory of "non-alcoholic beverages" in UNSPSC and "docking stations" a subcategory of "computers" in eCl@ss. Now, there exists at least one context in which the hierarchy relation can be read as a taxonomic relation in the sense of "`rdfs:subClassOf`", i.e. each instance of "ice" is also an instance of "non-alcoholic beverages" and each instance of "docking station" is also an instance of "computers". Then, however, the intension of the class "computers" is no longer any computer, but the concept "computer" from e.g. the perspective of cost accounting or spend analysis, where an incoming invoice for a docking station can be treated as an incoming invoice for a computer. Similarly will "non-alcoholic beverages" no longer represent all non-alcoholic beverages, but the union of non-alcoholic beverages and related commodities.

The negative consequence of interpreting the hierarchy as being equivalent to `rdfs:subClassOf`, is obvious: We can no longer use the resulting classes e.g. for buying processes, because a search for all instances of "computers" will also return docking stations, and ordering the cheapest available instance of non-alcoholic beverages will very likely return just ice cubes. One could argue that exactly this narrow definition of the classes is the original semantics of the categories. However, this is not true, since the plain text descriptions for these classes in UNSPSC and eCl@ss define the categories in the generic sense.

In a nutshell, most hierarchical categorization schemas are used with varying semantics in multiple contexts, and depending on the respective context, the hierarchy relations may constitute a subsumption hierarchy or just "narrower then/broader then" relations. Our claim is that by restricting the interpretation of a categorization scheme to a particular context, we can derive more useful ontologies, even if that means that the hierarchical order of the original schema does not constitute a subsumption hierarchy in this particular context.

2.2 Formal Definition

We view a hierarchical categorization schema as a directed graph where nodes represent categories and edges represents the "narrower term" or "has subcategory" relation. Depending on the context, a set is related to each category. This set represents the items associated with the category in a particular context.

Formally, a hierarchical categorization schema S is a 6-tuple $S = \langle V, E, C, J, l^V, l^C \rangle$ with:

- V a set of categories,
- E a binary relation over V: $E : V \times V$ reflecting the original edges in the hierarchy,
- C a set of contexts,
- J a partial function which assigns to every context $c \in C$ a partial function which assigns to every category $v \in V$ a set of items such that $J(c)(v)$ is the set of items associated with category v in context c,
- l^V a function which associates labels with categories: $l^V : V \rightarrow string$, and
- l^C a function which associates labels with contexts: $l^C : C \rightarrow string$.

We can see from the definition that a category is interpreted differently depending on the context of usage. We say that the *interpretation of a category* $v \in V$ *in a context* $c \in C$, denoted $S^c(v)$, is the set $S^c(v) = J(c)(v)$. The *interpretation of a category* $v \in V$, denoted *S(v)* is the union of the interpretations of *v* at every context in *C*: $S(v) = \cup\{S^c(v) \mid c \in C\}$.

Using the formal model, we can specify a number of properties which a categorization schema may have. First of all, it is not clear whether a particular hierarchical classification is a consistent taxonomy or rather a thesaurus.

We would call a classification *S* a *taxonomy*, if the hierarchy is a valid subsumption hierarchy, i.e., for all pairs of concepts v_a, v_b holds that if v_a is a descendant of v_b then v_a is also a *subclass* of v_b. Formally, *S* is a taxonomy if, and only if, for all $v_a, v_b \in V$ holds:

$$\text{if } \langle v_b, v_a \rangle \in E \text{ then } S(v_a) \subseteq S(v_b).$$

We call *S* a *taxonomy with respect to context c* if the interpretations of all categories in context *c* form a valid subsumption hierarchy. Formally, *S* is a taxonomy with respect to a context $c \in C$ if, and only if, for all $v_a, v_b \in V$ holds:

$$\text{if } \langle v_b, v_a \rangle \in E \text{ then } S^c(v_a) \subseteq S^c(v_b)$$

Several of the hierarchical classification schemas we looked at are taxonomies only for some contexts.

We say a categorization is *cyclic* if there is a $v \in V$ such that $\langle v, v \rangle \in tr(E)$ with *tr(E)* the transitive closure of *E*. For the remainder of this paper, we assume the input categorization not to be cyclic.

3 Deriving OWL and RDF-S Ontologies from Hierarchical Categorization Schemas

In this section, we describe a novel approach of deriving consistent OWL and RDF-S ontologies from hierarchical categorization schemas. Our approach allows for the semi-automatic creation of meaningful ontology classes for a particular context while preserving the original hierarchy, even if the latter is not a consistent subsumption hierarchy in this particular context. The basic idea of our *GenTax* methodology is to derive two ontology classes from each category: one **gen**eric concept in a given context and one broader **tax**onomic concept which allows preserving the original hierarchy.

The input required is minimal and limited to (1) an informal specification of a hierarchical categorization schema as defined in section 2.2 and (2) a notion of the context in which the ontology should be used. In particular, we do not need instance data or any additional information, as most ontology learning approaches usually would. The transformation itself is semi-automatic in the sense that human intervention is limited to checking some conceptual properties and identifying frequent anomalies. In other words, the actual generation of the ontology can be done by a script that only needs to be configured properly by a human.

The resulting ontologies can be either RDF-S or OWL DLP; in fact, they require no reasoning support beyond `rdfs:subClassOf`, which allows for the use of lightweight, scalable reasoners, while still being able to merge an OWL DLP variant with OWL DL data without leaving the boundaries of OWL DL (which would be the case if e.g. RDF-S meta-modeling was used).

3.1 Overview

The basic idea of our approach is as follows:

- We derive meaningful, generic ontology classes from the categories in the original classification by narrowing them down to their meaning in one particular context.
- We define taxonomic concepts for the categories of the original schema so that they form a consistent taxonomy when the edges in the schema are interpreted as a subsumption hierarchy.

Our algorithm depends on a number of external functions.

- The function **genS** takes as input a categorization hierarchy and returns a formal specification, as defined in Section 2.2. This function takes care of all the pre-processing, disambiguating labels of categories, etc.
- The function **getContextInfo** takes as input a formal categorization and returns a formal categorization which includes an interpretation for every category at every context, except for possibly c_{cat}.
- The function **genURI** takes as argument a context label and a category label and returns a URI based on this information.

The input to the algorithm is some categorization and a set of contexts *C*. The output of the algorithm is an RDF-S or OWL DLP ontology. There is a special context c_{cat} with $l^C(c_{cat})$ = *"Category"*. If this context is included in *C*, then the algorithm will create special category classes in the output ontology for each of the categories in the categorization.

Step 1: Pre-processing and creating a formal representation of the model

The input to this step is an arbitrary hierarchical categorization schema *H*. The output is $S = \langle V, E, C, J, l^C, l^V \rangle$ with V the set of categories, *E* such that $\langle v_{1,} v_2 \rangle \in E$ if there is an arc $\langle v_{1,} v_2 \rangle$ in the original categorization, *C* the set of input contexts, and *J* is not defined for any context. S is obtained as follows: *S* = **genS**(*H*).

Step 2: Deriving context information

This step defines the function *J* in *S* for each context $c \in C$ with $c \neq c_{cat}$. The (external) algorithm finds an interpretation $S^c(v)$ for each category $v \in V$. The output is *S'* = **genContextInfo**(*S*), where *J(c)(v)* is defined for every $c \in C, v \in V$ such that $c \neq c_{cat}$.

Step 3: Category context
If $c_{cat} \notin C$, proceed to the next step. Otherwise, choose $S^{c_{cat}}$ as follows: (a) for any $v \in V$, $S^{c_{cat}}(v) \supseteq S(v)$, (b) for all $v_1, v_2 \in V$ such that $\langle v_1, v_2 \rangle \in E$, $S^{c_{cat}}(v_1) \subseteq S^{c_{cat}}(v_2)$, and for every $v \in V$, $S^{c_{cat}}(v)$ is the smallest set such that the conditions (a) and (b) hold. Obviously there is such a $S^{c_{cat}}$.

Step 4: Generating the ontology
Start with an empty set G.

Step 4.1: Generating ontology classes
For each category $v \in V$ and relevant context $c \in C$, add the triple
$\langle genURI(l^C(c), l^V(v)), \text{rdf : type}, \text{rdfs : Class} \rangle$ (for an RDF-S ontology) or
$\langle genURI(l^C(c), l^V(v)), \text{rdf : type}, \text{owl : Class} \rangle$ (for an OWL DLP ontology)
to *G*.

Step 4.2: Generating subclassOf relations
Since many categorization schemas that are so large that it is infeasible to determine individually whether $S^{c_1}(v_a) \subseteq S^{c_2}(v_b)$ holds, we use the following approximations for creating subclassOf relations:

If $c_{cat} \in C$, then for any $v_a, v_b \in V$, $\langle v_a, v_b \rangle \in E$, add the triple
$\langle genURI(l^C(c_{cat}), l^V(v_a)), \text{rdfs : subClassOf}, genURI(l^C(c_{cat}), l^V(v_b)) \rangle$
to G, and for any $c \in C, c \neq c_{cat}$, $v \in V$, add the triple
$\langle genURI(l^C(c), l^V(v)), \text{rdfs : subClassOf}, genURI(l^C(c_{cat}), l^V(v)) \rangle$

If for **all** $v_a, v_b \in V$ such that $\langle v_a, v_b \rangle \in E$, $c \in C, c \neq c_{cat}$ holds
$S^c(v_a) \subseteq S^c(v_b)$, add the triple
$\langle genURI(l^C(c), l^V(v_a)), \text{rdfs : subClassOf}, genURI(l^C(c), l^V(v_b)) \rangle$
for any $v_a, v_b \in V$ such that $\langle v_a, v_b \rangle \in E$.

As a simplification, we may use a representative sample and statistic inferencing to determine whether the hierarchy would be a valid subsumption hierarchy for the categories in this particular context. In other words, instead of manually checking this property for the whole input schema, we draw a representative sample from the categories in the original schema and determine manually whether for this set of categories, the above mentionend conditions hold.

The output of step 4 is the ontology G (apart from the ontology header etc.). Figure 1 illustrates this for an ontology that contains classes for one context c and the category context c_{cat}. In this example, the original hierarchy would not be a valid subsumption hierarchy in the context c. If this was the case, there would be an additional `rdfs:subClassOf` relation from $S^c(v_2)$ to $S^c(v_1)$.

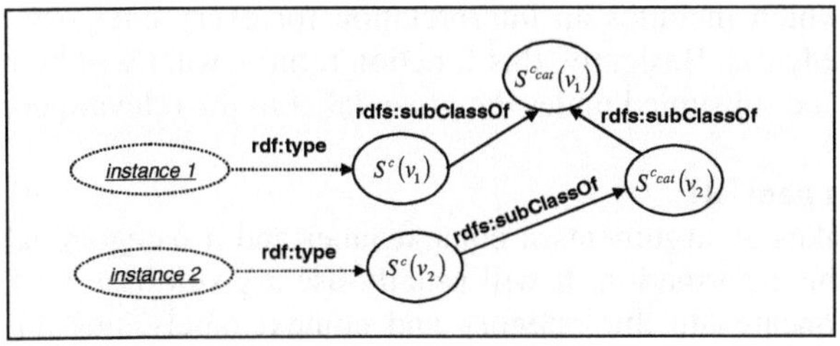

Fig. 1. Example of the representation of two categories v_1, v_2 as four ontology classes

3.2 Implementation

Our algorithm depends, as said, on a number of external functions, which we explain in this section.

3.2.1 Function Gens

This function takes as input a categorization hierarchy and returns a formal specification, as defined in Section 2.2. In particular, it handles all pre-processing and for disambiguating local labels. Local labels are such that are unique only in their particular position in the hierarchy, e.g. "Portable" in the following example.

```
Computer Equipment
     !- Laser Printers
                    !- Portable
```

One approach to handle such cases is by representing each node by a logical formula that takes into account the label of the node and its position in the hierarchy, as proposed by Giunchiglia, Marchese, and Zaihrayeu [9]. The simplest approach (and often sufficient for our purpose) is to disambiguate local names by concatenating the local name with the label of the path of parent nodes (with a suitable way of escaping colons). This would turn the label "Portable" in our example into:

```
Computer Equipment: Laser Printers: Portable
```

Since most classifications that we found were very limited with regard to the depth of branching, the growth in length created no problems.

A related anomaly is that of a varying semantics of the hierarchical relation by depth of branching. In UNSPSC, for example, the last level of the hierarchy reflects so called "Business Functions" for the next higher level:

```
Computer Equipment
     !- Laser Printers
             !- Sales
             !- Lease
```

We can handle this in the same way as local labels; however, this will usually make it impossible to use the hierarchy as a subsumption hierarchy in this context, since the lease of laser printers is not a subclass of laser printers etc.

3.2.2 Function getContextInfo

This function takes as input a formal categorization and returns a formal categorization which includes an interpretation for every category at every context, except for possibly c_{cat}. Basically, this function returns what will be relevant instances of the classes to be subsumed under the given label in the relevant contexts.

3.2.3 Function genURI

This function takes as arguments a context label and a category label and returns a URI based on this information. It will usually use a given base URI for the resulting ontology and concatenate the category and context labels, possibly separated by a slash. If any of the labels contains extra characters, the function will also rewrite them so that the result is a valid URI. Since we have disambiguated local labels, we can assume that the category labels are unique. In practice, we can often also assume the category labels to be unique.

As an example, the category "TV Set" in the two contexts "Product or Service" and "Category" could be transformed into e.g.

```
http://www.foo.org/myontology/TV_Set_ProductOrService
http://www.foo.org/myontology/TV_Set_Category
```

3.3 Statistical Diagnosis of Conceptual Properties and Relevant Anomalies

Depending on the size of the schema and our knowledge of its properties, we may not know a priori whether the categories in our selected context build a proper subsumption hierarchy. Also, we might need to check for the anomalies outlined in section 3.2, since they will require additional preprocessing.

We advocate the use of representative random samples and deductive statistics for these diagnosis tasks.

When taking a random sample, we should include only such categories v that are not top-level nodes. A nice property of this approach is that we can calibrate the test depending on our needs and thus deal with the unavoidable trade-off decisions between the value of an additional subsumption hierarchy vs. the risk of an undetected inconsistency, which is naturally domain-dependent.

It should be stressed that it can be attractive to make such decisions for one context as a whole, since only this allows for quick and cheap script-based creation of derived ontologies without substantial human intervention and engineering effort.

3.4 Example

We want to built a products and services ontology based on a fictive hierarchical schema for electronic-related categories, as shown in Fig. 2. In this case, the relevant target context is "Products or Services". We create two ontology classes for each category, one reflecting the category concept (e.g. "Radio and TV (Category)"), and one reflecting respective types of electronic equipment (e.g. "Radio and TV (Product or Service)"). We see that the original hierarchy is not a consistent subsumption hierarchy in the context of products or services, since "TV Maintenance", read as the actual type of services, is not a subclass of "TV Set", and "Radio Antenna" is not a subclass of "Radio". Thus, we arrange the category concepts in a subsumption hierarchy that represents the original edges, but do not arrange the products and services classes in such a hierarchy. All products and services classes are just subclasses of the respective category concepts. Fig. 3 shows the resulting ontology. Elipses represent ontology classes (`rdfs:Class` or `owl:Class`) and arrows represent `rdfs:subclassOf` relations.

If our target context was "cost accounting branch", then we could additionally arrange the context-specific ontology classes in a subsumption hierarchy, since invoices accounting for radio antennas are usually also regarded as invoices accounting for radios.

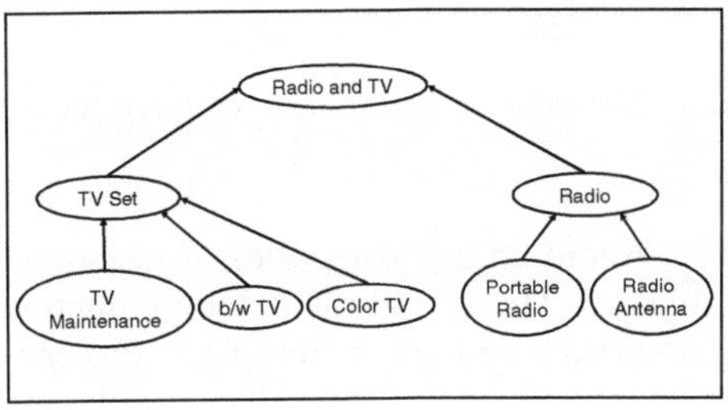

Fig. 2. Example of a hierarchical categorization for electronics

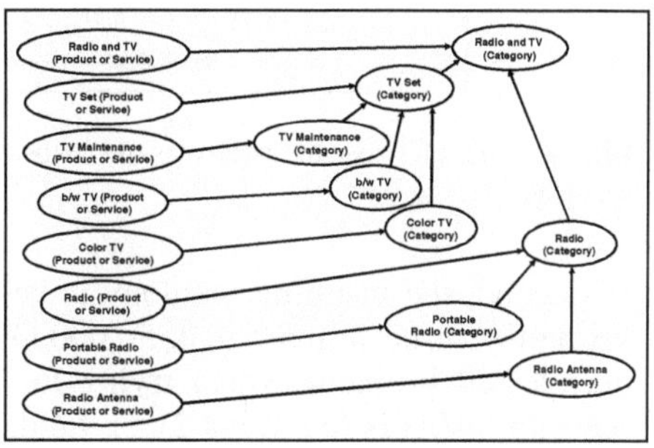

Fig. 3. The resulting ontology for the context "Pro-duct or Service"

4 Evaluation: eClassOWL and unspscOWL

In order so evaluate our approach, we tried to derive useful e-business ontologies in OWL-DLP from eCl@ss 5.1de [2] and UNSPSC [1]. Our goal was to generate, with minimal human intervention, one eCl@ss ontology that can be used to annotate products and services that unting purposes, i.e. aggregating incoming invoices by spend categoriare available on the Web, and another UNSPSC ontology to be used for cost accoes. This exercise is also practically relevant, since the existing prototypes of UNSPSC and eCl@ss ontologies are not very useful in these application domains as has been detailed in [6]. Both categorization schemas contain more than 20,000 categories, which renders manual steps in the transformation infeasible.

We have implemented preliminary tooling support for our methodology. Our prototype consists of a Java program that expects the informal categorization schema to be stored in a RDBMS. The program accesses the categories via an ODBC link. The reason why we use an RDBMS is that we needed nested queries. Also, it proved to be handy to import the various source formats into the RDBMS using standard tooling instead of developing proprietary import interfaces.

4.1 eCl@ss as a Products and Services Ontology

The eCl@ss standard is available at http://www.eclass.de in the form of separate CSV files containing categories, properties, values, class-property recommendations, property-value recommendations, and keywords. For evaluating our methodology, it was sufficient to import the categories.

The application of our methodology to eCl@ss creates only minor problems. First, the original hierarchy does not constitute a correct subsumption hierarchy if the categories are interpreted as products and services categories. Fig. 4 gives an example of how services of repairing assembly and maintenance technology are subnodes of machine. Thus, the structure of the resulting ontology is as in the example in Fig. 3.

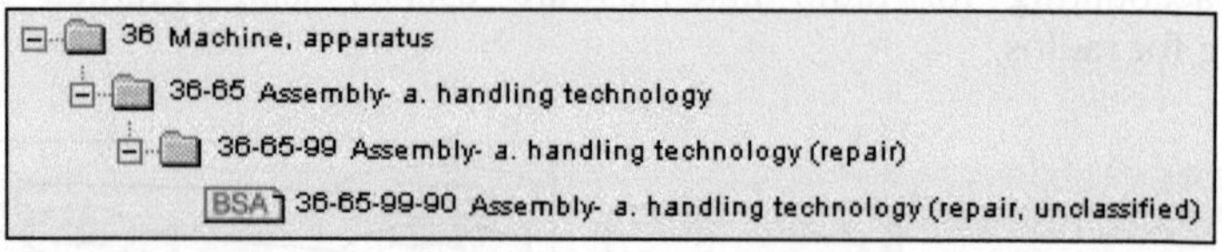

Fig. 4. The eCl@ss hierarchy is no subsumption hierarchy in the context of products and services

Second, the resulting ontology is very big: About 25,000 categories in the source taxonomy result in more than 50,000 OWL classes. The size of the ontology imposes unexpected problems when trying to use standard ontology editors (e.g. Protégé), repositories/APIs (e.g. Jena 2), or validators (e.g. vowlidator). They all exit with error messages when trying to process the full ontology. It was possible, though, to validate and use a restricted version of the ontology that contains only a small subset of the actual eCl@ss concepts. Also, we were able to load the full ontology into an OWLIM [10] configuration.

As compared to our early approaches described in [6] and [7], our new approach requires only two ontology classes instead of three per category, while the old evaluation results still hold.

While we started our experiments with version 5.0 of eCl@ss, we were able to generate new versions of our ontology based on new releases of eCl@ss in a fully automated fashion. The only manual steps required were importing the new CSV files into our RDBMS and updating the namespace for the new release. The total time for creating the new ontology was less than two hours.

Generating ontologies in other ontology languages than OWL (e.g. WSML) was also successful and just required expressing the OWL ontology patterns used in the respective target ontology language.

4.2 UNSPSC as a Cost Accounting Ontology

UNSPSC [1] is similar to eCl@ss in its structure, but has more top-level categories and is limited to the hierarchy of labels, while eCl@ss also includes properties and other elements.

Same as eCl@ss, UNSPSC contains hierarchical relations that do not constitute a correct subsumption hierarchy in some contexts, in particular when reading the labels in the literal sense. For example, we can find the following two candidate inconsistencies:

a) Non-dairy creamers are neither coffee nor tea, and not even a true beverage.

```
-family-[50.20.00.00] Beverages
     -class-[50.20.17.00] Coffee and tea
            -commodity-[50.20.17.14] Non-dairy creamers
```

b) Ice is not a beverage.

```
-family-[50.20.00.00] Beverages
     -class-[50.20.23.00] Non-alcoholic beverages
            -commodity-[50.20.23.02] Ice
```

However, in this second example, the target context of our ontology is “Cost Accounting Categories”. If we interpret the labels in this sense, then it is acceptable for “Beverages” to subsume “Ice”, since anything spend on ice may be correctly regarded as a beverage-related expenditure. Thus, other than in the example in section 4.1, the ontology classes in the target context “Cost Accounting Categories” can also be arranged in a subsumption hierarchy, which reflects the original order. If we wanted to create a products and services ontology from UNSPSC, the situation would be the same as with eCl@ss, i.e. the classes in this context cannot be arranged in a subsumption hierarchy.

We expect that running our script on other hierarchical categorization schemas, e.g. eOTD or XBRL standard reporting taxonomies should require only slight modifications in the embedded SQL.

5 Discussion

There exists a substantial amount of publications on the analysis of the meaning of taxonomic relationships, especially the fundamental work of [11]. This yielded the insight that there are multiple types of taxonomic relationships, which should be represented separately. In this paper, we have presented a generic methodology for deriving consistent ontologies in a script-based fashion from hierarchical categorization schemas, and successfully applied it to eCl@ss and UNSPSC. While the resulting ontologies are rather lightweight, the cost/benefit ratio of our ontologies seems very convincing, since the amount of human intervention is limited to importing source data into an RDBMS and determining some parameters in a script.

Related work to ours can be classified into the following main groups:

- Methodologies for and experiences with the reuse of consensus in classifications, thesauri, and taxonomies for the creation of ontologies. This is the most related field of work. [12] discusses the transformation of tangled hierarchies, as e.g. such derived from ambiguous “broader than / narrower than” taxonomies in library science, into formal ontologies. [13] presents the experiences gained while transforming the constructs of an existing semantic net in the medical domain into an OWL ontology. [14] describe how machine learning approaches can be used to integrate objects from taxonomies available on the Web into a consolidated master taxonomy. [6] is a detailed description of creating products and services ontologies based on UNSPSC and eCl@ss, but requires three classes per category and is also not generically applicable. [15] shows the reuse and semantic enrichment of an existing hierarchical standard, and demonstrates this for the Art and Architecture Thesaurus (AAT). [16] and [8] are consequent works of this stream of research. An important characteristic of [16] and [8] is that the authors leave the limits of OWL DL in order to capture semantics contained in the original thesaurus, namely to be able to treat classes as instances and vice versa. [9] presents a formal theory of classifications; [17] is an extension of this work and proposes how lightweight ontologies can be derived from such specifications.

- Prototypes of products and services ontologies in standard ontology languages derived from UNSPSC. To our knowledge, there are currently two examples of UNSPSC transformations into ontology representation languages: The DAML+OIL and RDF-S variants created by [5] and the DAML+OIL variant from the Knowledge Systems Laboratory at Stanford University [4]. For eCl@ss, there exists one early prototype by Bizer and Wolk [18] and the official release of eCl@ssOWL [19], which is based on our previous work [6].
- Ontology engineering methodologies, implicitly or explicitly focusing on the manual creation of ontologies based on knowledge engineering principles. A comprehensive discussion of all approaches in this field is beyond the scope of this paper, for an overview see e.g. [20] and [21]. The main difference between our work and traditional ontology engineering is that we advocate the script-based transformation without involving an ontology engineer for revising the modeling in every single case.

Our approach is different from previous work in that it allows for the script-based creation of meaningful ontology classes (1) for a particular context while (2) preserving the original hierarchy, even if the latter is not a real subsumption hierarchy in this particular context. The resulting ontologies can be either RDF-S or OWL DLP; in fact, they require no reasoning support beyond `rdfs:subClassOf`, which allows for the use of lightweight, scalable reasoners, while still being able to merge an OWL DLP variant with OWL DL data without leaving the boundaries of OWL DL (which would be the case if e.g. RDF-S meta-modeling would be used).

Our proposal comes not without cost. First, the resulting ontology provides quite limited reasoning support. Second, we create at least two ontology classes per each category, which increases the size of the ontology. However, the unwanted ontology growth has to be set in relation to the low costs of reasoning and to fact that the ontology building process requires almost no human labor.

In general, we agree that a greater amount of e.g. axioms would be desirable. On the other hand, we see no lightweight way of automatically adding more semantics because it cannot be easily derived from the input schemas. Also, we will have to set the resources necessary for the respective enrichment in relation to the gain in automation and the resulting economies.

Acknowledgements. Parts of the work presented in this paper have been supported by the European Commission under the projects DIP (FP6-507483), SUPER (FP6-026850), and MUSING (FP6-027097), and by the Austrian BMVIT/FFG under the FIT-IT project myOntology (grant no. 812515/9284).

References

[1] United Nations Development Programme, "United Nations Standard Products and Services Code (UNSPSC)," available at http://www.unspsc.org/, retrieved March 15, 2007.

[2] eClass e.V., "eCl@ss: Standardized Material and Service Classification," available at http://www. eclass-online.com/, retrieved March 15, 2007.

[3] M. Hepp, J. Leukel, and V. Schmitz, "A Quantitative Analysis of Product Categorization Standards: Content, Coverage, and Maintenance of eCl@ss, UNSPSC, eOTD, and the RosettaNet Technical Dictionary," *Knowledge and Information Systems*, (forthcoming).

[4] D. L. McGuinness, "UNSPSC Ontology in DAML+OIL," available at http://www.ksl.stanford.edu/projects/DAML/UNSPSC.daml, retrieved March 15, 2007.

[5] M. Klein, "DAML+OIL and RDF Schema representation of UNSPSC," available at http://www.cs.vu.nl/~mcaklein/unspsc/, retrieved March 15, 2007.

[6] M. Hepp, "Products and Services Ontologies: A Methodology for Deriving OWL Ontologies from Industrial Categorization Standards," *Int'l Journal on Semantic Web & Information Systems (IJSWIS)*, vol. 2, pp. 72-99, 2006.

[7] M. Hepp, "Representing the Hierarchy of Industrial Taxonomies in OWL: The gen/tax Approach," Proceedings of the ISWC Workshop Semantic Web Case Studies and Best Practices for eBusiness (SWCASE05), Galway, Irland, 2005.

[8] M. van Assem, M. R. Menken, G. Schreiber, J. Wielemaker, and B. J. Wielinga, "A Method for Converting Thesauri to RDF/OWL," Proceedings of the ISWC'04, Hiroshima, Japan, 2004.

[9] F. Giunchiglia, M. Marchese, and I. Zaihrayeu, "Towards a Theory of Formal Classification," Proceedings of the AAAI-05 Workshop on Contexts and Ontologies: Theory, Practice and Applications (C&O-2005), Pittsburgh, Pennsylvania, USA, 2005.

[10] A. Kiryakov, D. Ognyanov, and D. Manov, "OWLIM – a Pragmatic Semantic Repository for OWL," Proceedings of the International Workshop on Scalable Semantic Web Knowledge Base Systems (SSWS 2005), New York City, USA, 2005.

[11] R. J. Brachman, "What IS-A Is and Isn't: An Analysis of Taxonomic Links in Semantic Networks," *IEEE Computer*, vol. 16, pp. 30-36, 1983.

[12] A. L. Rector, C. Wroe, J. Rogers, and A. Roberts, "Untangling Taxonomies and Relationships: Personal and Practical Problems in Loosely Coupled Development of Large Ontologies," Proceedings of the K-CAP'01, Victoria, British Columbia, Canada, 2001.

[13] V. Kashyap and A. Borgida, "Representing the UMLS Semantic Network using OWL," Proceedings of the 2nd International Semantic Web Conference 2003 (ISWC 2003), Sanibel Island, Florida, USA, 2003.

[14] D. Zhang and W. S. Lee, "Learning to integrate web taxonomies," *Journal of Web Semantics*, vol. 2, pp. 131-151, 2004.

[15] B. J. Wielinga, A. T. Schreiber, and J. A. C. Sandberg, "From Thesaurus to Ontology," Proceedings of the First International Conference on Knowledge Capture (K-CAP 2001), Victoria, British Columbia, Canada, 2001.

[16] B. J. Wielinga, J. Wielemaker, G. Schreiber, and M. van Assem, "Methods for Porting Resources to the Semantic Web," Proceedings of the First European Semantic Web Symposium (ESWS'04), Heraklion, Greece, 2004.

[17] F. Giunchiglia, M. Marchese, and I. Zaihrayeu, "Encoding Classifications into Lightweight Ontologies," Proceedings of the 3rd European Semantic Web Conference (ESWC 2006), Budva, Montenegro, 2006.

[18] C. Bizer and J. Wolk, "RDF Version of the eClass 4.1 Product Classification Schema," available at http:////www.wiwiss.fu-berlin.de/suhl/bizer/ecommerce/eClass-4.1.rdf, retrieved March 15, 2007.
[19] M. Hepp, "eCl@ssOWL. The Products and Services Ontology," available at http://www.heppnetz.de/eclassowl/, retrieved March 15, 2007.
[20] M. Fernández-López and A. Gómez-Pérez, "Overview and analysis of methodologies for building ontologies," *The Knowledge Engineering Review*, vol. 17, pp. 129-156, 2002.
[21] J. de Bruijn, "Using Ontologies. Enabling Knowledge Sharing and Reuse on the Semantic Web," *DERI Technical Report DERI-2003-10-29, October 2003*, pp. 1-49, 2003.

SPARQLeR: Extended Sparql for Semantic Association Discovery*

Krys J. Kochut and Maciej Janik

Department of Computer Science, University of Georgia
415 Boyd Graduate Studies Research Center
Athens, GA 30602-7404
{kochut,janik}@cs.uga.edu

Abstract. Complex relationships, frequently referred to as semantic associations, are the essence of the Semantic Web. Query and retrieval of semantic associations has been an important task in many analytical and scientific activities, such as detecting money laundering and querying for metabolic pathways in biochemistry. We believe that support for semantic path queries should be an integral component of RDF query languages. In this paper, we present SPARQLeR, a novel extension of the SPARQL query language which adds the support for semantic path queries. The proposed extension fits seamlessly within the overall syntax and semantics of SPARQL and allows easy and natural formulation of queries involving a wide variety of regular path patterns in RDF graphs. SPARQLeR's path patterns can capture many low-level details of the queried associations. We also present an implementation of SPARQLeR and its initial performance results. Our implementation is built over BRAHMS, our own RDF storage system.

1 Introduction

The size of ontologies in the Semantic Web has grown significantly within the last few years. The vision of ontologies containing millions of entities interconnected by meaningful relationships presented in [22] has become reality. The current query languages for RDF bases, such as SPARQL [24], RQL [14] and RDQL [21], support defining graph patterns and expressing various restrictions on entities and relationships participating in the defined patterns. However, all of them lack the necessary constructs that directly support the discovery of semantic associations, which cannot be explicitly defined by fully specified structure of a graph pattern.

We believe that querying for semantic associations is an important feature missing in the current RDF query languages, most notably in SPARQL. This paper presents SPARQLeR, a novel extension of SPARQL that enables the discovery of semantic associations among entities in RDF knowledge bases.

Semantic association is an undirected path that connects two entities in the knowledge base using named relationships, which represent its meaning. Discovery of

* This research has been partially supported by the National Science Foundation Grant No. IIS-0325464 entitled "SemDIS: Discovering Complex Relationships in the Semantic Web".

E. Franconi, M. Kifer, and W. May (Eds.): ESWC 2007, LNCS 4519, pp. 145–159, 2007.

semantic associations is the process of finding paths of possibly unknown length that connect the given entities and have a specific semantics. Therefore, the search for paths must focus on the semantics of both the entities and the properties on the path. Moreover, the order of the relationships in the path and their directionality is crucial in expressing the semantics of the associations. To fulfill these requirements SPARQLeR uses regular expressions over properties for specifying the required semantics of the queried paths. The paths are treated as RDF meta-resources represented as sequences. They can be used in other patters, specifying the required properties of the individual path elements. This approach gives the user a detailed control over each of the elements on the path, as well as its overall semantics.

The paper is organized as follows. In Sec. 2, we give a motivation for adding semantic association discovery in RDF query languages. In Sec. 3, we discuss semantic associations and different types of paths in RDF bases. Sec. 4 introduces the concept of a path in SPARQLeR, shows the syntax and semantics of the language, and describes its prototype implementation. In Sec. 5, we discuss the initial performance results of our implementation.

2 Motivation

An important discovery in medicine made by Dr. D.R. Swanson of a dependency between Magnesium and Migraine [25] is a clear example of finding meaningful semantic associations. He manually searched through papers in PubMed [17] to establish a sequence of facts, supported by co-occurrence of significant terms in papers, that Magnesium may alleviate Migraine. With the suitable biomedical knowledge base extracted from PubMed and stored in RDF, as proposed in [19], finding such associations can be accomplished with the use of regular path queries.

Many interesting examples of semantic associations can be found in biological sciences. Metabolic pathways, composed of sequences of chemical reactions occurring within a cell, involve a gradual modification of the initial substance into the final product with the desired chemical structure.

N-Glycan Biosynthesis pathway [12] is an example of a well known metabolic pathway (presented later in Fig. 5). It starts from *dolichol phosphate* and ends with the production of *glyco peptide G00009*. It contains 15 chemical reactions and, even though this pathway may not be regarded as very long among the biochemical pathways, it is considered long for a path in the area of the Semantic Web.

Locating and retrieving metabolic pathways is a difficult problem. Regular path queries can be used searching for metabolic pathways. Using such queries, scientists should be able to query for and retrieve ordered sequences of specific reactions that lead from a given substance to a desired final product.

Additional interesting applications of semantic association discovery include BioPatentMiner [16] and Insider Threat [1]. We believe that there is a clear need for an RDF query language capable of semantic association retrieval.

Introduced in this paper SPARQLeR offers a variety of constructs for easy formulation of regular path queries which are suitable for solving the above problems.

3 Background

Path queries have been a focus of formal studies as well as practical applications. The complexity of finding regular paths in graphs was investigated in [15] and [7]. The authors showed that in general case finding all simple paths matching a given regular expression is NP-Complete, whereas in special cases it can be tractable. The complexity of various types of path queries, such as linear, regular and context-free was also described in [27]. Another approach was proposed in [6]. Here, the authors focused on finding paths in labeled graphs. In this case, a regular language is defined beforehand and a special index is maintained for all edge inserts and deletes.

Some of the query languages created for semi-structured databases support defining regular path queries. Among the well known languages are: G [10] and G+ [9], and Graphlog [8]. The relationship between the chain programs with recursive predicates and regular path queries is described in [4]. For RDF data, partial support for path queries, but not regular paths, can be found in SeRQL [5], TRIPLE [23], and Versa [18]. Versa introduced the *traverse* keyword which allowed querying for variable-length paths using a set of specified transitive properties. In [2], the authors present only the initial work on PSPARQL, a language supporting regular expressions in SPARQL. However, the regular expressions were to be used in place of properties in triple patterns, which limited the ability of testing individual path elements. It also significantly altered the syntax of SPARQL.

3.1 Semantic Associations in RDF Description Bases

Paths in RDF description bases represent a variety of explicit and implied semantic relationships among the participating resources (entities). This is based on the assumption that entities are semantically related if there exists a path connecting them. In [3], Anayawu and Sheth proposed a ρ-path (and related concepts) as a way of expressing semantic associations between entities in RDF bases. A ρ-path has been defined as a directed path connecting two entities.

While directed paths naturally capture semantic associations between entities, we also believe that undirected paths also capture important semantic associations which should not be ignored. Therefore, we view semantic associations as implied by the presence of either directed, undirected paths, or undirected paths with specific directionality of the included properties. A good illustration of this observation is an RDF graph, shown in Fig. 1, describing a part of a well known Glycan biosynthesis pathway (we discuss it further later in this paper). The shown fragment includes 3 reactions, represented by the entities *R05972, R05973,* and *R06238* and 4 glycans *(G00002-G00005)* as their reactants and products. For clarity of presentation, other properties have not been included in the shown graph.

The glycan *G00002* is a *predecessor* of *G00005*. Clearly, they are semantically associated, even though there is no directed path connecting them. In fact, the whole pathway links the starting substance, *dolichol phosphate*, and the final product, *peptide G00009*, using a sequence of reactions similar to the ones above. Again, a directed path connecting *dolichol phosphate*, and *peptide G00009* does not exist, but the two molecules are semantically related by this important pathway.

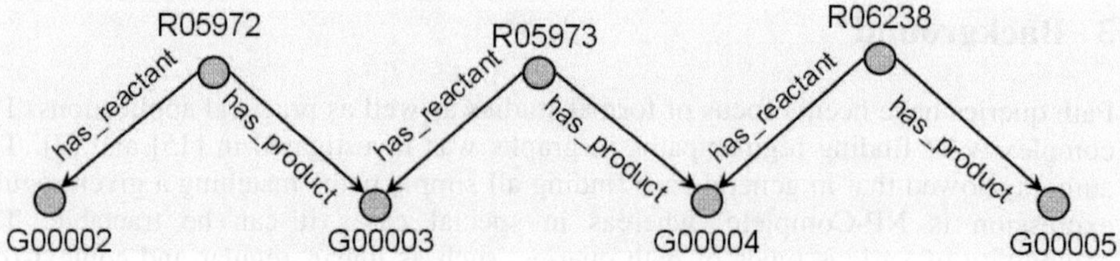

Fig. 1. An example of a chemical reaction graph

Below, we define semantic associations taking into account any type of connection between two entities. In what follows, we will interchangeably use the terms *triple* and *RDF statement*. We will assume that $\mathcal{R}$ is an RDF description base.

Def. 1. A *directed path* between resources r_0 and r_n in $\mathcal{R}$ is a sequence $r_0\ p_1\ r_1\ p_2\ r_2$, ... $p_{n-1}\ r_{n-1}\ p_n\ r_n$ $(n>0)$ if $r_0\ p_1\ r_1$, $r_1\ p_2\ r_2$, ... $r_{n-2}\ p_{n-1}\ r_{n-1}$, $r_{n-1}\ p_n\ r_n$ $(n>0)$ are triples in $\mathcal{R}$. The *length* of the path is *n*. Moreover, we require that all of the resources r_i $(0 \leq i \leq n)$ in the path be distinct (we will only consider *simple paths*). □

Def. 2. An *undirected path* between the resources r_0 and r_n in $\mathcal{R}$ is a sequence $r_0\ p_1\ r_1\ p_2\ r_2$, ... $p_{n-1}\ r_{n-1}\ p_n\ r_n$ $(n>0)$ if for each property and the two neighboring resources $r_{i-1}\ p_i\ r_i$ $(0 < i \leq n)$ in the path, either $r_{i-1}\ p_i\ r_i$. or $r_i\ p_i\ r_{i-1}$. is a triple in $\mathcal{R}$. We will consider only *simple undirected paths*. □

Def. 3. Two resources *r* and *s* in $\mathcal{R}$ are *semantically associated* if there exists an undirected path in $\mathcal{R}$ connecting the two resources. □

3.2 Defined Directionality Paths

While searching for semantic associations between two given entities we may be interested in paths in which properties follow a specific *defined directionality pattern*, according to the desired semantics of the connection between the entities. Creating such patterns requires an inverse property operator, not present in SPARQL. In SPARQLeR, we will use the '–' (minus) character to denote the inverse of a property.

Spatial relationships, such as *A is inside B*, offer illustrative examples for defined directionality paths. Let us consider the following three path queries with regular patterns (SPARQLeR's path patterns are defined later, in section 4.2):

1. *spatial:inside** - when used in a search for directed paths, it locates semantic associations illustrated by a diagram shown in Fig. 2a.
2. *spatial:inside** - when used in a search for undirected paths, it locates semantic associations illustrated by diagrams shown in Fig. 2a, 2b and 2c.
3. *(spatial:inside –spatial:inside)** [read as: concatenation of *inside* with *inverse of inside*] - when used in a search for directed paths, it locates semantic associations illustrated by a diagram shown in Fig. 2c, showing very specific, a chain-like inclusion structure.

Following the above observation, we believe that semantic associations require more than directed or undirected paths and should be treated as *defined directionality paths.*

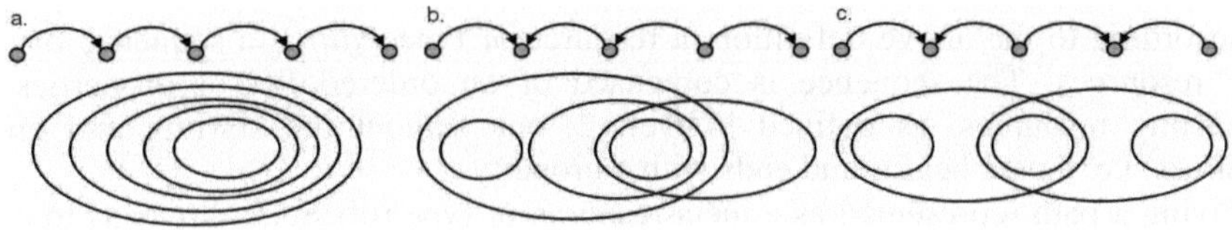

Fig. 2. Example results of spatial path queries

From a graph theoretical perspective, a path that matches a defined directionality pattern is an undirected path, and its implied semantics is set by the specific directionality of its member properties.

4 SPARQLeR

SPARQLeR (SPARQL ***e***xtended with ***R***egular paths) is an extension of SPARQL designed for querying for semantic associations. Our intension was to introduce minimal changes to SPARQL's syntax and semantics. Querying in SPARQLeR focuses on building *path patterns* involving undirected and directed paths as well as paths with defined directionality of the participating properties. Note, that since all properties have their inverses, the expressiveness of directed path queries is sufficient, as it enables us to construct undirected path patterns with the use of properties and their inverses. Nevertheless, to simplify the creation of path patterns, undirected path patterns are also supported. Syntax of proposed extensions fit seamlessly into current SPARQL language grammar. The new constructs in SPARQLeR are designed for the discovery of the semantic associations and, in particular, allow the user to:

- search for undirected paths or for paths with specific directionality of properties,
- filter located paths with the use of regular expressions formed over properties included in the path (use of inverse properties is also allowed),
- filter located paths by imposing constraints on the length of paths,
- filter located paths by requiring the presence of specific resources on the path, possibly even at a specific position,
- specify if located paths can include instance entities, schema classes and/or literals,
- indicate if the hierarchy of sub-properties should be used in property matching.

4.1 Path as RDF Meta-resource and Path Patterns

We will treat paths in RDF description bases as RDF *meta-resources*. In order to place these new meta-resources within the RDF vocabulary, we have created a new class *Path* defined in the new vocabulary *rdf-meta-schema*. The class Path has been defined as the sub-class of both rdf:Property and rdf:Seq as follows:

```
<rdf:Class rdf:about="http://meta.org/rdf-meta-schema#Path">
  <rdfs:isDefinedBy rdf:resource="http://meta.org/rdf-meta-schema#"/>
  <rdfs:subClassOf
    rdf:resource="http://www.w3.org/1999/02/22-rdf-syntax-ns#Property"/>
  <rdfs:subClassOf
    rdf:resource="http://www.w3.org/1999/02/22-rdf-syntax-ns#Seq"/>
  <rdfs:label>Path</rdfs:label>
  <rdfs:comment>The class of RDFMS paths.</rdfs:comment>
</rdf:Class>
```

According to the above definition, a resource of type *Path* is a sequence of other RDF resources. The sequence is composed of an ordered list of properties and connecting resources, as defined in Def. 2, but without the starting and ending resources, i.e. a path begins and ends with a property.

Having a path represented as a meta-resource of type rdfs:Seq, allows us to create patterns for inspecting the path elements with the use of the properties rdf:Member, and rdf:_*N*, which will be illustrated later in this paper.

A semantic association path between two resources *r* and *s* in an RDF graph forms a natural extension of a regular RDF property connecting the two resources. As such, the two resources and the path can be regarded as a *meta-triple* of the form *r path s*, where *path* is a meta-resource of type *Path*.

4.2 Path Patterns

We have extended SPARQL's triple patterns to include path patterns. Paths can be matched only by path variables. A *path variable* is a name beginning with the % character, for example *%connection.* A *path pattern* is a triple pattern created with the use of a path variable in place of the property. A *path query* is any SPARQL query involving at least one path pattern. Instead of formally defining SPARQLeR's syntax, we will present a number of examples illustrating path patterns.

The following SELECT path query, involving a *two-source path pattern*

```
SELECT %path WHERE {<r> %path <s>}
```

matches any path between the resources r and s. By default, the matched paths must be directed. As expected, for every matched path, the variable %path is bound to the located path, represented as a sequence (*rdf:Seq*) of properties and the connecting resources (the first and last elements of the sequence are properties). Therefore, the above query returns a list of blank nodes representing the matched paths (sequences).

In order to list the resources on each matched path, the *list* operator (applicable only to the path variables) must be applied to the path variable, as shown below:

```
SELECT list(%path) WHERE {<r> %path <s>}
```

Path patterns allow for searching for any resources reachable from a given one. The following *single source* SELECT query locates resources reachable from resource r:

```
SELECT %path, ?res WHERE {<r> %path ?res}
```

For every match, the reached resource is bound by the variable `?res`, and both the path leading to it and the resource are returned by the query. The analogous form of the above query relies on the inverse path pattern of the form *{?res %path <r>}*. This pattern matches all resources (and paths) from which the resource r is reachable.

Since a matched path is a meta-resource of type *Path*, and therefore also of type rdf:Seq, resources on the path may be examined with the use of patterns involving the container membership properties. For example, the following query:

```
SELECT %path WHERE
{<r> %path <s> .  %path rdfs:Member <e>}
```

matches any semantic path between the resources r and s, provided the path includes the resource e. Even though it involves a path variable, the second triple is not a path triple, since the path variable is not used in place of the property. Similarly, we can formulate queries examining resources at specific positions on the path. For example, the following query:

```
SELECT %path WHERE {<r> %path <s> . %path rdfs:_1 <p>}
```

matches any semantic path between the resources r and s, provided the path begins with the property p (a SPARQLeR path always begins with a property). Similarly, the property *rdfs:_2* can be used to examine the first connecting resource on the path. In addition, two meta-properties *rdfms:entityResource* and *rdfms:propertyResource* (not described here) access the connecting resources and relationships, respectively (*rdfms* is the namespace prefix of the meta-schema, discussed in Sec. 4.1).

Path patterns may be also used in *construct, describe* and *ask* queries. As expected, a path variable used in a CONSTRUCT query returns all triples forming the paths matched by the querie's graph pattern. For example, the query

```
CONSTRUCT {<r> %path <s>} WHERE {<r> %path <s>}
```

returns all triples forming all paths between the resources r and s. The *list* operator cannot be used within the CONSTRUCT expression.

It is interesting to note that CONSTRUCT queries can be used to extract interesting sub-graphs, satisfying certain specific semantic properties. Combination of multiple path queries with use of CONSTRUCT, possibly with common intersecting points, may lead to creating semantically highly informative sub-graphs [20].

The ASK query functionality for path patters is defined as testing for existence of at least one specified path. The DESCRIBE query returns the description of all resources included in the found paths. DESCRIBE and ASK queries have not been included yet in presented implementation.

4.3 Testing Paths

Testing of the located paths can be performed with the use of special expressions used within the FILTER clause. A path can be tested if it matches a given regular expressions, or if its length is within certain bounds.

The *regex* operator in SPARQLeR has been extended to specify regular path expression filters. Syntactically, it is identical to the usual *regex* operator, but the first argument must be a path variable. The second argument must be a path expression, while the optional third argument specifies the path matching flags:

regex(pathvar, pathexpr, pathflags)

The path expressions can be formed with the use of property names, their inverses, classes of properties, and the usual collection of regular expression operators. They are intended to specify the semantics of the path between a pair of resources.

Def. 4. SPARQLeR's *path expressions* are defined recursively as follows (p, p_1, p_2,..., and p_n denote property names, while x and y denote path expressions). We also define paths between resources r and s which are matched by the defined path expressions.

- *p* matches a path between *r* and *s* of length 1 if a triple *r p s* exists;
- *-p* (the inverse of *p*) matches a path between *r* and *s* of length 1 if a triple *s p r* exists;
- [p_1 p_2 … p_n] (class of properties) matches a path between *r* and *s* of length 1 if a triple *r* p_i *s* exists for some *i* $(1 \leq i \leq n)$;
- -[p_1 p_2 … p_n] matches a path between *r* and *s* of length 1 if a triple *s* p_i *r* exists for some *i* $(1 \leq i \leq n)$; inverse operator is not allowed for properties inside the set;
- [^p_1 p_2 … p_n] matches a path between *r* and *s* of length 1 if a triple *r p s* exists and $p \neq p_i$ $(1 \leq i \leq n)$;
- -[^p_1 p_2 … p_n] matches a path between *r* and *s* of length 1 if a triple *s p r* exists and $p \neq p_i$ $(1 \leq i \leq n)$; inverse operator is not allowed for properties inside the set;
- . (wildcard) matches a path between *r* and *s* of length 1 if either triple *r p s* or *s p r* exists for some property *p*;
- also supported: *x* | *y* (alternative); *xy* (concatenation); *x** (Kleene star); *x+* (one or more repetition); *(x)* (match a path matched by *x*). □

For example, the following query matches paths between resources r and s that use only property *foo:prop*:

```
SELECT list(%path) WHERE
{<r> %path <s>
 FILTER(regex(%path,"foo:prop+")}
```

In order to keep the size of the path expressions manageable, only the prefix-abbreviated names of properties are allowed. The type of the located path (directed or undirected) can be requested as part of the *regex* expression and is indicated in the (optional) path flags of the *regex* expression. For example, the query

```
SELECT list(%path) WHERE
{<r> %path <s>
 FILTER(regex(%path,"(foo:prop|foo:rel)+","u")}
```

allows the matched path to be undirected. When the path directionality is left unspecified, the path is assumed to be directed (the flag "d" is assumed). Also, the path expression may be omitted, as in *regex(%path,,"u")*. Here, each path bound to variable %path may be undirected and be composed of arbitrary properties. A regex with no path expression is equivalent to *regex(%path,".*","u")*. Note, that *regex(%path,".*")* matches only directed paths, even though the wildcard expression (.) matches both a property and its inverse.

The other path flags include *i*, *s*, *l*, and *h*. The flags *i*, *s*, and *l* specify that the path is restricted to resources which are instances (entities), schema classes, and literals, respectively. The last flag, *h* (hierarchy), indicates that when matching properties,additionally their ancestor properties (following the *subPropertyOf* property) may be used. The path flags may be combined. For example,

```
regex(%path,".*foo:prop.*","uis")
```

specifies that the path must involve property *foo:prop*, may be undirected, and can only involve connecting resources which are instances or schema classes. The default path flags string is *"di"*.

The new *length* operator returns the length of the path and can be used as part of a FILTER expression. For example,

```
SELECT list(%path) WHERE
{<r> %path <s>
 FILTER(length(%path)<5)}
```

restricts the matched paths to be of length less than 5. Due to implementation optimization, the length of a path may be compared only to constant values. Path filtering expressions may be combined, and mixed with any other filter tests, involving other variables and resources. As discussed in Sec. 0, the located paths may not be required to be fully directed, but with a specified directionality of individual properties. This may be requested by a suitable path expression, as in the following select query:

```
SELECT list(%path) WHERE
{<r> %path <s>
 FILTER(length(%path)<=6 && length(%path)>=4 &&
        regex(%path,"(foo:prop -foo:rel)+")}
```

which requires that the matched paths be composed of sequences of pairs of properties: *foo:prop* followed by the inverse of the property *foo:rel.*

4.4 Prototype Implementation of SPARQLeR

Our implementation of SPARQLeR uses BRAHMS, our own RDF storage system [13]. The implementation relies on BRAHMS's low level API to iterate over triples, depending on whether the subject, property, object, or their combination has already been fixed (by bound variables or explicit resources). The graph pattern included in a SPARQL query is converted into a composition of such iterators, according to a created query plan.

The path iterator, necessary for path pattern matching, has been implemented as a hybrid of a bidirectional breadth-first search and a simulation of a deterministic finite automaton (DFA) created for a given path expression. During our previous experiments [13], a bidirectional breadth first search proved to be the most efficient method in practice for finding all simple paths up to certain hop limit. For each instance of the iterator created for a path pattern, two DFAs are constructed. The first one accepts the regular language defined by the original path expression, while the second one accepts the reversed language, which is also regular. The path search uses the steps from the bidirectional BFS to grow the frontiers of entities used to connect paths. Before an entity is placed on the frontier for the next expansion, a check is performed if the partial path leading to it is not rejected by the appropriate DFA. This guarantees that the partial results, which are not accepted by DFA, will not be further expanded. Making this check for each node before adding it to a frontier causes the frontiers to grow very slowly for some regular expressions. From the practical point of view, it significantly increases the possibility of finding longer paths in an acceptable amount of time and of not exhausting the memory used by the search.

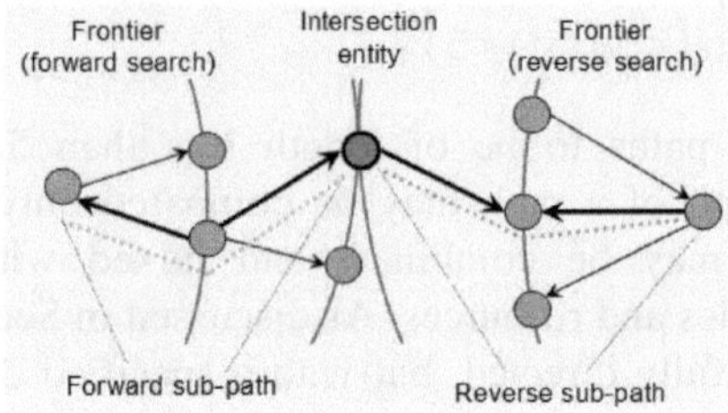

Fig. 3. Path finding and sub-paths in breadth-first search

A candidate path is located when an entity from the forward frontier matches an entity from the reverse frontier. At this point, it is only known that the "forward" sub-path has not been rejected by the forward DFA and that the "reverse" sub-path has not been rejected by the DFA accepting the reverse language. Before the concatenated path is returned, it must be accepted by the forward DFA, created from the original path expression.

A similar solution is used for single source path patterns. In this case, only one DFA in conjunction with a standard breadth first search is used to grow a single frontier of entities.

5 Experiment Design and Results

We have tested our implementation of SPARQLeR using a collection of path queries against a modified DBLP dataset [11]. We also performed path queries locating metabolic pathways in the Glycomics domain, using the GlycO ontology [26].

Tests were performed on machine with 2 Intel(R) Xeon(TM) 3.06GHz CPUs and 4Gb memory, running Red Hat 9.0 Enterprise Linux. C/C++ code was compiled using *gcc* (GCC) 3.2.3 20030502 (Red Hat Linux 3.2.3-56) with '–O6' optimization flag.

5.1 Data Sets

In our searching for metabolic pathways, we used the GlycO ontology. It represents information about glycans and includes a comprehensive schema as well as instances. GlycO is still under development and many new instances representing theoretical as well as experimental data are being added. Currently, the ontology has 362 classes (mainly glycans classification taxonomy) and 84 specialized properties.

Our scalability experiments required a much bigger data set. For this purpose we used a modified version of the DBLP ontology generated from the data available in September, 2006. It contains information about authors, published papers, articles, year of publication, etc. Unfortunately, the citations have been assigned to very few documents, rendering this set unsuitable for scalability test purposes. To be able to search for long, meaningful paths, we have replaced the current (few) citations with a list of randomly created citations (1 to 10 random citations to papers selected from all of the previous years in the knowledge base, using a normal distribution). The total number of randomly inserted citations in the full dataset reached almost 4.3 million.

The full DBLP dataset contains 790,635 publications with set publish year. For scalability testing, we used a subset of publications published in or after 1981. It contains 760,369 publications and has been subdivided it into 26 subsets, each one including publications from an increasingly wider time range, starting with 2006 and ending with 1981 (the smallest set included only 2006 publications and the largest one included publications from years 1981-2006). The smallest test dataset contained almost 300,000 instance statements, while the largest one had over 6.6M instance statements. Fig. 4 presents numbers of publications in full DBLP (starting from year 1936) and sizes of used test datasets in statements.

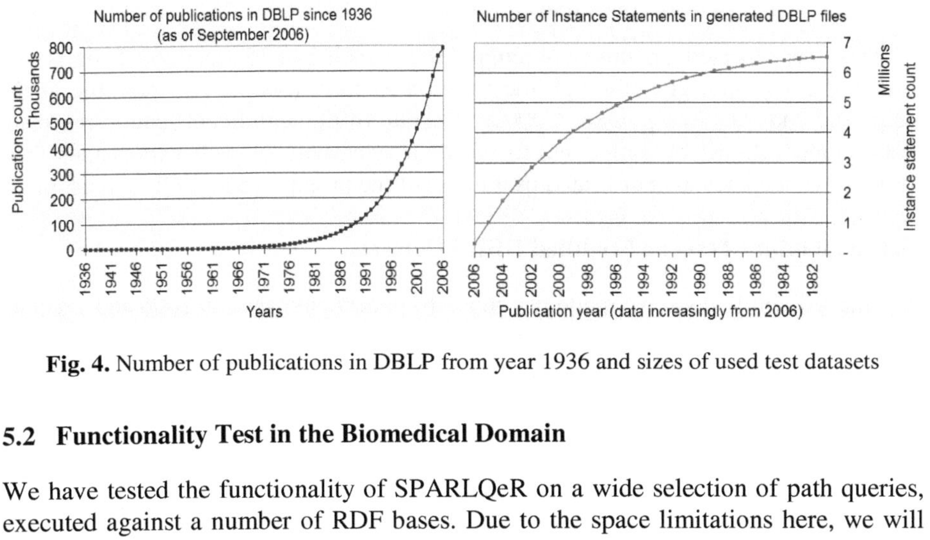

Fig. 4. Number of publications in DBLP from year 1936 and sizes of used test datasets

5.2 Functionality Test in the Biomedical Domain

We have tested the functionality of SPARLQeR on a wide selection of path queries, executed against a number of RDF bases. Due to the space limitations here, we will

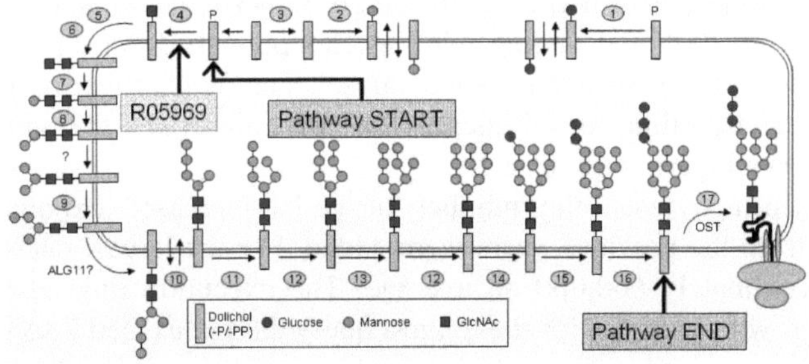

Fig. 5. N-Glycan biosynthesis pathway with query start and end points *(courtesy of Dr. Alison Vandersall-Nairn, University of Georgia)*

only discuss a particularly representative query in the biochemistry domain, retrieving a major part of the well known *N-Glycan biosynthesis pathway* [12]. The pathway is shown in Fig. 5 on the next page, where each arrow represents one reaction. The pathway is represented in GlycO, with the reactions represented as illustrated by the RDF graph in Fig 1.

We chose this pathway for its high regularity and a significant length. It enabled us to test if specifying paths using path expressions would help to find long, semantically relevant paths within an acceptable time. For this test we used the GlycO ontology and the SPARQLeR query used is presented below.

```
SELECT list(%path) WHERE {
  glyco:dolichol_phosphate %path glyco:glyco_peptide_G00009 .
  %path  rdfs:member  enzyo:R05969
  FILTER ( length(%path) <= 30 &&
    regex(%path, "((-glyco:has_acceptor_substrate|
      -glyco:has_reactant) glyco:has_product)*" ) ) }
```

This query located a pathway of length 30, consisting of 15 reactions. It starts with *dolichol phosphate*, goes through the reaction *R05969* (one of two possible at this step) and ends at *glyco peptide G00009*. Despite of the significant length, the result was retuned almost instantly, due to the high selectivity of path expressions. This proof of concept test demonstrated usefulness of the proposed SPARQL extension.

5.3 Scalability Tests on Modified DBLP Datasets

For the scalability tests, we randomly chose 14 papers published in 2006 and executed single-source queries to find all paths leading to papers they cited, using the relation *cites_publication*. A sample SPARQLeR query in presented below:

```
PREFIX opus: <http://lsdis.cs.uga.edu/projects/semdis/opus#>
SELECT ?end_publication WHERE {
  <http://dblp.uni-trier.de/rec/bibtex/journals/ai/Huber06>
  %path ?end_publication
  FILTER ( length(%path)<=26 &&
    regex(%path, "(opus:cites_publication)*" ) ) }
```

The queries were performed on increasingly larger datasets, starting with articles published only in 2006 and ending with articles published during 1981-2006. Each query was executed 4 times against each dataset. The plots in Fig. 6 on the next page present the execution time for all queries for each dataset and the number of located paths plotted on a logarithmic scale.

In the performed tests, the number of paths increased exponentially as the publications from the previous years were added. For the largest dataset, each query returned approximately 660,000 on average. The execution time also followed the exponential growth, but even for the longest query did not exceed 7 seconds.

Additionally, we performed tests for finding semantic associations between two given entities. We identified 4 early publications that were reachable by a relatively large number of paths from all previously chosen 14 starting publications. These 4

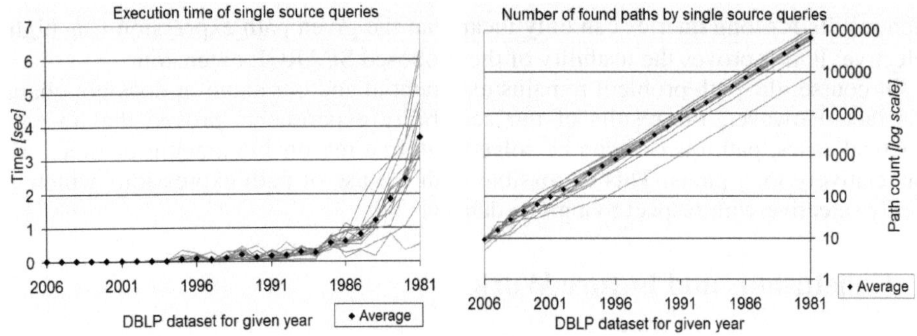

Fig. 6. Query execution times and number of found paths for single-source path queries

entities become endpoints for path queries between two resources. A sample SPARQLeR query is presented below:

```
SELECT list(%path) WHERE {
  <http://dblp.uni-trier.de/rec/bibtex/journals/ai/Huber06>
  %path
  <http://dblp.uni-trier.de/rec/bibtex/conf/programm/BarbutiM80>
  FILTER ( length(%path)<=26 &&
    regex(%path, "(opus:cites_publication)*" ) ) }
```

The queries were performed on increasingly larger datasets, while the length limit was increasing from 1 to 26, according to number of covered years in the datasets. For each of the 14 start entities we ran the path query to the 4 previously selected publications and averaged the results. The plots in Fig. 7 present the execution time and numbers of located paths for 14 start entities (each queried with 4 endpoints) for each dataset.

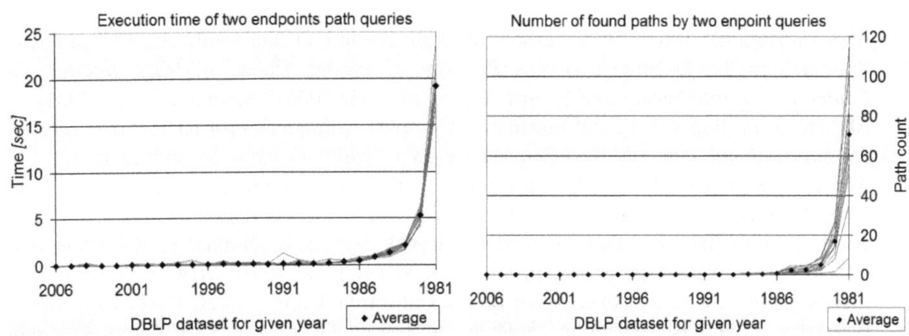

Fig. 7. Query execution times and number of found paths for path queries with set endpoints

In these tests, due to specificity of the dataset, although the number of results is a small fraction of previous ones, the search space became significantly larger than for the single-source queries. Nevertheless, the execution time did not exceed 25 seconds, which we think is a reasonable result for searching paths of length up to 26 hops. For shorter paths, the execution time drops drastically to below 1 second. In both cases, such results for long queries can only mean that the given path expression was highly selective. It also proves the usability of the proposed SPARQL extension.

Of course, the path problem remains exponential and our solution does not change this fact. However, the results of our scalability experiments proved that in some practical cases, path queries can be solved within a reasonable amount of time, even for relatively long paths. This is possible with the use of path expressions which are highly selective with respect to a given dataset.

6 Conclusions and Future Work

We have presented SPARQLeR, a novel extension of SPARQL designed for finding semantic associations in RDF bases, and described its working implementation. SPARQLeR's path patterns have been seamlessly incorporated within SPARQL's graph patterns and allow for capturing both structural and semantic requirements of semantic association queries. Our experiments with path pattern queries have demonstrated the expressive power of SPARQLeR, effectiveness of its implementation, as well as its practical value in the presented examples.

Our future plans involve the optimization of regular path queries and incorporation of regular context into SPARQLeR. Despite the presented good timing results, we think that the optimization of path queries is very important for the practical use of the proposed language. This line of research involves not only optimization of simple queries, but of complex expressions and queries spanning multiple paths, as well.

We plan to base the notion of a context on our path patterns inducing RDF sub-graphs that will allow us to semantically specify a sub-graph of interest within an RDF description base. Consequently, this would support the idea that the same query executed in different contexts should return different results.

References

1. Aleman-Meza, B., Burns, P., Eavenson, M., Palaniswami, D. and Sheth, A., An Ontological Approach to the Document Access Problem of Insider Threat. in *IEEE International Conference on Intelligence and Security Informatics (ISI-2005)*, (Atlanta, Georgia, 2005).
2. Alkhateeb, F., Baget, J.-F. and Euzenat, J. Complex path queries for RDF *Poster paper in 4th International Semantic Web Converence (ISWC2005)*, Galway, Ireland, 2005.
3. Anyanwu, K. and Sheth, A., r-Queries: Enabling Querying for Semantic Associations on the Semantic Web. in *12th International World Wide Web Conf.*, (Budapest, Hungary, 2003).
4. Beeri, C., Kanellakis, P., Bancilhon, F. and Ramakrishnan, R., Bounds on the propagation of selection into logic programs. in *6th ACM SIGACT-SIGMOD-SIGART Symposium on Principles of Database Systems*, (San Diego, California, United States, 1987), 214 - 226.
5. Broekstra, J. and Kampman, A. SeRQL: A Second Generation RDF Query Language *SWAD-Europe Workshop on Semantic Web Storage and Retrieval*, Amsterdam, Netherlands, 2003.
6. Buchsbaum, A.L., Kanellakis, P.C. and Vitter, J.S., A data structure for arc insertion and regular path finding. in *1st annual ACM-SIAM symposium on Discrete algorithms*, (San Francisco, California, United States, 1990), 22-31.
7. Calvanese, D., Giacomo, G.D., Lenzerini, M. and Vardi, M.Y., Containment of Conjunctive Regular Path Queries with Inverse. in *7th International Conference on the Principles of Knowledge Representation and Reasoning (KR 2000)*, (2000), 176-185.

8. Consens, M. and Mendelzon, A.O., Graphlog: a visual formalism for real life recursion. in *ACM Symposium On Principles of Database Systems*, (1990), 404-416.
9. Cruz, I.F., Mendelzon, A.O. and Wood, P.T., G+: Recursive queries without recursion. in *2nd International Conference on Expert Database Systems*, (1988), 355-368.
10. Cruz, I.F., Mendelzon, A.O. and Wood, P.T., A graphical query language supporting recursion. in *ACM SIGMOD International Conference on Management of Data*, (San Francisco, California, United States, 1987), ACM Press, 323-330.
11. Hassell, J., Aleman-Meza, B. and Arpinar, I.B. Ontology-Driven Automatic Entity Disambiguation in Unstructured Text *5th International Semantic Web Conference (ISWC-2006)*, Athens, GA, 2006.
12. Helenius, A. and Aebi, M. Roles of N-Linked Glycans in the Endoplasmic Reticulum. *Annual Review of Biochemistry, 2004, 73*. 1019-1049.
13. Janik, M. and Kochut, K., BRAHMS: A WorkBench RDF Store And High Performance Memory System for Semantic Association Discovery. in *4th International Semantic Web Conference*, (Galway, Ireland, 2005).
14. Karvounarakis, G., Alexaki, S., Christophides, V., Plexousakis, D. and Scholl, M., RQL: A Declarative Query Language for RDF. in *11th International World Wide Web Conference*, (Honolulu, Hawaii, USA, 2002), ACM.
15. Mendelzon, A.O. and Wood, P.T., Finding Regular Simple Paths In Graph Databases. in *15th Conference on Very Large Databases*, (Amsterdam, The Netherlands, 1989), Morgan Kaufman pubs. (Los Altos CA).
16. Mukherjea, S. and Bamba, B., BioPatentMiner: An Information Retrieval System for Biomedical Patents. in *13th International Conference on Very Large Data Bases (VLDB 2004)*, (Toronto, Canada, 2004), Morgan Kaufmann.
17. NLM. PubMed The National Library of Medicine, Bethesda MD.
18. Ogbuji, U. RDF Query using Versa Thinking XML: Basic XML and RDF techniques for knowledge management, Part 6, 10 April 2002.
19. Ramakrishnan, C., Kochut, K. and Sheth, A., A Framework for Schema-Driven Relationship Discovery from Unstructured text. in *5th International Semantic Web Conference (ISWC 2006)*, (Athens, Georgia, USA, 2006).
20. Ramakrishnan, C., Milnor, W.H., Perry, M. and Sheth, A.P. Discovering Informative Connection Subgraphs in Multi-relational Graphs. *SIGKDD Explorations*, *7* (2). 56-63.
21. Seaborne, A. RDQL - A Query Language for RDF, 2004.
22. Sheth, A., From Semantic Search & Integration to Analytics. in *Dagstuhl Seminar Proceedings 04391*, (Dagstuhl, Germany, 2005).
23. Sintek, M. and Decker, S. TRIPLE - An RDF Query, Inference, and Transformation Language *Deductive Databases and Knowledge Management*, Tokyo, Japan, 2001.
24. SPARQL. Query Language for RDF. Prud'hommeaux, E. and Seaborne, A. eds., 2005.
25. Swanson, R.D. Migraine and Magnesium: Eleven Neglected Connections. *Perspectives in Biology and Medicine*, *31* (4). 526-557.
26. Thomas, C.J., Sheth, A.P. and York, W.S., Modular Ontology Design Using Canonical Building Blocks in the Biochemistry Domain. in *International Conference on Formal Ontology in Information Systems (FOIS)*, (November 2006), IOS Press.
27. Yannakakis, M., Graph-theoretic methods in database theory. in *9th ACM SIGACT-SIGMOD-SIGART symposium on Principles of database systems*, (Nashville, Tennessee, United States, 1990), ACM Press, 230-242.

Simple Algorithms for Predicate Suggestions Using Similarity and Co-occurrence

Eyal Oren, Sebastian Gerke, and Stefan Decker

Digital Enterprise Research Institute
National University of Ireland, Galway
Galway, Ireland
firstname.lastname@deri.org

Abstract. When creating Semantic Web data, users have to make a critical choice for a vocabulary: only through shared vocabularies can meaning be established. A centralised policy prevents terminology divergence but would restrict users needlessly. As seen in collaborative tagging environments, suggestion mechanisms help terminology convergence without forcing users. We introduce two domain-independent algorithms for recommending predicates (RDF statements) about resources, based on statistical dataset analysis. The first algorithm is based on similarity between resources, the second one is based on co-occurrence of predicates. Experimental evaluation shows very promising results: a high precision with relatively high recall in linear runtime performance.

1 Introduction

The Semantic Web is decentralised in terms of autonomy, allowing everyone to make any statement, but centralised in terms of vocabulary: others can only understand statements that use familiar terminology. Given this situation, we consider the following problem: how to ensure that individuals, free to use arbitrary terminology, converge towards shared vocabularies?

As a particular use case we consider authoring in Semantic Wikis [12, 14, 18]. These enhanced Wikis allow users to describe information both in free text and through semantic descriptions. Allowing users to make arbitrary statements is important, since it ensures domain-independence of the Wiki.

Without further considerations, the authoring freedom in Semantic Wikis would result in statements with different vocabularies, defying the purpose of the Semantic Wiki. A terminology policy could be enforced but that would highly restrict users. A suggestion mechanism, recommending terminology based on the dataset, would help converge terminology without forcing users, as demonstrated in collaborative tagging [9, 10].

In collaborative data entry, participants construct a dataset by continuously and independently adding further statements to existing data. Each participant faces the question: *when creating Semantic Web data, which vocabulary to use*? To ensure convergence, the answer is: use the most relevant and frequently occurring vocabulary.

E. Franconi, M. Kifer, and W. May (Eds.): ESWC 2007, LNCS 4519, pp. 160–174, 2007.

Finding the most *frequent* vocabulary is straightforward: one can simply count the occurrences. We therefore focus on finding the *relevant* vocabulary. Datasets typically contain heterogeneous data. Finding the vocabulary that is relevant for one resource therefore means: finding similar resources and use their vocabulary.

Problem statement. Our problem is thus to suggest relevant and frequent terminology for extending a resource in an RDF dataset based on similarity with other resources and our question is *how well simple algorithms solve this problem*?

We present two algorithms that address this problem, based on the following hypotheses that simple algorithms do well enough: (a) computing resource similarity based only on *outgoing arcs* yields good results; (b) approximating resource similarity through *pairwise predicate co-occurrence* yields good results.

We will present the two algorithms in sections 2 and 3, and their implementation in section 4. We verify our hypotheses and the performance of these algorithms empirically in section 5. We conclude with a discussion of related work in section 6.

2 Classification-Based Algorithm

The task of the suggestion algorithm is to find, for a certain resource in focus, predicates to further describe that resource. The general idea of the classification-based algorithm is to divide the knowledge base in two groups, those similar to the current resource and those not similar, and to suggest the frequently occurring predicates from the similar group.

For example, figure 1 shows a simple knowledge base with three resources: the person "John", with his name, some friends, and homepage, the book "The Pelican Brief", with its title and author, and the person "Stefan", with his name. We want to suggest relevant predicates for "Stefan" based only on the given graph.

The algorithm consists of two steps, as shown in listing 1.1. In the first step, we divide all existing resources in the knowledge base into two sets, the similar

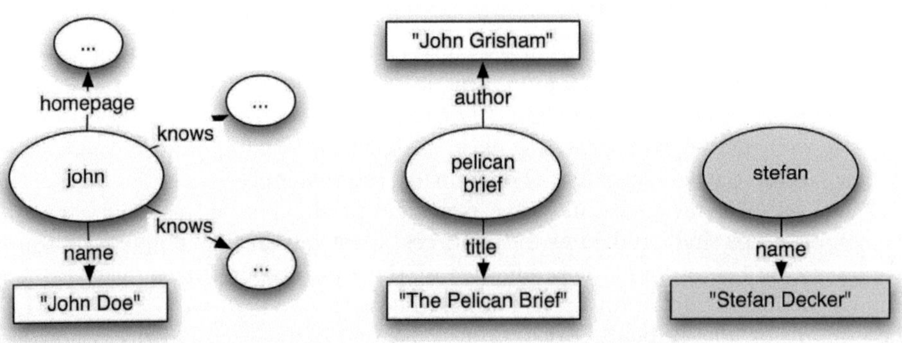

Fig. 1. Example knowledge base

Listing 1.1. Classification-based algorithm

```
def suggest(r, resources)
  # select similar resources
  similar_resources = resources.select { |r'| similarity(r,r') > threshold }

  # then collect all predicates from similar resources
  candidates = similar_resources.collect { |r'| r'.predicates }

  # then rank all candidate predicates
  return rank(candidates)
end
```

and unsimilar ones. In the second step, we look at all predicates from the similar group and rank them using a ranking function. In the remainder of this section, we explore each step in more detail: how to define similarity between resources, and how to rank the selected predicates.

2.1 Preliminaries

First we introduce some necessary definitions:

Definition 1 (RDF Graph). *An RDF graph G is defined as $G = (V, E, L, l)$ where V is the set of vertices (subjects and objects), E is the set of edges (predicates), L is the set of labels, $l : E \rightarrow L$ is the labelling function for predicates. The projection $source : E \rightarrow V$ and $target : E \rightarrow V$ return the source and target nodes of a given edge.*

Definition 2 (Outgoing edges). *The set of outgoing edges $E_o(v)$ of a vertex is defined as: $E_o(v) = \{e \in E | source(e) = v\} \subseteq E$. The bag of labels $L(E)$ of a set of edges is defined as $L(E) = [l(e) | e \in E]$. The bag of labels $L_o(v)$ of outgoing edges of a vertex v is defined as $L_o(v) = L(E_o(v))$. The set of outgoing edges of v whose label is l is defined as $E_o(v, l) = \{e \in E_o(v) | l(e) = l\}$.*

2.2 Classification Step

In the first step, we classify resources into those similar to the current one, and those not similar. The main requirement for the similarity metric is domain-independence: the algorithm should not rely on domain-specific knowledge. We use two well-known, widely used generic similarity metrics [2, 3]: the *containment* of one resource in another and their mutual *resemblance*.

Since we are interested in suggesting new predicates, we use these metrics to compare existing predicates of resources. Containment thus defines resource similarity as the amount of predicates of the first resource that are also contained in the second resource, as shown in equation (1). Resemblance measures how many of all predicates used in at least one of the two resources are used in both resources, as shown in equation (2). For example, in figure 1, the resource

"Stefan" uses the predicate "name" and the resource "John" uses "name", "knows" and "homepage", resulting in a containment value (of "Stefan" in "John") of 1 and a resemblance of $\frac{1}{3}$.

$$s_c(v', v) = \frac{|O(v) \cap O(v')|}{|O(v)|} \tag{1}$$

$$s_r(v', v) = \frac{|O(v) \cap O(v')|}{|O(v) \cup O(v')|}. \tag{2}$$

Since predicates can have multiple values, when computing this containment or resemblance metrics we need to decide whether to count multiple predicate occurrences once or several times.

In the example, the resource "John" uses the "knows" predicate twice with different values; we can either count these two occurrences only once, thus using $O(v)$ as a set, as shown in equation (3). The resemblance between "Stefan" and "John" would then be $\frac{1}{3}$. But we could also count each occurrences separately, using $O(v)$ as a bag as shown in equation (4), yielding a resemblance of $\frac{1}{4}$.

$$O_s(v) = \{l(e) | e \in E_o(v)\} \tag{3}$$

$$O_b(v) = [l(e) | e \in E_o(v)] \tag{4}$$

If we generalise from these two choices, the result of the first phase is the set of similar resources $V_s(v)$, as defined in equation (5), where s_t is some similarity threshold and $s(v, v')$ is either resemblance or containment measure. For example, with a threshold of 0.9 the set of similar resources to "Stefan" would consist only of the resource "John".

$$V_s(v) = \{v' \in V : s(v, v') \geq s_t\} \tag{5}$$

2.3 Ranking Step

After classifying all resources into two groups we collect all predicates from the set of similar resources $V_s(v)$ and use them as candidates for the suggestion. Since there might be many candidates, we need to rank these candidates and suggest the more useful predicates first. The most straightforward ranking function is based on the occurrence frequency of these predicates in the set of similar resources.

In this example, since only the resource "John" is similar to "Stefan", the candidates would be "knows" and "homepage", ignoring the predicates that "Stefan" uses already. Out of these two candidates, "knows" would be ranked first since it appears most frequently.

But again, since predicates in RDF can be multi-valued, we can define the (relative) occurrence frequency of a label l in the set of similar resources $V_s(v)$ in two ways. We can either count each predicate occurrence, as shown in equation (6). Or we can count each occurrence only once, or stated differently, count the set X of resources that use l in their outgoing edges and divide them by the total number of resources, as shown in equations (7). In the latter case, "knows" and "homepage" would be ranked the same since they are both used by one resource.

$$r_v(e) = f_s^p(v,l) = \frac{\sum_{v' \in V_s(v)} |E_o(v',l)| \cdot w(v,v')}{\sum_{v' \in V_s(v)} |E_o(v')| \cdot w(v,v')} \tag{6}$$

$$r_v(e) = f_s^r(v,l) = \frac{\sum_{v' \in X} w(v,v')}{\sum_{v' \in V_s(v)} w(v,v')}$$
$$X = \{v \in V_s(v) | l \in O(v)\} \tag{7}$$

In both methods of counting, we could allow for a weighting factor $w(v, v')$. The reason for this is that even in the set of similar resources $V_s(v)$, some are more similar than other: in ranking the predicates, it would be natural to "promote" the predicates from similar resources over those from less similar resources. If we choose to prefer predicates from resources more similar to v, the weight factor could be given by the resource similarity, shown in equation (8). A simpler approach would not to weigh the predicates, as shown in equation (9). In our example, these methods would yield the same ranking since both candidates originate from the same resource "John".

$$w_s(v,v') = s(v,v') \tag{8}$$

$$w_c(v,v') = \begin{cases} 1 & : \quad v' \in V_s(v) \\ 0 & : \quad v' \in V_n(v) \end{cases} \tag{9}$$

2.4 Qualitative Results

To investigate our hypothesis, we have evaluated the performance and quality of the algorithm using various different datasets. We are interested in the quality of the basic algorithm (using containment, counting multi-valued predicates only once, and without weighting) and in whether the various parameters, while reducing simplicity, improve the basic algorithm. We present and discuss these results in section 5.

2.5 Performance

Regarding the runtime performance of the algorithm, we can analyse the description in listing 1.1. We see that, ignoring data access, the overall algorithm should run linearly to the number of resources: The first phase, classifying the similar resources, runs linear to the number of resources r and the average number of predicates per resource p: comparing the similarity of each resource against the one resource in focus by comparing all their predicates. The second phase, ranking, is linear in the number of candidates c. The complete algorithm would therefore run in $O(r \cdot p + c)$, which is linear in r, since p will be constant on average and c is presumably smaller than r.

However, in practice we cannot ignore lookup performance on large datasets. To compute similarity, we need to lookup all predicates of each resource. Depending on the lookup performance of the used datastore, this could cause the whole algorithm to run logarithmic or even quadratic to the size of the dataset, rendering the algorithm impractical for reasonably large datasets.

A simple solution would be to materialise the similarity between resources in memory, obliterating the need for data lookup during suggestion time. Direct materialisation however has two problems: the required memory space would be quadratic in the size of the dataset, and updating one resource (prone to happen often in a data entry scenario) would require recalculation of all similarity values with respect to this resource.

The next algorithm remedies exactly this problem and allows materialisation without large memory requirements.

3 Co-occurrence-Based Algorithm

The general idea of the co-occurrence-based algorithm is to approximate resource similarity through the co-occurrence of predicates. Since usually datasets contain far less predicates than resources, predicate co-occurrence requires far less space than resource similarity. We then further reduce the required space by not considering the complete power set over all predicates, but instead approximate full co-occurrence through binary co-occurrences. We thus consider only pairwise occurrences of predicates, suggest predicate candidates for each pairwise occurrence, and combine these candidates through intersection.

We therefore make two assumptions on the probabilistic model of the dataset: (1) that predicate co-occurrence correlates with resource similarity, and (2) that considering binary predicate co-occurrences to be independent events (which they are not) yields acceptable predictions. The latter allows us to pairwise consider binary co-occurrences instead of all permutations.

The algorithm is based on association rule mining [1, 17] used for recommendations in e.g. online stores: when buying one book, other books that are often bought together with this book are recommended. In our case, books are replaced by predicates and shopping transactions by resources.

3.1 Pre-computation Step

To better show the details of the algorithm, we extend our earlier example, adding the person "Sebastian" and some more statements about John, as shown in figure 2. Again, we want to suggest further predicates to the resource "Stefan".

In the first step we calculate usage statistics of predicates in the knowledge base. We count for each predicate, the resources that use this predicate, defined in equation (10). Secondly, we count for each pair of predicates, the number

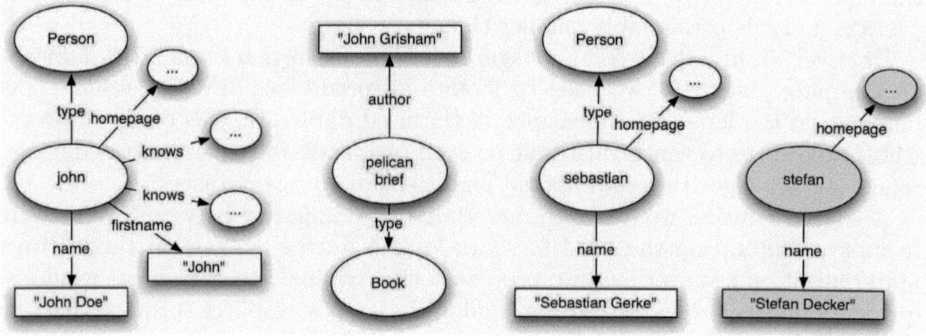

Fig. 2. Extended Knowledge Base

of times they co-occur together in the same resource, as defined in equation (11). The particular statistics for the example in figure 2 are given in table 2a and table 1.

$$occ(p) = |\{v \in V | p \in L_o(v)\}| \tag{10}$$

$$coocc(p_1, p_2) = |\{v \in V | p_1 \in L_o(v) \wedge p_2 \in L_o(v)\}| \tag{11}$$

Table 1. Predicate occurrence and co-occurrence frequency

predicate	*freq.*
type	3
name	2
knows	1
homepage	2
firstname	1
author	1

(a) occurrence

	type	name	knows	homepage	firstname	author
type	3	2	1	2	1	1
name	2	2	1	2	1	0
knows	1	1	1	1	1	0
homepage	2	2	1	2	1	0
firstname	1	1	1	1	1	0
author	1	0	0	0	0	0

(b) co-occurrence

3.2 Suggestion Step

In the second step, we compute suggestions for a given resource. We consider all predicates in the knowledge base that occur more than once with each of the predicates from "Stefan" as suggestion candidates, as defined in equation (12). In our example, the predicates "type", "knows", and "firstname" are candidates for the resource "Stefan".

$$cooccurring(p_1) = \{p_2 : coocc(p_1, p_2) > 1\} \tag{12}$$

For each candidate we calculate our confidence in suggesting it. As shown in equation (13), the confidence for suggesting a predicate p for a selected resource r, is formed by combining the confidence for p from each of r's predicates p_i. In the earlier example, the total confidence for suggesting "type" is computed by combining $confidence(name \Rightarrow type)$ and $confidence(homepage \Rightarrow type)$.

$$confidence(p, r) = \prod_{p_i \in cooccurring(p) \cap L_o(r)} confidence(p_i \Rightarrow p) \qquad (13)$$

Each constituent is computed as shown in equation (14): the confidence for suggesting any p_2 based on the existence of a p_1 is given as the co-occurrence frequency of p_1 and p_2 relative to the occurrence frequency of p_1 by itself. In our example, p_2, the candidate, would be "type", "knows", or "firstname", and p_1, the existing predicates, would be "name" and "homepage". Intuitively, we consider a relatively frequent co-occurrence as evidence for predicting p_2.

$$confidence(p_1 \Rightarrow p_2) = \frac{coocc(p_1, p_2)}{occ(p_1)} \qquad (14)$$

In our example, as shown in table 2, "type" co-occurs with both predicates of "Stefan" 100% of the time, whereas the two other candidates ("knows" and "firstname") co-occur only 50% of the time with each of the predicates of "Stefan". We rank each candidate by the combined (unweighted) confidence: in this example, "type" will be ranked first, with a combined confidence of 100%, and the other two second, with a combined confidence of 25%.

Table 2. Relative co-occurrence ratios for Stefan

candidate	*name*	*homepage*	*confidence*
type	1.0	1.0	1.0
knows	0.5	0.5	0.25
firstname	0.5	0.5	0.25

4 Implementation

We have implemented both algorithms in Ruby. We use the ActiveRDF [13] datastore abstraction layer which allows us to run this algorithm on various RDF datastores. The implementations are distributed as part of the ActiveRDF. We have also implemented the co-occurrence algorithm as a wrapper for an RDF datastore, in particular for the rdflite[1] RDF store.

Since rdflite uses a relational database with one table, `triple(s,p,o)`, we have implemented the (co)occurrence statistics as views on this database, comparable to [7]. Depending on the relational database, these views can be materialised or computed for each suggestion. The views, shown in listing 1.2, are a straightforward translation of the equations (10) and (11) given before.

[1] `http://wiki.activerdf.org/rdflite/`

Listing 1.2. Co-occurrence as database views

```
create view occurrence as
  select p, count(distinct s) as count
  from triple
  group by p;

create view cooccurrence as
  select t0.p as p1, t1.p as p2, count(distinct t0.s) as count
  from triple as t0 join triple as t1 on t0.s = t1.s and t0.p != t1.p
  group by t0.p, t1.p
```

4.1 Example Suggestions

Figure 3 shows an example of our suggestion system, for a randomly chosen resource from a dataset[2] about the Mindswap research group. The resource (a blank node representing Dan Connolly) and its predicates, such as name and email address, are listed on the left-hand side. Our suggestions, based on the other resources in this dataset, are listed in ranked order on the right-hand side.

```
_:#1 foaf:name Dan Connolly .
_:#1 owlweb:name Dan Connolly .
_:#1 foaf:mbox mailto:connolly@w3.org .
_:#1 owlweb:email mailto:connolly@w3.org .
_:#1 owlweb:homepage
        http://owl.mindswap.org/~danC/ .
_:#1 rdf:type owlweb:FamilyFriend .
```

1. foaf:workInfoHomepage
2. foaf:homepage
3. owlweb:personalHomepage
4. owlweb:bachelorsFrom
5. owlweb:mastersFrom
6. owlweb:homepage
7. foaf:nick
8. owlweb:phdFrom

Fig. 3. Suggested predicates (right) for example resource (left)

5 Evaluation

A predicate suggestion system is a kind of recommender system, using the opinions of a community to help individuals decide between a potentially overwhelming set of choices [6, 16]. In our case, this "potentially overwhelming set of choices" is formed by the terminology (ontologies or schemas) available.

Evaluations of recommender systems can be divided into two categories [5, 6]: when regarding recommendations as an information retrieval problem (selecting the interesting predicates from all possible predicates), evaluation is usually performed off-line, focused on accuracy, and measured using precision and recall. When, on the other hand, recommendation is approached as a machine learning regression problem (learning and predicting user's annotation preferences), evaluation is commonly performed online, focused on utility and usefulness, and measured using a training set and a test set.

[2] `http://www.cs.umd.edu/~hendler/2003/MindPeople4-30.rdf`

5.1 Evaluation Approach

Our evaluation combines both the information-retrieval and the machine-learning approach: we show both precision and recall ratings and evaluate our approach using training/testing datasets through a commonly applied technique of evaluating prediction of deleted values from existing data [6].

Because the distribution of data can alter the performance of the algorithms quite severely, we evaluated on five existing RDF datasets: a webcrawl[3] of arbitrary RDF, the Mindswap research group[4], a FOAF dataset[5], a terror dataset[6] augmented with terrorist data, and the ontoworld.org Semantic Wiki[7]. These datasets have differing characteristics, as shown in Table 3: both large and small, with homogeneous and heterogeneous data, and both highly structured and highly unstructured distribution.

Table 3. Evaluation datasets

dataset	*classes*	*resources*	*triples*
webcrawl	2	112	6766
mindpeople	14	273	1081
foaf	4	3123	10020
terror	25	1553	16632
ontoworld	42	4467	28593

Our primary evaluation technique is prediction of deleted values: we pick a random resource from the dataset as a candidate for which further predicates should be suggested. We then randomly remove one ore more statements about this candidate and analyse if and at which rank position the removed predicates are re-suggested. Repeated over n random resources this yields the *average resuggestion rate* (how often was the deleted predicate resuggested), the *empty suggestion rate* (how often were no suggestions given), and the *average rank* of the resuggested predicate. Since in practice not all suggestions can be displayed or will be considered by the user, we also show how many of the predicates were resuggested within the top-k of suggestions.

Secondly, we measure suggestion precision (how many suggestions are valid) and recall (how many valid suggestions have we missed) based on the schema definition: we define "valid" predicates as those predicates that, according to the schema, fall within the domain of the selected candidate. For recall computation, we consider only predicates that are actually used in the dataset; since the algorithm considers only instance data, unused predicates are unattainable.

[3] `http://www.activerdf.org/webcrawl_10k.nt`
[4] `http://www.cs.umd.edu/~hendler/2003/MindPeople4-30.rdf`
[5] `http://rdfweb.org/2003/02/28/cwm-crawler-output.rdf`
[6] `http://reliant.teknowledge.com/DAML/TerroristActs.owl`
[7] `http://ontoworld.org/RDF/`

Table 4. Results per dataset for classification-based algorithm

dataset	*resugg.*	*empty*	*rank*	*top-5*	*top-10*	*top-20*	*prec.*	*recall*	F_1
webcrawl	0.95	0.04	1.06	0.94	0.94	0.94	0.96	0.73	0.83
mindpeople	0.80	0.19	1.30	0.79	0.80	0.80	0.81	0.83	0.83
foaf	0.92	0.06	1.30	0.92	0.93	0.93	0.94	0.80	0.87
terror	0.98	0.02	1.10	0.97	0.97	0.98	0.98	0.91	0.95
ontoworld	0.85	0.13	1.39	0.84	0.84	0.85	0.87	0.72	0.79
average	0.90	0.08	1.22	0.89	0.90	0.90	0.92	0.80	0.85

Table 5. Results per dataset for co-occurrence-based algorithm

dataset	*resugg.*	*empty*	*rank*	*top-5*	*top-10*	*top-20*	*prec.*	*recall*	F_1
webcrawl	1.00	0.00	1.18	0.99	0.99	1.00	1.00	0.74	0.85
mindpeople	1.00	0.00	1.23	1.00	1.00	1.00	1.00	0.76	0.87
foaf	1.00	0.00	1.51	0.95	1.00	1.00	1.00	0.59	0.74
terror	1.00	0.00	1.15	0.98	1.00	1.00	1.00	0.95	0.97
ontoworld	1.00	0.00	1.14	0.98	1.00	1.00	1.00	0.78	0.88
average	1.00	0.00	1.24	0.98	1.00	1.00	1.00	0.77	0.87

5.2 Results

All tests were run on an AMD Opteron 1993MHz machine with 2GB of RAM. The similarity algorithm was run 300 times over five random samples (n=100, n=150, n=200, n=250, n=300) since its performance prevented us from using the full datasets; the co-occurrence algorithm was run 20.000 times over the complete datasets. In each run, we randomly selected a resource and deleted between one and ten of its existing predicates. We then let the algorithms suggest additional predicates and compare these to the randomly deleted predicates.

We first show the results of the two primary algorithms for each dataset: table 4 shows the results of the classification-based algorithm, table 5 the results of the co-occurrence-based algorithm. The tables show, for each dataset and for all datasets combined, the resuggestion rate, empty suggestion rate and average rank. It also shows the resuggestion rate when only considering the top-k results, and the precision, recall, and the F_1-measure for each algorithm.

We can see that in general the co-occurrence performs better than any of the classification-based variants, especially when looking at the top-5 results. We can see that the co-occurrence algorithm has very high precision (100% on average). The co-occurrence algorithm has a slightly lower recall than the classification-based ones, due to the intersection of candidates which results in only high-confidence candidates. The F_1-measure (harmonic mean of precision and recall) shows that co-occurrence has the highest quality over all datasets.

Table 6 shows (again) the results for the primary algorithms and then lists the results for each classification variant, averaged over all five datasets. We see that using resemblance instead of containment yields very low results, which is

Table 6. Results of algorithm variants (averaged over all datasets)

algorithm	*resugg.*	*empty*	*rank*	*top-5*	*top-10*	*top-20*	*prec.*	*recall*	F_1
co-occurrence	1.00	0.00	1.24	0.98	1.00	1.00	1.00	0.77	0.87
similarity (default)	0.90	0.08	1.22	0.89	0.90	0.90	0.92	0.80	0.85
resemblance (s_r)	0.10	0.86	1.01	0.11	0.11	0.11	0.14	0.99	0.24
similarity weigh (w_s)	0.90	0.09	1.24	0.89	0.90	0.90	0.91	0.80	0.85
count predicates (f_s^p)	0.91	0.08	1.48	0.89	0.90	0.91	0.92	0.80	0.85
threshold (s_t=0.8)	0.93	0.06	1.29	0.91	0.92	0.93	0.94	0.79	0.86

Table 7. Runtime performance with n resources

algorithm	*n=100*	*n=150*	*n=200*	*n=250*	*n=300*	*n=1555*	*n=3123*	*n=4467*
sim. (rdflite)	1.64s	4.02s	8.51s	15.10s	30.22s	–	–	–
sim. (Sesame)	0.71s	1.40s	2.74s	4.33s	7.88s	–	–	–
co-occ. (view)	0.63s	0.0.78s	1.46s	1.00s	0.93s	7.72s	9.65s	10.10s
co-occ. (constr.)	0.21s	0.27s	0.46s	0.47s	0.70s	2.73s	4.71s	6.34s
co-occ. (query)	0.01s	0.01s	0.01s	0.01s	0.01s	0.01	0.01	0.01

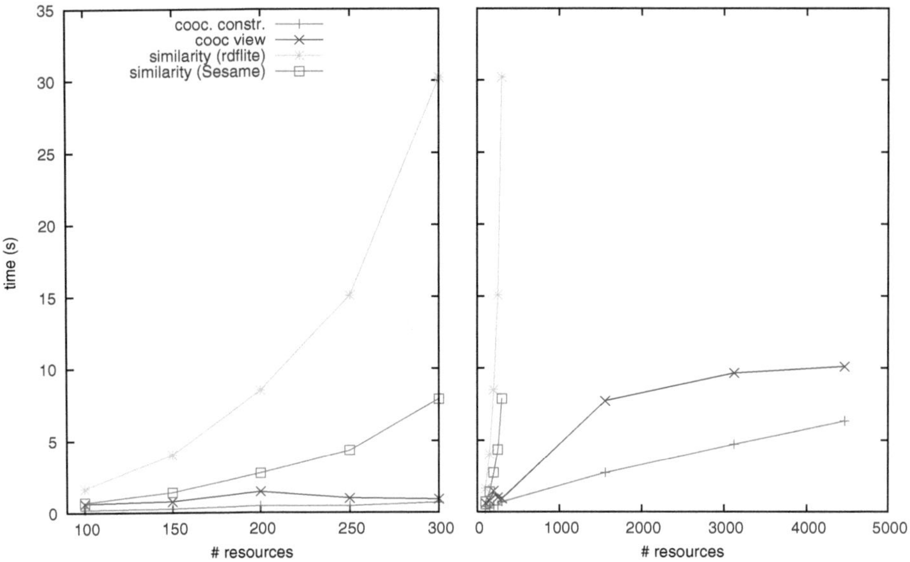

Fig. 4. Runtime performance

most probably due to a too high threshold value. The other variations do not seem to affect the results much.

Finally, Table 7 shows the performance times for the algorithms (only one classification variant is shown since runtime is similar for all). Figure 4 shows two graphs for these results; the left graph is zoomed for up-to 300 resources, the right graph shows the full results.

Timing for the co-occurrence algorithm is divided in matrix construction and query answering. We evaluated the classification on two different datastores, rdflite and Sesame[8], to evaluate scaling independent of a particular datastore implementation. We can see that the classification algorithm scales quadratic, which is due to the linear lookup times of the used datastores, although the Sesame datastore performs much better than rdflite.

Both variants of the co-occurrence algorithm perform well and scale linearly. The materialised co-occurrence implementation performs better than the view-based, which is due to the fact that the sqlite database does not support view materialisation; as mentioned earlier, both approaches have their advantages.

The classification algorithm was too slow to include tests with more than 300 resources but that was again due to data lookups on the underlying datastore: the algorithms themselves scale linearly when ignoring data-access. The materialised co-occurrence implementation shows that we can circumvent data access, leading to very good performance, without requiring large memory space.

6 Related Work

Annotation tools such as OntoMat [4] support semi-automatic annotation of documents; they suggest semantic annotation based on natural language analysis of the annotated resources, but do not take existing semantic descriptions into account. Annotea [8] supports collaborative annotations but annotations are made manually without a suggestion mechanism. Semantic Wikis such as SemperWiki [14] or Semantic MediaWiki [18] allow arbitrary Semantic Web authoring but do not guide users in selecting appropriate terminology.

Automatic schema mapping techniques [15] consider a similar problem (automatically finding relations between elements of a schema) but typically operate on class-level as opposed to instance level and use e.g. concept correlation to unify schema elements [11] whereas we try to discover combined usage patterns of predicates.

Our co-occurrence algorithm is based on association rule mining [1] but our techniques for memory conservation differ: Agrawal *et al.* [1] focus on advanced pruning techniques, whereas we approximate n-ary interdependencies using pairwise binary relations (resulting in a much simplified implementation). Furthermore, our technique allows online processing with incremental updates, whereas their algorithms are iterative and need to run over the complete database.

7 Conclusion

We have discussed the problem of choosing vocabulary during Semantic Web data entry; a crucial bottleneck, since only through shared vocabularies can

[8] `http://www.openrdf.org`

meaning be established. We introduced two algorithms for suggesting possible predicates based on statistical data analysis.

The first algorithm is based on a simple intuitive principle of resource classification: we suggest predicates from similar resources. We have discussed parametric variations that differ in the definition of similarity. We showed that the quality is good (F_1 : 85%) and that variations in similarity computation do not lead to much better results.

The second algorithm approximates resource similarity through pairwise predicate co-occurrence, treating predicate occurrences as independent events (which they are not). This simplifies computation and allows for memory-efficient materialisation, while still resulting in high-quality suggestions (F_1 : 87%). Runtime performance of the co-occurrence algorithm is good, scales linearly with the size of the dataset, and is constant in the presence of materialisation.

We conclude that suggesting predicates based on resource similarity works well and that, for this task, similarity based on outgoing arcs seems a "good-enough" metric. Seeing that co-occurrence suggestion quality is even better than in the classification algorithm, our second hypothesis on similarity approximation using predicate co-occurrence seems to hold as well.

Acknowledgements. This material is based upon works supported by the Science Foundation Ireland under Grants No. SFI/02/CE1/I131 and SFI/04/BR/CS0694.

References

[1] R. Agrawal, T. Imielinski, and A. N. Swami. Mining association rules between sets of items in large databases. In *Proceedings of the ACM SIGMOD International Conference on Management of Data*, pp. 207–216. 1993.

[2] A. Z. Broder, S. C. Glassman, M. S. Manasse, and G. Zweig. Syntactic clustering of the web. *Computer Networks*, 29(8-13):1157–1166, 1997.

[3] D. Dhyani, W. K. Ng, and S. S. Bhowmick. A survey of web metrics. *ACM Computer Surveys*, 34(4):469–503, 2002.

[4] S. Handschuh. *Creating Ontology-based Metadata by Annotation for the Semantic Web*. Ph.D. thesis, University of Karlsruhe, 2005.

[5] C. Hayes, P. Massa, P. Avesani, and P. Cunningham. An on-line evaluation framework for recommender systems. In *Workshop on Personalization and Recommendation in E-Commerce*. 2002.

[6] J. L. Herlocker, J. A. Konstan, L. G. Terveen, and J. T. Riedl. Evaluating collaborative filtering recommender systems. *ACM Transactions on Information Systems*, 22(1):5–53, 2004.

[7] M. Houtsma and A. Swami. Set-oriented data mining in relational databases. *Data and Knowledge Engineering*, 17(3):245–262, 1995.

[8] J. Kahan, M. Koivunen, E. Prud'Hommeaux, and R. Swick. Annotea: An open RDF infrastructure for shared web annotations. In *Proceedings of the International World-Wide Web Conference*, pp. 623–632. 2001.

[9] B. Lund, T. Hammond, M. Flack, and T. Hannay. A case study – Connotea. *D-Lib Magazine*, 11(4), 2005.

[10] C. Marlow, M. Naaman, D. Boyd, and M. Davis. HT06, tagging paper, taxonomy, Flickr, academic article, to read. In *Proceedings of the ACM Conference on HyperText and Hypermedia*. 2006.
[11] N. F. Noy and M. A. Musen. PROMPT: Algorithm and tool for automated ontology merging and alignment. In *Proceedings of the National Conference on Artificial Intelligence (AAAI)*, pp. 450–455. 2000.
[12] E. Oren, J. G. Breslin, and S. Decker. How semantics make better wikis. In *Proceedings of the International World-Wide Web Conference*. 2006.
[13] E. Oren, R. Delbru, S. Gerke, A. Haller, *et al.* ActiveRDF: Object-oriented semantic web programming. In *Proceedings of the International World-Wide Web Conference*. 2007.
[14] E. Oren, M. Völkel, J. G. Breslin, and S. Decker. Semantic wikis for personal knowledge management. In *Proceedings of the International Conference on Database and Expert Systems Applications (DEXA)*. 2006.
[15] E. Rahm and P. A. Bernstein. A survey of approaches to automatic schema matching. *The VLDB Journal*, 10(4):334–350, 2001.
[16] P. Resnick and H. R. Varian. Recommender systems. *Communications of the ACM*, 40(3):56–58, 1997.
[17] R. Srikant and R. Agrawal. Mining generalized association rules. In *Proceedings of the International Conference on Very Large Data Bases (VLDB)*, pp. 407–419. 1995.
[18] M. Völkel, M. Krötzsch, D. Vrandevic, H. Haller, *et al.* Semantic wikipedia. In *Proceedings of the International World-Wide Web Conference*. 2006.

Learning Disjointness

Johanna Völker[1], Denny Vrandečić[1], York Sure[1], and Andreas Hotho[2]

[1] Institute AIFB, University of Karlsruhe, Germany
[2] University of Kassel, Germany
{voelker,vrandecic,sure}@aifb.uni-karlsruhe.de,
hotho@cs.uni-kassel.de

Abstract. An increasing number of applications benefits from light-weight ontologies, or to put it differently *"a little semantics goes a long way"*. However, our experience indicates that more expressiveness can offer significant advantages. Introducing disjointness axioms, for instance, greatly facilitates consistency checking and the automatic evaluation of ontologies. In an extensive user study we discovered that proper modeling of disjointness is a difficult and very time-consuming task. We therefore developed an approach to automatically enrich learned or manually engineered ontologies with disjointness axioms. This approach relies on several methods for obtaining syntactic and semantic evidence from different sources which we believe to provide a solid base for learning disjointness. After thoroughly evaluating the implementation of our approach we think that in future ontology engineering environments the automatic discovery of disjointness axioms may help to increase the richness, quality and usefulness of any given ontology.

1 Introduction

An increasing number of applications benefits from light-weight ontologies, or, to put it differently, *"a little semantics goes a long way"* (Jim Hendler). Our experience in building ontology-based systems indicates, however, that adding more expressivity in a controlled manner can reap further benefits. Introducing disjointness axioms, for example, greatly facilitates consistency checking and the automatic evaluation of individuals in a knowledge base with regards to a given ontology.

In description logics two classes are considered as disjoint *iff* their *taxonomic overlap*, i.e. the set of common individuals, *must* be empty. This does not include classes with actual extensions that coincidentally do not have common individuals, for instance *Woman* and *US President*, but only those where the common subset must be empty in all possible worlds – like, for example, *Woman* and *Car*.

Disjointness allows for far more expressive and meaningful ontologies, as shown exemplary in the following. An ontology language with the expressivity of RDFS does not constrain the possible assertions in any way. Even after we set up an ontology defining terms like *Book*, *Student* and *University*, stating that *John* is both a *Student* and a *University* is logically perfectly viable, and would not be recognized as an error by a reasoner. Only if we define these classes as being disjoint, the reasoner will be able to infer the error in the above ontology, guaranteeing that particular constraints are met by

E. Franconi, M. Kifer, and W. May (Eds.): ESWC 2007, LNCS 4519, pp. 175–189, 2007.

the knowledge base and a certain quality of facts is achieved – thus raising the quality of the whole ontology-based system [17].

Despite the obvious importance of stating disjointness among classes, many of today's ontologies do not contain any disjointness axioms. In fact, a survey of 1,275 ontologies [22] recently found only 97 of them to include disjointness axioms. We can only speculate about the reasons, but it is very likely that ontology engineers often forget to introduce disjointness axioms, simply because they are not aware of the fact that classes which are not explicitly declared to be disjoint will be considered as overlapping. Particularly, inexperienced users usually assume the semantics of partitions, or even complete partitions, when they build a subsumption hierarchy (see [15]). Also, as the size of an ontology is a major cost driver for ontologies [2], the manual engineering and addition of the axioms actually costs more time, and thus money.

Therefore, we believe that an approach to automatically introduce disjointness axioms into an ontology would be a valuable addition to any ontology learning or engineering framework. The principle feasibility of learning disjointness based on simple lexical evidence has already been shown by [9]. However, our experiments indicate that a single heuristic is not suitable for detecting disjointness with sufficiently high precision, i.e. better than an average human could do.

For this paper, we performed an extensive survey in order to collect experience with modeling disjoint classes, and identified several problems frequently encountered by users who try to introduce disjointness axioms. Based on the results of our survey we developed a variety of different methods in order to automatically extract lexical *and* logical features which we believe to provide a solid basis for learning disjointness. These methods take into account the structure of the ontology, associated textual resources, and other types of data sources in order to compute the likeliness of two classes to be disjoint. The features obtained from these methods are used to build an overall classification model which we evaluated against more than 10,000 disjointness axioms provided by 30 human annotators. Due to the encouraging evaluation results we are confident that our implementation can be used, for example, to extend state-of-the-art ontology learning systems, to support ontology debugging [17], or to evaluate manually added disjointness axioms.

The survey also showed that deciding if two classes are disjoint is far from trivial. Although experts have a higher agreement on disjointness than non-expert users, their agreement is still lower than we expected. Discussing these problematic formalizations, we uncovered a number of problems humans have with formal disjointness.

In this paper, we will, in Section 2, first present the features we have used in order to automatically learn disjointness axioms. Section 3 describes the set up and execution of the experiments we conducted in order to train a classifier and evaluate the results of our implementation (Section 4). We close with an overview of related work in Section 5 and a summary of the key contributions and remaining open questions in Section 6.

2 Features for Learning Disjointness

Assuming that there is not the one and only approach to determine the disjointness of two classes in an ontology, we developed a variety of different methods to obtain evidence for or against disjointness from different sources. The features delivered by these methods will help us to train a classifier which is able to distinguish between disjoint and non-disjoint classes.

Preliminaries: In this paper we adopt the OWL ontology model, although we do not restrict our approach to OWL. Any ontology model that allows to state disjointness between two classes can be used with all the methods described in this paper.

The methods are provided with an unsorted list of all the pairs previously tagged by human annotators. In the following the set of pairs will be denoted by $P = \{p_1, ...p_n\}$ for $0 \leq n \leq |C|^2$, where C is the set of all classes in the ontology. Each pair $p_k = (c_{k_1}, c_{k_2})$ consists of two classes $c_{k_1}, c_{k_2} \in C$ and $c_{k_1} \neq c_{k_2}$. The confidence of the system in c_{k_1} and c_{k_2} being (not) disjoint is denoted by $conf(p_k, +)$ or $conf(p_k, -)$ respectively.

All methods are allowed to look up these classes within their semantic context, i.e. the domain **ontology** they have been extracted from (see Section 3.1). And finally, as additional sources of background knowledge, the methods may make use of a **corpus** of textual resources associated with the ontology. We automatically selected a subset of 957 documents from the Reuters corpus[1] [16]. For efficiency reasons we only chose those documents with at least 20 occurrences of any of the classes in the ontology.

It is important to mention, that we assume 'meaningful' labels for all classes in the ontology, i.e. labels which may be understood by humans even without knowing the whole taxonomy. This assumption is particularly relevant for all methods which make use of textual resources such as the pattern-based disjointness extraction (cf. Section 2.4), the computation of extensional overlap with respect to Del.icio.us[2] and the algorithms for learning taxonomic relationships (see Section 2.1).

2.1 Taxonomic Overlap

In description logics two classes are disjoint *iff* their taxonomic overlap, i.e. the set of common individuals, *must* be empty. Because of the open world assumption in OWL, these individuals do not necessarily have to *exist* in the ontology. The taxonomic overlap of two classes is considered not empty as long as there *could* be common individuals within the domain of interest which is modeled by the ontology.

We developed three methods which determine the likeliness for two classes to be disjoint by considering their overlap with respect to (i) **individuals** and **subclasses** in the ontology – or learned from a corpus of associated textual resources – and (ii) **Del.icio.us documents** tagged with the corresponding class labels.

Ontology. Both, individuals and subclasses can be imported from an ontology (see Section 3.1) or from a given corpus of text documents. In the latter case, `subclass-of` and `instance-of` relationships are extracted by different algorithms provided by the Text2Onto[3] ontology learning framework. A detailed description of these algorithms can be found in [4]. All taxonomic relationships – learned and imported ones – are associated with rating annotations $r_{subclass-of}$ (or $r_{instance-of}$ respectively) indicating

[1] `http://trec.nist.gov/data/reuters/reuters.html`
[2] `http://del.icio.us/`
[3] `http://ontoware.org/projects/text2onto/`

the certainty $x > 0$ of the underlying ontology learning framework in the correctness of its results. For imported relationships the confidence is 1.0.

$$r_{subclass-of}(c_1, c_2) = \begin{cases} x & c_1 \texttt{ subclass-of } c_2 \\ 0 & \text{otherwise} \end{cases} \quad (1)$$

The following formula defines the confidence $conf(p, -)$ for a pair $p = (c_1, c_2)$ to be **not** disjoint based on the taxonomic overlap of c_1 and c_2 with respect to common **subclasses** (the same for **instance**):

$$conf(p, -) = \frac{\sum_{c \in sub_1 \cap sub_2} (r_{subclass-of}(c, c_1) \cdot r_{subclass-of}(c, c_2))}{\sum_{c \in sub_1} r_{subclass-of}(c, c_1) + \sum_{c \in sub_2} r_{subclass-of}(c, c_2)} \quad (2)$$

where sub_i denotes the set of subclasses of c_i.

Del.icio.us. Del.icio.us is a server-based system with a simple-to-use interface that allows users to organize and share bookmarks on the internet. It associates each URL with a description, a note, and a set of tags (i.e. arbitrary class labels). For our experiments, we collected $|U| = 75,242$ users, $|T| = 533,191$ tags and $|R| = 3,158,297$ resources, related by in total $|Y| = 17,362,212$ triples. The idea underlying the use del.icio.us in this case is that two labels which are frequently used to tag the same resource are likely to be disjoint, because users tend to avoid redundant labeling of documents.

$$conf(p, -) = \frac{|\{d|c_1 \in t(d), c_2 \in t(d)\}|}{\sum_{c \in C} |\{d|c_1 \in t(d), c \in t(d)\}| + \sum_{c \in C} |\{d|c_2 \in t(d), c \in t(d)\}|} \quad (3)$$

where $t(d)$ is the set of del.icio.us *tags* associated with document d. The normalized number of co-occurrences of c_1 and c_2 (their respective labels to be precise) as del.icio.us tags aims at capturing the degree of association between the two classes.

2.2 Subsumption

If one class is a subclass of the other we assume the two classes of a pair $p = (c_1, c_2)$ to be **not** disjoint with a confidence equal to the likeliness associated with the `subclass-of` relationship (cf. Section 2.1).

$$conf(p, -) = \max(r_{subclass-of}(c_1, c_2), r_{subclass-of}(c_2, c_1)) \quad (4)$$

2.3 Semantic Similarity

The assumption that a direct correspondence between the semantic similarity of two classes indicates their likeliness to be disjoint led to the development of three further methods: The first one implements the similarity measure described by [24] to compute the semantic similarity sim of two classes c_1 and c_2 with respect to **WordNet** [6]:

$$conf(p, -) = sim(s_1, s_2) = \frac{2 * depth(lcs(s_1, s_2))}{depth(s_1) + depth(s_2)} \quad (5)$$

where $s_i = first(c_i)$ denotes the first sense of c_i, $i \in \{1, 2\}$ with respect to WordNet, and $lcs(s_1, s_2)$ is the least common subsumer of s_1 and s_2. The depth of a node n in WordNet is recursively defined as follows: $depth(root) = 1$, $depth(child(n)) = depth(n) + 1$.

The second method measures the distance of c_1 and c_2 with respect to the given background **ontology** (see Section 3.1) by computing the minimum length of a path q of `subclass-of` relationships connecting c_1 and c_2.

$$conf(p, +) = \min_{p \in paths(c_1, c_2)} length(q) \quad (6)$$

And finally, the third method computes the similarity of c_1 and c_2 based on their **lexical context**. Along with the ideas described in [5] we exploit Harris' distributional hypothesis [10] which claims that two words are semantically similar to the extent to which they share syntactic contexts.

For each occurrence of a class label in a corpus of textual documents (see prelimaries of this section) we consider all the lemmatized tokens in the same sentence (except for stop words) as potential features in the context vector of the corresponding class. After the context vectors for both classes have been constructed, we assign weights to all features using a modified version of the tf-idf formula:

Let $v_i = (f_1^i ... f_n^i)$ be the context vector of class c_i where each f_j^i, $n \geq 1$ is the frequency of token j in the context of c_i. Then we define $TF(f_j^i) = \sum_{d \in doc(c_i)} freq(f_j^i, d)$ and $N = |doc(c_i)|$ and $DF = |doc(c_i) \cap doc(f_j^i)|$, where $doc(t)$ is the set of documents containing term t and $freq(t, d)$ is the frequency of term t in document d. And finally, we get $TFIDF(f_j^i) = TF(f_j^i) \cdot \log(\frac{N}{DF})$.

Given the *weighted* context vectors v_1' and v_2' the confidence in c_1 and c_2 being **not** disjoint is defined as $conf(p, -) = \cos(v_1', v_2')$.

2.4 Patterns

Since we found that disjointness of two classes is often reflected by human language, we defined a number of lexico-syntactic patterns to obtain evidence for disjointness relationships from a given corpus of textual resources. The first type of pattern is based on enumerations as described in [9]. The underlying assumption is similar to the idea described in section 2.1, i.e. terms which are listed separately in an enumeration mostly denote disjoint classes. Therefore, from the sentence

The pigs, cows, horses, ducks, hens and dogs all assemble in the big barn, thinking that they are going to be told about a dream that Old Major had the previous night.

we would conclude that *pig*, *cow*, *horse*, *duck*, *hen* and *dog* are disjoint classes. This is because we believe that – except for some idiomatic expressions it would be rather unusual to enumerate overlapping classes such as *dogs* and *sheep dogs* separately which would result in semantic redundancy. More formally:

Given an enumeration of noun phrases NP_1, NP_2, $\ldots$, $(and|or)$ NP_n we conclude that the concepts c_1, c_2, $\ldots$, c_k denoted by these noun phrases are pairwise disjoint, where the confidence for the disjointness of two concepts is obtained from the number of evidences found for their disjointness in relation to the total number of evidences for the disjointness of these concepts with other concepts.

The second type of pattern is designed to capture more explicit expressions of disjointness in natural language by phrases such as *either* NP_1 *or* NP_2 or *neither* NP_1 *nor* NP_2. For both types of patterns we compute the confidence for the disjointness of two classes c_1 and c_2 as follows:

$$conf(p,+) = \frac{freq(c_1,c_2)}{\sum_{j\neq 1} freq(c_1,c_j) + \sum_{i\neq 2} freq(c_i,c_2)} \tag{7}$$

where $freq(c_i, c_j)$ is the number of patterns providing evidence for the disjointness of c_i and c_j with $0 \leq i, j \leq |C|^2$ and $i \neq j$.

2.5 OntoClean

In [20] we introduced AEON, an approach to automatically evaluate ontologies according to the OntoClean methodology [8]. The basic idea is to use a pattern-based approach on top of the Web (and other textual data sources) for annotating classes of a given ontology with the OntoClean properties such as unity, identity and rigidity. Parts of the approach can be reused for learning disjointness axioms.

Two classes are disjoint if they have incompatible unity or identity criteria. This implies that a class carrying anti-unity (~U) must be disjoint of a class carrying unity (+U) – and similarly for identity. Since we use the same subset of the PROTON ontology as in our AEON experiments, we can rely on the manual OntoClean taggings we collected earlier for the evaluation of AEON.

$$conf(p,+) = \begin{cases} 1 & \text{if } c_1 \text{ tagged with } \phi\Omega,\ c_2 \text{ tagged with } \psi\Omega, \\ & \text{for } \Omega \in \{U, I\},\ \phi,\psi \in \{\sim,+\},\ \phi \neq \psi \\ 0 & \text{otherwise} \end{cases} \tag{8}$$

2.6 Meta Algorithm

The meta algorithm considers superclasses known to be disjoint (from previously computed confidence values) and propagates this information downwards in the taxonomic hierarchy[4]. For $p = (c_1, c_2)$ the confidence for c_1 and c_2 being disjoint is computed as follows:

$$conf(p,+) = \frac{\sum_{p^s}(conf(p^s,+) - conf(p^s,-))}{|super(c_1)| \cdot |super(c_2)|} \tag{9}$$

where $p^s = (c_1^s, c_2^s)$ with $c_i^s \in \{c|subclass - of(c_i, c)\}$ for $i \in \{1, 2\}$ and $subclass - of(c_i, c_j)$ being the `subclass-of` relationship between c_i and c_j. Moreover, $super(c)$ denotes the set of superclasses of c.

[4] This algorithm was not used in the final evaluation, since early experiments indicated that it introduces too much noise. However, we report on it for reasons of completeness. And we still believe that it constitutes a potentially interesting direction of future work, because it allows for integrating subsumption information into any other algorithm.

3 Experiment: Human Annotation of Disjointness

We thoroughly evaluated our approach by performing a comparison of learned disjointness axioms against a large number of manually created ones to calculate (among other things) the degree of overlap. This section describes the generation of the evaluation dataset consisting of 2000 pairs of classes tagged by 30 annotators and discusses methodological aspects related to the manual creation of disjointness axioms. The complete dataset is available from `http://www.aifb.uni-karlsruhe.de/WBS/jvo/data/disjointness-111206.zip`.

3.1 Ontology

As a basis for the creation of the evaluation datasets and as background knowledge for the ontology learning algorithms we took a subset (*system*, *top* and *upper* module) of the freely available PROTON ontology (PROTo ONtology)[5]. In total our subset of PROTON contains 266 classes, 77 object properties, 34 datatype properties and 1388 siblings.

PROTON is a basic upper-level ontology to facilitate the use of background or pre-existing knowledge for automatic metadata generation. PROTON covers the general concepts necessary for a wide range of tasks, including semantic annotation, indexing, and retrieval of documents. The design principles can be summarized as follows (as described in [19]) (i) domain-independence; (ii) light-weight logical definitions; (iii) alignment with popular standards; (iv) good coverage of named entities and concrete domains (i.e. people, organizations, locations, numbers, dates, addresses).

3.2 Evaluation Setting: Manual Taggings

To be able to compare the results of our trained model with the results generated by manual annotation we created a dataset consisting of 2000 pairs of classes as follows: First, we manually selected 200 (potentially) *non-disjoint* pairs from the ontology, since we assumed the set of non-disjoint pairs to constitute a weak minority class (which would have hampered the construction of a good model for our classifier). Then, we randomly chose 500 *siblings* – which constitute a subset of the data, which is of particular interest from a practical and theoretical aspect. And finally, we added another 1300 pairs chosen *randomly* without any selection criteria.

Once the dataset was complete, each pair was randomly assigned to 6 different people – 3 from each of two groups, the first one consisting of PhD students from our institute (all of them professional "ontologists"), the second being composed of undergraduate students without profound knowledge in ontological engineering. Each of the annotators was given between 385 and 406 pairs along with natural language descriptions of the classes whenever those were available. Possible taggings for each pair were $+$ (disjoint), $-$ (not disjoint) and ? (unknown). The result were two datasets A and B for "ontologists" and "students". A third dataset C was created by merging A and B (cf. table 1a). Dataset D is a subset of C consisting of all siblings, whereas E contains all those pairs of classes which were randomly selected.

[5] PROTON is available from `http://proton.semanticweb.org`.

Table 1.

a) Evaluation Datasets

ID	Dataset	Annotators	Tags per Pair	Pairs
A	Experts	15	3	2000
B	Students	15	3	2000
C	All	30	6	2000
D	Siblings	30	6	541
E	Random	30	6	1300

b) Tagged Pairs (Individual)

Dataset	Individual Taggings					
	$+$	$-$	?	all	$-/+$	avg. agree.
A	3849	2007	144	6000	0.521	0.869
B	3881	2106	13	6000	0.543	0.858
$avg.$	3865.0	2056.5	78.5	6000	0.532	0.864
C	7730	4113	157	12000	0.532	0.824
D	1362	1822	62	3246	1.338	0.754
E	6166	1554	80	7800	0.252	0.853

In order to get cleaner and less ambiguous training data for our classification model (see Section 4) we computed the *majority votes* for all the above mentioned datasets by considering the individual taggings for each pair (3 in the case of A and B, and 6 for C). If at least 50% (or 100% respectively) of the human annotators agreed upon $+$ or $-$ this decision was assumed to be the majority vote for that particular pair. In case of equally many positive and negative taggings, the majority vote was defined as ? or *unknown*. These pairs were not used for training purposes. In this way we reduced the noise the classifier had to deal with in the training phase, and obtained a better overall model. Some statistical properties of the majority vote datasets are given by table 2.

3.3 Analysis of Human Annotations

In order to determine how difficult it is for humans to tag pairs of classes as being disjoint or not we measured the human agreement within and across the different subsets of the data. Table 1b shows the average agreement among the individual taggers, i.e. the average maximum ratio of annotators who agreed upon the same tag for a pair of classes. By analysing the figures we find that the average agreement for D is significantly lower than the agreement for any of the other datasets – which seems to imply that pairs of siblings (classes with a common direct superclass) are much more difficult to tag for human annotators than randomly chosen pairs of classes. This might be due to the fact that it is comparably hard to determine the differences between the intension and extension of classes which are semantically very close.

In addition to the computation of the agreement within each of the datasets, we also tried to capture commonalities and differences between the taggings of people from the two groups of annotators – ontologists (A) and students (B).

First, we measured the average agreement of the individual taggings of the experts with the majority vote 100% of the students and vice versa. The figures – 0.852 for the agreement between A and the majority vote of B, and a slightly lower value of 0.834 for the agreement between B and the majority vote of A – indicate that, maybe due to the relatively higher disagreement among the students (see table 1b), those tend to agree mainly on very evident cases of disjointness.

The hypothesis that there is a considerable number of pairs which are comparably easy to tag, thus provoking a high agreement, is supported by the figures we get for the agreement among the majority votes 100% (0.964) and 50% (0.793) of A and B.

Table 2. Tagged Pairs (Majority Vote)

Dataset	Majority Vote 50%					Majority Vote 100%				
	+	−	?	all	−/+	+	−	?	all	−/+
A	1297	649	54	2000	0.500	931	330	739	2000	0.354
B	1346	648	6	2000	0.481	846	307	847	2000	0.363
$avg.$	1321.5	648.5	30.0	2000	0.490	888.5	318.5	793.0	2000	0.359
C	1276	537	187	2000	0.421	616	194	1190	2000	0.315
D	188	274	79	541	1.457	28	96	417	541	3.429
E	1072	140	88	1300	0.131	588	35	677	1300	0.060

Table 3. Differences between Majority Votes 100% of A (Experts) and B (Students)

Vote		Disjoint Classes ?		Vote		Disjoint Classes ?	
A	B			A	B		
−	+	RailroadFacility	Pipeline	+	−	Canal	Harbor
−	+	Order	Abstract	+	−	OfficialPoliticalMeeting	Parliament
−	+	Newspaper	HomePage	+	−	Week	Month
−	+	School	MineSite	+	−	Mountain	Peninsula
−	+	TelecomFacility	Monument	+	−	Island	Valley
−	+	ReligiousLocation	Canal	+	−	Government	Parliament
−	+	InternationalOrganization	StockExchange	+	−	Service	Telecom
−	+	WaterRegion	PoliticalRegion	+	−	Park	Festival
−	+	InternetDomain	EntitySource	+	−	OilField	Province
−	+	ReligiousOrganization	Airline	+	−	Patent	AirplaneModel
−	+	RecreationalFacility	Capital	+	−	Ministry	Location
−	+	City	Archipelago	+	−	Delta	River
−	+	Pipeline	LaunchFacility	+	−	TVCompany	Movie
−	+	AstronomicalObject	Mountain				
−	+	GovernmentOrganization	AmusementPark				
−	+	AmusementPark	Galaxy				
−	+	LaunchFacility	Bridge				

And finally, we completed our analysis of the annotation results by inspecting concrete examples of differently tagged pairs. Table 3 shows the listing of all pairs of classes which were assigned different tags by the majority votes 100% (which means that all 3 annotators of A or B agreed upon each tag) of experts and students. An extensive discussion of the differences which tries to explain some of the problems the human annotators encountered can be found in the following section.

3.4 Discussion

During the creation of the human annotations, we had the chance to study the problems humans face when using disjointness. Even in the taggings of the experts group – consisting of post-graduates all involved in Semantic Web research – the overlap of the taggings was lower than expected (cf. Section 3.3). Table 3 shows all pairs where all experts agreed on one tagging, and all students agreed on the other. Based on an analysis of the taggings and subsequent discussions with the taggers, we identified several types of problems regarding disjointness:

1. The label and comment of a class often do not provide an unambiguous idea of what is meant with this class.

2. Some disjointness axioms may depend on the context: whereas *Dog* and *Livestock* may be disjoint in most parts of Europe, in the Chinese Wordnet[6] the latter is actually a hypernym of the former.
3. Classes can have abstract individuals, like *Money*, *Message* or *Idea*.
4. Often the extension of two classes are disjoint, although their intension is not, e.g. *US President* and *Woman*. Annotators struggle with this difference.
5. Also, the extensions of two classes might be not disjoint, even though their intensions are: although *Weapon* and *Pitchfork* are disjoint intensionally (in the literal sense), their extensions do not need to be.
6. Roles and so called basic classes are often mixed, e.g. the role *Professor* and the *Person* itself that plays the role, which may be defined disjoint (depending on how roles are modeled [11]).
7. Mereological and instantiation relations can be mixed: a *Week* is part of a *Month*, so are these two classes disjoint? What about *Delta* and *River*?
8. Mixing other types of relations with instantiation relations may lead to misunderstandings: see for example the pairs *Movie/TVCompany*, *Government/Parliament*, or *Patent/AirplaneModel*, where the instances have close relations and thus seem to confuse the annotators.
9. Instantiations can occur at different levels of abstraction. E.g., when describing animals, *Eagle* may be the label of both an individual (e.g. of the class *Species*) and of a class itself. Are then the two classes *Species* and *Eagle* disjoint? Note that the individual *Eagle* is not the same as the class *Eagle*, but they may be connected via an axiom like $\mathit{Class{:}Eagle} \equiv \exists \mathit{species}.\{\mathit{Individual{:}Eagle}\}$.
10. Sometimes, lexical information is mixed with ontological one. The PROTON ontology contains concepts like *Alias* that form lexical information. Is a *JobTitle* disjoint from a *Job* or the *Person* having the *Job* or *JobTitle*?

Note that this list does not speak about problems of disjointness with regards to its definition in description logics, but rather with the problems our annotators had when they had to decide if two classes are disjoint or not. Many of the above problem types have a well-defined answer with regards to the formal semantics of disjointness, e.g. #7, where *Week* and *Month* are disjoint as they don't have common instances (since a week consists of seven days, and months consist of around 28-31 days. Note that the definition of week and month can change, but this basically means that we introduce new concepts which may or may not have the same name).

Recognizing the problem type would allow an ontology development environment to offer much more appropriate help than just a general description of the meaning of the disjointness axiom, which can be hard to apply at times.

Often the decision, if two classes are disjoint or not, will uncover underspecified or ambiguous classes, i.e. moot points in the description of one or both classes. Instead of simply adding (or, which is far harder to tract, *not* adding) a disjointness axiom, the rationale behind this decision should also be documented, following an ontology lifecycle methodology like DILIGENT [21] for the continuous evolution and refinement of the ontology.

[6] `http://www.keenage.com/`

4 Evaluation: Learning a Classifier

In this section we present the evaluation procedure and analyse the results of the comparison between the classifier which has been trained on the features described in Section 2 and the sets of manual annotations (see Section 3).

4.1 Experimental Settings

To train the classifier we skipped pairs of classes tagged with ? since the definition of disjointness only distinguishes between disjoint and not disjoint classes. For the rest of the evaluation we will consider this two-class problem. We evaluate our learned classifier against two baseline: the random and majority baseline.

Random Baseline: The idea of the random baseline is to randomly choose the target class of the classifier. As we have a two-class problem we will distribute the pairs equally over the two classes. This will result in a 50% baseline for accuracy as 50% of the $+$ examples will be classified in $+$ which means that these examples are classified correctly. The same holds for the $-$ class.

Majority Baseline: The majority baseline is determined by taking the largest class as default classification. This way, we will get a high accuracy if the classes are unequally distributed. In this case, of course, the majority baseline is much more difficult to beat than the random baseline. Nevertheless, since in the experiments at hand we only have to deal with two classes ($+$ or $-$) which are not equally distributed, the majority baseline should be considered as more realistic than the random baseline.

Classifier settings: In order to be able to classify each pair of classes as being disjoint ($+$) or not ($-$), we trained a classifier based on the manual taggings created by human annotators. The features for the classifier are the confidence values obtained from various sources as described in section 2.

We tested a couple of different classifiers made available by the Weka package[7]. In general, decision trees outperformed all other classifiers – maybe, because of the highly selective character of our features – while the performance of different types of decision trees was more the less comparable. Therefore, we finally chose the *ADTree* classifier [7] with default settings for our experiments which shows very good performance while at the same time providing interpretable results.

First, we performed a 10-fold cross-validation against the majority votes 100% and 50% of the datasets A (ontologists), B (students), C (all) and E (random) (cf. table 1a). The results for the random dataset are included to show the performance of our approach for an unbiased dataset (E contains examples chosen randomly from the set of all possible pairs without any selection criteria). To get the results for dataset D (siblings), we split dataset C into two independent parts - one for evaluation and one for training. The training set for the evaluation with dataset D consists of all manually tagged pairs except for the siblings.

[7] `http://www.cs.waikato.ac.nz/ml/weka/`

4.2 Results

Table 5 and 4 list the results of our evaluation experiments by means of Precision (P), Recall (R), F-Measure (F) and Accuracy (Acc) (for definitions cf. [23]). From the tables it becomes evident that we easily beat the baselines for the datasets A (experts), B (students) and C in both cases majority vote 50% and 100%. With an accuracy of over 90% the performance of our system for dataset C is remarkable, especially in the case of the total majority vote. These results are comparable with the human inter-annotator agreement for experts and students – and even better for dataset C (90.9%) in comparison to the human agreement of 86.4%.

Dataset D, which only contains pairs of siblings, is certainly the most difficult to handle – for the classifier, but also for the human annotators – because, as explained in Section 3.3, siblings are semantically close, so that differences between their intensions and extensions may often be hard to grasp. As dataset D shows a relatively low average agreement compared to the other datasets (cf. table 1b) the classifier seems to have more difficulties to learn it. This is also expressed by the very bad classification accuracy with 37% for majority vote 100%.

An investigation of the learned classifier revealed that the rather important taxonomic feature (see Section 2.2) is not well populated in the siblings part of the dataset. To analyse the influence of this feature we constructed a dataset without this feature. As expected the accuracy for the training dataset drops, whereas for the evaluation set it is improved considerably from 37.9% to 74.2%. Moreover, the results for the majority vote 50% rise to 76.6% which can be interpreted as an indication to the noise insert by this feature.

Our approach seems to work very well also for the random dataset E as we got a better accuracy in both cases. The difference to the majority baseline is much smaller than for A, B, and C but the baseline of around 90% is very difficult to beat. To conclude, the results – not only for the random dataset – are very promising and allow us to setup a competitive classifier to support ontology engineering.

In order to find out which classification features contributed most to the overall performance of the classifier we performed an analysis of our initial feature set with respect to the gain ratio measure [14]. The ranking produced for data set C clearly indicates an exceptionally good performance of the features taxonomic overlap (Section 2.1), similarity based on WordNet and lexical context (Section 2.3), and del.icio.us (Section 2.1). The contribution of other features such as the one presented in Section 2.4 relying on

Table 4. Evaluation against Majority Vote 50% (ADTree)

Dataset	P			R			F			Acc	Acc_{random}	$Acc_{majority}$
	+	−	avg.	+	−	avg.	+	−	avg.			
A	0.815	0.638	0.727	0.823	0.626	0.725	0.819	0.632	0.726	0.757	0.500	0.666
B	0.807	0.642	0.725	0.844	0.580	0.712	0.825	0.609	0.717	0.758	0.500	0.675
$avg.$	0.811	0.640	0.726	0.834	0.603	0.719	0.822	0.621	0.722	0.758	0.500	0.671
C	0.854	0.682	0.768	0.874	0.644	0.759	0.864	0.663	0.764	0.806	0.500	0.704
D	0.558	0.628	0.593	0.255	0.861	0.558	0.350	0.726	0.538	0.615	0.500	0.593
E	0.910	0.761	0.836	0.990	0.250	0.620	0.948	0.376	0.662	0.904	0.500	0.884

Table 5. Evaluation against Majority Vote 100% (ADTree)

Dataset	P			R			F			Acc	Acc_{random}	$Acc_{majority}$
	+	−	avg.	+	−	avg.	+	−	avg.			
A	0.896	0.720	0.808	0.903	0.703	0.803	0.899	0.712	0.806	0.851	0.500	0.738
B	0.866	0.790	0.828	0.942	0.599	0.771	0.903	0.681	0.792	0.851	0.500	0.734
$avg.$	0.881	0.755	0.818	0.923	0.651	0.787	0.901	0.697	0.799	0.851	0.500	0.736
C	0.934	0.823	0.879	0.946	0.789	0.868	0.940	0.805	0.873	0.909	0.500	0.760
D	0.237	0.806	0.522	0.786	0.260	0.523	0.364	0.394	0.379	0.379	0.500	0.774
E	0.977	0.955	0.966	0.998	0.600	0.799	0.987	0.737	0.862	0.976	0.500	0.944

lexico-syntactic patterns seems to be less substantial. However as the classification accuracy tested on every single feature is always below the overall performance the combination of all features is necessary to achieve a very good overall result.

5 Related Work

Several ontology learning frameworks have been designed and implemented in the last decade. The Mo'K workbench [1], for instance, basically relies on unsupervised machine learning methods to induce concept hierarchies from text collections. In particular, the framework focuses on agglomerative clustering techniques and allows ontology engineers to easily experiment with different parameters. OntoLT [3] is an ontology learning plug-in for the Protégé ontology editor. It is targeted at end users and heavily relies on linguistic analysis, i.e. it makes use of the internal structure of noun phrases to derive ontological knowledge from texts. JATKE[8] is a Protégé based unified platform for ontology learning which allows for inclusion of modules for ontology learning. The OntoLearn framework [13] mainly focuses on the problem of word sense disambiguation, i.e. of finding the correct sense of a word with respect to a general ontology or lexical database. TextToOnto [12] is a framework implementing a variety of algorithms for diverse ontology learning subtasks. In particular, it implements diverse relevance measures for term extraction, different algorithms for taxonomy construction as well as techniques for learning relations between concepts. The recent RelExt approach [18] focusses on the extraction of triples, i.e. classes connected by a relation. None of the mentioned approaches deals with disjointness.

6 Conclusion and Future Work

Learning of disjointness axioms is an intuitive and useful extension of existing ontology learning frameworks. We have motivated the need for richter ontologies which include disjointness axioms and presented an approach consisting of a number methods to extract expressive feature for learning disjointness from different sources of evidence. In a thorough evaluation our learning approach behaved competitive to human annotators.

[8] `http://jatke.opendfki.de/`

As a by-product we captured lessons learned from human annotators with respect to their difficulties when modeling disjointness axioms.

Future work includes a combination with ontology evaluation approaches for richly axiomatized ontologies such as [17]. Moreover, we want to integrate the novel methods into the Text2Onto [4] framework for ontology learning from texts.

Acknowledgments. Research reported in this paper has been partially financed by the EU in the IST project SEKT (IST-2003-506826) (`http://www.sekt-project.com`). We would like to thank our colleagues, especially Peter Haase, for fruitful discussions, and all the students and colleagues at the AIFB and the University of Kassel for providing us with more than 10,000 taggings.

References

1. G. Bisson, C. Nedellec, and L. Canamero. Designing clustering methods for ontology building - The Mo'K workbench. In *Proc. of the ECAI Ontology Learning Workshop*, pages 13–19, 2000.
2. E. P. Bontas, C. Tempich, and Y. Sure. ONTOCOM: A cost estimation model for ontology engineering. In I. Cruz et al., editors, *Proceedings of the 5th International Semantic Web Conference (ISWC 2006)*, volume 4273 of *LNCS*, pages 625–639. Springer-Verlag Berlin Heidelberg, 2006.
3. P. Buitelaar, D. Olejnik, and M. Sintek. OntoLT: A protégé plug-in for ontology extraction from text. In *Proc. of the 2nd Int. Semantic Web Conference (ISWC2003)*, 2003.
4. P. Cimiano and J. Völker. Text2onto – a framework for ontology learning and data-driven change discovery. In *Proc. of the 10th Int.l Conf. on Applications of Natural Language to Information Systems (NLDB'05)*, June 2005.
5. P. Cimiano and J. Völker. Towards large-scale, open-domain and ontology-based named entity classification. In G. Angelova, K. Bontcheva, R. Mitkov, and N. Nicolov, editors, *Proc. of the International Conference on Recent Advances in Natural Language Processing (RANLP)*, pages 166–172, Borovets, Bulgaria, September 2005. INCOMA Ltd.
6. C. Fellbaum. *WordNet, an electronic lexical database*. MIT Press, 1998.
7. Y. Freund and L. Mason. The alternating decision tree learning algorithm. In *ICML*, pages 124–133, 1999.
8. N. Guarino and C. A. Welty. A formal ontology of properties. In *Knowledge Acquisition, Modeling and Management*, pages 97–112, 2000.
9. P. Haase and J. Völker. Ontology learning and reasoning - dealing with uncertainty and inconsistency. In *Proc. of the Workshop on Uncertainty Reasoning for the Semantic Web (URSW)*, pages 45–55, 2005.
10. Z. Harris. Distributional structure. In J. Katz, editor, *The Philosophy of Linguistics*, pages 26–47, New York, 1985. Oxford University Press.
11. K. Kozaki, E. Sunagawa, Y. Kitamura, and R. Mizoguchi. Fundamental considerations of role concepts for ontology evaluation. In *Proc. of the Workshop EON – Evaluation of Ontologies for the Web*, 2006.
12. A. Maedche and S. Staab. Ontology learning for the semantic web. *IEEE IS*, 16(2), 2001.
13. R. Navigli, P. Velardi, A. Cucchiarelli, and F. Neri. Extending and enriching WordNet with OntoLearn. In *Proc. of the GWC 2004*, pages 279–284, 2004.
14. J. R. Quinlan. *C4.5 Programs for Machine Learning*. Morgan Kaufmann, California, 1993.

15. A. Rector, N. Drummond, M. Horridge, J. Rogers, H. Knublauch, R. Stevens, H. Wang, and C. Wroe. OWL pizzas: Practical experience of teaching OWL-DL – common errors & common patterns. In *Proc. of EKAW 2004*, pages 63–81, 2004.
16. T. Rose, M. Stevenson, and M. Whitehead. The reuters corpus volume 1-from yesterdays news to tomorrows language resources. *Proc. of the Third International Conference on Language Resources and Evaluation*, pages 29–31, 2002.
17. S. Schlobach. Debugging and semantic clarification by pinpointing. In *Proc. of the 2nd European Semantic Web Conference (ESWC2005)*, volume 3532 of *LNCS*, pages 226–240. Springer, 2005.
18. A. Schutz and P. Buitelaar. RelExt: A tool for relation extraction in ontology extension. In *Proc. of the 4th International Semantic Web Conference (ISWC2005)*, 2005.
19. I. Terziev, A. Kiryakov, and D. Manov. Base upper-level ontology (BULO) guidance. SEKT deliverable 1.8.1, Ontotext Lab, Sirma AI EAD (Ltd.), 2004.
20. J. Völker, D. Vrandecic, and Y. Sure. Automatic evaluation of ontologies (AEON). In *Proc. of the 4th International Semantic Web Conference (ISWC2005)*, volume 3729 of *LNCS*, pages 716–731. Springer, 2005.
21. D. Vrandečić, H. S. Pinto, Y. Sure, and C. Tempich. The DILIGENT knowledge processes. *Journal of Knowledge Management*, 9(5):85–96, 2005.
22. T. D. Wang. Gauging ontologies and schemas by numbers. In *Proc. of the Workshop EON – Evaluation of Ontologies for the Web*, 2006.
23. I. H. Witten and E. Frank. *Data Mining: Practical Machine Learning Tools and Techniques*. Morgan Kaufmann Series in Data Management Sys. Morgan Kaufmann, 2nd edition, June 2005.
24. Z. Wu and M. Palmer. Verb semantics and lexical selection. In *32nd. Annual Meeting of the Ass. for Computational Linguistics*, pages 133–138, New Mexico, 1994.

Developing Ontologies for Collaborative Engineering in Mechatronics

Violeta Damjanović, Wernher Behrendt, Manuela Plößnig, and Merlin Holzapfel

Salzburg Research, Jakob Haringer Strasse 5/II,
5020 Salzburg, Austria
{violeta.damjanovic, wernher.behrendt, manuela.ploessnig, merlin.holzapfel}@salzburgresearch.at
http://www.salzburgresearch.at

Abstract. Creating a coherent set of ontologies to support a collaborative design process amongst different firms which develop mechatronic products is a challenge due to the semantic heterogeneity of the underlying domain models and the amount of domain knowledge that needs to be covered. We tackle the problem of semantic heterogeneity by employing the DOLCE foundational ontology and by aligning our models to it. We approach the problem of scale, i.e. the amount of knowledge modeled by keeping the models at a descriptive level which is still granular enough to connect them with domain and task specific engineering tools. In order to manage the complexity of the modeling task we separate the models into the foundational layer, the mechatronic layer consisting of three domain ontologies, one process model and one cross-domain model, and the collaborative application layer. For the development process, we employ a methodology for dynamic ontology creation, which moves from taxonomical structures to formal models.

1 Introduction

The mechatronic engineering process covers an interdisciplinary combination of different domains comprising of mechanical engineering, electrical engineering, and software engineering. For each of these engineering domains there exist diverse knowledge models, mostly in the form of documents or glossaries, but hardly as comprehensive ontologies. Furthermore each domain covers a specific mechatronic field, so that the intersection of knowledge models between these different engineering domains remains relatively small.

The focus of the ImportNET[1] project lies in this intersection, specifically in the collaboration of the three mainstream mechatronic domains, i.e. mechanical, electrical, and software engineering. For this reason, the process of ontology modelling in ImportNET is considered from two perspectives:

Firstly - in the research perspective - we employ a methodology for dynamic creation of ontologies (i.e. moving from less formalised models to more rigorous models).

---

[1] The ImportNET project is co-funded by the European Commission within the Sixth Framework Programme under Contract 033610, in the area of ICT for Networked Businesses.

E. Franconi, M. Kifer, and W. May (Eds.): ESWC 2007, LNCS 4519, pp. 190–204, 2007.

Using the DynamOnt methodology [1] we model the reference ontologies for mechatronics on the basis of the DOLCE foundational ontology. This includes a generic mechatronic process model which will also be used to describe the usage scenarios for the modeling.

The second perspective concerns the actual use of ImportNET tools: there needs to be a methodology to modify the reference ontologies in order to adapt them to the requirements of concrete companies and their products. The reference ontologies must be tailored to the requirements of the actual, planned collaboration. This will be done by the Ontology Integration Tool (OIT) which allows to modify and to expand the reference ontologies.

The paper describes the early stages of work in a European research project and addresses collaborative design processes that are used for development of mechatronic products as follows: The introduction section gives a brief overview of the state of the art and existing research gaps in collaborative engineering, mechatronic engineering and mechatronic domain modeling. Section 2 introduces the ImportNET approach to mechatronic domain modeling. This section firstly discusses the ontology landscape in ImportNET, and then illustrates the alignment of the mechatronic ontology with the DOLCE foundational ontology. Furthermore the DynamOnt methodology, which is used for the evolutionary creation and development of the mechatronic ontology, is explained in more detail. Section 3 discusses the main objectives of ImportNET, the possible system architectures, as well as usage and validation scenarios. Preliminary conclusions are drawn in Section 4.

1.1 State of the Art and Research Gaps in Collaborative Engineering

In recent years, collaboration not only between engineers but also across organisational boundaries has become a key research issue for the development of flexible engineering processes. Collaborative engineering aims at providing the main concepts, solutions, as well as technologies for development of products by multiple engineering teams. We found the following main research challenges and gaps in the domain of collaborative engineering:

- technical aspects: Web-based electronic design environments; architectures and technologies for knowledge sharing; standards for exchange formats/protocols; security aspects;
- social aspects: handling multi-cultural issues in collaborative design; knowledge sharing; collaborative learning; collaborative engineering; distributed engineering work; social aspects of collaboration teams;
- organizational and economic aspects: benefits of using collaboration approaches; validation scenarios.

At the same time, there is a number of unsolved problems from the industrial perspective, including application integration e.g. how can Web Services contribute to closing this gap?; knowledge integration e.g. how can Semantic Web technologies contribute?; and process integration e.g. how can approaches like Enterprise Modeling answer to this challenge?

1.2 State of the Art and Research Challenges in Mechatronic Engineering

Mechatronic engineering is one of the most recent branches of engineering and it has increasing impact on many sectors of the economy and on society overall. The competitive use of mechatronic engineering will soon require more model-driven development using design repositories of mechatronic components. We have found two notable metamodels which address this issue: Thramboulidis describes a four-layer model of Integrated Mechatronics distinguishing mechanical, resource, application and mechatronic layers [2]. The model is the basis for "Archimedes, a system platform that supports the engineering through a methodology, a framework and a set of tools to automate the development process of agile mechatronic manufacturing systems" [2]. The problem of ontological modelling was addressed by Yoshioka [3] in a layered knowledge structure for the Knowledge Intensive Engineering Framework (KIEF). They also introduce the concept of plug-in models to specialise and refine the metamodel into concrete models. Their paper indicates that there is at least a proof-of-concept prototype in which some of the proposed concepts are validated. Unfortunately, the actual implementation is not in the public domain. Each of the two frameworks has a particular angle: Thramboulidis focuses on the mechatronic process whereas Yoshioka emphasises the modelling of mechatronic artefacts. Both models will have to be considered as frameworks for our collaboration-centered approach to mechatronics.

1.3 State of the Art and Research Challenges in Mechatronic Domain Modeling

Ontological engineering covers a whole range of topics such as the basic philo sophical and metaphysical issues as well as knowledge representation formalisms, me thodology for ontology development, business process modelling, commonsense knowledge, systematisation of domain knowledge, Internet information retrieval, standardisation, evaluation, and many more [4].

If we put ontological engineering in the context of other disciplines, then many similarities and analogies arise. They allow us to make connections between ontological engineering and the other disciplines, to bridge potential comprehension gaps, and to shed a different light on already known concepts and practices. For example, when applying the Unified Modeling Language (UML) to a mechatronic system it turns out that some additional concepts are needed to model the mechatronic system [5]. Such concepts can be added by introducing stereotypes, e.g. a special stereotype called the Function Block Adapter (FBA) is described in [5]. The FBA stereotype can be used to specify the mapping from UML signals to the function block signals.

2 ImportNET Approach to Mechatronic Domain Modeling

A review of the literature about mechatronics rapidly results in a number of definitions, each of which emphasises a slightly different aspect of the mechatronics concept, ranging from design to precision engineering and from sensors to actuators [6]. Most of the definitions do manage to agree that mechatronics is concerned with the integration of its core engineering themes to generate novel technological solutions in the form of products and systems whose functionality is integrated across those core technologies.

The design of an ontology for mechatronics can be approached using a variety of scientific methods, such as the following paradigms [7]:

- empirically-based research (cognitive models),
- axiom-based research (computational models); and
- conjecture-based research (computational models):
  - conjectures based on an analogy with cognitive processes; and
  - conjectures based on an analogy with computational processes.

Empirically-based research involves the development of experimental studies of designers that result in cognitive models of designing. Axiom-based research involves the identification of a set of axioms and their consequences to derive a logic-based computational model of designing. Conjecture-based research involves an analogy between a cognitive or computational process that leads to a computational model specific to designing.

The approach taken by ImportNET is to move from a cognitive model to a computational model, with the help of a foundational ontology which could be seen as a compromise between cognitive conjectures (the concepts of the ontology) and axiom-based computational models (the axiomatic framework defined by the DOLCE foundational ontology).

2.1 Ontology Landscape in ImportNET

Ontologies provide the vocabulary for referring to the terms in a subject area, as well as the logical statements that describe what the terms mean, how they are related to each other, as well as the rules for combining terms and relations to define extensions to the vocabulary.

Figure 1 provides a landscape of reference ontologies employed in ImportNet. The DOLCE ontology represents the foundational layer which gives us a useful structure for building novel knowledge based architectures. Aligned to DOLCE, we place the domain ontologies for mechanical, electronics and software engineering. The new cross-domain engineering ontology is built as a result of the integration of these contributing ontologies, while the mechatronic engineering lifecycle ontology has to be linked ultimately, to distributed service execution and orchestration processes.

The ImportNET ontologies are created in support of a collaborative engineering process for developing mechatronic products. The process for development of the mechatronic products requires some ontology integration and configuration based on the overall set of ontologies. The ontology landscape and its configuration via the Ontology Integration Tool (OIT) is shown in Figure 1. The resulting Collaboration Ontology is a meaningful subset of concepts from the ontology landscape. Any collaboration between organisations developing a specific product will be based on such a specialised collaboration ontology. The design and detailed functionality of OIT are outside the scope of this paper.

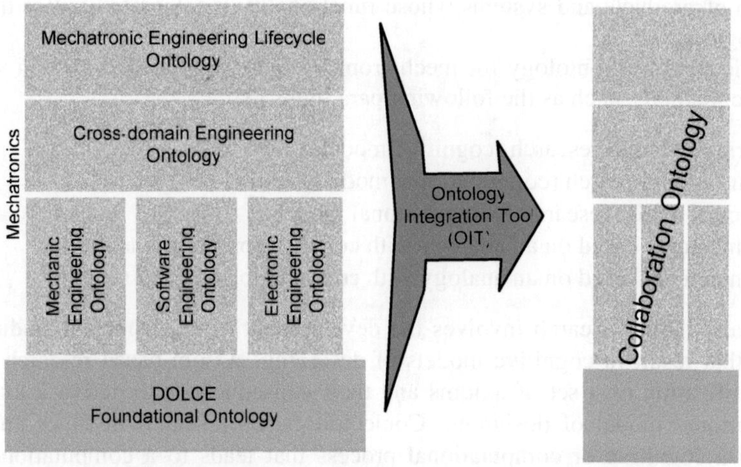

Fig. 1. Ontology landscape and configuration of a collaboration ontology

The design of the cross-domain engineering ontology is considered to be an essential theme for mechatronics since it attempts to bring together concepts and ideas in relation to a product or system [6]. Furthermore, the design of a flexible mechatronic engineering lifecycle ontology to support the collaborative development of mechatronic products amongst various communities of practice and virtual organizations is the main challenge in ImportNET.

A partial taxonomy of mechatronic ontologies is represented in Tables 1-5.

Table 1. A partial taxonomy of the mechanical engineering ontology

Criteria	Explanation
Spatial	Spatial description of mechanical components
Composing	Aggregation / assembly
Properties	Physical properties, e.g. liquid
Process	- domain specific workflow - mechanical behaviour of components, e.g. rotation or movement along a trajectory
Role	Roles of agents in the domain of mechanical engineering, e.g. material stress tester
Methods	Methods of mechanical engineering

Table 2. A partial taxonomy of the electronic engineering ontology

Criteria	Explanation
Spatial	Spatial description of electronic components
Composing	Aggregation / assembly
Properties	Physical properties
Process	- Domain specific workflow - Electro magnetic behaviour
Role	Roles of agents in the domain of electronic engineering
Methods	Methods of electronic engineering

Table 3. A partial taxonomy of the software engineering ontology

Criteria	Explanation
Functions	Architecture of the runtime environment, hardware drivers
Composing	Aggregation / assembly
Properties	Description of design, documentation, code, APIs...
Process	- Software life cycle - Behaviour of software components
Role	Roles of agents in the domain of software engineering
Methods	Methods of software engineering

Table 4. A partial taxonomy of the mechatronic engineering lifecycle (process) ontology

Criteria	Explanation
Composing	Sub-processes at different levels of granularity, requiring input/output parameters to be modeled at corresponding levels of detail
Properties	Characterising different instantiations of a process model (e.g. waterfall, V-model, etc), order of sub-processes, duration, pre- and postconditions
Role	Roles of agents in particular those engaged in coordinating and resolving conflicts between the engineering domains
Methods	E.g. conflict resolution between roles

Table 5. A partial taxonomy of the Cross-domain Ontology

Criteria	Explanation
Spatial	Runtime environment, hardware drivers
Composing	Aggregation / assembly
Process	Electro magnetic behaviour, software/ hardware execution...
Role	Union of roles defined in the other domains
Methods	E.g. conflict resolution between roles

Current work is addressing the relationship between the initial taxonomies and the frameworks proposed by Thramboulidis [2] and Yoshioka [3]. One of the main issues in combining the knowledge of these other models with ImportNET is that once we have made a commitment to a foundational ontology we need to also align external models to that foundation. For example, each mereological element of an external ontology needs to be mapped into the corresponding primitives of the foundational ontology. Whether or not there is a specific ontological bias in any of the external models can only be determined once we have access to the full models.

2.2 Ontology Alignment to the DOLCE Foundational Ontology

The Descriptive Ontology for Linguistic and Cognitive Engineering (DOLCE) was originally developed in the EU WonderWeb project [8] and has been extended in a number of other projects since then. The design philosophy of DOLCE is modularity in order for ontology projects to be able to pick and choose thus making only as much "ontological commitment" as needed. The typical process of developing an ontology is then to either "align" existing knowledge models to the DOLCE model or to develop the ontology from scratch, by using the conceptual primitives defined by DOLCE.

Despite the ambition to capture some "common sense" DOLCE constitutes a strictly formal approach to ontology modeling, which is a necessary condition if we want computational services or agents to make autonomous use of the ImportNET knowledge models while remaining "accountable" for their activities. Such semantic accountability is an important requirement for future work spaces where some of the decision making in cross-organisational processes will be delegated to machines and where there will be a need at least for boundary conditions to be defined explicitly in order to safeguard against unwanted behaviour of partly autonomous systems. Furthermore, the axiomatisation is a prerequisite for any logic based inferencing done by such machines.

The fundamental difference between current "semantic" terminologies as used in annotations and "proper" semantic models as envisaged for the ImportNET Semantic Application Server (SAS) is that the latter will have to implement a partly autonomous inference module in order to manage the cross-organisational work processes, which will be context-sensitive to the mechatronic design artefacts which will be exchanged between the engineers (i.e. the users of the system).

Since ImportNET will focus on cross-organisational processes it will be necessary to add the capability for modeling tasks to the basic model. The process of aligning the set of mechatronic ontologies to DOLCE is shown in Figure 2. In the recent EU METOKIS project, DOLCE was extended by an ontology called "Descriptions and Situations" (D&S), which includes a representation language for tasks or processes [9]. D&S shows its practical value when applied to ontology design patterns for (re)structuring application ontologies that require contextualization [10]. Figure 3 represents the process of aligning the mechatronic ontologies with respect to the basic categories of DOLCE, as well as using of Semantic Web Services approach to support the collaborative design process.

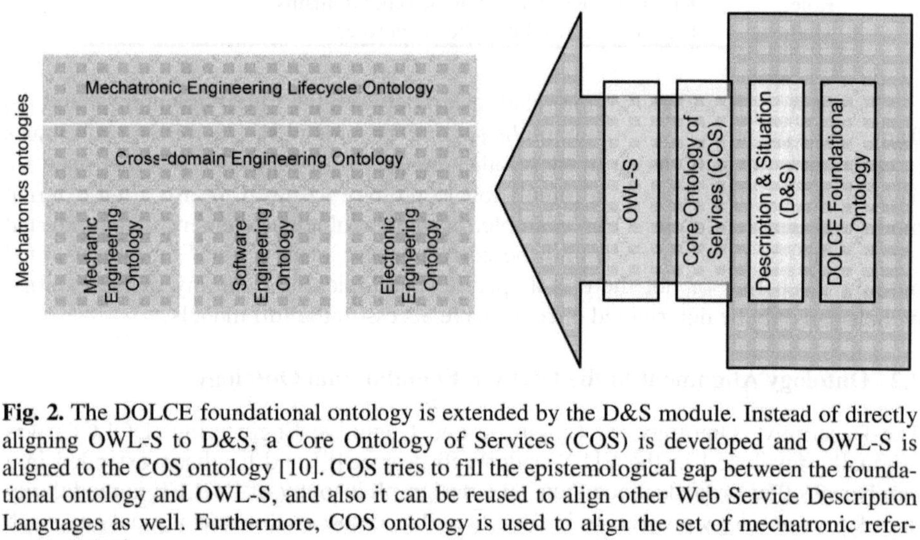

Fig. 2. The DOLCE foundational ontology is extended by the D&S module. Instead of directly aligning OWL-S to D&S, a Core Ontology of Services (COS) is developed and OWL-S is aligned to the COS ontology [10]. COS tries to fill the epistemological gap between the foundational ontology and OWL-S, and also it can be reused to align other Web Service Description Languages as well. Furthermore, COS ontology is used to align the set of mechatronic reference ontologies.

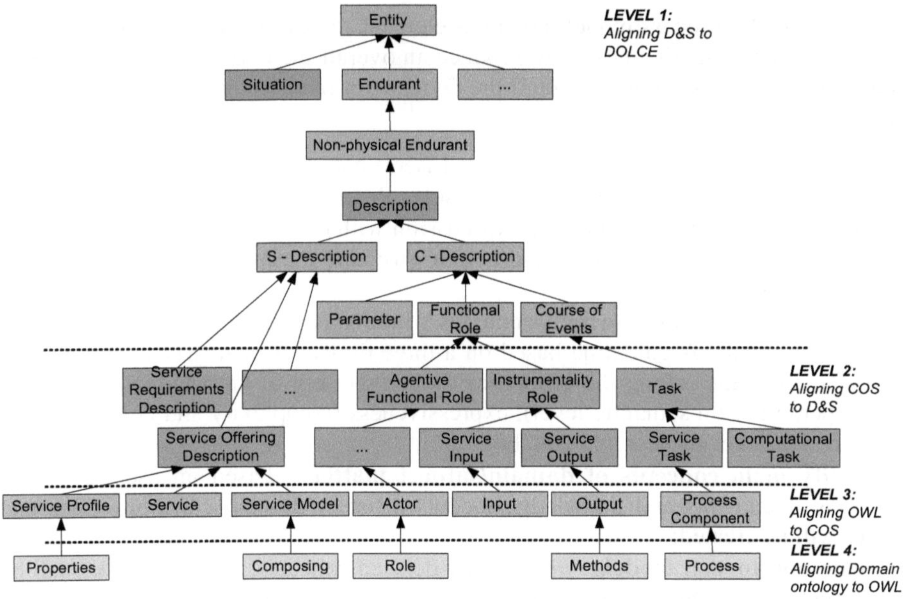

Fig. 3. Indirectly aligning mechatronic ontologies to the DOLCE foundational ontology

To summarise, the extended DOLCE foundational ontology for which a full implementation in OWL-DL exists has been chosen as the working hypothesis from which the modeling of the ImportNET ontologies start. Achieving such a combined representation in the area of mechatronics would be a significant result because to our knowledge, no other foundational model has a comparable degree of coherence and formalisation.

2.3 Methodology for the Development of the Mechatronic Ontology

Ontology development methodologies are intended to help with the complex process of ontology building and managing. They help knowledge engineering projects to successfully reach the main goals in time, especially when it comes to knowledge sharing in dynamic environments due to frequent changes of user needs.

There are two general ontology engineering approaches, *centralized* and *decentralized* methodologies. On-To-Knowledge (OTK) [11] and METHONTOLOGY [11], [12] are mostly centralized, while DILIGENT [13] and the recently proposed DynamOnt [1] methodology can be seen as decentralized and distributed approaches for ontology engineering where a community of ontology users and developers converges towards a shared view.

For the development of the contributing ImportNET ontologies, we use the DynamOnt methodology. The DynamOnt methodology enables the dynamic creation of ontologies based on communication and experience exchange amongst different communities of practice - in our specific case those communities which are concerned with the development of mechatronic products.

The DynamOnt process model integrates elements of known knowledge and ontology-engineering methods in order to produce an overall methodology for engineering of knowledge-based systems. In detail the DynamOnt model comprises the following phases [1]:

- **Identify the problem** – domain experts (users) could describe the situation when the problem occurs or they have ideas to solve the problem;
- **Structure the problem** – a broader discussion with domain experts (users) and the description of user scenarios would help to structure the problem in order to get a broader view of the topic;
- **Identify concrete purpose and scenarios** – the focus is a mutual understanding of the project goals. A guideline based on a three dimensional matrix is proposed to classify ontologies along the properties scope (stability of knowledge models and interoperability on semantic level), expressiveness (complexity and costs), and acceptance (market success and collaboration);
- **Identify main concepts of domain/subject matter** – based on user scenarios, existing documents and knowledge models a list of domain concepts, roles and tasks will be created;
- **Create non-formal models** – the already defined concepts, roles and tasks will be interrelated through attributes and relations. This will be supported by guided questions;
- **Create formal models (knowledge design)** – the classification according to the expressiveness dimension of the three dimensional matrix helps to decide which parts of the ontologies has to be formalised to a certain degree. Based on the non-formal model and maybe other available models, a conceptual (formal) model will be defined and the output will be machine readable (e.g. OWL, RDFS, XML);
- **Create acceptance (community design)** - the acceptance within the main user communities (e.g. developers, the domain experts, external user communities of the system) is an important factor for the success of the model and the system. The acceptance can be raised by trainings (e.g. workshops) and by adapting existing business processes according to inputs of the resulting formal model;
- **Create system (software design)** – based on software engineering methods and techniques the software will be specified and designed;
- **Implement Target System** - the scope of this phase is to provide a fully developed knowledge-driven application.

In the formalisation steps, DynamOnt uses the following ontological design patterns (based on DOLCE) to guide domain experts in creating conceptualisations of their domain knowledge [1]:

- the Participation pattern;
- the Description-Situation pattern;
- the Role-Task pattern;
- the Design-Artefact pattern;
- the Agent-Activities pattern;
- the Information-Object pattern.

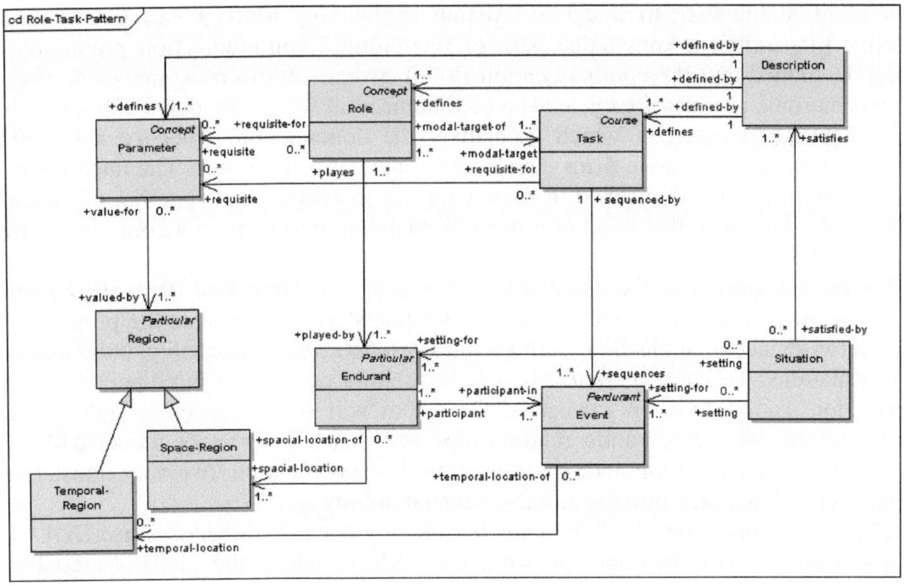

Fig. 4. Role-Task ontological design pattern defined in DynamOnt using DOLCE concepts

Each of these patterns acts as a modelling template to describe how agents in various roles, participate in situations and use information objects for communication. The use of these patterns is similar to the design patterns in object oriented programming and it should lead to a more homogeneous way of modelling intelligent agents, roles and activities in any environment. Figure 4 shows the Role-Task pattern, which is defined in DynamOnt methodology by using the DOLCE concepts.

One of the early lessons of our ontology work is that a common set of knowledge engineering methods would be desirable. Methontology appears to be the most straightforward approach to semiformal modelling, but lacks the foundational rigour of DOLCE which is better supported by the ontological patterns of DynamOnt. Methontology on the other hand, offers good ways to express axioms and rules which are absolutely necessary constructs for designing real-world semantic applications.

3 Objectives, ImportNET System Architecture, Usage and Validation Scenario

The ImportNET project is addressing on the one hand, the issue of creating a support environment for virtual enterprises in cross-domain engineering and on the other hand the problem of cultural differences and misunderstandings which may lead to communication failures between engineers who try to collaborate with each other.

3.1 Objectives and Initial Findings

The technical approach is to first create the collaboration environment by integrating the knowledge models of the three engineering domains and by creating a layer of supporting middleware to integrate existing engineering tools (CAE, CAD/CAM, CASE). In parallel, a knowledge base of intercultural communication problems is being developed and the communication flow between engineers is analyzed, along the mechatronic product life cycle. The communication will be modeled explicitly, in the collaboration ontology which specialises the domain ontologies for a specific collaboration between some firms developing some defined product. The intercultural knowledge base will be indexed in such a way as to enable the triggering of "warnings" when there is a likelihood of a misunderstanding occurring in a communication act along the lifecycle.

For the integration of the engineering tools into a collaborative lifecycle support environment it will be necessary to create "wrappers" which translate the proprietary or otherwise incompatible data formats into semantically comparable intermediate representations. To automate some of this translation process an Intelligent Adapter Generation Tool (IAGT) is envisaged. This tool will use compiler-compiler techniques to specify the semantic relationships between a proprietary model and the intermediate representation and to create from this specification, two-way translators which can be integrated into the collaboration environment.

The integration of the three domains has already been described: we use DOLCE as a foundational ontology and specialize the D&S module to the needs of modeling processes in cross-domain engineering. In order to make it easier for organisations as well as for technology integrators, to specify a workflow for a new collaboration, we make use of the OIT. This tool will offer semantic templates ("ontological patterns") to the integrator, which can be specialised for the needs of a new collaborative engineering project.

3.2 ImportNET System Architecture and Issues Around Semantic Modelling

The system comprises of a knowledge based back-end called SAS (Semantic Application Server), and a client front-end application called MDET (Multi Domain Engineering Tool). The MDET offers different engineers their preferred view of the overall system and it mediates potential misunderstandings by being aware of the communication acts between the participants of the collaboration. One of the roles of MDET will be to mediate between mechanical engineering views (which are typically 3D) and electronic views (normally 2D). The challenge lies in merging the internal representations of external engineering design tools (eCAD and mCAD) into a common one with uniform semantics. This resolution will be done in the SAS with the help of the tool adapters (i.e. semantic wrappers) constructed with the help of the IAGT. The SAS plays the role of a semantics-based middleware which connects the external tools to the ImportNET communication and collaboration processes. We have identified three issues that such a system needs to address: a) the role of inference support; b) the need for semantic web services; c) the degree of interoperation between current engineering tools.

Role of inference support: current CAD tools are based on object-oriented, often proprietary database back-ends. Similarly, even most of the open research systems in the field of engineering are based on object-oriented data models. Any semantic interoperation approach is faced with the dilemma that one has to either replicate the data in a Semantic-Web enabled knowledge base in order to use inference engines or, to reimplement some inferencing capability on top of the existing OO datastore. This is a general problem facing Semantic Web applications when they need to interoperate with software in the commercial domain.

The need for semantic web services in ImportNET: the implementation architecture of ImportNET could be envisioned as an open, yet collaborative lifecycle support environment in which different Semantic Web Services can find each other automatically. This kind of ImportNET system architecture could be based on the Web Service Execution Environment (WSMX) core architecture, which enables discovery, selection, mediation, invocation and interoperation of Semantic Web Services [14]. However, it is not yet clear whether this kind of spontaneous semantic service integration is really needed for ImportNET, because the philosophy behind the system is a planned collaboration between known organisations and systems.

Degree of interoperation between current engineering tools: we see a major hurdle for the envisaged system still, in the complex yet proprietary solutions that are currently prevalent in engineering domains. This necessitates firstly, the approach of building an external semantic application server with its associated problem of inference engines versus object-model. Secondly, it also bears the danger of "research at a dead end" because we cannot research semantically interoperable models when the actual target application software is *designed to hinder or defeat*, interoperation, for reasons of market protection. One such example is that object structures are based on OIDs which are generated afresh each time a design is loaded and there are only limited ways of exchanging typed structural (schematic) information between different tools. This leads to a need for effectively reverse-engineering some of the functionalities of the target tools which is neither a worthwhile research question nor strictly legal in some cases. There is, however, an interesting side effect to this issue: semantic modelling points directly at methods by which commercial players are trying to protect their intellectual property and market share. The legal system may one day employ semantic modelling to determine what kinds of protection are fair and which methods of protection are detrimental to a competitive market.

3.3 Usage and Validation Scenario for Collaborative Mechatronic Design

As described above mechatronic engineering deals with collaboration across different domains. Each of these engineering domains is well supported by a range of engineering tools which cover at most the domain itself, but typically focus on a specific aspect e.g. the design of physical artefacts or the specification of automated tests for an electronic device. The focus of ImportNET and consequently of the case studies in ImportNET is on the design phases of the mechatronic lifecycle and the cross domain cooperations. The mechatronic life cycle coordinates the different tasks of the mechatronic engineering domains and the engineering tasks of one domain often influence the engineering tasks of another domain. As a result, the precise hand-over points of these tasks are sometimes not clear and coordination conflicts may occur. Against that background several aspects have to be studied and validated through use cases:

- During design many documents (e.g. output of CAD tools) need to be exchanged between the mechanic and the electronic engineering domains. Most of the documents are in a proprietary format and are therefore not easily imported by other tools. Based on known exchange formats such as DXF[2] and STEP[3] ImportNET analyses where data can be automatically exchanged during cross-domain collaboration.

– Designing a mechatronic product involves engineering experts from different domains and conflicts can occur for several reasons. Often conflicts have simply a factual basis where e.g. a mechanic and an electronic engineer have to clarify technical issues. The mechatronic design process comprises in these cases the coordination of cross-domain issues with respect to spatial, temporal or causal relationships. The coordination between mechanical and electronic engineering can be very intricate because of interactions in space and in behaviour (e.g. thermal or electromagnetic dependencies).
– Companies are often from different countries and conflicts can also be caused by different cultural backgrounds (e.g. different time conceptualisations or communications habits). This may lead to misunderstandings when messages or behaviours are being interpreted in different ways.

ImportNET is developing two use cases where engineering experts (mechanic, electronic, software, testing) from different companies and different countries are involved. The basis for the description of the use cases is a general mechatronic life-cycle model which will be tailored firstly to the needs of the participating companies and secondly to requirements of the target mechatronic product which will be designed between these companies.

4 Conclusions

Creating a cross-domain engineering environment requires - irrespective of whether one uses a Semantic Web based approach or not - some understanding of the underlying domains and also an understanding of the maturity of the field. In the case of mechatronics, we found a mixed situation: each of the domains has relatively mature software tools for the design of new artefacts and the domain of manufacturing overall, has relatively mature standards such as STEP for the description of products. What is clearly missing is the integration of the design tools along the product life cycle and in the case of cross-domain engineering, the ability to transform the representations of one design tool into semantically equivalent representations for the perspective of a corresponding tool in the other engineering domain. Initial interviews with senior engineers revealed that up to a third of the development cost originates in the area of testing and that there is large scope for improvement in this phase of the product life cycle.

A first analysis of candidate ontologies revealed a good number of conceptual models not only at varying levels of generalisation but also with varying angles on the purpose of the system and hence, the choice of concepts.

There are at least three challenges in the project: defining a coherent set of partial ontologies, integrating a knowledge base of intercultural communication conflicts into the workflow model and integrating a Semantic Web Service architecture with the

---

[2] Drawing Exchange Format.

[3] STandard for the Exchange of Product model data.

process model of the mechatronic domain to ensure interoperation during the mechatronic design phase. In this paper, we have described our approach and methodological choices with respect to the development of the ontologies and we have outlined the implementation architecture for the case of Semantic Web Services. We have not addressed the integration of the intercultural issues yet. Another issue which is still to be addressed concerns the suitability of DOLCE as a foundational ontology for domains such as mechanical engineering and electronics. The current view is that DOLCE is a good choice as long as we do not need to engage in "deep modelling" i.e. the modeling of causes and effects or of constraints in physical systems. However, this begs the question whether current Semantic Web modelling is capable of integrating well, with any models that offer "analogue", i.e. numerical or function-based simulation of system behaviour.

Acknowledgments. This work has been supported by the ImportNET project (IST-2006-033610). We would like to thank the referees for their helpful comments. We also acknowledge fruitful discussions with our colleagues Miklós Szőts, Alexander Mahl, Oliver Hornberg, Ron Jamieson and Rupert Westenthaler.

References

1. DynamOnt Deliverable 201: First specification of Methodology and Workbench for Dynamic Ontology Creation (2006)
2. Thramboulidis, K.: Model Integrated Mechatronics – Towards a new Paradigm in the Development of Manufacturing Systems. IEEE Transactions on Industrial Informatics, Vol. 1, No. 1 (2005) 54-61
3. Yoshioka M.: Physical Concept Ontology for the Knowledge Intensive Engineering Framework. In Advanced Engineering Informatics, Vol. 18, No. 2 (2004) 95-113
4. Devedzić, V.: Understading Ontological Engineering. Communications of the ACM, Vol. 45, No. 4 (2002) 136–144
5. Heverhagen, T., Tracht, R.: Using Stereotypes of the Unified Modeling Language in Mechatronic Systems. In Proceedings of the First International Conference on Information Technology in Mechatronics, ITM'01, Istanbul, Turkey (2001) 333–338
6. Bradley, D.: What is Mechatronic and Why Teach It? International Journal of Engineering Education (2004) Online: http://findarticles.com/p/articles/mi_qa3792/is_200410/ai_n10298146/g_1 (Last access: 2007-03-13)
7. Gero, J.S., Maher, M.L.: A Framework for Research in Design Computing. In Martens, B., Linzer, H., and Voigt, A. (eds), ECAADE'97, Osterreichischer Kunst und Kulturverlag, Vienna (1997) Online: http://people.arch.usyd.edu.au/~john/ publications/1997/ecaade/ index.html (Last Access: 2007-03-13)
8. Masolo, C., Borgo, S., Gangemi, A., Guarino, N.: Ontology Library. WonderWeb Deliverable D18 (2004) Online: http://wonderweb.semanticweb.org/deliverables/documents/ D18.pdf (Last access: 2006-10-23)
9. Gangemi, A., Borgo, S., Catenacci, C., Lehmann, J.: Task Taxonomies for Knowledge Content, METOKIS Deliverable D07 (2004)
10. Mika, P., Oberle, D., Gangemi, A., Sabou, M.: Foundations for Service Ontologies: Aligning OWL-S to DOLCE, In Proceedings of the 13th International Conference on World Wide Web 2004 (2004) 563-572

11. Nagypál, G.: Methodology for Building SWS Ontologies in DIP. DIP Deliverable D3.11 (2005)
12. Fernández M., Gómez-Pérez, A., Juristo, N.: METHONTOLOGY: From Ontological Art towards Ontological Engineering. In Proceedings of the AAAI Spring Symp. Series, AAAI Press, Menlo Park, California (1997) 33-40
13. Sofia Pinto, H., Staab, S., Tempich, C.: DILIGENT: Towards a Fine-Grained Methodology for Distributed, Loosely-controlled and Evolving Engineering of Ontologies. In Proceedings of the ECAI 2004 (2004) 393-397
14. Zaremba, M., Moran, M., Haselwanter, T., Lee, H-K.: WSMX Architecture, WSMX Deliverable D13.4 (2005)

Media, Politics and the Semantic Web
An Experience Report in Advanced RDF Usage*

Wouter van Atteveldt, Stefan Schlobach, and Frank van Harmelen

Department of Artificial Intelligence
Free University Amsterdam (The Netherlands)
De Boelelaan 1071, 1071 HV Amsterdam
{wva,Frank.van.Harmelen}@cs.vu.nl, schlobac@few.vu.nl

Abstract. The media play an important role in the functioning of our society. This role is extensively studied by Communication Scientists, requiring a systematic analysis of media content. The methods developed in this field utilize complex data models and background knowledge. This data is generally represented ad hoc, making it difficult to analyze, combine and share data sets.

In this paper we present our work on formalizing this representation using RDF(S). We discuss the requirements for a good representation, highlighting a number of non-trivial modeling decisions. We conclude with a description of the resulting system and the benefits for a recent investigation of the 2006 Dutch parliamentary campaign. This case study shows concrete improvements for annotating, querying, and analyzing data, but also indicates a number of aspects that were more difficult to model in RDF(S), contributing to the discussion on modeling with and improving RDF(S) and associated tools.

1 Introduction

The media play an important role in our society. Citizen access to information about world events is almost exclusively mediated, making the press a vital part of our democracy. This underscores the need for the systematic study of the media done by Communication Science [1]. *Relational Content Analysis (RCA)* conducts this systematic analysis by representing news content as a graph linking the relevant actors and issues, which can then be used as input for further analysis [2]. This representation is currently mostly informal forcing answers to complex queries to be composed in a procedural way. Moreover, there are differences in the information captured by various RCA methods and in the vocabulary used for existing data sets. These aspects make it difficult to combine and reuse data.

In this paper we describe a recent case-study on how current Semantic Web technology can help to overcome these limitations. Since 1994, Communication

* The authors would like to thank Mark van Assem for his insightful contributions to the discussions leading to this paper and for his comments on the final version. We also thank the reviewers for their thorough reading and useful suggestions.

E. Franconi, M. Kifer, and W. May (Eds.): ESWC 2007, LNCS 4519, pp. 205–219, 2007.

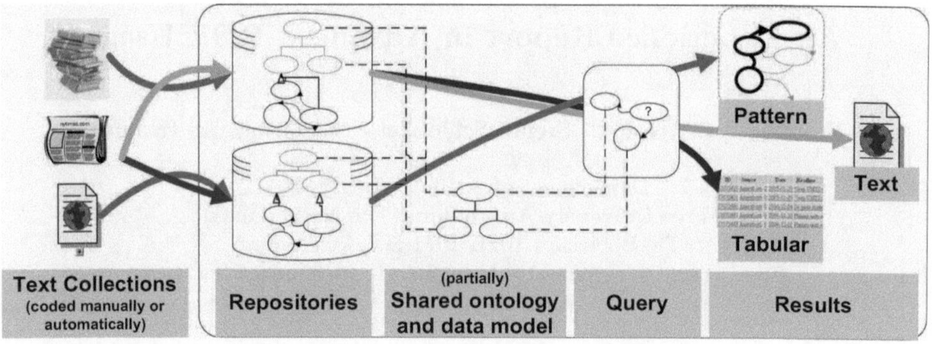

Fig. 1. A Framework for Querying Heterogeneous News Sources

Scientists at the Free University and the University of Amsterdam have conducted an extensive study of the influence of the media coverage of Dutch electoral campaigns on public opinion, most recently in Nov. 2006 [3]. These studies were based on the *NET method*, an RCA framework introduced in [4].

As visualized in Fig. 1, in our framework relational content data is formalized in a (partially) shared data model and vocabulary based on RDF and RDFS, and stored in Sesame repositories. Standard RDF query languages (such as SeRQL or SPARQL) can be used to express queries using the least common denominator of the models and vocabularies of the used data sets. The results of these queries can be used as input for quantitative analysis, to retrieve the original articles for qualitative review, and visualized within the graphs that represent these articles.

This paper describes a use case, and provides a detailed experience report, of an intricate data analysis in a highly complex domain, with many non-trivial modeling decisions. It discusses the literature on a number of aspects where the representational and inferential requirements stretch the possibilities of the current standards. This case study serves both to underscore the use of Semantic Web technology to practically formalize a complex domain and to point out a number of issues on which there is still room for improvement.

In Sect. 2, we will briefly describe Relational Content Analysis and the NET method, and give a list of desiderata for a respresentation and inference mechanism for this domain. Section 3 is the main section of this paper, containing an overview of the existing options and work in progress on each point, and describes the modeling choices made to satisfy the requirements. Section 4 outlines the system that was created and the benefits that it has brought. +

2 Content Analysis as the Domain of Formalization

One of the goals of this paper is to find a suitable framework for formalizing the data produced by Relational Content Analysis (RCA). This section will outline what Content Analysis and RCA are, and then describe a particular method, the NET method, in more detail. Finally, a number of requirements for formalizing this method will be listed, which will serve to guide the discussion in the next section.

2.1 (Relational) Content Analysis

Content Analysis is a social science method to analyze textual content by determining the occurrence of social scientific phenomena [1]. These phenomena are generally complex and subjective in nature, making the extraction a difficult task to automate (but see [5]). For this reason Content Analysis often uses human coders to read the text and directly indicate the presence of these phenomena.

Relational Content Analysis works by identifying smaller concepts, such as individual actors and issues, and coding the relations between these concepts as a graph [2]. The social scientific concepts are subsequently extracted as patterns or metrics defined over this graph. This two-step approach makes the data less dependent on the specific research question, creating a greater potential for sharing and reusing data.

Realizing this potential is difficult because we are dealing with data sets with heterogeneous data models and vocabularies. A formal representation that allows us to standardize both syntax and semantics of these data collections while remaining flexible enough to allow for different methods would be of great value in building the large data sets needed for statistically analyzing the complicated interaction between media and politics.

2.2 Relational Content Analysis Using the NET Method

The NET method [4] is the Relational Content Analysis method used in our case study. It has a fairly complex data model compared to other Relational Content Analysis methods [6]. Moreover, it includes a set of rules to make inferences about triples such as a form of transitivity. Furthermore, normal practice in NET is to annotate using very detailed concepts (such as 'Balkenende') and then aggregate to more general concepts (such as 'Politician') using a taxonomy.

In NET, sentences are coded as <*subject*, *predicate*, *object*> triples. The *subject* and *object* are drawn from a predefined hierarchy of concepts. The predicate is complex, consisting of a *type*, and *quality*. The *type* indicates the kind of sentence code; possible types include 'causative', 'action', and 'affinitive.' *Quality* is a number that indicates the strength and direction of association between subject and object and ranges from -1 to 1.

Additionally, each triple can be augmented with two pieces of information. An *angle* can be specified for some statements, which captures the reason of a disagreement or action in sentences such as "Blair and Brown disagreed *about Iraq.* " Also, some sentences in a newspaper are quoted or paraphrased sources: "Blair stated that it was certain Saddam had WMD". In such sentences, the optional *quoted source* argument captures the source of the statement.

As an example, consider the newspaper excerpt in Fig. 2. The headline is coded as a reciprocal negative relation between the political blocks Left and Right. The first sentence of the lead is more complicated: The main message is that incumbent prime minister Balkenende (CDA / Christian Democrats) and

Hard confrontation Right and Left

The champions for the premiership, Labor leader Bos and Christian Democrat leader Balkenende, attacked each other over poverty and health care. Bos is needlessly scaring people, according to the prime minister. [..] Bos: "Good health care costs money, so we should invest more."

	Source	Subject	Relation	Object	Angle
1		Left	$\xleftrightarrow{-1\ \text{affinitive}}$	Right	
2		Bos	$\xleftrightarrow{-.7\ \text{affinitive}}$	Balkenende	Poverty
2		Bos	$\xleftrightarrow{-.7\ \text{affinitive}}$	Balkenende	Healthcare
3	Balkenende:	Bos	$\xrightarrow{-.7\ \text{acting}}$	Citizens	
4		Bos: Invest in Health	$\xrightarrow{+.5\ \text{causative}}$	Healthcare	
4		Bos	$\xrightarrow{+1\ \text{affinitive}}$	Invest in Health	

Source: De Telegraaf, 22 November 2006 (tr.auth). Reading: sentence 3 means that according to Balkenende, Bos is acting against the good of the citizens

Fig. 2. Example article with NET annotation

the challenger Wouter Bos (PvdA / Labor Party) are fighting, but it is also stated what they are fighting about: the issues Poverty and Health Care. In the next sentence, Balkenende states that Bos is scaring people, which is coded as Bos acting against the Dutch citizens with Balkenende as source. The final sentence expresses two relations: according to Bos, investing more money would be good for the Health Care, and Bos wants to invest money in Health Care, here coded as an affinity (issue position) relation between Bos and Health Care Investments.

Currently, NET-encoding of sentences is performed manually; research is performed to automate this but due to the complex nature of the information this has yet to lead to satisfying accuracy [7]. Given the cost of manual annotation, this difficulty in extraction only underscores the need to share and reuse existing data.

2.3 Requirements

This section will list a number of aspects of NET that we need to be able to capture in a formalization framework.

Representational Requirements. Relational Content Analysis methods use triples as the main primitive, but these triples are enriched in various ways.

R1: Background Knowledge A central feature of the NET method is that data is coded at a very detailed level, and aggregated to higher level theoretical concepts. This aggregation requires background knowledge, for example party memberships (Bos member-of Labor), political functions (Balkenende leads Cabinet), and is-a relations (FreeHealthCare is-a HealthIssue). This needs to be encoded and the concrete annotations need to be expressed in terms of this background knowledge. Moreover, although the taxonomies currently used in NET are purely hierarchical it would be useful if this could be relaxed. For example, Balkenende is both a member of the CDA and the prime minister, and depending on the research question we want to use either fact for the aggregation.

R2: Statement types NET and other Relational Content Analysis methods use qualitatively different relations. For example, the statement "Bos and Balkenende attacked each other" is affinitive while "Good health care costs money" is causative. In Social Networks terminology these are called multiplex networks [8].

R3: Quantitative value In addition to different statement types, Relational Content Analysis often includes a quantitative indicator of the strength and direction of a relation. The statement ".. we should invest more" from the example is positive (+0.7) while the statement "Hard confrontation Left and Right" is strongly negative (-1). In Social Networks terms, graphs labeled with values are called signed and/or valued networks.

R4: Article Metadata To trace the evidence for an analysis and for time based analyses it is necessary to attach metadata to annotations, including publisher and publishing date, location in the newspaper, and a link to the original article.

R5: Extra Arguments Sometimes we need to code certain additional aspects of a relation. For example, in the sentence ".. attacked each other over poverty", we want to capture the topic of the disgreement as well as the fact that they disagree.

R6: Quoted Sources The example sentence "Bos: 'good health care costs money, so we should invest more" contains a positive causal relation between Investing and Health Care, but this relation is not directly stated by the newspaper but rather by a quoted source. In order to analyze such sentences correctly, it is necessary that the contained triples are accessible for analysis, but they should be kept separate from the main graph.

Usage Requirements. The requirements above all specify what kind of information we need to be able to represent. Next to these requirements there are also a number of things we need to be able to do, mainly while analyzing the annotated media material. These 'usage' requirements are outlined below.

R7: Shareability One of the purposes for formalizing NET is to make it easier to share and combine data sets. Therefore, it has to be possible to combine data sets that differ both in exact data structure and in the used vocabulary.

R8: Time-bound roles In Content Analysis, the social and political roles played by actors are generally considered background knowledge and remain static during a project. However, social roles are dynamic and especially for longitudonal analyses we need to be able to represent the temporal validity of political roles

R9: Disjoint Categorization Often, we want to aggregate the nodes in our media data to higher-level categories. These categorizations generally have to be disjoint and exhaustive. This is necessary to avoid counting one instance twice and is also assumed by many statistical analyses. Therefore, we want to be able to express disjoint categorizations and to check or prove that a categorization scheme is disjoint and exhaustive given the structure of the background knowledge.

R10: Extraction As stated above, it is often useful to categorize the nodes in the relational network into higher level categories. Therefore, it is necessary to have a formalization that allows extracting the network data using such categorizations.

3 Formalizing the NET Method

As described above, the NET method is a Relational Content Analysis (RCA) method with a fairly complex data model and usages. The relational nature of NET and the need to combine and share data sets with different structure and vocabulary make Semantic Web technologies a logical choice for formalizing this domain. This section will describe the modeling choices we made to meet the requirements listed in the section above.

3.1 R1-2, R7: Low Hanging Fruit

Due to their relational nature, Semantic Web technologies seem a natural match for the formalization of RCA. This was confirmed by a number of requirements that were fulfilled easily and elegantly.

Background Knowledge (R1) Background knowledge can be expressed elegantly using RDF(S), using either *rdfs:SubClassOf* statements or custom vocabulary. RDF(S) places no restriction on the types and amount of relations between concepts, allowing for multiple inheritance and different relation types.

Statement types (R2) In RDF, the predicative part of triples consists of an RDF resource that can be described in the same way as other resources. This means that it is natural to express multiplex networks in RDF.

Shareability (R7) RDF(S) does not solve the conceptual and substantive problems of combining heterogeneous data sets. It does, however, remove a number of technical difficulties. Globally unique names using URI's minimize vocabulary clashes while the subclass and subproperty mechanisms facilitate mapping specific vocabulary onto more general terms.

3.2 R3-6: Enriching Triples with Extra Information

Requirements R3-6 boil down to a single wish: enriching triples by adding extra information. This is difficult, as RDF is meant to describe resources, not triples: triples do not have URIs and hence cannot be part of triples. We are not the first to signal this difficulty: [9] cites the need for enriching triples to describe event data, and a number of authors want to use RDF for describing RDF documents, for example for reasoning about provenance and trust [10].

RDF(S) allows some form of adding information to existing triples. Trivially, we can replace each of the nodes in a triple by a node carrying more information and point to the original node. In RDFS, it is possible to do so transparently by making the new node a subclass or subproperty of the original node. Additionally, the RDFS specification includes a reification mechanism [11]. Essentially, an anonymous instance is made to represent the statement, and standardized vocabulary is used to define the subject, object, and predicate of the relation. The anonymous instance, being a first class citizen, can then be used in further statements. According to the definition, a reified statement does not necessarily imply the original statement: it is describing a hypothetical event.

Another solution is using the n-ary relation design patterns described in [12]. This is similar to reification in that a new node is created that represents the relationship, but the reification vocabulary is eschewed since "in n-ary relations [..] additional arguments in the relation do not usually characterize the statement but rather provide additional information about the relation instance itself" [12]. This has the same disadvantage as reification (the original triple semantics are lost) but additionally it has no formal meaning or standardized vocabulary.

To overcome these problems, a number of authors have suggested extending the notion of a triple to include a fourth place, often seen as a context marker [9,10,13,14,15]. For example, [14] propose a context mechanism that explicitly assumes the context marker to indicate provenance and they include a complicated system of lifting and aggregating mechanisms to combine RDF documents from different sources. On the other extreme, [9,13] support replacing triples by quadruples without restricting the interpretation of such triples.

A proposal that seems to be gaining ground is Named Graphs [10]. This proposal also adds a fourth places to the triple and defines the semantics of this added element but does not prescribe the interpretation in the way [14] does. Named Graphs semantics allow for nested graphs and they propose a predicate for indicating nesting. The main disadvantage of this method is that it is not standardized, leaving tool support and declarative semantics to be desired. Also, as the intended meaning of the context is the containing graph, Named Graphs add extra information to the whole statement rather than to the predicate much like reification does.

The proposals for adding information to triples in the literature are diverse. Part of the reason for this diversity is that the problem they are trying to solve is diverse. We think that there are two main factors on which the proposed solutions diverge: the *meaning* of the extra information with respect to the original triple; and the *opacity* of the enrichment. In terms of meaning, we distinguish four possible relations of the new information x to the existing triple Rab:

- $\boldsymbol{R^x ab}$: Adding information about the predicate of the triple;
- $\boldsymbol{Ra^x b}$: Adding information about the subject or object of the triple;
- $\boldsymbol{(Rab)^x}$: Adding information about the whole triple; and
- $\boldsymbol{Rabx}$: Adding an extra argument to the triple on equal footing with the subject and object.

In terms of opacity, we distinguish between transparent and opaque additions:

- **Transparant** additions preserve the original meaning of the triple in the graph, meaning that applications that do not interpret the richer relation can still see the original relation; while
- **Opaque** additions remove the original triple from the graph, meaning that it will not be visible to an application that does not (or cannot) interpret the enrichment technique.

Depending on the modeling requirements, we want to add certain information to a triple in a certain way. For example, a quoted source in a newspaper should be an opaque statement about the whole triple, while quality should be a transparant addition to the predicate. Thus, rather than looking for a single 'correct' solution we think that multiple options are needed to express these differences in enrichment. Table 1 categorizes the discussed proposals in these terms and serves as the basis for making the appropriate modeling choices. The proposal by [14] is left out of this table because its purpose is strictly describing graphs rather than enriching triples

Table 1. Suitability of discussed mechanisms for expressing different triple enrichments

	Transparent				Opaque			
	Enriching an argument Ra^xb	Enriching the predicate R^xab	Enriching the triple $(Rab)^x$	Extra argument $Rabx$	Enriching an argument Ra^xb	Enriching the predicate R^xab	Enriching the triple $(Rab)^x$	Extra argument $Rabx$
RDF					$\pm^1$	+		
RDFS	+	$\pm^1$			$\pm^1$	+		
N-ary							$\pm^2$	+
Reification							+	$\pm^2$
Quadruples			$\pm^3$	$\pm^3$				
Named Graphs			+				+	

[1]Adding a dicrete categorization is possible, but adding quantitative information is very difficult. [2]N-ary patterns are explicitly intended to express an extra argument to a statement, while reification is intended to express information about a statement, making other use of these solutions difficult to interpret. [3]Since there is no specified interpretation of the extra argument it is not possible to distinguish between these two cases.

We will now reconsider the requirements from the previous section in terms of Table Table 1. As listed in the previous sections, the basic unit of information is a triple representing a media relation (eg. Bos dislikes Balkenende). To this triple we add information to *quantify* (R3) the predicate, add *extra arguments* (R5) to the relation, *specify the source of a quoted statement* (R6), and link the media statement to *metadata* (R4) such as publisher and publishing date. As stated above, quoted sources should be opaque as the quoted statement is not directly asserted by the newspaper. The other additions should all be transparent: the original triple is a valid part of the graph with or without the extra information. The *quantification* is an enrichment of the predicate, but very difficult to represent using subproperties because of the quantitative and unrestricted nature of the information. The *extra arguments* and *quoted source* both add extra arguments that are subordinate to the main triple, falling somewhere between the intended meaning of reification (statements about triples) and n-ary relations (multiple arguments of equal weights). The *metadata* is adding information to the whole triple, and fits in the use case of reification and named graphs.

Surveying the table above, there is no perfect method for adding information to triples. Named graphs have the desired transparency but offer no solution for distinguishing between extra arguments and metadata. Quadruples allow extra arguments in a natural way but this comes at the expense of flexibility and semantic clarity. Within the existing standards, reification covers adding metadata and a case could be made for using reification to represent additional arguments. N-ary relations are better suited for the additional arguments but suffer from the lack of a standard vocabulary. It is possible to mix and match mechanisms, but this comes at the expense of increasing complexity and if multiple non-standard mechanisms are mixed it will be difficult for third parties to understand what we mean.

Solution. For the current application, we decided to stick to one representation for all enrichments. Since tool support for the proposed extensions is still limited, and the intended meaning of our enrichment is closer to meta-statements than to adding arguments, we decided to use RDFS reification.

3.3 R8: Dynamic Roles

Social Roles, such as being a party member or fulfilling a political function, are a complex topic that has received extensive attention in the literature [16,17,18,19, 20]. As described by [16], two defining characteristics of roles are that they are anti-rigid[1], and dynamic[2]. In this definition, background knowledge such as party membership and political office can be classified as knowledge on role memberships of actors.

[17] surveys a large body of literature and notes that there are three main approaches to representing roles. The first approach is calling the places in a predicate roles, i.e. in a predicate $memberOf(member, group)$ the roles *member* and *group* are implied. This corresponds to creating a simple RDF relation between the member and the group. Using this mechanism, it is impossible to represent temporal constraints on roles, and such statements should be considered snapshots of a dynamic relation rather than descriptions.

The second approach is to make the role a subtype of the natural type corresponding to it. This means that playing the role of being a PvdA member means creating a subclass of politician, the PvdAMemberPolitican. As described in [17], this leads to a number of complications and does not solve the representation problem inherent in the first approach.

The third approach is creating an *adjunct instance* representing the relation, which is an instance of the role type but unique for each instantiation. Since this promotes the role membership to first class citizen, it allows for further specification such as temporal aspects. In RDF, this can be described as a (blank) node representing the membership, with relations to the two role players and the role type, which is also the approach taken by in [21]. Interestingly, if the RDF reification vocabulary is used to denote the integral aspects of the role, this is equivalent to reifying a simple statement expressing the relation directly.

Solution. In the terminology introduced in section 3.2, we want the enrichment of the original $memberOf$ predicate to be opaque since the roles are invalid outside their (temporal) context. This makes creating *adjunct instances* by using reification a natural choice for representing social roles.

---

[1] The role players do not depend on their playing the role for their existence.

[2] Roles change over time and there is no 1-on-1 relation between roles and players.

3.4 R9: Disjoint Categorization

As described above, adding background knowledge and using this knowledge to link 'data level concepts' to 'theory level concepts' can be done elegantly using RDF. A frequent use case in Content Analysis is to define a set of categories on the media data, for example statements with an opposition politician as subject, with a coalition politician as subject, and statements with a societal actor as subject. Counts of such statements per period are then used either in statistical analysis or presented in a table. Both uses require the categories to be disjoint and exhaustive with respect to the higher category, in this case 'actor statements'. In other words, the higher category should be *partitioned* by the proposed categories.

Using model checking (e.g. SPARQL queries), it is trivial to *check* whether a categorization, expressed as a set of requirements, is a partitioning given a concrete data set. By pair-wisely conjoining the requirements disjointness can be checked, and by negating the whole conjunction exhaustiveness can be checked.

In some cases, such as presenting data real-time on a web page, we would like to be able to *prove* that such a categorization will always be a partitioning. In RDF this is impossible due to the fact that cardinality, disjointness, and negation cannot be asserted, so it is impossible to express the constraint that a politician is a member of exactly one party or that societal actors are all non-political actors. In OWL these restrictions can be expressed, and proving disjointness boils down to proving that each pairwise conjunction of the categories is unsatisfiable. Exhaustiveness can be shown by proving that the higher category implies membership of one of the lower categories. More formally, proving that the categories $\{A_1 \ldots A_n\}$ partition B in the ontology $\mathcal{O}$ means proving $\mathcal{O} \models A_i \sqcap A_j = \bot$ for all $i \neq j$, $i, j \leq n$ and $\mathcal{O} \models B \sqsubseteq A_1 \sqcup \ldots \sqcup A_n$.

Solution. For the current application, we decided to stay within RDF and only use query-based model checking of the disjointness.

3.5 R10: Categorizing and Extracting Data

As described above, it is useful to define categorization schemes and aggregate the media data to a higher level using such schemes. A scheme will generally consist of a high level category and all its instances and parts and members of these instances. However, as described in Sect. 3.3, these part-of and member-of relations will often be dynamic and represented using adjunct instances. Therefore, we need to check whether a role is actually valid at the publishing date of the article. Moreover, as described in Sect. 3.4, it is often necessary to include negations in the definition of categorization to prevent actors with multiple roles from being counted twice.

This leads to a complex definition for these categories. Since they will often include negation, they cannot be represented in RDF(S). It would be possible to represent them in OWL, but this requires complex concrete domain reasoning for the date comparisons. Practically, it is possible to conduct such categorizations using closed-world model checking in RDF, for example using a SeRQL query.

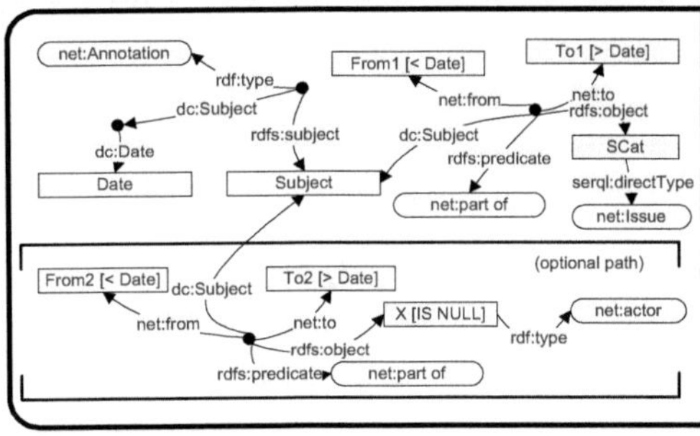


```
Select PublishDate, Subject, SCat FROM
{} rdf:type       {net:Annotation};
   rdfs:Subject {Subject};
   dc:subject   {} dc:date {Date},
{} rdf:type       {net:partOf};
   rdfs:Subject {Subject};
   net:from     {From1}; net:to {To1};
   rdfs:Object  {SCat}
         serql:directType {net:Issue}
[{} rdf:type      {net:partOf};
   rdfs:Subject {Subject};
   net:from     {From2}; net:to {To2};
   rdfs:Object  {X}
          serql:directType {net:Actor}
   WHERE From2 < Date
     AND To2 > Date]
WHERE From1 < Date
  AND To1 > Date
  AND X IS NULL
```

Fig. 3. The (partial) extraction query, represented visually and in SeRQL

This results in a query such as shown in Fig. 3, where the Subject of a statement is categorized as an issue but only if it is not categorizable as an Actor. To create the real query, this has to be duplicated for subject and object and a UNION query has to be created joining all category definitions.

Solution. For the applications described below we used SeRQL queries to extract data, using query rewriting to hide some of the query complexity from the user. If we describe the categorizations in OWL it should be possible to automatically rewrite the OWL definitions into RDF queries (assuming a closed world), or cache the categorization results from the DL reasoner.

4 Implementation

The sections above outlined the challenges encountered while modeling the Relational Content Analysis domain and the possible solution for these challenges. This section will briefly describe the actual systems that were built around the RDF representation, especially the annotation tool, the browser/visualizer and the extractor.

4.1 Data Model and Ontology

This section will describe the data model that resulted from the choices described in Sect. 3. Fig. 4 visualizes this model. The main element of the data model, the original triple of the relational method, is now a reified triple, making the annotation (a subclass of `rdf:Statement`) the central element. Annotations have a subject, predicate and object as required for reification, and also have the quantitative value ‘connection’ and an angle. On the left hand side, annotations are connected to textual units (sentences) from an article using the `dc:subject`, and metadata about this article are recorded. Additionally, the coder, the creator of the annotation, is recorded.

On the right hand side the subject, object, and predicate are all drawn from the ontology, having `net:entity` as its base. This ontology contains an IS-A hierarchy of (political) actors and issues together with role information such as party

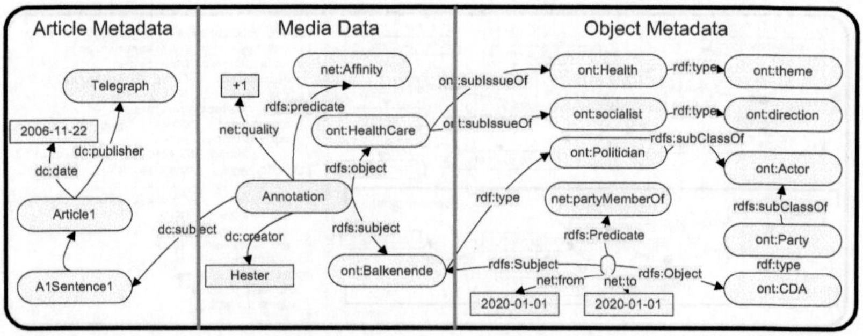

Fig. 4. The data model used

membership. As described in Sect. 3.3, these roles are made dynamic by reifying the role membership statement, creating an adjunct instance, and adding from and to dates. The ontology is a formalization of an existing taxonomy, containing 478 actors in 32 (nested) categories and an issue hierarchy of 103 issues.

4.2 The iNET Annotation Tool

A new version of the existing iNET tool was created in Java Eclipse for annotating newspaper articles in this framework. As can be seen in Fig. 5, iNET shows

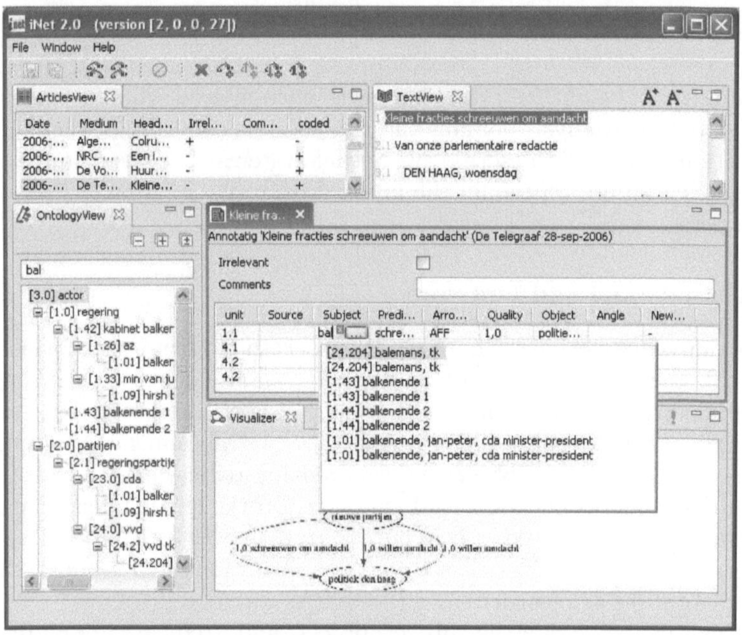

Fig. 5. iNET: RDF-based annotation with Autocomplete and Visualization

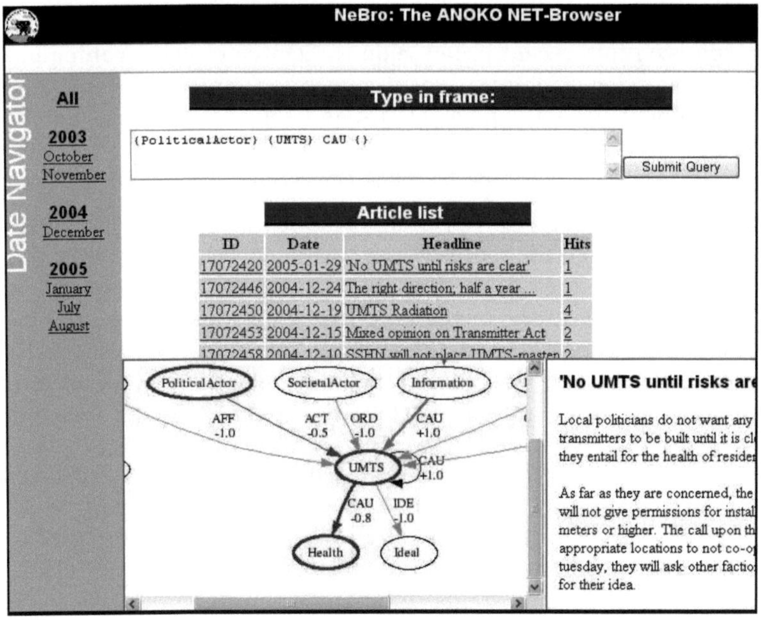

Fig. 6. NeBro: Browsing, Querying, and Visualizing the RDF repository

the content of the current article and the ontology, and assists annotators by offering autocomplete functionality and by allowing search through the ontology. A coder logs into a (relational) article database, which gives him or her a list of jobs, consisting of the to-be-annotated articles and the annotation scheme and a pointer to the ontology in the Sesame repository. After coding an article, results are immediately posted to that repository. To help coders check their annotations, a visual representation of their coding is also presented. Using this tool, a group of 14 coders annotated over 13,000 articles and tv news broadcasts, resulting in 30,000 triples.

4.3 The NeBro Browser / Visualizer

To browse through these results, a web application was created that allows a user to browse through articles and view a visualization of the annotation. As shown in Fig. 6, a user is able to enter queries to look for specific relations. Since the internal representation became quite complex, these queries can be posted in terms of the original NET relations and are translated to SeRQL queries on the Sesame Repository. In the visualization of the retrieved article, the relations matching the query are highlighted.

5 Conclusions

In this paper we reported our experiences in designing a formal representation for Relational Content Analysis, a method used in Communication Science to conduct quantitative media analysis. This domain, and particularly the NET methodology used in a recent use-case on the Dutch election campaign in 2006, highlighted a number of different requirements on representation and querying which could only partly be implemented seamlessly using current Semantic Web technology.

The main contribution of this paper is threefold:

1. The paper describes a real world, large size case-study where Semantic Web technology was used for representing and querying highly complex data on media coverage on the Dutch election campaign 2006, and thus exemplifies a state of matureness of the technology. Based on a detailed requirement analysis we designed workable and flexible data models to code the Relational Content Analysis data. On some aspects we identified conceptual problems that require more general solutions which should lead to extensions of existing standards in the future.
2. The paper identifies a number of such requirements to a formal representation framework, which are domain and application specific at first instance. Nevertheless, we expect the majority of these requirements to be recurrent in many other practical applications, and although we do not claim exhaustiveness the collection of items in Sect. 2.3 will be indicative for problems faced in similar applications.
3. For each of the requirements, this paper provides an overview of the state of the art in practical Semantic Web research. We surveyed the literature on extending RDF triples with additional information, and presented a categorization of existing and proposed mechanisms in terms of the meaning of the extra information and the transparency of the original triple. We hope that such a study (from an application perspective) can pinpoint relevant open research questions for the Semantic Web community.

The general conclusion is that Semantic Web technology offers a useful set of standards and tools for formalizing this background knowledge and storing and inferencing with the combination of background knowledge, metadata about the annotated articles, and the annotations themselves. Having all this data within one representation enabled us to develop the set of tools presented in this paper for efficiently annotating, visualizing, querying, and extracting from the data set.

References

1. Krippendorff, K.: Content Analysis: An Introduction to Its Methodology (second edition). Sage Publications (2004)
2. Carley, K.: Network text analysis: The network position of concepts. In Roberts, C., ed.: Text Analysis for the Social Sciences. Lawerence Erlbaum Associates, Mahwah, NJ (1997) 79–100
3. Kleinnijenhuis, J., Scholten, O., van Atteveldt, W., van Hoof, A., Krouwel, A., Oegema, D., de Ridder, J.A., Ruigrok, N., Takens, J.: Nederland vijfstromenland: De rol van media en stemwijzers bij de verkiezingen van 2006. Bert Bakker, Amsterdam (2006)

4. Van Cuilenburg, J.J., Kleinnijenhuis, J., De Ridder, J.A.: Towards a graph theory of journalistic texts. European Journal of Communication **1** (1986) 65–96
5. Wiebe, J.M., Wilson, T., Bruce, R.F., Bell, M., Martin, M.: Learning subjective language. Computational Linguistics **30(3)** (2004) 277–308
6. Van Atteveldt, W., Kleinnijenhuis, J., Carley, K.: Rcadf: Towards a relational content analysis standard. In: Presentated at the International Communication Association (ICA), Dresden (2006)
7. Van Atteveldt, W., Oegema, D., van Zijl, E., Vermeulen, I., Kleinnijenhuis, J.: Extraction of semantic information: New models and old thesauri. In: Proceedings of the RC33 Conference on Social Science Methodology, Amsterdam (2004)
8. Wasserman, S., Faust, K.: Social Network Analysis. CUP, Cambridge (1994)
9. MacGregor, R., Ko, I.Y.: Representing contextualized data using semantic web tools. In: Practical and Scalable Semantic Web Systems (workshop at second ISWC). (2003)
10. Carroll, J., Bizer, C., Hayes, P., Stickler, P.: Named graphs, provenance and trust. In: Proceedings of the Fourteenth International World Wide Web Conference (WWW2005), Chiba, Japan. Volume 14. (2005) 613–622
11. Brickley, D., Guha, R.: Rdf vocabulary description language 1.0: Rdf schema. W3C Recommendation (http://www.w3.org/TR/rdf-schema/) (2004)
12. Noy, N., Rector, A.: Defining n-ary relations on the semantic web. Working Draft for the W3C Semantic Web best practices group (2005)
13. Dumbill, E.: Tracking provenance of rdf data. Technical report, ISO/IEC (2003)
14. Guha, R., McCool, R., Fikes, R.: Contexts for the semantic web. In: Proceedings of the Third International Conference on the Semantic Web (ISWC'04). (2004)
15. Sintek, M., Decker, S.: Triple - a query, inference, and transformation language for the semantic web. In: Proceedings of ISWC02. (2002)
16. Masolo, C., Vieu, L., Bottazzi, E., Catenacci, C., Ferrario, R., Gangemi, A., Guarino, N.: Social roles and their descriptions. In Dubois, D., Welty, C., Williams, M., eds.: Proceedings of the Ninth International Conference on the Principles of Knowledge Representation and Reasoning (KR2004), Whistler, Canada (2004) 267–277
17. Steimann, F.: On the representation of roles in object-oriented and conceptual modelling. Data and Knowledge Engineering **35** (2000) 83–106
18. Sowa, J.: Using a lexicon of canonical graphs in a semantic interpreter. In Evens, M., ed.: Relational models of the lexicon. Cambridge University Press, Cambridge UK (1988)
19. Sowa, J.: Knowledge Representation: Logical, Philosophical, and Computational Foundations. Brooks/Cole, Pacific Grove, CA (2000)
20. Guarino, N.: Concepts, attributes and arbitrary relations: Some linguistic and ontological criteria for structuring knowledge bases. Data and Knowledge Engineering **8** (1992) 249–261
21. Mika, P., Gangemi, A.: Descriptions of Social Relations. In: Proceedings of the 1st Workshop on Friend of a Friend, Social Networking and the (Semantic) Web. (2004)

SEEMP: An Semantic Interoperability Infrastructure for e-Government Services in the Employment Sector

E. Della Valle[1], D. Cerizza[1], I. Celino[1], J. Estublier[2], G. Vega[2], M. Kerrigan[3], J. Ramírez[4], B. Villazon[4], P. Guarrera[5], G. Zhao[5], and G. Monteleone[6]

[1] CEFRIEL – Politecnico of Milano, Via Fucini 2, 20133 Milano, Italy
[2] Equipe Adele, LSR, Université Joseph Fourier,F-38041 Grenoble Cedex 9, France
[3] DERI, University of Innsbruck, Technikerstraße 21a, 6020 Innsbruck, Austria
[4] Universidad Politecnica de Madrid, 28660 Boadilla del Monte, Madrid, Spain
[5] Le Forem, Boulevard Tirou 104, 6000 Charleroi, Belgium
[6] TXT e-solutions, via Frigia 27, 21126 Milano, Italy

Abstract. This paper presents SEEMP, a marketplace to coordinate and integrate public and private employment services (ESs) around the EU Member States. The need for flexible collaboration in the marketplace gives rise to the issue of interoperability in both data exchange and share of services. SEEMP proposes a mixed approach that relies on the concepts of services and semantics. SEEMP approach combines Software Engineering and Semantic Web methodologies/tools in an infrastructure that allows for a *meaningful service-based communication among ESs.*

1 Introduction

SEEMP project[1] aims at designing and implementing in a prototype an Interoperability infrastructure for e-government. More specifically, SEEMP is developing an EIF-compliant [1] to allow interoperability among the hundreds of public and private Employment services (ESs) that exist in Europe. The resulting European Employment Marketplace will overcome the national barriers complying, at the same time, with the local policies of each Member States.

Thanks to SEEMP, which promotes increased partnership between labour market actors and the development of closer relations between private and public employment services, job-seekers and employers will have better services that operate at European scale. For instance, the matching process between job offers and CVs across all Europe will become possible increasing, eventually, labour hiring and workforce mobility. In order to fulfil such an ambitious goal several problems must be solved at organizational and technical level.

At an organizational level, the business model of SEEMP has to be catchy for all ESs. The main reason for a ES to buying in is creating added value for its local users (both job seekers and employers) by offering interconnections with

[1] SEEMP is funded by EU (IST-4-027347); please visit http://www.seemp.org/.

E. Franconi, M. Kifer, and W. May (Eds.): ESWC 2007, LNCS 4519, pp. 220–234, 2007.

other ESs. Today it is normal for all users to insert CV and Job Offers in many ESs and collect, laboriously, the results personally. When SEEMP will be in place each ES will be able to collaborate with other ESs. From the perspective of the end user, the add-value is the outreach to other niches of the job market without 'being stretched out". End users could insert the CV and Job Offers once and collect pan-European results. From ESs perspective it will results in increase both the number of users and their faithfulness to each ES, thus an increase in transaction volume.

The need for such flexible collaboration between ESs, gives rise to the issue of interoperability in both data exchange and share of services. The technical approach of SEEMP relies on the concepts of Web Services and semantics. Web Services, exploited in a Software Engineering manner, enable an easier maintenance of the integration. Semantics, encoded in the systems by the means of ontologies and mediators, allows for reconciliation of the hundreds local professional profiles and taxonomies.

SEEMP solution will expose, following the well established Software Engineering approach of Mélusine [2], a single consistent set of abstract services each ES can invoke. Such abstract services will provide a multilateral interoperability solution that delegates the execution of the services to the local ESs (in accordance with the subsidiarity principle) and aggregates the results before sending the response back to the invoker. Moreover, following the innovative Web Service Modeling Ontology [3] approach, we will capture the semantics shared among ESs in a single consistent model. Such model includes a reference ontology in which and from which the local semantics is mapped as well as semantic description of the local Web Services for their automatic use. A set tools will be provided to each ES for modeling its local ontology and for aligning the local ontology with the reference one. As a technical result *SEEMP will enable a meaningful service-based communication among ESs.*

The paper is structured as follows: Section 2 presents a running example that will be discussed throughout the paper; Section 3 explains the interoperability issue that arises in SEEMP project; Section 4 presents the SEEMP approach to support meaningful service-based communication among ESs; Sections 5 and 6 outline SEEMP solution architecture and components rooting them to the related work in Web Services and Semantic Web community; Section 7 briefly discusses the approach of SEEMP comparing it with already implemented approaches; and finally Section 8 presents future work.

2 A e-Employment Running Example

European Member States have introduced major reforms to make the labour market more flexible, transparent and efficient. Such major reforms include decentralization, liberalization of the mediation market (competition between public and private), and quality monitoring of ES staff and services. As an effect ESs understood the need for making available on-line a one-stop shop for the employment. This results in an increased used of ICT and in a boost in differentiating and personalizing the services they offer (e.g., Borsa Lavoro Lombardia, Le FOREM[2]). For the discussion of this paper we will consider a running example derived by the user requirements of the SEEMP project (for a more detailed description of the socio-economic constrains please refer to [4]):

> *Job seekers (companies) put their CVs (Job Offers) on a local ES and ask to match them with the Job Offers (CVs) other users put in different ESs through SEEMP.*

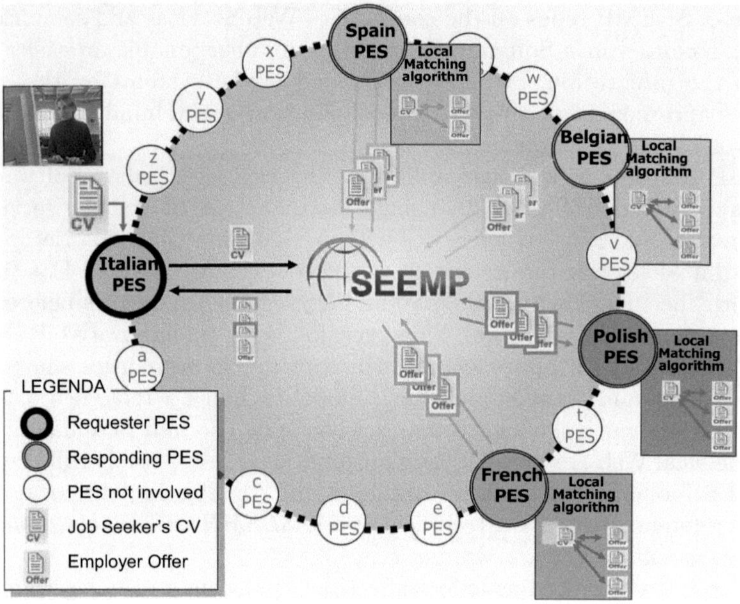

Fig. 1. The running example of distributed matching of CVs and job offers

It may look like a fairly simple example, but to reach its potential EU-wide audience, this e-Employment running example (see figure 1) needs to fulfill a wider set of requirements than the respective local ES services. A local matching service is designed for national/regional requirements only (i.e., central database, single professional taxonomy, single user language, etc.). SEEMP has to be able to send the request, which an end-user submits to the local ES (the Italian ES on left in the figure), to all the other ESs in the marketplace. In order to avoid asking "all" ESs, SEEMP has to select those ESs that most likely will be able to provide an answer and send the request only to them (the four ES on the right in the figure). Moreover, the answers should be merged and ranked homogeneously by SEEMP before they are sent back.

[2] Respectively http://www.borsalavorolombardia.net/ and http://www.leforem.be/

3 Interoperability Issues

The running example presented in section 2 highlights the need for a system that covers the whole EU and subsumes hundreds of real **heterogeneous systems** existing in EU countries and regions. It implies by-passing:

- *language* heterogeneity, e.g., an Italian Java Analyst Programmer may be looking for job offers written in all the different European languages;
- *CVs and Job Offers structur*al heterogeneity, i.e., the use of standards like HR-XML[3] is not wide spread and a multitude of local formats exists;
- *CVs and Job Offers content* heterogeneity, i.e., European level occupation classifications like ISCO-88[4] exist, but they do not reflect legitimate differences and perspectives of economic, cultural and political environments;
- heterogeneity of *service interface and behavior*, i.e., no standard exists for e-employment services thus each ES implemented them differently.

All those are typical interoperability issues. The need for interoperability at European Level among e-Government services has been perceived since 1999 [5] with the adoption of a series of actions and measures for electronic interchange of data between administrations, businesses and citizens (IDABC) [6].

The main results of IDABC is the European Interoperability Framework (EIF) [1]. EIF follows the principle of subsidiarity[5] in addressing the interoperability problem at all levels: organizational, semantic and technical. One crucial aspect, deriving from the principle of subsidiarity, is to keep responsibility decentralized; in other words each partner should be able to keep its own business process almost unchanged[6] and to provide externally point of exchange for its processes. EIF names these points "business interoperability interfaces" (BII).

EIF does not prescribe any solution, but it rather recommends the principles to be considered for any e-Government service to be set up at a pan-European level: accessibility, multilingualism, security, privacy, use of open standards and of open source software and use of multilateral solutions. SEEMP proposes itself as an implementation of EIF in the domain of e-employment.

4 The SEEMP Approach

SEEMP relies on the concept of Service. Following the EIF, each ES locally must expose its BII as Web Services. All these Web Services differ but they are fairly similar. SEEMP, as marketplace, models a single consistent set of

[3] http://www.hr-xml.org/

[4] http://www.warwick.ac.uk/ier/isco/isco88.html

[5] The principle of subsidiarity recommends not to interfere with the internal workings of administrations and EU Institutions.

[6] Quoting from IDABC: "it is unrealistic to believe that administrations from different Member States will be able to harmonize their business processes because of pan-European requirements".

Web Service out of those exposed by the ESs. Therefore the services exposed by SEEMP become the standard for the distributed independent service providers.

For instance, for the running example provided in section 2 each ES should expose two Web Services: match an external CV against the job offers stored locally and match an external job offer against the CVs stored locally. Therefore a service that subsume all the local heterogeneous ones can be modeled in SEEMP for each of the two families of similar Web Service. These two abstract service are those that SEEMP, as a marketplace, offers to the ESs.

SEEMP uses Mélusine [2] as tool for modeling those abstract services and orchestrating the process of delegating the execution to the distributed independent service providers.

SEEMP relies on the concept Semantics (both ontologies and mediators. As for the service, each local ES has its own *local ontology* for describing at a semantic level the Web Services it exposes, and the structure/content of the messages it exchanges. All these ontologies differ but they are fairly similar, because a common knowledge about employment exists as well as the needs for exchange (i.e., you don't exchange on things with no equivalence calculus). So, SEEMP, as marketplace, models a single consistent ontology out of those exposed by the ESs. Therefore the reference ontology of SEEMP becomes the actual standard for the ESs that should provide the *mediators* for translating from the local ontologies to the reference one and vice versa.

For instance, for the running example each ES should model in a local ontology the structure/content of its CV/job offers and the way they are exchanged via Web Services. A reference ontology that subsumes all the local heterogeneous ones can be modeled in SEEMP and it becomes the source of shared understanding. Each ES has to provide its mediator for local-reference semantic mapping.

SEEMP adopts WSMO [3] a way to semantically describe Web Service, ontologies and mediators, WSML [7] as concrete syntax for encoding those descriptions and METHONTOLOGY [8] as methodology for developing and maintaining those semantic descriptions.

Minimal shared commitment. In implementing SEEMP approach particular attention is paid in keeping a "win-win" situation among all ESs. The commitment (both at services and semantics level) of each ES should be minimal. ESs care about being able to share while maintaining all the necessary and unnecessary disagreements. It may appear counter-intuitive, but the most suitable set of services and ontologies is the one that enables ESs to "agree while disagreeing". Both the reference set of services and the reference ontology must cover the various aspects of the employment market with an *acceptable level of details that leaves leeways of disagreement.*

Considering the running example of section 2, the minimal shared commitment the SEEMP consortium agreed upon consists in sharing a subset of the CV named *candidacy* and a subset of job offer named *vacancy*. Candidacy (vacancy) enables matching without reveling how to contact the job seeker (employer) that

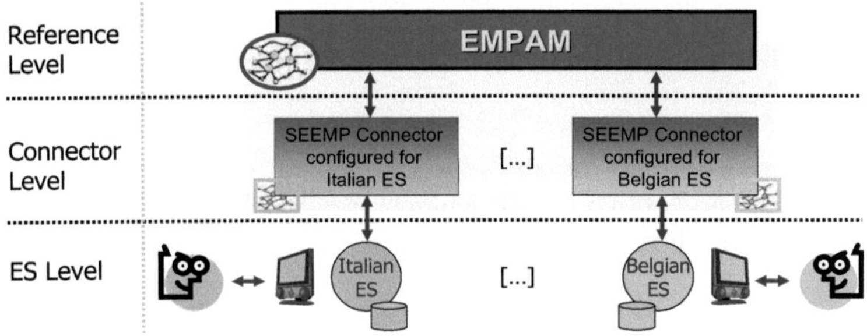

Fig. 2. An overview of the SEEMP solution

is, instead, in the CV (job offer) stored in the ESs and that can be contacted by invoking a service of the ES responsible for the CV (job offer). In this way SEEMP approach also takes into consideration privacy issues of collaborative network, which is a technical issue as well as a business constraint.

5 The SEEMP Solution Architecture

SEEMP solution is composed of *a reference part* (all the dark components in figure 2), which reflects the "minimal shared commitment" both in terms of services and semantics, and by *the connectors* toward the various local actors (the components in shading colors in figure 2).

5.1 Structural Overview

The reference part of SEEMP solution. is made up of the central abstract machine, named EMPAM (Employment Market Place Abstract Machine) and a set of SEEMP services.

The EMPAM is an "abstract machine", in that it does not perform directly any operation, but rather offers abstract services that are made concrete by delegation: when the abstract service is invoked, the EMPAM delegates its execution to the appropriate ES by invoking the correspondent concrete services. It acts as a multilateral solution (as request by EIF), in which all the services connected to the EMPAM are made available to all other ESs, i.e. they ensure a pan-European level of services without interfering with the Business processes of each ES.

The SEEMP services are meant to support EMPAM execution. The running example requires two SEEMP services: discovery and ranking. The *discovery* service is offered by Glue [9]. The EMPAM invokes Glue Discovery Engine before delegating the execution to the concrete services exposed by the ESs. Glue analyzes the CV sent by the invoking ES and it selects among all ESs those that most likely would be able to return relevant job offers. The ranking service is invoked by the EMPAM after all the concrete services have answered and it merges the results providing an homogeneous ranking of the returned job offers. It also delete possible duplicated job offers, which different ESs may have returned.

The SEEMP connectors. enables all communication that occurs between the EMPAM and a given ES. A SEEMP connector will exist for each of the ESs that are connected to the EMPAM and has two main responsibilities:

- *Lifting and Lowering*: when communicating with the ES any outgoing (or incoming) data which is exchanged by the means of Web Services must be lifted form XML to WSML in terms of the local ontologies of the ES (or lowered back to the XML level from WSML).
- *Resolving Heterogeneity*: each ES has its own local ontology that represents its view on the employment domain. The SEEMP connector is responsible for resolving these heterogeneity issues by converting all the ontologized content (the content lifted from the XML received from the ES) into content in terms of the reference ontology shared by all partners and vice versa.

5.2 Functional Overview

By combining the EMPAM and the connectors SEEMP solution enables a *meaningful service-based communication* among ESs. Figure 3 illustrates how such meaning communication takes place in running the example of section 2:

1. the user inserts a CV into the Italian ES and requests job offers,
2. the Italian ES invokes the marketplace matching service passing the CV encoded in the Italian ontology,
3. the SEEMP connector translates the CV from the Italian ontology to the reference one,
4. the discovery SEEMP service analyzes the CV and selects the ESs to be contacted,
5. the EMPAM invokes in parallel the local matching service of selected ESs,
6. the SEEMP connectors of the selected ESs translate the CV from the reference ontology to the local ontology (i.e., the Belgian and the French ESs) and invoke the ES's Web service,
7. the Belgian and the French ES compute the matching locally and returns a set of job offers,
8. each of the connector translates the job offers from each local ontology to the reference one,
9. the ranking SEEMP service merges the responses and ranks the job offers,
10. the job offers are sent back in the reference ontology to the Italian connector that translates them in the Italian ontology,
11. the connector responds to the Italian ES, and
12. finally the Italian ES displays the job offers to the user.

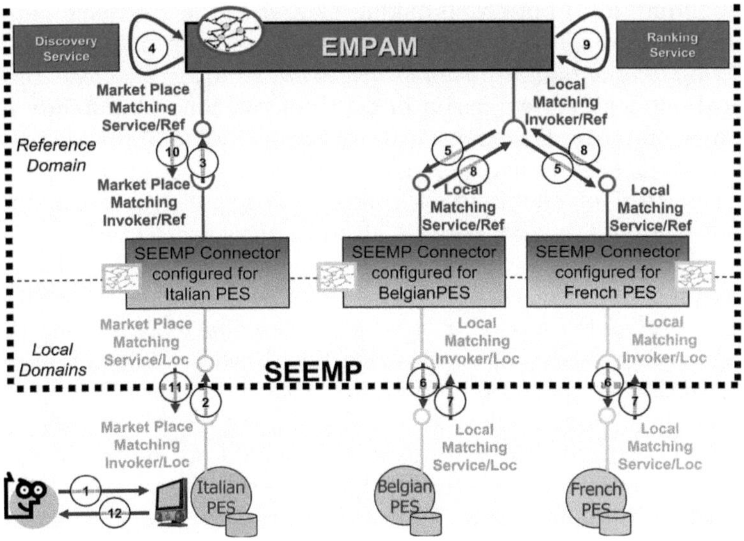

Fig. 3. How SEEMP solution enables meaningful service-based communication

6 The SEEMP Solution Components

6.1 Reference and Local Ontology for e-Employment

The Reference Ontology is a core component of the system. It acts as a common "language" in the form of a set of controlled vocabularies to describe the details the employment sector. The Reference Ontology has to be rich enough to support the semantic needs of all the ES (Local Ontologies) involved currently and in the future. The Reference Ontology also has to be a scalable, adaptable and maintainable ontology. For all those reason SEEMP follows the METHONTOLOGY [8] approach:

1. specifying competency questions and identified necessities;
2. selecting the standards that cover most of them;
3. semantic enrichment of the chosen standard; and
4. evaluating the Ontology content.

In order to build the Reference Ontology, the standards identified are: ISO 4217 for currencies, the 12 levels of driving license recognized by the European legislation, NACE Rev. 1.1 for economic activities, ISCO-88 (COM) and ONET taxonomy of occupations, FOET and ISCED97 for education, ISO 3166 country codes, LE FOREM classifications of contract types and work rule types, ISO 6392 for languages, and EDS classification for skills.

Local Ontologies. Based on the proposed SEEMP architecture, the possible options for building the local ontologies in SEEMP ranges between to extreme options: building local ontologies taking as a seed the reference ontology and building local ontologies as a reverse engineering process from ES schema sources.

In building local ontologies taken as a seed the reference ontology, the concepts in the local ontology are extension in depth of the concepts already present in the reference ontology. By extension we mean including application dependent concepts that appear in each ES schema source.

The exchange of job offers and CV (once ontologized), required by the running example of section 2, is easy because all local ontologies extends the same reference vocabulary. On the contrary mappings complexity are on local ontology and schema mappings (cf. section 6.3 for more details).

Building local ontologies as a reverse engineering process from ES schema sources, is the easiest way for ontologizing ESs. Each concept in the local ontology is the semantic expression of a relevant concept in the ES. In this way ESs becomes ontology-based applications that are more efficient because mappings between local ontologies and schema sources should not be complex, but complex mappings appear between the local and reference ontology. Therefore data exchange will require more time, in comparison to the previous option, due to the execution of two complex mappings.

The SEEMP way is keeping close to the first option in the beginning, when few ESs are in the marketplace and the union of ES semantics is the most straight forward solution. Then while more and more ESs would be added to the marketplace, the solution would move toward the second option. The equilibrium between the two extreme solutions is related to the need for a "minimal shared commitment" explained in section 4.

6.2 An Employment Market Place Abstract Machine

The EMPAM machine is implemented as a Mélusine application, which means it is structured following the Mélusine approach in three layers (cf. Figure 4).

Layer 1: The abstract machine. The higher EMPAM machine layer is a java abstrct program where abstract classes represent the concepts present in SEEMP Employment Marketplace Platform (EMP). EMP acts as a ES covering completely EU, i.e. it acts as if all the CV and vacancies were present in its repositories. However the EMP is abstract since it does not have any information locally, but delegates to real ESs the duty to process part of the job. The EMP program defines functions like matching CV and job offers that are, indeed, not implemented at all, or only sketching what they are supposed to do.

Layer 2: The adapters. The second layer duty is to do in such a way that empty or dummy methods, found in the abstract layer, really perform what they are supposed to perform. To that end this layer is itself structured in three layers:

- *The injection machine*, whose duty is to capture those methods that need to be implemented, and to transfer the call to the following layer.
- *The mediation and orchestration layer* which is in charge of transforming a single abstract method call into a potentially complex orchestration of real lower level services that together will perform the required function.

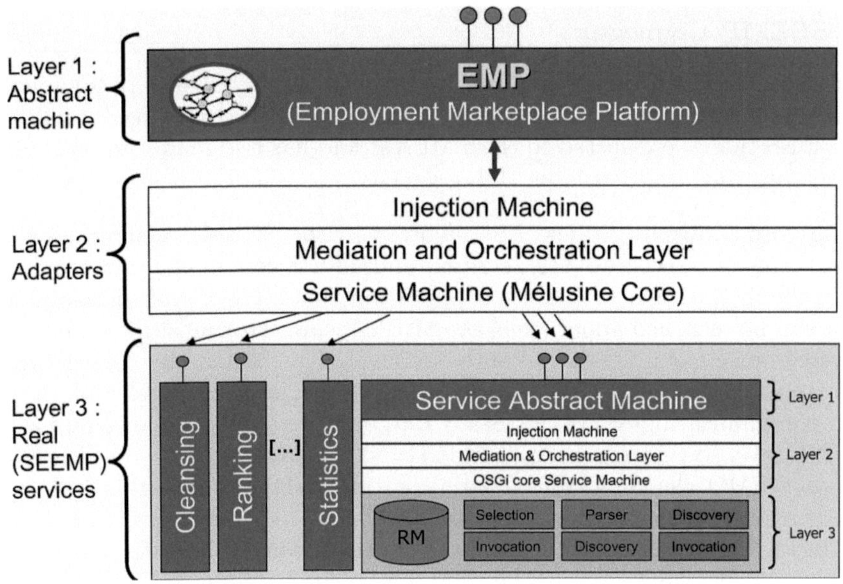

Fig. 4. The levels that made up the EMPAM as a Mélusine application

- *The Service machine*, whose duty is to transparently find and load the required SEEMP service and to call them. In SEEMP, this service machine is the core Mélusine service machine (an implementation of OSGi[7]).

Layer 3: The SEEMP services, which are OSGi services, and are called accordingly by Mélusine. This solution ensures optimal performance to the EMPAM, while allowing large facilities to future extensions (new SEEMP service) and even dynamic changes (dynamic loading/unloading of services).

Two classes of services have been identified:

- those dedicated to calling a Web Service exposed by a ES through the Service Abstract Machine (SAM). Most of the issues raised by EMP are related to discovering, selecting, parsing, and finally invoking a remote service; more exactly the SEEMP connector of each ES. SAM is itself a Mélusine application and therefore contains itself an abstract layer in which are defined the fundamental concepts and functions of a service machine. This layer is captured and delegated to an orchestration layer that calls local services which, in the scope of SEEMP, are WSMX components [10] wrapped as OSGi services.
- the other service; currently these services, in SEEMP, include the cleansing, ranking and statistic functions, and will include, in the future, the implementation of the specific functions and repositories of the EMPAM machine i.e. those functions and information not available in the ESs. Functions and information available in ESs are available calling the SAM service.

[7] http://www.osgi.org/

6.3 SEEMP Connectors

The SEEMP connectors behave as the mechanism through which all communication occurs between the EMPAM and a given ES. A SEEMP connector will exist for each ES that is connected to the EMPAM and has two main responsibilities: Lifting/Lowering and Resolving Heterogeneity.

Lifting and Lowering: The ESs involved in the SEEMP marketplace only deal in terms of structured XML content and do not deal in terms of ontologies. Within the SEEMP marketplace it is important that all content is ontologized so that it can be reasoned about, thus the SEEMP connector must lift all messages received from a given ES to the ontology level. This is done by converting the XML content received to WSML in terms of the local ontologies of the ES. When communicating with the ES any outgoing data must be lowered back to the XML level so that the ES can understand the content.

Since WSMO elements can be serialized in a RDF format, this task could be done by converting XML content to RDF first, and then converting RDF to WSML. In SEEMP this task is achieved by the means of an extension to R_2O language [11], which enables to describe mappings between XML schemas and ontologies, and to its related processor ODEMapster [12].

Resolving Heterogeneity: Each ES talks in its own language, essentially having its own local ontology that represents its view on the employment domain. The SEEMP connector is responsible for resolving these heterogeneity issues by converting all the ontologized content (the content lifted from the XML received from the ES) into content in terms of the SEEMP reference ontology. By doing this all the ESs in the marketplace talk in the same language, and thus heterogeneity issues are resolved. Similar to the lowering back to XML, when communicating with a given ES the SEEMP connector is also responsible for converting back from the reference ontology to the local ontology of the given ES.

As described in section 6.1 the reference ontology represents the bridge, or common vocabulary, that the ESs will communicate through. Rather than managing mappings between every possible ontology pair, which essentially becomes unmanageable once a number of ESs have joined the marketplace, each ES need only maintain mappings to and from the reference ontology. These mappings represent a set of instructions (or rules) on how to convert an instance from the local ontology to an instance of the reference ontology (and vice versa).

Technologically this is achieved using the WSMX Data Mediation [13]. This work is made up of two components, the first being the design time component, within which the ES will describe the mappings between their local ontology and the reference ontology, and the second being the run time component, which is responsible for executing the mappings at run time to transform the messages between ontologies.

A reusable SEEMP connector is built by bringing together the functionality described above. The architecture of the SEEMP Connector outlined in the

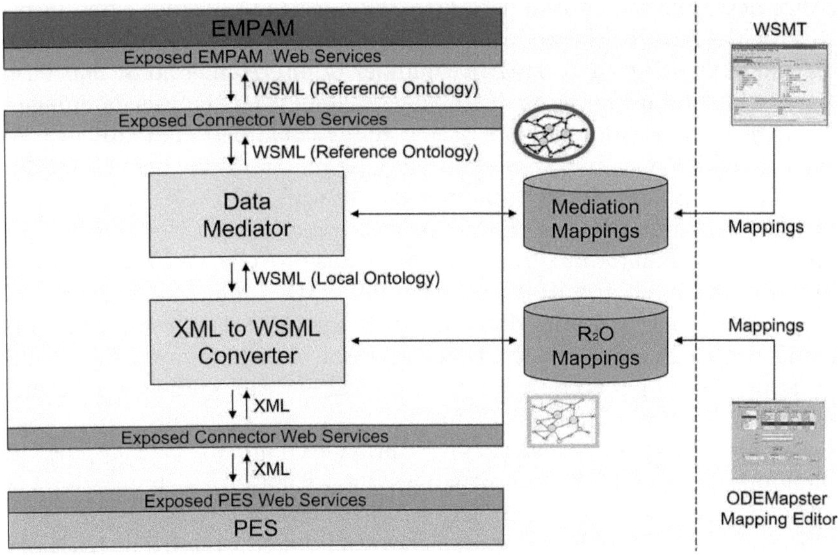

Fig. 5. The SEEMP Connector Architecture

figure 5 shows the ES communicating with the connector using XML via the exposed web services. This XML is then lifted to the local ontology using the R_2O mappings stored in the repository and furthermore converted to the reference ontology using the data mediation mappings. Ultimately the EMPAM is invoked using messages in the reference ontology via its exposed web services. Communication also occurs in the opposite direction.

Each of the ESs joining the marketplace will require its own SEEMP connector, however the only difference between any two connectors is the code for executing the ESs exposed Web Services as each ES will expose services in a different way. The need for individual SEEMP connectors could be removed through the use of WSMO Choreography [3] to describe the interfaces of the ES services and the integration of the WSMX choreography engine [14] and invoker into the SEEMP Connector, however this was not considered for the first prototype of the SEEMP solution.

7 Comparing SEEMP with Other Approaches

In order to draw a comparison between SEEMP and other approaches we selected two case studies: private employment networks (e.g. Adecco) and hierarchical network (e.g. Borsa Lavoro Lombardia, EURES). Moreover we consider the differences both from the point of view of CEO (the decision makers) and CTO (the IT experts).

Compared to other approaches SEEMP solution offers CEO a way to enforce subsidiarity principle, therefore valuing each ES contribution in the marketplace.

In private networks the subsidiarity principle is not applicable, while in hierarchical networks most of the nodes are passive actors. Moreover the marketplace creates added value by increasing the number of interconnections, hence resulting in more faithful users (more JO/CV accessible using the user language) and in more transactions. Many job offers that today could be found only at the cost of inserting the CV multiple times and merging manually the results of different ESs, becomes available through the interface of each ES.

For CTO SEEMP solution enable an easier maintenance of the integration with other ESs and minor integration costs. It was proved that Web Services used in a Service Oriented Architecture easies integration and maintenance. Moreover semantics make mapping different terminology easier because tools (such as WSMT [15]) can analyzed local and reference ontology (e.g., by comparing sub-structures and by searching for synonymies) and can guide the IT Administrator in drawing the mappings. Thank to this support, the mapping definition process requires less time or, eventually, it provides more precise mappings in the same amount of time. That support comes out with a minor integration costs.

In order to achieve this benefit CEO has to develop "partnership", i.e., the ability to collaborate with other peers, ES or staffing industries. The partnerships are different in the two case studies. In private network everything is agreed in advance. In hierarchical network partnership are necessary, but no peer to peer decision taking is possible. Decisions are institutionally imposed top-down. Moreover SEEMP supports CTO with comprehensive set of tools and methodologies for service and semantic interoperability.

Concerning services CTO has to expose ES APIs as Local Web Services and has to provide support for invoking EMPAM services. However, they don't have to understand interfaces and behavior of other ES (as in hierarchical solutions) because the connector presents the market place as if the ES was invoking its own services.

Concerning semantics, CTO has to model data structure and content and has to defining mappings with the reference ontology, but, as discussed above, this is easier and more precise than it is nowadays without ontologies and mediators.

What has to be built, and SEEMP alone won't be able to do, is a comprehensive reference ontology and abstract service machine that encompasses several employment domains. Developing and maintenance this reference part of SEEMP more then a ICT problem; it is a matter of reaching agreement at organizational level. As already discussed in section 4 the goal of SEEMP is reaching a "minimal shared commitment" in which ESs *agree* on high-level aspects, allowing for collaboration among them, *while disagreeing* on minor details that differentiate one ES from the others.

8 Conclusions and Future Work

This paper presented the SEEMP approach in supporting *meaningful service-based communication* among public and private employment services. The following results have been shown:

- **services and semantics** are the key concepts for abstracting from the hundreds of heterogeneous systems already in place that are evolving separately. They provide a straight forward way to implement the subsidiarity principle.
- **the combination of an abstract service machine with a reference ontology** is a technically sound approach to multi-laterality for marketplace implementation. Each actor in the marketplace has to care only about integrating with the marketplace. The marketplace will offer services to support the interaction with the other actors.
- **a mix of Software Engineering and Semantic approach** is required to achieve flexibility. The two approaches nicely complement each other. By means of "conventional" software engineering design SEEMP build an abstract machine that can run on "conventional" technology and at the same time embeds semantics both in the form of ontology/mediator and in the form of semantic-aware components (i.e. ODE mapster, WSMX data mediation, Glue Discovery Engine).

Currently SEEMP consortium is running a pilot that shows the integration of EURES and Borsa Lavoro Lombardia. This integration has allowed for so far testing the functional aspects of SEEMP approach. The next step is integrating Le FOREM ES as a validation case. We expect, the reference ontology and the abstract machine to be so well designed that Le FOREM introduction would have no impact on them.

Future work includes extending the number of abstract services included in the EMPAM and the respective concepts in the reference and local ontologies. For instance, one essential service of SEEMP should be the possibility to have regularly (monthly, weekly, daily,) a set of key indicators regarding labour market in all participant regions (job seekers, job offers, training, etc.), in a common and comparable language, both in terms of methods (definitions, calculation of indicators, etc.) and in terms of technical requirements.

Acknowledgments

The research has been supported by the SEEMP (IST-4-027347-STREP) EU funded project. We thank all the colleagues for the fruitful discussion we have and in particular Flavio De Paoli, Mirko Cesarini and Maria Grazia Fugini.

References

1. European Communities: European interoperability framework for pan-european egovernment services. Technical report, Office for Official Publications of the European Communities (2004)
2. Estublier, J., Vega, G.: Reuse and variability in large software applications. In: ESEC/SIGSOFT FSE. (2005) 316–325
3. Fensel, D., Lausen, H., Polleres, A., de Bruijn, J., Stollberg, M., Roman, D., Domingue, J.: Enabling Semantic Web Services – The Web Service Modeling Ontology. Springer (2006)

4. Della Valle, E., Cerizza, D., Celino, I., Estublier, J., Vega, G., Kerrigan, M., Ramírez J., Villazon, B., Guarrera, P., Zhao, G., Monteleone, G.: SEEMP: Meaningful service-based collaboration among labour market actors. In: proceedings of BIS 2007, LNCS 4439, Poznan, Poland, Springer-Verlag (2007)
5. 1720/1999/EC: Decision of the European Parliament and of the Council of 12 July 1999 (1999)
6. 2004/387/EC: Decision of the European Parliament and of the Council on Interoperable Delivery of pan-European Services to Public Administrations, 2004 (2004)
7. de Bruijn, J., Lausen, H., Polleres, A., Fensel, D.: The web service modeling language: An overview. In: Proceedings of the 3rd European Semantic Web Conference (ESWC2006), Budva, Montenegro, Springer-Verlag (2006)
8. Gómez-Pérez, A., Fernández-López, M., Corcho, O.: Ontological Engineering. Springer Verlag (2003)
9. Della Valle, E., Cerizza, D.: The mediators centric approach to automatic web service discovery of glue. In: MEDIATE2005. Volume 168 of CEUR Workshop Proceedings., CEUR-WS.org (2005) 35–50
10. Haller, A., Cimpian, E., Mocan, A., Oren, E., Bussler, C.: WSMX - A Semantic Service-Oriented Architecture. In: ICWS. (2005) 321–328
11. Barrasa, J., Corcho, O., Gómez-Pérez, A.: R2O, an extensible and semantically based database-toontology mapping language. In: Second International Workshop on Semantic Web and Databases. (2004)
12. Rodriguez, J.B., Gómez-Pérez, A.: Upgrading relational legacy data to the semantic web. In: WWW '06: Proceedings of the 15th international conference on World Wide Web, New York, NY, USA, ACM Press (2006) 1069–1070
13. Mocan, A., Cimpian, E., Kerrigan, M.: Formal model for ontology mapping creation. In: International Semantic Web Conference. (2006) 459–472
14. Cimpian, E., Mocan, A.: WSMX Process Mediation Based on Choreographies. In: Business Process Management Workshops. (2005) 130–143
15. Mocan, A., Cimpian, E.: Mappings creation using a view based approach. In: MEDIATE2005. Volume 168 of CEUR Workshop Proceedings., CEUR-WS.org (2005) 97–112

Combining RDF Vocabularies for Expert Finding*

Boanerges Aleman-Meza[1], Uldis Bojārs[2], Harold Boley[3], John G. Breslin[2], Malgorzata Mochol[4], Lyndon JB Nixon[4], Axel Polleres[2,5], and Anna V. Zhdanova[5]

[1] LSDIS Lab, University of Georgia, USA
[2] DERI, National University of Ireland, Galway
[3] University of New Brunswick and National Research Council, Canada
[4] Free University of Berlin, Germany
[5] Universidad Rey Juan Carlos, Madrid, Spain
[6] University of Surrey, UK

Abstract. This paper presents a framework for the reuse and extension of existing, established vocabularies in the Semantic Web. Driven by the primary application of expert finding, we will explore the reuse of vocabularies that have attracted a considerable user community already (FOAF, SIOC, etc.) or are derived from de facto standards used in tools or industrial practice (such as vCard, iCal and Dublin Core). This focus guarantees direct applicability and low entry barriers, unlike when devising a new ontology from scratch. The Web is already populated with several vocabularies which complement each other (but also have considerable overlap) in that they cover a wide range of necessary features to adequately describe the expert finding domain. Little effort has been made so far to identify and compare existing approaches, and to devise best practices on how to use and extend various vocabularies conjointly. It is the goal of the recently started ExpertFinder initiative to fill this gap. In this paper we present the ExpertFinder framework for reuse and extension of existing vocabularies in the Semantic Web. We provide a practical analysis of overlaps and options for combined use and extensions of several existing vocabularies, as well as a proposal for applying rules and other enabling technologies to the expert finding task.

1 Introduction

The Semantic Web has arrived! A growing number of people and institutions provide metadata on their personal or institutional Webpages in vocabularies based on RDF. Microformats provide another way to embed metadata directly within XHTML documents. The GRDDL working group [1], recently founded by the W3C, provides ways to derive semantically richer RDF from the structured data such as that of microformats. Additionally, the Semantic Web Best Practices and Deployment Working Group[2] provides guidelines on how to publish and syntactically combine RDF/XML or OWL

* This work has been partially supported by the EU NoE Knowledge Web (FP6-507482), the Spanish MEC (projects TIC2003-9001, TIN2006-15455-CO3), the German BMBF funded project Knowledge Nets and the Science Foundation Ireland (SFI/02/CE1/I131).

[1] `http://www.w3.org/2001/sw/grddl-wg/`

[2] `http://www.w3.org/2001/sw/BestPractices/`

E. Franconi, M. Kifer, and W. May (Eds.): ESWC 2007, LNCS 4519, pp. 235–250, 2007.

data and ontologies. Furthermore, browser extensions such as Semantic Radar[3] allow the detection of RDF data on web pages, and generic RDF browsers such as Tabulator[4] make exploration of Semantic Web data easier. While the syntactical issues are being solved by others, we take a closer look here at the actual vocabularies and ontologies that can be used for capturing metadata about persons and organisations for publishing on the Semantic Web. We propose a frequently-addressed, but still challenging, key application for the take-up of Semantic Web technologies: automating the task of finding experts (individuals, teams, and organisations), which is a daunting manual effort at the moment. Our assumption is that when persons, institutions, projects, and events are described in Web pages using agreed-upon machine readable formats, the automatic location of experts/expertise in a particular area or for a particular task will become feasible. To achieve this goal, and similarly for other applications of Semantic Web search, we identify three critical success factors:

- *Common machine readable formats* (syntax and semantics) supported by a
- *critical mass of users* (low entry barrier, tool support, reuse) as well as
- *enabling technologies* in place to solve practical use cases.

To this end, the following contributions are made in the remainder of this paper: In Sec. 2 we discuss each of these critical success factors. Sec. 3 describes the ExpertFinder initiative and we identify specific use cases which shall be covered within our chosen application domain. Sec. 4 contains the core of this paper, namely, we identify several quasi-standard ontologies relevant to expert finding, their relations to our domain, and propose how to combine them in the ExpertFinder Vocabulary Framework. We discuss related initiatives and approaches in Sec. 5 and conclude in Sec. 6.

2 Critical Success Factors

Common machine readable formats. As mentioned before, we will consider syntactical issues to be solved for the moment and assume that people know how to publish semantic annotations with their webpages and that there is proper tool support. We therefore focus on semantic aspects, by which we mean common existing ontologies and RDF vocabularies that can be used. The vocabularies that need to be considered comprise areas such as descriptions of personal and institutional data (including curriculum vitae and addresses), actual ontologies for modeling areas of knowledge/expertise, business sectors and communities, events, and publications. In section 4 we provide an analysis of existing vocabularies, ontologies and business standards for each of these fields.

Critical mass of users. Independent of whichever vocabulary we finally decide on, in order to really enable expert finding on a (Semantic) Web scale we have to either (i) convince a critical mass of users and content publishers to support the chosen vocabulary, (ii) translate/import existing content automatically into the chosen vocabulary, or (iii) provide mechanisms within the chosen vocabulary that can facilitate reuse of the vocabularies already utilised on the Web (e.g. using `owl:equivalentClass` and

[3] `http://sioc-project.org/firefox/`

[4] `http://www.w3.org/2005/ajar/tab`

owl:equivalentProperty). As for (i), creating a new ontology from scratch and disseminating it within a closed community is difficult. Within the wider Web community we can expect that enforcing the take-up of a single vocabulary is not only unlikely, but probably even impossible. Individual users and organisations will choose portions of ontologies, add extensions for their own purposes and will use different URIs to describe the same or similar concepts. In the context of (ii), text retrieval methods or wrapper technologies facilitated by approaches like PiggyBank[14] have become more stable and successful over the last few years. However, they still do not offer 100% precision or recall depending on how structured the underlying data source is, nor solve the problem of the right ontology/vocabulary to use for the generated annotations. Hence, there is nothing that can be called *the* right ontology for our domain and we will therefore focus on (iii) and try to reuse and effectively combine all existing and actually *used* formats. For this purpose, we can analyse and formally define their overlaps, and provide best practices on how to apply them together. This is the approach we focus on in the ExpertFinder initiative. A practical side effect of reusing and extending existing vocabularies is that we can rely on existing tools. For example, iCal[9], vCard[23], or BibTeX[19] provide vocabularies that are supported by tools (such as calendars, address book software) and online citation indexes, and are used already as de facto standards for data exchange.

Enabling technologies. The previous two success factors were concerned with how to get the necessary metadata on the Web. In order to solve practical use cases, such as the ones listed in Sec. 3, we have to consider several additional technologies. For example, recommendation algorithms, rules, strategies, collaborative filtering techniques, statistical methods, etc. Such methods will help to rate the value of metadata, but also allow support in annotation tools and search engines in order to find related ontological terms. Moreover, security and trust mechanisms that allow restricting access to certain metadata by encryption or authentication are gaining importance as the Web gets populated with personal information which should not be public. As a final key enabling technology, rules will play a crucial role in several respects. First, rules (together with expressive ontology languages) allow us to formally define the exact relationships between the existing vocabularies. For example, some vocabularies (e.g., FOAF, SIOC) already formalise their structure to some degree in OWL but usually do not incorporate many more features beyond simple taxonomies expressible in RDFS alone. We expect that expressive features beyond OWL[10] might be needed to define the exact relations between overlapping vocabularies. As there is no standard yet for defining such mappings rules, one could imagine SPARQL CONSTRUCT[5] along with built-in predicates. For instance imagine a mapping from vCard:homeTel to foaf:phone where the former is a datatype property and the latter an object property. Basically, a mapping needs a conversion function, generating a URI from the source RDF literal value:

```
CONSTRUCT { ?X foaf:phone ?T . } WHERE { ?X vCard:tel ?T1 .
  FILTER (fn:str(?T)=fn:concat("tel:",fn:encode-for-uri(?T1)))}
```

[5] Lacking a standard language to define complex mappings as this one, we admittedly "abuse" SPARQL FILTER expressions here for data conversion with XPath 2.0 Functions, which is likely not (yet) supported by existing engines.

Secondly, rules, published together with RDF metadata, can serve to "link" or define implicit metadata[26]. This would enable us to link to metadata published elsewhere involving (possibly negative) dependencies[20]. Sample rules could be for instance: `http://www.w3.org/People/Berners-Lee/card#i` *is an expert in* `http://en.wikipedia.org/wiki/Semantic_web`[6].

All persons listed at `http://www.rdfweb.org/topic/ExpertFinder_2fMembers` *but not those working for companies listed at* `http://www.myCompetitors/` *are my friends.* `http://www.w3.org/People/Berners-Lee/card#i` *is author of all publications listed at* `http://dblp.uni-trier.de/Berners=Lee:Tim.html`

Syntactically, one possibility is again to adopt CONSTRUCT queries from SPARQL as a view/link definition language[21], but also here a dedicated standard is still missing. We expect that W3C efforts like the RDF Data Access (DAWG)[7] and Rule Interchange Format (RIF)[8] working groups will soon provide adequate solutions in this direction.

3 Practical Use Cases from the ExpertFinder Initiative

ExpertFinder[9] is an international collaborative initiative with the aim of devising vocabulary and rule extensions (e.g. FOAF and SIOC) and best practices and recommendations towards standardisation in order to annotate personal homepages, pages of institutions, conferences, publication indexes, etc. with adequate metadata to enable computer agents to find experts on particular topics. The initiative was founded in 2006 with the goal to align related research efforts and to tackle precisely those critical factors outlined in Sec. 2. Among others, the following use cases were identified as potential early adopters of properly aligned Semantic Web data and vocabularies.

Automatic generation of institutional and personal webpages from metadata and rules. RDF metadata itself is an excellent source for content management systems separating content from any layout related issues, but in a more flexible manner than current solutions. Members of institutions could be allowed to provide their own metadata as extended FOAF files, but, if missing, the institution itself could specify standard policies for generating implicit metadata by means of default rules. Such rules could allow to aggregate metadata from some 3rd party sources. For instance, imagine that your office colleague is too lazy to generate his own homepage/metadata file. No problem: basic data can be aggregated from metadata available at the university personnel database, a default publication list can be generated by the metadata extracted from DBLP, and so on, by means of rules such as the ones in Sec. 2 or other techniques. By relying on common metadata formats natively, exporting, querying and combining information aggregated from different sources into annotated pages becomes trivial. Preliminary examples of this use of FOAF and other metadata are already proposed in[11].

Human Resource Management. People searching for a job could publish their CV and profiles as metadata on the Web or employees in a company could make their skill set

[6] Linking to Wikipedia terms is one of many possibilities to give areas of expertise an identifier.

[7] `http://www.w3.org/2001/sw/DataAccess/`

[8] `http://www.w3.org/2005/rules/`

[9] `http://www.rdfweb.org/topic/ExpertFinder`

and experiences available on the intranet in agreed metadata formats. Job agencies can deploy agents which they feed with their preferred profile crawling the web to identify suitable candidates for a given vacancy, or vice versa job vacancies could be published in the same formats. Team building within companies can be partly automated by selecting the right set of employees to successfully complete a given project through semantic matching and rules. ExpertFinder shall enable such scenarios and decentralise the process of expert and job finding, as opposed to current central recruitment or corporate portals, just as FOAF itself was aimed to decentralise social networks.

Public Semantic Research Portals. The ExpertFinder idea is fruitfully applicable to R&D community portals such as EU's successful CORDIS [10] which enable institutions to find and contact each other for joint research projects. Semantic enrichment of such portals may enable refined search down to the level of individual researchers, or allow decentralised publication by the institutions themselves. In the resulting scenario additional requirements on assessing trustworthy information arise: Instead of providing central portals, public bodies like the EU could assess/certify published content.

Semantic Reviewer Selection. In the academic realm, finding good reviewers remains a daunting task. However, many publications already provide pre-classified keywords such as ACM categories. Now, using citation indexes, committees of previous conferences etc. published in the agreed metadata format, one could define in a declarative rule language (possibly with priorities) some selection criteria to find appropriate expert reviewers, or adapt the selection criteria of previous workshops, if it is published by the organisers. Using common agreed vocabularies for categories, publications etc. mock-up examples using a combination of declarative rules and OWL such as presented at[10] could become a practical reality.

Trust and security for privacy-relevant meta-data. In all of the above scenarios it is desirable that (parts of) metadata can be protected, for instance by provision of time-restricted keys for decryption during a process of rule based negotiation (cf.[6] for pointers). Rules, similar to the ones mentioned above, can guide such negotiation. For example, the person who wants my phone number needs to invoke a service to get my phone number where all persons I know are registered. In the simplest case the service could work per email and check the sha1-sum of people in my FOAF file and send a mail back to that address with a (temporarily valid) decryption key for an encrypted telephone number also provided in my FOAF file. For different versions of this scenario, with different credentials, more involved negotiation processes are imaginable.

4 The ExpertFinder Vocabulary Framework

Instead of proposing a new ontology for tackling the challenges of semantic expert finding we rather suggest a framework of existing vocabularies which shall be fruitfully combined. As shown in Fig. 1 we identify the following "components" to describe experts (persons, organisations or communities):

- *general descriptions* of persons, communities and organisations,
- *relations* between persons, communities and organisations,

[10] http://cordis.europa.eu/en/home.html

- *educational aspects*,
- past and present *activities and projects*, and
- *skills*

Additional fields not uniquely connected to particular persons or organisations that we want to cover are *events* and *publications*, opinions and ratings, endorsements as well as recommendations and references. Our goal is to pick some of the most widely used vocabularies in these areas check how far they are formalised, identify what overlaps exist between these formats and how they can be reused and combined[11].

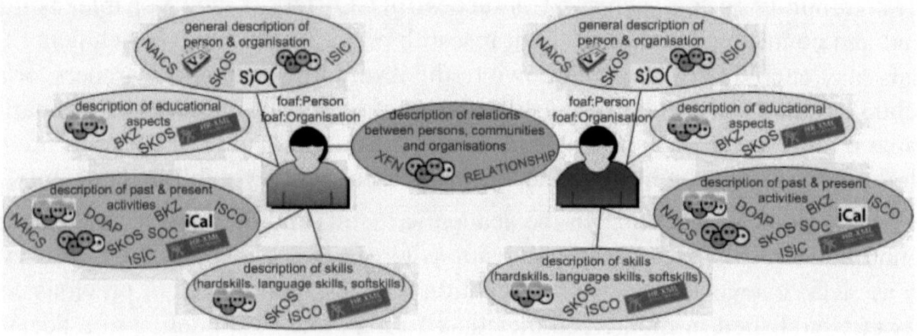

Fig. 1. How to describe an expert?

4.1 Starting Points: FOAF, SIOC and SKOS

The Friend of a Friend (FOAF)[8] and Semantically-Interlinked Online Communities (SIOC)[7] ontologies mark the starting points of our work as, on one hand, they already cover the description of much of the "components" mentioned above, and, on the other hand, they are being adopted by a steadily increasing user community.

The FOAF ontology was developed to create machine readable information/metadata for people, groups, organisations and other related concepts - basically, to describe people, what they do and how they interact with each other. One of the most used properties of the FOAF ontology is the "knows" property: a simple way to create social networks through the addition of knows relationships for each individual that a person knows. Aggregations of FOAF data from many individual homepages are creating distributed social networks; this can in turn be connected to FOAF data from larger online social networking sites such as LiveJournal[12] or Tribe.

In terms of definitions of expertise the FOAF ontology has a number of relevant properties, e.g. (i) the `foaf:interest` property defines topics of interest to a person, and can be used directly to find those with an interest a particular domain (e.g. `foaf:interest` has been used to match music preferences[13]), (ii) people can create `foaf:publications` or other `foaf:Documents` (via `foaf:made/maker`)

[11] We do not aim at providing an exhaustive list of all ontologies developed in all related areas.

[12] `http://rdfweb.org/topic/LiveJournal`

[13] `http://foafing-the-music.iua.upf.edu/`

which may have an associated `foaf:topic` or `foaf:primaryTopic` that can again be used to determine a person's domains of interest, and (iii) `foaf : current Project/pastProject` gives information on "some collaborative or individual undertaking" that a person may be (or have been) involved in.

There have been a number of extensions or modules for the FOAF ontology that are of interest to the expert finding scenarios previously mentioned. FOAFRealm[16] is a user profile management system based on FOAF that provides authentication, access control and social networking features such as "semantic social collaborative filtering". The system allows users to share and annotate their personal taxonomies across a social network using WordNet, DDC[14] and DMoz[15] as base classifications. When implemented in document exchange systems such as JeromeDL[16], a semantic digital library, users can classify their documents or bookmarks and allow others to access these resources using FOAFRealm's social networking functionality. Each user's collection is assigned an expertise value that reflects the quality of the information that they provide; this value is calculated based on a PageRank calculation of their social network. Users are then also aware of the expertise level of others on given topics.

The SIOC project[17], founded by one of the authors, aims to provide a framework for the connection and interchange of information from internet-based discussions and community portals. Such communities are primarily made up of users, the posts that they create, and the discussion forums that they subscribe to across a multitude of sites and discussion platforms. The basis for SIOC is the SIOC ontology, an RDF-based schema which describes the main concepts found in online communities[7]. While there are many classes and properties in SIOC, the main notion is that `sioc:Users` create `sioc:Posts` that are contained in `sioc:Forums` that are hosted on `sioc:Sites`. With respect to finding experts in a social network, the main SIOC property of interest is `sioc:topic` defining a resource that a particular discussion post is related to; by aggregating all the `sioc:topics` that are associated with a particular user's posts across a number of sites, a picture emerges as to where their topics of interest and related expertises lie. `sioc:Forums` or sites may also have associated `sioc:topics`, and a user with an interest in a particular topic may be a `sioc:subscriber_of` a certain discussion channel.

The Simple Knowledge Organisation System (SKOS)[17] completes the base we want to build on. It allows to describe general terms and concepts and define many useful properties of such terms such as declaring whether a concept is broader/narrower than another, preferred and alternative labels in multiple languages for terms, as well as related terms. SKOS facilitates sharing and representing terminologies that may not extensively require the expressive power of other languages such as OWL and where a strict hierarchy such as definable by `rdfs:subClassOf` cannot be imposed. In the context of ExpertFinder, we can view SKOS as the basis to define and relate skills, areas of expertise/interest (via the `foaf:interest` property) or topics people discuss in online communities described by SIOC.

---

[14] `http://www.oclc.org/dewey/`
[15] `http://dmoz.de/`
[16] `http://www.jeromedl.org/`
[17] `http://sioc-project.org/`

The SIOC ontology developers have worked with the authors of FOAF and SKOS to align concepts and avoid any unnecessary duplication or term conflicts. The concept of `sioc:User` has been defined to be a sub-type of `foaf:onlineAccount`, so that existing properties from FOAF can be reused and so that new properties for users can be defined in SIOC without directly impacting on the FOAF ontology. As shown in Fig. 2, a `foaf:Person` can own many `sioc:User` profiles (via the `foaf : holds OnlineAccount` relationship). Similarly, content that a `sioc:User` creates on a particular Forum (e.g., a Weblog, Mailing List, Bulletin Board) can be linked using `sioc : topic` to a `skos:Concept` (e.g., in Fig. 2 one post is talking about clouds and another post is referring to a narrower concept, that of rain clouds). Using SKOS to define topics under discussion and of interest combined with additional rule extensions, which we plan as a next step in the ExpertFinder framework, facilitates flexible definitions of relationships between the various skills formalised using SKOS concepts.

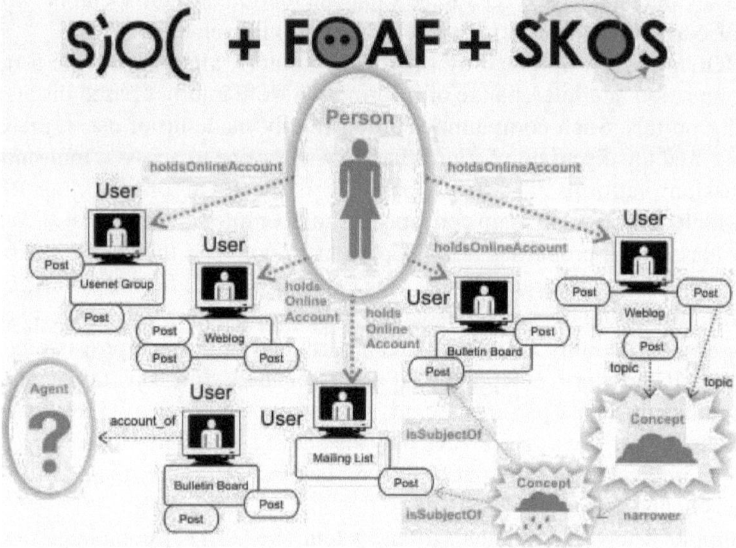

Fig. 2. Connections between SIOC, FOAF and SKOS

4.2 ExpertFinder Framework Extensions for the Core Vocabularies

FOAF, SIOC and SKOS largely cover general descriptions of, as well as relations between persons, communities and organisations. However, still some pieces are missing in order to obtain a complete picture of Fig. 1:

- FOAF misses detailed information about address data, but complementary standards such as vCard close this gap.
- The only relation between persons in FOAF is `foaf:knows`, but as we want to support more fine-grained relations, we propose the RELATIONSHIP & XFN vocabularies to close this gap.

- Projects can be linked to persons and groups by `foaf:currentProject` and `foaf:pastProject`. However, this might again be too coarse-grained if e.g. we want to know the exact timing of a project and we want to provide guidelines on how to annotate projects. We propose the use of DOAP here.
- More detailed CV information can be provided by the DOAC vocabulary.
- In the scientific context, publications are an important measure of expertise. These can be linked by the `foaf:maker` and `foaf:publications` properties to the `foaf:Document` class, but details on how to describe publications are missing. A de facto standard in the scientific community in this area is BibTeX.
- As for describing skills and topics of interest, SKOS defines a general framework, but details of concrete classifications to use for annotation are missing.
- Finally, events and their participants are not yet describable in a sufficient manner. iCal as a de facto standard for sharing events is a natural candidate to fill this gap.

A preliminary list of mappings for overlapping concepts and attributes in the mentioned vocabularies is omitted here for lack of space, but available at `http : //www.rdfweb.org/topic/ExpertFinder`$_2$`fmappings`

Refining Personal Data: vCard vCard is a standard for representing personal data such as business cards. Although there are various forms in which vCard data can be written, our interest is on the RDF-based representation [18]. Contact information, such as phone numbers or email-addresses can be expressed more fine-grained in vCard than in FOAF by means of distinguishing properties such as `vCard:homeTel` and `vCard:workTel`. Note, however, that vCard phone numbers are not directly mappable to `foaf:phone` as a subclass, as vCard uses RDF literal values whereas foaf uses URIs using the fully qualified `tel:` scheme. A workaround we propose is to adopt FOAF's representation in vCard/RDF and make the respective properties subproperties of `foaf:phone`, or otherwise define straightforward mapping rules for conversion. Such mappings can not necessarily be bidirectional, e.g., `vCard:email` may not be simply mapped to `foaf:mbox` as a `foaf:mbox` is supposed to be unique for a person, which is not necessarily the case for vCard. Affiliation information or role information in vCard can indicate knowledge areas or particular expertise aspects, which again should be linkable to SKOS concepts.

Refining Relations: RELATIONSHIP & XFN RELATIONSHIP[19] and XFN[20] (XHTML Friends Network) are two vocabularies used for describing interpersonal relationships. Since `foaf:knows` describes relationships between people rather sketchily, these vocabularies are deployed to fill the gap and assert such relationships in more detail by defining different subproperties.

Refining Project Descriptions: Description of a Project (DOAP) DOAP[21] is an XML/RDF vocabulary mainly conceived to describe open source projects. Its initial goals include: (i) internationalisable description of a software project and its resources; (ii) data exchange between software directories; (iii) automatic configuration for resources

[18] `http://www.w3.org/TR/vcard-rdf,http://www.w3.org/2006/vcard/ns`
[19] `http://vocab.org/relationship/`
[20] `http://gmpg.org/xfn/join`
[21] `http://usefulinc.com/doap/`

such as shared CVS repositories; (iv) interoperability with other popular Web metadata projects (RSS, FOAF, DC) and (v) the ability to extend the vocabulary for specialist purposes. DOAP describes the current state of a project but it does not highlight changes and updates. However, to keep the repository up to date with releases, the CodeZoo with an Atom[22] feed containing embedded DOAP can be used. Nevertheless, even if a feed to keep older versions can be used for DOAP a way to transform the information into RDF and to distinguish between current release and past releases is still needed.

DOAP uses `foaf:Person` to describe the corresponding contributors of each project part (e.g. project maintainer, developer). From the FOAF side, the use of `foaf:project`, `foaf:currentProject` and `foaf:pastProject` properties do not really allow to define the duration of participation in a certain project. Neither are project durations definable properly in DOAP alone. We suggest to remedy this problem by either adding new attributes for start and end (possibly subclassing iCal events) or using temporal RDF as mentioned below.

Refining CV information: Description of a Career (DOAC), Resume RDF Schema and BIO Vocabulary: DOAC[23] is a RDF metadata vocabulary to describe professional capabilities of workers gleaned for example from CVs or resumes. The metadata enhances specific description and facilitate the search to locate suitable (regarding the given position requirements) job candidates. DOAC has been designed to be compatible with the European CV (known as Europass), which can be generated from a FOAF+DOAC file. It includes information about education, work experience, publications, spoken languages and other skills that can be shared and processed by applications. As an alternative, one of the authors[5] proposes the Resume RDF schema[24] for extending FOAF profiles with curriculum vitae information. This schema includes terms for work and academic experience, skills, courses and certifications, publications, references, etc. Again, we propose to link to SKOS concepts to describe the respective concepts. BIO[25] describes biographical information about (living and dead) people and has been designed to be compatible with both RDF and non-RDF XML formats.

DOAC uses the `foaf:Person` class to general descriptions of job seekers and `foaf:Organisation` to define which schools and institutions the individual attended. Furthermore, the `foaf:pastProject` concept could be added as a subclass to the `doac:Experience` class. This would allow description of not only a job seeker's general experiences in a company but also their experiences in different projects. Next, the `doac:publication` property which establishes a connection between the `foaf:Person` and `doac:Publication` can be defined as a `foaf:publications` linking `foaf:Person` with `foaf:Document`.

Refining Bibliographic Descriptions: BibTeX, DC and others BibTeX was designed by Patashnik and Lamport in 1985 as the LaTeX bibliographic format[19] and has established itself as a de facto standard format for publishing bibliographic information in several online citation indices. Several RDF versions of BibTeX (e.g. bibtex2rdf,

[22] `http://www.codezoo.com/about/doap_over_atom.csp`
[23] `http://ramonantonio.net/doac/`
[24] `http://purl.org/captsolo/semweb`
[25] `http://vocab.org/bio/0.1/`

bib2rdf, bibTeX [26]) reuse existing formats in the same spirit as ExpertFinder. This could be directly adopted or combined with more comprehensive ontologies for digital libraries such as MarcOnt[27]. As MarcOnt also allows to import/export BibTeX, we currently suggest the RDF vocabulary supported by bibtex2rdf (cf.[11]). The Dublin Core (DC) initiative, started 10 years ago by librarians in order to provide a metadata standard for describing documents, may be viewed as a subset of bibtex, and is actually reused by bibtex2rdf.

Classifications & Standards for Skills and Topics. In the following we describe a few selected standards and classifications of occupations, competencies and economic activities as possible schemes which could be used for defining skills and topics of interest. Some of these standards are used for example as an instrument for assembling and presenting statistics of education/training on national as well as international level, some others are developed for fostering international comparability of data in studying economic phenomena. While the national and international types of classifications are used for example by federal agencies for education and training statistics, the international standards should also facilitate international communication.

The *Occupational Classification (SOC)*[28] system is used by federal US statistical agencies to classify workers into over 820 occupational categories, grouped into23 major groups, 96 minor groups, and 449 broad occupations. Each broad occupation includes detailed occupation(s) requiring similar job duties, skills, education, or experience. The *Profession Reference Number Classification (BKZ)*[29] is a German version SOC, detailing 5597 occupations. The *International Standard Classification of Occupations (ISCO-88)* is developed to facilitate international communication regarding occupations and occupational groups. Persons are classified by occupation through their relationship to a past, present or future job. The *International Standard Industrial Classification of All Economic Activities (ISIC)* [30] is a standard classification of economic activities arranged to classify entities according to the type of activity they carry out. *North American Industry Classification System (NAICS)*[31] provides common industry definitions for Canada, Mexico, and the US to facilitate economic analyses. Further standards to classify products and services like *eCl@ss*, *eOTD*, or the *RosettaNet Technical Dictionary*, or *UN-SPSC* could also partly serve to describe skills and topics. All these are however, with few exceptions such as e.g. eCl@ssOWL[13], not (yet?) available in "ontologised" versions. Mappings of concepts into SKOS terms is an open issue on our agenda. A key issue here is to assign proper URIs usable in RDF to these concepts:

However, apart from special suitable classification systems which yet need to be "webised", a simpler possibility to define topics which we already used in an example of Sec. 2, was simply referring to an online encyclopedia such as Wikipedia. Recent efforts towards semantically structuring wikis (cf.[15]) support such an approach. For

[26] `http://www.l3s.de/~siberski/bibtex2rdf/`, **http : //www.cs.vu. nl/ mcaklein/bib2rdf/**, `http://zeitkunst.org/bibtex/0.1/`

[27] `http://www.marcont.org/`

[28] `http://www.bls.gov/soc/`

[29] `www.arbeitsamt.de/hst/markt/news/BKZ_alpha.txt`

[30] `http://unstats.un.org/unsd/cr/registry/regcst.asp?Cl=17&Lg=1`

[31] `http://www.census.gov/epcd/www/naics.html`

some smaller domains such as computer science, ACM categories [32] or the WWW Conference Archive areas [33] already provide URI-addressable categories of topics, usable as SKOS terms, e.g. using the latter, we refine the example of Sec. 2 to:
`http://www.w3.org/People/Berners-Lee/card#i` *is an expert in* `http://wwwconf.ecs.soton.ac.uk/view/subjects/2004-C4.html`
where we can further add:
`http://wwwconf.ecs.soton.ac.uk/view/subjects/2004-C4.html` *is* `skos:narrower` *than* `http://en.wikipedia.org/wiki/Semantic_Web`.

Events and termporal information iCal[9] as a de facto standard for calendar information supported by many applications is a natural starting point for ExpertFinder to refer to events. RDF formats and conversion tools for iCal are available[34]. Still, iCal alone might not be sufficient to denote e.g. the validity duration of certain RDF information such as participation duration in a project. Exploring the use of RDF extensions by temporal information[12] to express the validity duration of triples would be an interesting option, but standardisation of such extensions does not yet seem to be in sight.

5 Related Projects, Initiatives and Approaches

Many other projects and initiatives overlap with or are relevant to the ExpertFinder initiative. As an umbrella initiative involving several organisations, some of these efforts are continued among the ExpertFinder participants and results will be exchanged both from these to ExpertFinder and vice versa, giving ExpertFinder the opportunity to already impact through its work into present activities as well as being open to these activities to impact upon the broader ExpertFinder efforts.

5.1 Related Projects

Several projects in the Semantic Web realm have already created their own ontologies for describing persons, organisations and activities. For instance, the *KnowledgeWeb platform ontologies*[35], the *AKT portal ontology*[36], the *SWRC portal ontology*[37], and the *DERI Semantic Web Portal* (SWP) working group's ontology[38] cover many aspects of the expert finding domain, and could thus be arguably seen as equally valid starting points. However, so far these approaches seem to have experienced little take-up outside the projects where they have been developed and were developed from scratch rather than being based in pre-existing vocabularies or de facto standards. Only the DERI SW-portal ontology reuses FOAF, RSS and BibTeX to some extent, while such reuse is a central rationale in our approach.

[32] `http://www.acm.org/class/1998/`
[33] `http://wwwconf.ecs.soton.ac.uk/view/subjects/subjects.html`
[34] `http://www.w3.org/2002/12/cal/`, `http://www.kanzaki.com/courier/ical2rdf` or `http://torrez.us/ics2rdf/`
[35] `http://knowledgeweb.semanticweb.org/semanticportal/OWL/`
[36] `http://www.aktors.org/ontology/portal`
[37] `http://swrc.ontoware.org/ontology`
[38] `http://sw-portal.deri.org/ontologies/swportal`

Some related projects conducted by ExpertFinder initiative members have already followed this rationale, like Knowledge Nets, SemDis, FindXpRT and SIOC.

The *Knowledge Nets*[39] project explores the potential of Semantic Web from a business and a technical viewpoint by means of pre-selected use case scenarios. For this purpose, a prototype for the e-Recruitment domain containing the online job seeking and job procurement processes has been developed[2,18,22]. The requirements analysis revealed the necessity of aligning with commonly used domain standards and classifications (SOC, BKZ, WZ2003, NAICS, HR-XML[40] and Skill Ontology) in order to integrate job seeker profiles and job postings as well as to support common practices from industry. Reusing these standards, the HR-ontology contributes to the realisation of more powerful and flexible e-Recruitment solutions which include advanced search and presentation facilities based on knowledge about the application domain.

The *SemDis* project addresses development of query/discovery techniques for semantic relationships. For example, `dblp:co-authorship` and `foaf:knows` were used to detect possible conflicts of interest between reviewers and authors in a peer-review process[1]. An extension on such work aims at determining possible reviewers by comparing their expertise to the topics of a paper. The expertise of a person on different topics or areas is described using the SwetoDblp dataset[41], a large populated ontology of computer science publications based on the DBLP[42] bibliography database.

The *FindXpRT* (Find an eXpert via Rules and Taxonomies)[26] focuses on the important aspect of rules by combining FOAF facts and RuleML[4] rules. This implemented system[43] allows users to derive FOAF data by deploying person-centric rules, either before FOAF publication or, on demand, from published (RuleML FOAF) pages.

Finally, the SIOC initiative, which forms a central part of the ExpertFinder framework is being developed by and in collaboration with initiative members with cross-fertilizing effects between the two initiatives.

5.2 Community-Driven Approaches

In this paper we focused on the specific domain of expert finding and explored "established" vocabularies in this domain. Other Web and Semantic Web application areas show the dynamics and need for alignment even more drastically: A recent trend in many popular non-academic portals is to allow communities to create their own vocabularies and tag the items/information they want to share with others with arbitrary tags from their vocabularies: The *del.icio.us* portal[44] allows communities to tag and share their bookmarks, and search others bookmarks on the basis of these tags. The *43Things* and *43Places* community portals[45] allow describing and sharing by community-created tags information about the things people do and places they travel or want to travel.

[39] `http://wissensnetze.ag-nbi.de`
[40] Developed by the HR-XML Consortium, `http://www.hr-xml.org`
[41] `http://lsdis.cs.uga.edu/projects/semdis/swetodblp/`
[42] `http://www.informatik.uni-trier.de/~ley/db/`
[43] `http://www.ruleml.org/usecases/foaf/JieLiMCSThesis.pdf`
[44] `http://del.icio.us`
[45] `http://www.43things.com` and `http://www.43places.com`

flickr[46] allows members to share, search and tag photos, again with arbitrary tags. *GoogleBase*[47] is a community application which allows Web users to share and search arbitrary items (pictures, text, ads, web-sites) and annotate these items using arbitrary attribute-value pairs. Most popular/shared attributes and attribute values come up in the upper level of Google search interfaces and are proposed to be used for searching and browsing the available items. None of these sites is based directly on Semantic Web technologies. However, the offered functionality is reminiscent significantly of earlier academic proposals in the Semantic Web realm, e.g. the People's Portal[24]. The examples reveal a trend of the Web becoming more structured and annotated in a community-driven manner via social processes and contributions of Web users. Reuse and adoption of already existing broadly used formats as we propose here could accelerate this process on the one hand, and on the other hand extensions for existing vocabularies could be developed in a community-driven process.

A common problem in completely unguided community-driven approaches is that entities and tags are different, yet semantically similar. This tendency brings difficulties for the community members in reuse of the community-contributed knowledge contained in the system. Defining mappings and finding an agreement on a meta-level upon which tags might become superfluous/deprecated by enforcing best practices are crucial for applications in an open web environment. Existing community-driven proposals ignore this problem to a large extent, or, in the most advanced cases, users are proposed to create ad-hoc, non-reusable alignments to achieve a specific task. Minimal support for reuse, such as auto-completion of tag names or suggestion of related search terms in annotation tools and search engines is partly supported by the above-mentioned platforms but mappings are not definable themselves in a community-driven process.

Community-driven ontology management and ontology matching extends conventional ontology matching by involving end users, knowledge engineers, and developer communities in the processes of establishing, describing and reusing vocabularies and inter-ontology mappings[25]. We believe that easy to use mapping and rules languages and tools as the next logical step. However, as we mentioned already in Sec. 2, a standard format for defining these mappings is still missing.

6 Conclusions and Outlook

We described the integration of efforts of members of the ExpertFinder initiative towards a common goal: combining commonly-agreed vocabularies including but not limited to describing information of people and their expertise, organisations, contact information, social and collaborative networks, etc. As members of this initiative, we have described various practical use cases for the task of expert finding which we identified as promising applications for actual take-up of Semantic Web technologies. We described three key success factors for bringing agreement and facilitating the take off of a joint vocabulary for expert finding. Based on this, we proposed the ExpertFinder vocabulary framework which stresses reuse and cautious extension of existing and established vocabularies in the Semantic Web. In this framework, we described how FOAF,

[46] `http://www.flickr.com`

[47] `http://base.google.com`

SIOC and SKOS mark the starting points. We also discussed how to use these together extended with various existing vocabularies, pointing out the necessity for formal mappings between overlapping terms which we provide at the initiative's Web page. Along the way, we have given a survey and analysis of the related vocabularies and classifications which, although restricted to the particular domain of expert finding, we hope to be useful as such also for other related Semantic Web applications. Although we deem the core defined so far a useful start which can already be used to cover several of our proposed use cases, we have to leave some extensions towards security, reputation and trust mechanisms (e.g., referencing endorsements or trust ontologies), which we only treated superficially so far, for future work.

References

1. B. Aleman-Meza, et al. Semantic Analytics on Social Networks: Experiences in Addressing the Problem of Conflict of Interest Detection. *15th Intl. WWWW Conference 2006,*, 2006.
2. C. Bizer, R. Heese, M. Mochol, R. Oldakowski, R. Tolksdorf, and R. Eckstein. The Impact of Semantic Web Technologies on Job Recruitment Processes. *7th Internationale Tagung Wirtschaftsinformatik 2005*, 2005.
3. U. Bojars. Extending FOAF with Resume Information. In *Proc. of the 1st Workshop on FOAF, Social Networks and the Semantic Web*, 2004.
4. H. Boley, S. Tabet, and G. Wagner. Design Rationale of RuleML: A Markup Language for Semantic Web Rules. *Semantic Web Working Symposium (SWWS'01)*, 2001.
5. U. Bojars. Extending FOAF with Resume Information. *1st Workshop on FOAF, Social Networks and the Semantic Web*, 2004.
6. P.A. Bonatti and D. Olmedilla. Semantic web policies: Where are we and what is still missing? Tutorial at *3rd European Semantic Web Conference (ESWC'06)*, 2006.
7. J.G. Breslin, A. Harth, U. Bojars, and S. Decker. Towards Semantically-Interlinked Online Communities. *2nd European Semantic Web Conference (ESWC'05)*, 2005.
8. D. Brickley and L. Miller. Friend of a Friend Vocabulary Specification. `http://xmlns.com/foaf/0.1/`, 2001.
9. F. Dawson and D. Stenerson. Internet Calendaring and Scheduling Core Object Specification (iCalendar). `http://www.ietf.org/rfc/rfc2445.txt`, 1998.
10. T. Eiter, G. Ianni, A. Polleres, and R. Schindlauer. Answer set programming for the semantic web. Tutorial at *3rd European Semantic Web Conference (ESWC'06)*, 2006.
11. G. AAstrand Grimnes, S. Schwarz, and L. Sauermann. RDFHomepage or "Finally, a use for your FOAF file". *2nd Workshop on Scripting for the Semantic Web (SFSW '06)*, 2006.
12. C. Gutierrez, C. Hurtado, A. Vaisman. Temporal RDF. *2nd European Semantic Web Conference (ESWC'05)* 2005.
13. M. Hepp. Products and Services Ontologies: A Methodology for Deriving OWL Ontologies from Industrial Categorization Standards, Intl. Journal on Semantic Web & Information Systems, 2(1):72–99, 2006.
14. D. Huynh, S. Mazzocchi, and D. Karger. Piggy Bank: Experience the Semantic Web Inside Your Web Browser. *Intl. Semantic Web Conference 2005 (ISWC2005)*, 2005.
15. M. Krötzsch, Denny Vrandecić, and M. Völkel. Wikipedia and the Semantic Web - The Missing Links. *Proc. of WikiMania2005*, 2005.
16. S.R. Kruk and S. Decker. Semantic Social Collaborative Filtering with FOAFRealm. *Semantic Desktop Workshop colocated with Intl. Semantic Web Conference (ISWC2005)*, 2005.
17. A. Miles and D. Brickley (eds.). SKOS Core Vocabulary Specification, 2 November 2005. W3C Working Draft, `http://www.w3.org/TR/swbp-skos-core-spec`.

18. M. Mochol, R. Oldakowski, and R. Heese. Ontology-based Recruitment Process. *GI 2004*.
19. O. Patashnik. BIBTeXing, 1998.
20. A. Polleres, C. Feier, and A. Harth. Rules with contextually scoped negation. *3rd European Semantic Web Conference (ESWC'06)*, 2006.
21. A. Polleres. SPARQL Rules! Tech. Report, `http://www.polleres.net/publications/GIA-TR-2006-11-28.pdf`, 2006.
22. R. Tolksdorf, M. Mochol, R. Heese, R. Eckstein, R. Oldakowski, and C. Bizer. Semantic-Web-Technologien im Arbeitsvermittlungsprozess. *Wirtschaftsinformatik: Internetoekonomie*, 48(1):17–26, 2006.
23. A versit Consortium. vCard: The Electronic Business Card. `http://www.imc.org/pdi/vcardwhite.html`, 1997.
24. A.V. Zhdanova. An Approach to Ontology Construction and its Application to Community Portals, PhD thesis, 2006.
25. A.V. Zhdanova and P. Shvaiko. Community-Driven Ontology Matching. *3rd European Semantic Web Conference (ESWC'06)*, 2006.
26. J. Li, H. Boley, V.C. Bhavsar, and J. Mei. Expert Finding for eCollaboration Using FOAF with RuleML Rules. 2006 Conference on eTechnologies. Montreal, Canada, 2006. .

Extracting Social Networks Among Various Entities on the Web

Yingzi Jin[1], Yutaka Matsuo[2], and Mitsuru Ishizuka[1]

[1] University of Tokyo, Hongo 7–3–1, Tokyo 113-8656, Japan
eiko-kin@mi.ci.i.u-tokyo.ac.jp, ishizuka@i.u-tokyo.ac.jp
[2] National Institute of Advanced Industrial Science and Technology
y.matsuo@aist.go.jp

Abstract. Social networks have recently attracted much attention for their importance to the Semantic Web. Several methods exist to extract social networks for people (particularly researchers) from the web using a search engine. Our goal is to expand existing techniques to obtain social networks among various entities. This paper proposes two improvements, i.e. *relation identification* and *threshold tuning*, which enable us to deal with complex and inhomogeneous communities. Social networks among firms and artists (of contemporary) are extracted as examples: Several evaluations emphasize the effectiveness of these methods. Our system was used at the International Triennale of Contemporary Art (Yokohama Triennale 2005) to facilitate navigation of artists' information. This study contributes to the Semantic Web in that we increase the applicability of social network extraction for several studies.

1 Introduction

Social networks explicitly exhibit relationships (called *ties* in social sciences) among individuals and groups (called *actors*). They have been studied in social sciences since the 1930s. To date, vastly numerous studies using social network analysis have been conducted [22]. In the context of the Semantic Web, social networks are crucial to realize a web of trust that facilitates estimation of information's credibility and its provider's trustworthiness [10]. Ontology construction is also related to social networks: P. Mika discusses the relation between the community and emergent ontology from a social network perspective [18]. Information sharing and recommendation [19,9] on social networks are other applications that are served by the Semantic Web. Our lives are influenced strongly by social networks without our knowledge of their implications. For that reason, many applications are relevant to social networks [23].

Social networks are obtained from various sources, such as e-mail archives, FOAF documents, and DBLP. For example, T. Finin et al. extract a social network from the web by collecting FOAF documents [7]. Particularly, several studies have been undertaken to use a search engine to extract social networks from the entire web [11,16,17]. Co-occurrence of names on the web, which is basically obtained by posing a query including two names to a search engine, is commonly used as proof of relational strength. Using a search engine to recognize the relation of two entities (or two words) has increasingly gained attention in the field of natural language processing [5,12,24].

E. Franconi, M. Kifer, and W. May (Eds.): ESWC 2007, LNCS 4519, pp. 251–266, 2007.

This study is intended to expand current social-network mining techniques using a search engine to obtain a social network among various entities. Specifically in this paper, two improvements are proposed in order to apply our method to complex and inhomogeneous communities: *relation identification* and *threshold tuning*. We extract two social networks as examples: artists of contemporary art, and famous firms in Japan. We must identify the relation types such as alliances and lawsuits; consequently, we can make elaborate queries and apply text processing to extract a social network among firms. Our algorithm adds a *relation keyword* to the search query to emphasize a specific relationship. Extracting a social network of artists, on the other hand, requires adaptive tuning of thresholds because the appearance of each artist on the web is completely different. Optimal thresholds are sought to invent appropriate edges between entities.

Our contributions are summarized as follows: First, through the two improvements, i.e. relation identification and threshold tuning, which respectively focus on complex and inhomogeneous communities on the web, social network extraction becomes more generally applicable to various entities. We argue the general social network extraction in the last part of the paper, which can cultivate existing studies using social networks in the Semantic Web. Second, because our method can extract relations from among entities, it can output machine-processable knowledge about the relations automatically from the information on the current web. Although some approaches exist to generate RDF statements by web mining, our study provides an alternative; our intuition is that extracting a social network might provide information that is only recognizable from the network point of view. For example, the *centrality* of each firm is identified only after generating a social network.

The next section introduces related studies. Section 3 describes the investigation of different appearance of entities on the web and addresses our ideas to obtain various social networks from the web. Sections 4 and 5 introduce our case studies, which specifically investigate two types of networks: those of firms and artists. In Section 6, before we conclude the paper, we propose a general architecture of social network extraction and discuss applications of the extracted social networks to the Semantic Web.

2 Related Works

Numerous studies have obtained and analyzed social networks on the web: L. Adamic collects relations among students from web link structure and text information, and characterizes the social networks among Stanford students and MIT students [1]. T. Finin describes a large collection of FOAF documents (over 1.5 million) from the web and analyzes the structure of friendship networks in the Semantic Web [7]. Trust calculation [10] is a major application of social networks. Some studies seek other applications: A. McCallum and his group present an end-to-end system that automatically integrates both e-mail and web content to help users maintain large contact databases [6]. Aleman-Meza et al. use relational data from both FOAF and DBLP to detect relationships among potential reviewers and authors of scientific papers [2].

Several studies have particularly addressed use of a search engine for social network extraction. In the mid-1990s, H. Kautz and B. Selman developed a social network extraction system called the *Referral Web* [11]. The system uses a search engine to retrieve web documents that include a given personal name. Recently, P. Mika developed

Flink, a system for extraction, aggregation, and visualization of online social networks for the Semantic Web community [17]. A social network of 608 researchers from both academia and industry is extracted and analyzed. The web-mining component of Flink, similarly to that used in Kautz's work, employs co-occurrence analysis. The strength of relevance of two persons, X and Y, is estimated by putting a query X AND Y to a search engine: If X and Y share a strong relation, we can usually find much evidence on the web such as links found on home pages, lists of co-authors in technical papers, organizational charts, and so on. In Flink, the strength of relations among individuals is calculated using the Jaccard coefficient $n_{X \cap Y}/n_{X \cup Y}$, where $n_{X \cap Y}$ represents the number of hits yielded by the query X AND Y and $n_{X \cup Y}$ represents the number of hits by the query X OR Y. The two researchers are considered to share a relation if the value is greater than a certain threshold. The term "*Semantic Web* OR *ontology*" is added to the query for name disambiguation.

Matsuo et al. developed a system called *POLYPHONET*, which also uses a search engine to measure the co-occurrence of names [15,16]. In their study, several co-occurrence measures [13] have been compared, including the matching coefficient ($n_{X \cap Y}$), mutual information, Dice coefficient, Jaccard coefficient, and overlap coefficient. The overlap coefficient $n_{X \cap Y}/\min(n_X, n_Y)$ performs best according to the experiments. In addition, POLYPHONET was operated at several AI conferences in Japan and a couple of international conferences to promote participants' communication. For disambiguating personal names, key phrases such as affiliations are added to queries.

We regard the two studies by Mika and Matsuo as relevant precedent studies, and propose some improvements to increase the applicability of that approach.

3 Extraction of Social Networks

3.1 Problem of Existing Methods

The fundamental idea underlying the existing studies by Mika and Matsuo is that *the strength of a relation between two entities can be estimated by co-occurrence of their names on the web*. The criteria to recognize a relation, such as the measure of co-occurrence and a threshold, are determined beforehand. An edge will be invented when the relation strength by the co-occurrence measure is higher than the predefined threshold. Although the approach is effective for extracting a social network of researchers, our preliminary study indicates that it does not perform well for various entities on the web.

As the first reason, co-occurrence-based methods become ineffective when two entities co-occur universally on numerous web pages. For example, when we want to infer two firms' relations from the web, we submit a query "*Matsushita* AND *JustSystem*"[1] to a search engine. Consequently, we are referred to as many as 425,000 pages, for which the Jaccard coefficient is 0.031. However, this figure is unreliable considering the media effect on the web. In the domain of firms, many relations are published in news reports and on news releases that are distributed on the web. Many web pages describe and comment on the relation if the news is given attention by media services or people. Conversely, if it were not attention given, only a small number of pages

[1] Both are names of famous Japanese corporations.

would describe the relations. Considering that media effects influence the number of web pages, co-occurrence of names on the web is not always available to represent the relational strength of two entities.

For the second reason, co-occurrence-based methods function ineffectively when applied to *inhomogeneous* communities. An inhomogeneous community means, in this paper, a community that includes people in different fields, different nations, or different cultures, where a relation is difficult to obtain using a single criterion. The researchers' communities (of the same research field) usually present a homogeneous character; for that reason, using a single criterion to calculate the relation works well. In contrast, the international artist community is more inhomogeneous. For example, two Japanese artists, "*Taisuke Abe*" and "*Jun Oenoki*", have no prior relationship, but their Jaccard coefficient is high: 0.024. Two international artists "*Beat Streuli*" from Switzerland and "*Nari Ward*" from Jamaica have co-participated in several exhibitions, but their coefficient is low: 0.0009. This happens because the community consists of many people from different contexts. For that reason, it is difficult to precisely recognize the relation using a single criterion.

We consider that the precedent studies on the research domain implicitly use the following two assumptions:

Assumption 1. Generally, web pages are created according to results of two actors' co-participation in events. Therefore, the number of web pages is assumed to show a useful correlation to the strength of two actors.

Assumption 2. A community to be extracted as a social network is assumed to be homogeneous.

In the following section, we will introduce our improvements, *relation identification* (in Section *3.2*) and *threshold tuning* (in Section *3.3*), which respectively mitigate violations of these assumptions. Furthermore, to emphasize the effectiveness of our methods, we apply each method to our case studies: Extracting social networks of firms (in Section *4*) and artists (in Section *5*). A general extraction model bundling these different extraction methods will be described in Section *6*.

3.2 Relation Identification

In social sciences, the definition of a weak or strong tie might vary among contexts [14]. For example, the frequency or degree of relations affects that strength; multiple relations between two actors also can imply a stronger tie. In the firm case, the types of relations define the strength: For example, a capital alliance relation is stronger than a business alliance relation. Consequently, to present a tie among firms, it is appropriate that we identify the concrete relations of firms. As a solution, we add some word or combination of words to a search query. Using this strategy, we can efficiently identify relations among firms. For example, when we wish to extract lawsuit relations, we add a term "*lawsuit*". We issue a query "*Matsushita* AND *JustSystem* AND *lawsuit*" so that the search engine will return the lawsuit pages that are associated with the two firms. Then we can conduct text processing to these pages to validate the relation's existence. This idea is similar to keyword spices [20], which extend queries for domain-specific web searches. Question answering systems also construct elaborate queries for using search engines [21].

We call the keyword to be added a *relation keyword*. By adding relation keywords, we can extract particular relations among entities, which can be a solution for validation of **Assumption 1**. Below, we explain some issues about relation types and extraction of relation keywords.

Relation Types. It is considered that a pair of entities has multiple relations. For example, two firms share alliance and lawsuit relations. Each relation is typed in a more detailed way. Alliance relations between firms include capital alliances and business alliances, where the former usually represents a stronger relation than the later. A lawsuit relation has multiple stages: at some time, it will be settled by mutual accommodation or by final judgement. Consequently, the relation can be typed into the claim phase and the accommodation phase. For dynamic and complex relational networks, it is important to distinguish such typical and temporal relations for detailed analyses of social networks [14,22].

Relation Keyword Extraction. To extract particular types of relations between firms, we need some relation keywords. The intuitive method for finding relation keywords is to select terms that appear often in target pages (where the target relation is described) and which do not appear in other pages. Therefore, as a training corpus, we must collect annotated web pages that describe specific relations of the firms. Once we find appropriate relation keywords, we can extract the relations among many firms.

Collecting and annotating the training corpus requires many hours of tedious work. In our study, we also try to use a search engine to extract relation keywords. This method is identical to that of Mori's work [19], in which a specific word w_c is assigned, which can represent the relation most precisely. If we want to retrieve an alliance relation, we add "*alliance*" (denoted as w_c) to a search query; words that co-occur frequently with it also become good clues to discern the relation. We use the Jaccard coefficient $n_{w_c \cap w}/n_{w_c \cup w}$ to measure relevance of word w to word w_c. The words w with large Jaccard coefficients are also used as relation keywords aside from w_c. It would saves costs of annotating training data with relevance or non-relevance manually.

3.3 Threshold Tuning

In studies of social network analysis, network questionnaires have traditionally been conducted. Typically, participants are asked "Please name your four closest friends." The respondents would then list the relations that are personally important. In other words, the relation is recognized by a subjective criterion for each participant. We propose to use this subjective criterion for the solution against **Assumption 2**. For example, even if the relation between "*Beat Streuli*" and "*Nari Ward*" is weaker than the objective standard, it is important to "*Beat Streuli*" if there are no other persons with a stronger relation. Consequently, we might add an edge between them.

We employ two criteria that correspond to objective and subjective importance of relations for actors. We first invent edges using objective criteria with a consistent threshold T. Then we invent edges using subjective criteria for actors who have no certain number M of edges. This procedure alleviates the problem of some nodes having too many edges and some nodes being isolated. The combination of two criteria enables

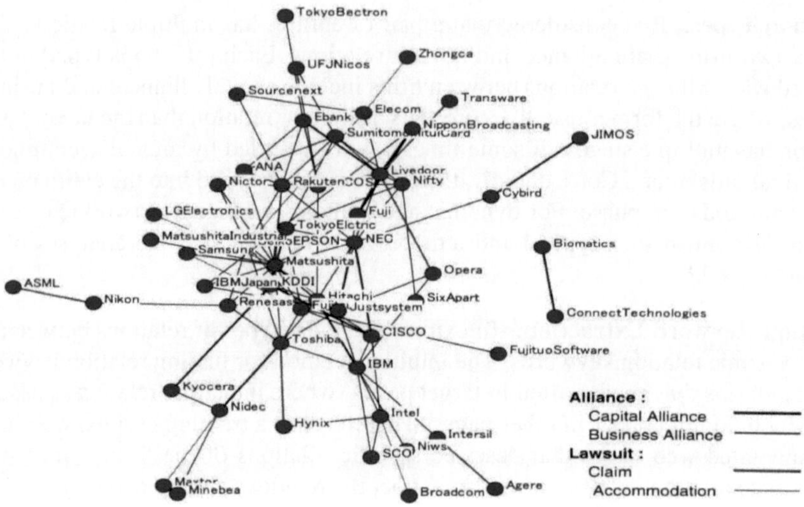

Fig. 1. Social network of 60 firms in Japan

more exhaustive extraction for every node than the previous method, although it sometimes yields low precision. For that reason, we must find the appropriate parameters so that the target network is extracted as precisely as possible.

Setting Parameters for Each Community. Parameters vary according to the domain of a community. For example, T in the researcher community might be higher than that in artist community, simply because researchers' names are more likely appear on the web than artists' names. Therefore, some training data are necessary to learn the appropriate values for each target community. Simply, the parameters are tuned so that the performance of relation identification is maximized: We maximize the F-value. More effective ways to determine the parameters are bootstrapping or user interaction. For the bootstrapping method, we can repeat the sampling and estimation process to determine parameters; for the user interaction method, we can use the users' feedback to reconstruct the network dynamically using the best parameters that can maximize the F-value.

4 Social Network Extraction for Firms

We describe the extraction of a firm network as a case study of *relation identification* (mentioned in Section *3.2*). Many relationships among firms are published in news

Table 1. Relation keywords extracted from the web using Jaccard coefficient

Alliance relation	t_w	Capital alliance	t_w	Business alliance	t_w
alliance AND *corporate*	1.000	*operation* AND *capital*	1.000	*alliance* AND *business*	1.000
alliance AND *stock*	0.878	*capital* AND *manage*	0.553	*alliance* AND *company*	0.475
alliance AND *company*	0.704	*capital* AND *company*	0.548	*alliance* AND *operation*	0.459
alliance AND *system*	0.565	*capital*	0.543	*alliance* AND *develop*	0.437
alliance AND *business*	0.534	*capital* AND *manage*	0.533	*alliance* AND *company*	0.432

Lawsuit relation	t_w	Claim phase	t_w	Accommodation phase	t_w
violate AND *lawsuit*	1.000	*violate* AND *sue*	1.000	*lawsuit* AND *accommodate*	1.000
violate AND *claim*	0.514	*patent* AND *sue*	0.533	*accommodate* AND *company*	0.648
violate AND *judge*	0.490	*sue* AND *technology*	0.486	*accommodate* AND *announce*	0.646
violate AND *court*	0.458	*sue* AND *develop*	0.483	*accommodate* AND *develop*	0.641
violate AND *indemnify*	0.444	*sue* AND *relevance*	0.469	*accommodate* AND *product*	0.640

articles and on news releases that are distributed on the web. In our work, we extract alliance and lawsuit relations as respective representatives of positive and negative relations among firms. We further distinguish these relations into two detailed relations: capital and business alliance relations, and claims and accommodation of lawsuit relations. A social network of 60 firms in Japan is extracted; it includes IT, communication, broadcasting, and electronics firms. We will describe details of our system and experimental results.

4.1 System Flow

Our system has two major procedures: an online procedure and an offline procedure. In the offline procedure, relation keywords for each relation are obtained beforehand using the methods introduced in Section *3.2*. We gathered 456 pages and 165 pages for alliance and lawsuit relations, respectively, from Nikkei Net and IP News site[2]. As preprocessing, we first eliminate all html tags and scripts; then we extract the body text of pages and apply a part-of-speech tagger Chasen[3] to choose nouns and verbs (except stop words). These words are candidates of relation keywords. We also use combinations of two words as candidates. We measure the score of

```
function RelationExtraction (D, x, y, W)
  score_xy ← 0
  S ← GetSentences(D, x, y)
  for each s ∈ S do
    if s contains "x" and s contains "y" then
    score_s ← Σ_{w_i(∈W) contained in s} t_{w_i}
    if score_s > score_xy then
      score_xy ← score_s
  done
  if score_xy > score_thre then
    do set an edge between x and y in G
  done
```

Fig. 2. A procedure to extract relations by text processing

[2] Nikkei Net (http://release.nikkei.co.jp/) is a famous online business newspaper. IP News (http://news.braina.com/judge.html) is an online news archive on intellectual property issues.

[3] http://chasen.naist.jp/hiki/ChaSen/

each candidate word / phrase by calculating the Jaccard coefficient with specific relation keywords w_c[4]. Candidates with the highest scores are recognized as relation keywords. Table 1 shows the top five relation keywords and their Jaccard scores denoted as t_w[5].

In the online procedure, a list of firms and specific relation types is given as input; the output is a social network of firms. Three steps exist: making queries, Google search, and network construction. First, we make queries by adding relation keywords to each pair of firms. We use top n_q relation keywords from Table 1. Then, we put these queries into the Google search engine to collect top-n_p web pages. (In this experiment, we set $n_q = 2$ and $n_p = 5$.) Lastly, for each downloaded document D, we conduct text processing to judge whether or not the relation actually exists. A simple pattern-based heuristic (as described in Fig. 2) is useful in our experience: We first pick up all sentences S that include the two firm names (x and y), and assign each sentence the sum of relation keyword scores t_w in the sentence. The score of firms x and y is the maximum of the sentence scores. If $score_{xy}$ is greater than a certain threshold (in other words, if the two firms seem to have the target relation with high reliability), an edge is invented between the two firms.

4.2 Results and Evaluation

The obtained network for 60 firms in Japan is shown in Fig. 1. Black lines represent alliances (bold ones are capital alliances and thin ones are business alliances) and red lines represent lawsuits (bold ones are in the claim phase and thin ones are in the accommodation phase).

The precision and recall of our system are shown in Table 2. For ${}_{60}C_2 = 1770$ pairs of firms, 113 pairs actually show alliance relations. Our system extracted 70 pairs correctly. There were actually 21 and 100 pairs of capital and business alliances; our system extracted 9 and 60, respectively. Compared with alliances, the lawsuit relations have higher recall, probably because lawsuit relations are described in rather common formats using words such as *judgment*, *lawsuit*, or *accommodate*.

Table 2. Precision and Recall of the System

Target relation	Precision	Recall
Alliance	60.9% (70/115)	62.0% (70/113)
capital alliance	75.0% (9/12)	42.9% (9/21)
business alliance	67.4% (60/89)	60.0% (60/100)
Lawsuit	61.5% (16/26)	100% (16/16)
claim phase	63.6% (14/22)	87.5% (14/16)
accommodation	72.7% (8/11)	88.9% (8/9)

Although they are not comparable technically, we obtained alliance and lawsuit relations from Nikkei Net and IP News, and compared the precision and recall to our results. The precision values at these sites are 100%, but the recall of alliance and lawsuit relations among 60 firms are low: 22.8% and 68.8%, respectively. This is true because these sites deal little with information on small companies and corporations that are capitalized with foreign capital (i.e. foreign companies).

[4] We used *alliance* AND *corporate* as w_c for alliance relations. Furthermore, we use the word appearing in the first lines in Table 1 as w_c for each relation: We determine these words through preliminary experiments.

[5] In our experiment, we mainly used web pages that had been composed in Japanese. For that reason, relation keywords are translated from Japanese.

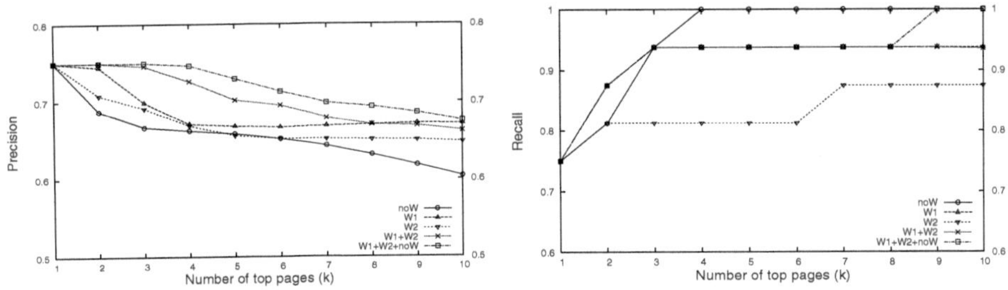

(a) Precision of retrieved pages (b) Recall of relations

Fig. 3. Evaluation of relation keywords for lawsuit relations

Some detected relations are wrong: As one example, Hitachi and IBM are shown to be embroiled in a lawsuit relation, but they actually are not. Our algorithm took the sentence *"Hitachi and HDD, a subsidiary of IBM have been sued a Chinese HDD maker for patent violations"* as spurious proof of a lawsuit relation. Some relations are described using uncommon phrases (such as *trouble* and *uproar*) that do not appear often in the training corpus. More sophisticated text processing might improve the results in these cases.

4.3 Effectiveness of Relation Keywords

The effectiveness of relation keywords is shown in this section. We compared the information contained in retrieved pages merely by using a pair of names as a search query to add relation keywords to the query. We compared the five methods described below:

noW: A firm pair (without relation keywords) is used as a query.
W1: A firm pair and the top-weighted relation keyword (w_1) are used as a query.
W2: A firm pair and the second-weighted relation keyword (w_2) are used as a query.
W1+ W2: It generates two queries – W1 and W2.
W1+W2+noW: It generates three queries – W1, W2, and noW.

The **noW** is considered to be the existing method (i.e. Mika and Matsuo's method). The others are variations of the proposed method. In all cases, we downloaded the same number of web pages. All other conditions are identical.

Figure 3 shows the results. Overall, the proposed methods perform better than the existing method (**noW**) with respect to precision. The precision and recall are respectively 65.7% and 95.0% if we do not use any relation keywords. Relation keywords improve the precision using the same number of downloaded documents. By integrating multiple queries (as **W1+W2+noW** case), we can achieve the highest precision as 71.9% while retaining high recall (92.5%).

5 Social Network Extraction for Artists

In this section, we describe the algorithm of *threshold tuning* (described in Section 3.3) for extracting a social network of artists of contemporary art.

```
/* First, we invent edges using two objective criteria: T_ov and T_co. */........(step 1)
for each x ∈ L and y ∈ L
    if (overlap(x, y)> T_ov AND cooc(x, y)> T_co)
        do set an edge between x and y in G

/* Then, invent edges using two subjective criteria M_1 and M_2 (≤ M_1). */......(step 2)
for each x ∈ L
    do Y_x ← ConnectedNodes(x),      /* Y_x are nodes set connected with x. */
       Ȳ_x ← L \ Y_x, Ȳ'_x ← L \ Y_x
    while |Y_x| < M_1 and Ȳ_x ≠ φ      /* |Y_x| is the number of nodes in Y_x. */
       y ← argmax_{y_j ∈ Ȳ_x} overlap(x, y_j), Ȳ_x ← Ȳ_x \ {y}
       if overlap(x, y)> T_ov OR cooc(x, y)> T_co ......................(step 2a)
          do set an edge between x and y in G, Y_x ← Y_x ∪ {y}
    done
    while |Y_x| < M_2 and Ȳ'_x ≠ φ
       y ← argmax_{y_k ∈ Ȳ'_x} overlap(x, y_k), Ȳ'_x ← Ȳ'_x \ {y}
       if overlap(x, y)> 0 AND cooc(x, y)> 0 .........................(step 2b)
          do set an edge between x and y in G, Y_x ← Y_x ∪ {y}
    done
done
```

Fig. 4. Detailed Algorithm of threshold tuning used at the Yokohama Triennale 2005

5.1 System Flow

This system includes online and offline procedures. In the offline procedure, we tune four parameters: T_{ov}, T_{co}, M_1, and M_2. For them, T_{ov} and T_{co} are thresholds to invent edges by the overlap coefficient and matching coefficient, and M_1 and M_2 are the minimum numbers of edges for each node. We sample 1000 pairs of artists as training data: 146 positive examples and 854 negative examples. We change the values of parameters, classify every pair of artists into positive and negative using the parameters, and find the optimal values where the F-value is maximized: $T_{ov} = 0.82$, $T_{co} = 20$, $M_1 = 5$ and $M_2 = 1$. We try different settings for the four parameters; T_{ov} is changed from 0 to 1 at every 0.01, and T_{co} is changed from 0 to 60 in steps of 5, M_1 and M_2 are incremented from 0 to 5[6].

For the online procedure, a list of artists' names are given as input; the output is a social network of artists. Three steps exist: making queries, Google search, and network construction. First, we make queries for each pair of names. Then we put them into the Google search engine to obtain the hit counts. Finally, we construct a social network after tuning the parameters.

A detailed algorithm to generate a social network is shown in Fig. 4. Edges are added using an objective criterion (in step 1): An edge is added between the nodes if

[6] We might use more sophisticated algorithms such as hill-climbing searches. However, we do not specifically examine the optimization method in this paper. For that reason, we employed a simple (but reliable) approach.

Table 3. Maximized precision, recall, and F-value using the precedent approach

Cases	T_{ov}	T_{co}	Precision	Recall	F-value	Extracted number*	Correct number*
case (a)	0.24	30	92.9%	26.7%	0.41	42 (42, 0, 0)	39 (39, 0, 0)
case (b)	0	0	14.6%	100%	0.25	1000 (1000, 0, 0)	146 (146, 0, 0)
case (c)	0.05	20	76.4%	37.7%	0.50	72 (72, 0, 0)	55 (55, 0, 0)

*: Numbers in brackets are numbers of edges invented in step 1, step 2a, and step 2b.

Table 4. Maximized precision, recall, and F-value using the proposed approach

Cases	T_{ov}	T_{co}	M_1	M_2	Precision	Recall	F-value	Extracted number	Correct number
case (a)'	0.24	30	3	2	34.4%	65.1%	0.45	277 (42, 227, 8)	95 (39, 54, 2)
case (b)'	0	0	0	0	14.6%	100%	0.25	1000 (1000, 0, 0)	146 (146, 0, 0)
case (c)'	0.05	20	1	0	55.4%	49.3%	0.52	130 (72, 58, 0)	72 (55, 17, 0)
case (d)	0.82	20	5	1	43.4%	74.0%	0.55	249 (23, 212, 14)	108 (19, 84, 5)

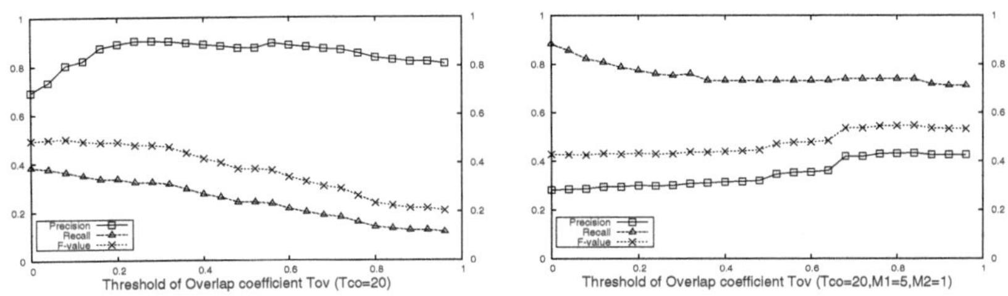

(a) Existing algorithm (b) Proposed algorithm

Fig. 5. Precision, recall and F-value for different T_{ov}

the overlap coefficient and the matching coefficient are both over the thresholds. Then subjective criteria are used to add edges (in step 2): If node x has less then M_1 edges, we choose nodes that have the strongest relations with node x. Node x is connected to the other nodes until the number of edges reaches M_1 (in step 2a). After that, if node x has no M_2 edges yet, we add edges in descending order of overlap coefficient (in step b).

Although the algorithm is highly customized for dealing with web information, the concept is simple. We use the objective criteria (using T_{ov} and T_{co}) first, and the subjective criteria (using M_1 and M_2) subsequently. It is important to combine multiple criteria to infer the relations among artists correctly from the available web information.

5.2 Evaluation

The existing approach by Mika and Matsuo generates a social network based on an objective criterion with a predefined threshold. It corresponds to the case where $M_1 = 0$ and $M_2 = 0$ in our algorithm. To compare the existing method with our method, we tune T_{ov} and T_{co} so that precision, recall, and F-value are maximized, respectively. The results are shown in Table 3. The maximal recall is 100% by setting T_{ov} and T_{co} as zero (which means the algorithm recognizes all the pairs having a relation), which yields precision as low as 14.6%. Conversely, the maximal precision is 92.9% when the recall is as low as 26.7%. The precision is 76.4% and the recall is 37.7% when the F-value is maximized.

(a) The whole network. (b) Centering artist *Curatorman.*

Fig. 6. System Interface for Yokohama Triennale 2005

Our algorithm can achieve better performance in either case. Table 4 shows results of our algorithm using four parameters. Even if we set T_{ov} and T_{co} as identical to those in Table 3, we can achieve better results by adjusting M_1 and M_2. The most balanced parameters achieve F-value of 0.55, which is more than 0.05 points better than the proposed algorithm. Figure 5 shows a notable difference: the proposed algorithm produces high recall while maintaining modest precision. It is useful when the purpose is to promote navigation and communication using a social network.

In this section, we emphasize detection of relationships using only the hit number of search engine. This is treated as a first step in the Yokohama Triennale system. As second step, we further identify concrete relation types from web pages retrieved by names of artists who are considered as related; we also filter out noisy edges to improve the precision. Details about the relation type identification are available from [16].

5.3 Navigation Site for Yokohama Triennale

Our system was put into operation on the official support site for Yokohama Triennale 2005 (http://mknet.polypho.net/tricosup/) to provide an overview of the artists (133 artists with 71 projects) along with informational navigation for users. At exhibitions, it is usual for participants to enjoy and evaluate each work separately. However, our supposition was that if participants knew the background and relations of the artists, they might enjoy the event more. For that purpose, the system provided relations of artists and evidential web pages for users.

The system interface is shown in Fig. 6. It was implemented using Flash display software to facilitate interactive navigation. The system provides a retrieval function. Information about the artist is shown on the left side if a user clicks a node. In addition, the edges from the nodes are highlighted in the right-side network. The user can proceed to view the neighboring artists' information sequentially, and can also jump to the web pages that show evidence of the relation.

6 General Extraction of a Social Network Using a Search Engine

Based on the two case studies described in the preceding sections, this section presents and explains an architecture to support general social network extraction from the web using a search engine. The types of social networks depend on their purpose [22]. A "good" social network should represent a target domain most appropriately.

We consider that social network extraction is generally written as

$$f(\mathbb{S}_r(X,Y),\Theta) \rightarrow \{0,1\} \quad (1)$$

where $\mathbb{S}_r(X,Y)$ is an m-dimensional vector space $(S_r^{(1)}(X,Y), S_r^{(2)}(X,Y), \ldots, S_r^{(m)}(X,Y))$ to represent various measures for X and Y in relation r. For example, $S_r^{(i)}(X,Y)$ can be either $n_{X\cap Y}$ (matching coefficient), $n_{X\cap Y}/n_{X\cup Y}$ (Jaccard coefficient), or $n_{X\cap Y}/min(n_X,n_Y)$ (overlap coefficient). It can possibly be a score function based on sentences including both mentions of X and Y (as the algorithm in Section 4). The parameter Θ is an n-dimensional vector space $(\theta^{(1)},\theta^{(2)},\ldots,\theta^{(n)})$. For example, Θ can be as a combination of T_{ov}, T_{co}, M_1, and M_2 as the algorithm in Section 5. The function f determines whether an edge should be invented or not based on multiple measures and parameters.

A social network should represent the particular relations of entities depending on purposes. Therefore, function f should not always be the same. We must have a method to infer an appropriate function f, thus the algorithm inevitably consists of an offline module and an online module. Function f is learned from the training examples and provides good classification to other examples.

In the online phase, it is important to extract a social network from the web in an efficient manner. We must consider how to use a search engine better and how to process web documents efficiently and correctly. Generally, the procedure consists of three steps:

Making queries. Two entities are used to generate a query. Basically, we put a query X AND Y to a search engine. In this paper, we add relation keywords to extract a particular type of relation efficiently. A combination of multiple queries might improve the result, as explained in Section 4. Entity disambiguation is another important issue that has already been addressed in several studies [3,4].

Google search. We put the queries into a search engine. Sometimes the counts are used to infer relational strength. In other cases, we download some documents (or snippets) and investigate the mentions of X and Y. A good combination of Google counts and text analysis would make the search more efficient and scalable, as discussed in [16].

Network construction. We use Google counts and downloaded text as evidence to construct a social network. The value of function f is calculated and the existence of an edge is determined. Usually, the obtained social network is visualized and reviewed. Sometimes we must change settings of the algorithm (or increase the training data) and repeat the entire process to improve the quality.

Previous studies have emphasized how to calculate the strength of two names on the Web in the **Google search** step, simply using X AND Y as query and construct networks based on objective criteria. Our method, i.e., *relation identification* and *threshold tuning*

Table 5. Centrality of firms in the extracted social network

(a) Eigenvector centrality.

Rank	Name	Value
1	Matsushita	0.366
2	Hitachi	0.351
3	NEC	0.289
4	Fujitsu	0.275
5	Toshiba	0.263
6	Rakuten	0.257
7	Just System	0.241
8	KDDI	0.208
9	Tokyo Electric	0.207
10	Seiko Epson	0.204

(b) Betweenness centrality.

Rank	Name	Value
1	Matsushita	168.981
2	IBM	149.192
3	NEC	144.675
4	Hitachi	136.978
5	Toshiba	113.239
6	Rakuten	109.887
7	Just System	77.175
8	Livedoor	74.141
9	CISCO	64.558
10	Fujitsu	56.081

are proposed for **Making queries** and **Network construction** steps respectively for complex and inhomogeneous communities. All of these methods are combined into our architecture of general extraction of social networks for various entities.

The obtained network is useful for Semantic Web studies in several ways. For example (inspired by [2]), we can use a social network of artists for detecting COI among artists when they make evaluations and comments on others' work. We might find a cluster of firms and characterize a firm by its cluster. Business experts often make such inferences based on firm relations and firm groups, so the firm network might enhance inferential abilities in the business domain. As a related work, F. Gandon et al. built a Semantic Web server that maintains annotations about the industrial organization of Telecom Valley to partnerships and collaboration [8].

We present a prototypical example of applications using a social network of firms. We calculate the *centrality*, which is a measure of the structural importance of a node in the network, for each firm on the extracted social network (on alliance relations). Table 5(a) shows the top ten firms by eigenvector centrality. These firms have remained large and reliable corporations in Japan for decades. Table 5(b) shows the top ten by betweenness centrality. Interestingly, IBM, Livedoor, and Cisco are on the list. These firms might bridge two or more clusters of firms: IBM and Cisco are United States firms and form alliances with firms in multiple clusters; Livedoor is famous for its aggressive M & A strategy in Japan. Such information can only be inferred after extracting a social network. There seem to be many potential applications that can make use of social networks in the Semantic Web.

7 Conclusion

This paper describes methods of extracting various social networks from the web. To date, numerous studies have addressed the researcher domain to estimate extraction methods. It is an important test-bed. Nevertheless, the next step must be taken to depart from the domain of researchers. This paper steps further to show that researcher networks might be an easy domain for social network extraction from the web. Our method, equipped with relation identification and threshold tuning that specifically focus on complex and inhomogeneous communities respectively, can extract other types of social networks: those of firms and artists. We show various evaluations of the methods along with discussions of the application of social network in the context of the Semantic Web. The proposed architecture toward general extraction of social networks, which bundles these different extraction methods, will enable us to extract various social networks from available information on the web.

In addition to some direct applications of social networks, we believe that a network point of view is important for knowledge integration and articulation and for (lightweight) ontology emergence. The combination of social networks and ontology emergence might prepare a fertile ground for Semantic Web research.

Acknowledgements. This research was supported by the New Energy and Industrial Technology Development Organization (NEDO) as project ID 04A11502a.

References

1. L. Adamic and E. Adar. Friends and neighbors on the web. *Social Networks*, 25(3):211–230, 2003.
2. B. Aleman-Meza, M. Nagarajan, C. Ramakrishnan, A. Sheth, I. Arpinar, L. Ding, P. Kolari, A. Joshi, and Tim Finin. Semantic analytics on social networks: Experiences in addressing the problem of conflict of interest detection. In *Proc. WWW2006*, 2006.
3. N. Aswani, K. Bontcheva, and H. Cunningham. Mining information for instance unification. In *Proc. ISWC2006*, 2006.
4. R. Bekkerman and A. McCallum. Disambiguating web appearances of people in a social network. In *Proc. WWW2005*, 2005.
5. H. Chen, M. Lin, and Y. Wei. Novel association measures using web search with double checking. In *Proc. COLING-ACL2006*, pages 1009–1016, 2006.
6. A. Culotta, R. Bekkerman, and A. McCallum. Extracting social networks and contact information from email and the web. In *CEAS-1*, 2004.
7. T. Finin, L. Ding, and L. Zou. Social networking on the semantic web. *The Learning Organization*, 2005.
8. F. Gandon, O. Corby, A. Giboin, N. Gronnier, and C. Guigard. Graph-based inferences in a semantic web server for the cartography of competencies in a telecom valley. In *Proc. ISWC2005*, 2005.
9. S. Ghita, W. Nejdl, and R. Paiu. Semantically rich recommendations in social networks for sharing, exchanging and ranking semantic context. In *Proc. ISWC2005*, 2005.
10. J. Golbeck and B. Parsia. Trust network-based filtering of aggregated claims. *International Journal of Metadata, Semantics and Ontologies*, 2006.
11. H. Kautz, B. Selman, and M. Shah. The hidden Web. *AI magazine*, 18(2):27–35, 1997.
12. F. Keller and M. Lapata. Using the web to obtain frequencies for unseen bigrams. *Computational Linguistics*, 29:459–484, 2003.
13. C. D. Manning and H. Schütze. *Foundations of statistical natural language processing*. The MIT Press, London, 2002.
14. P. Marsden. Measuring tie strength. *Social Forces*, 63:482–501, 1984.
15. Y. Matsuo, M. Hamasaki, H. Takeda, J. Mori, D. Bollegala, Y. Nakamura, T. Nishimura, K. Hasida, and M. Ishizuka. Spinning multiple social networks for semantic web. In *Proc. AAAI-06*, 2006.
16. Y. Matsuo, J. Mori, M. Hamasaki, H. Takeda, T. Nishimura, K. Hasida, and M. Ishizuka. POLYPHONET: An advanced social network extraction system. In *Proc. WWW2006*, 2006.

17. P. Mika. Flink: Semantic web technology for the extraction and analysis of social networks. *Journal of Web Semantics*, 3(2), 2005.
18. P. Mika. Ontologies are us: A unified model of social networks and semantics. In *Proc. ISWC2005*, 2005.
19. J. Mori, M. Ishizuka, T. Sugiyama, and Y. Matsuo. Real-world oriented information sharing using social networks. In *Proc. ACM GROUP2005*, 2005.
20. S. Oyama, T. Kokubo, and T. Ishida. Domain-specific web search with keyword spices. *IEEE TKDE*, 16(1):17–27, 2004.
21. G. Ramakrishnan, S. Chakrabarti, D. Paranjpe, and P. Bhattacharyya. Is question answering an acquired skill? In *Proc. WWW2004*, 2004.
22. J. Scott. *Social Network Analysis: A Handbook (2nd ed.)*. SAGE publications, 2000.
23. S. Staab, P. Domingos, P. Mika, J. Golbeck, L. Ding, T. Finin, A. Joshi, A. Nowak, and R. Vallacher. Social networks applied. *IEEE Intelligent Systems*, pages 80–93, 2005.
24. P. Turney. Mining the web for synonyms: PMI-IR versus LSA on TOEFL. In *Proc. ECML2001*, pages 491–502, 2001.

Towards Semantic Social Networks

Jason J. Jung[1] and Jérôme Euzenat[2]

[1] Department of Computer and Information Engineering, Inha University
Incheon, Republic of Korea 402-751
[2] INRIA Rhône-Alpes & LIG, Montbonnot, France
j2jung@intelligent.pe.kr, Jerome.Euzenat@inrialpes.fr

Abstract. Computer manipulated social networks are usually built from the explicit assertion by users that they have some relation with other users or by the implicit evidence of such relations (e.g., co-authoring). However, since the goal of social network analysis is to help users to take advantage of these networks, it would be convenient to take more information into account. We introduce a three-layered model which involves the network between people (social network), the network between the ontologies they use (ontology network) and a network between concepts occurring in these ontologies. We explain how relationships in one network can be extracted from relationships in another one based on analysis techniques relying on this network specificity. For instance, similarity in the ontology network can be extracted from a similarity measure on the concept network. We illustrate the use of these tools for the emergence of consensus ontologies in the context of semantic peer-to-peer systems.

1 Introduction

Social networks, i.e., networks based on the relation between people is our common environment. Many have realized that social networks are a key facilitator for collaborating. Social scientists have been analyzing the effectiveness of of these networks by helping characterizing the key individuals that must be touched in order to achieve some goals [1]. Recently, on the semantic web, social networks are very often described through FOAF or FOAF-like schemata. They could as well be built upon shared address books.

However, when communities are really different or when people do not want to describe explicitly their relationships, social networks have to be inferred from secondary sources. These secondary sources are often documents: bibliometrics has for long made a speciality of inferring social networks and many other clues from databases of co-authoring and citations. The web is a rich source of documents that can be used as secondary sources for inferring social networks from analyzing the hyperlinked structure on the web [2] (along the way used by google for inferring most authoritative web pages) to exploiting the more intimate personal infospheres made of web pages, weblogs, and so on.

E. Franconi, M. Kifer, and W. May (Eds.): ESWC 2007, LNCS 4519, pp. 267–280, 2007.

Most of these sources are based on explicit links that still need from document authors (or web page designers) to know each others. In this paper, we investigate the dual principles of:

- using the knowledge structure used by individuals in order to infer some of their social relationships, and
- taking advantage of this inferred social structure in order to help people sharing their knowledge.

For that purpose, we introduce a structure made of three superposed networks that are assumed to be strongly linked:

Social network relating people on the basis of common interest;
Ontology network relating ontologies on the basis of explicit import relationships or implicit similarity;
Concept network relating concepts on the basis of explicit ontological relationships or implicit similarity.

We call this stack of interlinked networks a semantic social network.

The top-down links between these networks are obvious: people use ontologies that defines or refer to concepts. The less obvious aspect that is illustrated here is bottom-up inference of relationships from one network to another. These relationships can be inferred by analyzing the structure of knowledge.

Once the structure of knowledge allows to infer "intellectual" relationships between people, it is possible to use it for helping people managing their knowledge. For instance, finding that similar people use similar ontologies or the same ontology makes them candidates to be standardized. In addition, when starting some collaboration between partners in order to merge their knowledge or to define one common ontology, it is better to start from the ontology used by considering social features such as centrality and authority.

The remainder of the paper is organized as follows. Sect. 2 provides a description of a target application which uses network analysis in order to help people sharing ontology annotated resources. Sect. 3 provides the basic knowledge in social network analysis. As the main contribution, Sect. 4 and 5 explain the three-layered semantic social network structure. Sect. 4 considers each layer, the kind of relationships it expresses and the tools that can be used for manipulating these networks. Sect. 5 investigates the relationships between these layers and the opportunity this provides to lift network structure from one layer to another. Sect. 6 presents an experiment we have conducted. Finally, in Section 7, this work is compared with previous studies related to social networks on semantic web.

2 Emerging Collaboration in Peer to Peer Networks

We apply semantic social networks in the context of peer-to-peer (P2P) sharing of annotations in which individuals can use ontologies for annotating resources (in our case, photographs). Because, these individuals can have different needs and different standpoints, they are not constrained to use the same ontology (contrary to BibSter [5], for instance). They can take advantage of standard ontologies on the web (like FOAF or EXIF), but they can extend these ontologies to their needs.

However, when people want to query other peers, for benefiting from better annotation or from other pictures, the network of heterogeneous ontologies must be dealt with. This ontology mismatch is solved by offering an alignment infrastructure to the peers that allows them to find some alignments between their ontologies and to process queries through mediators [6].

Current work on ontology matching consists of either automatically matching two ontologies without regard to its context or environment, or of interactively helping some knowledge engineer to match two ontologies. Our goal is to take into account the social networking context in order to improve the alignment results and, in turn, the social networking expertise.

In this context, semantic social network analysis has two purposes:

- helping people to find other peers with similar interests, and
- helping peers to find the best company for starting designing consensus ontologies.

For the first purpose, there are two options. On the one hand, the peers can use social network analysis (SNA) to find in their explicit personal network of peers with some the same interests. Such people would certainly use ontologies that are not totally remote and trying to match these ontologies promise to be easier and more rewarding. On the other hand, if one peer has no relation with known peers sharing the same interest, but is looking for some pictures, it will be more convenient that he looks for peers having similar ontologies. The peers can use metrics on the ontology network in order to find ontologies which are close to theirs. The underlying assumption is that these peers will have similar interest to theirs even if they are not explicitly recorded as connected. Moreover, if the metric is compatible with some matching algorithm, the matching will be easier to perform. So, in this situation, the social network is used for satisfying the need of some user (finding interesting annotated resources) and this action will contribute strengthening the social network by exhibiting new links between peers. As a simple example, Fig. 1 illustrates the potential connection between unknown users by ontology alignment in a semantic social network.

Another contribution of SNA to this application is to identify the central (authoritative) ontologies at the ontology level such that aligning with these ontologies allows to be connected with the maximum number of peers (and thus retrieve the maximum number of answers).

Matching algorithms can take advantage of SNA in order to help find the closest ontologies with regards to the connectivity of users and to identify which users are more prone to the use of a pivotal ontology.

Beside matching ontologies in an ad-hoc manner, there is time when peers can imagine benefiting from putting their experience together and building some consensus ontologies. To that extent, they can also use SNA in order to identify the community (cohesive subgroups on the social level), the best peer for proposing consensus (centrality on the social network), or the ontologies which could be the seed for this collaboration process (centrality on the ontology layer). Of course, when both coincide (i.e., when the central individuals use a central ontology), the chances of success are higher.

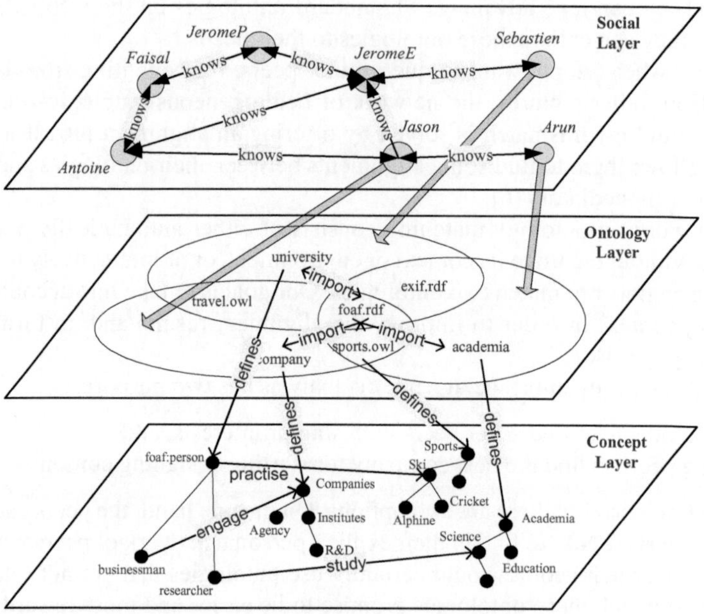

Fig. 1. A three-layered social semantic network

3 Network Analysis

Network analysis is based on the analysis of the relationships within a population. We here consider networks with several different relations between individuals. So, a network is characterized as a set of objects (or nodes) and a set of relations.

Definition 1 (Network). *A network $\langle N, E^1, \ldots E^n \rangle$ is made of a set N of nodes and n sets of object pairs $E^i \subseteq N \times N$ the set of relations between these nodes.*

We will below freely use the relations as functions ($E^i(n) = \{n' \in N | \langle n, n' \rangle \in E^i\}$). As usual, a path p between node e and e' in the graph of E^i is a sequence of edges $\langle e_0, e_1 \rangle, \langle e_1, e_2 \rangle, \ldots, \langle e_{k-1}, e_k \rangle$ from E^i in which $e_0 = e$ and $e_k = e'$. The length of a path is its number of edges (here k), $sp^i(e, e')$ is the set of shortest paths between e and e' and the shortest path distance $spd^i(e, e')$ between two nodes e and e' is the length of the shortest paths between them, when it exists. By convention, $spd(e, e) = 0$. Thus, $spd(Arun, Sebastien)$ in Fig. 1 is three via $Jason$ and $JeromeE$.

Social network analysis [1] has considered various measures on the networks between people (note that these measures apply only if the network is connected)[1]:

Closeness. The inverse of average length of the shortest path between a node e and any other node in the network:

$$Closeness^i(e) = \frac{|N - 1|}{\sum_{e' \in N} spd^i(e, e')} \tag{1}$$

[1] These measures are often normalized (between 0 and 1) but we present their simplest form.

Betweenness. [7] The proportion of shortest paths between two nodes which contains a particular node (this measures the power of this node):

$$Betweenness^i(e) = \sum_{e',e'' \in N} \frac{|\{p \in sp^i(e',e), p' \in sp^i(e,e'') | p \cdot p' \in sp^i(e',e'')\}|}{|sp^i(e',e'')|} \tag{2}$$

Hub and authority. There are different but interrelated patterns of power: Authorities that are referred to by many and hubs that refers to many. The highest authorities are those which are referred to by the highest hubs and the highest hubs that those which refers to the highest authorities. Kleinberg [2] proposes an iterative algorithm to measure authority and hub degree of each node in interlinked environment. Given initial authority and hub degrees of 1, the degrees are iteratively computed by

$$Hub^i_{t+1}(e) = \sum_{e':\langle e,e'\rangle \in E^i} Auth^i_t(e') \text{ and } Auth^i_{t+1}(e) = \sum_{e':\langle e',e\rangle \in E^i} Hub^i_t(e') \tag{3}$$

Similarly to betweenness, the hub weight indicates the structural position of the corresponding user. It is a measure of the influence that people have over the spread of information through the network.

Our purpose being the identification of individual that are prone to collaborate with each others, we would like to find these clusters of individuals in the network. There are no standard method for extracting so-called cohesive subgroups in social network analysis. Many different methods are proposed based on graph-theoretic terms (e.g., cliques [1]) or clustering methods (e.g., [8]).

Definition 2 (Distance network). *A distance network $\langle N, E^1, \ldots, E^n\rangle$ is made of a set N of nodes and n sets of distance functions $E^i : N \times N \longrightarrow [0\ 1]$ defining the distance between nodes (so satisfying symmetry, positiveness, minimality, and triangular inequality).*

Distance values can also be seen a weights or costs. It is clear that any network is a distance network which attributes either 0 or 1 as a distance. The definitions of SNA mentioned above can be adapted to distance networks if each time the cardinality of a set of edges if used, it is replaced by the sum of its distances. The distance of a path is obtained by summing the distances of its edges. One extension that must be made is to use the distance between nodes to reduce their influence to others in the computation of authority and hub degrees:

$$Hub^i_{t+1}(e) = \sum_{e' \in N} \frac{Auth^i_t(e')}{E^i(e,e')} \text{ and } Auth^i_{t+1}(e) = \sum_{e' \in N} \frac{Hub^i_t(e')}{E^i(e',e)} \tag{4}$$

4 Three-Layered Architecture for Semantic Social Networks

In order to uncover the links between people from those that can be found from their knowledge, we propose the three-layered architecture for constructing the semantic social network. As shown in Figure 1, it consists of i) a social network ($\mathcal{S}$), ii) an ontology network ($\mathcal{O}$), and iii) a concept network ($\mathcal{C}$). The characteristics of each layer and the relationships between layers are described below.

4.1 Social Layer

In the social layer ($\mathcal{S}$), nodes are representing people, and relations are the connections between peoples. A social network $\mathcal{S}$ is a directed graph $\langle N_S, E_S^{knows} \rangle$, where N_S is a set of person and $E_S^{knows} \subseteq N_S \times N_S$ the set of relations between these persons. In most current applications, the relation used by SNA is the $knows$ relation that can be found in FOAF.

Table 1. Closeness, authoritative and hub weights in Fig. 1

Weights	$Arun$(AS)	$Antoine$(AZ)	$Faisal$(FAK)	$JeromeE$(JE)	$Jason$(JJ)	$JeromeP$(JP)	$Sebastien$(SL)
Closeness	0.5	**0.67**	0.5	0.6	**0.67**	0.46	0.4
Authoritative	0.21	0.45	0.37	**0.69**	0.243	0.236	0.13
Hub	0.01	**0.52**	0.27	0.32	**0.54**	0.42	0.275

From the social network in Fig. 1, the authoritative and hub weights of three users are shown in Table 1. Obviously, the highest hub weight is assigned to $Jason$ because he is an important and unavoidable role of bridging between the rest of users.

4.2 Ontology Layer

The ontology network $\mathcal{O}$ is a network $\langle N_O, E_O^i \rangle$, in which N_O is a set of ontologies and $E_O^i \subseteq N_O \times N_O$ the relationships between these ontologies. There can be two main kinds of relations at this stage:

import when some ontology explicitly import another ontology;
refer when some ontology uses some concept defined in another ontology.

The objective relationship from the $\mathcal{S}$ to the $\mathcal{O}$ is through the explicit usage of an ontology by a user which can be expressed by a relation: $Use \subseteq N_S \times N_O$.

We can easily interpret the hubs as being the ontologies that combine a large number of other ontologies. These would be an interesting starting point for any newcomer willing to annotate a similar set of objects as his friend. Likewise, authorities will be ontologies that are extended and imported by many different actors (i.e., *de facto* standard ontologies).

There is a difference between ontology networks and social networks though: while in social networks it is normal to be connected to several authorities, an ontology will only import one ontology on some topic. It would thus be useful to recognize those hubs that connects authorities on the same topics, these "ontologies" are likely to be the expression of an alignment between the two authorities.

4.3 Concept Layer

In the concept layer ($\mathcal{C}$), nodes are concepts, and links are the numerous kinds of links that can be found in ontologies. The concept network $\mathcal{C}$ is a network $\langle N_C, E_C^i \rangle$, in which N_C is a set of entity of an ontology (classes, properties, individuals) and $E_C^i \subseteq$

$N_C \times N_C$ the relationships between these entities. These relationships are far more numerous than in the other layers and depends on the kind of entity considered. If we restrict our attention to classes, the:

subClass linking a class to its subclasses;
superClass (=subClass^{-1}) linking a class to its super classes;
sibbling linking a class to its siblings;
disjoint linking a class to the classes it is explicitly disjoint with;
property (=domain^{-1}) linking a class to its properties;
range^{-1} linking a class to the properties that refer to it.

The objective relationship from the $\mathcal{O}$ to the $\mathcal{C}$ is through the definition of concept in an ontology which can be expressed by a relation: $Defines \subseteq N_O \times N_C$. However, this notion of definition is not easy to catch: it could be based on either the assertion of a constraint on some ontology entity or the namespace in which entity belongs. We will consider the namespace in the following.

We are here further away from social networking. As noted in [4], the notions of hub and authority cannot be understood in the same way for all the relations expressed in $\mathcal{C}$.

5 Inferring Relationships

This three-level semantic social network does not bring in itself new improvement for our P2P sharing application. In order to provide new insight in the possible collaborations it is necessary to analyze these networks and to propagate information from one layer to another. We explain how, starting from the lower concept layer, it is possible to enrich the upper ontology and social layers with new relations from which SNA helps finding relevant peers.

5.1 Similarity on the Concept Layer

Beside the numerous relationships that can be found by construction of the concept layer, new relationships can be inferred between the entities. One particular relationship that will be interesting here is similarity. In order, to find relationship between concepts from different ontologies, identifying the entities denoting the same concept is a very important feature. As a matter of fact, most of the matching algorithms use some similarity measure or distance in order to match entities.

Similarity on the concept layer can be obtained by various means [9]. Some distances can be established from the local features of entities. For instance, the name of entities can be the basis for matching them. Many techniques have been developed for comparing strings, based on their structures (like edit distance), their morphology (through lemmatization), their entry in lexicons (using WordNet). Another kind of similarity can be established based on set of shared instances like in [3].

Some other distances, more in the spirit of network analysis, can be defined from the structure of the network. For instance, [10], defines possible similarities (e.g., Sim_C, Sim_R, Sim_A) between classes, relationships, attributes, and instances. Given a pair of classes from two different ontologies, the similarity measure Sim_C is assigned in $[0, 1]$. The similarity (Sim_C) between c and c' is defined as

$$Sim_C(c, c') = \sum_{E \in \mathcal{N}(C)} \pi_E^C MSim_Y(E(c), E(c')) \tag{5}$$

where $\mathcal{N}(C) \subseteq \{E^1 \dots E^n\}$ is the set of all relationships in which classes participate (for instance, subclass, instances, or attributes). The weights π_E^C are normalized (i.e., $\sum_{E \in \mathcal{N}(C)} \pi_E^C = 1$).

If we consider class labels (L) and three relationships in $\mathcal{N}(C)$, which are superclass (E^{sup}), subclass (E^{sub}) and sibling class (E^{sib}), Equ. 5 is rewritten as:

$$\begin{aligned} Sim_C(c, c') = \ & \pi_L^C sim_L(L(A_i), LF(B_j)) \\ & + \pi_{sup}^C MSim_C(E^{sup}(c), E^{sup}(c')) \\ & + \pi_{sub}^C MSim_C(E^{sub}(c), E^{sub}(c')) \\ & + \pi_{sib}^C MSim_C(E^{sib}(c), E^{sib}(c')) \end{aligned} \tag{6}$$

where the set function $MSim_C$ computes the similarity of two entity collections.

As a matter of fact, a distance between two set of classes can be established by finding a maximal matching maximizing the summed similarity between the classes:

$$MSim_C(S, S') = \frac{\max\left(\sum_{\langle c,c' \rangle \in Pairing(S,S')} Sim_C(c, c')\right)}{\max\left(|S|, |S'|\right)} \tag{7}$$

in which $Pairing$ provides a matching of the two set of classes. The OLA algorithm is an iterative algorithm that compute this similarity [10]. This measure is normalized because, if Sim_C is normalized, the divisor is always greater or equal to the dividend.

A normalized similarity measure can be turned into a distance measure by taking its complement to 1 ($E_C^{dist}(x, y) = 1 - Sim_C(x, y)$). Such a distance introduces a new relation E_C^{dist} in the concept network $\mathcal{C}$. This relation indeed defines a distance network as introduced above.

5.2 From Concept Similarity to Ontology Similarity

Once such a distance has been introduced at the concept level, it can be used for computing a new distance at the ontology level. Again, a distance between two ontologies can be established by finding a maximal matching maximizing similarity between the elements of this ontology and computing a global measure which can be further normalized:

Definition 3 (Ontology distance). *Given a set of ontologies N_O, a set of entities N_C provided with a distance function $E_C^{dist} : N_C \times N_C \longrightarrow [0\ 1]$ and a relation $Defines : N_O \times N_C$, the distance function $E_O^{dist} : N_O \times N_O \longrightarrow [0\ 1]$ is defined as:*

$$E_O^{dist}(o, o') = \frac{\max(\sum_{\langle c,c' \rangle \in Pairing(Defines(o), Defines(o'))} E_C^{dist}(c, c'))}{\max(|Defines(o)|, |Defines(o')|)}$$

Of course, even with these heavy computations, $\forall o \in N_O, E_O^{dist}(o, o) = 0$.

This is the measure that is used in the OLA algorithm for deciding which alignment is available between two ontologies [10]. However, other distances can be used such as the well known single, average and multiple linkage distances.

This ontology distance introduces a new relation on the ontology layer. This measure provides a good idea of the distances between ontologies. These distances, in turn, provide hints of the difficulty to find an alignment between ontologies. It can be used for choosing to match the closest ontologies with regard to this distance. This can help a newcomer in a community to choose the best contact point: the one with who ease of understanding will be maximized. This will be further developed in Section 5.4.

5.3 From Concept Similarity to Alignment

It can however happen that people have similar but different ontologies. In order for them to exchange their annotations, they use alignments existing within the ontology network. Alignments, in turn, are the results of applying matching algorithms based on the correspondence between ontologies.

As a result, from concept similarity these algorithms will define a new relation E^{align} at the ontology level.

Definition 4 (Alignment relation). *Given a set of ontologies N_O, a set of entities N_C provided with a relation $E_C^{dist} : N_C \times N_C$, a matching algorithm $Match$ based on E_C^{dist} and a relation $Defines : N_O \times N_C$, the alignment relation E^{align} is defined as:*

$$\langle o, o' \rangle \in E^{align} \textit{ iff } Match(o, o') \neq \emptyset$$

If one has a measure of the difficulty to use an alignment or of its quality, this network can also be turned into a distance network on which all these measures can be performed.

This new relation in the ontology layer allows the agents to choose the ontology that they will align with first. Indeed, the ontologies with maximal hub centrality and closeness are those for which the benefit to align to will be the highest because they are aligned with more ontologies at the ontology level. In the P2P sharing application, choosing such an ontology will bring the maximum answers to queries. For example, in the concept layer of Fig. 1, two alignments between i) po_{Arun} and po_{Jason} and ii) $po_{Sebastien}$ and po_{Jason} enable *Arun* and *Sebastien* to share information, even though they are not explicitly linked with each other.

This is the occasion to note the difference between the relations in the same network: in the ontology network, the hub ontologies for the import relation are rather complete ontologies that cover many aspects of the domains, while hub ontologies for the E^{align} relation are those which will offer access to more answers.

Of course, when an alignment exists between all the ontologies used by two peers, there is at least some chance that they can talk to each others. This can be further used in the social network.

5.4 From Ontology Similarity to People Affinity

Once these measure on ontologies are obtained, this distance can be further used on the social layer. As we proposed it is possible to think that people using the same ontologies should be close to each other. We can consider measuring the affinity between people from the similarity between the ontology they use.

Definition 5 (Affinity). *Given a set of people N_S, a set of ontologies N_O provided with a distance $E_O^{dist} : N_O \times N_O \longrightarrow [0\ 1]$ and a relation $Uses : N_S \times N_O$, the affinity is the similarity measure defined as*

$$E^{aff}(p,p') = \frac{\max\left(\sum_{(o,o')\in Pairing(Use(p),Use(p'))} 1 - E_O^{dist}(o,o')\right)}{\max(|Use(p)|, |Use(p')|)} \quad (8)$$

Since this measure is normalized, it can be again converted to a distance measure through complementation to 1.

Introducing the distance corresponding to affinity in the social network allows to compute the affinity relationships between people with regard to their knowledge structure. Bottom-up inference from $\mathcal{C}$ allows to find out the semantic relationships between users based on this space.

For completing the P2P application, the last step consists of identifying the subgroups of users, according to the various social characteristics, as follows:

1. The subgroup whose *members are assigned very high semantic authoritative weight* (or *semantic hub weight*) can be identified by comparing the weights computed with Equ. 4. These peers have the social power to control and select semantic information to distribute.
2. The subgroups of *people with very similar personal ontologies* can be obtained by computing the cohesive subgroup of the $\mathcal{S}$ network using affinity (E^{aff}).
3. The subgroup of *people which are interested in the same topics* extends the previous subgroups, depending on a particular topic. Their members can efficiently share information about that topic.

6 Experimental Results

As a first evaluation of our framework, we want to differentiate social affinity E^{aff} (p,p') from the previous measures for social features. Our experimentation scenario follows these steps (i) building personal ontologies during photo annotation, (ii) aligning these personal ontologies for measuring social affinity between people, and (iii) discovering the most powerful person for semantic interoperability.

For collecting data, we invited seven members of our team to select a set of photographs and annotate them by using a specific annotation tool (Picster[2]). Table 2 shows the specification of personal ontologies.

From co-occurrence patterns between the annotated photos, Mika's social centrality C_M [3] can be formulated by

$$C_M(U_i) = \frac{\sum \frac{\cap_{k=1,k\neq i}^{|U|}(\mathcal{R}_{U_k},\mathcal{R}_{U_i})}{\mathcal{R}_{U_i}}}{|U|-1} \quad (9)$$

where $|U|$ is the total number of people in the social network. The results are shown in Table 3. We found out that the number of annotated resources are barely related to the

[2] http://gforge.inria.fr/projects/elster

Table 2. Specification of personal ontologies as test bed

	AS	AZ	FAK	JE	JJ	JP	SL
Number of annotated photographs ($\mathcal{R}_{User}$)	47	47	37	49	47	30	25
Number of used ontologies ($\mathcal{O}_{User}$)	3	5	2	6	1	1	2

Table 3. Experimental results of a) closeness and centrality by co-occurrence patterns b) people affinity E^{aff} and centrality in the semantic social network

(a/b)	AS	AZ	FAK	JE	JJ	JP	SL	C_M	C_{aff}
AS	-	0.98/0.65	0.62/0.33	0.94/0.73	1.00/0.26	0.60/0.32	0.23/0.62	0.73	0.49
AZ	0.98	-	0.62/0.49	**0.94/0.825**	0.98/0.31	0.62/0.3	0.26/0.52	0.73	0.52
FAK	0.78	0.78	-	0.70/0.57	0.78/0.28	0.54/0.22	0.30/0.32	0.65	0.37
JE	0.90	**0.90**	0.53	-	0.90/0.46	0.57/0.49	0.16/0.75	0.66	**0.64**
JJ	1.00	0.98	0.62	0.94	-	0.60/0.72	0.23/0.39	0.73	0.40
JP	0.93	0.97	0.67	0.93	0.93	-	0.13/0.51	**0.76**	0.43
SL	0.44	0.48	0.44	0.32	0.44	0.16	-	0.38	0.52

social centrality. SL annotated the least number of resources, so that his centrality also lowest among people. But, even though JE's annotations were the largest one, JP has shown the most powerful centrality.

We measured semantic affinity on the semantic social network (Eq. 8). For doing so, the ontology distances E_O^{dist} between personal ontologies are measured. We used string edit distance between class labels. For instance, E_O^{dist} between JE and AZ is shown in Table 4. Then, we can measure $E^{aff}(\text{AZ}, \text{JE}) = \frac{\max(\sum_{\langle o,o' \rangle} 1 - E_O^{dist}(o,o'))}{\max(5,6)} = \frac{4.95}{6} = 0.82$ where ontology distance between non aligned ontologies ('-') are regarded as 1. We computed this for all pairs of people on the social network, as shown in Table 3. The matrix for social affinity is symmetric. We found out that the number of ontologies are playing an important role in social affinity. JE has shown the highest centrality in the given social network, while JP annotated the most common resource with other people. This means that collaborations can be effectively be provided with JE. Meanwhile, $E^{aff}(\text{JJ}, \text{JP})$ was relatively high (0.72). We found that they were using the same large ontology (i.e., SUMO). So the number of found correspondences where very high. To deal with this problem, we have to consider preprocessing personal ontologies.

Table 4. Ontology distance E_O^{dist} between JE and AZ; Mark '-' means no alignments between two ontologies

	JE_foaf.owl	JE_Meteo.owl	JE_Picster.owl	JE_space.owl	JE_UrbanLand.owl	JE_World.owl
az_support-ontology.owl	0.03	-	-	0.17	-	-
az_hasSupplyLineOnt.owl	0.46	-	0.09	0.05	0.04	0.49
az_office.owl	0.47	-	0.04	0.05	0.06	0.04
az_people+petsB.owl	0.06	-	-	0.16	-	-
az_space-basic.owl	0.18	-	-	0.5	-	0.01

Another issue is the discovery of potential collaborators (or like-minded people) by comparing the social distances in Table 1. While the social distance between AS and SL is 3 on the social layer of Figure 1 (i.e., 1. once normalized), their social affinity is measured as 0.62 which is relatively high (the corresponding distance would be .38). We can expect that they can share common interests.

7 Related Work

Many semantic systems on distributed environments like P2P networks have been introduced to efficiently share information and knowledge between heterogeneous sources. Some have studied the relevance of peers (or users) by analyzing topology and interactions like message passing. In [11], for selecting the expert peers, semantic topology analysis is exploited. But they make assumptions that every peers have to use the same ontology for calculating semantic similarity. Practically, [12] applies the similar idea to multi-agent architecture, and as an example of application, Jung has introduced a social communication framework for collaborative web browsing [13]. Meanwhile, Alani and colleagues introduced a system, called *Ontocopi*, for ontology-based network analysis (ONA) [14]. This system can identify the communities by using informal data.

More closely related to this work, Mika proposes a three-layered space which is composed of a social network and a knowledge network relying on concepts and instances [15]. However, the knowledge network is simply based on sets for co-occurrence analysis, and in this networks, the super/sub-relationships are retrieved based on statistical overlapping. This approach does not deeply consider semantic relationships between concepts and ontologies, so it is hard to use the structure of knowledge for structuring the social network.

For measuring the relevance of ontologies, in [4], the AKTiveRank system ranks the ontologies applying a number of classical metrics such as class match, centrality, density, and semantic similarity. In [16], network analysis methods (in particular, Hermitian matrices-based eigensystem analysis) are used for analyzing ontologies as concept graphs in the same way social network are analyzed. This is used for contrasting the "style" of ontologies but not for social networking, though this last application could be investigated.

8 Concluding Remarks and Future Work

We have focused on using the structure of knowledge used by people in order to extract meaningful relations at the social level. Moreover, the extraction of these new relations is used to further improve the collaborative sharing and exploitation of this knowledge.

We propose a three-layered architecture for constructing semantic social network, which is composed of a social layer, an ontology layer, and a concept layer. This space not only supports the relations within a layer, but also the propagation of relations between layers. We have provided the principles for extracting similarity between concepts and propagating this similarity to a distance and an alignment relation between ontologies. This distance relation can be used for discovering affinity in the social network. In return, users can take advantage of these newly established relations to find people closer to them on the basis of the structure of their knowledge. For that purpose, we have only used classical SNA measures (that we extended to distance networks).

The basic assumption of this work is that these newly discovered relations between people will facilitate mutual understanding as well as ontology matching and resource sharing. This remains to be demonstrated experimentally. To that extent we are developing and experimenting the presented P2P sharing system. We will make measure and generate alignment and ask users to rate and correct the provided alignments.

There remains important issues to be investigated: all these networks are not equal and their exploitation with classical SNA tools can be meaningless (in the same sense that considering the "loves" and "hates" relations as the same would lead to problems). It is thus important to characterize the various relations that were provided with regard to the measures that can be used on them. We also plan to extend this work by defining and extracting meaningful (and useful) clusters among people. We expect to apply this information to build social and semantic grid environment [17]. Moreover, the cost of computing these networks can become important. We have not considered this issue here because most of the given measures are examples, but this can become an important factor if the networks and measures have to be computed on-line.

References

1. Wasserman, S., Faust, K.: Social network analysis: methods and applications. Cambridge University Press, Cambridge (UK) (1994)
2. Kleinberg, J.M.: Authoritative sources in a hyperlinked environment. J. of the ACM **46**(5) (1999) 604–632
3. Mika, P.: Ontologies are us: A unified model of social networks and semantics. In Gil, Y., Motta, E., Benjamins, V.R., Musen, M.A., eds.: Proc. of the 4th Int. Semantic Web Conf. Volume 3729 of Lecture Notes in Computer Science., Springer (2005) 522–536
4. Alani, H., Brewster, C.: Ontology ranking based on the analysis of concept structures. In: Proc. of the 3rd Int. Conf. on Knowledge capture (K-CAP '05), New York, NY, USA, ACM Press (2005) 51–58
5. Broekstra, J., Ehrig, M., Haase, P., van Harmelen, F., Menken, M., Mika, P., Schnizler, B., Siebes, R.: Bibster - a semantics-based bibliographic peer-to-peer system. In: Proc. of the 2nd Work. on Semantics in Peer-to-Peer and Grid Computing (SemPGRID '04), New York, USA (2004) 3–22
6. Euzenat, J.: Alignment infrastructure for ontology mediation and other applications. In: Proc. of the 1st Int. Work. on Mediation in Semantic Web Services, Amsterdam (NL). (2005) 81–95
7. Freeman, L.: Centrality in social networks: Conceptual clarification. Social Networks **1** (1979) 215–239
8. Fortunato, S., Latora, V., Marchiori, M.: Method to find community structures based on information centrality. Physical review E. **70** (2004) 056104
9. Shvaiko, P., Euzenat, J.: A survey of schema-based matching approaches. J. on data semantics **4** (2005) 146–171
10. Euzenat, J., Valtchev, P.: Similarity-based ontology alignment in OWL-Lite. In de Mántaras, R.L., Saitta, L., eds.: Proc. of the 16th Euro. Conf. on Artificial Intelligence, IOS Press (2004) 333–337

11. Haase, P., Siebes, R., van Harmelen, F.: Peer selection in peer-to-peer networks with semantic topologies. In Bouzeghoub, M., ed.: Proc. of the Int. Conf. on Semantics in a Networked World (ICNSW'04). Volume 3226 of Lecture Notes in Computer Science., Springer-Verlag (2004) 108–125
12. Valencia, E., Sansonnet, J.P.: Building semantic channels between heterogeneous agents with topological tools. In: Proc. of the 2nd Euro. Work. on Multi-Agent Systems. (2004)
13. Jung, J.J.: Collaborative web browsing based on semantic extraction of user interests with bookmarks. J. of Universal Computer Science **11**(2) (2005) 213–228
14. Alani, H., Dasmahapatra, S., O'Hara, K., Shadbolt, N.: Identifying communities of practice through ontology network analysis. IEEE Intelligent Systems **18**(2) (2003) 18–25
15. Staab, S., Domingos, P., Mika, P., Golbeck, J., Ding, L., Finin, T., Joshi, A., Nowak, A., Vallacher, R.R.: Social networks applied. IEEE Intelligent Systems **20**(1) (2005) 80–93
16. Hoser, B., Hotho, A., Jäschke, R., Schmitz, C., Stumme, G.: Semantic network analysis of ontologies. In: Proc. of the 3rd Euro. Semantic Web Conf. (ESWC). Volume 4011 of Lecture Notes in Computer Science (2006) 514–529
17. Jung, J.J., Ha, I., Jo, G.: BlogGrid: Towards and efficient information pushing service on blogspace. In Zhung, H., Fox, G., eds.: Proc. of the Int. Conf. Grid and Cooperative Computing (GCC 2005). Volume 3795 of Lecture Notes in Computer Science., Springer-Verlag (2005) 178–183

Knowledge Sharing on the Semantic Web

Nicholas J. Kings[1], Caroline Gale[2], and John Davies[1]

[1] Next Generation Web Research Group, BT Group, Adastral Park, Ipswich, UK
{nick.kings,john.nj.davies}@bt.com
[2] Chimera – Institute for SocioTechnical Research and Innovation, Essex University, UK
cgale@essex.ac.uk

Abstract. This paper details the design, implementation and evaluation of an ontology-based knowledge sharing tool. The system, "Squidz", automatically classifies browsed web pages against an ontology, and allows users to share comments made about those pages to members of a community. As the user browses web pages, recommendations of relevant documents which have already been shared are produced, based upon both the user's social network as well as the semantic content of the page currently in view. Key to the design of the system has been the requirement, evidenced by earlier studies, that sharing should be easily effected as a side-effect of browsing rather than comprising a separate and distracting task. Another feature of the system is the linkage of a formal ontology with user-provided tags of shared information, thus combining the proven popularity of folksonomy-based systems with the shared and formal domain model provided by an ontology.

Keywords: Semantic Web, Knowledge Sharing, Tagging, Social Software, Communities of Practice.

1 Introduction

In general, knowledge sharing tools combine the functions of searching for and distributing information. As a user requires information to undertake a task, information relevant to that task can be located. Underpinning knowledge sharing tools is the premise that someone in the user's wider community has already created or accessed relevant information (explicit knowledge transfer) or someone is able to provide help or advice (tacit knowledge transfer).

Knowledge management can be defined as the "systematic application of actions to ensure that an organisation obtains greatest benefit from the information that is available to it" [1]. Knowledge sharing software supports the activities of collating, categorising and distributing information [2], which creates a group memory and improves team awareness [3, 4].

The application of a knowledge sharing tool has a direct impact on the community's behaviour; any interactive digital technology has embedded implicit cultural assumptions. Raybourn *et al.* [5] suggest that there is no recipe or standard format for encouraging participation, nor should any one cultural perspective be forced on such a community. However, functions to support the community may be just as important

E. Franconi, M. Kifer, and W. May (Eds.): ESWC 2007, LNCS 4519, pp. 281–295, 2007.

as other functions supported by a system [5-10]. Kings *et al.* [11, 12] further suggest that a sense of history and a user's reputation are prerequisites for the development of a shared community purpose.

1.1 Models of Knowledge Sharing

The Semantic Web [13] can provide enhanced information access based on the exploitation of machine-processable meta data. Central to the vision of the Semantic Web are ontologies, which are seen as facilitating knowledge sharing and re-use between agents, be they human or artificial [14]. They offer this capability by providing a consensual and formal conceptualisation of a given domain. As such, the use of ontologies and supporting tools offer an opportunity to significantly improve knowledge management capabilities on the intranets of organisations and on the wider web. Furthermore, Mika [15] suggests that although the Semantic Web has been defined to facilitate machine understanding of the World Wide Web, the process of creating and maintaining that shared ontology is a purely social activity. Each ontology is created in a process that requires a group, or community, to build and share an agreed understanding of the important concepts and objects for that self same community. Mika further proposes that the understanding of social presence is crucial in understanding how an ontology evolves and gains acceptance.

Recently, we have seen the emergence of a number of very popular community-based systems on the Web. Typically in systems such as flickr[1] and delicious[2], instead of using a centralized form of classification, users are encouraged to assign freely chosen keywords, called tags, to pieces of information or data, a process known as tagging. As a community of users generate a series of tags for overlapping and common items, a "folksonomy" can been seen to emerge. Since folksonomies are incremental and end user-generated and therefore inexpensive to implement, advocates of folksonomy believe that it provides a useful low-cost alternative to more traditional controlled vocabularies or classification schemes.

The combination of social networking systems such as FOAF[3] and XFN[4] with the development of tag-based folksonomies, implemented as blogs and wikis initally were seen as a basis for the semantic environment. This has been seen as part of a change from accessing static web pages to the use of the web as an application platform [16]:

- The change from centralised information sources to an approach of creating and distributing Web content itself, characterised by open communication, the willingness to share and re-use information, and "the market as a conversation" [17].
- The change from using web sites as point sources of information, stored within static Web pages, to sources of remotely accessible information, through the use of network accessible APIs or services.

In contrast to defined, formal taxonomies, categories in a folksonomy may appear to be arbitrary and idiosyncratic. However, a particular tag is chosen for a particular Web page based upon an individual's own understanding of the content being tagged,

---

1 http://www.flickr.com/
2 http://del.icio.us/
3 http://xmlns.com/foaf/0.1/
4 http://gmpg.org/xfn/

which combines the personal, social, and technical understanding of that content [15]. By publishing a tag and tagged content to a wider audience, other users are subtly encouraged to explore other tagged content and other users' interests.

Pind [18] suggests that there is anecdotal evidence to show that the sets of tags used within a community tend to converge upon a common set of agreed meanings and usage. Pind further suggests, however, that tagging software could be improved by the addition of the following five features: the software should suggest appropriate tags; the software should display related tags and topics; suggest tags that others have used to describe the same items; infer topic hierarchies from the way tags are used; and, allow a user to quickly edit and change tags that have already been applied. The Squidz tool addresses all of these issues.

In Section 2, we describe the Squidz tool. Two key features of the system are, firstly, the requirement, as evidenced by earlier studies, that sharing should be easily effected as a side-effect of browsing rather than comprising a separate and distracting task [19-21]; and, secondly, the linkage of a formal ontology with user-provided tags of shared information, thus combining the proven popularity of folksonomy-based systems with the shared and formal domain model provided by an ontology. Section 3 describes the operation of the software in more detail. Section 4 details the evaluation of Squidz.

2 Software Overview

The main function of Squidz is to share knowledge in the form of textual annotations about web pages, within a community of business users. The secondary purpose of Squidz is to allow a user to discover new social contacts or sources of information, through discovery of shared interests. The important features to be implemented by the software are:

- To improve software adoption, Squidz must be useful to an individual user without relying on any other's contributions. Squidz should be able to be used as an advanced book marking tool.
- All users can view and comment upon any other's annotations.
- An annotation is made in a technical and a social context. The technical context is represented by the topics associated with the annotation; the social context is represented by the community where the annotation is posted.
- The software filters and presents appropriate annotations, based upon each user's technical and social context. In effect, a relevant annotation is one made by a closely related person within a topic area related to the current page.
- Squidz models a user's social network in simple manner [2]: sharing pages to oneself; sharing to close work or team colleagues; sharing to members of a community of practise; or, sharing to all users of the system

Squidz is implemented as a browser plug-in, making use of a number of web services to classify browsed web pages and to retrieve annotations made by other members of the user community; Squidz is just one potential interface to the meta-data generated within the community. Squidz is being developed for communities of interest, or practice, within a corporate environment. In common with other tools being developed within the SEKT[5] project, Squidz is utilising the PROTON ontology [22, 23]; PROTON[6] is an ontology for knowledge management, modelling entities such as Documents with associated Topics as well as Users with Interests and Communities of Users.

As a user browses the internet or their intranet, the web page is classified against the formal ontology, and that classification is used to retrieve related pages. By clicking on the icons and words in the plug-in, separate information windows are created, rather than attempting to merge information within the browsed page.

Fig. 1. Pages classified during browsing

Figure 1 shows the main user interface to Squidz, after a user has been browsing: web page's main topic is "copyright" and that six related annotated pages have been previously shared by other users; the main topic for a web page is either the highest ranked topic returned by the classifier or, if this user has already annotated this page, the main topic is taken from that annotation.

2.1 System Architecture

Squidz has been implemented as a number of server side components and the user interface is supplied as an Internet Explorer Browser Helper Object (or plug-in). The server side components write to a database which is mapped to an OWL ontology (PROTON) held in KAON. Information generated by using Squidz is thus fully accessible to other semantic web applications. Thus, other applications developed within SEKT, such as Semantic Search and Browse, are enable to share information through a semantic repository [22, 24-26]. Squidz has been envisaged as being used with a corporate setting, for use within communities of practise or communities of interest. Squidz monitors and tracks a user's browsing behaviour; privacy and secrecy were seen as less relevant within an enterprise setting. For use of Squidz within a

5 http://www.sekt-project.com/

6 http://proton.semanticweb.org/

community of users across the wider internet, further work would be required to satisfy concerns about user security and privacy.

For use within British Telecom (BT), BT procures the Inspec and ABI bibliographic record databases, giving access to over 4 million bibliographic records. The format for each bibliographic record is based on ISO 2709 (which is based on the Library of Congress MARC format[7]).

A classifier has been developed, which assigns subject categories from the ABI and Inspec controlled vocabularies to content retrieved from the Web [27]. The classifier identifies the occurrence of controlled indexing terms in the text, and selects those terms for classification that are deemed most significant. The significance of each controlled indexing term is not only dependent on its frequency of occurrence in the text, but also on its inter-dependencies with other controlled indexing terms. The classifier web service returns the four highest ranking subject topics, for each web page, as well as up to ten of the most significant words and phrases identified in the text.

An inspection of the classifications presented by the classifier for a number of web pages suggests that it is capable of producing sensible classifications for previously unseen Web content. However, a full evaluation of the classification technique is currently being undertaken.

2.2 Semantic Annotations

When sharing a web page in Squidz, the user has the opportunity to provide some folksonomy-style tags which describe the content of the page, along with a comment about the page. Squidz models an annotation as the following: the user's comment; a set of formal topics, generated by the page classifier; a set of informal user-supplied tags; the target audience (individual, team, community or world as chosen by the user); and, a set of keywords and phrases, also generated by the page classifier. For each user tag, Squidz derives an associated set of keywords; each tag is semantically characterised by that keyword vector rather that the character string the user chose to represent the concept denoted by that particular tag.

As web pages are annotated with tags, the set of words for each tag is re-calculated automatically based upon the keywords stored within the page annotation. The keywords are associated with each tag in order to find related pages even if pages have been tagged with different terms; the derived keywords allow pages to be tagged with similar words, even though each user has their own understanding of that word. For example, one user's tag of "project" may represent the same concept as another user's tag of "SEKT". Adding an annotation forces the system to recalculate the keywords associated with the user's tags and subsequently all pages that are now related to the current page.

Though a folksonomy allows a community to evolve their own vocabulary, at a low cost for each contributor, there are associated problems of synonomy and polysemy [28], as there is no central defined meaning for each keyword or phrase used. Using the approach proposed, the semantic annotation uses the information available from the tagged pages themselves to provide a common basis for understanding the community's concepts.

---

[7] http://www.loc.gov/marc/

Annotations are also made to a particular context, such as "Self", "Team", "Community" or "Everyone". This is used as a recommendation as to whom the annotation may be most relevant; by taking into account the suitability of an annotation, this may reduce the "cost" of a user understanding the importance or relevance of that particular web page [2].

As pages are shared in Squidz, a number of OWL metadata elements are generated and stored in the PROTON ontology: a Squidz annotation is represented as a PROTON document with an associated set of ontology topics and informal user-generated tags. Thus, the community generates relevant annotated content for its own use and this content can be accessed via other Semantic Web applications, such as the Semantic Search and Browse tool in BT's Digital Library [22, 26].

3 Squidz in Action

Section 2 presented an overview of Squidz, whereas this section examines each of the main software functions in more detail.

3.1 Fetching Related Pages

Squidz provides the user with "peripheral vision" of previously shared web pages and annotations related to the current page. As a web page is viewed by a user, a request for related pages and their associated annotations is made to the remote web service. Related pages are determined in the following manner:

```
Initialise candidate topic and tag set, CTT

Add topics returned by classifier web service, for the
current page, to CTT

Add all topics from previous annotations made on cur-
rent page to CTT

Calculate similarity of page content to users' tags,
using Dice coefficient

        If similarity of tag exceeds threshold, add tag
        to CTT

Initialise set of candidate annotations, CA

Add annotations to CA, for each annotation which has
been annotated with at least one topic in CTT

Rank each web page in CA, based on the number of anno-
tations stored for that page and who made the annota-
tion
```

The similarity of each tag in the system to the current page content is calculated with the Dice coefficient, as shown in Equation 1 [29], where KWC and KWT are sets of keywords: KWC, the set of keywords derived for a viewed page; and, KWT, the set of keyword associated with a particular user tag.

$$Dice(KWC, KWT) = \frac{|KWC \cap KWT|}{\frac{1}{2}\left(|KWC| + |KWT|\right)} \quad (1)$$

The similarity measure is taken between the currently viewed page's content and a tag's keyword vector rather than between the current page contents and the keywords associated with each previously stored page for two reasons. Firstly, this decision improves the performance: as there are fewer user tags than annotations, fewer similarity calculations need to be under taken. Secondly, by matching user tags, Squidz is returning pages that are associated with a user's concepts: one user may be interested another user's "project" pages, without actually knowing of that other person's interest.

In effect, each annotation "votes" for a web page to be brought to the user's attention; the weight associated to the vote varies based on who made the annotation. Thus, there is a chance that annotations made by people outside of a user's direct social network will be brought to a user's attention. In this manner, the technical or topical context of web page and well as a user's social context is taken into account to account to rank a related page.

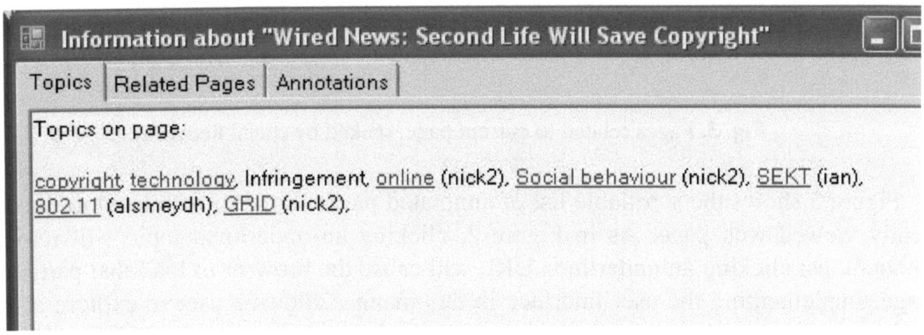

Fig. 2. Topics related to current page

Figure 2 shows that eight topics have been identified that are related to the current page: three topics from the defined topic ontology, and, five user created tags: user- generated tags are identified by placing the name of the user who generated the tag in parentheses. As described above, the defined topics have been identified by the classifier.

The list of related user tags, shown in Figure 2, is formed from the tags found within the returned list of related pages. Each underlined topic or tag can be clicked to cause all of the annotations made with that particular topic to be retrieved and displayed. Related topics are calculated in the following manner:

```
Initialise related topic set, RTL
Add all library topics, as calculated by page classi-
fier, to RTL
For each related page (as calculated above), RP
    For each annotation, stored for RP
        Add user tag, UT, to RTL
```

Thus, the list of related topics contains ontology topics, user tags that are similar to the current page and user tags that have been explicitly added to those related pages. Only if an annotation has been created with a particular topic or tag, will the name be clickable; Figure 2 shows no annotations have been made with the topic of "Infringement".

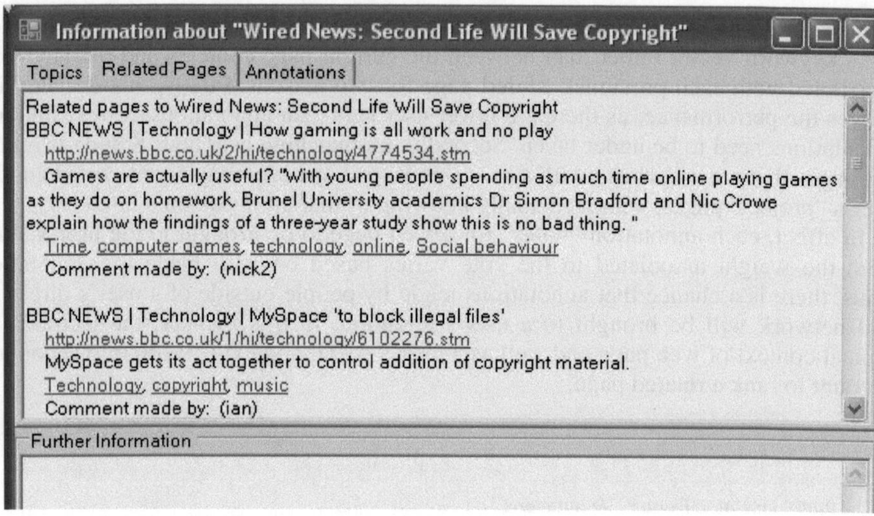

Fig. 3. Pages related to current page, ranked by social network

Figure 3 shows the scrollable list of annotated pages found to be related to the currently viewed web page. As in Figure 2, clicking an underlined topic will retrieve annotations; clicking an underlined URL will cause the browser to load that particular page. Implementing the user interface in this manner allows a user to explore across web pages, rather than having to visit a particular website and then start exploring tags and relations from that point onwards.

3.2 Sharing Annotations

By clicking on "Self", "Team", "Community" or "Everyone" (as shown Figure 1), a user can choose to share an annotation about the currently viewed web page. This is implemented via a less intrusive pop-up window, rather than by changing the contents of the main browser widow. Figure 4 shows that comments about a particular web page about to be shared to members of this user's community.

Sharing to a particular set of people does not preclude that others outside of group will not see the annotation, as having the ability to view every person's contribution is a crucial way of building an active set of users. However, the user-chosen target group of the annotation is used in the ranking algorithm when calculating related pages: a page shared to "Team" by one of my team members is ranked higher than a page shared by the same person but shared to "Community".

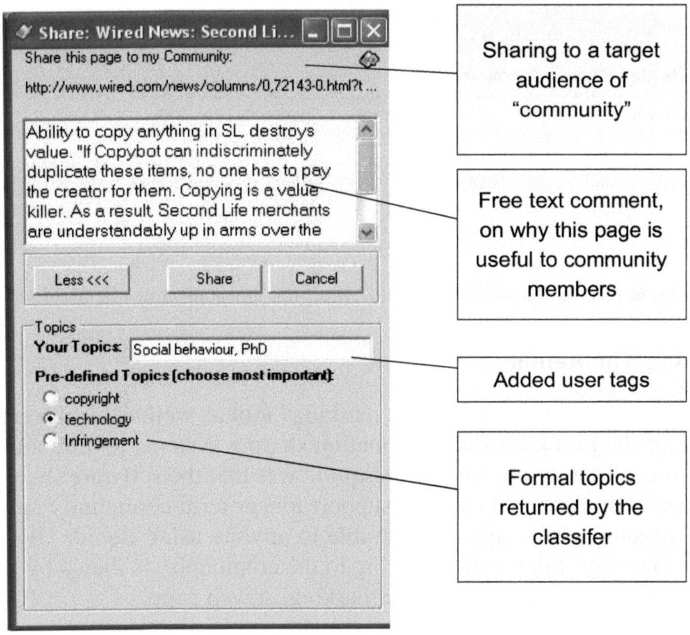

Fig. 4. Storing an annotation with the user choosing to add two tags and changing the main topic

Figure 5 shows that after the annotation has been made, the web page is now related to nine pages, and the main formal topic for the page is shown as "technology".

By making an annotation the topics related to this page have changed, as shown by comparing Figures 2 and 6. As the added annotation explicitly mentions the user's tag of "PhD", this tag would be expected to be shown for this page. However, by annotating, the algorithm has calculated that there is also a similarity between this page and the user's tags of "Tagging" and "Classification techniques". Figure 6 also shows that "Infringement" can now be clicked, as there has been an annotation stored for that topic.

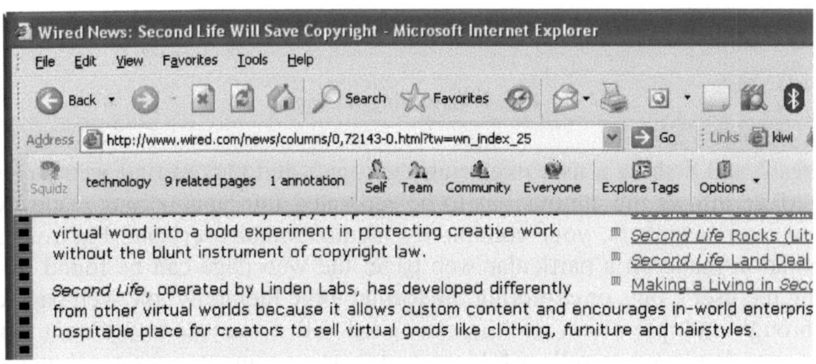

Fig. 5. Further related pages after adding an annotation

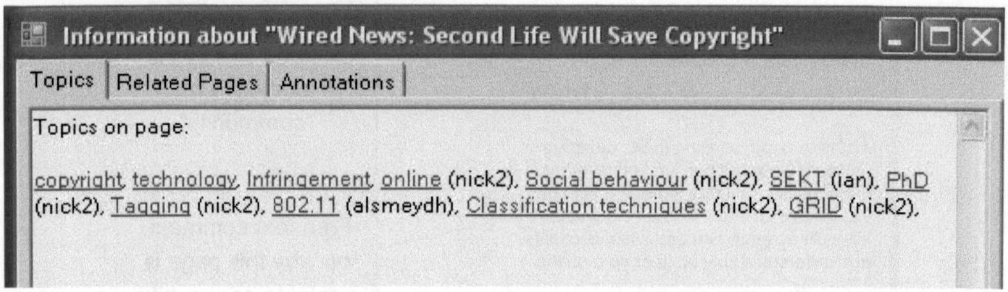

Fig. 6. Recalculated related topics, after the annotation has been added

3.3 Exploring Annotations

Squidz can be used as a personal "book marking" tool as well as for information sharing. Because of the problems that information sharing systems require a critical mass of users [4], one of the requirements for Squidz was that the software should be useful for a single user. However, in order to support longer term community spirit, it is important that all annotations should be visible to anyone using Squidz. By developing Squidz in this manner, information sharing to the community is then a by-product of a simple, well understood action of "book marking" a web page.

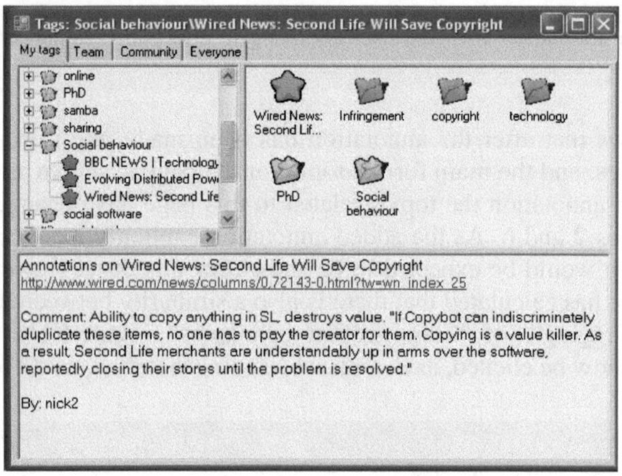

Fig. 7. Exploring my tags

Figures 7 and 8 show a user exploring the topics and tags to find web pages. The user interface allows the annotations to be separated into annotations made by your self, your team members, your community members and everyone. Figure 9 shows the annotation made on a particular web page: the web page can be found by either exploring the user's tags or exploring the formal topic hierarchy; the web page can be found through multiple routes, as it is associated with a number of different topics. A user' tags are displayed as yellow folders, and their own annotations are shown as yellow stars, while formal topics are shown as red folders.

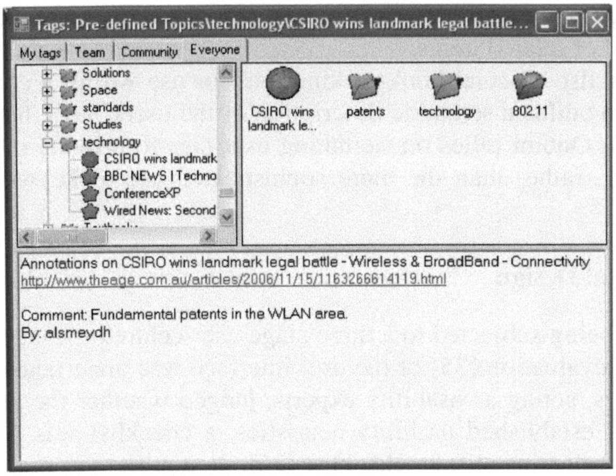

Fig. 8. Exploring another user's annotations and tags

Figure 8 shows a user exploring all the annotations made by everyone. Here annotations made by other people are denoted by blue globes, and their tags are denoted by blue folders. Figure 8 also shows that four annotations have been made with the topic of "technology": three have been by made by the user himself, and a fourth has been made by user "alsmeydh". From this screen, it can be seen that there is a user tag of "802.11", and that tag can be subsequently explored.

4 Evaluation

4.1 Related Work

The design for Squidz has been influenced by lessons learnt from the Jasper [30] and OntoShare [20, 21] knowledge sharing tools. In those tools, a profile of user interests is built to filter information, based upon which web pages have been shared by various users. Notifications, about pages being stored, were sent by email which caused, however this caused a disconnect between sharing and commenting and there was little dialogue encouraged between users. Sharing information should be easily effected as a side-effect of browsing rather than comprising a separate and distracting task.

Piggy Bank [31] presents a tool integrated into a web browser, coupled with the use of associated RDF collection utilities. Like Squidz, Piggy Bank allows a community of users to share and collaborate over items of information found. The approach taken by Piggy Bank, however, is to present the RDF metadata directly in the user interface; Squidz simplifies the user interface and only exposes parts of the metadata associated with document sharing, such as topic and user.

Annotea [32, 33] is a framework to support collaboration and sharing of semantically marked objects. The toolset allows users to tag Web pages with concepts, and allow the metadata to be repurposed through various XSLT style sheets. Annotea provides a flexible interface to explore bookmarks and topics. Squidz, however, is aimed at providing page recommendations, through the classification of the current web page and retrieval of related items.

Onomi [34] is also a social bookmarking tool for use within a corporate environment. Onomi also builds a semantic description of the users' tags, however, the technique used within Onomi relies on stemming user tags to provide the common basis for understanding, rather than the more sophisticated use of keyword sets, as used within Squidz.

4.2 Experimental Design

Squidz has been being subjected to a three-stage user-centred evaluation. For the first stage, a heuristic evaluation [35] of the user interface was undertaken. A small group of five researchers, acting as usability experts, judged whether the user interface adhered to a list of established usability heuristics; a checklist was adapted from the Xerox heuristic evaluation system checklist [36]. A number observations were made, most of which were concerned with minor interface problems and system performance. The results of the evaluation were collated and discussed with the development team. Squidz was then modified in accordance with these observations.

The second stage comprised a cognitive walkthrough evaluation [37]. Users were asked to use Squidz in order to complete a number of knowledge sharing tasks, where the user's actions and behaviours will be recorded. Users were encouraged to talk through their actions and their concerns as they undertake each task, as this provided additional information about the usability of Squidz and the user's thought processes as they used the application. At the end of each task, users were also asked to complete a short questionnaire. The findings were again discussed with the Squidz development team. By the completion of this stage, all major interface issues had been resolved.

At the time of submitting this paper, the third and final stage of evaluation is underway: preliminary results are given in the next section. Squidz is being rolled out for use within an intranet, to users with a range of technical experience. This stage consists of a series of semi-structured interviews to find out how useful people have found Squidz [38]. The purpose of the interviews is to validate the following hypotheses: knowledge sharing requires both a technical and social context; sharing is more likely to occur if the costs of sharing are reduced; and, sharing allows the knowledge of the community to built and enhanced.

4.3 Results and Analysis

Currently, twenty users have registered and downloaded the software: nine of those are regular users and a further eight users have made at least one annotation. All of the users are familiar with information sharing with tools such as email and delicious. Longer term user acceptance will be an important aspect to measure, since Squidz constantly monitors web page access. As the current trial has only been running a relatively short time at the time of writing, and further longitudinal data is required to determine the users' long term attitude to sharing information in this manner.

Nevertheless, data from the current valuation has validated the Squidz approach: users have commented on the simplicity and ease of use for the tool. The user interviews have also identified a number of new functions that could be incorporated into later versions of the software. For example, one requested function would be ability to "notify immediately" other people that a page has been annotated, as well still having the ability to be made aware of that page while browsing.

The users found the Squidz ranking of pages taking into account the content of the page and the original sharer of the page natural and intuitive. Annotations were typically tagged with only a few keywords or phrases. The characterization of tags with a keyword vector, however, gives a richer semantic representation of tags than in typical user-tagging applications. Semantic tags allowed users to name topics closer to the user's own higher level concepts, rather than each user having to add explicitly the complete set of keywords describing each web page.

5 Discussion and Conclusions

In this paper we have presented the design, implementation of a semantic tool for annotating and sharing information about web pages. Squidz is intended to test the hypothesis that information sharing is more effective when the software is aware of both the social and technical context of that information. The approach taken is to associate a formal topic ontology alongside an informal folksonomy, through the ability to annotate web pages. Presenting information on related pages and topics, through the toolbar, allows a user to browse and explore when convenient to the user, rather than forcing a particular mode of usage; knowledge sharing occurs as a result of normal user activity (browsing). Though further improvements to the tool are planned, Squidz has already gained positive user feedback and acceptance.

Acknowledgements. The work described in this paper was developed as part of the SEKT project, under EU funding (IST IP 2003506826). Further project details may be found at http://www.sekt-project.com/.

References

1. Marwick, A.D., *Knowledge management technology.* IBM Systems Journal, (2001). **40**(4): p. 814-830.
2. Shneiderman, B., *Leonardo's Laptop: Human Needs and the New Computing Technologies.* (2003), London, England: MIT Press.
3. Udel, J., *The New Social Enterprise.* InfoWorld, (2004). **Vol. 26**(No. 13): p. 47,50-52.
4. Rafaeli, S. and D.R. Raban, *Information sharing online: a research challenge.* International Journal of Knowledge and Learning, (2005). **1**(2): p. 62-79.
5. Raybourn, E.M., N. Kings, and J. Davies, *Adding cultural signposts in adaptive community-based virtual environments.* Interacting with Computers, (2003). **15**(1): p. 91-107.
6. Preece, J., *Online Communities: Designing Usability, Supporting Sociability.* (2000), Chichester, England: John Wiley & Sons.
7. Merali, Y. and J. Davies. *Knowledge capture and utilization in virtual communities.* in *International Conference On Knowledge Capture.* (2001). Victoria, British Columbia, Canada.
8. Kim, A.J., *Community Building on the Web.* (2000), Berkeley, Ca: Peachpit Press.

9. Crossley, M., N.J. Kings, and J.R. Scott, *Profiles – Analysis and Behaviour*, in *Location and Personalisation: Delivering Online and Mobility Services*, D. Ralph and S. Searby, Editors. (2004), IEE: London.
10. Sharratt, M. and A. Usoro, *Understanding Knowledge-Sharing in Online Communities of Practice.* Electronic Journal of Knowledge Management, (2003). **1**(2): p. 187-196.
11. Kings, N. *Knowledge sharing using Semantic Web technologies.* in *1st Workshop on Friend of a Friend, Social Networking and the Semantic Web.* (2004). Galway, Ireland.
12. Kings, N., D. Alsmeyer, and F. Owston, *Libraries as Shared Spaces*, in *Universal Access in HCI: Inclusive Design in the Information Society. Proceedings of HCI International, 2003, Volume 4.* (2003), Lawrence Erlbaum Associates.
13. Berners-Lee, T., J. Hendler, and O. Lassila, *The Semantic Web.* Scientific American, (2001). **May, 2001**.
14. Fensel, D., *Ontologies: Silver Bullet for Knowledge Management and Electronic Commerce.* (2001): Springer-Verlag, Berlin.
15. Mika, P. *Ontologies Are Us: A Unified Model of Social Networks and Semantics.* in *The Semantic Web - ISWC 2005.* (2005). Galway, Ireland: Springer.
16. O'Rielly, T., *What Is Web 2.0: Design Patterns and Business Models for the Next Generation of Software.* (2005), O'Reilly Media, Inc.
17. Levine, R., et al., *The Cluetrain Manefesto: The End of Business as Usual.* (2000), London: Pearson Education.
18. Pind, L., *Folksonomies: How we can improve the tags.* (2005), Pinds.com.
19. Davies, J., S. Stewart, and R. Weeks. *Knowledge Sharing over WWW.* in *WebNet '98.* (1988).
20. Davies, J., A. Duke, and A. Stonkus, *OntoShare: Evolving Ontologies in a Knowledge Sharing System*, in *Towards The Semantic Web*, J. Davies, D. Fensel, and F.v. Harmelen, Editors. (2003), John Wiley & Sons, Ltd.: Chichester. p. 161-177.
21. Davies, J., A. Duke, and Y. Sure, *OntoShare: a knowledge management environment for virtual communities of practice*, in *Proceedings of the international conference on Knowledge capture.* (2003), ACM Press: Sanibel Island, FL, USA.
22. Bontcheva, K., et al., *Semantic Information Access*, in *Semantic Web Technologies: Trends and Research in Ontology-based Systems*, J. Davies, R. Studer, and P. Warren, Editors. (2006), John Wiley & Sons Ltd: Chichester, England.
23. Kirakov, A., *Ontologies for Knowledge Management*, in *Semantic Web Technologies: Trends and Research in Ontology-based Systems*, J. Davies, R. Studer, and P. Warren, Editors. (2006), John Wiley & Sons Ltd: Chichester, England.
24. Davies, J., et al., *Next generation knowledge management.* BT Technology Journal, (2005). **23**(3): p. 175-190.
25. Glover, T. and J. Davies, *Integrating device independence and user profiles on the Web.* BT Technology Journal, (2005). **23**(1): p. 239-248.
26. Warren, P., I. Thurlow, and D. Alsmeyer, *Applying Sematic Technology to a Digital Library*, in *Semantic Web Technologies: Trends and Research in Ontology-based Systems*, J. Davies, R. Studer, and P. Warren, Editors. (2006), John Wiley & Sons Ltd: Chichester, England.
27. Thurlow, I. *Classifying Web content for a corporate digital library.* in *London Communications Symposium.* (2006).
28. Golder, S. and B.A. Huberman, *Usage Patterns of Collaborative Tagging Systems.* Journal of Information Science, (2006). **Vol. 32**(No. 2): p. pp198-208.
29. Grossman, D.A. and O. Frieder, *Information Retrieval: Algorithms and Heuristics*, ed. T.I.R. Series. (2006): Springer.

30. Davies, J., S. Stewart, and R. Weeks, *Knowledge Sharing Agents over the WWW.* BT Technology Journal, (1998). **Vol. 16**(No. 3, July).
31. Huynh, D., S. Mazzocchi, and D. Karger. *Piggy Bank: Experience the Semantic Web Inside Your Web Browser.* in *International Semantic Web Conference 2005.* (2005). Galway, Ireland: Springer.
32. Koivunen, M.-R. *Annotea and Semantic Web Supported Collaboration.* in *ESWC 2005, UserSWeb workshop.* (2005).
33. Koivunen, M.-R. *Semantic Authoring By Tagging with Annotea Social Bookmarks and Topics.* in *SAAW2006 - 1st Semantic Authoring and Annotation Workshop.* (2006). Athens, GA, USA.
34. Damianos, L., J. Griffith, and D. Cuomo. *Onomi: Social Booking on a Corporate Intranet.* in *Collaborative Web Tagging Workshop.* (2006). WWW2006, Edinburgh, UK.
35. Neilsen, J. and R. Molich, *Heuristic Evaluation of User Interfaces*, in *Proceedings of the SIGCHI conference on Human factors in computing systems.* (1992): Monterey, California, United States. p. 373 – 380.
36. Christiansson, P., *Heuristic Evaluation - A System Checklist.* (2000), Xerox Corporation.
37. Wharton, C., et al., *Applying Cognitive Walkthroughs to more Complex User Interfaces: Experiences, Issues, and Recommendations*, in *Proceedings of the SIGCHI conference on Human factors in computing systems.* (1992): Monterey, California, United States. p. 381-388.
38. Reynolds, T.J. and J. Gutman, *LADDERING THEORY, METHOD, ANALYSIS, AND INTERPRETATION.* Journal of Advertising Research, (1988)(February/March).

Real-World Reasoning with OWL

Timo Weithöner[1], Thorsten Liebig[1], Marko Luther[2], Sebastian Böhm[2], Friedrich von Henke[1], and Olaf Noppens[1]

[1] Inst. of AI, Ulm University, Ulm, Germany
firstname.lastname@uni-ulm.de

[2] DoCoMo Communications Laboratory Europe GmbH, Munich, Germany
lastname@docomolab-euro.com

Abstract. This work is motivated by experiences in the course of developing an ontology-based application within a real-world setting. We found out that current benchmarks are not well suited to provide helpful hints for users who seek for an appropriate reasoning system able to deal with expressive terminological descriptions, large volumes of assertional data, and frequent updates in a sound and complete way. This paper tries to provide some insights into currently available reasoning approaches and aims at identifying requirements to make future benchmarks more useful for application developers.

1 On Benchmarking OWL Reasoners

Having sufficiently exhaustive knowledge about the influence of the underlying reasoning approach on the practical tractability of a particular ontology is of fundamental importance when selecting an inference engine for a real-world application. By real-world we mean an ontology-based application with an expressivity at least beyond $\mathcal{ALC}$, containing more than thousands of individuals, and an inference response time of less than a second, even in a dynamical setting of frequent ontology updates. For instance, context-aware applications want to offer services to users based on their actual situation. Experiences in the course of operating a context-aware application for mobile users [1] clearly have shown that the quality of such an application hosted on a server significantly depends on the availability of reliable and scalable reasoning systems able to deal with constantly changing data. In order to meet real-world needs a reasoning system also has to offer a sufficiently expressive query language as well as a flexible and efficient communication interface.

Unfortunately, current benchmarks or system comparisons neither draw a clear picture of the landscape of practically tractable language fragments with respect to large amounts of instance data, give valuable insights into pros and cons of different reasoning approaches, identify performance penalties caused by certain language features, nor consider issues such as updates, incremental query answering, or interfaces.

For instance, many benchmarks consist of synthetical generated and sparsely interrelated data using inexpressive ontology languages such as the widely used

E. Franconi, M. Kifer, and W. May (Eds.): ESWC 2007, LNCS 4519, pp. 296–310, 2007.

Lehigh University Benchmark (LUBM) [2]. In case of the LUBM an incomplete query answering procedure exploiting told information about the individuals from a classified TBox is sufficient to answer the given queries correctly. The published RacerPro results [3] heavily rely on this property of the LUBM. The bottom line is that RacerPro shows an impressive performance for solving this ABox benchmark by switching off ABox reasoning. Obviously, this cannot be considered as a meaningful benchmark and it is not surprising that this test suite has let to exceptional performance for almost all inherently incomplete reasoning systems. On the other hand, the University Ontology Benchmark (UOBM) [4], a direct extension of the LUBM in terms of expressiveness, turned out to be much too difficult for most systems to answer correctly within reasonable time. This well known trade-off between tractable and effectively un-tractable ontologies, the so called *computational cliff*, is caused by an increase in language expressivity [5]. A more fine-grained map of the border of effectively tractable ontologies still needs to be practically explored in order to be helpful for developers.

The discussion of inherent drawbacks and advantages of different approaches with respect to diverse application tasks has been largely neglected in recent benchmarks or system comparisons. However, application developers need to be aware of potential trade-offs and a serious benchmark should discuss its results with respect to alternative reasoning approaches.

Another performance related issue deals with the way of feeding the systems with large amounts of data. Our selective tests have shown that for some systems not only the transmission format (RDF/XML or DIG [6]) is of importance, but also the way data is encoded (e. g. deep vs. flat serialization).

A real-world requirement which has not been taken into account in any benchmark so far is concerned with dynamic data. The ABox is not only expected to be the largest part of an ontology but is also subject to frequent changes. In order to serve as an efficient reasoning component within a realistic setting it is necessary to perform well under small ABox updates. First results in this research direction, e. g. [7], need to be evaluated by appropriate benchmarks.

Finally, all benchmark results need to be weighted with respect to soundness and completeness of the underlying inference procedure. Assuming that soundness and completeness is an indispensable requirement for knowledge-based applications — of which we think it is — many of the existing benchmark results are not helpful at all. Some of our randomly selected tests showed that even systems assumed to implement a sound and complete calculus fail on a number of OWL Lite test cases.

Our overall goal is to qualitatively analyze various benchmark suites and results in order to identify requirements for a comprehensive benchmark suite suitable to allow ontology-based application developers to pick the right system for their individual task. In the following section, we compare alternative reasoning approaches. We then (Section 3) analyze existing benchmark suites, discuss corresponding results and compare them with our own tests. As a result we compiled a collection of requirements (Section 4) to make future benchmarks more useful for application developers. Section 5 summarizes our experiences and suggestions.

2 System Analysis

Understanding and interpreting benchmarking results correctly requires to have some insights into alternative processing methods of different system implementations. In the context of reasoning with OWL, or fractions thereof, one can roughly distinguish between four different approaches.

Due to its historical origins the inference calculus implemented in **tableaux-based provers** for DLs is an obvious choice and available via systems like Pellet [8], RacerPro [9], or FaCT++ [10]. They implement a conceptually sound as well as complete approach for which many optimizations are known so far. Unfortunately, complete instance reasoning still requires expensive computations but recent research on elaborated reduction methods [11] show enormous optimization possibilities in this respect.

An alternative, equally sound and complete, approach is to transform an OWL ontology into a disjunctive datalog program and to utilize a **disjunctive datalog engine** for reasoning as implemented in KAON2 [12]. This allows for fast query answering due to well-known optimization techniques from deductive databases such as magic set transformation. A drawback is that this approach does not support nominals and has some performance problems with cardinality restrictions in presence of certain other axioms.

Other systems like OWLIM [13] or OWLJessKB use a standard **rule engine** to reason with OWL. This is fast and easily tunable to different language fragments just by manipulation the rule set. However, this procedure is known to be incomplete and resource consumptive when filled with large amounts of implicit knowledge because of their materialization strategy.

A couple of more or less **hybrid approaches** such as QuOnto [14], Minerva [15], Instance Store [16], or LAS [17] combine an external reasoner (often a tableaux-based system) with a Database system. This enables to process large data volumes due to secondary storage mechanisms. On the other hand, this combination only allows for a very limited language expressivity.

3 Benchmarking Experiences

This section tries to roughly draw a picture of practically tractable OWL repositories with current reasoning systems. This is done by gathering data from different existing as well as own benchmarks. The collected results are reviewed with respect to the system, i. e. the underlying approach, as well as the kind of test ontologies.

A common benchmark for today's DL reasoners is to measure the performance in processing huge ontologies. Ontologies with relatively small and inexpressive TBoxes and ABoxes containing hundreds of thousands or even millions of individuals and relations are predominantly used in benchmarking tests. It is assumed that real world applications will also exhibit the described characteristics.

A selection of such "real world ontologies" (e.g. Gene Ontology[1] or Airport Codes Ontology[2]) which are used for benchmarking can be found in [18].

Nowadays, the Lehigh University Benchmark (LUBM) is the de facto standard when it comes to reasoning with large ontologies [3,19,8,20,21]. But as mentioned before, many reasoners that achieved excellent results when benchmarked with LUBM, failed to reason about other ABox or TBox tests (cf. results for OWLIM and KAON2 from Sections 3.1 and 3.2).

The University Ontology Benchmark (UOBM) [4], extends the LUBM by adding extra TBox axioms making use of all of OWL Lite (UOBM Lite) and OWL DL (UOBM DL). In addition, the ABox is enriched by interrelations between individuals of formerly separated units, which then requires ABox reasoning to answer the given UOBM queries. Not surprisingly, it turned out that incomplete systems now can only answer a fraction even of the OWL Lite queries completely. Only one theoretically sound and complete approach, namely Pellet [8], was able to handle about a tenth of the number of individuals compared to the LUBM. The others failed either due to a timeout or the lack of memory.

These shortcomings motivated us to experiment with a set of tests of a different kind using both existing and newly created benchmarks. In the following, we will present some of these measurements, whereas the aim of these tests was not to simply nominate the fastest or most reliable reasoner. Also, instead of overloading this report with a complete set of all of our measurements we will highlight some results that demonstrate the necessity and requirements for a comprehensive benchmark that goes beyond the LUBM.

In the following, we will present selected results for KAON2 (built 05-12-2005), Pellet 1.3, OWLIM 2.8.3, and RacerPro 1.9.0 which were selected from three out of four different reasoning approaches mentioned in Section 2. FaCT++ was dropped due to a missing query language and all hybrid systems were not appropriate because of their limited language expressivity. We divided the whole process of loading an ontology, any preprocessing as applicable and processing of a query into two separate measurement stages:

Loading and preprocessing the ontology. This stage summarizes the measurements for the process of feeding the ontology into the reasoner and any parsing as required by the interface. Also any preprocessing that is either done automatically or can be started manually is included into this measure. For most of the benchmarks presented in this report this measurement is dominated by the time needed to load the ABox as TBoxes tend to be very small and incomplex.

Query processing. This stage measures the time and resources needed to process a given query and for some systems might also include preprocessing efforts.

Loading of ontologies was repeated three times (discarding them after the first two passes, keeping them after the third). Then the respective queries were repeated

[1] `http://archive.godatabase.org/`

[2] `http://www.daml.ri.cmu.edu/ont/AirportCodes.daml`

ten times each. For both stages time and memory consumption were measured and maximum, minimum, as well as average measurements were recorded. Subsequently the machine was rebooted after each test case. All diagrams in this report show the average of three measurement turns as described above.

The benchmarking tests were conducted on a Windows XP Workstation (3.0 GHz Intel Pentium 3 Processor, 1 GB physical RAM). KAON2, OWLIM, and Pellet were run in the same Java virtual machine (JVM) as the benchmarking application itself. RacerPro was running in a separate process and was connected using JRacer[3]. For all systems the JVM was set to initial and maximum heap size of 700 MB.

3.1 Starting Point: Existing ABox Benchmarks

Figure 1 shows the time needed to load different ontologies from LUBM. While RacerPro shows the worst performance and Pellet not being able to load the largest ontology, KAON2 turned out to be the fastest system directly followed by OWLIM. These two systems show a linear relationship between ontology size and load time.

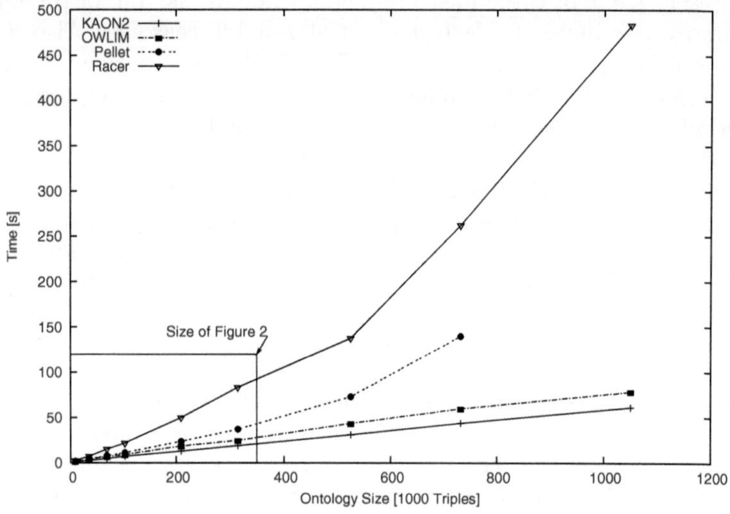

Fig. 1. Comparing LUBM load times for KAON2, Pellet, OWLIM, and RacerPro

We compared these results with the Semintec Benchmark which is based on a benchmark suggested by [20][4]. The Semintec ontology[5] consists of an even simpler TBox that even does not contain existential quantifiers or disjunctions.

[3] `http://www.racer-systems.com/products/download/nativelibraries.phtml`

[4] We only used the second of the two queries suggested in [20] since the concept `Person` (referenced in query one) is not present in the ontology.

[5] The Semintec ontology was originally created by the Semintec project: `http://www.cs.put.poznan.pl/alawrynowicz/semintec.htm`

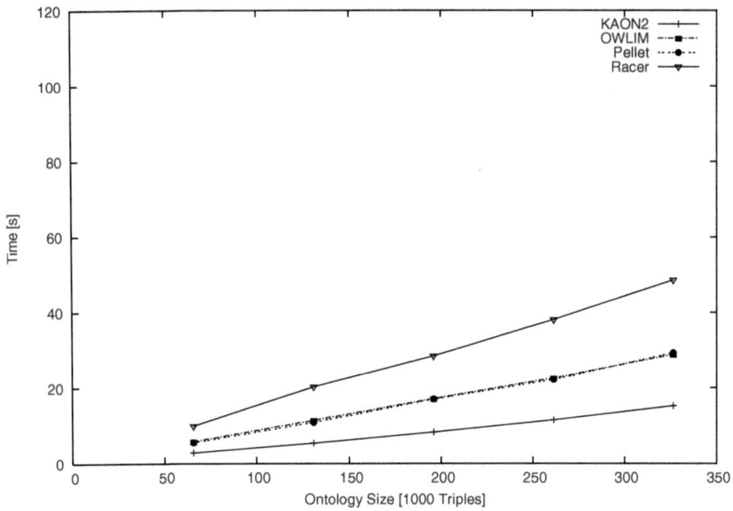

Fig. 2. Semintec Benchmark load times for KAON2, RacerPro, and OWLIM

Again we measured a somewhat linear relationship between the size of the ontology and the loading time for the named systems (cf. Figure 2). But we also noticed that RacerPro is only marginally slower than the other systems, in contrast to the LUBM. It seems that the lower expressivity also reduces the complexity of index creation during preprocessing.

3.2 Implicit Knowledge - A Stumbling Block?

The results from LUBM and the Semintec Benchmark were in general unspectacular, even though the small difference between the benchmarks could not be explained definitely. Thus we designed a Benchmark consisting of a very simple TBox for the next tests.

The TBox of the so called "Exquant Benchmark" consists of a transitive property and a concept defined as existential quantification upon this transitive property (`someValueFrom` restriction). The ABox consists of a chain of individuals related via the transitive property. This individual chain is of different length for every ontology in the benchmark, where 100.000 instances marks the maximum length. The query collects the extension of the restriction. The layout of this benchmark reflects one aspect of the social network ontology (part of our application scenario), which heavily uses transitive properties.

Suddenly, the picture changes. OWLIM, performing very well for LUBM and Semintec, is unable to load an ontology consisting of a chain of 1.000 individuals linked by a transitive property (all tests interrupted after 1 hour). In contrast RacerPro and KAON2 never needed longer than 3.5 seconds. Obviously OWLIM's total forward chaining and materialization approach to compute all implicit knowledge on loading the ontology causes this performance deficit[6]. In the Exquant Benchmark the amount of implicit knowledge grows quadratic with the size of the ontology.

This also influences KAON2. Even though the system performs slightly better than RacerPro on loading the ontology, KAON2 was unable to answer the mentioned query, even for some of the smallest ontologies (a 500 element individual chain) within the given time limit of 10 minutes.

3.3 Influence of Serialization

Our next benchmark (the List Benchmark) consists of head|tail lists modeled in OWL. The biggest ontology contains a list of 100.000 elements. All ontologies in this benchmark are present in two different OWL serializations. One serialization follows a "flat" approach in the sense that all list elements are defined one after the other, referencing their respective tail. In the alternative "deeply nested" serialization, list elements are defined at the place where they are used.

An interesting result, when processing the list benchmark was that RacerPro is sensitive to the OWL serialization of the ontology loaded. We found that RacerPro easily loads the flat serialization of the List Benchmark, while the system fails to load deeply nested serializations with more then 6.400 list elements.

This emphasizes that reasoners should not be reduced to the performance of their core reasoning component when selecting a system for an application. Weaknesses might appear at unexpected places.

3.4 TBox Complexity

We are convinced that an ABox benchmark can not be conducted without scaling the TBox size, too. Inevitably this will also increase TBox reasoning complexity which again might influence ABox reasoning performance. Thus as a first test set we created the Unions Benchmark which checks the influence of TBox complexity on ABox reasoning. The benchmark primarily consists of a set of ontologies with gradually increasing TBox complexity. For every TBox, a set of ontologies with a growing number of ABox individuals is created.

The different TBoxes all consist of a concept hierarchy tree, in which every concept (except for leaf concepts) is defined as a union of further concepts modeled the same way. The TBox size is controlled by the number of concepts per union and the depth of the hierarchy tree. We then scale the size of the ABoxes by instantiating different amounts of individuals per concept. The query collects the extension of the root concept of the concept hierarchy, representing the superset of all ABox individuals.

Once again different reasoning techniques show different performance characteristics in this benchmark. While RacerPro's performance when querying the Unions Benchmark seems to solely depend on the size of the ABox, KAON2 mainly depends on the complexity of the TBox. Figures 3 and 4 depict these findings (pls. observe the direction of the level curves on the base of the graphs).

[6] Reportedly OWLIM v2.8.5 will feature optimized handling of transitive properties.

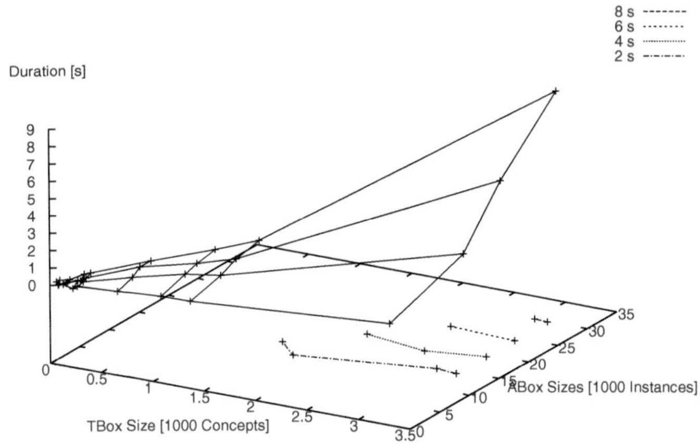

Fig. 3. RacerPro's query times for Unions Benchmark

In an additional test we introduced TBox Axioms irrelevant for the actual reasoning task. We suspected that some reasoners might switch off some optimizations in the presence of TBox axioms of higher complexity. Initial tests suggest that we were too censoriously regarding this assumption as we could not measure any differences.

3.5 Query Repetition and Query Specialization

We introduced the Query Specializing Benchmark to determine whether reasoners do profit from previously calculated results or not. If so, the executing of a specialization of a previous query would perform better than the execution of the specialized query alone.

We defined a set of five queries in selecting publications and their respective authors from the LUBM ontologies. Thereby, we restricted the possible authors from `Person` over `Faculty`, and `Professor` to `FullProfessor`. The last query additionally restricted the possible authors to `FullProfessors` working for a given department. The queries were processed against a "3 universities" ontology from most specific to most general and vice versa.

Unfortunately, we were not able to measure any significant speed up in comparison to the independent execution of the queries. Curious enough we were even unable to measure effects for RacerPro using its "query repository" [22] which is designed to make use of previously calculated answers.

Even if the same query is repeated several times, the query times do not necessarily decrease after the first execution. Considering all measurements we were not able to detect a significant speed up for KAON2 and only minor improvements (under 15%) for Pellet and RacerPro without query repository. OWLIM saved approximately one third of the initial query time, while the biggest speed

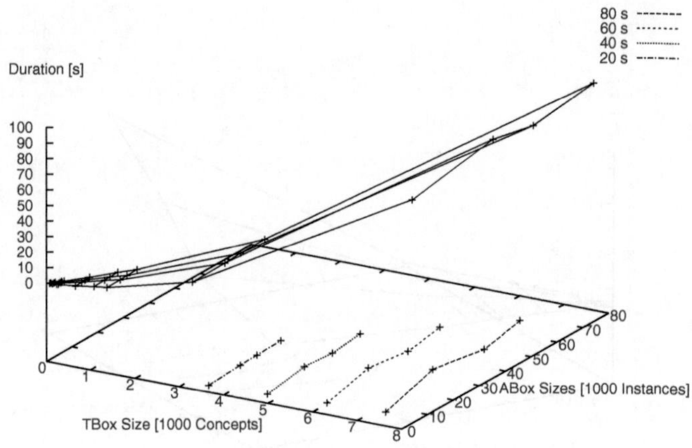

Fig. 4. KAON2's query times for Unions Benchmark

up was measured for RacerPro with activated query repository. Only in this configuration, repeated query executions in average were seven times faster compared to the first execution.

3.6 Dynamic Behavior

Existing performance results of DL reasoners are often limited to the classification of static ontologies. However, in the case of frequent updates (a KB submission, discarding, and re-submission cycle) the communication overhead introduced on loading the ontology can easily dominate the overall performance. In this respect, the delay caused by ontology-based inferencing easily becomes a major obstacle for its use in context-aware applications [23]. One approach to realize high-level situational reasoning for this type of application is to apply dynamic assertional classification of situation descriptions represented as concrete ABox individuals. Each situation individual is assembled of a set of ABox entities representing qualitative context information such as the location (e.g., office), the time (e.g., afternoon) and the persons in proximity (e.g., friends). Finally, the direct subsuming concepts of the situation individual determine the user's abstract situation. The whole process of determining the situation of a user (including the gathering and transformation of the relevant context data) is limited to about 2 seconds per classification. Retraction improves the performance for this type of application drastically, since only a small fraction of the ontology changes between two requests.

The standard DL interface DIG 1.1 [6] does not support the removal of specific axioms, making it necessary to re-submit the complete ontology for each request. As active members of the informal DIG 2.0 working group[7] we therefore propose

[7] http://dig.cs.manchester.ac.uk/

a modular extension to the interface that supports incremental reasoning and retraction [24]. Unfortunately, current reasoners only provide some kind of batch-oriented reasoning procedure. A notable exception is RacerPro, which offers low-level retraction support for most of its statements.

We compared different retraction strategies implemented in Racer. Reloading of ontologies from a local Web server can be accelerated by either loading from a image file (up to 3 times faster) or by cloning an ontology in memory (up to 70 times faster). For small ABoxes, cloning the ontology outperformed even the retraction of single axioms with forget statements (usually 80 times faster). However, it turned out that the fastest strategy was to keep situation individuals up to a certain number (about 20 in our case) within the ABox before cloning a fresh pre-loaded ABox.[8] Due to the lack of incremental classification algorithms, RacerPro still initiates a complete reclassification after each change in the ontology. Initial empirical results from [7], performed with an experimental version of Pellet, indicate that such algorithms for $\mathcal{SHOIN}(\mathcal{D})$ can be quite effective.

Without retraction support, the time needed to compute simple reasoning problems, is easily dominated by the communication overhead caused by the reasoner interface. For example, accessing RacerPro via its native API using TCP is about 1,5 times faster then via HTTP/DIG and even 2 times faster than the access realized with the triple-oriented framework Jena2 [25]. The best performance can be achieved by using the Pellet reasoner running in the same Java virtual machine as the application itself, this way without the need for any external communication.

Another problematic issue we observed was that some reasoners tend to allocate more and more memory over time. This leads to a considerable decrease in performance and makes it necessary to restart the reasoning component after a certain amount of transactions.

3.7 Completeness Versus Performance

Reasoning with OWL Lite as well as OWL DL is known to be of high worst-case complexity. By using the "right" combination of costly expressions, one can intentionally or incidentally create even a very small ontology whose complexity will make practical reasoning impossible. Therefore, in case of taking the whole vision of the Semantic Web literally as the domain for reasoning-aware applications, one obviously has to give up soundness and completeness [26]. However, besides some preliminary empirical evaluation [27], there are currently no attempts to reason with all ontologies found on the Web in parallel. Instead, when assuming the currently more realistic application range in which applications need to reason about information represented via distributed ontologies, soundness and completeness typically do matter. It seems very unlikely that users of large scale ontologies in the context of industrial or scientific research such as SWEET or GO, or defense critical approaches such as the "Profiles in Terror"

[8] Keeping individuals and axioms in the ABox is only possible if they do not influence later classifications.

ontology will accept incomplete reasoning results. Note that in the presence of full negation, as in OWL, one can not really distinguish between completeness and correctness anymore. Because the answers you miss due to incompleteness will be your incorrect answers of the complementary problem.

As a consequence we tested our systems with help of an empirical evaluation using spot tests which are intentionally designed to be hard to solve but small in size. They try to meter the correctness of the reasoning engines with respect to inference problems of selected language features.[9] Surprisingly, only RacerPro and KAON2 were able to solve those tests which lay within the language fragment (above $\mathcal{ALC}$) they claim to support. Others such as Pellet and FaCT++ even failed on some OWL Lite test cases (not to mention OWLIM and related systems). Besides this semantical errors we also found a couple of parsing problems. For instance, all of the systems failed to parse either an empty intersection, union, or enumeration via XML/RDF or DIG 1.1. We also experienced that there is an unpredictable scatter of runtime from case to case even within one system implementation. Actually we discovered random runtime behavior for Pellet for one test case ranging from less than a second up to effectively non-termination. Finally, an expressive all-embracing test case with less than 50 classes and individuals overextended almost all systems.

The discovered failures have been communicated to the system developers and a more detailed description of our test suite can be found at [29].

In addition, we found out that the given answer sets of UOBM are wrong in the DL part of the benchmark suite. Their approach of importing all statements into a RDBMS and manually build SQL queries for answer set computing failed for query 11 with five universities for example. The presumably correct number of answers is 6230 (as opposed to the official 6225) and was computed by Pellet. At least the additional retrieved individuals from Pellet represent correct answers. This can easily be seen by manually collecting the information which makes them a legal candidate. As far as we see the official result set does not take into account that `isHeadOf` is an inverse functional property.

4 Requirements for a Comprehensive ABox Benchmark

In the above sections we demonstrated the impact of some important influencing factors neglected by today's standard ABox benchmarks. This weakness renders the named benchmarks useless when choosing a reasoner for a real-world application. We thus suggest to build future benchmarks along the lines of the following requirements.

R1 Separate measurements should be taken for each processing stage (loading and querying) as described in Section 3.

R2 The benchmark should investigate query performance when processing a set of ontologies with gradually growing ABoxes while size and complexity of

[9] Very similar to the system evaluation of [28] and the system comparisons conducted at various DL workshops.

the TBox remains constant. It must thereby be ensured that the ABox can't be split into disjoint and unconnected parts.

R3 The benchmark should also pinpoint the influence of TBox complexity on ABox reasoning. Thus TBox complexity should gradually be increased. In one setting this increase in complexity should influence the ABox reasoning task while in a separate setting TBox axioms which are unrelated to the actual benchmark should be added. The second setting is to trick the reasoner into switching off optimizations even if this would not have been necessary for the actual reasoning task.

R4 Include benchmarks, that comprise TBoxes modeled in a way such that adding explicit knowledge to the ABox also adds large quantities of implicit knowledge (e.g. transitive properties). This is to reveal the possibly negative influences of materialization approaches or maintenance of index structures.

R5 OWL allows for different serializations of the same ontology. The benchmark should check the influence of different serializations on the process of loading these ontologies. A well implemented reasoner should be agnostic to such differences.

R6 A reasoner with well implemented query caching should answer a repetition, a specialization, or a sub query of a previous query almost instantly. Thus tests should be included which disclose the reasoners capabilities with respect to query caching.

R7 Most reasoners support different interfaces, like a proprietary API and a DIG interface. Since these interfaces might exhibit different performance the benchmark should compare loading and processing of ontologies through the varying interfaces. Clearly results from this benchmark can be disregarded if only very time consuming reasoning tasks are triggered. In such cases the communication overhead is negligible.

R8 Real world applications will be subject of constant change. These changes will appear most frequently in the ABox. Thus additional benchmarks should be available measuring the performance of ABox modifications like addition or retraction of axioms and the time required for subsequent reasoning tasks.

A future comprehensive ABox benchmark should consist of a set of specialized benchmarks tackling a variety of different requirements. Though, we won't suggest to define a procedure to reduce the various results of the different benchmarks to a single metric (as done in [2]). Because, we do not believe that a single score would be of any help when selecting a reasoner for a particular application scenario. For instance, consider the case where two reasoners claim to support the same expressivity but one of them is not sound/incomplete. Then, strictly speaking, they are not comparable at all. Therefore, as a kind of meta requirement, comparisons should carefully interpret all measured results with respect to the different underlying approaches and their theoretical properties.

From a practical point of view, it is advisable to analyze the specific requirement of the planned application and then choose the relevant benchmarks for comparison of potential reasoners. In this respect a set of special purpose benchmarks will be of great help. As a starting point we compiled Table 1, that lists

Table 1. Requirements covered by the benchmarks presented in this paper

Benchmark	Description	Meets Requirements
LUBM	The original Lehigh University benchmark	R2
UOBM	Extended LUBM which introduces an OWL Lite and an OWL DL Version of the benchmark	R2, R3 partially
Semintec	Based on a real-word TBox, modeling the financial domain. ABox size is increased in five steps.	R2
List	Synthetic ontology modeling a head|tail list in OWL. Amount of implicit knowledge rises exponentially with the number of list elements.	R2, R4, R5
Exquant	Another synthetic ontology heavily using transitive property instances.	R2, R4
Unions	Benchmark that increases ABox size as well as TBox complexity	R2, R4, R5, R3 partially
Query Specializing	Based on LUBM. Consists of increasingly specialized queries. Checks for query caching capabilities.	R2, R6

the benchmarks presented in this paper together with the covered requirements (requirements R0 and R1 are not mentioned there as they are independent of the concrete benchmark).

5 Summary

We showed that today's ABox benchmarks fall short on providing comprehensive and meaningful information in order to support users in selecting a reasoner for real world applications. We highlighted and discussed some benchmarking results gained from well known as well as newly created benchmarks. These benchmarks cover traditional aspects like ABox size but also measure influences due to ontology serialization, TBox complexity, query caching, and dynamic ontology changes. The results clearly show that there is still no single benchmark suite which covers all of the issues above and that there is no reasoner able to deal with large and complex ABoxes in a robust manner. As a consequence we suggest a set of general benchmarking requirements which will be helpful when designing future OWL reasoner benchmarking suites.

References

1. Luther, M., Böhm, S., Wagner, M., Koolwaaij, J.: Enhanced Presence Tracking for Mobile Applications. In: Proc. of 4th Int. Semantic Web Conference (ISWC'05), Galway, Ireland. Volume 3729., Galway, Ireland, Springer (2005)
2. Guo, Y., Pan, Z., Heflin, J.: An Evaluation of Knowledge Base Systems for Large OWL Datasets. In: Proc. of the 3rd Int. Semantic Web Conference (ISWC'04), Hiroshima, Japan (2004) 274–288

3. Möller, R., Haarslev, V., Wessel, M.: On the Scalability of Description Logic Instance Retrieval. In: Proc. of the 29th German Conf. on Artificial Intelligence. LNAI, Bremen, Germany, Springer (2006) 171–184
4. Ma, L., Yang, Y., Qiu, Z., Xie, G., Pan, Y., Liu, S.: Towards a Complete OWL Ontology Benchmark. In: Proc. of the 3rd European Semantic Web Conference (ESWC'06). Volume 4011 of LNCS., Budva, Montenegro, Springer (2006) 125–139
5. Brachman, R., Levesque, H.: The Tractability of Subsumption in Frame-based Description Languages. In: Proc. of the 4th Nat. Conference on Artificial Intelligence (AAAI'84). (1984) 34–37
6. Bechhofer, S., Möller, R., Crowther, P.: The DIG Description Logic Interface. In: Proc. of the Int. Workshop on Description Logics (DL'03). Volume 81 of CEUR., Rome, Italy (2003)
7. Halaschek-Wiener, C., Parsia, B., Sirin, E., Kalyanpur, A.: Description Logic Reasoning for Dynamic ABoxes. In: Proc. of the Int. Workshop on Description Logics (DL'05), Edinburgh, Scotland. Volume 147 of CEUR. (2006)
8. Sirin, E., Parsia, B., Grau, B., Kalyanpur, A., Katz, Y.: Pellet: A Practical OWL DL Reasoner. Journal of Web Semantics (2006) To Appear.
9. Haarslev, V., Möller, R.: Racer: A core inference engine for the Semantic Web Ontology Language (OWL). In: Proc. of the 2nd Int. Workshop on Evaluation of Ontology-based Tools. (2003) 27–36
10. Tsarkov, D., Horrocks, I.: FaCT++ Description Logic Reasoner: System Description. In: Proc. of the Int. Joint Conference on Automated Reasoning (IJCAR'06). (2006) To Appear.
11. Fokoue, A., Kershenbaum, A., Ma, L., Schonberg, E., Srinivas, K.: The Summary Abox: Cutting Ontologies Down to Size. In: Proc. of the 5th Int. Semantic Web Conference (ISWC'06), Athens, GA, USA, Springer Verlag (2006) 343–356
12. Motik, B., Studer, R.: KAON2 – A Scalable Reasoning Tool for the Semantic Web. In: Proceedings of the 2nd European Semantic Web Conference (ESWC'05), Heraklion, Greece (2005)
13. Kiryakov, A., Ognyanov, D., Manov, D.: OWLIM — a Pragmatic Semantic Repository for OWL. In: Proc. of the Int. Workshop on Scalable Semantic Web Knowledge Base Systems (SSWS'05), New York City, USA, Springer (2005) 182–192
14. Acciarri, A., Calvanese, D., Giacomo, G.D., Lembo, D., Lenzerini, M., Palmieri, M., Rosati, R.: QuOnto: Querying ontologies. In: Proc. of the 20th Nat. Conf. on Artificial Intelligence (AAAI 2005), Pittsburgh, USA, The MIT Press (2005) 1670–1671
15. Zhou, J., Ma, L., Liu, Q., Zhang, L., Yu, Y., Pan, Y.: Minerva: A Scalable OWL Ontology Storage and Inference System. In: Proc. of 1st Asian Semantic Web Conference (ASWC 2006), Beijing, China, Springer Verlag (2006) 429–443
16. Bechhofer, S., Horrocks, I., Turi, D.: The OWL Instance Store: System Description. In: Proc. of the 20th International Conference on Automated Deduction (CADE 2005), Tallinn, Estonia, Springer Verlag (2005) 177–181
17. Chen, C., Haarslev, V., Wang, J.: LAS: Extending Racer by a Large Abox Store. In: Proc. of the Int. Workshop on Description Logics (DL'05), Edinburgh, Scotland, UK (2005) 200–2007
18. Gardiner, T., Horrocks, I., Tsarkov, D.: Automated Benchmarking of Description Logic Reasoners. In: Proc. of the Int. Workshop on Description Logics (DL'06), Lake District, UK. Volume 189 of CEUR., Lake District, UK (2006) 167–174
19. Haarslev, V., Möller, R., Wessel, M.: Querying the Semantic Web with Racer + nRQL. In: Proc. of the 3rd Int. Workshop on Applications of Description Logics (ADL'04). CEUR, Ulm, Germany (2004)

20. Motik, B., Sattler, U.: A Comparison of Techniques for Querying Large Description Logic ABoxes. In: Proc. of the 13th Int. Conf. on Logic Programming Artificial Intelligence and Reasoning (LPAR'06). Volume 4246 of LNCS., Phnom Penh, Cambodia, Springer (2006)
21. Fokoue, A., Kershenbaum, A., L., M., Schonberg, E., Srinivas, K.: The Summary Abox: Cutting Ontologies Down to Size. Technical Report TR-404, IBM Research – Intelligent Application Analysis, Hawthornem, NY (2006)
22. Wessel, M., Möller, R.: A High Performance Semantic Web Query Answering Engine. In: Proc. of the Int. Workshop on Description Logics (DL'05), Edinburgh, Scotland, UK (2005) 84–95
23. Luther, M., Fukazawa, Y., Souville, B., Fujii, K., Naganuma, T., Wagner, M., Kurakake, S.: Classification-based Situational Reasoning for Task-oriented Mobile Service Recommendation. In: Proc. of the ECAI'06 Workshop on Contexts and Ontologies. (2006)
24. Bechhofer, S., Liebig, T., Luther, M., Noppens, O., Patel-Schneider, P., Suntisrivaraporn, B., Turhan, A., Weithöner, T.: DIG 2.0 – Towards a Flexible Interface for Description Logic Reasoners. In: Proc. of the OWL Experiences and Directions Workshop (OWLED'06) at the ISWC'06. (2006)
25. Carroll, J., Dickinson, I., Dollin, C., Reynolds, D., Seaborne, A., Wilkinson, K.: Jena: Implementing the Semantic Web Recommendations. Technical Report HPL-2003-146, HP Labs (2004)
26. van Harmelen, F.: How the Semantic Web will change KR: challenges and opportunities for a new research agenda. The Knowledge Engineering Review **17** (2002) 93–96
27. Guo, Y., Qasem, A., Heflin, J.: Large Scale Knowledge Base Systems: An Empirical Evaluation Perspective. In: Proc. of the 21st National Conf. on Artificial Intelligence (AAAI 2006), Boston, USA (2006) to appear.
28. Heinsohn, J., Kudenko, D., Nebel, B., Profitlich, H.J.: An Empirical Analysis of Terminological Representation Systems. Technical Report RR-92-16, German Research Center for Artificial Intelligence (DFKI), Kaiserslautern, Germany (1992)
29. Liebig, T.: Reasoning with OWL – System Support and Insights –. Technical Report TR-2006-04, Ulm University, Ulm, Germany (2006)

How to Design Better Ontology Metrics

Denny Vrandečić and York Sure

Institut AIFB, Universität Karlsruhe (TH), Germany
{vrandecic,sure}@aifb.uni-karlsruhe.de

Abstract. You can only control what you can measure. Measuring ontologies is necessary to evaluate ontologies both during engineering and application. Metrics allow the fast and simple assessment of an ontology and also to track their subsequent evolution. In the last few years, a growing number of ontology metrics and measures have been suggested and defined. But many of them suffer from a recurring set of problems, most importantly they do not take the semantics of the ontology language properly into account. The work presented here is a principal approach to facilitate the creation of ontology metrics with the clear goal to go beyond structural metrics to proper semantic-aware ontology metrics. We have developed guidelines and a set of methodological tools based on the notions of "normalization" and "stable metrics" for creating ontology metrics. These guidelines allow the metric author to decide which properties metrics need to fulfil and to appropriately design the desired metric. A discussion of an exemplary metric (taken from literature) illustrates and motivates the issues and suggested solutions.

1 Introduction

Did you ever dare to raise the issue of ontology quality assurance? How did you control the process of improvement? As in many other related fields, you can only control what you can measure [4]. Measuring ontologies is necessary to evaluate ontologies both during engineering and application and is a necessary precondition to perform quality assurance and control the process of improvement. Metrics allow the fast and simple assessment of an ontology and also to track their subsequent evolution. In the last years, many ontology metrics and measures have been suggested and some principal work has been done to study the nature of metrics and measures for ontologies in general. We are extending this work.

There is a recurring set of problems with existing ontology metrics and measures, whereby we focus on the W3C standardized ontology language OWL [12]. We argue that most metrics are based on *structural notions* without taking into account the *semantics* which leads to incomparable measurement results. First, most ontology metrics are defined over the RDF graph that represents an OWL DL ontology and thus are basically graph metrics which take only structural notions into account. Second, only a very small number of metrics is taking the semantics of OWL DL into account (subsumption etc.). Third, almost no metric is taking the open world assumption into account. We believe that foundational work addressing these issues will substantially facilitate the definition of proper ontology metrics in the future.

In this paper we will study these issues, describe how they can be avoided, and under what circumstances they have to be avoided, and under which they are acceptable,

E. Franconi, M. Kifer, and W. May (Eds.): ESWC 2007, LNCS 4519, pp. 311–325, 2007.

will outline the foundations for a novel set of metrics and measures, and discuss the advantages and problems of the given solutions. Our approach is based on two notions, first "normalization" of an ontology, and second "stable metric".

Normalization consists of the five steps (i) name anonymous classes, (ii) name anonymous individuals, (iii) materialize the subsumption hierarchy and unify names, (iv) propagate instances to deepest possible class or property within the hierarchy, and (v) normalize property instances. We argue that such a normalization is useful as a kind of pre-processing in order to apply known structural metrics *in a semantics-aware way*. For instance, a known structural metric is the depth of of the class-hierarchy. However, the current measures of ontology depth depend on a number of structural parameters such as whether subsumption reasoning has been performed and whether the results have been materialized before measurement. Performing the normalization steps before measuring ensures that the value for the maximum depth of an ontology is comparable to the maximum depth of another ontology.

Stable metrics are metrics that take the open world assumption properly into account, that means that they are stable with regards to possible additions of further axioms to the ontology. Stable metrics allow us to make statements about the behaviour of an ontology in the context of a dynamic and changing world wide web, where ontologies may frequently be merged together in order to answer questions over integrated knowledge. We give an exemplary extension of the depth metric towards a stable metric in order to demonstrate how a classic metric can be turned into a stable one.

In this paper we assume the term to include both axioms and facts (as well as annotations and ontology properties, although those are not taken into regard for normalization), i.e. the TBox and the ABox. Here, ontology does not mean only the axioms (as it is assumed in many other works), but also a knowledge base, and any of them could be empty. Thus we follow the definition of ontology in the OWL standard [12].

The paper is structured as follows. In Section 2 we will examine existing metrics and measures, and thus survey related work. Section 3 contrasts the underlying notions of semantic metrics with structural metrics, and discusses which scenario will require what kind of metric. Section 4 introduces the notion of "normalization" of an ontology which forms the heart of our approach. In Section 5 we illustrate the practical application of normalization on examples. Section 6 addresses the issue of stable metrics with regards to the open world assumption. We conclude in Section 7, where we also discuss future work.

2 Current Metrics and Measures

In this paper we will concentrate on some foundational aspects that form the base for automatically acquirable measures. Therefore we will not define a number of metrics and measures, but rather take a step back and discuss conditions that measures have to adhere to in order to be regarded as semantically aware ontology metrics. This also helps to understand clearly what it means for a metric to remain on a a structural level.

Thus the scope of this work compares best to other metric frameworks, like the QOOD (quality oriented ontology description) framework [7] and the O^2 and *oQual* models [6]. The authors created semiotic models for ontology evaluation and validation, and thus describe how measures should be built in order to actually assess quality. They also describe the relation between the ontology description, the ontology graph, and the conceptualization that is expressed within the graph, and they define measures for the structural, functional, and usability dimension. In [5] they introduce further measures that can be applied within that framework. We will take one of the measures introduced in [5] as an example in Section 5, and show some shortcomings of the actual descriptions of such a measure (not of the framework as a whole!). We think that the work described here fits well into the QOOD framework by making the assumptions underlying such measures explicit.

A framework for metrics in the wider area of ontology engineering is provided by OntoMetric [11]. The authors name and sort a long list of metrics into several different areas, like tools, languages, methodologies, costs, and content. They define the relations between the different metrics, their attributes, and the quality attributes they capture. Within the OntoMetric framework, the work presented in this paper is based solely in the area of content metrics. It extends the discussions around content metrics, and elaborates properties of such metrics in more detail. Whereas OntoMetric regards all kind of metrics, we gear the results described here towards automatically measurable metrics.

OntoQA is a tool that implements a number of metrics [14], and thus it allows for the automatic measurement of ontologies. They define metrics like richness, population, or cohesion. Whereas all these metrics are interesting, they fail to define if they are structurally or semantically defined – which is a common lapse. Most of the metrics in OntoQA actually can be applied both before and after normalization (as described in the following section). We suppose that comparing these two measures will yield further interesting results.

Often metrics are defined purely structural. An example is given by [1], where the authors describe metrics for ranking ontologies, like the class match or the density measure. Interestingly even the so called semantic similarity measure is not a semantic measure in the sense described here, since they apply all these measures on the graph that describes the ontology, not on the ontological model.

OntoClean [9], currently the most well-known ontology evaluation approach, is a philosophically inspired approach for the evaluation of formal properties of a taxonomy. Some tools offer support for the manual tagging with OntoClean properties (OntoEdit [13] and WebODE [2]), a recent work deals with the automation of OntoClean [15]. From a practical perspective OntoClean provides means to derive measurable mismatches of a taxonomy with respect to an ideal structure which takes into account the semantics of the "is-a" relationship. Such mismatches have a structural nature, e.g. one is able to derive that a certain concept should not be the subconcept of another concept. OntoClean provides an explanation of why mismatches occur which subsequently might help to improve the taxonomical structure. For many people the philosophical notions of OntoClean are subject of long discussions, however, strictly speaking, this is not part of the evaluation but of the ontology engineering because deciding the proper nature of a class forces the ontology to commit itself to a more specified meaning, which in turn allows for a more object evaluation technique.

Measures applied to ontologies from the Semantic Web are usually still in a very simple state [17] (unsurprising due to the overall bad quality of ontologies in the wild, and the high costs on resources for providing reasoning on a big number of ontologies). We think that the most prevalent hurdle towards applying more semantic measures on ontologies on the web is an actual lack of some foundational work towards defining such measures, and a subsequent lack of implementation. The work presented here is a step towards such an implementation, that will allow to measure the web in several new dimensions.

3 Ontological Metrics

As shown in the previous section, current metrics often measure structural properties of the ontology. In the case of OWL DL, this often means that they measure the structure of the RDF graph that describes the ontology with well-known graph measures. Another approach is to measure the explicitly stated facts and axioms. Within these paper, we regard both approaches as structural. Structural metrics are often useful, and this paper does not suggest to replace them. It rather offers a way to extend the possibilities available to the ontology engineer with truly ontological metrics.

We define ontological, or semantic, metrics to be those who do not measure the structure of the ontology, but rather the models that are described by that structure. In a naïve way, we could state that we base our metrics not on the explicit statements, but on every statement that is entailed by the ontology.

But measuring the entailments is much harder than measuring the structure, and we definitively need a reasoner to do that. We also need to make a difference between a statement X that is entailed by an ontology O to be true ($O \models X$), a statement that is not entailed by an ontology ($O \not\models X$), and a statement that is entailed not to be true ($O \models \neg X$). To properly regard this difference leads us to so called stable metrics that can deal with the open world assumption of OWL DL. We will return to them in Section 6.

Note that measuring the entailments is more an intuitive description of how to describe ontological metrics than the actual approach. In many cases – for example for a measure that simply counts the number of statements in an ontology – measuring all entailed statements instead of measuring all explicit statements often leads to an infinite number of statements. Just to give one example, the ontology $\exists R.\top \sqsubseteq C$ also entails the statements $\exists R.\exists R.\top \sqsubseteq C$, $\exists R.\exists R.\exists R.\top \sqsubseteq C$, and so on, an endless chain of existentials. But only terminating measures are of practical interest, and thus we need approaches that allow us to capture ontological metrics in a terminating way.

In order to gain the advantage of the simple and cheap measurement of structural features, we can transform the structure of the ontology. These transformation need to preserve the semantics of the ontology, that is, they need to describe the same models. But they also need to make certain semantic features of the ontology explicit in their structure – thus we can take structural measures of the transformed ontology and interpret them as ontological measures of the original ontology. We call this kind of transformations normalization. The following section describes five steps of normalization.

With these tools we will be enabled to define ontological metrics in a simpler and less error prone way than in current practice. We will show this on an exemplary metric in Section 5.

4 Normalization of an Ontology

This section describes several steps of normalization. Their goal is to explicate some features of the semantics of an ontology within its structure, so that the structural metrics actually capture the semantics they are supposed to capture.

The following normalization steps are defined here:

1. name all relevant classes, so no anonymous complex class descriptions are left
2. name anonymous individuals
3. materialize the subsumption hierarchy and normalize names
4. instantiate the deepest possible class or property
5. normalize property instances

Notice that if we speak of names, we mean, in the context of OWL DL, the URI of the class, property, or individual, not the human readable label.

In the **first normalization** our aim is to get rid of anonymous complex class descriptions. After the first normalization, the TBox will contain two kind of axioms: class definitions of the form $A \equiv C$, where A is a class name and C a class description (or class name), and subsumption axioms of the form $A \sqsubseteq B$, where both A and B are class names. The ABox will consist of property instantiations of the form $R(i, j)$, and of facts of the form $A(i)$, with A being a class name.

The first normalization can be done as follows:

1. in all axioms of the form $C \sqsubseteq D$ where C (or D) is a complex class description, add a new axiom $A \equiv C$ ($B \equiv D$) with A (B) being a new class name. Replace the axiom $C \sqsubseteq D$ with $A \sqsubseteq D$ ($C \sqsubseteq B$, or even $A \sqsubseteq B$)
2. in all axioms of the form $C \equiv D$ where both C and D are complex class descriptions, replace that axiom with the two axioms $A \equiv C$ and $A \equiv D$, with A being a new class name
3. in all axioms of the form $C \equiv A$ where C is a complex class descriptions and A an atomic class name, replace that axiom with $A \equiv C$
4. in all axioms of the form $C(i)$ where C is a complex class description, replace that axiom with the axioms $A(i)$ and $A \equiv C$ with A being a new class name

None of these structural changes change the possible models, that means, that they are semantically equivalent. They do introduce new class names to the ontology, which may not be desirable in all cases (for example for presentation purposes, for counting the classes, and so on).

Note that it is possible to introduce named classes that are unsatisfiable. This does not mean that the ontology becomes unsatisfiable, but solely these newly introduced classes. Instead of introducing new names for unsatisfiable classes though, we could simply use the name $\perp$.

The **second normalization** gets rid of anonymous individuals. This means that every blank node that is of the (asserted or inferred) type individual needs to be replaced with an URI reference. Especially in FOAF [3] files this occurs regularly since, for some time, it was regarded as good practice *not* to define URIs for persons. Integration of data was not done via the URI, but with inverse functional properties. This practice is problematic, since the semantics of blank nodes in RDF are rather often not fully understood, and should thus be avoided. The second normalization as defined here captures the semantics most users wanted to express anyway.

It is possible that these newly introduced individual names give a further name to already existing (or other newly introduced) individuals. But since OWL DL does not adhere to a unique name assumptions, this is no problem. Furthermore, the next step of normalization will take care to resolve such synonyms.

The **third normalization** will materialize the subsumption hierarchy and normalize the names. The first step requires a reasoner.

1. for all pairs of simple class names (A, B) in the ontology, add the axiom $A \sqsubseteq B$ if the ontology entails that axiom (that is, materialize all subsumptions between simple named classes).
2. detect all cycles in the subsumption structure. For each set of classes $A_1 \ldots A_n$ that participate in a cycle, remove all subsumption axioms from the ontology where both classes are members of this set. In subsumption axioms where only one class is a member of this set, replace the class with B in the axioms. Add the axioms $B \equiv A_1 \ldots B \equiv A_n$ to the ontology. B is a new class name for each cycle. If B is unsatisfiable, take $\bot$ instead of B. If B is equal to $\top$, take $\top$.
3. regarding solely the subontology H_3 that consists of all subsumption axioms of an ontology O, remove all redundant ones (that is, remove all subsumption axioms that are redundant due to the transitivity of the subsumption relation alone).

The subsumption structure now forms a directed acyclic graph that represents the complete subsumption hierarchy of the original ontology. We define a set of normal classes of an ontology as follows: every class that participates in an subsumption axiom after the third normalization of an ontology is a normal class of that ontology.

Since we got rid of facts with complex class descriptions in the first normalization, we do not need a reasoner in order to take care of fact normalization. We still have to replace every class name that is not normal with its normal equivalent within the facts.

Note that instead of creating a new class name for each detected cycle, often it will make more sense to choose a name from the set of classes involved in that cycle, based on some criteria (like the class name belonging to a certain namespace, the popularity of the class name on the web, etc.). For many ontology metrics, this does not make any difference, so we disregard it for now, but we expect the normalizations to have beneficial effects in other scenarios as well, in which case some steps of the normalization need to be revisited in more detail. We will further discuss this in Section 7.

Since in OWL DL it is not possible to make complex property descriptions besides inverse properties, property subsumption, and transitivity, (extensions towards enabling more complex property descriptions are suggested in the OWL 1.1 proposal [8]) no heavy reasoning is involved for property normalization in most cases. In case a property has more than one name, we choose one (or introduce a new name and state the equality). All normal property names have to be stated explicitly to be equivalent to all other property names they are equal to (that is, we materialize the equality relations between the normal property names and the non-normal ones). All occurrences of non-normal property names (besides within the axiom stating equality with the normal property name, and besides within annotation property instances) are replaced with the normal property name.

The same holds true for individuals. In case an individual has more than one name, we decide on or introduce a normal one and state explicitly equality to the normal name, and then replace all occurrences of the non-normal individual names with the normal one (besides within the axiom stating equality with the normal individual name, and besides within annotation property instances).

We disregard annotation property instances since they may be used to state annotations about the URI, and not about the actual concept, property, or individual. There could be annotations that describe when a certain URI was introduced, who created it, its deprecation state, or that point to a discussion related to the introduction of the URI. Some annotations on the other hand may be useful for the normal name as well – especially labels, or sometimes comments. Since annotations do not have impact on the DL semantics of the ontology anyway, they may be dropped for the purpose of measuring semantic metrics. Nevertheless, if the normalization is done for some other purpose, and it is planned to further use the normalized version of the ontology in some scenario, than the possible replacement of names within annotation property instances depends both on the scenario and the instantiated annotation property (for example, it may be useful to normalize the label when the ontology will be displayed on the user interface, but it may be bad to normalize versioning information that is captured within annotations).

The **fourth normalization** aims towards moving the instantiations to the deepest possible level, as this conveys the most information explicitly (and deriving instantiations of higher levels is very cheap because of the asserted explicitness of the hierarchy due to third normalization). This does not mean that every instance will belong to only one class, multiple instantiations will still be necessary in general.

Here is a possible (though not efficient) algorithm to perform the fourth normalization of an ontology O.

1. for each normal class C and each normal individual i in O, add $C(i)$ to O if it is entailed by the ontology.
2. for each normal object property instance $R(i, j)$ and each object property S so that $S \sqsubseteq R$ is an explicit axiom in O, add $S(i, j)$ if it is entailed by the ontology. Check this also for the property instances added this way (this step will terminate since the subsumption hierarchy is finite).
3. for each normal data property instance $T(i, d)$ and each data property U, proceed as in the previous step.
4. create a subontology H_4 out of O including only the facts (that is, the ABox), and the explicitly stated subsumption hierarchy of the classes and properties (after third normalization)
5. remove all facts from O that are redundant in H_4.

We do not want to remove all redundant facts from the ontology at this step, since there may be some facts that are redundant due to an interplay of different other axioms in the TBox. For example, in the following ontology:

$Person(Adam).$
$likes(Adam, Eve).$
$Person \sqsubseteq \exists likes.\top$

the first statement is actually redundant, but would not be removed by the above algorithm (the third statement states that the domain of *likes* is *Person*). This is because we only remove axioms that are redundant within the subontology H_4, and the axiom stating the domain of *Person* would not be part of it. This is due to the fact that after first normalization, the ontology would look like this:

$Person(Adam).$
$likes(Adam, Eve).$
$Person \sqsubseteq A$
$A \equiv \exists likes.\top$

So H_4 would not include the last axiom, and thus the first axiom would not be redundant within H_4.

The **fifth normalization** finally normalizes the properties: we materialize property instances of symmetric and inverse properties, and we clean the transitivity relationship. This can be done similar to the creation of the subsumption hierarchy in the third normalization: after materializing all property instances, we remove all that are redundant in the subontology H_5, which contains only the property instances of all transitive properties, and the axioms stating the transitivity of these properties.

It is important to mention that normalization does not lead to a canonic normalized version. This means that there may be many different ontologies that result from the normalization of an ontology. Often normalizations do not result in canonical, unique results (think about conjunctive normal forms). The normalization as described here can be extended in order to result in canonic normalized forms, but the benefit of such an extension is not clear. Considering that common serializations, like the RDF/XML serialization of OWL ontologies [12], lack a canonic translation anyway, and thus ontologies cannot be compared on a character by character base, for example as some version control systems like CVS or SVN would require.

Also, normalization is not an applicable solution for every metric. For example, if we want to know the number of atomic classes in an ontology, first normalizing it and then calculating the number actually will return the wrong result in the general case. The goal of normalization is to actually provide the metric designer some tools in order to simplify the description of his metric. In the following section we describe an example of how to apply the normalization for the description of a metric.

5 Examples of Normalization

The metric we will regard in this example is the *depth of the ontology*. What we want to measure is intuitively described as the length of the subsumption hierarchy, or else the number of levels the class hierarchy has. In [5], this is the measure (M3), called *Maximal depth*, and the definition is given as follows:

$$m = N_{j \in P}$$

$$\forall i \exists j (N_{j \in P} \geq N_{i \in P})$$

where $N_{j \in P}$ is the set of all nodes in the path j from the set of all paths through the digraph g that represents the ontology, that is, the definition is the length of the longest succession of explicitly stated subsumption relations.

Let us regard the following ontology:

$C \equiv \geq 1R.\top$
$D \equiv \geq 2R.\top$
$E \equiv \geq 3R.\top$

By the definition of (M3), the depth of the ontology is 1 (since there are no explicitly stated subsumption axioms, every path has one node). But after normalization the ontology gets transformed to this:

$C \equiv \geq 1R.\top$
$D \equiv \geq 2R.\top$
$E \equiv \geq 3R.\top$
$D \sqsubseteq C$
$E \sqsubseteq D$

Now the very same metric, applied to the normalized ontology, actually captures the intuition of the depth of the ontology and returns 3.

As discussed earlier, this example also shows us that some metrics will not work with normalization. In [5], metric (M30) is the *axiom/class ratio*. On the original ontology it is 1, but raises to 5/3 in the normalized version. In case the original ontology is being distributed and shared, (M30) – if stated as metadata of the ontology, for example in some kind of ontology repository [10] – should be 1, and not calculated on the normalized version.

Let us regard another example. In the following ontology

$D \sqsubseteq C$
$E \sqsubseteq D$
$D \sqsubseteq E$
$F \sqsubseteq E$

(M3) will be ∞ due to the subsumption cycle between D and E. The cycle can be resolved by rewriting the axioms in the following way:

$D \sqsubseteq C$
$D \equiv E$
$F \sqsubseteq E$

But due to the definition, (M3) would yield 2 here – there are two explicit subsumption paths, C, D and E, F, both having two nodes, and thus the longest path is 2.

The structural measure again does not bring the expected result. After normalization, though, the ontology will look like this:

$A \sqsubseteq C$
$A \equiv D$
$A \equiv E$
$F \sqsubseteq A$

We have introduced a new class name A that replaces the members of the cycle, D, E. Now the depth of the ontology is 3, as we would have expected from the start, since the cycle is treated appropriately.

Existing structural metrics, as discussed in Section 2, often fail to capture what they are meant for. Normalization is a tool that is easy to apply and that can easily repair a number of such metrics. Even seemingly simple metrics, as demonstrated here with the ontology depth, are defined in a way that makes too many assumption with regards to the structure of the measured ontologies.

As we can see in this section, simple structural measures on the ontology do yield values, and often these values may be highly interesting. If we know that (M3) resolves to ∞, then this tells us that we have a cycle in the subsumption hierarchy. Also a high number of classes and complex axioms, but a low (M3) may indicate an expensive to reason about ontology, since the major part of the taxonomy seems to be implicitly stated (but such claims need to be evaluated appropriately). But both results do not capture what the measure was meant to express, that is, the depth of the class hierarchy.

But this leads us to the possibility of creating measures by combining structural metrics on the original ontology and on its normalized version, for example to calculate ratios like $M_3(O)/M_3(N(O))$ (with $M_3(O)$ returning measure (M3) as described above, and $N(O)$ being a function that returns the normalized version of the ontology O). This could describe the *explicitness of the subsumption hierarchy*. Further work needs to investigate and evaluate such measures, and to assess their usefulness for evaluating ontologies.

6 Stability of Metrics

Often metrics intend to capture features of the ontology that are independent of the actual representation of the ontology. But as we have seen, structural transformations of the ontology description often lead to differences in the metrics even though the semantics remained untouched. Normalization offers a way to overcome these problems in many cases.

One aspect of metrics, that are not touched upon by normalization, is the issue of how stable the metrics are with regards to the open world assumption of OWL DL ontologies. In order to illustrate this issue let's take a look at a simple example. Imagine an ontology with the following three facts:

$author(paper, York).$
$author(paper, Denny).$
$author(paper, Zdenko).$

Now let us ask the simple question: how many authors does the *paper* have? It seems that the answer should be 3. But now, if you knew that *Zdenko* is just another name for

Denny, and thus state *Zdenko* $\approx$ *Denny*, then you suddenly would change your answer to 2, or even, becoming more careful, giving an answer like "I am not sure, it is either 1 or 2". So finally we can state that *York* $\not\approx$ *Denny* and thus arrive at the answer that the paper indeed has 2 authors (and even that is possibly wrong if we consider that we could add statements any time in an open world that add further authors to the paper – all we know *as of now* is that the paper has *at least* two authors).

When creating a metric, we have to ask ourselves the following, similar question: how does the metric behave when additions to the ontology happen? Since ontologies are meant to be smushed and integrated constantly and dynamically, can we predict how certain properties of the ontology will behave, that is, if $M(O_1)$ and $M(O_2)$ for a metric M and two ontologies O_1 and O_2 are known, what can we state about $M(O_1 \cup O_2)$? Or even, can we give a function f_M so that $f_M(M(O_1), M(O_2)) = M(O_1 \cup O_2)$ without having to calculate $M(O_1 \cup O_2)$ (which may be much more expensive)?

In the previous section we have discussed the simple example of ontology depth. Let us return to this example again. We define the function $M_3(O)$ that returns the measure (M3) as described in [5], and already described above. If we have an ontology O_1:

$D \sqsubseteq C$
$E \sqsubseteq D$

And a second ontology O_2:

$C \sqsubseteq D$
$E \sqsubseteq D$

In this case, $M_3(O_1) = 3, M_3(O_2) = 2$. We would expect $M_3(O_1 \cup O_2)$ to be 3, since M(3) is defined as the maximal depth, but since the union of both ontologies actually creates a cycle in the subsumption hierarchy, (M3) is ∞ – or, after normalization, just 2, and thus even smaller than the maximal depth before the union.

We can avoid such behaviour of the metrics by carefully taking the open world assumption into account when defining the metric. But this leads us to three possibilities for defining metrics,

1. to base the value on the ontology as it is,
2. to measure an upper bound, or
3. to measure a lower bound.

We need a more complicated example to fully demonstrate these metrics:

$C \equiv D \sqcup E$
$D \sqcap E \sqsubseteq \bot$
$F \sqsubseteq E$
$G \equiv \neg C$
$H \sqsubseteq C$
$F(i).$
$D(j).$
$G(k).$

This ontology says that D and E form a complete partition of C (the first two axioms), that E has the subclass F, that there are elements that are not in C, and it states the existence of three individuals, i, j and k, and the classes they belong to.

The normalized version of this ontology looks like this (shortened slightly for readability):

$C \equiv D \sqcup E$
$\bot \equiv D \sqcap E$
$D \sqsubseteq C$
$E \sqsubseteq C$
$F \sqsubseteq E$
$G \equiv \neg C$
$H \sqsubseteq C$
$F(i).$
$D(j).$
$G(k).$

(M3) of this ontology is 3 (C, E, F). But besides the actual depth, we can also calculate the minimal depth of this ontology, that is, no matter what axioms are added, what is the smallest number of levels the ontology will have (under the condition that the ontology remains satisfiable)?

In the given example, if we add the axiom $F \equiv E$, (M3) will decrease to 2. But on the other hand, no matter what axiom we further add, there is no way to let C collapse with D and E, therefore C is a proper superset of both (that is, it contains more individuals than D or E alone). And because C cannot become $\top$ (due to k being outside of C), the minimum depth of the ontology is 2.

The maximum depth of an ontology is usually ∞ (since we can always add axioms about an arbitrarily long class hierarchy). Only in the case of an ontology with a closed domain, that is, if we have an axiom like $\top \equiv \{a, b, c\}$, then the maximum depth is set (to $|\top| - 1$, since there may be a class C with one element, a class D with two elements that subsumes C, and then $\top$ with three elements, but since $\top$ is not counted, the longest path would be (C, D), and every further class in this path would become equivalent to an already existing class or be empty). But we expect such axioms to usually appear only in theoretical musings and hardly be of any practical relevance.

Therefore we need to define a maximum depth in a slightly different way in order to be of practical value. In the following, we will discuss two possible definitions.

Instead of allowing for arbitrary axioms that may be added, we only allow to add axioms of the form $A \sqsubseteq B$ with A and B being normal class names of the normalized ontology. In the above example, we may add the axiom $H \sqsubseteq F$ to the ontology in order to increase (M3) from 3 to 4. No longer subsumption path is possible, since all the other named classes would become unsatisfiable when added to an existing path. So this metric will provide with a maximum depth of the ontology, assuming no new class names are added.

Another possibility to constrain the axioms to be added, is to allow only for axioms that do not relate to the existing ontology, that is, the intersection of the signatures of the two ontologies is empty. The signature of an ontology is the set of all names used in the ontology (besides the names from the OWL, RDF, RDFS, and XSD namespaces). In this case, (M3) of the merged ontology is the maximal (M3) of the single ontologies, since no interaction between the axioms happen that may increase or reduce (M3). We can thus define $f_{M_3}(M_3(O_1), M_3(O_2)) = \max(M_3(O_1), M_3(O_2))$, which is much cheaper to calculate than $M_3(O_1 \cup O_2)$.

Stable metrics are metrics that take the open world assumption into account. Stable metrics will help us to evaluate ontologies for the wide wild web. Since we expect ontologies to be merged on the web dynamically, stable metrics allow us to state conditions that the ontology will fulfil in any situation. The depth of an ontology may be a too simple example to demonstrate the advantages of stable metrics, but imagine a dynamic, ontology-based graphical user interface. Having certain guarantees with regards to the future development of the properties of the ontology may help the designer of the user interface tremendously, even if it is such a seemingly trivial statement like "the depth of the ontology is never less than 3".

There is no simple recipe to follow in order to turn a metric into a stable metric, but the question outlined at the beginning of this section, and then discussed throughout the rest – how does the ontology behave when axioms are added? – can be used as a guideline in achieving a stable metric.

We expect that the ready availability of metrics that take the open world assumption into account will lead to more robust ontologies. Since ontology engineers will have these numbers available at engineering and maintenance time, they will learn easier how to achieve their actual goals. For example, ontology engineers that want to create a class hierarchy that will not collapse to less levels can always check if the minimum depth as described above corresponds to the asserted depth. Tools could guide the ontology engineer towards achieving such goals. Ontology engineers get more aware of such problems, and at the same time get tools to measure, and thus potentially control them.

7 Conclusion and Future Work

We have discussed the properties of ontology metrics. Sometimes simple structural metrics are sufficient for the task at hand, and many structural metrics exist today. Our goal in this paper was to raise the awareness for the difference between structural and ontological metrics, and to provide principle means for the simple definition of metrics that take the semantics of the ontology appropriately into account.

Ontology normalization was introduced as a preprocessing step in order to align structural measures with intended semantic measures. Further properties, like the stability of a metric towards ontology extension and merges, and the non-dichotomous nature of ontologies were discussed, and an approach towards encapsulating these problems was suggested by introducing stable metrics.

In addition to offering the theoretical tool of normalization, we are planning to implement it as an extension to the OWL tools[1]. This will allow to access these metrics both from the command line as well as from a Java API. Besides making normalization available to tools and metric suites, this will also allows us to evaluate if there are further benefits to normalized ontologies. We assume such benefits with regards to query answering performance, usability, and ontology maintenance. Some properties of normalization suggest advantages in these and other areas, but we expect some parts of

[1] `http://owltools.ontoware.org/`

the normalization process to be adapted based on differing requirements by these other use cases.

Based on the foundational work provided in this paper, we plan to adapt, extend, and implement several metrics already known in literature. We hope that a thorough evaluation of these metrics will allow to correlate quality attributes to these metrics, and thus to finally lead to viable sets of metrics and measures for the whole ontology life cycle. We don't think that there is one single such set, but the ideas presented here make several design decision when creating metrics more explicit and point to common problems and pitfalls when creating metrics and measures in this field. This will help in deciding which metrics to choose for a given scenario.

We expect that future work will continue on this basis in order to create a bigger tool set for everybody dealing with ontologies to allow them to evaluate ontologies during every step of the ontology life cycle. This will lead to an overall higher quality of ontologies, and thus to a stronger foundation on which the Semantic Web is being built.

Acknowledgments. Research reported in this paper has been partially financed by the EU in the IST project Knowledge Web (FP6-507482, `http://knowledgeweb.semanticweb.org`). We want to thank our colleagues for the fruitful and interesting discussions about the ideas presented in this paper, especially Aldo Gangemi, Marta Sabou, Markus Krötzsch, Jos Lehmann, and Malvina Nissim.

References

1. H. Alani and C. Brewster. Metrics for ranking ontologies. In Vrandečić et al. [16].
2. J. C. Arpírez, O. Corcho, M. Fernández-López, and A. Gómez-Pérez. WebODE: a scalable workbench for ontological engineering. In *Proceedings of the First International Conference on Knowledge Capture (K-CAP) Oct. 21-23, 2001, Victoria, B.C., Canada*, 2001.
3. D. Brickley and L. Miller. The friend of a friend (FOAF) vocabulary specification, July 2005. Namespace Document 27 July 2005 ('Pages about Things' Edition).
4. T. DeMarco. *Controlling Software Projects: Management, Measurement & Estimation*. Yourdon Press, New York, 1982.
5. A. Gangemi, C. Catenacci, M. Ciaramita, and J. Lehmann. Ontology evaluation and validation: an integrated formal model for the quality diagnostic task. Technical report, Laboratory of Applied Ontologies – CNR, Rome, Italy, 2005. available at http://www.loa-cnr.it/Publications.html.
6. A. Gangemi, C. Catenaccia, M. Ciaramita, and J. Lehmann. Modelling ontology evaluation and validation. In Y. Sure and J. Domingue, editors, *Proceedings of the 3rd European Semantic Web Conference (ESWC2006)*, number 4011 in LNCS, Budva, Montenegro, June 2006. Springer-Verlag.
7. A. Gangemi, C. Catenaccia, M. Ciaramita, and J. Lehmann. Qood grid: A metaontology-based framework for ontology evaluation and selection. In Vrandečić et al. [16].
8. B. C. Grau(ed.). OWL 1.1 web ontology language, November 2006. Available at http://owl1_1.cs.manchester.ac.uk/.
9. N. Guarino and C. Welty. Evaluating ontological decisions with OntoClean. *Communications of the ACM*, 45(2):61–65, February 2002.

10. J. Hartmann, Y. Sure, P. Haase, R. Palma, and M. del Carmen Suarez-Figueroa. OMV – ontology metadata vocabulary. In C. Welty and A. Gangemi, editors, *ISWC 2005 - In Ontology Patterns for the Semantic Web*, Galway, Ireland, November 2005.
11. A. Lozano-Tello and A. Gómez-Pérez. OntoMetric: A method to choose the appropriate ontology. *Journal of Database Management, Special Issue on Ontological analysis, Evaluation, and Engineering of Business Systems Analysis Methods*, 15(2), April-June 2004.
12. M. K. Smith, C. Welty, and D. McGuinness. OWL Web Ontology Language Guide, 2004. W3C Recommendation 10 February 2004.
13. Y. Sure, J. Angele, and S. Staab. OntoEdit: Multifaceted inferencing for ontology engineering. *Journal on Data Semantics*, 1(1):128–152, NOV 2003. LNCS 2800.
14. S. Tartir, I. B. Arpinar, M. Moore, A. P. Sheth, and B. Aleman-Meza. OntoQA: Metric-based ontology quality analysis. In *Proceedings of IEEE Workshop on Knowledge Acquisition from Distributed, Autonomous, Semantically Heterogeneous Data and Knowledge Sources*, 2005.
15. J. Völker, D. Vrandečić, and Y. Sure. Automatic evaluation of ontologies (AEON). In Y. Gil, E. Motta, V. R. Benjamins, and M. A. Musen, editors, *Proceedings of the 4th International Semantic Web Conference (ISWC2005)*, volume 3729 of *LNCS*, pages 716–731. Springer Verlag Berlin-Heidelberg, NOV 2005.
16. D. Vrandečić, M. del Carmen Surez-Figueroa, A. Gangemi, and Y. Sure, editors. *Proceedings of the 4th International Workshop on Evaluation of Ontologies for the Web (EON2006) at the 15th International World Wide Web Conference (WWW 2006)*, Edinburgh, Scotland, May 2006.
17. T. D. Wang. Gauging ontologies and schemas by numbers. In Vrandečić et al. [16].

Measuring Inconsistencies in Ontologies

Xi Deng, Volker Haarslev, and Nematollaah Shiri

Department of Computer Science and Software Engineering
Concordia University, Montreal, Canada
{xi_deng,haarslev,shiri}@encs.concordia.ca

Abstract. In this paper, we propose a novel approach to measure inconsistencies in ontologies based on Shapley values, which are originally proposed for game theory. This measure can be used to identify which axioms in an input ontology or which parts of these axioms need to be removed or modified in order to make the input consistent. We also propose optimization techniques to improve the efficiency of computing Shapley values. The proposed approach is independent of a particular ontology language or a particular reasoning system used. Application of this approach can improve the quality of ontology diagnosis and repair in general.

1 Introduction

Ontologies play an important role in the Semantic Web as they provide a common shared model to represent a domain and to reason about the objects in the domain. As the size of the ontologies grows and applications developed become more complex, inconsistency becomes inevitable in the design and development of ontologies. According to the classical *ex contradictione quodlibet* (ECQ) principle, anything follows from an inconsistent ontology is useless. In order to help users to resolve the inconsistencies in ontologies, several approaches to identify and explain the cause of these inconsistencies have been proposed [1,2,3]. An assumption often made in these approaches is that all inconsistencies are equally "bad". However, as shown in real world applications, it is possible for an ontology to contain two or more sources of inconsistencies and they may have different impact towards the inconsistencies. They may not necessarily contain the same contradiction and the same information, and may have overlapping content. The following are just two possible scenarios.

- Non-overlapping: there are more than one set of axioms that are needed to produce an inconsistency in an ontology and they are independent of one another.

  Suppose K' and K'' are two inconsistent subsets of ontology $\mathcal{O}$. When we say $\mathcal{O}$ has non-overlapping sources of inconsistencies, it means that $K' \cap K'' = \emptyset$. Note that inconsistency might occur at different levels: in the level of a single axiom, and in the level of sets of axioms. For example, consider $K' = \{C \sqcup \neg C \sqsubseteq C \sqcap \neg C\}$ and $K'' = \{\top \sqsubseteq \exists R.B, \top \sqsubseteq \forall R.\neg B\}$.[1] The

[1] Please refer to Section 2 for the definition of the notations.

E. Franconi, M. Kifer, and W. May (Eds.): ESWC 2007, LNCS 4519, pp. 326–340, 2007.

axiom in K' is inherently inconsistent, while both of the two axioms in K'' are responsible to produce a contradiction.

- Overlapping: there are more than one set of axioms that are needed to produce an inconsistency in an ontology and they are interweaved with one another.

  $\mathcal{O}$ having overlapping sources of inconsistencies means that $K' \cap K'' \neq \emptyset$. When two sets of inconsistent axioms are overlapping, it indicates that certain axioms contribute more to the inconsistencies and these axioms are possibly more problematic than others. It is most likely the case that removing one of these axioms from $\mathcal{O}$ will result in resolving other inconsistencies as well.

In this paper we propose a quantitative measure of inconsistencies in ontologies which gives users guidelines on priorities of the axioms to be removed and their consequences. Our approach borrows some ideas from [4], which presents an approach of using the Shapley value to obtain an inconsistency measure for propositional logic. It can be applied to clauses instead of axioms. Since clauses are more fine-grained than axioms, it allows us to take a deeper look inside the axioms and find out which proportion of the axiom contributes to the inconsistency. We also discuss the relationship between the inconsistency measure and the minimal inconsistent subsets of ontologies. The computational complexity of calculating the Shapley value is at least Exp-time, which shows that it does not scale well in general [5]. Therefore we propose to optimize the calculation based on the structural relevance of the axioms and properties of the defined inconsistency measure.

The main contribution of this paper is twofold: we combine previously known game theory strategies into ontology reasoning and present a measure to systematically evaluate the inconsistencies in ontologies. To the best of our knowledge, this is the first work in Description Logics towards providing a quantitative measure of inconsistencies. This approach is independent of a particular species of ontology languages or a particular reasoning system used. Moreover, we illustrate how the application of this method can improve the quality of ontology diagnosis and repair.

The remainder of the paper is organized as follows: Section 2 presents a brief introduction of Description Logics, the underlying logics of the Web Ontology Language OWL. Section 3 introduces the basic concepts used in our approach. Section 4 describes the algorithms. The paper closes with a summary and also a discussion of future work.

2 Preliminaries

In this section, we first review some basic concepts and terms related to the ontology languages in the Semantic Web, and their relationships to Description Logics (DLs). We also define the notion of inconsistency in an ontology.

2.1 Ontologies in the Semantic Web

The Web Ontology Language (OWL) has been recommended as the standard web ontology language by the World Wide Web Consortium (W3C) [6]. It is a machine-readable language for sharing and reasoning information on the Internet. OWL is an extension of the Resource Description Framework (RDF) and a revision of the DAML+OIL Web Ontology Language.

OWL represents the domain of interest by defining hierarchies of classes and properties. An OWL ontology consists of axioms and facts. Axioms define intensional knowledge by building relationships between classes and properties. Facts describe the extensional knowledge about individuals. OWL currently has three flavors: OWL Lite, OWL DL, and OWL Full. OWL Full contains OWL DL, which in turn contains OWL Lite. OWL DL and OWL Lite correspond semantically with certain Description Logic languages. Reasoning tasks are undecidable in OWL Full and currently there is no reasoner that supports reasoning of every feature of OWL Full.

2.2 Description Logics

We shall not give a detailed introduction of Description Logics here, but refer the interested reader to [7]. Description logics are a family of concept-based knowledge representation formalisms. It represents the knowledge of a domain by first defining the relevant concepts of the domain. These concepts are used to specify properties of the objects in the domain. Typically a DL language has two parts: *terminology* (TBox) and *assertion* (ABox). The TBox includes intensional knowledge in the form of axioms whereas the ABox contains the extensional knowledge that is specific to elements in the domain, called individuals. The TBox together with the ABox is called a *knowledge base* in DL. In the TBox, basic descriptions are *atomic concepts*, designated by unary predicates, and *atomic roles*, designated by binary predicates to express relationships between individuals. Concept descriptions can be built on atomic concepts by iteratively applying constructors such as intersection, union, negation, value restriction and existential quantification. Axioms express how concepts and roles are related to each other. Generally, it is a statement of the form $C \sqsubseteq D$, read as "concept C is subsumed by concept D", or $C \equiv D$, indicating that $C \sqsubseteq D$ and $D \sqsubseteq C$, where C and D are concept descriptions. An ABox is a set of assertions that of the form $C(a)$ and $R(a, b)$, where R is a role, and a, b are individuals.

An interpretation $\mathcal{I}$ defines the formal semantics of concepts, roles, and individuals. It consists of a non-empty set $\Delta^{\mathcal{I}}$, called the domain. The interpretation function $\mathcal{I}$ maps every atomic concept A to a subset $A^{\mathcal{I}}$ of $\Delta^{\mathcal{I}}$, and maps every atomic role R to a binary relation $R^{\mathcal{I}} \subseteq \Delta^{\mathcal{I}} \times \Delta^{\mathcal{I}}$. In addition, $\mathcal{I}$ maps each individual name a to an element $a^{\mathcal{I}} \in \Delta^{\mathcal{I}}$. An interpretation $\mathcal{I}$ satisfies $C \sqsubseteq D$ if $C^{\mathcal{I}} \subseteq D^{\mathcal{I}}$. It satisfies $C \equiv D$ if $C^{\mathcal{I}} = D^{\mathcal{I}}$. It satisfies $C(a)$ if $a^{\mathcal{I}} \in C^{\mathcal{I}}$ and it satisfies $R(a, b)$ if $(a^{\mathcal{I}}, b^{\mathcal{I}}) \in R^{\mathcal{I}}$.

The basic inference services in TBoxes include satisfiability, subsumption, equivalence, and disjointness. A concept C in a TBox $\mathcal{T}$ is said to be *satisfiable* w.r.t $\mathcal{T}$ if there exists a model of $\mathcal{T}$ (that is an interpretation $\mathcal{I}$ that satisfies the axioms of $\mathcal{T}$), such that $C^{\mathcal{I}}$ is nonempty. The other three inference services can be reduced to (un)satisfiability. Another important reasoning service in TBoxes is to check whether a TBox $\mathcal{T}$ is *consistent*, i.e., whether there exists a model for $\mathcal{T}$. The basic reasoning tasks in ABoxes include instance checking, realization, and retrieval. The instance check verifies if a given individual is an instance of a specified concept. The realization finds the most specific concept that an individual is an instance of. The retrieval finds the individuals in the knowledge base that are instances of a given concept. An ABox $\mathcal{A}$ is *consistent* w.r.t a TBox $\mathcal{T}$, if there is an interpretation that is a model of both $\mathcal{A}$ and $\mathcal{T}$ (that is an interpretation $\mathcal{I}$ that both satisfies the axioms of $\mathcal{T}$ and the assertions of $\mathcal{A}$). Similar to the inference services in TBoxes, the other three inference services in ABoxes can also be reduced to the consistency problem of ABoxes. In this paper, we use the term "consistency" to refer to consistency problems in both TBoxes and ABoxes. We will also discuss the measure regarding unsatisfiable concepts.

Roughly speaking, a concept in DL is referred to as a class in OWL. A role in DL is a property in OWL. The terms axioms and individuals have the same meaning in DL and OWL. OWL DL is based in part on the DL $\mathcal{SHOIN}(\mathcal{D})$, which includes special constructors such as oneOf, transitive properties, inverse properties and datatype properties, and its subset OWL Lite which is based on the less expressive DL $\mathcal{SHIF}(\mathcal{D})$, is $\mathcal{SHOIN}(\mathcal{D})$ without the oneOf constructor and with the number restriction constructors limited to 0 and 1. Due to the close connection between OWL and DLs, in this paper, we will make no distinction between ontologies and knowledge bases in DL, and the examples are given mainly in DL syntax. The DL languages that we work on are those for which consistency checking is decidable.

3 Inconsistency Measures

In this section, we study methods of measuring inconsistencies in DL knowledge bases. The idea is to first define an inconsistency value, and then take it as the characteristic function to compute the Shapley value.

3.1 Motivating Example

The following example is adapted from [3], which we modified to be inconsistent. In the example, A, B, C, D, $A1$ to $A6$ denote concepts, and a and b are individuals.

1. $A1 \sqsubseteq A2 \sqcap \neg A \sqcap A3$	**2.** $A2 \sqsubseteq A \sqcap A4$
3. $A3 \sqsubseteq A5 \sqcap A4$	**4.** $A4 \sqsubseteq C \sqcap \forall S.B$
5. $A5 \sqsubseteq \exists S.\neg B$	**6.** $D \sqcup \neg D \sqsubseteq D \sqcap \neg D$
7. $A1(a)$	**8.** $A3(b)$
9. $A6 \equiv D$	

A complete DL reasoner, such as FaCT++ [8], RACER [9] or Pellet [10], reports this knowledge base to be inconsistent. However, they can not provide crucial information, e.g., that there are four inconsistent subsets (3, 4, 5 and 8; 1, 2, and 7; 1, 3, 4, 5 and 7; 6) in this knowledge base, and that one axiom (axiom 6) is inherently inconsistent. We will use this as our running example throughout the paper to show how the hidden information can be unraveled using our method.

3.2 Definitions

In this section we review some definitions of the Shapley value in game theory [4], which we tailor for use in DL in this work.

Definition 1. *Given a set of axioms and assertions in a knowledge base K, a characteristic function $v : 2^K \rightarrow \mathbb{R}$ assigns a value to each coalition K', where $K' \subseteq K$.*

An example of the characteristic function is the drastic inconsistency value, which assigns 1 to a set of axioms if it is inconsistent, and 0 to the set if it is consistent.

Definition 2. *For a set of axioms and assertions $K' \subseteq K$, the drastic inconsistency value of K' is defined as:*

$$I_d(K') = \begin{cases} 0 & \textit{if } K' \textit{ is consistent or } K' \textit{ is empty} \\ 1 & \textit{otherwise} \end{cases} \tag{1}$$

Example 1. Some drastic inconsistency values of the running example are as follows, where we only show some of those with the inconsistency value 1, as well as some consistent ones (with this value being 0).[2]

$I_d(\{1\}) = 0$ $\quad I_d(\{2\}) = 0$ $\quad I_d(\{3\}) = 0$
$I_d(\{1, 2\}) = 0$ $\quad I_d(\{3, 4, 5\}) = 0$ $\quad I_d(\{1, 2, 6\}) = 1$
$I_d(\{3, 4, 5, 8\}) = 1$ $\quad I_d(\{3, 4, 5, 6\}) = 1$
$I_d(\{1, 3, 4, 5, 7\}) = 1$

As shown above, the coalition {1, 2, 6} has the inconsistency value 1. Axiom 6 is often of a great value for a coalition it joins. For example, it can bring 1 to {2, 6} for making the coalition {1, 2, 6}. And it can also bring 1 to the coalition {3, 4, 5, 6}.

Similarly, we can define another characteristic function which assigns 0 to a set of axioms if a concept A is satisfiable and 1 otherwise.

Definition 3. *The concept-related inconsistency value of a set of axioms K' w.r.t. a concept A (A occurs in K) is defined as:*

$$I_A(K) = \begin{cases} 0 & \textit{if } A \textit{ is satisfiable w.r.t. } K' \\ 1 & \textit{otherwise} \end{cases} \tag{2}$$

[2] For the sake of simplicity, we refer to the axioms and assertions by their numbers.

Example 2. Some of the concept-related inconsistency values of the working example are:

$$I_{A1}(\{1, 2\}) = 1 \qquad I_{A1}(\{1, 3, 4, 5\}) = 1$$
$$I_{A3}(\{3, 4, 5\}) = 1 \qquad I_{A3}(\{3, 4, 9\}) = 0$$

An inconsistent measure is to quantify the contribution of each axiom or assertion in the knowledge base to the overall inconsistencies. The higher the measure is, the more weight an axiom carries in contributing to the inconsistencies. Shapley [11] proposed such a measure, known as Shapley values, in the context of game theory in 1953, which describes a fair allocation of gains obtained by the cooperation among several agents.

The Shapley value is defined for a game that has n agents. In the game, the agents can form coalitions, which is a subset of the n agents. Each coalition has a gain when all its members work together as a team. A question which may arise here is "which agent contributes the most to different coalitions?" A solution to this problem can help determine which agent values more to the game than the others. The Shapley value is proposed to tackle this problem. The basic idea is as follows. Suppose the agents join a coalition according to a certain order, and the payoff of an agent in this coalition is its marginal contribution to the gain of the coalition. The Shapley value takes all the possible orders of the coalition formation into account and averages the agent's marginal contribution over them.

As the inconsistency checking can be deemed as a game, each axiom (or assertion) in the knowledge base can be deemed as an agent. Analogously, the contribution of each axiom (or assertion) to the inconsistencies can be measured using the Shapley value.

Let K be a knowledge base, σ_K be the set of all permutations on K, and $n = |K|$ be the cardinality of K. Given an order $\sigma \in \sigma_K$, we use p_σ^α to denote the set of all the axioms and assertions in σ that appear before an axiom (or an assertion) α.

Definition 4. *The Shapley value for an axiom (or an assertion) α in an knowledge base K is defined as:*

$$S_\alpha(K) = \frac{1}{n!} \sum_{\sigma \in \sigma_K} v(p_\sigma^\alpha \cup \{\alpha\}) - v(p_\sigma^\alpha)$$

The Shapley value can be directly computed from the possible coalitions without considering the permutations, with the following expression:

$$S_\alpha(K) = \sum_{C \subseteq K} \frac{(c-1)!(n-c)!}{n!} (v(C) - v(C \backslash \{\alpha\}))$$

where C is any coalition of the axioms and assertions K, $n = |K|$, and $c = |C|$.

3.3 Inconsistency Measure Based on the Shapley Value

We can take the drastic inconsistency measure defined in Definition 2 (the same computation can be applied to the concept-related measure defined in Definition 3) as the characteristic function, and then use the Shapley value to compute to what extent an axiom or an assertion is concerned with the inconsistency.

For example, suppose K is a knowledge base and α is an axiom (or an assertion) in K, then the Shapley value of α based on the drastic inconsistency value I_d is defined as:

$$S_\alpha(K) = \sum_{C \subseteq K} \frac{(c-1)!(n-c)!}{n!}(I_d(C) - I_d(C \backslash \{\alpha\})) \tag{3}$$

where $n = |K|$ and $c = |C|$.

The Shapley value of a knowledge base K is a vector with each element denoting the Shapley value of each axiom (or assertion) in K.

Example 3. The Shapley value of the knowledge base $K = \{C \sqcup \neg C \sqsubseteq C \sqcap \neg C, \top \sqsubseteq \exists R.B, \top \sqsubseteq \forall R.\neg B, A \sqsubseteq D\}$ is $(\frac{4}{6}, \frac{1}{6}, \frac{1}{6}, 0)$.

This shows that $\{C \sqcup \neg C \sqsubseteq C \sqcap \neg C\}$ is the most problematic.

Example 4. The Shapley value of the working example is $(\frac{268}{7!}, \frac{250}{7!}, \frac{120}{7!}, \frac{120}{7!}, \frac{120}{7!}, \frac{3774}{7!}, \frac{268}{7!}, \frac{120}{7!}, 0)$.

This shows that axiom 6 is the one that causes the most problems. 3, 4, 5 and 8 are equally responsible for the inconsistencies, so are 1 and 7. Axiom 9 has a value of 0, which means it does not contribute to the inconsistency. Axiom 2 has a higher value than 3, 4 or 5, which shows that the inconsistency is more equally distributed among 3, 4 and 5.

3.4 Properties of the Inconsistency Measures

We make a few observations and remarks regarding the inconsistency measures.

Definition 5. *The characteristic function v is increasing if $X \subseteq Y$, $v(X) \leq v(Y)$.*

An increasing function indicates that adding more agents to the coalition will never decrease the value. In the worst case, they contribute nothing to the coalition. Due to the monotonic nature of DL reasoning, we can prove the following.

Proposition 1. *The drastic (concept-related) inconsistency value (see Definition 2 and 3) is increasing.*

A set of axioms and assertions is called *convergent* if its inconsistency value is convergent, i.e., it is the same as the inconsistency values of its super sets, and adding any other axioms or assertions does not change its inconsistency value.

Definition 6. *The convergent subset K' of a knowledge base K is defined as a set of axioms and assertions that satisfies the following two properties:*

1. $I_d(K') = 1$ *(or* $I_A(K') = 1$*), and*
2. $I_d(K'') = 0$ *(or* $I_A(K'') = 0$*), for all* $K'' \subset K'$.

Intuitively, K' is a convergent point, in which the inconsistency value flips from 0 to 1. There is a direct relation between the convergent subsets and minimally inconsistent subsets of a knowledge base, defined as follows.

Definition 7 (MIS). *We say that $\mathcal{T}'$ is the minimally inconsistent subset (MIS) of a knowledge base $\mathcal{T}$ if the following two conditions hold:*

1. $\mathcal{T}'$ *is inconsistent, and*
2. $\mathcal{T}''$ *is consistent, for every* $\mathcal{T}''$ *such that* $\mathcal{T}'' \subset \mathcal{T}'$.

Proposition 2. *K' defined in Definition 6 is a minimally inconsistent subset.*

Another inconsistency measure of a coalition K' that can be defined is based on the number of minimally inconsistent subsets that would be removed if we remove K' from the knowledge base. In other words, this measures the impact of K' on the knowledge base, formalized as follows.

Definition 8. *The impact inconsistency measure of a subset K' in a knowledge base K can be defined as follows:*

$$I_i(K') = |MIS(K)| - |MIS(K - K')|$$

where $|MIS(K')|$ denotes the number of minimally inconsistent subsets of K'.

Example 5. There are four convergent subsets, i.e., four MISs in the working example.

$MIS_1 = \{1, 2, 7\}$, $MIS_2 = \{3, 4, 5, 8\}$, $MIS_3 = \{1, 3, 4, 5, 7\}$, $MIS_4 = \{6\}$
$I_i(\{1\}) = I_i(\{7\}) = I_i(\{3\}) = I_i(\{4\}) = I_i(\{5\}) = 2$
$I_i(\{2\}) = I_i(\{6\}) = I_i(\{8\}) = 1$
$I_i(\{9\}) = 0$

Obviously, the removal of 1, 7, 3, 4, or 5 will remove the most number of inconsistencies. The removal of axiom 9 will not affect the inconsistencies at all.

3.5 Apply the Inconsistency Measures to Clauses

The inconsistency measures discussed in this paper so far are applied to axioms in ontologies. It excludes the possibility of a more fine-grained inspection of the content of the axioms. In particular, if the inconsistency is in the level of a single axiom, then only two values can be obtained: consistent or inconsistent. In our previous work [1,12], we have developed a resolution based technique to explain inconsistency in ontologies. We found that clauses work on a more fine-grained level than DL axioms, so an approach based on clauses can identify specific parts of axioms that are responsible for an inconsistency. Consequently, the inconsistency measures can also be applied to clauses and this allows us to look inside the axiom and identify which proportion of the axiom is contributing to the inconsistency.

4 Computational Complexity Concerns

The most important source of computational complexity in calculating the Shapley value is the difficulty to obtain the inconsistency value of an axiom. It is directly dependent on the complexity of consistency checking in DL reasonings. One of the possible optimizations is to reduce the number of consistency checks. Besides, the complexity is also related to the computation process itself. The computation of the Shapley value considers all the subsets of the axioms/assertions in the knowledge base, and hence it results in Exp-time. However, we do not really need to compute the inconsistency values of all the subsets. In the following sections we will discuss such optimizations.

4.1 Partition Based on Structural Relevance

In DLs, axioms[3] can be related to each other through structure relevance. For example, the axiom $A \sqsubseteq B$ is structurally related to $\neg B \sqsubseteq \top$ but not to $\neg C \sqsubseteq D$. It is clear that adding a structurally unrelated axiom to a coalition will not change the relative inconsistency value. Structural relevance is an equivalence relation, hence it can be exploited to induce a partitioning of the axioms, which as shown below, can be used as an optimization to speed up the computation of the Shapley inconsistency value.

Definition 9. *We say an axiom is directly structurally related to another axiom if the intersection of their signature (the set of all (negated) concept names and role names occurring in the axiom) is not empty. The structural relevance is the transitive closure of direct structural relevance.*

Example 6. In the motivating example, $A1 \sqsubseteq A2 \sqcap \neg A \sqcap A3$ is directly structurally related to $A2 \sqsubseteq A \sqcap A4$. It is structurally related to $A4 \sqsubseteq C \sqcap \forall S.B$ (because $A3 \sqsubseteq A5 \sqcap A4$), but it is not related to $D \sqcup \neg D \sqsubseteq D \sqcap \neg D$.

Example 7. There are two partitions in the motivating example: $\{A1 \sqsubseteq A2 \sqcap \neg A \sqcap A3, A2 \sqsubseteq A \sqcap A4, A3 \sqsubseteq A5 \sqcap A4, A4 \sqsubseteq C \sqcap \forall S.B, A5 \sqsubseteq \exists S.\neg B, A1(a), A3(b)\}$, $\{D \sqcup \neg D \sqsubseteq D \sqcap \neg D, A6 \equiv D\}$.

The following result suggests that partitioning according to structural relevance can be used to reduce the computational complexity of the Shapley value in our context.

[3] For the sake of simplicity, we only refer to the axioms, assertions can be considered in the same way.

Lemma 1. *If $K = \sum_{i=1}^{T} K_i$ is a partitioning of a knowledge base, and all K_i have the same inconsistency value function I, then for any axiom (or assertion) $\alpha \in K_i$,*

$$S_\alpha(K_i) = \sum_{C \subseteq K_i} \frac{(c-1)!(n-c)!}{n!}(I(C) - I(C \backslash \{\alpha\}))$$

where $n = |K_i|$ and $c = |C|$.

After the partitioning, the Shapley value of an axiom (or an assertion) can be computed in $O(\sum_{i=1}^{T} 2^{|K_i|})$.

The partitioning based on structural relevance preserves the total ordering on the Shapley values of the axioms (and assertions) inside the same partition. In other words, if an axiom has a higher Shapley value than another axiom in the partition, then it will also have a higher Shapley value in the knowledge base.

Theorem 1. *If $K = \sum_{i=1}^{T} K_i$ is a partitioning of a knowledge base, and all K_i have the same inconsistency value function I, then for any $\alpha, \beta \in K_i$, if $S_\alpha(K_i) > S_\beta(K_i)$, then $S_\alpha(K) > S_\beta(K)$.*

Proof. For each partition $K_i \subseteq K$ and $\alpha, \beta \in K_i$, if $S_\alpha(K_i) > S_\beta(K_i)$, then $\sum_{C \subseteq K_i} \frac{(c-1)!(n-c)!}{n!}(I(C) - I(C \backslash \{\alpha\})) > \sum_{C \subseteq K_i} \frac{(c-1)!(n-c)!}{n!}(I(C) - I(C \backslash \{\beta\}))$. Let us prove $S_\alpha(K) > S_\beta(K)$ by cases: for any $C_i \subseteq K$, if $C_i \subseteq C$, as previously indicated, $\sum_{C_i \subseteq K} \frac{(c-1)!(n-c)!}{n!}(I(C_i) - I(C_i \backslash \{\alpha\})) > \sum_{C_i \subseteq K} \frac{(c-1)!(n-c)!}{n!}(I(C_i) - I(C_i \backslash \{\beta\}))$. Otherwise, if $I(C_i - C) = 1$, then $I(C_i) - I(C_i \backslash \{\alpha\}) = I(C_i) - I(C_i \backslash \{\beta\})$, if $I(C_i - C) = 0$, then $I(C_i) - I(C_i \backslash \{\alpha\}) > I(C_i) - I(C_i \backslash \{\beta\})$. So $\sum_{C_i \subseteq K} \frac{(c-1)!(n-c)!}{n!}(I(C_i) - I(C_i \backslash \{\alpha\})) \geq \sum_{C_i \subseteq K} \frac{(c-1)!(n-c)!}{n!}(I(C_i) - I(C_i \backslash \{\beta\}))$. So $\sum_{C \subseteq K} \frac{(c-1)!(n-c)!}{n!}(I(C) - I(C \backslash \{\alpha\})) > \sum_{C \subseteq K} \frac{(c-1)!(n-c)!}{n!}(I(C) - I(C \backslash \{\beta\}))$, considering all the possible cases. Hence $S_\alpha(K) > S_\beta(K)$.

4.2 Optimization Based on Properties of the Inconsistency Measure

Partitioning the knowledge base according to structural relevance can work very well when the sizes $|K_i|$ are small, especially when the $|K_i|$ are bounded by a constant. This, however, is largely dependent on the particular knowledge base considered. In what follows, we propose an algorithm to calculate the Shapley value based on the properties of the inconsistency measure, especially the convergent property. This method aims to reduce the number of consistency checks. The basic idea is quite simple: according to Definition 6 in Section 3.4, a convergent subset of a knowledge base is the maximal subset whose inconsistency value has to be calculated. Any superset of a convergent knowledge base can have its inconsistency value derived to be 1 due to the monotonicity of DLs. Hence once a subset is convergent, there is no necessity to compute its supersets. The detailed algorithm is shown in Figure 1.

Example 8. To see how this algorithm works, we can apply the optimization to one of the partitions in the motivating example: $\{A1 \sqsubseteq A2 \sqcap \neg A \sqcap A3, A2 \sqsubseteq$

Input: an axiom (or assertion) α in K
Output: the Shapley value of α

for all the subsets $K' \subseteq K$, sorted by the cardinality of K'
    while $I(K' \cup \alpha) = 0$
        move to the next unvisited K'
    for all supersets K'' of K'
        $I(K'') = 1$
        tag K'' as visited
    move to the next unvisited K'
Compute the Shapley value of α

Fig. 1. Computing Shapley values

$A \sqcap A4, A3 \sqsubseteq A5 \sqcap A4, A4 \sqsubseteq C \sqcap \forall S.B, A5 \sqsubseteq \exists S.\neg B, A1(a), A3(b)\}$. The algorithm will first compute the inconsistency value of $A1(a)$, and then the coalition of $A1(a)$ and any one of the other axioms, and then the coalition of $A1(a)$ and any two of the other axioms, since the coalition of $A1 \sqsubseteq A2 \sqcap \neg A \sqcap A3, A2 \sqsubseteq A \sqcap A4$ and $A1(a)$ has an inconsistency value of 1, we will skip computing the inconsistency value of its supersets. It is the same case with the coalition of $A1 \sqsubseteq A2 \sqcap \neg A \sqcap A3, A3 \sqsubseteq A5 \sqcap A4, A4 \sqsubseteq C \sqcap \forall S.B, A5 \sqsubseteq \exists S.\neg B$ and $A1(a)$.

We can either compute the convergent sub-terminologies on the fly during the computation of the Shapley value, or use algorithms for MIS in [13] to compute them in advance.

5 Experimental Results

To evaluate whether our proposed method is useful for ontologies, we have implemented the algorithm with the suggested optimizations in the previous section, and have performed some preliminary experiments. It was run on two ontologies: the University ontology with 32 concepts and 24 axioms, the Koala ontology with 14 concepts and 19 axioms.[4] Tests were performed on a Windows XP system with a 3.0GHz Intel Pentium 4 processor, and 2GB memory. The implementation was developed in Java (JDK 1.5.0). The original University and Koala ontologies are consistent and they have 15 and 3 unsatisfiable concepts respectively. They are provided as sample ontologies to test the repair service of the OWL ontology editor SWOOP 2.3 beta 3, which uses Pellet as the default DL reasoner. In our experiments, we asserted fresh individuals for these unsatisfiable concepts in order to make the ontologies inconsistent. After the modification, repair plans can no longer be generated for these inconsistent ontologies in SWOOP.

There were four sources of inconsistencies in the University ontology. Calculating the measure took about 35 seconds. In the Koala ontology, there were

[4] The University and Koala ontologies were obtained from the website of SWOOP: http://www.mindswap.org/2005/debugging/ontologies/.

three sources of inconsistencies, and it took 12 seconds to calculate the inconsistency measure. If optimization techniques discussed in Section 4 are adopted, then the running time are 1.6 seconds and 467 ms, respectively.

Let's look at part of the Koala ontology as follows:

$$
\begin{array}{l}
1.Koala \sqsubseteq Marsupials \\
2.Quokka \sqsubseteq Marsupials \\
3.Person \sqsubseteq \neg Marsupials \\
4.Koala \sqsubseteq \exists isHardworking\{false\} \\
5.KoalaWithPhD \sqsubseteq \exists hasDegree\{PhD\} \sqcap Koala \\
6.Quokka \sqsubseteq \exists isHardworking\{true\} \\
7.\exists isHardworking.\top \sqsubseteq Person \\
8.\exists hasDegree.\top \sqsubseteq Person \\
9.Koala(Suzy) \\
10.KoalaWithPhD(Lizzy) \\
11.Quokka(Pan)
\end{array}
$$

There are three sources of inconsistencies in this ontology. A koala named Suzy is forced be to a member of the disjoint classes marsupial and person. She is a marsupial because koala is a subclass of marsupial and she is a person because person is the domain of isHardWorking and every koala must have at least one isHardWorking property. A koala with a PhD named Lizzy is also forced to a member of the disjoint classes marsupial and person. She is a person because person is the domain of hasDegree and every koala with a PhD must have at least one hasDegree property. Similarly, a Quokka named Pan is also causing inconsistency problems.

After the computation, axiom 3 gets the highest Shapley value, followed by axiom 7 and 1. In addition, axiom 3 has an impact inconsistency measure of 3. For the naive user of these ontologies, this information can be very helpful. Removing Axiom 3 will render the ontology consistent, and therefore enables most of the reasoning services that users have expected from their ontology development tools.

For the purpose of this paper, the implementation uses Racer as a black box to check the satisfiability. This black box approach comes with the advantage of independence of any particular reasoner or DL language. However, even with the optimization techniques, the worst case computational complexity is still exponential time. If the convergent ontologies are computed in a preprocessing step using algorithms presented in [14], the computational time can be further reduced.

6 Related Work

Measures of inconsistency haven been studied in [15,16,4]. These proposals for measuring inconsistency can be classified in two approaches. The first involves counting the minimal number of formulae needed to produce the inconsistency. The more formulae needed to produce the inconsistency, the less inconsistent the set [16]. The second approach involves inspecting the proportion of the language that is affected by the inconsistency. [4] is the closest to our work. Our work is inspired by their proposal to use the Shapley value to obtain an inconsistency measure for propositional logic. However, our work differs from theirs in two aspects: our approach works for a much more expressive logic and we also consider the optimizations of this method from a practical point of view.

In the DL community, several approaches have been proposed to deal with diagnosis of a knowledge base. [17] provides a general theory for diagnosis of Description Logics knowledge bases, based on Reiter's diagnosis theory from first principles. [13] uses a modified tableau algorithm for $\mathcal{ALC}$ to first find a minimally inconsistent subset of a knowledge base, and then use Reiter's hitting set algorithm to find the maximally consistent subsets. They also propose to choose and eliminate axioms that most frequently participate in the underlying logical contradictions. [14] modifies the tableau rules and extends the DL language to $\mathcal{SHOIN}$. These approaches focus on the incoherence problem, i.e., if there exists an unsatisfiable concept in the ontology and they cannot render a repair plan if the ontology contains more than one inconsistency. Our approach also differs from these proposals in that they assume that all inconsistencies are equally bad, while we present a quantitative way to differentiate these inconsistencies.

7 Conclusion and Future Work

With the development of more expressive ontologies in the Semantic Web community, inconsistency has become an increasing problem that can seriously hamper the construction and application of web ontologies. In this paper, we have presented a technique for measuring inconsistencies which uses the Shapley value in game theory. Since the Shapley value aims to distribute the gains from cooperation in a fair manner, it can be used to impute the inconsistency to each axioms in the problematic ontology. The idea is to first define an inconsistency value, and then take it as the characteristic function, using the Shapley value to compute the contribution of each axiom or assertion to the inconsistencies in the ontology. This technique is conceptually flexible in the sense that although we focus on inconsistency problems in this paper, by choosing different inconsistency value functions, we can also address incoherence problems. It should be an easy extension to adopt other consistency values in the future. The measure associated with an axiom shows the degree of its responsibility for the inconsistency, therefore it can give guidelines for repairing the ontologies. We have also proposed and implemented some optimization techniques based on the structural relevance of the axioms and properties of the defined inconsistency measure, in order to reduce the computational complexity. An implementation of the proposed algorithm to be applied to clauses is underway to be incorporated as part of our explanation module developed earlier for DL reasoning.

Acknowledgements

This work was supported in part by Natural Sciences and Engineering Research Council (NSERC) of Canada, by Genome Quebec, and by Faculty of ENCS, Concordia University. We also thank anonymous reviewers for their valuable comments.

References

1. Deng, X., Haarslev, V., Shiri, N.: A framework for explaining reasoning in description logics. In: Proceedings of the AAAI Fall Symposium on Explanation-aware Computing, Washington, DC, USA, AAAI Press (2005) 189–204
2. Parsia, B., Sirin, E., Kalyanpur, A.: Debugging owl ontologies. In: Proceedings of the 14th International World Wide Web Conference (WWW 2005), Chiba, Japan, ACM Press (2005) 633–640
3. Schlobach, S., Cornet, R.: Non-standard reasoning services for the debugging of description logic terminologies. In: Proceedings of the eighteenth International Joint Conference on Artificial Intelligence (IJCAI'03), Acapulco, Mexico, Morgan Kaufmann (2003) 355–362
4. Hunter, A., Konieczny, S.: Shapley inconsistency values. In: Proceedings of the International Conference on Knowledge Representation (KR'06), Windermere, UK, AAAI Press (2006) 249–259
5. Conitzer, V., Sandholm, T.: Computing shapley values, manipulating value division schemes, and checking core membership in multi-issue domains. In McGuinness, D.L., Ferguson, G., eds.: Proceedings of the Nineteenth National Conference on Artificial Intelligence, Sixteenth Conference on Innovative Applications of Artificial Intelligence (AAAI 2004), San Jose, California, USA, AAAI Press/The MIT Press (2004) 219–225
6. In: OWL Web Ontology Language Overview. (2004) `http://www.w3.org/TR/owl-features/`.
7. Baader, F., Nutt, W.: Basic description logic. In Baader, F., Calvanese, D., McGuinness, D., Nardi, D., Patel-Schneider, P.F., eds.: The Description Logic Handbook: Theory, Implementation, and Applications. Cambridge University Press (2003) 5–44
8. Tsarkov, D., Horrocks, I.: FaCT++ description logic reasoner: System description. In: Proceedings of the International Joint Conference on Automated Reasoning (IJCAR 2006). Volume 4130 of Lecture Notes in Artificial Intelligence., Seatle, Washingtion, USA, Springer (2006) 292–297
9. Haarslev, V., Möller, R.: Racer system description. In R. Gori, A. Leitsch, T.N., ed.: Proceedings of International Joint Conference on Automated Reasoning (IJCAR 2001), Siena, Italy, Springer-Verlag (2001) 701–705
10. Sirin, E., Parsia, B.: Pellet: An owl dl reasoner. In: Proceedings of the 2004 International Workshop on Description Logics (DL2004), Whistler, British Columbia, Canada (2004)
11. Shapley, L.: A value for n-person games. In Kuhn, H., Tucker, A., eds.: Contributions to the Theory of Games. Volume 2. Princeton University Press (1953) 307–317
12. Deng, X., Haarslev, V., Shiri, N.: Resolution based explanations for reasoning in the description logic $\mathcal{ALC}$. In: Proceedings of the Canadian Semantic Web Working Symposium, Quebec City, Canada, Springer (2006) 55–61

13. Schlobach, S.: Diagnosing terminologies. In: Proceedings, The Twentieth National Conference on Artificial Intelligence and the Seventeenth Innovative Applications of Artificial Intelligence Conference (AAAI 2005), Pittsburgh, Pennsylvania, USA, AAAI Press (2005) 670–675
14. Kalyanpur, A., Parsia, B., Sirin, E., Grau, B.C.: Repairing unsatisfiable concepts in owl ontologies. In Sure, Y., Domingue, J., eds.: Proceedings of the 3rd European Semantic Web Conference (ESWC 2006). Volume 4011 of Lecture Notes in Computer Science., Budva, Montenegro, Springer (2006) 170–184
15. Grant, J.: Classifications for inconsistent theories. Notre Dame Journal of Formal Logic **19**(3) (1978) 435–444
16. Knight, K.: Measuring inconsistency. Journal of Philosophical Logic **31**(2) (2002) 77–98
17. Friedrich, G., Shchekotykhin, K.M.: A general diagnosis method for ontologies. In Gil, Y., Motta, E., Benjamins, V.R., Musen, M.A., eds.: Proceedings of 4th International Semantic Web Conference (ISWC 2005). Volume 3729 of Lecture Notes in Computer Science., Galway, Ireland, Springer (2005) 232–246

Squirrel: An Advanced Semantic Search and Browse Facility

Alistair Duke, Tim Glover, and John Davies

Next Generation Web Research Group, BT Group, Adastral Park, Ipswich, UK
{alistair.duke,tim.glover,john.nj.davies}@bt.com

Abstract. Search is seen as a key application that can benefit from semantic technology with improvements to recall and precision over conventional Information Retrieval techniques. This paper describes Squirrel, a search and browse tool that provides access to semantically annotated data. Squirrel provides combined keyword based and semantic searching. The intention is to provide a balance between the speed and ease of use of simple free text search and the power of semantic search. In addition, the ontological approach provides the user with a much richer browsing experience. Squirrel builds on and integrates a number of semantic technology components. These include machine learning and information extraction components which generate, extract and manage semantic metadata contained within and about textual documents at index time. A number of run-time components have also been integrated to deliver an enhanced user experience which goes beyond merely presenting a list of documents as a query response. The tool has been trialled and evaluated in two case studies and we report early results from this exercise, revealing promising results.

1 Introduction

Search engines based on conventional IR techniques (employing keyword and phrase matching between the query and index) alone tend to offer high recall and low precision. The user is faced with too many results and many results that are irrelevant. A major reason for this is the failure to handle polysemy and synonymy, or more generally, the view of a document purely as a bag of words upon which no semantic analysis is carried out. Although it is possible to disambiguate queries by adding more terms and by making use of more sophisticated functions (e.g. Boolean operators) there is evidence to suggest that the majority of users do not do this [8]. Algorithms such as Google's PageRank offer significant improvement over previous approaches, however, the approach tends to attribute a low rank to new pages about popular topics and the rank stays low if people cannot find the site and then link to it. Also, the result set continues to be too large and context is not considered which means results from queries about unpopular topics are still hard to pick out. Moreover, there are doubts about the suitability of the algorithm for ranking pages on corporate intranets due to their more regular structure [15]. A further point is that people require information rather than just a list of links where they might find what they are looking for.

E. Franconi, M. Kifer, and W. May (Eds.): ESWC 2007, LNCS 4519, pp. 341–355, 2007.

This paper describes Squirrel, a search and browse tool based on semantic technology. Squirrel aims to address the issues stated above through combined keyword based and semantic searching by allowing the user to initially enter free text terms and see results immediately but following this to allow them to refine their queries with the use of ontological support e.g. by selecting from a set of returned topics or matching entities. The intention is to provide a balance between the speed and ease of use of simple free text search and the power of semantic search. In addition, the ontological approach provides the user with a much richer browsing experience.

Squirrel has been developed in the EU SEKT project[1]. One of the main aims of SEKT has been to address the metadata bottleneck by providing semi-automatic semantic annotation and ontology generation tools. Squirrel builds upon the efforts of SEKT in this area to allow users to more easily find relevant knowledge with the support of semi-automatically derived semantic metadata about unstructured textual documents. The metadata describes both documents themselves and their content. Squirrel is targeted a large enterprises where it aims to offer a single search interface over the heterogeneous data sources that are common in that setting.

Squirrel provides a number of novel features to enhance the search and browse experience. These include natural language generation which provides natural language summaries of knowledge held in formal (ontological) structures; device independence which allows the tool to be run on multiple devices; and result consolidation which presents the most relevant textual content of result documents rather than a simple list of results.

The paper is structured as follows. The next section introduces a scenario which illustrates both the requirements and benefits of enhanced search and browse. Section 3 describes the salient features of the supporting components that comprise Squirrel and presents an architecture for the system. In Section 4 further detail is provided about the user experience with a description of the important features of the interface. Early results from an evaluation exercise are described in Section 5. Section 6 discusses related work and outstanding issues whilst we conclude in Section 7 with a view of the way forward.

2 Scenario

Squirrel has been trialled within the SEKT project in a case study which is developing an improved Digital Library. The following scenario describes a Digital Library user carrying out a knowledge seeking task, making use of the features provided by Squirrel and its supporting components.

The user has an initial high level goal to seek information about the field of Home Health Care and has little idea about what is contained in the library which might be of use. They first enter 'Home Health Care' into the search box and hit the 'Go!' button. The first result screen includes a summary of the sorts of resources that have been found that might be able to meet their needs. These include the textual resources from the library i.e. that there are a number of journal articles, conference papers,

[1] http://www.sekt-project.com

periodicals and web pages (that have been shared by library users and indexed as library resources) that match their query. In addition, there are a number of library topics (subject areas against which relevant documents are classified) that also match their query including one which is itself called 'Home Health Care'. Further matches include a number of organizations from a domain knowledge base whose description includes the search term.

In addition to this summary, the user is also presented with the top ranked textual resources (in typical search engine style). This simple list of documents is augmented with the ability to refine the search based on the properties of the documents in the result set, including a hierarchical display of the topics of the results documents, date of publish, author name, etc.

Our user decides that they are interested in current affairs related to their topic rather than scientific papers. They choose to refine the search to the periodicals section of the library. The topic hierarchy is rebuilt to accommodate the documents in the refined results set. Furthermore, our user is interested in the financial aspects of home health care and notices that a topic called 'Economic Conditions' has been presented. They then choose this topic to further refine their results set.

A short taster of each result together with its title and other metadata is presented. This is enhanced by highlighting the named entities that have been recognized in the text such as people, organization, locations, etc. Our user places their mouse of one such highlighted term, 'United Health Products' and is shown a popup which tells them that this is a company. Intrigued about the company, they click on the term and are taken to a screen giving further information about the company including a natural language summary about the company as well as related entities such as the people who work for it. There is also a link to all of the documents where this company (including any of its aliases such as UHP, etc.) is mentioned. Our user clicks on this link but then decides they'd rather view the results as a consolidated summary rather than a list of discrete results. In this view the most relevant portions of the result documents are shown to the user meaning they can quickly view the available material rather than having to navigate to several different pages. The user again refines the contents of the summary by selecting one of more of the topics that have been assigned to the presented material.

3 Architecture

Squirrel integrates a number of components developed in the SEKT project. This section briefly describes the features of the components and then presents an architecture illustrating how they have been integrated.

3.1 PROTON

PROTON is a lightweight general purpose ontology developed in SEKT which includes a module specific to the knowledge management domain. PROTON is used by SEKT components for metadata generation and as a basis for knowledge modelling and integration.

A world knowledge base (originally developed as part of the KIM platform [11]) has been expressed in PROTON. The knowledge base is comprised of more than 200,000 entities. These are gathered semi-automatically from a range of high quality public data sources. It includes around 36,000 locations, 140,000 companies and other organisations and the world's pre-eminent politicians, businesspeople and technologists

3.2 Full-Text Index

The Lucene full-text indexing facility is used to provide an index over the various forms of textual resource but also over the labels and literal properties of the ontological instances and classes. This allows metadata e.g. authors' names, topic names, knowledge base entities to be discovered and presented to the user as a way to refine their search or as an alternative way to find documents, since the metadata is always related to a set of documents in some way. The index of ontological instance will, given a search term, provide a set of matching URIs which can then be used to query the ontology repository to extract further details of the concept they refer to.

3.3 KAON2

The KAON2 [7] ontology management and inference engine provides an API for the management of OWL-DL and an inference engine for answering conjunctive queries expressed using the SPARQL[2] syntax. KAON2 also includes a module for extracting ontology instances from relational databases via a mapping facility. This greatly increases the number of instances that can be held in the ontology as compared to a file based version.

KAON2 also supports the Description Logic-safe subset of the Semantic Web Rule Language[3] (SWRL). This allows knowledge to be presented against concepts that goes beyond that provided by the structure of the ontology. For example, one of the attributes displayed in the document presentation is 'Organisation'. This is not an attribute of a document in the PROTON ontology; however, affiliation is an attribute of the Author concept and has the range 'Organisation'. As a result, a rule was introduced into the ontology to infer that the organisation responsible for a document is the affiliation of its lead author.

3.4 Natural Language Generation

Natural Language Generation (NLG) takes structured data in a knowledge base as input and produces natural language text, tailored to the presentational context and the target reader [12]. In the context of the semantic web and knowledge management, NLG is required to provide automated documentation of ontologies and knowledge bases and to present structured information in a user-friendly way [1].

When a Squirrel search is carried out and certain people, organisations or other entities occur in the result set, the user is presented with a natural language summary of information regarding those entities. In addition, document results can be enhanced

[2] http://www.w3.org/TR/rdf-sparql-query/

[3] http://www.w3.org/Submission/SWRL/

with brief descriptions of key entities that occur in the text. For example, the knowledge base contains company related metadata, which can be presented to the user as natural language. Squirrel uses the ONTOSUM component described in [1]. An example is shown in Figure 3.

3.5 DIWAF

The SEKT Device Independence Web Application Framework is a server side application, which provides a framework for presenting structured data to the user [5]. The framework does not use a mark-up language to annotate the data. Instead, it makes use of templates, which are "filled" with data, rather like a mail merge in a word processor. These templates allow the selection, repetition and rearrangement of data, interspersed with static text. The framework can select different templates according to the target device, which present the same data in different ways.

Squirrel has been built as a DIWAF application. This enables the applications to be easily adapted to alternative device or context requirements (such as different languages). Currently, templates are available to deliver the application to users via a WAP-enabled mobile device, via a PALM PDA and via a standard web browser.

3.6 Ontology Generation

The Ontology Generation (OG) component can automatically identify an ontology from the content of the documents and then classify the documents according to the ontology [4]. In SEKT this process is generally carried out at index-time which allows the knowledge engineer to modify the generated ontology if required. The process results in hierarchical topic ontology and a set of document classifications into that ontology (in accordance with PROTON). These results are used by Squirrel to present the topic hierarchy thus allowing the user to refine their search by topic. The OG component can also be used by Squirrel to generate clusters of documents at query-time. This could be used in the general case where some or all of the documents being searched have not been classified at index-time.

3.7 User Profile Construction and Usage

The PROTON ontology includes concepts allowing the interests of users to be stored and modelled as a profile. Such a profile is an important step in providing relevant knowledge to people and is used within Search and Browse to personalise the search experience. The profile allows the search tool to predict what the user is interested in based upon their observed interests. Profiles can be manually or automatically created. An automatic approach places less burden on the user than a manual one but relies on the assumption that the techniques used to collect the profile are accurate (which is why a combination of automatic and manual approaches is often adopted). The approach adopted in SEKT has been to observe the web browsing habits of the individual users and to extract the topics associated with the pages that have been accessed. A level of interest for each of the topics is also determined.

In PROTON, the profile consists of an expression of both long-term and short-term interest in topics from a PROTON topic ontology. The Squirrel search tool makes use of these profiles to modify the way the results are presented to the user i.e. providing context to the user's search query. When ranking documents, the topics of the documents in the result set can be considered against the level of interest in topics in the user's profile. Thus documents with a topic recorded in the profile with a strong short-term interest would be presented first. Topics are related to other topics (either by a sub or super topic or some other weaker relation) so these relationships can also be considered to support the application of profiles.

3.8 Massive Semantic Annotation

Squirrel makes use of the results of the Massive Semantic Annotation process, carried out at index time. The function uses KIM [11], which employs GATE's ontology-based information extraction module to identify named entities in texts. For each of these it carries out instance disambiguation i.e. it either identifies an existing PROTON instance of which the entity is an occurrence or creates a new instance. An annotation (which consists of the instance URI, the document it occurs in and its location within the text) is stored using KIM's highly scalable (due to the very large number of annotations) knowledge store.

In order to support user responsive entity presentation and browsing, Squirrel makes use of the OWLIM semantic repository [9]. OWLIM is considered to be the fastest OWL repository available. The dual repository approach adopted by Squirrel is justified by the benefits that each repository provides. KAON2 offers relational database mapping and rule support whilst the scalability and speed of OWLIM are suited to handling the volume of data resultant from the semantic annotation process.

3.9 Segmentation

During the indexing process, documents are segmented at topical boundaries in order to support the presentation of consolidated results to the user. The segmenter used is a C99 segmenter [2] which is enhanced to consider the distribution of named entities and key phrases identified in the text in addition to the distribution of words. Following segmentation the subdocuments are classified using the OG classifier described above. These classifications are used in two ways by Squirrel. Firstly, to allow the user to refine their summaries based upon topics of interest and secondly to reorder the presentation of the summary based upon the topics in the user's profile.

3.10 Integration

The components described above have been integrated as a complete end-to-end architecture for indexing and querying

The sources of textual data are stored together with their associated metadata in a database. These sources may be e.g. records of bibliographic data (as is the case in the SEKT Digital Library case study), webpages identified by a crawler, or legal judgements (as is the case in the SEKT Legal case study). At this stage, the metadata can be augmented by one or more entity extraction or classification components.

Once these steps have been carried out, the contents of the database can then be indexed by Lucene which indexes all the textual fields in the database, allowing subsequent keyword based retrieval. Entities within the PROTON world knowledge base are described by one or more aliases. These aliases are also added to the index allowing the entities to which they refer to be retrieved at query time.

A parallel activity to free-text indexing is to load the contents of the database into KAON2. KAON2 provides a facility to map database schema to an ontological representation. This allows KAON2 to represent the contents of the database as a PROTON ontology. In addition to the database, KAON2 also reads in the PROTON ontology, its world knowledge base and user profiles.

User queries are first passed to Lucene in order to retrieve a set of matching documents and entities by their URI. Squirrel then queries KAON2 to extract further details about these concepts as described in Section 3.2

Following this, Squirrel is able to present an initial result to the user. The user interface and the various options provided by Squirrel for refining and presenting results are described in Section 4.

4 Interface Description

4.1 Initial Search

Users are permitted to enter terms into a text box to commence their search. This initially simplistic approach was chosen based on the fact that users are likely to be comfortable with it due to experience with traditional search engines. If they wish, the user can specify which type of resource (from a configurable list) they are looking for e.g. publications, web articles, people, organisations, etc. although the default is to search in all of these.

The first task Squirrel carries out after the user submits a search is to call the Lucene index and then use KAON2 to look up further details about the results, be they textual resources or ontological entities. In addition to instance data, the labels of ontological classes are also indexed. This allows users to discover classes and then discover the corresponding instances and the documents associated with them without knowing the names of any of the instances e.g. a search for 'Airline Industry' would match the 'Airline' class in PROTON. Selecting this would then allow the user to browse to instances of the class where they can then navigate to the documents where those instances are mentioned. This is an important feature since with no prior knowledge of the domain it would be impossible to find these documents using a traditional search engine.

Textual content items can by separated by their type e.g. Web Article, Conference Paper, Book, etc. Squirrel is then able to build the meta-result page based upon the textual content items and ontological instances that have been returned.

4.2 Meta-result

The meta-result page is intended to allow the user to quickly focus their search as required and to disambiguate their query if appropriate. The page presents the different types of result that have been found and how many of each type. In order not to introduce unnecessary overhead on the user, the meta-result page also lists a default set of results allowing the user to immediately see those results deemed most relevant to their query by purely statistical means.

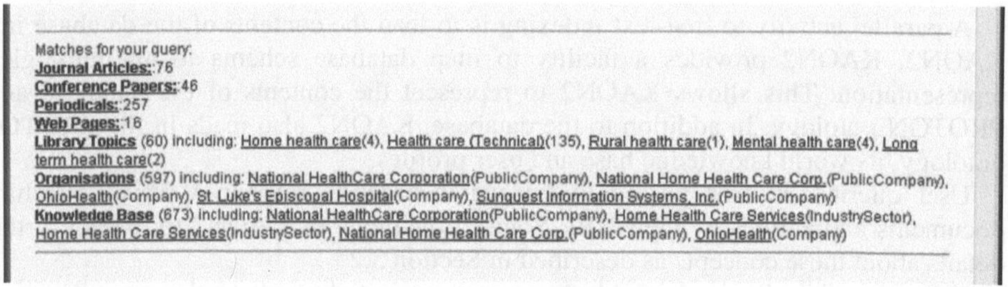

Fig. 1. Meta-result

The meta-result for the 'home health care' query is shown in Figure 1 under the sub-heading 'Matches for your query'. The first items in the list are the document classes. Following this is a set of matching topics from the topic ontology. In each case, the number in brackets is the number of documents attributed to each class or topic. Following the topics, a list of other matching entities is shown. The first 5 matching entities are also shown allowing the user to click the link to go straight to the entity display page for these. Alternatively they can choose to view the complete list of items. Squirrel can be configured to separate out particular entity types (as is the case with topics and organisations as shown in Figure 1) and display them on their own line in the meta-result.

The returned documents can be ranked according to the user's profile if one has been specified. The algorithm to perform this ranking checks to see which documents are attributed to topics that exist in the profile and then adjusts the ranking using the degree of interest for those topics. For each document, the degree of relevance provided by Lucene is added to the cumulative degree of interest from the topics in the profile that are also attributed to the document. The documents are then re-ranked based upon these new relevance figures and displayed to the user. The algorithm can be formulised as shown in Equation 1.

$$R = r + \sum_{i=1..k.} \begin{cases} w(t_i) & t_i \in U \\ 0 & t_i \notin U \end{cases} \tag{1}$$

Where R is the degree of relevance for a document and r is the (statistical) degree of relevance provided by Lucene. A topic attributed to the document is denoted by t_i where i is the index of a topic in the set of attributed topics. $w(t_i)$ denotes the level of interest in the topic t_i if it is a member of the set of topics in the user's profile, U. If the topic is not in the profile then no adjustment is made.

The algorithm attempts to maintain a balance between results with a high relevance ranking mainly due to the Lucene result and those that are principally deemed to be of interest according to the profile. Thus, if the user has just switched their focus, the results should not skew too far in favour of the profile which might be slightly biased towards their previous focus.

4.3 Refining by Topic

Alongside the results, the user is presented with the topics associated with the documents in the result set. Not all topics are shown here since in the worst case where each document has many distinct topics the list of topics presented to the user would be unwieldy. Instead an algorithm takes the list of topics that are associated with the collection of documents in the result set, and generates a new list of topics representing the narrowest "common ancestors" of the documents' topics.

Having selected a topic and viewed the subset of documents from the result set, the user can switch to an entity view for the topic. The user can also reach this view by selecting a topic from the meta-result section. Instead of showing the documents of the topic, the meta-data for the topic is shown, which includes the broader, narrower and related topics. This allows the user to browse around the topic ontology. Each topic is shown with the number of documents it contains in brackets after it. Two links to document results are also shown. The first takes the user to a list of documents that have been attributed to the topic itself. The second takes the user to a list of documents that include all subtopics of the current topic. The layout of entity views in Squirrel are defined by templates. This allows the administrator to determine what meta-data is shown and what is not. The use of these templates is discussed further in Section 4.6 where a 'Company' entity view is described.

4.4 Attribute-Based Refinement

Any document result list has a link, which opens a refiner window. This allows the user to refine the results based upon the associated metadata. The metadata shown and the manner in which it is displayed are configurable through the use of entity type specific templates that are configured by an administrator or knowledge engineer. Documents can be refined by the user based upon their authors, date of publication, etc. The approach adopted has been to allow the user to enter free text into the attribute boxes and to re-run the query with the additional constraints. An alternative would be to list possible values in the result set. However, the potential size of this list is large and is difficult to present to the user in a manageable way. The downside to the free-text approach is that the user can reduce the result set to zero by introducing an unsatisfiable constraint – which is obviously undesirable. Squirrel attempts to address this by quickly showing the user the size of the result set once constraints have been set. The user can then modify them before asking Squirrel to build the result page.

4.5 Document View

The user selects a document from the reduced result set, which takes them to a view of the document itself. This shows the meta-data and text associated with the document and also a link to the source page if appropriate – as is the case with web-pages. Since web-pages are published externally with specific formatting data, the text of the page is extracted at index-time. Any semantic mark-up that is applied at this stage can then be shown on the extracted text at query-time. However, the user should always be given the option to navigate to the page in its original format. A screenshot of the document view is shown in Figure 2.

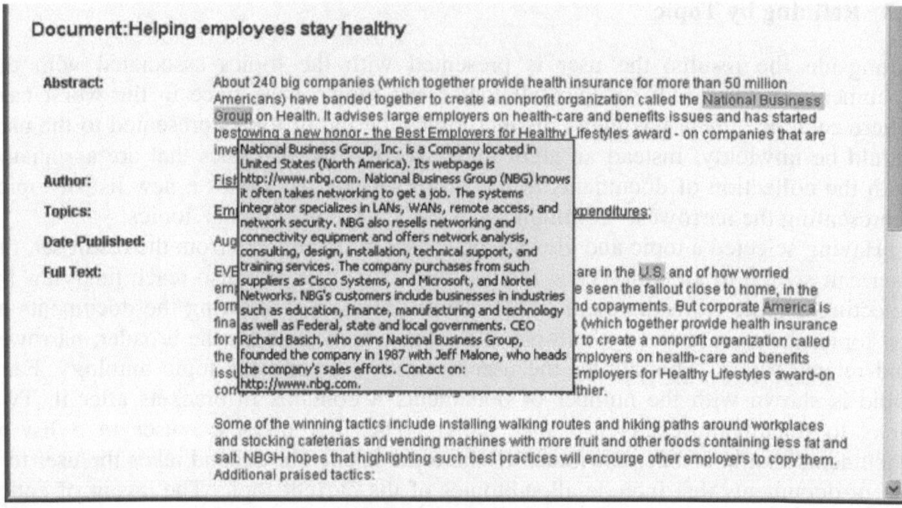

Fig. 2. Document View

The document view also shows the whole document abstract, marked-up with entity annotations. 'Mousing-over' these entities provides the user with further information about the entity extracted from the ontology. Clicking on the entity itself takes the user to the entity view.

4.6 Entity View

The entity view for 'Sun Microsystems' is shown in Figure 3. It includes a summary generated by the NLG Web Service described in Section 3.4. The summary displays information related not only to the entity itself but also information about related

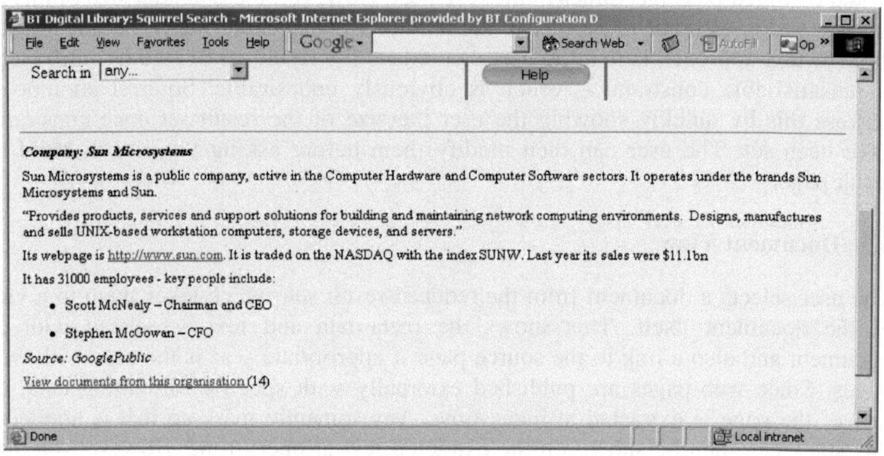

Fig. 3. Company Entity View

entities such as people who hold job roles with the company. This avoids users having to browse around the various entities in the ontology that hold relevant information about the entity in question. The relationship between people and companies is made through a third concept called JobPosition. Users would have to browse through this concept in order to find the name of the person in question.

4.7 Consolidated Results

Users can choose to view results as a consolidated summary of the most relevant parts of documents rather than a discrete list of results. This view is appropriate when a user is seeking to gain a wider view of the available material rather than looking for a specific document. The view allows them to read or scan the material without having to navigate to multiple results. Figure 4 shows a screenshot of a summary for a query for 'Hurricane Katrina'. For each subdocument in the summary the user is able to view the title and source of the parent document, the topics into which the subdocument text has been classified or navigate to the full text of the document. A topic tree is built which includes a checkbox for each topic. These allow the user to refine the summary content by checking or unchecking the appropriate checkboxes.

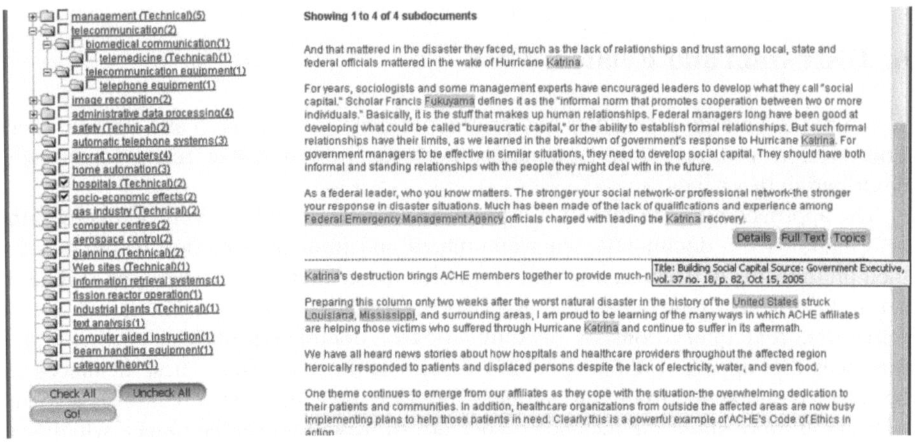

Fig. 4. Consolidated Results

5 Evaluation

Squirrel has been being subjected to a three-stage user-centred evaluation with users of BT's Digital Library. Firstly, a heuristic evaluation [10] of the user interface was undertaken. A small group of researchers, acting as usability experts, judged whether the user interface adhered to a list of usability heuristics; a checklist was adapted from the Xerox heuristic evaluation system [3]. A number of observations were made, most of which were concerned with minor interface problems and system performance. The results of the evaluation were collated and discussed with the development team. Squirrel was then modified in accordance with these observations.

The second stage comprised a cognitive walkthrough evaluation [14]. Users were asked to use Squirrel in order to complete a number of tasks, where the user's actions and behaviours were recorded. Users were encouraged to talk through their actions and their concerns as they undertook each task, since this provided additional information about usability and the user's thought processes. At the end of each task, users were also asked to complete a short questionnaire. The findings were again discussed with the Squirrel development team in order to resolve interface issues.

The final stage involved a set of field tests. Subjects were provided with set of information seeking tasks which they were invited to complete using both the existing keyword based search system and Squirrel. They then provided feedback on the perceived quality of results and progress in search. Qualitative feedback was sought via rating forms and a Software Usability Measurement Inventory (SUMI) questionnaire.

Initial results reveal promising results in the perceived information quality (PIQ) of search results obtained by the subjects. From 20 subjects, using a 7 point scale the average (PIQ) using the existing library system was 3.99 vs. an average of 4.47 using Squirrel – an 12% increase. The SUMI assessment showed that users rate the application positively and believe that it has attractive properties, but were concerned about its performance and speed.

6 Discussion and Related Work

A number of emerging products and prototypes exist in the Search and Browse space. This section will now briefly consider how the best of these relate to the work described in this paper

The approach adopted by ISX Corporation's Concept Object Web [13] is to gather information from documents using automated and human extraction techniques and then present this to users as answers to queries rather than leaving the user to read the set of documents in the query response in order to gain the answer. The response to an initial free-text query consists of summaries of documents together with a list of instances that appear in them split into customisable categories. These instances can be selected and added to the initial query to refine the results set. The user is also able to find out more about the instances. They are shown a knowledge object, which is an aggregation of the semantic information about the entity with links to the source documents and how each fact was obtained. Users can navigate through the knowledge base by selecting the relation values of the current object. The approach is similar in many respects to that adopted by Squirrel. It provides better support for refinements based upon entities but does not allow the user to refine their search based on the topic of documents or browse a topic ontology in order to find related documents. Additional facilities offered by Squirrel include NLG and tailoring results according to a user profile.

/facet [6] adopts a navigational model or facet browsing approach where a user can navigate to heterogeneous resources by making selections on the properties of other semantically related types. This benefits users who do not have a clear idea of what they are searching for or know how it is described. Only those attribute values that will lead to at least one result being shown are made available. This ensures the user does not take any 'blind alleys'. They can also back-up by removing previously chosen attributes from their search. One issue with facet browsing is that where there are many possible selections the interface becomes unwieldy. /facet overcomes this by allowing the user to enter terms in a textbox and by suggesting possible terms as they type. Again, only keywords that produce actual results are suggested. An impressive feature of /facet is its ability to build facet selection directly from the metadata with no configuration and at query time. Whilst /facet does not contain many of the features of Squirrel such as named entity recognition, result consolidation or NLG, the navigation approach and its ability to cope with a large number of facets with no configuration indicate how Squirrel could be extended in this fashion.

The Excalibur[4] product from Convera provides a search interface that allows the user to enter queries and then refine based on topics which tailors the results indicating that documents have been attributed to topics at index time. Further topics are offered to broaden or narrow the search which indicates that there is some sort of topic hierarchy behind the system, however, it is not clear that this extends to a full ontology akin to PROTON. The system highlights named entities in the text, but the user is not able to select these and find out more about them, browse to related entities or refine a search based on the values of properties that particular entities have. Furthermore, the lack of an ontology behind the entities makes it harder to apply rules of the nature described in Section 3.

7 Conclusion and Future Work

We have described Squirrel, a tool for searching and browsing semantically annotated textual resources. The tool provides a hybrid approach to search allowing the user to enter simple search terms and then refine their search or browse to related material through the presentation of appropriate metadata. The tool integrates a number of state-of-the-art components to provide ontology management, named entity recognition on ontology generation and classification. It also includes a set of novel features such as result consolidation, natural language generation, ontology based user profiling and device independence.

The tool has been trialled in two case studies where evaluation shows promising results in terms of Perceived Information Quality. Performance concerns were raised by users and this certainly requires further attention with respect to the choice and use of the reasoner. Whilst KAON2 is more flexible and includes relational database support, OWLIM is faster and may prove more appropriate for this application were it to be extended to provide similar database facilities.

A number of potential areas for further development include the ability to identify the relationships between entities that are discovered in the query response. For example, a query such as 'Java Microsoft' would match a number of different entities including both a topic and a company. The appropriate entities could be chosen based upon a combination of the popularity in the knowledge base i.e. how many documents contain them as annotations, the user's profile and lastly an order of precedence that is specified for the domain e.g. in the digital library, if a topic is identified then this

---

[4] http://www.convera.com

might be chosen over an entity with the same name. In this case the topic Java and the company entity Microsoft might be chosen and as such it might be appropriate to initially show documents from the topic where annotations of the company have been identified. Similarly a query for 'Bill Gates Microsoft' would identify the person and company from the knowledge base. Here, as well as showing documents where annotations of both occur, it might be appropriate to show how they are related, which in this case is via the JobPosition instance relating the person to the company.

The display of entity results could also be improved with the use of user profiles. Where a user searches using a term that closely matches two or more entities there is a need for disambiguation i.e. predicting which of the entities the user is searching for by matching the results against the profile. For example, if a user was searching for documents written by an author called Davies (of which there are many different separate instances) the correct author can be chosen by matching the topics of the documents the individual authors have written against topics in the user's profile.

A second area for development is concerned with adopting a reduced configuration, faceted browsing approach such as that offered by /facet and described in the previous section.

Finally, there is a need for better and optimised integration between ontology management and full-text search. Currently the two are separate components and must be queried separately either sequentially (as in Squirrel) or in parallel with a subsequent intersection step. This has inherent performance issues which could be addressed by a combined approach that is able to employ optimisation techniques.

Acknowledgements. The work described in this paper was developed as part of the SEKT project, under EU funding (IST IP 2003506826). Further project details may be found at http://www.sekt-project.com/.

References

1. Bontcheva, K. 'Generating Tailored Textual Summaries from Ontologies'. Proceeding of the 2nd European Semantic Web Conference, Crete, Greece, June (2005), p. 531-545
2. Choi, F. Y. Y., :'Advances in domain independent linear text segmentation'. In Proceedings of NAACL, Seattle, USA, April, (2000), p. 26-33.
3. Christiansson, P., Heuristic Evaluation - A System Checklist. (2000), Xerox Corporation.
4. Fortuna, B.., Grobelnik, M. & Mladenic, D. :'Semi-automatic Construction of Topic Ontology', Semantics, Web and Mining, Joint International Workshop, EWMF 2005 and KDO 2005, Porto, Portugal, October 3-7, (2005)
5. Glover, T. and Davies, J.: 'Integrating Device Independence and User Profiles on the Web'. BT Technology Journal Vol 23. No. 3 July (2005).
6. Hildebrand, M., van Ossenbruggen, J. & Hardman, L. :'/facet: A Browser for Heterogeneous Semantic Web Repositories', Proceedings of the 5th International Semantic Web Conference, Athens, USA, November (2006).
7. Hustadt, U., Motik, B., Sattler, U. : Reducing SHIQ Description Logic to Disjunctive Datalog Programs. In Dubois, D., Welty, C., Williams, M.A., eds.: Proc. of the 9th Int. Conf. on Knowledge Representation and Reasoning (KR2004), Menlo Park, California, USA, AAAI Press (2004) 152–162

8. Jansen, B. J., Spink, A., and Saracevic, T. : 'Real life, real users, and real needs: A study and analysis of user queries on the web', Information Processing and Management. 36(2), (2000), 207-227.
9. Kiryakov, A., Ognyanov, D. & Manov, D. :'OWLIM – a Pragmatic Semantic Repository for OWL', In Proc. of Int. Workshop on Scalable Semantic Web Knowledge Base Systems (SSWS 2005), WISE 2005, 20 Nov, New York City, USA (2005).
10. Neilsen, J. and R. Molich, Heuristic Evaluation of User Interfaces, in Proceedings of the SIGCHI conference on Human factors in computing systems. (1992): Monterey, California, United States. p. 373 – 380.
11. 11.Popov, B., Kiryakov, A., Ognyanoff, D., Manov, D. and Kirilov, A. : 'KIM - a semantic platform for information extaction and retrieval', Journal of Natural Language Engineering, Vol. 10, Issue 3-4, Sep 2004, pp. 375-392, Cambridge University Press.
12. Reiter, E. & Dale, R.: 'Building Natural Language Generation Systems'. Cambridge University Press, Cambridge, (2000).
13. Starz, J., Kettler, B., Haglich, P., Losco, J., Edwards, G. and Hoffman, . :'The Concept Object Web for Knowledge Management' Proceedings of the 4th International Semantic Web Conference, Galway, Ireland, November (2005).
14. Wharton, C., et al., Applying Cognitive Walkthroughs to more Complex User Interfaces: Experiences, Issues, and Recommendations, in Proceedings of the SIGCHI conference on Human factors in computing systems. (1992): Monterey, California, United States.
15. Xue, G., Zeng, H., Chan, Z., Ma, W., Zhang, H., and Lu, C. : 'Implicit link analysis for small Web search', SIGIR'03, July 28-August 1, (2003), Toronto, Canada.

User-Centric Faceted Search for Semantic Portals

Osma Suominen, Kim Viljanen, and Eero Hyvönen

Semantic Computing Research Group (SeCo),
Helsinki University of Technology (TKK), Laboratory of Media Technology
University of Helsinki, Department of Computer Science
firstname.lastname@tkk.fi
http://www.seco.tkk.fi/

Abstract. Many semantic portals use faceted browsing, where the facets are based on the underlying indexing ontologies of the content. However, in many cases, like in medical applications, the ontologies may be very large and complex, and do not provide the end-user with intuitive facet hierarchies for conceptualizing the content, for formulating queries, and for classifying the search results. We argue that in such cases end-user facets should be separated from the annotation ontologies, and show how to generalize the semantic view-based search paradigm to take into account this fact. A user-centric card sorting method is proposed for designing intuitive views for the end-users and a method for mapping its facets onto the indexing ontologies and search items is presented. The system has been implemented in a prototype of the semantic portal TerveSuomi.fi, a national health promotion portal in Finland.

1 Introduction

Faceted search (i.e., faceted browsing and view-based search) [1,2,3], is a search paradigm developed originally in the field of information retrieval. The idea of the scheme is to analyze and index search items along multiple orthogonal taxonomies that are called subject *facets* or *views*. From the end-users viewpoint searching is then reduced to the selections of categories along the facets. This idea has later been developed into *semantic faceted search* [3,4,5], where the facets are based on ontological structures, such as subclass and part-of hierarchies. The usefulness of faceted search has been demonstrated in many applications [5,6] and the first commercial products are already on the market[1].

A limitation of semantic faceted search is that facets based on indexing ontologies do not always provide the end-user with natural categorizations of the content for formulating queries or for organizing search result lists. In many domains, very large and complex ontologies are used for indexing by domain professionals. The point of view of indexing may differ substantially from the point of view of the end-user, who also may not be familiar with the professional terminology. The ontologies may also be too general or too specific for her needs.

[1] http://www.siderean.com/, http://www.express.ebay.com/, http://endeca.com/

E. Franconi, M. Kifer, and W. May (Eds.): ESWC 2007, LNCS 4519, pp. 356–370, 2007.

The main hypotheses underlying this paper is that end-users, such as ordinary citizens, often conceptualize the domain of discourse in terms of categories that are different from the ontological representations used by the domain specialist and content indexers. To bridge the semantic gap between end-users and professionals, we need 1) a method for finding out the end-user facets and search categorizations about the domain, and 2) a method for mapping the facet categories onto the content indexed along ontologies. To solve the facet creation problem, we present a user-centric card sorting method [7,8,9] for creating intuitive search facets for the end-users, and present results of applying the method in creating the facets for the prototype of the national semantic health promotion portal TerveSuomi.fi [10,11] in Finland. To address the mapping problem, we show how the user-centric facets can be mapped onto the ontologies used for describing the content, and be used for answering queries.

The paper is organized as follows. We first present a general architecture and a method for answering faceted queries based on user-centric facets and indexing ontologies. After this the method is applied to our case study by describing ontological metadata and the indexing ontologies which are used to describe the content. Based on this, card sorting methods for creating user-centric facets are discussed and applied for the TerveSuomi.fi portal, and a prototype implementation of the system is presented. Finally, contributions of the work are summarized, related work discussed, and future research suggested.

2 Extending Faceted Search Architechure

In semantic faceted search the documents are annotated along different facets that typically correspond to the elements (fields) of the metadata (annotation) schema used. For example, in MuseumFinland [5] the collection artifact metadata schema has nine resource-valued properties such as *Artifact type* and *Material* whose values are taken from a set of seven orthogonal *indexing ontologies*. A single ontology, such as "Places", can be used for expressing values of several different elements, such as "Place of Manufacture" and "Place of Usage". When the same place is selected in the "Place of Manufacture" or "Place of Usage" facet for querying, different result sets are obtained. Although facets share hierarchical structures of the indexing ontologies, the facet categories are different from the corresponding ontology concepts and have different facet-wise URIs. The resulting facet hierarchies will be called *faceted ontologies.*

By using a set of logical projection rules each facet-URI can be associates with a set of related search items [12]. The *extension* $ext(S)$ of a facet-URI S is the set of all search items associated with it and any of its subcategories. The faceted search query in semantic faceted search with n facets is a conjuctive Boolean expression $Q = S_1 \wedge ... \wedge S_n$ where each S_i is a Boolean expression of the facet ontology category (URIs). The result set is obtained by interpreting the logical operations of Q as set operations over the search item set E in the following way: $S \wedge T \equiv ext(S) \cap ext(T)$, $S \vee T \equiv ext(S) \cup ext(T)$, $S - T \equiv S \wedge \neg T \equiv ext(S) - ext(T)$, and $\neg S \equiv E - ext(S)$.

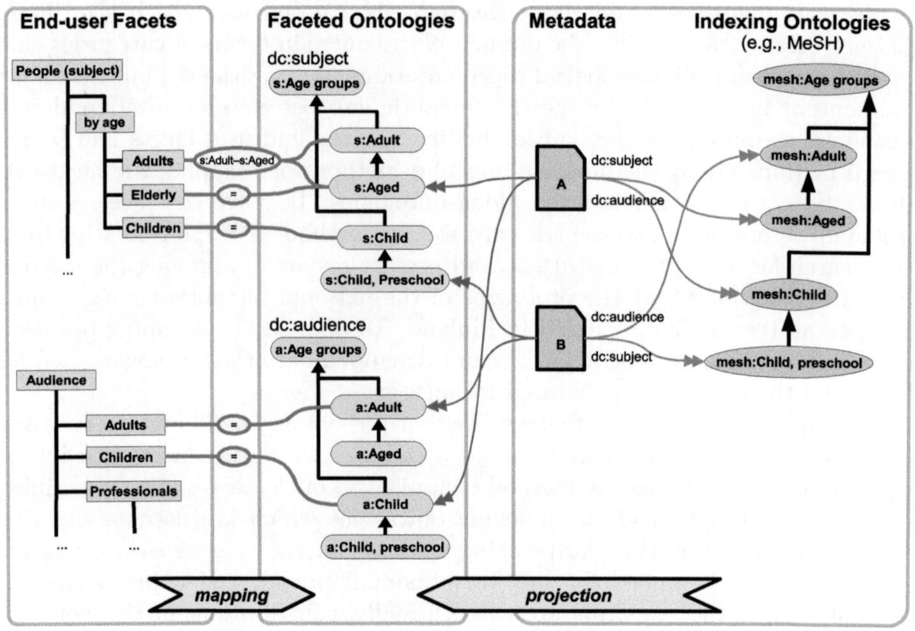

Fig. 1. Components of an end-user-centric view-based semantic search framework

In MuseumFinland, faceted ontologies are used directly as querying facets (with some filtering). In the case of TerveSuomi.fi this is not feasible but a set of m user-centric facets is created for formulating the query $Q_U = U_1 \wedge ... \wedge U_m$, where U_i is a category in a user-centric facet i. In order to map such queries onto queries at the faceted ontology level, each user-centric facet category can be defined as a Boolean expression of the faceted ontology categories. This means that queries expressed in terms of end-user facet categories can be reduced into faceted ontology queries that can be processed with a semantic faceted search engine such as Ontogator [13].

Figure 1 depicts how the search documents can be found using user-centric facets, indexing facets and indexing ontologies in our view-based semantic search scheme. The documents are indexed with ontologies using resource-valued metadata fields `dc:subject` and `dc:audience` whose values are taken from the indexing ontology (e.g., MeSH). The document A tells about (dc:subject) aged people (elderly people in layman terms) and is intended for children (dc:audience). Similarly, the document B is about preschool children and is intended for adults. The corresponding two faceted ontologies, projected from the indexing ontologies, are seen in the middle left in different namespaces s (subject) and a (audience). Some of the extension of the categories are: ext(s:Aged)=ext(s:Adult)={A}, ext(s:Child)={B}, ext(a:Child)={A} and ext(a:Adult)=ext(a:AgeGroup)={B}. The extensions can be projected by simple logic rules based on the metadata. If faceted ontologies were used for searching (like in MuseumFinland), then the selection *s:Adult* would return the document A and *a:Adult* would return the document B. By introducing end-user facets, the queries can be expressed in terms of new facet categories. For example, the end-user facet category *Adults* is defined as $Adults \equiv ext(s : Adult) - ext(s : Aged)$, and the document A would not be returned with the facet selection *Adults*, because the document is about aged people which has been excluded from the end-user facet category *Adults*.

This model can be extended to deal with uncertainty and relevance by using the "fuzzy view-based search" approach presented in [10] in the following way. Fuzzy annotations can be used if exact classifications are not appropriate. For example, the boundary between children and adults is fuzzy. Therefore we could say that the dc:audience of the document B in figure 1 is not the crisp set $\{mesh : Adult\}$ but rather the fuzzy set $\{(mesh : Adult, 0.8), (mesh : Child, 0.2)\}$ indicating that the document is targeted to some degree also to children. In [11] a method for determining such fuzzy annotations based on ontologies and the *tf-idf* method is presented. In this way the extensions of the indexing facets can be seen as fuzzy sets where the membership values are based on fuzzy annotations and are interpreted as a measure of relevance. By defining the Boolean operators used in faceted search as fuzzy Boolean operators, the relevance of hits in the search results can be determined—an important feature missing in the traditional faceted search paradigm. In the example above, the document B would be less relevant when looking for material targeted to children than to adults. It is also possible to generalize the mappings into fuzzy mappings by attaching a membership function value to them, indicating only partial match between an end-user category and its definition in terms of the facet ontology categories, and by interpreting the mapping as a fuzzy inclusion.

3 Ontological Metadata for a Health Promotion Portal

When building a semantic portal, one key decision is to choose which ontologies are used for indexing the content and whether existing ontologies can be used compared to building custom ontologies. Typically suitable ontologies exist, which have been created for a similar domain but a different purpose. In such cases the decision has to be made between using them as-is, modifying them to suit the purposes of the portal, or creating a new ontology from scratch.

Using an existing, established ontology has several advantages. Creating a new ontology requires substantial amounts of manual work, which can be avoided by reusing an existing ontology. An existing and established, large ontology is also more likely to have broad and deep coverage of concepts within its domain, which will allow documents to be annotated to very specific concepts, whereas a custom-made ontology might only cover the topic areas and concepts that are relevant for the need of the semantic portal. Finally, reusing a shared ontology furthers the vision of the Semantic Web [14] by allowing semantic interoperability between different systems, as long as they use the same ontology (or a compatible one, interlinked by semantic mappings). On the other hand, existing ontologies may differ from the goals of the portal in their scope or the point of view. These and other problems, such as licensing and technical issues, involved in reusing existing ontologies have to be balanced against the benefits.

In the case of the health promotion portal TerveSuomi.fi, the information items of interest are web-accessible publications such as web pages and PDF documents, and they are described using Dublin Core[2] metadata such as `dc:subject` which contains the subject topic(s) of the given document. We decided to use the following ontologies[3] as indexing ontologies for the subject field: the Finnish General Upper Ontology (YSO)[4] [15] which is based on the General Finnish Thesaurus YSA[5] that is widely used in Finland for indexing contents of various kinds, Medical Subject Headings (MeSH)[6] and the European Multilingual Thesaurus on Health Promotion (HPMULTI)[7]. The reason for combining them (see also Table 1) was that none of them was alone adequate for describing the topics of the whole variety of content documents in the portal. YSO is broad but too general with regard to medical content. On the other hand, MeSH is too focused on clinical healthcare while HPMULTI has very narrow coverage, focusing exclusively on health promotion terminology.

Table 1. Core subject ontologies of the TerveSuomi.fi portal

Name	YSO	MeSH	HPMULTI
Publisher	National Library of Finland & FinnONTO	National Library of Medicine, USA	European Commission
# concepts	23 000	23 000	1 200
Languages	Finnish; Swedish and English under construction	English; Finnish and Swedish translations available	Multilingual, including Finnish, Swedish and English
Intended use	Cataloging of material published in Finland	Cataloging of biomedical documents	Cataloging of material on health promotion
Intended user group	Librarians	Medical professionals; librarians within the field of medicine	Professionals involved in health promotion
Examples of concepts	Travellers Water pipes Cities Vegetables	Metabolic Syndrome X Endocrine Disruptors Biopsy, Fine-Needle DNA Damage	Traffic accidents Behavioural change Voluntary work Sunburn

To prevent the creation of internal semantically incompatible islands within the portal, corresponding concepts in each ontology had to be mapped to each other. Mapping the ontologies was done using three complementary approaches:

---

[2] http://dublincore.org

[3] The term *ontology* is used in a broad sense, covering also thesauri in the sense that they are formal, explicit classifications for describing the content of documents.

[4] http://www.seco.tkk.fi/ontologies/yso/

[5] http://vesa.lib.helsinki.fi

[6] http://www.nlm.nih.gov/mesh/

[7] http://www.hpmulti.net

a) using available, existing mappings between MeSH and HPMULTI; b) creating automatic mappings between MeSH and YSO based on textual matching of concept labels; and c) manually mapping HPMULTI to YSO. The result was a interlinked combination ontology, where YSO provides the upper concepts and MeSH and HPMULTI the more exact concepts.

When analysing the ontologies, we noticed that these ontologies were created for use by professionals and their intended use is somewhat different from their use within the portal. This disparity between the points of view of the ontologies and the users of the portal manifests itself in many ways. First, the concept hierarchies are often inappropriate for the portal. MeSH e.g. uses deep hierarchies with complex subclassification criteria, and YSO contains many generic concepts that would probably only be confusing as facet categories. Second, concepts in professional classifications have typically been labelled using professional terminology instead of layman terms used by the end-users. On the other hand, there are many terms in everyday use that are not used by professionals due to their ambiguity. A portal must also be able to deal with queries based on such terminology even if it is not considered appropriate from a professional perspective. As an example of the problems involved, consider the MeSH top-level categories *Anthropology, Education, Sociology and Social Phenomena*, *Biological Sciences* and *Technology and Food and Beverages*. Such categories in a facet would not be very good starting points for a person looking for information about dieting. Another user might not realize that information about breast cancer can be found under the MeSH concept *Breast Neoplasms*.

4 Creating User-Centric Search Facets with Card Sorting

To solve the mismatch between the indexing ontologies and the expectations of the end-users, we propose using user-centric design practices to construct a custom classification system for the portal based on the users' expectations and their mental models, with the intent of later mapping it to the underlying ontological concepts to provide semantically sound faceted browsing and other functionalities.

A practical method for gathering information about the end-user's mental models of an information space, i.e., how users of a website tacitly group, sort and label tasks and content, is the *card sorthing* method [7,8,9]. A card sorting study is typically performed using index cards, with each card bearing the title and possibly a short description of an individual document. The study is then performed on volunteers that are asked to sort the cards into piles based on intuitive feeling of similarity or relateness of the given cards, and to give the piles descriptive names. This variation of card sorting where the categories (piles) are not given beforehand but are created by the participants is called *open card sorting* [9,16]. Card sorting doesn't directly give the designer a finished categorisation structure, but provides insight into the design choices for creating such a structure.

4.1 Selecting Card Contents

When using card sorting for creating user-centric facets for organizing ontologically indexed content, we propose using the ontological concepts as values of the index cards. To avoid overwhelming the participants of the study, only the most frequently used indexing concepts (based on a sample of indexed content) excluding overly general concepts should be used in the cards.

In our case, we created a list of all the concepts that occurred in our annotated content items (n=523). These 1722 concepts were then ranked by their frequency of occurrence. Concepts with only a few documents were pruned, as well as overly general concepts such as *health* and *health promotion*. Finally, concepts judged to be uninteresting or unnecessary from the point of view of the study were eliminated. These included, e.g., geographical locations, individual organizations, abstract concepts such as *Development* and *Promotion* as well as concepts that were considered very similar to others that were already on the list[8]. The pruning brought down the number of concepts to 177, which was deemed acceptable for the card sorting exercise. The labels of the concepts were printed on index cards together with numeric identifiers for ease of analysis.

4.2 Performing the Card Sorting

The study participants should be representative of the expected users of the system. Nielsen recommends using 15 participants [17] while Maurer & Warfel recommends seven to ten individuals [16]. Each participant is advised to group the cards into piles according to their meaning or topical similarity. Participants are asked to think aloud, especially when facing difficult decisions. During the exercise, the facilitator takes notes of important events and insightful comments made by participants during the experiment. If a pile becomes very large, the participants are instructed to split it into smaller parts. After sorting the cards into piles, they are asked to write down a descriptive label for each pile.

In our case, the card sorting study was performed on volunteers that were chosen to represent potential users of the system. A total of ten individuals of varying ages and backgrounds participated, with three of them doing the exercise as a group while the others performed the study alone. Thus, a total of eight rounds were performed. The raw data obtained during the card sorting study is a set of labeled piles of cards such as those shown in Figure 2.

4.3 Creating the Result Categories Based on the Card Piles

When analyzing card sort results, both qualitative and quantitative aspects need to be considered [9]. When card sorting is used for getting input into the design of website navigation, gaining insight from the data is of foremost importance; whether that requires a rigorous statistical analysis depends on the situation at hand. In many projects, simply "eyeballing" the data may provide enough

[8] E.g., only a sample of the dozens of food items such as *Meat* and *Cheese* were kept.

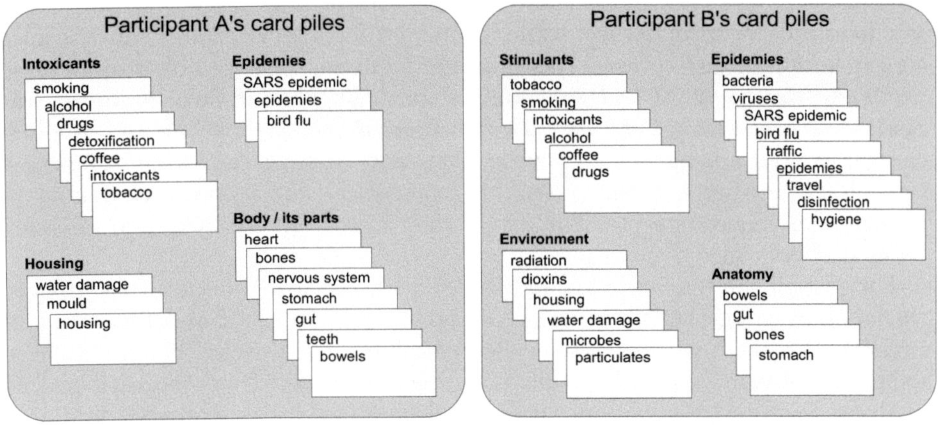

Fig. 2. Examples of card piles created by study participants

insight to create a workable design [8]. In our case, we used a spreadsheet template [18] to calculate some metrics such as card co-occurrence and the average number of cards in each category, but did not perform a full-fledged statistical cluster analysis. However, automatic tools have been created that perform cluster analysis and create tree diagrams that might be used directly as a basis for constructing web site navigation [19].

To create the result categories, we used the following processing steps: First, the categories created by individual participants were manually clustered to create a standardized set of categories for the purposes of analysis. As an example of the clustering process, the category *Body and its parts* created by one participant was considered the same as category *Anatomy* created by another participant, and these were both mapped to the standard category *Body part*. The clustering resulted in 29 standard categories and a mapping of each participant's categories to these. Sometimes similar categories created by a single participant were mapped to the same standard category for the purposes of analysis (e.g., *Body and its parts* and *Teeth* were both mapped to *Body part*), and not all standard categories were present in all user categories (e.g., not all participants had included a category for *Weight control*).

The second step of analysis was to enter the raw data about user-created categories and their contents (individual concepts) into the spreadsheet, using the above defined mappings.

The third step was to actually analyze the data, looking for patterns of interest. The analysis spreadsheet revealed, for example, that there was a high agreement about the existence of a category for body parts (all participants had included such a category) as well as the contents of that category. E.g., all participants had placed the concepts *Stomach* and *Skeletal system* in that category. On the other hand, while three participants had included a category for well-being, there was low agreement about the contents of that category. The interpretation for these results was that participants (and, by extension, users of the portal) have a clear mental model of body parts as a category distinct from, say, food and nutrition issues, while a category for lifestyle doesn't invoke such a clear notion of distinctness. Based on the analysis, we were able to pick good candidates for top-level facets as well as construct part of their contents. The analysis revealed that *Body part*, *Group of people* and *Life event* were popular categories. Perhaps more importantly, they were at least somewhat orthogonal towards each other and the rest of the categories, so they were chosen to be presented as separate facets. The rest of the categories were then used to create a fourth facet called *Topic*.

For each of the remaining categories, we had to decide whether to a) discard the category altogether (in cases of low agreement), b) use it as a top-level category or c) place it below a top-level category in the hierarchy. The hierarchical relations between categories could not directly be seen from the card sorting analysis. However, hints about these could be found in the notes made during the card sorting sessions, e.g., situations where a participant had split a large pile into smaller components.

Some of the discarded categories included *Oversensitivities*, *Disease prevention and self-help* and *Health problems*. The resulting hierarchy is quite shallow; additional levels may need to be added using other methods such as *laddering* [20]. However, for the purposes of our portal we have expanded the hierarchy simply by examining the underlying ontologies and building up the hierarchy by mirroring their structure, while trying to make sure that the terminology and groupings are suitable for end-users. We feel that while the design of lower hierarchy levels could benefit from user-centric design methods, the issues here are not as critical as the choice of facets and their topmost categories.

4.4 Finalizing and Evaluating the Categorisation

When an initial version of the facets is created using the card sorting method described above, the result should be evaluated and possibly reviewed both by additional user testing and by domain experts. One way for evaluating the result categories with users is to do new rounds of card sorting using the closed card sort method where the categories are given beforehand and the study participants are asked to sort the cards into those piles. The intuitiveness of the categories can be estimated based on how well the results of the closed card sorting matches the initial facets.

In our case, the user testing of the *Topic* facet (see Figure 3) using the closed card sorting method was done with two volunteers who were asked to sort the ontological concepts into the suggested top-level categories. The results were encouraging: participants placed nearly all concepts in the category that was intended by the designer. More user evaluations would probably need to be done to find subtler errors.

Additionally, an expert review of the initial facets was done by health promotion experts, which revealed some problems. The category *Catastrophes & Epidemies* was considered problematic: the two concepts are not very closely related and lumping them together may send false signals to users of the portal.

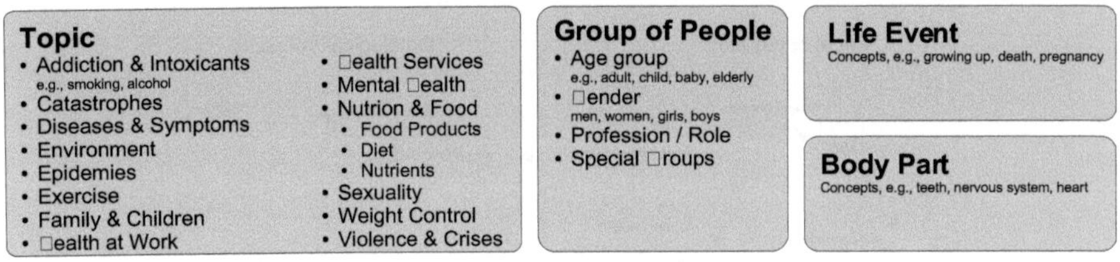

Fig. 3. The finalized end-user facets with some examples of the concepts

A new look into the analysis process revealed that these two concepts had been paired together during the clustering phase, and in fact only one user had created a category where both aspects were present – a clear mistake in the clustering. Thus, separating the two aspects into their own top-level categories was an easy decision. Another problem discovered by domain experts was that there was no category for issues related to occupational health, such as the hazards of dangerous chemicals used at work. Such concepts were not very well represented in the set used for the card sort, possibly because the initial set of documents lacked documents specific to occupational health. A new top-level category for occupational health issues was created, with subcategories taken from ontologies as well as classifications used on existing websites on the topic. The finalized facets are presented in Figure 3.

More generally, the lesson learned was that skewed initial data and errors during the analysis may cause subtle errors in the hierarchy. However, using user evaluations and expert reviews helps alleviate at least some of the problems.

5 Mapping User-Centric Facets to Ontologies

When the initial user-centric facets have been created using the method above, the facets should be logically mapped to the ontological facets as described in section 2. Since the card sorting is done using a selection of typically used ontological concepts, and since the relations between the standard categories and these concepts are known, these relations can be directly used as mappings between the facets and the facet ontologies. To make the mapping comprehensive, additional concepts are needed, which must be added manually. E.g., if the concept *Food* was used in the card sorting but the concept *Nutrition* was not, the latter might be relevant also for a facet category of *Nutrion and Food.*

We have decided to represent the facet hierarchies described in the previous chapter in RDF using the SKOS Core vocabulary[9] and their connections to the underlying ontologies using the SKOS Mapping vocabulary[10]. Each facet is described as a `skos:ConceptScheme` and each facet category is represented as a `skos:Concept` with a human-understandable `skos:prefLabel`. The facet

[9] http://www.w3.org/2004/02/skos/core/ (URI prefix `skos`)

[10] http://www.w3.org/2004/02/skos/mapping/ (URI prefix `skosmap`)

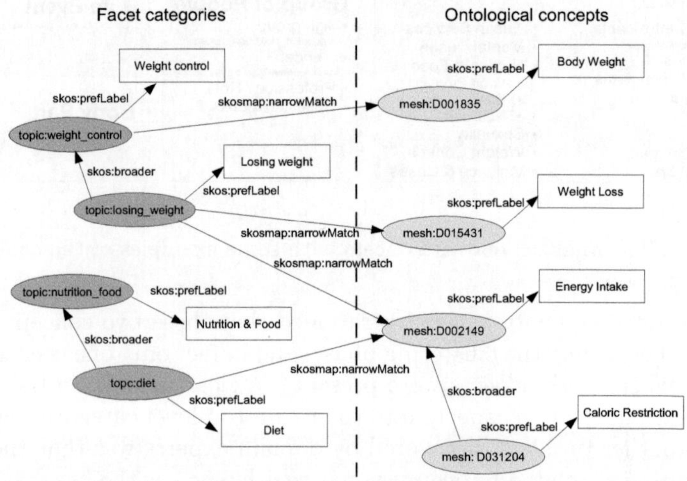

Fig. 4. Examples of mappings between facet and ontology concepts. The URI prefix `topic` refers to the *Topic* facet and `mesh` to the indexing ontology MeSH.

hierarchies are represented using `skos:broader` and `skos:narrower` relationships. Mappings to the underlying ontological concepts are represented using `skosmap:narrowMatch`. This mapping is a subset of the mappings described in Section 2; such more complex mappings can also be expressed using the SKOS Mapping vocabulary and will likely be used in the future. An example of mappings between facet categories and ontological concepts is shown in Figure 4.

This representation implies, by the SKOS inference rules, that a category within a facet contains (is the subject of) all documents that are annotated with one of the ontological concepts that the category has been mapped to. In addition, the category subsumes its child categories, and the ontology concept is the subject of all its narrower concepts. Thus, in Figure 4, the category `topic:weight_control` will contain all documents indexed against any of the MeSH concepts in the figure due to the `skos:broader` and `skosmap:narrowMatch` relations present.

6 Prototype Implementation

As a proof of concept, the methods discussed have been implemented in the prototype of TerveSuomi.fi[11] where the faceted search functionality has been created using the faceted search engine Ontogator [13]. Figure 5 shows the user interface, where the user has selected the category *Diet* from the *Topic* facet and the category *Pregnancy* from the *Life event facet.* The result of this faceted search query is the list of links to web pages. The user could now either visit

[11] http://www.seco.tkk.fi/applications/tervesuomi/

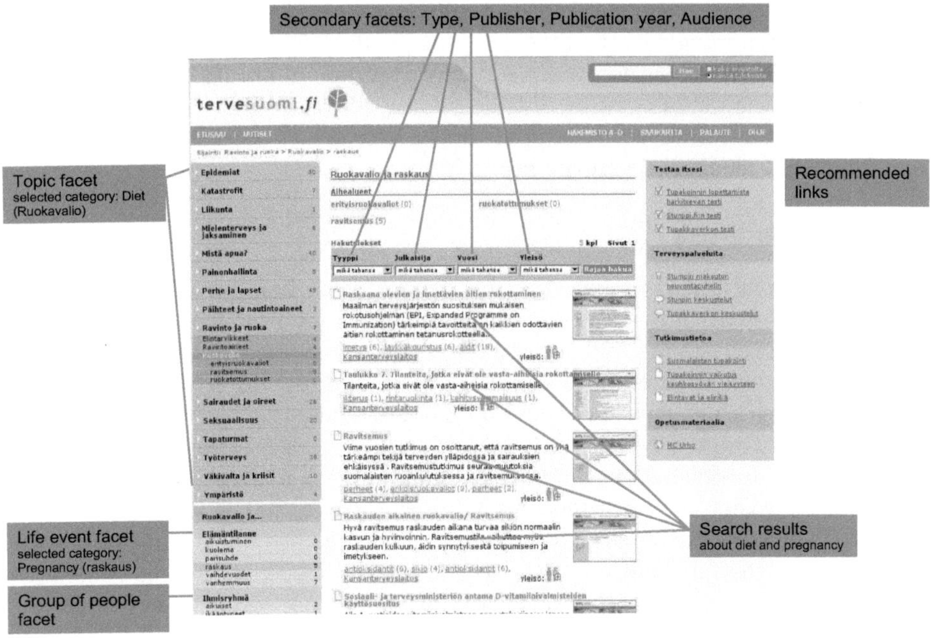

Fig. 5. TerveSuomi.fi portal user interface

some of the resulted web pages or modify the query by selecting, e.g., additional facets or by clicking on context based semantical recommend links on the right.

7 Discussion

This paper argued that card-sorting combined with mappings provides a promising approach for designing and implementing semantic view-based search based on user-centric facets.

7.1 Contributions

The main benefits of separating end-user facets from content indexing ontologies are: First, more intuitive and useful user interfaces can be provided. Second, the same ontologically annotated metadata can be re-used for different use cases and interfaces without changing the metadata or the content by defining new alternative user-centric facets and mappings. This flexibility would not be achieved if the metadata were described using application-specific or user-centric categorizations directly. For example, the same metadata could be used to create both a professional facet and a citizen's facet to the same content, where the professional facet is more directly based on the indexing ontologies and the citizen facets more on the various information needs of ordinary life.

The downside of using user-centric facets is the extra work needed in creating them and in mapping search categories onto annotation ontologies. Also, if the card sorting is based on non-representative example annotations, the resulting user-centric facets might not be optimally designed when more content is added to the portal. Therefore, readjustments to the user-centric facets might be needed based on, e.g., feedback from the users.

7.2 Related Work

In earlier semantic portals based on the faceted browsing paradigm, the facets have been automatically created from the underlying ontological hierarchies using projection rules (e.g. [12]). A distinction can be made between systems where the ontologies are created to become facets in the user interface and systems that use pre-existing general purpose ontologies. The former group includes MuseumFinland [5] and SWED[12], whereas /facet[13] [6] is an example of the latter approach. The problems of matching the hierarchical structure of the ontology with user needs and expectations only become apparent in the latter case, as the point of view of the original ontology may differ a lot from the end-users' mental models of the information space. In /facet, the automated facet generation sometimes results in a user interface that is hard to use [6].

Another approach for creating a navigational hierarchy based on an ontology is presented by Stoica & Hearst [21,22]. Their system uses the WordNet lexical ontology as a basis for creating a hierarchical classification which can then be used in faceted browsing. The Castanet algorithm simplifies the WordNet IS-A hierarchy by eliminating branches that aren't represented in the document collection as well as unnecessary levels of the hierarchy. The resulting taxonomies can be used either as-is or after some manual adjustments. However, the relationship of Stoica & Hearst's work with ontological metadata is weak: WordNet is only used as a basis for creating the navigational hierarchies, and the document metadata is later assumed to reference the newly created taxonomy directly.

Card sorting has been previously used in the construction of ontologies as a means of knowledge elicitation. While card sorting is usually performed manually outside the ontology engineering process, a computerized card sorting plugin has been developed for the Protégé[14] ontology editor [23]. However, the focus of this work is on the ontology creation process itself; there is no direct intent of using the resulting ontology in a search-oriented user interface.

7.3 Future Work

We are currently implementing a more finalized prototype of the semantic portal TerveSuomi.fi. After this, user tests should be done to evaluate the prototype and the underlying hypotheses such as the end-user-centric facets. We are currently also investigating how ontologies could be used to model health care services

[12] http://www.swed.org.uk

[13] http://slashfacet.semanticweb.org

[14] http://protege.stanford.edu

using methods presented in [24]. In the future, the portal may be extended to incorporate access to personal medical records and health care services.

Acknowledgements

We thank Markus Holi and Petri Lindgren for their input to the work reported in this paper. This work is part of the FinnONTO project[15], funded mainly by the National Funding Agency for Technology Innovation (Tekes) and the Ministry of Social Affairs and Health. The TerveSuomi.fi project is co-ordinated by the National Public Health Institute in Finland (Project Coordinator Eija Hukka).

References

1. Pollit, A.S.: The key role of classification and indexing in view-based searching. Technical report, University of Huddersfield, UK (1998) http://www.ifla.org/IV/ifla63/63polst.pdf.
2. Hearst, M., Elliott, A., English, J., Sinha, R., Swearingen, K., Lee, K.P.: Finding the flow in web site search. CACM **45**(9) (2002) 42–49
3. Hyvönen, E., Saarela, S., Viljanen, K.: Application of ontology techniques to view-based semantic search and browsing. In: The Semantic Web: Research and Applications. Proc. of the 1st European Semantic Web Symposium (ESWS 2004). (2004)
4. Mäkelä, E., Hyvönen, E., Sidoroff, T.: View-based user interfaces for information retrieval on the semantic web. In: Proceedings of the ISWC-2005 Workshop End User Semantic Web Interaction. (Nov 2005)
5. Hyvönen, E., Mäkelä, E., Salminen, M., Valo, A., Viljanen, K., Saarela, S., Junnila, M., Kettula, S.: Museumfinland – finnish museums on the semantic web. Journal of Web Semantics **3**(2) (2005) 25
6. Hildebrand, M., van Ossenbruggen, J., Hardman, L.: /facet: A browser for heterogeneous semantic web repositories. In Cruz, I.F., Decker, S., Allemang, D., Preist, C., Schwabe, D., Mika, P., Uschold, M., Aroyo, L., eds.: International Semantic Web Conference. Volume 4273 of Lecture Notes in Computer Science., Springer (2006) 272–285
7. Rugg, G., McGeorge, P.: The sorting techniques: a tutorial paper on card sorts, picture sorts and item sorts. Expert Systems **14**(2) (1997) 80–93
8. Nielsen, J., Sano, D.: Sunweb: User interface design for sun microsystem's internal web. In: Proceedings of the 2nd World Wide Web Conference, Chicago, IL. (Oct 17-20 1994) 547–557
9. Rosenfeld, L., Morville, P.: Information Architecture for the World Wide Web. second edn. O'Reilly (2002)
10. Holi, M., Hyvönen, E.: Fuzzy view-based semantic search. In: Proceedings of the 1st Asian Semantic Web Conference (ASWC2006), Beijing, China, Springer-Verlag (September 3-7 2006)
11. Holi, M., Hyvönen, E., Lindgren, P.: Integrating tf-idf weighting with fuzzy view-based search. In: Proceedings of the ECAI Workshop on Text-Based Information Retrieval (TIR-06). (Aug 2006)

[15] http://www.seco.tkk.fi/projects/finnonto/

12. Viljanen, K., Känsälä, T., Hyvönen, E., Mäkelä, E.: Ontodella - a projection and linking service for semantic web applications. In: Proceedings of the 17th International Conference on Database and Expert Systems Applications (DEXA 2006), Krakow, Poland, IEEE (September 4-8 2006) 370–376
13. Mäkelä, E., Hyvönen, E., Saarela, S.: Ontogator — a semantic view-based search engine service for web applications. In: Proceedings of the 5th International Semantic Web Conference (ISWC 2006). (Nov 2006)
14. Berners-Lee, T., Hendler, J., Lassila, O.: The semantic web. Scientific American **284**(5) (May 2001) 34–43
15. Hyvönen, E., Valo, A., Komulainen, V., Seppälä, K., Kauppinen, T., Ruotsalo, T., Salminen, M., Ylisalmi, A.: Finnish national ontologies for the semantic web - towards a content and service infrastructure. In: Proceedings of International Conference on Dublin Core and Metadata Applications (DC 2005). (Nov 2005)
16. Maurer, D., Warfel, T.: Card sorting: a definitive guide. Boxes and Arrows (Apr 7 2003) http://boxesandarrows.com/S1937.
17. Nielsen, J.: Card sorting: How many users to test (Jul 19 2004) Alertbox column, http://www.useit.com/alertbox/20040719.html.
18. Lamantia, J.: Analyzing card sort results with a spreadsheet template. Boxes and Arrows (Aug 26 2003) http://boxesandarrows.com/S1708.
19. Dong, J., Martin, S., Waldo, P.: A user input and analysis tool for information architecture. In: CHI'01 extended abstracts on Human factors in computing systems. (March 31 - Apr 05 2001)
20. Rugg, G., Malcolm, E., Mahmood, A., Rehman, N., Andrews, S., Davies, S.: Eliciting information about organizational culture via laddering. Journal of Information Systems **12**(3) (2002) 215–230
21. Stoica, E., Hearst, M.: Nearly-automated metadata hierarchy creation. In: Proceedings of HLY-NAACL'04, Boston. (May 2004)
22. Stoica, E., Hearst, M.: Demonstration: Using wordnet to build hierarchical facet categories. In: Proceedings of the International ACM SIGIR Workshop on Faceted Search, Seattle, WA. (Aug 2006)
23. Wang, Y., Sure, Y., Stevens, R., Rector, A.: Knowledge elicitation plug-in for Protégé: Card sorting and laddering. In: Proceedings of the 1st Asian Semantic Web Conference, Beijing, China. (Sep 3-7 2006)
24. Laukkanen, M., Viljanen, K., Apiola, M., Lindgren, P., Hyvönen, E.: Towards ontology-based yellow page services. In: Proceedings of WWW2004 Workshop, Application Design, Development, and Implementation Issues. (May 2004)

An Approach for Identification of User's Intentions During the Navigation in Semantic Websites

Rafael Liberato Roberto and Sérgio Roberto P. da Silva

Universidade Estadual de Maringá, Av. Colombo 5790, zona 07,
Maringá – PR – Brasil
{liberato,srsilva}@din.uem.br

Abstract. The growing need for content customization in websites has fostered the development of systems which try to identify the user's navigation patterns. These may be, normally, identified by means of log file analysis. However, this solution does not identify the semantic intention behind user's navigation. This paper provides an approach to incorporating semantic knowledge to the process of identifying the user's intentions in the navigation of a website with semantic support. The capture of the user's intentions is achieved by the semantic enrichment of the log files and the use of and approach that takes into account the linguistic and cognitive aspects in the development of the user model.

Keywords: User Model, Semantic Web, Web Personalization.

1 Introduction

A user's website navigation is strongly related to his interests and necessities. However, most of today's websites do not take this into account. A solution to this problem would use of personalization mechanisms. A personalized website may include new index pages, provide personalized search results, dynamically create recommendations (such as new links), or even define new layouts for a webpage. Several projects have come out aiming to improve the user's interaction with the website such as, for instance, Letizia [14], WebMate [5], PersonalWebWatcher [17] and OBIWAM [11], which construct a user model analyzing the webpages visited by the user and making adaptations on the pages he has visited.

Today's personalization mechanisms, in general, utilize a navigational behavior analyses to create a user model [4] [10] [13], extracting behavioral patterns by means of machine learning techniques. These models can be represented in several different ways. Letizia [14] produce a list of webpages, while QuickStep [16] creates a list of concepts of interests, and Persona [21] creates ontologies. It is interesting to note that these models are constructed by using a great variety of learning techniques such as, vectorial space models [5], probabilistic models [17] or clustering [18]. However, as seen in [17], the exclusive use of the navigation data can be problematic, because there may be difficulties when there is not enough data to extract patterns related to certain categories of the domain, or when new webpages, which have not been visited already by the users yet, are added to the website.

E. Franconi, M. Kifer, and W. May (Eds.): ESWC 2007, LNCS 4519, pp. 371–383, 2007.

The incorporation of information related to the content of the webpages or the website's structure, i.e., data concerning its semantic relations, provides a way of overcoming the problems mentioned above, improving the personalization process [8] [10]. A common approach in this context is the integration of the characteristics of the webpages contents with a classification defined by the users [6] [19]. Generally, in this approach, keywords are extracted from the website's content and are used to index or classify its webpages in several categories of content. Thus, these approaches would permit the recommendation engines to indicate pages to a user based not only in the similarity among the users' navigation patterns, but (or alternatively) on the similarity of the content of the webpages as well. Some projects have adopted the integration of the similarity of the webpages' content to enrich the process of user model construction [6] [19] [9]. SEWeP [9], for example, has as main characteristic the creation of C-logs (concept-logs), a semantically enriched log file based on the extraction of keywords of the webpages. After they have been found, the keywords are mapped to the concept of a taxonomy created to model the domain concepts. Each register of log file is improved with the concepts representing the semantic of the respective URI. However, even these systems may not be able to capture more complex relations among the information as, for example, relations originating from a deeper semantic level, which are based on attributes and properties of the concepts involved. To do so, a richer model is necessary, one which is able to represent the semantically richer relations as, for example, a domain ontology.

In the context of the semantic web [3], webpages must be understood not only by people, but also by the machines, in the form of computational agents. One way of making this idea viable is through the usage of ontologies associated to the websites [12]. These ontologies would permit the software agents to reason about the relations of the websites' content and, by doing so, to make it possible to improve the attention given to the necessities of the user.

This article suggests an approach to associate the benefits provided by the semantic structuring offered by the semantic web proposal, by means of ontologies associated to the websites, along with the analyses of the user's navigational behaviors, creating what is called a semantic log. In this way, it proposes a process of user model creation based on the identification of possible interests of a user when s/he is navigating a website with semantic support. This process is based on the analyses of a semantic log, using the domain ontology available for the website. To do so, an algorithm was developed in order to classify the user's intentions. This algorithm is based on the idea that all the concepts involved on the website are candidates to be the user's intentions. To determine the real relevancy of each concept, a set of parameters was defined to ponders the influence of the linguistic control over the user's power of expression in the form of a segmentation of the domain ontology, and the influence of a concept in the user's interaction applying the idea of cognitive force, derived from the theory of spreading activation [1] [2], and developed under the cognitive psychology to explain how human memory works.

This article is organized as follows. In section 2 we describe the algorithm responsible for identifying the possible user's interests. In section 3 we discusses a comparative analyses done by means of a simulation of the proposed algorithm with algorithms based on the frequency and the classic algorithm of the Naïve Bayes classifier. Finally, section 4 describes the results and our conclusions.

2 The User Modeling

The creation a user model is always a very complex task, for the set of parameters to be considered is always large. The choice of these parameters along with the machine learning techniques which are applied makes the difference in the quality of the model.

To define which parameters should be used in our project we take into account the linguistic and cognitive aspects which affect the users in the expression of their interests. In this way, we are initially going to discuss the effect that the vocabulary available to the user has on his form of expression. Then, we are going to discuss the effect that the semantic relationship among the concepts has on the human memory and how that can affect the user's expression and the identification of his intention when navigating a website.

2.1 The Linguistic Aspects That Affect the User Model

A domain ontology consists of a set of concepts and relationships that describe the knowledge about a domain of interest. It can also be seen as a vocabulary definer for a controlled language that permits the users to express themselves about the domain.

However, when developing a software application, be it for the desktop or for the web, not all the concepts present in the ontology domain needs to be involved. This partition of domain ontology, which we can call the **application model,** has indirect influence on the expression of the user's interests about the domain.

In the case of a website, the problem goes further, for each webpage uses only a part of the knowledge contained in the application ontology, limiting even more the user's power of expression. To reflect this new reduction in the user's vocabulary, we use a **presentation model**, which represents the segmentation of the application model, which is presented in each webpage. Thus, the partition of the knowledge for a website is structured as illustrated on Fig. 1.

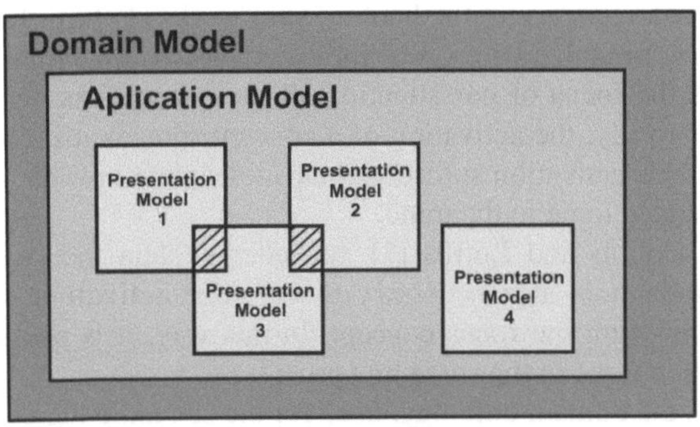

Fig. 1. Segmentation of knowledge in a web application

It is important to emphasize that, in this project, the presentation model is related, mainly, to the information content present in the webpages. It does not take into account the effect that the esthetic presentation has on the user's attention and that can also direct his intentions. The introduction of the presentation model tries to contemplate the fact that a person can only express something for which he has a vocabulary. This fact has an indirect influence on the creation of a model of the user's interests, but that should be considered. Thus, the presentation model is necessary to model the limitations imposed by the language to the user's power of expression.

Trying to consider, partially, the effect that the layout has on the presentation model, we have defined a **status (S)** parameter, in which we isolate three main components in a webpage, based on its level of prominence, which are: the "Main Menu", the "Secondary Menu" and the "Body" of the webpage, as illustrated in Fig. 2. As each component possesses a level of importance in the webpage, drawing the user's attention in different forms, we attribute differentiated weights for the concepts involved to each component.

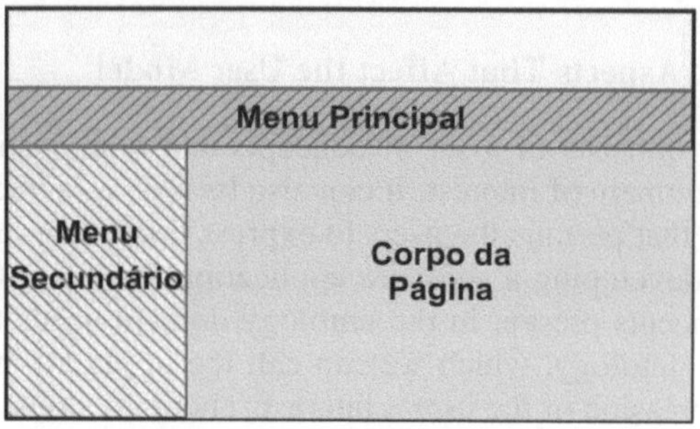

Fig. 2. Parts of the page with different status

2.2 The Cognitive Aspects That Affect the User Model

One of the dominant theories to explain the semantic processing in the cognitive psychology is known as spreading activation [1] [2] [20]. This theory tries to explain how the information recovery of the human brain works [1] [20]. It considers that the human memory is organized in a semantic network form, proposing that when a concept becomes the **focus** of our attention all the concepts associated to it are also activated. That is to say, the activation of a concept spreads itself to all the concepts associated to it. This activation spreading helps to explain how the remembrance of a topic can bring related topic to the mind.

According to Collins and Loftus [7], to better explain the cognitive process of spreading activation it is also necessary to consider **activation strength** for each existent association with the focus concept. In this way, it is possible to amplify or reduce the cognitive force in the spreading process.

In this project, the domain ontology, used for the semantic log construction as well as the models of application and presentation, is represented by a semantic network expressed in the OWL language [15]. If we considerer that by choosing a link in a semantic website page the user will be activating a concept (the **focus concept**) in the semantic network which composes it, it is very reasonable to apply the concept of cognitive strength and spreading activation to evaluate the real interest of the user.

2.3 Linking Navigational Patterns and Semantic Content

Aiming at aiding the user's intentions modeling, in the development of semantic website pages we have adopted a strategy to monitor in a transparent way, the user's interactions with the website, creating a proper log file. Thus, each website page visited inserts a register in the log file storing the date and time of access, and its address. However, in the context of websites personalization, as seen in [17], using data originating from the user's navigation can bring difficulties, especially when there is not enough data to extract patterns related to certain domain categories, or when new website pages are added which have not been visited by the user yet. In this way, due to the fact that all information on the website pages we are interested are based on the domain ontology, the concepts involved in the information of the pages visited, also are inserted in the log file, creating what we call a **semantic log**. Thus, the construction of the semantic log associates the semantic information of website's content with the user's navigation behavior.

It is important to emphasize that, in the scope of this project; we will be limiting our discussion to websites constituted of intranets portals which have been constructed with technology for semantic support since the beginning. We know that the treatment of the websites constructed with modern technology is very relevant, but it will be the target of this project at the moment.

2.4 An Identification Algorithm of a User's Interests

The algorithm proposed here takes into consideration the linguistics and cognitive aspects that influence the process user model creation. Thus, it uses the presentation model and the idea of spreading activation, considering the cognitive strength of each concept, in order to select the concept that expresses the present user's interest.

The algorithm has as entrance the *semantic log* and the presentation model of the page that the user is in. The presentation model has two functions. The first is taking into account the linguistic limitation imposed by the content presented on the webpage. As the concepts present in it are more strongly activated in the user's memory, focusing its scope of expression, they are pondered more strongly than the other concepts of application model, when determining the probability of a concept expresses the user's intention. The second is to take into account the cognitive aspect of the spreading activation in the human memory. The cognitive strength increases or decreases the strength of the activation of the concept in the spreading process. To determine this cognitive strength for each concept is necessary to determine the following parameters: ***R***, which define the number of relations that the concept in question possesses, and ***S***, that defines the status of the concept in the presentation model, as mentioned before in section 2.1. The relations that a concept possesses were classified in two groups: $\#\boldsymbol{R}_{aplic}$ provides the number of relations with other concepts in the application model (concepts that do not appear in the presentation model), $\#\boldsymbol{R}_{apress}$ provides the number of relations with other concepts of the presentation model. Thus, the parameter ***R*** is defined by the formula $\boldsymbol{R}=\mathbf{P}_{\mathbf{aplic}}\ \#(\mathbf{R}_{\mathbf{aplic}}).+\ \mathbf{P}_{\mathbf{apres}}\ \#(\mathbf{R}_{\mathbf{apres}})$, were $\mathbf{p}_{\mathbf{x}}$ is a specific weight for each type of relation, respecting the first function of the presentation model. The parameter ***S*** is defined by the position in which the concept occupies in the webpage layout, as seen in Fig. 2. The cognitive strength of a concept is defined by the formula $\boldsymbol{F = S * R}$.

Fig. 3 illustrates the parameters that must be determined for any concept of presentation model during the process of spreading activation. Besides the cognitive strength, differentiated weights were defined for three different types of concepts: the **focus concept** – associated to the *link* chosen by the user and that, therefore, has a high possibility of being his real interest, has a higher value; the **directly connected concepts** to the focus concept – which have a good probability of being the user's real interest, have a value dependent on its cognitive strength, as previously defined; and the **disconnected concepts** of the focus concept, which are not involved in the spreading process and are considered **noise** (once they do not share the user's attention) have a residual value, since they must not be ignored, as it will be shown in the Section 3.

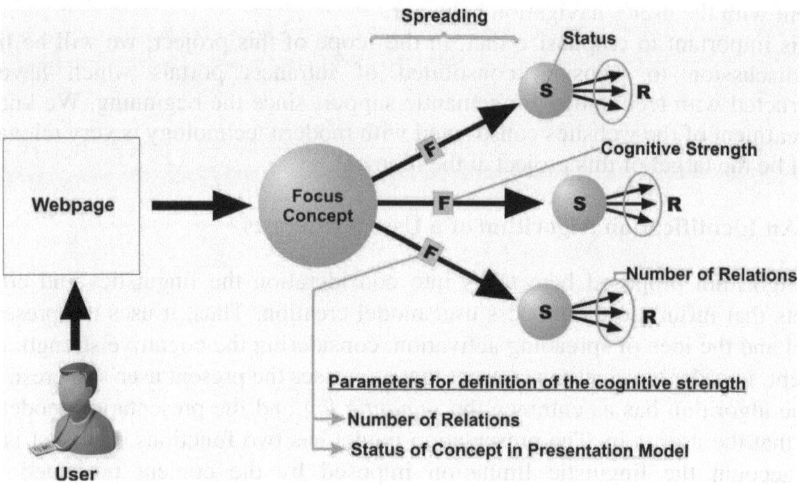

Fig. 3. Parameters in the spreading process

Thus, at each user's interaction, represented by a register of the semantic log, the algorithm identifies the focus concept, the connected concept and the noises in the presentation and application mode and attributes the weights correspondent to each of them and defines the probabilities of interest of each concept. This probability is defined by the formula:

$$P(C_i \mid I_1 \ldots I_j) = \sum_{k=1}^{j} (Tc_i)_k + (F_i)_k \tag{1}$$

where:

- P = probability of user's interest to a determined concept;
- C = the concept in question among all the domain ontology concepts;
- I = the present interaction in the semantic log;

- Tc = the value correspondent to the type of concept (Focus Concept, Directly Connected Focus and Disconnected Concept);
- F = the cognitive strength of concept ($F = 0$, if the type of concept is a Disconnected Concept).

To obtain the concept that represents the greater interest the following formula is used:

$$\arg\max_i (P(C_i \mid I_1 ... I_j)) \tag{2}$$

where $\boldsymbol{I}$ varies for all the concepts and $\boldsymbol{j}$ is the total of interactions. In this way, the algorithm keeps an updated list of the concepts with their respective probabilities of interest.

3 An Evaluation of the Proposed Algorithm

In this section we present an experimental study based on simulated empiric tests, in which navigations are defined with pre-defined interests. The objective of these tests is to verify the validity of the approach of semantic knowledge integration to navigation data in the identification process of the user's interests, considering the linguistics and cognitive aspects of the process of user model creation.

To compare our results, tests were also performed using classification algorithms based on frequency and on a Bayesian approach. It is important to emphasize that the parameters used on the proposed approach were defined in an empiric form. However, as future work, we intended to develop a learning algorithm to identify their ideal values.

Thus, in the Bayesian approach, we have used the Naïve Bayes Classifier algorithm, using just positive learning examples, since it is not possible to obtain negative examples in this case without an explicit intervention of the user. The positive examples were extracted by means of the simulated semantic log, in which each entrance was considered as a positive example. In the classification approach based on frequency, just the frequency of visits to the webpage is considered. However, as the frequency is obtained by means of semantic log, it will be considered in fact the frequency of concepts involved on the webpage.

For the execution of the tests a semantic website was developed for the Computer Department of the State University of Maringá, in which the navigations were applied. For each navigation, one hundred interactions with their respective interest concepts were defined. As the result of navigating in the semantic website, a semantic log was generated and departing from it, and from the presentation model, the three approach previously mentioned were applied.

In the tests, each navigation was defined to characterize a determined interest. As an example, the objective of one of the pre-defined navigation was to search for "publications" and "events" in which some "professors" are involved. To find the desired professor it is necessary to access the research projects in which s/he takes part. When the professor is found, her/his publications as well the events in which s/he took part can be visualized. In this example, the greatest concentration of accesses will be found in the process of searching the professor, that is to say, most of

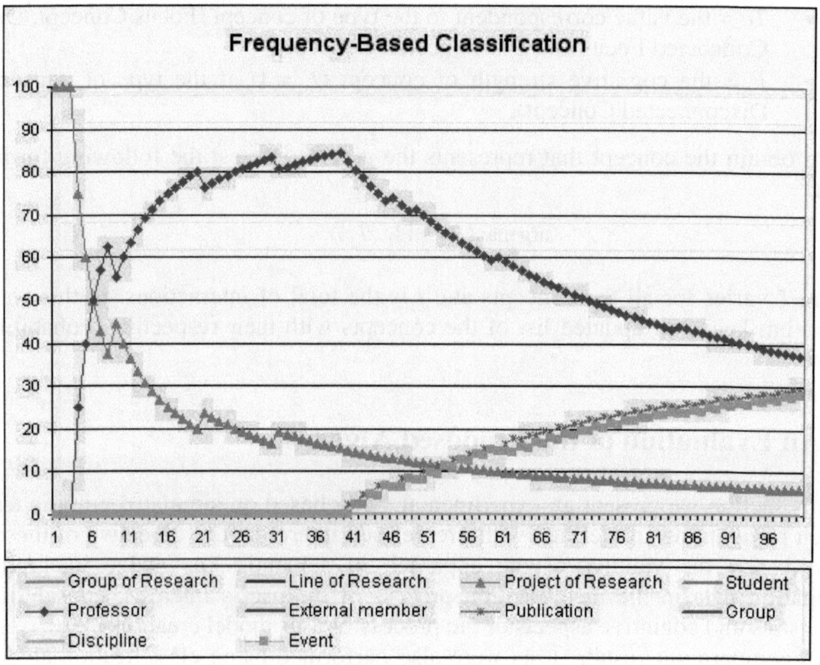

Fig. 4. Interests Percentages progress in the navigation for the algorithm based on frequency

the accessed pages belong to the professors. In this way, the concepts pre-defined as of greater navigation interest were "Professor", "Publication" and "Event".

The Bayesian classification approach and the frequency-based approach, presented similar results. In the frequency-based classification, the concepts identified as of greater interest are the concepts involved on the most accessed webpages. Thus, initially we had a very high value (100%) for the concept "Research Project" and a null value for the others, with a better convergence for the real value with the increase of interactions number, as illustrated in **Fig. 4**. In the Bayesian classification, this total preference for just one concept is corrected, as illustrated in **Fig. 5**, for the Naïve Bayes Classifier considers a similar probability for all concepts in the beginning, but with the increase in the interactions number its classification converge to the same result.

Thus, both approach identified the same pre-defined concept for the navigation, that is: "Professor", "Publication" and "Event", as being the ones of the user's greatest interests. The "Professor" concept becomes the user's greatest interest. This happens due to the high access to the professors' pages. On the other hand, the frequency-based classification do not take into account all the concepts involved in the webpage, which were not accessed, attributing a null value to the percentage of interest. The same happens in the Bayesian classification, however it attributes a very small value to the percentage of interest, as illustrated in Fig. 6.

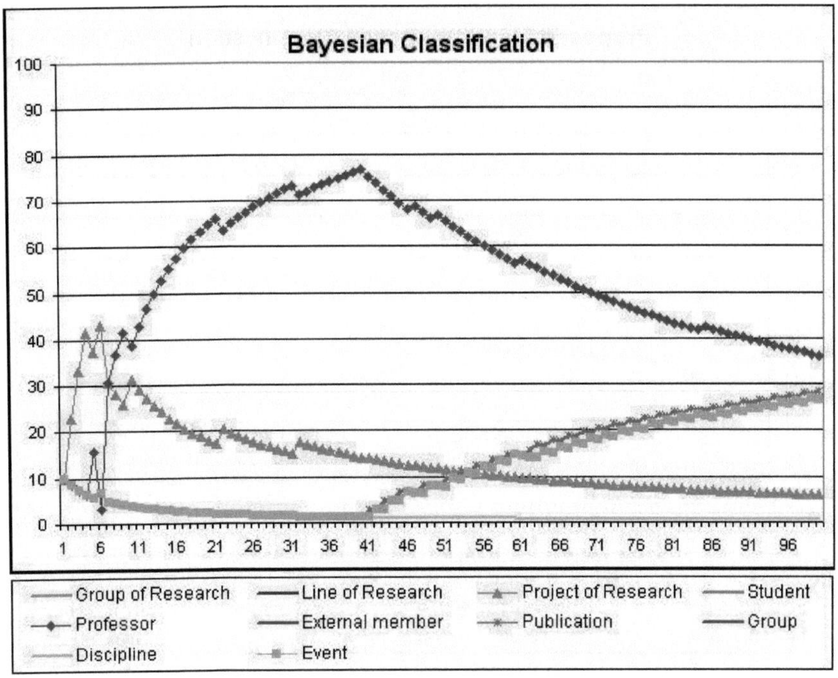

Fig. 5. Interest Percentage Progress in the navigation for the Bayesian algorithm

Bayesian Classification		Frequency-based Classification		Proposed Algorithm	
Concepts		**Concepts**		**Concepts**	
Professor	- 36,22%	Professor	- 37%	Publication	- 22,54%
Publication	- 28,40%	Publication	- 29%	Professor	- 20,54%
Event	- 26,45%	Event	- 28%	Event	- 14,27%
Project of Research	- 5,99%	Project of Research	- 06%	Project of Research	- 10,89%
Group	- 0,49%	Group	- 0%	Group of Research	- 6,32%
Discipline	- 0,49%	Discipline	- 0%	Student	- 5,76%
Line of Research	- 0,49%	Line of Research	- 0%	Group	- 5,70%
Student	- 0,49%	Student	- 0%	External Member	- 5,62%
External Member	- 0,49%	External Member	- 0%	Line of Research	- 4,30%
Group of Research	- 0,49%	Group of Research	- 0%	Discipline	- 4,05%

Fig. 6. Preferences resulted from the Bayesian, the Frequency-based and the Proposed algorithm

One characteristic, which we consider negative in these two approaches, is that they do not take into account the concepts involved on the webpages that were not accessed yet. These concepts have a great semantic relation to the present user's interest concepts and, due to that semantic proximity, can be good candidates to become the user's interest focus in a posterior navigation.

Our algorithm also identified as main user's interest the concepts that were predefined for the navigation, as illustrated in **Fig. 7**, however we have found some interesting differences.

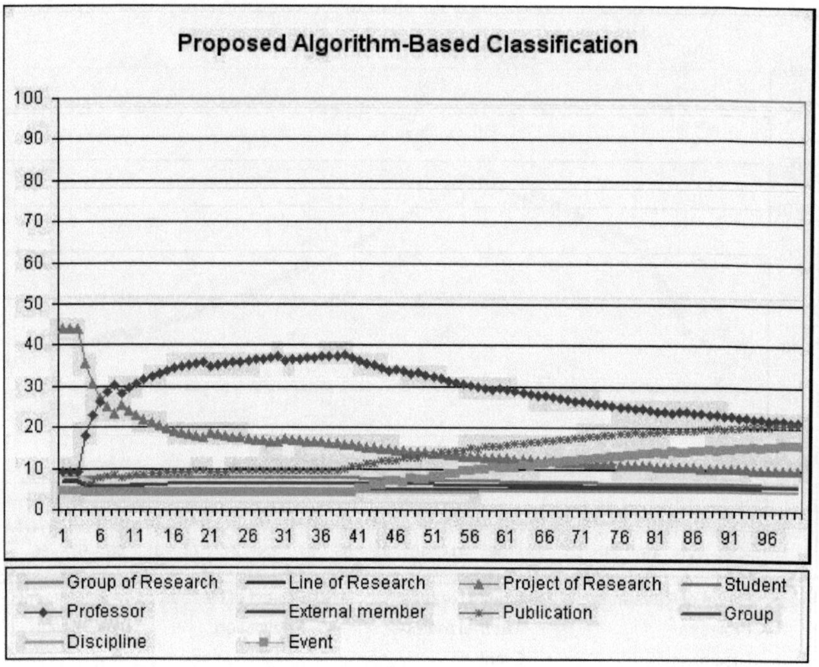

Fig. 7. Progress in the percentage of interest in the navigation when applied to the proposed algorithm

First, the "Publication" concept was identified as being of the user's greater interest, even having the less access number in the navigation. This fact is very relevant, considering that during the navigation, the process of searching the professor is secondary, for s/he is only the way to find the "Publications" and "Events" of the desired professor. In this way, taking in to account the linguistics and cognitive aspects of the process of user model creation, the proposed algorithm was able to identify this semantic relation and ponder on the "Publication" concept (according to its cognitive strength) each time that the "Professor" concept was accessed. As can be seen in **Fig. 6**, the difference between the percentage of interest of the "Publication" and "Professor" concepts became small which confirms an equivalent interest between the concepts. The same do not happen in the other approaches, in which there is a considerable difference between the "Publication" and "Professor" concepts.

Secondly, we can observe that the percentages graphic suffers a flattening. This phenomenon can be explained by the fact that in our algorithm, the concepts involved on the non-accessed pages were also considered, this occurs due to their semantic relations with the concepts of greater interest and their cognitive strength. This consideration of the concepts which do not appear directly on the pages raises the percentage of interest for these concepts and permits a faster response to the changes of the user's future interests.

Finally, we can consider that the results of this experiment were satisfactory, conforming our expectations that the use of the cognitive and linguistics aspects make a great difference in determining the user's intention

4 Conclusion and Future Projects

In this article, we present an approach to integrate semantic knowledge and navigation data in the process of identification of possible user's interests. For this, we have created a semantic log associating navigational patterns to concepts defined in a domain ontology. We have developed an algorithm using the linguistics and cognitive aspects that affect the process of user models creation. We identified that the form with which the concepts are disposed in the webpages, limit the user's capacity of choice, influencing indirectly the expression of her/his interests. In this way we consider the effect that the segmentation of the application model, i.e., the presentation model, has over the concepts presented in the webpages. The presentation models along with the semantic log act as the basis for the proposed algorithm. In order to take into account the way the concepts are organized in the human memory, and how the remembrance of a certain concept can bring to the memory the related concepts, we apply the idea of spreading activation and cognitive strength, to consider the degree of relevance of the concepts in the knowledge model.

In our analysis of the proposed approach, our algorithm presented very positive aspects in relation to other well known approaches (the Frequency-based classification and the Bayesian classification and the). One of the relevant points was to make it possible to identify a concept that has not had the largest number of access as being it the concept of greatest interests by the user. This behavior shows that the spreading activation considered with the different status of the accessed concepts, defined by the presentation model, permits the compensating the deviation caused by greater number of access resulted from a secondary path used to achieve the main objective. This does not happen in the other approaches, in which these concepts have a considerable difference of percentage of interest, in which the concept with greater number of access prevails. Another important fact is the determination of greater values for the probability of the non-visible concepts, which makes possible to obtain a better answer in the change of the user's interests possible.

This article also shows interesting research problems to be discussed and future projects. Most urgent among them is the development of learning techniques to identify ideal values of the parameters for the cognitive strength consideration and of the individual values of the different types of concepts. We also envisage a great deal of work to be done to consider the influence of the aesthetic aspects of the webpages.

Acknowledgements

Special thanks to CNPq – (Scientific and Technological Development National Counsel for financial support to the Post Graduation in Computer Science Program of the State University of Maringá - UEM).

References

1. Anderson, J. R. (1983a). A spreading activation theory of memory, *Journal of Verbal Learning and Verbal Behavior*, 22.
2. Anderson, J. R. (1983b). *The architecture of cognition. Cambridge*, MA: Harvard University Press.
3. Berners – Lee, t., Hendler, J., Lassila O.(2001). The Semantic Web. *Scientifc American*, May 2001.
4. Brusilovsky, P., (2001). *Adaptive Hypermedia, User Modeling and User-Adapted Interaction*, Kluwer Academic Publishers, Netherlands, pp. 87-110.
5. Chen L., Sycara K. (1998). Web Mate: A Personal Agent for Browsing and Searching. In *Proceedings of the 2nd International Conference on Autonomous Agents and Multi Agent Systems, AGENTS '98*, ACM, pp. 132 – 139.
6. Claypool, M., Gokhale, A., Miranda, T., Murnikov, P., Netes, D. and Sartin, M.(1999). Combining Content-based and Collaborative Filters in an Online Newspaper. In *Proceedings of the ACM SIGIR '99 Workshop on Recommender Systems: Algorithms and Evaluation.* University of California, Berkeley, Aug.
7. Collins, A. M., & Loftus, E. F. (1975). A spreading activation theory of semantic priming. *Psychological Review*, 82, 407-428.
8. Dai, H. and Mobasher, B. (2003). A Road map to More Effective Web Personalization: Integrating Domain Knowledge with Web Usage Mining. *Proc.of the International Conference on Internet Computing 2003 (IC'03)*, Las Vegas, Nevada, June 2003.
9. Eirinaki, M.; Vazirgiannis, M., Varlamis, I. (2003) SEWeP: using site semantics and a taxonomy to enhance the Web personalization process. *KDD 2003*: 99-108.
10. Eirinaki, M. and Vazirgiannis, M. (2003). Web Mining for Web Personalization, *ACM Transactions on Internet Technology (TOIT)*, February 2003/ Vol.3, No.1, 1-27.
11. Gauch S., Chaffee J., Pretschner A. Ontology Based Personalized Search. *Web Intelligence and Agent Systems* (in press).
12. Hendler, J., Berners-Lee, T. and Miller, E. (2002). Integrating Applications on the Semantic Web, *Journal of the Institute of Electrical Engineers of Japan*, Vol 122(10), October, 2002, p. 676-680. http://www.w3.org/2002/07/swint.
13. Lei, Y., Motta, E. and Domingue, J. (2004). Modelling Data- Intensive Web Sites with OntoWeaver, in *International Workshop on Web Information Systems Modelling (WISM 2004)*, Riga, Latvia.
14. Lieberman, H. (1995). Letizia: An Agent That Assists Web Browsing. *In Proceedings of the International Joint Conference on Artificial Intelligence*, Montreal, CA.
15. MCGUINNESS, D. L.; VAN HARMELEN, F.(2006). OWL Web Ontology Language Overview. W3C Recommendation 10 February 2004. Available in: http://www.w3.org/TR/2004/REC-owl-features-20040210/, Acess in: jan. 2006.
16. Middleton, S., De Roure, D., Shadbolt, N. (2001). Capturing knowledge of user preferences: ontologies in recommender systems. *In Proceedings of the 1st International Conference on Knowledge Capture (K-Cap2001)*, Victoria, BC Canada.
17. Mladenic, D. (1999). Text-learning and related intelligent agents. Revised version In *IEEE Expert*, special issue on Applications of Intelligent Information Retrieval.
18. Mobasher, B., Daí, H., Luo, T., Sung, Y. and Zhu, J. (2000). Integrating Web Usage and Content Mining for More Effective Personalization, in *Proc. of the International Conference on E-Commerce and Web Technologies (ECWeb2000)*, Greenwich, UK, September 2000.

19. Pazzani, M. A (1999). Framework for Collaborative, Content-Based and Demographic Filtering. *Artificial Intelligence Review*, Dec. 1999, pp. 393-408.
20. Quillian, M. R. (1968). Semantic memory. In M. L. Minsky (Ed.), *Semantic Information Processing*. Cambridge, MA: MIT Press.
21. Tanasa, D. and Trousse, B. (2004). Advanced data preprocessing for intersites web usage mining. *IEEE Intelligent Systems*, 19(2):59-65, March-April 2004.

A Novel Combination of Answer Set Programming with Description Logics for the Semantic Web

Thomas Lukasiewicz*

Dipartimento di Informatica e Sistemistica, Università di Roma "La Sapienza"
Via Salaria 113, I-00198 Rome, Italy
lukasiewicz@dis.uniroma1.it

Abstract. We present a novel combination of disjunctive logic programs under the answer set semantics with description logics for the Semantic Web. The combination is based on a well-balanced interface between disjunctive logic programs and description logics, which guarantees the decidability of the resulting formalism without assuming syntactic restrictions. We show that the new formalism has very nice semantic properties. In particular, it faithfully extends both disjunctive programs and description logics. Furthermore, we describe algorithms for reasoning in the new formalism, and we give a precise picture of its computational complexity. We also provide a special case with polynomial data complexity.

1 Introduction

The *Semantic Web* [4,14] aims at an extension of the current World Wide Web by standards and technologies that help machines to understand the information on the Web so that they can support richer discovery, data integration, navigation, and automation of tasks. The main ideas behind it are to add a machine-readable meaning to Web pages, to use ontologies for a precise definition of shared terms in Web resources, to use knowledge representation technology for automated reasoning from Web resources, and to apply cooperative agent technology for processing the information of the Web.

The Semantic Web consists of several hierarchical layers, where the *Ontology layer*, in form of the *OWL Web Ontology Language* [41,22], is currently the highest layer of sufficient maturity. OWL consists of three increasingly expressive sublanguages, namely *OWL Lite*, *OWL DL*, and *OWL Full*. OWL Lite and OWL DL are essentially very expressive description logics with an RDF syntax [22]. As shown in [19], ontology entailment in OWL Lite (resp., OWL DL) reduces to knowledge base (un)satisfiability in the description logic $\mathcal{SHIF}(\mathbf{D})$ (resp., $\mathcal{SHOIN}(\mathbf{D})$). As a next important step in the development of the Semantic Web, one aims at sophisticated representation and reasoning capabilities for the *Rules*, *Logic*, and *Proof layers* of the Semantic Web.

In particular, there is a large body of work on integrating rules and ontologies, which is a key requirement of the layered architecture of the Semantic Web. Significant research efforts focus on hybrid integrations of rules and ontologies, called *description logic programs*[1] (or *dl-programs*), which are of the form $KB = (L, P)$, where L is a

* Alternate address: Institut für Informationssysteme, Technische Universität Wien, Favoritenstraße 9-11, A-1040 Vienna, Austria; e-mail: lukasiewicz@kr.tuwien.ac.at.

[1] Note that we use the notion of "description logic programs" in a generic way, that is, to denote a class of different formalisms, similarly to the notion of "description logics".

E. Franconi, M. Kifer, and W. May (Eds.): ESWC 2007, LNCS 4519, pp. 384–398, 2007.

description logic knowledge base and P is a finite set of rules involving either queries to L in a loose integration (see especially [11,12,9,10]) or concepts and roles from L as unary resp. binary predicates in a tight integration (see especially [36,37]).

However, especially the tight integration of rules and ontologies presents many semantic and computational difficulties [37]. Since many expressive description logics are very close to the decidability / undecidability frontier (such as $\mathcal{SHOIN}(\mathbf{D})$, which is only decidable when number restrictions are limited to simple abstract roles [23]), developing decidable extensions of them by rules turns out to be a naturally hard task, and often comes along with strong syntactic restrictions on the resulting language (such as syntactic safety conditions and/or syntactic partitionings of the vocabulary).

Nevertheless, in rule-based systems in the Semantic Web, we would like to use vocabulary from formal ontologies, and we would like to do it without syntactic restrictions. In this paper, we show that the main difficulties with the above tight integrations of rules and ontologies lies actually in the *perspective of the integration*. That is, they all look from the *perspective of description logics* at the integration of rules and ontologies. However, for extending certain kinds of rule-based systems by vocabulary from formal ontologies, we actually do not need the full power of a rule-based extension of description logics. This is the main idea behind this paper. More precisely, we look at the integration of rules and ontologies from the *perspective of rule-based systems*.

The main contributions of this paper can be summarized as follows:

- We present a new combination of disjunctive logic programs under the answer set semantics with description logics. In detail, we present a new form of tightly integrated disjunctive dl-programs $KB = (L, P)$ under the answer set semantics, which allows for decidable reasoning, without assuming any syntactic restrictions (see Section 8 for a detailed comparison to previous approaches to dl-programs).
- Intuitively, the main idea behind the semantics of such dl-programs $KB = (L, P)$ is to interpret P relative to Herbrand interpretations that also satisfy L, while L is interpreted relative to general interpretations over a first-order domain. That is, we modularly combine the standard semantics of disjunctive programs P and of description logics L, via a well-balanced interface between P and L.
- We show that the new approach to disjunctive dl-programs under the answer set semantics has very nice semantic features. In particular, the cautious answer set semantics faithfully extends both disjunctive programs and description logics, and its closed-world property is limited to explicit default-negated atoms in rule bodies. Furthermore, the new approach does not need the unique name assumption.
- We also analyze the computational aspects of the new formalism. We describe algorithms for deciding answer set existence, brave consequences, and cautious consequences. This shows in particular that these decision problems are all decidable. We also draw a precise picture of the complexity of all these decision problems.
- Finally, we delineate a special case of stratified normal dl-programs where the above decision problems all have a polynomial data complexity.

The rest of this paper is organized as follows. Sections 2 and 3 recall disjunctive programs under the answer set semantics resp. the description logics $\mathcal{SHIF}(\mathbf{D})$ and $\mathcal{SHOIN}(\mathbf{D})$. In Section 4, we introduce our novel approach to disjunctive dl-programs under the answer set semantics, and in Section 5, we analyze its semantic properties.

Sections 6 and 7 focus on the computational properties. In Section 8, we discuss related work in the literature. Section 9 summarizes our main results and gives an outlook on future research. Some selected proofs are given in the appendix. Note that detailed proofs of all results are given in the extended report [31].

2 Disjunctive Programs Under the Answer Set Semantics

In this section, we recall disjunctive programs (with default negation) under the answer set semantics; see especially [26] for further details and background.

Syntax. Let Φ be a first-order vocabulary with nonempty finite sets of constant and predicate symbols, but no function symbols. Let $\mathcal{X}$ be a set of variables. A *term* is either a variable from $\mathcal{X}$ or a constant symbol from Φ. An *atom* is of the form $p(t_1, \ldots, t_n)$, where p is a predicate symbol of arity $n \geqslant 0$ from Φ, and $t_1, \ldots, t_n$ are terms. A *literal* l is an atom p or a negated atom *not* p. A *disjunctive rule* (or simply *rule*) r is of the form

$$\alpha_1 \vee \cdots \vee \alpha_k \leftarrow \beta_1, \ldots, \beta_n, \mathit{not}\, \beta_{n+1}, \ldots, \mathit{not}\, \beta_{n+m}\,, \tag{1}$$

where $\alpha_1, \ldots, \alpha_k, \beta_1, \ldots, \beta_{n+m}$ are atoms and $k, m, n \geqslant 0$. We call $\alpha_1 \vee \cdots \vee \alpha_k$ the *head* of r, while the conjunction $\beta_1, \ldots, \beta_n, \mathit{not}\, \beta_{n+1}, \ldots, \mathit{not}\, \beta_{n+m}$ is its *body*. We define $H(r) = \{\alpha_1, \ldots, \alpha_k\}$ and $B(r) = B^+(r) \cup B^-(r)$, where $B^+(r) = \{\beta_1, \ldots, \beta_n\}$ and $B^-(r) = \{\beta_{n+1}, \ldots, \beta_{n+m}\}$. A *disjunctive program* P is a finite set of disjunctive rules of the form (1). We say P is *positive* iff $m = 0$ for all disjunctive rules (1) in P. We say P is a *normal program* iff $k \leqslant 1$ for all disjunctive rules (1) in P.

Example 2.1. An online store (such as *amazon.com*) may use the subsequent set of rules P to express that (1) pc_1 and pc_2 are personal computers, and obj_3 is either a personal computer or a laptop, (2) pc_1 and obj_3 are brand new, (3) *dell* is the vendor of pc_1 and pc_2, (4) a customer avoids all cameras not on offer, (5) all electronic products that are not brand new are on offer, (6) each vendor of a product is a provider, (7) each entity providing a product is a provider, (8) all related products are similar, and (9) the binary similarity relation on products is transitively closed.

(1) $pc(pc_1)$; $pc(pc_2)$; $pc(obj_3) \vee laptop(obj_3)$;
(2) $brand\_new(pc_1)$; $brand\_new(obj_3)$;
(3) $vendor(dell, pc_1)$; $vendor(dell, pc_2)$;
(4) $avoid(X) \leftarrow camera(X), not\ offer(X)$;
(5) $offer(X) \leftarrow electronics(X), not\ brand\_new(X)$;
(6) $provider(V) \leftarrow vendor(V, X), product(X)$;
(7) $provider(V) \leftarrow provides(V, X), product(X)$;
(8) $similar(X, Y) \leftarrow related(X, Y)$;
(9) $similar(X, Z) \leftarrow similar(X, Y), similar(Y, Z)$.

Semantics. The answer set semantics of disjunctive programs is defined in terms of finite sets of ground atoms, which represent Herbrand interpretations. Positive disjunctive programs are associated with all their minimal satisfying sets of ground atoms, while the semantics of general disjunctive programs is defined by reduction to the minimal model semantics of positive disjunctive programs via the Gelfond-Lifschitz reduct [15].

More concretely, the *Herbrand universe* of a disjunctive program P, denoted HU_P, is the set of all constant symbols appearing in P. If there is no such constant symbol, then $HU_P = \{c\}$, where c is an arbitrary constant symbol from Φ. As usual, terms, atoms, literals, rules, programs, etc. are *ground* iff they do not contain any variables. The *Herbrand base* of a disjunctive program P, denoted HB_P, is the set of all ground atoms that can be constructed from the predicate symbols appearing in P and the constant symbols in HU_P. Hence, in the standard answer set semantics, the Herbrand base is constructed from all constant and predicate symbols in a given disjunctive program, and thus the Herbrand base is finite. A *ground instance* of a rule $r \in P$ is obtained from r by replacing every variable that occurs in r by a constant symbol from HU_P. We denote by $ground(P)$ the set of all ground instances of rules in P.

An *interpretation* I relative to a disjunctive program P is a subset of HB_P. Informally, every such I represents the Herbrand interpretation in which all $a \in I$ (resp., $a \in HB_P - I$) are true (resp., false). An interpretation I is a *model* of a ground atom $a \in HB_P$, or I *satisfies* a, denoted $I \models a$, iff $a \in I$. We say I is a *model* of a ground rule r, denoted $I \models r$, iff $I \models \alpha$ for some $\alpha \in H(r)$ whenever $I \models B(r)$, that is, $I \models \beta$ for all $\beta \in B^+(r)$ and $I \not\models \beta$ for all $\beta \in B^-(r)$. We say I is a *model* of a disjunctive program P, denoted $I \models P$, iff $I \models r$ for every $r \in ground(P)$. An *answer set* of a positive disjunctive program P is a minimal model of P relative to set inclusion. The *Gelfond-Lifschitz reduct* of a disjunctive program P relative to $I \subseteq HB_P$, denoted P^I, is the ground positive disjunctive program obtained from $ground(P)$ by (i) deleting every rule r such that $B^-(r) \cap I \neq \emptyset$, and (ii) deleting the negative body from each remaining rule. An *answer set* of a disjunctive program P is an interpretation $I \subseteq HB_P$ such that I is an answer set of P^I. A disjunctive program P is *consistent* iff P has an answer set.

Hence, under the answer set semantics, every disjunctive program P is interpreted as its grounding $ground(P)$. Note that the answer sets of any disjunctive program P are also minimal models of P. An equivalent definition of the answer set semantics is based on the so-called *FLP-reduct* [13]: The *FLP-reduct* of a disjunctive program P relative to $I \subseteq HB_P$, denoted P^I, is the set of all $r \in ground(P)$ such that $I \models B(r)$. An interpretation $I \subseteq HB_P$ is an *answer set* of P iff I is a minimal model of P^I.

We finally recall the notions of *cautious* (resp., *brave*) *reasoning* from disjunctive programs under the answer set semantics. A ground atom $a \in HB_P$ is a *cautious* (resp., *brave*) *consequence* of a disjunctive program P under the answer set semantics iff every (resp., some) answer set of P satisfies a. Observe that for positive disjunctive programs P, since the set of all answer sets of P is given by the set of all minimal models of P, it holds that $a \in HB_P$ is a cautious consequence of P under the answer set semantics iff a is a logical consequence of the propositional positive disjunctive program $ground(P)$. Note that, more generally, this result holds also when a is a ground formula constructed from HB_Φ using the Boolean operators $\wedge$ and $\vee$. This means that the *closed-world property* (that is, the derivation of negative facts from the absence of derivations of positive facts) of the above notion of cautious reasoning under the answer set semantics is actually limited to the occurrences of default negations in rule bodies.

3 Description Logics

In this section, we recall the description logics $\mathcal{SHIF}(\mathbf{D})$ and $\mathcal{SHOIN}(\mathbf{D})$, which stand behind the web ontology languages OWL Lite and OWL DL [19], respectively. Intuitively, description logics model a domain of interest in terms of concepts and roles, which represent classes of individuals and binary relations between classes of individuals, respectively. A description logic knowledge base encodes especially subset relationships between concepts, subset relationships between roles, the membership of individuals to concepts, and the membership of pairs of individuals to roles.

Syntax. We first describe the syntax of $\mathcal{SHOIN}(\mathbf{D})$. We assume a set of *elementary datatypes* and a set of *data values*. A *datatype* is either an elementary datatype or a set of data values (called *datatype oneOf*). A *datatype theory* $\mathbf{D} = (\Delta^{\mathbf{D}}, \cdot^{\mathbf{D}})$ consists of a *datatype domain* $\Delta^{\mathbf{D}}$ and a mapping $\cdot^{\mathbf{D}}$ that assigns to each elementary datatype a subset of $\Delta^{\mathbf{D}}$ and to each data value an element of $\Delta^{\mathbf{D}}$. The mapping $\cdot^{\mathbf{D}}$ is extended to all datatypes by $\{v_1, \ldots\}^{\mathbf{D}} = \{v_1^{\mathbf{D}}, \ldots\}$. Let $\mathbf{A}$, $\mathbf{R}_A$, $\mathbf{R}_D$, and $\mathbf{I}$ be pairwise disjoint (nonempty) denumerable sets of *atomic concepts*, *abstract roles*, *datatype roles*, and *individuals*, respectively. We denote by $\mathbf{R}_A^-$ the set of inverses R^- of all $R \in \mathbf{R}_A$.

A *role* is an element of $\mathbf{R}_A \cup \mathbf{R}_A^- \cup \mathbf{R}_D$. *Concepts* are inductively defined as follows. Every $\phi \in \mathbf{A}$ is a concept, and if $o_1, \ldots, o_n \in \mathbf{I}$, then $\{o_1, \ldots, o_n\}$ is a concept (called *oneOf*). If ϕ, ϕ_1, and ϕ_2 are concepts and if $R \in \mathbf{R}_A \cup \mathbf{R}_A^-$, then also $(\phi_1 \sqcap \phi_2)$, $(\phi_1 \sqcup \phi_2)$, and $\neg\phi$ are concepts (called *conjunction*, *disjunction*, and *negation*, respectively), as well as $\exists R.\phi$, $\forall R.\phi$, $\geqslant nR$, and $\leqslant nR$ (called *exists*, *value*, *atleast*, and *atmost restriction*, respectively) for an integer $n \geqslant 0$. If D is a datatype and $U \in \mathbf{R}_D$, then $\exists U.D$, $\forall U.D$, $\geqslant nU$, and $\leqslant nU$ are concepts (called *datatype exists*, *value*, *atleast*, and *atmost restriction*, respectively) for an integer $n \geqslant 0$. We write $\top$ and $\bot$ to abbreviate the concepts $\phi \sqcup \neg\phi$ and $\phi \sqcap \neg\phi$, respectively, and we eliminate parentheses as usual.

An *axiom* has one of the following forms: (1) $\phi \sqsubseteq \psi$ (called *concept inclusion axiom*), where ϕ and ψ are concepts; (2) $R \sqsubseteq S$ (called *role inclusion axiom*), where either $R, S \in \mathbf{R}_A$ or $R, S \in \mathbf{R}_D$; (3) $\mathrm{Trans}(R)$ (called *transitivity axiom*), where $R \in \mathbf{R}_A$; (4) $\phi(a)$ (called *concept membership axiom*), where ϕ is a concept and $a \in \mathbf{I}$; (5) $R(a, b)$ (resp., $U(a, v)$) (called *role membership axiom*), where $R \in \mathbf{R}_A$ (resp., $U \in \mathbf{R}_D$) and $a, b \in \mathbf{I}$ (resp., $a \in \mathbf{I}$ and v is a data value); and (6) $a = b$ (resp., $a \neq b$) (*equality* (resp., *inequality*) *axiom*), where $a, b \in \mathbf{I}$. A *knowledge base* L is a finite set of axioms. For decidability, number restrictions in L are restricted to simple abstract roles [23].

The syntax of $\mathcal{SHIF}(\mathbf{D})$ is as the above syntax of $\mathcal{SHOIN}(\mathbf{D})$, but without the oneOf constructor and with the atleast and atmost constructors limited to 0 and 1.

Example 3.1. The subsequent description logic knowledge base L expresses that (1) textbooks are books, (2) personal computers and laptops are mutually exclusive electronic products, (3) books and electronic products are mutually exclusive products, (4) objects on offer are products, (5) every product has at least one related product, (6) only products are related to each other, (7) the relatedness between products is symmetric, (8) *tb_ai* and *tb_lp* are textbooks, (9) which are related to each other, (10) *pc_ibm* and *pc_hp* are personal computers, (11) which are related to each other, and (12) *ibm* and *hp* are providers for *pc_ibm* and *pc_hp*, respectively.

(1) *textbook* $\sqsubseteq$ *book*; (2) *pc* $\sqcup$ *laptop* $\sqsubseteq$ *electronics*; *pc* $\sqsubseteq \neg$*laptop*;
(3) *book* $\sqcup$ *electronics* $\sqsubseteq$ *product*; *book* $\sqsubseteq \neg$*electronics*; (4) *offer* $\sqsubseteq$ *product*;
(5) *product* $\sqsubseteq\ \geqslant 1$ *related*; (6) $\geqslant 1$ *related* $\sqcup \geqslant 1$ *related*$^{-}$ $\sqsubseteq$ *product*;
(7) *related* $\sqsubseteq$ *related*$^{-}$; *related*$^{-}$ $\sqsubseteq$ *related*;
(8) *textbook*(*tb_ai*); *textbook*(*tb_lp*); (9) *related*(*tb_ai*, *tb_lp*);
(10) *pc*(*pc_ibm*); *pc*(*pc_hp*); (11) *related*(*pc_ibm*, *pc_hp*);
(12) *provides*(*ibm*, *pc_ibm*); *provides*(*hp*, *pc_hp*).

Semantics. An *interpretation* $\mathcal{I} = (\Delta^{\mathcal{I}}, \cdot^{\mathcal{I}})$ w.r.t. a datatype theory $\mathbf{D} = (\Delta^{\mathbf{D}}, \cdot^{\mathbf{D}})$ consists of a nonempty (*abstract*) *domain* $\Delta^{\mathcal{I}}$ disjoint from $\Delta^{\mathbf{D}}$, and a mapping $\cdot^{\mathcal{I}}$ that assigns to each atomic concept $\phi \in \mathbf{A}$ a subset of $\Delta^{\mathcal{I}}$, to each individual $o \in \mathbf{I}$ an element of $\Delta^{\mathcal{I}}$, to each abstract role $R \in \mathbf{R}_A$ a subset of $\Delta^{\mathcal{I}} \times \Delta^{\mathcal{I}}$, and to each datatype role $U \in \mathbf{R}_D$ a subset of $\Delta^{\mathcal{I}} \times \Delta^{\mathbf{D}}$. We extend $\cdot^{\mathcal{I}}$ to all concepts and roles, and we define the *satisfaction* of an axiom F in an interpretation $\mathcal{I} = (\Delta^{\mathcal{I}}, \cdot^{\mathcal{I}})$, denoted $\mathcal{I} \models F$, as usual [19]. We say $\mathcal{I}$ *satisfies* the axiom F, or $\mathcal{I}$ is a *model* of F, iff $\mathcal{I} \models F$. We say $\mathcal{I}$ *satisfies* a knowledge base L, or $\mathcal{I}$ is a *model* of L, denoted $\mathcal{I} \models L$, iff $\mathcal{I} \models F$ for all $F \in L$. We say L is *satisfiable* (resp., *unsatisfiable*) iff L has a (resp., no) model. An axiom F is a *logical consequence* of L, denoted $L \models F$, iff each model of L satisfies F.

4 Disjunctive DL-Programs Under the Answer Set Semantics

In this section, we present a novel integration between disjunctive programs under the answer set semantics and description logics. The basic idea behind this integration is as follows. Suppose that we have a disjunctive program P. Under the answer set semantics, P is equivalent to its grounding $\mathit{ground}(P)$. Suppose now that some of the ground atoms in $\mathit{ground}(P)$ are additionally related to each other by a description logic knowledge base L. That is, some of the ground atoms in $\mathit{ground}(P)$ actually represent concept and role memberships relative to L. Thus, when processing $\mathit{ground}(P)$, we also have to consider L. However, we only want to do it to the extent that we actually need it for processing $\mathit{ground}(P)$. Hence, when taking a Herbrand interpretation $I \subseteq HB_{\Phi}$, we have to ensure that the ground atoms of I represent a valid constellation relative to L.

In other words, the main idea behind the semantics is to interpret P relative to Herbrand interpretations that also satisfy L, while L is interpreted relative to general interpretations over a first-order domain. Thus, we modularly combine the standard semantics of disjunctive programs and of description logics as in [11,12,10], which allows for building on the standard techniques and the results of both areas. However, our new approach here allows for a much tighter integration of L and P.

Syntax. We assume a function-free first-order vocabulary Φ with nonempty finite sets of constant and predicate symbols, as in Section 2. We use Φ_c to denote the set of all constant symbols in Φ. We also assume pairwise disjoint (nonempty) denumerable sets $\mathbf{A}$, $\mathbf{R}_A$, $\mathbf{R}_D$, and $\mathbf{I}$ of atomic concepts, abstract roles, datatype roles, and individuals, respectively, as in Section 3. We assume that Φ_c is a subset of $\mathbf{I}$. This assumption guarantees that every ground atom constructed from atomic concepts, abstract roles, datatype roles, and constants in Φ_c can be interpreted in the description logic component. We do not assume any other restriction on the vocabularies, that is, Φ and $\mathbf{A}$ (resp., $\mathbf{R}_A \cup \mathbf{R}_D$) may have unary (resp., binary) predicate symbols in common.

A *disjunctive description logic program* (or simply *disjunctive dl-program*) $KB = (L, P)$ consists of a description logic knowledge base L and a disjunctive program P. It is *positive* iff P is positive. It is a *normal dl-program* iff P is a normal program.

Example 4.1. A disjunctive dl-program $KB = (L, P)$ is given by the description logic knowledge base L and the disjunctive program P of Examples 2.1 and 3.1, respectively. Another disjunctive dl-program $KB' = (L', P')$ is obtained from KB by adding to L the axiom $\geqslant 1\, \mathit{similar} \sqcup \geqslant 1\, \mathit{similar}^- \sqsubseteq \mathit{product}$, which expresses that only products are similar. Observe that the predicate symbol *similar* in P' is also a role in L', and it freely occurs in both rule bodies and rule heads in P' (which is not possible in [11]).

Semantics. We now define the answer set semantics of disjunctive dl-programs via a generalization of the FLP-reduct of disjunctive programs (see Section 2).

In the sequel, let $KB = (L, P)$ be a disjunctive dl-program. A *ground instance* of a rule $r \in P$ is obtained from r by replacing every variable that occurs in r by a constant symbol from Φ_c. We denote by $\mathit{ground}(P)$ the set of all ground instances of rules in P. The *Herbrand base* relative to Φ, denoted HB_Φ, is the set of all ground atoms constructed with constant and predicate symbols from Φ. Observe that we now define the Herbrand base relative to Φ and not relative to P. This allows for reasoning about ground atoms from the description logic component that do not necessarily occur in P. Observe, however, that the extension from P to Φ is only a notational simplification, since we can always make constant and predicate symbols from Φ occur in P by "dummy" rules such as $\mathit{constant}(c) \leftarrow$ and $p(\mathbf{c}) \leftarrow p(\mathbf{c})$, respectively. We denote by DL_Φ the set of all ground atoms in HB_Φ that are constructed from atomic concepts in $\mathbf{A}$, abstract roles in $\mathbf{R}_A$, concrete roles in $\mathbf{R}_D$, and constant symbols in Φ_c.

An *interpretation* I is any subset of HB_Φ. We say I is a *model* of a description logic knowledge base L, denoted $I \models L$, iff $L \cup I \cup \{\neg a \mid a \in HB_\Phi - I\}$ is satisfiable. Note that the former defines the truth of description logic knowledge bases L in Herbrand interpretations $I \subseteq HB_\Phi$ rather than first-order interpretations $\mathcal{I}$. Note also that a negative concept membership $\neg C(a)$ can be encoded as the positive concept membership $(\neg C)(a)$. The following theorem shows that also negative role memberships $\neg R(b, c)$ can be reduced to positive concept memberships and concept inclusions.

Theorem 4.1. *Let L be a description logic knowledge base, and let $R(b, c)$ be a role membership axiom. Then, $L \cup \{\neg R(b, c)\}$ is satisfiable iff $L \cup \{B(b), C(c), \exists R.C \sqsubseteq \neg B\}$ is satisfiable, where B and C are two fresh atomic concepts.*

An interpretation $I \subseteq HB_\Phi$ is a *model* of a disjunctive dl-program $KB = (L, P)$, denoted $I \models KB$, iff $I \models L$ and $I \models P$. We say KB is *satisfiable* iff it has a model.

Given a disjunctive dl-program $KB = (L, P)$, the *FLP-reduct* of KB relative to an interpretation $I \subseteq HB_\Phi$, denoted KB^I, is the disjunctive dl-program (L, P^I), where P^I is the set of all $r \in \mathit{ground}(P)$ such that $I \models B(r)$. An interpretation $I \subseteq HB_\Phi$ is an *answer set* of KB iff I is a minimal model of KB^I. A disjunctive dl-program KB is *consistent* (resp., *inconsistent*) iff it has an (resp., no) answer set.

We finally define the notions of *cautious* (resp., *brave*) *reasoning* from disjunctive dl-programs under the answer set semantics as follows. A ground atom $a \in HB_{\Phi}$ is a *cautious* (resp., *brave*) *consequence* of a disjunctive dl-program KB under the answer set semantics iff every (resp., some) answer set of KB satisfies a.

5 Semantic Properties

In this section, we summarize some semantic properties (especially those relevant for the Semantic Web) of disjunctive dl-programs under the above answer set semantics.

Minimal Models. The following theorem shows that, like in the ordinary case (see Section 2), every answer set of a disjunctive dl-program KB is also a minimal model of KB, and the answer sets of a positive KB are given by the minimal models of KB.

Theorem 5.1. *Let $KB = (L, P)$ be a disjunctive dl-program. Then, (a) every answer set of KB is a minimal model of KB, and (b) if KB is positive, then the set of all answer sets of KB is given by the set of all minimal models of KB.*

Faithfulness. An important property of integrations of rules and ontologies is that they are a faithful [33,34] extension of both rules and ontologies.

The following theorem shows that the answer set semantics of disjunctive dl-programs faithfully extends its ordinary counterpart. That is, the answer set semantics of a disjunctive dl-program $KB = (L, P)$ with empty description logic knowledge base L coincides with the ordinary answer set semantics of its disjunctive program P.

Theorem 5.2. *Let $KB = (L, P)$ be a disjunctive dl-program such that $L = \emptyset$. Then, the set of all answer sets of KB coincides with the set of all ordinary answer sets of P.*

The next theorem shows that the answer set semantics of disjunctive dl-programs also faithfully extends the first-order semantics of description logic knowledge bases, which here means that a ground atom $a \in HB_{\Phi}$ is true in all answer sets of a positive disjunctive dl-program $KB = (L, P)$ iff a is true in all first-order models of $L \cup ground(P)$. The theorem holds also when a is a ground formula constructed from HB_{Φ} using $\wedge$ and $\vee$. Observe that the theorem does not hold for all first-order formulas a, but we actually also do not need this, *looking from the perspective of answer set programming*, since we actually cannot refer to all general first-order formulas in P.

Theorem 5.3. *Let $KB = (L, P)$ be a positive disjunctive dl-program, and let a be a ground atom from HB_{Φ}. Then, a is true in all answer sets of KB iff a is true in all first-order models of $L \cup ground(P)$.*

As an immediate corollary, we obtain that $a \in HB_{\Phi}$ is true in all answer sets of a disjunctive dl-program $KB = (L, \emptyset)$ iff a is true in all first-order models of L.

Corollary 5.1. *Let $KB = (L, P)$ be a disjunctive dl-program with $P = \emptyset$, and let $a \in HB_{\Phi}$. Then, a is true in all answer sets of KB iff a is true in all first-order models of L.*

Closed-World Assumption. It is often argued that the closed-world assumption is not very desirable in the open environment of the Semantic Web [20]. The notion of cautious reasoning from disjunctive dl-programs under the answer set semantics also has some closed-world property. However, as also shown by Theorem 5.3, this closed-world property is actually limited to the explicit use of default negations in rule bodies, and thus we can actually control very easily its use in disjunctive dl-programs.

Unique Name Assumption. Another aspect that may not be very desirable in the Semantic Web [20] is the *unique name assumption* (which says that any two distinct constant symbols in Φ_c represent two distinct domain objects). It turns out that we actually do not have to make this assumption, since the description logic knowledge base of a disjunctive dl-program may very well contain or imply equalities between individuals.

This result is included in the following theorem, which shows an alternative characterization of the satisfaction of L in $I \subseteq HB_\Phi$: Rather than being enlarged by a set of axioms of exponential size, L is enlarged by a set of axioms of polynomial size. This characterization essentially shows that the satisfaction of L in I corresponds to checking that (i) the ground atoms in $I \cap DL_\Phi$ satisfy L, and (ii) the ground atoms in $I \cap (HB_\Phi - DL_\Phi)$ do not violate any equality axioms that follow from L. Here, an equivalence relation $\sim$ on Φ_c is *admissible* with an interpretation $I \subseteq HB_\Phi$ iff $p(c_1, \ldots, c_n) \in I \Leftrightarrow p(c'_1, \ldots, c'_n) \in I$ for all n-ary predicate symbols p, where $n > 0$, and constant symbols $c_1, \ldots, c_n, c'_1, \ldots, c'_n \in \Phi_c$ such that $c_i \sim c'_i$ for all $i \in \{1, \ldots, n\}$.

Theorem 5.4. *Let L be a description logic knowledge base, and let $I \subseteq HB_\Phi$. Then, $L \cup I \cup \{\neg b \mid b \in HB_\Phi - I\}$ is satisfiable iff $L \cup (I \cap DL_\Phi) \cup \{\neg b \mid b \in DL_\Phi - I\} \cup \{c \neq c' \mid c \not\sim c'\}$ is satisfiable for some equivalence relation $\sim$ on Φ_c admissible with I.*

Conjunctive Queries. It is often argued that the processing of conjunctive queries is important for the Semantic Web [37]. As for this issue, observe that (Boolean unions of) conjunctive queries in our approach can be reduced to atomic queries. A *Boolean union of conjunctive queries* Q is of the form $\exists \boldsymbol{x}(\gamma_1(\boldsymbol{x}) \vee \cdots \vee \gamma_n(\boldsymbol{x}))$, where $\boldsymbol{x}$ is a tuple of variables, $n \geqslant 1$, and each $\gamma_i(\boldsymbol{x})$ is a conjunction of atoms constructed from predicate and constant symbols in Φ and variables in $\boldsymbol{x}$. We call Q a *conjunctive query* when $n = 1$. If we assume that $\boldsymbol{x}$ ranges over all constant symbols in Φ_c (which is sufficient for our needs, *looking from the perspective of answer set programming*, since in P we can refer only through Φ_c to elements of a first-order domain), then Q can be expressed by adding the rules $q(\boldsymbol{x}) \leftarrow \gamma_i(\boldsymbol{x})$ with $i \in \{1, \ldots, n\}$ to P and thereafter computing the set of all entailed ground instances of $q(\boldsymbol{x})$ relative to Φ_c (see also Section 6).

6 Algorithms and Complexity

In this section, we describe algorithms for deciding whether a disjunctive dl-program has an answer set, and for deciding brave and cautious consequences from disjunctive dl-programs under the answer set semantics. Furthermore, we also draw a precise picture of the complexity of all these decision problems.

Algorithm consistency

Input: vocabulary Φ and disjunctive dl-program $KB = (L, P)$.
Output: *Yes*, if KB has an answer set; *No*, otherwise.

1. **if** there exists $I \subseteq HB_\Phi$ such that I is a minimal model of $KB^I = (L, P^I)$
2. **then return** *Yes*;
3. **else return** *No*.

Fig. 1. Algorithm consistency

Algorithms. The problem of deciding whether a disjunctive dl-program $KB = (L, P)$ has an answer set can be solved by a simple guess-and-check algorithm, which guesses a subset I of the finite Herbrand base HB_Φ, computes the FLP-reduct $KB^I = (L, P^I)$, and then checks that I is in fact a minimal model of KB^I (see Fig. 1).

The problem of deciding brave and cautious consequences can be reduced to deciding answer set existence (like in the ordinary case), since a ground atom $a \in HB_\Phi$ is true in some (resp., every) answer set of a disjunctive dl-program $KB = (L, P)$ iff $(L, P \cup \{\leftarrow \mathit{not}\, a\})$ (resp., $(L, P \cup \{\leftarrow a\})$) has an (resp., no) answer set.

Complexity. We now show that the problems of deciding consistency and brave / cautious consequences have the same complexity in disjunctive dl-programs under the answer set semantics as in ordinary disjunctive programs under the answer set semantics.

The following theorem shows that deciding the consistency of disjunctive dl-programs is complete for NEXP^{NP} (combined complexity). The lower bound follows from the NEXP^{NP}-hardness of deciding the consistency of ordinary disjunctive programs [6]. The upper bound follows from the result that deciding knowledge base satisfiability in $\mathcal{SHIF}(\mathbf{D})$ (resp., $\mathcal{SHOIN}(\mathbf{D})$) is complete for EXP (resp., NEXP) [40,19].

Theorem 6.1. *Given Φ and a disjunctive dl-program $KB = (L, P)$ with L in $\mathcal{SHIF}(\mathbf{D})$ or $\mathcal{SHOIN}(\mathbf{D})$, deciding whether KB has an answer set is complete for* NEXP^{NP}.

The next theorem shows that deciding cautious (resp., brave) consequences from disjunctive dl-programs is complete for co-NEXP^{NP} (resp., NEXP^{NP}) in the combined complexity. This result follows from Theorem 6.1, since the two problems of consistency checking and cautious (resp., brave) reasoning can be reduced to each other.

Theorem 6.2. *Given Φ, a disjunctive dl-program $KB = (L, P)$ with L in $\mathcal{SHIF}(\mathbf{D})$ or $\mathcal{SHOIN}(\mathbf{D})$, and a ground atom $a \in HB_\Phi$, deciding whether a holds in every (resp., some) answer set of KB is complete for* co-NEXP^{NP} *(resp.,* NEXP^{NP}*).*

7 Tractability Results

In this section, we describe a special class of disjunctive dl-programs for which the problems of deciding consistency and of query processing have both a polynomial data complexity. These programs are normal, stratified, and defined relative to *DL-Lite* [5], which allows for deciding knowledge base satisfiability in polynomial time.

We first recall *DL-Lite*. Let $\mathbf{A}$, $\mathbf{R}_A$, and $\mathbf{I}$ be pairwise disjoint sets of atomic concepts, abstract roles, and individuals, respectively. A *basic concept in DL-Lite* is either an atomic concept from $\mathbf{A}$ or an exists restriction on roles $\exists R.\top$ (abbreviated as $\exists R$), where $R \in \mathbf{R}_A \cup \mathbf{R}_A^-$. A *literal in DL-Lite* is either a basic concept b or the negation of a basic concept $\neg b$. *Concepts in DL-Lite* are defined by induction as follows. Every basic concept in *DL-Lite* is a concept in *DL-Lite*. If b is a basic concept in *DL-Lite*, and ϕ_1 and ϕ_2 are concepts in *DL-Lite*, then $\neg b$ and $\phi_1 \sqcap \phi_2$ are also concepts in *DL-Lite*. An *axiom in DL-Lite* is either (1) a concept inclusion axiom $b \sqsubseteq \psi$, where b is a basic concept in *DL-Lite* and ϕ is a concept in *DL-Lite*, or (2) a *functionality axiom* $(\mathsf{funct}\ R)$, where $R \in \mathbf{R}_A \cup \mathbf{R}_A^-$, or (3) a concept membership axiom $b(a)$, where b is a basic concept in *DL-Lite* and $a \in \mathbf{I}$, or (4) a role membership axiom $R(a, c)$, where $R \in \mathbf{R}_A$ and $a, c \in \mathbf{I}$. A *knowledge base in DL-Lite* L is a finite set of axioms in *DL-Lite*.

Every knowledge base in *DL-Lite* L can be transformed into an equivalent one in *DL-Lite* $\mathit{trans}(L)$ in which every concept inclusion axiom is of form $b \sqsubseteq \ell$, where b (resp., ℓ) is a basic concept (resp., literal) in *DL-Lite* [5]. We then define $\mathit{trans}(P) = P \cup \{b'(X) \leftarrow b(X) \mid b \sqsubseteq b' \in \mathit{trans}(L), b' \text{ is a basic concept}\} \cup \{\exists R(X) \leftarrow R(X,Y) \mid R \in \mathbf{R}_A \cap \Phi\} \cup \{\exists R^-(Y) \leftarrow R(X,Y) \mid R \in \mathbf{R}_A \cap \Phi\}$. Intuitively, we make explicit all the rule-based relationships between the predicates in P that are implicitly encoded in L. We define stratified normal dl-programs as follows. A normal dl-program $KB = (L, P)$ is *stratified* iff (i) L is defined in *DL-Lite* and (ii) $\mathit{trans}(P)$ is (locally) stratified.

It can be shown that stratified normal dl-programs $KB = (L, P)$ have either no or a unique answer set, which can be computed by a finite sequence of fixpoint iterations (relative to $\mathit{trans}(P)$), as usual. This implies immediately that for such programs consistency checking and query processing have both a polynomial data complexity.

Theorem 7.1. *Given Φ and a stratified normal dl-program KB, (a) deciding whether KB has an answer set, and (b) deciding whether a given ground atom $a \in HB_\Phi$ is true in the answer set of KB (if it exists) have both a polynomial data complexity.*

8 Related Work

There is a large body of related works on combining rules and ontologies, which can essentially be divided into the following three lines of research: (a) loose integration of rules and ontologies, (b) tight integration of rules and ontologies, and (c) reductions from description logics to logic programming formalisms. In this section, we discuss only the works that are most closely related to the framework of this paper.

Representatives of the loose integration of rules and ontologies are in particular the dl-programs in [11,12], their extension to HEX-programs [9,10], to probabilistic dl-programs [28,29], and to fuzzy dl-programs [30]. The combination of defeasible reasoning with description logics in [3], the calls to description logic reasoners in TRIPLE [38], and the hybrid MKNF knowledge bases in [33,34] are also close in spirit. More concretely, compared to the present paper, the dl-programs $KB = (L, P)$ in [11] also consist of a description logic knowledge base L and a normal program P. However, P may also contain classical negations, and rather than using concepts and roles from L as predicates in P, rule bodies in P may only contain queries to L, which may also contain facts as additional input to L. Like in this paper, P is interpreted relative to

Herbrand interpretations under the answer set semantics, while L is interpreted relative to first-order interpretations under the classical model-theoretic semantics. However, differently from the concepts and roles in P here, the queries in P in [11] are evaluated independently from each other. HEX-programs [9,10] extend the approach to dl-programs in [11] by multiple sources of external knowledge, with possibly different semantics, while probabilistic dl-programs [28,29] and fuzzy dl-programs [30] are extensions by probabilistic uncertainty and fuzzy vagueness, respectively. Closely related to the dl-programs in [11] are also the hybrid MKNF knowledge bases in [33,34]. They essentially allow for querying a description logic knowledge base L via the operators $\mathbf{K}$ and **not**, which can be used more flexibly than the queries in [11] (the operators can also occur in rule heads, while the queries are restricted to rule bodies), but which do not allow for passing facts to L in the form of query arguments. Note that closely related to the hybrid MKNF knowledge bases in [33,34] is also the embedding of non-ground logic programs into autoepistemic logic in [7]. The following example shows that our novel dl-programs here generally do not have the same meaning as the dl-programs in [11] (note that a similar example can be constructed for the approach in [33,34]).

Example 8.1. The normal dl-program $KB = (L, P)$, where

$$\begin{aligned} L &= \{person(a),\ person \sqsubseteq male \sqcup female\} \text{ and} \\ P &= \{client(X) \leftarrow male(X),\ client(X) \leftarrow female(X)\} \end{aligned}$$

implies $client(a)$, while the normal dl-program $KB' = (L', P')$ as in [11]

$$\begin{aligned} L' &= \{person(a),\ person \sqsubseteq male \sqcup female\} \text{ and} \\ P' &= \{client(X) \leftarrow DL[male](X),\ client(X) \leftarrow DL[female](X)\} \end{aligned}$$

does *not* imply $client(a)$, since the two queries are evaluated independently from each other, and neither $male(a)$ nor $female(a)$ follows from L'. To obtain the conclusion $client(a)$ in [11], one has to directly use the rule $client(X) \leftarrow DL[male \sqcup female](X)$.

Some representatives of tight integrations of rules and ontologies are in particular the works due to Donini et al. [8], Levy and Rousset [27], Grosof et al. [16], Motik et al. [35], Heymans et al. [17], and Rosati [36,37]. SWRL [21] and WRL [2] also belong to this category. Closest in spirit to this paper among the above works is perhaps Rosati's approach [36,37]. Like here, Rosati's hybrid knowledge bases also consist of a description logic knowledge base L and a disjunctive program (with default negations) P, where concepts and roles in L may act as predicate symbols in P. However, differently from this paper, Rosati partitions the predicates of L and P into description logic predicates and logic program predicates, where the former are interpreted under the classical model-theoretic semantics, while the latter are interpreted under the answer set semantics (and thus in particular default negations of concepts and roles are not allowed in P). Furthermore, differently from this paper, he also assumes a syntactic restriction on rules (called *weak safeness*) to gain decidability, and he assumes the standard names assumption, which includes the unique name assumption.

Finally, the works reducing description logics to logic programming are less closely related to the framework of this paper. Some representatives are in particular the works by Alsaç and Baral [1], Swift [39], Heymans and Vermeir [18], and Motik et al. [24].

9 Summary and Outlook

We have presented a novel combination of disjunctive programs under the answer set semantics with description logics for the Semantic Web. The combination is based on a well-balanced interface between disjunctive programs and description logics, which guarantees the decidability of the resulting formalism without assuming any syntactic restrictions on the resulting language (such as syntactic safety conditions and/or syntactic partitionings of the vocabulary). We have shown that the new formalism has very nice semantic properties. In particular, it faithfully extends both disjunctive programs and description logics. We have also provided algorithms and precise complexity results for the new formalism, as well as a special case of polynomial data complexity.

The presented mechanism of integrating rules and ontologies is of general importance, since it can actually also be used for the decidable integration of other reasoning techniques (such as reasoning about defaults, probabilistic uncertainty, and fuzzy vagueness) with description logics, since it applies to all reasoning techniques that are based on interpretations over finite Herbrand bases (or also finite sets of propositional symbols). It thus paves the way for decidable reasoning formalisms on top of description logics for the Semantic Web. Note that a companion paper [32] explores the use of this novel integration mechanism in fuzzy description logic programs.

An interesting topic of future research is to develop more sophisticated algorithms for reasoning from the new disjunctive dl-programs, and to implement the approach. Another interesting issue is to extend disjunctive dl-programs by classical negation.

Acknowledgments. This work has been supported by a Heisenberg Professorship of the German Research Foundation (DFG). I thank the reviewers for their constructive comments, which helped to improve this work.

References

1. G. Alsaç and C. Baral. Reasoning in description logics using declarative logic programming. Report, Department of Computer Science and Engineering, Arizona State University, 2001.
2. J. Angele, H. Boley, J. de Bruijn, D. Fensel, P. Hitzler, M. Kifer, R. Krummenacher, H. Lausen, A. Polleres, and R. Studer. Web Rule Language (WRL), Sept. 2005. W3C Member Submission. http://www.w3.org/Submission/WRL/.
3. G. Antoniou. Nonmonotonic rule systems on top of ontology layers. In *Proc. ISWC-2002*, pp. 394–398, 2002.
4. T. Berners-Lee. *Weaving the Web*. Harper, San Francisco, CA, 1999.
5. D. Calvanese, G. De Giacomo, D. Lembo, M. Lenzerini, and R. Rosati. *DL-Lite*: Tractable description logics for ontologies. In *Proc. AAAI-2005*, pp. 602–607, 2005.
6. E. Dantsin, T. Eiter, G. Gottlob, and A. Voronkov. Complexity and expressive power of logic programming. *ACM Comput. Surv.*, 33(3):374–425, 2001.
7. J. de Bruijn, T. Eiter, A. Polleres, and H. Tompits. Embedding non-ground logic programs into autoepistemic logic for knowledge-base combination. In *Proc. IJCAI-2007*, pp. 304–309, 2007.
8. F. M. Donini, M. Lenzerini, D. Nardi, and A. Schaerf. $\mathcal{AL}$-log: Integrating datalog and description logics. *J. Intell. Inf. Syst.*, 10(3):227–252, 1998.

9. T. Eiter, G. Ianni, R. Schindlauer, H. Tompits. A uniform integration of higher-order reasoning and external evaluations in answer-set programming. In *Proc. IJCAI-2005*, pp. 90–96.
10. T. Eiter, G. Ianni, R. Schindlauer, H. Tompits. Effective integration of declarative rules with external evaluations for semantic web reasoning. In *Proc. ESWC-2006*, pp. 273–287, 2006.
11. T. Eiter, T. Lukasiewicz, R. Schindlauer, and H. Tompits. Combining answer set programming with description logics for the Semantic Web. In *Proc. KR-2004*, pp. 141–151, 2004.
12. T. Eiter, T. Lukasiewicz, R. Schindlauer, and H. Tompits. Well-founded semantics for description logic programs in the Semantic Web. In *Proc. RuleML-2004*, pp. 81–97, 2004.
13. W. Faber, N. Leone, and G. Pfeifer. Recursive aggregates in disjunctive logic programs: Semantics and complexity. In *Proc. JELIA-2004*, pp. 200–212, 2004.
14. D. Fensel, W. Wahlster, H. Lieberman, and J. Hendler, editors. *Spinning the Semantic Web: Bringing the World Wide Web to Its Full Potential*. MIT Press, 2002.
15. M. Gelfond and V. Lifschitz. Classical negation in logic programs and disjunctive databases. *New Generation Comput.*, 9(3/4):365–386, 1991.
16. B. N. Grosof, I. Horrocks, R. Volz, and S. Decker. Description logic programs: Combining logic programs with description logics. In *Proc. WWW-2003*, pp. 48–57, 2003.
17. S. Heymans, D. V. Nieuwenborgh, D. Vermeir. Nonmonotonic ontological and rule-based reasoning with extended conceptual logic programs. In *Proc. ESWC-05*, pp. 392–407, 2005.
18. S. Heymans and D. Vermeir. Integrating semantic web reasoning and answer set programming. In *Proc. ASP-2003*, pp. 194–208, 2003.
19. I. Horrocks and P. F. Patel-Schneider. Reducing OWL entailment to description logic satisfiability. In *Proc. ISWC-2003*, pp. 17–29, 2003.
20. I. Horrocks and P. F. Patel-Schneider. Position paper: A comparison of two modelling paradigms in the Semantic Web. In *Proc. WWW-2006*, pp. 3–12, 2006.
21. I. Horrocks, P. F. Patel-Schneider, H. Boley, S. Tabet, B. Grosof, and M. Dean. SWRL: A Semantic Web rule language combining OWL and RuleML, May 2004. W3C Member Submission. Available at http://www.w3.org/Submission/SWRL/.
22. I. Horrocks, P. F. Patel-Schneider, and F. van Harmelen. From $\mathcal{SHIQ}$ and RDF to OWL: The making of a web ontology language. *J. Web Sem.*, 1(1):7–26, 2003.
23. I. Horrocks, U. Sattler, and S. Tobies. Practical reasoning for expressive description logics. In *Proc. LPAR-1999*, pp. 161–180, 1999.
24. U. Hustadt, B. Motik, and U. Sattler. Reducing SHIQ-description logic to disjunctive datalog programs. In *Proc. KR-2004*, pp. 152–162, 2004.
25. L. Iocchi, T. Lukasiewicz, D. Nardi, and R. Rosati. Reasoning about actions with sensing under qualitative and probabilistic uncertainty. In *Proc. ECAI-2004*, pp. 818–822, 2004.
26. N. Leone, G. Pfeifer, W. Faber, T. Eiter, G. Gottlob, S. Perri, and F. Scarcello. The DLV system for knowledge representation and reasoning. *ACM TOCL*, 7(3):499–562, 2006.
27. A. Y. Levy and M.-C. Rousset. Combining Horn rules and description logics in CARIN. *Artif. Intell.*, 104(1–2):165–209, 1998.
28. T. Lukasiewicz. Stratified probabilistic description logic programs. In *Proc. URSW-2005*, pp. 87–97, 2005.
29. T. Lukasiewicz. Probabilistic description logic programs. In *Proc. ECSQARU-2005*, pp. 737–749, 2005. Extended version in *Int. J. Approx. Reasoning*, in press.
30. T. Lukasiewicz. Fuzzy description logic programs under the answer set semantics for the Semantic Web. In *Proc. RuleML-2006*, pp. 89–96, 2006.
31. T. Lukasiewicz. A novel combination of answer set programming with description logics for the Semantic Web. Report 1843-06-08, Institut für Informationssysteme, TU Wien, 2006.
32. T. Lukasiewicz and U. Straccia. Tightly integrated fuzzy description logic programs under the answer semantics for the Semantic Web. In *Proc. RR-2007*, 2007. To appear.
33. B. Motik, I. Horrocks, R. Rosati, and U. Sattler. Can OWL and logic programming live together happily ever after? In *Proc. ISWC-2006*, pp. 501–514, 2006.

34. B. Motik and R. Rosati. A faithful integration of description logics with logic programming. In *Proc. IJCAI-2007*, pp. 477–482, 2007.
35. B. Motik, U. Sattler, and R. Studer. Query answering for OWL-DL with rules. *J. Web Sem.*, 3(1):41–60, 2005.
36. R. Rosati. On the decidability and complexity of integrating ontologies and rules. *J. Web Sem.*, 3(1):61–73, 2005.
37. R. Rosati. $\mathcal{DL}$+*log*: Tight integration of description logics and disjunctive datalog. In *Proc. KR-2006*, pp. 68–78, 2006.
38. M. Sintek and S. Decker. TRIPLE - A query, inference, and transformation language for the Semantic Web. In *Proc. ISWC-2002*, pp. 364–378, 2002.
39. T. Swift. Deduction in ontologies via ASP. In *Proc. LPNMR-2004*, pp. 275–288, 2004.
40. S. Tobies. *Complexity Results and Practical Algorithms for Logics in Knowledge Representation*. PhD thesis, RWTH Aachen, Germany, 2001.
41. W3C. OWL web ontology language overview, 2004. W3C Recommendation (10 Feb. 2004). Available at http://www.w3.org/TR/2004/REC-owl-features-20040210/.

Appendix: Selected Proofs

Proof of Theorem 5.2. Observe first that $I \subseteq HB_\Phi$ is a model of $KB^I = (L, P^I)$ iff (i) $I \models L$ and (ii) $I \models r$ for every $r \in P^I$. Since $L = \emptyset$, this is equivalent to $I \models r$ for every $r \in P^I$. Thus, $I \subseteq HB_\Phi$ is a minimal model of KB^I iff I is a minimal model of P^I. That is, $I \subseteq HB_\Phi$ is an answer set of KB iff I is an ordinary answer set of P. □

Proof of Theorem 5.3. Observe first that, by Theorem 5.1, since P is positive, the set of all answer sets of KB is given by the set of all minimal models $I \subseteq HB_\Phi$ of KB. Observe then that $a \in HB_\Phi$ is true in all minimal models $I \subseteq HB_\Phi$ of KB iff a is true in all models $I \subseteq HB_\Phi$ of KB. It thus remains to show that a is true in all models $I \subseteq HB_\Phi$ of KB iff a is true in all first-order models of $L \cup ground(P)$:

($\Rightarrow$) Suppose that a is true in all models $I \subseteq HB_\Phi$ of KB. Let $\mathcal{I}$ be any first-order model of $L \cup ground(P)$. Let $I \subseteq HB_\Phi$ be defined by $b \in I$ iff $\mathcal{I} \models b$. Then, $\mathcal{I}$ is a model of $L^\star = L \cup I \cup \{\neg b \mid b \in HB_\Phi - I\}$, and thus $L^\star$ is satisfiable. Hence, I is a model of L. Since $\mathcal{I}$ is a model of $ground(P)$, also I is a model of $ground(P)$. In summary, this shows that I is a model of KB. Hence, a is true in I, and thus a is true in $\mathcal{I}$. Overall, this shows that a is true in all first-order models of $L \cup ground(P)$.

($\Leftarrow$) Suppose that a is true in all first-order models of $L \cup ground(P)$. Let $I \subseteq HB_\Phi$ be any model of KB. Then, $L^\star = L \cup I \cup \{\neg b \mid b \in HB_\Phi - I\}$ is satisfiable. Let $\mathcal{I}$ be a first-order model of $L^\star$. Then, $\mathcal{I}$ is in particular a model of L. Furthermore, since I is a model of $ground(P)$, also $\mathcal{I}$ is a model of $ground(P)$. In summary, $\mathcal{I}$ is a model of $L \cup ground(P)$. It thus follows that a is true in $\mathcal{I}$, and thus a is also true in I. Overall, this shows that a is true in all models $I \subseteq HB_\Phi$ of KB. □

Algorithms for Paraconsistent Reasoning with OWL*

Yue Ma[1,2], Pascal Hitzler[2], and Zuoquan Lin[1]

[1] Department of Information Science, Peking University, China
[2] AIFB, Universität Karlsruhe, Germany
{mayue,lz}@is.pku.edu.cn, {yum,hitzler}@aifb.uni-karlsruhe.de

Abstract. In an open, constantly changing and collaborative environment like the forthcoming Semantic Web, it is reasonable to expect that knowledge sources will contain noise and inaccuracies. Practical reasoning techniques for ontologies therefore will have to be tolerant to this kind of data, including the ability to handle inconsistencies in a meaningful way. For this purpose, we employ paraconsistent reasoning based on four-valued logic, which is a classical method for dealing with inconsistencies in knowledge bases. Its transfer to OWL DL, however, necessitates the making of fundamental design choices in dealing with class inclusion, which has resulted in differing proposals for paraconsistent description logics in the literature. In this paper, we build on one of the more general approaches which due to its flexibility appears to be most promising for further investigations. We present two algorithms suitable for implementation, one based on a preprocessing before invoking a classical OWL reasoner, the other based on a modification of the KAON2 transformation algorithms. We also report on our implementation, called ParOWL.

1 Introduction

Real knowledge bases and data for Semantic Web applications will rarely be perfect. They will be distributed and multi-authored. They will be assembled from different sources and reused. It is unreasonable to expect such realistic knowledge bases to be always logically consistent, and it is therefore important to study ways of dealing with inconsistent knowledge. This is particularly important if the full power of logic-based approaches like the Web Ontology Language OWL [1] shall be employed, as classical logic breaks down in the presence of inconsistent knowledge.

The study of inconsistency handling in Artificial Intelligence has a long tradition, and corresponding results are recently being transferred to description logics, which underly OWL. Two fundamentally different approaches can be distinguished. The first is based on the assumption that inconsistencies indicate erroneous data which is to be repaired in order to obtain a consistent knowledge base, e.g. by selecting consistent subsets for the reasoning process [2,3]. The other approach yields to the insight that inconsistencies

* We acknowledge support by the German Federal Ministry of Education and Research (BMBF) under the SmartWeb project (grant 01 IMD01 B), by the EU under the IST project NeOn (IST-2006-027595, `http://www.neon-project.org/`), by the Deutsche Forschungsgemeinschaft (DFG) in the ReaSem project, and by China Scholarship Council, and partially by NSFC (grant numbers 60373002 and 60496322) and by NKBRPC (2004CB318000).

E. Franconi, M. Kifer, and W. May (Eds.): ESWC 2007, LNCS 4519, pp. 399–413, 2007.

are a natural phenomenon in realistic data which are to be handled by a logic which tolerates it [4,5,6]. Such logics are called paraconsistent, and the most prominent of them are based on the use of additional truth values standing for *undefined* (i.e. neither true nor false) and *overdefined* (or *contradictory*, i.e. both true and false). Such logics are appropriately called *four-valued logics* [7]. We believe that either of the approaches is useful, depending on the application scenario.

In this paper, we contribute to the paraconsistency approach. We indeed extend on the preliminary work in [6], which has the following features.

- It is grounded in prominent research results from Artificial Intelligence [8].
- It is very flexible in terms of design choices which can be made when developing a paraconsistent description logic. This concerns the issues arising from the fact that there are different ways of defining the notion of logical implication in four-valued logics. The approach which we follow allows the full and simultaneous use of the different notions of implication.
- It does not increase worst-case computational complexity of reasoning if compared to standard reasoning methods for consistent knowledge bases.

In this paper, we present two algorithms for practical paraconsistent reasoning based on this approach. The first one is based on a transformation from a paraconsistent ontology O to a classical two-valued ontology $\overline{O}$ in such a way that paraconsistent reasoning on O can be simulated by classical reasoning on $\overline{O}$. The second algorithm is based on an adaptation of the algorithms underlying the KAON2 OWL Reasoner[1] [9] by realizing a resolution-based decision procedure for paraconsistent reasoning. We spell out the details for the description logic $\mathcal{ALC}$, which is considered to be the most foundational one and comprises a large fragment of OWL DL.

The paper is structured as follows. We first review briefly preliminaries in Section 2. In Section 3 we then describe the syntax and semantics of the paraconsistent description logic which we will use. Sections 4 and 5 describe the two reasoning procedures, respectively based on a transformation for preprocessing and on an adaptation of the KAON2 algorithms. In Section 6, we describe the prototype of our paraconsistent approach to reasoning with a possible inconsistent ontology, and discuss future work and conclude in Section 7.

This paper is a substantial continuation of work presented as preliminary results in [6]. Due to space limitations, proofs are omitted. They can be found in a technical report available from http://www.aifb.uni-karlsruhe.de/WBS/phi/pub/parowltr.pdf.

2 Preliminaries

2.1 The Description Logic $\mathcal{ALC}$

We briefly review notation and terminology of the description logic $\mathcal{ALC}$, but we basically assume that the reader is familiar with description logics. For comprehensive background reading, please refer to [10].

[1] http://kaon2.semanticweb.org

Table 1. Syntax and semantics of $\mathcal{ALC}$

Constructor Name	Syntax	Semantics
atomic concept A	A	$A^I \subseteq \Delta^I$
abstract role R_A	R	$R^I \subseteq \Delta^I \times \Delta^I$
individuals I	o	$o^I \in \Delta^I$
top concept	$\top$	Δ^I
bottom concept	$\bot$	$\emptyset$
conjunction	$C_1 \sqcap C_2$	$C^I \cap D^I$
disjunction	$C_1 \sqcup C_2$	$C^I \cup D^I$
negation	$\neg C$	$\Delta^I \setminus C^I$
exists restriction	$\exists R.C$	$\{x \mid \exists y, (x, y) \in R^I \text{ and } y \in C^I\}$
value restriction	$\forall R.C$	$\{x \mid \forall y, (x, y) \in R^I \text{ implies } y \in C^I\}$

Axiom Name	Syntax	Semantics
concept inclusion	$C_1 \sqsubseteq C_2$	$C_1^I \subseteq C_2^I$
concept assertion	$C(a)$	$a^I \in C^I$
role assertion	$R(a, b)$	$(a^I, b^I) \in R^I$

We assume that we are given a set of atomic concepts (or concept names), a set of roles (or role names), and a set of individuals. With the symbols $\top$ and $\bot$ we furthermore denote the top concept and the bottom concept, respectively.

Complex concepts in $\mathcal{ALC}$ can be formed from these inductively as follows.

1. $\top$, $\bot$, and each atomic concept are concepts;
2. If C, D are concepts, then $C \sqcup D$, $C \sqcap D$, and $\neg C$ are concepts;
3. If C is a concept and R is a role, then $\forall R.C$ and $\exists R.C$ are concepts.

An $\mathcal{ALC}$ ontology consists of a set of assertions, called the $ABox$ of the ontology, and a set of inclusion axioms, calld the $TBox$ of the ontology. Assertions are of the form $C(a)$ or $R(a, b)$, where a, b are individuals and C and R are concepts and roles, respectively. Inclusion axioms are of the form $C \sqsubseteq D$, where C and D are concepts. Informally, an assertion $C(a)$ means that the individual a is an instance of concept C, and an assertion $R(a, b)$ means that individual a is related with individual b via the property R. The inclusion axiom $C \sqsubseteq D$ means that each individual of C is an individual of D.

The formal definition of the (model-theoretic) semantics of $\mathcal{ALC}$ is given by means of interpretations $I = (\Delta^I, \cdot^I)$ consisting of a non-empty domain Δ^I and a mapping $\cdot^I$ satisfying the conditions in Table 1, interpreting concepts as subsets of the domain and roles as binary relations on the domain. An interpretation satisfies an $\mathcal{ALC}$ ontology (i.e. is a model of the ontology) iff it satisfies each axiom in both the $ABox$ and the $TBox$. An ontology is called satisfiable (unsatisfiable) iff there exists (does not exist) such a model. In $\mathcal{ALC}$, reasoning tasks, i.e. the derivation of logical consequences, can be reduced to satisfiability checking of ontologies [10,11].

Table 2. Truth table for 4-valued connectives

α	f	f	f	f	t	t	t	t	$\ddot{\top}$	$\ddot{\top}$	$\ddot{\top}$	$\ddot{\top}$	$\ddot{\bot}$	$\ddot{\bot}$	$\ddot{\bot}$	$\ddot{\bot}$
β	f	t	$\ddot{\top}$	$\ddot{\bot}$	f	t	$\ddot{\top}$	$\ddot{\bot}$	f	t	$\ddot{\top}$	$\ddot{\bot}$	f	t	$\ddot{\top}$	$\ddot{\bot}$
$\neg\alpha$	t	t	t	t	f	f	f	f	$\ddot{\top}$	$\ddot{\top}$	$\ddot{\top}$	$\ddot{\top}$	$\ddot{\bot}$	$\ddot{\bot}$	$\ddot{\bot}$	$\ddot{\bot}$
$\alpha \wedge \beta$	f	f	f	f	f	t	$\ddot{\top}$	$\ddot{\bot}$	f	$\ddot{\top}$	$\ddot{\top}$	f	f	$\ddot{\bot}$	f	$\ddot{\bot}$
$\alpha \vee \beta$	f	t	$\ddot{\top}$	$\ddot{\bot}$	t	t	t	t	$\ddot{\top}$	t	$\ddot{\top}$	t	$\ddot{\bot}$	t	t	$\ddot{\bot}$
$\alpha \mapsto \beta$	t	t	t	t	f	t	$\ddot{\top}$	$\ddot{\bot}$	$\ddot{\top}$	t	$\ddot{\top}$	t	$\ddot{\bot}$	t	t	$\ddot{\bot}$
$\alpha \supset \beta$	t	t	t	t	f	t	$\ddot{\top}$	$\ddot{\bot}$	f	t	$\ddot{\top}$	$\ddot{\bot}$	t	t	t	t
$\alpha \rightarrow \beta$	t	t	t	t	f	t	f	$\ddot{\bot}$	f	t	$\ddot{\top}$	$\ddot{\bot}$	$\ddot{\bot}$	t	$\ddot{\bot}$	t

2.2 Four-Valued Logic

The major studies of four-valued logics have been carried out in the setting of propositional logic. We will very briefly review the preliminaries which set the state for the four-valued version of $\mathcal{ALC}$ which we will present later in Section 3.

The idea of four-valued logic is based on the idea of having four truth values, instead of the classical two. The four truth values stand for *true, false, unknown* (or *undefined*) and *both* (or *overdefined, contradictory*). We use the symbols $t, f, \ddot{\bot}, \ddot{\top}$, respectively, for these truth values, and the set of these four truth values is denoted by FOUR. The truth value $\ddot{\top}$ stands for contradictory information, hence four-valued logic lends itself to dealing with inconsistent knowledge. The value $\ddot{\top}$ thus can be understood to stand for *true and false*, while $\ddot{\bot}$ stands for *neither true nor false*, i.e. for the absence of any information about truth or falsity.

Syntactically, four-valued logic is very similar to classical logic. Care has to be taken, however, in defining meaningful notions of implication, as there are several ways to do this. Indeed, there are three major notions of implication in the literature, all of which we will employ in our approach. The logical connectives we allow are thus negation $\neg$, disjunction $\vee$, conjunction $\wedge$, material implication $\mapsto$, internal implication $\supset$, and strong implication $\rightarrow$. We will discuss them in detail later on as the presence of all three implications is crucial for our approach.

Four-valued interpretations for formulae (i.e. 4-interpretations) are obviously mappings from formulae to (the set of four) truth values, respecting the truth tables for the logical connectives, as detailed in Table 2.

Four-valued models (4-models) are defined in the obvious way, as follows, where t and $\ddot{\top}$ are the designated truth values.

Definition 1. *Let I be a 4-interpretation, let Σ be a theory (i.e. set of formulae) and let φ be a formula in four-valued logic. Then I is a 4-model of φ if and only if $I(\varphi) \in \{t, \ddot{\top}\}$. I is a 4-model of Σ if and only if I is a 4-model of each formula in Σ. Σ four-valued entails φ, written $\Sigma \models_4 \varphi$, if and only if every 4-model of Σ is a 4-model of φ.*

Proposition 1. *We note the following general properties.*

- *The language* $L = \{\neg, \vee, \wedge, \supset, \ddot{\bot}, \ddot{\top}\}$ *is functional complete for the set* **FOUR** *of truth values, i.e. every function from* $\textbf{FOUR}^n$ *to* **FOUR** *is representable by some formula in L [8, Theorem 12].*
- *Any formula containing only connectives from* $\{\neg, \vee, \wedge, \supset\}$ *always has a four-valued model.*

Some general remarks about the different notions of implication are in order. They are the major notions of implication used in the literature, and are discussed in detail in [8,12]. The basic rationales behind them are the following: Material implication can be defined by means of negation and disjunction as known from classical logic. However, it does not satisfy Modus Ponens or the deduction theorem, and is thus of limited use as an *implication* in the intuitive sense. Internal implication satisfies Modus Ponens and the deduction theorem, but cannot be defined by means of other connectives. Furthermore, internal implication does not satisfy contraposition. Strong implication is stronger than internal implication, in that it additionally satisfies contraposition. Indeed, an alternative view on the truth tables for the implication connectives is as follows.

$$\varphi \mapsto \psi \text{ is definable as } \neg\varphi \vee \psi. \qquad \text{(Material Implication)}$$

$$\varphi \supset \psi \quad \text{evaluates to} \quad \begin{cases} \psi & \text{if } \varphi \in \{t, \ddot{\top}\} \\ t & \text{if } \varphi \in \{f, \ddot{\bot}\} \end{cases} \qquad \text{(Internal Implication)}$$

$$\varphi \rightarrow \psi \text{ is definable as } (\varphi \supset \psi) \wedge (\neg\psi \supset \neg\varphi) \qquad \text{(Strong Implication)}$$

Further properties of the implication connectives are summarised in the following proposition (as shown in [8, Corollary 9] and [12]).

Proposition 2. *The following claims hold, where* Γ *is a theory and* ψ, ϕ *are formulae.*

- *Internal implication is not definable in terms of the connectives* $\neg, \vee, \wedge$.
- $\Gamma, \psi \models_4 \phi$ *iff* $\Gamma \models_4 \psi \supset \phi$.
- *If* $\Gamma \models_4 \psi$ *and* $\Gamma \models_4 \psi \supset \phi$ *then* $\Gamma \models_4 \phi$.
- $\psi \rightarrow \phi$ *implies that* $\neg\phi \rightarrow \neg\psi$.

Apart from the formal properties of the different notions of implication, it is obviously important to consider their intuitive meaning and their usefulness for knowledge base modelling. We will discuss this in detail in the next section.

3 The Four-Valued Description Logic $\mathcal{ALC}4$

We describe the syntax and semantics of our four-valued description logic $\mathcal{ALC}4$. The approach is fairly standard apart from the fact that we allow the simultaneous use of all three notions of implication introduced in Section 2.2. We will thus devote significant space to a detailed discussion of the intuitions behind these different implications.

Syntactically, $\mathcal{ALC}4$ hardly differs from $\mathcal{ALC}$. Complex concepts and assertions are defined in exactly the same way. For class inclusion, however, the question arises how to interpret the underlying implication connective in the four-valued setting. We thus allow

Table 3. Semantics of $\mathcal{ALC}4$ Concepts

Constructor Syntax	Semantics
A	$A^I = \langle P, N \rangle$, where $P, N \subseteq \Delta^I$
R	$R^I = \langle R_P, R_N \rangle$, where $R_P, R_N \subseteq \Delta^I \times \Delta^I$
o	$o^I \in \Delta^I$
$\top$	$\langle \Delta^I, \emptyset \rangle$
$\bot$	$\langle \emptyset, \Delta^I \rangle$
$C_1 \sqcap C_2$	$\langle P_1 \cap P_2, N_1 \cup N_2 \rangle$, if $C_i^I = \langle P_i, N_i \rangle$ for $i = 1, 2$
$C_1 \sqcup C_2$	$\langle P_1 \cup P_2, N_1 \cap N_2 \rangle$, if $C_i^I = \langle P_i, N_i \rangle$ for $i = 1, 2$
$\neg C$	$(\neg C)^I = \langle N, P \rangle$, if $C^I = \langle P, N \rangle$
$\exists R.C$	$\langle \{x \mid \exists y, (x, y) \in \text{proj}^+(R^I) \text{ and } y \in \text{proj}^+(C^I)\}$, $\{x \mid \forall y, (x, y) \in \text{proj}^+(R^I) \text{ implies } y \in \text{proj}^-(C^I)\}\rangle$
$\forall R.C$	$\langle \{x \mid \forall y, (x, y) \in \text{proj}^+(R^I) \text{ implies } y \in \text{proj}^+(C^I)\}$, $\{x \mid \exists y, (x, y) \in \text{proj}^+(R^I) \text{ and } y \in \text{proj}^-(C^I)\}\rangle$

three kinds of class inclusions, corresponding to the three implication connectives we have discussed. They are as follows, $C \mapsto D$, $C \sqsubset D$, and $C \to D$, called material inclusion axiom, internal inclusion axiom, and strong inclusion axiom, respectively.

Semantically, interpretations map individuals to elements of the domain of the interpretation, as usual. For concepts, however, we need to make modifications to the notion of interpretation in order to allow for reasoning with inconsistencies.

Intuitively, in four-valued logic we need to consider four situations which can occur in terms of containment of an individual in a concept: (1) we know it is contained, (2) we know it is not contained, (3) we have no knowledge whether or not the individual is contained, (4) we have contradictory information, namely that the individual is both contained in the concept and not contained in the concept. There are several equivalent ways how this intuition can be formalised, one of which is described in the following.

For a given domain Δ^I and a concept C, an interpretation over Δ^I assigns to C a pair $\langle P, N \rangle$ of (not necessarily disjoint) subsets of Δ^I. Intuitively, P is the set of elements known to belong to the extension of C, while N is the set of elements known to be not contained in the extension of C. For simplicity of notation, we define functions $\text{proj}^+(\cdot)$ and $\text{proj}^-(\cdot)$ by $\text{proj}^+\langle P, N \rangle = P$ and $\text{proj}^-\langle P, N \rangle = N$.

Formally, a four-valued interpretation is a pair $I = (\Delta^I, \cdot^I)$ with Δ^I as domain, where $\cdot^I$ is a function assigning elements of Δ^I to individuals, and subsets of $(\Delta^I)^2$ to concepts, such that the conditions in Table 3 are satisfied. Note that the conditions in Table 3 for role restrictions are designed in such a way that the logical equivalences $\neg(\forall R.C) = \exists R.(\neg C)$ and $\neg(\exists R.C) = \forall R.(\neg C)$ are retained – this is the most convenient way for us for handling role restrictions, as it will allows for a straightforward translation from $\mathcal{ALC}4$ to classical $\mathcal{ALC}$. Note also that for roles we actually require only the positive part of the extension – we nevertheless require interpretations to assign pairs of sets to roles, which is a technical formality to retain consistency of notation with possible extensions to more expressive description logics (see [6]).

Obviously, under the constraints $P \cap N = \emptyset$ and $P \cup N = \Delta$, four-valued interpretations become just standard two-valued interpretations.

Table 4. Semantics of inclusion axioms in $\mathcal{ALC}4$

Axiom Name	Syntax	Semantics
material inclusion	$C_1 \mapsto C_2$	$\Delta^I \setminus \text{proj}^-(C_1^I) \subseteq \text{proj}^+(C_2^I)$
internal inclusion	$C_1 \sqsubset C_2$	$\text{proj}^+(C_1^I) \subseteq \text{proj}^+(C_2^I)$
strong inclusion	$C_1 \rightarrow C_2$	$\text{proj}^+(C_1^I) \subseteq \text{proj}^+(C_2^I)$ and $\text{proj}^-(C_2^I) \subseteq \text{proj}^-(C_1^I)$
concept assertion	$C(a)$	$a^I \in \text{proj}^+(C^I)$
role assertion	$R(a, b)$	$(a^I, b^I) \in \text{proj}^+(R^I)$

The correspondence between truth values from FOUR and concept extensions is the obvious one: For instances $a \in \Delta^I$ and concept name C we have

- $C^I(a) = t(\ddot{\top})$, iff $a^I \in \text{proj}^+(C^I)$ and $a^I \notin (\in)\text{proj}^-(C^I)$,
- $C^I(a) = f(\ddot{\bot})$, iff $a^I \notin \text{proj}^+(C^I)$ and $a^I \in (\notin)\text{proj}^-(C^I)$,

When defining the semantics as we just did, we ensure that a number of useful equivalences from classical logic hold, as follows.

Proposition 3. *For any four-valued interpretation I and concepts C, D, the following claims hold.*

$$(C \sqcap \top)^I = C^I, (C \sqcup \top)^I = \top^I, (C \sqcap \bot)^I = \bot^I, (C \sqcup \bot)^I = C^I,$$
$$(\neg\neg C)^I = C^I, (\neg\top)^I = \bot^I, (\neg\bot)^I = \top^I, (\neg(C \sqcup D))^I = (\neg C \sqcap \neg D)^I,$$
$$(\neg(C \sqcap D))^I = (\neg C \sqcup \neg D)^I, (\neg(\forall R.C))^I = (\exists R.\neg C)^I, (\neg(\exists R.C))^I = (\forall R.\neg C)^I.$$

We now come to the semantics of the three different types of inclusion axioms. It is formally defined in Table 4 (together with the semantics of concept assertions). We say that a four-valued interpretation I satisfies a four-valued ontology $\mathcal{O}$ (i.e. is a model of it) iff it satisfies each assertion and each inclusion axiom in $\mathcal{O}$. An ontology $\mathcal{O}$ is satisfiable (unsatisfiable) iff there exists (does not exist) such a model.

With the formal definitions out of the way, it remains to address the intuitions underlying the different inclusion axioms. These intuitions are evidenced by the formal properties of the underlying implications as discussed in Section 2.2 as well as the behaviour of the implications in practice. We actually foresee a possible workflow for handling inconsistent ontologies, as follows. In a first step, inclusion axioms are classified into the three types of four-valued inclusion axioms available. Then four-valued reasoning is performed based on the classification, in order to arrive at a meaningful 4-valued conclusion. The question, how such a classification can be performed, will not be addressed in this paper. It constitutes a seperate substantial piece of work which is under investigation by the authors. A combination of automated detection and a user-interaction process may be the most workable solution, where the user-interaction process may be guided by the intuitive explanations which we will now give for the three types of inclusion. An example which displays the effects of the different inclusions is given in Section 6.

Strong inclusion respects the deduction theorem and contraposition reasoning. In a paraconsistent context, it is thus the inclusion to be used for universal truth, such as Square $\rightarrow$ FourEdged.

Internal inclusion propagates contradictory information forward, but not backward as it does not allow for contraposition reasoning. It can thus be characterized as a *brave* way of handling inconsistency. It should be used whenever it is important to infer the consequent even if the antecedent may be contradictory. To give an example, consider a robot fault diagnosis system and an axiom stating that oil leakage is indicative of a robot malfunction. Obviously, it is important to check on a possible malfunction even in case there is contradictory information about an oil leakage. In a paraconsistent context, the axiom is thus best modeled by means of internal inclusion, i.e. as OilLeakage $\sqsubset$ RobotMalfunction.

Material inclusion is *cautious* in the sense that contradictory information is not propagated. The intuition behind material inclusion becomes apparent by studying the truth table for material implication: $a \mapsto b$ indicates that the only way for b to be *not true* (i.e. to be f or $\ddot{\bot}$) is if there is information of falsity of a (i.e. it is f or $\ddot{\top}$). This kind of modeling becomes important if an inclusion has to be second-guessed e.g. after a merging of knowledge bases. Consider, for example, an ontology about marathon runs containing the axiom Healthy $\sqsubset$ eqMarathonParticipant which is supposed to say that somebody (i.e. a person who has signed up for a run) participates in a marathon if he checks out to be healthy. The axiom is reasonable if the domain is for the management of marathon participants' data only. Now imagine that this ontology is merged with other sports knowledge bases, e.g. a boxing domain. It is wrong to infer that every healthy boxer will participate in the marathon, so the original axiom will likely lead to contradictions. We propose to handle this kind of information by modelling the axiom as material inclusion, i.e. as Healthy $\mapsto$ MarathonParticipant, which will indeed not infer participation from a positive health status. However, the weak form of contraposition reasoning featured by material inclusion results in the following situation: If an individual is not known to be contained in MarathonParticipant, then it is known to be not Healthy, resulting in a possible contradiction on health status while avoiding contradiction in terms of marathon participation, which may be preferred in the domain. Material inclusion may thus propagate contradictory information backwards (to the antecedent), while internal inclusion may propagate contradictory information forward (to the consequent).

We remark here that different inclusion axioms provide ontology engineers with a flexible way to define different ontologies according to the intuition explained above. In case only one kind of inclusion shall be used, we recommend to use strong inclusion, as it should serve the ontology engineer's original intention most closely. To give an example, consider the inconsistent subontology of BuggyPolicy[2] (with additional assertions) which says "*GeneralReliabilityUsernamePolicy* (*G* for short) is a subset of *Reliable*, *G* and *Messaging* are disjoint, *Reliable* is a subset of *Messaging*, p_1 is an individual of *G* and p_2 is an individual of *Reliable*". Using strong inclusion results in the ontology $\{R \to M, G \to R, M \to \neg \mathrm{G}, G \to \neg M, G(p_1), R(p_2)\}$, where we use obvious abbreviations for the class names. Under the semantics of strong inclusion, $M(p_1), R(p_1), M(p_2), \neg G(p_1)$, and $\neg M(p_1)$ hold, but $G(p_2)$ does not hold. This example shows that our four-valued semantics can give meaningful answers when an ontology is inconsistent, while classical semantics fails to do so.

[2] http://www.mindswap.org/2005/debugging/ontologies/

4 Transforming $\mathcal{ALC}4$ to $\mathcal{ALC}$

It is a pleasing property of $\mathcal{ALC}4$, that it can be translated easily into classical $\mathcal{ALC}$, such that paraconsistent reasoning can be simulated using standard $\mathcal{ALC}$ reasoning algorithms. We briefly present the translation, a preliminary report has appeared in [6].[3]

For any given concept C, its transformation $\overline{C}$ is the concept obtained from C by the following inductively defined transformation.

- If $C = A$ for A an atomic concept, then $\overline{C} = A^+$, where A^+ is a new concept;
- If $C = \neg A$ for A an atomic concept, then $\overline{C} = A^-$, where A^- is a new concept;
- If $C = \top$, then $\overline{C} = \top$;
- If $C = \bot$, then $\overline{C} = \bot$;
- If $C = E \sqcap D$ for concepts D, E, then $\overline{C} = \overline{E} \sqcap \overline{D}$;
- If $C = E \sqcup D$ for concepts D, E, then $\overline{C} = \overline{E} \sqcup \overline{D}$;
- If $C = \exists R.D$ for D a concept and R is a role, then $\overline{C} = \exists R.\overline{D}$;
- If $C = \forall R.D$ for D a concept and R is a role, then $\overline{C} = \forall R.\overline{D}$;
- If $C = \neg\neg D$ for a concept D, then $\overline{C} = \overline{D}$;
- If $C = \neg(E \sqcap D)$ for concepts D, E, then $\overline{C} = \overline{\neg E} \sqcup \overline{\neg D}$;
- If $C = \neg(E \sqcup D)$ for concepts D, E, then $\overline{C} = \overline{\neg E} \sqcap \overline{\neg D}$;
- If $C = \neg(\exists R.D)$ for D a concept and R is a role, then $\overline{C} = \forall R.\overline{\neg D}$;
- If $C = \neg(\forall R.D)$ for D a concept and R is a role, then $\overline{C} = \exists R.\overline{\neg D}$;

Based on this, axioms are transformed as follows, where C_1, C_2 are concepts.

- $\overline{C_1 \mapsto C_2} = \neg\overline{\neg C_1} \sqsubseteq \overline{C_2}$
- $\overline{C_1 \sqsubset C_2} = \overline{C_1} \sqsubseteq \overline{C_2}$;
- $\overline{C_1 \rightarrow C_2} = \{\overline{C_1} \sqsubseteq \overline{C_2}, \overline{\neg C_2} \sqsubseteq \overline{\neg C_1}\}$.
- $\overline{C(a)} = \overline{C}(a)$, $\overline{R}(a, b) = R(a, b)$, where a, b are individuals, C a concept, R a role.

The following theorem shows that paraconsistent reasoning can indeed be simulated on standard reasoners by means of the transformation just given.

Theorem 1. *For any ontology O we have $O \models_4 \alpha$ if and only if $\overline{O} \models_2 \overline{\alpha}$, where $\models_2$ is the entailment in classical $\mathcal{ALC}$.*

We note that the transformation algorithm is linear in the size of the ontology. This implies that paraconsistent reasoning in our paradigm is not more expensive than classical reasoning.

5 Resolution-Based Reasoning with $\mathcal{ALC}4$

There exist two fundamentally different approaches to reasoning with description logics. The first, historic approach is based on an adaptation of the tableaux method from first-order predicate logic (see [10]), and is implemented in most current reasoners. The second approach is based on resolution and has been realised in the KAON2 reasoner [9]. While the first method invokes a classical reasoner as a black-box by a

[3] In [6], it was actually spelled out for $\mathcal{SHOIN}$.

preprocessing spelled out in Section 4, the paraconsistent resolution given in this section views the classical reasoner KAON2 as a glass-box, thus avoiding the preprocessing step. We basically follow [9, Chapter 4], and indeed we have to assume that the reader is familiar with the KAON2-approach because space restrictions do not allow us to spell everything out in detail.

We first note that resolution relies heavily on the *tertium non datur*, and thus does not lend itself easily to a paraconsistent setting. In particular, resolution cannot be based on the negation present in paraconsistent logics, as in this case $A \vee B$ and $\neg A \vee C$ does not imply $B \vee C$. We thus start by introducing a second kind of negation, called the *total negation*, denoted by $\sim$. In order to avoid confusion, we will refer to the standard negation as *paraconsist negation*.

Definition 2. *The* total negation $\sim$ *on* $\{\langle P, N\rangle \mid P, N \subseteq \Delta\}$ *is defined by*

$$\sim\langle P, N\rangle = \langle \Delta \setminus P, \Delta \setminus N\rangle.$$

The intuition behind total negation is to reverse both the information of being true and of being false. Notice that we do not extend our four-valued DLs to have the total negation as a concept constructor. We rather use it only to provide a resolution-based decision procedure for four-valued DLs.

Proposition 4. *For total negation, the following hold for all concepts C, D and roles R. For any four-valued interpretation I,*

$$(\sim\sim C)^I = C^I, (\sim\top)^I = \bot^I, (\sim\bot)^I = \top^I, (\neg\sim C)^I = (\sim\neg C)^I$$
$$(\sim(C \sqcup D))^I = (\sim C \sqcap \sim D)^I, (\sim(C \sqcap D))^I = (\sim C \sqcup \sim D)^I,$$
$$(\sim(\forall R.C))^I = (\exists R.\sim C)^I, (\sim(\exists R.C))^I = (\forall R.\sim C)^I.$$

A second issue which we have to address when adjusting resolution to the paraconsistent setting, is to obtain a representation of internal implication (i.e. of internal inclusion) in terms of clauses. We have already remarked in Section 2.2 that internal implication cannot be represented by means of the connectives conjunction, disjunction and paraconsistent negation. However, with total negation a representation of $C \sqsubset D$ as $\sim C \sqcup D$ is possible. The representation is actually not logically equivalent, but it is equisatisfiable, which suffices for setting up a resolution procedure. We indeed have the following theorem.

Theorem 2. *Let O be a four-valued $\mathcal{ALC}4$ ontology, C, D be concepts, I be an interpretation and ι be a new individual not occurring in O. Then the following hold.*

1. $(C \sqsubset D)^I \in \{t, \ddot{\top}\}$ *if and only if* $(\sim C \sqcup D)^I \in \{t, \ddot{\top}\}$.
2. $O \models_4 C(a)$ *if and only if* $O \cup \{\sim C(a)\}$ *is four-valued unsatisfiable.*
3. $O \models_4 C \mapsto D$ *if and only if* $O \cup \{\sim(\neg C \sqcup D)(\iota)\}$ *is four-valued unsatisfiable.*
4. $O \models_4 C \sqsubset D$ *if and only if* $O \cup \{(C \sqcap \sim D)(\iota)\}$ *is four-valued unsatisfiable.*
5. $O \models_4 C \rightarrow D$ *if and only if* $O \cup \{(C \sqcap \sim D)(\iota), (\neg D \sqcap \sim\neg C)(\iota)\}$ *is four-valued unsatisfiable.*

5.1 Translating $\mathcal{ALC}4$ into Clauses

Resolution-based calculi operates on sets of clauses in normal form, so we introduce next clausal forms for $\mathcal{ALC}4$ expressions. We were inspired by [13]. We first define a negation normal form for $\mathcal{ALC}4$ concepts, which we call quasi-NNF.

Definition 3. *A concept C is a quasi-atom, if it is an atomic concept, or in form $\neg A$ where A is an atomic concept. A concept C is a quasi-literal, if it is a quasi-atomic concept, or in form $\sim L$ where L is a quasi-atomic concept. A concept C is in quasi-NNF, if the total negation $\sim$ occurs only in front of quasi-literals.*

To give an example, let A, B, and C be atomic concepts. Then $(A \vee \sim\neg B) \sqcup \forall R.(\sim C)$ is in quasi-NNF. By propositions 3 and 4, the following is obvious.

Theorem 3. *All $\mathcal{ALC}4$ concepts can be transformed into equivalent expressions in quasi-NNF.*

We next define the Definitorial form of $\mathcal{ALC}4$ concepts, which is a technicality to control the size of clauses (see to [9] for details). If C is a concept, then we set

$$\mathrm{Def}(C) = \begin{cases} \{C\} & \text{if } C \text{ is a literal concept,} \\ \{\sim Q \sqcup C|_p\} \cup \mathrm{Def}(C[Q]|_p) & \text{if } p \text{ is eligible for replacement in } C. \end{cases}$$

where $C|_p$ is the position p in concept C, as defined in [14,9].

As an example, we have $\mathrm{Def}(A \sqcap \exists R.(A \sqcap B)) = \{A \sqcap \exists R.Q, \sim Q \sqcup (A \sqcap B)\}$. Note that $\sim Q \sqcup (A \sqcap B)$ can be interpreted as internal inclusion $Q \sqsubset (A \sqcap B)$, which allows us to use Q as a new name for $(A \sqcap B)$ in $\exists R.(A \sqcap B))$.

The following proposition is as expected.

Proposition 5. *For an $\mathcal{ALC}4$ concept C in quasi-NNF, $C(x)$ has information to be true for all individuals x if and only if all concepts $D_i(x)$ with $D_i \in Def(C)$ have the information to be true. Formally, $\{\top \sqsubset C\}$ is four-valued satisfiable iff $\{\top \sqsubset Def(D_i) \mid D_i \in Def(C)\}$ is.*

We next translate the concepts into predicate logic. This is done by the standard translation as e.g. spelled out in [9] in terms of the function π_y – we just have to provide for the total negation. This is done by allowing the total negation to occur in the predicate logic formulae as well, and by translating total negation in the same way as paraconsistent negation. We make one exception, namely for unversal restriction, where we set $\pi_y(\forall R.C, x) = \forall y.(\sim R(x, y) \sqcup C(y))$.

Following the above transformations step by step, any $\mathcal{ALC}4$ concept can be translated into a set of first order predicate logic clauses (with total negation) in polynomial size of the original concepts.

The obtained predicate logic formulae (with total negation) can now be translated into clauses in the standard way, i.e. by first casting them into Skolem form, and then into conjunctive normal form by exhaustive application of well-known logical equivalences (see e.g. [14]), which are adjusted for total negation in the obvious straightforward way.

Table 5. Clause Types

1	$\bigvee(\sim)(\neg)C_i(x) \vee R(x, f(x))$
2	$\bigvee(\sim)(\neg)C_i(x) \vee (\sim)(\neg)D(f(x))$
3	$\bigvee(\sim)(\neg)C_i(x)$
4	$\bigvee(\sim)(\neg)C_i(x) \vee R(x, y) \vee (\sim)(\neg)D(y)$
5	$(\sim)(\neg)C(a)$
6	$(\sim)(\neg)R(a, b)$

If C is a concept (where the additional use of total negation is allowed), then we denote by Cls(C) the set of clauses which is obtained by the just mentioned transformation. These clauses are predicate logic formulae (with total negation).

We finally translate an $\mathcal{ALC}4$ knowledge base KB into a set $\Xi(KB)$ of predicate logic clauses (with total negation), as follows. The knowledge base $\Xi(KB)$ is the smallest set satisfying the following conditions:

- For each $ABox$ axiom α in $ABox$, Cls(α) $\subseteq \Xi(KB)$
- For each $TBox$ axiom $C \mapsto D$ in $TBox$, Cls($\neg C \sqcup D$) $\subseteq \Xi(KB)$
- For each $TBox$ axiom $C \sqsubset D$ in $TBox$, Cls($\sim C \sqcup D$) $\subseteq \Xi(KB)$
- For each $TBox$ axiom $C \rightarrow D$ in $TBox$, Cls($\sim C \sqcup D, \sim\neg D \sqcup \neg C$) $\subseteq \Xi(KB)$

Theorem 4. *Let KB be an $\mathcal{ALC}4$ ontology. Then, the following hold.*

- *KB is satisfiable iff $\Xi(KB)$ is satisfiable.*
- *$\Xi(KB)$ can be computed in time polynomial in $|KB|$.*
- *Each clause in $\Xi(KB)$ is of one of the syntactic forms listed in Table 5. We refer to these clauses as* 4-valued clauses.

5.2 Ordered Resolution with Selection Function $O4_{DL}$ for $\mathcal{ALC}4$

The KAON2 approach for $\mathcal{ALC}$ is based on a modification of the original resolution calculus, known as ordered resolution (see [9] for the necessary preliminaries). We will now define the corresponding notions of clause ordering and of selection function, which we require for this. We assume in the sequel that all 4-valued clauses are of the form described in theorem 4.

Given any fixed ordering $\succ$ on ground quasi-atoms which is total and well-founded, we can obtain an *ordering on sets of clauses* as follows.

1. Extend $\succ$ to an ordering $\succ_L$ on ground literals by setting $\sim A \succ_L A$ for any A, and $[\sim]A \succ_L [\sim]B$, if $A \succ B$.
2. Extend $\succ_L$ to an ordering $\succ_C$ on ground clauses by setting $\succ_C = (\succ_L)_{mul}$ to be the multi-set extension of $\succ_L$ (see [9] for formal definition). The intuition is that $C_1 \succ_C C_2$ iff the maximal quasi-literal in clause C_1 is greater then that in clause C_2 w.r.t. $\succ_L$.

By a slight abuse of notation, we use $\succ$ also for $\succ_L$ and $\succ_C$ where the meaning is clear from the context.

By a *selection function* we mean a mapping S that assigns to each clause C a (possibly empty) multiset $S(C)$ of literals with the prefix $\sim$ in C. For example, both $\{\sim\neg A\}$ and $\{\sim\neg A, \sim D\}$ can be selected in clause $\sim\neg A \vee \sim D \vee B \vee \neg C$.

An ordered resolution step with selection function can now be described by the inference rule

$$\frac{C \vee A \quad D \vee \sim B}{C\sigma \vee D\sigma}$$

where

- $\sigma = MGU(A, B)$ is the most general unifier of the quasi-atoms A, B, and C, D are quasi-clauses.
- $A\sigma$ is strictly maximal in $C\sigma \vee A\sigma$, and no literal is selected in $C\sigma \vee A\sigma$;
- $\sim B\sigma$ is either selected in $D\sigma \vee \sim B\sigma$, or it is maximal in $D\sigma \vee \sim B\sigma$ and no literal is selected in $D\sigma \vee \sim B\sigma$.

The corresponding ordered factorization rule is

$$\frac{C \vee A \vee B}{(C \vee A)\sigma}$$

where $\sigma = MGU(A, B)$ and $A\sigma$ is maximal in $C\sigma \vee A\sigma$ and nothing is selected in C.

Theorem 5. *(Soundness and Completeness of $O4_{DL}$) Let N be an $\mathcal{ALC}4$ knowledge base. Then $\Xi(N) \vdash_{O4_{DL}} \Box$ iff N is four-valued unsatisfiable.*

We can now select suitable parameters in order to arrive at a decision procedure based on $O4_{DL}$. This can be done as follows.

- The literal ordering $\succ$ is defined such that $R(x, f(x)) \succ \sim C(x)$ and $D(f(x)) \succ \sim C(x)$, for all function symbols f, and predicates R, C, and D.
- The selection function selects every binary literal which is preceeded by $\sim$.

Theorem 6. *(Decidability) For an $\mathcal{ALC}4$ knowledge base KB, saturating $\Xi(KB)$ by $\mathcal{O}4_{DL}$ decides satisfiability of KB and runs in time exponential in $|KB|$.*

6 Implementation

ParOWL[4] is a prototype implementation of our paraconsistent reasoning approach. It realises the algorithm from Section 4 by means of a command line tool based on KAON2. In order to allow the use of standard OWL syntax, the tool expects four input files as parameters, all of which are standard OWL documents. In the first file, class inclusion is interpreted as material inclusion, in the second as internal inclusion, and in the third as strong inclusion. The fourth file is expected to contain an ABox. ParOWL outputs an OWL file which contains the translation.

To display the usage of the different inclusion axioms in ParOWL, consider the following example ontology O, which consists of the axioms Bird $\sqsubseteq$ FlyAnimal,

[4] http://logic.aifb.uni-karlsruhe.de/wiki/Paraconsistent_reasoning

Table 6. Paraconsistent reasoning examples

Ontology	Bird	FlyAnimal	Penguin	notBird	notFlyAnimal	notPenguin
O_1	$\emptyset$	$\emptyset$	tweety	$\emptyset$	$\emptyset$	$\emptyset$
O_2	tweety	tweety	tweety	$\emptyset$	tweety	$\emptyset$
O_3	tweety	tweety	tweety	tweety	tweety	tweety
O_4	tweety	$\emptyset$	tweety	$\emptyset$	tweety	$\emptyset$

Penguin $\sqsubseteq$ Bird, Penguin $\sqsubseteq$ $\neg$FlyAnimal, and Penguin(tweety), which is obviously inconsistent. We compare the following different four-valued ontologies which can be derived from O: For O_1 all inclusions are material, for O_2 all inclusions are internal, for O_3 all inclusions are strong. For O_4, the first inclusion is material, the second is internal, and the third is strong. Table 6 shows the extensions of the concepts in these ontologies as computed with ParOWL. The desired result may be O_4, and indeed the choice of inclusion axioms in this case follows the intuitions laid out in Section 3.

7 Conclusions and Further Work

We have motivated and formally described an approach for paraconsistent reasoning with ontologies, which is based on the simultaneous use of different kinds of paraconsistent inclusion. We have provided guidelines for the use of these different inclusions. We have provided algorithms for implementing our approach and presented a publicly available tool which realises it.

Concerning the two algorithms provided in Sections 4 and 5, it is rather apparent that all the benefits from the KAON2 system – like the ability to handle large ABoxes – can also be achieved by invoking KAON2 after employing the transformation algorithm from Section 4 in a preprocessing manner using ParOWL. However, although the transformation algorithm is polynomial, it may be time consuming for large ontologies. This can possibly be improved by the direct algorithm from Section 5. Most of the technical details of the KAON2 implementation can indeed be carried over to our algorithm from Section 5.

In the literature, there are basically two other approaches to four-valued description logics, namely [4] and [5]. Our approach differes from theirs in two important aspects. The first is that we allow for simultaneous usage of different inclusions. The second is that we propose a translation from our logic into standard description logic such that established reasoners can be used. We thus benefit directly from the highly optimised systems currently available. The logics in [4,5] have neither of these features.

Obviously, much work remains to be done to make our approach fit for practice. Besides the obvious task of providing a better implementation than just a prototype, we also have to address in more detail the question, which kinds of paraconsistent inclusion are to be chosen when translating paraconsistent ontologies to standard ontologies. We envision a combination of system recommendations with user interactions. Alternatively, inclusions could be weakened gradually from strong inclusion to weaker versions – probably involving even further notions of inclusion – until a reasonable answer to a query is found. These issues are currently under investigation by the authors.

References

1. Hayes, P., Horrocks, I., Patel-Schneider, P.F.: OWL Web Ontology Language Semantics and Abstract Syntax. W3C Recommendation 10 February 2004 (2004)
2. Schlobach, S., Cornet, R.: Non-standard reasoning services for the debugging of description logic terminologies. In Gottlob, G., Walsh, T., eds.: IJCAI, Morgan Kaufmann (2003) 355–362
3. Haase, P., van Harmelen, F., Huang, Z., Stuckenschmidt, H., Sure, Y.: A framework for handling inconsistency in changing ontologies. In Gil, Y., Motta, E., Benjamins, V.R., Musen, M.A., eds.: International Semantic Web Conference. Volume 3729 of Lecture Notes in Computer Science., Springer (2005) 353–367
4. Patel-Schneider, P.F.: A four-valued semantics for terminological logics. Artificial Intelligence **38** (1989) 319–351
5. Straccia, U.: A sequent calculus for reasoning in four-valued description logics. In Galmiche, D., ed.: TABLEAUX. Volume 1227 of Lecture Notes in Computer Science., Springer (1997) 343–357
6. Ma, Y., Lin, Z., Lin, Z.: Inferring with inconsistent OWL DL ontology: A multi-valued logic approach. In Grust, T., et al., eds.: EDBT Workshops. Volume 4254 of Lecture Notes in Computer Science., Springer (2006) 535–553
7. Belnap, N.D.: A useful four-valued logic. Modern uses of multiple-valued logics (1977) 7–73
8. Arieli, O., Avron, A.: The value of the four values. Artif. Intell. **102** (1998) 97–141
9. Motik, B.: Reasoning in description logics using resolution and deductive databases. PhD theis, University Karlsruhe, Germany (2006)
10. Baader, F., Calvanese, D., McGuinness, D., Nardi, D., Patel-Schneider, P., eds.: The Description Logic Handbook: Theory, Implementation, and Applications. Cambridge University Press (2003)
11. Horrocks, I., Patel-Schneider, P.F.: Reducing OWL entailment to description logic satisfiability. J. Web Sem. **1** (2004) 345–357
12. Arieli, O., Avron, A.: Reasoning with logical bilattices. Journal of Logic, Language and Information **5** (1996) 25–63
13. Kamide, N.: Foundations of paraconsistent resolution. Fundamenta Informaticae **71** (2006) 419–441
14. Fitting, M.: First-Order Logic and Automated Theorem Proving, 2nd Edition. Texts in Computer Science. Springer (1996)

Vague Knowledge Bases for Matchmaking in P2P E-Marketplaces

Azzurra Ragone[1], Umberto Straccia[2], Tommaso Di Noia[1], Eugenio Di Sciascio[1], and Francesco M. Donini[3]

[1] SisInf Lab - Politecnico di Bari, Via Re David, 200, I-70125, Bari, Italy
{a.ragone,t.dinoia,disciascio}@poliba.it
[2] ISTI - CNR, Via G. Moruzzi 1, 56124 Pisa, Italy
straccia@isti.cnr.it
[3] Università della Tuscia, via San Carlo, 32, I-01100, Viterbo, Italy
donini@unitus.it

Abstract. In this paper we propose an approach to semantic matchmaking that exploits various knowledge representation technologies to find most promising partners in peer-to-peer e-marketplaces. In particular we mix in a formal and principled way the semantic expressiveness of DLR-lite Logic Programs, fuzzy logic and utility theory. We adopt DLR-Lite Logic Programs to obtain a reasonable compromise between expressiveness and complexity to ensure the scalability of our approach to large e-marketplaces, and Fuzzy Logic to model logical specifications as soft constraints. Furthermore, fully exploiting the peer-to-peer paradigm, we consider in the matchmaking process preferences and corresponding utilities of both parties.

1 Introduction

The distinguishing characteristic of a peer-to-peer (P2P) e-marketplace is that basically peer users – both buyers and sellers – can submit their advertisements, browse through available ads, and be assisted in finding the best available counterparts to meet their needs and initiate a commercial transaction. Furthermore, in such marketplaces, there is often the need to negotiate not only on single numerical features such as price, quantity, etc., but also on some good's characteristics. Then there is the need to represent advertisements in a machine understandable way, using languages able to model the background domain knowledge. Also, descriptions could have logical implications, *e.g.*, *If a car has leather seats then it is also provided with air conditioning* or bundles *e.g.*, *Sports car with optional package including both GPS system and alarm system*, and some kind of logical theory, able to let users express their needs/offers, could surely help. Finally, while performing a matchmaking process between two peers advertisements, we should take into account user preferences – soft constraints – and distinguish them from mandatory – hard – ones.

In an e-commerce setting, matchmaking can be defined as the process of finding "good" counterparts for a given entry in the marketplace. Of course, the evaluation of how "good" a counterpart is constitutes most of the effectiveness of a matchmaking system. Currently, many commercial sites force the buyer to enter her request browsing

E. Franconi, M. Kifer, and W. May (Eds.): ESWC 2007, LNCS 4519, pp. 414–428, 2007.

a predefined classification that may be completely unsuitable for the characteristics the buyer might have in mind *e.g.*, they require to enter a brand first, then a model of that brand, etc. while a buyer may be not interested in a specific brand, but only on some limitations on price and color. In this respect, one may say that they provide no matchmaking assistance: the matchmaker is the buyer herself. Even the use of textual search engine (as in eBay) does not help a lot, since the result is a (sometimes very long and tedious) list to browse.

To assist buyers and sellers in marketplaces, several research proposals on matchmaking systems were issued. They either try to compute a score of possible counterparts, based on textual information [25], or to compare the logical representations of supply and demand [10], or combine both scores and logic in some way [7,16]. Our proposal falls in this last category, mixing in a formal way ontologies in DLR-Lite, Datalog rules, Fuzzy sets, and Utility Theory. While all the above logic-based proposals suffer on the scalability side, our resort on DLR-Lite and Datalog ensures the scalability of our approach w.r.t. to large datasets.

Furthermore, following the economical approach to negotiation, the matchmaker computes a score as the maximum value of the *product* of the weighted utility of the buyer u_β times the weighted utility of the seller u_σ over all possible agreements between the buyer and the seller. In this way both buyer's and seller's preferences are taken into account ruling out of the match list those counteroffers that, although seemingly appealing for the buyer, would probably lead to failure due to contrasting preferences of the seller, that we take already into account.

The remaining of the paper is as follows: Section 2 introduces basics of languages and technologies we adopt. In Section 3 requirements for matchmaking process, including preferences, utility functions are illustrated and vague knowledge bases are introduced. The description of the matchmaking process over vague knowledge bases follows in Section 4. Next, in Section 5 rules for item classification in P2P marketplaces are outlined. An illustrative example is presented in Section 6, discussion about relevant related work and conclusions close the paper.

2 Basic Technologies

As our representation and query tool, we use a specific combination of Description Logics [3] (DLs), Logic Programming [6] (LPs) and Fuzzy Theory [26].

A *Knowledge Base* is a triple $\mathcal{K} = \langle \mathcal{F}, \mathcal{O}, \mathcal{P} \rangle$, where $\mathcal{F}$ is a set of facts stored into a relational database, $\mathcal{O}$ is the DL component and $\mathcal{P}$ is the rule component. The DL component is used to model the application domain's ontology (*e.g.*, the intentional knowledge). Specifically, we use a description logic of the family of DLs, *DLR-Lite* [5]. Concerning the rule and query component, we use an extension of *Datalog* (cf. [6] among others), in which we allow soft constraint predicates to appear in rules and queries [22]. Basically, we allow vague/fuzzy predicates to occur in rule bodies, which have the effect that each tuple in the answer set of a query has now a score in [0,1]. The main problem to be addressed in the resulting language is the problem to compute the top-k answers in case the set of facts is huge, without evaluating all the tuples' score. As matching a buyer's request with a seller's offer is a matter of degree, our purpose is to find the top-k matches only, rather than all matches.

In the following, consider a knowledge base $\mathcal{K} = \langle \mathcal{F}, \mathcal{O}, \mathcal{P} \rangle$.

Facts component. Concerning the facts component, $\mathcal{F}$ is simply a finite set of formulae of the form $R(c_1, \ldots, c_n)$, where R is an n-ary predicate and c_i are constants. Facts, representing extensional information, are stored in relational tables of an underlying database. For instance,

$$CarTable(544, FiatPunto, 2004, 15000)$$

is a fact stating that item 544 is a Fiat Punto, built in 2004 and having 15000 kilometers.

An *interpetation* $\mathcal{I} = \langle \Delta, \cdot^{\mathcal{I}} \rangle$ consists of a *fixed infinite domain* Δ and an *interpretation function* $\cdot^{\mathcal{I}}$ that maps an n-ary predicate R into an n-ary relation $R^{\mathcal{I}}$ over Δ and maps constants into constants of Δ such that $a^{\mathcal{I}} \neq b^{\mathcal{I}}$ if $a \neq b$ (unique name assumption). We assume to have one object for each constant, denoting exactly that object. In other words, we have standard names, and we will not distinguish between the alphabet of constants and the objects in Δ.

DL component. Concerning the DL component, $\mathcal{O}$ is a finite set of DLR-Lite axioms [1]. A DLR-Lite *axiom* has the form $C_1 \sqsubseteq C_2$ (*concept inclusion*) or has the form $(disjoint\ C_1, \ldots, C_n)$ (*disjointness axiom*), where C_i is a concept expression. Informally, $C_1 \sqsubseteq C_2$ says that the set denoted by C_1 is a subset of the set denoted by C_2, while $(disjoint\ C_1, \ldots, C_n)$ declares that the concepts are pairwise disjoint. Concepts expressions are constructed starting from a set of atomic concepts and relations by applying suitable constructs. In DLR-Lite we distinguish between constructs that are allowed in the left-hand side (Cl) and those in the right-hand side (Cr) of concept inclusions, according the following syntax:

$$\begin{aligned} Cl &\longrightarrow A \mid \exists i : R \mid Cl_1 \sqcup Cl_2 \\ Cr &\longrightarrow A \mid \exists i : R \mid Cr_1 \sqcap Cr_2 \end{aligned}$$

where R is an n-ary predicate and $i \in \{1, \ldots, n\}$.

An interpretation $\mathcal{I} = \langle \Delta, \cdot^{\mathcal{I}} \rangle$ maps a concept C into subsets $C^{\mathcal{I}}$ of Δ. In the following, R denotes an n-ary predicate, and we use $\mathbf{c}$ to denote an n-tuple of constants, and $\mathbf{c}[i]$ to denote the i-th component of $\mathbf{c}$. Then $\cdot^{\mathcal{I}}$ has to satisfy:

$$\begin{aligned} A^{\mathcal{I}} &\subseteq \Delta \\ (C_1 \sqcap C_2)^{\mathcal{I}} &= {C_1}^{\mathcal{I}} \cap {C_2}^{\mathcal{I}} \\ (C_1 \sqcup C_2)^{\mathcal{I}} &= {C_1}^{\mathcal{I}} \cup {C_2}^{\mathcal{I}} \\ (\exists i : R)^{\mathcal{I}} &= \{\mathbf{c}[i] \mid \mathbf{c} \in R^{\mathcal{I}}\} \end{aligned}$$

An interpretation *satisfies* (*is a models of*) $C_1 \sqsubseteq C_2$ iff ${C_1}^{\mathcal{I}} \subseteq {C_2}^{\mathcal{I}}$, while an interpretation *satisfies* (*is a models of*) $(disjoint\ C_1, \ldots, C_n)$ iff $\forall i \neq j . {C_i}^{\mathcal{I}} \cap {C_j}^{\mathcal{I}} = \emptyset$.

[1] Here we refer to DLR-Lite$_{core}$ but any DL of the DL-Lite or DLR-Lite family can be as well considered.

Example 1. The following set of axioms is an excerpt of the encoding for the web directory behind the car selling site www.autos.com:

$$\begin{array}{ll} Cars \sqsubseteq Vehicles & CompactCars \sqsubseteq PassengerCars \\ Trucks \sqsubseteq Vehicles & Vehicles \sqsubseteq \exists 1 : hasMaker \sqcap \exists 1 : hasPrice \\ Vans \sqsubseteq Vehicles & \exists 1 : hasPrice \sqcup \exists 1 : hasMaker \sqsubseteq Vehicles \\ LuxuryCars \sqsubseteq Cars & \exists 2 : hasMaker \sqsubseteq CarMaker \\ PassengerCars \sqsubseteq Cars & Cars \sqsubseteq \exists 1 : hasKmWarranty \sqcap \exists 1 : hasFuel \\ (disjoint\ Cars, Trucks, Vans) & \exists 2 : hasFuel \sqsubseteq FuelType \end{array}$$

Given a DLR-Lite ontology related to a relation model, in [5] it is shown how to rewrite a conceptual query over the ontological model in a conjunctive query over the relational model in the form

$$q(\mathbf{x}) \leftarrow \exists \mathbf{y} body(\mathbf{x}, \mathbf{y})$$

Here *body* is the conjunction of n-ary predicates representing the information modeled by concepts and roles in the DLR-Lite ontology.

Notice that even apparently simple, DLR-Lite family languages are expressive enough to represent RDFS[2] ontologies (at least their DL subset).

LP component. Concerning the rule component of a knowledge base, $\mathcal{P}$ is a finite set of vague Datalog rules[22], which are defined as follows.

A Datalog *rule* is a Horn clause of the form

$$P(\mathbf{t_0}) \leftarrow R_1(\mathbf{t_1}), \ldots, R_n(\mathbf{t_n}) ,$$

where $P(\mathbf{t_0})$ is the *head* of the rule, and $R_1(\mathbf{t_1}), \ldots, R_n(\mathbf{t_n})$ is the *body* of the rule. $P(\mathbf{t_0})$ and all $R_i(\mathbf{t_i})$ are atoms, $\mathbf{t_i}$ are arrays of *terms*. A *term* is either a *variable* or a *constant*. Here we also provide a set of built-in predicates, like $+, *, -, /, \geq, \leq, =$, with (obvious) fixed interpretation, which may appear in a rule body.

A Datalog *query predicate* q is a designated n-ary predicate symbol appearing in the head of a Datalog rule in $\mathcal{P}$. For instance,

$$q(x,p) \leftarrow Cars(x), hasPrice(x,p), p \leq 15000$$

is a query asking for cars whose price is less or equal than 15000.

The interpretation of n-ary predicates, terms and the notions of satisfiability (is model of) and logical consequence are as usual.

Vague Datalog rules are as Datalog rules except that we additionally allow fuzzy predicates to occur in Datalog rule bodies (see [22]). Specifically, let r be an n-ary predicate symbol. We add an additional position to r, making it $n+1$-ary. Then a vague Datalog *rule* is of the form

$$r(\mathbf{x}, s) \leftarrow \exists \mathbf{y} body(\mathbf{x}, \mathbf{y}), s = f(p_1(\mathbf{z}_1), \ldots, p_n(\mathbf{z}_n))$$

where

1. $\mathbf{x}$ are the n *distinguished variables*;
2. s is the *score variable*, taking values in $[0, 1]$, and r is functional on s;

[2] http://www.w3.org/TR/rfs-schema/

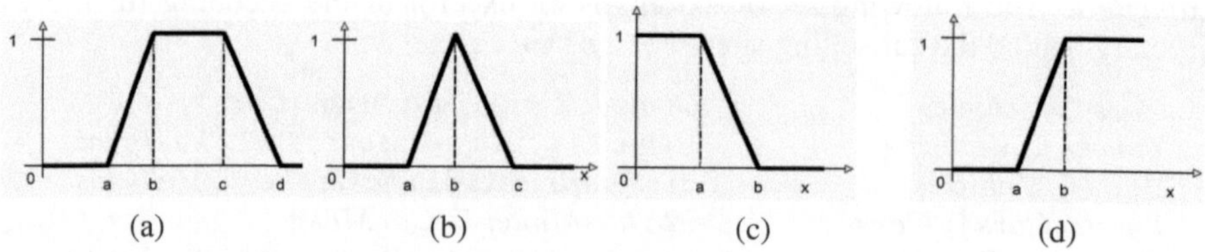

Fig. 1. (a) Trapezoidal function; (b) Triangular function; (c) L-function; (d) R-function

3. $\mathbf{y}$ are so-called *non-distinguished variables* and are distinct from the variables in $\mathbf{x}$;
4. $body(\mathbf{x}, \mathbf{y})$ is a conjunction of Datalog atoms;
5. $\mathbf{z_i}$ are tuples of constants or variables in $\mathbf{x}$ or $\mathbf{y}$;
6. p_i is an n_i-ary *fuzzy predicate* assigning to each n_i-ary tuple $\mathbf{c}_i$ as *score* $p_i(\mathbf{c}_i) \in [0, 1]$;
7. f is a *scoring* function $f\colon [0, 1]^n \to [0, 1]$, which combines the scores of the n fuzzy predicates p_i into and overall *query score* to be assigned to the score variable s. We assume that f is *monotone*, *i.e.*, , for each $\mathbf{v}, \mathbf{v}' \in [0, 1]^n$ such that $\mathbf{v} \leq \mathbf{v}'$, $f(\mathbf{v}) \leq f(\mathbf{v}')$ holds, where $(v_1, \ldots, v_n) \leq (v_1', \ldots, v_n')$ iff $v_i \leq v_i'$ for all i; we assume that the computational cost of f and all fuzzy predicates p_i is bounded by a constant.

We call $s = f(p_1(\mathbf{z}_1), \ldots, p_n(\mathbf{z}_n))$ a *scoring atom*. For instance,

$$\begin{aligned} CheapCar(x, p, s) &\leftarrow NewCar(x), hasPrice(x, p), \\ &\quad\; s = max(0, 1 - p/15000) \\ CheapCar(x, p, s) &\leftarrow SecondHandCar(x), hasPrice(x, p), \\ &\quad\; s = max(0, 1 - p/7500) \end{aligned}$$

are vague Datalog rules that can be used to look for cheap cars, assigning to each car a score depending on its price. If the price of a new car is above 15,000€ the car is not considered as a cheap one, while the scoring function is increased as the price lowers. Hence, it is quite natural that if we are looking for cheap cars one wants that the retrieved cars are sorted in decreasing order with respect to its score, *i.e.*, degree of cheapness. Furthermore, as the database may contain thousands of tuples, one usually wants to retrieve just the top-k ranked ones.

Concerning fuzzy predicates involved in scoring atoms, we recall that in fuzzy set theory and practice there are many membership functions for fuzzy sets membership specification. However, the *trapezoidal* $trz(x; k_1, k_2, a, b, c, d)$, the *triangular* $tri(x; k_1, k_2, a, b, c)$, the *L-function* (left shoulder function) $LS(x; k_1, k_2, a, b)$ and the *R-function* (right shoulder function) $RS(x; k_1, k_2, a, b)$ are simple, yet most frequently used to specify membership degrees (see Figure 1) [3].

From a semantics point of view (see [22]), we have to take into account the additional scoring atom $s = f(p_1(\mathbf{z}_1), \ldots, p_n(\mathbf{z}_n))$.

Informally a vague Datalog rule is interpreted in an interpretation $\mathcal{I}$ as the set $r^{\mathcal{I}}$ of tuples $\langle \mathbf{c}, v \rangle$, such that when we substitute the variables $\mathbf{x}$ and s with the constants $\mathbf{c}$

[3] k_1, k_2 is the domain of the functions.

and the score value $v \in [0,1]$, the formula $\exists \mathbf{y} body(\mathbf{x},\mathbf{y}), s = f(p_1(\mathbf{z}_1),\ldots,p_n(\mathbf{z}_n))$ evaluates to true in $\mathcal{I}$.

Due to the existential quantification $\exists \mathbf{y}$, for a fixed $\mathbf{c}$, there may be many substitutions $\mathbf{c}'$ for $\mathbf{y}$ and, thus, we may have many possible scores for the tuple $\mathbf{c}$. Among all these scores for $\mathbf{c}$, we select the highest one, *i.e.*, the sup.

In case that the atom $r(\mathbf{x}, s)$ is the head of multiple rules, for each tuple $\mathbf{c}$ there may be a score v_i computed by each of these rules. In that case, we assume that the overall score for $\mathbf{c}$ is the *maximum* among the scores v_i.

Now, let $\theta^{\mathbf{cc}'v}_{\mathbf{xy}s} = \{\mathbf{x}/\mathbf{c}, \mathbf{y}/\mathbf{c}', s/v\}$ be a substitution of the variables $\mathbf{x}, \mathbf{y}$ and s with the tuples $\mathbf{c}, \mathbf{c}'$ and score value $v \in [0,1]$. Let $\psi(\mathbf{x},\mathbf{y},s)$ be $body(\mathbf{x},\mathbf{y}), s = f(p_1(\mathbf{z}_1),\ldots,p_n(\mathbf{z}_n))$. With $\psi(\mathbf{x},\mathbf{y},s)\theta^{\mathbf{cc}'v}_{\mathbf{xy}s}$ we denote the ground formula obtained by applying the substitution $\theta^{\mathbf{cc}'v}_{\mathbf{xy}s}$ to $\psi(\mathbf{x},\mathbf{y},s)$.

We say that an interpretation $\mathcal{I}$ is a *model* of $\psi(\mathbf{x},\mathbf{y},s)\theta^{\mathbf{cc}'v}_{\mathbf{xy}s}$ iff $\psi(\mathbf{x},\mathbf{y},s)\theta^{\mathbf{cc}'v}_{\mathbf{xy}s}$ evaluates to true in $\mathcal{I}$, *i.e.*, all ground atoms and the grounded scoring atom occurring in $\psi(\mathbf{x},\mathbf{y},s)\theta^{\mathbf{cc}'v}_{\mathbf{xy}s}$ are true. We will write $\mathcal{I} \models \psi(\mathbf{x},\mathbf{y},s)\theta^{\mathbf{cc}'v}_{\mathbf{xy}s}$ in this case.

Then, the interpretation $\mathbf{r}^{\mathcal{I}}$ of a set of rules with same head $\mathbf{r} = \{r_1,\ldots,r_n\}$ in $\mathcal{I}$ is

$$\mathbf{r}^{\mathcal{I}} = \{\langle \mathbf{c}, v\rangle \mid v = \max(v_1,\ldots,v_n), v_i = \sup_{\mathbf{c}'}\{v' \mid \mathcal{I} \models \psi_i(\mathbf{x},\mathbf{y},s)\theta^{\mathbf{cc}'v'}_{\mathbf{xy}s}\}\}\,, \quad (1)$$

where each rule $r_i \in \mathbf{r}$ is of the form $r(\mathbf{x}, s) \leftarrow \exists \mathbf{y}\psi_i(\mathbf{x},\mathbf{y},s)$, $\sup\emptyset$ is undefined, and $\max(v_1,\ldots,v_n)$ is undefined iff all its arguments are undefined.

Note that some tuples $\mathbf{c}$ may not have a score in $\mathcal{I}$ and, thus, $\langle \mathbf{c}, v\rangle \notin \mathbf{r}^{\mathcal{I}}$ for no $v \in [0,1]$. Alternatively we may define $\sup\emptyset = 0$ and, thus, all tuples $\mathbf{c}$ have a score in $\mathcal{I}$, *i.e.*, $\langle \mathbf{c}, v\rangle \in \mathbf{r}^{\mathcal{I}}$ for some $v \in [0,1]$. We use the former formulation to distinguish the case where a tuple $\mathbf{c}$ is retrieved, though the score is 0, from the tuples which do not satisfy the query and, thus, are not retrieved. Finally, for all $\mathbf{c}$ and for all $v \in [0,1]$, we say that $\mathcal{I}$ is a *model* of $r(\mathbf{c}, v)$ (denoted $\mathcal{I} \models r(\mathbf{c}, v)$) iff $\langle \mathbf{c}, v\rangle \in \mathbf{r}^{\mathcal{I}}$.

We say that a vague knowledge base $\mathcal{K} = \langle \mathcal{F}, \mathcal{O}, \mathcal{P}\rangle$ *entails* $q(\mathbf{c}, v)$, written $\mathcal{K} \models q(\mathbf{c}, v)$, iff for all models $\mathcal{I}$ of $\mathcal{K}$, $\mathcal{I} \models q(\mathbf{c}, v)$ holds.

Linking **Facts component** $\mathcal{F}$, **DL component** $\mathcal{O}$ and **LP component** $\mathcal{P}$ with each other in $\mathcal{K}$ we have that:

– Atoms, representing unary predicates, and predicates occurring in $\mathcal{O}$ may appear in rules in $\mathcal{P}$.

– Predicates occurring in $\mathcal{F}$ do not occur in the head of rules in $\mathcal{P}$ — essentially, we do not allow that the fact predicates occurring in $\mathcal{F}$ can be redefined by $\mathcal{P}$.

2.1 Top-k Retrieval

The basic inference services that concerns us is the top-k retrieval problem, where this latter is defined as:

Top-k retrieval: Given a vague knowledge base $\mathcal{K} = \langle \mathcal{F}, \mathcal{O}, \mathcal{P}\rangle$, retrieve the top-$k$ ranked tuples $\langle \mathbf{c}, v\rangle$ that instantiate the query q and rank them in decreasing order w.r.t. the score v, *i.e.*, find the top-k ranked tuples of the answer set of q, denoted

$$ans_k(\mathcal{K}, q) = \mathtt{Top}_k\{\langle \mathbf{c}, v\rangle \mid \mathcal{K} \models q(\mathbf{c}, v)\}\,.$$

For instance,

$$q(x, p, s) \leftarrow CheapCar(x, p, s)$$

is a query asking for cheap cars. The top-k ranked cars, according to the score (that depends on their price), is obtained by $ans_k(\mathcal{K}, q)$.

From a reasoning point of view, [23] shows a that the top-k problem for DL-Lite knowledge bases can be solved in LogSpace data complexity. The result holds also for the top-k problem in DLR-Lite, using the results described in [5]. On the other hand, [22] shows that the top-k problem for vague Datalog can also be solved in LogSpace data complexity, if the set of vague Datalog rules is not recursive. Both solutions rely on a query rewriting method which is conceptually based on the same idea as for [5]. In fact, it can be shown that for a vague knowledge base $\mathcal{K} = \langle \mathcal{F}, \mathcal{O}, \mathcal{P} \rangle$, where $\mathcal{P}$ is not recursive, the top-k problem can be solved in LogSpace data complexity. Essentially, we combine [22] and [23] to rewrite a query into a set R of new queries and then we apply relational top-k database technology (see, *e.g.*, [14]) to solve the queries in R.

3 Matchmaking Scenario

In this section we outline the matchmaking scenario and show how to model it as a top-k retrieval problem over a vague knowledge base $\mathcal{K}$. Without loss of generality, in the examples we refer to an automobile marketplace, and motivate our work in this domain. In such domain features as look, comfort, optionals, type have to be modeled, as well as numerical features as price, warranty or delivery time. In fact, based on these features both the buyer is able to formulate her request and the seller to describe the good to be sold *i.e.*, a buyer can specify conditional preferences, such as "***If** it is a luxury car, **then** it has to be provided with leather seats*" or "*I want a cheap car, yet if the car has an alarm system I'm ready to pay up to 18,000 €*", conversely the seller can offer "*a compact car with 4 years warranty or 12,0000 km warranty*". Constraints can involve only numerical features,or non numerical ones, as well as both of them.

Hard and Soft Constraints. In a typical e-marketplace scenario, the issues within both the buyer's request and the seller's offer can be split into *strict requirements* and *preferences*. Strict requirements represent what the buyer and the seller want to be necessarily satisfied in order to accept the final agreement – in our framework we call strict requirements *hard constraints*. Preferences denote issues they are willing to negotiate on – this is what we call *soft constraints*. Hence, the matchmaker has to be able to handle both hard and soft specifications of both the buyer and the seller. Let us now introduce an example request, we will use to explain some aspects of our approach:

Example 2. Suppose to have a buyer's request like "*I want a passenger car black or gray. Preferably I would like to pay less than 14,000 € furthermore I'm willing to pay up to 17,000 € if warranty is greater or equal than 100000 km.* ". In this example we identify:

Hard Constraints. *Body Type:* Passenger car; *Color:* Black or Grey.

Soft Constraints. *Price:* $\leq$ 14,000; *Warranty-Price:* if Warranty $\geq$ 100,000 then Price $\leq$ 17,000 €.

Utility. Given a request and several supplies, in the final agreement, both buyer's and seller's *hard constraints* have to be satisfied. Nevertheless, how should the matchmaker find –and rank– the most *suitable* or *promising* agreements to be proposed to both parties? In the ranking process, *soft constraints* are key information the matchmaker should use to evaluate the match degree. The final agreement is computed in order to maximize both buyer's and seller's preferences satisfaction. For instance, w.r.t. Example 2 suppose to have three supplies[4]:

σ' = *Body Type:* Compact car; *Color:* Grey; *Price:* 16,000 €; *Warranty:* 200,000 km.
σ'' = *Body Type:* Passenger Car; *Color:* Black; *Price:* 13,000 €; *Warranty:* 50,000 km.
σ''' = *Body Type:* Van; *Color:* Brown; *Price:* 19,000 €.

Comparing these supplies with buyer's request we note that σ''' will be discarded because its hard constraints are in conflict with the buyer's one. For σ' and σ'' we have that: σ' satisfies the preference β_2 = {**Warranty-Price:** if Warranty $\geq$ 100,000 then Price $\leq$ 17,000 €}; σ'' the preference β_1 = {**Price:** $\leq$ 14,000}. Now the question is: *how to evaluate the best one?*

In order to provide an answer to such a question, we take into account utility values assigned by the buyer and representing the preference relevance to sub-parts of *soft constraints*. In this case we assume utility values — $u(\beta_1)$ and $u(\beta_2)$ — both for β_1 and β_2. [5] Notice that actually the same holds from the seller's side. In a P2P e-marketplace the seller may express his preferences — *soft constraints e.g.*, on selling price, warranty, delivery time — with corresponding utilities $u(\sigma_j)$, as well as his *hard constraints* (*e.g.*, color, model, engine fuel, etc.). The only constraint on utility values is that both seller's and buyer's ones are normalized to 1 to eliminate outliers, and make them comparable [12].

$$\sum u(\beta_i) = 1 \ , \sum u(\sigma_j) = 1 \tag{2}$$

Since we assume utilities on preferences as additive, here we can write the global utility of the buyer u_β and of the seller u_σ as just a sum of the utilities of preferences satisfied in the agreement. Let s_i and s_j be a score representing the degree of preference satisfaction, then the global utility will be:

$$u_\beta = \sum s_i * u(\beta_i) \ , u_\sigma = \sum s_j * u(\sigma_j) \tag{3}$$

Matchmaking Steps. Now we can outline the steps of the matchmaking process:

1: Every time a seller enters the marketplace, he proposes his supply expressing both *hard* and *soft constraints* (preferences). Eventually, for each preference σ_j (if any) he expresses the corresponding utilities $u(\sigma_j)$.
2: Similarly the buyer who enters the marketplace will express *hard* and *soft constraints* as well as the utility $u(\beta_i)$.

[4] Without loss of generality, for the sake of simplicity in this example we consider supplies where only *hard constraints* have been set.

[5] It is not in the scope of this paper to investigate on how to compute $u(\beta_1)$ and $u(\beta_2)$; we might assume, without loss of generality, they are determined in advance by means of either direct assignment methods (Ordering, Simple Assessing or Ratio Comparison) or pairwise comparison methods (like AHP and Geometric Mean) [20].

3: Based on buyer's and seller's specifications, the matchmaker returns a ranked list of agreements such that: [a] they satisfy both the *hard constraints* in the request and conversely their *hard constraints* are satisfied by the request; [b] the rank is evaluated taking into account preferences and utility functions u_β and u_σ as defined by equations (3).

In a P2P e-marketplace, the aim is to maximize both buyer's and seller's utilities in the final agreement, so the matchmaker has to propose agreements mutually beneficial for both of them. Such agreements are computed considering the higher values of u_β and u_σ **utilities product** [18].

4 Matchmaking with Vague Knowledge Bases

E-Marketplaces are typical systems where the notion of fuzziness is often involved. It is usual to find, among others, concepts like $Cheap$ or $Expensive$. Similarly, numerical variables involved in a commercial transaction expose a fuzzy behavior. For instance, suppose to have a buyer looking for a car provided with a warranty greater than 100,000 kilometers and a supplier selling his car with a 80,000 kilometers warranty. If the buyer's warranty specification is modeled as preference, we can not say they do not match at all. Instead we can say they match with a certain degree.

Also notice that in both cases — conceptual and numerical information — the fuzziness is: (1) strongly dependent from the user point of view. The idea of "cheapness" changes moving from a user to another one; (2) allowed only in *soft constraints*. If the user expresses the willingness of selling a car with 80,000 kilometers warranty within *hard constraints*, it means that he does not want to negotiate on it at all.

A logical language able to allow the user to express also her fuzzy *soft constraints* would be then a good choice to model matchmaking. Given a DLR-Lite ontology $\mathcal{O}$ and a set of facts $\mathcal{F}$ we model a vague knowledge base $\mathcal{K}_{match} = \langle \mathcal{F}, \mathcal{O}, \mathcal{P} \rangle$ for matchmaking in e-marketplaces where:

- $\mathcal{P}$ represents user's (both buyer and seller) *soft constraints*. In $\mathcal{P}$'s rules, also roles and concepts form $\mathcal{O}$ can be used.

In this case, $\mathcal{F}$ represents the tuples of the relational database storing P2P advertisements. Notice that since we model a P2P marketplace, then advertisements can be either supplies or demands, depending on the "searcher" point of view. The former are used in case a buyer enters the marketplace the latter in case a seller decides to find promising requests.

To represent requester requirements in a vague knowledge base setting, we model *hard constraints* as a conceptual query over $\mathcal{O}$ and *soft constraints* as a query over a vague Datalog program $\mathcal{P}$.

Hereafter we will use the following notation:

$$\textit{hard constraints} = \begin{cases} \beta(x, \mathbf{y}) \ , \ buyer's\ strict\ requirements \\ \sigma(x, \mathbf{y}) \ , \ seller's\ strict\ requirements \end{cases}$$

$$\textit{soft constraints} = \begin{cases} \beta_i(x, \mathbf{y}, s) \ or \ \beta_i(x, s) \ , \ buyer's\ preferences \\ \sigma_j(x, \mathbf{y}, s) \ or \ \sigma_j(x, s) \ , \ seller's\ preferences \end{cases}$$

x is a single variable. It is usually instantiated with the key value of a database tuple;

y (if present) will be instantiated by numerical values. It is used whenever an agreement on numerical variables (price, km/years warranty, etc.) has to be reached;

s is the score variable as defined in Section 2 [22]. It represents the score associated to fuzzy predicates involved in the body of the rules;

Notice that since a score is associated to each fuzzy predicate, we can compute the global utility based on the two utility functions in Section 3. Furthermore, the two queries σ and β model the minimal requirements the buyer and the seller want to be satisfied in order to accept the final agreement. Notice that, if seller and buyer set hard constraints in conflict with each other, the corresponding supply will not be retrieved. *Soft constraints* are modeled via Datalog predicates β_i for the buyer and σ_j for the seller, where each of them represents a sub-part of the buyer/seller preferences.

The use of x, **y** and s should be clearer looking at how buyer's request in Example 2 is formalized:

$$\begin{aligned}
\beta_A(x) &\leftarrow PassengerCars(x)\\
\beta_B(x) &\leftarrow hasColor(x,y), Gray(y)\\
\beta_B(x) &\leftarrow hasColor(x,y), Black(y)\\
\beta(x) &\leftarrow \beta_A(x), \beta_B(x)
\end{aligned}$$

$$\begin{aligned}
\beta_1(x,p,s) \leftarrow\ & hasPrice(x,p),\\
& LS(0, 100000, 14000, 16000, p, s)
\end{aligned}$$

$$\begin{aligned}
\beta_2(x,p,kmw,s) \leftarrow\ & KmWarranty(x,kmw), hasPrice(x,p)\\
& RS(0, 400000, 80000, 100000, kmw, s_1),\\
& LS(0, 100000, 17000, 19000, p, s_2),\\
& s = max(1 - s_1, s_2)
\end{aligned}$$

With respect to the previous encoding we notice that defining β_1 and β_2 here we can use two different *L-functions* to specify membership degrees for the variable p.

5 Top-*k* Retrieval for Matchmaking in Vague Knowledge Bases

Given a DLR-Lite ontology $\mathcal{O}$ and a set of facts $\mathcal{F}$ in a relational database, we can detail the matchmaking framework in a vague knowledge base $\mathcal{K}_{match} = \langle \mathcal{F}, \mathcal{O}, \mathcal{P} \rangle$. In order to formulate a query and rank all the retrieved results, what is still missing is how to put together both buyer's and seller's requirements. Then:

1. for each buyer's preference β_i, write the corresponding rule in vague Datalog where the head contains the predicate $\beta_i(x, \mathbf{y}, s)$ as shown in Section 4; add the rule to $\mathcal{P}$; set the utility value $u(\beta_i)$ as shown in Section 3; The same is for the seller where the head of each rule is the predicate $\sigma_j(x, \mathbf{y}, s)$ and for each of them utility values are $u(\sigma_j)$
2. add to $\mathcal{P}$ the rules:

$$Buyer(x, \mathbf{y}, u_\beta) \leftarrow \beta_1(x, \mathbf{y_1}, s_1), \beta_2(x, \mathbf{y_2}, s_2), \ldots, u_\beta = u(\beta_1) \cdot s_1 + u(\beta_2) \cdot s_2 + \ldots$$
$$Seller(x, \mathbf{y}, u_\sigma) \leftarrow \sigma_1(x, \mathbf{y_1}, s_1), \sigma_2(x, \mathbf{y_2}, s_2), \ldots, u_\sigma = u(\sigma_1) \cdot s_1 + u(\sigma_2) \cdot s_2 + \ldots$$

where for each variable in **y** in the head of one of the two previous rules, the same variable occurs in at least one of the arrays of the corresponding body: $\mathbf{y_1}, \mathbf{y_2}, \ldots$;

3. encode buyer's *hard constraints* requirements as a conceptual query over $\mathcal{O}$. Rewrite the query as a conjunctive query where the query is denoted with $\beta(x, \mathbf{y}_\beta)$. The same is for the seller and his hard constraints $\sigma(x, \mathbf{y}_\sigma)$;
4. solve the **Top-*k* retrieval** problem:

$$ans_k(\mathcal{P}, Match) = \mathtt{Top}_k\{\langle x, \mathbf{y}, u\rangle \mid \langle \mathbf{y}, u\rangle \in \mathtt{Top}_1\{\langle x, \mathbf{y}', u'\rangle \mid \mathcal{P} \models Match(x, \mathbf{y}', u')\}\}.$$

where $Match$ is the conjunctive query

$$Match(x, \mathbf{y}, u) \leftarrow \beta(x, \mathbf{y}_\beta), Buyer(x, \overline{\mathbf{y}_\beta}, u_\beta), \sigma(x, \mathbf{y}_\sigma), Seller(x, \overline{\mathbf{y}_\sigma}, u_\sigma), u = u_\beta * u_\sigma$$

and for each variable in the array $\mathbf{y}$, the same variable occurs in $\mathbf{y}_\beta, \overline{\mathbf{y}_\beta}, \overline{\mathbf{y}_\sigma}$ or $\mathbf{y}_\sigma$.

Basically, for each key value x of the database, we compute the best match $\langle \mathbf{y}, u\rangle$ for it, *i.e.*, $\langle \mathbf{y}, u\rangle \in \mathtt{Top}_1\{\langle x, \mathbf{y}', u'\rangle \mid \mathcal{P} \models Match(x, \mathbf{y}', u')\}$, and then rank the top-$k$ key values.

Notice that the rank is computed considering the product of buyer's and seller's utilities as stated at the end of Section 3 in order to reach an agreement appealing both for the buyer and for the seller.

6 An Illustrative Example

Let us better clarify the approach with the aid of a tiny example. In Table 1 two possible offers stored in an e-marketplace database are presented. We call $CarTable$ the relation representing Table 1.

Table 1. The *CarTable* relation with a sample set of offers

ID	MODEL	TYPE	PRICE	DISCOUNT	KM	COLOR	AIRBAG	INTERIOR TYPE	AIR COND	ENGINE FUEL
34	ALFA 156	Sedan	12000	20%	25000	Black	1	LeatherSeats	0	Diesel
1812	FORD FOCUS	StationVagon	13000	20%	20000	Gray	1	LeatherSeats	1	Gasoline

Based both on these information and some specific domain knowledge, we will write the corresponding vague knowledge base $\mathcal{K}_{match}$ as explained is Section 4 and Section 5. In Figure 3 tuples related to the relation $CarTable$ are represented together with a DLR-Lite ontology $\mathcal{O}$ extending the one presented in Example 1. Auxiliary rules in Figure 2 are introduced and used only for the sake of clarity and conciseness.
Now, suppose to have the following buyer's request:

[Hard Constraint] (β) I want a *sedan* or a *station wagon*.
[Soft Constraint] (β_1) I would like *air conditioning* if the car has *leather seats*. (β_2) Preferably I would like to pay *less than 11,000 €*. (β_3) The car should have *less than 15,000 km*.

$$\begin{aligned}
hasPrice(x,p) &\leftarrow hasPossibleCarPrice(x,p)\\
Kilometers(x_1,x_6), &\leftarrow CarTable(x_1,\ldots,x_{11})\\
CataloguePrice(x_1,x_4), &\leftarrow CarTable(x_1,\ldots,x_{11})\\
MinimalPrice(x_1,y), &\leftarrow CarTable(x_1,\ldots,x_{11}), y = x_4 - x_4 \cdot x_5\\
LS(k_1,k_2,a,b,p,1) &\leftarrow k_1 \le p \le a\\
LS(k_1,k_2,a,b,p,0) &\leftarrow b \le p \le k_2\\
LS(k_1,k_2,a,b,p,s) &\leftarrow a < p < b, s = (b-p)/(b-a)\\
RS(k_1,k_2,a,b,p,0) &\leftarrow k_1 \le p \le a\\
RS(k_1,k_2,a,b,p,1) &\leftarrow b \le p \le k_2\\
RS(k_1,k_2,a,b,p,s) &\leftarrow a < p < b, s = (p-a)/(b-a)
\end{aligned}$$

Fig. 2. Auxiliary rules

$CarTable(34, \texttt{ALFA 156}, 2002, 12{,}000, 20\%, 25{,}000, \texttt{Black}, 1, \texttt{LeatherSeats}, 0, \texttt{Diesel})$
$CarTable(1812, \texttt{FORD FOCUS}, 2001, 13{,}000, 20\%, 20000, \texttt{Gray}, 1, \texttt{LeatherSeats}, 1, \texttt{Gasoline})$
$hasPossiblePrice(34, p), p \in \{12000, 11900, 11800, \ldots, 9700, 9600\}$
$hasPossiblePrice(1812, p), p \in \{13000, 12900, 12800, \ldots, 10500, 10400\}$

$\exists 1 : CarTable \sqsubseteq Cars$	$(disjoint\ Mazda, AlfaRomeo, Ford)$
$Sedan \sqcup StationWagon \sqsubseteq Cars$	$(disjoint\ Sedan, StationWagon)$
$\exists 9 : CarTable \sqsubseteq Seats$	$(disjoint\ AirConditioning, NoAirConditioning)$
$Mazda \sqcup AlfaRomeo \sqcup Ford \sqsubseteq CarMake$	$(disjoint\ LeatherSeats, VelvetSeats)$
$LeatherSeats \sqcup VelvetSeats \sqsubseteq Seats$	...

Fig. 3. A part of the vague knowledge base $\mathcal{K}_{match}$ used in the example

[Preferences Utilities] $u(\beta_1) = 0.05$; $u(\beta_2) = 0.5$; $u(\beta_3) = 0.45$

Then we add to the vague Datalog program $\mathcal{P}$ in $\mathcal{K}_{match}$ the rules:

$$\begin{aligned}
\beta_A(x) &\leftarrow Sedan(x)\\
\beta_A(x) &\leftarrow StationWagon(x)\\
\beta(x) &\leftarrow \beta_A(x)\\
\beta_1(x,1) &\leftarrow NoLeatherSeats(x)\\
\beta_1(x,1) &\leftarrow LeatherSeats(x), AirConditioning(x)\\
\beta_1(x,0) &\leftarrow LeatherSeats(x), NoAirConditioning(x)\\
\beta_2(x,p,s) &\leftarrow hasPrice(x,p),\\
&\quad\ LS(0, 100000, 11000, 13000, p, s)\\
\beta_3(x,s) &\leftarrow Kilometers(x,k),\\
&\quad\ LS(0, 400000, 15000, 20000, k, s)\\
Buyer(x,p,u_\beta) &\leftarrow \beta_1(x,s_1), \beta_2(x,p,s_2), \beta_3(x,s_3),\\
&\quad\ u_\beta = 0.05 \cdot s_1 + 0.5 \cdot s_2 + 0.45 \cdot s_3
\end{aligned}$$

Since we are in a P2P e-marketplace, also the seller can express hard and soft constraints. Looking at the information modeled in Table 1 we see that a soft constraint is expressed on price: the seller prefers to sell the car at the catalogue price, furthermore he may apply a discount. In this case also a constraint on price is set; in fact, he does not want to go down such defined discount (hard constraint). Without loss of generality, for the sake of clarity in this example we consider the same hard and soft constraints for all the sellers within the e-marketplace. Seller's requirements are then encoded in $\mathcal{K}_{match}$ as:

$$\sigma(x,p) \leftarrow hasPrice(x,p), CataloguePrice(x,catP), \\ MinimalPrice(x,minP), minP \leq p \leq catP$$

$$\sigma_1(x,p,s) \leftarrow hasPrice(x,p), CataloguePrice(x,catP), \\ MinimalPrice(x,minP), minP \leq p \leq catP, \\ RS(0,100000,minP,catP,p,s)$$

$$Seller(x,p,u_\sigma) \leftarrow \sigma(x,p), \sigma_1(x,p,s_1), u_\sigma = s_1$$

Notice that, in this particular case, the specification of $u(\sigma_1)$ is not necessary because of equation (3).

According to the encoding in Section 5 the query is:

$$Match(x,p,u) \leftarrow \beta(x), Buyer(x,p,u_\beta), \sigma(x,p), Seller(x,p,u_\sigma), u = u_\beta \cdot u_\sigma$$

Solving $ans_2(\mathcal{K}_{match}, Match)$ retrieval problem with respect to $\mathcal{K}_{match}$ defined in this example, the ranked list of agreements is:

x	p	u
34	11300	0.3010
1812	11800	0.1885

Then, the agreement between `ALFA 156` and the request is ranked better then `FORD FOCUS`.

7 Related Work

Recently, the problem of matchmaking has been investigated under different perspectives and many approaches have been proposed. An initial approach to matchmaking can be dated back to vague query answering [17] where the need to go beyond pure relational databases was addressed using weights attributed to several search variables. More recently similar approaches have been proposed extending SQL with "preference" clauses, in order to allow relaxed queries in structured databases [11] where only buyer's preferences are taken into account while retrieving promising supplies: no agreement is proposed as a result of the query process. In our framework we model the matchmaking process in a P2P marketplace, taking into account not only the buyer's preferences, but also the seller's ones, finding the most promising agreements w.r.t. preferences of them both. Classified-ads matchmaking, at a syntactic level, was proposed in [21] and [25] to perform a matchmaking between semi-structured descriptions. Approaches to matchmaking using LOOM as description language can be found, among others, in [2] and [9]. Due to the growing interest in the Semantic Web initiative many approaches to matchmaking have been proposed in the framework of DAML+OIL, OWL and their grounding logical languages in particular Description Logics (DL). Matchmaking as satisfiability of concept conjunction in DLs was first proposed in [10]. In the framework of Retsina Multiagent infrastructure [24], a specific language was defined for agent advertisement, and matchmaking engine was developed [19], which carries out the process on five possible match levels. The approach in [19] was later extended in [15], where two new levels for matching classification were introduced. A similar classification was proposed — in the same venue — in [8], along with properties that a matchmaker should have in a DL based framework, and algorithms to classify and semantically rank matches within classes. An initial DL-based approach, adopting penalty functions ranking, has been proposed in [4], in the framework of dating systems. An extended matchmaking approach, with negotiable and strict constraints in a DL framework has been proposed in [7], using both concept contraction and concept abduction. The need to work in someway with approximation and ranking in DL-based approaches to matchmaking has also recently led to adopting fuzzy-DLs, as in sMART [1] or hybrid approaches, as in the OWLS-MX matchmaker [13]. sMART is a semantic matchmaking portal, based on fuzzy-DLs, able to deal with approximation in the requests description handled by crisp DL-reasoners. Nevertheless in such approaches the matchmaking process is defined according to buyer's perspective. In [16] a language able to express conditional preferences is proposed to perform a matchmaking in Description Logics. Also in this case nothing is said on how to compute a agreement — as needed in P2P scenarios. Furthermore, the notion of fuzzy/vague requirements is not addressed.

8 Conclusion

In this work we propose a semantic matchmaking approach that mixes various knowledge representation technologies to find the most promising agreements in a P2P e-marketplace. In particular, by exploiting ontologies in DLR-Lite and fuzzy rules we are able to model both *hard constraints* and *soft constraints*, while taking into account both buyer's and seller's preferences and utilities to find matches mutually beneficial for them both. The information needed for the P2P matchmaking process are modeled as a vague knowledge base, taking into account domain knowledge, while keeping the approach effective and scalable. A prototype is currently being implemented to further validate the approach through large scale experiments.

References

1. S. Agarwal and S. Lamparter. smart - a semantic matchmaking portal for electronic markets. In *Proc. of 7th Int.IEEE Conference on E-Commerce Technology*, 2005.
2. Y. Arens, C. A. Knoblock, and W. Shen. Query Reformulation for Dynamic Information Integration. 6:99–130, 1996.
3. F. Baader, D. Calvanese, D. McGuinness, D. Nardi, and P. F. Patel-Schneider, editors. *The Description Logic Handbook: Theory, Implementation, and Applications*. Cambridge University Press, 2003.
4. A. Calì, D. Calvanese, S. Colucci, T. Di Noia, and F. M. Donini. A description logic based approach for matching user profiles. In *Proc. of DL'04*, volume 104 of *CEUR Workshop Proceedings*, 2004.
5. D. Calvanese, G. De Giacomo, D. Lembo, M. Lenzerini, and R. Rosati. DL-Lite: Tractable description logics for ontologies. In *Proc. of AAAI'05)*, 2005.

6. S. Ceri, G. Gottlob, and L. Tanca. *Logic Programming and Databases.* Springer Verlag, 1990.
7. S. Colucci, T. Di Noia, E. Di Sciascio, F. Donini, and M. Mongiello. Concept Abduction and Contraction for Semantic-based Discovery of Matches and Negotiation Spaces in an E-Marketplace. *Electronic Commerce Research and Applications*, 4(4):345–361, 2005.
8. T. Di Noia, E. Di Sciascio, F. M. Donini, and M. Mongiello. A system for principled matchmaking in an electronic marketplace. *International Journal of Electronic Commerce*, 8(4):9–37, 2004.
9. Y. Gil and S. Ramachandran. PHOSPHORUS: a Task based Agent Matchmaker. In *Proc. International Conference on Autonomous Agents '01*, pages 110–111. ACM, 2001.
10. J. Gonzales-Castillo, D. Trastour, and C. Bartolini. Description Logics for Matchmaking of Services. In *Proc. of ADL-2001*, volume 44. CEUR Workshop Proceedings, 2001.
11. B. Hafenrichter and W. Kießling. Optimization of relational preference queries. In *In Proc. of ADC'05*, pages 175–184, Newcastle, Australia, Jan. 2005.
12. R. L. Keeney and H. Raiffa. Decisions with multiple objectives - preferences and value trade-offs. *Cambridge University Press*, 1993.
13. M. Klusch, B. Fries, M. Khalid, and K. Sycara. Owls-mx: Hybrid owl-s service matchmaking. In *Proc. of 1st Int. AAAI Fall Symposium on Agents and the Semantic Web*, 2005.
14. C. Li, K. C.-C. Chang, I. F. Ilyas, and S. Song. RankSQL: query algebra and optimization for relational top-k queries. In *Proc. of SIGMOD-05*, pages 131–142, New York, NY, USA, 2005. ACM Press.
15. L. Li and I. Horrocks. A Software Framework for Matchmaking Based on Semantic Web Technology. In *Proc. of WWW '03*, 2003.
16. T. Lukasiewicz and J. Schellhase. Variable-strength conditional preferences for matchmaking in description logics. In *Proc. of KR 2006*, 2006.
17. A. Motro. VAGUE: A User Interface to Relational Databases that Permits Vague Queries. *ACM Trans. Office Inf. Syst.*, 6(3):187–214, 1988.
18. J. F. Nash. The bargaining problem. *Econometrica*, 18 (2):155–162, 1950.
19. M. Paolucci, T. Kawamura, T. Payne, and K. Sycara. Semantic Matching of Web Services Capabilities. In *Proc. of ISWC'02*. 2002.
20. J. Pomerol and S. Barba-Romero. *Multicriterion Decision Making in Management.* Kluwer Series in Operation Research. Kluwer Academic, 2000.
21. R. Raman, M. Livny, and M. Solomon. Matchmaking: distributed resource management for high throughput computing. In *Proc. of IEEE High Performance Distributed Computing Conf.*, 1998.
22. U. Straccia. Towards top-k query answering in deductive databases. In *Proc. of SMC-06*, pages 4873–4879. IEEE, 2006.
23. U. Straccia. Towards top-k query answering in description logics: the case of DL-Lite. In *Proc. of JELIA-06*, 2006.
24. K. Sycara, M. Paolucci, M. Van Velsen, and J. Giampapa. The RETSINA MAS infrastructure. *Autonomous agents and multi-agent systems*, 7:29–48, 2003.
25. D. Veit, J. Muller, M. Schneider, and B. Fiehn. Matchmaking for Autonomous Agents in Electronic Marketplaces. In *Proc. International Conference on Autonomous Agents '01*, pages 65–66. ACM, 2001.
26. L. A. Zadeh. Fuzzy sets. *Information and Control*, 8(3):338–353, 1965.

Symbol Grounding for the Semantic Web

Anne M. Cregan[1,2]

[1] National ICT Australia (NICTA)
[2] CSE, University of New South Wales, Australia
Anne.Cregan@nicta.com.au

Abstract. A true semantic web of data requires dynamic, real-time interoperability between disparate data sources, developed by different organizations in different ways, each for their own specific purposes. Ontology languages provide a means to relate data items to each other in logically well-defined ways, producing complex logical structures with an underlying formal semantics. Whilst these structures have a logical formal semantics, they lack a pragmatic semantics linking them in a systematic and unambiguous way to the real world entities they represent. Thus they are intricate "castles in the air", which may certainly have pathways built to link them together, but lack the solid foundations required for robust real-time dynamic interoperability between structures not mapped to each other in the design stage. Current ontology interoperability strategies lack such a meaning-based arbitrator, and depend instead on human mediation or heuristic approaches. This paper introduces the symbol grounding problem, explains its relevance for the Semantic Web, illustrates how inappropriate correspondence between symbol and referent can result in logically valid but meaningless inferences, examines some of the shortcomings of the current approach in dealing effectively at the level of meaning, and concludes with some ideas for identifying effective grounding strategies.

Keywords: Ontology Alignment, Semantic Interoperability, Semantic Web, Symbol Grounding.

1 Introduction

The purpose of the World Wide Web is to share and leverage information. But information is only ultimately useful if it produces some result in the real world, either in the physical environment, or in someone's state of understanding. Raw unprocessed data is not very helpful in this regard, as it requires significant human effort, and the application of implicit human knowledge to understand it and process it appropriately to produce tangible benefits. It is generally agreed that machines should be doing more of the work of turning data into knowledge in a way that supports the production of results for human benefit. The purpose of the Semantic Web is to address this; its stated objective being to make information more easily shared and applied, by making its meaning explicit [1]. The implicit assumption is that once meaning is represented explicitly, machines will be able to align and process data according to its meaning, thus turning it into knowledge, and supporting web services and intelligent agents to produce real-world results on our behalf.

E. Franconi, M. Kifer, and W. May (Eds.): ESWC 2007, LNCS 4519, pp. 429–442, 2007.

However, this implicit assumption has not yet been thoroughly investigated. To date, information processing has been based on a symbolic processing paradigm, and to process information at a semantic level requires a fundamental paradigm shift. New methodologies, processes, and criteria for judging success are needed. Many of the techniques for aligning or reconciling meaning are already known from programming, but not at a mature level where meaning is made explicit and machine processing does the rest: it requires a human being to analyze the meaning and devise and implement appropriate code to do the necessary transformations.

How are we to start making inroads into this new semantic territory? As an initial step, taking a good look at the really hard questions should help focus the effort, and provide foundations for this new information processing paradigm.

These hard questions include but are not limited to the following:

1. What is meaning?
2. WhaZt do we need to do to make meaning explicit?
3. What is the appropriate way to process meaning?
4. How can we judge whether we have been successful in representing and processing meaning at a semantic level?
5. Will the current Semantic Web approach, based on the Web Ontology Language OWL [10], produce the right kind of representations, and support the right kinds of processes, to achieve the results being sought, or is a key component of the solution missing?

Spanning from the very philosophical to the very practical is necessary because the issue of meaning is a fundamental philosophical issue, whilst the goals of the Semantic Web are very practical. By their very nature as a "specification of a conceptualization" [3], creating ontologies involves bridging between the realm of IT/ Engineering, and the realm of Cognitive Science/Philosophy. It is hoped that such as investigation can uncover the foundations for such a bridge, providing a basis not only for the Semantic Web but for the Pragmatic Web it will ultimately support.

Organization
The paper is organized as follows:

- **Section 1** introduces the challenge being undertaken
- **Section 2** relates meaning to both entailment and designation, looks at symbolize-tion and introduces the symbol grounding problem
- **Section 3** explains why symbol grounding is relevant for the Semantic Web in its aim to achieve dynamic real time interoperability, and extensional approaches and URIs are not sufficient in themselves to provide adequate symbol grounding.
- **Section 4** considers next steps in identifying suitable symbol grounding strategies for the Semantic Web and concludes.

2 Meaning and Symbol Grounding

What is meaning? The greatest philosophers and thinkers have considered this question for the last several thousand years, but as yet there seems to be no definitive answer. What are the implications for the Semantic Web, which is being built around the keystone of making meaning explicit and machine-processable? Is it ever really going to get off the ground, or perhaps do so initially but quickly collapse under its own weight for lack of good foundations? It seems somewhat foolhardy to attempt to devise explicit well-defined procedures for operating at the level of meaning, without attempting to lay good foundations by stating what meaning is taken to be.

Whilst a conclusive answer to the question is unlikely (isn't that what makes a good philosophical question after all?) and Semantic Web researchers are, generally speaking, practical people who want results in reasonable timeframes and certainly don't want to get bogged down in the vagaries of philosophy, I believe that as part of the construction of Semantic Web technologies, for purely practical reasons, there should be some attempt to state what we take meaning to be for the purposes of the Semantic Web.

A clear conception of meaning for the purposes of the Semantic Web should, at the very least, assist researchers in devising appropriate and precise procedures and methods for making meaning explicit, which then has the flow-on effect of supporting practitioners to build appropriate semantic models representing their respective domains, and will make such models better suited for interoperability. It also provides a theoretical basis for semantically processing the information captured by such models.

Whilst many modeling errors have been identified and are well understood e.g. [8], there is still quite a spectrum of "correct" models available for modeling any given domain. The ontology builder has considerable discretion in make design choices. Are some of the resulting models better than others? Intuitively the answer is yes, and depends on the intended function of the ontology. However, we are still seeking a more precise understanding of the nature of this dependence, and at the moment there is no one clear guiding methodology for building domain models. Whilst there are obviously several factors at play, the model's effectiveness in making meaning explicit should certainly be considered a key criterion.

Beyond this, an analysis of meaning also offers insights into the overall Semantic Web approach and whether it will ultimately be able to deliver on its promises. Capturing meaning is clearly a fundamental component, but are the current suite of Semantic Web standards and technologies adequate to the task of capturing machine-processable meaning to produce the outcomes being sought, or will they ultimately fall short? If we want to ultimately build a "Pragmatic Web", that delivers tangible real-world benefits, we need to make sure the foundations are firm enough to support this.

2.1 What Is Meaning?

Without getting too bogged down in philosophy, let's take a practical approach to home in on what meaning is, by identifying what it is that we really want when we ask the meaning of something. In everyday life, we generally don't ask the meaning of concrete things like a chair, or a train, or a person, or a pet. Such things have no meaning: they just are. We ask the meaning of actions and events, policies and such like, in which case we are generally trying to identify the relevant entailments, **or** we ask the meaning of symbols, in which case we want to know what they designate, or stand for. When we ask about meaning, we are usually asking for one of two things: either for entailment, or for designation.

2.2 Entailment and Designation

Entailment: What are the logical consequences of some action, event or state?
Examples:

- *If I take this promotion does it mean I will be able to afford the house?*
- *If Serena wins this point, does that mean she wins the match?*
- *If my business is registered as a public company, does that mean we are required to have annual audits conducted?*

Designation: What is being referred to? What does the symbol symbolize?
Examples:

- *What's the meaning of "verisimility"?*
- *What does that sign mean?*
- *What do you mean by giving me that wink?*
- *What does the green line on the graph mean?*

Designation gives the referent being represented by some kind of symbol: a word, a street sign, a gesture, a line on a graph. It uses symbols to point to something; a convenience originating from the need to identify and communicate something that does not have a local physical existence, is abstract, or is an internal state and not directly accessible. Designation is the back end of symbolization: it establishes the referent, or what the symbols symbolize.

Symbolization
Note that there are (at least) two related senses of symbolization. In the first sense, a recognizable concrete thing is used to stand for a more abstract intangible thing e.g. a dove is used to symbolize peace. It is usually chosen as a symbol because it has some kind of real-world historical or mythological relationship with the abstract thing, or evokes it through some other kind mental or perceptual association, or it can simply be a matter of convention. In this sense, a non-verbal meaning relation is pre-established and the symbolization makes use of it to evoke the intended referent. This should not be confused with symbolization as used in this setting, which is being referred to as "designation" for the purposes of clarity within this paper.

In our setting, symbolization refers to the scenario where a mark, character, sound, avatar or some such arbitrary thing is used to designate some physical or conceptual thing. In this case, the symbol is an arbitrary physical token, designed by humans specifically for the purpose of representation, and does not usually have a meaning in and of itself (Although in the case of avatars, some recognizable topographical resemblance may exist, and thus their form may be argued not to be completely arbitrary). In this kind of symbolization (designation), the essential question is how the relationship between an arbitrary symbol and its intended referent is to be established. This question has been identified in Artificial Intelligence Research as the "Symbol Grounding Problem".

2.3 Denotation and Connotation

Designation itself has two aspects: denotation and connotation, a distinction introduced by J.S. Mill [5]. To illustrate by example, the denotation of a term such as 'woman' refers to all the individuals to which may correctly be applied, whilst the connotation consists of the attributes by which the term is defined e.g. being human, adult and female. Connotation determines denotation, and in J.S. Mill, is taken to be meaning, whereas terms like proper names e.g. 'Mary', which have denotation if there is someone so called, are taken to lack meaning as they have no connotation, as no attributes define 'Mary'.

2.4 Relevance to the Semantic Web

Both entailment and designation have relevance for the Semantic Web: entailment relating to what can be concluded from what is already known, and designation relates to establishing the connection between symbols in a formal system and what they represent. There is already a very significant body of work around entailment for the Semantic Web [10], based on description logics providing an underlying formal semantics for the various flavours of OWL.

However, designation has had less attention to date. OWL's formal semantics have a set-theoretic basis, where a set ('concept' or 'class' in DLs) is essentially defined by its extension - clearly a denotational approach. However, meaning based on denotation is less than adequate for the needs of the Semantic Web, as will be explained below. The consequence is that the entailment parts of the Semantic Web have no theoretical basis for anchoring to anything in the real world, and are thus floating castles in the air.

To explain: an OWL ontology is made up of a set of logical axioms, themselves composed of primitive objects, predicates and operators, combined via formation rules into well-formed formulae. Unless some kind of faithful and appropriate correspondence is established between the primitives and whatever they are intended to represent outside the formal logical system, any entailment produced by the system will not result in reliable conclusions that correspond to the actual state of affairs in the real-world domain of interest. Establishing a correspondence between the primitives (which are effectively just symbols or symbol strings once they are inside the logical system), and the domain is an extra-logical consideration. The question of how the relationship between the symbol and the referent is to be established has been identified in Artificial Intelligence Research as the "Symbol Grounding Problem".

2.5 The Symbol Grounding Problem

The Symbol Grounding Problem, as described, for instance, by Harnad [4] relates to the inadequacy of defining symbols using only other symbols, as is commonly done in a dictionary or a formal logical system. In his exposition, Harnad takes Searle's [9] famous Chinese Room scenario, originally used by Searle to illustrate the difference between mechanical symbol manipulation, which merely simulates mind, and a true understanding of intrinsic meaning, which necessarily involves processing at the semantic level.

The scenario involves a machine hidden inside a room, which is given a set of Chinese language inputs and produces a set of Chinese language outputs. Searle points out that a machine using only symbolic manipulation to match a list of pre-defined inputs with a list of pre-defined outputs may be capable of simulating conversation with a Chinese speaker well enough to pass the Turing test. However, Searle argues, such a machine cannot be said to understand Chinese in any sense, any more than a human who uses such a list to produce statements in Chinese can be said to understand Chinese. Searle ultimately concludes that meaning is in the head, not in the symbols, and furthermore that cognition cannot be just symbol manipulation, as it clearly requires some activity to take place at the semantic level.

Harnad puts an alternate spin on Searle's Chinese Room scenario, asking the reader to imagine having to learn Chinese as a second language, where the only source of information available is a Chinese/Chinese dictionary. He observes that "*The trip through the dictionary would amount to a merry-go-round, passing endlessly from one meaningless symbol or symbol-string (the definientes) to another (the definienda), never coming to a halt on what anything meant.*" He then presents a second variant, where one has to learn Chinese as a first language, and again the only source of information available is a Chinese/Chinese dictionary. He argues that if the first variant is difficult, then the second must be impossible, relating it to the task faced by a purely symbolic model of the mind, and asking *"How can you ever get off the symbol/symbol merry-go-round? How is symbol meaning to be grounded in something other than just more meaningless symbols? This is the symbol grounding problem."*

How indeed, are we to get off the Symbol/Symbol merry-go-round? Firstly though, let us consider in detail how the symbol grounding problem is relevant for the Semantic Web.

3 Why the Semantic Web Needs Symbol Grounding

The ultimate vision of the Semantic Web is a web of data connected by meaning which is machine processable. The idea is to get meaning out of the technologists and domain expert's heads, and into some explicit, machine processable representation which defines how to link it up appropriately, in real-time, without reference to human mediators. But the current Semantic Web building blocks are a long way from achieving this vision. Let's take a look at why this is.

In building an ontology, the designer chooses terms for classes, instances and properties, and builds axioms/structure linking them. The terms are usually chosen for their meaning in some natural or domain-specific language. Additional annotations may explain the meaning of the term, using more natural and/or domain specific language. But natural language is notoriously ambiguous and slippery. Its symbols/semantic units are imperfectly grounded, as we will explain in the following section. And whilst domain specific-terminology may be unambiguous within the domain, it is not necessarily unambiguous when linking across domains.

If the basic terms used for ontologies are ambiguous, then having a well-defined structure that supports entailment is of dubious benefit. The structure by itself is not the meaning: as discussed, meaning requires both logical structure for the purposes of entailment, and grounding for the purpose of establishing correspondence between the domain and the logical structure. Only then can entailments made by virtue of the logical structure be guaranteed to be an accurate reflection of the real-world state. Garbage in, garbage out, as the old saying goes.

3.1 Meaningfulness

As an example of this principle, there is a considerable body of work e.g. [6] in Mathematical Psychology around determining which kinds of variables can be subjected to which kinds of mathematical operations, in order to produce only meaningful results and avoid meaningless conclusions. Considerable effort has gone into investigating "meaningfulness" to avoid the inappropriate use of statistics. In a classic example, the school football team are assigned numbers to wear on their football jerseys. This numerical assignment is simply to give each football player a unique label for the purpose of identification. However, this assignment does not support taking the average of those numbers, and asserting that this average reflects some meaningful property of the football team. This is because the numbers have no numerical properties attached to them - they are just labels, and could equally well be any other arbitrary symbol (letters of the alphabet, pictures of animals) - the only important factor is that each player is designated by a unique symbol. The underlying variable being represented (identity) is not a quantitative variable, so any mathematical inferences derived from the football jersey numbers are simply meaningless. This is not the case for a quantitative variable like the heights of the football players, where it is perfectly appropriate to represent heights as real numbers and calculate average and standard deviations in the height of this population.

Note that the "meaningfulness" criteria is not necessary because of any problem to do with numbers themselves, or with mathematical reasoning. The problem is that the real-world dimension being represented does not have the same properties as the chosen representation. The representation is richer and more structured than what is being represented, and thus permits reasoning and inferences which have no correspondence with the real-world. Inferences which are perfectly valid inside the representation symbol are thus meaningless when we attempt to map them back to the domain of reference. As a formal logical system without appropriate grounding strategies to connect it to the real-world, the Semantic Web faces a similar problem.

3.2 Semantic Interoperability Problems

Pollock and Hodgson's analysis of types of semantic conflicts [7] identified eleven kinds of semantic level clashes: DataType, Labeling, Aggregation, Generalization, Value Representation, Impedance Mismatch, Naming, Scaling & Unit, Confounding, Domain, and Integrity. This analysis has been adapted and re-organized to fit the Semantic Web and the focus of the current investigation.

Semantic Conflict:	Manifests As:	Example:
Terminology	The same term is used to mean different things (homonyms) or different terms are used to mean the same thing (synonyms).	- "mouse" as a hardware peripheral vs a rodent - "Holiday" vs "Vacation" : different terms, same meaning
Representation: Instance Level	The same information is being referred to at the meaning level but is being represented differently.	- Fahrenheit vs Celsius temperature scales.
Representation: Concept Level	Concepts have been abstracted differently	- StartTime and Duration vs StartTime and EndTime
Representation: Structural Level	Different choices about the division of the domain into instances, classes and properties, and/or different choice of axioms	- Modeling a Person's educational institution as an ObjectProperty connected to a Class, or as a DataProperty connected to string values.
Representation: Superstructure Level	Different modeling constructs or paradigms used, based on fundamentally different representation methodologies.	- Relational DB vs Object-Oriented (Impedance Mismatch) - Entity-Relationship model vs first order logic model
Granularity	The same information is represented at different levels of granularity.	- Daily Sales vs Monthly Sales - Temperature information: 39.3 degrees Celsius vs "hot"

Semantic Conflict:	Manifests As:	Example:
Perspective/ Context	The information may be from the point of view of a particular part of the supply chain, a particular business process or application and does not apply universally.	- Whilst related, "Cost" from a supplier's point of view is the cost of production, whilst to the consumer it is the cost of purchasing the finished product.
Underlying conceptualization	The information may be based on a different kind of conceptualization, theory or ideology	- Linnaeism vs Cladism: different criteria for classification of classes of animals - A commonsense classification vs a technical one (eg tomato as a vegetable vs a fruit)
Origin	Whilst the information is ostensibly the same, the purpose the data was collected for, or the way it was collected is different, creating bias.	- Information about income collected for tax purposes vs collected in a credit application.

As semantic conflicts, every one of these problems requires reference to the meaning level of the information for its resolution. The following table looks at the resolution method for each class of semantic conflict and indicates how Symbol Grounding and Meaning is relevant in each case.

Semantic Conflict:	Resolution Method:	How Symbol Grounding / Meaning is relevant:
Terminology	Merge or align the two if the two terms have the same meaning Separate or treat separately if the two terms have distinct meanings.	Joining or separating based on meaning is determined by the identity or divergence of the real-world entity the terminology designates ie what the symbol is symbolizing.
Representation	Transformation of content or structure, or mapping relations based on meaning	Need to determine real-world relationships between entities being represented to identify equivalences if any and determine which transformations would be valid.
Granularity	Identify how the two fit together and create some transformation and/or mapping of values if possible	Need to know what the underlying real-world dimension is? What is the base/lowest level of granularity? If aggregated, from what base and by what criteria?

Semantic Conflict:	Resolution Method:	How Symbol Grounding / Meaning is relevant:
Perspective/ Context	Make perspective explicit and align into a superstructure	Meaning is relative to perspective and context but how does this affect it? How do we specify the context and how it affects the meaning?
Underlying conceptualization	Make underlying assumptions explicit and align into superstructure that states these explicitly	What are the underlying assumptions and methodology? How do we represent these?
Origin	Make underlying assumptions explicit and align into superstructure that states these explicitly	What is the source and how does it affect the data? How do we represent this and adjust for it?

3.3 Current Support for Semantic Interoperability Conflict Resolution

As the analysis of the previous section shows, symbol grounding and meaning is at the heart of these interoperability problems. Resolving these kinds of problems are common occurrences in mapping and aligning ontologies as they exist today. Whilst tools and heuristics are available to humans to assist the process, it is a problem essentially being addressed by human beings, rather than machines. Somehow, the underlying meaning needed to resolve these interoperability problems is not being explicitly represented in the Semantic Web, and/or there are not sufficient tools and techniques available for resolving it through automated processing. The remainder of this section makes an initial pass at identifying where the shortcomings are, starting with specifics of the OWL language and broadening out from there.

Mapping Constructs Provided by OWL

OWL provides only very limited constructs for mapping ontologies, and essentially none for transformations. The OWL language has four constructs specifically for use in mapping ontologies for the purposes of merging. These constructs are intended for use when two or more ontologies have been built independently of each other and later need to be merged or linked, i.e. they are being mapped after the initial design phase has taken place. They are:

`owl:equivalentClass`	asserts that two or more classes have exactly the same instances
`owl:equivalentProperty`	asserts that two or more properties are equivalent
`owl:SameIndividual`	asserts that two or more individuals are identical
`owl:DifferentIndividuals`	asserts that two or more individuals are identical

Whilst these constructs provide the means to specify that two or more classes, properties or individuals are equivalent / identical, or that two or more individuals are different, they are only useful once the equivalence, identity or difference is determined. This determination is outside of OWL, and the Semantic Web technologies do not provide any formal basis for a machine to determine this without human guidance (albeit supported by tools and heuristics). The use of class extensions and URIs is insufficient, as explained below.

Extensive vs Intensive Class Definitions

Some may argue that if two classes contain the same set of individuals, they must be the same, and machine processing can determine this. However, having the same extension does not necessarily prove that two classes are the same: they may happen to contain the same individuals, but have different intensive meanings: that is, the criteria for membership of the class is different. Philosophically, this is the denotation vs connotation distinction.

For example, the extension of the members of the school basketball team in a Sports Ontology and the extension of the school's Grade A students in an Academic Ontology may conceivably, at some point in time, consist of exactly the same set of individuals, and machine processing may determine on the basis of these equal extensions that the two classes must be equivalent, and map them using `owl:equivalentClass`.

The possible ramifications of such an ill-advised mapping are obvious: if a student drops his grade, he will find he is no longer classified as a member of the basketball team. If she drops out of the basketball team, she will no longer be classified as a Grade A student. If a new student is added to the class of Grade A students, he will find himself automatically in the basketball team. A new member of the basketball team will find she is automatically classified as a Grade A student.

Clearly, the extensive approach to class definition and mapping is inadequate, especially when changes in the variables used for classification occur, or new, unclassified instances are encountered, simply because the classification criteria are not adequately captured. A complete specification of meaning needs to support a decision procedure which determines whether a previously unseen instance qualifies for membership of the class or not, by making the membership criteria explicit. We also note that it is the nature of some classes to have fuzzy boundaries [2] (e.g. the colour red), and support may be needed for graded class membership in such cases.

Representing Mappable Differences/Transformations

OWL also lacks the means to specify semantic differences and the transformations needed in a way that support interoperability. Programming code can get around this, but has to be written for each specific situation, straying from the Semantic Web ideal of explicit representation enabling automated processing at the meaning level.

As a simple example, a temperature in Fahrenheit and a temperature in Celsius can easily be converted either way via a simple arithmetic equation. The underlying measurement scales are interoperable, but there is no support for representing such a relationship within an ontology. Assuming that both ontologies map temperature as a DataProperty to a real number value, the numerical values of the respective properties differ, but have a well-defined arithmetic conversion. However, OWL provides no mechanism for specifying the scale, the properties of the scale or for doing such a transformation. Meaning that is available to the designers, and could be made explicit in building the respective ontologies, subsequently supporting interoperability via machine processing, is currently not able to be represented within the ontology.

Independent and Dependent Ontologies

The mapping constructs in OWL for mapping independently designed ontologies were considered above. However, because OWL ontologies are built from URIs, the components can reside anywhere. The principle of composability in OWL means that when an ontology is being built or revised, the designer can freely use any constructs from any other available ontology. This results in one ontology having logical dependence on another. Interoperability is an issue for both dependent and independent ontologies, and some of the broader interoperability concerns are addressed below.

Dependent Ontologies: If a base ontology changes, other dependent ontologies are affected. This is potentially much more serious than just a "broken link" because it can potentially change the inferences made. In the World Wide Web, broken links and web page changes are not critical, they are simply a dead-end and one can always look for information elsewhere. However, when an external URI is an essential part of a logical structure, a change or deletion can have serious real-world consequences, such as an incorrect classification as an illegal alien for example.

Independent Ontologies: In the case where two independent ontologies need to be aligned or mapped and have no common constructs other than the Ontology language itself, currently the options for resolution are either heuristic approaches with varying success rates, or the human designers of the respective ontologies can communicate with each other or check other sources to establish the meanings of terms devise an appropriate mapping. The problem here is not only that this mapping problem is potentially in the order of N^2 to achieve interoperability across the semantic web, but the problem is an N^2 human-to-human problem. The resulting reward to effort ratio causes pause for reflection.

3.4 Why Using URIs Is Not a Sufficient Grounding Strategy

A commonly encountered argument is that Unique Resource Identifiers (URIs) can be used for any disambiguation needed for the semantic web. This is Sir Tim Berners-Lee's own view (conversation with author, November 2006). After all, anything can have a URI, why not simply use that as a unique identifier? If two concepts, individuals or properties have the same URI, they must therefore be the same. Problem solved!

Whilst this approach has its merits, it is not sufficient in itself to resolve all the kinds of semantic interoperability problems. Granted, it can identify cases where two ontologies are referring to the same thing, but it cannot identify what that same thing is that they both refer to, or that either of them are representing or processing it appropriately based on its meaning. This is because the URI does not have a grounding mechanism to connect it to anything outside the information system: A textual description residing at the URI or within the URI itself is natural language and thus subject to ambiguity and vagueness.

The exception, of course, is when the thing being referred to is, exactly, the information that resides at the URI. For instance, an identifier for a particular tax law can be grounded to a URI that contains the exact text of the tax law, and thus there is no need to go further. But if the URI is intended to reference a real world physical thing, like a person or a building, or something else outside the information system itself, it needs a symbol grounding strategy.

A further consideration when considering an information system on the intended scale of the Semantic Web, is who is going to check that every URI maps to one and only one real-world referent? No doubt many things will have more than one URI, in which case we still have the human problem discussed earlier, of determining that the URIs have the same referent and mapping them, which obviously cannot be resolved using the URI itself. And on the flip side, there will no doubt be many real-world things that have no corresponding URI, so the system is incomplete.

4 Next Steps and Conclusions

If Searle is right, cognition cannot be reduced to symbol manipulation. Semantic processing therefore requires an understanding of cognition in regard to meaning.

The next steps underway in this line of research are the investigation of a wide range of grounded symbol systems, such as Musical Notation, Cartography, Chess Notation, Circuit Diagrams, Barcodes and even Knitting Patterns. These are being analyzed to determine the grounding strategies used, and how and why they are effective or ineffective. The analysis will identify the kinds of grounding strategies available, and determine appropriate criteria for assessing them, and it is hoped it will provide a theoretical basis for constructing Symbol Grounding strategies for the Semantic Web will be identified. Following this, the question of devising appropriate processing and procedures to produce meaningful results will be addressed.

In conclusion, this paper has put forward some of the hard questions the semantic Web needs to answer, examined some of the pitfalls that may occur if they are not addressed, and explained the relevance of the symbol grounding problem for the kinds of semantic interoperability issues commonly encountered. Some insights from measurement theory in Mathematical Psychology were briefly covered to illustrate how inappropriate correspondence between symbol and referent can result in logically valid but meaningless inference. Some of the shortcomings of the current Semantic Web technologies in dealing effectively at the level of meaning level were investigated. The arguments that set extensions and URIs can provide an appropriate basis for grounding the Semantic Web were considered and found wanting. Finally, next steps for identifying effective grounding strategies and doing meaning-level processing were briefly discussed.

Acknowledgments. To my infant nephew, Gaelen Raphael Bonenfant, for providing a beautiful example of how the grounding process gets started.

Research reported in this paper has been partially financed by NICTA (http://www.nicta.com.au). NICTA is funded by the Australian Government's Department of Communications, Information Technology and the Arts, and the Australian Research Council through Backing Australia's Ability and the ICT Centre of Excellence program. It is supported by its members the Australian National University, University of NSW, ACT Government, NSW Government and affiliate partner University of Sydney.

References

1. Berners-Lee, T., & Fischetti, M. (1999). *Weaving the Web.* SanFrancisco, Harper.
2. Gärdenfors, P. *Conceptual Spaces.* (2000) MIT Press, Cambridge MA.
3. Gruber, T. R. (1993) A translation approach to portable ontologies. *Knowledge Acquisition,* 5(2), 199-220.
4. Harnad, S. (1990). The Symbol Grounding Problem. *Physica D*, 42: 335-346.
5. Mill, J.S. (1843) *A System of Logic*, London.
6. Narens, L.E (2002) *All you ever wanted to know about meaningfulness.* Volume in the Scientific Psychology Series, Lawrence Erlbaum Associates, Mahwah, NJ.
7. Pollock, J.T. and Hodgson, R. (2004). *Adaptive information: Improving business through semantic interoperability, grid computing, and enterprise integration.* Wiley Series in Systems Engineering Management, John Wiley & Sons, Inc.

8. Rector, A., Drummond, N., Horridge, M., Rogers, J., Knublach, H, Stevens,R., Wang, H. and Wroe, C (2004). OWL Pizzas: Practical Experience of Teaching OWL-DL: Common Errors & Common Patterns. *EKAW2006 Proceedings*, (pp. 63-81). Available at www.co-ode.org/resources/papers/ekaw2004.pdf
9. Searle, J. (1980). Minds, brains and programs. *Behavioral and Brain Sciences*, 3: 417-457.
10. Smith, M.K., Welty, C., & McGuinness, D.L (eds) (2004). *OWL Web Ontology Language Guide*, W3C Recommendation, 10 February 2004. Available at http://www.w3.org/TR/2004/REC-owl-guide-20040210/. Latest version available at http://www.w3.org/ TR/owl-guide/

Ontology-Driven Semantic Ranking for Natural Language Disambiguation in the OntoNL Framework

Anastasia Karanastasi and Stavros Christodoulakis

Laboratory of Distributed Multimedia Information Systems / Technical University of Crete (MUSIC/TUC)University Campus, Kounoupidiana, Chania, Greece
{allegra,stavros}@ced.tuc.gr

Abstract. The measurement of the semantic relatedness has many applications in natural language processing, and many different measures have been proposed. Most of these measures use WordNet as their central resource and not domain ontologies of a particular context. We propose and evaluate a semantic relatedness measure for OWL domain ontologies that concludes to the semantic ranking of ontological, grammatically-related structures. This procedure is used to disambiguate in a particular domain of context and represent in an ontology query language, natural language expressions. The ontology query language that we use is the SPARQL. The construction of the queries is automated and also dependent on the semantic relatedness measurement of ontology concepts. The methodology has been successfully integrated into the OntoNL Framework, a natural language interface generator for knowledge repositories. The experimentations show a good performance in a number of OWL ontologies.

Keywords: natural language interfaces, ontologies, semantic relatedness, query representation.

1 Introduction

The need to determine semantic relatedness between two lexically expressed concepts is a problem that concerns natural language processing. Measures of relatedness or distance are used in applications of natural language processing as word sense disambiguation, determining the structure of texts, information extraction and retrieval and automatic indexing.

It is also well known that a problem with the natural language interfaces to information repositories is the ambiguities of the requests, which may lead to lengthy clarification dialogues. Due to the complexity of natural language, reliable natural language understanding is an unaccomplished goal in spite of years of work in fields like Artificial Intelligence, Computational Linguistics and other. The natural language understanding could be approached by applying methods for consulting knowledge sources such as domain ontologies. Ontologies are usually expressed in a formal knowledge representation language so that detailed, accurate, consistent, sound, and meaningful distinctions can be made among the classes (general concepts), properties (those concepts may have), and the relations that exist among these concepts. A

E. Franconi, M. Kifer, and W. May (Eds.): ESWC 2007, LNCS 4519, pp. 443–457, 2007.

module dealing with ontologies can perform automated reasoning using the ontologies, and thus provide advanced services to intelligent applications such as: conceptual/semantic search and retrieval, software agents, decision support, speech and natural language understanding and knowledge management.

Knowing the context in which an ambiguity occurs is crucial for resolving it. This observation leads us to try to exploit domain ontologies that describe the domain of use of the natural language interface. The methodology that we have developed is reusable, domain independent and works with input only from the OWL ontology that was used as a reference schema for constructing a knowledge repository.

This methodology is integrated in the OntoNL Framework [3], a natural language interface generator to knowledge repositories. In comparison with natural language interfaces that focus either on developing methodologies only for syntactic analysis or for a specific application, the OntoNL Framework is able to address uniformly a range of problems in sentence analysis each of which traditionally had required a separate computational mechanism. In particular a single architecture handles both syntactic and semantic ambiguities, handles ambiguity at both a general and a domain specific environment and uses semantic relatedness measures on the concepts of the ontology to provide better ranked results. The communication is done through APIs. Note that different domain ontologies may be just imported in the system, provided that they are expressed in the same knowledge representation language (OWL). The Framework is therefore reusable with different domain ontologies.

We examine how consulting domain ontologies can help to do semantic language processing and disambiguation, not just syntactic. To this end, we have developed and evaluated a semantic relatedness measure for domain ontologies that concludes to semantic ranking. The semantic ranking is a methodology for ranking related concepts based on their commonality, related senses, conceptual distance, specificity and semantic relations. This procedure concludes to the natural language representation for information retrieval using an ontology query language, the SPARQL. The SPARQL queries are ranked based on the semantic relatedness measure value that is also used for the automatic construction of the queries.

An application of the OntoNL Framework that addresses a semantic multimedia repository with digital audiovisual content of soccer events and metadata concerning soccer in general, has been developed and demonstrated in the 2nd and 3rd Annual Review of the DELOS II EU Network of Excellence (IST 507618) (http://www.delos.info/).

2 Related Work

The known methodologies for measuring semantic relatedness are based on lexical resources or WordNet [2] and other semantic networks or computing taxonomic path length. All approaches that we are aware of measuring semantic relatedness that use a lexical resource construe the resource, in one way or another, as a network or directed graph, and then base the measure of relatedness on properties of paths in this graph [4], [5].

Most of the methods use the WordNet [1], a broad coverage lexical network of English words, as a semantic network. Nouns, verbs, adjectives, and adverbs are organized into synonym sets (synsets), each representing one underlying lexical concept, that are interlinked with a variety of relations. A simple way to compute semantic relatedness in a taxonomy such as WordNet is to view it as a graph and identify relatedness with path length between the concepts [9]. This approach was followed in other networks also, like the MeSH (Medical Subject Headings) (http://www.ncbi.nlm.nih.gov/entrez/query.fcgi?db=mesh), a semantic hierarchy of terms used for indexing articles in the bibliographic retrieval system MEDLINE, by Rada et al., [7], [8]. The principal assumption of Rada and colleagues was that “the number of edges between two terms in the MeSH hierarchy is a measure of conceptual distance between the terms”.

Despite its apparent simplicity, a widely acknowledged problem with the edge-counting approach is that it typically “relies on the notion that links in the taxonomy represent uniform distances”, which is typically not true. Sussna’s approach to scaling [10], Wu and Palmer’s Conceptual Similarity [11] and Leacock and Chodorow’s normalized path length [5] are efforts in WordNet to overcome the problem of the edge-counting approach.

One last approach for measuring semantic relatedness attempts to counter problems inherent in the structures of a general ontology by incorporating an additional, and qualitatively different, knowledge source, namely information from a corpus [1], as was first proposed in [9]. The key idea underlying Resnik’s approach [9] is the intuition that one criterion of similarity between two concepts is “the extent to which they share information in common”, which in an IS-A taxonomy can be determined by inspecting the relative position of the most-specific concept that subsumes them both. In order to overcome the information loss in Resnik’s method, Lin presented a universal similarity measure [6]. Noticing that all of the similarity measures known to him were tied to a particular application, domain, or resource, Lin attempted to define a measure of similarity that would be both universal (applicable to arbitrary objects and “*not presuming any form of knowledge representation*”) and theoretically justified (“*derived from a set of assumptions*”, instead of “*directly by a formula*”, so that “*if the assumptions are deemed reasonable, the similarity measure necessarily follows*”). Lin’s measure also referred to similarities and he took into account only the commonality and differences of two terms. The objective was to compute the similarity of ordinal values and words.

All the research results presented in the literature so far [5], [6], [7], [9], [10], [11] were tested in specific ontologies like the WordNet and the MeSH ontology, they are not general and have not been tested in different domain ontologies that refer to different contexts. The WordNet and MeSH ontologies are well formed hierarchies of terms and the methodologies that have used them examined basically similarity between terms and not relatedness between concepts.

In a framework like the OntoNL that needs to preserve its generality we could not rely on a general hierarchy of terms like the WordNet to disambiguate user expressions or the MeSH ontology a semantic hierarchy of terms used for indexing articles in the medical domain. We propose a method that can be used for computing semantic relatedness between concepts that constitute domains of context and are described by OWL domain ontologies.

The semantic ranking procedure proposed here is designed to clarify sense ambiguities. The procedure uses information from the ontologies and the specific clusters of context inside an ontology. Given an OWL ontology, weights are assigned to links based on certain properties of the ontology, so that they measure the level of relatedness between concepts. In this way we can identify related concepts in the ontology that guide the semantic search procedure. The semantic relatedness is used for the determination of the optimum, most related path that leads from the source concept-subject part to the target concept-object part of a natural language expression.

An important issue that we also attack is the need of an asymmetric measure, since all the previous approaches are based on symmetric measures. Asymmetric relatedness denotes that the relatedness between A and B is not necessarily the same as the relatedness between B and A. This is an important aspect for natural language processing since relations that are described with natural language do not indicate mathematical rules. Also, in a domain ontology we need to take into account the total of information loss in an IS-A taxonomy between the nodes that we want to test their similarity or relatedness and their common subsumer (common root node). The semantic information each concept inherits from the root node may be the same but its specialization defined by new properties that it carries is not the same for all the concepts.

All these parameters modulated the proposed semantic relatedness measure described in Section 4. We also need to point that this measure was developed to help the natural language disambiguation process when the use of domain ontologies is not enough to determine the sense words are used in an utterance for a specific domain of context, as it is described in Section 3.

3 The OntoNL Semantic Disambiguation Algorithm

The purpose of semantic disambiguation in natural language processing, based on a particular domain is to eliminate the possible senses that can be assigned to a word in the discourse, and associate a sense which is distinguishable from other meanings (WordNet gives only generic categories of senses and not domain specific. This domain specific disambiguation is much more powerful).

In particular, the common types of ambiguity encountered in the OntoNL Framework are:

1. The natural language expression contains general keywords that can be resolved by using only the ontology repository (ontological structures and semantics).
2. One of the subject or object part of the language model cannot be disambiguated by using the ontology repository.
3. Neither the subject nor the object part contains terms disambiguated using the ontological structures.

Next, we describe the entire semantic disambiguation algorithm based on the different levels of ambiguities using a UML Activity Diagram (Fig. 1). It is a general approach where the disambiguation is based on an OWL repository.

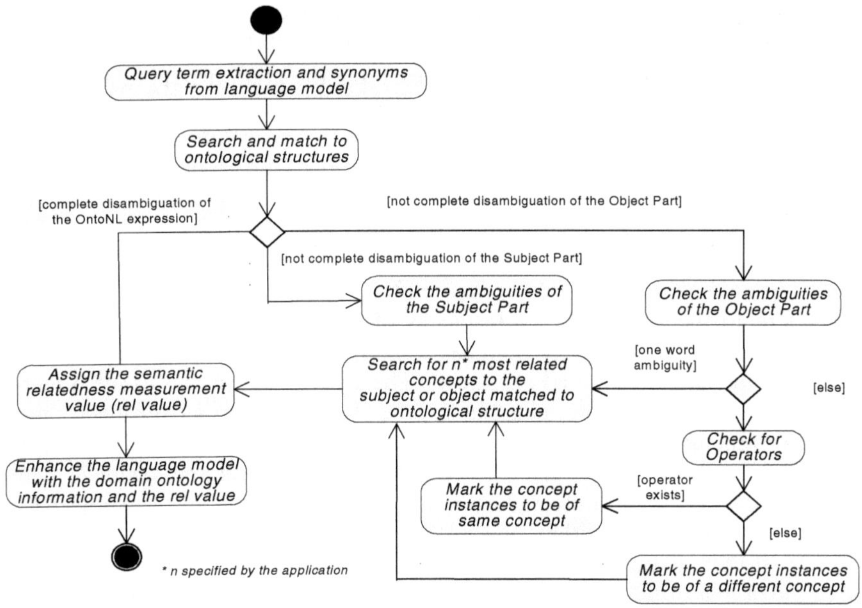

Fig. 1. The OntoNL Semantic Disambiguation procedure

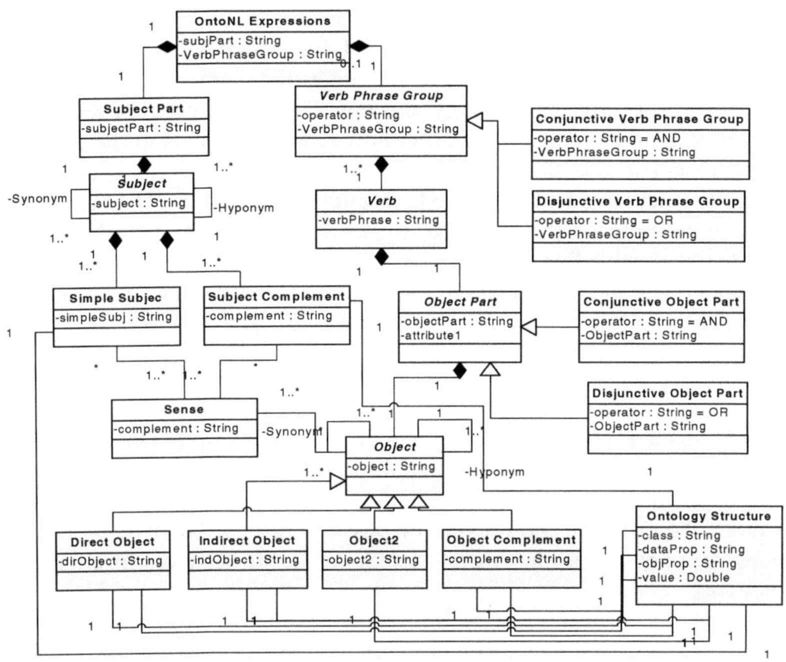

Fig. 2. The OntoNL Language Model that derives from the syntactic and semantic analysis, based on the OntoNL Natural Language Expressions

The language model that is referred in the last activity of the semantic disambiguation procedure of Fig. 1is described by the UML Class Diagram of Fig. 2. In this model diagram there are classes representing the grammatical relations that are connected with associations. There are lists of words that constitute the basic sentence structures, like the subject and the object and there are complements and special cases of objects that predicate them. The OntoNL Expressions is the general class that summarizes the cases of possible grammatical dependencies inside an utterance. It consists of a Subject Part and possibly of a Verb Phrase Group. The classes of the OntoNL language model are enhanced with the parsed information from the OntoNL syntactic disambiguation phase described in [3].

The input to the algorithm are instances of the language model, which include terms extracted from the natural language input, their synonyms, and their tagging according to the language model constructs. The algorithm searches to see if there is a correspondence between the naming of the language model instance and the ontological structures. If there is a complete match, a Relatedness Value measure is assigned with value 1 to indicate the complete relevance of the sentence with the specific domain. If the disambiguation is not complete (either in the Subject Part or the Object Part) the algorithm checks for the number of the terms that show ambiguity. If there is only one term with an ambiguity then the algorithm checks and retrieve the output of the OntoNL Ontologies Processor for a number, specified by the application, of the most related concepts to the concept that comprise the subject or the object part (if the ambiguity is in the object or the subject part respectively) of the expression. If in the Object Part are more than one terms with ambiguities the algorithm checks for operators (or/and). In the existence of an operator the algorithm considers the terms to be concept instances of the same concept of the domain ontology. In the absence of an operator the algorithm considers the terms to be concept instances of a different ontology concept. Then the algorithm searches for a number, specified by the application, of the most related concepts to the concept that found a correspondence to the ontological structures and assigns the relatedness measure, already calculated by the OntoNL Ontologies Processor. The last activity of the algorithm is to enhance the Ontology Structure class of the OntoNL Language Model with the corresponding ontology concepts to natural language terms in the class attribute and with the relatedness measurement value the value attribute.

4 The OntoNL Ontology-Driven Semantic Ranking

When a query cannot be disambiguated completely from the OntoNL Semantic Disambiguation procedure, OntoNL returns all the possible results ranked according to a value computed by the system that represents the possibility that the user has requested them. To compute the ranking of possible results, OntoNL borrows ideas and develops new ones from the research of Semantic Relatedness of concepts in a semantic network.

The output of the measurement of the relatedness between concepts of domain ontologies is a matrix containing a weight of relatedness between any two concepts. It is crucial to identify more specific domains inside the domain, based on concepts and relationships of those concepts. Consider an nth row in this matrix and a function $F_n(i)$

which takes the nth row and returns the set of the largest values. Then this function defines a local association cluster around the concept C_n. The clustering has the effect of reducing the size of a domain by creating groups of more specific information from one or more ontologies to search for semantic information.

The relatedness is also a metric that depends on the semantic relations defined by properties in OWL. Properties can be used to state relationships between individuals (named **ObjectProperties**) or from individuals to data values (named **DatatypeProperties**). Based on the semantic relations when we detect that a source concept-class is related via an **ObjectProperty** with the target concept, the relatedness value is 1 independently from their commonality or common senses or conceptual distance.

The algorithm also takes into account the semantic relation of **EquivalentClass**. The EquivalentClass of the source class has a **similarity (not relatedness) value 1** with the source class in order to also consider the relatedness measurement value of the equivalent class with the remaining classes of the Ontology.

The **commonality** depends on the amount of the common information two concepts share. We cannot use commonality like it was used by Resnik [9] when we consider domain ontologies other than WordNet because there are no senses to count the frequency of a word. We can accept partly that the distance from the most specific common subsumer of the two concepts is a criterion that must be taken into account but we have also to consider the number of common relations. We do need to keep the measure asymmetric so it will depend on the reference concept of which the relatedness to another concept we calculate. The measure that we developed has two factors: The position of the concepts relatively to the position of their most specific common subsumer (how far is their common root node) and the relativity of their properties (OWL ObjectProperties):

To measure the relativity of the properties of any two concepts we first count the number of the common properties that the two concepts share.

$$rel_{C_1}(c_1,c_2)=\frac{\sum_{i=1}^{n} p_{i12}}{\sum_{i=1}^{n} p_{i1}}. \tag{1}$$

The value p_{ij} represents the fact that concept c_1 is related to concept c_i. The value p_{i12} represents the fact that both concepts c_1 and c_2 are related to concept c_i. This measure takes into account that concepts share more common properties with other concepts that relate.

We then count the number of the common properties the two concepts share that are **inverseOf** properties:

$$rel_{C_2}(c_1,c_2)=\frac{\sum_{i=1}^{n} p_{invi12}}{\sum_{i=1}^{n} p_{i12}}. \tag{2}$$

where the p_{invi12} represents the fact that both concepts are inversely related.

The motivation to measure the common **inverseOf** properties is to release the relatedness measure from the similarity dimension. If we only counted the common **ObjectProperties** then we would assign a great value of relatedness between siblings (subclasses with common superclass) which are similar but not semantically related as the OntoNL Framework defines.

The measures rel_{C1} and rel_{C2} are combined with relative weights that show the relative importance of these two factors (*f* values):

$$\forall f_1 \geq f_2, f_2 > 0, f_1 + f_2 = 1:$$

$$rel_{prop}(c_1,c_2) = (f_1 \times \frac{\sum_{i=1}^{n} p_{ijk}}{\sum_{i=1}^{n} p_{ij}}) + (f_2 \times \frac{\sum_{i=1}^{n} p_{invijk}}{\sum_{i=1}^{n} p_{ijk}}), \quad (3)$$

The factors f_1 and f_2 in general depend on the ontologies used, and we assume that they are experimentally determined for a given ontology. A systematic algorithm for the quantification of the factors is ongoing.

The **conceptual distance measure** is based on two factors; the path distance and the specificity. The **specificity** of the concepts is based on their position in the ontology (the leaf nodes are the most specific concepts in the hierarchy). The path distance counts the edges in the minimal path of edges from a concept to another. Within one conceptual domain, the relatedness of a concept (C_1) to another concept (C_2) is defined by how closely they are related in the hierarchy, i.e., their structural relations (IS-A relation). In the OntoNL, the IS-A relations are implemented through the rdfs:subClassOf syntax of OWL. The parameter that differentiates our measure from the classic measures of distance counting is the change of direction that is combined with the specificity factor. We claim that when the change of direction (from superclassing to subclassing) is close to the initial concept-c_1 (that is the **subject** of the natural language expression) of the pair we test the relatedness; the two concepts are more related. When the direction of the path changes far from the first concept then the semantics change quite as well (more specialization). Also we take into account the place of the concepts in the hierarchy. The terms located higher in the hierarchy have higher values of relatedness than located terms lower in the hierarchy.

The value of distance can be measured with the following measure

$$pathDist(c_1,c_2) = \frac{d_{C1} + d_{C2}}{2 * D} \in (0,1]. \quad (4)$$

where d_{C1} is the number of edges to go from the concept 1 to the closer common superconcept (subsumer) and d_2 the number of edges to go from the concept 2 to the closer common superconcept (subsumer). With D we count the maximum depth of the ontology. The OntoNL disambiguation algorithm uses the relatedness of concepts of the domain ontologies and not the similarity, so the measure excludes the cases were $d_{C1} = 0$ and $d_{C1} + d_{C2} = 2$. So, the path distance measure becomes

$$\forall d_{C1} \geq 1, d_{C2} \geq 1, d_{C1} + d_{C2} > 2 : pathDist(c_1,c_2) = \frac{d_{C1} + d_{C2}}{2 * D} \in (0,1]. \quad (5)$$

We need a factor to determine the specificity of the concepts inside the ontology. As we have already stated is the value of d_{C1} is close to the value of $(d_{C1}+d_{C2})/2$ then the relatedness must be decreased, because the initial concept c_1 is specialized a lot in comparison with the subsumer concept.

$$w1_{spec\,C1} = \begin{cases} -\log\dfrac{2\times d_1}{d_1+d_2} \in (0,1], \text{if } d_1 < \dfrac{d_1+d_2}{2} \\ 0, \quad \text{if } d_1 \geq \dfrac{d_1+d_2}{2} \end{cases} \tag{6}$$

We, also use a method of counting the specialization of the concept – c_1 based on the object properties of the subsumer (root OWL Class), by the factor:

$$spec_{C1} = \frac{\#ObjP_{C1} - \#ObjP_S}{\#ObjP_S} \in [0,\infty)\,. \tag{7}$$

were $ObjP_{C1}$ is the number of Object Properties of the concept c_1 and $ObjP_S$ is the number of ObjectProperties of the subsumer concept. If the factor becomes 1 or greater then the specialization is so big that we cannot count the relatedness based on the specificity. The range of the $spec_{C1}$ is $[0,\infty)$. To limit the range in $[0,1]$ we need to restrict the number of ObjectProperties of the concept c_1. We normalize the factor and we subtract it from 1, with the restriction that the number of the ObjectProperties of the concept – c_1 is at most 10 times the number of the ObjectProperties of the subsumer.

$$\forall \#ObjP_{C1} \leq 10\times\#ObjP_S : w2_{spec\,C1} = 1 - \log\frac{\#ObjP_{C1}}{\#ObjP_S} \in [0,1]\,. \tag{8}$$

The conceptual distance measure then becomes

$$rel_{CD} = (w1_{specC1} + w2_{specC1} + 1 - pathDist(c_1,c_2))/3\,. \tag{9}$$

The **amount of related senses measure** is a measure that concerns the domain ontology and the WordNet Ontology. From the WordNet Ontology we exploit the noun glosses. Glosses are descriptions of a word's sense and it consists of a descriptive part and an example of use case. From the domain ontologies we exploit the concept descriptions that are expressed in the <owl:label> and <owl:comment> constructs. The measure is based on sets of each concept that contain synonyms and nouns extracted from the descriptive part of the glosses of each concept:

$$rel_{RS}(c_1,c_2) = \frac{|S_1 \cap S_2|}{|S_1 \cap S_2| + |S_1 \setminus S_2|}\,. \tag{10}$$

were S_1 is the description set of senses for concept C_1 and S_2 the description set of senses for concept C_2.

The overall relatedness measure is the following:

$$\begin{aligned}&\forall w_1 + w_2 + w_3 = 1, (w_1, w_2, w_3) > 0,\\&rel_{PROP}(c_1, c_2), rel_{CD}(c_1, c_2), rel_{RS}(c_1, c_2) \in [0,1]:\\&rel_{OntoNL} = w_1 \times rel_{PROP} + w_2 \times rel_{CD} + w_3 \times rel_{RS} .\end{aligned} \quad (11)$$

The three factors w_1, w_2 and w_3, can help of choosing which parameter can better express the semantic relatedness. We test the values of the factors in the evaluation section.

The measure is applied in all concepts of the ontology in the preprocessing phase and constructs a NxN matrix, were N is the total number of concepts, with the relatedness values of each concept with all the other concepts inside the disambiguation ontology.

5 Representation of Natural Language Interactions

After the syntactic and semantic disambiguation, we have concluded to the subject of the query, specialized by additional description that forms the object part or possible object parts of the query. We need a formal way to represent the query, a standardized query language that will meet the specification of the ontology language (OWL) and will be easily mapped to various forms of repository constructions. Although we could in principle use an internal representation of the preprocessed NL interactions, we opted to use a representation that is near to the languages used in the Semantic Web, so that when the repository is based on OWL or RDF to be able to directly use it to access the repository. We choose SPARQL as the query language to represent the natural language queries since SPARQL is defined in terms of the W3C's RDF data model and will work for any data source that can be mapped into RDF.

To provide an automatic construction of SPARQL queries we need at any point to define the path that leads from the subject part to the object part of the natural language expression by taking into account the constraints that are declared from the keywords and the relatedness value between the related classes of the ontology. The path connecting the classes directed from the user expression is given by an algorithm solving the problem:

Given a connected graph G = (V,E), a weight d:E->R+ and a fixed vertex s in V, find a optimized path from s to each vertex v in V. The optimized path is determined by the highest normalized sum value of the weights of the related concepts.

In the OntoNL Framework the edges linking the classes of the ontology graph are the objectProperties of the OWL syntax and the weight values are specified by the relatedness measure calculation described earlier in this chapter.

The general algorithm of the OntoNL query representation of domain-ontology disambiguated natural language expression in SPARQL is shown in Fig. 3:

```
Program String SPARQLRepr (List, List, DoubleList)
 List subjOper, objOper, Values, OptPath;
 Double relVal;
 DoubleList SemRelMeas, ListNLStoOnto, ListNLOtoOnto;
 String Query, QueryTemplate, OntoSubjTerm, OntoObjTerm, value, value1,
 value2, val1, val2;
Begin
QueryTemplate=" PREFIX ins:<ontology_path> SELECT ?OntoSubjTermIDs WHERE
{?OntoSubjTermIDs rdf:type ?OntoSubjTerm ."
If ListNLOtoOnto.size()=0 && subjOper.size()=0
 OntoSubjTerm = ListNLStoOnto.get(term)
 Query = QueryTemplate + "}";
ElseIf ListNLOtoOnto.size()=0 && subjOper.size()!=0
 For all terms i of ListNLStoOnto
 OntoSubjTerm(i) = ListNLStoOnto.getTerm(i)
 Query = QueryTemplate + "}";
Else
 relVal = ListNLOtoOnto.get(relatedness value)
 value = Values.get(not_Disambiguated_Term)
 If objOper.size()=0 && relVal=1
  OntoObjTerm = ListNLOtoOnto.get(term)
  Query = QueryTemplate +
  "{{?OntoSubjTerm ins:hasObjPropTo ?OntoObjTerm . "
  "?OntoObjTerm ins:hasDataProp "value"}"
 ElseIf objOper.size()=0 && relVal!=1
  OntoObjTerm = ListNLOtoOnto.get(term)
  OptPath = findOptPath(OntoSubjTerm, OntoObjTerm)
  Query = QueryTemplate + "
  For all ObjProperties of OptPath
  "{{?OntoSubjTerm ins:OptPath.get(hasObjProp) ?OntoObjTerm . "
  "?OntoObjTerm ins:hasDataProp "value"}"
 Else
  For all terms of OntoObjTerm
   OntoObjTerm = ListNLOtoOnto.get(term)
  If Values.size() = 1
   If relVal=1
    Query = QueryTemplate +
    "{{?OntoSubjTerm ins:hasObjPropTo ?OntoObjTerm1."
    "?OntoObjTerm1 ins:hasDataProp ?val1}UNION"
    "{{?OntoSubjTerm ins:hasObjPropo ?OntoObjTerm2."
    "?OntoObjTerm2 ins:hasDataProp ?val2}"
    "FILTER(?val1 = "value" || ?val2 = "value")"
   Else
    Query = QueryTemplate +
    For all ObjProperties of OptPath
    "{{?OntoSubjTerm ins:Opt.get(hasObjProp) ?First_Rel_Class ."
    "?First_Rel_Class ins:hasDataProp ?val1} UNION"
    "{{?OntoSubjTerm ins: OptPath.get(hasObjProp) ?Sec_Rel_Class."
    "?Sec_Rel_Class ins:hasDataProp ?val2}"
    "FILTER(?val1 = "value" || ?val2 = "value")"
  Else
   For all terms of Values
    If relVal=1
     Query = QueryTemplate +"
     "{{?OntoSubjTerm ins:hasObjPropTo ?First_Rel_Class."
     "?First_Rel_Class ins:hasDataProp "value1"}UNION"
    "{{?OntoSubjTerm ins:hasObjPropTo ?Sec_Rel_Class."
    "?Sec_Rel_Class ins:hasDataProp "value2"}"
   Else
    Query = QueryTemplate +"
    For all ObjProperties of OptiPath
     "{{?OntoSubjTerm ins: OptPath.get(hasObjProp) ?First_Rel_Class."
     "?First_Rel_Class ins:hasDataProp "value1"}UNION"
     "{{?OntoSubjTerm ins: OptPath.get(hasObjProp) ?Sec_Rel_Class."
     "?Sec_Rel_Class ins:hasDataProp "value2"}"
 End
```

Fig. 3. The OntoNL query representation of domain-ontology disambiguated natural language expression in SPARQL

6 Evaluation

A complete evaluation framework has been designed. Such a framework takes into account a large number of parameters regarding the characteristics of the ontologies involved and the types of users.

Our objective so far was to integrate all the components involved, to test interoperability, to integrate with a knowledge repository and to experiment with the performance of the disambiguation component. The integration among the components is complete and serves all the needs currently anticipated. A complete scenario utilizing all the components of the system with semantic MPEG-7 descriptions of soccer games which utilize an extensive soccer ontology [10] was tested successfully.

We have focused now our attention to the performance experimentation in a generic way utilizing readily available ontologies in the web, not carefully constructed by hand ontologies. Our objective was to analyze the semantic disambiguation process, to see if it works satisfactorily, and which components can be improved with different algorithms.

To assess our relatedness measure's usefulness, we needed to evaluate it against a "gold standard" of object relatedness. To that end we designed a detailed experiment in which human subjects were asked to assess the relatedness between two objects. As Budanitsky and Hirst [1] found in a study comparing WordNet similarity measures human judgments give the best assessments of the "goodness" of a measure.

We have obtained relatedness judgments from 20 human subjects, 10 from the computer science field that had knowledge of the domain ontologies and 10 from the liberal arts field, that were used for the evaluation, for 25 pairs of concepts that we meet in 3 OWL domain ontologies freely available on the web, for the domains of **soccer**, **wine** and **people with pets**. The pairs ranged from "highly related" to "semantically unrelated", and the subjects were asked to rate them, on the scale of 0.0 to 1.0, according to their "relatedness of meaning". After calculating the mean ratings from the experiments on the concept pairs produced by the human ratings and the ratings the equations 3, 4, 10 and 11 produced for the three ontologies, we present the absolute values of the coefficients of correlation between the ratings in **Table 1**. We also present in this table, the overall satisfaction of the users after presenting them the results of the OntoNL Semantic Ranking procedure for the pairs of concepts used for the experimentation. The users were showed the semantically related concepts to the source-initial concept accompanied with the value of relatedness and the users could evaluate if the ranking was correct to their sense of the domain.

We have observed that the ratings from human subjects that come from the liberal arts field were closer to the ratings from the Properties sub-measure (rel_{PROP}) and the Related Senses sub-measure (rel_{RS}). On the contrary, the human subjects that were aware of the structures of the tested domain ontologies (from the Computer Science field) came closer to the ratings from the Conceptual Distance sub-measure (rel_{CD}) and to the ratings from the Properties sub-measure (rel_{PROP}). The impact of each of the sub-measures expressed by the factors w_1, w_2 and w_3 of the eq. 11 can generally be tuned after experimentation of each specific application in a particular domain, that is expressed by an OWL domain ontology.

The values of the factors f_1 (for rel_{C1}), f_2 (for rel_{C2}) of equation 3, w1(for rel_{PROP}), w2 (for rel_{RS}) and w3 (for rel_{CD}) of equation 11 are shown in **Table 2**. The experimentation pointed to some first conclusions that are the basis for the relative weights value calculation algorithm extraction. The parameters that we take into account are described next but are not limited to them since the evaluation tests and the methodology for the relative weights values extraction are ongoing:

Table 1. The values of the coefficients of correlation between human ratings of relatedness and four computational measures; the three submeasures that constitute the OntoNL Semantic Relatedness Measure and the overall OntoNL measure with relative weights of Table 2

Measure	Humans LibArts Field			Humans CompSc Field			User Satisfaction over Ranking (%)		
Ontology	Soccer	Wine	PP	Soccer	Wine	PP	Soccer	Wine	PP
rel_{RS}	0,938	0,967	0,953	0,908	0,925	0,917	80%	85%	83%
rel_{CD}	0,935	0,947	0,927	0,929	0,961	0,982	82%	90%	89%
rel_{PROP}	0,964	0,948	0,954	0,945	0,943	0,963	87%	85%	89%
rel_{OntoNL}	0,978	0,981	0,969	0,968	0,972	0,987	92%	95%	93%

Table 2. The values of the relative weights f1 and f2 of eq. 3 and w1 (for rel_{PROP}), w2 (for rel_{RS}) and w3 (for rel_{CD}) of eq. 11 for each one of the ontologies used for the specific experimentation

Ontology	rel_{PROP}		rel_{OntoNL}		
	f_1	f_2	w_1	w_2	w_3
Soccer	0,5	0,5	0,7	0,1	0,2
Wine	0,8	0,2	0,25	0,2	0,55
P 'n' P	0,8	0,2	0,45	0,2	0,35

The language the ontology uses for its terminology. When ontologies are used directly from their source (web) a major factor of the rel_{RS} parameter's performance is the names that are used to describe the ontologies. If the names for the concepts and the logical relationships among the concepts used are near the "natural language" names the performance of the system is significantly better.

The number of the properties over the concepts. When the concepts of the ontology have a number of properties that specialize them over other concepts (the semantic network has a significantly greater number of edges over nodes) then the parameter rel_{PROP} can participate with a great value of influence in the overall OntoNL semantic relatedness measure calculation.

The depth of the domain ontology. When the ontology is of a great depth then the conceptual distance needs to be assigned with a big relative weight because the information loss is significant over the inheritance.

To summarize the observations over the experimentations, the application of the semantic relatedness measure on a number of OWL ontologies produced some first conclusions. We have observed that when ontologies are used directly from their source (web) a major factor in the performance of the natural language interaction system is the names that are used to describe the ontologies. If the names for the concepts and the logical relationships among the concepts used are near the "natural language" names the performance of the system is significantly better. This may imply that for ontologies that do not utilize "natural language" names for their concepts and relationships we have to provide a mapping to more natural language expressed ontologies. Alternatively, algorithms for automatic mappings should also be investigated.

When the concepts of the ontology have a great number of properties that specialize them over other concepts, like in the 'Soccer' Ontology then if the parameter rel_{PROP} takes a great weight of influence in the overall OntoNL measure, the user satisfaction over the OntoNL Semantic Ranking procedure increases up to almost 10% of the average satisfaction using the three parameters of the measure individually.

Also, the conceptual distance is a measure that has a great influence if the ontology depth is big because this means that there are several paths that lead from the source concept (that is the subject part of a natural language expression) to the target concept (that is the object part of a natural language expression). This observation was applied in the evaluation of the measure over the 'Soccer' and the 'People with Pets' Ontologies and produced very good results.

7 Conclusions

We have presented the OntoNL ontology-driven semantic ranking methodology for ontology concepts used for natural language disambiguation. The methodology uses domain specific ontologies for the semantic disambiguation. The ontologies are processed offline to identify the strength of the relatedness between the concepts. Strongly related concepts lead to higher ranked pairs of results during disambiguation. The disambiguation procedure is automatic and quite promising, since it is linguistically as complete as possible in an automatic environment [3] and it is enhanced with information based on the domain that the request refers to. It is easily reusable in many domains since the only restrictions are the used language (English) and OWL as the standard language for representing ontologies of a specific domain.

The OntoNL semantic ranking methodology depends on the OntoNL semantic relatedness measure for OWL domain ontologies. The measure is based on commonality of two concepts, the related senses that may share, their conceptual distance in the ontology, their specificity in comparison with their common root concept and the semantic relations to other ontological concepts.

The motivation of this work came from the absence of a general, domain-independent Natural Language Interface Generator with good results in the Natural Language Disambiguation process. The disambiguation process depends on the domain ontologies and when necessary, the OntoNL Semantic Relatedness Measure is used to rank ontological, grammatically-related concepts. We have developed a semantic relatedness measure over OWL ontologies that is general, domain independent and covers the lack of a systematic way for calculating asymmetric semantic relatedness of concepts.

Overall, we state that the semantic relatedness measure that leads to the ontology-based semantic ranking of concepts for natural language disambiguation is quite complete and shows very good results. For future improvements, we may need to investigate the influence of more complex structures of OWL vocabulary to the performance.

References

1. Budanitsky, A., Hirst, G. 2006. Evaluating WordNet-based Measures of Lexical Semantic Relatedness. Computational Linguistics 32(1): 13-47
2. Fellbaum, C. editor. 1998. WordNet: An Electronic Lexical Database. The MIT Press, Cambridge, MA.
3. Karanastasi, A., Zwtos, A., Christodoulakis, S. 2006. User Interactions with Multimedia Repositories using Natural Language Interfaces - OntoNL: an Architectural Framework and its Implementation, in Journal of Digital Information Management - JDIM, Volume 4, Issue 4, December 2006
4. Kozima, H., Furugori, T. 1993. Similarity between words computed by spreading activation on an English dictionary. In Proceedings of 6th Conference of the European Chapter of the Association for Computational Linguistics (EACL-93), pages 232–239, Utrecht.
5. Leacock, C., Chodorow, M. 1998. Combining local context and WordNet similarity for word sense identification. In Christiane Fellbaum, editor, WordNet: An Electronic Lexical Database, chapter 11, pages 265–283. The MIT Press, Cambridge, MA.
6. Lin, D. 1998. An information-theoretic definition of similarity. In Proceedings of the 15th International Conference on Machine Learning.
7. Rada, R., Bicknell, E. 1989. Ranking documents with a thesaurus. JASIS, 40(5):304–310, September.
8. Rada, R., Mili, H., Bicknell, E., Blettner, M. 1989. Development and application of a metric on semantic nets. IEEE Transactions on Systems, Man, and Cybernetics, 19(1):17–30, February.
9. Resnik, P. 1995. Using information content to evaluate semantic similarity. In Proceedings of the 14th International Joint Conference on Artificial Intelligence, pages 448–453, Montreal, Canada.
10. Sussna, M. 1993. Word sense disambiguation for free-text indexing using a massive semantic network. In Proceedings of the Second International Conference on Information and Knowledge Management (CIKM-93), pages 67–74, Arlington, Virginia.
11. Wu, Z., Palmer, M. 1994. Verb semantics and lexical selection. In Proceedings of the 32nd Annual Meeting of the Association for Computational Linguistics, pages 133–138, Las Cruces, New Mexico

Web-Annotations for Humans and Machines

Norbert E. Fuchs[1] and Rolf Schwitter[2]

[1] Department of Informatics & Institute of Computational Linguistics
University of Zurich, Switzerland
fuchs@ifi.unizh.ch

[2] Department of Computing & Centre for Language Technology
Macquarie University, Australia
rolfs@ics.mq.edu.au

Abstract. We propose to manually annotate web pages with computer-processable controlled natural language. These annotations have well-defined formal properties and can be used as query relevant summaries to automatically answer questions expressed in controlled natural language, and as the basis for other forms of automated reasoning. Last, but not least, the annotations can also serve as human-readable summaries of the contents of the web pages. Arguably, annotations written in controlled natural language can bridge the gap between informal and formal notations and leverage true collaboration between humans and machines. This is a position paper that proposes a solution combining existing methods and techniques to achieve a highly relevant practical goal, namely how to effectively access information on the web. However, our solution introduces a "chicken and egg" problem: a critical mass of web annotations will be necessary that people perceive the value of these annotations and start annotating web pages themselves. Only the future will show whether this – basically non-technical – problem can be solved.

Keywords: Annotations, Controlled Natural Languages, Semantic Web, Question Answering, Automated Reasoning.

1 Getting Your Questions Answered – Or Perhaps Not

1.1 The Problem

When visiting restaurants in Sydney you notice that many dishes contain capers, and you may ask yourself "Does this Mediterranean plant perhaps grow in Australia?" Asking the service personnel remains inconclusive, and you eventually turn to Google with the query

Are capers grown in Australia?

and get more than 74'000 references. Realising that Google gives you references to all web pages that contain the keywords of your query in any order and any context, you let Google search for the exact phrase

E. Franconi, M. Kifer, and W. May (Eds.): ESWC 2007, LNCS 4519, pp. 458–472, 2007.

"Are capers grown in Australia?"

and receive no answer at all. Recalling that Google orders its results according to their rank you select the top result of the 74'000 references found for your first query. This top result refers to an interview of the Landline[1] program of the Australian Broadcasting Corporation. Perusing the complete interview of 2000 words you find that the text of the interview nowhere explicitly states

Capers are grown in Australia.

which would exactly answer your question. Instead there are text snippets containing variants of and references to the words "capers", "grow" and "Australia" like

- *... a young South Australian couple decided they could grow capers in this country. ...*
- *... So they sought out some plants and now boast Australia's first and only commercial caper crop. ...*
- *... "No-one grows capers in Australia" ...*
- *... as Australia's first and only commercial growers of capers. ...*
- *... because they've never been grown in Australia before ...*
- *... that we can grow capers in Australia ...*
- *... Australia's first home-grown capers and caper-berries. ...*
- *.... they could be grown in Australia ...*

These text snippets do not readily help you to answer your question, are rather confusing, perhaps even contradictory. Absorbing the complete contents of the interview, and applying unspecified world-knowledge, you eventually infer that capers are experimentally grown in South Australia.

A little frustrated, you wished that a search engine would be able to automatically find a satisfying answer to your query without you having to extract the answer from lengthy documents.

So what can be done to automatically find an answer to your question on the web? Note that a solution to this problem is intimately related to another problem. Which answer do you actually expect? Which answer would you accept as satisfactory?

1.2 Question Answering

One approach to answer questions has been investigated by researchers of the language engineering community. To this community question answering (QA) systems are of great interest because they combine information retrieval (IR), natural language processing (NLP), and often machine learning (ML) within the same task. QA systems

- receive natural language queries as input – not keywords,
- process large unstructured document collections – usually not web pages,
- return precise answers as output – not (references to) documents.

Though the fields IR, NLP and ML have seen spectacular progress in recent years, a sobering realisation must be made – there seems to be a ceiling of what can be achieved. Here is a representative current result. The best values cited in the

[1] www.abc.net.au/landline/content/2006/s1602940.htm

"Overview of the TREC 2005 Questions Answering Track"[2] are 71% accuracy, 64% precision and 53% recall for answering mainly simple factoid questions. Though new methods may bring some improvements, we believe that no real breakthrough can be expected, and that eventually automatic methods must be complemented by human intervention to get better results. This echoes our experience that interpreting the Landline interview required additional knowledge not found in the interview itself.

1.3 Automatic Summarisation

An alternative approach has been to automatically summarise documents, and – among other things – to use summaries of documents instead of the documents themselves to answer questions. Summaries can be generated simply by extraction, i.e. by copying relevant information of the document into the summary, or by abstraction, i.e. by paraphrasing and condensing the contents of the document. Though there does not seem to be a consensus on evaluation methods, the results of automatic summarisation are not more encouraging than those of question answering[3], and again we have to realise that human intervention would eventually be required to improve the results. Interestingly, Hovy [1] writes in this context

> *... Since the result [of summarisation] is something new not explicitly contained in the input this stage [of summarisation] requires that the system have access to knowledge separate from the input. ...*

Again, we encounter the situation that additional knowledge is required to understand a text.

1.4 Semantic Web

Another approach – aimed directly at web pages – is taken by the semantic web[4] that states as its goals

> *The Semantic Web is about two things. It is about common formats for interchange of data, where on the original Web we only had interchange of documents. Also it is about language for recording how the data relates to real world objects. That allows a person, or a machine, to start off in one database, and then move through an unending set of databases which are connected not by wires but by being about the same thing.*

These goals are to be achieved by languages like RDF[5] and OWL[6]

> *... the World Wide Web Consortium released the Resource Description Framework (RDF) and the OWL Web Ontology Language (OWL) as W3C Recommendations. RDF is used to represent information and to exchange knowledge in the Web. OWL is used to publish and share sets of terms called ontologies, supporting advanced Web search, software agents and knowledge management.*

---

[2] trec.nist.gov/pubs/trec14/papers/QA.OVERVIEW.pdf
[3] acl.ldc.upenn.edu/E/E99/E99-1011.pdf
[4] www.w3.org/2001/sw
[5] www.w3.org/RDF
[6] www.w3.org/2004/OWL

Like Katz and Lin [2] we see two main problems of the semantic web. The first problem is

... to transform existing sources (stored in HTML pages, in legacy databases etc.) into a machine-understandable form (i.e. XML/RDF/OWL) ...

which is hard to do automatically since the transformation involves hurdles similar to those encountered in automatic question answering and automatic summarisation.

The second problem is that this transformation

... is sometimes at odds with a human-based natural language view of the world.

concretely, that languages like RDF and OWL are intended for computers, not for humans.

To solve the second problem natural language front-ends have been proposed. Within the Metalog[7] project of W3C, Marchiori and collaborators developed the language PNL ("Pseudo Natural Language") that they describe as follows

The goal of the Metalog's PNL ("Pseudo Natural Language") is to define a technology that is very close to the people, even if this possibly means sacrificing part of the expressive power of the underlying tower (in other words, to start filling up the upper parts of the P axis). The PNL, as the name says, aims to use a very colloquial form of communication, that is very close to humans: natural language.

and give the example

JOHN and MARY OWN a "red house".

that – capitalising some constituents and putting others in quotes – can hardly be called natural. Incidentally, the example is also ambiguous as to whether John and Mary own together one house, or individually two houses.

Alternative approaches to bridge the gap between the languages of the semantic web and natural language are offered within the Attempto project [3]. There is a bidirectional translation between Attempto Controlled English (ACE) – a subset of standard English equivalent to full first-order logic – and OWL DL that allows users to interface OWL ontologies in ACE without having to know the languages OWL, RDF or XML.

A slightly different approach is proposed by Schwitter and Tilbrook [4] who use a controlled natural language to directly describe knowledge of the semantic web without taking recourse to RDF.

1.5 Augmenting RDF by Natural Language Annotations

To render RDF friendlier to humans and to facilitate question answering from web pages and data bases, Katz and his collaborators [2] propose to augment RDF with natural language annotations. Using these techniques they have developed the Natural Language Question Answering System START[8] that uses pattern matching to answer natural language questions from a variety of sources.

---

[7] www.w3.org/RDF/Metalog

[8] start.csail.mit.edu/

Unlike information retrieval systems (e.g., search engines), START aims to supply users with "just the right information," instead of merely providing a list of hits. Currently, the system can answer millions of English questions about places (e.g., cities, countries, lakes, coordinates, weather, maps, demographics, political and economic systems), movies (e.g., titles, actors, directors), people (e.g., birth dates, biographies), dictionary definitions, and much, much more.

While the START system is rather impressive and answers questions from a wide range of sources, we believe that the fixed patterns of the RDF annotations employed by START are too rigid and too restrictive to anticipate the diversity of questions users may ask.

1.6 Annotations in Controlled Natural Language

Realising that attempts to automatically answer questions from legacy texts and web pages have encountered fundamental problems, we propose a radical solution, namely to have humans annotate web pages in a way that facilitates question answering. Concretely, we propose to augment web pages with annotations in a controlled natural language [4]. This proposal offers the following advantages:

- annotations in computer-processable controlled languages permit formal reasoning, specifically question answering by deduction,
- question answering from annotations in controlled natural languages can easily be supported by the necessary linguistic and domain-specific background knowledge,
- annotations in controlled natural languages are readable by anybody, and thus can also serve as a summary of the respective web page,
- annotations in controlled natural languages can be written according to standard guidelines of good summary writing, for instance Wikipedia's guidelines for lead sections[9].

In a similar approach [5] propose to annotate scientific publications with summaries in controlled natural languages, and point out that these annotations can be used for question answering and a number of additional reasoning tasks.

Here is a possible annotation to the above-mentioned Landline interview on capers that contains just a small amount of the factual knowledge of the interview.

A couple grows capers in Australia.

This annotation is written in a controlled natural language [6, 7] similar to ACE or PENG[10], and allows us to answer our question by deduction

Are capers grown in Australia?

provided that we take into account the linguistic knowledge relating the active and the passive forms of the transitive verb *grow*.

So we are already done, are we?

---

9 en.wikipedia.org/wiki/Wikipedia:Lead_section

10 www.ics.mq.edu.au/~rolfs/peng/

In fact, we are not done at all since we skillfully crafted the annotation in a way that allowed us to more or less immediately deduce our question, and we carefully sidestepped a number of serious problems of this approach.

Foremost, there is a "chicken and egg" problem [2] that similarly affects our approach, the semantic web and the START project, namely

> *... people will not spend extra time marking up their data unless they perceive a value for their efforts, and metadata will not be useful until a "critical mass" has been achieved.*

Only the future will show whether this – basically non-technical – problem can be solved, and we will not discuss it further here.

Furthermore, there are technical problems that are specific to our approach. We will address these problems in the remainder of this paper. In section 2 we describe controlled natural languages and in section 3 how these languages can be used to annotate web pages. Section 4 discusses guidelines for writing good annotations. In section 5 we outline how annotations can be added to web pages in a way that benefits both humans and machines. Section 6 is dedicated to question answering on the basis of annotations, to the need for linguistic and other background knowledge, and to the problems that arise when annotations are inconsistent, incomplete, on different levels of detail, using different conceptualisations, or are plainly not understandable. Here we also discuss possible solutions to these problems. Section 7 suggests some alternative uses of annotations. In section 8 we summarise the main ideas and the advantages of our approach.

2 Controlled Natural Languages

Formal languages such as RDF and OWL have exclusively been designed for machines and are hard to write and understand for humans. There is an urgent need for an expressive high-level interface language to the semantic web that allows humans to write annotations in a familiar notation which is unambiguous and offers the same precision as a formal language.

A promising candidate for such a high-level interface language is the use of a controlled natural language. In general, a controlled natural language can be defined as a subset of a natural language with explicit constraints on grammar, lexicon, and style. These constraints usually have the form of writing rules and help to reduce both ambiguity and complexity of a full natural language [6, 7].

The probably most successful controlled natural language is ASD Simplified Technical English [8] that has been designed to improve the readability of aircraft maintenance documentation for non-native readers. However, readability of the language is only one important characteristic which needs to be combined with machine-processability to make the language useful in the context of question answering.

There are some relatively new controlled natural languages such as Attempto Controlled English [9], Common Logic Controlled English [10], and Processable English [11] that combine and balance readability and processability in such a way that makes it easier for humans and machines to work in cooperation. These highly expressive controlled natural languages are equivalent to large – in fact undecidable – fragments of first-order predicate logic, and have already been used as specification and knowledge representation languages in various application domains.

With some instruction, or supported by an intelligent authoring tool [12], even non-specialists can use these machine-oriented controlled natural languages to write annotations in a familiar notation without the need to formally encode their knowledge, and without a steep learning curve.

3 Controlled Natural Languages for Web Annotations

Traditionally, RDF-based languages and technologies have been promoted to semi-automatically generate annotations for web pages with machine-processable information. These annotations are usually not very expressive, and – once generated – are difficult to read and modify by humans.

For example, in the Friend of a Friend (FOAF) project[11], individual web pages are linked to machine-readable RDF documents which describe people, the links between them and the things these people create and are interested in. FOAF makes it easy to share and transfer information and to automatically extend, merge and reuse this information online. In the Annotea project[12], RDF-based annotation schemata are used for describing annotations as metadata and XPointer for locating the annotations in the annotated document. The annotations are stored locally, or in one or more annotation servers. When a document is browsed, a client such as Amaya[13] queries each of these servers, requesting the annotations related to that document.

RDF has been criticized as the formal underpinning for the semantic web [13, 14]. In particular the current RDF-based architecture for the semantic web has severe problems when more expressive rule languages are incorporated. An alternative approach is to use first-order logic as the semantic underpinning [13]. First-order logic is well-established and numerous state-of-the-art tools exist for processing first-order axiomatisations. There are various subsets of first-order logic that offer different tradeoffs with respect to expressivity, complexity and computability. Moreover, the direct mapping of subsets of first-order logic languages – for example between Horn logic and description logic – provides immediate semantic interoperability [15].

The controlled natural language we promote for annotating web pages is first-order equivalent, but we have shown that subsets thereof can be translated automatically into OWL DL [16, 17]. However, exclusively relying on description logic would considerably reduce the expressive power of the controlled natural language [5].

4 How to Compose Web Annotations

To compose meaningful annotations for web pages let us have a brief look at what we can learn from the field of news writing and from existing guidelines for well-designed web pages.

[11] www.foaf-project.org/
[12] www.w3.org/2001/Annotea/
[13] www.w3.org/Amaya/Amaya.html

4.1 Inverted Pyramid Style

Information in news reports is usually presented in an inverted pyramid style which begins with the conclusion, expressed as a single sentence. The subsequent paragraphs will then convey the most important and interesting information, leaving details and background information to further paragraphs in an order of diminishing importance. This format has the advantage that a reader can leave a report at any time without missing the most important facts. It also allows less important information to be more easily removed to fit a fixed size in a print medium.

Web usability experts recommend the inverted pyramid style for presenting textual information on web pages[14]. Putting the most important information into a lead section at the beginning of a web page better supports scanning of web pages by the human eye, and additionally minimizes the need for scrolling. Usability studies show that 79% of users scan a new web page and only 16% read it word-by-word[15].

4.2 The Lead Section

The lead section is the most important structural element of a well-designed web page and should convey the conclusion in a succinct form, usually in not more than 20-25 words. The importance of this lead section is also eminent in the manual of style of Wikipedia. This manual recommends that a Wikipedia article should be introduced by a lead section before the first headline and should summarize the most important information[16]:

> *The* ***lead section*** *should briefly summarize the most important points covered in an article in such a way that it could stand on its own as a concise version of the article. It is even more important here than for the rest of the article that the text be accessible, and consideration should be given to creating interest in reading the whole article.*

It is obvious that such a lead section needs to be easy to read and write by humans and that machine-processability would add enormous benefits for various reasoning tasks such as question answering, consistency checking and information fusion. Representing this information in an RDF-based formal language is not very helpful, since this language is probably not expressive enough, and its syntax is a slap in the face of any human author (specialists or non-specialists alike). At first glance, writing a lead section looks like a challenging optimization problem, but a machine-oriented controlled natural language can bridge the gap here and the field of news writing can give us some valuable guidelines how to do this in a clever and informative way.

4.3 The Five W's (plus H)

The lead section not only encompasses specific constraints on sentence structure but also promotes a particular way in which the content is presented. The basic idea is that the lead section should attempt to answer all the fundamental questions about a

---

[14] www.useit.com/alertbox/9606.html
[15] www.useit.com/alertbox/9710a.html
[16] en.wikipedia.org/wiki/Wikipedia:Lead_section

peculiar event and this can be memorised as: *who* did *what when where* and *why*, and occasionally also *how*.

Let us illustrate how the most important information of the Landline web page[17] can be represented in a lead section using these five W's (plus H) as a guiding principle. For this purpose, we will use the following linguistic schema for sentences

subject + predicate + object + {modifiers}

where the subject answers the question about *who* is involved in a specific situation, the predicate states a particular event or state, the object answers a *what* question, and optional modifiers answer a *when, where, why* or *how* question.

Of course, users can freely compose lead sections following this schema. Alternatively, composing a lead section can be supported by an intelligent authoring tool that displays predictive information while the lead section is being written (cf. [4]).

Here is the step-wise construction of a possible lead section of the Landline web page with the help of a predictive authoring tool.

In our case, the transitive verb *cultivates* as predicate requires both a subject and an object.

[subject: who]
A couple ... **[predicate]**
A couple cultivates ... **[object: what]**
A couple cultivates capers ... **[modifiers: how | where | when | why]**

The *how*, *where*, *when*, and *why* can all be expressed as a sequence of modifiers terminated by a period.

A couple cultivates capers experimentally ... **[modifiers: where | when | why]**
A couple cultivates capers experimentally in South Australia ... **[modifiers ...]**
...
A couple cultivates capers experimentally in South Australia since 1999 for economical benefit.

Please note that the information expressed in each constituent can directly be queried by questions in controlled natural language. For example:

Who cultivates capers? → *a couple*
Where does a couple cultivate capers? → *in South Australia*

but to answer our original question

Are capers grown in Australia?

we need additional linguistic background knowledge in the form of a lexical derivation rule (*if somebody cultivates something then somebody grows something*), the linguistic knowledge relating the active and passive forms of the transitive verb *grow*, and domain specific knowledge that specifies that *South Australia* is part of *Australia*.

As we will see in section 6.2, in general much more background information will be needed that has to be provided by external knowledge sources or explicitly by statements in controlled natural language.

---

17 www.abc.net.au/landline/content/2006/s1602940.htm

5 How to Attach Web Annotations to a Web Page

Under the proposed model annotations function as lead sections of web pages. Therefore they need to be directly embedded into a web page by the author. Internally, the lead section is marked up as a paragraph and labeled with the help of an XHTML language attribute ("lang") together with an experimental language tag[18] ("x-cnl"). A search engine supporting this tag could then recognise that a paragraph is written in controlled natural language. In our case the result looks as follows:

<p lang="x-cnl"><strong>A couple cultivates capers experimentally in South Australia since 1999 for economical benefit .</strong></p>

In this example the *lang* attribute's value *cnl* stands for an experimental language tag and indicates that the following snippet is written in controlled natural language. Figure 1 illustrates how the lead section can add value to a web page for both humans and machines.

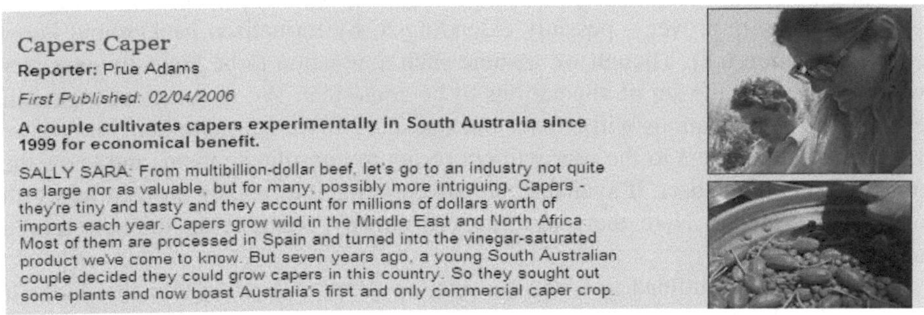

Capers Caper
Reporter: Prue Adams
First Published: 02/04/2006

A couple cultivates capers experimentally in South Australia since 1999 for economical benefit.

SALLY SARA: From multibillion-dollar beef, let's go to an industry not quite as large nor as valuable, but for many, possibly more intriguing. Capers - they're tiny and tasty and they account for millions of dollars worth of imports each year. Capers grow wild in the Middle East and North Africa. Most of them are processed in Spain and turned into the vinegar-saturated product we've come to know. But seven years ago, a young South Australian couple decided they could grow capers in this country. So they sought out some plants and now boast Australia's first and only commercial caper crop.

Fig. 1. Landline Article with Lead Section in Controlled Natural Language

The annotation being part of the web page, it will be indexed by search engines, and will also be available for any ranking that the search engine performs.

6 Deductions from Web Annotations

Question answering on the basis of annotations is done by a two-step process that in general needs linguistic and domain-specific background knowledge, and has to cope with the problems arising from inconsistent, incomplete, or differently conceptualised annotations.

Though the proposed annotations have a simple structure, background knowledge is complex, and in general involves quantification, negation, and disjunction. Thus question answering cannot be reduced to mere pattern matching, but requires first-order theorem proving.

[18] www.w3.org/TR/html4/struct/dirlang.html

6.1 A Two-Step Process to Answer Questions

Assuming that web pages are annotated by a lead section in controlled natural language, we suggest a two-step process to concisely answer questions in controlled natural language. This two-step process again reflects our decision to split the work between humans and machines according to their abilities, and thus complements our proposal to have web pages manually annotated.

In a first step, the question expressed in controlled natural language is automatically split into keywords that are then submitted together with the XHTML language attribute *lang="x-cnl"* to a ranking search engine that supports the language attribute. Since annotations are part of the respective web pages, the search engine will only return web pages containing the tag "x-cnl". Furthermore, the returned web pages are ranked with respect to the keywords of the question.

In the second step, we select the *N* top-ranked web pages and then try to automatically deduce the answer to our question separately from each of the *N* annotations. Deduction is done by converting the question *Q* and the annotations A of the selected web pages into their logical representations, Q' respectively A' and submitting $A' \cup \neg Q'$ to a theorem prover – possibly extending A' by formalised background knowledge (cf. section 6.2). Though we assume each annotation to be logically consistent, we cannot expect the set of annotations to be consistent. We also cannot expect that each of the *N* annotations will answer our question. If we get more than one answer, we present all answers to the user without trying to consolidate them, and leave their interpretation to the user. If available, we also provide information on the trustworthiness of the source. Note that page ranking already provides an implicit level of trustworthiness.

To support the outlined two-step process we propose a query tool that hides the computational details from the users and that contains a predictive editor to formulate questions in controlled natural language (cf. [4] for details).

6.2 Background Knowledge

No system can answer real world questions and make inferences without additional knowledge, i.e. knowledge that is not contained in the input. This applies also to our case: annotations written in controlled natural language require additional linguistic and domain specific background knowledge to serve as a complete knowledge base. However, an attractive feature of our approach is that much of the linguistic knowledge and all of the domain knowledge can be expressed in controlled natural language and is thus accessible for both man and machine.

Linguistic background knowledge is already needed in the first step of our approach when we split a question into keywords. If the annotation is

A couple cultivates capers experimentally in South Australia since 1999 for economical benefit.

and the question is

Are capers grown in Australia?

we cannot expect to get the question answered. However, we increase the probability to find adequate answers if we do a query expansion by allowing for synonyms of the content words of the question. For instance, linguistic resources like WordNet[19] provide for the verb *grow* the synonyms *cultivate*, *develop*, *increase*, *mature*, *originate*, *change* that we can add as alternatives to the keyword *grow* when we submit the keywords to the search engine. This will allow us to retrieve the above annotation as the basis for question answering.

More linguistic – and also domain-specific – background knowledge is required for the second, deductive, step of our approach. Assuming that the word *grown* of the question has been replaced by *cultivated* then we need linguistic knowledge to relate the active *somebody cultivates* and the passive *something is cultivated.* This relation can be expressed in controlled natural language, for example

If somebody cultivates something then something is cultivated by somebody.

Domain-specific knowledge needed for the second, deductive, step can conveniently be expressed in controlled natural language, for instance the geographic fact

South Australia is a part of Australia.

Now the question can be positively answered on the basis of the annotation. Other questions, like

Who grows capers in Australia?

What grows in South Australia?

Where do capers grow?

Since when are capers cultivated in South Australia?

Why are capers cultivated in South Australia?

can similarly be answered provided the required background knowledge is made available.

Where does the background knowledge come from, where is it stored, and how is it applied?

Linguistic knowledge can be extracted from linguistic resources such as WordNet, expressed in controlled natural language, and directly be converted to the logical representation of the controlled natural language. Domain knowledge can be composed by the user in controlled natural language, or (semi-) automatically extracted from existing ontologies or knowledge bases such as Cyc[20], and then converted into controlled natural language.

Since writers of annotations cannot anticipate the variety of questions asked, it seems natural to associate the background knowledge with the question, concretely to incorporate it in a suitable representation into the query tool. Alternatively, it may turn out to be more convenient to split the background knowledge into a user-independent part that is associated with the annotation and stored on some server, and a user-specific part that is associated with the question.

---

[19] wordnet.princeton.edu/perl/webwn

[20] www.cyc.com/

6.3 Missing and Inconsistent Answers

We cannot expect that each retrieved annotation will answer our question since annotations can violate the principles of good writing presented in section 4. One should rather assume that some annotations are incomplete, conceptualised differently to the question, or expressed in a way that no satisfying answer can be deduced.

Another issue is inconsistency. Though each annotation is expected to be consistent, the set of retrieved annotations is not necessarily consistent, and thus answers to our question can be inconsistent. Some researchers [18] have suggested to replace standard first-order logic by paraconsistent logic. Though this might be applicable in some cases, we believe that the enormous range of information available on the web simply does not allow for a coherent solution[21]. Instead, we leave it to the user to interpret the validity and the trustworthiness of the answers.

7 Other Uses of Web Annotations

Since web annotations in computer-processable controlled natural language have a logical foundation they can be used for many other purposes involving deduction, for instance comparing annotations of different web pages or checking annotations for compliance with respect to ontologies and knowledge bases.

If required by an application, annotations written in controlled natural language can be exported in RuleML[22], or in non-XML notations.

The annotations can also be exported as news feeds, for instance in our capers example, to inform Australian exporters of fruit and vegetable of an opportunity to expand their business with a new product.

Last, but not least an annotation in controlled natural language is a human-readable summary of the respective web page, and fulfills similar functions to the lead section of Wikipedia articles.

8 Conclusions

We propose to manually augment web pages with annotations in a controlled natural language. Our approach offers the following advantages:

- annotations in computer-processable controlled languages permit formal reasoning, specifically question answering by deduction,
- question answering from annotations in controlled natural languages can easily be supported by the necessary linguistic and domain-specific background knowledge,
- annotations in controlled natural languages are readable by anybody, and thus can also serve as a summary of the respective web page,
- annotations in controlled natural languages can be written – preferably with the support of an authoring tool – according to standard guidelines of good summary writing, for instance Wikipedia's guidelines for lead sections.

[21] www.w3.org/DesignIssues/Inconsistent.html
[22] www.ruleml.org

Arguably, annotations written in controlled natural language can bridge the gap between informal and formal notations and leverage true collaboration between humans and machines. However, our solution introduces a "chicken and egg" problem: a critical mass of web annotations will be necessary that people perceive the value of these annotations and start annotating web pages themselves. Only the future will show whether this – basically non-technical – problem can be solved.

Acknowledgements

The reported work was performed while N. E. Fuchs was visiting the Centre for Language Technology at Macquarie University in Sydney, Australia. The authors are grateful to Macquarie University's Visiting Scholars Scheme 2006 that provided the major part of the necessary funding. The authors would also like to thank their colleagues Mark Dras, Kaarel Kaljurand, Tobias Kuhn, Diego Mollá, Luiz Augusto Pizzato and three anonymous ESWC'07 reviewers for helpful hints and comments.

References

1. Hovy, E.: Text Summarization, in: R. Mitkov (ed.), The Oxford Handbook of Computational Linguistics, Oxford University Press (2003)
2. Katz, B., Lin, J.: Annotating the Semantic Web Using Natural Language, in: Proceedings of the 2nd Workshop on NLP and XML at COLING 2002, Taipei, Taiwan (2002)
3. Attempto Project; www.ifi.unizh.ch/attempto and attempto.ifi.unizh.ch
4. Schwitter, R., Tilbrook, M.: Annotating Websites with Machine-processable Information in Controlled Natural Language, in: M. A. Orgun and T. Meyer (eds.), Advances in Ontologies 2006, Proc. of AOW 2006, Hobart, Australia, Australian Computer Society, Conferences in Research and Practice in Information Technology, Vol. 72, (2006), 75-84
5. Kuhn, T., Royer, L., Fuchs, N. E., Schroeder, M.: Improving Text Mining with Controlled Natural Language: A Case Study for Protein Interactions, in: U. Leser, B. Eckman, and F. Naumann, editors, Proceedings of the 3rd International Workshop on Data Integration in the Life Sciences 2006 (DILS'06), Lecture Notes in Bioinformatics, Springer (2006)
6. Huijsen, W. O.: Controlled Language - An Introduction, in: Proceedings of CLAW 1998, Pittsburgh (1998) 1-15
7. O'Brien, S.: Controlling Controlled English – An Analysis of Several Controlled Language Rule Sets, in: Proceedings of EAMT-CLAW 03, Controlled Language Translation, Dublin City University (2003), 105-114
8. Simplified Technical English. Specification ASD-STE100, A Guide or the Preparation of Aircraft Maintenance Documentation in the International Aerospace Maintenance Language, Issue 3, January (2005)
9. Fuchs, N. E., Kaljurand, K., Schneider, G.: Attempto Controlled English Meets the Challenges of Knowledge Representation, Reasoning, Interoperability and User Interfaces, in: FLAIRS'2006, 2006.
10. Sowa, J. F.: Common Logic Controlled English Draft, 24 February 2004, available at: http://www.jfsowa.com/clce/specs.htm, (2004)

11. Schwitter, R.: English as a Formal Specification Language, in: Proceedings of the Thirteenth International Workshop on Database and Expert Systems Applications (DEXA 2002), W04: Third International Workshop on Natural Language and Information Systems - NLIS, 2-6 September 2002, Aix-en-Provence, France, (2002) 228-232
12. Schwitter, R., Ljungberg, A., Hood, D.: ECOLE – A Look-ahead Editor for a Controlled Language, in: Controlled Translation, Proceedings of EAMT-CLAW03, Joint Conference combining the 8th International Workshop of the European Association for Machine Translation and the 4th Controlled Language Application Workshop, May 15-17, Dublin City University, Ireland, (2003) 141-150
13. Horrocks, I., Patel-Schneider, P. F.: Three Theses of Representation in the Semantic Web, in: Proc. of WWW 2003, (2003) 39-47
14. Patel-Schneider, P. F.: A Revised Architecture for Semantic Web Reasoning, in: Proceedings of Third Workshop on Principles and Practice of Semantic Web Reasoning (PPSWR'05), Dagstuhl, Germany (11-16th September 2005), Organization: REWERSE, LNCS 3703, (2005) 32-36
15. Grosof, B. N., Horrocks, I., Volz, R., Decker, S.: Description Logic Programs: Combining Logic Programs with Description Logic, in: Proceedings of the 12th International Conference on the World Wide Web (2003) 48-57
16. Schwitter, R., Tilbrook, M.: Let's Talk in Description Logic via Controlled Natural Language, in: Proc. of the Third International Workshop on Logic and Engineering of Natural Language Semantics (LENLS2006) in Conjunction with the 20th Annual Conference of the Japanese Society for Artificial Intelligence, Tokyo, Japan, June 5-6, (2006) 193-207
17. Kaljurand, K., Fuchs, N. E.: Bidirectional mapping between OWL DL and Attempto Controlled English, in: Fourth Workshop on Principles and Practice of Semantic Web Reasoning, Budva, Montenegro, (2006)
18. Schaffert, S., Bry, F., Besnard, P., Decker, H., Decker, S., Enguix, C., Herzig A.: Position Paper: Paraconsistent Reasoning for the Semantic Web. Technical Report PMS-FB-2005-42. Institut für Informatik der Ludwig-Maximilians-Universität München, (2005)

PANTO: A Portable Natural Language Interface to Ontologies

Chong Wang, Miao Xiong, Qi Zhou, and Yong Yu

APEX Data and Knowledge Management Lab,
Department of Computer Science and Engineering,
Shanghai JiaoTong University, Shanghai, 200240, P.R. China
{wangchong,xiongmiao,jackson,yyu}@apex.sjtu.edu.cn

Abstract. Providing a natural language interface to ontologies will not only offer ordinary users the convenience of acquiring needed information from ontologies, but also expand the influence of ontologies and the semantic web consequently. This paper presents PANTO, a Portable nAtural laNguage inTerface to Ontologies, which accepts generic natural language queries and outputs SPARQL queries. Based on a special consideration on nominal phrases, it adopts a triple-based data model to interpret the parse trees output by an off-the-shelf parser. Complex modifications in natural language queries such as negations, superlative and comparative are investigated. The experiments have shown that PANTO provides state-of-the-art results.

1 Introduction

Ontology[1], which both explicitly represents the taxonomy of a domain (classes and properties) and stores a lot of knowledge (instances and instance relations), plays a key role in the semantic web by enabling knowledge sharing and exchanging [1]. However, in order to acquire the formal knowledge in ontologies, users have to be familiar with:

- the ontology syntax, such as RDF and OWL;
- some formal query language, such as RDQL[2] and SPARQL[3];
- the schema (structure) and vocabulary of the target ontology.

Consequently, as Bernstein et al. stated in [2], there is a *gap* between *the logic-based semantic web* and *real-world users*. In order to bridge the *gap*, this paper presents PANTO, a Portable nAtural laNguage inTerface to Ontologies, which offers users the convenience of acquiring needed information from formally defined ontologies. Specifically, users can express their information needs in natural language even without considering the syntax of RDF or OWL, the formal query language, or the schema and vocabulary of ontologies.

[1] In this paper, the term *ontology* refers to a knowledge base (KB) that includes concepts, relations, instances and instance relations that together model a domain.
[2] http://www.w3.org/Submission/RDQL/
[3] http://www.w3.org/TR/rdf-sparql-query/

E. Franconi, M. Kifer, and W. May (Eds.): ESWC 2007, LNCS 4519, pp. 473–487, 2007.

1.1 Background

Although the first natural language interface system came out more than three decades ago (LUNAR [3]), a fully portable and widely used system for formalized knowledge bases is still unavailable. As mentioned in [4], two major obstacles lie in the way:

Firstly, the ambiguity and complexity make it difficult for a machine to understand arbitrary natural language. The NLP community have been keeping on paying efforts in this area. The state-of-the-art statistical parsers [5] can reach about 90% in precision and recall. However, it is still well regarded as an "AI Complete" problem and far from totally resolved.

Secondly, even with correctly parsed natural language queries, a lot of challenges remain in translating them to correct formal queries:

- Vocabularies of the knowledge bases are controlled and lean compared with users', so it is a challenge to correctly map users' words to vocabularies of the knowledge bases.
- Together with the lexical and syntactic information of the parsed queries, semantic information in the knowledge bases can also be utilized to help formulate the formal query, but how to accomplish this is still an open problem.
- Different knowledge representations require different techniques to interpret the semantics of users' queries. For example, how to deal with queries containing negations in ontologies is different from that in databases. SPARQL has been recommended as the standard query language for the semantic web community, but few researches have been carried out to investigate the new opportunities and challenges in translating natural language queries into it.

1.2 Features and Contributions

PANTO focuses on the second obstacle mentioned above. It utilizes an off-the-shelf statistical parser StanfordParser [5] to deal with the first major obstacle. Multiple existing techniques and tools are integrated to interpret parse trees of natural language queries into SPARQL. The following are the major features and contributions:

Firstly, PANTO is designed to be ontology portable and no assumption is made about any specific knowledge domain. To help make sense of the words in the NL queries and map them to the entities (concepts, instances or relations) in the ontology, existing tools such as WordNet [6] and string metrics algorithms [7] are integrated.

Secondly, nominal phrase constituents in the parse trees are specially considered. We extract nominal phrases in the parse trees as pairs to form an intermediate representation called *QueryTriples*. Then, by utilizing knowledge in the ontology, PANTO maps *QueryTriples* to *OntoTriples* which are represented with entities in the ontology. Finally, together with *targets* and *modifiers* extracted from the parse trees, *OntoTriples* are interpreted as SPARQL.

Thirdly, we investigate certain problems in the process of translating natural language queries into SPARQL queries. The translation of advanced semantic features such as negation, comparative and superlative modifications are supported.

The rest of this paper is organized as follows. Section 2 introduces the system architecture. Section 3 describes how the ontologies and other resources are wrapped to construct a lexicon. Section 4 presents the key process to interpret parse trees into SPARQL queries. Section 5 elaborates on the experiments. Section 6 discusses about limitations and directs the future work. Section 7 describes related work and section 8 concludes this paper.

2 PANTO Architecture

Fig. 1 depicts the architecture of PANTO. It takes ontologies and natural language queries as input, and finally returns SPARQL queries as output. When an ontology is selected as the underlying knowledge base, the *Lexicon Builder*

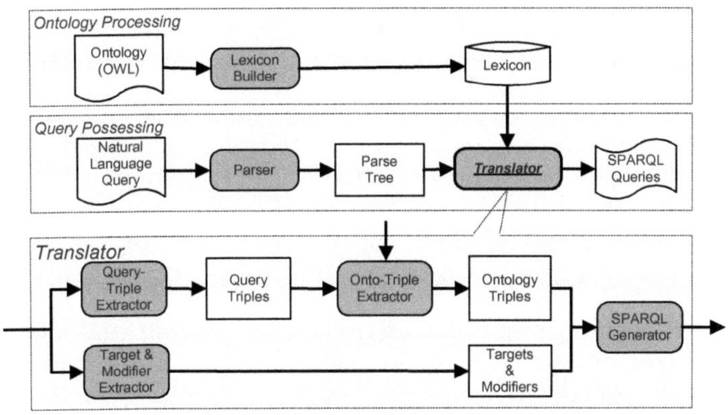

Fig. 1. Architecture of PANTO

automatically extracts entities out of the ontology to build the *Lexicon*. WordNet is utilized to find synonyms to enlarge the vocabulary of the ontology. Once the user inputs a natural language query, PANTO first invokes the *Parser* to produce the parse tree, which is then transfered to the *Translator*. The *Translator* is the core processing engine of PANTO. Upon receiving a parse tree, the *Translator* processes it in two ways in parallel:

- The parse tree is first transformed into *QueryTriples*, which are less restricted, for their predicates can be prepositions, verbs, phrases, or even omitted. Then *QueryTriples* are mapped to *OntoTriples* with the help of the *Lexicon*.
- The *Target and Modifier Extractor* extracts the potential words for *targets* (variables after "SELECT") and *modifiers* (information related to "FILTER", "UNION", etc.) of the target SPARQL queries from the parse tree.

Finally the domain-compliant *OntoTriples*, *targets* and *modifiers* are sent to the *SPARQL Generator* to produce SPARQL queries.

3 The Lexicon

The *Lexicon* is mainly designed for making sense of words in natural language queries and mapping them to entities in the ontology. It is composed of the following contents:

Ontology Entities. This is the most important part of the *Lexicon*. Entities in the ontology, including classes (concepts), properties (relations), and instances (individuals), are extracted and stored for fast access and matching. Since different ontology entities may have the same name (e.g. "Mississippi" river and "Mississippi" state) and one ontology entity may have different names (e.g. "US", "United States", and "USA" denote the same entity "United States of America" in the ontology), ontology entities and their names are put into a special hash table, in which a key maps to a set of ontology entities and an ontology entity can be obtained by different keys. Given a word from the natural language query, the *Lexicon* will acquire a set of possible entities. Proper nouns (e.g. "New Mexico") are also extracted from the ontology for fast access and matching.

General Dictionaries. In order to help bridge the *gap* between user vocabulary and ontology vocabulary, general dictionary WordNet is utilized. The synsets in the dictionary defined by linguists enable PANTO to match "work" in user's query to concept "Job" in the ontology. Also with the help of the dictionaries, we are able to retrieve the property "length" when the user asks "how *long* ...". This module is open and other thesauri can also be adopted if available.

User-Defined Synonyms. Since users may use jargons and abbreviations to denote entities, words from general dictionaries only may not be enough. Therefore, the *Lexicon* allows users to define their own "synonymy words" (a set of words that match the same entity in the ontology). This will be helpful when PANTO is adopted to a certain domain.

Note that, the user-defined synonyms are not mandatory for the *Lexicon*, and all the mandatory contents are extracted in a totally automatic way. Therefore, the construction of the *Lexicon* is portable.

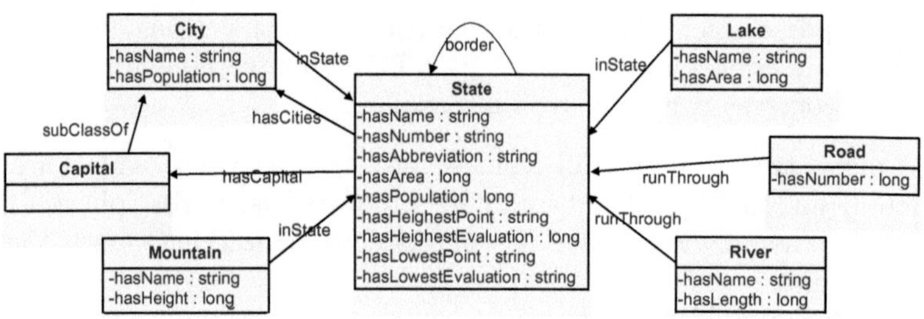

Fig. 2. The schema definition of *geography* ontology (the following examples are based on this ontology)

4 Translator: Translating Parse Tree to SPARQL

The translator is the backbone of PANTO. Our basic idea for translating natural language queries into formal ones is to specially consider nominal phrases. We observe that a natural language query can usually be viewed as the combination of multiple nominal-phrase pairs. From the parse trees by a deep parser, such pairs are easily recognized. Inside each pair is a verb phrase, a preposition, a conjunction or the like to represent the relationship between the two nominal phrases. At the same time, nominal phrases or words also play an important role in ontologies which store facts to model a domain. The facts are explicitly or implicitly stated in the triple form <*subject*, *predicate*, *object*>. The *subject* and the *object* may be classes, instances or literal values and usually should be named with nominal words or phrases [8]. The *predicate* may be prepositions, verbs, verb phrases and so on, and sometimes may also be nominal phrases. Because a nominal-phrase pair represents some kind of semantic relationship between the two nominal phrases, we expect it to be mapped to a triple in the ontology. Fig. 2 presents the schema definition of an example ontology and Fig. 3 depicts the processing steps of an NL query to this ontology. In the following subsections, we will detail the whole process of translating a parse tree of a natural language query into SPARQL based on the above idea.

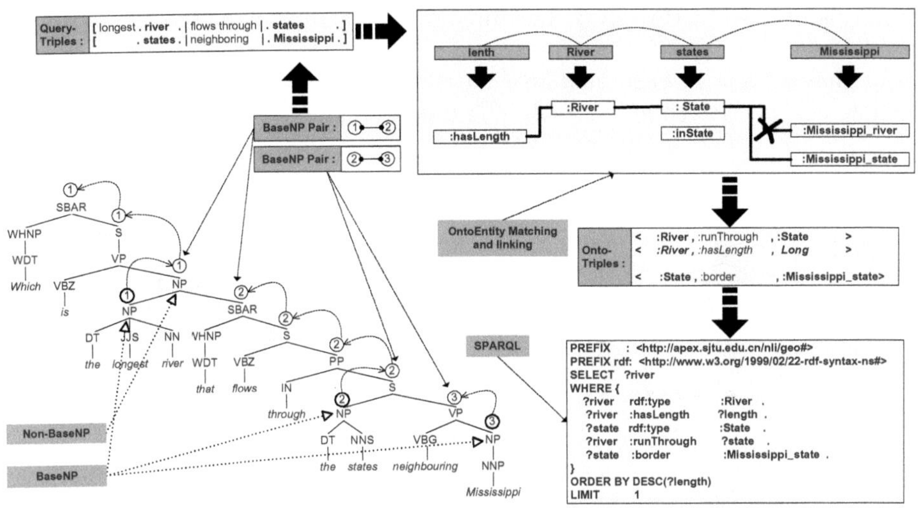

Fig. 3. Parse tree of query "which is the longest river that flows through the states neighboring Mississippi" by StanfordParser (with the default root node "ROOT" removed for brevity). *BaseNP*, *Head BaseNP* propogation, *BaseNP pairs*, *QueryTriples*, *OntoEntity* matching, linking *OntoEntities* as *OntoTriples*, *OntoTriples* and final SPARQL query are illustrated.

4.1 QueryTriple Extractor

In this part, we extract the nominal-phrase pairs out of the parse tree to form the intermediate representation *QueryTriples*. This includes the following steps:

Identify and propagate Head BaseNP. In the parse tree, a nominal phrase is tagged as NP[4]. Since NPs may be nested, we need to find those that can be mapped to entities in the ontology. We introduce the notion *BaseNP* (Fig. 3). **BaseNP** is formally defined as *NonrecursiveNP* (and *BaseNP* is called for short) in [9], where it refers to an NP that does not directly dominate an NP itself, unless the dominated one is a possessive NP (i.e. it directly dominates a POS-tag "POS"). Here we slightly extend the concept of *BaseNP* to also include a constituent (some WHNP) if it contains nouns but does not dominate an NP. We make this extension to capture all nominal words, which may not be parsed to be contained in a *BaseNP* by a statistical parser. For example, StanfordParser parses "how many rivers" in the query "how many rivers does alaska have" as *(WHNP(WHADJP((WRB how)(JJ many))(NNS rivers))*. There is a noun "rivers" inside WHNP, but there is no NP.

With this definition, we first identify all the *BaseNPs* from a parse tree. After that, we prepare for identifying and linking those related NPs as pairs. We follow the rules used by Michael Collins [9] for mapping from trees to dependency structures (StanfordParser also uses these rules and this is one of the reasons why it is utilized in PANTO). The difference is: Collins used those rules to find *head words* (see [9] for details), and we use a subset of the rules to propagate *head BaseNPs*. The only modification to the rules is: if there is a node containing conjunctive children nodes, the *head BaseNPs* of all the children will be

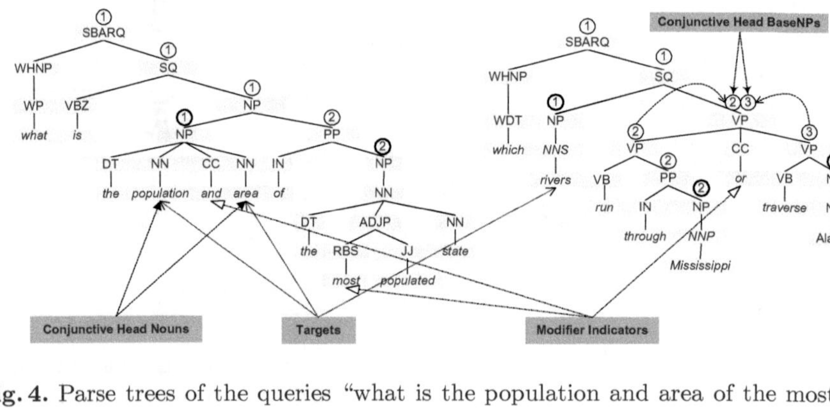

Fig. 4. Parse trees of the queries "what is the population and area of the most populated state" and "which rivers run through Mississippi or traverse Alaska" by StanfordParser (with default root node "ROOT" removed for brevity). *Targets, modifier indicators*, conjunctive *Head Nouns* and conjunctive *Head BaseNPs* are illustrated.

[4] NP, short for "Nominal Phrase", is a syntactic tag in parse tree. In the following, we will use such syntactic tags as well as POS tags without explanations.

propagated to it as its *head BaseNPs* (Fig. 4). All these operations can be done via a bottom-up traversal along the parse tree.

Link BaseNP Pair to form QueryTriple. After the basic constituent *BaseNPs* are identified and propagated on the parse tree, we link them one another where there is modification relationship to form *BaseNP pairs*. The process is: *for each BaseNP, traverse up towards the root node and link it with the first different head BaseNP(s) as BaseNP pair(s).* The two *BaseNPs* in such a pair, together with the words which syntactically connect them, form a *QueryTriple* (Fig. 3). A **QueryTriple** is a triple in the form of [*subject* |*predicate* |*object*]. The *subject* and the *object* are the two *BaseNPs* and the modifiers to them extracted from the original query. The *predicate* is composed of the words (may be a preposition, a verb phrase, etc) that connect the *subject* and the *object* together.

Specify internal structure for QueryTriple. A linked *BaseNP* pair is only a raw *QueryTriple*. Since a *BaseNP* may contain more than nominal words, we need to separate different contents. First of all, the *head nouns* for the *BaseNPs* are identified. A **Head Noun** is a nominal word acting as *head word* [9] in a *BaseNP*. The rules to find *head noun* again follows [9], with the exception that when conjunction exists, treat all nominal words of conjunctive relations as *head nouns* (Fig. 4). As the next step, the internal structure of *QueryTriples* are specified. For the *subject* and the *object* of each *QueryTriple*, the internal structure is in the form of [*pre-modifier . head noun . post-modifier*]. Here, only the *head noun* is mandatory and the *pre/post-modifiers* represent the words modifying the *head noun* or the whole *BaseNP*.

4.2 OntoTriple Extractor

QueryTriples are only the intermediate forms of the user's query. We need to map them to the semantic content in the ontology. This is carried out as below. First, the *OntoTriple Extractor* matches the words (especially nominal words) with the *OntoEntities* (**OntoEntity**, short for *Ontology Entity*, represents concepts, instances, relations/properties in the ontology). Then, it makes use of lexical and syntactic information extracted with *QueryTriples* to find the semantic relationships among *OntoEntities*, which will be explicitly represented with *OntoTriples*. The detail is as follows:

Map user words to OntoEntities. The first is to find the corresponding *OntoEntities* for each word in the query. For each *QueryTriple*, we retrieve the matching *OntoEntities* for the *head nouns* of the *subject* and the *object*, by invoking the *Lexicon*. The *Lexicon* employs a number of matching methods, which can be classified as two types: (1) *semantic matching* mainly uses general dictionaries like WordNet to find synonyms of words; (2) *morphological matching* uses WordNet, string metrics or heuristic rules (e.g. algorithms to find abbreviations, which may also be separately designed when PANTO is adopted to a particular domain) to find morphologically similar matchings. Different matching methods are sometimes combined to find matching entities for a word. For example, the word "states" gets the matching list {State, inState} (Fig. 3). "State" is a *class* entity retrieved by morphological matching and "inState" is a *property* entity matched by a heuristic rule based on naming conventions for ontology and string metrics algorithms.

Map QueryTriples to OntoTriples. An **OntoTriple** (short for *Ontology Triple*) is a triple that is compatible with some statements in the ontology, in the form of <*subject, predicate, object*>, where the *predicate* can be a relation (property) and the *subject* or the *object* can be a concept (class) or an instance. When the *predicate* is a *datatype* property, the *object* must be a literal value or the value type. A nominal word may also be mapped to the *predicate*, besides the *subject* and the *object*. Therefore, the *subject* or the *object* in the *QueryTriple* does not necessarily be mapped to that in the *OntoTriple*. In PANTO, 11 cases are enumerated for OWL ontology to generate *OntoTriple*(s) from two *OntoEntities*. For example, if one *OntoEntity* is a *property* and the other is a *class* which is the domain of the *property*, one *OntoTriple* is generated; if both are *properties* and can be related by a *class*, two *OntoTriples* are generated. With different combinations of the matching *OntoEntities* of the *head nouns* in the *subject* and the *object* of the *QueryTriple*, multiple *OntoTriples* will be generated. When the *predicate* of a *QueryTriple* is not empty, we use it to verify the generated *OntoTriples*. For example, from the *QueryTriple* [river | flows through | mississippi], we get two *OntoTriples* <:Mississippi, rdf:type, :River> and <:River, :runThrough, :Mississippi>. Since "flows through" can be mapped to ontology property ":runThrough", we discard the first *OntoTriple*. On the other hand, if the two *OntoTriples* are generated from *QueryTriple* [river | | mississippi], we discard the second one. Because this pattern is usually used by people to indicate an entity [10], in such a case we believe with high confidence that "mississippi" should be mapped to the entity "Mississippi_river" rather than "Mississippi_state". Thus we remove all the other matching *OntoEntities* for the word "mississippi", and finally also discard the triple <:Mississippi, rdf:type, :River> and only hold that *OntoEntity*. Besides the *OntoTriples* generated above, this module also generates *OntoTriples* from inside a *BaseNP*. Take the *BaseNP* "the longest river" in Fig. 3 as an example, the *OntoTriple Extractor*, with the help of WordNet, transforms "longest" to "long" and then to the nominal form "length". Now "length" is used to match *OntoEntities*. These matching *OntoEntities* are then used to generate *OntoTriples* with those matching *OntoEntities* of "river". Finally, a valid *OntoTriple* <:River, hasLength, Long> is generated.

Link OntoTriples. A natural language query represents the semantic relationships and constraints among different concepts and individuals in the domain. When multiple *BaseNP pairs* are available, there are *BaseNPs* shared by two or more *QueryTriples*. Therefore, a valid *OntoTriple* set should be linked one another to form a tree (Fig. 3). Since one word may match multiple *OntoEnties*, there may be different combinations and multiple valid *OntoTriple* result sets.

4.3 Target and Modifier Extractor

To translate a natural language query into a SPARQL query, we must find the *targets*, i.e. the words that correspond to the variables after "SELECT" in the resultant SPARQL query. This process is as follows: first find the allowed wh-word [10] like "what", "who" and "how" or an enumerated set of command words like "list", "give me" and "return"; then take the nouns in the same or the directly followed constituent as *targets* (Fig. 4). Detailed rules vary for different question/command words, and are usually common for different domains.

Filter is provided in SPARQL to enable the users to specify constraints on relations and variables. *Solution Modifier* is provided for the users to carry out operations (*Order by*, *Limit*, *Offset*) on the result. In natural language queries, both of these are expressed through certain types of words, which we call *modifier indicators* (Fig. 4). In the current version of PANTO, we mainly deal with the following: (1)*negation*, including "not" and "no"; (2)*superlative*, superlative words will be tagged as "JJS" (superlative adjective) or "RBS" (superlative adverb); (3)*comparative*, comparative words will be tagged as "JJR" (comparative adjective) or "RBR" (comparative adverb); (4)*conjunctive/disjunctive*, including "and" and "or".

The extractor records the positions and types of *targets* and *modifier indicators*, and then send them as input to the SPARQL Generator.

4.4 SPARQL Generator

The *targets* and *modifiers* are extracted from the parse tree, and it is straightforward to relate them with corresponding *OntoEntities* and *OntoTriples* extracted by the *OntoTriple Extractor*. After that, we interpret them into SPARQL:

SELECT. *OntoEntities* matched with *target* words are interpreted as variables after "SELECT" according to the following rules: (1) If it is a *class* entity ":SomeClass", interpret it as the variable "?someclass", and add a *triple pattern*[5] <?someclass, rdf:type, :SomeClass> to the "WHERE" clause. (2) If it is an *RDF Literal Type* entity, such as "Long", "String" or the like, directly interpret it as a variable, e.g. "?long", "?string", etc.

WHERE. An ordinary *OntoTriple* is directly interpreted as a *triple pattern* after "WHERE". *Instance* and *property* entities are directly interpreted as their URIs in the ontology. As for *class* entities or *RDF Literal Type* entities in each *OntoTriple* (except those with the *property* "rdf:type"), interpret them as variables according to the above rules for *targets*. For some complex queries, there may be multiple words mapped to the same *class* entity. For example, the two "state" in the query "what state borders the state that has the largest population?" are both mapped to the *class* entity ":State". However, they should obviously be interpreted as different variables. Our current solution is to always interpret such entities as different variables. [11] mentioned a technique to check whether such two words mean the same or not, by comparing their local contexts in the query according to some heuristics, but it requires ad hoc rules. According to the current experiments, our approach works fine. More investigations will be carried out as future work. *RDF Literal Type* entities are treated similarly.

---

[5] http://www.w3.org/TR/rdf-sparql-query/

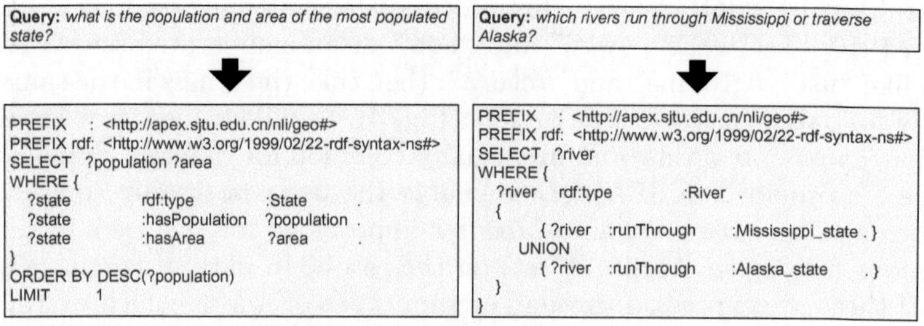

Fig. 5. SPARQL queries for the two example queries in Fig. 4

FILTER and Modifiers. Corresponding to the *modifier indicators* mentioned above, this part currently includes the following: (1)*negation.* Negations are interpreted by utilizing the operators "!" and "bound" of SPARQL. For example, "not" in the query "which river's length is not 5000" is interpreted as a FILTER clause "(FILTER(?length != 5000))". Negation on *property* (relation) are interpreted by a combination of "OPTIONAL" and "FILTER". Take the query "which rivers do not run through Alaska" as an example, "do not" is interpreted as "{OPTIONAL {{?river :runThrough ?state.} FILTER (?state = :Alaska)} FILTER (!bound(?state))}". However, some kinds of negations can not be interpreted, e.g. queries like "which rivers do not run through states bordering Alaska". (2)*superlative.* Superlative modification on *datatype* can be interpreted with *Order By* and *Limit* (*Limit* can be removed if there are multiple results) of SPARQL (e.g. the NL query and its resultant SPARQL query shown in Fig. 3), but superlative modifications that require the functionality of *count* are not supported (e.g. "most" in query "which river runs through the most states" can not be expressed with the current version of SPARQL). (3)*comparative.* Similar as above, currently only comparative modification on *datatype* is supported. (4)*conjunctive/disjunctive. OntoTriples* related to conjunctive constituents (conjunctive *BaseNPs* or conjunctive *head nouns*) indicated by "and" are interpreted as conjunctive *triple patterns* to the "WHERE" clause by default. Since "and" may sometimes be used in disjunctive relation (e.g. in the query "list all the cities in California and Maine"), we need to link their *triple patterns* with a "UNION" in such a situation. Currently, when the two conjunctive words are both mapped to the same *class* entity (or *instances* of the same *class*), multiple SPARQL queries are produced for different interpretations of "and". Those *OntoTriples* related to "or" are always interpreted with a linking "UNION".

As an example, Fig. 3 depicts the final SPARQL query. Fig. 5 depicts the final SPARQL queries for the two example queries used in Fig. 4.

5 Experiments and Evaluation

The goal of the experiments is to quantitatively assess the performance and effectiveness of our approach in the current implementation.

Table 1. Performance of PANTO. Row 2 shows the number of original Mooney queries and row 3 shows that of the selected testing queries (duplicated ones are removed).

Domain	*Geography*	*Restaurant*	*Job*
Original Mooney Queries#	880	250	641
Selected Testing Queries#	877	238	517
Precision	88.05%	90.87%	86.12%
Recall	85.86%	96.64%	89.17%

5.1 System Implementation and Experiment Setup

PANTO was implemented as a stand-alone web application. In the current version, it adopts Protégé API[6] to access the underlying ontologies. The version of WordNet used in the system is 2.1. The lexicalized StanfordParser[7], version 1.5.1 without additional training, is adopted as the statistical parser. It produces one parse tree for each input. The experiments were performed on a PC with 2.4GHz P4 CPU and 1G Memory.

Test Data. The test data are based on those provided by Mooney[8] which have been widely used to evaluate natural language interfaces [2,12,13]. There are test data and queries on three domains: one about geography data in the United States, one about restaurant information and the third about job announcements. We translated the original three Prolog databases into OWL as the testing ontologies. There are several hundreds of natural language queries for each domain. We removed the duplicated ones in each query set, Table 1 presents the numbers of queries for experiments.

5.2 Results and Discussion

Performance. With the initialization time (i.e. loading ontologies and parsers) excluded, the total average processing time of a query was less than one second (0.787s) and the running time is related with the scale of the ontology and the complexity of the query (the length and the number of clauses).

Correctness. In order to assess the correct rate that how many of the translated queries correctly represent the semantics of the original natural language queries, we compare the output with the manually generated SPARQL queries. The metrics we used are *precision* and *recall*. For each domain, *precision* means the percentage of correctly translated queries in the queries that PANTO produced an output; *recall* refers to the percentage of queries that PANTO produced an output in the total testing query set. Since the queries are originally prepared for evaluating natural language interface to database, some features in the queries are not supported by the current version of SPARQL, but we also count them as correct if they match the manually generated ones.

[6] http://protege.stanford.edu/download/registered.html

[7] http://nlp.stanford.edu/software/lex-parser.shtml

[8] http://www.cs.utexas.edu/users/ml/nldata.html

To the best of our knowledge, the only existing system that also translates generic natural language queries into SPARQL is Querix [14], which provides preliminary results on the geography dataset with 215 selected queries. Querix claimed to achieve a precision of 86.08% and recall 87.11%. From Table 1 we can see that PANTO provides comparable results.

Coverage. In this part we analyze the effectiveness of the approach to interpreting the queries in a triple-based model. The experiments on the Mooney data and queries show that all the correctly parsed natural language queries can be correctly translated into *QueryTriples* and then be mapped to *OntoTriples* at a high accuracy. What's more, we also parsed and analyzed the 170 sample queries presented on the AquaLog web site[9]. AquaLog claims to be able to parse these kinds of queries, and with PANTO, all of them can generate pretty good *QueryTriples*. Hence we can expect that the PANTO approach can cover the query scope supported by AquaLog.

6 Limitations and Future Work

The limitations with the current version of PANTO mainly include the following:

Restrictions on Query Scope. Though the triple-based analysis is effective to cover a broad range of natural language queries, the supported query scope is still limited. First, PANTO depends on an off-the-shelf parser to correctly parse the NL queries and thus is limited by the NLP techniques. But we can continually utilize the new achievements in NLP community. Second, it can not totally interpret semantics that is beyond the expressiveness of SPARQL (e.g. queries involving *count* on instances). When more features are added to SPARQL, these limitations are expected to be resolved.

Weakness in User Interaction. At present, PANTO is not a full-fledged system. It focuses on the query processing step and is currently weak in supporting complex user interactions. However, a well-designed interaction model can enable users to paraphrase the query and guide them correctly express their information need with system processable input. In future, we will investigate more on user interaction to make PANTO more effective.

Scalability. The ontologies used for evaluation is relatively small and all processing operations are carried out in memory. An investigation on the system performance with larger ontologies will be part of the future work. Database and indexing techniques will also be involved.

7 Related Work

Natural language interface to knowledge bases, which can help ordinary users express their information needs in natural language that they are familiar with and

[9] http://kmi.open.ac.uk/technologies/aqualog/examples.html

can consequently populate the knowledge bases, has been studied for decades [15, 16]. According to the ability of processing the natural language input, such natural language interfaces can be classified into two categories, namely, full natural language interface and restricted natural language interface.

Masque [17] is a typical natural langue interface to databases. It first transforms the natural language query into an intermediate logic representation and then translates the logic query into SQL. Systems which employ machine learning methods for transforming natural language query into formal logic representation or formal query have been studied for many years too. With the training on domain specific sentences, these systems [18,19,20] gain a good result (precision can be higher than 95% and recall can reach 80%). However, they need a lot of domain specific training. In order to avoid the defects brought up by NLP parsers, many systems only use the shallow parsing result of NLP tools, e.g. POS tags, chunks, etc. Rodrigo et al. tried to break down the query into words and form a formal query with words in their semantic search engine for the international relation sector [21]. Kang et al. have also formed the SQL query with keywords in [22]. As it is summarized by the authors in [21], "...the query construction is at present the weakest link in the chain...". Since PANTO combines the *OntoTriples* and the modifiers generated from the parse tree and the *Lexicon*, it can easily constructs the SPARQL query. NaLIX [4,11] is a generic natural language interface to XML Database. It focuses on correct parse trees output by a dependency parser and translated them into XQuery. Querix [14] is a domain-independent natural language query interface to Ontologies based on clarification dialogs. It is the only known system that also translates full natural language queries into SPARQL. Similar to PANTO, Querix also adopts an off-the-shelf parser (also StanfordParser), but it differs from PANTO in analyzing the parse trees. Querix directly uses the POS tags and a set of heuristic rules to extract skeletons (e.g. Q-V-N for "what are the population sizes"), while PANTO identifies *BaseNPs* and utilizes structure information inside and among the *BaseNPs*.

Because of the complexity and ambiguity of full natural language, many systems only accept queries which are in a subset of natural language. Controlled natural language, which restricts the terminology and grammar and is equivalent to First Order Logic [23], avoids the ambiguity of full natural language. Bernstein et al. [2] have adopted Attempto Controlled English to query ontologies and Nelken et al. [24] have employed controlled language to query historical databases. The queries are in the form of natural language, but users have to first learn the syntactic restrictions to make sure they are in the "controlled" set. PRECISE [13] defines a notion of "semantic tractable" questions on database and can translate them into SQL queries. However, all tokens must be distinct and questions with unknown words are not semantically tractable and cannot be handled. In contract, with the *Lexicon*, PANTO can deal with questions even some of them are not "semantic tractable". Pattern based methods are also widely used in natural language interfaces. Lopez et al. [25] have classified questions into 23 categories in AquaLog. If the input query is classified into some category, AquaLog will process it correctly. However, due to the limited coverage of the patterns, many queries will be left unresolved. Comparing to other systems, AquaLog is more similar to PANTO. Its underlying knowledge base is ontology, it also adopts a triple-based intermediate presentations and it is the one that introduces the notion of *query-triple* and *onto-triple*. The difference is, AquaLog is based on a shallow parser and depends on handcrafted grammars to identify *terms*, *relations* for composing *query-triples*, while the parse tree by the deep parser provides PANTO more modification information between nominal phrases. What's more, different from the *query-triple* of Aqualog, our *Query-Triple* are always formed with two nominal phrases (the *BaseNP pair*). Bernstein et al. have also proposed a guided natural language search engine [12,26] to help users form the query and avoid ambiguity, but the processing ability of the system is also limited to the defined grammar.

8 Conclusion

This paper presents PANTO, a portable natural language interface to ontologies. Based on the observation that nominal words or phrases play an important role in both natural language query and ontology triples, PANTO adopts a triple-based data model as the intermediate representation to translate natural language queries into SPARQL. The experiments on three different ontologies have shown that the PANTO approach produces promising results. Our approach helps bridge the *gap* between *the logic-based semantic web* and *real-world users.*

Acknowledgments

This work is carried out as part of a university joint research project between IBM China Research Lab and Department of Computer Science and Engineering, Shanghai JiaoTong University. The authors sincerely thank the anonymous reviewers for their valuable comments. We also would like to thank Zhangmei Yao for his participation in this research.

References

1. Chandrasekaran, B., Josephson, J.R., Benjamins, V.R.: What Are Ontologies, and Why Do We Need Them? IEEE Intelligent Systems **14**(1) (1999) 20–26
2. Bernstein, A., Kaufmann, E., Göhring, A., Kiefer, C.: Querying Ontologies: A Controlled English Interface for End-Users. In: International Semantic Web Conference. (2005) 112–126
3. Woods, W., Kaplan, R., Webber, B.: The Lunar Sciences Natural Language Information System: Final Report. Technical report, Bolt Beranek and Newman Inc., Cambridge, Massachusetts (1972)
4. Li, Y., Yang, H., Jagadish, H.V.: NaLIX: an interactive natural language interface for querying XML. In: SIGMOD Conference. (2005) 900–902

5. Klein, D., Manning, C.D.: Accurate Unlexicalized Parsing. In: ACL. (2003) 423–430
6. Fellbaum, C. In: Wordnet: An Electronic Lexical Database. Cambridge: MIT Press (1998)
7. Cohen, W.W., Ravikumar, P., Fienberg, S.E.: A Comparison of String Distance Metrics for Name-Matching Tasks. In: IIWeb. (2003) 73–78
8. Noy, N.F., McGuinness, D.L.: Ontology Development 101: A Guide to Creating Your First Ontology. Technical Report SMI-2001-0880, Stanford University School of Medicine (2001)
9. Collins, M.: Head-driven Statistical Models for Natural Language Parsing. PhD thesis, University of Pennsylvania (1999)
10. Quirk, R., et al. In: A Comphrehensive Grammar of the English Language. Longman, London (1985)
11. Li, Y., Yang, H., Jagadish, H.V.: Constructing a Generic Natural Language Interface for an XML Database. In: EDBT. (2006) 737–754
12. Bernstein, A., Kaufmann, E., Kaiser, C.: Querying the Semantic Web with Ginseng: A Guided Input Natural Language Search Engine. In: 15th Workshop on Information Technology and Systems (WITS 2005). (2005) 45–50
13. Popescu, A.M., Etzioni, O., Kautz, H.A.: Towards a Theory of Natural Language Interfaces to Databases. In: Intelligent User Interfaces. (2003) 149–157
14. Kaufmann, E., Bernstein, A., Zumstein, R.: Querix: A Natural Language Interface to Query Ontologies Based on Clarification Dialogs. In: 5th International Semantic Web Conference (ISWC 2006), Springer (2006) 980–981
15. Androutsopoulos, I., Ritchie, G., Thanisch, P.: Natural Language Interfaces to Databases - An Introduction. Natural Language Engineering **1**(1) (1995) 29–81
16. Copestake, A., Jones, K.S.: Natural Language Interfaces to Databases. Knowledge Engineering Review **5**(4) (1990) 225–249
17. Androutsopoulos, I., Ritchie, G., Thanisch, P.: An Efficient and Portable Natural Language Query Interface for Relational Databases. In: 6th International Conference on Industrial and Engineering Applications of Artificial Intelligence and Expert Systems. (1993) 327–330
18. Zelle, J.M., Mooney, R.J.: Learning to Parse Database Queries Using Inductive Logic Programming. In: AAAI/IAAI, Vol. 2. (1996) 1050–1055
19. Thompson, C.A., Mooney, R.J.: Automatic Construction of Semantic Lexicons for Learning Natural Language Interfaces. In: AAAI/IAAI. (1999) 487–493
20. Zhang, L., Yu, Y.: Learning to Generate CGs from Domain Specific Sentences. In: ICCS. (2001) 44–57
21. Rodrigo, L., Benjamins, V.R., Contreras, J., Patón, D., Navarro, D., Salla, R., Blázquez, M., Tena, P., Martos, I.: A Semantic Search Engine for the International Relation Sector. In: International Semantic Web Conference. (2005) 1002–1015
22. Kang, I.S., Na, S.H., Lee, J.H., Yang, G.: Lightweight Natural Language Database Interfaces. In: NLDB. (2004) 76–88
23. Fuchs, N.E., Schwertel, U., Torge, S.: Controlled Natural Language Can Replace First-Order Logic. In: ASE. (1999) 295–298
24. Nelken, R., Francez, N.: Querying Temporal Databases Using Controlled Natural Language. In: COLING. (2000) 1076–1080
25. Lopez, V., Pasin, M., Motta, E.: AquaLog: An Ontology-Portable Question Answering System for the Semantic Web. In: ESWC. (2005) 546–562
26. Bernstein, A., Kaufmann, E., Kaiser, C., Kiefer, C.: Ginseng: A Guided Input Natural Language Search Engine for Querying Ontologies. In: 2006 Jena User Conference. (2006)

Mining the Web Through Verbs: A Case Study

Peyman Sazedj and H. Sofia Pinto

Inesc-ID
Rua Alves Redol 9, Apartado 13069 1000-029 Lisboa, Portugal
{psaz,sofia}@algos.inesc-id.pt

Abstract. Mining non-taxonomic relations is an important part of the Semantic Web puzzle. Building on the work of the semantic annotation community, we address the problem of extracting relation instances among annotated entities. In particular, we analyze the problem of verb-based relation instantiation in some detail and present a heuristic domain independent approach, based on verb chunking and entity clustering, which doesn't require parsing. We also address the problem of mapping linguistic tuples to relations from the ontology. A case study conducted within the biography domain demonstrates the validity of our results in contrast to related work, whilst examining the complexity of the extraction task and the feasibility of verb-based extraction in general.

1 Introduction

Unsupervised annotation tools have promoted a new perspective on harvesting data from the web, whilst presenting a convenient solution for populating knowledge bases with concept instances. Concept instances alone are not very expressive, unless accompanied by relations, therefore a lot of effort is being directed towards unsupervised extraction of non-taxonomic relations. There is an enormous amount of data on the web, requiring efficient extraction methods. Moreover the data is error-prone, requiring fault tolerant systems. Shallow extraction techniques are therefore clearly advantageous over more knowledge intensive methods, specially when they produce similar or even better results. Thus, the aim of this work is to present a shallow heuristic approach for mining relation instances through verbs, which achieves satisfactory precision and recall. The approach has been implemented within the FactBox framework [1], specifically designed for prototyping relation extraction algorithms and requires no parsing. Instead, it relies on verb-chunking and clustering to extract verb and entity pairs. A novel algorithm is presented, to match extracted relation candidates with relations from the ontology, whenever applicable. Being more efficient than other approaches based on dependency or full syntactic parsing, it is a more suitable candidate for mining large and error-prone data from the web. It also improves over the algorithms presented in [1] and compares to other state of the art algorithms which follow more knowledge intensive approaches.

Section 2 presents an overview of related work. Section 3 and 4 provide a generic problem description and a formal analysis of the task at hand. Section 5 discusses

E. Franconi, M. Kifer, and W. May (Eds.): ESWC 2007, LNCS 4519, pp. 488–502, 2007.

our algorithms. Section 6 presents the framework which served as a testbed for their evaluation, followed by a specific case study in section 7, the results of which are analyzed in section 8. Section 9 summarizes our main conclusions.

2 Related Work

Relation Instantiation falls within the category of Ontology Population, a newly emerging field within the Semantic Web community. Nevertheless, underlying techniques are often one and the same with those of the Information Extraction [2] and Question Answering [3,4] communities. A sister field, Ontology Learning [5,6], also has similar purpose and methods: to learn relations among concepts, instead of extracting relation instances among concept instances. One of the differences is that ontology learning puts a strong emphasis on statistical analysis, whereas in relation instantiation the balance is more inclined towards the linguistic component. In the following we discuss some relevant contributions for mining non-taxonomic relation instances from text.

One class of extraction systems are pattern based systems, with varying degrees of linguistic analysis, where patterns may be hand-crafted or automatically learned. A famous example of hand-crafted patterns are the Hearst patterns for extracting is-a relations [7]. Instead of using lexical patterns which hardly generalize over different domains, a popular alternative is to apply patterns to deeper syntactic or semantic structures. Some recent approaches use hand-crafted patterns to extract verb-based relations over dependency parse trees [8,9]. In [10] a generalization of the Hearst patterns is presented, based on an automatic pattern induction system. Similarly, and equally interesting, are pattern induction systems over dependency trees [3]. Even though the approach in [3] is targeted at question answering, it resembles relation instantiation, since the answer to a question often consists in querying relations among entities. DIPRE [11] and Snowball [12] are two relation extraction systems which learn patterns from seed tuples. Similarly, Armadillo [13] applies adaptive information extraction to learn extraction patterns.

Machine learning systems use a set of training examples, alike pattern induction systems, to train a classifier. A popular set of classifiers are based on kernel methods, and in particular, support vector machines. Kernel methods for relation extraction have been presented in [14]; [15] defines dependency tree kernels for learning relations from dependency parse trees. LEILA is a recent learning system which trains a classifier on link grammar structures [16]. A mixed approach is followed in [17], where probabilistic parse trees are augmented with entities and relations in order to train a classifier.

Yet another approach consists of transforming text into logical formula [18], where deduction can be used to extract more complex relations.

3 Generic Problem Description

Our goal is to develop a shallow method for extracting relation instances from web pages. Throughout the remainder of this paper, we assume that named entities of a text have been previously annotated with concepts from an ontology. In practical terms, annotations are produced by a third-party tool and are then read by the FactBox framework [1], whose main features will be reviewed in section 6. Consequently, the challenge lies in extracting those relations among entities, which have been modeled within the ontology. The problem is a very complex one, since different relations materialize through an unnumbered variety of linguistic expressions. Even a single relationship may often be expressed in so many ways, that we know of no single method for extracting all relations of a kind. Recent work [5,8] has shown that verbs are invaluable sources of evidence for retrieving relationships among entities, and methods such as dependency parsing have become popular for extracting verbs and their arguments. This work further explores the trend of verb-based extraction methods and aims to address the problem of relating verb arguments, without parsing, and selecting those verb-entity pairs which match the semantics of the ontology.

3.1 Divide and Conquer

Ontology-based extraction of relation instances from verbs can be divided into two sub-problems: (1) the selection of verb arguments (e.g. entity pairs in case of binary relations) and (2) the mapping of verbs with appropriate relations from the ontology. Both sub-problems are dependent parts of one whole, resembling a constraint satisfaction problem, where the solution of one restricts the solution space of the other. Nevertheless, they can be solved in any order, and the choice is relevant as we will show. Before we analyze each of the problems in greater detail, consider the example sentence "John works at IBM and likes Jill", where John and Jill are instances of `person` and IBM of `company`. Assume that the ontology connects the classes `person` and `company` with two relations, namely `employee` and `investor`. Problem (1) consists of selecting the correct arguments of each verb and would result in the two potential relation candidates `works(John, IBM)` and `likes(John, Jill)`. Problem (2), on the other hand, consists of mapping verbs to relations from the ontology, in this case, to map the verb `works` to the relation `employee`. This two-step process would result in the extraction of `employee(John, IBM)`. If we start out by determining that John and IBM are related, the problem is simplified to that of discovering whether any of `employee` or `investor` relations hold. On the other hand, if we had started by discovering that the verb `works` matches the `employee` relation, the argument selection process would have become restricted to choosing whether John or Jill work at IBM. In the following we analyze each of the problems in greater detail.

3.2 Selecting Verb Arguments

The most naive solution for this problem is to randomly map each verb with all entities in some vicinity. If the vicinity is sufficiently large, at maximum having the size of the whole text, it is guaranteed that all potential matches will be found. This approach makes little sense of course, and yields factorial complexity $O(n!)$[1]. To simplify argument selection, two restrictions are often

[1] To simplify we assume there are no reflexive relations.

implicitly adopted: **a locality restriction**, only considering arguments in a local vicinity of the verb (e.g. a part of the sentence); **an arity restriction**, focusing on binary relations (since n-ary relations can be decomposed into n binary relations). The first restriction substantially reduces the search space of potential entities, whereas the second restriction reduces the complexity of the task to $O(n^2)$. In an ontology-based setup, as in many schema-based setups, a third restriction applies: **a type restriction** where the arguments of a relation must be of a specified type, in other words, only some entity types potentially match to form relation candidates, filtering out incompatible entity pairs (those which are not related within the ontology). Complexity is reduced to $O(|i|\,|j|)$, where the relation holds among instances of classes i and j. Nevertheless, the worst case complexity is still of quadratic order.

3.3 Mapping Verbs to Relations

This problem is sometimes solved by using a lexicon which directly maps lexical elements (such as verbs) to relations from the ontology. Although convenient in situations where such a lexicon is available, it requires a large amount of work to build one, specially when mining heterogeneous data from different domains. We aim to develop a domain-independent strategy which tries to map verbs to relations based on linguistic evidence. For that purpose, we may roughly distinguish three degrees of linguistic analyses: (1) a lexical analysis which tries to establish a lexical equivalence among the verb and the relation name; (2) a shallow semantic analysis which aims to establish an intentional equivalence among the verb and the relation name (e.g. by considering synonyms); and (3) a deep semantic analysis, which tries to match the selectional restrictions of the verb with the constraints of a relation. In this work we focus on the first two degrees - a lexical and a shallow semantic approach implemented within the FactBox framework. The approach consists of locating a verb and a relation name within the WordNet hierarchy, and deciding whether they are sufficiently related or not.

4 Formal Analysis

We define an ontology as a set of concepts C ordered by a subsumption relation within a taxonomy.[2] We consider a set of labeled non-taxonomic binary relations R among concepts of C, denoted as $r(c, d)$, $r \in R \wedge c, d \in C$. Entities in the text are considered instances of concepts and denoted by a set I. For an entity pair $i, j \in I$ and two classes $c, d \in C$ of which i and j are instances, let us consider a relation selector $S_{rel} : C \times C \rightarrow R \cup \{\perp\}$, which, given two concepts c and d, returns a subset of R, augmented with the empty relation $\perp$, which can be interpreted as the set of potential relations among c and d which are not modeled within the ontology.

[2] We have omitted the formal definition of the subsumption relation for the sake of simplicity.

As explained in section 3.2, the argument selection task, S, has a worst case time complexity of $O_S(n^2)$ for each verb. This translates into a total complexity of $O_S(n^2 \, |V|)$, considering V the set of all verb occurrences of a corpus. The mapping task, M, has a worst case complexity of $O_M(|V| \, |R|)$. In order to understand how the two sub-problems affect each other and the total problem complexity, let us consider both solutions SM and MS, depending on the order in which the problems are solved. In SM (M is solved after S), only a limited subset of relations will have to be considered and final complexity is $O_{SM}(n^2 \, |V| \, |S_{rel}|)$. If the tasks are solved in inverted order, we obtain a complexity of $O_{MS}(|V| \, |R| \, |i| \, |j|)$, where $|i|$ and $|j|$ are the number of entities of type i and j, applicable to the relations picked by M.

A few observations follow. The locality restriction of verbs (section 3.2) says that only arguments within some local vicinity of a verb are plausible candidates. Let E be the set of entities within the vicinity of a verb, then $|E| < \epsilon_1$, for some constant ϵ_1. Likewise, $|S_{rel}| < \epsilon_2 < |R| + 1$. A particularity is that both ϵ_1 and ϵ_2 are constants, whereas $|V|$ grows with the size of the corpus. It follows that $|V| >> \epsilon_1, \epsilon_2$, for a large enough corpus, therefore the time complexity can be rewritten as $O(m\epsilon_1^2\epsilon_2)$, where $m = |V|$. Thus, in reality the problem is of linear complexity. In a similar way it can be shown that the same holds for O_{MS}.

The question remains which one of the two is more efficient. Since the number of entities within the vicinity of a verb is always less than a threshold ϵ_1, the answer lies within the complexity of the ontology. For a large ontology, $|R| \, / \, |S_{rel}| > \epsilon_1^2$, and SM is the preferred method. The more specific the relations of the ontology are, the smaller is $|S_{rel}|$ in contrast to $|R|$, and the better the performance of SM.

5 Algorithms

Based on the division of the general problem into two sub-problems (section 3.1), we present distinct algorithms, for the selection of verb arguments and for mapping verbs to relations.

5.1 Selecting Verb Arguments

Regarding the selection of verb arguments, we used a simple verb chunker to identify verb groups. For each verb group, the challenge consists of discovering the syntactic arguments of the verb, without parsing. As a first approach (S_1) we consider all entities of a sentence as potential arguments of a verb, thus every possible entity pair of a sentence is matched with each verb of the sentence. We are implicitly restricting the vicinity of a verb to the lexical elements of a sentence, while ignoring verb anaphora, where the arguments of the same verb instance can span several sentences. Since S_1 yields a 100% recall in finding verb arguments, it is a good baseline to compare with more complex approaches.

$$S_1 : E \rightarrow \mathcal{P}(V) = V_s, V_s \subseteq V$$

V_s denotes the set of verbs of a sentence and is an element of the powerset of V. Since a sentence often contains several verbs, we developed a second approach (S_2) which further restricts the vicinity of verbs by dividing the sentence into segments and centering each segment around a verb. In other words, we employ clustering, where the number of clusters is equivalent to the number of verbs within the sentence and each verb is the center of a cluster. For each entity $e \in E$ of the sentence, the challenge consists of assigning it to the correct verb cluster $v \in V$. [3]

$$S_2 : E \rightarrow V = \arg\min_{v \in V} distance(e, v)$$

We define a distance function, $distance : E \times V \rightarrow \mathbb{N}$, which represents the numerical distance between an entity and the center of a verb cluster. Several metrics are possible for such a distance function, based on the linguistic evidence that is considered. We experiment with two metrics (short d_1 and d_2): (1) the distance between two words is given by the number of characters in between them, and (2) the distance is given by the number of words in between them. For the sentence "John works for IBM but loves Sisco.", $d_1(IBM, works) = 5$ and $d_1(IBM, loves) = 11$, whereas $d_2(IBM, works) = 1$ and $d_2(IBM, loves) = 2$. Moreover $d_2 \leq d_1$, because the number of words is always less than the number of characters. The main characteristic of d_2 over d_1 is that it disregards the length of words and is expected to have a more constant behavior in presence of both short and lengthy words.

5.2 Mapping Verbs to Relations

Let us recall that the problem consists of mapping verbs to relations from an ontology (for a more detailed description see section 3.3). We define a mapping function as a function M which maps a verb $v \in V$ to a relation $r \in R$ from the ontology.

$$M : V \rightarrow R \cup \{\bot\} = \begin{cases} \arg\max_{r \in R} similarity(v, r) \geq \theta, \theta \in [0, 1] \\ \\ \bot, \text{otherwise.} \end{cases}$$

M returns the relation that maximizes a normalized similarity function defined as $similarity : V \times R \rightarrow [0, 1]$ and returns no mapping ($\bot$), if the maximum similarity between a verb and a relation is below a threshold θ.

As a first approach (M_1), we try to establish a mapping by means of string matching, based on a normalized Smith-Waterman distance [19].

We also define a shallow semantic approach (M_2), which queries WordNet in order to detect whether lexically dissimilar verbs and relation names are similar or equivalent in meaning, such as synonyms. For example, we may need to match the compound verb "was born" with a relation from the ontology which may

[3] Within a verb cluster, the semantic restrictions of the ontology allowed us to select entities in the correct order, without the need for additional criteria.

have been named "born", "birthday" or "date of birth". Likewise, suppose we are mining death dates. It is desirable to match the verbs "deceased" or "passed away" with a relation such as "date of death". More concretely, we are looking for a strategy to assess the similarity of verbs with verbs and verbs with nouns, whether simple or compound. We start with simple verbs and nouns, then we generalize for the compounds.

For matching two simple verbs, we simply check whether they belong to the same synset. We don't use a criterion based on the subsumption hierarchy of verbs as commonly done with nouns, because the verb hierarchy is very shallow and could potentially cause many false positives. For matching a simple verb with a simple noun, we measure the similarity among the derivationally related nouns of the verb with the noun. Thus, the task remains of measuring the relatedness of simple nouns, a well studied problem which may be solved by calculating the path from the root to the lowest common node that subsumes both nouns. Finally, the compounds, namely compound verbs and nouns remain. The main verb of a compound stands always in last position, therefore it is straightforward to filter out the main verb.[4] A compound noun, on the other hand, can be split into a set of simple nouns. Then, we filter nouns which do not exist in the WordNet hierarchy and those which characterize the domain of the relation and not the relation itself. For example a relation named "death date" has a date as its domain, therefore the noun "date" can be filtered out. We obtain a final set of nouns, each of which is compared individually with its counterpart, whether verb or noun. The efficiency of the previous method is not of great concern, because relation names within the ontology only need to be processed once. This can be done conveniently as a pre-processing step, before the extraction commences.

To summarize, consider V the set of simple verbs and N the set of simple nouns within WordNet. Consider V' and N' the sets of all possible compound verbs and nouns, respectively. A compound verb has a main verb $v \in V$, so that $\exists v_{\in V}, last(v') = v, v' \in V'$ returns the main verb. Further, consider the functions $synset(v), v \in V$, which returns the synset of a verb and $deriv(v), v \in V$, which returns the derivationally related nouns of a verb. Finally, consider the following functions: $sim(n_i, n_j) \in [0, 1]$, calculates the normalized similarity among two nouns, $split(n) = X, X \cap N \neq \{\}$, splits a compound noun into its simple parts and $filter(X) = X', X' \subseteq X \cap N$, filters irrelevant nouns.

Our WordNet based similarity algorithm (Sim_{Wn}) is hence defined as follows:

$$Sim_{Wn} : x \times y \rightarrow [0,1] = \begin{cases} 1 & x \in synset(y) \wedge x, y \in V \\ sim(x,y) & x, y \in N \\ sim(last(x), y) & x \in V' \\ \max_{\forall n, n \in deriv(x)} sim(n,y) & x \in V \wedge y \in N \\ \max_{\forall n, n \in filter(split(y))} sim(x,n) & y \in N' \\ 0 & x, y \notin V \cup V' \cup N \cup N' \end{cases}$$

[4] This observation applies, at least, to the verb chunker we used.

6 FactBox Framework

FactBox is a testbed for relation instantiation algorithms and supports ontology population. It is composed of three plugin-based components: (1) an Extraction Component, (2) a Relation Base and (3) a Similarity Matcher. The Extraction Component is divided into an entity and a relation extraction component. Annotations of named entities can be produced by a third-party tool, given that a plugin is provided which reads the annotations and populates the Entity Base of the system. Entities are then read by relation extraction algorithms which populate the Relation Base with relation candidates. The Relation Base contains a rule chain which is capable of adding, retracting and transforming relation candidates by applying deductive rules (similar to implications of FOL). Finally, the Similarity Matcher tries to match instances of the Relation Base with relations from an ontology and populates a knowledge base with relation instances. Given a gold standard, the system performs automatic evaluation of the relation extraction algorithms over a corpus.

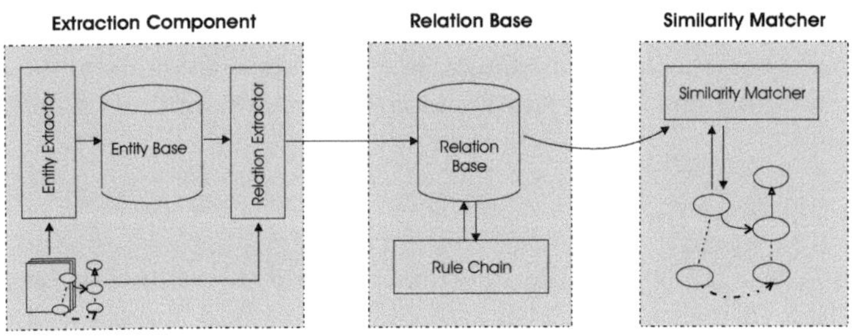

Fig. 1. The conceptual architecture of FactBox and its three main components

Figure 1 illustrates the architecture of the framework. The algorithms for selection of verb arguments (section 5.1) have been implemented as relation extraction plugins and the mapping algorithms (section 5.2) as plugins of the Similarity Matcher. For additional information on the framework, see [1].

7 Case Study: Biographies from IMDb

One of the challenges of ontology based relation extraction is to find appropriate corpora and ontologies for testing systems. This limitation has led some researchers to evaluate their systems on different domains, making comparison of results difficult; it has led others to work on a very small set of relations of a particular domain. In choosing a domain of interest, we attempted to pick a domain which is generally interesting for the public and can be reused in other experiments. We also tried to pick a domain which had already been tackled by other researchers. As another requirement, it should present the researcher with a wealth of possible relations. The last requirement was that the corpus should be as large as possible, preferably having been written by different people, in order to cover as many forms of expressing a relation as possible. The aforementioned conditions led us to create a corpus of biographies.[5] In this case study, we aim to explore the practical complexity of the extraction task, the performance of our approach, and the coverage of verbs with regard to other part of speech that may indicate the same relationship. In the following, we present the corpus, explain how it was acquired, describe the relations to be extracted and present some useful statistics on the corpus.

7.1 The IMDb Biography Corpus

The Internet Movie Database (IMDb) is a popular site for looking up information on movie reviews.[6] The site also offers a wealth of information about people involved with movies, including biographies of actors, directors and producers. Information on IMDb is a result of social collective effort. Having been written by thousands of people, it is reasonable to expect that biographies reflect different writing styles, rendering this corpus particularly interesting for the proposed analysis. Additionally, the IMDb site also offers some information in semi-structured form, such as date and place of birth or death. This information is very useful for automatically mining a golden standard for those relations. Crawling the IMDb site for biographies, we obtained a corpus of 2695 documents.[7]

7.2 Relations of Interest

From among the many possible relations of interest within the biography domain, we had to pick a few. Although our corpus is domain-specific, covering biographies of the movie domain, we aim to extract relations which are general enough to be applied to any biography. This choice is based on two reasons: (1) to ease comparison with other works, (2) because the annotation tool we used for marking-up named entities, by default, only recognizes `people`, `organizations`, `dates` and `locations`, making it difficult to work with domain-specific relations.

Thus, we came up with the following relations of interest: `birthday`, `birthplace`, `death date`, `death place`, `spouse` and `study`. Our ontology containing the aforementioned concepts and relations is illustrated in figure 2. The ontology was formalized in OWL. The `birthday` and `death date` relations were designed as functional relations, to ensure that each person had unique birth and death dates.

7.3 Statistics

Table 1 summarizes useful statistical estimates for the corpus, which were extracted from a random sample of 50 biographies. The first row summarizes the

---

[5] http://www.inesc-id.pt/~psaz/

[6] http://www.imdb.com

[7] At some point we ended the crawling process; 2695 is *not* the total number of biographies available at IMDb.

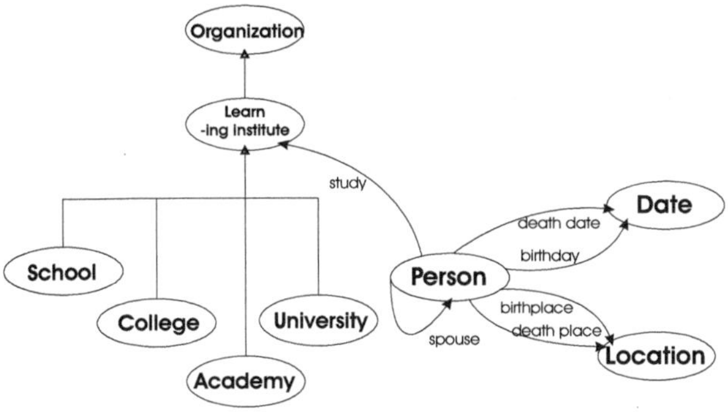

Fig. 2. The conceptual model of the biography ontology

number of instances of each relation without anaphora resolution, that is, disregarding pronouns as possible verb arguments. The second row considers only those relations where pronoun anaphora occurred, and row three sums up the total number of verb instances for each kind of relation. The next row is an estimate of the number of relation instances that occur per document, given in percentages. Finally, we show how often a relation occurs in other part of speech, apart from verbs, and calculate the coverage of verbs within the last row.

Table 1. Sample statistics and rough estimates for the corpus. VI stands for Verb Instances.

	birthday	birthplace	study	spouse	death date	death place
VI	26	53	7	3	4	4
VI with anaphora	1	5	21	14	10	4
Total VI	27	58	28	17	14	8
VI p/ document	0.52	1.1	0.14	0.06	0.08	0.08
Non VI	0	0	0	16	1	0
VI coverage (%)	100	100	100	52	93	100

A number of conclusions follows from the data in table 1. Only the `birthday` and `birthplace` relations exist abundantly, without considering anaphora. This is justified by the fact that a person's name is usually referred to in the beginning of a biography, jointly with date and place of birth, whereas future references to the person are made with pronouns. With anaphora resolution, the `study`, `spouse` and `death date` relations also occur sufficiently often. Regarding dates and places of death, they appear with low frequency, since most of the covered actors are still alive.

A major point under study is the coverage of verbs with regard to other part of speech. The data clearly shows the selected relations are mostly covered by verbs, except for the `spouse` relationship. The estimates hint that verb-based extraction methods promise to perform well on many domains such as the one under evaluation.

Table 2. Different modes of expressing the same relation

birthday	**study**
<person> was born on <date>	<person> was educated at <school>
Born on <date>, <person>	<person> majored at <school>
<person> was born ... on <date>	<person> attended <school>
birthplace	<person> was transferred to <school>
<person> was born in <location>	during ... at <school>, <person>
Born in <location>, <person>	After graduation at <school>, <person>
<person> was born ... in <location>	<person> graduated from <school>
<person> was born ... in <location>, <location>	**death date**
spouse	<person> died on <date>
After <person>'s divorce from <person>	On <date>, <person> was killed by
<person> married <person>	<person> was found dead ... <date>
<person> remarried <person>	**death place**
his wife <person>	<person> died in <location>
her husband <person>	<person> passed away ... <location>

From among the different verbal expressions we encountered in the sample, we selected a few to illustrate different ways of expressing each relation. The patterns are summarized in table 2.

8 Evaluation

It is extremely difficult to evaluate extraction systems on large corpora. Either a gold standard is somehow available for the entire corpus, or a sample has to be selected for which a reference is manually created. The work in [12] follows an interesting approach, automatically mining a reference standard from structured data, whilst manually creating a reference standard for a smaller subset of their corpus. They argue that both solutions are interesting. The former allows to gain valuable insight into large-scale extraction, which is often the actual aim of the work, while the latter allows to fine-tune results by detecting named-entity tagging errors. For obvious reasons, the two approaches may differ considerably in their results.

An overall important detail is to distinguish among *a priori* and *a posteriori* evaluations [5,20]. In an *apriori* evaluation, the gold standard is created before hand and the results of the system are measured against the reference. In an *a posteriori* evaluation, results are presented to an evaluator, who then decides which of them are correct and which are not. Comparing the strict a priori method with the more relaxed a posteriori, the work in [5] reports a degradation of about 10% in the precision of the former over the latter.

In our case, the FactBox framework is provided with a gold standard in digital format and the system automatically produces an *a priori* evaluation without human intervention. Following the approach of [12], we attempted to create two gold standards, one manually over a small subset of documents (section 8.1), and the other automatically over a larger subset (section 8.2).[8] Both experiments are described in the following. It is important to note that we used an unsupervised

[8] Structured information for automatically creating a gold standard was only available for a subset of the corpus.

named entity tagger and verb chunker, plugins of the Gate architecture [21], therefore a degradation of 10-20% is expected in all experiments due to tagging errors.

In both experiments, evaluation scores are presented separately for each relation, in order to convey deeper insight on the performance of each. Standard Precision, Recall and F1 metrics are defined as

$$P = \frac{|R_{es} \cap G|}{|R_{es}|}, R = \frac{|R_{es} \cap G|}{|G|}, F1 = \frac{2PR}{P+R}$$

where R_{es} is the result set (the set of instances that were mined) and G is the gold standard set.

8.1 Experiment 1

We randomly selected a sample of 50 documents from the corpus and manually created a gold standard for the sample. Recall that in section 5.1 we defined three algorithms for selecting verb arguments, a baseline S_1 and an improvement over the baseline which was defined based on two different distance metrics, $S_{2.d1}$ and $S_{2.d2}$. We compare the performance of the three algorithms with the `birthday` and `birthplace` relations. Since these two relations are those which occur more frequently, they are appropriate candidates. Additionally, both relations always materialize through the verb "born", meaning that the performance of the verb mapping algorithm can be discarded and the results do accurately reflect the performance of the verb argument selection algorithms. Table 3 summarizes our results.

Table 3. Comparison of different verb selection algorithms defined in section 5.1

	S_1	$S_{2.d1}$	$S_{2.d2}$
Precision (%)	58.5	75.4	75.8
Recall (%)	77.5	69.0	70.4
F1 (%)	66.7	72.1	73.0

The results are quite interesting and a few conclusions follow. First, we remark that the baseline which was supposed to obtain 100% recall, only obtains 77.5% recall. It is highly likely that the 22.5% decline in recall is due to name entity and verb chunking tagging errors. Thus, we should bear in mind that all results presented in the remainder of this paper may actually be improved up to approximately 20% if better taggers are used. Moreover, both distance metrics improve over the baseline and algorithm $S_{2.d2}$, based on a word distance metric, works best. In the remainder of our evaluations we use this algorithm.

In the following we inspect the performance of our approach for each kind of relation. Evaluation scores are presented in table 4.

The `birthday` and `birthplace` relations obtain very high precision and recall, given that the values include up to 20% of tagging errors. Our results are similar to those obtained by the machine learning approach in [16], who trained

Table 4. Evaluation scores for each kind of relation

	birthday	birthplace	study	spouse	death date	death place	study(2)	marry
Precision (%)	77.3	75.0	0	0	50.0	21.4	66.7	25.0
Recall (%)	73.9	68.8	0	0	50.0	75.0	57.1	33.3
F1 (%)	75.6	71.7	0	0	50.0	33.3	61.5	28.6

a classifier on the `birthday` relation and obtained precision and recall in the range of 70-80%. It is unclear whether their approach included tagging errors or not.

Regarding the `study` and `spouse` relations, we found that both of them had unhappy designations, since our mapping algorithm failed to match them with corresponding verbs. In table 2 one can see that the `study` relation is not expressed in a consistent way; very loose semantic associations between people and schools are sufficient to express this relation. Therefore, we implemented a transformation rule within FactBox, which transformed relations between people and schools into `study` relations. Regarding the `spouse` relation, we changed its name to `marry`. The last two columns of table 4 reflect the improvements that followed from these simple adaptations.

8.2 Experiment 2

In a second experiment, we evaluated the scalability of our approach by using larger samples. For that purpose, we automatically created a gold standard by mining structured data from the IMDb web site. This data is available for the `birthday`, `birthplace`, `death date`, `death place` and `spouse` relations. The mined data for each relation was only considered suitable, to be included in the gold standard of a biography, if it actually occured within the biography. This is so, because data available in structured form was often not mentioned within the biographical text. Unfortunately, this approach has a pitfall, since it does not exclude relations whose entities are referred to by pronouns. Since we do not have an automatic anaphora resolution module yet, only the `birthday`

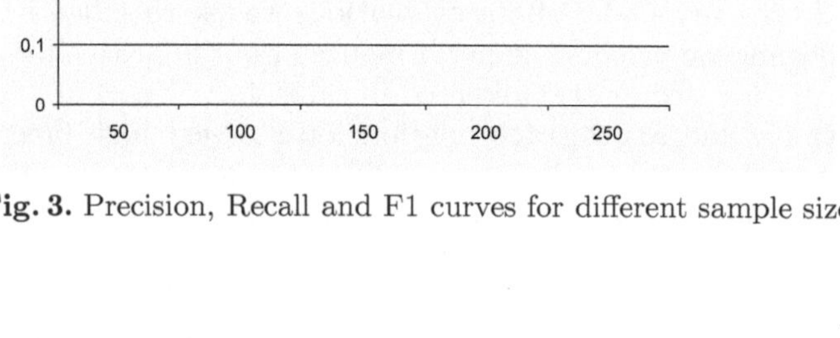

Fig. 3. Precision, Recall and F1 curves for different sample sizes

and `birthplace` relations are considered in this experiment, since they do not frequently occur with pronouns. Results are shown in figure 3.

We started with a sample of 50 documents and gradually increased the sample size. The recall values are very low in this experiment due to a faulty construction of the gold standard. In particular, having automatically obtained the place of birth of an actor, we check whether it is mentioned in the biography in order to decide whether to include it in the gold standard or not. This seems to be an unreliable approach, since the place of birth is often mentioned with some other intention, not expressing the actual `birthplace` relationship. Nevertheless, the experiment shows that the output of our approach remains stable over large samples, indeed results improved with larger samples until they converged.

9 Concluding Remarks

This work is a contribution towards large scale relation extraction from the web. We analyzed the problem of verb-based relation extraction and divided it into two sub-problems, leading to two novel algorithms: an unsupervised heuristic approach which performs verb-entity clustering, and an algorithm for mapping verbs to relations from an ontology. In a first experiment we showed that verb-based relation extraction is a feasible solution for mining relations from the web, and that, our approach in particular, compares to other state of the art algorithms that require more complex techniques. In a second experiment we tested the scalability of our approach, showing that it converges over larger samples. We also described the problems we encountered, namely, that verb arguments are often referred to by pronouns, requiring anaphora resolution to increase recall up to 5 times for of some relations. Future work includes the development of an efficient anaphora resolution module to be integrated within the FactBox framework and an improvement of current algorithms.

References

1. P. Sazedj and H. S. Pinto, FactBox - a Framework for Instantiating Ontological Relations from Text, in *Workshop on Web Content Mining with Human Language Technologies at ISWC*, 2006.
2. E. Marsh and D. Perzanowski, MUC-7 Evaluation of IE Technology: Overview of Results, http://www.itl.nist.gov/iaui/894.02/related_projects/muc/index.html.
3. D. Lin and P. Pantel, DIRT - Discovery of Inference Rules from Text, in *Proceedings of KDD*, pp. 323–328, 2001.
4. D. Ravichandran and E. H. Hovy, Learning surface text patterns for a Question Answering System, in *ACL*, pp. 41–47, 2002.
5. A. Schutz and P. Buitelaar, RelExt: A Tool for Relation Extraction from Text in Ontology Extension, in *Proceedings of ISWC*, pp. 593–606, 2005.
6. A. Maedche and S. Staab, Discovering Conceptual Relations from Text, in *Proceedings of ECAI*, pp. 321–325, 2000.
7. M. A. Hearst, Automatic acquisition of hyponyms from large text corpora., in *COLING*, pp. 539–545, 1992.

8. M. Ciaramita, et al, Unsupervised Learning of Semantic Relations between Concepts of a Molecular Biology Ontology, in *Proceedings of IJCAI*, pp. 659–664, 2005.
9. L. Specia and E. Motta, A Hybrid Approach for Relation Extraction Aimed at the Semantic Web, in *Proceedings of FQAS*, pp. 564–576, 2006.
10. R. Snow, D. Jurafsky, and A. Y. Ng, Learning Syntactic Patterns for Automatic Hypernym Discovery, in *NIPS*, 2004.
11. S. Brin, Extracting Patterns and Relations from the World Wide Web, in *WebDB*, pp. 172–183, 1998.
12. E. Agichtein, *Extracting Relations From Large Text Collections*, Ph.D. thesis, Columbia University, 2005.
13. F. Ciravegna, et al, Learning to Harvest Information for the Semantic Web, in *Proceedings of ESWS*, pp. 312–326, 2004.
14. D. Zelenko, C. Aone, and A. Richardella, Kernel Methods for Relation Extraction. *Journal of Machine Learning Research*, volume 3, pp. 1083–1106, 2003.
15. A. Culotta and J. S. Sorensen, Dependency tree kernels for relation extraction., in *ACL*, pp. 423–429, 2004.
16. F. M. Suchanek, G. Ifrim, and G. Weikum, Combining linguistic and statistical analysis to extract relations from web documents, in *KDD*, pp. 712–717, 2006.
17. S. Miller, et al, A Novel Use of Statistical Parsing to Extract Information from Text, in *ANLP*, pp. 226–233, 2000.
18. D. I. Moldovan and V. Rus, Logic Form Transformation of WordNet and its Applicability to Question Answering, in *ACL*, pp. 394–401, 2001.
19. T. F. Smith and M. S. Waterman, Identification of common molecular subsequences. *Journal of Molecular Biology*, volume 147, pp. 195–197, 1981.
20. P. Cimiano, M. Hartung, and E. Ratsch, Finding the Appropriate Generalization Level for Binary Relations Extracted from the Genia Corpus, in *LREC*, pp. pp. 161–169, 2006.
21. H. Cunningham, R. J. Gaizauskas, and Y. Wilks, A General Architecture for Language Engineering (GATE) - a new approach to Language Engineering R&D, Technical Report, Dept. of Computer Science, University of Sheffield, 1996.

What Have Innsbruck and Leipzig in Common? Extracting Semantics from Wiki Content

Sören Auer[1,2] and Jens Lehmann[1]

[1] Universität Leipzig, Department of Computer Science, Johannisgasse 26, D-04103 Leipzig, Germany
{auer,lehmann}@informatik.uni-leipzig.de
[2] University of Pennsylvania, Department of Computer and Information Science Philadelphia, PA 19104, USA
auer@seas.upenn.edu

Abstract. Wikis are established means for the collaborative authoring, versioning and publishing of textual articles. The Wikipedia project, for example, succeeded in creating the by far largest encyclopedia just on the basis of a wiki. Recently, several approaches have been proposed on how to extend wikis to allow the creation of structured and semantically enriched content. However, the means for creating semantically enriched structured content are already available and are, although unconsciously, even used by Wikipedia authors. In this article, we present a method for revealing this structured content by extracting information from template instances. We suggest ways to efficiently query the vast amount of extracted information (e.g. more than 8 million RDF statements for the English Wikipedia version alone), leading to astonishing query answering possibilities (such as for the title question). We analyze the quality of the extracted content, and propose strategies for quality improvements with just minor modifications of the wiki systems being currently used.

1 Introduction

Wikis are established means for the collaborative authoring, versioning and publishing of textual articles. Founded on Ward Cunninghams design principles[1], wikis dramatically simplify the process of creating and maintaining content by a community of readers and at the same time authors.

A large variety of wiki systems for all possible technical environments and application domains emerged, ranging from lightweight personal wikis focusing on personal information management to full-fledged enterprise wiki systems with integrated groupware functionality. Services based on wikis, such as the provision of wiki farms, support knowledge base wikis, or the maintenance of wikis as intranet sites are provided and employed by startup companies and established global players. Countless special interest wikis on the Web build enormous content collections from travel information for the global village (e.g. on Wikitravel) to local news and gossip on city wikis (such as stadtwiki.net).

[1] http://c2.com/cgi/wiki?WikiDesignPrinciples

E. Franconi, M. Kifer, and W. May (Eds.): ESWC 2007, LNCS 4519, pp. 503–517, 2007.

However, the most famous and successful wiki project probably is Wikipedia[2]. It succeeded in creating the by far largest encyclopedia authored by a globally distributed community just on the basis of a wiki. Wikipedia editions are available in over 100 languages with the English one accounting for more than 1.5 million articles.

To be able to 'tap' this knowledge by machines, recently, several approaches have been proposed on how to extend wiki systems to allow the creation of structured and semantically enhanced content. Modulo minor variations, all of them suggest to enrich the textual wiki content with semantically interpretable statements. The project Semantic Wikipedia [15,24] for example proposes to integrate typed links and page attributes into Wiki articles in a special syntax. It is a straightforward combination of existing wiki systems and the Semantic Web knowledge representation paradigms.

Unfortunately, this approach has several drawbacks: Wikipedia authors have to deal with another means of syntax within wiki texts (in addition to many existing ones). Adding more and more syntactic possibilities counteracts ease of use for editors, thus antagonizing the main advantage of wikis - their unbeatable simplicity. In addition to that, existing (possibly already structured) content in Wikipedia may have to be manually converted or even duplicated. Finally, the approach requires fairly deep changes and additions to the MediaWiki software with unknown effects on its scalability. Scalability is, due to persistently enormous growth in access rates, presently the most pressing technical issue Wikipedia has to face.

However, the means for creating semantically enhanced structured content are already available and used (although unconsciously) by Wikipedia authors. More precisely, MediaWiki, the wiki software behind Wikipedia, enables authors to represent structured information in an attribute-value notation, which is rendered inside a wiki page by means of an associated template. Many Wikipedia pages contain templates, often for layout purposes, but still approximately a quarter to one third of the Wikipedia pages already today contain valuable structured information for querying and machine interpretation.

To be able to query, recombine and reason about this structured information, we present in this paper methods (a) to separate valuable information from less important one, (b) to extract this information from templates in wiki texts and convert it into RDF under usage of unified data types, (c) to query and browse this information even though its schema is very large and partly rudimentary structured. Further, we analyze the quality of the extracted content, and propose strategies for quality improvements with just minor modifications of the wiki systems being currently used.

2 Knowledge Extraction from Wikipedia Templates

We are not aware of any general approaches for extracting information from all templates in Wikipedia. We will first explain templates and then show how their

[2] http://www.wikipedia.org

inherent structure can be used to extract meaningful information. In contrast to the Semantic Wikipedia approach, this has the advantage that we can use already existing information, i.e. we do not need to modify the MediaWiki software (on which Wikipedia is based) to enrich it with support for expressing RDF statements. Hence, our approach can be of immediate use and overcomes the obstacles outlined in Section 1. Semantically enriching wiki templates has been discussed in [15, Section 3.1]. However, we will show that even with the template mechanisms currently available in many wikis, in particular Wikipedia, it is possible to accurately extract useful information.

2.1 MediaWiki Templates

MediaWiki supports a sophisticated template mechanism to include predefined content or display content in a determined way. A special type of templates are infoboxes, aiming at generating consistently-formatted boxes for certain content in articles describing instances of a specific type. An example of the application of an infobox template for the city Innsbruck and the generated HTML table representation on the resulting Wikipedia page is visualized in Figure 1. Such infobox templates are used on pages describing similar content. Other examples include:

- *Geographic entities:* countries, cities, rivers, mountains, ...
- *Education:* university, school, ...
- *Plants:* trees, flowers, ...
- *Organizations:* companies, sports teams, ...
- *People:* politicians, scientists, presidents, athletes ...

More information can be found in the Wikipedia infobox template article[3].

2.2 Extraction Algorithm

To reveal the semantics encoded in templates we developed an extraction algorithm operating in five stages:

*Select all Wikipedia pages containing templates.*Wikipedia pages are retrieved by an SQL query searching for occurrences of the template delimiter "`{{`" in the `text` table of the MediaWiki database layout. The SQL query can be adopted to select only pages for particular Wikipedia categories or containing specific templates to generate fragments of the Wikipedia content for a certain domain.

Extract and select significant templates. All templates on a Wikipedia page are extracted by means of a recursive regular expression. Since templates serve different needs, we extract those with a high probability of containing structured information on the basis of the following heuristic: templates with just one or two template attributes are ignored (since these are templates likely to function as shortcuts for predefined boilerplates), as well as templates whose usage count is below a certain threshold (which are likely to be erroneous). The extraction

[3] `http://en.wikipedia.org/wiki/Wikipedia:Infobox_templates`

```
1  {{Infobox Town AT |
2    name = Innsbruck |
3    image_coa =  InnsbruckWappen.png |
4    image_map = Karte-tirol-I.png |
5    state = [[Tyrol]] |
6    regbzk = [[Statutory city]] |
7    population = 117,342 |
8    population_as_of = 2006 |
9    pop_dens = 1,119 |
10   area = 104.91 |
11   elevation = 574 |
12   lat_deg = 47 |
13   lat_min = 16 |
14   lat_hem = N |
15   lon_deg = 11 |
16   lon_min = 23 |
17   lon_hem = E |
18   postal_code = 6010-6080 |
19   area_code = 0512 |
20   licence = I |
21   mayor = Hilde Zach |
22   website = [http://innsbruck.at] |
23 }}
```

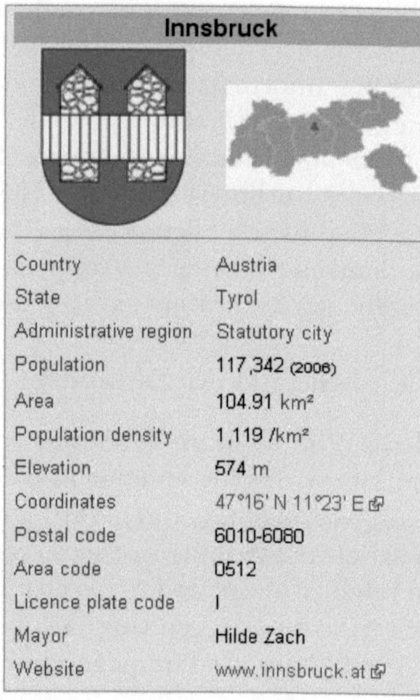

Fig. 1. Example of a Wikipedia template and rendered MediaWiki output for Austrian towns applied for Innsbruck

algorithm can be further configured to ignore certain templates or groups of templates, based on specific patterns.

Parse each template and generate appropriate triples. A URL derived from the title of the Wikipedia page the template occurs in is used as subject for templates which occur at most once on a page. For templates occurring more than once on a page, we generate a new identifier being used as subject. Each template attribute corresponds to the predicate of a triple and the corresponding attribute value is converted into its object. MediaWiki templates can be nested, i.e. the attribute value within a template can again be a template. In such a case, we generate a blank node linking the attribute value with a newly generated instance for the nested template.

Post-process object values to generate suitable URI references or literal values. For MediaWiki links (e.g. "`[[Tyrol]]`" in line 4 of Figure 1) suitable URI references are generated referring to the linked Wikipedia article. (Currently, we ignore the special case that the link denoting brackets could be part of the template definition.) Typed literals are generated for strings and numeric values. Common units (such as m for meter, kg for kilogram, s for seconds) are detected and encoded as special datatypes (cf. Table 1). However, a conversion between different scales (e.g. between mm, cm, m, km) is not performed. Furthermore, comma separated lists of the form `[[Jürgen Prochnow]]`, `[[HerbertGrönemeyer]]`, `[[Martin Semmelrogge]]` are detected and, depending on configuration options, converted into RDF lists or individual statements.

Table 1. Detection of literal datatypes (excerpt)

Attribute type	Example	Object data type	Object value
Integer	7,058	xsd:integer	7058
Decimals	13.3	xsd:decimal	13.3
Images	[[Image:Innsbruck.png|30px]]	Resource	c:Innsbruck.png
Links	[[Tyrol]]	Resource	w:Tyrol
Ranks	11th	u:rank	11
Dates	[[January 20]] [[2001]]	xsd:date	20010120
Money	$30,579	u:Dollar	30579
Large numbers	1.13 [[million]]	xsd:Integer	1130000
Big money	$1.13 [[million]]	u:Dollar	1130000
Percent values	1.8%	u:Percent	1.8
Units	73 g	u:Gramm	73

Determine class membership for the currently processed Wikipedia page. Wikipedia pages are organized in categories. In some cases, these can be interpreted as classes subsuming instances described by Wikipedia pages in the corresponding category. Furthermore, the name of the template can be an indicator for a certain class membership. The categories itself are Wikipedia pages and are often organized into super-categories. Unfortunately, here the sub-category super-category relationship often refers more to being "related-to" than constituting a subsumption relation. We are currently working on improving class membership detection.

2.3 Extraction Results

We tested the extraction algorithm with the English Wikipedia content (available from `http://dumps.wikimedia.org/enwiki`). The overall time needed to extract template instances and convert them to RDF for the approx. 1.5 Mio English Wikipedia articles (accounting for roughly 10GB raw data) was less than one hour on a computer with Xeon 2.80GHz CPU and 1GB of main memory. The raw extraction results as well as the source code of the extraction algorithm are available from `http://wikipedia.aksw.org/`.

Table 2 shows some extraction statistics. The first column contains general information about extracted quantities. Overall, more than 8 Mio triples were obtained from the English Wikipedia. Each triple belongs to one of about 750,000 templates, which can be grouped in approx. 5,500 template types. This means a template is used 137 times on average. In the extracted ontology, 650,000 individuals are connected by 8,000 properties and the class hierarchy consists of 111,500 classes (all numbers approximated).

The second column displays the most frequently used templates and how much instances of them exist in Wikipedia. The third column shows similar information for attributes. Table 3 exhibits the properties extracted from some frequently used templates.

Table 2. Extraction results: overall statistics, most used templates, and most used attributes

Statistics:		**Templates:**		**Attributes:**	
Template types	5,499	succession_box	72262	name	301020
Template instances	754,358	election_box	48206	title	143887
Templates per type	137.18	infobox_album	35190	image	110939
Attributes per instance	8.84	taxobox	29116	years	89387
Categories	106,049	fs_player	25535	before	79960
Classes	111,548	nat_fs_player	15312	after	78806
Instances	647,348	imdb_title	15042	genre	77987
Properties	8,091	infobox_film	12733	type	74670
Triples	8,415,531	imdb_name	12449	released	74465
		fs_squad2_player	10078	votes	59659
		infobox_cvg	7930	reviews	58891
		infobox_single	7039	starring	57112
		runway	6653	producer	53370

2.4 Obstacles

Since there are not many restrictions on the design of Wikipedia templates, there are a number of obstacles, which can lead to undesired extraction results in some cases.

First of all, templates are not yet used everywhere in Wikipedia, where they are appropriate. Sometimes tables or other means are used to display structured information.

For certain content (e.g. planets) infobox templates are not used to separate content from presentation, but for each content object a separate template containing attributes is created. Similarly, layout information is sometimes encoded directly in templates (e.g. color information) and templates for certain content are made up of from many small element boxes (e.g. chemical elements), even when this is not necessary.

Table 3. Extracted properties for specific templates

Template/ Class	**No. of Instances**	**Used properties**
Music album	35190	name, artist, cover, released, recorded, genre, length, label, producer, reviews, last album, next album
Species	29116	binomial, genus, genus_authority, classis, phylum, subfamilia, regnum, species, subdivision
Film	12733	starring, producer, writer, director, music, language, budget, released, distributor, image, runtime
Cities	4872	population_total, population_as_of, subdivision_type, area_total, timezone, utc_offset, population_density, leader_name, leader_title
Book	4576	author, genre, release_date, language, publisher, country, media_type, isbn, image, pages, image_caption

Attributes sometimes contain (from a knowledge representation viewpoint) redundant information, whose purpose is more intuitive visual representation as for example: `[[Innsbruck]], [[Austria]]`. In other cases, multiple values are encoded in one attribute instead of cleanly separating them in different attributes, e.g. `foundation = [[California]] ([[April 1]], [[1976]])`. Furthermore, duplicate information is sometimes present in attribute values, e.g. `height = 5'11" (180cm)`. The last example also shows that different units are used as attribute value, often depending on the locality of the intended audience of an article.

2.5 Guide for Designing Semantically Rich Templates

Despite all the obstacles described in the previous section, we were still surprised by the enormous amount of meaningful and machine interpretable information we were able to extract from Wikipedia templates. In order to improve extraction, we want to suggest some guidelines for defining templates in this section. We mention them here to outline how the extraction results could be improved further. Please note, that following these guidelines is not only good for semantic extraction, but usually also improves the corresponding template in general, i.e. it becomes more convenient to use by article authors.

- Do not define attributes in templates, which encode layout information. Rather, let the template handle the representation. (This corresponds to the well known principle of separating content and its presentation.) Along the same line, HTML markup should be used in attribute values only when necessary.
- Use only one template for a particular item of interest, instead of using one template for each attribute. Currently, the later version of templates is still present in many Wikipedia articles, although the former is emerging as the standard.
- Each attribute should have exactly one value within an article template. This value can be a list of values. However, one should not mix several statements (from an RDF point of view) within one attribute value.
- Currently, images in the English Wikipedia are retrieved from two places, depending on the definition of a template: Wikipedia Commons and the media library for the English Wikipedia . Thus, it is not possible to determine the location of an image without analyzing the definition of a template, which is an unnecessary complication. However, Wikipedia Commons is emerging as a standard, so the number of problematic cases is already decreasing.
- Do not use different templates for the same purpose, e.g. there are currently template infoboxes for "Infobox_Film", "Infobox Film" and "Infobox film". This problem is already tackled by the Wikipedia community[4].
- Do not use different attribute names for the same kind of content and do not use the same attribute name for different kinds of content. Support for this can be added to the wiki software (see below).
- Use standard representations for units, such that they can be detected by the extraction algorithm.

[4] `http://en.wikipedia.org/wiki/Wikipedia:Infobox_templates`

Furthermore, the following improvements could be made to the MediaWiki software, which is used for Wikipedia, and other software to make the design of clean templates easier:

- Offer the possibility to fix the data type of an attribute value, e.g. by attribute templates as mentioned above. For instance, the template designer could specify in the template definition that the attribute value of the attribute `budget` has to be a number. Although this greatly improves the extraction process, we are aware that sometimes verbal descriptions are necessary to explain attribute values, e.g. a value for budget could be "estimated to be between 20 and 30 million dollars".
- Offer the possibility of language and unit tags if one attribute value can be given in different languages or units, i.e. the budget can be specified in Euro and Dollar. A name of a French city can be given in English and French.
- If an attribute is defined in a template, the MediaWiki software could list templates, where this attribute already exists and show other characteristics of the attribute, to give the template designer the possibility to check whether these attributes have the same intended meaning.

One of the aims of these proposals is to improve extraction without putting the burden on the user (in this case the article author). In many cases, following the guidelines makes it clearer for the article writer how to use templates and many of these guidelines are common sense. They enable a clean extraction of information without the need to recreate the content of Wikipedia or dramatically change the way templates are currently defined.

3 Browsing and Querying Extracted Knowledge

Compared to most of the other Semantic Web knowledge bases currently available, for the RDF extracted from Wikipedia we have to deal with a different type of knowledge structure – we have a very large information schema and a considerable amount of data adhering to this schema. Existing tools unfortunately mostly focus on either one of both parts of a knowledge base being large, schema *or* data.

If we have a large data set and large data schema, elaborated RDF stores with integrated query engines alone are not very helpful. Due to the large data schema, users can hardly know which properties and identifiers are used in the knowledge base and hence can be used for querying. Consequently, users have to be guided when building queries and reasonable alternatives should be suggested. In this section, we present with the facet-based browsing in OntoWiki and our graph pattern builder two approaches to simplify navigation and querying of knowledge bases possessing a large information schema.

3.1 OntoWiki

To allow browsing of the extracted information in an intuitive manner, we adopted our tool for social semantic collaboration – OntoWiki [2]. It facilitates the visual presentation of a knowledge base as an information map, with different views on the instance data. It enables intuitive authoring of semantic content and fosters social collaboration aspects (however these features are not of primary interest for browsing the extracted Wikipedia content). Furthermore, OntoWiki enhances the browsing and retrieval by offering semantically enhanced search strategies. In particular the facet-based browsing implemented in OntoWiki allows to intuitively explore the extracted Wikipedia content.

Taxonomic structures give users only limited ways to access the information. Also, the development of appropriate taxonomic structures requires significant initial efforts. Only a very restricted taxonomic structure can be extracted from Wikipedia content by means of categories. As a pay-as-you-go strategy, facet-based browsing allows to reduce the efforts for a knowledge structuring, while still offering efficient means to retrieve information. To enable users to select objects according to certain facets, all property values (facets) of a set of selected instances are analyzed. If for a certain property the instances have only a limited set of values, those values are offered to restrict the instance selection further. Hence, this way of navigating through data will never lead to empty results. The analyzing of property values as well as the appropriate filtering and sorting of instances though can be very resource demanding. Since the respective optimizations in OntoWiki are not yet finished we deployed just an excerpt of the extraction for the film domain for demonstration at `http://wikipedia.aksw.org`. However, we aim to adopt and optimize OntoWiki further to function as easy to use browser for the complete semantic Wikipedia content.

3.2 Graph Pattern Builder

In addition to support browsing with OntoWiki we specifically developed a graph pattern builder for querying the extracted Wikipedia content. Users query the knowledge base by means of a graph pattern consisting of multiple triple patterns. For each triple pattern three form fields capture variables, identifiers or filters for subject, predicate and object of a triple. While users type identifier names into one of the form fields, a look-ahead search proposes suitable options. These are obtained not just by looking for matching identifiers but by executing the currently built query using a variable for the currently edited identifier and filtering the results returned for this variable for matches starting with the search string the user supplied. This method ensures, that the identifier proposed is really used in conjunction with the graph pattern under construction and that the query actually returns results. In addition, the identifier search results are ordered by usage number, showing commonly used identifiers first. All this is executed in the background, using the Web 2.0 AJAX technology and hence completely transparent for the user. Figure 2 shows a screenshot of the graph pattern builder.

3.3 Example Queries

In the previous sections, we have shown how to extract information from templates. We gave an impression about the sheer volume of knowledge we obtained

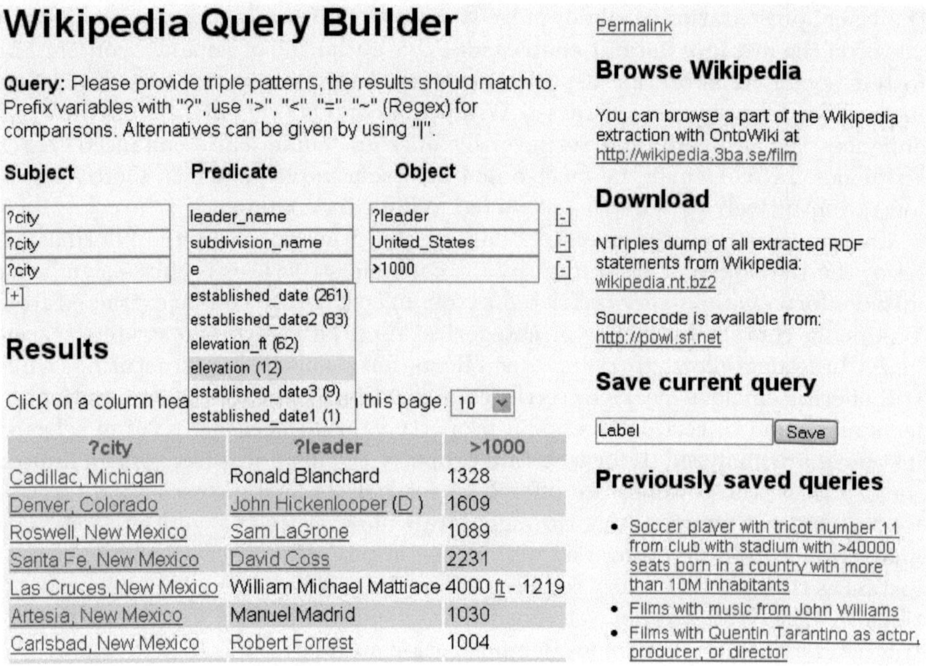

Fig. 2. Query interface of the Wikipedia Query Builder with AJAX based look-ahead identifier search

and suggested ways to improve templates to ease the conversion to RDF triples. The aim of this subsection is to present some example queries to justify our claim that we can obtain a huge amount of semantically rich information from Wikipedia – even in its current form.

Of course, reasonable queries can only involve a very small fraction of the information we obtained. We invite the reader to browse the obtained knowledge and pose example queries by visiting `http://wikipedia.aksw.org`.

The first query uses the film template. We ask for films starring an Oscar winner (as best actor) with a budget of more than 10 million US dollars[5].

```
SELECT ?film ?actor ?budget
WHERE {
   ?film p:starring ?actor;
         p:budget ?budget.
   ?actor a c:Best_Actor_Academy_Award_winners.
   FILTER regex(str(?budget),"1[0123456789]{6,}")
}
```

[5] The examples make use of the namespace prefixes `rdf` for RDF, `xsd` for XML-Schema data types, `p` for generated property identifiers, `w` for Wikipedia pages, `c` for Wikipedia categories and `u` for units.

The following table summarizes the result:

film	actor	budget
w:The_Da_Vinci_Code_(film)	w:Tom_Hanks	”125000000”^^u:Dollar
w:Ghost_Rider_(film)	w:Nicolas_Cage	”120000000”^^u:Dollar
w:Apocalypse_Now	w:Robert_Duvall	”31500000”^^u:Dollar
w:Jackie_Brown_(film)	w:Robert_De_Niro	”12000000”^^u:Dollar
w:Bobby_(2006_film)	w:Anthony_Hopkins	”10000000”^^u:Dollar
w:Confidence_(film)	w:Dustin_Hoffman	”15000000”^^u:Dollar
w:Apocalypse_Now	w:Marlon_Brando	”31500000”^^u:Dollar
w:The_Mission_(film)	w:Robert_De_Niro	”17218000”^^u:Dollar
w:The_Silence_of_the_Lambs	w:Anthony_Hopkins	”19000000”^^u:Dollar

Note the automatic detection of the budget unit (all US dollars in this case). The next query is more complex, involving different kinds of templates, namely soccer players, soccer clubs, and countries. We ask for soccer players with number 11 (on their jersey), who play in a club whose stadium has a capacity of more than 40000 people and were born in a country with more than 10 million inhabitants.

```
SELECT ?player ?club ?country ?capacity
WHERE {
   ?player p:currentclub ?club .
   ?player p:clubnumber 11 .
   ?player p:countryofbirth ?country .
   ?club p:capacity ?capacity .
   ?country p:population_estimate ?population .
   FILTER (?capacity > 40000) .
   FILTER (?population > 10000000)
}
ORDER BY DESC(?capacity) LIMIT 1000
```

The following table shows the result of the query:

player	club	country	capacity
w:Mehrzad_Madanchi	w:Persepolis_FC	w:Iran	90000
w:Cicinho	w:Real_Madrid	w:Brazil	80354
w:Ram%C3%B3n_Morales	w:Chivas_de_Guadalajara	w:Mexico	72480
w:Lukas_Podolski	w:FC_Bayern_Munich	w:Poland	69901
w:Gonzalo_Fierro	w:Colo-Colo	w:Chile	62000
w:Robin_van_Persie	w:Arsenal_F.C.	w:Netherlands	60432
w:Michael_Thurk	w:Eintracht_Frankfurt	w:Germany	52000
w:Stein_Huysegems	w:Feyenoord_Rotterdam	w:Belgium	51177
w:Mark_Gonz%C3%A1lez	w:Liverpool_F.C.	w:South_Africa	45362

Both queries are interesting and realistic. In both cases the results are probably not complete, because templates are still not used everywhere where appropriate or are badly designed (see Section 2.5). You can find more queries and build your own ones at `http://wikipedia.aksw.org`.

4 Related Work

The free encyclopedia Wikipedia has been tremendously successful due to the ease of collaboration of its users over the Internet [16]. The Wikipedia wiki is the representative of a new way of publishing [4] and currently contains millions of articles.

It is a natural idea to exploit this source of knowledge. In the area of Machine Learning [17] the Information Retrieval community has applied question answering [14], clustering, categorization and structure mapping to Wikipedia content. An XML representation of (most of) the Wikipedia corpus has been extracted [6] to support these tasks. Within the Semantic Web community this corpus has been used to extract common sense knowledge from generic statements [22] by mapping them to RDF.

In a different ongoing project the link structure and basic metadata of Wikipedia articles are mapped to a constantly updated RDF dataset, which currently consists of approximately 47 million triples [19]. Amongst other uses, the link structure has been exploited to build semantic relationships between articles to analyze their connectivity [5], which can improve the search capabilities within Wikipedia.

In contrast to the information extraction approaches mentioned above, i.e. full text extraction and link structure extraction, we focused on the extraction of Wikipedia templates. The corresponding algorithm, presented in Section 2.2, uses standard pattern matching techniques [1] to achieve this. We argued that the resulting RDF statements are a rich source of information contributing to existing results of knowledge extraction from Wikipedia. The goal of a related, but much more focused project is to extract personal data from the "Personendaten" template in the german Wikipedia[6].

Our research also fits within the broader scope of integrating Semantic Web standards with different sources of information like LDAP servers [7] and relational databases [3]. The integration of different data sources is considered an important issue in Semantic Web research [8].

Additionally, there are strong links to knowledge extraction from table structures. A general overview of work on recognizing tables and drawing inferences from them can be found in [27]. [21] is an approach for automatic generation of F-Logic frames out of tables, which subsequently supports the automatic population of ontologies from table-like structures. Different approaches for interpreting tables [12,13,25], i.e. deriving knowledge from them, have been tested for plain text files, images and HTML tables. Naturally, an additional difficulty of these approaches compared to the template extraction we perform, is to properly recognize tables (see e.g. [9,11,18,26] for table recognition techniques) and the relationships between entries in these tables using techniques like row labeling and cell classification [11,20]. Amongst other target formats for extraction, there has also been work on ontology extraction from tables [23], in particular for HTML tables [10]. Since templates in Wikipedia have a predefined structure,

---

6 http://de.wikipedia.org/wiki/Hilfe:Personendaten/Datenextraktion

our results are most likely more accurate than those one would obtain using more general table extraction approaches. (We are working on measuring the quality of the information we have extracted.) For this reason, we did not consider to apply general-purpose extraction of HTML tables in Wikipedia, but focused on the already structured templates.

5 Conclusions

As we outlined in the introduction, we created an approach for extracting information from Wikipedia and similar wiki systems, which is usable right now without further modifications on the MediaWiki software or expensive updates of Wikipedia content. We showed that we obtained a vast amount of useful machine processable information. Obstacles of our approach were discussed and suggestions for solving them have been given. They mostly involve common sense rules for template authors and reasonably small modifications of existing wiki software.

Further, we discussed the problem of querying and browsing the extracted knowledge. A simple and easy-to-use query engine for our extraction was developed and some example queries were presented to give the reader an intuition of the extracted knowledge. As with textual Wikipedia content the templates might contain incomplete or even wrong information. This cannot be detected by the extraction algorithm and prospective users should be aware of it. However, by providing a platform exhibiting the structured and interlinked content in Wikipedia, we were able to create one of the largest existing ontologies and hopefully a useful contribution to the Semantic Web in general. The presented approach was implemented for MediaWiki and evaluated with Wikipedia content. However, it can be similarly applied to different Wiki systems supporting templates and other means for handling frequently occurring structured content.

Finally, we want to solve the question from the title of this article for a commonality between Innsbruck and Leipzig: A fairly simple query on our extraction results reveals that both share Kraków as a twin town. However, this information can not be found on either of the Wikipedia pages and we are not aware of knowledge bases able to answer similar unspecific and domain-crossing queries.

Acknowledgments

This research was supported in part by the following grants: BMBF (SE2006 #01ISF02B), NSF (SEIII #IIS-0513778). We are grateful to Jörg Schüppel, who participated in the implementation of the extraction algorithm, and the anonymous reviewers for their suggestions.

References

1. A. Apostolico and Z. Galil, editors. *Pattern Matching Algorithms.* OUP, 1997. SEP.
2. Sören Auer, Sebastian Dietzold, and Thomas Riechert. Ontowiki - A tool for social, semantic collaboration. In Isabel F. Cruz, Stefan Decker, Dean Allemang, Chris Preist, Daniel Schwabe, Peter Mika, Michael Uschold, and Lora Aroyo, editors, *International Semantic Web Conference*, volume 4273 of *Lecture Notes in Computer Science*, pages 736–749. Springer, 2006.
3. Christian Bizer. D2R MAP - A database to RDF mapping language. In *WWW (Posters)*, 2003.
4. Bryant, Susan L., Andrea Forte, and Amy Bruckman. Becoming wikipedian: transformation of participation in a collaborative online encyclopedia. In *GROUP'05: International Conference on Supporting Group Work*, Net communities, pages 1–10, 2005.
5. S. Chernov, T. Iofciu, W. Nejdl, and X. Zhuo. Extracting semantic relationships between wikipedia categories. In *1st International Workshop: "SemWiki2006 - From Wiki to Semantics" (SemWiki 2006), co-located with the ESWC2006 in Budva, Montenegro, June 12, 2006*, 2006.
6. L. Denoyer and P. Gallinari. The Wikipedia XML Corpus. *SIGIR Forum*, 2006.
7. S. Dietzold. Generating rdf models from ldap directories. In C. Bizer S. Auer and L. Miller, editors, *Proceedings of the SFSW 05 Workshop on Scripting for the Semantic Web, Hersonissos, Crete, Greece, May 30, 2005.* CEUR Workshop Proceedings Vol. 135, 2005.
8. Dimitre A. Dimitrov, Jeff Heflin, Abir Qasem, and Nanbor Wang. Information integration via an end-to-end distributed semantic web system. In Isabel F. Cruz, Stefan Decker, Dean Allemang, Chris Preist, Daniel Schwabe, Peter Mika, Michael Uschold, and Lora Aroyo, editors, *The Semantic Web - ISWC 2006, 5th International Semantic Web Conference, ISWC 2006, Athens, GA, USA, November 5-9, 2006, Proceedings*, volume 4273 of *Lecture Notes in Computer Science*, pages 764–777. Springer, 2006.
9. Shona Douglas and Matthew Hurst. Layout and language: lists and tables in technical documents. In *Proceedings of ACL SIGPARSE Workshop on Punctuation in Computational Linguistics*, pages 19–24, jul 1996.
10. David W. Embley, Cui Tao, and Stephen W. Liddle. Automatically extracting ontologically specified data from HTML tables of unknown structure. In Stefano Spaccapietra, Salvatore T. March, and Yahiko Kambayashi, editors, *Conceptual Modeling - ER 2002, 21st International Conference on Conceptual Modeling, Tampere, Finland, October 7-11, 2002, Proceedings*, volume 2503 of *Lecture Notes in Computer Science*, pages 322–337. Springer, 2002.
11. J. Hu, R. S. Kashi, D. P. Lopresti, and G. T. Wilfong. Evaluating the performance of table processing algorithms. *International Journal on Document Analysis and Recognition*, 4(3):140–153, 2002.
12. M. Hurst. Layout and language: Beyond simple text for information interaction – modelling the table. In *Proceedings of the 2nd International Conference on Multimodal Interfaces, Hong Kong*, 1999.
13. M. Hurst. *The Interpretation of Tables in Texts.* PhD thesis, University of Edinburgh, 2000.

14. B. Katz, G. Marton, G. Borchardt, A. Brownell, S. Felshin, D. Loreto, J. Louis-Rosenberg, B. Lu, F. Mora, S. Stiller, O. Uzuner, and A. Wilcox. External knowledge sources for question answering. In *Proceedings of the 14th Annual Text REtrieval Conference (TREC2005), November 2005, Gaithersburg, MD*, 2005.
15. Markus Krötzsch, Denny Vrandecic, and Max Völkel. Wikipedia and the Semantic Web - The Missing Links. In Jakob Voss and Andrew Lih, editors, *Proceedings of Wikimania 2005, Frankfurt, Germany*, 2005.
16. Bo Leuf and Ward Cunningham. *The Wiki Way: Collaboration and Sharing on the Internet.* Addison Wesley, Reading, Massachusetts, apr 2001.
17. Thomas Mitchell. *Machine Learning.* McGraw Hill, New York, 1997.
18. Hwee Tou Ng, Chung Yong Lim, and Jessica Li Teng Koo. Learning to recognize tables in free text. In *ACL*, 1999.
19. System One. Wikipedia3. http://labs.systemone.at/wikipedia3, 2006.
20. David Pinto, Andrew McCallum, Xing Wei, Croft, and W. Bruce. Table extraction using conditional random fields. In *Proceedings of the 26th Annual International ACM SIGIR Conference on Research and Development in Information Retrieval*, IR theory, pages 235–242, 2003.
21. Aleksander Pivk, Philipp Cimiano, and York Sure. From tables to frames. *Journal of Web Semantics*, 3(2-3):132–146, 2005.
22. S. Suh, H. Halpin, and E. Klein. Extracting common sense knowledge from wikipedia. In *Proceedings of the ISWC-06 Workshop on Web Content Mining with Human Language Technologies*, 2006.
23. Yuri A. Tijerino, David W. Embley, Deryle W. Lonsdale, and George Nagy. Ontology generation from tables. In *WISE*, pages 242–252. IEEE Computer Society, 2003.
24. Max Völkel, Markus Krötzsch, Denny Vrandecic, Heiko Haller, and Rudi Studer. Semantic wikipedia. In Les Carr, David De Roure, Arun Iyengar, Carole A. Goble, and Michael Dahlin, editors, *Proceedings of the 15th international conference on World Wide Web, WWW 2006*, pages 585–594. ACM, 2006.
25. Xinxin Wang. *Tabular abstraction, editing, and formatting.* PhD thesis, Waterloo, Ont., Canada :University of Waterloo, Computer Science Dept.,, 1996.
26. Yalin Wang, Ihsin T. Phillips, and Robert M. Haralick. Table structure understanding and its performance evaluation. *Pattern Recognition*, 37(7):1479–1497, 2004.
27. R. Zanibbi, D. Blostein, and J. R. Cordy. A survey of table recognition: Models, observations, transformations, and inferences. *International Journal on Document Analysis and Recognition*, 7(1):1–16, mar 2004.

SALT - Semantically Annotated LaTeX for Scientific Publications

Tudor Groza, Siegfried Handschuh, Knud Möller, and Stefan Decker

DERI, National University of Ireland, Galway,
IDA Business Park, Lower Dangan, Galway, Ireland
{tudor.groza,siegfried.handschuh,knud.moeller,stefan.decker}@deri.org
http://www.deri.ie/

Abstract. Machine-understandable data constitutes the foundation for the Semantic Web. This paper presents a viable way for authoring and annotating Semantic Documents on the desktop. In our approach, the PDF file format is the container for document semantics, being able to store both the content and the related metadata in a single file. To achieve this, we provide a framework (SALT - Semantically Annotated LaTeX), that extends the LaTeX writing environment and supports the creation of metadata for scientific publications. SALT allows the author to create metadata concurrently, i.e. while in the process of writing a document. We discuss some of the requirements which have to be met when developing such a support for creating semantic documents. In addition, we describe a usage scenario to show the feasability and benefit of our approach.

1 Introduction

The vision of the Semantic Web, as well as the personal Semantic Desktop aims at integrated personal information management, at information distribution and collaboration. This will be enabled by the use of ontologies, semantic metadata (machine-understandable data) and Semantic Web protocols. Hence, semantic metadata constitutes the foundation for Semantic Web and Desktop. Authoring and annotating semantic documents on the desktop is one of the possible means to create semantic metadata.

This paper introduces a new way for authoring and annotating Semantic Documents on the Desktop. In our approach, the PDF file format is used as the container for document semantics, being able to store both the content and the related metadata in a single file. To achieve this, we provide a framework (SALT - Semantically Annotated LaTeX[1]) together with an associated ontology, that extends the LaTeX writing environment and supports the creation of metadata for scientific publications. SALT allows the author to create metadata while in the process of writing the content of a research paper.

Previous work in the creation of semantic metadata and annotation of documents has been mainly focused on the annotation of HTML documents for the

[1] Not to be confused with the SALT KA system by Marcus and McDermott.

E. Franconi, M. Kifer, and W. May (Eds.): ESWC 2007, LNCS 4519, pp. 518–532, 2007.

Semantic Web. Most of these HTML annotation tools [1,2,3] are following an *a posteriori* annotation approach. In order to provide metadata about the content of a web page, the author must first create the content and then annotate it as an additional, *a posteriori* step. This approach is reasonable, when the annotator is not the creator of the web document, as it is a common use case in the web. However, if author and annotator are the same person, the possibility arises to easily combine authoring of a document with the creation of the metadata describing its content. We will call this approach *concurrent* annotation. First steps towards this approach for HTML documents in a web context are described in Handschuh et al. [4], or for blogs in Möller et al. [5].

HTML is the document format for the web, and thus research on semantic annotations is mainly centered around it. However, another important format is PDF – the Portable Document Format. PDF can be seen at the moment as the *de facto* standard in terms of electronic publishing, especially in the research area. We observed that there exist a small number of solutions for creating semantic annotations on PDF documents, most of them following the *a posteriori* approach ([6]). In the case of *concurrent* annotations – to our knowledge – there is no clear defined approach. Also, when it comes to embedding the semantic annotations in the document itself, the existing support is poor and rarely used.

Adobe has defined the Extensible Metadata Platform[2] (XMP), a platform (methodology, schemas, tools, ...) for embedding RDF metadata in data files. XMP supports metadata in a broad variety of file formats, among them PDF. However, even though it is possible to embed arbitrary metadata in a PDF's XMP field, in practice only shallow DublinCore[3] descriptions are used. As a result, neither the inherent structure nor the semantic content of a document are reflected in the metadata.

Our approach proposes to extend the shallow metadata schemas currently used with a set of three ontologies which are able to capture the structural information of the document as well as the semantics of its content. The three ontologies are (i) the *Document ontology*, (ii) the *Rhetorical ontology* and (iii) the *Annotation ontology*. All three will be discussed in more detail in Sect. 3.1.

We support our proposal with a method for creating *concurrent* semantic annotations for PDF documents, by exploiting the rich environment provided by LATEX. The annotation process takes place while writing and the actual integration is realized at syntax level by exploiting regular LATEX commands plus a series of newly introduced special annotation commands. The final result is a semantically enriched PDF document encapsulating instances of the afore-mentioned ontologies together with associated visual annotations. We believe that the ontologies presented in our proposal can be used independently of the format used for the scientific publications. Therefore, we intend to use the current approach as a proof of concept and extend our investigations to other formats in the near future.

---

[2] Adobe Systems Incorporated - XMP. `http://www.adobe.com/products/xmp/`

[3] DublinCore Metadata Initiative. `http://dublincore.org/`

In Sect. 2 we will present the automatic creation of online-proceedings as a use case for our framework. Then, we describe a modularization of the used ontologies and define the support for creating annotations, i.e. the annotation syntax (Sect. 3). In Sect. 4 we give an overview of the annotation process and revisit the proposed use case from the implementation point of view. Before concluding, we present a discussion of the proposed solution in Sect. 5, give an overview of the related work in Sect. 6 and discuss some aspects of our solution in Sect. 7.

2 Use Case

An increasing number of applications make use of metadata contained in PDF documents, or otherwise analyze the document's content. The type of functionality offered by such applications is varying from Personal Information Management (e.g. Gnowsis [7]) or searching (e.g. Beagle++ [8]) to digital libraries (e.g. JeromeDL[9]). All these applications have in common: (i) that they either use the limited metadata captured by the DublinCore elements present in the XMP field – which offers only shallow information about the document, or (ii) they perform full-text indexing in order to maximize the searching capabilities. Even though richer semantic annotations are in theory possible (and also in practice, as we show in this paper), they are currently not used. As a result, none of the applications mentioned can make use of them.

We believe that by using semantic PDF documents (i.e. PDF documents encapsulating rich RDF, e.g. instances of our ontologies), all the afore-mentioned applications would bring more value to the user: more accurate results, better visualization, etc. In order to provide an example, we will describe how such semantic documents enable an easy, low-effort information distribution, collaboration and integration for the purpose of an innovative online workshop proceedings. The goal is not only to ease the process of creation of the online proceedings, but also provide added value to the reader of these proceedings.

The process for the online publication of accepted workshop papers is usually done manually. The editor typically creates a list containing the authors and the titles and afterwards they link the corresponding PDF document to it. However, additional information can easily be retrieved given that each author would use our framework while writing the scientific publication. SALT enables a combination of automatically retrieved annotations based on i) the analysis of the used LaTeX commands, ii) the rhetorical structure of the document and iii) the arbitrary annotations included in the document.

For our use case, we took the following approach: we first create an individual HTML page for each annotated paper (cf. Fig. 1). The rich annotations in each paper can be visualized and exploited for navigation in many different ways. In our example, we chose to present each paper in such a way that the focus is on the linear structure, including information regarding the rhetorical structure. In addition, the page also contains some simple metadata associated with the publication (such as title, authors, etc), the link to the PDF document and if desired even the original instances of the ontologies associated with the publication.

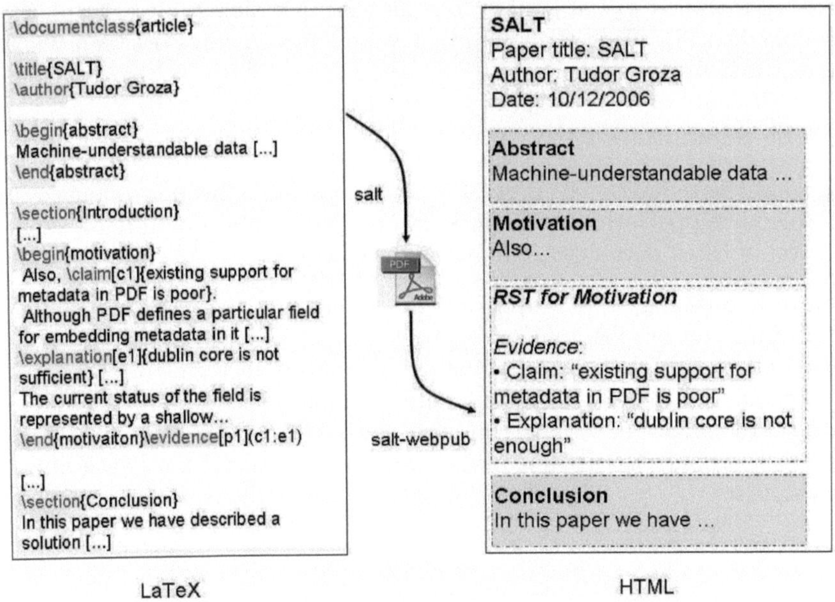

Fig. 1. HTML creation from annotated paper

The second phase of the process iterates over all the created pages and generates an entry point in the form of an index page. The index page gives a short overview of all papers, but more information — generated from the metadata — is available. Readers can quickly glance through the contribution and skip to the section they are interested in.

3 Ontological Foundation and Syntactical Support

There are two types of annotations that can be embedded in PDF documents: (i) visual annotations in the form of notes, bookmarks or markups and (ii) arbitrary metadata in the XMP field. Our proposal for the creation of Semantic Documents is exploiting and extending both possibilities. In the following, we will present both a semantic foundation — a set of three ontologies — and a means to express those semantics in an extended LaTeX syntax.

The semantic layer consists of three ontologies: document ontology, annotation ontology and rhetorical ontology. Instances of these ontologies will be placed in the XMP field, thus extending typical current use of PDF XMP, and providing a much richer environment for capturing the document's semantics.

The syntactical implementation proposes an enrichment of the LaTeX syntax. This is done by considering the existing commands and performing analysis and metadata extraction on them, and by introducing a series of new commands. These commands provide the support for creating rhetoric elements, creating implicit and explicit visual annotations and for inserting arbitrary annotations in the document. In effect, the semantic layer creates a bridge between the actual document and its metadata.

3.1 The Semantic Layer

The goal of the semantic layer (see Fig. 2) is to define a proper semantic framework able to support the entire annotation process. As a result, we created a federation of three ontologies, enumerated as follows:

Document ontology [4] – Capturing the internal structure of the document (sections, paragraphs, sentences, etc).

Rhetorical ontology [5] – Modelling the document in terms of rhetorical elements and rhetorical structure (claims, evidence, etc).

Annotation ontology [6] – Creating the bridge between the rhetorical structure and the ordinary structure. It also captures additional metadata about the document.

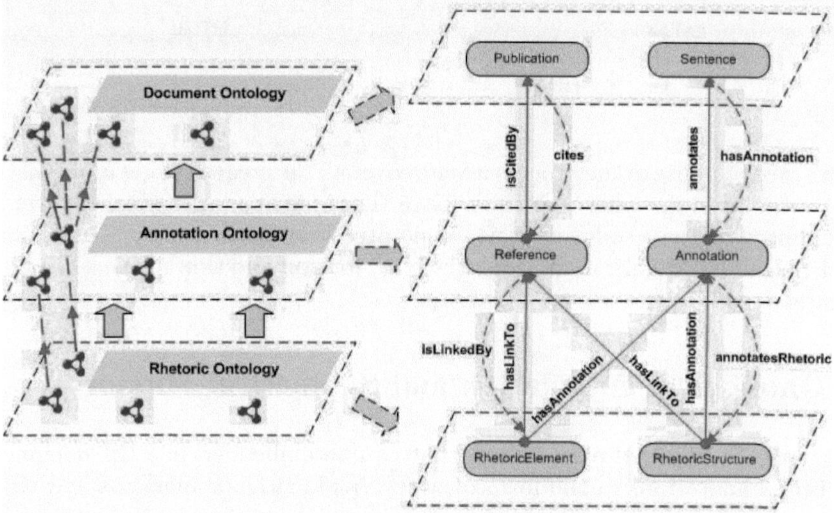

Fig. 2. Ontology Layers

The Document Ontology. The document ontology, depicted in see Fig. 3, captures the structural layout of the document and provides hooks to its annotated parts. The motivation behind the current level of decomposition is given by the need of instantiating the annotated parts of the text. The sentence

[4] http://salt.semanticauthoring.org/onto/2006/12/document-ontology.rdfs
[5] http://salt.semanticauthoring.org/onto/2006/12/rhetoric-ontology.rdfs
[6] http://salt.semanticauthoring.org/onto/2006/12/annotation-ontology.rdfs

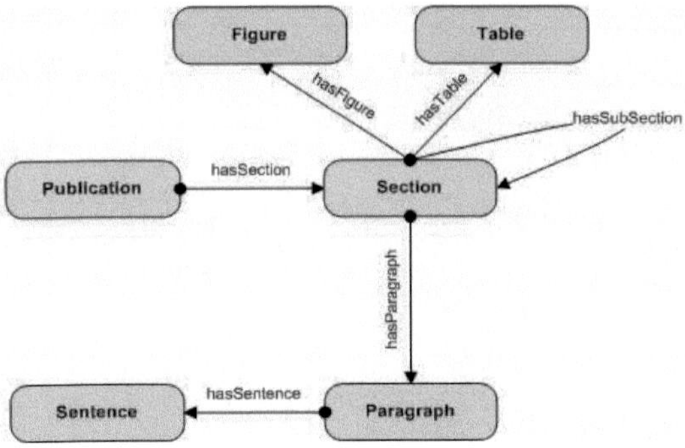

Fig. 3. The Document Ontology schema

currently represents the finest granularity of physical structure. However, they are mapped to specific substrings of a document using the `Annotation` class, which also allows to map arbitrary sub-phrases of a sentence.

The Rhetorical Ontology. The rhetorical structure ontology (see Fig. 4) represents a union of (i) the knowledge captured by the rhetorical relations within the text, (ii) the rhetorical structure modeling the positioning of the contained information chunks and (iii) the argumentative support providing the mean for building a stable foundation for the rhetoric elements. In the following, we will analyze the three parts of the ontology.

The first part of the ontology (*Rhetorical Relations*) deals with modeling the information chunks present in the document as rhetoric elements. This approach has its roots in the Rhetoric Structure of the Text (RST) theory [10], which describes the text in terms of the rhetoric relations existing between a Nucleus (modeled by us as the *Claim*) and a Satellite (in our case, the *Explanation*). Although the theory contains around 30 such relations, we currently only consider those that seem most relevant when annotating scientific documents (e.g. *Antithesis*, *Concession* or *Means*). The main role of these rhetoric relations (modeled as concepts) is to provide a reason for the existence of claims and explanations in the document. Furthermore, we considered their placement in the frame created by the rhetorical structure (captured by the second part of the ontology) as a natural integration and thus we introduced a relation between the rhetorical relation concept and rhetorical structure concept.

The second part of the ontology (*Rhetorical Structure*) takes care of capturing the rhetorical structure of the document. It represents an extension of the ABCDE format for the annotation of scientific papers [11]. ABCDE stands for: **A**nnotation, **B**ackground, **C**ontribution, **D**iscussion, **E**ntities. In SALT, we build on ABCDE, but propose a more comprehensive and fine-grained set of concepts. The simple metadata like title and authors is covered by the ***A**nnotation* concept in ABCDE. In SALT, this is covered elsewhere (see the next section), which is why our **A** is the ***A**bstract* of the document. Furthermore, we extend ABCDE with the concepts *Motivation*, *Scenario* and *Conclusion*. Finally, the

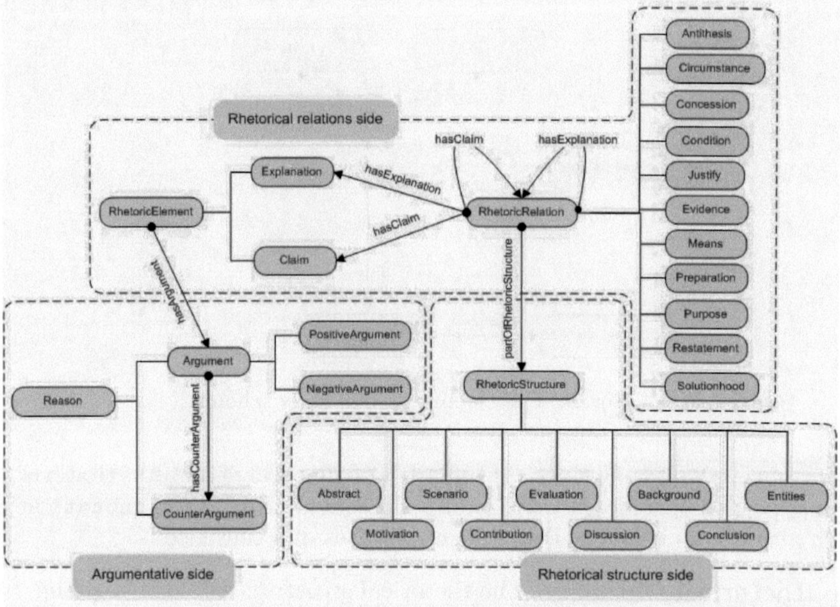

Fig. 4. The Rhetorical Ontology schema

Argumentative part of the ontology allows the further modeling of scientific discourse in the form of *Arguments* and *Counter Arguments.*

The Annotation Ontology. The main role of the annotation ontology (see Fig. 5) is to create the link between the document ontology and the rhetorical ontology. Conceptually, the rhetorical structure represents an annotation of the physical structure. Thus, one is able to enrich the document with rhetoric elements by attaching semantic annotations to it. In ontological terms, this would translate to creating instances of the *Annotation* concept and attaching them to the appropriate parts of the text.

A second role of the ontology is to provide metadata about the publication as a whole. This part can be seen as an alignment to the DublinCore initiative and to the SWRC ontology [12], as each of the concepts corresponds directly to a DublinCore element or to a concept of the SWRC ontology.

3.2 Syntactical Support

Providing a syntactical implementation of the ontologies discussed above is done in two ways: the extraction of metadata from existing LaTeX commands and the introduction of new commands. The new set of commands was kept small in order to avoid a steep learning curve for new SALT users. Functionality is extended in three directions:

Insertion of arbitrary annotations. This possibility was introduced in order to allow the authors to freely insert arbitrary metadata about the publication

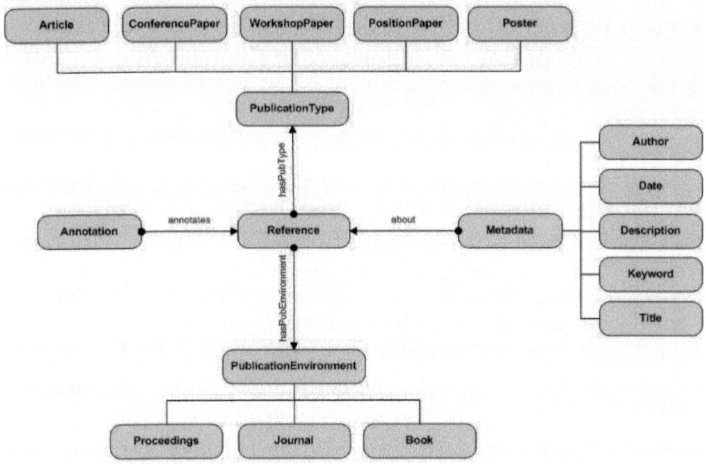

Fig. 5. The Annotation Ontology schema

by using the N3 notation[7]. As body of the *N3* command, one can insert valid N3 statements, for example by referring to a specific domain ontology.

Creation of rhetoric elements. We allocated a special command for each type of rhetorical relation and a special environment for each type of rhetorical structure. As a foundation for all these, there exist also the commands for creating the basic rhetorical elements, i.e. the **Claim** and the **Explanation**. Here are some command examples: *claim*, *explanation*; rhetorical relations: *antithesis*, *concession*; rhetorical environments: *begin*{*motivation*} ...\end{motivation}.

Explicit creation of visual annotations. The author themselves can create visual annotations by using the *note* command, which has three parameters: the subject, the author and the content of the visual annotation. These annotations will e.g. show up as a little post-it in the PDF rendering.

A more detailed description of all the concepts present in the ontology, as well as of the annotation syntax can be found on the SALT web page: `http://salt.semanticauthoring.org/`

4 Annotation and Publishing

We implemented SALT and the workshop online proceedings publication scenario as two independent applications. SALT itself can be used stand-alone from the command line, or can be integrated in different LaTeX editors. For example, we integrated it in Kile[8]. However, one can integrate it in any editor which provides the flexibility of choosing a custom LaTeX - PDF compiler, not the implicit

[7] http://www.w3.org/DesignIssues/Notation3

[8] http://kile.sourceforge.net/

one. The second application, called SALT-WebPub, is a stand-alone application with an easy to use graphical user interface. In the following we will detail both applications separately.

4.1 The SALT Process

The SALT application is responsible for analyzing the annotations and embedding the ontology instances into the resulting PDF document. In order to create the final document, a series of processing steps need to be performed:

Syntactic analysis and annotation extraction. As a first step, the syntax tree of the LaTeX document is searched for elements which will add to the document metadata (both ordinary commands and new commands defined by SALT). The result are two separate metadata graphs representing both the physical document structure (according to the document ontology) and the rhetorical structure (according to the rhetorical ontology).

Annotation analysis and ontology population. In this step, both metadata graphs are joined. Also in this step the arbitrary RDF triples extracted from the N3 command are added to the graph.

PDF document compilation. In the final step, the PDF document is created using an ordinary PDFLatex compiler (the user can choose which compiler to use). Afterwards, the complete metadata graph is added in the document's XMP field, and visual annotations (notes, etc) are added.

4.2 The Publishing Process

SALT-WebPub, the publishing application, takes as input a list of semantic PDF documents and generates a set of corresponding HTML files, together with the associated index. In order to provide flexibility to the format of the resulting HTML files, we let the user specify a template for the page associated with each publication and a template for the index file. This way, it is possible to customize the presentation of the online proceedings without affecting the web page content generation.

The process of generating generating the proceedings from the PDF documents is split into a series of steps. The first step is to extract the ontology instances out of the document. The second step is to interpret the extracted metadata and to prepare it for the final output format. The last step creates the associated HTML page by taking the user's template and filling it with the output from the previous step. Finally, the index page is created, based on the information extracted from each individual document.

5 First Experiences

To get a better idea of how SALT works "in real life", we performed a test with a group of six authors from this the SAAW2006 workshop[9]. Together with the

[9] `http://saaw2006.semanticweb.org`

authors, we annotated their LaTeX source code and generated semantic PDF documents using our tools. Taking all documents, we produced a richly annotated online proceedings[10]. Both the author's feedback and our own observations are summarized in the following discussion.

Ontological foundation – The three ontologies discussed in Sec. 3.1 did undergo small modifications as a result of the evaluation. For a more comprehensive capturing of the shallow metadata about scientific publications we felt the necessity of adding several concepts in the *Annotation Ontology*, like *PublicationType* or *PublicationEnvironment*. Also, in order to build the support for creating semantic network between the annotated documents, we had to introduce the *Reference* concept in the same ontology and link the rhetorical elements and the rhetorical structure to it. Regarding the *Rhetorical Ontology*, the only necessary modification needed was to allow the rhetorical relations to act as rhetorical elements in more complex relations. This modification provides a better degree of flexibility and allows the creation of rhetorical structure of text trees (one of the possible views over a Semantic Document – see Sect. 2). The ontology layer as a whole was seen as comprehensive enough to capture both shallow metadata and the content's semantics.

Annotation syntax – The proposed LaTeX syntax was very well received. The number of newly introduced commands was small enough not to create any significant extra workload on the authors. Based on this, we intend to leave the syntax untouched. Some small modifications were necessary, in order to reflect the actual status of the ontologies.

A general issue that we discovered was that we need to take more into consideration the semantics of the existing LaTeX commands and the overall structure of the LaTeX documents. Therefore, our framework will support automatic information extraction from bibliographical items and from citing commands. In this way it will automatically create possible relations between instances present in the currently annotated scientific publication and the cited ones.

6 Related Work

As already mentioned, our ontologies have their roots in the Rhetorical Structure of Text (RST) Theory [10]. The paper provides the underlying semantics of the concepts modelled by the theory together with their definitions. A second publication by the same authors [13] provides a deep analysis of the application domains in which RST was used until a certain point in time. It is interesting to observe that the mentioned range of domains varies from computational linguistics, cross-linguistic studies and dialogue to multimedia presentations.

A similar approach is presented by Tempich et. al in [14]. The DILIGENT *Argumentation Ontology* was designed in line with the terminology proposed by

[10] `http://salt.semanticauthoring.org/experiment/saaw2006/`

the IBIS methodology [15] and captures the argumentative support for building discussions in DILIGENT processes. In DILIGENT, the argumentative support is equivalent to one of the three parts of our Rhetorical Ontology and so is less expressive. Uren et. al [16] describe a framework for sensemaking tools in the context of the Scholarly Ontologies Project. Their starting point is represented by the requirements for a discourse ontology, which has its roots in the CCR (Cognitive Coherence Relations) Theory and models the rhetorical links in terms of similarity, causality and challenges. Although the ontological foundation is very similar, the application approach is different (see below).

In terms of applications, we found the approach by Peter et al. [17] to be one of the most interesting ones in terms of similarity with our research. Their goal is to extract semantics from a LaTeX document content based on the references and index present in the document (for example *see* and *see also* references). We have a similar approach when it comes to extracting the structural information, but our focus is more oriented on the rhetorical structure of the text and the semantic links between claims placed in different documents.

MMISS [18] fits into the category of using LaTeX as a development environment. The project aims at building an internet-based, adaptive multimedia educational system. Based on a series of custom LaTeX commands they are able to build ontologies and semantically link the resulting lecture slides (via a central repository). SALT also uses custom LaTeX commands to semantically annotate the document, the difference being that we embed the annotations in their natural environment (i.e. the resulting PDF document) and not in a central storage place. This is also one of the main differences when compared to the system developed by Uren et. al [16]. Their goal is to create and visualize claim networks using scholarly documents (represented as HTML files) using a central knowledge server. One of our goals is also to create such knowledge networks, but using active reference embedded in the semantic document.

Another interesting system is described by Geurts et al. [19]. It models the process of transforming semantic graphs into multimedia presentations, using domain knowledge and discourse analysis. Their work is focussing more on using parts of the text for presentation purposes. SALT on the other hand provides a method for enriching the normal documents with semantic annotations, based also on discourse analysis. However, in their approach it is not very clear how the knowledge base is structured and how they chose the domain knowledge effectively.

7 Discussion

In this section we will discuss a number of relevant issues that appeared while researching the concepts presented before focussing on: i) annotation instance generation and maintenance and ii) object identification and reference.

Annotation instance generation and maintenance refers to the mechanism of generating and mapping the ontology instances which annotate information chunks to the actual content present in the document. This issue is especially

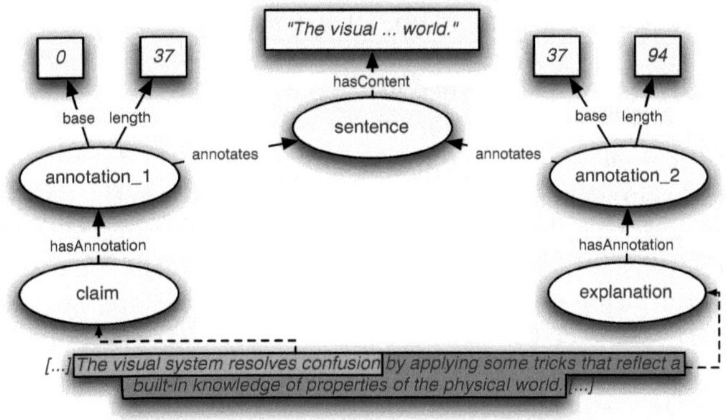

Fig. 6. Metadata Graph for a Sentence

sensitive when it comes to the document structure objects. In order to have a clear view over the subject, we will compare possible solutions for HTML and PDF documents.

In the case of HTML, if we would like to create an annotation instance and connect that instance to the piece of text being annotated, we could do this directly by referencing an element within the documents DOM tree. Thus, a simple pointer solves the problem, which is not the case for PDF documents. Although the internal organization of the document is represented by a tree of complex objects and streams, referencing inside this tree is not straightforward. The reasons are mainly related to accessing rights, image analysis or text retrieval algorithms' accuracy. At the same time, we also have to consider the fact that we're dealing with *concurrent* annotations, which makes the situation even more complex. Because the annotation process is interleaved with the writing process in the LaTeX environment, the targeted PDF document does not even exist yet.

Our current approach solves this issue by creating an instance for every annotated information chunk, the finest granularity being part of sentence. To give an example, consider the sentence in Fig. 6: The first part of the sentence is annotated as a claim, and the second part as an explanation. SALT will now (i) create a *Sentence* instance for the entire sentence, having the *hasContent* property set to the sentence's content, (ii) create a *Claim* instance, (iii) an *Explanation* instance and (iv) two *Annotation* instances connecting the claim and the explanation to the sentence. Finally (v) the *base* and *length* of the annotation instances are set to reflect their position in the sentence.

Obviously this solution presents two disadvantages: on one side, it increases the space of the document (linearly by the number of annotated sentences), and on the other side, it generates redundancy. In order to correct this issue, we intend to implement the XPointer Framework [20] applied for our case. The result will replace the actual content of a sentence with pointers in the PDF document to it, and thus the redundancy will be eliminated and the document space decreased, but it will increase the complexity of the metadata analysis process, since it will introduce a pre-PDF-creation and a post-PDF-creation analysis step.

The second discussion issue which we would like to raise is the object identification and reference. One of our goals is to be able to create references between different rhetorical elements placed in different semantic documents, and to be able to provide arguments and counter-arguments based on the ontology support. In order to achieve this, the first step that we took is to impose the definition of a unique identifier for each rhetorical element present in the document. Thus, the identification inside one document is solved. There are two problems that appear now: (i) how can the actual reference between rhetorical elements be realized and (ii) what happens if there are two versions of the same document between which the element identification is different.

A possible solution for the first issue could be to impose the presence of a valid URL pointing to the original document for each cited publication. Thus, when referencing a rhetoric element in a particular publication, for example *[Handschuh2006]#claim1*, it will be possible to create a valid reference by resolving *[Handschuh2006]* to e.g. *http://example.org/handschuh2006.pdf*.

This solution brings us to the second problem. Usually a publication resides in more than one place, and there are cases in which the version of the publication differs from one place to another. The simplest solution to solve the issue would be not to care about the version. When citing a document, the author would provide a direct link to that document, and thus, all the references will be created based on that document, and presuming that the author realizes the referencing correctly. Of course, one could continue the discussion and raise another issue, i.e. what happens with journal articles and copyright issues regarding them, but we will tackle this point in future.

8 Conclusion and Future Work

In this paper we have described a solution for authoring and annotation of semantic documents. SALT leaves the semantic data where it can be handled best: within the document. Also, it provides a means to create Semantic Documents in a simple and intuitive way for LaTeX authors.

To attain this objective, we have defined the SALT process, the appropriate ontologies and the architecture of the application. We have incorporated the means for rhetorical markup of a document that allows the scientific authors to explicitly markup their contribution, the claims they made and the support for their claims. The framework brings added value to the applications using PDF documents and to the users, as shown in our online proceedings scenario, where we used it to automate the presentation and improve the navigation of scientific publications. Used in this way, SALT could also be integrated in the workflow of generating the semantic metadata for conferences such as ESWC or ISWC, a process that can be long and tedious if performed manually.

For the future, there is a list of open issues concerning the authoring of semantic PDF documents that we will consider: (i) PDF referencing or creation of semantic knowledge networks by means of PDF documents and using active references, as we described it in Section 7, (ii) integrating the framework with existing (semantic) digital libraries and semantically-interlinked online communities and (iii) automatic derivation of markup. We believe that these options make SALT a good approach for the authoring of scientific semantic documents.

Our ultimate goal is to convince the community about the value that semantic documents bring and to transform the ontological framework into a *de facto* standard for annotating scientific publications. Thus, the next step that we will take is to perform an intensive evaluation phase with researchers coming from different backgrounds. This phase will provide us with both a better understanding of the weak points in our framework and with a perfect environment for improvement.

Acknowledgements. This work has been funded by the European Commission 6th Framework Programme in the context of the NEPOMUK IP - The Social Semantic Desktop, FP6-027705. We would like to thank Alexander Schutz for the fruitful discussions.

References

1. Handschuh, S., Staab, S., Maedche, A.: CREAM — creating relational metadata with a component-based, ontology-driven annotation framework. In: The First International Conference on Knowledge Capture (K-Cap 2001), Victoria, B.C., Canada (2001) 76–83
2. Vargas-Vera, M., Motta, E., Domingue, J., Lanzoni, M., Stutt, A., Ciravegna, F.: MnM: Ontology Driven Semi-Automatic and Automatic Support for Semantic Markup. In: The 13th International Conference on Knowledge Engineering and Knowledge Management, Sigüenza, Spain (2002) 379–391
3. Ciravegna, F., Dingli, A., Petrelli, D., Wilks, Y.: User-system cooperation in document annotation based on information extraction. In: The 13th International Conference on Knowledge Engineering and Knowledge Management. (2002) 122+
4. Handschuh, S., Staab, S.: Authoring and annotation of web pages in CREAM. In: 11th International World Wide Web Conference, WWW 2002, Honolulu, Hawaii, ACM Press (2002) 462–473
5. Möller, K., Bojārs, U., Breslin, J.G.: Using semantics to enhance the blogging experience. In: The third European Semantic Web Conference (ESWC2006), Budva, Montenegro (2006)
6. Eriksson, H.: A PDF storage backend for Protege. In: Proceedings of the 9th Protege International Conference, Stanford, California, USA (2006)
7. Sauermann, L.: The Gnowsis Semantic Desktop for information integration. In: The 1st Workshop on Intelligent Office Appliances: Knowledge-Appliances in the Office of the Future (IOA 2005), at WM 2005, Kaiserslautern, Germany (2005)
8. Brunkhorst, I., Chirita, P.A., Costache, S., Gaugaz, J., Ioannou, E., Iofciu, T., Minack, E., Nejdl, W., Paiu, R.: The Beagle++ toolbox: Towards an extendable desktop search architecture. Technical report, L3S Research Centre, Hannover, Germany (2006)
9. Kruk, S.R., Decker, S., Zieborak, L.: JeromeDL - adding semantic web technologies to digital libraries. In: DEXA 2005, Copenhagen, Denmark (2005)

10. Taboada, M., Mann, W.C.: Rhetorical structure theory: looking back and moving ahead. Discourse Studies **8, No. 3** (2006) 423–459
11. de Waard, A., Tel, G.: The ABCDE format - enabling semantic conference proceeding. In: Proceedings of 1st Workshop: "SemWiki2006 - From Wiki to Semantics" at ESWC2006, Budva, Montenegro (2006)
12. Sure, Y., Bloehdorn, S., Haase, P., Hartmann, J., Oberle, D.: The SWRC ontology - semantic web for research communities. In: Proceedings of the 12th Portuguese Conference on Artificial Intelligence (EPIA 2005), Covilha, Portugal (2005)
13. Taboada, M., Mann, W.C.: Applications of rhetorical structure theory. Discourse Studies **8, No. 4** (2006) 567–588
14. Tempich, C., Pinto, H.S., Sure, Y., Staab, S.: An Argumentation Ontology for Distributed, Loosely-controlled and evolvInG Engineering processes of oNTologies (DILIGENT). In: The Second European Semantic Web Conference, (ESWC 2005), Heraklion, Crete, Greece (2005) 241–256
15. Kunz, W., Rittel, H.: Issues as elements of information system. Working paper 131, Institute of Urban and Regional Development, University of California (1970)
16. Uren, V., Shum, S.B., Li, G., Bachler, M.: Sensemaking tools for understanding research literatures: Design, implementation and user evaluation. Int. Jnl. Human Computer Studies **64, No.5** (2006) 420–445
17. Peter, H., Sack, H., Beckstein, C.: Document indexing - providing a basis for semantic document annotation. In: XML-Tage 2006, Berlin (2006)
18. Krieg-Brückner, B., Lindow, A., Lüth, C., Mahnke, A., Russell, G.: Semantic interrelation of documents via an ontology. In: Proceedings of DeLFI 2004: Die 2. e-Learning Fachtagung Informatik, Paderborn, Germany (2004)
19. Geurts, J., Bocconi, S., van Ossenbruggern, J., Hardman, L.: Towards ontology-driven discourse: From semantic graphs to multimedia presentations. Technical report, Centrum voor Wiskunde en Informatica (INS-R0305) (May 31, 2003)
20. DeRose, S., Maler, E., Jr., R.D.: XPointer xpointer() scheme (2002) `http://www.w3.org/TR/xptr-xpointer/`.

Annotating Relationships Between Multiple Mixed-Media Digital Objects by Extending Annotea

Ronald Schroeter, Jane Hunter, and Andrew Newman

School of ITEE, The University Of Queensland
{ronalds,jane,anewman}@itee.uq.edu.au

Abstract. Annotea provides an annotation protocol to support collaborative Semantic Web-based annotation of digital resources accessible through the Web. It provides a model whereby a user may attach supplementary information to a resource or part of a resource in the form of: either a simple textual comment; a hyperlink to another web page; a local file; or a semantic tag extracted from a formal ontology and controlled vocabulary. Hence, annotations can be used to attach subjective notes, comments, rankings, queries or tags to enable semantic reasoning across web resources. More recently, tabbed browsers and specific annotation tools, allow users to view several resources (e.g., images, video, audio, text, HTML, PDF) simultaneously in order to carry out side-by-side comparisons. In such scenarios, users frequently want to be able to create and annotate a link or relationship between two or more objects or between segments within those objects. For example, a user might want to create a link between a scene in an original film and the corresponding scene in a remake and attach an annotation to that link. Based on past experiences gained from implementing Annotea within different communities in order to enable knowledge capture, this paper describes and compares alternative ways in which the Annotea Schema may be extended for the purpose of annotating links between multiple resources (or segments of resources). It concludes by identifying and recommending an optimum approach which will enhance the power, flexibility and applicability of Annotea in many domains.

Keywords: Annotea, Annotation, Semantic Web, Relationships.

1 Introduction

Simple Web annotation tools for annotating individual web objects have existed for over ten years [1, 2].

They began with annotation tools for attaching comments to web pages and textual documents, but then expanded to images and video, audio and 3D objects as more multimedia content was published on the Web. More recently, as many communities have formed online collaborative groups, annotation tools have transformed from asynchronous to synchronous - enabling real-time online discussions about resources. Figure 1 illustrates the evolution of annotation tools over the past ten or so years.

E. Franconi, M. Kifer, and W. May (Eds.): ESWC 2007, LNCS 4519, pp. 533–548, 2007.

In the past year, we have observed yet a new phase in the demands of users with respect to annotation tools. Our observation is related to the establishment of more online communities, who have established a consensus on exchangeable data standards, terminologies (defined through mark-up languages and machine-processable ontologies) and who want to be able to share and compare overlapping and related resources of many types. These resources may be of many media types (images, video, audio, multimedia, 3D), associated with specific disciplines (e.g., scientific models) or may comprise XML files used to represent shareable, exchangeable objects (e.g., scientific workflows).

To summarize, communities have been voicing a demand for annotation tools that enable a combination of the following:

1. The specification of links between whole objects or segments within objects and annotation of these links [3];
2. Support for annotating links between objects of the following types: images, video, audio, text, HTML, PDF, 3D objects and XML files;
3. Viewing of more than one object simultaneously to enable side-by-side comparison and association;
4. Annotations that are based on domain-specific terms from either controlled vocabularies or (OWL) ontologies. This enhances the ability for other application programs to process the annotations;
5. The ability to share these comparative interpretations and associations amongst communities of users through shared annotation servers, using a protocol such as Annotea.

Some specific examples that we have seen through our eResearch collaborations with different scientific communities include:

- In the humanities, film/media researchers want to link, compare and annotate segments between books, screenplays and different films and film versions;
- In molecular biology, researchers want to be able to relate and compare 3D protein structures – to discuss protein-protein docking interactions and protein function;
- In the geosciences, geologists want to be able to compare and annotate different types of computational models with still photos and videos of earth quakes.

The remainder of this paper is structured as follows. Section 2 provides an overview of previous related work and a description of the Vannotea tool developed by the authors, which has been the driver for the work described in this paper. Section 3 describes the existing Annotea protocol and the advantages of extending this to support new user demands. Section 4 describes extensions to Annotea to support machine-processable annotations (based on ontologies). Section 5 describes different possible approaches to extending Annotea to support the annotation of links between multiple objects. Section 6 concludes with a recommendation for the optimum approach and future work.

2 Previous Work

Significant previous work has focussed on the development of annotation tools. Fig. 1 provides a 3D classification of existing annotation systems – classified according to:

- Annotation level (x-axis) – from simple free-text annotations and tagging, to the attachment of local files and URLs, and more controlled annotations based on simple vocabularies and ontologies.
- Content type (y-axis) – text, HTML, images, video, audio and 3D objects.
- Number of simultaneous resources (z-axis) - the ability to compare multiple files and annotate links between them.

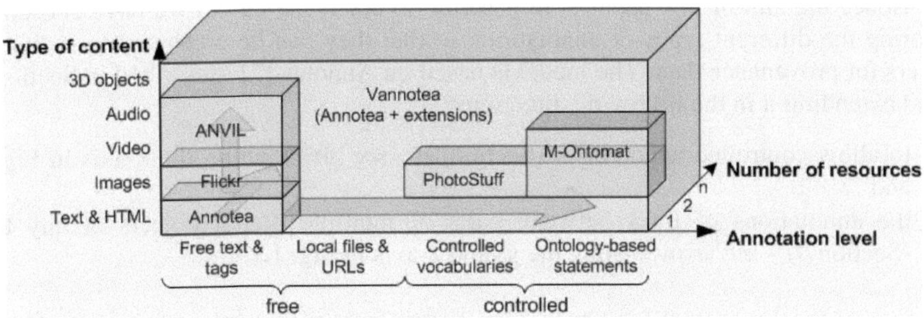

Fig. 1. Annotation tools

2.1 Free Annotation Tools

Examples of simple free-text annotation tools are depicted in the left column of Fig. 1. They include tools that are based on Annotea without extensions such as Annozilla[1] and Amaya[2]. Flickr[3] is an online photo management and sharing application. It allows users to upload their photos and freely annotate them. ANVIL [4] is a stand-alone tool which allows free-text annotations of audio and video files. Many more tools could be mentioned here, but the focus of this paper is on annotation systems that support ontology-based annotation of links between multiple web-accessible digital resources.

2.2 Semantic Annotation Tools

Systems that support controlled vocabulary-based and ontology-based annotations of multimedia objects include the following:

PhotoStuff [5] is a tool that allows users to highlight regions within images, create instances from any ontology through sophisticated forms and link the instance to the region of the image. The users are able to perform the semantic annotation locally and then upload the RDF instance to a central database, where the RDF file - and

[1] http://annozilla.mozdev.org
[2] http://www.w3.org/Amaya
[3] http://www.flickr.com

therefore the whole graph including multiple instance statements - is then attributed to the user through his/her user account and time stamped for provenance data.

The M-Ontomat-Annotizer [6] provides ontology-based image and video frame (and region) annotation. This tool also supports initialization and linking of RDF/S domain ontologies with low-level MPEG-7 visual descriptors.

Vannotea [3] is a collaborative tool that enables fine-grained annotation of objects of any media type, where the annotations themselves can be free-text, files or URLs or from a controlled vocabulary (e.g., WordNet) or ontology. As a result of user demand, Vannotea was recently extended to enable the viewing of multiple related objects simultaneously. Users in geographically distributed locations can share the Vannotea application and simultaneously view two or more videos or 3D objects through a user interface that allows side-by-side comparisons.

Hence the aim of this paper is to describe in detail the model we have chosen for storing the different types of annotations so that they can be attributed to individual users for provenance data. The model is based on Annotea [7] (described in Section 0) and extending it in the following directions:

- to allow controlled annotations (Section 0) - see arrow along the x-axis in Fig. 1; and
- the annotations of links between parts of multiple digital objects of any type (Section 0) – see arrows along the y- and z-axis in Fig. 1.

3 Annotea

Through earlier work [8], we identified Annotea [7] as an ideal approach for implementing an annotation server. Annotea, in its original sense, is a Web-based annotation system that uses the Resource Description Framework (RDF) to model free annotations as a set of statements or assertions made by the author about a particular webpage. These annotations are then stored in a HTTP-enabled server, which enables clients to query, update, post, delete and reply to annotations.

A key strength of the Annotea protocol is that it uses open W3C standards such as RDF, XPointer, XLink and HTTP. The use of machine-processable RDF descriptions enables easy search, retrieval and linking of the annotations to related resources and services using Semantic Web technologies (e.g., OWL, SPARQL).

Fig. 2 illustrates the RDF Schema of Annotea and an RDF instance – an Annotea object – separated by the dividing dotted line. The Annotea Schema introduces properties that point to the annotated Web document (`annotates`) and to a specific location within a structured Web document, thus describing the `context` of an annotation, for which Annotea uses the XPointer technology. Furthermore, the specification provides a `related` property, which relates the resource representing the 'content' of an Annotation to the annotated resource.

Developers are encouraged to create new types of annotations by sub-classing from the `Annotation` class and creating sub-properties of the `related` property. Fig. 2 shows such a new type: the `Comment` class and the `body` property. In essence, the

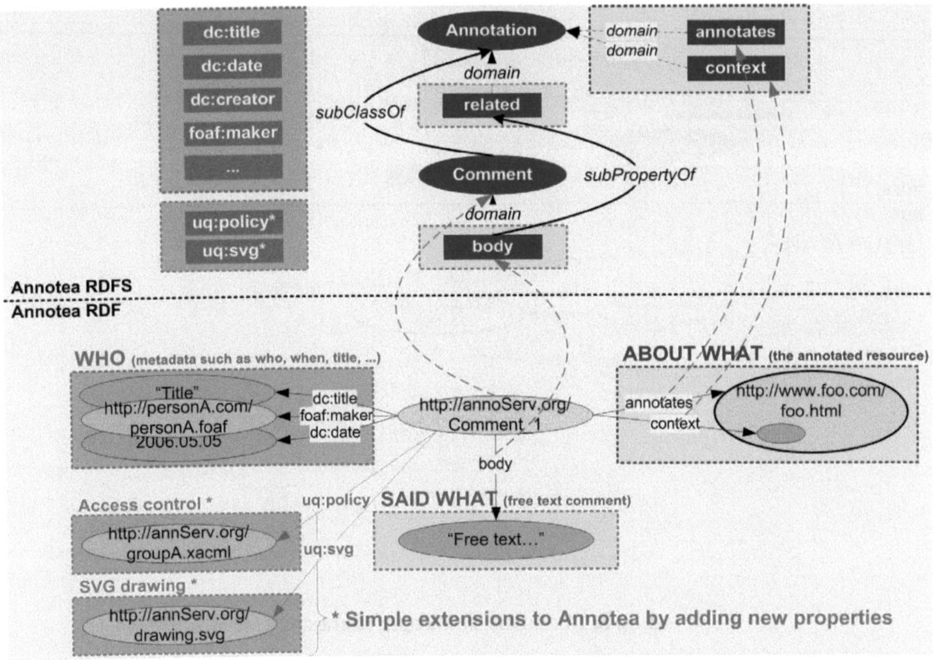

Fig. 2. Annotea Schema and instance with access control and SVG extension

RDF instance of a simple Annotea object will carry information about "*Who said what about which resource?*"

RDF/S allows easy addition of metadata properties from other schemas such as the Dublin Core (`title`, `date`, `creator`) and FOAF (Friend-of-a-Friend) namespaces, which are used to describe the provenance of an annotation.

We have also made our own, application-specific extensions. For example, we have extended Vannotea [3] to allow annotations in the form of drawings on top of media types such as images or videos, through the use of SVG.

Furthermore, we have added functionality to enable users to apply fine-grained access control to their Annotations through XACML policies and implemented the Annotation Server as a Shibboleth Service Provider [9]. A survey of current Web-based annotation systems [10] reveals that they vary in the way in which annotations may be attached, the way in which they are presented and in the access control mechanisms. Some systems are designed for private use only, whilst others permit sharing amongst groups and/or public access. None of the surveyed systems provide the kinds of fine-grained access control mechanisms that are achieved by our implementation and are required by collaborative teams of scientists engaging in eResearch.

Koivunen [11] introduced new Bookmark and Topic objects to Annotea. These social bookmarks and topics can be used for semantic authoring by allowing ordinary users tag interesting web documents with their *own personal* concepts or folksonomies.

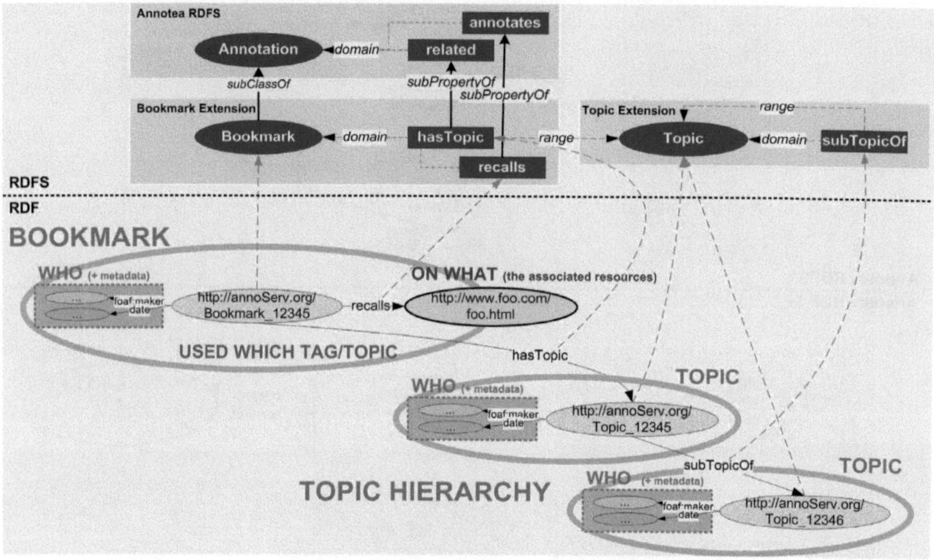

Fig. 3. Bookmark and Topic Hierarchy

Currently, the Boomark class is a separate, new class within Annotea. In our view however, bookmarks are just a special type of annotation. Rather than a user attaching a free-text comment to a specific resource, the user can build their own folksonomy using the topic hierarchies and attach those topics to the resource. Semantically we are basically describing, "*who used which topic about which resource?*", rather than "*who said what about which resource?*" as mentioned in the previous section.

Therefore, we suggest to subclass the `Bookmark` class from the `Annotation` class as shown in Fig. 3, which includes making the `recall` property a sub-property of the `annotates` property, and `hasTopic` a sub-property of `related`.

As a result, we will be able to query the Annotation Server to return any Annotation that is attached to a specific resource, whether it is of type Comment, Bookmark, or any other types that will follow in this paper.

4 Ontology-Based Annotations Using Annotea

As mentioned earlier, various communities - especially within the field of eScience - are creating their own domain-specific ontologies through group consensus. These ontologies can be hierarchical, controlled vocabularies modelled in RDFS, or more complex knowledge representations in OWL. This section illustrates how Annotea can be extended to allow users to take advantage of these formal concepts in order to create subjective semantic annotations. These formal annotations can aid in bridging the semantic gap between automatic recognition techniques that extract different low-level visual or audio features and highly subjective free-text annotations by humans. The line between objectivity and subjectivity is not always clear. When does a subjective annotation become objective and in whose eyes? As the provenance data for these controlled annotations is recorded, machines will be able to evaluate the objectivity of the individuals' semantic statements, based on trust relationships between the users and statistical calculations.

4.1 Controlled Vocabularies

One extension to Annotea using controlled vocabularies is to allow users to attach pre-defined ranking information to a specific web resource (e.g., the controlled terms "strong accept", "accept" and "reject" for a collaborative review process of scientific papers). Another example is using terms from an ontology such as the WordNet Ontology[4].

In any case, the controlled vocabulary or ontology is modelled in RDFS or OWL and publicly available over the web, so that an Annotea client can access and present it to the user when he/she wants to attach a controlled term to the resource.

In Fig. 4, the user searched for the term "animal" and then browses through the WordNet Ontology to navigate to a controlled vocabulary of a specific animal. The controlled term can then be attached to the resource (or part of the resource) that the user is currently viewing in his browser or Vannotea client.

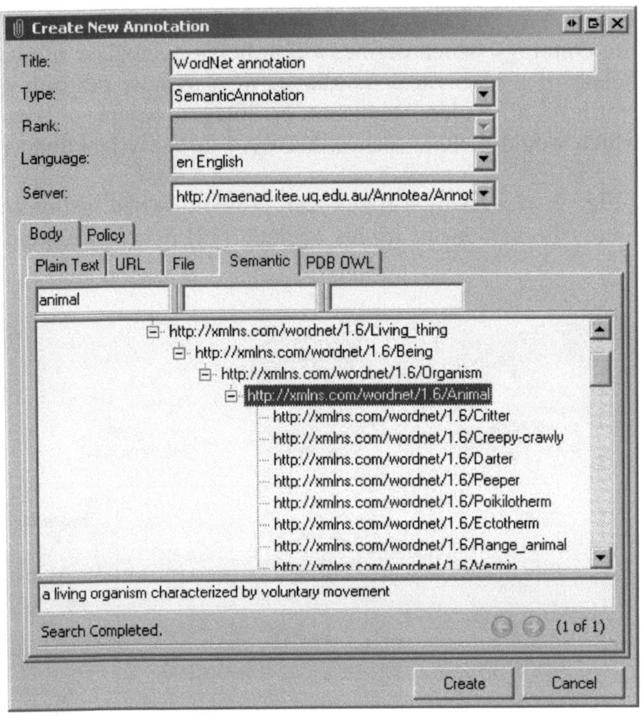

Fig. 4. Screenshot of creating a link to a controlled term

[4] http://xmlns.com/2001/08/wordnet

This is very similar to creating bookmarks (see Fig. 3), except that the topic is being replaced by a predefined controlled vocabulary. The benefit of using these controlled vocabularies is that we can perform searches using these terms, taking advantage of the ontology to infer that a "fish", for example, is a subclass of an "animal", and therefore returning all resources about a "fish" when querying for resources about "animals". Since we store provenance data about who created the annotation, we envisage taking definitions of trust relationships inside a user's FOAF profile into consideration when querying the Annotation Server, e.g., "*retrieve the ranking information about a particular resource from all users that I trust and calculate an average rating*".

4.2 Simple Formal Statements

Using the same interface depicted in Fig. 4, users can also attach formal triple statements based on ontologies to a resource, i.e., relate a formal statement to the annotated resource and context. A statement is a more complex instance of an ontology compared to the controlled terms in the previous section. Fig. 5 illustrates the schema of the `FormalStatement`. We introduce a `states` property which is a sub-property of `related` and has a range of `rdf:Statement`. The statement itself consists of a subject, predicate and object from an Ontology.

The example in Fig. 5 shows a statement that says "*lion eats gazelle*" from a simple Wildlife Ontology which defines a lion being a subclass of a carnivore and a gazelle a subclass of a herbivore. As above, we can now perform ontology based searches to retrieve all video segments or images where a "*carnivore eats a herbivore*", which would include scenes that were formally labelled as a "*lion eating a gazelle*".

However, Fig. 5 also demonstrates the problem when using reification with `rdf:Statement`: the amount of triples explode [12] as every statement carries the

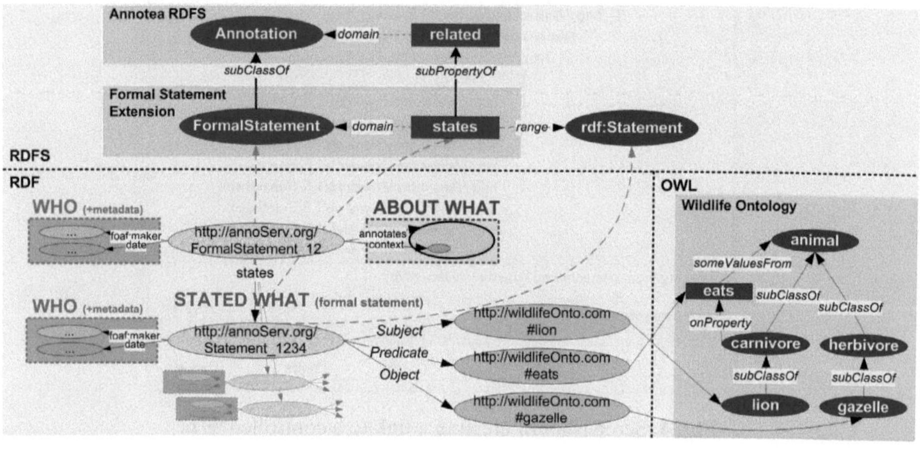

Fig. 5. Formal Statement

same metadata (as indicated by the greyed out statements). Therefore, we are currently investigating the use of named graphs as a mechanism for reasoning about provenance [13].

5 Comparisons and Associations Using Annotea

As mentioned earlier, tabbed browsers and tools like Vannotea allow users to view several objects (images, video, audio, text, HTML, PDF) simultaneously and carry out side-by-side comparisons. In such scenarios, users want to be able to annotate the link between two or more objects or between segments of multiple objects. For example, a user might want to annotate the link

- between a scene in an original film and the corresponding scene in a remake;
- between an image and a location (through Google maps URLs);
- between regions within several images; or,
- between structural components of two different 3D protein structures.

This section will investigate several approaches to model the annotation of such comparisons and associations within Annotea, in a way that follows the best practices described earlier. A *comparison* annotates multiple resources and describes their similarities or dissimilarities, whereas an *association* describes a user's mental connection between the resources. Although comparisons and associations are semantically different, they are both conceptually similar in the fact that they are about multiple resources, where the order of the resources is irrelevant and the description applies to the collection of resources.

As Annotea is based on RDF, it is very flexible with regards to adding/extending it to other properties as we have demonstrated earlier. Furthermore, since there is no cardinality defined for properties of an Annotea object, we can add multiple properties of the same type to the object. Therefore, the most convenient way to relate an `Association` (or `Comparison`) to multiple resources would be to use multiple `annotates` properties.

However, if we want to create an `Association` object between two parts (`contexts`) of two resources, we run into the following problem: According to the Annotea Protocol [7], the `context` property is supposed to include an XPointer, e.g.:

```
<context>
http://mydomain.com/foo.html#xpointer(id("Main")/p[2])
</context>
```

As we have identified in our earlier work with Vannotea, XPointer does not suit time-continuous media. Instead, temporal fragment identifiers, which have been discussed for URIs[5], could be used to refer to a time segment as follows:

```
<context>http://mydomain.com/foo.mpg#?t=15.2-
18.7</context>
```

[5] http://www.annodex.net/TR/draft-pfeiffer-temporal-fragments-02.html

The two examples above show that the `context` includes information about the resource it refers to. Unfortunately, we cannot always assume this to be the case. The `context` for images might be a definition of regions in SVG or some other format:

```
<context><svg id="SvgGdi_output"><g id="root_group">
  <rect height="102" id="77" width="79" x="95" y="125" />
</g></svg></context>
```

The context within 3D models might be an application-specific string describing the zoom factor, position, angle, selected polygons or even selected atoms and molecules within JMOL[6] models, e.g. using a JMOL script string:

```
<context>SAH 21.OXT number:46,moveto 1 58 -16 93 83.7 58
</context>
```

This leads to the problem illustrated in Fig. 6, where the literals denoting the `context` information have no formal connection to the resource they refer to.

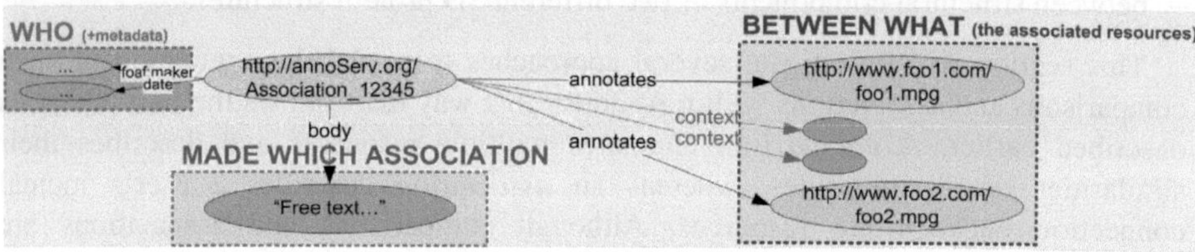

Fig. 6. The context property

The following sections will illustrate several attempts to bypass this problem - each attempt with its own advantages and disadvantages - before providing a recommended solution.

5.1 Attempt 1: The `isLinkedTo` Property

Attempt 1 is illustrated in Fig. 7, in which every resource that is part of an association is annotated as a separate `Association` object, and all `Association` objects are then linked together by a new `isLinkedTo` property, which has a domain `Association` and a range `Association`.

The advantage of this approach is that it is easy to implement as the additions don't require any changes to the current Annotation Server implementation. A general query such as "*Give me all annotations that annotate this resource*" (`?Annotation annotates "http://www.foo1.com/foo1.mpg"`) will retrieve `AssociationItem_1` and its `isLinkedTo` property is automatically returned as part of it. However, to retrieve the resource `AssociationItem_2` annotates, we would have to perform a nested query.

Another disadvantage is that the association as a whole cannot be addressed. In the example above it would have several addresses. This has several implications, not

[6] http://jmol.sourceforge.net/

being able to delete or reply to the association are just a few. Furthermore, managing an update (modification) of the association becomes very cumbersome to implement, as there would be many linked statements to fix. This might not be apparent in the above example, but an association might involve more than two resources, in which case deleting one `AssociationItem` would involve cleaning up n-1 links. Finally, the retrieval of all the links requires a recursive query, which is not supported by SPARQL.

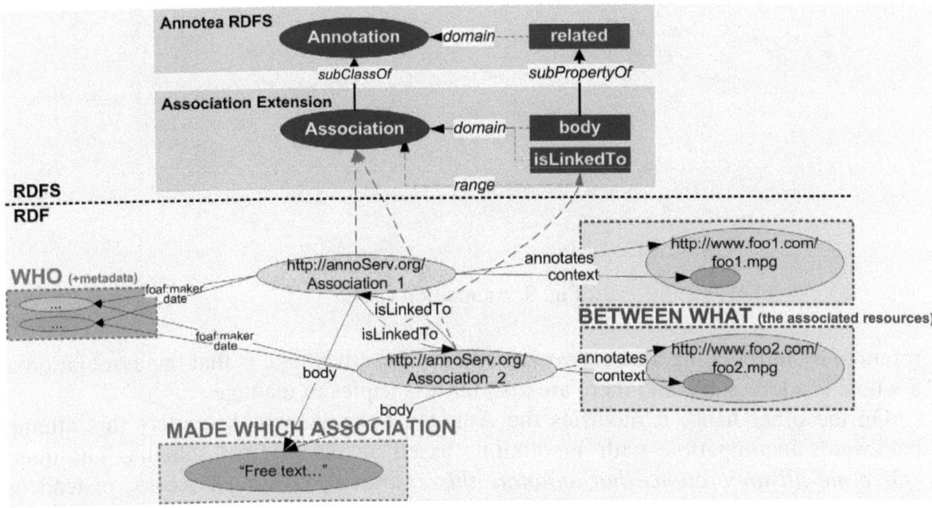

Fig. 7. The isLinkedTo property

5.2 Attempt 2: The `AnnotationGroup` Object

Fig. 8 illustrates Attempt 2, in which a newly introduced `Group` class (an rdf:Bag) links to all the `AssociationItems`. This means that the association as a whole can now be addressed through the `AssociationGroup` object. However, since the `Group` class is not a subclass of the `Annotation` class, it is not very useful. The query "*Give me all annotations that annotate this resource*" still returns an `AssociationItem` and a fairly complex and expensive nested query will need to retrieve the other AssociationItems through the `AssociationGroup` object.

Although this is likely to be supported by SPARQL using inferencing[7], it should be avoided if possible.

5.3 Attempt 3: The `Target` Object

Fig. 9 shows the third attempt, which unlike the previous two, views an Association as a subclass of an Annotation, and tries to combine the `context` with the resource it

[7] Question 3.3. of the SPARQL FAQ (http://thefigtrees.net/lee/sw/sparql-faq)

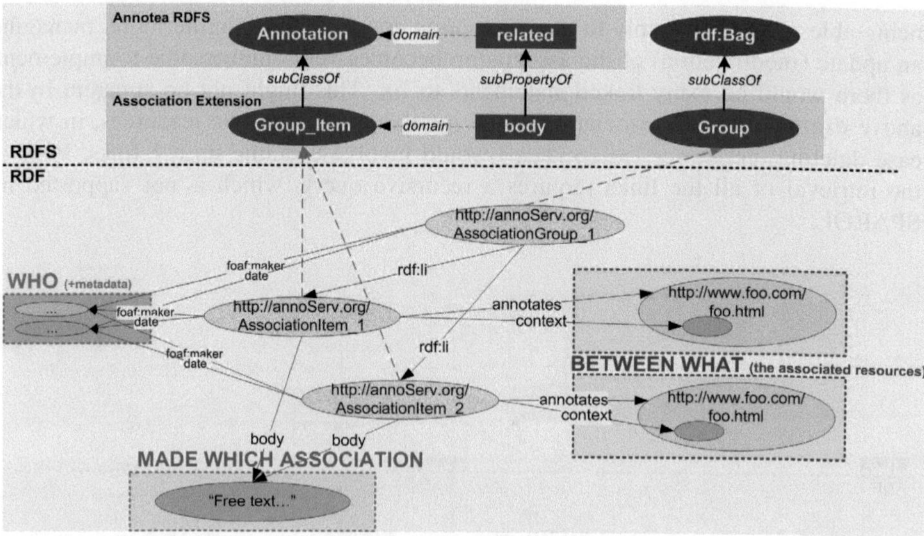

Fig. 8. Annotation Group

refers to by introducing a new `Target` object. The advantage is that the association as a whole is addressable and there are also far less triples to manage.

On the other hand, it modifies the Annotea Schema, which renders this attempt backwards incompatible with annotations based on the original schema. The query "*Give me all annotations that annotate this resource*" no longer works, instead we have to extend it to follow the graph through the `Target` object (a blank node), e.g.:

```
?Annotation annotates _targetObject,
_targetObject hasResource
"http://www.foo1.com/foo1.mpg"
```

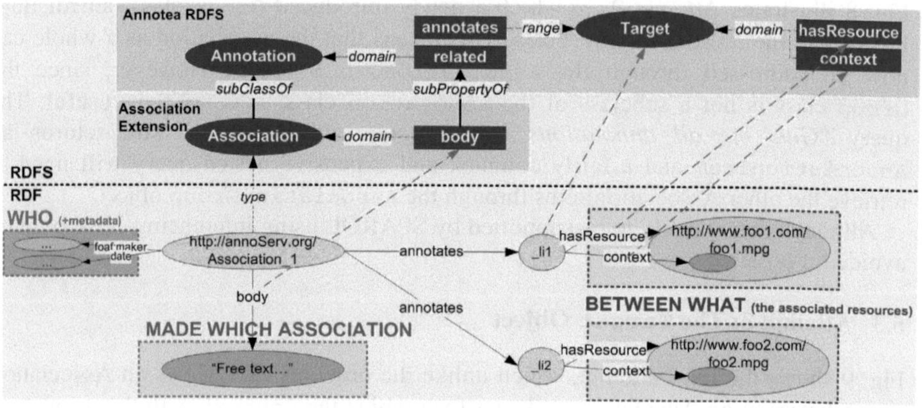

Fig. 9. Target Class

5.4 Recommended Solution: The `Context` Object

Finally, the recommended solution is illustrated in Fig. 10. The `context` property has a range `Context` class, which is the domain of two new properties, `hasResource` and `contextDescription`. Additionally, content-type specific `Context` classes and `contextDescription` properties can be subclassed, e.g., `VideoContext` and `mediaTime` for resources that are videos.

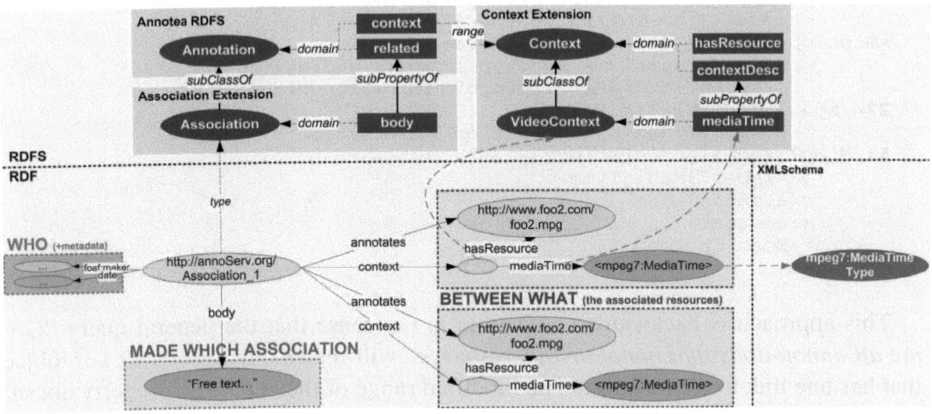

Fig. 10. Context Class

The RDF instance in Fig. 10 shows how the context property is pointing to a `Context` object (a blank node) which links to the same resource (`hasResource`) as the `annotates` property of the `Association` object. Additionally, it can contain any formalized or standardized description to represent the context, e.g., using the XML-Schema datatype `MediaTimeType` from the Multimedia Description Scheme[8] (MDS) of the MPEG-7 standard [14]:

```
<rdf:RDF
xmlns:rdf="http://www.w3.org/1999/02/22-rdf-syntax-ns#"
xmlns:a="http://www.w3.org/2000/10/annotation-ns#"
xmlns:dc="http://purl.org/dc/elements/1.0/"
xmlns:foaf="http://xmlns.com/foaf/0.1/">
  <rdf:Description rdf:about="http://mydomain/Anno/10894">
    <rdf:type rdf:resource= "http://www.w3.org/2000/10/annotation-
ns#Association" />
    <dc:creator>ronalds</dc:creator>
    <foaf:maker rdf:resource="http://www.my/~ronalds#ron" />
    <dc:date>2006-11-09T14:28:27Z</dc:date>
    <a:body rdf:resource="http://mydomain/Anno/body/10894" />
    <a:annotates>http://foo1.org/foo1.mpg</a:annotates>
    <a:annotates>http://foo2.org/foo2.mpg</a:annotates>
    <a:context>
      <a:hasResource>http://foo1.org/foo1.mpg</a:hasResource>
      <a:mediaTime rdf:parseType="XmlLiteral"
```

[8] http://m7itb.nist.gov/mds-2001.xsd

```
xmlns:mpeg7="urn:mpeg:mpeg7:schema:2001#">
        <mpeg7:MediaTime xsd:type="mpeg7:MediaTimeType">
          <mpeg7:MediaRelTimePoint mediaTimeUnit="PT1S">
10</Mpeg7:MediaRelTimePoint>
          <mpeg7:MediaIncrDuration mediaTimeUnit="PT1S">
5</Mpeg7:MediaIncrDuration>
        </mpeg7:MediaTime>
      </a:mediaTime>
    </a:context>
    <a:context>
      <a:hasResource>http://foo2.org/foo2.mpg</a:hasResource>
      <a:mediaTime rdf:parseType="XmlLiteral"
xmlns:mpeg7="urn:mpeg:mpeg7:schema:2001#">
        <mpeg7:MediaTime xsd:type="mpeg7:MediaTimeType">
          <mpeg7:MediaRelTimePoint mediaTimeUnit="PT1S">
22</Mpeg7:MediaRelTimePoint>
          <mpeg7:MediaIncrDuration mediaTimeUnit="PT1S">
5</Mpeg7:MediaIncrDuration>
        </mpeg7:MediaTime>
      </a:mediaTime>
    </a:context>
  </rdf:Description>
</rdf:RDF>
```

This approach is backwards compatible in the sense that the general query "*Give me all annotations that annotate this resource*" will return any `Association` object that has one link to the resource. The modified range of the `context` property doesn't have to be defined within the Annotea Schema, but could be defined as a new sub-property of the `context` property within the Context extension.

6 Conclusion

In this paper we have demonstrated various ways of extending the Annotea Schema to enable annotation of links between segments of multiple objects. We have shown that careful considerations need to be made as the flexibility of RDF and the ease to add and extend new classes and properties might have wide-reaching implications if not approached and considered cautiously.

When extending Annotea, we recommend the following best practices:

- Reuse existing Annotea classes and properties as well as RDF and XML Schema (built-in) types where possible;
- A general query such as "*Give me all Annotations that annotate this resource*" should always return *all* objects that are sub-classes of `Annotation`, e.g. `Comment`, `Bookmark`, `Ranking`, `FormalStatement`, `Association`. It is then up to the client to display the different objects accordingly;
- A general query such as the one above should return all the information needed to enable clients to display an appropriate overview/list, i.e., avoiding nested queries on the server-side to retrieve additional information where possible;
- Avoid unnecessary explosion of triples in the triple-store;
- Investigate SPARQL implications.

7 Future Work

In the future, we will investigate the following:

- The use of named graphs instead of triple-hungry reification where possible to avoid triple explosion in the RDF store.
- Consider definitions of trust relationships inside a user's FOAF profile when querying the Annotation Server, e.g. filter out annotations that are not trusted.
- Combine Comparisons/Associations with ontologies, which should be straightforward, based on the work presented in this paper.
- Add generic HTML-based ontology browsers and forms to create more complex ontology instances and embed these into the user interface of our Annotea Sidebar and Vannotea.

References

1. Roscheisen, M., Mogensen, C., and Winograd, T.: Shared Web Annotations as a Platform for Third-Party Value-Added, Information Providers: Architecture, Protocols, and Usage Example. Computer Science Dept., Stanford University. Technical Report CSDTR/DLTR STAN-CS-TR-97-1582 (1994).
2. Davis, J. and Huttenlocher, D.: The CoNote System for Shared Annotations. (1995) http://www.cs.cornell.edu/home/dph/annotation/annotations.html.
3. Schroeter, R., Hunter, J., Guerin, J., Khan, I., and Henderson, M.: A Synchronous Multimedia Annotation System for Secure Collaboratories. In 2nd IEEE International Conference on E-Science and Grid Computing (eScience 2006). Amsterdam, Netherlands (2006)
4. Kipp, M.: Anvil - A Generic Annotation Tool for Multimodal Dialogue. In 7th European Conference on Speech Communication and Technology (Eurospeech). Aalborg (2001) 1367-1370
5. Halaschek-Wiener, C., Golbeck, J., Schain, A., Grove, M., Parsia, B., and Hendler, J.: Annotation and provenance tracking in semantic web photo libraries In International provenance and annotation workshop. Chicago (2006)
6. Petridis, K., Kuehn, K., Handschuh, S., Bloehdorn, S., Saathoff, C., Avrithis, Y., Kompatsiaris, Y., and Staab, S.: Semantic Annotation of Images and Videos for Multimedia Analysis and Retrieval. In Lecture Notes in Computer Science, vol. 3532: Lecture Notes in Computer Science (2005) 592-607.
7. Koivunen, M.-R. and Kahan, J.: Annotea: an open RDF infrastructure for shared Web annotations. In Proceedings of the 10th international conference on World Wide Web. Hong Kong. ACM Press (2001)
8. Schroeter, R., Hunter, J., and Kosovic, D.: Vannotea - A Collaborative Video Indexing , Annotation and Discussion System For Broadband Networks. In Knowledge Markup and Semantic Annotation Workshop, K-CAP 2003. Sanibel, Florida (2003)
9. Khan, I., Schroeter, R., and Hunter, J.: Implementing a Secure Annotation Service. In International Provenance and Annotation Workshop (IPAW2006). Chicago (2006)
10. Uren, V., Cimiano, P., Iria, J., Handschuh, S., Vargas-Vera, M., Motta, E., and Ciravegna, F.: Semantic annotation for knowledge management: Requirements and a survey of the state of the art. Web Semantics: Science, Services and Agents on the World Wide Web, vol. 4 (2006) 14-28

11. Koivunen, M.-R.: Semantic Authoring by Tagging with Annotea Social Bookmarks and Topics In The 5th International Semantic Web Conference (ISWC2006) - 1st Semantic Authoring and Annotation Workshop (SAAW2006). Athens, GA, USA (2006)
12. Carroll, J. J. and Stickler, P.: RDF triples in XML. In 13th international World Wide Web conference. New York, NY, USA (2004) 412 - 413
13. Watkins, E. R. and Nicole, D. A.: Named Graphs as a Mechanism for Reasoning about Provenance. In APWeb 2006: 8th Asia-Pacific Web Conference. Harbin, China (2006) 943-948
14. Salembier, P. and Smith, J. R.: MPEG-7 multimedia description schemes. IEEE Transactions on Circuits and Systems for Video Technology, vol. 11 (2001) 748-759

Describing Ontology Applications

Thomas Albertsen and Eva Blomqvist

Jönköping University, Jönköping, Sweden
{alth|blev}@ing.hj.se

Abstract. Semantic Web technologies are finally, after a few years of infancy, truly entering the business world to support the growing needs of computer aided information selection and processing. There are already quite well-defined development processes and methods in the software engineering field to handle the construction of large scale and complex enterprise systems, and to reuse knowledge in different software domains patterns are considered to be common practise. Patterns can be described on different levels of abstraction, but the patterns in the focus of this paper are on the software architecture level. In this paper we present a definition of the notion "ontology application pattern", as a special form of software architecture patterns describing an ontology-based system. We also show how such patterns, as well as the description of the pattern instantiations, can be described using a modified architecture description language.

1 Introduction

In recent years the area of semantic applications has grown from small web applications, to large scale enterprise systems showing real benefits in the business applications domain. Technologies intended for the web are now also applied on company intranets and in enterprise information systems etc. As the technology development increases system capabilities, the systems keep getting larger and more complex. Semantic applications are comparable to any other software systems in terms of size and complexity, and thereby require the same rigorous development process as any other system.

Complexity of a system to be built can be managed through established development processes and knowledge reuse (through for example patterns), this is common practise in software development. When constructing semantic applications ontologies are usually used as the core knowledge component of the system. This paper addresses the issue of specifically incorporating ontologies in the architecture design of a software system through a specialisation of software architecture patterns, and extending an architecture description language to better suit software architectures including ontologies.

The following section presents motivation and background of our approach, including related work on patterns and architecture description languages. In section 3 the term ontology application pattern is defined and described in more

E. Franconi, M. Kifer, and W. May (Eds.): ESWC 2007, LNCS 4519, pp. 549–563, 2007.

detail. A way of describing the specific kinds of software architectures are discussed in section 4. Finally, the paper is concluded with a discussion and some future work possibilities.

2 Background

This section describes the general background, including definitions, and some previous work on patterns.

2.1 Ontologies and Patterns

Ontology is a popular term today, used in many areas and defined in many different ways. In our research we adopt the definition from [1], stating that an ontology is a formal explicit specification of a shared conceptualisation. In our view this means that an ontology is a hierarchically structured set of concepts describing a specific domain of knowledge, that can be used to create a knowledge base. An ontology contains concepts, a subsumption hierarchy, arbitrary relations between concepts, and possibly other axioms.

Even using this definition, ontologies can be used for many different purposes and applications, and they can be constructed and structured in many different ways. Our research focuses mainly on the abstraction level of domain and application ontologies (as defined in [2]) within enterprises, so called enterprise application ontologies, to be used in enterprise applications. When considering the complexity of the ontologies, our research is mainly focused on terminological ontologies, to structure and retrieve information.

In a previous paper [3] we have attempted to describe and classify different kinds of patterns concerning ontologies. In general, patterns are here used in the sense of a generalised solution to a recurring problem that can be reused (instantiated and adapted) in the specific engineering task at hand. In [3] the patterns concerning ontology engineering are divided into five intuitive levels of abstraction. The four lowest levels deal with the internal structure of the ontologies, so called ontology patterns, while the fifth level deals with complete ontologies used within a system, thereby making the connection to software engineering and software architectures. This level of abstraction is in [3] denoted "ontology application patterns" and will be the focus of the rest of this paper.

2.2 Ontology Applications

There are a lot of applications that use ontologies today. In [4] a description of four classical ontology application scenarios is made: Neutral authoring, common access to information, ontology as specification and semantic search. Another description of ontology-based applications, more focused on the business domain, is the classification in [5]. Four slightly different scenarios are described: Knowledge management, information retrieval, portals and web communities and finally e-commerce.

The trend is now to move away from traditional heavily knowledge based systems, using a single ontology, to more open semantic applications (as for example stated in [6]) exploiting and handling the huge size of the web or company intranets and the diversity of both ontologies and information sources. As stated already in the introduction of this paper, these kinds of semantic applications are entering also the business area, through for example intranets, company portals and e-commerce. As technologies develop, more advanced applications become possible, which also increases the complexity and size of the software systems that use the ontologies.

To describe a whole ontology and its use in an application there is a need to describe the characteristics of the ontology. A formulation of such characteristics has been described in connection with the Ontology Definition Metamodel presented by the Object Management Group (OMG) in [7], the dimensions described there are:

- Level of Authoritativeness: This is a measure of how authoritative the ontology is of the area it describes. If the author of the ontology is the organisation that is responsible for specifying the conceptualisation then this is clearly a highly authoritative ontology.
- Source of Structure: Either the ontology is developed separately from the application that will use it or else the structure comes directly from the intended application.
- Degree of Formality: Degree of formality refers to the level of formality of the specification of the conceptualisation. This could range from highly informal ontologies, via semantic networks, to highly formal ontologies that include complex logical axioms.
- Model Dynamics: This concerns the rate of change in the ontology, from the extreme where the ontology is stable and rarely or ever changes, to very volatile ontologies that change continuously.
- Instance Dynamics: This dimension is closely related to Model Dynamics but concerns the instances of the ontology.
- Control and Degree of Manageability: This dimension considers who decides when and how to change an ontology. One extreme is that the author of the ontology has the sole decision on changes, and the other extreme is of course that the ontology changes are based on decisions of outside parties.
- Application Changeability: The applications that use the ontology might be on one hand developed once and for all, and not frequently changed, and on the other hand they might be changing dynamically during run time.
- Coupling: This dimension describes how closely coupled applications committed to shared ontologies are.

To further describe the ontology and what content it should have, competency questions could be used. Competency questions, as described in [8], are used to exactly define what information the ontology should provide, in terms of constraints on its content. A competency question can be expressed as a constraint, but it also contains information on the content of the ontology. Competency questions can be expressed in formal logic or more informally, as requirements on the ontology. In this paper we mainly consider informally expressed competency questions.

Other requirements that should be specified according to most common ontology development methodologies, as described in [9], concern:

- Purpose and Goal: The intended use of the ontology.
- Scope: Describing major subject areas that need to be covered in the ontology, and possible boundaries to what is not to be included.
- Level of Detail: Specifying the lowest level of detail considered important (the instance level).
- Representation Suggestions: Suggestions of naming conventions, representation language appropriateness etc.

Often a system could exploit not only one ontology but several, perhaps even exchangeable during runtime. Then these ontologies might need a connection, in order to for example translate information expressed in one ontology to the other. This process of translation requires an ontology alignment to be present, which can be obtained through an ontology matching process as defined in [10].

2.3 Software Architectures and Patterns

A description of a software architecture is the description of the high level structure of a software system. In [11] the term software architecture is defined as *"the description of elements from which systems are built, interactions among those elements, patterns that guide their composition and constraints on these patterns. [...] a particular system is defined in terms of a collection of components and interactions among those components"*.

This definition describes the overall design of the system that includes global control structure, communication protocols, data access, and the system's major components and the behaviour of the components. The essential idea of architecture is abstraction, to hide some of the details of the system in order to make it easier to understand the properties of the system, and to connect the functionality in the requirements to elements of the high level design. If the system is complex there can be several levels of abstraction, and the elements on each level can be decomposed into new architectures.

The elements of an architecture can as noted above be divided into components, connectors and their configuration. Both components and connectors provide interfaces that act as connection points to that entity. Sometimes also the data that is exchanged between components is included in the architectural view (see for example [12]). The component is classically an abstract unit of software instructions and internal state that provides a transformation of data. This can be transformations as computation or as simply loading data to memory from secondary storage. The connectors are an abstract mechanism that facilitates communication between the components. A connector may have a subsystem inside it, in order to make this communication possible. Data is the information that is transferred between the components through the connectors of the architecture.

An architecture style describes a generalised architectural organisation of software systems, which means that it describes how components and connectors can be configured into a certain kind of software system with certain properties. In [11] the notion of architecture style is defined as follows: *"... an architectural style defines a vocabulary of components and connector types, and a set of constraints on how they can be combined."* It is also stated that many styles might in addition include a semantic model that defines how to determine the system's overall properties from its parts.

Examples of common architecture styles are "Pipes and Filters" where the components are filters while the connectors are the pipes. Another architecture style is the "Object Oriented" style. The components encapsulate the data and the operations, as connectors, make it possible for the components, the objects, to communicate. The connectors are usually procedure calls. If an architecture style is more formally described it constitutes and architecture pattern, such as the patterns in [13] and [14].

2.4 Architecture Description Languages

In order to support the description and communication of software architectures the use of architecture description languages (ADLs) have been proposed. An ADL is usually described as a language designed for describing high level architectural notions, including components, connectors, and their interfaces, and that treats connectors as first-class entities. ADLs were originally developed when existing formal languages and programming languages were found to be insufficient or inappropriate to describe a software architecture (see discussion in [11]).

For software ADLs the components in the language are software processes or modules and the ADL is used to define and model the software architecture. There are numerous ADLs existing today, examples are Rapide, UniCon, Darwin and xADL. In order to transfer an architecture description from one ADL to another the ADL ACME has been constructed to facilitate mapping from one ADL to another. For a survey and comparison of ADLs see [15].

xADL (see [16]) has been developed by the University of California Irvine and is defined as a set of XML schemas. The language has a core model with four elements: Components, connectors, interfaces and configurations. These are mainly the same elements as discussed in section 2.3. The language has a modular design that makes it easy to extend with new structures.

3 Ontology Application Patterns

In this section we attempt to define the notion of ontology application patterns, as an extension of software architecture patterns. As software architecture patterns are usually described using a special template, one such template is described and detailed to better fit ontology applications.

3.1 Definition

Ontology application patterns are very much related to software architecture patterns, although ontology application patterns include ontologies as components. Our definition of ontology application patterns therefore build on the definition of software architecture patterns stated in section 2.3.

We define an ontology application pattern as *"a software architecture pattern describing a software system that utilises ontologies to create some of its functionality. The pattern also describes properties of the ontology or ontologies in the system, and the connection between the ontology, other ontologies and the rest of the system"*.

3.2 Specific Characteristics

The difference between software architecture patterns and ontology application patterns is that there always exist ontology components in the latter and that the pattern should give insight in what capabilities these ontology components should have. First we consider the topology of the complete architecture, to address the nature of ontology components. How are the ontology components used in the complete system? There can be several ways to view this, on several levels of abstraction. If the abstraction of the system is high, the ontology component might include a lot of "intelligence" such as an inference engine and additional software functionality. If on the other hand the abstraction level is low, the component might represent only the ontology, as a data source. These two views imply different interfaces and connectors of the ontology component, but both views might be used for describing patterns as well as for describing a specific software architecture.

To give a simple example, we use a scenario from a research project at Jönköping University called MediaILOG. The project aims at introducing semantic technologies to address information processing problems within companies of the media industry. Specifically a local newspaper of the Jönköping region is used as an application case of the project. One part of a proposed framework in the context of this project is the possibility to rank incoming documents (whether they are e-mails, articles from a news agency or update reports from a website) with respect to the company's interests (as represented in their enterprise ontology) and the individual interests or job descriptions of the employees of the company (also described in the enterprise ontology). This process should be supported by a software system for evaluating and ranking incoming documents. A simplified version of such an architecture can be depicted as in figure 1, where part A illustrates the architecture on a high level of abstraction, the enterprise ontology is an internal part of the profile matching component, and part B illustrates the case when the ontology is viewed as a component in itself.

In both views described above the internal structure and content of the ontology itself needs to be considered. This leads to the question, what characteristics should be described concerning an ontology in an architecture or even an ontology application pattern? In section 2.2 a set of ontology characteristics are given, along with the requirements usually used when constructing ontologies

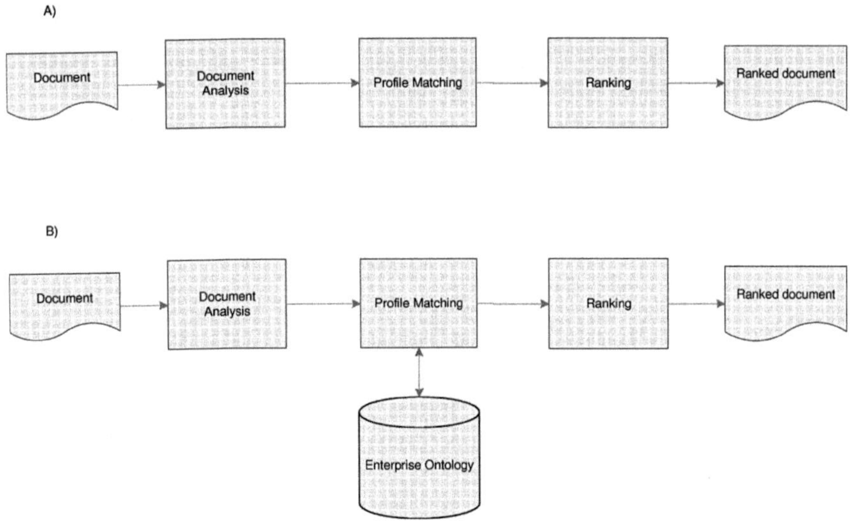

Fig. 1. Examples of architecture alternatives

(like scope, level of detail and a set of competency questions). Together this can be viewed as a description characterising the ontology as well as requirements for its construction. In the example architecture such information would include the scope of the enterprise ontology, the required level of detail and a set of competency questions describing the ontology capabilities.

There can be competency questions on several levels; general abstract questions stated for a pattern, and more specific ones developed when implementing the pattern. An ontology pattern used for ontology construction might describe the content and internal structure of a part of the ontology in general terms, like the ontology design patterns of [17], and could be used as a way to describe the content requirements of an ontology. If for example the ontology application pattern would suggest the use of a structure similar to the DOLCE roles and tasks pattern in [17] (describing how objects and events in a certain situation are described by roles and tasks) it could also include suggestions for competency questions, representing an abstract interface to the ontology. An example is to be able to "ask" the ontology "what role has a certain object that participates in this specific event". When the pattern is instantiated such a question can be specified into "what role has this person participating in this meeting". An ontology application pattern might suggest ontology patterns for realising the ontology component and competency questions for realising its interface.

In the example of figure 1 such description of the ontology interface (of alternative B) could include competency questions like i) "what are the related topics of a topic t" or ii) "what are the interests of employee e". This should be part of the component description of an ontology component in a software architecture and thereby also, in more general terms, might be addressed in an ontology application pattern. Such a more general version of the competency question ii could for example be "what are the concepts considered relevant from the perspective of instance i". In addition, if the more general view (alternative A) is adopted, including software structures in the ontology component, also the functionality of the surrounding software and its interface should be described.

Another issue when considering interfaces to ontology components are the possible and intended connections. The above discussed example component interfaces support mainly the connection of an ontology component to a pure software component. Another possibility is to connect an ontology component directly to another ontology component. Then the interface might constitute either only the competency questions (for dynamic connections) or the complete internal structure of the ontology (for more static approaches). Interfaces can be described both as general types, for ontology application patterns, and as specific instances of those types for the instantiation of the patterns.

When connecting two ontology components directly the connector between these probably contains an ontology alignment (see section 2.2). The interface of the ontology components are in the case of a static alignment the complete ontology, since the alignment can operate on any internal part of the ontologies. In the case of a dynamically constructed alignment the interface is a set of competency questions and the connector a set of queries to construct the alignment. On a pattern level this can be described as an abstract alignment, an alignment between components, but on the level of the specific pattern instantiation the nature of the alignment might be further specified.

An example of this situation could also be illustrated through a simplified part of the framework envisioned in the MediaILOG project previously mentioned. When collecting relevant information from news agency services the enterprise ontology of the newspaper can be aligned to ontologies describing the structure and general content of the information provided by different news agencies, as illustrated in figure 2. In this way when the newspaper wants to retrieve information from a news agency a wrapper interpreting the alignment between the enterprise ontology and the news agency ontology is used, to translate the information need of the newspaper into a query to be sent to an appropriate news agency. In this simplified example, there exists a static connection between the enterprise ontology and several news agency ontologies. Each ontology acts as a component of this architecture and the alignments act as the connector component between ontology components. But in practise it may also be used by the wrapper components, as illustrated in figure 2, which gives the wrappers access to the aligned ontologies.

3.3 Pattern Template

In order to make it easier to create ontology application patterns a template, corresponding to the commonly used template in [18] is detailed to better fit ontology application patterns. The suggested template in [18] has the following headings:

- Name of pattern
- Aliases: Aliases, if there are any.

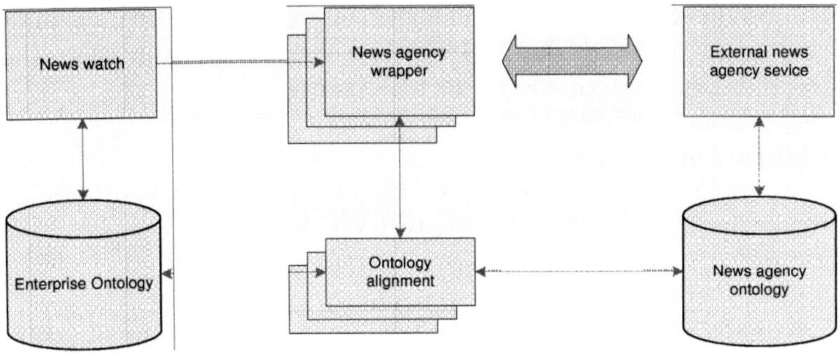

Fig. 2. Example of an architecture involving static alignments

- Problem: Gives a short description of the problem. This could be a statement or a question.
- Context: The context in which the pattern is valid.
- Forces: Lists and describes each relevant force. A force is a factor or attribute relevant to the problem, the solution is the result of the "tension" between the forces. Examples of forces can be factors such as efficiency or robustness. Forces can be conflicting and the solution should balance the tension between the forces present in the problem.
- Solution: Gives a statement of the solution to the problem, can include diagrams.
- Resulting Context: Describes the context after the solution is applied.
- Rationale: Explains the logic behind the solution. The user will understand why this solution is a good solution.
- Known Uses: Gives examples on where the pattern have been used.
- Related Patterns: Lists any related patterns.
- Sketch: A sketch to describe the pattern and a description of the sketch, if needed.
- Author(s): Names of the authors.
- Date: Date created.
- Contact Details: Authors' emails
- Pattern Source: The source of the pattern, for example the affiliation of the authors.
- References: A list of references cited in the pattern.
- Keywords: A string of comma delimited terms used for searching.
- Example: Referenced in the pattern. This could be pseudo code, skeleton for classes etc.

To further detail this template for ontology application patterns, information about the ontology components need to be specified in the template sections. In the following list subheadings illustrate where new information should be added:

- Context
  - Application Context

  - Knowledge Context: Describing the context, especially regarding level of authoritativeness, model and instance dynamics, application changeability and coupling, for the pattern to be applicable.
- Forces
  - General Forces
  - Ontology Forces: Describing forces related to the use of ontologies and the requirements set by the pattern on the ontology.
- Solution
  - Software Solution
  - Ontology Solution: Describing the intention and general requirements set on the ontology by the pattern, like purpose, scope, source of structure, level of detail, degree of formality, representation suggestions (and reasoning capabilities) and a set of general competency questions.
- Resulting Context
  - Application Context
  - Knowledge Context: Describing the model and instance dynamics, application changeability and control, and degree of manageability.
- Example: The examples might include parts of ontologies and references to reusable ontologies.

4 Architecture Description

When using the above described template much of the information regarding an ontology application pattern can be captured, but only in the ambiguous manner of arbitrary textual descriptions, simple illustrations and examples. To more formally describe a pattern, or a pattern instance implementation, a structured language (such as an ADL) is needed. Existing ADLs focus on software components and the interactions between them. We feel there is a need for focusing more on knowledge components to be able to define an ontology application pattern or an application architecture containing ontologies properly. Many of the existing ADLs are possible to use for describing ontology components, but to capture the above noted characteristics of ontology components and connectors the languages have to be extended.

4.1 Suggested ADL

Architecture description languages usually contain the elements components and connectors. If ontologies are treated as another component there are several issues to be considered, as described in section 3. If the component description in an ADL were to be extended by the characteristics of the ontology this could be a way to show what the ontologies must be able to do within the component and the complete system.

As described in section 2.4 there are several ADLs available. In order extend an ADL to accommodate the use of ontologies an ADL with the property of being easily extendible were chosen, namely xADL 2.0. xADL has a modular organisation, where new modules can be added to include more specific expressions in the language. This is a good way to create a description of a system using ontologies i.e. creating an ADL for ontology applications.

Ontologies can be seen as components in the system with some sort of information inside and given interfaces. Other components within the system should be able to access and manipulate the ontology. With this in mind, a component with the additional properties of an ontology should be added to the language. The characteristics mentioned in 2.2 are used as additional properties of the ontology component and thereby a new element called ontology is created in xADL. The ontology element has a description element called the Ontology Description, which has all the ontology characteristics as attributes. This is illustrated in figure 3 through a graph based notation provided by xADL 2.0.

The interfaces are the connection points to the ontology component and should describe what services the ontology can provide to the other components through some connector. In section 3.2 examples of such services are given, and those are examples of what can be considered ontology interfaces. Mainly there are two kinds of interfaces; interfaces directly to another ontology component and software interfaces (possibly constrained by some competency questions). The interface between ontologies also has two variants, either constituting the complete ontology or just the set of competency questions. These variants are added as specialised interface description attributes as illustrated in figure 4.

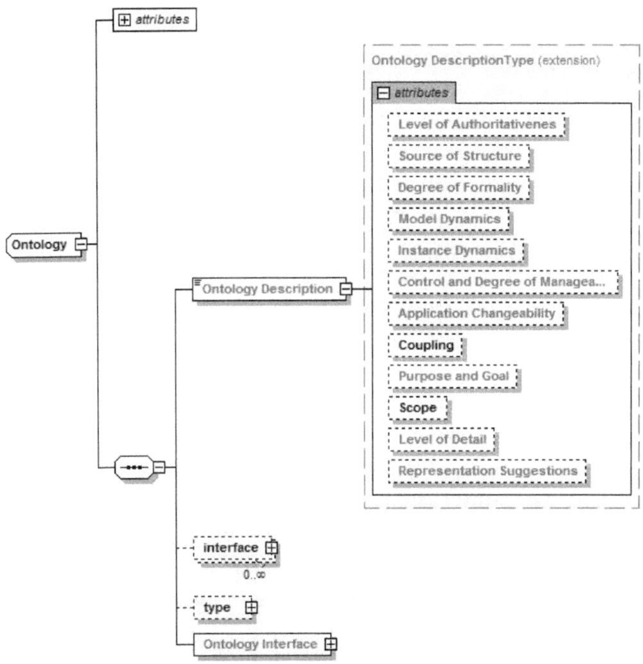

Fig. 3. Ontology component

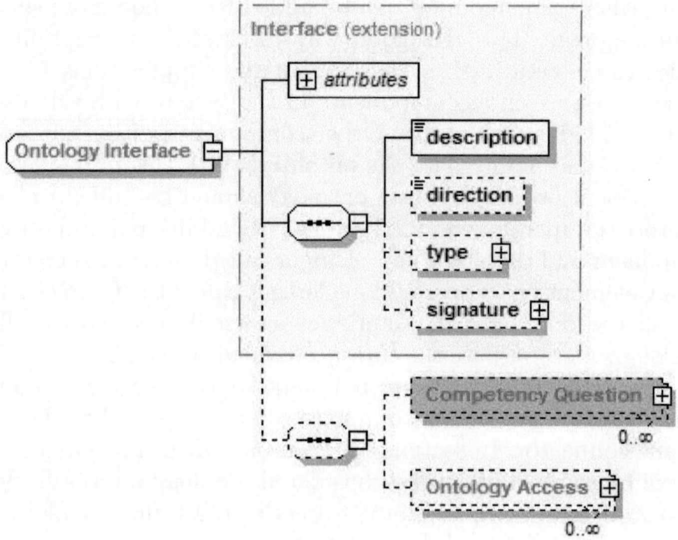

Fig. 4. Ontology interface

The last of the added elements is the connector, that will make communication to and from the ontology components possible. Connectors can also be of several kinds, but if the ontology is connected to a software component there is no major difference to an ordinary software connector, only the possible addition of constraints (competency questions). If the ontology is on the other hand connected to another ontology component directly, the connector might be a set of queries or a set of competency questions, or both. These are added as a specialised attribute in the ontology connector as illustrated in figure 5.

4.2 Architecture Example

When considering the simplified project scenarios discussed earlier in section 3.2, the first architecture (version B of figure 1) could for example be described using the suggested additions to xADL. Due to space limitation we can here only give a brief idea of what this description could contain (using the added attributes of ontology components, connectors and interfaces).

The ontology component is in this example architecture considered as a data store with no software functionality of its own. The description of this enterprise ontology using the component template depicted in figure 3 should include information of all the attributes of an ontology component. Level of authoritativeness would in this case describe the level of authoritativeness required from the ontology to support accurate ranking of incoming documents. The level of authoritativeness should be high in order for the rankings to be useful, so the ontology needs to be tailored to the media company at hand. Source of structure would include information of the sources for ontology construction, in the case

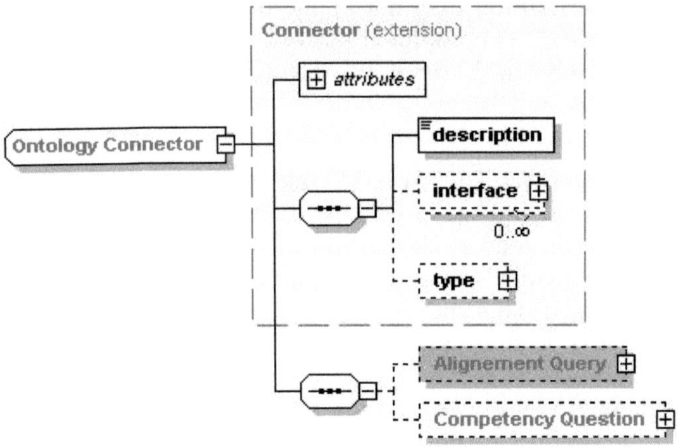

Fig. 5. Ontology connector

of a local newspaper this might be the paper sections and the actual work division structure between departments and individual employees. The structure of the ontology needs to be tailored to the task it should perform. The degree of formality states if informal definitions are acceptable or if each concept needs a formal logical definition. In this application case a semantic net representation of the ontology is sufficient. Model dynamics describes how dynamic the enterprise ontology will be, in this case a traditional newspaper with fixed topics probably has a quite static enterprise ontology. On the other hand, the level of detail needs to be high, which gives a higher level of instance dynamics. This kind of reasoning can be performed for each of the attributes of the ontology component, based on the application case requirements.

As an interface description the competency questions of the ontology should be one part, in addition the representation language of the ontology needs to be specified. For the example case the following two competency questions were suggested in a previous section: "what are the related topics of a topic t" and "what are the interests of employee e". The ontology representation is needed to support the implementation of the connector interface. Finally, the connector between the ontology component interface and the software component (in the example denoted "profile matching") needs to be specified. This will be the reasoning software used to query the ontology, for example using some ontology query language, allowing queries corresponding to the competency questions of the interface. This reasoner will then connect to the software interface of the profile matching component, thereby letting that component use the services corresponding to the competency of the ontology.

So far we have not generalised these example architectures into patterns, but we envision that several architecture patterns can be created in each application field (as discussed in section 2.2) simply by looking at existing systems. The extended version of xADL presented above can equally well be used to describe those patterns. All attributes may not apply for every pattern, but generally the attributes are still valid. The modified pattern template suggested in section 3.3 can be viewed as the natural language correspondence to the added component templates in xADL.

5 Conclusions and Future Work

In this paper we have defined the notion of ontology application patterns, as a specialisation of software architecture patterns. Then we have continued to study the characteristics that differentiate these patterns from ordinary software architecture patterns. Mainly we can see that engineering ontology applications may benefit from descriptions and patterns more focused on ontologies and their use in the system. Therefore we propose a more detailed template to guide the future description and documentation of ontology application patterns. A next step will be to construct a catalogue of ontology application patterns according to our definition and defined using the extended template.

Both pattern descriptions and architecture descriptions in themselves can also benefit from being described in a more formal manner. Therefore we have extended the general architecture description language xADL 2.0 with specific ontology component templates, together with specialised connectors and interfaces. Future work will involve to evaluate the extended ADL through applying it in a set of cases and determine the completeness of the attributes. It will also be used to describe future system architectures of ontology applications developed by our research group, in projects like MediaILOG.

Acknowledgements

Examples used in this paper were provided by the MediaILOG research project at Jönköping University, financed by the Swedish Hamrin Foundation. Also special thanks to the three anonymous reviewers for their valuable comments on how to improve this paper.

References

1. Gruber, T.: A translation approach to portable ontology specifications. In: Knowledge Acquisition. Volume 5. (1993) 199–220
2. Guarino, N.: Formal Ontology and Information Systems. In: Proceedings of FOIS'98. (1998) 3–15
3. Blomqvist, E., Sandkuhl, K.: Patterns in Ontology Engineering: Classification of Ontology Patterns. In: Proc. of ICEIS2005, Miami Beach (2005)
4. Jasper, R., Uschold, M.: A Framework for Understanding and Classifying Ontology Applications. In: Proceedings of the Twelfth Workshop on Knowledge Acquisition Modeling and Management KAW'99. (1999)
5. OntoWeb: D21 Successful Scenarios for Ontology-based Applications v1.0. OntoWeb Deliverable, D2.1, Available at: http://www.ontoweb.org/About/Deliverables/ (2002)

6. Motta, E., Sabou, M.: Next Generation Semantic Web Applications. In: Proceedings of the First Asian Semantic Web Conference - ASWC 2006, Springer (2006)
7. Object Management Group: Ontology definition metamodel. IBM and Sandpiper Software Inc, Available at: http://www.omg.org/docs/ad/06-05-01.pdf (2006)
8. Gruninger, M., Fox, M.S.: The Role of Competency Questions in Enterprise Engineering. In: IFIP WG5.7 Workshop on Benchmarking - Theory and Practice, Trondheim, Norway (1994)
9. Öhgren, A., Sandkuhl, K.: Towards a methodology for ontology development in small and medium-sized enterprises. In: IADIS Conference on Applied Computing, Algarve, Portugal (2005)
10. Shvaiko, P., Euzenat, J.: A Survey of Schema-based Matching Approaches. Journal on Data Semantics (2005)
11. Shaw, M., Garlan, D.: Software Architecture - Perspectives on an Emerging Discipline. Prentice-Hall, Upper Saddle River (1996)
12. Fielding, R.T.: Architectural Styles and the Design of Network-based Software Architectures. PhD thesis, University of California, Irvine, Irvine, California (2000)
13. Buschmann, F., Meunier, R., Rohnert, H., Sommerlad, P., Stal, M.: Pattern-oriented Software Architecture - A System of Patterns. John Wiley & Sons (1996)
14. Fowler, M.: Patterns of Enterprise Application Architecture. Addison-Wesley (2003)
15. Medvidovic, N., Taylor, R.: A Classification and Comparison Framework for Software Architecture Description Languages. IEEE Transactions on Software Engineering **26** (2000) 70–93
16. Dashofy, E.M., van der Hoek, A., Taylor, R.N.: A Highly Extensible, XML-Based Architecture Description Language. In: Proc. of WICSA 2001. (2001)
17. Gangemi, A.: Ontology Design Patterns for Semantic Web Content. In: Proceedings of ISWC2005, Springer (2005) 262–276
18. Meszaros, G., Doble, J.: Metapatterns: A pattern language for pattern writing. In: The 3rd Pattern Languages of Programming conference, Monticello, Illinois (1996)

The SPARQL Query Graph Model for Query Optimization

Olaf Hartig and Ralf Heese

Humboldt-Universität zu Berlin
Department of Computer Science
{hartig|rheese}@informatik.hu-berlin.de

Abstract. The Semantic Web community has proposed several query languages for RDF before the World Wide Web Consortium started to standardize SPARQL. Due to the declarative nature of the query language, a query engine should be responsible to choose an efficient evaluation strategy. Although all RDF repositories provide query capabilities, some of them require manual interaction to reduce query execution time by several orders of magnitude.

In this paper, we propose the SPARQL query graph model (SQGM) supporting all phases of query processing. On top of the SQGM we defined transformations rules to simplify and to rewrite a query. Based on these rules we developed heuristics to achieve an efficient query execution plan. Experiments illustrate the potential of our approach.

1 Introduction

With introducing the RDF data model researchers have investigated approaches to manage RDF data efficiently. As a part of these efforts, researchers as well as developers are looking for approaches to reduce query execution time. Current RDF repositories often rely on existing database technologies, e.g., relational databases [1,2,3] or Berkley-DB [4]. A main reason can be seen in the experience with efficient query processing gained over the past decades.

However, a posting in a newsgroup[1] illustrates that there is still room for improvement. A user of the Jena Semantic Web Framework [3] asked in a posting why his program containing only simple queries on a small RDF database runs so slowly. The answer was quite surprising: the user should put the more specific part of the query first, because it made a significant difference. Rearranging the queries resulted in a reduction of the execution time by a factor of 220, i.e., $33000ms \rightarrow 150ms$.

The development of RDF repositories came along with several proposals for query languages. Combining concepts of these languages, the World Wide Web Consortium currently standardizes a query language for RDF, namely SPARQL. Although the specification is still in the status of a working draft, it has already been adopted by recent implementations of RDF repositories. Due to its declarative nature, a query engine has to choose an efficient way to evaluate a query.

[1] http://groups.yahoo.com/group/jena-dev/message/21436 (posted on Mar 8, 2006)

E. Franconi, M. Kifer, and W. May (Eds.): ESWC 2007, LNCS 4519, pp. 564–578, 2007.

As shown in the above example, the user has still to choose the right order of triple patterns to minimize query execution time – contradicting to the nature of a declarative query language.

In this paper, we make a first step to consider all phases of query optimization in RDF repositories. We adopted the well-known query graph model developed for the Starburst database management system [5] to represent SPARQL queries and propose the SPARQL query graph model (SQGM). In our approach, this model forms the key data structure for all phases of query processing and is used to store information about the query being processed. Figure 1 depicts the main phases of query processing in database systems. The small arrows depict the information flow and the large arrows depict the control flow. See [6] for a description of the phases. We defined transformation rules on top of the SQGM to provide means for rewriting and simplifying the query formulation.

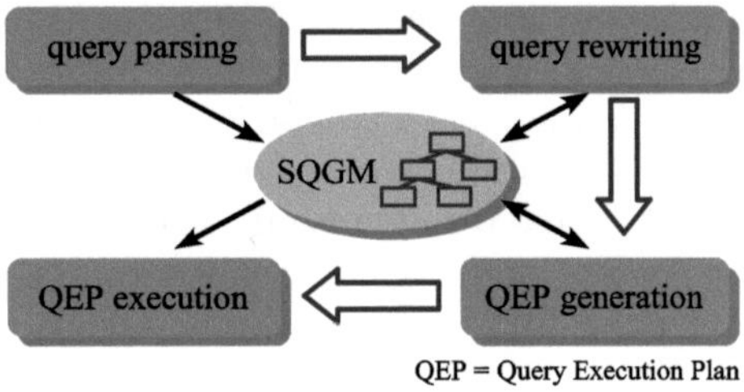

Fig. 1. Phases of the query processing

1.1 Running Example

Throughout this paper we use the same SPARQL query (Figure 2) as a running example. The query asks for the names of all graduate students taking some course. Due to the optional clause in line 7, the result of this query may also include students taking no courses at all.

```
1 PREFIX ub: <http://www.lehigh.edu/.../univ-bench.owl#>
2 PREFIX rdf: <http://www.w3.org/1999/02/22-rdf-syntax-ns#>
3 SELECT ?n ?c
4 FROM <http://example.org/University0.owl>
5 WHERE {
6       ?s rdf:type ub:GraduateStudent .
7       OPTIONAL { ?s ub:takesCourse ?c . }
8       ?s ub:name ?n .
9 }
```

Fig. 2. Example SPARQL query

Except for the prologue containing prefix declarations (lines 1–2), the structure of a SPARQL query is similar to the query language SQL for relational database systems. A SPARQL query consists basically of three parts: (a) the result specification part including solution modifiers (line 3), (b) the dataset definition part (line 4), and (c) the restriction definition part (lines 5–9). We refer to [7] for further explanation of the syntax .

1.2 Structure and Goals of This Paper

In the next section, we define the SPARQL query graph model and its graphical representation. Section 3 describes transformation rules based on this model which rewrite a query into a semantically equivalent one. The goal of rewriting a query is to achieve an efficient query execution plan. To evaluate our approach, we implemented it on top of the Jena Semantic Web Framework. We present and discuss some results of our experiments in Section 4. Section 5 discusses related work; Section 6 concludes the paper.

2 SPARQL Query Graph Model

In [5] Pirahesh et al. developed the *query graph model* (QGM) which defines a conceptually more manageable representation of an SQL query. We adapted this model to represent SPARQL queries – the SPARQL query graph model (SQGM). The basic elements of an SQGM are operators and dataflows – an operator processes data and a dataflow connects the output and input of two operators. In this section we describe the adaption of the query graph model to represent SPARQL queries. We begin with the description of the basic elements of the model and then explain the translation of SPARQL queries into the model.

2.1 Fundamentals

An SQGM can be interpreted as a directed labeled graph with vertices and edges representing operators and dataflows, respectively. Figure 3 shows the graphical representation of the SQGM for our example (cf. Figure 2). Operators are depicted as boxes consisting of a head, a body, and additional annotations. The Definitions 1 and 2 give a basic definition of an operator and a dataflow. In reference to the SPARQL specification, we refine the first definition and introduce operators having special properties and graphical representations below.

Definition 1. *An* operator *performs operations on its input data to generate output data.* □

An edge symbolizes the dataflow between two operators indicating that an operator consumes the output of another. Edges are directed and point to the data consumer.

Definition 2. *A* dataflow *connects two operators and transfers the data provided by one of them and consumed by the other.* □

An operator processes and generates either an RDF graph (a set of RDF triples), a set of variable bindings, or a Boolean value. Any operator has the properties *input* and *output*. The property *input* specifies the dataflow(s) providing the input data for an operator and *output* specifies the dataflow(s) pointing to another operator consuming the output data. We call the operator providing the

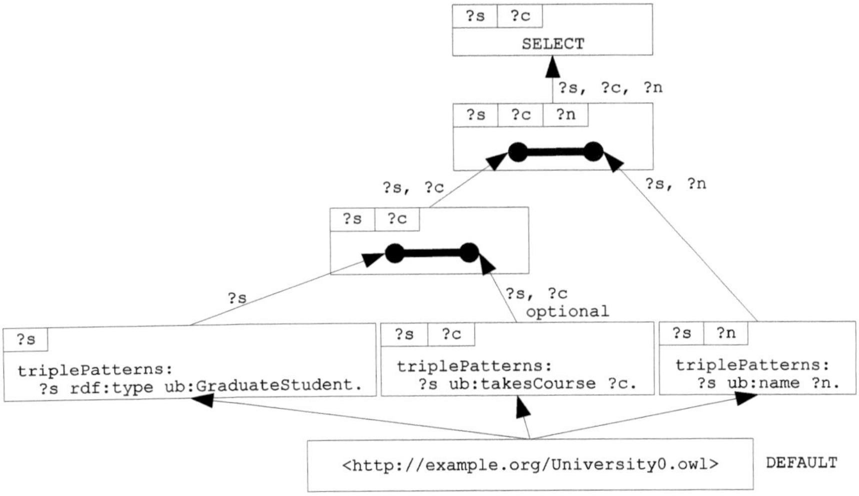

Fig. 3. Graphical representation of the SQGM for the SPARQL query in Figure 2

data of a dataflow the *providing operator* of this dataflow. The operator consuming the data of a dataflow is called the *consuming operator*. The mapping UsingOp : $DF \rightarrow OP$ assigns the consuming operator to every dataflow and the mapping ProvOp : $DF \rightarrow OP$ assigns the providing operator to every dataflow.[2] The following expressions formally define these mappings; please note that we use a dot notation to access a property.

$$\text{UsingOp}(d) := o \Leftrightarrow d \in o.\text{input}$$
$$\text{ProvOp}(d) := o \Leftrightarrow d \in o.\text{output}$$

According to the produced output of an operator we distinguish between *V-operator* and *G-operator*. A V-operator creates a set of variable bindings. The variables that are bound by an operator are specified in the property *provVars*. In the graphical representation, their names are listed in the head of the operator box. For example, in Figure 3 the right-most operator provides bindings for the variables `?s` and `?n`. The output of a G-operator is an RDF graph, i.e., a set of RDF triples. Since these operators do not bind any variables, they do not have the property *provVars* and the head is omitted in their graphical representation.

Dataflows are also divided into two categories: V-dataflow and G-dataflow. We denote with *GF* the set of all G-dataflows and with *VF* the set of all V-dataflows. V-dataflows are dataflows that originate in a V-operator, i.e., variable bindings are transferred. A V-dataflow has the property *vars* containing all variables that are used by subsequent operators. In the graphical representation a V-dataflow is annotated with the names of the variable contained in *vars*. It holds

[2] We denote with *OP* the set of all operators of an SQGM and with *DF* the set of all dataflows.

$\forall d \in VF : d.\text{vars} \subseteq \text{ProvOp}(d).\text{provVars}$ because not every consuming operator processes the bindings for all variables offered by the providing operator.

G-dataflows are dataflows that originate in a G-operator, i.e., an RDF graph is transferred. That is why they do not have the property *vars*.

Operator Types. We defined a set of operator types to cover the language structures of the SPARQL specification. Table 1 gives an overview of all defined operator types, their meaning, and their specific properties. If it is not noted otherwise, the values of the properties of an operator are listed in the body part of its box in the graphical representation.

Table 1. Overview of all SQGM operator types

Operator Type	Meaning and Properties
Graph Operator	Accesses an RDF graph *iri*: IRI of the RDF graph
Graph Merge Operator	Merges a set of RDF graphs
Graph Selection Operator	Accesses a set of named RDF graphs *var*: a variable bound to the IRI of the selected RDF graph
Graph Pattern Operator	*See detailed description below*
Join Operator	Joins two sets of variable bindings
Union Operator	Calculates the set union of two sets of variable bindings
Solution Modifier Operator	Applies solution modifiers to a set of variable bindings *distinct*: indicates duplicate elimination *orderBy*: determines the order of the result set *limit*: restricts the size of the provided set of variable bindings *offset*: an offset within the provided set of variable bindings
Select Result Operator	Returns only the bindings for the given variable names
Describe Result Operator	Creates an RDF graph describing a set of IRIs and the resources that are bound to given variable names *describedResources*: IRIs and variable names to be considered
Construct Result Operator	Creates an RDF graph by instantiating a template *template*: template graph pattern for constructing the query result
Ask Result Operator	Returns TRUE if the input is not empty

Due to the limited space we cannot define all operators in this paper. Instead, we selected one operator type, the graph pattern operator, which we describe in detail. The graph pattern operator corresponds to the basic graph pattern defined in the SPARQL specification. It is the main building block to specify the part of RDF dataset that is of interest. In the SPARQL query graph model the graph pattern operator is defined as follows:

Definition 3. *A* graph pattern operator *is a V-operator which takes an RDF graph as input and returns the variable bindings for a set of triple patterns and value constraints as defined in [7].The property* input *of a graph pattern operator* *is restricted to G-dataflows. Furthermore, the following specific properties are defined:*

- triplePatterns*: a list of triple patterns to be matched*
- constraints*: value constraints to be satisfied by the variable bindings*
- contr*: indicates a provable contradiction in the value constraints* □

In the graphical representation of a graph pattern operator, the head of the box lists the variable names bound by the operator and the body contains the properties of the operator (see Figure 3).

SPARQL Query Graph Model. Before we describe the translation of a SPARQL query into an SQGM in the following section, we present the definition of SQGM.

Definition 4. *A* SPARQL query graph model (SQGM) *represents a SPARQL query. It is a tuple* $(OP, DF, r, dflt, NG)$ *where*

- OP *denotes the set of all operators necessary to model the query,*
- DF *denotes the set of all dataflows necessary to model the query,*
- r *is an operator responsible for generating the result of the query* $(r \in OP)$*,*
- $dflt$ *is an operator providing the default RDF graph of the queried RDF dataset* $(dflt \in OP)$*,*
- NG *is the set of graph operators that provide the named graphs* $(NG \subset OP)$*.* □

2.2 Translating a SPARQL Query to an SQGM

Our process for constructing an SQGM from a SPARQL query is described in Algorithm 1. It takes a query as input and returns the corresponding SQGM. The SPARQL query is given as a tuple (DS, GP, SM, R) where DS is the queried RDF dataset, GP is a graph pattern, SM is a set of solution modifiers, and R is the result form [7]. In the remainder of this section, we describe each step separately.

Algorithm 1. Translating a SPARQL query q into an SQGM Q

INPUT: $q := (DS, GP, SM, R)$ – a SPARQL query
OUTPUT: $Q := (OP, DF, r, dflt, NG)$ – an SQGM representing q

1. Generate operators for the RDF dataset DS;
2. Generate operators for the graph pattern GP;
3. Generate operators for the set of solution modifiers SM;
4. Generate operators for the result form R;

Generate Operators for the RDF dataset. In the first step of the algorithm, the RDF dataset is modeled. An RDF dataset DS consists of a default RDF graph G and a set of named graphs $(\langle u_j \rangle, G_j)$. We use *graph operators* and *graph merge operators* to represent these parts. Although RDF graphs can be stored differently, e.g., in secondary storage or main memory, they are modeled in the same way. To model the RDF dataset the algorithm creates a graph operator for the default graph and each named graph (cf. Table 1). The property *iri* is set to the IRI of the respective RDF graph. The operators for the named graphs $(\langle u_j \rangle, G_j)$ are added to the sets OP and NG, the operator providing the default graph G is added to OP and is assigned to the *dflt* element. In the case that the default graph consists of a multiple RDF graphs, a graph merge operator is additionally created which provides access to the merge of these RDF graphs.

In the graphical presentation, the operator providing the default graph is additionally annotated with the keyword `DEFAULT`. In our running example (see Figure 3), the box at the bottom models the access to the default RDF graph.

Generate Operators for the Graph Pattern. The second step of query translation models the graph pattern GP of a SPARQL query. GP is either a basic graph pattern, a group graph pattern, value constraints, an optional graph pattern, an union graph pattern, or an RDF dataset graph pattern. Some of them (group graph pattern, optional graph pattern, union graph pattern, and RDF dataset graph pattern) contain other graph patterns. Thus, we use a tree of connected operators to represent GP and the algorithm traverses the graph pattern depth first to generate the operators. While traversing the graph pattern GP the algorithm creates the corresponding SQGM operators and dataflows bottom-up and adds them to the sets OP and DF, respectively. Furthermore, the properties of the operators and dataflows such as *input*, *output*, *provVars*, and *vars* are set accordingly.

In the following, we describe which operators are generated with respect to the different graph pattern types. For each *basic graph pattern* and *value constraint* a graph pattern operator is created and added to the set OP. Its properties *triplePatterns* and *constraints* are initialized according to the graph pattern at hand. Its property *provVars* contains all variables occurring in the basic graph pattern and value constraint. While constructing the SQGM the algorithm keeps track of the data source needed to set the *input* property of the new operators.

The example SQGM (Figure 3) contains three graph pattern operators – boxes containing a *triplePattern* property – representing the three basic graph patterns in our example query. The incoming edges represent G-dataflows indicating that these operators process the default RDF graph.

A *union graph pattern* $U(P_1, \cdots, P_n)$ operates on a set of graph patterns P_i. It is modeled in two steps. First, the operators of all graph patterns P_i are created recursively. The result is a set of operator trees, one for each P_i. Second, the root operators of these trees are connected using union operators. Unions involving three or more graph patterns are modeled as a binary tree of union operators.

A *group graph pattern* $G(P_1, \cdots, P_n)$ containing multiple graph patterns P_i is modeled similarly to a union graph pattern. The only difference is that the root operators of the trees are connected by join operators. The example query in Figure 2 contains a group graph pattern consisting of two graph patterns: a optional graph pattern (line 6–7) and a basic graph pattern (line 8). It translates into a single join operators, i.e., the upper one. The second join operator is a result of translating the optional graph pattern which we discuss next.

An *optional graph patterns* $O(P_1, P_2)$ consists of two graph pattern P_1 and P_2. It is basically modeled in the same way as a group graph pattern with two patterns, except that the dataflow between the operator representing P_2 and the join operator is initialized with the property *optional* set to TRUE. In the graphical representation this is reflected by the keyword `optional` (see Figure 3).

A *RDF dataset graph pattern* GRAPH(g,P) matches a pattern P on one or more named graphs depending on whether g is an IRI or a variable. In case g is an IRI, the data source of the operators representing P is the graph operator that provides access to the RDF graph with the IRI g. Otherwise, if g is a variable, the algorithm creates a graph selection operator providing access to all named RDF graphs. Depending on its type the graph pattern P is modeled as described before.

Generate Operators for the Solution Modifiers. In the third step, the algorithm generates the operators representing the solution modifier set SM of the SPARQL query. Solution modifiers such as `order by` or `limit` manipulate their input data to change the order of the result set or to select a subset of it. If the set of solution modifiers SM of the SPARQL query is not empty, a solution modifier operator is created and its specific properties are initialized according to values in SM. The new operator is added to set OP of the SQGM and connected to the root operator representing the graph pattern GP.

Generate Operators for the Result Form. The last step of Algorithm 1 generates the operators for the result form R. The authors of the SPARQL specification [7] distinguish the four result forms `SELECT`, `DESCRIBE`, `CONSTRUCT`, and `ASK`. According to the result form of the query, the algorithm creates the appropriate operator and connects it to the SQGM generated so far, i.e., it constructs a dataflow to either the solution modifier operator or the operator representing the graph pattern. Furthermore, the operator specific properties are set. For example, if the result form is `CONSTRUCT`, then the template graph pattern is assigned to the property *template* of the construct result operator.

3 Query Rewriting Based on SQGMs

While our query graph model for SPARQL is intended to support all phases of query processing (cf. Figure 1), we currently focus on query rewriting. In the query rewriting phase, the generated SQGM is transformed into a semantically equivalent one to achieve a better execution strategy when processed by the plan optimizer. For instance, rules may aim at simplifying complexly formulated queries by merging graph patterns, e.g., avoiding join operations, and eliminating redundant or contradicting restrictions. In this section, we first define semantical equivalence of two SQGMs and, thereafter, specify transformation rules. These rules are finally combined to heuristics which promise to result in efficient query execution plans.

It is essential for query rewriting that applying transformation rules to a query has no impact on the query result. We define semantical equivalence of two SQGMs as follows:

Definition 5. *Two SQGMs q and q' are* semantically equivalent, *if the equation*

$$\text{Result}_D(q) = \text{Result}_D(q')$$

holds for any RDF dataset D, where $\text{Result}_D(q)$ *denotes the result set of evaluating q on D.* □

We distinguish two categories of rules for SQGM restructuring: transformation rules and rewrite rules. While transformation rules change an SQGM only locally, i.e., an operator and its immediate neighbors are affected, rewrite rules are more complex and affect the complete SQGM. Since transformation rules are the building blocks of rewrite rules, we introduce them first.

Definition 6. *A* transformation rule *is a tuple (n, D, P, I) where n is the* name *of the rule, D is a set of operators for which the rule is applicable* (domain), *P is an optional set of Boolean valued expressions being the* preconditions *for applying the rule, and I is a non-empty list of* instructions *for changing the SQGM.* □

Given a transformation rule, if an operator is contained in D and all preconditions P are fulfilled for this operator then the transformation rule can be applied to the SQGM, e.g., the instructions in I are executed. In the context of query rewriting, we are interested only in transformation rules which transform an SQGM into an semantical equivalent one. In the following, we give an example for the definition of a transformation rule.

Example 1. Figure 4(a) depicts a part of an SGQM. The two graph pattern operators can be merged with the join operator without affecting the semantics of the represented query. Merging would be beneficial in several cases, e.g., offering more options for index application or potentially facilitating further simplification. Therefore, we developed the transformation rule (MergeJoinedGPOs, D, P, I). It merges the join of two graph pattern operators to a single operator. The rule is applicable to join operators, i.e., $D = JO$, where JO denotes the set of all join operators.

Let $o \in D$ be a join operator, $(i_L, i_R) := o.\text{input}$ be the left and right input dataflows originating from the two operators $o_L := ProvOp(i_L)$ and $o_R := ProvOp(i_R)$, and GPO be the set of all graph pattern operators in the SQGM. Then the set of preconditions P contains the following expressions:

i) $o_L \in GPO$
ii) $o_R \in GPO$
iii) $i_L.\text{optional} = \text{FALSE}$

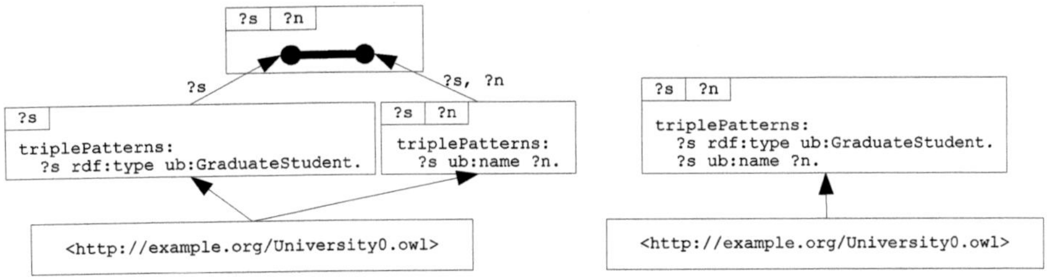

Fig. 4. Part of an SQGM before (a) and after (b) applying the transformation rule MERGEJOINEDGPOS

iv) i_{R}.optional = FALSE
v) $|o_{\mathrm{L}}.\mathrm{output}| = 1$
vi) $|o_{\mathrm{R}}.\mathrm{output}| = 1$
vii) $\mathrm{ProvOp}(o_{\mathrm{L}}.\mathrm{input}) = \mathrm{ProvOp}(o_{\mathrm{R}}.\mathrm{input})$

Hence, the preconditions of MERGEJOINEDGPOS are the following: both operators are graph pattern operators (i and ii), none of them provides optional data (iii and iv), they are not used in another context (v and vi), and the input of both operators originates from the same RDF graph (vii).

The rule transforms $q = (OP, DF, r, dflt, NG)$ to $q' = (OP', DF', r, dflt, NG)$ using the following instructions in I:

i) $o_{\mathrm{L}}.\mathrm{triplePatterns} := o_{\mathrm{L}}.\mathrm{triplePatterns} \sqcup o_{\mathrm{R}}.\mathrm{triplePatterns}$
ii) $o_{\mathrm{L}}.\mathrm{constraints} := o_{\mathrm{L}}.\mathrm{constraints} \wedge o_{\mathrm{R}}.\mathrm{constraints}$
iii) $o_{\mathrm{L}}.\mathrm{contr} := o_{\mathrm{L}}.\mathrm{contr} \vee o_{\mathrm{R}}.\mathrm{contr}$
iv) $o_{\mathrm{L}}.\mathrm{output} := o.\mathrm{output}$
v) $o.\mathrm{output} := ()$
vi) $DF' := DF \setminus \{i_{\mathrm{L}}, i_{\mathrm{R}}, o_{\mathrm{R}}.\mathrm{input}\}$
vii) $OP' := OP \setminus \{o, o_{\mathrm{R}}\}$

The operator $\sqcup$ used in the instructions i) denotes the merge of two triple patterns. This merge of triple patterns is similar to the merge of RDF graphs except that the triple patterns may contain variables. Especially, implementations of this operation have to consider the scope of blank nodes.

The operators in Figure 4(a) satisfy the preconditions of the transformation rule MERGEJOINEDGPOS. Thus, the rule can be applied to this part of the SQGM. Figure 4(b) shows the part after executing the instructions of the transformation rule. The three operators have been merged to a single graph pattern operator containing all triple patterns of the original graph pattern operators. □

Rewrite rules are similar to transformation rules, but consider the complete SQGM. Every rewrite rule follows a certain *goal*, e.g., merge as many graph pattern operators as possible. To reach a goal it may be necessary to apply a single or a sequence of transformation rules several times. A goal of a rewrite rule is *reached* if no further steps are possible.

Definition 7. *A* rewrite rule *is a tuple* $(n, G, S, A_{check}, A_{compile})$ *where* n *is the* name *of the rule,* G *specifies a goal,* S *defines steps to reach the goal,* A_{check} *is an algorithm to determine if another step is executable, and* $A_{compile}$ *is an algorithm that compiles a sequence of transformation rules to actually execute the next step.* □

In the case that multiple rewrite rules are applied to an SQGM, it may happen that the goal of a rewrite rule is contrary to the goal of another. Our current approach to solve this problem is to choose manually the set of rewrite rules to be applied.

Example 2. As mentioned in Example 1, merging joined graph pattern operators is beneficial in some cases. Following this assumption, we developed the rewrite rule MERGEALLJOINEDGPOs. Its goal is to merge as many graph pattern operators as possible.
In every step, the rewrite engine selects two graph pattern operators and merges them if possible. The algorithm A_{check} searches for a pair of candidate operators and checks if another step is executable. This is not trivial. As already mentioned, the translation algorithm constructs a join tree. Thus, graph pattern operators that could be merged by the transformation rule MERGEJOINEDGPOs, may occur at any place in the join tree. In order to merge these operators nevertheless, the join tree has to be restructured, so that the candidates become children of the same join operator. Due to the limited space, we do not discuss the transformation rules to restructure a join tree in this paper. However, the algorithm A_{compile} compiles a sequence of these additional rules being suitable to restructure the join tree appropriate. After restructuring the transformation rule MERGEJOINEDGPOs finally performs the merge.

Considering Figure 3 as an example, the compiled sequence contains transformation rules to switch the places of the left-most and the right-most graph pattern operators and ends with the rule MERGEJOINEDGPOs. □

Having rewrite rules defined, we are able to specify *heuristics*. A heuristic consists of a set of preconditions and a set of rewrite rules. If the preconditions are fulfilled and the rewrite rules are applied to an SQGM then the heuristic promises that in most cases the resulting SQGM meets a certain efficiency criterion, e.g., reducing query execution time. The following example illustrates the idea of one heuristic.

Example 3. We developed a heuristic that supports the fast path algorithm implemented in the Jena Semantic Web Framework [3]. The fast path algorithm detects triple patterns that can be executed as a single query within the underlying relational database system, e.g., the database can optimize the joins. These triple patterns have to be part of the same execution stage, e.g., contained in the same basic graph pattern. We believe, that merging graph pattern operators of an SQGM reduces the query execution time, because larger sets of triple patterns are created and the fast path algorithm can push the evaluation of larger sets of triple patterns into the relational database. The heuristic suggests to apply rewrite rule MERGEALLJOINEDGPOs (cf. Example 2) if (a) the SQGM contains joined graph pattern operators, (b) the query RDF dataset is stored in a database, and (c) query execution is performed by Jena. □

4 Implementation and Evaluation

We prototypically implemented the SPARQL query graph model on top of the Jena Semantic Web Framework and the ARQ query processor (Version 1.3). Afterwards, we run experiments to evaluate our approach. In this section, we outline our implementation, describe the testing environment, and present some results of our experiments.

The SPARQL query graph model is implemented as an extension to ARQ. While our query engine is derived from the classes of ARQ, the query model has been implemented from scratch. Additionally, we developed a rule engine being responsible for transforming query graph models.

Using our extension, a SPARQL query is processed as shown in Figure 5: First, the ARQ query processor parses the query and generates a query model specific to ARQ. We chose the indirection over the ARQ query model to reuse the SPARQL parser of ARQ. However, this model has the disadvantage that it is very close to the syntax of the query. For example: in contrast to ARQ, our model considers basic graph patterns and filter expressions as a single operation. Because the dataflows are not explicitly defined in the ARQ model, it is not possible to distinguish between provided and actually used variables. After parsing the query, the generated query model is translated to an SQGM and heuristics are applied to provide a good basis for an efficient query execution plan. Then the restructured SQGM is translated back into an ARQ query model. ARQ generates a query execution plan which is executed finally.

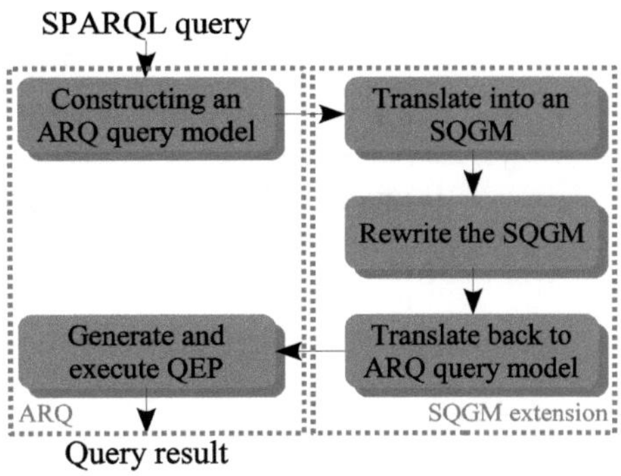

Fig. 5. Processing of a SPARQL query using the SQGM extension

Based on the Lehigh University Benchmark [8], we generated three sets of RDF data with increasing size using the scaling factors 1, 5, and 10, e.g., the sets contained about 100k, 624k, and 1272k triples, respectively. The data was managed by a relational database system and stored on secondary storage. Furthermore, we developed 41 queries which combined basic graph patterns, OPTIONAL, FILTER and UNION clauses in various ways. For each query we measured the query execution time two times: with and without applying heuristics.

The diagram in Figure 6 illustrates the results of applying the heuristic presented in Example 3 to our example query. We see that the execution of the reformulated query is about 2.4 times faster than the original query. The reason for the better performance of our approach is that Jena implements the fast path algorithm. Considering our example, the algorithm can not facilitate combined pattern matching, because the correlation between the lines 6 and 8 in Figure 2 is not detected by Jena. The previously presented heuristic (cf. Example 3) transforms the query so that the fast path algorithm can facilitate to push down the pattern evaluation into the database.

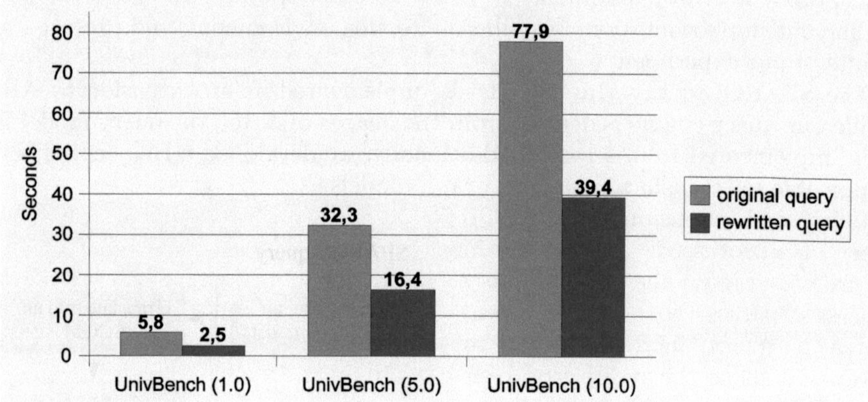

Fig. 6. Query execution time of our running example (Figure 2)

Further results of our experiments are the following. Parsing and transforming the query took less than 1 ms. Thus, these operations have little impact on the query execution time. The average savings were about 87%. Partly, this high savings are caused by a query which contained a contradicting value constraint. In contrast to our implementation, ARQ did not detect the contradiction. Instead of immediately returning an empty result, ARQ executed the query nevertheless.

5 Related Work

Although many approaches have been investigated how to store and to query RDF data, less attention has been paid to query optimization as a whole. However, some phases of query processing have already been considered. Cyganiak [9] and Frasincar et al. [10] considered query rewriting and proposed algebras for RDF. Derived from the relational algebra they allow to construct semantically equivalent queries. Furthermore, Serfiotis et al. developed algorithms for containment and the minimization of RDF/S query patterns in [11]. This approach examines only hierarchies of concepts and properties. All these approaches have in common that they consider only a small part of the query processing, while the proposed query graph model supports all phases of query processing.

Pérez et al. present an approach to formalize the semantics of the core of SPARQL in [12]. They also define a set of equivalence expressions that allow to transform any SPARQL query into a simple normal form consisting of unions of graph patterns. Currently, we investigate how their formal model can be used to formulate transformation rules on top of SQGMs.

To improve the query execution time, several ways of storing RDF data have been developed and evaluated, e.g., Jena [3], Sesame [2], Redland [4] and a path-based relational RDF database [13]. But as the introductory example demonstrated, we have reasons to believe that developers of current RDF repositories have to re-engineer the query engines to declaratively deal with queries.

Christophides et al. focused on indexing RDF data, e.g., in [14] they developed labeling schemes to access subsumption hierarchies efficiently. Again, developing index algorithms is only a small part of query processing. It is also important to enable the query processor to transform the query such that an effective index can be invoked.

6 Conclusion and Future Work

Over the last years several approaches for storing and querying RDF data have been developed and evaluated. Although some research has been undertaken to provide means for query rewriting, we think that query optimization as a whole has not been considered so far. In this paper, we proposed the SPARQL query graph model (SQGM) supporting all phases of query processing. This model forms the key data structure for storing information that is relevant for query optimization and for transforming the query. We presented transformation rules which enable the query processor to transform an SQGM. We combined sets of transformation rules to rewrite rules as the base for heuristics. These heuristics enabled the Jena-based query execution to exploit the fast path algorithm more often. Our experiments demonstrated the potential of our approach to reduce query execution time.

The SPARQL query graph model can easily be extended to represent new concepts. This is important since the current SPARQL specification defines only basic query structures. Widely used structures such as group by, subqueries, and views are currently not supported, but will certainly be added in near future.

In our future work, we will develop further heuristics and exploit additional information to decide on the transformation rules being applied to the SQGM, e.g., information about the RDF schema or statistical data about the RDF datasets. Furthermore, we currently work on the problem of selecting indexes to minimize the costs of query execution. In the future, we want to combine the transformation of SQGMs with the selection of indexes. For example, apply rules to an SQGM such that an index with a high selectivity becomes usable. As a long-term goal, we will investigate extensions to the current SPARQL specification, e.g., subqueries and views, and develop appropriate transformation rules.

References

1. Alexaki, S., Christophides, V., Karvounarakis, G., Plexousakis, D., Tolle, K.: The RDFSuite: Managing Voluminous RDF Description Bases. In Decker, S., Fensel, D., Sheth, A.P., Staab, S., eds.: Proceedings of the Second International Workshop on the Semantic Web. (2001) 1–13
2. Broekstra, J., Kampman, A., van Harmelen, F.: Sesame: A generic architecture for storing and querying RDF and RDF schema. In Horrocks, I., Hendler, J.A., eds.: Proceedings of the First International Semantic Web Conference. Volume 2342 of Lecture Notes in Computer Science., Springer (2002)
3. Wilkinson, K., Sayers, C., Kuno, H., Reynolds, D.: Efficient RDF Storage and Retrieval in Jena2. In Cruz, I.F., Kashyap, V., Decker, S., Eckstein, R., eds.: Proceedings of the First International Workshop on Semantic Web and Databases. (2003)
4. Beckett, D.: The Design and Implementation of the Redland RDF Application Framework. In: Proceedings of the Tenth International Conference on World Wide Web, New York, NY, USA, ACM Press (2001)
5. Pirahesh, H., Hellerstein, J.M., Hasan, W.: Extensible/rule based query rewrite optimization in Starburst. SIGMOD Records **21**(2) (1992) 39–48
6. Heese, R.: Query graph model for sparql. In: International Workshop on Semantic Web Applications: Theory and Practice. Proceedings of ER workshops. (2006)
7. Prud'hommeaux, E., Seaborne, A.: SPARQL Query Language for RDF. http://www.w3.org/TR/rdf-sparql-query/ (2006) W3C Candidate Recommendation.
8. Guo, Y., Pan, Z., Heflin, J.: Lubm: A benchmark for owl knowledge base systems. Journal of Web Semantics **3**(2) (2005) 158–182
9. Cyganiak, R.: A relational algebra for SPARQL. Technical Report HPL-2005-170, HP Laboratories Bristol (2005)
10. Frasincar, F., Houben, G.J., Vdovjak, R., Barna, P.: RAL: an Algebra for Querying RDF. In Feldman, S.I., Uretsky, M., Najork, M., Wills, C.E., eds.: Proceedings of the 13th International conference on World Wide Web, New York, NY, USA, ACM Press (2004)
11. Serfiotis, G., Koffina, I., Christophides, V., Tannen, V.: Containment and Minimization of RDF/S Query Patterns. In: International Semantic Web Conference. Lecture Notes in Computer Science, Springer (2005) 607–623
12. Pérez, J., Arenas, M., Gutierrez, C.: Semantics and Complexity of SPARQL. In Cruz, I., Decker, S., Allemang, D., Preist, C., Schwabe, D., Mika, P., Uschold, M., Aroyo, L., eds.: Proceedings of the 5th International Semantic Web Conference. Volume 4273 of Lecture Notes in Computer Science., Springer (2006)
13. Matono, A., Amagasa, T., Yoshikawa, M., Uemura, S.: A path-based relational RDF database. In: CRPIT '39: Proceedings of the sixteenth Australasian conference on Database technologies, Darlinghurst, Australia, Australia, Australian Computer Society, Inc. (2005) 95–103
14. Christophides, V., Karvounarakis, G., Scholl, D.P.M., Tourtounis, S.: Optimizing Taxonomic Semantic Web Queries Using Labeling Schemes. Web Semantics: Science, Services and Agents on the World Wide Web **1**(2) (2004) 207–228

A Unified Approach to Retrieving Web Documents and Semantic Web Data

Trivikram Immaneni and Krishnaprasad Thirunarayan

Department of Computer Science and Engineering, Wright State University,
3640 Colonel Glenn Highway, Dayton, OH 45435, USA
{immaneni.2,t.k.prasad}@wright.edu

Abstract. The Semantic Web seems to be evolving into a property-linked web of RDF data, conceptually divorced from (but physically housed in) the hyperlinked web of HTML documents. We discuss the Unified Web model that integrates the two webs and formalizes the structure and the semantics of interconnections between them. We also discuss the Hybrid Query Language which combines the Data and Information Retrieval techniques to provide a convenient and uniform way to retrieve data and documents from the Unified Web. We present the retrieval system SITAR and some preliminary results.

Keywords: Semantic Web, Information Retrieval, Data Retrieval, Hybrid Retrieval, Unified Web, Hybrid Query Language.

1 Introduction

Semantic Web [1] is a term used to describe the family of description languages, standards, and other technologies which aim at "extending" the current web by making its content machine accessible. Since the Resource Description Framework (RDF) forms the foundation of this "extension", we can visualize the Semantic Web (SW) as a labeled graph with resources as nodes and binary predicates as edges (web of data). This is in contrast to the Hypertext Web (HW) which is a graph with resources (usually documents) as nodes and hyperlinks as edges (web of documents).

An interesting question that arises is as to where the Web documents fit into the SW and how they can be retrieved. Intuitively, since the Web documents are resources that are identified by their URIs, we can view them as nodes in the SW graph. The document *content* can be explicitly incorporated into the SW as literals. We can then use RDF query languages such as SPARQL [2], which enable RDF graph traversal and support regular expression matching of strings (literals), to retrieve the documents based upon their neighborhood as well as their content. For example, we can pose queries such as *retrieve documents authored by Tragula and contain the string "Spectrographic"*. This is classic Data Retrieval (DR).

Arguably, for this method of retrieving Web documents to have any remote chance of out-performing current Information Retrieval (IR) techniques, each and every Web document should have highly useful semantic *descriptions*. Some technologies such as RDFa [3] enable embedding of semantic markup in a HTML document. Even if such technologies gain wide usage, unless we find a way to (automatically) create

E. Franconi, M. Kifer, and W. May (Eds.): ESWC 2007, LNCS 4519, pp. 579 – 593, 2007.

semantic descriptions of all of the existing HW documents, there will always be a large corpus of documents isolated from the SW. Another issue here is that query languages such as SPARQL require the users to have intimate knowledge of the underlying schema (exact URIs) to compose queries. The simple keyword-based interfaces that systems such as Yahoo! and Google expose to their users is another compelling reason to stick to IR techniques for retrieving Web documents. So, we seem to be better off retrieving data from the SW using DR techniques and retrieving documents from the HW using keyword-based IR techniques. In this sense, when seen from (data or document) retrieval perspective, the Semantic Web is, conceptually, a web of data that is estranged from the web of documents that is the Hypertext Web.

Our high level goal is to *view* the Semantic Web and the Hypertext Web as a unified whole and retrieve data and documents from this Unified Web (UW) [4]. This way, we can utilize the available semantic descriptions to enhance Web document retrieval and will also have the option of using the information from the (unstructured documents of) HW to improve the SW data retrieval.

The web documents can be broadly divided into the following three categories – those meant primarily for human consumption (HTML, plain text, jpg, etc.), those meant primarily for machine consumption (RDF, OWL, RDFS, etc.) and hybrid documents that are meant for both machine and human consumption (RDFa, microformats and other such technologies that allow embedding of semantic markup in HTML/XHTML documents [5]). Our goal is to facilitate the retrieval of all the above three types of documents while fully exploiting semantic markup/descriptions when available to increase retrieval effectiveness.

We want to enable lay users to retrieve human-consumable documents (first and third types) using the traditional keyword-based query mechanism (with minimal enhancements). We want to transparently use the available SW data to enhance the retrieval process. For example, if the user knows that she is looking for Jaguar the car, she should be able to communicate this disambiguating information to the system using a query such as "car::jaguar".

For more informed users, we want to provide a light-weight, keyword-based hybrid query language, that does not require knowledge of the underlying schema (exact URIs). These users should be able to use the query language to retrieve all three types of documents. That is, they should be able to (i) retrieve human-consumable documents by posing queries such as *retrieve documents authored by Tragula (the professor) and are about spectrography,* or *retrieve homepages of professors named John*, (ii) query for and retrieve SW documents (RDF, OWL, RDFS, etc.) by posing queries such as *retrieve documents that assert triples about the ventriloquist John Smith,* and (iii) retrieve hybrid documents (e.g., RDFa) by posing queries that combine the features of the above two types of queries. In addition, the users should be able to query and retrieve *data* by posing questions such as *what is professor Tragula's phone number* or *list all the elements in group 1 of the periodic table*, even in the absence of schema information.

We first describe the Unified Web model in Section 2, followed by the description of the Hybrid Query Language in Section 3. We discuss the implementation of our system SITAR and present some results in Section 4. We discuss related research in Section 5 and conclude with Section 6.

2 The Unified Web Model

The Unified Web model aims to integrate the SW and the HW into a single unified whole by encoding the two webs and the connections between them. The UW model is a graph of nodes and edges (*N, E*). A node is an abstract entity that is uniquely identified by its URI. There are two categories of nodes: (i) Natural nodes (*NN*) and (ii) System defined nodes (*SN*). The natural nodes can be further classified as plain (or non-document) nodes (*PN*) and document nodes (*DN*) based on whether or not a node has an associated document. The system defined nodes can be further classified as literal nodes (*LN*), triple nodes (*TN*) and blank nodes (*BN*). The system creates a URI and assigns it to each blank node, triple and literal that it encounters on the Web.

There are two categories of edges: (i) User defined edges (*UE*) and (ii) System defined edges (*SE*). The user defined edges come from the triples in the (Semantic Web) documents while the system defined edges are defined to make explicit the interconnections between the HW and the SW. The system defined edges are the following. The *asserts* edge exists from a node (document) to each of the RDF statements found in the associated document. The RDF statement itself has a subject, a property and an object. There is no restriction as to how a triple is obtained from the document. The *hasDocument* edge exists from a node to a literal. The literal is the string representation of the document associated with the node. A *hyperlinksTo* edge exists from a node A to another node B if there is a hyperlink from the document of node A to the document of node B. The *linksTo* edge exists from node A to node B if a *hyperlinksTo* relationship exists from node A to node B, or node B occurs in any of the triples asserted by node A (see Figure 1).

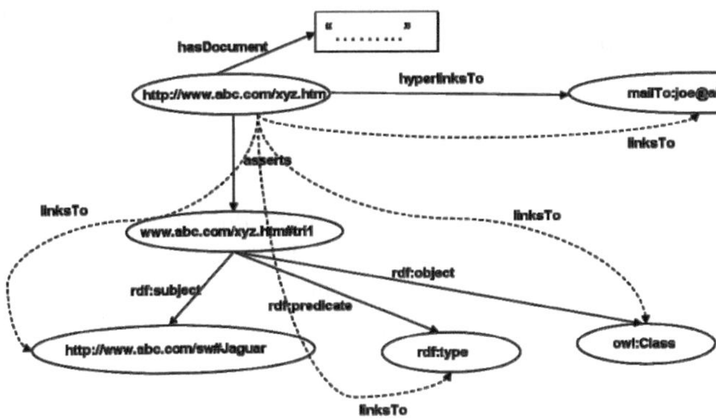

Fig. 1. Relationships

More formally, they can be specified as functions/relations in terms of their signatures (domains and ranges), and include:

$$hyperlinksTo \subseteq DN \times NN$$
$$linksTo \subseteq DN \times NN$$
$$asserts \subseteq DN \times TN$$
$$hasDocument: DN \rightarrow LN$$

These relations are not independent and cannot be assigned arbitrarily. They must satisfy at least the following constraints:

$\forall$n $\in$ *DN*, $\forall$m $\in$ *NN*: if [n, *hyperlinksTo*, m] $\in$ *SE* then [n, *linksTo*, m] $\in$ *SE*

The Unified Web model is not a simple super-imposition of the SW graph over the hypertext graph. The Semantic Web can be thought of as a global RDF graph constructed by gathering all possible RDF triples from documents that reside on the Hypertext Web. The UW reifies each of the SW triples by explicitly encoding the *asserts* relationship between a document and the triple that is extracted from it. The UW can be visualized as a meta – Semantic Web which in itself can be an RDF graph (one that subsumes all RDF graphs found on the web). In addition, this RDF graph also encodes the Hypertext Web (HTML documents and hypertext links between them). The aim is to encode the HW (*hyperlinksTo* and *hasDocument*) and the SW and the connections between the two (such as *asserts* which is not explicitly defined by the user but is rather constructed by the system) while allowing easy mapping of data retrieval queries meant for the "conventional" SW to those for the UW [4]. The *linksTo* tries to define a generic "connection" between two nodes. It seeks to establish a definite connection between a document node and a SW data node (which, of course, can be a document node as well) and to deliberately blur the distinction between such a connection and a hypertext connection. The *linksTo* edge is for the Unified Web what the hyperlink is for the HW and the property-link is for the SW. In our implementation, we use *linksTo* to view a document as being annotated by the URIs that it *linksTo* and use this information while retrieving documents.

The UW model can be specified and implemented using RDF [4]. Since all the "user triples" are present (in reified from) in the model, query languages like SPARQL can be used to retrieve the data – all we need is a straight-forward mapping of the SPARQL query for the SW to the SPARQL query for the UW.

3 Hybrid Query Language Specification

The Hybrid Query Language [4] enables convenient *navigation* and *extraction* of information from the UW. It enables formulation of precise queries involving URIs, and "approximate" word-based queries that capture context (e.g., wordset, wordset-pairs queries) and/or content (e.g., keyword queries). In other words, it enables access to both HW documents and SW data, incorporating indexing information from the neighboring nodes. Specifically, the wordset queries can use anchor text in the HW to retrieve SW nodes, and wordset pair queries can express disambiguation information using the ISA edges encoded in the SW for semantic search of HW documents.

Before we go into the details of the query language, let us first define some more utility functions/relations in addition to the four in the previous section:

homeURI: N → Set(URI)
externalTexts: NN → PowerSet(STRINGS)
indexWords : NN → PowerSet(STRINGS)
parameters: DN → PowerSet(STRINGS)
hasTriples: DN → PowerSet(NN x *NN* x *NN)*
hasLiteral: LN → STRINGS

URI denotes a string that must satisfy the URI syntax requirement (RFC 3986), while *STRINGS* denotes a set of words, phrases, and other fragments. *PowerSet* operator yields a set of all subsets. The members of *NN* x *NN* x *NN* are referred to as the triples (such as those found in RDF documents).

homeURI maps a node to its URI. The URI is what we use to refer to a node explicitly. *externalTexts* maps a node to a set of strings, possibly from its neighborhood, providing contextual information. *indexWords* maps a node to a set of strings that can serve as an index to it. These can be composed from the URI and the anchor text from the neighboring nodes among other things. *hasDocument* maps a document node to the associated document text string. *parameters* maps a document node to a set of attribute-value strings capturing OS/Server related book-keeping information on the document. *hyperlinksTo* relates a document node to a node that appears in a hyperlink in the corresponding document. *linksTo* relates a document node to a node that appears in the corresponding document. This can be in the form of a hyperlink or embedded in a triple. *hasTriples* maps a document node to the set of 3-tuples of nodes that appear in the corresponding document. *asserts* relates a document node to a triple node that reifies the triple that appears in the corresponding document. *hasLiteral* maps a literal node to the string it is associated with. It is possible to have multiple literal nodes associated with the same string. *Note that a specific instantiation of the framework can be obtained by defining how these functions/relations* (such as *externalText*, *IndexWords*, etc) *are obtained from the node's neighborhood.*

These functions must satisfy at least the following constraints:

$$\forall n \in NN,\ \forall [n1, n2, n3] \in NN \times NN \times NN:$$
$$[n1, n2, n3] \in hasTriples\ (n) \quad \text{only if}$$
$$[n, linksTo, n1] \in SE \ \wedge\ [n, linksTo, n2] \in SE \ \wedge\ [n, linksTo, n3] \in SE$$

$$\forall n \in N,\ \forall [n1, n2, n3] \in NN \times NN \times NN:$$
$$[n1, n2, n3] \in hasTriples\ (n) \quad \text{if and only if}$$
$$\exists\ tn \in TN : [n, asserts, tn] \ \wedge [tn, rdf{:}subject, n1] \in SE$$
$$\wedge [tn, rdf{:}predicate, n2] \in SE \ \wedge [tn, rdf{:}object, n3] \in SE$$

For convenience, we abuse the language and say that n1, n2, and n3 *appear in* tn in the context of the reification constraint.

In what follows, we motivate and specify the abstract syntax of the queries using a context-free grammar, and the semantics of the queries in terms of the Unified Web model, in sufficient detail to enable prototyping. Our presentation focuses on queries that yield a set of nodes. The "domain information bearing" strings such as the document text, literal, etc. can be easily obtained from a URI by calling corresponding system functions such as *hasDocument*, *hasLiteral*, etc. and from triples using *rdf:subject*, *rdf:predicate*, and *rdf:object*, etc.

TopLevelQuery ::= Nodes-ref | Triples-ref | ...

QUERY: *Nodes-ref ::= u*, where $u \in Set(URI)$.

ANSWER: $Result(u) = \{\ n \in N \mid HomeURI(n) = u\ \}$

SEMANTICS: The *URI-query* returns the set containing the unique node whose *HomeURI* matches the given URI. Otherwise, it returns an error.
EXAMPLE: *http://www.aifb.uni-karlsruhe.de/Personen/viewPersonenglish?id_db=20*

QUERY: *Nodes-ref::=ss*, where $ss \in PowerSet(STRINGS)$.
ANSWER: $Result(ss) = \{ n \text{ in } N \mid ss \subseteq IndexWords(\text{n}) \}$
SEMANTICS: The *wordset query*, *ss*, usually written as a set of strings delimited using angular brackets, returns the set of nodes whose *IndexWords* contain *ss*.
EXAMPLE: *<peter haase>*

QUERY: *Nodes-ref ::= pp::ss*, where $pp, ss \in PowerSet(STRINGS)$.
ANSWER: $Result(pp::ss) = \{ \text{n} \in N \mid ss \subseteq IndexWords(\text{n}) \wedge \exists \text{m} : \text{n ISA m} \wedge pp \subseteq IndexWords(\text{m}) \}$
SEMANTICS: The *wordset-pair query, pp::ss,* usually written as two wordsets delimited using colon, returns the set of nodes such that each node has *IndexWords* that contains *ss* and has an ISA ancestor whose *IndexWords* contains *pp*.
EXAMPLE: *<student>::<peter>*

QUERY: *Triples-ref ::=* u, where $\text{u} \in Set(URI)$.
ANSWER: $Result(\text{u}) = \{ \text{n} \in TN \mid HomeURI(\text{n}) = \text{u} \}$
SEMANTICS: The *triple node URI-query* returns the set containing the unique node whose *HomeURI* matches the given triple node URI. Otherwise, it returns an error. The triple nodes are system generated.
EXAMPLE: *http://www.aifb.uni-karlsruhe.de/Personen/viewPersonFOAF/foaf_80.rdf#tri52*

QUERY: *Triples -ref ::= Single-Var-Triples-ref*
Single-Var-Triples -ref ::= [?var *Nodes-ref Nodes-ref*]
Single-Var-Triples -ref ::= [*Nodes-ref* ?var *Nodes-ref*]
Single-Var-Triples –ref ::= [*Nodes-ref Nodes-ref* ?var]
where *?var* is a variable.
ANSWER: *Result*([?var *Nodes-ref1 Nodes-ref2*]) =

$$\{ \text{t} \in TN \mid \text{n1} \in Result(Nodes\text{-}ref1) \wedge \text{n2} \in Result(Nodes\text{-}ref2) \wedge \exists \text{m} \in N : [\text{m}, asserts, \text{t}] \wedge [\text{t}, rdf{:}predicate, \text{n1}] \wedge [\text{t}, rdf{:}object, \text{n2}] \}$$

Similarly, for the other two cases.
SEMANTICS: The *part triple query* [?var *Nodes-ref1 Nodes-ref2*] returns the set of (system generated) triple nodes that are related by a binary predicate denoted by *Nodes-ref1* to some node denoted by *Nodes-ref2*. Similarly, for the other two cases. Note that this query characterizes a node using its neighborhood.
EXAMPLE: *[<silver> <atomic weight> ?x]*

QUERY: *Triples-ref* ::= [*Nodes-ref, Nodes-ref, Nodes-ref*]
ANSWER: *Result*([*Nodes-ref1*, *Nodes-ref2*, *Nodes-ref3*]) =

$$\{ \text{t} \in TN \mid \text{n1} \in Result(Nodes\text{-}ref1) \wedge \text{n2} \in Result(Nodes\text{-}ref2) \wedge \text{n3} \in Result(Nodes\text{-}ref3) \wedge \exists \text{m} \in N : [\text{m}, asserts, \text{t}] \wedge [\text{t}, rdf{:}subject, \text{n1}] \wedge [\text{t}, rdf{:}predicate, \text{n2}] \wedge [\text{t}, rdf{:}object, \text{n3}] \}$$

SEMANTICS: The *full triple query* [*Nodes-Ref*, *Nodes-Ref*, *Nodes-ref*] returns the set of (system generated) triple nodes matching the node references.
EXAMPLE: *[http://www.aifb.uni-karlsruhe.de/Personen/viewPersonOWL/id2062instance <name> <peter>]*

QUERY: *Triples-ref* ::= *Double-Var-Triples-ref*
Double-Var-Triples-ref ::= [?var, ?var, *Nodes-ref*]
Double-Var-Triples-ref ::= [?var, *Nodes-ref*, ?var]
Double-Var-Triples-ref ::= [*Nodes-ref*, ?var, ?var]
where ?var is a variable.

ANSWER: *Result*([?var, ?var, *Nodes-ref*]) =

$$\{ t \in TN \mid m \in Result(Nodes\text{-}ref) \wedge [t, rdf{:}object, m] \wedge \exists n \in N : [n, asserts, t] \}$$

Similarly, for the other two cases.
SEMANTICS: The *part triple to triples query* [?var ?var *Nodes-ref*] returns the set of (system generated) triple nodes that are related to some node denoted by Nodes-ref. Similarly, for the other two cases. Note that this query characterizes the node neighborhood. Each variable occurrence is independent of the other occurrences.
EXAMPLE: *[?x <title> ?x]*

QUERY: *Nodes-ref* ::= *Nodes-ref AND Nodes-ref*
Nodes-ref ::= *Nodes-ref OR Nodes-ref*
Triples-ref ::= *Triples-ref AND Triples-ref*
Triples-ref ::= *Triples-ref OR Triples-ref*

SEMANTICS: "OR" and "AND" are interpreted as set-union and set-intersection respectively. Each variable occurrence is independent of the other occurrences.

3.1 Queries for Exploring the System-Generated Neighborhood of a Node

QUERY: *Nodes-ref* ::= *getAllTriples*(*Nodes-ref*)

ANSWER: *Result*(*getAllTriples*(*Nodes-ref*)) =

$$\{ t \in TN \mid n \in Result(Nodes\text{-}ref) \wedge n \text{ appears in } t \wedge \exists m \in N: [m, asserts, t] \}$$

SEMANTICS: This query retrieves the (system generated) triple nodes in which the queried node URI appears.
EXAMPLE: *getAllTriples(http://www.daml.org/2003/01/periodictable/PeriodicTable#group_11)*

QUERY: *Nodes-ref* ::= *getLinkingNodes*(*Nodes-ref*)

ANSWER: *Result*(*getLinkingNodes*(*Nodes-ref*)) =

$$Result(Nodes\text{-}ref) \cup \{ m \in N \mid \exists n \in Result(Nodes\text{-}ref) : [m, linksTo, n] \}$$

SEMANTICS: This query retrieves the nodes corresponding to *Nodes-ref* and the document nodes containing references to the nodes corresponding to *Nodes-ref*. Effectively, nodes and their neighborhoods are retrieved.
EXAMPLE: *getLinkingNodes(http://www.aifb.uni-karlsruhe.de/Personen/viewPerson?id_db=2023)*

QUERY: *Nodes-ref* ::= *getAssertingNodes*(*Triples-ref*)
ANSWER: *Result*(*getAssertingNodes*(*Triples-ref*)) =
$\{ m \in DN \mid \exists t \in Results(Triples\text{-}ref) : [m, asserts, t] \}$
SEMANTICS: This query retrieves document nodes containing the triples.
EXAMPLE: *getAssertingNodes([<peter haase> <publication> ?x])*

QUERY: *Nodes-ref* ::= *getDocsByKeywords*(*ss*), where *ss* ∈ *PowerSet(STRINGS)*
ANSWER: *Result*(*getDocsByKeywords*(*kws*)) =
$\{ m \in DN \mid hasDocument(m) = dt \wedge match(kws, dt) \}$
SEMANTICS: This query is analogous to the traditional keyword query that takes a set of keywords and retrieves document nodes that match the keywords. *match* embodies the criteria for determining when a document text is "relevant" to a keyword. It can be as simple as requiring verbatim occurrence, to as complex as requiring stemming, synonym generation, spelling correction, etc. *match* may be *compositional*, that is, *match* (kws, dt) = ∀w ∈ kws: *match*(w, dt), but it is not required.

QUERY: *Nodes-ref* ::= *getLiteralsByKeywords* (*ss*),
where *ss* ∈ *PowerSet*(*STRINGS*)
ANSWER: *Result*(*getLiteralsByKeywords*(kws)) =
$\{ m \in LN \mid hasLiteral(m) = dt \wedge match(kws, dt) \}$
SEMANTICS: This is analogous to the above query customized for literal nodes.
EXAMPLE: *getLiteralsByKeywords(semantic grid)*

3.2 Further Queries for Retrieving Documents

QUERY: *getDocsByContent*: *PowerSet*(*STRINGS*) → PowerSet(DN)
ABBREVIATION FOR: *getDocsByContent*(kws) =
getLinkingNodes(*getDocsByKeywords*(kws))
where kws ∈ *PowerSet*(*STRINGS*)
SEMANTICS: This query retrieves the document nodes with content matching keywords in kws and the neighboring document nodes that reference such nodes. Intuitively, we want to pursue both the "authorities" and the "hubs" [6], assisting both navigational searches and research searches [7].

QUERY: *getDocsByIndexOrContent*: *PowerSet*(*STRINGS*) → *PowerSet*(*DN*)
ABBREVIATION FOR: *getDocsByIndexOrContent* (kws) =

$$getDocsByKeywords(kws) \vee \bigvee_{kw \in kws} getLinkingNodes(kw)$$

where kws ∈ *PowerSet*(*STRINGS*)
SEMANTICS: This query retrieves the document nodes with content matching the keywords kws or in the neighborhood of nodes indexed by kws. Implicitly, the former captures syntactic retrieval and the latter enables semantic retrieval.
EXAMPLE: *getDocsByIndexOrContent(semantic web)*

QUERY: *getDocsByIndexAndContent*:
Nodes-ref x *PowerSet*(*STRINGS*) → *PowerSet*(*DN*)

ABBREVIATION FOR: *getDocsByIndexAndContent* (nr, kws) =
getLinkingNodes(*Result*(nr)) $\wedge$ *getDocsByKeywords*(kws)
where nr $\in$ *Nodes-ref*, kws $\in$ *PowerSet*(*STRINGS*)

SEMANTICS: This query retrieves the document nodes with content matching the keywords kws and in the neighborhood of nodes corresponding to nr. Implicitly, if nr is a URI of a document node containing the keywords kws, then the result will contain this document node. If nr is a URI and this URI and the keywords kws are contained in a document, then the result will contain the latter document node. Similarly, for nodes in *Result*(nr) when nr contains wordset and wordset-pairs.

QUERY: *getDocsByTriplesAndContent*:
Triples-ref x *PowerSet*(*STRINGS*) → *PowerSet*(*DN*)

ABBREVIATION FOR: *getDocsByTriplesAndContent*(tr, kws) =
getAssertingNodes(tr) $\wedge$ *getDocsByKeywords*(kws)
where tr $\in$ *Triples-ref*, kws $\in$ *PowerSet*(*STRINGS*)

SEMANTICS: This query retrieves (semantic web) document nodes that match the keywords and contain the referenced triples.

QUERY: *Single-Var-Triples-list* ::= *Single-Var-Triples-ref*
Single-Var-Triples-list ::= *Single-Var-Triples-ref* *Single-Var-Triples-list*
Nodes-ref ::= *getBindings*(*Single-Var-Triples-list*)

QUERY1: *Nodes-ref* ::= *getBindings*([?var *Nodes-ref* *Nodes-ref*])

ANSWER: *Result*(*getBindings*([?var *Nodes-ref1* *Nodes-ref2*]))
= { n $\in$ N | n1 $\in$ *Result*(*Nodes-ref1*) $\wedge$ n2 $\in$ *Result*(*Nodes-ref2*)
$\wedge$ $\exists$m $\in$ N : [m, *asserts*, t] $\wedge$ [t, *rdf:subject*, n]
$\wedge$ [t, *rdf:predicate*, n1] $\wedge$ [t, *rdf:object*, n2]}

Similarly, for the other two cases.

QUERY2: *Nodes-ref* ::= *getBindings*(*Single-Var-Triples-ref* *Single-Var-Triples-list*)

ANSWER: *Result*(*getBindings*(*Single-Var-Triples-ref* *Single-Var-Triples-list*)) =
Result(*getBindings*(*Single-Var-Triples-ref*)) $\cap$
Result(*getBindings*(*Single-Var-Triples-list*))

SEMANTICS: This query retrieves the bindings for the variables that satisfy all the triple references with single variable. All the variable occurrences are considered identical, that is, they must all be assigned the same value throughout the *getBindings*-argument.

EXAMPLE: *getBindings*([<phdstudent>::<peter> <name> ?x])

EXAMPLE: *getBindings*([?x <group> <group 1>] [?x <color> <white>])

QUERY: *getDocsByBindingsAndContent*:
Single-Var-Triples-list x *PowerSet*(*STRINGS*) → *PowerSet*(*DN*)

ABBREVIATION FOR: *getDocsByBindingsAndContent*(vtl, kws) =
getBindings (vtl) $\wedge$ *getDocsByKeywords*(kws)
where vtl $\in$ *Single-Var-Triples-list*,
kws $\in$ *PowerSet*(*STRINGS*)
SEMANTICS: This query retrieves document nodes that match the keywords and contain the matching triples.
EXAMPLE: *getDocsByBindingsAndContent([<phdstudent>::<peter> <homepage> ?x] "Semantic Grid")*

4 Implementation and Results

We have implemented an Apache Lucene [8] based retrieval system called SITAR (Semantic InformaTion Analysis and Retrieval system) based upon our model. The system can currently index HTML and RDF/OWL files in addition to RDF data. At present, the system does not support pdf, doc files etc. (we index their URIs but their content is not being analyzed).

Evaluating such a hybrid system is an extremely tricky process. The system has DR components (triple matching) which render the precision and recall criteria irrelevant. But at the same time, the system also has IR components such as keyword based retrieval of documents in which case precision and recall become important. In order to evaluate the system in terms of precision and recall, we would need a standard data set (such as MEDLINE dataset) which has documents, their semantic descriptions, some queries, and results of those queries (adjudged to be relevant by human experts). We are still looking for such data sets. Here, we present qualitative results obtained by experimenting with the invaluable AIFB SEAL [9] data.

The AIFB SEAL website has human-consumable XHTML documents (in English and German) along-side OWL documents. Some of the XHTML documents have explicit semantic descriptions (in the OWL documents). We crawled the SEAL website looking only for English versions of web pages and RDF/OWL files using heuristics. The crawler collected a total of 1665 files. Of these, we chose to deliberately ignore some large OWL files (multiple copies of the same file with each copy identified by a different URI) to simplify matters. Our system uses the CyberNeko HTML parser [10] to parse HTML documents and the Jena ARP [11] parser to parse RDF documents. The ARP parser could not parse some of the RDF documents - possibly because of problems with container elements. In the end, a total of 1455 (610 RDF files and 845 XHTML files) were successfully parsed and indexed. A total of 193520 triples were parsed and indexed though there is no guarantee that the same triple was not asserted by two different documents. Note that, all the index structures are persistently stored.

Every URI is analyzed (using several heuristics) to build a set of index words for it. More importantly, if the URI occurs in a HTML document as a hyperlink, we use the anchor text to add to its index words set. A HTML document is indexed by the URIs that it *linksTo* (or *hyperlinksTo*) as well as the words that are extracted from the URIs. In this sense, it can be seen as a bag of URIs and words. An RDF document is indexed by the URIs that it *linksTo* and by the triples (URIs) that it asserts. In this sense, an RDF document can be seen as a bag of URIs and triples. Note that a hybrid document such as an RDFa document would be indexed by all of the above. But as mentioned before, we are not experimenting with RDFa documents at present.

A user can use the HQL (Hybrid Query Language) described in the previous section to query for data and documents. A user searching for information about a person named *peter* can pose the query *<peter>*. This query, in effect returns all nodes (URIs) that have been indexed using the word *peter*. A total of 52 URIs were retrieved in response to the above query including OWL files (instance data of people) and HTML files. The user can convey to the system that she is looking specifically for Ph.D. students named *peter* using the query *<phdstudent>::<peter>*. The following URIs were retrieved in response to this query:

http://www.aifb.uni-karlsruhe.de/Personen/viewPersonOWL/id2023instance
http://www.aifb.uni-karlsruhe.de/Personen/viewPersonOWL/id2119instance
http://www.aifb.uni-karlsruhe.de/Personen/viewPersonOWL/id2062instance

These are apparently URIs (of OWL files) representing individuals and containing information about them. In order to find out the names of these individuals, the user can use the query *getBindings([<phdstudent>::<peter> <name> ?x])*. This query returned 125 literal nodes gathered from different RDF files (apparently FOAF files). Note that the above queries are keyword-based, and hence easy to formulate, and enable transparent traversal of the semantic web. The system finds the bindings for the variable from triples such as those shown below:

uri: http://www.aifb.uni-karlsruhe.de/Personen/viewPersonFOAF/foaf_80.rdf#tri52
sub: http://www.aifb.uni-karlsruhe.de/Personen/viewPersonOWL/id2023instance
pred : http://xmlns.com/foaf/0.1/name
obj (Literal): Peter Haase

uri: http://www.aifb.uni-karlsruhe.de/Personen/viewPersonFOAF/foaf_2127.rdf#tri27
subj: http://www.aifb.uni-karlsruhe.de/Personen/viewPersonOWL/id2119instance
pred: http://xmlns.com/foaf/0.1/name
obj (Literal): Peter Bungert

uri: http://www.aifb.uni-karlsruhe.de/Personen/viewPersonFOAF/foaf_2069.rdf#tri132
sub: http://www.aifb.uni-karlsruhe.de/Personen/viewPersonOWL/id2062instance
pred: http://xmlns.com/foaf/0.1/name
obj (Literal): Peter Weiß

These triples repeated themselves in different files (with different URIs) and so a lot of duplicate data has been indexed by the system. The user can search for the homepages of Ph.D. students named *peter* by posing the query, *getBindings ([<phdstudent>::<peter> <homepage> ?x]),* which returns the following results:

http://www.aifb.uni-karlsruhe.de/WBS/pha/
http://www.aifb.uni-karlsruhe.de/Forschungsgruppen/WBS
http://www.aifb.uni-karlsruhe.de/Personen/viewPerson?id_db=2023
http://www.aifb.uni-karlsruhe.de/Personen/viewPerson?id_db=2119
http://www.aifb.uni-karlsruhe.de/Personen/viewPerson?id_db=2062

The above URIs are, apparently, home pages of the above three individuals. The interesting thing is that all of the URIs except the first one points to a German page (whose content has not been indexed by our system). So, we cannot pose queries such as *get homepages of Ph.D. students named peter which talk about "semantic grid"* which translates into *getDocsByBindingsAndContent([<phdstudent>::<peter>*

<homepage> ?x] "semantic grid") , unless we can convey to the system that the German version of the page should be treated "same as" the English version. Now that the user has the names, she can use the names to query the system. The query *<peter haase>* retrieves the following URIs:

http://www.aifb.uni-karlsruhe.de/Personen/viewPersonOWL/id2023instance
http://www.aifb.uni-karlsruhe.de/Personen/viewPerson?id_db=2023
http://www.aifb.uni-karlsruhe.de/Personen/viewPersonenglish?id_db=2023
http://www.aifb.uni-karlsruhe.de/Publikationen/viewPublikationenPersonOWL/id2023.owl
http://www.aifb.uni-karlsruhe.de/Personen/viewPersonOWL/id2023.owl

These URIs are a mix of HTML (second and third URIs) and OWL documents. The second URI is the homepage of the individual named Peter Haase. It is almost synonymous with the individual [4] and so the pages that link to (*linksTo*) this page must be, arguably, relevant to the individual. The query
getLinkingNodes (http://www.aifb.uni-karlsruhe.de/Personen/viewPerson?id_db=2023)
retrieves a set of RDF and HTML documents most of which are pages of projects on which Peter Haase is working. Some of these results are shown below:

http://www.aifb.uni-karlsruhe.de/Personen/viewPersonDC/en/dc_2023.rdf
http://www.aifb.uni-karlsruhe.de/Personen/viewPersonFOAF/foaf_2023.rdf
http://www.aifb.uni-karlsruhe.de/Personen/Projekte/viewProjektenglish?id_db=78
http://www.aifb.uni-karlsruhe.de/Personen/Projekte/viewProjektenglish?id_db=80
http://www.aifb.uni-karlsruhe.de/Forschungsgruppen/Projekte/viewProjektenglish?id_db=51
http://www.aifb.uni-karlsruhe.de/Forschungsgruppen/Projekte/viewProjektenglish?id_db=71
http://www.aifb.uni-karlsruhe.de/Forschungsgruppen/Projekte/viewProjektenglish?id_db=81
http://www.aifb.uni-karlsruhe.de/Forschungsgruppen/Projekte/viewProjektenglish?id_db=42
http://www.aifb.uni-karlsruhe.de/Forschungsgruppen/Projekte/viewProjektenglish?id_db=54

The user can query for publications by Peter Haase that have the word "semantic" in the title by composing the query:
getBindings([<peter haase> <publication> ?x] [?x <title> <semantic>])
which retrieves the following URIs:

http://www.aifb.uni-karlsruhe.de/Publikationen/viewPublikationOWL/id399instance
http://www.aifb.uni-karlsruhe.de/Publikationen/viewPublikationOWL/id449instance
http://www.aifb.uni-karlsruhe.de/Publikationen/viewPublikationOWL/id748instance
http://www.aifb.uni-karlsruhe.de/Publikationen/viewPublikationOWL/id1003instance

All of the above are OWL files corresponding to publications. The user can query for documents asserting the triples used to find the above bindings by using a query such as *getAssertingNodes([<peter haase> <publication> ?x]).* The query *getDocByKeywords* corresponds to straight-forward keyword search of HTML documents. The query *getDocByKeywords(peter haase)* retrieves 251 HTML documents. A Google search for "*peter haase*" retrieves 325 documents (with omitted results) on the AIFB website. But note that we are not indexing all the AIFB web pages and that we are completely ignoring PDF documents and the like.

SITAR indexes and retrieves RDF files too. In other words, SITAR aims at treating HTML and RDF files with equal importance. SITAR allows users to simply enter a set of keywords which is then automatically plugged into the query *getDocsByIndexOrContent*. So, the query *peter haase* returns 299 documents which are a mix of HTML and OWL documents (it also retrieves a PDF document URI).

Note that in all of the above queries, the user is using intuitive keywords to explore the RDF data. She is not aware of the underlying schema and hardly ever needs to know the exact URIs of the resources. The user however is required to have an idea of the underlying model. The idea is to retrieve data and document nodes from the same unified whole. As can be imagined, this will especially be useful when dealing with those documents that have both text and semantic markup. Such documents can be indexed using URIs, triples and text, and the *getLinkingNodes* and *getAssertingNodes* will play a major role in retrieving those documents. We are currently looking for an RDFa like dataset to test this.

5 Related Research

Storing and retrieving RDF data is an area of research that has been well explored by researchers in the recent past [12,13,14,15]. Retrieving RDF data is typically viewed as a data retrieval problem and, not surprisingly, most of the query languages have the SQL flavor [14]. When seen purely from the perspective of querying the RDF data, HQL is unique because it allows the users to explore the RDF graph even without any knowledge about the underlying schema (namespaces, exact URIs, ontologies, etc.). The user can use HQL to quickly get a feel for the underlying data.

As far as document retrieval is concerned, there are several retrieval systems that retrieve documents based upon their annotations/descriptions [6,16,17,18,19,20,21], but none seems to aim at retrieving HTML, RDF *and* hybrid documents (that is, all the three types). We index a document based upon words, URIs, and triples that can be extracted from the document and give the user a light-weight query language to retrieve documents based upon this information. The query language is hybrid in the sense that it has both "formal" and keyword components but what is unique is that the "formal" component itself is expressible using keywords.

Unlike our unified approach, Semantic Search [6] treats the SW and the HW as two separate repositories and aims at retrieving documents from the HW (in fact they use Google to search for documents). It lets the user communicate the disambiguation information using the user interface. Like SITAR, quizRDF[16] indexes a document (URL) using words obtained from its body as well as from the literals of triples in which its URL participates. Like Semantic Search, quizRDF too uses a GUI to let the user communicate disambiguation information.

SITAR views the SW and the HW as a unified whole (unlike Semantic Search). One benefit of this, compared to quizRDF, is that the URL of a document is also indexed by the anchor text words. Further, SITAR indexes a document using any URIs (*linksTo*) or triples (*asserts*) that can be extracted from the document. This allows it to index and retrieve RDF documents (and hybrid RDFa kind of documents in the future). Also, unlike the above two systems, the user can specify the disambiguation information (the "class") using word-set pairs, and use it in conjunction with *linksTo* information to retrieve documents.

Swoogle[18] specializes in retrieving ontology documents and URIs. It doesn't seem to index HTML documents or support triple search or keyword-based querying of the RDF graph.

OWLIR's[17] approach of treating a triple (that appears in the document) as an indexing term corresponds to what we are doing. But the way the indexing information is used and the nature of the query language is quite different. HQL is keyword-based and so the users can retrieve an "asserting" document even when the exact URIs are not known. Also, we index a document based upon the component URIs of the triples and the hyperlinks that appear in the document (*linksTo*).

There are several other systems [19,20,21] that perform hybrid retrieval but our system is different due to the reasons discussed above and due to the fact that we view SW and HW as a single UW. We situate the SW data and the HW documents side by side and query the Unified Web using HQL which has both keyword and "formal" components. We also exploit existing hyperlink (*linksTo*) connections between HW documents and SW nodes while retrieving documents.

6 Conclusion and Future Work

We have discussed the Unified Web model that seeks to present a unified view of the SW and the HW, and the design and implementation of the Hybrid Query Language that can be used to retrieve data and documents from the UW. We have presented preliminary results obtained by experimenting with AIFB SEAL data.

HQL is a light-weight, keyword-based query language that allows the users to query and explore the RDF graph even when no schema information is available. This can then lead to composition of more involved queries using languages such as SPARQL. If, in the future, we expect lay users to pose queries such as *what is the phone number of Ph.D. student Peter Bungert,* to the Semantic Web and get back answers, query languages like HQL are a step in the right direction (though at present the user is still required to have knowledge of the RDF model).

One of the fundamental ideas behind HQL is to index a URI using a set of keywords, which is a common notion in the literature. But because we position the RDF data in a web of hypertext documents, we have the freedom to exploit information from the hypertext documents (such as the anchor text) to enrich a URI's index words. At this level, we again see natural language induced problems such as synonymy, polysemy, etc. (which only got pushed to a lower level). The resulting uncertainty necessitates ranking (not unlike what Swoogle [18] is doing). But, this is where the novel wordset pair queries such as *<phdstudent>::<peter>* enable disambiguation, stating that the user is only interested in URIs of Peter the PhD student. This, in essence, is how ontologies can help in document retrieval. And this is where the "Semantic Web enabled Information Retrieval" starts deviating from traditional IR. Otherwise, we are simply pushing the problem of keyword-based document retrieval to the level of URIs (we have simply reduced the size of a typical term vector) and there is nothing "semantic" about it – *jaguar* will retrieve both the car and animal URIs in spite of "meaningful" label-literals.

SITAR and HQL are both works in progress and are gradually evolving. The major piece of the puzzle missing from SITAR is ranking of URIs and documents. Even though Lucene does rank URIs (SITAR stores a URI in a Lucene document that is indexed by the index words), and of course, documents, we need a ranking algorithm that is based on *linksTo* relationship among others (especially to rank RDF and hybrid files). We are currently working on a ranking algorithm and its implementation.

References

1. Semantic Web Activity page, [Webpage], http://www.w3.org/2001/sw/.
2. E. Prud'hommeaux, A. Seaborne, Eds., "SPARQL Query Language for RDF," [W3C Working Draft], October 2006, http://www.w3.org/TR/rdf-sparql-query/.
3. B. Adida, M. Birbeck, Eds., "RDFa," [W3C Working Draft], 2006, http://www.w3 .org/ TR/xhtml-rdfa-primer/.
4. T. Immaneni and K. Thirunarayan, "Hybrid Retrieval from the Unified Web," *Proceedings of the 22nd ACM Symposium on Applied Computing, Semantic Web and Applications Track (ACM SAC 2007)*, Seoul, Korea, March 2007.
5. K. Thirunarayan, "On Embedding Machine-Processable Semantics into Documents," in *IEEE Transactions on Knowledge and Data Engineering*, Vol. 17, No. 7, pp. 1014-1018, July 2005.
6. J. Kleinberg, "Authoritative sources in a hyperlinked environment," *Proceedings of the 9th ACM-SIAM Symposium on Discrete Algorithms*, 1998.
7. R. Guha, R. McCool, and E. Miller, "Semantic search," in *Proceedings of the Twelfth International Conference on World Wide Web*, Budapest, Hungary, New York: ACM Press, May 2003.
8. Apache Lucene, [Webpage], http://lucene.apache.org/.
9. J. Hartmann, Y. Sure., "An Infrastructure for Scalable, Reliable Semantic Portals," IEEE Intelligent Systems 19 (3): 58-65. 2004.
10. CyberNeko HTML Parser, [Webpage], http://people.apache.org/~andyc/neko/doc/html/ .
11. Jena ARP, [Webpage], http://www.hpl.hp.com/personal/jjc/arp/.
12. D.Beckett, "SWAD-E Deliverable 10.2: Mapping Semantic Web Data with RDBMSes," [Online Document] 2003, http://www.w3.org/2001/sw/Europe/reports/ scalable_ rdbms_ mapping_report/
13. D. Beckett, "SWAD-Europe Deliverable 10.1: Scalability and Storage: Survey of Free Software / Open Source RDF storage systems," [Online Document] 2002, http://www.w3. org/2001/sw/Europe/reports/rdf_scalable_storage_report/.
14. J. Bailey, F. Bry, T. Furche, and S. Schaffert, "Web and Semantic Web Query Languages: A Survey," *Reasoning Web*, Eds., N. Eisinger and J. Maluszynski , Springer-Verlag, 2005.
15. P. Haase, J. Broekstra, A. Egerhart, and R. Volz, "A comparison of RDF query langauges," in *Proceedings of the Third International Semantic Web Conference*, Hiroshima, Japan, 2004.
16. J. Davies, R. Weeks, and U. Krohn, "QuizRDF: Search technology for the semantic web," *Workshop on Real World RDF and Semantic Web Applications*, 11th International World Wide Web Conference, Hawaii, USA, 2002.
17. J. Mayfield and T. Finin, "Information retrieval on the semantic web: Integrating inference and retrieval," in *Proceedings of the SIGIR 2003 Semantic Web Workshop,* 2003.
18. Li Ding et al., "Finding and Ranking Knowledge on the Semantic Web", in *Proceedings of the 4th International Semantic Web Conference*, November 2005.
19. C. Rocha, D. Schwabe, and M.P. Aragao, "A Hybrid Approach for Searching in the Semantic Web," in *Proceedings of the 13th International World Wide Web Conference*, New York, May 2004, pp. 374-383.
20. L. Zhang, Y. Yu, J. Zhou, C. Lin, Y. Yang, "An enhanced model for searching in semantic portals," in *Proceedings of the 14th International World Wide Web Conference*, Chiba, Japan, NY: ACM Press, May 2005.
21. D. Vallet, M. Fernández, and P. Castells, "An Ontology-Based Information Retrieval Model," in Proc. of 2nd European Semantic Web Conf. (ESWC 2005), Berlin Heidelberg, 2005.

Distributed Knowledge Representation on the Social Semantic Desktop: Named Graphs, Views and Roles in NRL

Michael Sintek[1], Ludger van Elst[1], Simon Scerri[2], and Siegfried Handschuh[2]

[1] Knowledge Management Department
German Research Center for Artificial Intelligence (DFKI) GmbH,
Kaiserslautern, Germany
firstname.surname@dfki.de
[2] DERI, National University of Ireland, Galway
firstname.surname@deri.org

Abstract. The vision of the *Social Semantic Desktop* defines a user's personal information environment as a source and end-point of the Semantic Web: Knowledge workers comprehensively express their information and data with respect to their own conceptualizations. Semantic Web languages and protocols are used to formalize these conceptualizations and for coordinating local and global information access. From the way this vision is being pursued in the NEPOMUK project, we identified several requirements and research questions with respect to knowledge representation. In addition to the general question of the expressivity needed in such a scenario, two main challenges come into focus: i) How can we cope with the heterogeneity of knowledge models and ontologies, esp. multiple knowledge modules with potentially different interpretations? ii) How can we support the tailoring of ontologies towards different needs in various exploiting applications?

In this paper, we present NRL, an approach to these two question that is based on named graphs for the modularization aspect and a view concept for the tailoring of ontologies. This view concept turned out to be of additional value, as it also provides a mechanism to impose different semantics on the same syntactical structure.

We think that the elements of our approach are not only adequate for the semantic desktop scenario, but are also of importance as building blocks for the general Semantic Web.

1 Motivation: The Social Semantic Desktop

The very core idea of the Social Semantic Desktop is to enable data interoperability on the personal desktop based on Semantic Web standards and technologies, *e. g.*, ontologies and semantic metadata. The vision [6] aims at integrated personal information management as well as at information distribution and collaboration, envisioning two expansion states: i) the *Personal Semantic Desktop* for personal information management and later ii) the *Social Semantic Desktop* for distributed information management and social community aspects.

E. Franconi, M. Kifer, and W. May (Eds.): ESWC 2007, LNCS 4519, pp. 594–608, 2007.

In traditional desktop architectures, applications are isolated islands of data—each application has its own data, unaware of related and relevant data in other applications. Individual vendors may decide to allow their applications to interoperate, so that, *e. g.*, the email client knows about the address book. However, today there is no consistent approach for allowing interoperation and a system-wide exchange of data between applications. In a similar way, the desktops of different users are also isolated islands—there is no standardized architecture for interoperation and data exchange between desktops. Users may exchange data by sending emails or upload it to a server, but so far there is no way of seamless communication from an application used by one person on their desktop to an application used by another person on another desktop. The problem on the desktop is similar to that on the Web.

The Social Semantic Desktop paradigm adopts the ideas of the Semantic Web (SW) paradigm [3], which offers a solution for the web. Formal ontologies capture both a shared conceptualization of desktop data and personal mental models. RDF (Resource Description Format) serves as a common data representation format. Together, these technologies provide a means to build the semantic bridges necessary for data exchange and application integration. The Social Semantic Desktop will transform the conventional desktop into a seamless, networked working environment, by loosening the borders between individual applications and the physical workspace of different users. By aligning the Social Semantic Desktop paradigm with the Semantic Web paradigm, a Semantic Desktop can be seen as source and end-point of the Semantic Web.

This viewpoint of the user comprehensively generating, manipulating and exploiting private as well as shared and public data has to be adequately reflected in the representational basis of such a system. While we think in general the assumptions of knowledge representation in the Semantic Web are a good starting point the Semantic Desktop scenario generates special requirements. We identified two core questions which we try to tackle in the knowledge representation approach presented in this paper:

1. How can we cope with the heterogeneity of knowledge models and ontologies, esp. multiple knowledge modules with potentially different interpretation schemes?
2. How can we support the tailoring of ontologies towards different needs in various exploiting applications?

The first question is rooted in the fact that with heterogeneous generation and exploitation of knowledge there is no "master instance" which defines and ensures the "interpretation sovereignty." The second question turned out to be an important prerequisite for a clean ontology design on the semantic desktop, as many applications shall use a knowledge worker's "personal ontology."

From these general questions, we specialized the following five main requirements for knowledge representation on the Social Semantic Desktop:

Epistemological adequacy of modeling primitives: In the Social Semantic Desktop scenario, knowledge modeling is not only performed offline (*e. g.*, by a distinguished knowledge engineer), but also by the end user, much like in the tagging systems of the Web 2.0 where a user can continuously invent new vocabulary for describing his information items. Even if much of the complexity of the underlying representation formalism can be hidden by adequate user interfaces, it is desirable that there is no big *epistemological gap* between the way an end-user would like to express his knowledge and the way it is represented in the system.

Integration of open-world and closed-world assumptions: The main principle of the SW is that it is an open world in which documents can add new information about existing resources. Since the Web is a huge place in which everything can link to anything else, it is impossible to rule out that a statement could be true, or could become true in the future. Hence, the global semantic web relies on a open-world semantic, with no unique-name assumption—the official OWL and RDF/S semantics. On the other hand, the main principle on the personal Semantic Desktop is that it is a closed-world as it mainly focuses on personal data. While most people find it difficult to understand the logical meaning and potential inferences statements of the open-world assumption, the closed-world assumption is easier to understand for the user. Hence, the Personal Semantic Desktop requires the closed-world semantics with a unique-name assumption or good smushing techniques to achieve the same effects. The next stage of expansion of the personal semantic desktop is the Social Semantic Desktop, which connects the individual desktops. This will require open-world semantics (in between desktops) with local closed-world semantics (on the personal desktop). Thus the desktop needs to be able to handle external data with open-world semantics. Therefore we require a scenario where we can always distinguish between data per se and the semantics or assumptions on that data. If these are handled analogously, the semantic desktop, a closed-world in theory, will also be able to handle data with open-world semantics.

Handling of multiple models: In order to adequately represent the social dimension of distributed knowledge generation and usage [12], a module concept is desirable which supports encapsulation of statements and the possibility to refer to such modules. The social aspect requires a support for provenance and trust information, when it comes to importing and exporting data. With the present RDF model, importing external RDF data from another desktop presents some difficulties, mainly revolving around the fact that there are no standard means of retaining provenance information of imported data. This means that data is propagated over multiple desktops, with no information regarding the original provider and other crucial information like the context under which that data is valid. This can result in various situations like ending up with outdated RDF data with no means to update it, as well as redundant RDF data which cannot be entirely and safely removed.

Multiple semantics: As stated before, the aspect of distributed (and independently created) information requires the support of the open-world assumption (as we have it in OWL and RDF/S), whereas local information created on a single desktop will have closed-world semantics. Therefore, applications will be forced to deal with different kinds of semantics.

Multiple views: Also required by the social aspect is the support for multiple views, since different individuals on different desktops might be interested in different aspects of the data. A view is dynamic, virtual data computed or collated from the original data. The best view for a particular purpose depends on the information the user needs.

In the next section, we will briefly discuss the state of the art which served as input for the NEPOMUK Representation Language (NRL). Sec. 3 gives an overview of our approach. The following sections elaborate on two important aspects of NRL, the *Named Graphs* for handling multiple models (Sec. 4) and the *Graph Views* for imposing different semantics on and application-oriented tailoring of models (Sec. 5). In Sec. 6, we present an example which shows how the concepts presented in this paper can be applied. Sec. 7 summarizes the NRL approach and discusses next steps.

2 State of the Art

The Resource Description Framework [8] and the associated schema language RDFS [4] set a standard for the Semantic Web, providing a representational language whereby resources on the web can be mapped to designated classes of objects in some shared knowledge domain, and subsequently described and related through applicable object properties. With the gradual acceptance of the Semantic Web as an achievable rather than just an ideal World Wide Web scenario, and adoption of RDF/S as the standard for describing and manipulating semantic web data, there have been many attempts to improve some RDF/S shortcomings to handling such data. Most where in the form of representational languages that extend RDF/S, the most notable of which is OWL [1]. Other work attempted to provide further functionalities on top of semantic data to that provided by RDF/S by revising the RDF model itself. The most successful idea perhaps is the named graph paradigm, where identifying multiple RDF graphs and naming them with distinct URIs is believed to provide useful additional functionality on top of the RDF model. Given that named graphs are manageable sets of data in an otherwise structureless RDF triple space composed of all existent RDF data, most of the practical problems arising from dealing with RDF data, like dealing with invalid or outdated data as well as issues of provenance and trust, could be addressed more easily if the RDF model supports named graphs. The RDF recommendation itself does not provide suitable mechanisms for talking about graphs or define relations between graphs [2,8,4,7]. Although the extension of the RDF model with named graph support has been proposed [5,11,9], and the motivation and ideas are clearly stated, a concrete extension to the RDF model supporting named graph has not yet materialized. So far, a basic syntax and semantics that models minimal manipulation of named graphs has been presented by participants of the Semantic Web Interest Group.[1] Their

[1] http://www.w3.org/2004/03/trix/

intent is to introduce the technology to the W3C process once initial versions are finalized. The SPARQL query language [9], currently undergoing standardization by the W3C, is the most successful attempt to provide a standard query language for RDF data. SPARQL's full support for named graphs has encouraged further research in the area. The concept of modularized RDF knowledge bases (in the spirit of named graphs) plus views that can be used to realize the semantics of a module (with the help of rules), amongst other things, has been introduced in the Semantic Web rule language TRIPLE [11].

Since the existing approaches are incomplete wrt. the needs of NEPOMUK and most Semantic Web scenarios in general, we propose a combination of named graphs and TRIPLE's view concept as the basis for NRL, the representational language we are presenting. In contrast to TRIPLE, we will add the ability to define views as an extension of RDF and named graphs at the ontological level, thus we are not dependent on a specific rule formalism as in the case of TRIPLE.

In the rest of this paper, we will give a detailed description of the named graphs and views features of NRL. Other features of NRL (which consist of some RDFS extensions mainly inspired by Protégé and OWL) will not be discussed.

3 Knowledge Representation on the Social Semantic Desktop: The NRL Approach

NRL was inspired by the need for a robust representational language for the Social Semantic Desktop, that targets the shortcomings of RDF/S. NRL was designed to fulfill requirements for the NEPOMUK Social Semantic Desktop project,[2] hence the particular naming, but it is otherwise domain-independent.

As discussed in the previous section, the most notable shortcoming of the RDF model is the lack of support for handling multiple models. In theory Named Graphs solve this problem since they are identifiable, modularized sets of data. Through this intermediate layer handling RDF data, *e. g.*, exchanging data and keeping track of data provenance information, is much more manageable. This has a great influence in the social aspect of the Social Semantic Desktop project, since the success of this particular aspect depends largely on how to successfully deal with these issues. All data handling on the semantic desktop including storage, retrieval and exchange, will therefore be carried out through RDF graphs. Alongside provenance data, more useful information can be attached to named graphs. In particular we feel that named graphs should be distinguished by their roles, *e. g.*, Ontology or Instance Base.

Desktop users may be interested in different aspects of data in a named graph at different times. Looking at the contents of an image folder for instance, the user might wish to see related concepts for an image, or any other files related to it, but not necessarily both concurrently even if the information is stored in the same graph. Additionally, advanced users might require to see data that is not usually visible to regular users, like additional indirect concepts related to the

[2] http://nepomuk.semanticdesktop.org/

file. This would require the viewing application to realize the RDF/S semantics over the data to yield more results. The desktop system is therefore required to work with extended or restricted versions of named graphs in different situations. However, we believe that such manipulations over named graphs should not have a permanent impact on the data in question. Conversely, we believe that the original named graph should be independent of any kind of workable interpretation executed by an application, which can be discarded if and when they are no longer needed.

For this reason, we present the concept of Graph Views as one of the core concepts in NRL. By allowing for arbitrary tailored interpretations for any established named graph, graph views fulfill our idea that named graphs should not innately carry any realized semantics or assumptions, unless they are themselves views on other graphs for exactly that purpose, and that they should remain unchanged and independent of any view applied on them. This means that different semantics can be realized for different graphs if required. In practice, different application on the semantic desktop will require to apply different semantics, or assumptions on semantics, to named graphs. In this way, although the semantic desktop operates in a closed-world, it is also possible to work with open-world semantic views over a graph. Importing a named graph with predefined open-world semantics on the semantic desktop is therefore possible. If required (and meaningful), closed-world applications can then work with a closed-world semantics view over the imported graph.

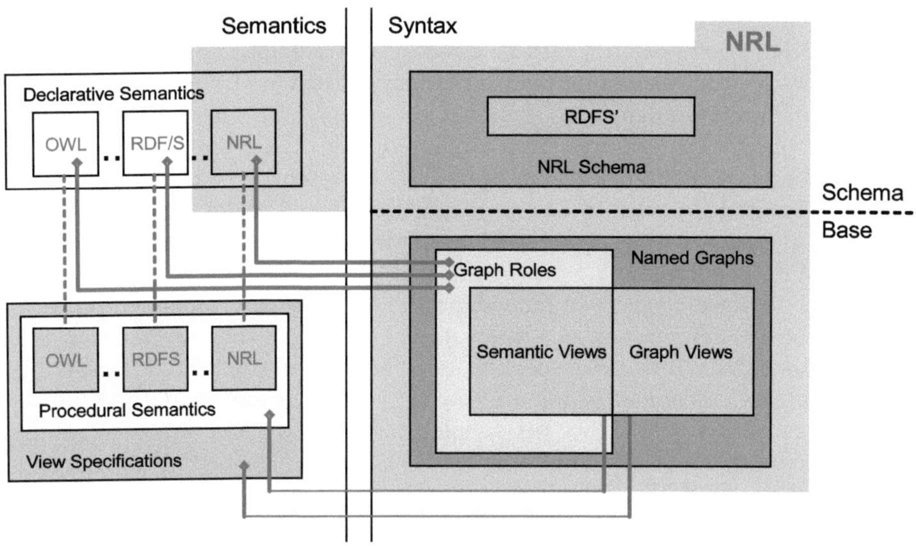

Fig. 1. Overview of NRL—Abstract Syntax, Concepts and Semantics

Fig. 1 gives an overview of the components of NRL, depicting both the syntactical and the semantic blocks of NRL. The syntax box contains, in the upper part, the NRL Schema language, which is mainly an extension of (a large subset of) RDFS. The lower part shows how named graphs, graph roles, and views are related, which will be explained in detail in the rest of this paper.

The left half of the figure sheds some light on the semantics of NRL, which has a declarative and a procedural part. Declarative semantics is linked with graph roles, *i. e.*, roles are used to assign meaning to named graphs (note that not all named graphs or views must be assigned some declarative semantics, *e. g.*, in cases when the semantics is (not) yet known or simply not relevant). Views are also linked to view specifications, which function as a mechanism to express procedural semantics, *e. g.*, by using a rule system. The procedural semantics has, of course, to realize the declarative semantics that is assigned to a semantic view.

4 Handling Multiple Models: NRL Named Graphs

Named graphs (NGs) are an extension on top of RDF, where every distinct RDF graph is identified by a unique name. NGs provide additional functionality on top of RDF particularly with respect to metametadata (metadata about metadata), provenance, and data (in)equivalence issues, besides making data handling more manageable. Our approach is based on the work described in [5] excluding however, the open-world assumption stated there. As stated earlier (*cf.* Sec. 3) we believe that named graphs should not innately carry any realized semantics or assumptions on the semantics. Therefore, despite being designed as a requirement for the Semantic Desktop, which operates under a closed-world scenario, NRL itself does not impose closed-world semantics on data. This and other semantics can instead be realized through designated views on graphs.

A named graph is a pair (n, g), where n is a unique URI reference denoting the assigned name for the graph g. Such a mapping fixes the graph g corresponding to n in a rigid, non-extensible way. The URI representing n can then be used from any location to refer to the corresponding set of triples belonging to the graph g. A graph g' consistent[3] with a distinct graph g named n cannot be assigned the same name n.

An RDF triple can exist in a named graph or outside any named graph. However, for consistency reasons, all triples must be assigned to some named graph. For this reason NRL provides a special named graph, `nrl:DefaultGraph`. Triples existing outside any named graph are considered part of this default graph. This ensures backward compatibility with triples that are not based on named graphs. This approach gives rise to the term RDF Dataset as defined in [9]. An RDF dataset is composed of a default graph and a finite number of distinct named graph, formally defined as the set $\{g, (n_1, g_1), (n_2, g_2), ..., (n_n, g_n)\}$ comprising of the default graph g and zero or more named graphs (n_i, g_i).

NRL distinguishes between graphs and graph roles, in order to have orthogonal modeling primitives for defining graphs and for specifying their role. A

[3] Two different datasets asserting two unique graphs but having the same URI for a name contradict one another.

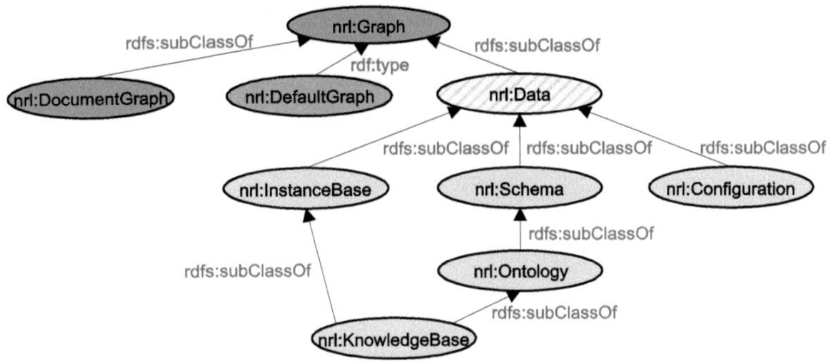

Fig. 2. NRL Named Graph Class Hierarchy

graph role refers to the characteristics and content of a named graph (*e. g.*, simple data, an ontology, a knowledge base, *etc.*) and how the data is intended to be handled. The NEPOMUK Graph Metadata vocabulary (NGM)[4] provides a vocabulary for annotating graph roles. Graph metadata will be attached to roles rather than to the graphs themselves, because its more intuitive to annotate an ontology, for example, rather than the underlying graph. Roles are more stable than the graphs they represent, and while the graph for a particular role might change constantly, evolution of the role itself is less frequent. An instantiation of a role represents specific type of graph and the corresponding triple set data.

Fig. 2 depicts the class hierarchy supporting NGs in NRL. Graph roles are defined as specialization of the general graph representation `nrl:Data`. A special graph, `nrl:DocumentGraph`, is used as a marker class for graphs that are represented within and identified by a document URL. We now present the NRL vocabulary supporting named graphs. General graph vocabulary is defined in Sec. 4.1 while Sec. 4.2 is dedicated entirely to graph roles.

4.1 Graph Core Vocabulary

nrl:Graph and nrl:DocumentGraph Instances of these classes represent named graphs. The name of the instance coincides with the name of the graph. The graph content for a `nrl:DocumentGraph` is located at the URL that is the URIref for the `nrl:DocumentGraph` instance. This allows existing RDF files to be re-used as named graphs, avoiding the need of a syntax like TriG[5] to define named graphs.

nrl:subGraphOf, nrl:superGraphOf, and nrl:equivalentGraph. These relations between named graphs have the obvious semantics: they are defined as $\subseteq$, $\supseteq$, and $=$ on the bare triple sets in these graphs.

nrl:imports is a subproperty of `nrl:superGraphOf` and models graph imports. Apart from implying the $\supseteq$ relation between the triple sets, it also requires

[4] NGM will not be described in this paper.

[5] http://sites.wiwiss.fu-berlin.de/suhl/bizer/TriG/

that the semantics of the two graphs is compatible if used on, *e. g.*, graphs that are ontologies.

nrl:DefaultGraph This instance of `nrl:Graph` represents the graph containing all triples existing outside any user-defined named graph. Since we do not apply any semantics to triples automatically, this allows views to be defined on top of triples defined outside of all named graphs analogously to the named-graph case.

4.2 Graph Roles Vocabulary

nrl:Data This subclass of `nrl:Graph` is an abstract class to make graph roles easy-to-use marker classes. It represents the most generic role that a graph can have, namely that it contains data.

nrl:Schema and nrl:Ontology are roles for graphs that represent data in some kind of conceptualization model. `nrl:Ontology` is a subclass of `nrl:Schema`.

nrl:InstanceBase marks a named graph to contain instances from schemas or ontologies. The properties `nrl:hasSchema` and `nrl:hasOntology` relate an instance base to the corresponding schema or ontology.

nrl:KnowledgeBase marks a named graph as containing a conceptual model plus instances from schemas or ontologies.

nrl:Configuration is used to represent technical configuration data that is irrelevant to general semantic web data within a graph. Other additional roles serving different purposes might be added in the future.

nrl:Semantics Declarative semantics for a graph role can be specified by referring to instances of this class via `nrl:hasSemantics`. These will usually link (via `nrl:semanticsDefinedBy`) to a document specifying the semantics in a human readable or formal way (*e. g.*, the RDF Semantics document [7]).

5 Imposing Semantics on Graphs: NRL Graph Views

A named graph consists only of the enumerated triples in the triple set associated with the name, and does not inherently carry any form of semantics (apart from the basic RDF semantics). However in many situations it is desirable to work with an extended or restricted interpretation of simple syntax-only named graphs. These can be realized by applying some algorithm (*e. g.*, specified through rules) which enhances named graphs with entailment triples, returns a restricted form of the triple set, or an entirely new triple set. To preserve the integrity of a named graph, interpretations of one named graph should never replace the original. To model this functionality and retain the separation between original named graph and any number of their interpretations, we introduce the concept of *Graph Views*.

Views are different interpretations for a particular named graph. Formally, a view is an executable specification of an input graph into a corresponding output graph. Informally, they can be seen as arbitrary wrappings for a named graph.

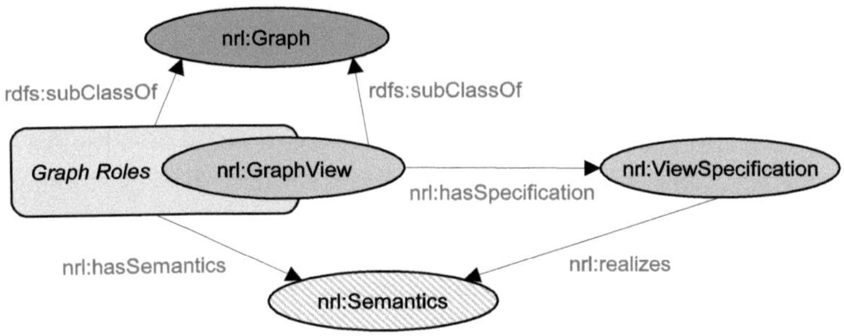

Fig. 3. Graph Views in NRL

Fig. 3 depicts graph view support in NRL. Views are themselves named graphs. Therefore one can have a named graph that is a different interpretation, or view, of another named graph. This modeling can be applied recurrently, yielding a view of a view and so on.

View specifications can execute the view realization for a view, via a set of queries/rules in a query/rule language (*e. g.*, a SPARQL query over a named graph), or via an external application (*e. g.*, an application that returns the transitive closure of `rdfs:subClassOf`). As in the latter example, view realizations can also realize the implicit semantics of a graph according to some language or schema (*e. g.*, RDFS, OWL, NRL *etc.*). We refer to these as *Semantic Views*, represented in Fig. 3 by the intersection of `nrl:GraphView` and graph roles. One can draw a parallel between this figure and Fig. 1. In contrast to graph roles, which have only declarative semantics defined through the `nrl:hasSemantics` property, semantic views also carry procedural semantics, since the semantics of these graphs are always realized, (through `nrl:realizes`) and not simply implied.

5.1 Views Vocabulary

In this section we briefly present the NRL vocabulary supporting graph view specifications.

nrl:GraphView represents a view, modeled as a subclass of named graph. A view is realized through a view specification, defined by an instance of `nrl:ViewSpecification` via `nrl:hasSpecification`. The named graph on which the view is being generated is linked by `nrl:viewOn`. The separation between different interpretations of a named graph and the original named graph itself is thus retained.

nrl:ViewSpecification This class represents a general view specification, which can currently take one of two forms, modeled as the two subclasses `nrl:RuleViewSpecification` and `nrl:ExternalViewSpecification`. As discussed earlier, semantic views realize procedural semantics and are linked to some semantics via `nrl:realizes`. This is however to be differentiated from `nrl:hasSemantics`, which states that a named graph carries (through a role) declarative semantics which is not necessarily (explicitly) realized via a view specification.

nrl:RuleViewSpecification Views can be specified by referring to a rule language (via `nrl:ruleLanguage`) and a corresponding set of given rules (via `nrl:rule`). These views are realized by executing the rules, generating the required output named graph.

nrl:ExternalViewSpecification Instances of this class map to the location of (via `nrl:externalRealizer`) an external application, service, or program that is executed to create the view.

6 Example: NRL in Use

In this section, we demonstrate the utilization of the various NRL concepts in a more complex scenario: Ella is a biologist and works as a senior researcher at Institute Pasteur in central Paris. She would like to compile an online knowledge base describing animal species for her students to access. She knows that a rather generic ontology describing the animal species domain, O_1, is already available (which, technically speaking, means it exists as a named graph). Someone else had also supplied data consisting of a vast amount of instances for the animals ontology as a named graph with the role of instance base, I_1. However this combined data does not provide extensive coverage of the animal kingdom as required by Ella. Therefore Ella hires a SW knowledge engineer to model another ontology that defines further species not captured in O_1, and this is stored as another named graph, O_2. Since Ella requires concepts from both ontologies, the engineer merges O_1 and O_2 in the required conceptualization by creating a named graph O as an ontology and defining it as supergraph of O_1 and O_2. Furthermore, a number of real instances of the new animal species defined in O_2 is compiled in an instance base, I_2.

Ella now requires to use all the acquired and generated data to power a useful service for the students to use. Schematic data from the graph O, and the instances from I_1 and I_2 are all imported to a new graph, KB, acting as a knowledge base. Ella would like the students to be able to query the knowledge base with questions like 'Are flatworms Deuterostomes or Platyzoa?'. Although by traversing the animals hierarchy it is clear that they are Platyzoa, the statement is not innately part of the graph KB. This can be discovered by realizing the semantics of `rdfs:subClassOf` as defined in the RDFS semantics. However KB might be required as is, with no assumed semantics, for other purposes. Directly enriching KB with entailment triples permanently would make this impossible.

Therefore the knowledge engineer creates a view over KB for Ella, consisting of the required extended graph, without modifying the original KB in any way. This is done by defining a view specification that computes the procedural semantics for KB. The specification uses a rule language of choice that provides a number of rules, one of which computes the transitive closure of `rdfs:subClassOf` for a set of RDF triples. Executing that rule over the triples in KB results in the semantic view $V_1(KB)$, which consists of the RDF triples in KB plus the generated entailment triples. The separation between the underlying model and the model with the required semantics is thus retained and through simple queries over $V_1(KB)$, students can instantly get answers to their questions.

Ella later on decides to provide another service for younger students by using 'Graph Taxonomy Extractor', a graph visualization API that generates an interactive graph depicting the animal hierarchy within $V_1(KB)$. However this graph contains other information in addition to that required (*e. g.*, properties attributed to classes). Of course, Ella does not want to discard all this useful information from $V_1(KB)$ permanently just to generate the visualization. The knowledge engineer is aware of a Semantic Web application that does exactly what Ella requires. The application acts as an external view specification and generates a view, consisting of only triples defining the class hierarchy, over an input named graph. The view generated by this application, $V_2(V_1(KB))$, is fed to the API to effectively generate the interactive graph for the students to explore.

It is worth to note that all seven named graphs on which this last view is generated upon are still intact and have not been affected by any of the operations along the way. If the knowledge engineer requires to apply some different semantics over KB, it may still be done since generating $V_1(KB)$ did not have an impact on KB. However, the content of KB needs to be validated, or generated, each time it is used since one of its subgraphs (O_1, O_2, I_1 and I_2) can change. Although from a practical point of view this might sound laborious, from a conceptual point of view it solves problems regarding data consistency and avoids other problems like working with outdated data that can't be updated because links to underlying models have been lost.

Fig. 4 presents the "dataflow" in our example scenario, demonstrating how the theoretical basis of NRL can be applied in practice to effectively model data for use in different scenarios in a clear and consistent way.

We now model the dataflow in Fig. 4 in TriG syntax.[6] TriG is a straightforward extension of Turtle.[7] Turtle itself is an extension of N-Triples[8] which carefully takes the most useful and appropriate things added from Notation3[9] while keeping it in the RDF model. TriG is a plain text format created for serializing NGs and RDF Datasets. Fig. 5 demonstrates how one can make use of the named graph paradigm and the syntax for named graphs:

- `[1]` namespace declarations
- `[2-5]` ontology graphs (`ex:o1` and `ex:o2` are defined and then imported into `ex:o`)
- `[6-8]` instance/knowledge base definitions
- `[9]` contents of ontology `ex:o2`, defining extended animal domain
- `[10]` contents of instance base `ex:i2`, defining instances of animals in (`ex:o2`

[6] http://sites.wiwiss.fu-berlin.de/suhl/bizer/TriG/
[7] http://www.dajobe.org/2004/01/turtle/
[8] http://www.w3.org/TR/rdf-testcases/#ntriples
[9] http://www.w3.org/DesignIssues/Notation3

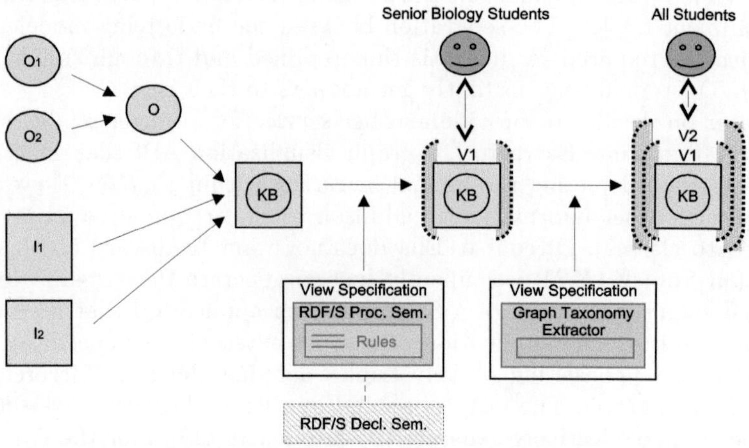

Fig. 4. NRL Dataflow Diagram

```
[1] @prefix nrl: <http://semanticdesktop.org/ontology/nrl-yyyymmdd#> .
    @prefix ex: <http://www.example.org/vocabulary#> .
[2] ex:o2 rdf:type nrl:Ontology .
[3] <http://www.domain.com/o1.rdfs> rdf:type nrl:Ontology ,
                     nrl:DocumentGraph .
[4] ex:o1 rdf:type nrl:Ontology ;
         nrl:equivalentGraph <http://www.domain.com/o1.rdfs> .
[5] ex:o rdf:type nrl:Ontology ;
         nrl:imports ex:o1, ex:o2 .
[6] ex:i2 rdf:type nrl:InstanceBase ;
         nrl:hasOntology ex:o2 .
[7] http://www.anotherdomain.com/i1.rdf> rdf:type nrl:InstanceBase ,
                                                   nrl:DocumentGraph .
[8] ex:kb rdf:type nrl:KnowledgeBase ;
      nrl:imports ex:o, ex:i2, <http://www.anotherdomain.com/i1.rdf> .
[9] ex:o2 {
      ex:Animal rdf:type rdfs:Class .
             ## further Animal Ontology definitions here ## }
[10]ex:i2 {
     ex:CandyCaneWorm rdf:type ex:Flatworm ;
             ## further Animal Instance definitions here ## }
[11] ex:v1kb rdf:type nrl:KnowledgeBase, nrl:GraphView ;
            nrl:viewOn ex:kb ; nrl:superGraphOf ex:kb ;
            nrl:hasSpecification ex:rvs .
[12] ex:rvs rdf:type nrl:RuleViewSpecification ;
            nrl:realizes ex:RDFSSemantics ; nrl:ruleLanguage "SPARQL" ;
            nrl:rule "CONSTRUCT {?s rdfs:subClassOf ?v} WHERE ..." ;
            nrl:rule "CONSTRUCT {?s rdf:type ?v} WHERE ..." .
[13] ex:RDFSSemantics rdf:type nrl:Semantics ; rdfs:label "RDFS" ;
            nrl:semanticsDefinedBy "http://www.w3.org/TR/rdf-mt/" .
[14] ex:v2v1kb rdf:type nrl:GraphView, nrl:KnowledgeBase ;
            nrl:viewOn ex:v1kb ; nrl:hasSpecification ex:evs .
[15] ex:evs rdf:type nrl:ExternalViewSpecification ;
            nrl:externalRealizer "GraphTaxonomyExtractor" .
```

Fig. 5. NRL Example—TriG Serialization

[11-13] `ex:v1kb` is defined as a view on `ex:kb` via the view specification `ex:rvs`; furthermore, `ex:v1kb` is a super graph of `ex:kb` as it realizes the RDFS semantics and thus contains the original graph plus the inferred triples; the view specification is realized (as an example) with some SPARQL-inspired `CONSTRUCT` queries (for this to work, a real rule language is required)

[14-15] similar to [11-13], but here we define `ex:v2v1kb` with the help of an external tool, the "GraphTaxonomyExtractor"

7 Summary and Outlook

Aligning knowledge representation on a Social Semantic Desktop with the general Semantic Web approaches (RDF, RDFS, OWL, ...) promises a comprehensive use of data and schemas and an active, personalized access point to the Semantic Web [10]. In such a scenario, ontologies play an important role, from very general ontologies stating which entities can be modeled on a Semantic Desktop (*e. g.*, people, documents, ...) to rather personal vocabulary to structure information items. One of the most important design decisions is the question of the *representational ontology*, constraining the general expressivity of such a system. In this paper, we concentrated on those parts of the NEPOMUK Representational Language (NRL) which are rooted in the requirements risen by the *distributed knowledge representation and heterogeneity aspects* of the Semantic Desktop scenario and which we think cannot satisfactorily be dealt with by the current state of the art. In a nutshell, the basic arguments and design principles of NRL are as follows:

- Due to the heterogeneity of the data creating and consuming entities in the social semantic desktop scenario, a single interpretation schema cannot be assumed. Therefore, NRL aims at a *strict separation between data (sets of triples, graphs) and their interpretation/semantics.*
- Imposing specific semantics to a graph is realized by generating *views* on that graph. Such a generation is directed by an (executable) *view specification* which may realize a declarative semantics (*e. g.*, the RDF/S or OWL semantics specified in a standardization document).
- Graph views cannot only be used for semantic interpretations of graphs, but also for application-driven tailoring of a graph.[10]
- Handling of multiple graphs (with different provenance, ownership, level of trust, ...) is essential. *Named graphs* are the basic means to this problem.
- Graphs can play different roles in different contexts. While for one application a graph may be an ontology, another one may see it as plain data. These *roles* can explicitly be specified.

While originally designed as a NEPOMUK internal standard for the Social Semantic Desktop, we believe that the arguments also hold for the general Semantic Web, especially when we review the current trends which more and more show a

[10] This corresponds to a database-like view concept.

development from the view of "the Semantic Web as one big, global knowledge base" to "a Web of (machine and human) actors" with local perspectives and social needs like trust, ownership, *etc.*

Within NEPOMUK, we are developing the approach technically, by complementing the NRL standard with tools that facilitate its use by the application programmer, as well as conceptually, by the development and integration of accompanying ontology standards, *e. g.*, an annotation vocabulary, an information element ontology, and an upper-ontology for *personal information models.*

Acknowledgements. This work was supported by the European Union IST fund (Grant FP6-027705, Project NEPOMUK) and by the German Federal Ministry of Education, Science, Research and Technology (bmb+f), (Grant 01 IW F01, Project Mymory: Situated Documents in Personal Information Spaces). The authors would especially like to thank all contributors to NEPOMUK's ontology taskforce.

References

1. S. Bechhofer, F. van Harmelen, J. Hendler, I. Horrocks, D. McGuinnes, P. Patel-Schneider, and L. Stein. OWL web ontology language reference, 2004.
2. D. Beckett. RDF/XML syntax specification (revised). W3C recommendation, W3C, February 2004. http://www.w3.org/TR/rdf-syntax-grammar/.
3. T. Berners-Lee, J. Hendler, and O. Lassila. The semantic web. *Scientific American*, 89, May 2001.
4. D. Brickley and R. Guha. RDF vocabulary description language 1.0: RDF Schema. Technical report, W3C, February 2004. http://www.w3.org/TR/rdf-schema/.
5. J. J. Carroll, C. Bizer, P. Hayes, and P. Stickler. Named graphs, provenance and trust. In *WWW '05: Proceedings of the 14th international conference on World Wide Web*, pages 613–622, New York, NY, USA, 2005. ACM Press.
6. S. Decker and M. Frank. The social semantic desktop. In *Proc. of the WWW2004 Workshop Application Design, Development and Implementation Issues in the Semantic Web*, 2004.
7. P. Hayes. RDF semantics. W3C recommendation, W3C, February 2004. http://www.w3.org/TR/rdf-mt/.
8. F. Manola and E. Miller. RDF primer. W3C recommendation, W3C, February 2004. http://www.w3.org/TR/rdf-primer/.
9. E. Prud'hommeaux and A. Seaborne. SPARQL query language for RDF. W3C working draft, W3C, 2005. http://www.w3.org/TR/rdf-sparql-query/.
10. L. Sauermann, A. Dengel, L. Elst, A. Lauer, H. Maus, and S. Schwarz. Personalization in the EPOS project. In M. Bouzid and N. Henze, editors, *Proceedings of the International Workshop on Semantic Web Personalization, Budva, Montenegro, June 12, 2006*, pages 42–52, 2006.
11. M. Sintek and S. Decker. TRIPLE—A query, inference, and transformation language for the Semantic Web. In *1st International Semantic Web Conference (ISWC2002)*, June 2002.
12. L. van Elst, V. Dignum, and A. Abecker. Towards agent-mediated knowledge management. In L. van Elst, V. Dignum, and A. Abecker, editors, *Agent-Mediated Knowledge Management International Symposium AMKM 2003, Stanford, CA, USA, March 24-26, 2003, Revised and Invited Papers*, volume 2926 of *LNAI*, pages 1–31. Springer, Heidelberg, 2004.

Semantic Process Retrieval with iSPARQL

Christoph Kiefer[1], Abraham Bernstein[1], Hong Joo Lee[2], Mark Klein[2], and Markus Stocker[1]

[1] Department of Informatics, University of Zurich, Switzerland
{kiefer,bernstein,stocker}@ifi.unizh.ch
[2] Center for Collective Intelligence, Massachusetts Institute of Technology, USA
{hongjoo,m_klein}@mit.edu

Abstract. The vision of semantic business processes is to enable the integration and inter-operability of business processes across organizational boundaries. Since different organizations model their processes differently, the discovery and retrieval of similar semantic business processes is necessary in order to foster inter-organizational collaborations. This paper presents our approach of using iSPARQL– our imprecise query engine based on SPARQL– to query the OWL MIT Process Handbook– a large collection of over 5000 semantic business processes. We particularly show how easy it is to use iSPARQL to perform the presented process retrieval task. Furthermore, since choosing the best performing similarity strategy is a non-trivial, data-, and context-dependent task, we evaluate the performance of three simple and two human-engineered similarity strategies. In addition, we conduct machine learning experiments to learn similarity measures showing that complementary information contained in the different notions of similarity strategies provide a very high retrieval accuracy. Our preliminary results indicate that iSPARQL is indeed useful for extending the reach of queries and that it, therefore, is an enabler for inter- and intra-organizational collaborations.

1 Introduction

One of the cornerstones of the Semantic Web services vision is to enable the design and execution of dynamic inter- and intra-organizational services (processes). A major prerequisite for fulfilling this vision is the ability to find services which have certain features (*i.e.*, the ability for adaptive service discovery/-matchmaking and/or mediation). Most approaches so far have relied on some type of logical reasoning [4,10]. In earlier works, we suggested that statistical methods based on a catalog of simple predefined similarity measures might be more suitable for this task [3]. Indeed, using the OWLS-TC matchmaking test collection, we showed that a straightforward method based on simple, off-the-shelf similarity metrics performed almost as well as the "best of bread" OWLS-MX matchmaker that was engineered to the task of matching OWL-S services.

While this success was remarkable it left open some important questions. First, the question of *which similarity measure* is applicable for a given problem needs to be answered. Findings from machine learning [8], information retrieval [1], and psychology [9] show that the best performing similarity measure might be both

E. Franconi, M. Kifer, and W. May (Eds.): ESWC 2007, LNCS 4519, pp. 609–623, 2007.

task (*e.g.*, OWL-S/WSML matchmaking, retrieval in ontologies, etc.) and domain dependent (*i.e.*, the ontologies involved). Indeed, finding the best similarity measure for any given task and domain can be mapped to an optimization problem, where the "No Free Lunch" theorem [15] has proven that no uniformly best solution exists. Hence, the choice of the best performing similarity measure for any given task given an application domain seems anything but straightforward.

Second, given that our similarity-based approach was still slightly outperformed by the human-engineered, task-optimized OWLS-MX matchmaker, gives rise to the question *if a (human-) engineered, task-optimized similarity measure would not perform better*? This question is especially important since in many practical applications a considerable amount of human (knowledge) engineering is expended to improve the performance of systems. Hence, the engineering effort would also go into similarity-based solutions.

Third, given that similarity is an inherently statistics-based notion almost begs *the use of statistical machine learning techniques for finding a similarity measure optimized for a given task and application domain.*

In this paper we use the *iSPARQL* framework to address exactly these questions. iSPARQL is an extension of official SPARQL that allows for similarity joins which employ any of about 40 different similarity measures implemented in *SimPack*[1] – our generic Java library of similarity measures for the use in ontologies. It, therefore, lends itself as a platform for any kinds of similarity-based retrieval experiments in ontologies. Specifically, the contributions of this paper are that it (i) shows the simplicity of designing similarity based Semantic Web applications with iSPARQL, (ii) analyzes the usefulness of human-engineered task- and domain-specific similarity measures in comparison to some off-the-shelf measures widely used in computer science and AI, and (iii) shows how similarity measures learned through supervised learning techniques outperform both the off-the-shelf as well as the human-engineered measures in a service retrieval task. Last, the paper introduces a new data set for (process/service) retrieval applications in ontologies based on the MIT Process Handbook [13] that provides a very rich structural and textual description of the provided processes.

The remainder of this paper is structured as follows. The next section succinctly summarizes the most important related work. Given the importance as an underlying framework, Section 3 introduces the relevant features of iSPARQL. Section 4 is the heart of the paper: it introduces the experimental setup including the data set used in the evaluations, provides some details on the experiments, and discusses the results. To close, Section 5 discusses the results in the light of the claims, related work, and limitations. We close the paper with our conclusions and some insight into future work.

2 Related Work

Several other studies focus on the comparison of semantic business processes either for retrieval, discovery, matchmaking, or process alignment. We introduced

[1] `http://www.ifi.unizh.ch/ddis/simpack.html`

in earlier works PQL – the Process Query Language to query the MIT Process Handbook [4]. PQL does not make use of similarity measures to retrieve similar query matches. However, PQL knows a "contains"-operator that can be roughly compared with the SQL "like"-operator performing string comparisons.

We are aware of two other studies that address the task of aligning semantic business processes using a similarity measure. Brockmans *et al.* and Ehrig *et al.* [5,7] propose an approach to semantically align business processes originally represented as Petri nets. After the nets have been transformed to OWL, similarity measures from different categories are employed to measure the affinity between elements of Petri nets. Since we are able to define similarity strategies (*i.e.*, compositions of several atomic similarity measurements and weighting schemes) with our iSPARQL system, we consider such an ontology alignment task as being in the range of tasks which could be perfectly carried out by iSPARQL.

With respect to matchmaking, Klusch *et al.* [10] present an approach to perform hybrid Semantic Web service matchmaking. Their OWLS-MX matchmaker uses both, semantic similarity measures, as well as logic-based reasoning techniques to discover similar web services to a given query service. Again, Semantic Web service/process matchmaking is a possible application for iSPARQL.

Last, imprecise RDQL (iRDQL) is the predecessor of iSPARQL [3]. In iRDQL, special keywords are used to specify the similarity strategy (and parameters) to measure the relatedness between resources in ontologies. We did not want to introduce new keywords in iSPARQL since this would break the official W3C SPARQL grammar. Hence, we decided to integrate imprecise statements as *virtual triples* allowing us to add similarity measures and parameters by simply extending the virtual triple ontology.

3 iSPARQL

This section succinctly introduces the relevant features of our iSPARQL framework that serves as the technical foundation to all evaluations.[2] iSPARQL is an extension of SPARQL [14] that allows to query by triple patterns, conjunctions, disjunctions, and optional patterns. iSPARQL extends the traditional SPARQL grammar but does not make use of additional keywords. Instead, iSPARQL introduces the idea of *virtual triples*. Virtual triples are not matched against the underlying ontology graph, but used to configure similarity joins: they specify which pair of variables (that are bound by SPARQL to resources) should be joined and compared using what type of similarity measure. Thus, they establish a *virtual relationship* between the resources bound to the variables describing their similarity. A similarity ontology defines the admissible virtual triples and links the different measures to their actual implementation in our library of similarity measures called *SimPack*. The similarity ontology also allows the specification of more sophisticated combinations of similarity measures, which we call *similarity strategies* (or simply *strategies*) in the rest of this paper. Note

[2] An online demonstration of iSPARQL is available at `http://www.ifi.unizh.ch/ddis/isparql.html`

```
1 PREFIX ph: <http://www.ifi.unizh.ch/ddis/ph/2006/08/ProcessHandbook.owl#>
2 PREFIX isparql: <java:ch.unizh.ifi.isparql.query.property.>
3
4 SELECT ?process1 ?name1 ?overallsimilarity
5 WHERE {
6   ?process1 ph:name ?name1 .
7   ?process1 ph:description ?description1 .
8   ?process2 ph:name ‘‘Sell’’ ; ph:name ?name2 .
9   ?process2 ph:description ?description2 .
10
11  # ImpreciseBlockOfTriples (lines 13-20, 22-24, and 26-33)
12
13  # NameStatement
14  ?strategy1 isparql:name ‘‘LoLN’’.
15  # ArgumentsStatement
16  ?strategy1 isparql:argument (?name1 ?name2) .
17  # IgnorecaseStatement
18  ?strategy1 isparql:ignorecase ‘‘true’’ .
19  # SimilarityStatement
20  ?strategy1 isparql:similarity ?sim1
21
22  ?strategy2 isparql:name ‘‘TFIDFD’’ .
23  ?strategy2 isparql:arguments (?description1 ?description2) .
24  ?strategy2 isparql:similarity ?sim2 .
25
26  ?strategy3 isparql:name ‘‘ScoreAggregator’’ .
27  # ScoresStatement
28  ?strategy3 isparql:scores (?sim1 ?sim2) .
29  # WeightsStatement
30  ?strategy3 isparql:weights (0.8 0.2) .
31  # AggregatorStatement
32  ?strategy3 isparql:aggregator ‘‘sum’’ .
33  ?strategy3 isparql:similarity ?overallsimilarity
34 } ORDER BY DESC(?overallsimilarity);
```

Listing 1.1. iSPARQL example query for the MIT Process Handbook

that the order of virtual triples is irrelevant since iSPARQL's query processor will inspect (reorder) them before the query is passed to the query engine. In the remainder of this section, we will briefly discuss the iSPARQL grammar and then introduce some of the similarity strategies employed in the evaluation.

3.1 The iSPARQL Grammar

The various additional grammar statements are explained with the help of the example query in Listing 1.1. This query aims at finding processes (or services) in a process ontology (we use the MIT Process Handbook introduced in Section 4.1) which are similar to the process "Sell" by comparing process names and descriptions. To implement our virtual triple approach we added an `ImpreciseBlockOfTriples` symbol to the standard SPARQL grammar expression of `FilteredBasicGraphPattern` [14]. Instead of matching patterns in the RDF graph, the triples in an `ImpreciseBlockOfTriples` act as *virtual triple* patterns, which are interpreted by iSPARQL's query processor

An `ImpreciseBlockOfTriples` requires at least a `NameStatement` (lines 14, 22, and 26) specifying the similarity strategy. iSPARQL has two kinds of strategies: *similarity strategies* and *aggregation strategies*. The former defines how the proximity of resources should be computed. The latter aggregates previously computed similarity scores to an overall similarity value. The example query in Listing 1.1 defines the two similarity strategies "LoLN" (lines 13–20) and "TFIDFD" (lines 22–24) as well as the aggregation strategy "ScoreAggregator" (lines 26–33; see Section 3.2 for a discussion of the available strategies).

In addition, an `ImpreciseBlockOfTriples` requires an `ArgumentsStatement` (lines 16 and 23) or a `ScoresStatement` (line 28), depending on whether it specifies a similarity or an aggregation strategy. An `ArgumentsStatement` specifies the resources under comparison to the iSPARQL framework. The `ScoresStatement` takes a list of previously calculated values (typically from similarity strategies) and summarizes the individual values in a user-defined way (*e.g.*, average, weighted sum, median, etc.). We found aggregators to be useful to construct overall (sometimes complex) similarity scores based on two or more previously computed similarity scores. The similarity ontology also allows the use of some additional triple patterns (statements) for most strategies to pass parameters to the strategies instructing them to, for example, ignore a string's case during a comparison operation (the `IgnorecaseStatement` on line 18) or to apply weights to the aggregated values (using the `WeightsStatement` on line 30).

Table 1. Selection of five iSPARQL similarity strategies

Strategy	Explanation
TFIDFD (simple)	TFIDF between process descriptions: the textual descriptions of two processes are compared by TFIDF, the standard information retrieval similarity measure. This measure makes use of pre-computed corpus of process descriptions which serves to retrieve statistics about words in the descriptions. The TFIDF measure extends the cosine measure with the traditional IR weighing scheme [1].
LevN (simple)	Levenshtein similarity of process names: two process names are compared with the Levenshtein string similarity measure [11]. The Levenshtein-based similarity measures are founded on the Levenshtein string edit distance that measures the relatedness of two strings (process names) in terms of the number of insert, remove, and replacement operations to transform one string into another string.
LoLN (simple)	Levenshtein Level 2 (Levenshtein of Levenshtein) similarity of process names: two process names such as "Buy over the internet" and "Sell via Internet" are compared string-by-string with the (inner) Levenshtein string similarity measure. If the similarity between two strings is above a user-defined threshold, the strings are considered as equal (*i.e.*, they match). These scores are used by the outer Levenshtein string similarity measure to compute an overall degree of similarity between the two process names (sequences of strings).
MITPH-LoLNTFIDFD (engineered)	Levenshtein Level 2 similarity between process names, TFIDF between process descriptions: this strategy is a combination of two atomic measures. An overall similarity score is computed by aggregating the individual scores.
MITPH-LoLNTFIDFD-JaccardAll (engineered)	Levenshtein Level 2 similarity between process names, TFIDF between process descriptions, Jaccard (Tanimoto) set-based similarity [6] between process exceptions, goals, resources, inputs, and outputs: a combination of six atomic measures; in addition to MITPH-LoLNTFIDFD, four single similarity scores are computed from two processes' goal, exception, resource, in- and output sets. An overall score is, again, determined by accumulating (and weighting) the individual scores.

3.2 Similarity Strategies

Currently, iSPARQL supports all of the about 40 similarity measures implemented in SimPack. The reference to the implementing class as well as all necessary parameters are listed in the iSPARQL ontology. It is beyond the scope of this paper to present a complete list of implemented strategies. Therefore, Table 1 summarizes the five similarity strategies we use to evaluate the performance of iSPARQL on the MIT Process Handbook (see Section 4). We distinguish between *simple* and *engineered* strategies: simple strategies employ a single, atomic similarity measure of SimPack, whereas engineered strategies are a (weighted) combination of individual similarity measures whose resulting similarity scores get aggregated by a user-defined aggregator. Table 1 lists in addition to the explanation of each strategy if it is considered as simple or engineered.

4 Experimental Analysis

The goal of our experimental analysis was to find some empirical evidence to answer the questions raised in the introduction: Which are the "correct" measures for a given task and domain? Do engineered measures outperform off-the-shelf measures? And, can an "optimal" measure be learned? To that end we constructed a large ontology retrieval data set and performed two sets of experiments: the pure *retrieval experiments* show a comparison of both off-the-shelf and domain/task-specific, engineered similarity strategies using iSPARQL; The *machine learning experiments* compare the performance of these predefined measures to learned strategies gained using supervised learning approaches. In the following, we first describe the experimental setup, explain the generation of the test set that is used to perform the aforementioned experiments, and present the results of our evaluations.

We conducted all our experiments on a two processor dual core AMD Opteron 270 2.0GHz machine with 4GB RAM, 7200rpm disks, using a 32Bit version of Fedora Core 5.

4.1 Test Set Generation – "Mutating" the MIT Process Handbook

In order to evaluate our ontology retrieval approach, we needed a substantial database of instances, which includes a sizable number of queries with its associated correct answers. The correct answers are crucial, as they allow the quantitative evaluation of the retrieval approach. But preparing manually a suitable database that is large enough to enable statistical analysis can be impracticably time-consuming. We, therefore, decided to bootstrap the data set generation process by a large existing knowledge base that describes business processes: The *MIT Process Handbook* is an electronic repository of best-practice business processes and the result of over a decade of development by over 40 researchers and practitioners centered around the MIT Center for Coordination Science.[3] The Handbook

[3] Now called the MIT Center for Collective Intelligence (`http://cci.mit.edu`).

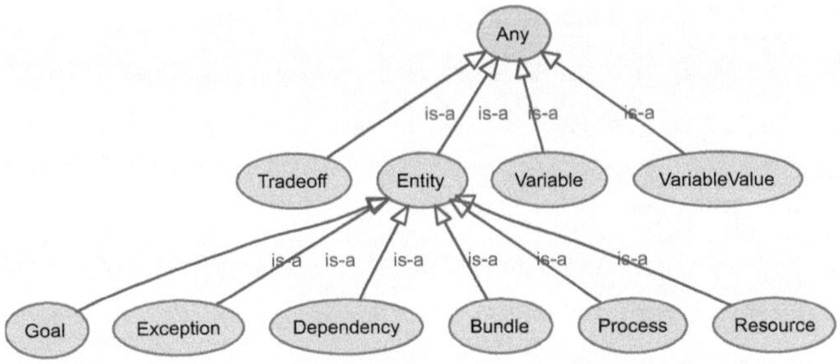

Fig. 1. Simplified structure of the OWL MIT Process Handbook Ontology

is intended to help people: (1) redesigning organizational processes, (2) inventing new processes, and (3) sharing ideas about organizational practices [13]. The Handbook includes a database of about 8000 business processes in addition to software tools for viewing, searching, and editing the database contents [12]. The Process Handbook is a process ontology: it provides a specialization hierarchy of processes (verbs) and their interrelationships in the form of properties, which connect the process to its attributes, parts, exceptions and dependencies to other processes. Note that specialization in the process handbook is non-monotonic. In other words, it is possible for a "child" process to overwrite or delete an inherited property. The Process Handbook, thus, has the advantage of being a sizable data set that was developed in a real-world setting (*i.e.*, by end-users and not by Semantic Web researchers).

In order to use the MIT Process Handbook for an evaluation, we had to *export it into an OWL-based format.* Given the non-monotonic inheritance structure the straight-forward translation of processes to concepts was not possible. We, therefore, decided to model the Process Handbook meta-model in OWL and export the processes in the Handbook as instances of the meta-model.[4] Hence, all major parts of the Handbook such as Process, Bundle, Goal, Exception, Resource, Dependency, and Trade-offs are represented as OWL classes (see Figure 1). With the ontology, we transformed the approximately 5000 business processes to OWL and stored them in their own files. Figure 2 shows a representative example of such a process.

Next, we had to find a sizable number of realistic queries and their corresponding correct answers in the Process Handbook. To that end *we adopted a novel approach for creating a test database that is based on semantics-preserving process mutation.* We began by *selecting 105 distinct process models* from within the Process Handbook repository. These models represent the target set. For each target process we then *created 20 variants of that process that are syntactically different but semantically equivalent* using mutation operators. These variants represent the "true positives" or correct answers (*i.e.*, the database items that should be returned when our retrieval algorithm is applied to find matches for

[4] In order to preserve the inherent semantics of the MIT Process Handbook, some additional rules in RuleML would be needed [2].

```
@prefix rdfs:             <http://www.w3.org/2000/01/rdf-schema#> .
@prefix daml:             <http://www.daml.org/2001/03/daml+oil#> .
@prefix rdf:              <http://www.w3.org/1999/02/22-rdf-syntax-ns#> .
@prefix owl:              <http://www.w3.org/2002/07/owl#> .
@prefix processHandbook: <http://www.ifi.unizh.ch/ddis/ph/2006/08/ProcessHandbook.owl#> .

<http://www.ifi.unizh.ch/ddis/ph/2006/08/E1024.owl#E1024> a processHandbook:Process ;
processHandbook:name "Determine cost" ;
processHandbook:description "This is a general activity to determine the cost
                             to the organization of purchase or production." ;
processHandbook:hasException <http://www.ifi.unizh.ch/ddis/ph/2006/08/E17159.owl#E17159> ;
processHandbook:hasSpecialization <http://www.ifi.unizh.ch/ddis/ph/2006/08/E6302.owl#E6302> ,
                                  <http://www.ifi.unizh.ch/ddis/ph/2006/08/E8007.owl#E8007> ;
processHandbook:hasGeneralization <http://www.ifi.unizh.ch/ddis/ph/2006/08/E3356.owl#E3356> .
```

Fig. 2. The figure shows process E1024 ("Determine cost") in Notation 3. E1024 has one exception, two process specializations, and one process generalization

the target process). All other items in the database are viewed as non-matches, and should not be returned by our retrieval algorithm if is it operating correctly.

Variants were created by applying semantics-preserving mutation operators to the target processes. Every variant represented the application of between 1 and 20 randomly selected operators to a target process. We used the following operators:

- a *process step* (*i.e.*, part of a process) is
  - split into two siblings (STEPSPLIT)
  - split into a parent/child (STEPCHILD)
  - merged with a (randomly selected) sibling (STEPMERGESIB)
  - merged with its parent (STEPMERGEPARENT)
  - deleted (STEPDELETE)
- a *word in the name* of a process is
  - deleted (NAMEDELETE)
- a *word in the description* of a process is
  - deleted (DESCRIPTIONDELETE)

The mutation operators were selected so that they produce a plausible alternative way of modeling the process they were applied to. If we were modeling a restaurant process, for example, some people might combine the "order" and "pay" actions into one substep (*e.g.*, for a fast food restaurant), while others might model the same process with separate substeps for "order" and "pay". These two approaches represent syntactically different, but semantically equivalent, ways of modeling the same process. The STEPMERGESIB operator could take a process model with distinct "order" and "pay" substeps and merge them into one. Conversely, the STEPSPLIT operator could take a process model where "order" and "step" are merged, and split them into two distinct substeps.

It should be noted, as a caveat, that there is a substantial random element in how the mutation operators work, since they do not perform a sophisticated semantic analysis of a step before, for example, deciding how to perform a split. Hence, the process variants may not look much like what a human might have generated, even though they are generated by a process that is similar to what a person might have used. It is our belief, however, that a semantics-preserving mutation approach represents a promising way for generating large query collections enabling rapid and useful evaluations of different retrieval algorithms. The algorithms that "rise to the top" as a result of this screening procedure can then be evaluated using hand-generated test sets that, presumably, will produce retrieval and precision figures that are closer to what we can expect in "real-world" contexts. The generated process retrieval test collection, including queries and variants (true positives) is available at our project web site.[5]

4.2 Retrieval Experiments – Off-the-Shelf vs. Engineered

In order to compare the performance of off-the-shelf versus specifically engineered similarity strategies, we first chose three simple strategies from SimPack: TFIDFD, LevN, and LoLN (see Section 3.2). Obviously, we did not choose them randomly but actually chose the off-the-shelf measures that we thought would perform well and then discarded the ones that were not performing sufficiently well to compete with the top-ranking ones. Second, we manually defined (or engineered) two task and domain specific complex similarity strategies that are both a combination of multiple similarity measures based on our experience with the Process Handbook: MITPH-LoLNTFIDFD and MITPH-LoLNTFIDFDJaccardAll. Note that while almost no domain knowledge is necessary to choose and define the off-the-shelf similarity strategies, some domain expertise is needed for the human-engineered strategies since specifying which measures should be used to determine the similarity between which elements of processes means to have a profound understanding of the structure of the data.

To compare the performance of the similarity strategies, we had to execute all 105 query processes with each of the five similarity strategies. Here, the capabilities of iSPARQL were very useful: since it was designed to run SPARQL queries with similarity joins, we could simply construct iSPARQL queries that would correspond to the retrieval operations. Consider the query depicted in Listing 1.2: it computes a similarity join between the process with the reference http://www.ifi.unizh.ch/ddis/ph/2006/08/ProcessHandbook.owl#E16056 and all other entries in the knowledge base using the TFIDFD strategy and returns them ordered descending by similarity. Actually, we were able to run all five strategies with one query by having an `ImpreciseBlockOfTriples` (see Section 3.1) for each strategy in the same query, exemplifying how iSPARQL simplifies the implementation of Semantic Web retrieval applications.

To evaluate the performance of the queries we chose the traditional information retrieval measures precision and recall. As a representative example, the results for process E16056[6] are shown in Figures 3(a) and 3(b), which depict precision and recall for the 100 most similar processes to the query process. As one can see, TFIDFD outperforms all other strategies in terms of precision

[5] `http://www.ifi.unizh.ch/ddis/ph-owl.html`
[6] `http://www.ifi.unizh.ch/ddis/ph/2006/08/E16056.owl`

```
PREFIX ph: <http://www.ifi.unizh.ch/ddis/ph/2006/08/ProcessHandbook.owl#>
PREFIX isparql: <java:ch.unizh.ifi.isparql.query.property.>

SELECT ?process2 ?name2 ?similarity
WHERE {
 ?process2 ph:name ?name2 .
 ?strategy isparql:name ``TFIDFD'' .
 ?strategy isparql:arguments (ph:E16056 ?process2) .
 ?strategy isparql:similarity ?similarity .
ORDER BY DESC(?similarity)
```

Listing 1.2. iSPARQL retrieval query

closely followed by MITPH-LoLNTFIDFD. Both, simple as well as engineered strategies start very high with precision=1, except for MITPH-LoLNTFIDFD that starts around 0.9. LoLN rapidly falls below 0.2 in precision (~25 returned processes), which expresses its low usefulness for this retrieval task. Considering recall (Figure 3(b)), MITPH-LoLNTFIDFD starts highest (recall ~0.7) but gets outperformed by TFIDFD (around 15 returns) for larger query result sets. Why does the standard TFIDF perform so well? We believe it is due to the large descriptions that are typically associated with Process Handbook entries. Given that the descriptions were not mutated in all cases and that mutation did essentially consist of deleting words, TFIDF, which has been found to be very useful in full text retrieval, may have an unfair advantage. Nonetheless, even disregarding TFIDF as a competitor, it is interesting to observe that neither of the engineered measures uniformly outperforms the off-the-shelf ones in terms of precision, but that they only gain with larger result sets. Why does the engineered measure MITPH-LoLNTFIDFDJaccardALL not perform equally well (LoLN initially outperforms it in terms of precision and almost uniformly outperforms it in terms of recall)? This might be due to badly chosen weights of the individual similarity strategies (*i.e.*, instead of giving the same weights to TFIDFD, LoLN, and Jaccard, TFIDFD should probably be weighted much higher as indicated by the simple strategies). We discuss an approach of how to learn such weights in the next subsection.

Figure 4 shows average precision and recall of the five employed similarity strategies across all 105 queries. As the figure illustrates the performance of all measures across all the queries is not as good as for the the single query. Nonetheless, we can see that the findings from the one query generalize qualitatively. Specifically, Figure 4(a) illustrates that the simple TFIDFD measure clearly outperforms all other strategies in terms of precision – it seems as if the descriptions across all the queries again are of much higher importance than other structure properties of a process. However note, that precision for all measures (including TFIDFD) on average is not as high as in the single query case. This due to the fact, that there are processes in the test collection which have shorter textual descriptions and/or fewer properties resulting in lower TFIDF similarity scores, which, in turn, leads to reduced average precision. In terms of *precision*, all three simple strategies outperform the engineered ones again for

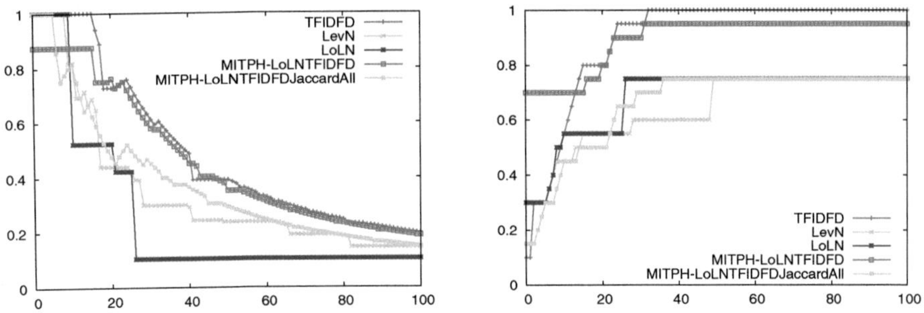

(a) Precision for E16056. (b) Recall for E16056.

Fig. 3. Precision and recall for a representative example process

few processes returned. For larger query result sets, the two engineered strategies MITPH-LoLNTFIDFD and MITPH-LoLTFIDFDJaccardAll perform better than LevN and LoLN but still worse than simple TFIDF. Inspecting *recall* (Figure 4(b)), the best performing similarity strategy is the engineered MITPH-LoLNTFIDFD until ∼40 returned processes. With larger result sets, it gets outperformed by TFIDFD that starts with about the recall of low (recall ∼0.3). We note, that also on average, similarity strategies incorporating TFIDF to measure the relatedness of processes of the MIT Process Handbook perform substantially better than strategies focusing on other modeling aspects. Thus, future strategies should probably use TFIDF as one of their component measures, assigning it a high enough weight in the overall similarity computation.

Summarizing, we can state, that the engineered measures do not uniformly outperform the off-the-shelf ones. Indeed, it seems that the simple ones that are heavily reliant on full-text (such TFIDF) are favored by this data set. Ignoring the description (and the TFIDF measure), however, we can see that the engineered measures perform better in terms of both precision and recall for large return sets. For small return sets the off-the-shelf measures are better in terms of precision and at least competitive for recall.

4.3 ML Experiments – Off-the-Shelf and Engineered vs. Learned

The last question raised in the introduction demands clarification on the performance on a learned measure in comparison to either the off-the-shelf or the engineered ones. To that end we decided to employ the widely used machine learning tool Weka[7] in conjunction with iSPARQL to learn a similarity measure based on the results obtained with the simple as well as engineered strategies. Specifically, for each of the 105 queries we took all the off-the-shelf but the TFIDF measures used so far. The rationale for not using the TFIDF measure was that we did not want the description to have too much influence in this evaluation. Together with the information if they were a correct or incorrect answer, we combined them

[7] `http://www.cs.waikato.ac.nz/ml/weka/`

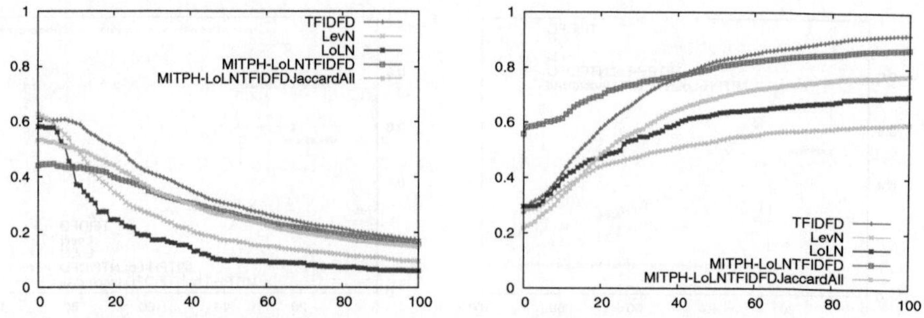

(a) Average precision. (b) Average recall.

Fig. 4. Average precision and recall for 105 queries and five similarity strategies

to a feature vector $\mathbf{shelf_{i,j}} = [LevN(i,j), LoLN(i,j), correct(i,j)]^T$, where i is the number of the target entity, j is the number of the entity from the Process Handbook and $correct(i,j)$ specifies if j is a correct answer to the query i. We then combined all the vectors $\mathbf{shelf_{i,j}}$ to the data set *shelf*. Analogously, we constructed the vector $\mathbf{engineered_{i,j}}$ that extended $\mathbf{shelf_{i,j}}$ with the engineered measures to the data set *engineered*.

For each of these two data sets we then learned a similarity measure using a logistic regression statistical learning algorithm performing an (almost) 10-fold cross validation.[8] We took 10% of the queries (always exactly 10, discarding the rest), learned the similarity measure using the logistic regression learner on the remaining 90% of the data, and then measured its effectiveness on these 10%. This approach is standard practice in machine learning. The averages of the results of the 10 runs are shown in Figure 5. As the Figures 5(a) and 5(b) show, the performance of the learned measures vastly outperforms both the engineered and the off-the-shelf measures (note the scale on the figures!). It, thus, seems that each of the measures employed contains some latent (potentially different) information about the similarity between the queries and its correct answers. Combined, they provide an excellent performance. Note also, that the similarity measure learned from the *engineered* data set (the upper line in Figures 5(a) and 5(b)) significantly outperforms the one learned from the *shelf* data set (lower curves). Since precision/recall curves are sometimes misleading when evaluating the performance of learning approaches, we also supply the average receiver operating characteristic (ROC) curves for both learned measures. The ROC curve graphs the true positive rate (y-axis) against the false positive rate (x-axis), where and ideal curve would go from the origin to the top left (0,1) corner, before proceeding to the top right (1,1) one. As Figure 5(c) clearly shows, the similarity measure learned from the *engineered* data set almost perfectly mimics a perfect prediction resulting in an accuracy of 99.469%; the one for the *shelf* data set being not much worse with an accuracy of 98.523%.

[8] We call it "almost" 10-fold cross validation because 105 queries cannot be divided into 10 equally sized groups, but 5 groups of 10 and 5 groups 11 queries.

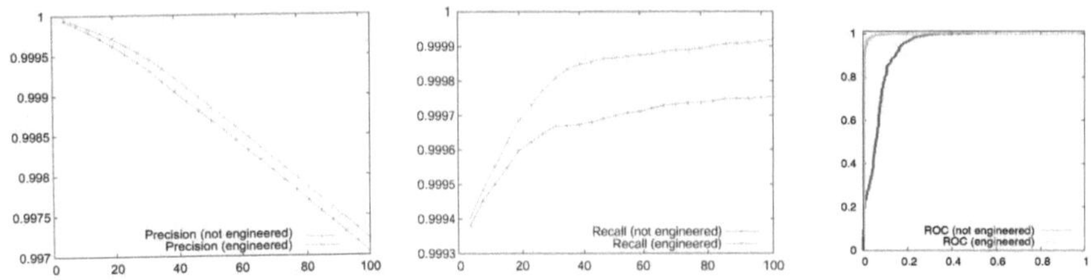

(a) Precision curves. (b) Recall curves. (c) ROC curves.

Fig. 5. Results for the learned similarity measure (logistic regression)

5 Discussion, Limitations, and Future Work

The findings of the preceding analysis are relatively clear. First, the ease of use of our iSPARQL framework for the presented semantic process retrieval task has been clearly shown. Evaluations, that previously would have had to be programmed tediously, could be effectuated by simply compose a query. The seamless integration of simple, off-the-shelf as well as human-engineered similarity strategies significantly simplified the implementation. We, therefore, believe that a declarative query language containing statistical reasoning elements like iSPARQL can significantly simplify the design and implementation of Semantic Web applications that include some element of similarity. Since such elements are included in many of the core Semantic Web applications (*e.g.*, matchmaking, retrieval in ontologies, ontology alignment, etc.), *tools such as iSPARQL can play an important role in simplifying the spread of the Semantic Web.*

Second, as our retrieval experiments showed, the human-engineered measures performed constantly better on large sets of processes than the off-the-shelf measures did. In contrast, the simple, off-the-shelf strategies turned out to be superior for smaller sets. Furthermore, strategies including the TFIDF measure, which heavily drew on the process descriptions performed better in terms of precision and recall. This indicates that the *off-the-shelf methods captured a different notion of the similarity between processes than the engineered ones.* This finding is further supported by the learned similarity measures. As Figure 5 shows the results of the algorithm learned with only the off-the-shelf data is somewhat less precise than the one learned with both the engineered and off-the-shelf methods. Hence, the information contained in the engineered measures is at least partially complimentary to the information contained in the off-the-shelf ones, which the learning algorithm can exploit. Arguably, this additional information is *the latent experience of the experts that was embedded in the engineered measures.*

Third, the learned measures clearly outperformed the designed or off-the-shelf ones. The learning algorithm's ability to combine the complimentary information contained in the different notions of similarity proved to provide an overall almost overwhelming accuracy. We can, therefore, clearly conclude that *the value of using learned similarity measures seems immense*, assuming that a sufficient number of examples is available: irrespective of whether we used off-the-shelf or expert measures, the learned measures performed close to perfect.

One major *limitation of our work is the choice of experimental data.* The generalizability of our findings across tasks and domains is limited by the fact that we (i) only used one data set, (ii) that this data set employed some generated data, (iii) we only ran one task, and (iv) that the test suite generation strategy might have influenced the results. Nonetheless, we believe that our findings are likely to hold across domains and task: First, extrapolating from information retrieval, where the choice of good similarity measures seem to permeate across both tasks and domains. Second, while our data set is not ideal, *it is one of the first ones* in the Semantic Web that contains a large data set with both queries and associated true answers. Such data sets are very costly to design and only their introduction to the community will allow comparative studies, which, ultimately, is the basis of science. Last, even though the true positives where generated (note that the data base itself was collected by domain experts), their generation process was guided by many years of experience with the type of data under study. We, therefore, believe that our findings will generalize at least across domains and possibly, given the ubiquity of similarity measures in computer science and AI, even across tasks.

We see a couple of *future research directions*: (1) extending our evaluation to other domains and tasks to ensure our findings' generalizability; (2) applying iSPARQL to different Semantic Web tasks such as service matchmaking and ontology alignment to shed some more light on its potential as a framework; and (3) investigating extensions to iSPARQL that will further improve its usefulness for additional tasks.

6 Conclusions

Our study investigated the use of similarity measures in a process ontology retrieval task using the iSPARQL framework. We found that the declarative nature of iSPARQL did significantly simplify the task prompting us to a deeper investigation of the applicability of iSPARQL to different Semantic Web tasks such as matchmaking and ontology alignment, beyond the presented process retrieval task. We also found that the combination of different notions of similarity string learning approaches significantly boosted the overall task performance. Therefore, as seen from our evaluations, the use of statistics, either directly employed by similarity strategies or by statistical learning algorithms, proved crucial for the performance in this task. For the Semantic Web in general, these findings raise the question whether the more wide-spread use of statistical reasoning elements would not improve the overall performance of its tools and applications.

Acknowledgment. One of the authors was supported by the Korea Research Foundation Grant of the Korean Government (KRF-2006-214-D00193). We would like to thank the anonymous reviewers for their valuable comments.

References

1. R. Baeza-Yates and B. Ribeiro-Neto. *Modern Information Retrieval.* Addison Wesley, 1999.
2. A. Bernstein, B. Grosof, and M. Kifer. Beyond Monotonic Inheritance: Towards Non-Monotonic Semantic Web Process Ontologies. In *W3C Ws. On Frameworks for Semantics in Web Services*, 2005.
3. A. Bernstein and C. Kiefer. Imprecise RDQL: Towards Generic Retrieval in Ontologies Using Similarity Joins. In *Proc. of the 2006 ACM Symp. on Applied Computing (SAC '06)*, pages 1684–1689, New York, NY, 2006.
4. A. Bernstein and M. Klein. Towards High-Precision Service Retrieval. In *Proc. of the 1st Int. Semantic Web Conf. on The Semantic Web (ISWC '02)*, pages 84–101, London, UK, 2002.
5. S. Brockmans, M. Ehrig, A. Koschmider, A. Oberweis, and R. Studer. Semantic Alignment of Business Processes. In *Proc. of the 8th Int. Conf. on Enterprise Information Systems (ICEIS '06)*, pages 191–196, Paphos, Cyprus, 2006.
6. W. W. Cohen, P. Ravikumar, and S. Fienberg. A Comparison of String Distance Metrics for Name-Matching Tasks. In *Proc. of the IIWeb Ws. (IJCAI '03)*, 2003.
7. M. Ehrig, A. Koschmider, and A. Oberweis. Measuring Similarity between Semantic Business Process Models. In *Proc. of the 4th Asia-Pacific Conf. on Conceptual Modelling (APCCM '07)*, Ballarat, Victoria, Australia, 2007. to appear.
8. L. Geng and H. J. Hamilton. Interestingness Measures for Data Mining: A Survey. *ACM Comp. Surv.*, 38(3), 2006.
9. D. Gentner and J. Medina. Similarity and the Development of Rules. *Cognition*, 65:263–297, 1998.
10. M. Klusch, B. Fries, and K. Sycara. Automated Semantic Web Service Discovery with OWLS-MX. In *Proc. of the 5th Int. Joint Conf. on Autonomous Agents and Multiagent Systems (AAMAS '06)*, pages 915–922, New York, NY, 2006.
11. V. I. Levenshtein. Binary Codes Capable of Correcting Deletions, Insertions and Reversals. *Soviet Physics Doklady*, 10:707–710, 1966.
12. T. W. Malone, K. Crowston, and G. A. Herman. *Organizing Business Knowledge: The MIT Process Handbook.* MIT Press, Cambridge, MA, 2003.
13. T. W. Malone, K. Crowston, J. Lee, B. Pentland, C. Dellarocas, G. Wyner, J. Quimby, C. S. Osborn, A. Bernstein, G. Herman, M. Klein, and E. O'Donnell. Tools for Inventing Organizations: Towards a Handbook of Organizational Processes. *Management Science*, 45(3):425–443, 1999.
14. E. Prud'hommeaux and A. Seaborne. SPARQL Query Language for RDF. Technical report, W3C, 2006.
15. D. Wolpert and W. Mcready. No Free Lunch Theorems for Optimization. *IEEE TOEC*, 1(1):67–82, 1997.

Integrating Folksonomies with the Semantic Web

Lucia Specia and Enrico Motta

Knowledge Media Institute – The Open University
Walton Hall, MK7 6AA, Milton Keynes, UK
{L.Specia,E.Motta}@open.ac.uk

Abstract. While tags in collaborative tagging systems serve primarily an indexing purpose, facilitating search and navigation of resources, the use of the same tags by more than one individual can yield a collective classification schema. We present an approach for making explicit the semantics behind the tag space in social tagging systems, so that this collaborative organization can emerge in the form of groups of concepts and partial ontologies. This is achieved by using a combination of shallow pre-processing strategies and statistical techniques together with knowledge provided by ontologies available on the semantic web. Preliminary results on the del.icio.us and Flickr tag sets show that the approach is very promising: it generates clusters with highly related tags corresponding to concepts in ontologies and meaningful relationships among subsets of these tags can be identified.

1 Introduction

Describing resources by means of a set of keywords is a very common way of organizing content for future use, including search and navigation. A collaborative form of this process for shared web-based resources, called "tagging", or "social tagging / annotation", has been gaining impressive popularity among web users. In the light of the Web 2.0 philosophy, the several social tagging systems available nowadays enable users to annotate their resources (web pages, images, videos, etc.) with a set of words, the so-called "tags", which they believe to be relevant to characterize the resource according to their own needs, without relying on a controlled vocabulary or a previously defined structure. The main goal of this annotation is to facilitate the access to resources and, since the systems allow users to share their resources and annotations, tags serve as links to resources annotated both by their owners and by other users. This allows the emergence of a shared and evolving classification structure, which is sometimes called "folksonomy"[1], i.e., a folk taxonomy, or a lightweight conceptual structure created by the users.

Social tagging systems such as Flickr (http://www.flickr.com/), for photo-sharing, and del.icio.us (http://del.icio.us/), for social bookmarking, are becoming more and more popular, covering nowadays a wide range of resources and communities, with a huge number participants sharing and tagging a large number of resources. For example, del.icio.us is said to have more than 1,000,000 registered users in September 2006, who have been posting more than 100,000 bookmarks each day (http://deli.ckoma.net/stats).

---

[1] As defined by T. Vander Wal (http://www.vanderwal.net/random/entrysel.php?blog=1750).

E. Franconi, M. Kifer, and W. May (Eds.): ESWC 2007, LNCS 4519, pp. 624–639, 2007.

Tagging systems are constituted by three main elements: users, resources and tags. Although in most of the systems tags are not mandatory, they are certainly a very important element. Besides establishing a relationship between a resource and a concept in the user's mind, tags can be thought of as the connecting element between resources and users, with these connections defining (even implicitly) relationships amongst users (several users may use the same tags) and amongst resources (resources can be tagged with the same words).

Taking subsets of the tags in Flickr and del.icio.us as examples, in this paper we focus on the relationships amongst tags themselves and their mapping into formal concepts in ontologies. We are therefore primarily interested in the collective purpose of the tags assigned to resources. In that sense, one of the greatest strengths of the social tagging systems, the fact that no pre-defined vocabulary is assumed, leads to a number of limitations and weaknesses in what concerns the use of the tags to retrieve content. As highlighted by Golder and Huberman (2005), the main problems of social tagging systems include *ambiguity*, *lack of synonymy* and *discrepancies in granularity*. An ambiguous word, e.g. `apple`, may refer to the `fruit` or the `computer company`, and this in practice can make the user retrieve undesired results for a certain query. Synonyms like `lorry` and `truck`, or the lack of consistency among users in choosing tags for similar resources, e.g., `nyc` and `new york city`, makes it impossible for the user to retrieve all the desired resources unless he/she knows all the possible variants of the tags that may have been used. Different levels of granularity in the tags may also be a problem: documents tagged `java` may be too specific for some users, but documents tagged `programming` may be too general for others.

We present an approach to minimize these problems by making explicit the semantics behind the tag space in social annotation systems. This is achieved by a pipeline of processes including the cleaning up of tags, the analysis of co-occurrence among tags, the clustering of tags based on the co-occurrence information, and finally the mapping of tags in a cluster into elements (concepts, properties or instances) in ontologies and the extraction of (taxonomic or non-taxonomic) semantic relations between them, using for that information from ontologies available on the semantic web, as well as resources like Wikipedia and Google. Although other attempts have been made to bring the semantics of folksonomies to the surface, as we discuss in Session 2, they do not go beyond finding groups of related tags - no assumption is made about the nature of the tags or the relationships within clusters. Moreover, a systematic evaluation on the quality of the clusters is not performed.

As result of our approach, we obtain groups of highly related tags corresponding to elements in ontologies, structured according to the relationships holding amongst those elements, which can be thought of as *faceted ontologies*, that is, partial ontologies conceptualizing specific facets of knowledge. In contrast with traditional "monolithic" ontologies, the resulting ontologies are constructed by putting together fragments derived from multiple ontologies on the semantic web2. These resulting ontologies can be used to enhance various tasks in the tagging systems (for users and semantic web applications):

a) **Query (tag) extension/disambiguation:** in searches for tags, query extension to all related tags (or a subset of them) can be offered to the user. Simple

[2] In (Motta and Sabou, 2006) we argue that this ability of dynamically combining and integrating information coming from multiple ontologies on the web is one of the key features of the emerging new generation of semantic web applications.

heuristics, based on the types of tags, can be used to extend the search to a subset of the related tags. Also, if the searched tag is ambiguous, the user can be given the related tags in each cluster it appears in order to choose the sense. The search can then be restricted to that sense by adding another tag (from the cluster) to the query.

b) **Visualization:** clusters of related tags (and the relationships among them) can be graphically presented to provide a better understanding on the way the searched tag is used in the system, which can also be used for query extension / disambiguation.

c) **Tag suggestion:** when tagging a resource, the user can be offered suggestions of "good" tags, based on other tags used by other people for that resource (like in del.icio.us), or related tags that are highly frequent in a given cluster.

The approach can also be used to support **ontology evolution and population:** the new and dynamic knowledge provided by users can complement the formal knowledge in ontologies by adding concepts (or instances of concepts) and relationships (or instances of relationships) between concepts in that ontology. Therefore, with our approach to integrating folksonomies and the semantic web we intend to show ultimately both (i) that the ontologies provided by the semantic web can be used to structure folksonomies semantically and (ii) that the dynamic knowledge provided by folksonomies can be used as a resource for bottom-up knowledge acquisition to support ontology evolution.

The rest of this paper is organized as follows. In Section 2 we describe a few approaches that are related to our work. In Section 3 we present our approach to integrate folksonomies with the semantic web. In Section 4 we show the results of initial experiments with Flickr and del.icio.us tag sets. We conclude with some remarks and future work in Section 5.

2 Related Work

Because one important step in our work is the identification of relations between tags, before comparing our work to other approaches that try to extract semantics from tagging systems, it is important to distinguish it from traditional approaches to relation extraction from texts (Schutz and Buitelaar, 2005; Specia and Motta, 2006). The main difference is that here we cannot count on the conventional notion of "context", i.e., surrounding words around the tag. Some attempts have been made to use information about the resources as context, but there is no guarantee that this *ad hoc* context will offer helpful clues. For example, (Aurnhammer et al., 2006) uses image content features as context to improve search in Flickr: an ordinary search by tag is accomplished and the user then selects a subset of the resulting images to perform another search by "similar" images according to two simple features - colour and texture. Images with other tags, not necessarily similar to the initial one, can therefore be retrieved. However, this image retrieval strategy is unlikely to work well with complex images. In our approach, we rely on no additional context except the tags themselves.

Aiming to induce faceted ontologies from Flickr tags, (Schmitz, 2006) uses a subsumption-based model, derived from the co-occurrence of tags, to find candidate subsumption relations: a tag x subsumes another tag y if the probability of x occurring given y is above a certain threshold and the probability of y occurring given x is below that same threshold. Given the resulting set of "candidate pairs of tags", a tree of possible parent-child relationships is built, with certain candidate pairs being filtered out according to their position and thus reinforcing the remaining relationships. For each leaf of the tree, the best path to the root is chosen accordingly and partial paths are merged into sub-trees. Some of the illustrated sub-trees show that common features hold amongst certain tags. For example, a resulting tree contains san `francisco` as the subsuming tag and a set of children like `civiccenter`, `cliffhouse`, `streetfair`, `muni`. From a semantic point of view this approach is however limited, as in the general case the identified relationships will vary considerably (e.g., these trees may mix type-of, hyponym, or part-of relationships), but these distinctions are not captured.

Other approaches that concentrate in finding groups of potentially related tags include those of (Begelman et al., 2006) and (Wu et al., 2006). In (Begelman et al., 2006), the tag space is first organized according to their co-occurrences in annotating different resources. A cutoff co-occurrence value is defined based on disruption points in frequency graphs. This new tag space is then represented as an undirected graph, having strongly related tags as vertices and edges with pairs of tags weighted according to the number of times they co-occur. This yields clusters of related tags, but since some clusters are very big, a spectral clustering algorithm is applied to refine them. Amongst the illustrated examples of clusters created for RawSugar data (http://www.rawsugar.com), some seem to group truly related tags (e.g., `{health, nutrition, food, diet}`), while tags in other clusters are less related (e.g., `{health, shopping, research}`). No assumption can be made about the nature of the relationships holding within a cluster.

Wu et al. (2006) present a probabilistic model, which aims to generate groups of semantically related tags based on the co-occurrence of tags, resources, and users. Entities (user, resource or tag) are represented as a multi-dimensional vector, a *conceptual space*, where each dimension represents a category of knowledge – whose meaning is unknown. The value in each dimension should measure the level of relationship between the entity and the corresponding category of knowledge. The log-likelihood of the dataset is estimated in order to determine the number of dimensions of that conceptual space and assign the relationship values of entities to each dimension. In experiments with a subset of del.icio.us data, 40-dimensions are estimated as sufficient to represent the major category of meanings. A small example taking 10 randomly selected dimensions and the top 5 closely related tags to each of those dimensions shows that relationships hold amongst the tags within each group (dimension), however, once more, the types of these relationships are not explored.

Mika (2005) extends the traditional bipartite model of ontology with a social dimension, yielding in a tripartite model involving users (actors), tags (concepts), and resources (instances of concepts). With a subset of del.icio.us' tags, based on the co-occurrence of tags with resources and users, the author builds graphs relating tags and users and also tags and resources. Techniques of network analysis, which are not discussed in the paper, are then applied on those graphs in order to discover emergent ontologies. For each graph, the result is a set of clusters of semantically related tags, but the relations are not made explicit.

Schmitz et al. (2006) extract association rules between projections of pairs of elements from the tripartite model of folksonomies, i.e., users, resources and tags. In experiments with data from del.ici.us, the authors illustrate two different projections, learning rules of the types: (i) users assigning certain tags to some resources often also assign another set of tags to those resources; and (ii) users labelling certain resources with a set of tags often also assign those tags to another set of resources. While these kinds of association rules make it possible to identify the existence of relationships among different tags, users or resources, they do not provide any information about the nature of these relationships.

Finally, most of the social systems, including Flickr and del.icio.us, provide facilities such as "clusters" and "related tags", which show groups of related tags to allow the user to tune the search to other (statistically) related tags. Del.icio.us also provides "recommended tags" and "popular tags" when a given resource is being tagged, based on tags previously used for the same resource. Apparently these facilities rely on co-occurrence information but the groups express nothing about the actual relationships between the tags.

Since none of the described approaches applies more sophisticated pre-processing of the tags than eliminating infrequent tags, tags like `Music` and `music` count as different elements. The same applies to tags with very little lexical variation, such as `blog` and `blogs`. As we describe in the next section, we use specific strategies for cleaning up tags, and, more importantly, besides identifying groups of related tags, we investigate the nature of these relationships by exploiting information available on the semantic web, in order to give semantics both to the tags themselves and to the relationships between tags.

3 Integrating Folksonomies with the Semantic Web

As we previously mentioned, the tag space in social tagging systems encompasses semantic aspects of the system that are not explicitly defined. By identifying formal elements corresponding to tags and relationships among them it is possible to make explicit a significant part of this underlying knowledge, which is crucial for the efficient use of these systems. In fact, only with a clear semantic structure the annotations in folksonomies can be useful not just to humans, but can be made available to software agents and applications on the semantic web. Our hypothesis is that this knowledge can be derived by means of a statistical analysis of the annotations combined with pragmatic information provided by the semantic web and additional clues given by external resources.

3.1 Datasets

We investigate the tag sets in Flickr and del.icio.us due both to their popularity (with a large number of resources, users, and tags) and availability. These datasets differ from each other in a series of features. In fact, Thomas Vander Wal[3] mentions these systems when distinguishing between broad and narrow folksonomies: in a broad

[3] http://www.personalinfocloud.com/2005/02/explaining_and_.html

folksonomy (e.g., del.icio.us) many users tag the same resource, while in a narrow folksonomy (e.g., Flickr) only the creator of the resource tags it. Other studies on the structure of both del.icio.us and Flickr, focusing on user activity, tag frequency, kinds and variability of tags, among other aspects, are presented in (Golder and Huberman, 2005) and (Marlow et al., 2006). In our experiments, we use the del.icio.us tags provided by Peter Mika, which were also used in (Mika, 2005), and Flickr tags for photos posted between 01-02-2004 and 01-03-2006. The total numbers of entries (i.e., a resource tagged by a user) and tags in both datasets, as well as the number of distinct users, resources and tags, are shown in Table 1.

Table 1. Number of tags (with their corresponding users and resources) from del.icio.us and Flickr

	Total		**Distinct**		
	# entries	*# tags*	*# users*	*# resources*	*# tags*
del.icio.us	19,605	89,978	7,164	14,211	11,960
Flickr	49,087	167,130	6,140	49,087	17,956

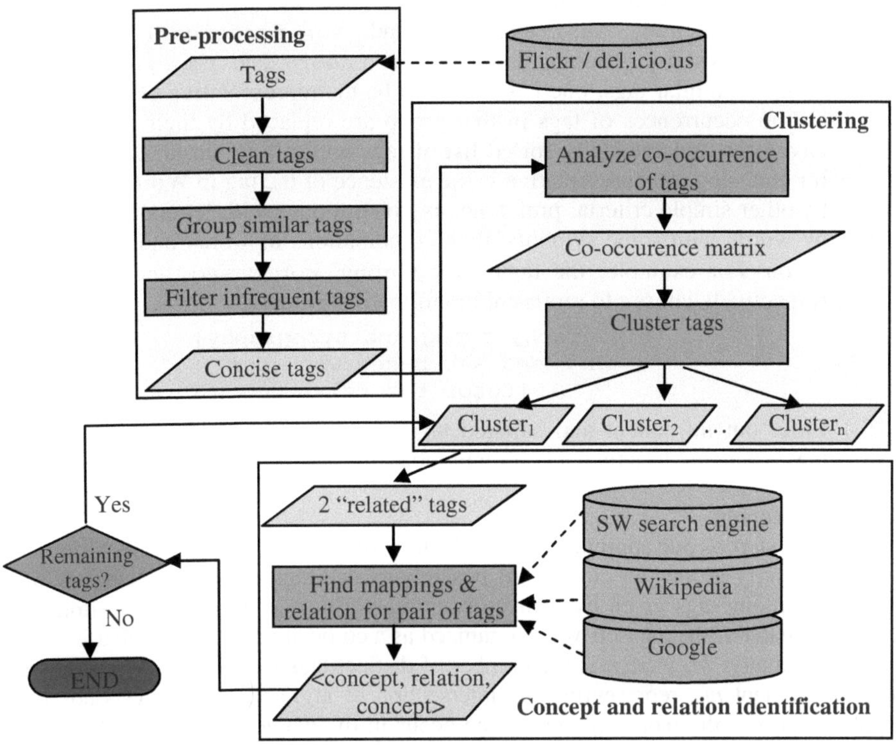

Fig. 1. System architecture

3.2 Methodology

Two very important features of our methodology are that it is unsupervised, i.e., it does not assume previously identified mappings or relationships to train the system, and it does not require any context besides the tags themselves, the resources being tagged and other tags used for those resources. The general approach, as given in Fig. 1, consists of three steps: pre-processing, clustering and concept / relation identification.

3.2.1 Pre-processing

The following shallow pre-processing steps were performed:

(1) Filter out unusual tags (and corresponding resources, if no other tag remains in that annotation). From a social perspective, all tags are relevant, even if they cannot be mapped to elements in ontologies. However, at this stage we are interested in tags with a more general applicability, which can be possibly found in ontologies, and therefore we define the following constraints: tags must start with a letter followed by any number of letters, numbers, and symbols like dash, dot, underscore, etc.

(2) Group morphologically very similar tags using the Levenshtein similarity metric[4] with a high threshold to determine "similar" words. This can tackle minor morphological variations (by grouping tags such as `cat` and `cats`, `san_francisco`, `sanfrancisco` and `san.francisco`) as well as misspellings (by grouping tags such as `theory` and `teory`). Within each group of similar tags, one is selected to be the representative of the group and all the occurrences of tags in that group are replaced by their representative. Given the alphabetically sorted list of tags within a group, the main criterion for choosing its representative is the existence of the tag in WordNet, followed by other simple criteria: preference is given to tags with letters only, followed by words with some symbols, then combinations of words and numbers, and so on. For example, the tags `typography`, `web-based`, and `tutor` were respectively chosen to represent the following groups:

```
{tipography typograph typography}
{web-based web_based webbased}
{tutor tutors}
```

(3) Filter out infrequent and isolated tags (and corresponding resource), that is, tags occurring less than a certain number of times or appearing only isolated.

3.2.2 Clustering

The second step of our approach is to perform a statistical analysis of the tag space in order to identify groups, or clusters, of possibly related tags. Clustering is based on the similarity among tags given by their co-occurrence. In order to find these similarities, the tags in each of the datasets were organized as a co-occurrence matrix, that is: a n x n symmetric matrix M, were n is the number of distinct tags in the dataset, and the value of each element m_{ij}, representing the intersection of tag_i and tag_j, corresponds to the number of times the pair tag_i and tag_j co-occur in the whole tagset (with the same or different resources/users). If $tag_i = tag_j$, then the intersection represents the frequency of the tag in the dataset.

---

[4] As implemented in the package SimMetrics in http://sourceforge.net/projects/simmetrics/

In this co-occurrence matrix, each line (or column) is a vector representing one of the tags. Therefore, several vector space statistics can be computed. We tried different metrics to calculate the similarity between the pairs of vectors, including Euclidian and Manhattan distance, angular separation (cosine), etc. Metrics computing absolute distance like Euclidian and Manhattan showed to be inappropriate, since they are much more sensitive to significant variations in a few elements than little variations in a large number of elements, which is relevant to our problem. We chose angular separation, illustrated in (1), which computes the cosine angle between two vectors and thus is more sensitive to small changes in various elements. It is also less complex than similar metrics such as correlation coefficient.

$$angular\_separation_{ij} = \frac{\sum_{k=1}^{n} x_{ik} . x_{jk}}{\left(\sum_{k=1}^{n} x_{ik}^2 . \sum_{k=1}^{n} x_{jk}^2 \right)^{\frac{1}{2}}} \quad (1)$$

As a result of computing the similarity between each possible pair of vectors in the co-occurrence matrix (i.e., n x n pairs), for each tag we obtain a list of its similarities to all the other tags. For example, Table 2 shows the top five similar words to the words `audio`, `semantic-web`, `adult`, `apple`, and `chat` in del.icio.us data.

As we can see in Table 2, this co-occurrence-based similarity computation already shows some semantics about the words. It goes beyond finding syntagmatic associations, since we do not simply check the pairs of tags that co-occur a significant number of times, such as in (Begelman, 2006). In our case, by using the co-occurrence matrix, we take into account all the other tags as context and state that to be considered similar to a certain tag_j, a tag_i has to co-occur not only with tag_j, but also with the other tags co-occurring with tag_j. That is, both tags must have a similar pattern of co-occurrence, which is given by their co-occurrence vectors. Similarities like these are sometimes called "paradigmatic associations".

Table 2. Top 5 similar words to some examples of tags

Top	audio	semantic-web	adult	apple	chat
1	mp3	rdf	girls	mac	aim
2	music	ontology	nude	macintosh	messenger
3	playlist	owl	babes	tiger	gtalk
4	streaming	semweb	pics	osx	msn
5	radio	daml	sex	macosx	icq

Certainly, a single tag can have two or more patterns of co-occurrence, representing different meanings or uses of the tag (e.g., `apple` as *computer brand* and as *fruit*). In that case, the most similar tags to a given tag will mix words referring to distinct domains. Therefore, although relevant, the information provided by the paradigmatic associations is limited to pairs of tags, i.e., it can only tell, for a given tag, that there is a set of other tags that are related to it, but this does not guarantee that a relationship also holds among the other tags in that set. For example, we could also have found a paradigmatic association between **apple** and **fruit** in Table 2, but clearly we should not include fruit in a group representing the *computer brand* sense of apple. Therefore, we extend the paradigmatic associations by defining a clustering algorithm on top of them.

In order to group the highly co-occurring tags, we first establish a similarity threshold to filter out pairs of tags that are not highly similar. Given the highly similar pairs of tags, the algorithm takes into account the mutual similarity amongst tags to identify the groups. It considers each pair of similar tags, for example, audio and mp3, as seeds constituting an initial cluster, and then tries to enlarge this cluster by looking for tags that are similar to both the initial tags. This procedure is recursively repeated for all the tags, i.e., each new "candidate" tag for a cluster must be similar to the whole (possibly enlarged) set of tags in that cluster. Once there are no more candidates for that cluster, a new pair of similar tags (e.g., audio and music) is taken as seed and this is repeated until all pairs of tags have been processed[5].

This procedure generates a set of clusters, including a number of identical clusters, resulting from distinct seeds that are in fact similar amongst each other. It also generates highly similar clusters, differing in only a few tags, which are in many cases a consequence of the threshold to filter out not so similar pairs of tags. We use two smoothing heuristics to avoid having a high number of these very similar clusters. For every two clusters:

1) If one cluster contains the other, that is, if the larger cluster contains all the tags of the smaller, remove the smaller cluster;
2) If clusters differ within a small margin, that is, the number of different tags in the smaller cluster represents less than a percentage of the number of tags in the smaller and larger clusters, add the distinct words from the smaller to the larger cluster and remove the smaller.

These heuristics make it possible to group two tags that are not sufficiently similar according to the established threshold, but are both similar to a large set of other tags. Therefore, we are able to eliminate redundancies but keep multiple clusters sharing a number of tags when those tags have multiple meanings, indicating that they are ambiguous tags. Good quality resulting clusters can already be used for several of the tasks described in Section 1, including tag extension/disambiguation, visualization and suggestion.

One important feature of our clustering technique is that it does not require establishing the number of clusters to be produced. The only parameters are the threshold to define the minimum co-occurrence for pairs of tags and the percentage of variation allowed for "similar clusters". Alternatively, we could have used traditional clustering algorithms. However, as we discuss in Section 5, this approach showed to be more appropriate for our problem.

---

[5] Our clustering strategy can be compared to the *Clustering by Committee* approach (Pantel, 2003), in the sense that multiple elements are chosen to be the cluster's centroids, as opposite to traditional partitional approaches in which only one centroid per cluster. However, our strategy is more strict, since all the elements within a cluster must be similar amongst each other, instead of being similar just to the centroids.

3.2.3 Concept and Relation Identification

Since our clusters are derived from co-occurrence information only, there is no indication of the relationships holding amongst subsets of the tags in each of them. Our goal is to use knowledge provided by different sources, including ontologies available on the semantic web, Wikipedia and Google, to discover if there are in fact relationships between tags in each cluster and, if they exist, categorize them. This process involves mapping the tags into concepts / instances / properties of ontologies and checking the possible relationships among the mapped tags. As a source for ontologies we use semantic web search engines such as Swoogle (Ding et al., 2004). The procedure within each cluster is the following:

1) Post each possible pair of tags to the semantic web search engine in order to retrieve ontologies that contain both tags. All combinations of pairs are tried, since it is not possible to know within which pairs a relation holds (look for matches with labels and identifiers).

2) If any of the tags is not found by the search engine, consider that they can be acronyms, misspellings or variations of known terms, and look for them in additional resources:

  2.1) Post the tag to Wikipedia in order to get (in the title field) the whole expression in case it is an acronym. For example, the query term `NYC` in Wikipedia returns as title `New York City`. Select the text in the title.

  2.2) If the tag is not in Wikipedia, consider it to be a misspelling or multi-word term. Post it to Google, looking for a suggestion of correct term. For example, in a search for `sanfrancisco` in Google, the system returns the following suggestion: "`Did you mean: san francisco?`" Select the suggested term.

3) If the two tags (or the corresponding terms selected from Wikipedia or Google) are not found together by the semantic web search engine, consider them not to be related and eliminate the pair from that cluster if they are not (possibly) related to any other tags, that is, all the combinations of pairs of tags must be searched.

4) Conversely, if ontologies are found containing the two tags:

  4.1) Check whether the tags were correctly mapped into elements of the ontologies. Tags can refer to the following elements: concepts, instances, or properties.

  4.2) Retrieve information about the tags in each of the ontologies: the type of tag (concept, instance, property), its parents (up to 3 levels) if it is a concept or an instance, and its domain and range or value if it is a property.

5) For each pair of tags for which the semantic web search engine retrieved information, investigate possible relationships between them:

  5.1) A tag is an ancestor of the other. For example, in the *Food* ontology[6], `apple` is a subclass of `fruit`.

  5.2) A tag is the range or the value of one of the properties of the other tag. For example, in the *Wine* ontology[7], the class representing the wine `Zinfandel` has a property `hasColor`, for which the value is `red`. Therefore, the relation `hasColor` holds between `Zinfandel` and `red`. This extends to properties defined in superclasses of the actual classes of the tags.

---

[6] http://lists.w3.org/Archives/Public/www-archive/2004Oct/att-0016/food.owl

[7] http://www.w3.org/TR/2003/CR-owl-guide-20030818/wine

5.3) Both tags have the ***same direct parent***. For example, `apple` and `pear` are concepts with the same parent (`fruit`) in the FOOD ontology.
5.4) Both tags have the ***same ancestors***, at the ***same level***. For example, in WordNet (Miller et al., 1994), `assembly` has as ancestors `building` (1st level) and `construction` (2nd level), while `formation` has the ancestors `fabrication` (1st level) and `construction` (2nd level).
5.5) Both tags have the ***same ancestors***, at ***different levels***. For example, in WordNet, `chapterhouse` has as ancestors `building` (1st level) and `construction` (2nd level), while `edifice` has as ancestor `construction` (1st level).

By looking for pairs of tags in single ontologies, instead of individual tags, we eliminate much of the ambiguity in those tags. For example, in the case (5.3) above, `apple` is also defined in other ontologies with different meanings, for example, the *computer brand* in the CLib-core-office ontology[8]. However, this ontology does not contain the tag `pear`, and thus it is not considered an information provider for the relation identification.

If more than one ontology contains both tags (and possibly relationships between those tags), we currently use a simple resolution strategy: we give preference to the ontology containing also other tags (and possibly relationships) in the cluster. If there are still multiple ontologies, the first fulfilling this constraint is chosen. If there are multiple relationships between a pair of tags in that ontology, we choose the first of them according to steps (5.1-5.5).

If the aforementioned procedure does not yield any relationship for a given pair of tags co-occurring in at least one single ontology, we assume they are not related and remove the pair from the current cluster (unless they are related to other tags). Obviously these strategies are rather simplistic and in the future we plan to use more elaborate strategies, in particular for deciding whether to merge information coming from multiple ontologies and how to do so (Sabou et al., 2006). It is important to notice that even when relationships are not found, if the tags can be correctly mapped to elements in ontologies, these mappings can already be used to support the various tasks discussed in Section 1. For example, related elements (concepts, instances, and properties) in the ontology, like subclasses of a concept representing a tag, can be used to extend searches. However, going one step ahead and finding self-contained structures within each cluster by merging knowledge from multiple ontologies can provide a much richer perspective on the underlying semantics of the tagging systems. In what follows we present initial experiments with our approach to find both meaningful clusters and the underlying relationships amongst their tags.

4 Experiments and Discussion

For both del.icio.us and Flickr datasets, in the first step of the approach, i.e., the **pre-processing** strategies, we empirically defined the parameters as follows: (1) to be "similar", a pair of words has to reach **0.83** or higher score (Levenshtein metric); (2) a tag has to occur at least **10** times. The resulting number of tags, resources, and users is shown in Table 3.

[8] http://www.cs.utexas.edu/users/mfkb/RKF/tree/CLib-core-office.owl

Table 3. Number of tags (and their corresponding users and resources) after the pre-processing steps

	Total		Distinct		
	# entries	*# tags*	*# users*	*# resources*	*# tags*
del.icio.us	18,882	70,194	7,090	13,579	1,265
Flickr	44,032	127,098	5,321	44,032	2,696

For the **clustering** step, after generating the co-occurrence matrix for each of the datasets and computing the similarity between all pairs of vectors (tags) in that matrix, we empirically established a similarity threshold of **0.5** for both datasets to filter out the pairs of tags that were not highly similar. This means eliminating a number of tags that are not similar enough to any other tag. Excluding symmetric pairs, this resulted in **2,298** pairs of tags (**847** distinct) in del.icio.us, and **4,983** (2140 distinct) in Flickr.

For the smoothing heuristics to avoid a high number of very similar clusters, we established the threshold of **0.3** as accepted difference to group two "similar" clusters, that is, the number of distinct tags cannot be greater than **30%** of the number of tags in each of the clusters. This resulted in **410** clusters for del.icio.us, and **882** for Flickr.

The high number of clusters is due to the existence of clusters with only two tags. This can mean simply that certain pairs of tags do not co-occur with other tags a significant number of times. However, it may also indicate that the tags in the pair constitute a compound word, co-occurring between each other much more than with any other tag. This happens mostly in Flickr data, where we found clusters containing pairs of words such as {el salvador}. If we discard clusters with two tags, the number of clusters decreases to **47** in del.icio.us and **206** in Flickr. Some examples of clusters are shown in Table 4.

Alternatively to our clustering strategy, we experimented with a traditional clustering algorithm, namely, k-means, using cosine as the similarity metric. K-means

Table 4. Examples of clusters found for del.icio.us and Flickr data

del.icio.us data
{author books literature}
{bicycle bike courier cycling}
{bingo blackjack casino gamble gambling keno poker roulette slots}
{bookmark bookmarking folksonomy social tagging tags}
{browse extension firefox mozilla thunderbird}
{aim chat gtalk icq instant jabber messaging messenger msn yahoo}
Flickr data
{activism anarchism banner brutality demonstration eu globalization gothenburg police protest riots summit syndicalism worker}
{backpacking hot humid iguacu lush rainforest waterfall wilderness}
{damage flooding hurricane katrina Louisiana}
{bosnia europe herzegovina Sarajevo}
{apple ibook mac ipod macintosh powerbook}

requires the number of clusters to be given a priori, and therefore we defined the following numbers: 30, 50, 100, 150 and 200. Amongst the generated clusters, some seem to be very good, but many of them contain noise and some are meaningless. One of the reasons for that is the need of defining the number of clusters, which can force clusters to be divided when it is not necessary (or vice-versa). Also, it is not possible to discard pairs of tags that do not co-occur a certain number of times and therefore all the tags will be in the resulting clusters. A final problem is that the random criterion to create seed clusters can yield completely different clusters at each run. In future we will experiment with a hybrid strategy involving hierarchical clustering, which does not require defining the number of clusters, and a means of establishing a similarity threshold to discard tags.

By looking at the clusters obtained for both tag sets, we found that clusters from del.icio.us express concepts that are apparently closer to formal categories than clusters from Flickr. This may be due to the distinct purposes of the two systems. In fact, more pre-processing steps seem to be necessary to allow identifying meaningful categories in Flickr. For example, dates in various formats could be mapped into a semantic category "date", while names of unknown people could be mapped to a category "person".

As we discussed in Section 3, meaningful clusters can be very useful to enhance certain tasks in folksonomies. However, systematically evaluating the quality of clusters is a very complex task, since it relies on subjective criteria. One way of carrying out this evaluation could be verifying whether the tags within each cluster are correctly mapped into elements in ontologies, as we show with a few examples in what follows while describing the next step of our experiments.

For the last step of the approach, i.e., the concept and relation identification, we used Swoogle, since it is the most comprehensive semantic web search engine available on the web right now. Although Swoogle can provide useful information, it does not take into account semantic particularities of the data when creating indexes, and this yields severe limitations for our purposes. Querying facilities are limited to keyword search, with a few modifiers allowed, while we need more fine-grained queries, which would allow distinguishing between concepts, instances, properties, etc. In the presentation of the results, only part of the information is returned in a structured way and thus in most of the cases it is necessary to download and parse the ontologies. In the near future, we will use a semantic web search engine under development in our group, Watson (d'Aquin et al., 2007), which aims to overcome these limitations.

Given the difficulties to find the information we need in Swoogle, we performed a few experiments considering the search for concepts only, with exact matching. Tags within a cluster were queried using the Ontology Dictionary facility in Swoogle 2005 and the retrieved ontologies were manually analyzed to find both mappings to concepts and relationships. The examples illustrated here aim to show the potential of the approach in finding mappings and relationships: a fully automated version will be implemented when our semantic web search engine is ready.

Starting from the previously obtained clusters (410 for del.icio.us: and 882 for Flickr), we posted all the possible pairs of tags within each cluster to Swoogle. Out of a total of 3152 pairs (847 distinct tags) in del.icio.us and 5031 pairs (2140 distinct tags) in Flickr, without using Wikipedia / Google to find unknown concepts in

Swoogle, the following numbers of pairs were found as concepts together in at least one single ontology: **569** pairs (**358** distinct tags) in del.icio.us and **309** pairs (**492** distinct tags) in Flickr.

Many of the pairs were found in WordNet only. WordNet, which is more like a dictionary than an ontology, and thus has a wide coverage over most of the concepts. In general, finding two concepts in any ontology does not mean that they are related, but this is even more critical with WordNet: concepts are more likely to be related via very generic semantic categories, such as "entity" or "thing". Moreover, WordNet is very limited in terms the relationships that are covered: mostly hierarchical. If we consider only the pairs of tags that were found in other ontologies than WordNet, the numbers decrease to **126** (**97** distinct tags) in del.icio.us and **67** (**94** distinct tags) in Flickr. This means that **97** tags in del.icio.us and **94** in Flickr could be mapped to concepts in ontologies (except WordNet). In order to assess the quality of these mappings, and therefore the quality of the clusters containing the corresponding tags, we manually verified whether the concepts identified in ontologies were in accordance with the knowledge represented by their cluster. For example, in *Cluster_1* in Fig. 2, all the tags except `sourcecode` where found in ontologies, i.e., there was a possible mapping for all the other tags, even if no relationship was found for many of them. The tags for which a valid relationship was found are considered correctly mapped. Out of the remaining 7 tags, namely `collection control dom form layout program repository`, only `dom` could not be correctly mapped to a concept in any ontology. We believe this extends to most of the clusters: there will not be a valid mapping for only a few tags in clusters. This may indicate that these tags do not belong to any of the clusters, but also may just reflect the low coverage of the current search engine over the ontologies available on the web. However, it is important to notice that the most frequent tags in the two datasets, which are supposedly some of the most relevant, are amongst the pairs of tags that were found in ontologies.

Regarding the identification of relationships, it is important to remember that not all the pairs of tags in a cluster are expected to be related. In fact, the most common situation is that a certain tag is related to a few others, which are, in turn, related to others, composing an incomplete directed acyclic or cyclic (possibly disconnected) graph. Without considering the pairs found in WordNet, in Fig. 2 we show examples of partial ontologies that were produced for some del.icio.us clusters by following the steps 5.1-5.5 (Section 3). Terms in italic are not part of the cluster, but were kept in the graphs to indicate indirect relationships. Arrows without explicit relationships represent "**subclass**" relations.

As we can see, not all the tags in each cluster are included in the graphs. This may be because either the tag was not found in Swoogle at all (e.g., `lms` in *Cluster_2*), the tag was not found in a single ontology together with at least one of the other tags (e.g., `sourcecode` in *Cluster_1*), or no relationship was found within an ontology (e.g., `layout` in *Cluster_1*). When a tag is not found in Swoogle, additional resources can be used. In *Cluster_3* these additional resources play a very important whole. Swoogle does not contain concepts referring to `distro`, `fedora`, `gentoo`, `kubuntu`, `mandriva`, or `rpm`. The only relationships that could be retrieved were: `suse` and `debian` are subclasses of `linux`. By using Wikipedia, we could infer the relationships between most of the other tags and `linux`.

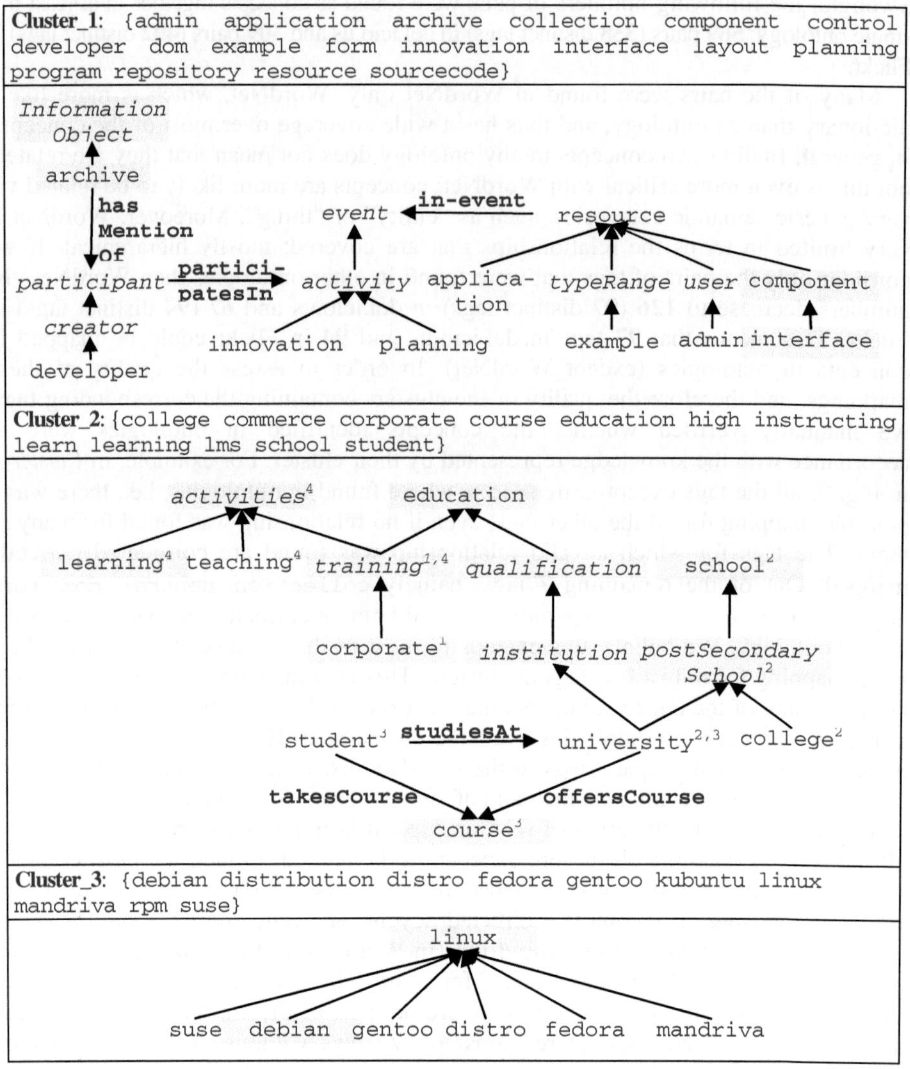

Fig. 2. Examples of relationships for del.icio.us clusters

5 Conclusions and Future Work

We have presented some initial work in the direction of integrating folksonomies with the semantic web, thus making explicit the semantics behind the tag space in folksonomies. Preliminary experiments with tags from del.icio.us and Flickr have shown that the approach is feasible and very promising: meaningful groups of tags corresponding to concepts in ontologies could be derived by means of co-occurrence analysis and clustering techniques, while relationships within tags in each cluster could be discovered by querying ontologies in Swoogle.

In the future we plan to improve the approach in several aspects. These include: (i) using a new clustering technique which combines hierarchical clustering with a threshold to discard tags that are not sufficiently similar to others, and (ii) implementing a fully automated version of the last step of the approach, that is, the process to map tags into ontology elements to build partial structures based on the knowledge provided by our new semantic web search engine. In order to achieve this goal we will also need to devise better strategies for ontology selection and matching, as well as strategies to extract information from external resources (Wikipedia / Google). We also plan to extrinsically assess the quality of our results by integrating them in the context of the various tasks discussed in the paper (tag disambiguation, result visualization, ontology evolution, etc).

References

Aurnhammer, M., Hanappe, P., Steels, L.: Augmenting Navigation for Collaborative Tagging with Emergent Semantics. 5th ISWC, Athens, GA, LNCS 4273 (2006) 58-71.

Begelman, G., Keller, P., and Smadja, F.: Automated Tag Clustering: Improving search and exploration in the tag space. Collaborative Web Tagging Workshop, 15th WWW Conference, Edinburgh (2006)

d'Aquin, M., Sabou, M., Dzbor, M., Baldassarre, C., Gridinoc, L., Angeletou, S. and Mottta, E.: WATSON: A Gateway for the Semantic Web. Poster Session at 4th ESWC (2007)

Ding, L., Finin, T., Joshi, A., Pan, R., Scott Cost, R., Peng, Y., Reddivari, P., Doshi, V.C., and Sachs, J.: Swoogle: A Search and Metadata Engine for the Semantic Web. 13th ACM Conference on Information and Knowledge Management, Washington D.C. (2004)

Golder, S., and Huberman, B.A.: The Structure of Collaborative Tagging Systems. HP Labs technical report. (available in http://www.hpl.hp.com/research/idl/papers/tags/) (2005)

Marlow, C., Naaman, M., Boyd, D., Davis, M.: Position Paper, Tagging, Taxonomy, Flickr, Article, ToRead. Collaborative Web Tagging Workshop, 15th WWW Conference, Edinburgh (2006)

Mika, P.: Ontologies are us: A unified model of social networks and semantics. 4th ISWC (2005)

Miller, A., Chorodow, M., Landes, S., Leacock, C., Thomas, R.G.: Using a Semantic Concordancer for Sense Identification. Arpa Human Language Technology Workshop (1994) 240-243

Motta, E and Sabou, M.: Next Generation Semantic Web Applications. In Mizoguchi et al. (eds.), The Semantic Web - ASWC 2006, LCNS 4185, Springer (2006)

Pantel, P. Clustering by Committee. Ph.D. Dissertation. University of Alberta (2003)

Sabou, M., d'Aquin, M., Motta, E.: Using the Semantic Web as Background Knowledge for Ontology Mapping. International Workshop on Ontology Matching, Athens, GA (2006)

Schmitz, C., Hotho, A., Jäschke, R. Stumme, G.: Mining Association Rules in Folksonomie. IFCS Conference, Ljubljana (2006) 261-270

Schmitz, P.: Inducing ontology from Flickr tags. Collaborative Web Tagging Workshop, 15th WWW Conference, Edinburgh (2006)

Schutz, A. and Buitelaar, P.: RelExt: A Tool for Relation Extraction from Text in Ontology Extension. 4th ISWC, Galway (2005) 593-606

Specia, L., Motta, E.: A hybrid approach for extracting semantic relations from texts. 2nd Workshop on Ontology Learning and Population at COLING/ACL 2006, Sydney (2006) 57-64

Wu, X., Zhang, L., Yu, Y.: Exploring social annotations for the semantic web. 15th WWW Conference, Edinburgh (2006) 417-426

IdentityRank: Named Entity Disambiguation in the Context of the NEWS Project

Norberto Fernández[1], José M. Blázquez[1], Luis Sánchez[1], and Ansgar Bernardi[2]

[1] Carlos III University of Madrid, Leganés, Madrid, Spain
{berto,jmb,luiss}@it.uc3m.es
[2] German Research Center for Artificial Intelligence, DFKI GmbH, Kaiserslautern, Germany
ansgar.bernardi@dfki.de

Abstract. In this paper we introduce the IdentityRank algorithm, developed as part of the EU-funded project NEWS to address the problem of named entity disambiguation in the context of semantic annotation of news items. The algorithm provides a ranking of the candidate instances within an ontology which can be associated to a certain entity. In order to do so, it uses as context the metadata available in a certain news item. The algorithm has been evaluated with promising results.

1 Introduction

The EU-IST funded project NEWS[1] (News Engine Web Services) [3], which has recently been completed, aimed at providing solutions which help news agencies to overcome limitations in their current workflows and increase their productiveness and revenues by using a Web Service based architecture and Semantic Web technologies.

In order to apply Semantic Web technologies to the news domain, in the NEWS project a set of components were developed. One of them is the NEWS ontology [4], a lightweight RDFS[2] ontology providing a formal model of the domain. Another one is an annotation component, developed by Ontology Ltd., which uses natural language processing techniques to provide capabilities such as categorization and named entity extraction.

Within the semantic annotation process, one of the key problems that we found in NEWS was the disambiguation of the entities detected by the natural language processing engine. This engine extracts named entities out of the news items, but, in order to allow a fine-grained semantic search for the user of the NEWS system, these entities have to be matched against instances of the NEWS ontology. That is, the natural language processing engine can detect that a certain occurrence of the piece of text *Bush* represents a person, but we also need to deduce that this person is represented in the NEWS ontology by a certain URI like *http://www.news-project.com/2005/1*.

[1] Contract number: FP6-001906. Web site: http://www.news-project.com
[2] http://www.w3.org/TR/rdf-schema/

E. Franconi, M. Kifer, and W. May (Eds.): ESWC 2007, LNCS 4519, pp. 640–654, 2007.

In this paper we describe the IdentityRank algorithm (a.k.a. IdRank) that we designed in order to address the entity disambiguation problem in the news domain. Our algorithm, inspired by PageRank [10], exploits the metadata currently provided by news agencies (like news item timestamp) and the information provided by the natural language processing engine (categories and entities) as a context for named entity disambiguation. Using all this information, IdRank allows to match news items' entities to ontology instances automatically.

The rest of this paper is organized as follows: section 2 describes with more detail the IdRank operational scenario within the NEWS workflow. Section 3 describes the algorithm. Section 4 shows the results of an experimental evaluation of the algorithm. Section 5 takes a deeper look at related work, and finally, section 6 gives concluding remarks and finalizes the paper.

2 Scenario

In order to give a clearer idea of the operational environment of IdRank, we describe in this section the NEWS workflow, which acts as an scenario for the entity disambiguation problem. The NEWS workflow design has taken into account that the journalists in the news agencies want to have control over all the content production process in order to ensure the quality of the results. This leads to a supervised solution, where the journalist can validate the results obtained in the different processing stages of a news items. These are:

1. The journalist creates a news item using the NEWS GUI. The news item is represented in XML and some metadata like author and timestamp are added to it.
2. The news item is processed by the natural language processing component. It annotates the news item with some entities and categories. The vocabulary used for categorization is taken from the NEWS ontology. Basically this vocabulary is an RDFS representation of the International Press Telecommunication (IPTC) standard Subject Codes NewsCodes[3]. These Subject Codes constitute a three level taxonomy that, at the moment, contains about 1300 different categories. In such taxonomy, each category is identified by a fixed eight decimal-digit string. The first two digits represent the first level of the taxonomy, which consists of 17 different categories. For instance, the Subject Code 01000000 represents the category *arts, culture and entertainment*, the Subject Code 01011000 represents the subcategory *music* and the Subject Code 01011006 represents the subsubcategory of news items talking about *rock music.*

   With respect to entities, these are also added to the news item. For each entity the natural language processing engine provides the tagged text and

[3] http://www.iptc.org/NewsCodes

the entity type, which in our case is one of the three possibilities: person, place or organization. For instance the following piece of XML:

```
<meta content="11000000" name="srs-category" />
<meta content="Gargano" name="entity-person" />
<meta content="Mexico" name="entity-places" />
<meta content="Liberal Party" name="entity-organization" />
```

would be added by the natural language processing engine to state that the news item belongs to category 11000000 *politics* and mentions the entities *Gargano*, a person, *Mexico*, a place, and *Liberal Party*, an organization.
3. The annotated document is sent back to the GUI and the journalist is allowed to check the annotations. The validated document is sent to other of the NEWS components: the Heuristic and Deductive Database (HDDB).
4. The HDDB stores the news item, indexes its textual content to allow keyword based search, stores the news item metadata, including the categories and entities, and then runs IdRank to disambiguate the entities to instances in the NEWS ontology.
5. The results of IdRank, a set of assignments (entity, instance), are then shown to the journalist. (S)he may confirm them, select a different instance for some entity (creating a new one if needed) or might simply drop the assignment and leave the entity without associated instance.
6. The results of the validation process are sent back to the HDDB, where are stored and used to train IdRank. All the information generated and stored in this process and the NEWS ontology are used by the HDDB to allow intelligent content distribution services.

3 The Algorithm

As we have seen in the previous section, the NEWS natural language processing engine is able to extract basic entities from text. But in order to allow fine-grained semantic search over the news item repository stored in the HDDB it is not enough to figure out, that the extracted text string *Alonso* represents a person, we need to know who is that person by mapping the entity to an instance in the NEWS ontology. For instance, for the entity (*Alonso,person*) there are the following candidates in the NEWS ontology:

```
Fernando Alonso, Airbus flight testing vice-president.
http://www.news-project.com/2005/11
Fernando Alonso, Formula 1 driver.
http://www.news-project.com/2005/12
Mikel Alonso, soccer player.
http://www.news-project.com/2005/13
Xabi Alonso, soccer player.
http://www.news-project.com/2005/14
Jose Antonio Alonso Suarez, Spanish politician.
http://www.news-project.com/2005/15
Alonso Cano, Spanish painter, architect and sculptor.
http://www.news-project.com/2005/16
Alonso de Ercilla y Zuniga, Spanish poet.
http://www.news-project.com/2005/17
```

So a problem of ambiguity arises: which is the best candidate instance to be assigned to a certain entity? Finding that instance is the main task of the IdRank algorithm, which is based on two principles:

Semantic coherence: Instances typically occur in news items of certain categories, e.g., the politician *Jose Antonio Alonso* in news items of *politics* category. Also the occurrence of a certain instance gives information about the occurrence of other instances. For example, the soccer player *Xabi Alonso* usually appears in news items in which the soccer team where he plays, *Liverpool*, is also mentioned.

News trends: Important events typically are described with several news items covering a certain period of time. For instance when the Formula 1 driver *Fernando Alonso* won the F1 world championship, several news items describing such event where composed, most of them including instances as *Fernando Alonso* and *Renault*, his F1 team.

In this section we will describe in detail the main processes involved in the IdRank algorithm. As we have said, IdRank is partially inspired by PageRank, so we will start by briefly describing PageRank before going into the IdRank details.

3.1 PageRank and Relation with IdRank

The PageRank algorithm [10] exploits the information in web links to compute the ranking of a certain web page. The basic idea is mentioned in [10]: *a page has high rank if the sum of the ranks of its backlinks is high.* So in PageRank the ranking or importance of a certain page depends on the ranking and number of the pages which point to it (backlinks). Mathematically this is represented by the following equation (see [10]):

$$R(u) = \lambda \sum_{v \epsilon B_u} \frac{R(v)}{N_v} + \lambda E(u) \tag{1}$$

Where:

- λ is a factor used for normalization.
- $R(u)$ represents the ranking of the web resource u. The L_1 norm of the vector R, composed of all $R(u)$, is such that $||R||_1 = 1$.
- B_u is the set of backlinks of u.
- N_v is the cardinality of F_v, the set of pages v points to (forward links of v).
- E is a vector that corresponds to a source of rank. As is indicated in [10], each component $E(u)$ can be used to adjust the rank of a certain resource u, for instance for personalization purposes (give more weight to certain pages).

This equation can be represented in a matricial manner:

$$R = \lambda A R + \lambda E \tag{2}$$

Where A is a matrix, $A_{uv} = 1/N_v$ if $v \epsilon B_u$ or 0 otherwise, $||A||_1 = 1$.

The relation between PageRank and IdRank comes from the application of one of the basic principles that inspire IdRank. The Semantic Coherence principle states that the appearance of an instance gives certain information about the occurrence of other instances. Paraphrasing the sentence in [10] we can say that: *an instance has high rank if the sum of the ranks in the news item of the instances that typically cooccur with it is high.* As PageRank does with web pages, the objective of IdRank is to obtain a ranking: the ranking of the possible *identities* (candidate instances) of a certain entity.

The next subsection will describe with more detail how IdRank works in practice.

3.2 IdRank

Three are the main steps needed to run IdRank on a certain news item: *finding* the candidates instances in the ontology for each entity in the news item, *ranking* that candidate instances using a modified version of PageRank and *retraining* the algorithm with the journalist feedback once the process is finished. The next subsections will describe each of these steps in more detail.

Find candidate instances. This process takes as input the entities detected by the natural language processing engine in a certain news item and produces as output a set of candidate instances for each of the input entities. For instance, given the entity (*Alonso,person*) the following steps are executed:

1. Given the entity type, the HDDB code is configured to decide which is the upper class in the NEWS ontology taxonomy which maps to the entity type. In our example the mapping is as follows: the entity type *person* maps to the ontology class *Human*. The other possible mappings are: the entity type *place* maps to the ontology class *Location* and the entity type *organization* maps to the ontology class of the same name.
2. The HDDB computes the transitive closure of the *subclassOf* property to find all the subclasses of the class of interest. For instance, in our example, the deductive part of the HDDB computes the transitive closure of the class *Human* finding the two subclasses of this class: *Man* and *Woman*.
3. An SQL query is automatically generated to query the database where the NEWS ontology A-box is stored. With this query we find the candidate instances that match the entity text and belong to the classes *Human, Man* or *Woman*. For instance, in the example introduced above, the SQL looks like:

```
SELECT DISTINCT(uri) from Instances
WHERE (
        label LIKE '% Alonso' OR label LIKE 'Alonso %' OR
        label LIKE '% Alonso %' OR label = 'Alonso'
)
AND (
        type IN (
                'http://www.news-project.com/Ontology/Content#Human',
                'http://www.news-project.com/Ontology/Content#Man',
                'http://www.news-project.com/Ontology/Content#Woman'
            )
);
```

The same process is repeated with all the entities detected in the news item by the natural language processing engine.

Rank the candidates using a modified version of PageRank. Once the candidate instances of all the entities are obtained, a semantic network with all these instances is defined. In such semantic network the nodes represent the different candidate instances for the entities in the news item. If an instance is candidate for more than one entity, it only appears once. The arcs between two nodes appear when the two instances have cooccurred in the past in at least one news item, that is, if at least one news item exists in the HDDB that is annotated with occurrences of both instances.

Then we apply a modified version of PageRank to the semantic network. In our algorithm, instead of dividing the importance of an instance among its forward links evenly, as PageRank does with the quotient $R(v)/N_v$ in equation (1), we will give weights to the links. That is, in IdRank, the occurrence of an instance can give more weight to certain instances than to others. These weights depend on the cooccurrence frequency of the involved instances.

Mathematically we have the following set of equations:

$$R(I_i) = \lambda \sum_{j \epsilon C_i} \alpha_{ij} R(I_j) \tag{3}$$

Where:

- λ is a factor used for normalization.
- $R(I_i)$ represents the ranking of the candidate instance I_i in the context of the news item.
- C_i is the set of candidate instances in the semantic network that cooccur with I_i in at least one news item apart from the one being analyzed.
- α_{ij} represent the weight of the link from I_j to I_i, that is, the proportional part of the I_j importance or ranking which is given to I_i. Mathematically, this can be expressed as:

$$\alpha_{ij} = \frac{f_{ij}}{\sum_{k \epsilon C_j} f_{kj}} \tag{4}$$

Where f_{ij} is the coocurrence frequency of I_i and I_j, that is, the number of news items where both I_i and I_j occur divided by the number of news items where I_j occurs. With this definition: $\alpha_{ij} \epsilon [0, 1]$ and:

$$\sum_{\forall i \epsilon C_j} \alpha_{ij} = 1 \tag{5}$$

Note that, as has been previously indicated, the weights α_{ij} and α_{ji} are, in general, not equal due to equation (4).

At this point, we have described how the coocurrence of instances is used by the algorithm, but still two contributions remain unclear: the *Semantic Coherence* principle is also dependent on the news item categories, and the *News Trends* principle, which uses the timestamp, has also not been exploited.

In order to exploit also that information, we will use the E component included in the original PageRank formula (equation (2)) where it was used to personalize the ranking. In our case, we are going to use it with a similar meaning, *personalizing* the ranking computation for the context of the concrete news item. Mathematically we will have now:

$$R(I_i) = \lambda \sum_{j \epsilon C_i} \alpha_{ij} R(I_j) + \lambda E(I_i) \tag{6}$$

In practice, the vector E, composed of all the $E(I_i)$, is computed as a normalized sum of contributions:

$$E = \sum_{\forall c} E_{cnorm} = \sum_{\forall c} \frac{E_c}{||E_c||_1} \tag{7}$$

At the moment, the set of contributions which are being considered in the context of NEWS are:

E_{tim}: **instance occurrence in last D days.** The value of each element $E_{tim}(I_i)$ of E_{tim} is computed taking into account the frequency of occurrence of the candidate instance I_i in the news items of the last D days, taking as time origin the timestamp of the news item being analyzed. D is a constant empirically determined (we worked with D=7).

E_{cat}: **instance occurrence in news items of certain category.** Takes into account the occurrence of the instance in news items belonging to a certain top level category (01000000-17000000). A news item belongs to a certain category if the annotation engine assigns it that category or one of its subcategories. As a news item can belong to several different top level categories, in practice E_{cat} is composed of the sum of several components.

For each top level category, *tlc*, in the news item, the value of each element of the vector $E_{cat}^{tlc}(I_i)$ is computed taking into account the frequency of occurrence of the candidate instance I_i in news items belonging to *tlc*. The final vector E_{cat} is just a linear combination of the different vectors E_{cat}^{tlc}. Less frequent categories have a higher weight in that linear combination, because they provide more information about the news item.

Taking into account these contributions and equation (7), we can represent the equation (6) in a matricial manner, as in 3.1:

$$R = \lambda AR + \lambda E = \lambda AR + \lambda E_{catnorm} + \lambda E_{timnorm} \tag{8}$$

Where, in the same way as in [10], A is a matrix, $A \in M_{nxn}$, $A_{ij} = \alpha_{ij}$ and $R, E_{catnorm}, E_{timnorm}$ are vectors, R, $E_{catnorm}$, $E_{timnorm} \in R^n$ and n is the total number of different candidate instances in the news item.

Due to equation (5) $||A||_1 = 1$ and due to equation (7) $||E_{catnorm}||_1 = 1$ and $||E_{timnorm}||_1 = 1$. The consequence of this fact is that if we use directly the equation (8), we give the same weight in the computation of the R vector to the components depending on instance coocurrence, A, depending on instance-category coocurrence, $E_{catnorm}$ and depending on the temporal information, $E_{timnorm}$. In order to control the effect in the final ranking of each contribution, we have assigned weights to each component, resulting the equation:

$$R = \lambda(k_a AR + k_{cat} E_{catnorm} + k_{tim} E_{timnorm}) \tag{9}$$

Where $k_a + k_{cat} + k_{tim} = 1$. As is indicated in [10], since $||R||_1 = 1$ the equation (9) can be rewritten as:

$$R = \lambda(k_a A + (k_{cat} E_{catnorm} * \mathbf{1}) + (k_{tim} E_{timnorm} * \mathbf{1}))R \tag{10}$$

Where $\mathbf{1}$ represents a row vector of all ones, $\mathbf{1} \in R^n$, and $*$ represents the matrix product.

Analyzing equation (10) we conclude that, as happens in the original PageRank algorithm, we can compute the vector R simply by determining the main eigenvector of a matrix. In our case the matrix is: $k_a A + (k_{cat} E_{catnorm} * \mathbf{1}) + (k_{tim} E_{timnorm} * \mathbf{1}) \in M_{nxn}$. In our implementation, that eigenvector is computed using a numerical method: the power method.

Once R is computed, we know the ranking of each candidate instance in the context of the news item being analyzed: the weight of the instance I_i is simply the component i of the vector R. For each entity in the news item, the algorithm returns a vector with all the pairs (candidate instance, weight) for such entity. This vector is sorted using the weight, so the candidate with the biggest weight is the one shown to the journalist as best candidate instance for the entity. If more than one candidate has the biggest weight, the algorithm randomly selects one of them as the first one.

Retraining the system. The results of the ranking process are shown to the journalist at the GUI. The journalist can check the suggestions of the system and correct the wrong ones. The resulting annotations are stored into the HDDB and used to retrain IdRank. Basically the retraining process consist in storing or updating into the relational database used by the algorithm information needed for the algorithm process. Concretely, for each instance in the news item we update the following information: the occurrence of the instance in a certain timestamp (used in computing E_{tim}), the counter of number of coocurrences between the instance and all the other instances detected in the new news item (used in computing A) and the counter of the number of occurrences of the instance in each of the top level categories of the news item (used in E_{cat}).

For each new news item, we store or update also the following information: the counter of the total number of news items, the counter of the number of news items belonging to a certain top level category (used in computing E_{cat}) and the association between news item and its timestamp (used in E_{tim}).

As IdRank reads on the fly from the database the information needed to perform its computations, the next time the algorithm is run, the new training information is taken into account.

4 Evaluation

The first step in the empirical evaluation of the algorithm was to define a theoretical threshold that we can take as reference to compare our algorithm. In our case, that theoretical threshold is provided by the average accuracy of a naive disambiguation algorithm that simply assigns randomly one of the possible candidate instances to each entity. We assume that all the candidates of a certain entity have the same probability of being chosen as the right one. We also assume that the decisions of that hypothetical algorithm are independent, that is, that it chooses the candidate instance for a certain entity independently of all the other elections in the corpus. Finally, we also assume that, for each entity in the corpus, there exists in the ontology at least one candidate: the right candidate instance to be mapped. Though this last assumption seems unrealistic, in practice in the NEWS scenario, as journalists are allowed to insert new instances into the knowledge base, entities without the right instance can get one as soon as they are detected.

With these assumptions, we get the following expression for the accuracy:

$$Av[Acc] = \frac{Av[right]}{total} = \frac{\sum_{\forall e} Occ(e)P(Right/e)}{N_{ent}} \tag{11}$$

That is, the average accuracy is defined as the average number of right assignments entity/instance of our naive algorithm divided by the total number of possible assignments. The total number of assignments coincides with the number of entities in the corpus (N_{ent}) due to the assumption that each entity has at least one candidate. The total average number of right decisions is the addition of the average number of right decisions for each entity e in the corpus. Due to the assumption of independent election, the average number of right decisions for a certain entity e can be computed as the number of occurrences of the entity e in the corpus $Occ(e)$ multiplied by the probability of making a right decision on that entity $P(Right/e)$. Due to the assumption of random uniform election between the candidates for a certain entity, $P(Right/e) = 1/Ncand_e$, where $Ncand_e$ represents the number of candidates for the entity e in the ontology. As N_{ent} is constant for the summatory in the fraction, we can reformulate the equation (11) as:

$$Av[Acc] = \sum_{\forall e} \frac{Occ(e)}{N_{ent}} \frac{1}{Ncand_e} \tag{12}$$

As can be seen, and not surprisingly, the final accuracy depends on the concrete corpus used for evaluation ($Occ(e)/N_{ent}$ component) and the ontology used as source of candidates instances in the evaluation ($1/Ncand_e$ component). This value of the average accuracy of the naive random election algorithm can also be interpreted as a measure of the degree of ambiguity of the pair corpus/ontology used in the evaluation: the bigger $Av[Acc]$ the lower the ambiguity.

Once the theoretical baseline that we use as reference for our algorithm is defined, we will describe in next subsections the concrete results of the empirical evaluation of the algorithm. We start by describing the corpus and ontology used in the evaluation process, after that, we describe the results of the accuracy evaluation and finally we include some measurements of the computation time of the algorithm.

4.1 Corpus and Ontology

As we have seen previously, the corpus and the ontology selected to perform the evaluation have direct influence on the results of the evaluation. Due to this, we have carried out our evaluation using two different corpora and two different ontologies, instead of only one.

Corpora. As we are evaluating an algorithm for ambiguity resolution, we are interested in having ambiguity in our corpora. In order to achieve this, the process of building our corpora started by selecting a possible ambiguous entity and querying the NEWS repository, which contains real news items of Spanish news agency EFE and Italian news agency ANSA, for news items where such entity appears. More in detail two entities were selected for the process, *Georgia, location* and *Alonso, person.* The query gave us 32 news items for the entity *Georgia, location* and 65 for the entity *Alonso, person.* All the entities appearing in the news items, where manually disambiguated using the NEWS ontology. The annotations were reviewed by two different persons to ensure as much as possible the quality of the evaluation corpora.

The results were 343 total entities in the Georgia corpus, 169 of them distinct (different pair entity text, entity type), and 742 entities in the corpus of Alonso, 229 of them distinct. The entity *Georgia, location* appeared with two different meanings in the Georgia corpus (Georgia as U.S. state -12 times- and Georgia as country -20 times-) and the entity *Alonso, person* appeared with three different meanings in the corpus of Alonso: Fernando Alonso, a Formula 1 driver (41 times), Jose Antonio Alonso a Spanish minister (23 times), and Xabi Alonso, a soccer player (only once).

The timestamps of the Alonso news items range from the 13/Oct/2005 to the 12/May/2006 and the ones of the Georgia corpus range from 17/Oct/2005 to 12/May/2006. Their distribution is shown in table 1. The distribution of the number of news items belonging to the top level categories (01000000 to 17000000) in both the Georgia and Alonso corpora is shown in table 2. Note that the total number of categories in each corpus is bigger than the total number of news items in such corpus, because a single news item can be categorized into different categories.

Ontologies. We have also used two different ontologies for our process. One of the ontologies was the NEWS ontology, the other one was built using the information on the annotated corpus. In order to build this second ontology, we considered as the only possible candidates for each different entity the ones that appear in the manually annotated corpus. For instance, as we have seen the entity *Alonso,person* has 7 different candidates in the NEWS ontology, but in the corpus it appears only with 3 different meanings, so the number of candidates for this entity is 7 in the NEWS ontology and 3 in the ontology built considering only the candidates that appear in the corpus.

Table 1. Number of news items in each month from Oct 2005 to May 2006

	Oct	Nov	Dec	Jan	Feb	Mar	Apr	May
Alonso	28	0	0	0	2	10	1	24
Georgia	27	0	0	0	0	2	0	3

Table 2. Number of news items in each of the top level categories

	01	02	03	04	05	06	07	08	09	10	11	12	13	14	15	16	17
Alonso	1	9	0	1	0	0	0	2	0	0	23	0	0	1	40	4	0
Georgia	2	7	0	3	0	0	2	3	0	0	16	0	0	0	3	4	1

4.2 Accuracy Evaluation

In our accuracy evaluation process, we were interested in measuring the effect of using different ontologies and corpora. Another aspect of interest was to measure the impact on the final results of the three components involved in the IdRank computation, instance coocurrences, categories and timestamp. Due to this, for each of the four possible combinations (ontology,corpus) we ran four experiments changing the values of the k_a, k_{cat} and k_{tim} parameters. So the total number of experiments performed was 16. As IdRank has a random component, each of these experiments was run 10 times and the average accuracy of IdRank was measured. We centered our attention on two aspects of the accuracy. One was the global accuracy, measured as total number of right assignments entity/instance divided by the total number of assignments. The other one was the relative accuracy for the entity used to construct the corpus, defined as the number of right assignments on the decisions of that entity divided by the number of decisions about the entity.

The results of this evaluation are shown in table 3. The first column, labeled as *Corp, Ont, Res* indicates the corpus (A, Alonso or G, Georgia) the ontology (N, NEWS ontology or C, Corpus dependent ontology) and the results (Tot, total accuracy or A/G, Alonso/Georgia, relative accuracy). So, for instance, the value *A,C,A* indicates that these results where obtained with the corpus of Alonso, using the ontology built taking as input such corpus and that only the relative accuracy for the entity *Alonso, person* is shown. The second column shows the theoretical results. These are computed with the equation 12 for the total accuracy case. For the relative accuracy case, the theoretical average accuracy is just the total number of occurrences of the entity, multiplied by the probability of choosing the right candidate for the entity and divided by the total number of occurrences of the entity, that is, just the probability of choosing the right candidate for the entity. As we have seen this depends on the concrete ontology, so for instance, the second column in the row *A,N,A* is 1/7*100 = 14.3%, because in the NEWS ontology, the one used in such experiments, the entity *Alonso, person* has 7 candidates. The rest of the columns of the table show the results of concrete experiences with the pair (ontology,corpus). The column A shows the results obtained when $k_a = 1$, the column E_{tim} the results when $k_{tim} = 1$, the column E_{cat} the results when $k_{cat} = 1$ and finally the column, *All* contains the results obtained when the three components of the algorithm were considered. In concrete we used: $k_a = 0.8$, $k_{tim} = 0.05$ and $k_{cat} = 0.15$. For each of the entries in the table, estimated by averaging the results of 10 executions of IdRank, the mean and the standard deviation are shown.

Table 3. Average accuracy results (percentages)

Corp, Ont, Res	Theo. (%)	A (%)	E_{tim} (%)	E_{cat} (%)	*All* (%)
A,N,Tot	82.89	96.44 (0.62)	95.21 (0.52)	96.27 (0.58)	96.48 (0.52)
A,N,A	14.3	93.69 (1.35)	74.62 (1.81)	93.23 (0.79)	95.54 (0.49)
A,C,Tot	92.07	97.91 (0.25)	96.35 (0.33)	97.78 (0.25)	98.07 (0.23)
A,C,A	33.3	95.38 (0.73)	74.46 (2.63)	93.23 (1.30)	95.73 (0.68)
G,N,Tot	88.56	97.32 (0.55)	93.67 (0.65)	93.09 (0.55)	96.24 (0.57)
G,N,G	33.3	93.13 (1.98)	57.81 (8.10)	54.69 (3.68)	85.00 (1.32)
G,C,Tot	95.04	98.89 (0.18)	95.92 (0.55)	95.66 (0.47)	98.22 (0.26)
G,C,G	50	94.06 (1.77)	61.25 (6.28)	57.50 (4.70)	85.62 (1.61)

Analyzing the results in table 3, we see that the A component, related with instances coocurrence, is more accurate in giving us the right candidate instance than the E_{tim}, E_{cat} components. The category-related component, works fine in the case of the entity *Alonso, person*, because most of the news items in category 15000000 (sports) talk about Fernando Alonso, the Formula 1 driver, whereas the news item in category 11000000 (politics) are mostly related with Jose Antonio Alonso, a Spanish minister. Nevertheless, the behavior of the category-based component, is worse in the case of the entity *Georgia, location*. This is due to the fact that locations usually are not directly related with a certain subject, so we can have news talking about very different events, and thus having completely different categories, mentioning the same location. In fact, due to the bad performance of the category-based component in the Georgia case, the results of the *All* test are worse than the ones obtained by using only the instance coocurrences information. With respect to the temporal component, its poor results can be explained by the fact that in our concrete corpora the occurrences of the different candidates are interleaved, and the temporal window is relatively long (D=7) to give good results. But, as the number of news items in the corpora is low and they are relatively disperse, we had to use long windows to get significative results for this component. So, more experiences with bigger corpora and with different values of the temporal window D must be accomplished to extract definitive conclusions.

4.3 Computation Time

As we have said in the initial sections, IdRank is designed to operate on the real production environment of a news agency. In such environment the time expended on creating and sending to the customers a new news item should be minimized, because the news agencies are interested in providing the relevant news items to the clients as soon as possible. In order to evaluate whether IdRank is adequate to operate on such an environment, we have conducted an evaluation of the time expended by the algorithm.

This evaluation consisted in running the algorithm 10 times with the Alonso corpus and the NEWS ontology, the case with bigger ambiguity, and compute the average time expended by the algorithm in each of its subprocess: candidate finding, ranking and retraining. The parameters of the evaluation were: $k_a = 0.80$, $k_{tim} = 0.05$. $k_{cat} = 0.15$. The tolerance and maximum number of iterations of the iterative method used to compute the matrix eigenvector where 0.0001 and 100 respectively. We conducted this experiment on a machine with Linux Debian 3.1 operative system, kernel 2.6.11, one Gigabyte of RAM memory and a Pentium(R) Mobile 1.60GHz processor.

Table 4. Average Computation Time

Nent	Ncat	Av[Find] (msec)	Av[Rank] (msec)	Av[Retrain] (msec)	Av[Total] (msec)
25	1	123.5	2705.7	3348.2	6177.4
26	1	270.4	1594.4	2015.6	3880.4
23	1	114.2	2246.3	2663.5	5024.0
23	1	197.1	2719.7	2347.6	5264.4
30	1	392.9	570.4	4868.4	5831.7

Table 4 shows the results for the five news items with worst total average execution times. For each news item, the number of distinct entities N_{ent}, the number of categories N_{cat} and the average time of finding, ranking and retraining, are shown. The last column shows the average total time needed by IdRank to process the news item.

As can be seen, the total time is in the order of seconds, which seems affordable for the proposed application scenario. Another conclusion is that the retraining time has a significative influence on the final results. On the positive side, we have to say that the retraining process does not have much effect on the time perceived by the journalist and the news agency client, because the retraining process is done when the edition process of the news item is finished.

5 Related Work

Named entity disambiguation or proper name disambiguation is a type of *word sense disambiguation* [8], in which the words to be disambiguated are named entities. There are lots of approaches in the state of the art dealing with word sense disambiguation and also with named entity disambiguation. These different approaches can be characterized according to a number of criteria:

- The *context* used to disambiguate the entity. Some approaches use the complete document where the entity is placed to disambiguate [2]. Others use as context a number of words before and after the entity. Those can be further classified on those that take a "bag of words" context (the position of the words taken as context is not considered) like [11] and those that try to use the role of each word in the context and their relation with the entity [9].

  Although some approaches use both common words and named entities as context [11], others suggest that better results can be obtained using as context only other named entities [9].
- The use of *knowledge sources* like lexical databases, etc., that define the instances that should be matched against the entities and can provide information that can be exploited to perform the matchings. There are of course several approaches that make use of such knowledge sources [1,7]. However, a remarkable number of approaches try to cluster the named entities without any reference to an available list of possible instances [11,9].
- The *disambiguation algorithms* employed can make use of a number of techniques or a combination of them: statistical procedures [6,11,9], morphosyntactic analysis [9,2], or exploiting ontologies that provide rich linguistic and semantic information about instances of interest [7].
- The *domain*: several approaches are oriented to a particular domain like biology [5] or bibliographic citations [1,6].

The usage of a semantic network ranking algorithm, which also takes into account the temporal component and the categorization system characteristic of the news domain, are the main differences of our approach compared with the ones in the state of the art.

6 Conclusions and Future Lines

In this paper we introduce the IdRank algorithm to address the problem of entity disambiguation in the context of semantic annotation of news items. The algorithm provides a ranking of the candidate instances within an ontology which can be associated to a certain entity. In order to do so the algorithm uses as context the metadata available in a certain news item. It is based on the principles of *Semantic Coherence* (instances typically occur in similar contexts) and *News Trends* (it is common to have temporal burst of news items talking about a certain event).

We have performed an empirical evaluation of the algorithm that shows its adequacy for the news domain, both for the quality of results and the computation time.

A possible future line of development that we want to explore is the possibility of using dynamic coefficients k_a, k_{cat} and k_{tim}, instead of the constant ones. In the training process we would decide the right coefficients for the next execution, depending on the quality of results obtained in the past ones. Evaluating the algorithm in bigger corpora and adapting it to other domains are also future lines of development.

References

1. N. Aswani, K. Bontcheva, H. Cunnigham. Mining Information for Instance Unification. In 5th International Semantic Web Conference. Ed. Springer, LNCS 4273, pp. 329-342. Athens, USA. November 2006.
2. A. Bagga, B. Baldwin. Entity-Based Cross-Document Coreferencing Using the Vector Space Model. In 17th International Conference on Computational Linguistics. Quebec, Canada. August 1998.
3. N. Fernández, J. M. Blázquez and J. Arias, L. Sánchez, M. Sintek, A. Bernardi, M. Fuentes, A. Marrara, Z. Ben-Asher. NEWS: Bringing Semantic Web Technologies into News Agencies. In 5th International Semantic Web Conference. Ed. Springer, LNCS 4273, pp. 778-791. Athens, USA. November 2006.
4. N. Fernández, L. Sánchez, J. M. Blázquez, J. Villamor. The NEWS Ontology for Professional Journalism Applications. A Handbook of Principles, Concepts and Applications in Information Systems. Ed. Springer, Integrated Series in Information Systems, Vol. 14. To appear in December 2006.
5. F. Ginter, J. Boberg, J.Ärvinen, T. Salakoski, New Techniques for Disambiguation in Natural Language and their Applications to Biological Text. Journal of Machine Learning Research, 5: 605-621, 2004.
6. H. Han, L. Giles, H. Zha, C. Li, K. Tsioutsiouliklis. Two Supervised Learning Approaches for Name Disambiguation in Author Citations. In Joint ACM/IEEE Conference on Digital Libraries. Tucson, USA. June 2004.
7. J. Hassell, B. Aleman-Meza, I. Budak Arpinar. Ontology-Driven Automatic Entity Disambiguation in Unstructured Text. In 5th International Semantic Web Conference. Ed. Springer, LNCS 4273, pp. 44-57. Athens, USA. November 2006.
8. N. Ide, J. Véronis. Word Sense Disambiguation: The State of the Art. Computational Linguistics, 24(1), 1998.
9. G. S. Mann, D. Yarowski. Unsupervised Personal Name Disambiguation. In 7th Conference on Natural Language Learning. Edmonton, Canada. June 2003.
10. L. Page, S. Brin., R. Motwani, T. Winograd. The PageRank Citation Ranking: Bringing Order to the Web. Stanford Technical Report available online at: http://dbpubs.stanford.edu/pub/1999-66, 1999
11. T. Pedersen, A. Purandare, A. Kulkarni. Name Discrimination by Clustering Similar Contexts. In 6th International Conference on Computational Linguistics and Intelligent Text Processing. Ed. Springer, LNCS 3406. Mexico City, Mexico. February 2005.

A Study in Empirical and 'Casuistic' Analysis of Ontology Mapping Results

Ondřej Šváb[1], Vojtěch Svátek[1], and Heiner Stuckenschmidt[2]

[1] Department of Information and Knowledge Engineering,
University of Economics, Prague, W. Churchill Sq. 4, 130 67 Praha 3, Czech Republic
{svabo,svatek}@vse.cz

[2] Universität Mannheim, Institut für Informatik, A5, 6 68159 Mannheim, Germany
heiner@informatik.uni-mannheim.de

Abstract. Many ontology mapping systems nowadays exist. In order to evaluate their strengths and weaknesses, benchmark datasets (ontology collections) have been created, several of which have been used in the most recent edition of the Ontology Alignment Evaluation Initiative (OAEI). While most OAEI tracks rely on straightforward comparison of the results achieved by the mapping systems with some kind of reference mapping created a priori, the 'conference' track (based on the *OntoFarm* collection of heterogeneous 'conference organisation' ontologies) instead encompassed multiway manual as well as automated analysis of mapping results themselves, with 'correct' and 'incorrect' cases determined a posteriori. The manual analysis consisted in simple labelling of discovered mappings plus discussion of selected cases ('casuistics') within a face-to-face consensus building workshop. The automated analysis relied on two different tools: the DRAGO system for testing the consistency of aligned ontologies and the *LISp-Miner* system for discovering frequent associations in mapping meta-data including the phenomenon of graph-based mapping patterns. The results potentially provide specific feedback to the developers and users of mining tools, and generally indicate that automated mapping can rarely be successful without considering the larger context and possibly deeper semantics of the entities involved.

1 Introduction

Ontologies can help integrate semantic views on real-world data. Unfortunately, designers of ontologies themselves apply different views of the same domain during ontology development. This yields semantic heterogeneity at ontology level, which is one of main obstacles to semantic interoperability. *Ontology mapping* (also called 'matching' or 'alignment') is the core component of approaches attempting to solve this problem. It consists in finding mappings (also called 'correspondences') among entities (classes, relations) from different ontologies. The set of mappings is called *alignment*. The process of mapping is followed by ontology merging, ontology transformation, data transformation etc. A survey of ontology mapping methods is e.g. in [11].

It is important to have means to *evaluate* the quality of mapping, and, consequently, the fitness of different methods and tools with respect to different domains and settings. Nowadays, the central approach to ontology mapping evaluation is based on the notion of *reference alignment* ('gold standard'), defined a priori, to which the results obtained

E. Franconi, M. Kifer, and W. May (Eds.): ESWC 2007, LNCS 4519, pp. 655–669, 2007.

by the matching systems are compared. This typically yields measures borrowed from the discipline of Information Retrieval, such as *precision* (the proportion of mappings returned by the matching system that are also present in the reference mapping) and *recall* (the proportion of mappings present in the reference mapping that are also returned by the matching system). The correspondences in both the reference and experimental alignments are most often expressed as simple concept-concept (or relation-relation) pairs, interpreted as *logical equivalence*. Sometimes, alignments interpreted as *logical subsumption* (analogously to the same notion as omnipresent in ontology design), and/or with a non-Boolean value of *confidence* are also considered. However, we might even be interested in more complex *alignment structures* (patterns), which could reveal interesting details about the relationship of the two ontologies—for example, the situation when an entity from one ontology can potentially be mapped on both a parent and a child from the other ontology.

In case there is no reference alignment (and providing it manually would be unacceptably tedious) or we are interested in more complex phenomena—say, *mapping patterns*—arising in ontology alignment, novel methods for mapping evaluation have to be devised. Let us outline four of them that are focal in this paper; while the first and the third are manual, the second and the fourth rely on automated procedures.

- Instead of formulating a reference alignment *a priori*, the mappings discovered by the system (which are often just a small fraction of the carthesian product of the sets of entities from two ontologies) can be *a posteriori* examined and *labelled* (as in/correct or possibly using a richer set of labels) by human evaluator/s. Such evaluation naturally lacks the rigour of 'blindfold' evaluation wrt. an (unbiased) reference alignment, and does not tell much about the *recall*. Still, the *precision* figure may be valuable[1]; and its subjective bias can be reduced via recourse to *multiple* (ideally, expert) evaluators.
- Automated *reasoning* over aligned ontologies. A complement to manual labelling of the mappings is the exploitation of an inference procedure that is usually considered as first step in exploiting ontologies: concept satisfiability testing. Clearly, mappings that incur inconsistency to ontologies (that have been consistent as long as standalone) are potentially inadequate.
- A side-product of both manual labelling and automated consistency checking can be a list of 'interesting' (ambiguous, dubious, surprising etc.) mappings. In order to get an overview of typical *reasons* (or 'arguments') for success/failure of automated mapping, some individual 'interesting' cases (not only the entities mapped but also their context within the ontologies and possibly some metadata about the mapping process) can be submitted to a *discussion board*. The outcome of discussion is definitely of different nature than that of quantitative evaluation, but can lead to complementary feedback to the developers of mapping tools. This discussion can also help identify candidate *mapping patterns* to be quantitatively evaluated in further analysis (see next item). We can view this approach as analogous to *casuistic* medical studies/literature, which is also sometimes used as complement to the nowadays dominant empirical (evidence-based) one. In the context of this paper, this 'discussion' approach is incarnated in the *consensus building workshop*.

[1] Cf. the discussion on prior and posterior precision in *ontology learning* [6].

- Large-scale *mining* over the mapping results with meta-data. The input to the mining process can be not only the name of the mapping system, name and nature of the ontologies mapped, the type of mapping (such as equivalence/subsumption) and the subjective posterior evaluation, but also the information whether the given mapping is part (and what part) of a certain mapping pattern. We believe that the hypotheses discovered via data mining (over ontology mapping data including information about mapping patterns), in particular, mining for *frequent associations*, can become useful feedback to the development and tuning of mapping tools, complementary to the feedback provided by Information Retrieval measures with respect to reference mapping.

The paper is structured as follows. Section 2 surveys the background of the current research: the underlying *ontology collection* and the *international initiative* (OAEI) within which the automated mapping experiments took place. Section 3 reports on the evaluation via *manual labelling*. Section 4 deals with *reasoning*-based evaluation, using the *Drago* distributed description logic (DDL) tool. Section 5 describes the *consensus building workshop* in which selected discovered mappings were discussed by humans. Section 6 first presents a simple typology of *mapping patterns* of interest; the rest of it is devoted to the *data mining* effort; the mining tool used is briefly presented and then the actual experiments in mining over ontology mappings (taking the mentioned patterns into account) are given. Finally, section 7 surveys some related research, and section 8 wraps up the paper.

2 Project Background

2.1 *OntoFarm* Collection

The motivation for initiating the creation of the *OntoFarm*[2] collection (in Spring 2005) was the lack of 'manageable' material for testing ontology engineering (especially, mapping) techniques. As underlying domain, we chose that of *conference organisation*—among other, for the following reasons:

- Most ontology engineers are academics who themselves submit and review papers and organise conferences: there is zero overhead of acquiring the *domain expertise*.
- Organisation of a conference shares some aspects with (heavier-weighted) *business activities*: access restrictions, hard vs. soft constraints, temporal dependencies among events, evolution of the meaning of concepts in time etc. There is also a wide range of supporting *software tools* covering various aspects of conference organisation. Their domain assumptions can also be captured using ontologies (specific for each system). The process of matching the requirements of conference organisers with the capacities of such tools is analogous with that of matching the requirements of a business with the capacities of an off-the-shelf enterprise information system.

[2] See `http://nb.vse.cz/~svabo/oaei2006`; the development of the collection is described in more detail in [12].

– In many cases, even the underlying *instance data* could be obtained, since legal restrictions are typically not as strong as e.g. in business or medicine.

The snapshot of the (constantly growing) collection used for the 2006 OAEI track, see below, consisted of ten OWL-DL ontologies, typically of the size of 30–80 concepts and 30–60 properties, some of them being endowed with DL axioms. The overview is in Table 1. Six among the ontologies were derived from different conference organisation *support tools* (for the review process, registration etc.), using their documentation and experiments with installed tools ('tool' ontologies); two of them are based on the experience of people with *personal participation* in conference organization ('insider' ontologies); finally, two of them are merely based on the content of *web pages* of concrete conferences ('web' ontologies). The ontology designers (partly students of a course on Knowledge Modelling and partly experienced knowledge engineers) did not interact among themselves. This should guarantee that, although the ontologies themselves are to some degree 'artificial' (their development not being drived by an application need), their *heterogeneity* was introduced in a 'natural' way, that possibly simulating the heterogeneity of ontologies developed by different communitites in the real world.

Table 1. Characteristics of ten *OntoFarm* ontologies

Name	Type	Number of Classes	Number of Properties	DL expressivity
EKAW	Insider	77	33	$\mathcal{SHIN}(\mathcal{D})$
SOFSEM	Insider	60	64	$\mathcal{ALCHIF}(\mathcal{D})$
SIGKDD	Web	49	28	$\mathcal{ELI}(\mathcal{D})$
IASTED	Web	140	41	$\mathcal{ALCIF}(\mathcal{D})$
Confious	Tool	57	57	$\mathcal{SHIN}(\mathcal{D})$
PCS	Tool	23	38	$\mathcal{ELUIF}(\mathcal{D})$
OpenConf	Tool	62	45	$\mathcal{ALCIO}(\mathcal{D})$
ConfTool	Tool	38	36	$\mathcal{SIF}(\mathcal{D})$
CRS	Tool	14	17	$\mathcal{ALCIF}(\mathcal{D})$
CMT	Tool	36	59	$\mathcal{ALCIF}(\mathcal{D})$

2.2 *OAEI 2006* Initiative

The Ontology Alignment Evaluation Initiative[3] (OAEI) is a coordinated international initiative that organizes the evaluation of the increasing number of ontology matching systems. The main goal of OAEI is to to compare systems and algorithms on the same basis and to allow anyone for drawing conclusions about the best matching strategies [3]. The first OAEI evaluation campaign was presented at the workshop on Integrating Ontologies held in conjunction with the International Conference on Knowledge Capture (K-Cap) 2005. The outcomes of the 2006 campaign were then presented at the Ontology Matching (OM-2006) workshop at ISWC, in Athens, Georgia, USA. There were six different test cases (ontology pairs/collections), related to different domains, which emphasised different aspects of the matching needs; each of them constituted

[3] http://oaei.ontologymatching.org

a specific track of evaluation. Four of the tracks ('benchmark', 'anatomy', 'jobs', 'directory') relied on some sort of pre-defined reference mappings to which those discovered by the systems could be compared (resulting in standard relevance measures such as precision/recall). The remaining ones ('food', 'conference') lacked such reference mappings, but the results were evaluated a posteriori. Here we concentrate on the 'conference' track, which was based upon the aforementioned *OntoFarm* collection and culminated at the OM-2006 *consensus building workshop*.

3 Initial Manual Empirical Evaluation

There were six participant groups to the 'conference' track, with mapping systems[4] named Automs, Coma++, OWL-CtxMatch, Falcon, HMatch and RiMOM. The alignments obtained were examined by the organizers, and each individual mapping was assigned a label. Results from the initial evaluation phase are on the *result report page*[5]; these consist in global statistics about the participants' results, which more-or-less reflect their quality.

The global statistics for each system amount to (among other):

- the distinction whether the mapping is true/false or is scaled between 0 and 1
- number of alignments (i.e. ontology pairs)
- number of individual mappings labeled as 'correct' vs. 'incorrect'
- number of 'interesting correct' mappings, namely, those that were subjectively 'not so easy to identify' at first sight (e.g. due to lack of string similarity)
- number of mappings that seemed to exhibit an interesting type of error (or problematic feature), specifically for: *subsumption* mistaken for equivalence, *sibling* concepts mistaken for equivalent ones, mutually *inverse* properties mapped on each other, *relation* mapped onto *class*
- *precision* as ratio of the number of all correct mappings to the number of all mappings, and *relative recall* as ratio of the number of all correct mappings to the number of correct mappings found by *any of the systems*

Additionally, some of the mappings that were retained as 'worth discussing' by both independent evaluators were then submitted to the consensus building workshop.

4 Empirical Evaluation Via Logical Reasoning

In addition to manual evaluation, we conducted an automatic analysis on a subset of the mappings. Mappings between class names in different ontologies were formalized in C-OWL [1] and the DRAGO system [10] was used to determine whether the mappings created by a particular system cause *logical inconsistencies* in one of the mapped ontologies. C-OWL was chosen as basis for the evaluation, as its semantics is tuned towards describing mappings between ontologies of the same domain; it solves some

[4] Descriptions of the systems are in the OAEI 2006 papers available from `http://om2006.ontologymatching.org/`, see the section *OAEI Papers*.

[5] `http://nb.vse.cz/~svabo/oaei2006/`

problems that occur when standard OWL is used for this purpose. A more detailed description of the approach can be found in [7]. The analysis was performed on six ontologies only, as SOFSEM, IASTED, Confious and OpenConf could not be processed by DRAGO. Further, we restricted the analysis to four matching systems, namely Falcon, OWL-CTXmatch, COMA++ and HMatch. The analysis can easily be extended to other two participating systems, though.

Table 2 shows the results of the reasoning-based analysis. Note that the precision only refers to mappings between *class* names and therefore naturally differs from the numbers at the result report page. The precision has been determined by a manual investigation of the mappings by three independent people (different from those doing almost the same task for the sake of the result report page). In cases of a disagreement the correctness of a correspondence was decided by a majority vote. It however turned out that there was little disagreement with respect to the correctness of correspondence. For only about 3% of the correspondences the result had to be determined by vote.

Table 2. Results of Reasoning-Based Evaluation

System	Inconsistent mappings[6]	Avg. number of inconsistent concepts	Overall Precision
Falcon	4	1,5	89,7 %
OWL-CTXmatch	6	9,6	85,67 %
Coma	12	2,2	67,7 %
HMatch	9	5,5	63,7 %

The results of this evaluation are useful in two ways. First of all, we can see from the numbers that a low number of inconsistent alignments is an indicator for the quality of mappings (we also see that the actual number of concepts that become unsatisfiable is less relevant). The second benefit of this evaluation is the fact that the information about inconsistent concepts and mappings that caused these inconsistencies reveal obvious and also non-obvious errors in mappings. Some examples of obviously incorrect mappings produced by matching systems in the experiments are the following:

$$Document = Topic$$
$$Decision = Location$$
$$Reception = Rejection$$

The real benefit of this evaluation is its ability to find non-obvious errors in mappings that can only be detected taking the position of the mapped concepts in the concept hierarchy into account. In our experiments, we found a number of such errors. Examples include the following mappings:

$$Regular\_Paper = Regular$$
$$Reviewing\_event = review$$
$$Main\_office = Location$$

In the case of the first correspondence, *Regular* actually denotes the regular participation fee as opposed to the early registration. The error in the second correspondence is caused by the fact that *Reviewing_event* represents the process of reviewing whereas review denotes the review document as such. The last correspondence is not correct, because the concept *Main_office* actually represents the main office as an organizational unit rather than a location. Such mappings are candidates for a closer inspection in terms of a committee of experts that analyze the reason for the inconsistency and decide whether the problem is in the mapping or in the ontologies.

5 'Casuistics' – Consensus Building Workshop

5.1 General Idea

The idea of consensus building workshop was to discuss some interesting mappings in detail. Such interesting mappings are determined as a result of the manual and the automatic evaluation of the matching results, as shown above. In the case of the *manual evaluation* mappings where the evaluators where in doubt or where they disagreed on the correctness of a mapping are candidates for a consensus workshop. In the *automatic evaluation*, mappings that have been shown to cause concepts in the mapped ontologies to become inconsistent are such candidates, especially if the mappings have been annotated as being correct in the manual evaluation. Often, a decision whether a mapping is correct or not can be made quite easily in a committee of experts. In some cases, however, it turns out that deciding whether a mapping is correct or not is far from being trivial. In particular, it turns out that sometimes a detailed analysis of the mapped ontologies is necessary to come to a decision.

As far as *arguments* against and for individual mappings are concerned, we experienced that *lexical* reasons of mapping were first considered by the workshop participants. Then followed arguments with regard to the *context* of elements in question. This means consideration of certain neighborhood, subclasses and superclasses (in the case of properties, we can consider subproperties and superproperties). This can disclose different extensions of classes (especially through their subclasses). Also, properties related to classes were considered. As a last resort, *axioms* (more complex restrictions) were taken into account if they were present.

5.2 Examples of Mappings Discussed

In the following, we focus on examples that illustrate the kinds of arguments used in the discussion and the insights gained.

Person vs. Human. At first sight the equivalence between the concepts person and human looks rather intuitive, it is however not obvious that the two concepts have the same intended meaning in different ontologies. First of all, the concept person can be interpreted in a legal context in which it also refers to organizations. Further, when we look at the hierarchies of the different ontologies, we see that the concepts have completely different sets of subconcepts depending on the scope of the ontology (compare figure 1.

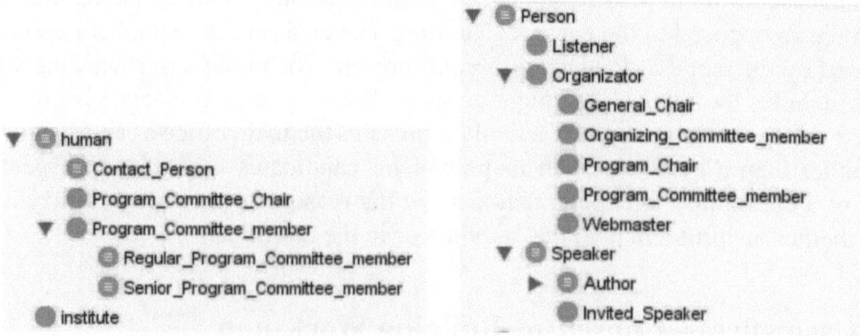

(a) IASTED (b) SIGKDD

Fig. 1. Subtrees rooted at the concepts *Human* and *Person*

As we can see, the notion of a person in SIGKDD also contains subclasses not subsumed under human in IASTED (e.g. speakers). As it is clear, however that both ontologies cover the same domain, it was decided that in this case the two concepts actually have the same intended meaning even though they do not share all subclasses.

PC_Member vs. Member_PC. The concepts *PC_member* and *member_PC* are another example of mappings that seem to be trivially correct at first sight. In this case the question is whether the ontologies assumes the same set of people to belong to the program committee. A look at the hierarchies reveals that the two ontologies use a different interpretation of the set of people belonging to the PC. In particular in one case the *PC_chair* is assumed to be a member of the committee, in the other case not (compare figure 2). This seems to imply that the notion of *PC_member* in EKAW is more general than that in *ConfTool*. However, this is only the case if we assume that the concepts *Chair_PC* und *PC_Chair* are equivalent. Another possible interpretation is that the concepts *PC_member* and *Member_PC* are equivalent but *Chair_PC* and *PC_Chair* are different concepts, namely one denoting PC chairs that are members of the PC and the other denoting PC chairs that are not member of the PC. While both interpretations are possible, the majority of workshop participants favored the first interpretation where PC chairs are the same concepts.

Rejection vs. Reject. Another mapping under discussion was the one between the concepts *Reject* and *Rejection*. It is clear that both are closely related to the outcome of the review of a submitted paper. Differences were only detected when looking at the subtrees of the superconcepts. While *Rejection* is a subconcept of *Decision*, *Reject* is defined as a subconcept of *Recommendation*. Understanding the difference between these two requires a deeper understanding of the process of reviewing, namely that a recommendation is the input for the final evaluation and the decision is the output.

Location vs. Place. A similar situation could be observed in connection with the concepts *Location* and *Place*. Both concepts are closely related as they refer to some geographical entity. A closer look however reveals that they are used in a very different

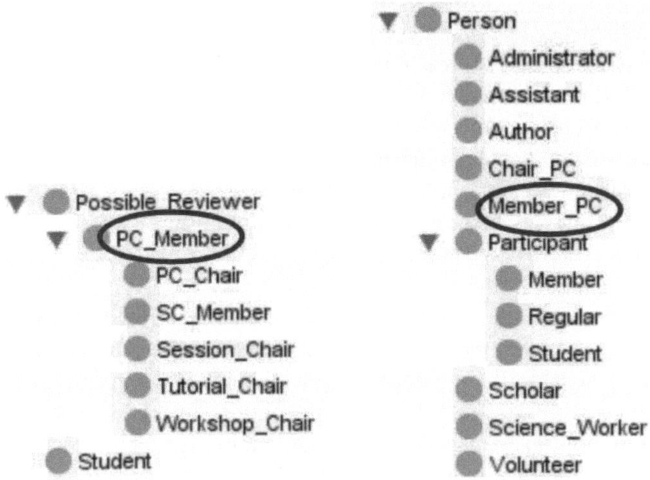

(a) EKAW (b) ConfTool

Fig. 2. Subtrees containing the concepts *PC_Member* and *Member_PC*

way. While *Location* refers to the country and city in which the conference is held, *Place* refers to buildings and parts of buildings in which certain conference-related events take place. The detection of this fundamental difference required a detailed analysis of the ontologies, in particular the range and domain restrictions of related properties in the ontologies.

5.3 Lessons Learned

The discussions at the consensus workshop revealed a number of insights about the nature of ontology matching and limitations of existing systems that provide valuable input for the design of matching tools. In the following we summarize the three most important insights gained.

Relevance of Context. Probably the most important insight of the consensus workshop was that in many cases it is not enough to look at the concept names to decide whether a mapping is correct or not. In all of the examples above, the position of the concept in the hierarchy and in some cases also the scope of the complete ontology had to be taken into account. In some cases, a decision actually requires deep ontological arguments, for instance to distinguish between a recommendation and the actual decision made on the basis of this recommendation. For existing matching tools this means that the use of lexical matching techniques and often even of local structure matching is not sufficient. Matchers rather have to take the complete ontology and its semantics or even background knowledge about basic ontological distinctions into account. This observation is also supported by the results of the reasoning-based evaluation where automatically created mappings often turned out to cause inconsistencies in the ontologies.

Semantic Relations. All of the systems participating in the evaluation were restricted to detecting equivalences between concepts or relations respectively. It turned out that this restriction is a frequent source of errors. Often ontologies contain concepts that are closely related but not exactly the same. In many cases one concept is actually a subclass of the other. Heuristics-based matching tools will often claim these concepts to be equivalent, because they have similar features and similar positions in the hierarchy. As a result, the corresponding mapping often becomes inconsistent. We believe that matching tools that are capable of computing subsumption rather than equivalence relations are able to produce more correct and suitable mappings.

Alternative Interpretations. The example of *PC_member* illustrates the fundamental dilemma of ontology matching, which tries to determine the intended meaning of concepts based on a necessarily incomplete specification. As a result, it is actually not always possible to really decide whether a mapping is correct or not. All we can do is to argue that a mapping is consistent with specifications in the ontologies and with the other mappings. In the example this leads to a situation where we actually have two possible interpretations each of which makes a different set of mappings correct. It is not completely clear how this dilemma can be handled by matching tools. The only recommendation we can give is in favor of using methods for checking the consistency of mappings as an indicator whether the mapping encodes a coherent view on the system.

6 Evaluation Via Pattern-Aware Data Mining

6.1 Introducing Mapping Patterns

Before starting to talk about mapping patterns, it could be useful to briefly discuss the notion of patterns as typically treated in ontological engineering research. We will consider three categories of patterns: content patterns, logical patterns and frequent errors. *Content patterns* [4] use specific non-logical vocabulary and describe a recurring, often domain-independent state of affairs. An example is the "Descriptions&Situations" pattern, which reflects the typical way a situation (with various entities and events involved) is described using some representation. *Logical patterns*, in turn, capture the typical ways certain modelling problems can be tackled in a specific ontological language. An example is the "Classes as Property Values" pattern[7], which defines multiple ways to satisfy the need for using a class in place of a value of an OWL property. Finally, *frequent errors* (though not usually denoted as patterns, they are clearly so) describe inadequate constructions that are often used by unexperienced modellers [9]. All three mentioned types of patterns are used to describe modelling behaviours that considered as either 'desirable' (content and logical patterns) or 'undesirable' (frequent errors). They can be qualified as *design* patterns; indeed, ontology building is essentially an activity carried out by human intellect (at least at the level of defining logical axioms, which are hard to obtain via automated ontology learning). In contrast, *mapping patterns* that will be discussed further are by themselves neither desirable nor undesirable;

[7] `http://www.w3.org/TR/swbp-classes-as-values/`

their desirability depends on the correctness of the mappings. They don't result from a deliberate activity by humans but can be detected in data output by automated mapping systems.

As opposed to ontology design patterns, which concern one ontology, mapping patterns deal with (at least) two ontologies. These patterns reflect the *structure of ontologies* on the one side, and on the other side they include *mappings* between elements of ontologies. A mapping pattern is a graph structure, where nodes are classes, properties or instances. Edges represent mappings, relations between elements (eg. domain and range of properties) or structural relations between classes (eg. subclasses or siblings).

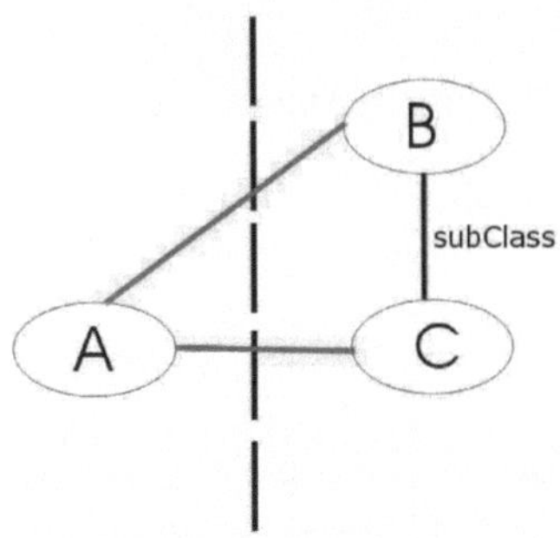

Fig. 3. Pattern 1 – 'Parent-child triangle'

The simplest (trivial) mapping pattern we do not consider here only contains one element from each of the two ontologies (let us call them O1 and O2), and a mapping between them. In our data mining experiments (described later) we employed three slightly more complex mapping patterns.

The first one is depicted in Figure 3. The left-hand side (class A) is from O1 and the right-hand side (class B and its subclass C) is from O2. There is a mapping between A and B and at the same time between A and C.

The second pattern is depicted in Figure 4. It is quite similar to the previous one, but now we consider a child and a parent from each ontology and simultaneous mappings between parents and between children.

The third mapping pattern we consider is depicted in Figure 5. It consists of simultaneous mappings between class A from ontology O1 and two sibling classes C and D from ontology O2.

A somewhat different kind of pattern could be that of mapping between a class and a property. Such 'heterogeneous mappings' are described in [5].

6.2 *4ft-Miner* Overview

The *4ft-Miner* procedure is the most frequently used procedure of the *LISp-Miner* data mining system [8]. *4ft-Miner* mines for association rules of the form $\varphi \approx \psi/\xi$, where φ, ψ and ξ are called *antecedent*, *succedent* and *condition*, respectively. Antecedent and succedent are conjunctions of *literals*. Literals are derived from attributes, i.e. fields

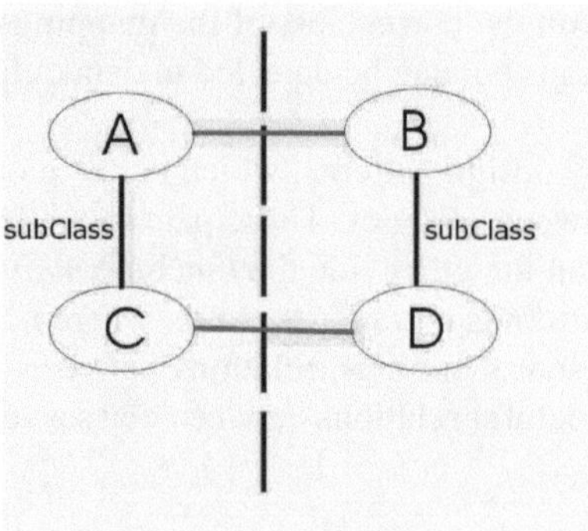

Fig. 4. Pattern 2 – 'Mapping along taxonomy'

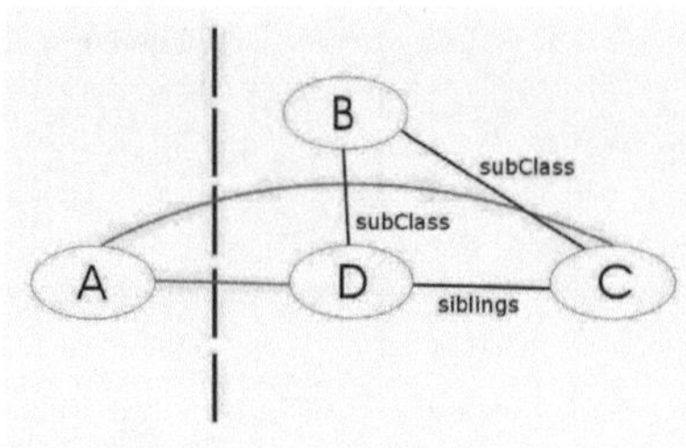

Fig. 5. Pattern 3 – 'Sibling-sibling triangle'

of the underlying data matrix; unlike most propositional mining system, they can be (at runtime) equipped with complex *coefficients*, i.e. value ranges. The association rule $\varphi \approx \psi/\xi$ means that on the subset of data defined by ξ, φ and ψ are associated in the way defined by the symbol $\approx$. The symbol $\approx$, called *4ft-quantifier*, corresponds to some statistical or heuristic test over the four-fold contingency table of φ and ψ.

The task definition language of *4ft-Miner* is quite rich, and its description goes beyond the scope of this paper. Let us only declare its two features important for our mining task: it is possible to formulate a wide range of so-called *analytic questions*, from very specific to very generic ones, and the underlying data mining algorithm is very fast thanks to highly optimised bit-string processing [8].

6.3 Using *4ft-Miner* for Mining over Mapping Results

For the purpose of data mining, a data matrix with each record capturing all information about one (occurrence of) correspondence was built[8]. This elementary information amounted to: name of mapping *system* that detected this (occurrence of) correspondence; *validity* assigned to the correspondence by the system; types of *ontologies* ('tool', 'insider', 'web') on both sides of the correspondence; 'correctness' *label* manually assigned to the correspondence (cf. section 3). In addition, there is information about *patterns*

[8] In sum, there are 5238 records.

(those from the previous section) in which the given correspondence participates. There are two data fields for each of the three patterns; the first one contains the 'correctness' *label* of the *other* mapping within the pattern (note that there are exactly two mappings in each of these simple patterns), and the second one contains the *validity* assigned to this *other* correspondence by the system.

The *analytic questions* (i.e. task settings) we formulated for *4FT-Miner* were for example as follows[9]:

1. Which systems give higher/lower validity than others to the mappings that are deemed 'in/correct'?
2. Which systems produce certain mapping patterns more often than others?
3. Which systems are more successful on certain types of ontologies?

Due to limited space we do not list complete nor detailed results of the data mining process. We only present some interesting association hypotheses discovered.

For the first question, we found for example the following hypotheses:

- Correspondences output by Falcon with medium validity (between 0,5 and 0,8) are almost twice more often 'incorrect' than such correspondences output by all systems (on average).
- Correspondences output by RiMOM and by HMatch with high validity (between 0,8 and 1,0) are more 'correct' than such correspondences output by all systems (on average).

For the second question, we found for example the following hypotheses:

- Correspondences output by HMatch with medium validity (between 0,5 and 0,8) are more likely to connect a child with a class that is also connected (with high validity) with a parent (Pattern 1) than such correspondences with all validity values (on average).
- Correspondences output by RiMOM with high validity (between 0,8 and 1,0) are more likely to connect class C with class D whose parent B is connected (with high validity) with A, which is parent of C (Pattern 2), than such correspondences with all validity values (on average).

These two hypotheses seem to have a natural interpretation (at the level of patterns, perhaps not so at the level of mapping systems). Pattern 1 represents a potential mapping conflict (aka love triangle with a father and a son competing for the same woman), i.e. increasing the validity of one may lead to decreasing the validity of the other. On the other hand, Pattern 2 seems to evoke positive feedback between the two mappings (as might be the case when a father is interested in the mother of his son's girlfriend).

A feature of the *OntoFarm* collection that was clearly beneficial for the 'data mining' approach to mapping evaluation was the fact that it contains (far) more than two ontologies that can be matched. Thanks to that, mapping patterns frequently arising because of the specific nature of some ontologi/es could be separated from mapping patterns that are frequent in general.

[9] Actually, the questions were even more generic; however, their generic form is less elegant when translated to natural language.

7 Related Work

To our knowledge, there has been no systematic effort in posterior analysis of ontolology mappings without reference alignment involving multiple methods like in our research. There are only projects with which we share some isolated aspects.

Mapping patterns are implicitly considered in [5]; however, they focus on 'heterogeneous mappings' (class to property) as special kind of pattern. We also considered this, but it appeared too infrequently (essentially, it was only output by the Coma++ system) to allow for meaningful data mining.

Data mining of a kind was also used for ontology mapping by Ehrig [2]. However, unlike our approach, this was supervised Machine Learning rather than mining data for frequent associations.

8 Conclusion and Future Work

The purpose of the current study was to examine multiple methods of posterior evaluation of ontology mappings, focussing on the situation when there is no reference mapping available and/or we want to get deeper insight into the nature of mappings. Our results could have at least two potential uses: to give the authors of individual *mapping systems* feedback on strong and weak points of the systems (going far beyond the usual precision/recall statistics), and to contribute to better insight of the whole *research community* into the possible argumentation used in the ontology mapping process.

Although the methods are principially different, they have certain dependencies. In particular, initial manual empirical evaluation is pre-requisite for selecting representative cases for the consensus building workshop (this role was also played by automated reasoning) as well as for subsequent data mining. Consensus workshop, in turn, helped refine the nature of mapping patterns. An outline of general methodology could easily be worked out from these dependencies.

In the future, we would like to more thoroughly compare the outcomes of the different methods used (manual labelling, board discussion, data mining, logical reasoning). We would also like to consider a richer variety of ontology mapping patterns. An important task is also to increase the size and improve the quality of *OntoFarm* collection, which would presumably be used in the next OAEI edition.

The authors cordially thank Jérôme Euzenat and Pavel Schvaiko for their cooperation in preparing the OM-2006 consensus building workshop, and Jan Rauch and Milan Šimůnek for their assistence with the LISp-Miner tool. The research leading to this paper was supported by the European Commission under contract IST FP6-507482, Knowledge Web Network of Excellence. Ondřej Šváb and Vojtěch Svátek are partially supported by the IGA VSE grant no.12/06 "Integration of approaches to ontological engineering: design patterns, mapping and mining", and by the grant no.201/05/0325 of the Czech Science Foundation, "New methods and tools for knowledge discovery in databases". Heiner Stuckenschmidt is partially supported by the German Science Foundation under contract STU 266/1 as part of the Emmy-Noeter Programme.

References

1. Bouquet, P., Giunchiglia, F., van Harmelen, F., Serafini. L., Stuckenschmidt, H.: C-OWL: Contextualizing ontologies. In: Proc. ISWC 2003, Springer 2003.
2. Ehrig M., Staab S., Sure Y.: Bootstrapping Ontology Alignment Methods with APFEL In: Proceedings of ISWC, Galway, Ireland, 2005.
3. Euzenat, J., Mochol, M., Shvaiko, P., Stuckenschmidt, H., Šváb, O., Svátek, V., van Hage, W. R., Yatskevich, M.: First Results of the Ontology Alignment Evaluation Initiative 2006. In: International Workshop on Ontology Matching collocated with the 5th International Semantic Web Conference ISWC-2006 , November 5, 2006: GA Center, Athens, Georgia, USA.
4. Gangemi, A.: Ontology Design Patterns for Semantic Web Content. In: The Semantic Web - ISWC 2005, 4th International Semantic Web Conference, Springer LNCS 3729.
5. Ghidini, C., Serafini, L.: Reconciling concepts and relations in heterogeneous ontologies. In: Proc. ESWC 2006, Budva, Montenegro, 2006.
6. Kavalec, M., Svátek, V.: A Study on Automated Relation Labelling in Ontology Learning. In: P.Buitelaar, P. Cimiano, B. Magnini (eds.), Ontology Learning and Population, IOS Press, 2005, 44-58.
7. Meilicke, C., Stuckenschmidt, H., Tamilin, A.: Improving Automatically Created Mappings Using Logical Reasoning. In: Ontology Matching 2006, Workshop at ISWC 2006.
8. Rauch, J., Šimůnek, M.: An Alternative Approach to Mining Association Rules. In: Lin, T. Y., Ohsuga, S., Liau, C. J., Tsumoto, S. (eds.), Data Mining: Foundations, Methods, and Applications, Springer-Verlag, 2005, pp. 211–232
9. Rector, A., et al.: OWL Pizzas: Practical Experience of Teaching OWL-DL: Common Errors and Common Patterns. Proc. EKAW 2004, LNAI3257, Springer-Verlag, 63-81.
10. Serafini, L., Tamilin, A.: DRAGO: Distributed Reasoning Architecture for the Semantic Web. In: Proc. of the Second European Semantic Web Conference (ESWC'05), Springer-Verlag, 2005.
11. Shvaiko P., Euzenat J.: A Survey of Schema-based Matching Approaches. Journal on Data Semantics, 2005.
12. Šváb O., Svátek V., Berka P., Rak D., Tomášek P.: OntoFarm: Towards an Experimental Collection of Parallel Ontologies. Poster Session at ISWC 2005.

Acquisition of OWL DL Axioms from Lexical Resources

Johanna Völker, Pascal Hitzler, and Philipp Cimiano

Institute AIFB, University of Karlsruhe, Germany

Abstract. State-of-the-art research on automated learning of ontologies from text currently focuses on inexpressive ontologies. The acquisition of complex axioms involving logical connectives, role restrictions, and other expressive features of the Web Ontology Language OWL remains largely unexplored. In this paper, we present a method and implementation for enriching inexpressive OWL ontologies with expressive axioms which is based on a deep syntactic analysis of natural language definitions. We argue that it can serve as a core for a semi-automatic ontology engineering process supported by a methodology that integrates methods for both ontology learning and evaluation. The feasibility of our approach is demonstrated by generating complex class descriptions from Wikipedia definitions and from a fishery glossary provided by the Food and Agriculture Organization of the United Nations.

1 Introduction

Knowledge modeling for semantic applications, particularly the creation of ontologies, is a difficult and time-consuming task. It usually requires to combine the knowledge of domain experts with the skills and experience of ontology engineers into a single effort with high demand on scarce expert resources. We believe that this bottleneck currently constitutes a severe obstacle for the transfer of semantic technologies into practice. In order to address this bottleneck, it is reasonable to draw on available data, applying automated analyses to create ontological knowledge from given resources or to assist ontology engineers and domain experts by semi-automatic means. Accordingly, a significant number of ontology learning tools and frameworks has been developed aiming at the automatic or semi-automatic construction of ontologies from structured, unstructured or semi-structured documents. However, both quality and expressivity of the ontologies which can be generated by the current state of the art in lexical ontology learning have failed to meet the expectations of people who argue in favor of powerful, knowledge-intensive applications based on ontological reasoning. Purely logical approaches on the other hand presuppose a large number of manually created ABox statements, and lack the scalability required by many application scenarios.

In this paper, we focus on text corpora as the source for automated ontology creation. Texts are available in abundance from the internet, and the knowledge expressed e.g. through defining sentences given by experts can be expected to be a good base for the creation of ontology axioms describing the same knowledge. Our approach is essentially based on a syntactic transformation of natural language definitions into description logic axioms. It hinges critically on the availability of sentences which have definitory character, like "*Enzymes are proteins that catalyse chemical reactions.*" Such sentences could be obtained e.g. from glossaries or software documentation related to

E. Franconi, M. Kifer, and W. May (Eds.): ESWC 2007, LNCS 4519, pp. 670–685, 2007.

the underlying ontology-based application. Here, we exemplify our approach by using definitions taken from Wikipedia[1] and a fishery glossary provided by the *Food and Agriculture Organization of the United Nations* (FAO), while also sketching a number of alternatives which could suggest themselves in different application scenarios. One of these alternatives, and a particularly interesting case in point is the exploitation of comments in less expressive ontologies given by means of annotation properties.

Consider, for example, an RDFS ontology (lying within OWL DL) essentially consisting of a class hierarchy with some simple use of properties. This ontology, e.g. modeling the scientific domain of Bioinformatics, could well describe a class *Enzyme* being annotated with a comment like the sentence given before, which documents the class for the ontology engineer – note that the sentence cannot be modeled properly in RDFS. In order to enhance the RDFS ontology, our automated tool LExO can be used to analyze this sentence and encode it as the OWL DL axiom

$$\text{Enzyme} \equiv (\text{Protein} \sqcap \exists \text{catalyse.}(\text{Chemical} \sqcap \text{Reaction})).$$

This OWL DL axiom might then be conveyed to the ontology engineer as a suggestion to enhance the ontology. We present here the initial work which can serve as an acquisition core for realizing such a semi-automated process. While we think that our approach has the necessary potential, several non-trivial obstacles remain to be addressed in forthcoming work. We will discuss these obstacles in detail and lay out a work plan for realizing our vision.

The paper is structured as follows. We first give some preliminaries in Section 2. In Section 3, we then describe our approach in detail, giving an ample supply of examples, and also describe our prototype implementation. This leads to a discussion, in Sections 4 and 5, of the obstacles which need to be overcome to realize the vision based on our initial work. We discuss related work and conclude in Section 6.

2 Preliminaries

The Web Ontology Language OWL [1] is being recommended as a web standard by the World Wide Web Consortium for modeling ontological knowledge which requires more expressivity than RDFS. Since then, OWL has found a multitude of uses, not only on the web, but for knowledge representation in general. In essence OWL, or more precisely its most important variant OWL DL, is a so-called *description logic*, or DL for short [2]. DLs combine a rigorous semantics based on first-order predicate logic with an intuitive way of structuring and encoding expressive conceptual knowledge.

In the following we give a formal introduction to the description logic underlying OWL DL. We will keep this introduction very brief, as an intuitive understanding of OWL DL suffices for understanding our approach. For the same reason, some finer details of the definitions will be omitted; the interested reader can find them in [2,1]. We will introduce a syntax which is known as *DL-syntax*, as it is most convenient (and the easiest to read) for the purpose of this paper.

OWL DL is essentially the description logic known as $\mathcal{SHOIN}(\mathbf{D})$, which we introduce below. Knowledge bases in OWL DL are called *ontologies*, and they express

[1] `http://en.wikipedia.org`

relationships between the basic entities in OWL DL, which are *concepts* (or *classes*, like *Enzyme* or *Bank*), *roles* (or *properties*, denoting relationships between things, like *provides* or *represents*), and *individuals* (or *instances*, like *Rudi* or *Lactase*). There are actually two types of roles, namely abstract roles, which relate individuals to individuals, like *fatherOf*, and concrete roles, like *hasAge*, which assign an element of a concrete datatype $\mathcal{D}$ – in this case a number – to an individual.

For describing a $\mathcal{SHOIN}(\mathbf{D})$ (i.e. an OWL DL) ontology, we thus require three sets: a set of concept names (called *atomic* or *named concepts*), a set of role names, and a set of individual names. $\mathcal{SHOIN}(\mathbf{D})$ now allows to combine concepts to (*complex*) *concepts* as defined by the following grammar, where A is an atomic concept, r is an abstract role, s is a concrete role, d is a concrete domain predicate, a_i are individuals, c_i are elements of a datatype, and n is a non-negative integer:

$$\begin{aligned} C &\rightarrow A \mid \bot \mid \top \mid \neg C \mid C_1 \sqcap C_2 \mid C_1 \sqcup C_2 \mid \exists r.C \mid \forall r.C \mid {\geq} n\, r \mid {\leq} n\, r \mid {=} n\, r \mid \\ &\quad \mid \{a_1, \ldots, a_n\} \mid \exists s.D \mid \forall s.D \mid {\geq} n\, s \mid {\leq} n\, s \mid {=} n\, s \\ D &\rightarrow d \mid \{c_1, \ldots, c_n\} \end{aligned}$$

Formally, the semantics of concepts is given by means of *interpretations*, which are mappings from the concept, role and individual names into sets, called *domains of interpretation* which are to be considered as their extensions. This is done essentially as in first-order predicate logic. We omit the details, which can be found in [1], and rather provide some examples to convey the intuition. *Female* $\sqcap$ *Human* stands for all human females, i.e. for all women. $\exists$ *hasChild.Male* denotes all things which have a male child. ≤ 3 *hasChild* is a so-called *cardinality restriction* and denotes all things which have at most 3 children. $\{a, b, c\}$ stands for the class containing exactly the instances a, b, c.

$\mathcal{SHOIN}(\mathbf{D})$ furthermore allows to specify relations between roles. In particular, two roles r_1 and r_2 can be in subrole relation ($r_1 \sqsubseteq r_2$), can be equivalent ($r_1 \equiv r_2$) or can be inverse to each other ($r_1 \equiv r_2^-$). Roles can also be specified as transitive.

Information about individuals is given in terms of the *ABox* of an ontology, which consists of statements of the form $C(a)$ and $r(a, b)$, where a, b are individuals, C is a (complex) concept and r is a role. Information about concrete roles like $s(a, d)$ can also be given, here d is an element of a concrete datatype, and the semantics is analogous.

Complex concepts can now be related in the following way: If C, D are complex concepts, then $C \sqsubseteq D$ and $C \equiv D$ are *inclusion axioms*, where the first denotes that C is a subconcept of D, and the second says that C and D are equivalent. The collection of all axioms of an ontology together with information about roles is called the *TBox* of the ontology. An ontology thus consists of a TBox and an ABox. By *OWL axioms* or *statements* we denote everything which can be contained in an ABox or a TBox. An *OWL element* is either an OWL axiom or a concept, role, or individual.

By means of the formal semantics which can be assigned to an ontology, it is possible to draw non-trivial logical inferences from an ontology. The ontology thus carries *implicit knowledge* by means of its semantics.

3 The LExO Approach and Examples

In this section we present a conceptual approach for transforming natural language definitions (e.g. annotation properties or associated dictionary entries) into sets of OWL

```
( E1 () (fin) C
   ( 3 is (be) VBE i
      ( 2 number (~) N s
         ( 1 A (~) Det det )
      )
      ( 6 entity (~) N pred
         ( E3 () (number) N subj 2 )
         ( 4 an (~) Det det )
         ( 5 abstract (~) A mod )
         ( E0 () (fin) C rel
            ( 7 that (~) THAT whn 6 )
            ( 8 represents (represent) V i
               ( E4 () (that) THAT subj 6 )
               ( 10 count (~) N obj
                  ( 9 a (~) Det det )
                  ( 11 or (~) U punc )
                  ( 12 measurement (~) N conj )
) ) ) ) ) )
```

Fig. 1. Dependency Tree (Minipar)

axioms. The technical feasibility of this approach is demonstrated by a prototypical implementation called LExO (*Learning EXpressive Ontologies*). However, the goal of our work is not to put forward our prototype, but rather to investigate the potential, limitations and challenges posed by attempting to acquire complex ontological knowledge from lexical evidence. We will thus abstract from some implementation details in favor of a critical discussion based on key examples, see Sections 3.1 and 4, and the development of a general methodology for putting our ideas to practical use, see Section 5.

The implementation of LExO basically relies on KAON2[2], an ontology management infrastructure for OWL DL, and the Minipar dependency parser [3]. Given a natural language definition of a class, LExO starts by analyzing the syntactic structure of the input sentence. The resulting dependency tree is then transformed into a set of OWL axioms by means of manually engineered transformation rules. In the following, we provide a step-by-step example to illustrate the complete transformation process. For more (and more complicated) examples please refer to Section 3.1.

Here, we assume that we would like to refine the description of the class *Number* which is part of the Proton ontology (see Section 3.1). The following definition of *Number* was taken from its corresponding Wikipedia article: *A **number** is an abstract entity that represents a count or measurement.*[3]

Initially, LExO applies the Minipar dependency parser wrapped by our own Java-based API in order to produce a structured output as shown in Figure 1. Every node in the dependency tree contains information about the token such as its lemma (base form), its syntactic category (e.g. *N* (noun)) and role (e.g. *subj*), as well as its surface position. Indentation in this notation visualizes direct dependency, i.e. each child node is syntactically dominated by its parent.

This dependency structure is now being transformed into an XML-based format (see Figure 2) in order to facilitate the subsequent transformation process, and to make LExO more independent of the particular parsing component.

[2] `http://kaon2.semanticweb.org`

[3] `http://en.wikipedia.org/wiki/Number`. In our experiments the word sense disambiguation required for identifying the correct article was done manually, although one could well imagine an automatic or semi-automatic solution depending on the requirements of the regarding application.

```
<?xml version="1.0" encoding="UTF-8"?>
<root>
  <C id="E1" pos="0">
    <VBE id="3" pos="3" role="i" phrase="is" base="be">
      <N id="2" pos="2" role="s" phrase="number">
        <Det id="1" pos="1" role="det" phrase="A"/>
      </N>
      <N id="6" pos="7" role="pred" phrase="entity">
        <N id="E3" pos="4" role="subj" base="number" antecedent="2"/>
        <Det id="4" pos="5" role="det" phrase="an"/>
        <A id="5" pos="6" role="mod" phrase="abstract"/>
        <C id="E0" pos="8" role="rel">
          <THAT id="7" pos="9" role="whn" phrase="that" antecedent="6"/>
          <V id="8" pos="10" role="i" phrase="represents" base="represent">
            <THAT id="E4" pos="11" role="subj" base="that" antecedent="6"/>
            <N id="10" pos="13" role="obj" phrase="count">
              <Det id="9" pos="12" role="det" phrase="a"/>
              <U id="11" pos="14" role="punc" phrase="or"/>
              <N id="12" pos="15" role="conj" phrase="measurement"/>
</N> </V> </C> </N> </VBE> </C> </root>
```

Fig. 2. XML Representation of Dependency Tree

```
rule: relative clause {
    arg_0:   //N
    arg_1:   arg_0 /C[@role='rel']
    arg_2:   arg_1 /V
    result:  [equivalent 0 [and 0-1 2]]
}
rule: verb and object {
    arg_0:   //V
    arg_1:   arg_0 /N[@role='obj']
    result:  [equivalent 0 [some 0-1 1]]
    result:  [subObjectPropertyOf 0 0-1]
}
```

Fig. 3. Transformation Rules

```
[equivalent lexo:a_number lexo:an_abstract_entity_that_represents_a_count_or_measurement]
[equivalent lexo:an_abstract_entity_that_represents_a_count_or_measurement
   [and lexo:an_abstract_entity lexo:represents_a_count_or_measurement]]
[equivalent lexo:represents_a_count_or_measurement [some lexo:represents lexo:a_count_or_measurement]]
[equivalent lexo:a_count_or_measurement [or lexo:a_count lexo:measurement]]
[equivalent lexo:abstract_entity [and lexo:entity lexo:abstract]]
```

Fig. 4. Resulting Axioms

The set of rules which are then applied to the XML-based parse tree make use of XPath expressions for transforming the dependency structure into one or more OWL DL axioms. Figure 3 shows a few examples of such transformation rules in original syntax. Each of them consists of several arguments (e.g. *arg_1:...*), the values of which are defined by an optional prefix, i.e. a reference to a previously matched argument (*arg_0*), plus an XPath expression such as */C[@role='rel']* being evaluated relative to that prefix. The last lines of each transformation rule define one or more templates for OWL axioms, with variables to be replaced by the values of the arguments. Complex expressions such as *0-1* allow for "subtracting" individual subtrees from the overall tree structure. A more complete listing of the transformation rules we applied can be found further below.

A minimal set of rules for building a complete axiomatization of the *Number* example could be, e.g. *Copula*, *Relative Clause*, *Transitive Verb Phrase*, *Disjunction* and *Subjective Adjective* (see Table 1). The resulting list of axioms (see Figure 4) in KAON2

```
[equivalent lexo:a_number [and [and lexo:entity lexo:abstract] [some lexo:represents [or lexo:a_count lexo:
    measurement]]]]
```

Fig. 5. Class Description (unfolded)

Table 1. Transformation Rules

Rule	Natural Language Syntax	OWL Axioms
Disjunction	NP_0 `or` NP_1	$X \equiv (\mathrm{NP}_0 \sqcup \mathrm{NP}_1)$
Conjunction	NP_0 `and` NP_1	$X \equiv (\mathrm{NP}_0 \sqcap \mathrm{NP}_1)$
Determiner	$\mathrm{Det}_0\ \mathrm{NP}_0$	$X \equiv \mathrm{NP}_0$
Intersective Adjective	$\mathrm{Adj}_0\ \mathrm{NP}_0$	$X \equiv (\mathrm{Adj}_0 \sqcap \mathrm{NP}_0)$
Subsective Adjective	$\mathrm{Adj}_0\ \mathrm{NP}_0$	$X \sqsubseteq \mathrm{NP}_0$
Privative Adjective	$\mathrm{Adj}_0\ \mathrm{NP}_0$	$X \sqsubseteq \neg\mathrm{NP}_0$
Copula	$\mathrm{NP}_0\ \mathrm{VBE}\ \mathrm{NP}_1$	$\mathrm{NP}_0 \equiv \mathrm{NP}_1$
Relative Clause	$\mathrm{NP}_0\ \mathrm{C}(\mathit{rel})\ \mathrm{VP}_0$	$X \equiv (\mathrm{NP}_0 \sqcap \mathrm{VP}_0)$
Number Restriction	$\mathrm{V}_0\ \mathrm{Num}\ \mathrm{NP(obj)}_0$	$X \equiv =\mathrm{Num}\ \mathrm{V}_0.\mathrm{NP}_0$
Negation (not)	`not` $\mathrm{V}_0\ \mathrm{NP}_0$	$X \sqsubseteq \neg\exists \mathrm{V}_0.\mathrm{NP}_0$
Negation (without)	NP_0 `without` $\mathrm{NP}(\mathit{pcomp\text{-}n})_1$	$X \equiv (\mathrm{NP}_0 \sqcap \neg\mathrm{with}.\mathrm{NP}_1)$
Participle	$\mathrm{NP}_0\ \mathrm{VP}(\mathit{vrel})_0$	$X \equiv (\mathrm{NP}_0 \sqcap \mathrm{VP}_0)$
Transitive Verb Phrase	$\mathrm{V}_0\ \mathrm{NP}(\mathit{obj})_0$	$X \equiv \exists \mathrm{V}_0.\mathrm{NP}_0$
Verb with Prep. Compl.	$\mathrm{V}_0\ \mathrm{Prep}_0\ \mathrm{NP}(\mathit{pcomp\text{-}n})_0$	$X \equiv \exists \mathrm{V}_0\_\mathrm{Prep}_0.\mathrm{NP}_0$
Noun with Prep. Compl.	$\mathrm{NP}_0\ \mathrm{Prep}_0\ \mathrm{NP}(\mathit{pcomp\text{-}n})_1$	$X \equiv (\mathrm{NP}_0 \sqcap \exists \mathrm{Prep}_0.\mathrm{NP}_1)$
Prepositional Phrase	$\mathrm{Prep}_0\ \mathrm{NP}_0$	$X \equiv \exists \mathrm{Prep}_0.\mathrm{NP}_0$
...	...	...

internal syntax is directly fed into KAON2 which interprets the textual representation of these axioms, and finally builds an unfolded[4] class description as shown in Figure 5.

In DL syntax, this final, unfolded axiomatization reads:

$$\mathrm{A\_number} \equiv ((\mathrm{Entity} \sqcap \mathrm{Abstract}) \sqcap \exists \mathrm{represents}.(\mathrm{A\_count} \sqcup \mathrm{Measurement}))$$

Obviously, all parts of this class description have to be normalized and possibly mapped to already existing content of the ontology before the results can be used to generate suggestions for ontology changes (cf. Section 5). As shown by the large body of research done in the domain of ontology mapping, this task is not trivial at all. Semantic ambiguities of labels (e.g. homonymy or polysemy), as well as the fact that a single entity or axiom in the ontology can have arbitrarily many lexicalizations – differing even in their syntactic category – make it necessary to consider a multitude of possible mappings. Moreover, idiomatic expressions, i.e. expressions the meaning of which cannot be directly derived from the meaning of their individual components, need to be treated properly. Therefore, in addition to integrating a state-of-the-art mapping framework, a significant degree of user involvement will be unavoidable in the end (see Section 5).

[4] By *unfolding*, a term borrowed from logic programming, we mean transformations like that of $\{A \equiv \exists R.B, C \equiv A \sqcap D\}$ to $\{C \equiv \exists R.B \sqcap D\}$. The specific for of output which we receive allows us to remove many of the newly generated class names by unfolding, in order to obtain a more concise output.

Table 1 gives an overview of the most frequently used transformation rules. Each row in the table contains the rule name (e.g. *Verb with Prepositional Complement*) and an expression describing the natural language syntax matched by that rule – like, for example, $V_0\ Prep_0\ NP(pcomp\text{-}n)_0$, where V_0 represents a verb, $Prep_0$ a preposition and $NP(pcomp\text{-}n)$ denotes a noun phrase acting as a prepositional complement. Please note that these expressions are very much simplified due to lack of space. The last column shows the OWL axioms generated in each case, where X denotes the atomic class name represented by the surface string of the complete expression matched by the regarding transformation rule.

It is important to emphasize that this set of rules is by no means exhaustive, nor does it define the only possible way to perform the transformation. In fact, there are many different modeling possibilities, and the choice of appropriate rules very much depends on the particular application or individual preferences of the user (see example *Tetraploid* in Section 3.1).

3.1 Examples

We exemplify our approach by giving a number of axiomatizations automatically generated by LExO. The example sentences are not artificial, but were selected from real sources. The first is a fishery glossary provided by the Food and Agriculture Organization (FAO) of the United Nations within the NeOn project[5] – these are examples 1 through 8 below. The remaining examples concern classes of the Proton ontology [4], which has been developed in the SEKT project.[6] While Proton is an OWL ontology, it is rather inexpressive, so we undertook to create more complex OWL axiomatizations for Proton classes. For this purpose, we checked whether for a given Proton class name there was a corresponding Wikipedia article, and took the first sentence of this article as definitorial sentence for the class name. This approach worked reasonably well, as we will see. We first list the example sentences together with their axiomatizations.

1. ***Data:*** *Facts that result from measurements or observations.*
   Data $\equiv$ (Fact $\sqcap$ $\exists$result_from.(Measurement $\sqcup$ Observation))
2. ***InternalRateOfReturn:*** *A financial or economic indicator of the net benefits expected from a project or enterprise, expressed as a percentage.*
   InternalRateOfReturn $\equiv$ ((Financial $\sqcup$ Economic) $\sqcap$ indicator $\sqcap$ $\exists$of.(Net $\sqcap$ Benefit $\sqcap$ $\exists$expected_from.(Project $\sqcup$ Enterprise)) $\sqcap$ $\exists$expressed_as.Percentage)
3. ***Vector:*** *An organism which carries or transmits a pathogen.*
   Vector $\equiv$ (Organism $\sqcap$ (carry $\sqcup$ $\exists$transmit.Pathogen))
4. ***Juvenile:*** *A young fish or animal that has not reached sexual maturity.*
   Juvenile $\equiv$ (Young $\sqcap$ (Fish $\sqcup$ Animal) $\sqcap$ $\neg\exists$reached.(Sexual $\sqcap$ Maturity))
5. ***Tetraploid:*** *Cell or organism having four sets of chromosomes.*
   Tetraploid $\equiv$ ((Cell $\sqcup$ Organism) $\sqcup$ $=4$ having.(Set $\sqcap$ $\exists$of.Chromosomes))
6. ***Pair Trawling:*** *Bottom or mid-water trawling by two vessels towing the same net.*
   PairTrawling $\equiv$ ((Bottom $\sqcup$ MidWater) $\sqcap$ Trawling $\sqcap$ $=2$ by.(Vessel $\sqcap$ $\exists$tow.(Same $\sqcap$ Net)))
7. ***Sustained Use:*** *Continuing use without severe or permanent deterioration in the resources.*
   SustainedUse $\equiv$ (Continuing $\sqcap$ Use $\sqcap$ $\neg\exists$with.((Severe $\sqcup$ Permanent) $\sqcap$ Deterioration $\sqcap$ $\exists$in.Resources))

[5] `http://www.neon-project.org`
[6] `http://sekt-project.com`

8. ***Biosphere**: The portion of Earth and its atmosphere that can support life.*
   Biosphere $\equiv$ (Portion $\sqcap$ $\exists$of.((Earth $\sqcup$ (Its $\sqcap$ Atmosphere)) $\sqcup$ $\exists$can_support.Life))
9. ***Vehicles** are non-living means of transportation.*
   Vehicle $\equiv$ ($\neg$Living $\sqcap$ Means $\sqcap$ $\exists$of.Transportation)
10. *A **minister** or a secretary is a politician who holds significant public office in a national or regional government.*
    (Minister $\sqcup$ Secretary) $\equiv$ (Politician $\sqcap$ $\exists$holds.((Office $\sqcap$ Significant $\sqcap$ Public) $\sqcap$ $\exists$in.(Government $\sqcap$ (National $\sqcup$ Regional))))
11. *A **currency** is a unit of exchange, facilitating the transfer of goods and services.*
    Currency $\equiv$ (Unit $\sqcap$ $\exists$of.Exchange $\sqcap$ $\exists$facilitate.(Transfer $\sqcap$ $\exists$of.(Good $\sqcap$ Service)))
12. *An **island** or isle is any piece of land that is completely surrounded by water.*
    (Island $\sqcup$ Isle) $\equiv$ (Piece $\sqcap$ $\exists$of.$Land$ $\sqcap$ $\exists completely\_surrounded\_by.Water$)
13. ***Days of the week** are: Sunday, Monday, Tuesday, Wednesday, Thursday, Friday, Saturday.*
    DayOfWeek $\equiv$ {Monday, Tuesday, Wednesday, Thursday, Friday, Saturday, Sunday}

Some critical remarks and observations on the examples:

1. This is a simple example, which works out very well.
2. This example shows the complex axiomatizations which can be obtained using our approach. Here (and in other examples) we note that adjectives are so far interpreted as being intersective – we discuss this in Section 4. Another recurring problem is the generic nature of the role *of*. Nevertheless, the output is a reasonable approximation of the intended meaning and would serve well as suggestion for an ontology engineer within an interactive process as we will draft in Section 5.
3. This is a Minipar parse error. The desired solution would be
   Vector $\equiv$ (Organism $\sqcap$ ($\exists$carry.Pathogen $\sqcup$ transmit.Pathogen)).
4. Take particular attention to the handling of negation and of the present perfect tense.
5. The natural language sentence is actually ambiguous whether the number should be read as *exactly four* or *at least four*, and the role name *having* is certainly not satisfactory. Even more difficult is how *set of chromosomes* is resolved. A correct treatment is rather intricate, even if modeling is done manually. The class name *Chromosomes* should probably rather be a nominal containing the class name as individual – which cannot be modeled in OWL DL, but only in OWL Full. Note also that the cardinality restriction is used as a so-called *qualified* one, which is not allowed in OWL DL but is supported by most DL reasoners.
6. *Same* is difficult to resolve. In fact, OWL DL is not expressive enough for properly modeling the sentence!
7. Apart from the very generic role *in* and the problem with adjectives already mentioned, this is a complex example which works very well.
8. The possessive pronoun *its* would have to be resolved.
9. Here, *means of transportation* should probably not be further broken down.
10. *Or* is ambiguous. Here it indicates that *minister* and *secretary* mean the same.
11. The word *and* actually indicates a disjunction in this example.
12. The handling of the adverb *completely* is insufficient.
13. This example would be easy to implement, but we have not done it yet because KAON2 currently cannot handle nominals.

4 Critical Discussion

The syntactic transformation proposed in Section 3 creates a set of OWL axioms which can be used to extend the axiomatization of any given class in an ontology. Our naive implementation of this approach is as simple as efficient, but obviously requires a significant amount of manual or automatic post-processing. This is to a major extent due to a number of problems which relate to limitations of the linguistic analysis and the transformation process, as well as fundamental differences between lexical and ontological semantics. In the following we will discuss some of these problems in more detail, and present possible solutions.

Semantic Aspects. Although the transformation takes into account some aspects of lexical semantics, it is certainly not capable of capturing much of the intension of the terms involved in the natural language expression that serves as an input for the transformation process. Much of the meaning of the resulting axioms is still brought in by the semantics of the underlying natural language terms. This does not necessarily constitute a significant problem as long as the semantics of the description logic expressions is sufficiently "in line" with the lexical semantics of the terms involved in their formation. Actually, the semantics of ontological elements – not of the constructs of the ontology language, but of the classes, properties and instances defined by means of these constructs – will always be grounded to some extent in natural language semantics.

As it is impossible to express all possible aspects of a concept's meaning by virtue of description logic axioms, natural language labels and comments undoubtedly play a key role in ontological knowledge representation. In fact, an ontology without natural language labels attached to classes or properties is almost useless, because without this kind of grounding it is very difficult, if not impossible, for humans to map an ontology to their own conceptualization, i.e. the ontology lacks human-interpretability.

However, a grounding of ontologies in natural language is highly problematic due to different semantics and the dynamic nature of natural language. It is important to mention that many problems linked to either of these aspects are not necessarily specific to ontology learning approaches such as the one we present in this paper. Since the way people conceive and describe the world is very much influenced by the way they speak and vice-versa (also known as the *Sapir-Whorf hypothesis*), ontology engineering is often subject to our intuitive understanding of natural language semantics.

Lexical Semantics. The semantics of lexical relations fundamentally differs from the model-theoretic semantics of ontologies. While lexical relations such as hyponymy, antinomy or synonymy are defined over lexemes, ontological relations are used for relating classes.[7] And it is not obvious in all cases how to map words – especially very abstract notions – to classes, as their extension often remains unclear.

[7] For example, each of these classes could be associated with one or more natural language expressions describing the intended meaning (intension) of the class. And still, since hyponymy is not "transitive" over (near-)synonymy it is not necessarily the case that all mutually synonymous words associated with a subclass are hyponyms of all synonymous words associated with its superclass.

For practical reasons it might be sensible to assume a correspondence between lexical relations and some types of axioms. Traditional ontology learning approaches often rely on information about hyponymy for creating subsumption hierarchies [5], or meronymy for identifying part-of relationships [6]. However, one has to be aware of the fact that a one-to-one mapping between lexical and model-theoretic semantics may affect the formal correctness of ontologies – even more, if ontology learning or engineering exclusively relies on clues by means of lexico-syntactic patterns for inferring lexical relationships. Due to the informal character of natural language it is no trouble to say, for instance, "*A person is an amount of matter*". But from the perspective of formal semantics this might be problematic as pointed out by [7].

Dynamics of Natural Language. Further problems with respect to the use of natural language in ontology engineering relate to the way in which semantics are defined. While ontologies have a clear model-theoretic semantics, the semantics of lexical relations is defined by so-called *diagnostic frames*, i.e. by typical sentences describing the context in which a pair of words may or may not occur given a certain lexical relation among them. This way of defining lexical relations does not guarantee for stable semantics, since natural languages, other than ontology representation languages, are dynamic. That means, each (*open-class*) word slightly changes its meaning every time it is used in a new linguistic context. These *semantic shifts*, if big enough, can affect the lexical relationships between any pair of words. And considering that natural language expressions are regularly used for the grounding of ontologies they can potentially lead to semantic "inconsistencies", i.e. conflicting intensional descriptions. This kind of inconsistencies can be avoided by more precise, formal axiomatizations of ontological elements. However, it is an open issue how many axioms are required to "pin down" the meaning of a given class or property.

Technical Problems. A significant objection one might have with respect to the technical implementation of our approach certainly refers to the rather sophisticated linguistic analysis which is required prior to the actual transformation process. And indeed, the Minipar[8] dependency parser we use sometimes fails to deliver a parse, particularly in the case of ill-formed or structurally complex sentences, or wrongly resolves **syntactic ambiguities** such as prepositional phrase attachments. However, dependency parsers are known to be much more robust than parsers using phrase structure grammar, for example. And as most of our transformation rules can be mapped to surface structure heuristics (see Table 1) in a relatively straightforward way, a chunker or shallow parser could complement Minipar in case of failure. Efficiency is not so much an issue as the parser is extremely fast, processing a few hundred sentences per second.

However, there are more severe problems apart from the quality or efficiency of the syntactic analysis. Many of them concern **semantic ambiguity** related to quantifier scope (e.g. "*any*", see example *Island*) or homonymy (e.g. "*net*"). Both types of problems are not appropriately handled at the moment. And our implementation also lacks an **anaphora resolution** step which would help to identify antecedents of pronouns (e.g. "*its atmosphere*") or nominal anaphora, for instance. Although some types

[8] http://www.cs.ualberta.ca/~lindek/minipar.htm

of coreference including relative pronouns can be handled by Minipar itself, the language we defined for describing the transformation rules is not expressive enough to deal with phenomena such as long distance dependencies or deictic expressions. Therefore, user intervention is still essential during the post-processing phase to replace pronouns and to map co-referring nominals to the same class. Similarly, depending on the desired degree of modeling granularity, user input might be required to support the semantic analysis of **compound nominals** (e.g. "*Pair Trawling*").

Moreover, the different semantics of **adjectives** are not taken into account by the translation rules. Ideally, one would have to distinguish between at least three types of adjectives – subsective ($Young\_Fish \sqsubseteq Fish$), intersective ($Sexual\_Maturity \sqsubseteq (Sexual \sqcap Maturity)$) and privative ($Fake\_Fish \sqsubseteq \neg Fish$). But since an automatic classification of adjectives into these classes as proposed by [8], for example, is a very challenging task, we currently assume intersective semantics for all adjectives. Even more difficult is the semantics of **adverbs** (e.g. "*completely surrounded*") and some types of auxiliary verbs which express a spatial, temporal or behavioral modality (e.g. "*can support life*"). And of course, temporal relationships expressed by past or future **verb tense** are also very difficult to handle without temporal reasoning.

Another problem which is not yet sufficiently handled by our transformation rules are so-called **empty heads**, i.e. nominals which do not contribute to the actual meaning of a genus phrase (cf. the "any" in the *Island* example). In particular, the rules relying on Hearst-style patterns [5] for the identification of hyponymy relationships may be mislead by expressions such as *one*, *any*, *kind*, *type*. This phenomenon has already been described [9,10] and could be handled by appropriate exception rules. An alternative solution to this and similar problems could be to increase the expressiveness of the **rule language** used in the transformation process. The language as it is defined by now does not permit the usage of regular expressions, for instance, which might be valuable means to generalize particular transformation rules. XSLT and *tgrep* could help to overcome these limitations.

Finally, our approach is restricted to texts with **definitory character** such as glossary entries or encyclopedic descriptions which have a universal reading and a more or less canonical form, i.e. including a genus category and additional information to distinguish the term from other members of the same category [11]. In order to extend the applicability of LExO to a greater variety of textual resources, one would need a component for the automatic identification of natural language definitions.

Further Remarks. Although we see a great potential in our approach (cf. Section 5), the discussion shows that there are still many open issues – technical, but also very fundamental questions. The most important ones according to our perception relate to the relationship of lexical and ontological semantics. Given a purely syntactical transformation such as ours, it will be crucial to investigate at which stage of the process and in which manner particularities of both semantics have to be considered. And finally, we will have to answer the question where the principal limitations of our approach with respect to the expressivity of the learned ontologies really are. It is reasonable to assume that at least some aspects of ontological semantics cannot (or not so easily) be captured by purely lexical ontology learning methods. However, we believe that a combination of lexical and logical approaches could help to overcome these limitations.

5 Realising the Vision

Despite the fact that our approach currently has a number of limitations as pointed out in Section 4, we believe that it has the potential to become a valuable component of a semi-automatic ontology engineering environment. Many of the technical drawbacks of the approach can be alleviated by integrating more sophisticated methods for natural language processing, ontology mapping or evaluation, and as a matter of course, by adding a human factor to the ontology acquisition process.

In this section we sketch our vision of a semi-automatic ontology engineering process involving a set of complementary methods for ontology engineering and evaluation along with an elaborate methodology. We describe the potential role of our approach within this scenario and identify the missing components.

Semi-automatic Ontology Engineering. The overall scenario we envision for engineering expressive OWL ontologies is a semi-automatic cyclic process of ontology learning, evaluation and refinement, see Figure 6. The process starts with a relatively inexpressive ontology, possibly a bare taxonomy given in RDFS, which is supposed to be enriched and refined to meet the requirements, e.g. of a reasoning-based application. In each iteration of the process, the user selects the class to be refined, and optionally specifies appropriate resources for the ontology generation phase (Step 1) such as

- manual user input,
- comments contained in the ontology,
- definitions extracted from ontology engineering discussions by email or Wiki,
- documentation of the underlying application and use cases,
- available glossaries and encyclopedias (e.g. Wikipedia), or
- textual descriptions of the domain which could be obtained by initiating a Google™search for definitions (e.g. "*define: DNS*").

A tool such as LExO can analyze the given resources to identify and extract definitory sentences, i.e. natural language descriptions of the class previously selected by the user. These definitions are parsed and transformed into OWL DL axioms (Step 2) that can be presented to the user, if she wants to intervene at this point. Otherwise, the system directly proceeds to the mapping phase which aims at relating the newly generated entities and axioms to elements in the initial ontology (Step 3). The outcome of this phase are a number of mapping axioms which can be added to the class axiomatization after being confirmed by the user.

Then, methods for ontology evaluation check for logical inconsistencies or potential modeling errors (Step 4). Based on the learned axiomatization and additional mappings the system now suggests ontology changes or extensions to the user (Step 5). The user now revises the ontology by modifying or removing some of the axioms (Steps 6 and 7), before the whole process starts over again. Further entities, e.g. those introduced by previous iterations, can be refined until the user or application needs are satisfied.

Ontology Evaluation. It is certainly necessary to add further functionalities to the interactive process. We point out two aspects which we judge to be of particular importance, namely how to aid the ontology engineer to ensure high *quality* of the ontology, and to ensure *completeness* of the modeling process in terms of the application domain.

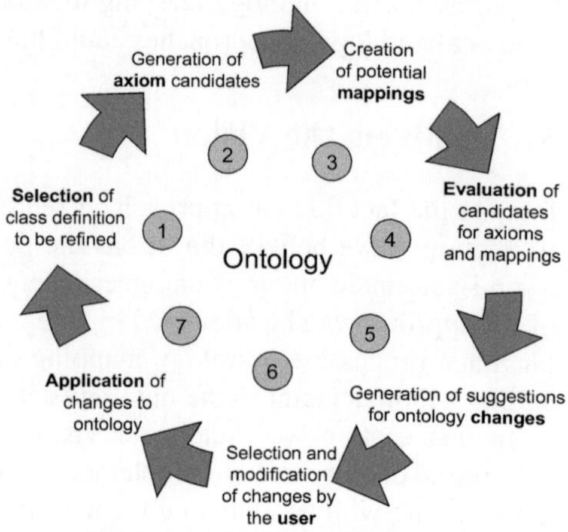

Fig. 6. Ontology Refinement Process

Quality insurance will have to be based on previous work on the field of ontology evaluation. Since the automatic generation of expressive ontologies can potentially lead to a substantial increase in complexity, a simple manual revision of the ontology generated by a system such as the one described here might be infeasible. Therefore, we believe that automatic techniques for ontology evaluation will play a crucial role in the ontology learning and engineering cycle. These techniques could check, for instance, the ontology's validity with respect to the OntoClean methodology [12], or assure the logical consistency of the ontology. In particular, debugging techniques like pinpointing [13] will be indispensable as soon as cardinality restrictions or any kinds of negation (e.g. class complement, disjointness) are introduced into learned ontologies.

Since this aspect has not emerged in lexical ontology learning up to now, the problem of integrating ontology evaluation and debugging into the learning process has not received much attention yet. As pointed out in [14] we see a great potential in exploiting metadata such as confidence and relevance values generated during the ontology learning process for resolving inconsistencies in learned ontologies. But still, the perfect synthesis of ontology learning and evaluation is a challenging problem.

In order to ensure completeness of the modeling process in terms of the application domain, a structured approach for an exhaustive exploration of complex relationships between classes is required. This can be realized e.g. by employing methods like *relational exploration* [15] which is an adaptation of attribute exploration from Formal Concept Analysis [16] to description logics. And finally, it might also be worthwhile to consider an integration of LExO with other learning approaches which could compensate for some of its limitations, e.g. with respect to the learnability of particular relations between roles [17], or disjointness axioms [18].

The issues discussed for creating an interactive ontology engineering tool are under investigation by the authors. In the medium term, we expect to develop a corresponding system as part of a powerful ontology engineering environment like OntoStudio, Protégé, or the forthcoming NeOn Toolkit[9]. It will also be worth investigating the use of LExO for automated question answering. Integrating LExO into any of these application scenarios will allow for a much more target-oriented evaluation of our approach.

[9] http://www.neon-project.org

6 Related Work and Conclusions

We have presented an approach which can support the semi-automatic engineering of ontologies by automatically processing dictionary definitions or ontology comments and translating these to axioms exploiting the expressive power of description logic languages, in particular $\mathcal{SHOIN}(\mathbf{D})$, i.e. OWL DL. An alternative to automatically processing definitory descriptions is to offer a natural language interface allowing users to interact with an ontology editor using natural language. Recently, several approaches relying on controlled language have been presented [19,20,21]. The drawback of such approaches is that users have to actually learn a restricted language which might sometimes even seem unnatural [22]. Though our approach also has limitations, it aims at processing language as used in dictionary definitions without notable restrictions.

There is also a large body of work on dictionary parsing reaching back to the 80s and 90s [23, Chapter 6]. Recent work has focused on extracting knowledge from online glossaries [24] and Wikipedia [25,26,27]. However, most of the work on processing machine-readable dictionaries up to now has mainly focused on extracting lexical relations, in particular hyponymy or meronymy relations. The aspect which distinguishes our approach from others is the fact that it aims at exploiting an expressive description logic language. In this line, our approach is related to the work of Gardent and Jacquey [28], who translate hypernyms, troponyms and antonyms as found in WordNet into DL axioms to be used within a question answering application.

Other work which indeed has aimed at inducing DL class descriptions from data are the ones of Lisi et al. [29] as well as Fanizzi et al. [30]. However, these approaches rely on extensional data, i.e. they assume the availability of an ABox from which a TBox is obtained by generalization. It remains an open issue how, and if these approaches can be applied to learning knowledge bases from texts.

Summarizing, the potential of our treatment lies in its flexibility and simpleness, as well as its suitability for an interactive process as spelled out in Section 5. We have reported on the decisive initial steps in realizing this vision, and accompanied it with a critical discussion of the obstacles which need to be overcome. We believe that these efforts will eventually result in an interactive system which will aid ontology engineers in the construction of expressive OWL DL ontologies.

Acknowledgments. This research has been partially supported by the European Commission under contracts IST-2003-506826 SEKT, IST-2006-027595 NeOn and IST-FP6-026978 X-Media, and by the German Federal Ministry of Education and Research (BMBF) under the SmartWeb project (grant 01 IMD01 B).

References

1. W3C: Web Ontology Language (OWL) (2004) `http://www.w3.org/2004/OWL/`.
2. Baader, F., et al., eds.: The Description Logic Handbook: Theory, Implementation, and Applications. Cambridge University Press (2003)
3. Lin, D.: Dependency-based evaluation of MINIPAR. In: Proceedings of the Workshop on the Evaluation of Parsing Systems. (1998)
4. Terziev, I., Kiryakov, A., Manov, D.: Base Upper-Level Ontology (BULO) guidance. SEKT deliverable 1.8.1, Ontotext Lab, Sirma AI EAD (Ltd.) (2004)

5. Hearst, M.: Automatic acquisition of hyponyms from large text corpora. In: Proceedings of the 14th International Conference on Computational Linguistics. (1992) 539–545
6. Poesio, M., Ishikawa, T., im Walde, S.S., Vieira, R.: Acquiring lexical knowledge for anaphora resolution. In: Proceedings of the 3rd Conference on Language Resources and Evaluation. (2002)
7. Guarino, N., Welty, C.A.: A formal ontology of properties. In: Knowledge Acquisition, Modeling and Management. (2000) 97–112
8. Amoia, M., Gardent, C.: Adjective based inference. In: Proceedings of the EACL Workshop on Knowledge and Reasoning for Answering Questions (KRAQ'06). (2006)
9. Guthrie, L., Slator, B., Wilks, Y., Bruce, R.: Is there content in empty heads? In: Proceedings of the 13th conference on Computational linguistics, Morristown, NJ, USA, Association for Computational Linguistics (1990) 138–143
10. Chodorow, M., Byrd, R., Heidorn, G.: Extracting semantic hierarchies from a large on-line dictionary. In: Proceedings of the 23rd annual meeting on Association for Computational Linguistics. (1985) 299–304
11. Klavans, J., Popper, S., Passonneau, B.: Tackling the internet glossary glut: Automatic extraction and evaluation of genus phrases. In: Proceedings of the SIGIR'03 Workshop on Semantic Web. (2003)
12. Völker, J., Vrandecic, D., Sure, Y.: Automatic evaluation of ontologies (AEON). In: Proc. of the 4th International Semantic Web Conference (ISWC2005). Volume 3729 of LNCS., Springer (2005) 716–731
13. Schlobach, S.: Debugging and semantic clarification by pinpointing. In: Proc. of the 2nd European Semantic Web Conference (ESWC'05). Volume 3532 of LNCS., Springer (2005) 226–240
14. Haase, P., Völker, J.: Ontology learning and reasoning – dealing with uncertainty and inconsistency. In: Proc. of the ISWC Workshop on Uncertainty Reasoning for the Semantic Web (URSW). (2005) 45–55
15. Rudolph, S.: Exploring relational structures via FLE. In Wolff, K.E., Pfeiffer, H.D., Delugach, H.S., eds.: Conceptual Structures at Work: 12th International Conference on Conceptual Structures. Volume 3127 of LNCS., Springer (2004) 196–212
16. Ganter, B., Wille, R.: Formal Concept Analysis – Mathematical Foundations. Springer, Berlin (1999)
17. Lin, D., Pantel, P.: DIRT–SBT–discovery of inference rules from text. In: Knowledge Discovery and Data Mining. (2001) 323–328
18. Völker, J., Vrandecic, D., Sure, Y., Hotho, A.: Learning disjointness. In: Proceedings of the 4th European Semantic Web Conference (ESWC'07). (2007)
19. Tablan, V., Polajnar, T., Cunningham, H., Bontcheva, K.: User-friendly ontology authoring using a controlled language. In: Proceedings of the 5th International Conference on Language Resources and Evaluation (LREC'06). (2006)
20. Bernstein, A., Kaufmann, E., Göhring, A., Kiefer, C.: Querying ontologies: A controlled english interface for end-users. In: Proceedings of the 4th International Semantic Web Conference (ISWC'05). (2005) 112–126
21. Pease, A., Murray, W.: An English to Logic Translator for ontology-based knowledge representation languages. In: Proceedings of the International Conference on Natural Language Processing and Knowledge Engineering. (2003) 777–783
22. Fuchs, N., Kaljurand, K., Schneider, G.: Attempto Controlled English meets the challenges of knowledge representation, reasoning, interoperability and user interfaces. In: Proceedings of FLAIRS'06. (2006)
23. Cimiano, P.: Ontology Learning and Population from Text: Algorithms, Evaluation and Applications. Springer Verlag (2006)

24. Hovy, E., Philpot, A., Klavans, J., Germann, U., Davis, P., Popper, S.: Extending metadata definitions by automatically extracting and organizing glossary definitions. In: Proceedings of the 2003 annual national conference on Digital government research, Digital Government Research Center (2003) 1–6
25. Ruiz-Casado, M., Alfonseca, E., Castells, P.: Automatic extraction of semantic relationships for WordNet by means of pattern learning from Wikipedia. In: Proceedings of the International Conference on Natural Language for Information Systems (NDLB'05). Number 3513 in LNCS, Springer Verlag (2005) 67–79
26. Suh, S., Halpin, H., Klein, E.: Extracting common sense knowledge from Wikipedia. In: Proceedings of the Workshop on Web Content Mining with Human Language Technologies at ISWC'06. (2006)
27. Weber, N., Buitelaar, P.: Web-based ontology learning with ISOLDE. In: Proc. of the ISWC Workshop on Web Content Mining with Human Language Technologies. (2006)
28. Gardent, C., Jacquey, E.: Lexical reasoning. In: Proceedings of the International Conference on Natural Language Processing (ICON'03). (2003)
29. Lisi, F., Esposito, F.: ILP meets knowledge engineering: A case study. In: Proc. of the International Conference on Inductive Logic Programming (ILP), Springer (2005) 209–226
30. Fanizzi, N., Iannone, L., Palmisano, I., Semeraro, G.: Concept formation in expressive description logics. In: Proc. of the 15th European Conference on Machine Learning (ECML'04), Springer Verlag (2004)

On Enriching Ajax with Semantics: The Web Personalization Use Case

Kay-Uwe Schmidt[1], Ljiljana Stojanovic[2], Nenad Stojanovic[2], and Susan Thomas[1]

[1] SAP Research, CEC Karlsruhe, Vincenz-Prießnitz-Str. 1, 76131 Karlsruhe, Germany
{Kay-Uwe.Schmidt,Susan.Marie.Thomas}@sap.com
[2] FZI Forschungszentrum Informatik, Haid-und-Neu-Straße 10-14, 76131 Karlsruhe, Germany
{Ljiljana.Stojanovic,Nenad.Stojanovic}@fzi.de

Abstract. With the dawn of Ajax the capabilities of tracking user behavior multiplied. The same holds for the capabilities of adapting the user interface in a Web browser. To provide meaningful adaptation, the events, context and elements of an Ajaxified Portal must be given meaning. We show the use of ontologies as a model for user-related context and portal-related content. Content-related concepts are used to annotate Ajax widgets to associate them with meaning. As a user navigates a portal and fires events related to the widgets, a semantically rich user model is built, enabling suitable adaptation. Both the user model and the adaptation are based on ontologies and logic rules. Since user tracking and portal adaptation in the era of Ajax, now takes place on the client-side we present a resource-saving approach to executing adaptation rules in the browser. The approach is applied in an e-Government case study.

Keywords: User Adaptivity, Ajax Portal, Semantic Web, e-Government.

1 Introduction

In most e-Government projects to date, technology was in the center of the project and not the user, although the user, e.g., the citizen or business person, is the one who shall in the end use all the new and exciting online e-Government services. As long as all efforts are technology driven, e-Government will not take off, and will not reach its full potential. To confront different citizens with a one-size-fits-all Web interface is not the optimum way to deliver public sector services because every person is an individual with different knowledge, abilities, skills and preferences. The conventional brick-and-mortar office has a more human face because the clerk can respond to different people in different manners. That is why people tend to use the conventional office rather than the e-Government services. To transfer some of the humanity to e-Government portals, it is necessary to build adaptive portals for public services. Such user-adaptive portals will increase the usability, and, thus, the acceptance of e-Government, enabling administrations to achieve the, as yet, elusive efficiency gains and user satisfaction that are the primary goals of e-Government projects.

E. Franconi, M. Kifer, and W. May (Eds.): ESWC 2007, LNCS 4519, pp. 686 – 700, 2007.

This paper describes an approach for adaptation that addresses these issues. This approach results in a system that is both user-adaptive and self-adaptive. By 'user-adaptive', we mean an interactive system that acquires a model of the individual user, and utilizes that model to adapt itself to the user. Such adaptation usually involves some form of learning, inference or decision making (paraphrased from [1]). By 'self-adaptive', we mean a system that observes the results of its actions and adapts itself to improve future performance. Such adaptation can be automatic or mediated by a human. This paper, however, concentrates on user-adaptivity.

Of the types of functions user-adaptation might fulfill – identified in [1] – the approach focuses on two: first, 'help with system use', in particular, help that enables a user to efficiently use an offered e-Government service; second, support for information acquisition, in particular, the information related to the offered e-Government services.

To achieve user-adaptivity we use a new approach that combines the power of Ajax, the underlying technology of Web 2.0, with Semantic Web technologies to create a client-side semantic framework for capturing the meaning of user behavior, recognizing the user's situation, and applying rules to adapt the portal to this situation.

Although, in this paper, the discussion centers around e-Government portals, the architecture is easily generalizable to other types of portals, since at an abstract level most portals can be said to offer some combination of services and associated information.

The rest of this paper is organized as follows. Section 2 presents some examples of e-Government services, and derives requirements on adaptive e-Government portals. Sections 3 and 4 explain the advantages of Ajax and Semantic Web technologies when it comes to meeting these requirements and achieving user-adaptation. Section 5 presents our approach, which combines Ajax and Semantic Web technologies. Section 6 compares the approach to related work. Section 7 indicates the direction of future work. Finally, Section 8 concludes the paper with acknowledgements.

2 Motivating Examples and Requirements

A typical service provided by an e-Government portal is submission of an application form related to a building project. Such a service is actually a complex process, subject to regulations that require the submission of different forms at different times, depending on the type of building project. For an inexperienced user the challenge starts here. With lack of background knowledge of the building regulations the user is confused and does not know which form to choose for her building project. Such users, unfamiliar with the portal and the specific service, need guidance to prevent them from getting stuck in the portal shallows. On the other hand, for an architect who works daily with the virtual building application within an e-Government portal, any guidance would only hinder her smooth sailing. Therefore, there is, among other things, a need to cater for different skill levels like novice, average and expert.

Moreover, adaptation, for example adaptation to skill level, is needed for each service, since an expert in one service may be a novice in another. For example, the architect, expert at building applications, may be a novice hen it comes to submitting an application for child support. In fact, many services will only rarely be used by any one user, so the majority of users will probably remain novices in their use.

Given this example we derive five basic requirements for adaptive e-Government portals. Firstly, a portal must provide guidance and information that matches the users, e.g., the different skill levels and interests of its citizens. Secondly, as citizens typically use e-Government services rarely they should not be bothered with providing and maintaining any user profiles.

The third important requirement is to observe the crucial usability principles such as responsiveness, predictability and comprehensibility, controllability and unobtrusiveness. Initial emphasis, in regard to usability, is placed on providing accurate, but unobtrusive, guidance, when and where it is needed by the user.

The fourth general requirement is that the portal should be subject to continual improvement. Explicit user feedback can be enormously helpful here, but the feedback requested should be relevant to the services that the user executed.

A fifth, and final, important requirement is related to the nature of e-Government services, and the fact that there are multiple units of e-Government at different levels, e.g., local, regional and national. Given the similarity of many of the services offered by these different units, there is an enormous potential for efficiency gains through sharing best practices. Therefore, the fifth general requirement is the ability to share successful adaptation strategies and rules.

In the rest of the paper, we describe an approach that meets these identified requirements.

3 Mashup of Ajax and the Semantic Web

As indicated by the definition of user-adaptivity given in Section 1, acquisition of a model of the user is the indispensable pre-requisite to adaptation. The second step is then to use this model to perform the adaptation. As shown by the requirements analysis of Section 2, it cannot be assumed that the system has any previous information about a user, so that in each user session the user model has to be acquired from scratch. Effectively, this means that the user model is based on the user actions during a session. Therefore, it is essential to be able to track and interpret these actions as accurately as possible. This section shows that Ajax enables the required fine-grained tracking of user behavior and that, additionally, it provides a richer set of adaptation options than standard HTML-based Web technology. The next section then shows how semantics enables interpretation of the user behavior that can be collected using Ajax.

In Adaptive Hypermedia Systems (AHS) adaptation strategies were already studied [2] and are well understood for conventional Hypermedia systems. Under conventional techniques the creation of HTML pages on a remote server and the typical request-response user paradigm of the Web are subsumed. With conventional techniques, the tracking of user clicks, the user modeling, as well as the adaptation all take place on the server. This limits the possibilities of user tracking to the user requests seen by the server [3], which is actually a subset of the user clicks. Furthermore adaptation can only take place when a user requests a new page, which then is adapted to her needs. On the fly adaptation without reloading the whole page in not obtainable.

With the dawn of Ajax in early 2005 [4] a new potential of tracking a user's browsing behavior, as well as new adaptation strategies arose. With Ajax the look and feel of Web pages can be transformed to that of desktop applications. This is the result of the seamless combination of three powerful technologies in Ajax: Asynchronous communication, JavaScript and XML. Using asynchronous communication, just the needed data can be obtained from the server without reloading the whole page. Additionally, JavaScript as a language to execute code in a browser, and XML for the ad-hoc manipulation of a Web page make it possible that a user can be supported without explicitly clicking a link on a page. Thus help and guidance can be already provided when the user behavior is recognized as searching for additional information without server communication explicitly originated by the user.

With Ajax the range of user actions that can be tracked is extended beyond just mouse clicks. For example, scrolling, mouse over and keystroke events can be tracked enabling the detailed recording of user actions on the client-side. In the world of the HTTP requests-response paradigm the Web server is not able to obtain such detailed information. A Web server can only track a subset of user clicks. It misses browser events, like the Back button and cached links. The well-known problem of assigning clicks to users is also solved on the fly, since user tracking takes place directly on the client-side. Additionally, the user's Web browsing behavior can be processed directly on the client and the browser can react immediately to recognized behavioral patterns.

The advanced user tracking possibilities are also accompanied by sophisticated adaptation techniques formerly only seen in desktop applications, like tool tips and fading help windows. However, this rich model of user actions and new adaptation options can only be leveraged if their meaning is machine readable as discussed in the next section.

4 Semantics-Based Adaptation

In this section we show how semantic technologies and in particular ontologies can be utilized for automatic adaptation of an e-Government portal to the individual requirements of the users. We firstly motivate the reasons for using ontologies and thereafter we introduce the semantic model of adaptive portals. Even though the paper is motivated by using e-Government examples, the proposed approach is general enough to be applied in any other domain.

4.1 Advantages of Using Ontologies for Adaptation

There are several reasons to build our approach upon the intensive use of semantic technologies. Firstly, ontologies enable semantic interpretation of user behavior in a portal, which enables meaningful, effective and context-aware adaptation.

The building permission example from Section 2 is elaborated next to show how Ajax and semantics together enable such context-aware adaptation. Assume that the user, who wants to apply for building permission, goes to the appropriate e-Government Web site. And, on this site, the user finds a list of hyperlinks to forms related to building permits. But, she does not know which one is appropriate for her building project. Being based on Ajax, the Web site implements mouse-over help for these hyperlinks. The user knows this, and places the mouse on a hyperlink for a time to make the help appear. Then the user does this for a second hyperlink, but still does not choose a form. Assuming that the hyperlinks have been associated with concepts in the ontology, the system can now make a semantic interpretation of the user's behavior. In this case, the conclusion would be that the user has a strong interest in the concepts associated with the two mouse-over hyperlinks, and that the user needs help choosing a form. In response to this context, the system can offer the user help. Not only that, this help can be tailored to the user by taking account of the concepts in which the user showed interest, concluded from her current navigation path and behavior. As explained later, adaptation such as this is based on using semantic annotation of a page and its structural elements (e.g. hyperlinks).

A second reason to use ontologies is that ontologies used in rules can make adaptation logic more explicit. This declarative representation, expressed as rules using concepts and relations from the ontology, helps the domain experts inspect, understand and even modify the rationales behind adaptive functionality. For example, the hierarchical organization of e-Government services allows the expert to model adaptation rules on a more abstract level, i.e., covering more than one concrete service (e.g. building permission service, independently of the type of building such as house, office, etc). This reduces significantly the number of rules and makes maintenance of the system much easier.

Finally, ontologies facilitate sharing knowledge between portals, especially for those offering similar services (e.g. two municipalities in one state are similar). For example, the best practices gathered in issuing building permits in one portal (e.g. inexperienced users need an additional explanation regarding the hyperlink "required documents") can be easily transferred to other portals that implement the same regulations for issuing building permits. This sharing is greatly facilitated by the fact that all of the terms used (e.g. additional explanation, hyperlinks, "required documents" etc.) are well defined. It is clear that the benefits for the users as well as for e-Government are enormous, since the public administration can improve its performance at much less expense.

4.2 Ontology-Based Model of Adaptive Portals

Since the data relevant for adaptation is rather sparse, or a great deal of interpretation must be done to turn it into actually useful information, we have developed the ontology-based model of adaptive portals. This model (the so-called Portal Adaptation Ontology) is used to decide if an adaptation should take place and how to do that. A part of the ontology is shown in Figure 1. The full version can be found in [5]. The ontology represents all aspects relevant for adaptation such as Web site structure (Web Portal Ontology), Web site content (Content Ontology), user profiles (User Ontology), and Web site usage data as well as knowledge about the adaptation process itself (Adaptation Ontology). Ajax-enabled Web pages as well as the UI elements contained by those pages will be annotated with individuals and concepts form the Portal Adaptation Ontology.

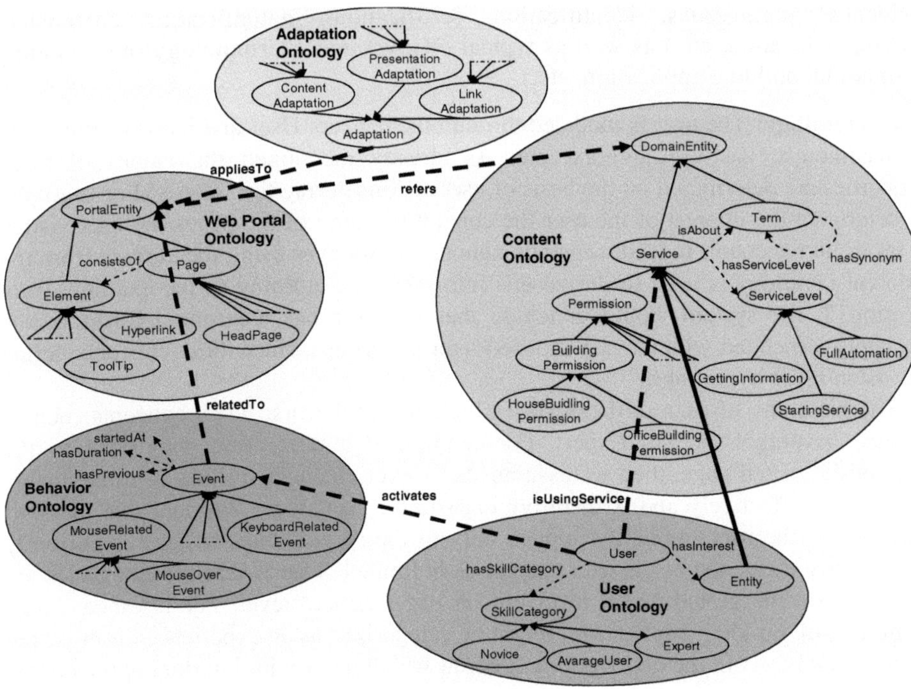

Fig. 1. A part of the Portal Adaptation Ontology showing several entities of the included ontologies as well as dependencies between them

Web Portal Ontology: The way that the Web site is physically laid out as well as the structure of each page can be useful toward understanding usage behavior and interpreting system suggestions. Additionally, the semantic information about the reasons why the structure exists in the way that it does may also be useful. Thus, the Web Portal Ontology contains entities representing the types of pages (such as Head Page, Navigation Page, FAQ, Combined Page etc.) and the structural elements of a page (e.g. Hyperlink, Figure, Table, Content, etc.). We note here that information about page structure can be used to derive or to verify the type of a page [6]. For example, a Navigation Page is a page with small content/link ratio; short time spent on page and is not a maximal forward reference.

Content (Domain) Ontology: The content[1] of Web pages themselves is essential to determining particular topical interests and understanding the relationships between pages. The Content Ontology consists of concepts and relations modeling the meaning of services/information offered by an e-Government portal. This includes already existing categorization[2] of e-Government services (such as residential affairs,

[1] By 'content' we assume the meaning and not the syntax of a page.

[2] It has been developed based on the existing standards for modeling life events such as the Swiss Standard eCH-001 that aims to give an overview over all relevant e-Government services in Switzerland and therefore to provide a consistent and standardized classification of the services.

residential permissions, identification, certifications, naturalization citizenship, moving, education, etc.) as well as typical e-Government terminology (e.g. building permission, building application, etc.).

User Ontology: The user is modeled through the concept User and its properties such as hasInterest, hasSkillCategory, etc. As already mentioned the values of these properties are determined on the basis of user actions during the session. For example, to determine the interest of the user the content/meaning of the pages the user visited is taken into account. Indeed, semantic annotation of pages using the entities from the Content Ontology is used to derive this information. Returning to the example from Section 2, the system would conclude that the user has a strong interest in the concepts associated with the two moused-over hyperlinks, since these concepts define the meaning of hyperlinks.

The hierarchy of user skill categories includes, at the first level, concepts such as Novice, AverageUser and Expert. For example if a user often goes back to the previously visited page, then we assume she is overwhelmed and has become unable to navigate effectively and is therefore classified as a novice. We note that the skill category of the user implicitly applies only to the service that the user is currently using, since the scope of the user categories is limited. That is that the categories are not valid on the global portal level but on page/service[3] level. For instance a user familiar with building applications might be categorized as an experienced user on the appropriate pages in the e-Government portal which deals with building applications. On the other hand, she might be a domain novice when trying to enroll her child in a public school.

Behavior Ontology: The most important data set is the recording of interactions of users with the Web site, in other words, the way that the Web site is used. Even though, this is by far the most abundant collection of data, provided by Ajax, it is, however, the least informative on its own and needs to be enriched with semantics and interpreted.

However, interpreting event data is difficult if the data is not normalized into a common, complete, and consistent model. This entails not only reformatting the data for better processing and for achieving readability, but also breaking it down into its most granular pieces. For example, the system has to be able to recognize all mouse-related events such as mouse-down, mouse-move, mouse-out, mouse-over, mouse-up, etc. Moreover, interpretation involves filtering out unwanted information to reduce analytical errors or misrepresentations. For example, the system should be able to condense the received events into a single event directly indicating a problem. Returning to the example from the beginning of this section, the adaptation should be generated only if two mouse-over events occur sequentially within a session. Finally, interpretation involves acquiring more information from outside the scope of the original event data, for example, from a page the event occurred on (e.g. replacing the meaningless information such as name and target of a hyperlink with the meaning of this hyperlink).

[3] A page must be annotated with the service it belongs to in order to enable the system to link the user with a service.

To cover all these requirements we have developed the Behavior Ontology that structures information about the user's interactions and relationships and/or dependencies between interactions. The main purpose of the ontology is to store all the interactions of the user which might help to identify her experience, actual context and goals.

The most important concept of this ontology is the concept Event that describes what happened, why it happened, when it happened, and what the cause was. The structure of the hierarchy of events reflects the underlying technology used for capturing events, i.e. Ajax. For example, events are decomposed at the first level into the event categories: keyboard, button, mouse, focus and general events. Each of these categories is further specialized. For example, keyboard-related events occurring when a user hits a key contain events such as key-down, key-up and key-press. The category of general events cover load, unload, submit, error handling, etc.

Adaptation Ontology: This ontology was derived from the taxonomy of adaptive hypermedia systems [2]. We distinguish between content, presentation and link adaptation. Each of these types can be further categorized. For example, adaptation of navigation which realizes adaptation by changing the links of the system (i.e. LinkAdaptation) can be realized by several techniques such as DirectGuidance, LinkSorting, LinkHiding, LinkAnnotation, LinkGeneration or MapAdaptation. Each technique might also be realized in several ways. For example, LinkHiding concerns links that are not considered relevant for a user (at the current time), and can be realized by hiding, disabling or removing links.

As shown in Figure 1, all the previously mentioned ontologies are combined in the Portal Adaptation Ontology that models adaptive functionality formally and explicitly. Moreover, it is enriched with rules[4], as discussed below, to enable automation of the adaptation process. In this way, we provide a logical characterization of self-adaptive e-Government systems. We note here that we use the OWL-DL ontology language to represent ontologies. Rules are encoded in the SWRL[5] language, and the KAON2[6] inference engine is used to perform ontology and rule-based reasoning.

We classify the rules into two types based upon their roles in the adaptation process:

Categorization Rules: These rules assign a current user to the predefined user categories. For example[7], a user is an expert for a service, if she uses a bookmark to load a page representing this service.

```
FORALL hasSkillCategory(U,"Expert") ←
  User(U) AND Service(S) AND isUsingService(U,S) AND
  Page(P) AND refers(P,S) AND bookmarkUsage(E) AND
  relatedTo(E,P) AND activates(U,E).
```

---

[4] Concepts and relations defined in the Portal Adaptation ontology directly or indirectly through included ontologies are used in rules.

[5] Semantic Web Rule Language: http://www.w3.org/Submission/SWRL

[6] KAON2: http://kaon2.semanticweb.org/

[7] Note that all terms used in these examples belong to the Portal Adaptation Ontology. Additionally, due to complexity of SWRL format, rules are represented using FLOGIC syntax.

Adaptation Rules: These rules automate corrective actions, i.e. adapt the content, structure or layout of a page to the current user based on the category to which she belongs. For example, if a user is not an expert, and if she spent more than 100ms reading a tool tip of some element on a page, then context-sensitive and content-sensitive help explaining the meaning of this element should be shown to this user.

```
FORALL D ShowAdditionalInformation(T,D) ←
  User(U) AND hasSkillCategory(U,"Expert") AND
  MouseOver(E) AND activates(U,E) AND hasDuration(E,t)
  AND greater(t,TimeConstant8) AND ToolTip(T) AND
  relatedTo(E,T) AND DomainEntity(D) AND refers(T,D).
```

5 Bringing Together Semantics and Ajax

In this section we relate the ontologies, introduced in the previous section to Ajax technology, and describe how their combination enables adaptivity. There are four key elements to achieving adaptivity: annotation, event interpretation and correlation, the user model and adaptation to the user. These four are explained in the initial part of this section. Then a more detailed description of the user model is given. After that we discuss the main challenges of integrating semantics and Ajax.

The fundamentally new in the idea to marry Ajax with the semantic Web is that JavaScript events are associated with concepts, thereby, becoming meaningful 'words' in the interaction with the user. By means of appropriate annotations, the context of JavaScript events can be recognized and the portal can react accordingly. The annotations come from our Portal Adaptation Ontology (see Section 4) and are stored in a knowledge base.

Semantics are indirectly associated to events by annotating the UI elements which fire events as the user interacts with the portal. The UI elements, also called Widgets, of an Ajax page can be annotated with concepts from the Content Ontology (see Section 4) e.g. the concept of a building application in an e-Government context. They can also be annotated with concepts related to the purpose of the widget e.g. with the concept of 'navigation' for a widget meant to navigate the user to a sought-after service.

Events on their own, even when coupled with semantics, are not enough. First, sequences of events have to be correlated into more meaningful units. Thus, simple JavaScript events like mouse over events are combined into compound events, which can, in turn, be a starting point for subsequent correlations. A compound event can consist in such a way of multiple simple events. Based on the list of the compound events, a model of the user can be derived, i.e. a proper instantiation of the Behavior Ontology (see Section 4) will be generated.

The user model is the basis for adaptation. One attribute of the user model is the user category. Categories are pre-defined, either on the basis of a priori knowledge, or on the basis of offline categories discovered by data mining. When possible a user is classified into one of these pre-defined categories. This is done on the basis of the

[8] TimeConstant is a numerical value that is dynamically changed based on the log information.

context information extracted from the semantic concepts related to the simple and compound events. The list of events and the context of the user derived from it, as well as the user category are the essential attributes of the user model that enable adaptation.

Adaptation rules evaluate the current user model and generate abstract actions, which can be interpreted by the portal to adapt to the user. Abstract actions must then be converted by the portal into concrete changes to the user interface. For example, from an adaptation rule, it might follow that the user is lost in the portal shallows, and needs an assistance window in order to reach her goal. The adaptation rule only specifies that an assistance window is needed. The content of the assistance window is derived from the semantics of the events and from UI-elements linked with those events. In this way, the exact conversion of the adaptation directives conforms to the style sheets used by the portal.

In these last few paragraphs of this section, we take a closer look at the user model and the advantages of SWRL rules. All significant events generated by user interactions are collected and stored in logical event queues. The chronological sequence of the events is guaranteed by the assignment of a time stamp to each event. The rules for the correlation of events are expressed in SWRL. One advantage of using SWRL is that the Portal Adaptation Ontology can be accessed by the rules directly. A further advantage is that SWRL can be serialized as an OWL ontology. Thus, reasoning support is available.

As discussed previously, SWRL rules are used to classify the user into a category. They are also used to derive abstract actions to adapt the portal to the user. Like the events, the actions are stored chronologically in a queue ordered by a time stamp.

Both logical queues are modeled in the Behavior Ontology during design time and serve as client-side data structures driving the portal adaptation. The event queue stores the JavaScript a.k.a. Ajax events and the action queue stores the resulting adaptation steps.

5.1 Challenges Integrating Semantics and Ajax

The combination of Semantics and Ajax brings many advantages for dynamic portal adaptation. But this does not come for free. While starting the implementation of our solution we were faced with a lot of tricky challenges, all resulting from moving user tracking from the server to the client-side.

The decision to react to user behavior at the level of JavaScript events leads to a rich and verbose user model. With every new JavaScript event, the user model, and thus the user context, may change. Every such change requires execution of rules. For this reason, rule execution on the server-side, by means of the asynchronous communication facility of Ajax is infeasible; it would overload the server. So, we needed to move rule execution to the client-side. Therefore the major challenge is to implement rule execution in the client, which has limited resources and limited programming libraries.

That is, we have to deal with adaptation on the client-side, adaptation based on JavaScript event streams, and JavaScript only programming capabilities. Taking these constraints into account the following concrete challenges arise.

- Resource saving ontology-based model of adaptive portals
- Annotating Ajax pages with semantic concepts from the ontologies
- Extracting semantic annotations on the client-side
- Rule-based portal adaptation and Execution of rules on the client-side
- Portal specifics, dynamic Web pages and self-adaptivity

5.1.1 Resource Saving Ontology-Based Model of Adaptive Portals

We have already solved this challenge by carefully designing and implementing the ontologies for portal adaptation with respect to the limited resources at the client-side and to the rich user model conditioned by the verbose Ajax events (see Section 4).

Because of the resource restrictions in typical browser environments we developed new and rather small user model and behavior ontologies and did not reuse already existing but large user model ontologies like GUMO [7]. In GUMO a rich user model is proposed with many concepts and properties not applicable to the domain of adaptive portals.

5.1.2 Annotating Ajax Pages with Semantic Concepts from the Ontologies

The challenge here is to establish a link between the UI elements of an Ajax page and the concepts of the Portal Adaptation Ontology that describe them. The annotation of HTML pages with RDF triples was already a topic of several investigations and there exist a couple of solutions [8]. However, because we wanted to avoid deep changes to the portal we decided to follow a different approach.

Our idea is to store the semantic descriptions in a knowledge base. The knowledge base is an extra ontology that is not directly integrated into the Web pages, but is left on the server together with the other ontologies. But the open question is how to link all relevant UI elements of an Ajax page to the concepts which provide the semantic context information. This can be done using the optional `id` attribute provided by nearly every HTML element. As all UI elements of an Ajax page are in fact HTML elements, we can add the optional `id` attribute to every UI element we want to annotate. In order to establish a link between the annotations stored in the knowledge base and the UI elements described by them, a special property was introduced in the ontology. This property carries the value of the `id` attribute of an UI element. Thus, whenever information is needed for a certain UI element, the `id` serves as a link to its semantic annotations.

However, this raises further challenges: How to guarantee the unambiguity of the identifiers, and how to access the ontology containing the annotations from the browser?

5.1.3 Extracting Semantic Annotations on the Client-Side

As discussed above we are in favor of using a separate ontology for storing the semantic annotations of the portal. That saves us from dealing with the awkward extraction of Metadata embedded directly in the HTML page. It also saves us from cumbersome XML and ontology processing. The challenge is how to access the Portal Adaptation ontologies stored on a Web server from the browser on the client-side.

Based on the work done in [9] we developed a prototypical Java library that translates ontologies to JavaScript objects. These objects can be directly accessed and evaluated within an Ajax page. A first promising candidate for the encoding of the ontologies in JavaScript is JSON [10]. The JSON string encoding the ontologies can be accessed by the Ajax page using its asynchronous communication facility.

Another issue for further investigation is to perform reasoning over the ontologies in the browser with JavaScript. Although, there exist at least one inference engine supporting JavaScript and backward-chaining reasoning [11] we decided not to use a reasoner on the client-side. There are mainly two reasons which caused this decision: Firstly, we already can perform the externalization[9] of the ontology at the server. This is only done once in advance on the server and thus has no negative implication on the portal adaptation at runtime. Secondly there is no explicit need for doing reasoning on the client-side because all HTML elements are also annotated in advance and thus well know before runtime.

Not covered by externalization are the JavaScript events and the adaptation actions because they are dynamically created at runtime. But events and actions are annotated via their targets, that is, the UI elements they are connected with. So we don't need reasoning at runtime. At this stage, having the Portal Adaptation ontologies and the link from the HTML elements to the ontologies, the open questions are: what is the most appropriate representation of the Portal Adaptation ontologies at the client, and what is the best way to synchronize the user model on the client with the user model on the server-side, in case there are rules to be executed at the server-side.

5.1.4 Rule-Based Portal Adaptation and Execution of Rules on the Client-Side

The challenge here is to develop easy to maintain rules taking into account the semantic knowledge of annotated JavaScript events. Four rule types have to be designed: Extraction, correlation, categorization and adaptation rules. Extraction rules add the semantic annotations from the knowledge base to the core JavaScript events. Since the events are only indirectly annotated by their target UI elements, some logic is necessary to combine the events with the semantics. Correlation rules combine simple JavaScript events and their semantics to an interpretable user behavior. Categorization rules evaluate the semantics of JavaScript events in order to categorize the current user properly. Based on the user categories, adaptation rules will propose appropriate adaptation strategies.

A promising rule language for OWL ontologies is SWRL. However, first prototypical implementations already show that SWRL might not be sufficient, as we also need production rules in order to fire adaptation actions. Thus, a further investigation of SWRL and other rule languages is necessary.

In order to tackle this challenge of dealing with the extraction, correlation, categorization and adaptation rules, we developed as an initial solution a server-side component that translates SWRL rules into JavaScript control statements, and into JavaScript objects or arrays, respectively. As JavaScript can be executed easily by any Web browser, the extraction, correlation, categorization, as well as the adaptation rules can now be executed at the client-side to guarantee instant portal adaptation.

To reduce the complexity of the client-side JavaScript rules, a two-stage rule-handling approach will be introduced. Simple rules are transformed into JavaScript by a server-side component, and are executed directly on the client-side. Complex rules,

---

[9] With externalization we mean the transformation of implicit knowledge of an ontology into explicit knowledge by reasoning.

with no time critical consequences for UI adaptation, remain on the server-side, in order to utilize the powerful features of ontology processing and reasoning, as well as rule execution frameworks at the server-side. Server-side rules are triggered and their results are evaluated by JavaScript callback functions encoded into the Ajax page, leveraging the XMLHttpRequest object for asynchronous communication.

5.1.5 Portal Specifics, Dynamic Web Pages and Self-adaptivity

So far, only static Web pages using the Ajax technology were examined. The open question is what effort must be made in order to support complex portals with dynamic Web pages. Another challenging area is portal self-adaptivity to achieve continual improvement. Some relevant open research questions are: Which collected data about the user and the user behavior should be recorded for the purpose of long-term adaptation? How can this data be used to check the effectiveness of adaptation rules and discover new adaptation needs?

6 Related Work

Related work to our approach includes standard models of adaptive hypermedia like [12], recent semantic-based personalization systems [13], [14] and Ajax-based personalized systems [15].

Comparing our work with standard models for adaptive hypermedia systems like e.g. AHAM [12], we observe that they use several models like conceptual, navigational, adaptational, teacher and learner models. Compared to our approach, these models correspond to ontologies presented in Section 4, but miss their formal representation. Moreover, we express adaptation functionalities as encapsulated and reusable OWL-DL rules, while the adaptation model in AHA uses a rule based language encoded into XML.

The Personal Reader [13] provides a framework for designing, implementing and maintaining Web content readers, which provide personalized enrichment of Web content for each individual user. The adaptive local context of a learning resource is generated by applying methods from adaptive educational hypermedia in a semantic Web setting. Similarly [14] focuses on content adaptation, or, more precisely, on personalizing the presentation of hypermedia content to the user. However, both approaches do not focus on the on-line discovery of the profile of the current user that is one of the main features of our approach. Another difference would be the self-adaptivity.

Recently some work has been done regarding the usage of Ajax for personalization, like [15]. However, our approach resolves the problem of the syntactical processing of the user's click stream by combines Ajax with semantic technologies. Indeed, our approach enables semantic interpretation of the user's behavior in a portal.

7 Conclusion and Outlook

This paper presented an approach to achieving user-adaptivity that combines Ajax with Semantic Web technologies. Three major advantages to this approach were discussed. First, with this approach, user-adaptation can be more accurate and more appropriate. Better accuracy is possible because Ajax enables finer-grained tracking of user behavior, and semantic annotation enables meaningful interpretation of this more accurate record of behavior. More appropriate adaptation is enabled by the richer set of options for adaptation offered by Ajax, which makes a Web application more like a desktop application.

A second advantage of the approach is that domain experts can inspect, understand and modify the adaptation logic, since it is expressed in the form of explicit rules. Moreover, hierarchical organization of rules makes them easier to maintain.

A third major advantage is that adaptation rules can be shared by groups, like public administrations, that agree to use the same ontologies, since the rules are formulated using concepts from ontologies.

Currently we are working on solving the open research challenges. We are implementing our ideas prototypically in an Ajax-enabled JBoss Portal[10]. After finishing this, we will first evaluate our prototype, and afterwards implement parts of it in real e-Government portals.

Acknowledgements

The work is based on research done within the FIT project – Fostering self-adaptive e-Government service improvement using semantic technologies. The FIT project is co-funded by the European Commission under the "Information Society Technologies" Sixth Framework Program (2002-2006).

References

1. Jameson, A.: Adaptive Interfaces and Agents. Chapter in Jacko, J. A., Sears, A. (eds.): Human-Computer Interaction Handbook (2nd ed.). Erlbaum, Mahwah, New Jersey (2006)
2. Brusilovsky, P.: Methods and techniques of adaptive hypermedia. In User Modelling and User-Adapted Interaction, 6(2-3):87–129, July 1996
3. Mobasher, B., Cooley, Srivastava, J.: Automatic Personalization Based on Web Usage Mining. Communication of ACM, 43(8):142–151, August 2000
4. Garrett, J. J.: Ajax: A New Approach to Web Applications. http://adaptivepath. com/ publications/essays/archives/000385.php, February 2005, retrieved on 2006-11-13
5. Stojanovic, L., et al., D2: Framework for self-adaptive e-Government. Available as Deliverable D2, EU/IST Project FIT, http://www.fit-project.org/index.htm, 2006
6. Thomas, S.M., et al., D4: Identification of typical problems in e-Government portals. Available as Deliverable D4, EU/IST Project FIT, http://www.fit-project.org/index.htm, 2006
7. Heckmann, D., Schwartz, T., Brandherm. B., Schmitz, M., von Wilamowitz-Moellendorf, M.: GUMO - the General User Model Ontology. In Proceedings of the 10th International Conference on User Modeling, Edinburgh, Scotland, Jun 2005
8. Palmer, S.: RDF in HTML: approaches. http://infomesh.net/2002/rdfinhtml/, 2002, retrieved on 2006-12-28.

[10] JBoss Portal: http://www.jboss.org/products/jbossportal

9. Kalyanpur, A. et al.: Automatic Mapping of OWL Ontologies into Java. In Proceedings of the 16th International Conference of Software Engineering and Knowledge Engineering, pages 98–103, 2004
10. Internet Engineering Task Force: The application/json Media Type for JavaScript Object Notation (JSON). RFC 4627, 2006, http://www.ietf.org/rfc/rfc4627.txt, retrieved on 2006-12-28.
11. Euler Proof Mechanism: http://eulersharp.sourceforge.net/, retrieved on 2006-12-28.
12. Bra, P. D., Aerts, A., Smits, D., Stash, N.: AHA! version 2.0: More adaptation flexibility for authors. In Proceedings of the AACE ELearn'2002 conference, pages 240-246, Oct. 2002
13. Dolog, P., Henze, N., Nejdl, W., Sintek, M.: The Personal Reader: Personalizing and Enriching Learning Resources using Semantic Web Technologies. AH 2004: International Conference on Adaptive Hypermedia and Adaptive Web-Based Systems, 2004
14. Frasincar, F., Houben, G.: Hypermedia presentation adaptation on the semantic Web. In Proccedings of the 2nd International Conference on Adaptive Hypermedia and Adaptive Web-Based Systems (AH 2002), Malaga, Spain, 2002
15. Köberl, K.: Erfassen von Benutzerkontextinformationen mit Ajax. MSc thesis, Technische Universitaet Graz, 2006

A Semantic Web Service Oriented Framework for Adaptive Learning Environments

Stefan Dietze, Alessio Gugliotta, and John Domingue

Knowledge Media Institute
Open University
Milton Keynes, MK7 6AA, UK
{s.dietze,a.gugliotta,j.b.domingue}@open.ac.uk

Abstract. The current state of the art in supporting e-learning objectives is primarily based on providing a learner with learning content by using metadata standards. Due to this approach, several issues have to be taken into account – e. g. limited re-usability across different standards and learning contexts and high development costs. To overcome these issues, this paper describes an innovative semantic web service-oriented framework aimed at changing this data- and metadata-based paradigm to a highly dynamic service-oriented approach. Instead of providing a learner with static data, our approach is based on fulfilling learning objectives based on a dynamic supply of services. Therefore, we introduce a semantic layer architecture to abstract from existing learning data as well as process metadata standards by using Semantic Web Service (SWS) technology. Furthermore, our approach is based on abstract and reusable learning process models describing a learning process semantically as a composition of learning goals. Based on the formal semantic descriptions of learning goals as well as web services, services appropriate to achieve a specific learning goal can be selected, composed and invoked dynamically. This supports a high level of re-usability since a dynamic adaptation to different learning contexts and requirements of individual learners is achieved while utilizing standard-compliant learning applications. To illustrate the application of our approach, we describe a prototypical implementation utilizing the introduced approach based on the SWS framework WSMO.

Keywords: Semantic Web Services, Service oriented Architecture, WSMO, E-Learning, IRS III, IMS Learning Design.

1 Introduction

Current approaches to support a learning objective are fundamentally based on providing a learner with appropriate learning content – the so called learning objects. Composite learning objects contain the learning resources - the physical data assets – as well as a description of the learning process to be followed by the learner. The latter usually is based on existing metadata standards - IEEE LOM [9], ADL SCORM [1] – based on IMS Simple Sequencing - or IMS Learning Design (IMS LD) [15]. Due to the approach of allocating learning resources – whether services or data - at

E. Franconi, M. Kifer, and W. May (Eds.): ESWC 2007, LNCS 4519, pp. 701–715, 2007.

design-time of a learning process model, the actual learning context – known at runtime only – cannot be considered. This means, a new learning content package has to be developed for every different learning scenario or individual needs of specific learners. For instance, a package suiting the needs of a learner with specific preferences – e. g. his native language or technological platform - can suit only this specific requirements and cannot be reused across different learning contexts. The identified limitations (cf. [2], [16], [6]) can be summarized as follows:

L1. *Limited appropriateness and dynamic adaptability to actual learning contexts.* It is assumed that every learning objective occurs in a specific context which is defined by e. g. the preferences of the actual learner. Learning data is allocated at design-time what limits the appropriateness of the data to the actual learning context. Moreover, the use of data excludes the dynamic adaptability a priori. In parallel to data-centric approaches, analogous issues can also be observed with service-oriented approaches. However, in that case, these issues are related to the allocation of services only.

L2. *Limited reusability across different learning contexts and metadata standards.* Due to L1, for every different learning context having distinct requirements or learner needs a new learning content package has to be developed. Since metadata is described based on standard-specific specifications, an individual content package cannot be reused across different standards. Besides that, the current approach limits opportunities for reusing available learning data and service repositories.

L3. *High development costs.* Due to L1 and L2, high development costs have to be taken into account when developing standard-compliant E-Learning packages.

To overcome these issues, the approach described in this paper changes this data- and metadata-based paradigm to a dynamic service-oriented approach based on Semantic Web Service (SWS) technologies.

SWS are aimed at enabling a automatic discovery, composition and invocation of available Web services. Based on semantic descriptions of functional capabilities of available Web services, a SWS broker automatically selects and invokes Web services appropriate to achieve a given goal.

IRS-III [5], the Internet Reasoning Service, is an implementation of a SWS broker environment. It provides the representational and reasoning mechanisms, which enable the dynamic interoperability and orchestration between services as well as the mediation between their semantic concepts. IRS-III utilizes a SWS library based on the reference ontology Web Service Modelling Ontology (WSMO) [26] and the OCML representation language [7] to store semantic descriptions of Web services and knowledge domains.

WSMO is a formal ontology for describing the various aspects of services in order to enable the automation of Web service discovery, composition, mediation and invocation. The meta-model of WSMO defines four top level elements: Ontologies, Goals, Web Services and Mediators. Whereas Ontologies describe the terminology and its semantics used by Web Services, Web Service descriptions describe the capabilities and interfaces of a particular service. Moreover, Goals describe a task from a user perspective and Mediators handle data and process interoperability issues that arise when handling heterogeneous systems. As a result, we enable the automatic allocation of the adequate services at runtime – not only data – and the integration of a wide variety of learning resources – whether data or services.

The following section of the paper outlines the issues of current learning technologies which are addressed with this paper. Section 3 then describes our approach of using a SWS oriented architecture to support learning processes followed by a section describing our ontological framework. The fifth section explains a SWS oriented architecture implemented as a first prototype and the used development principles. Finally, we summarize the contributions of our work, draw a conclusion and provide an outlook to future work related to our approach.

2 Related Work

Several approaches follow the idea of using semantic Web or Web service technologies to provide dynamic as well as personalized support for learning objectives.

To quote a few examples, [16] as well as [3] are concerned with bridging learning contexts and resources by introducing semantic learning context descriptions. This allows the adaptation to different contexts based on reasoning over the provided context ontologies, but does not provide solutions for building complex adaptive learning applications by reusing distributed learning functionalities. Moreover, [16] is entirely based on IMS LD.

[4] follows the idea of using a dedicated personalization web service which makes use of semantic learning object descriptions to identify and provide appropriate learning content. Integration of several distributed learning services or service allocation at runtime is not within the scope of this approach. The related research on a Personal Reader Framework (PRF) introduced in [8], [13] and [14] allows a mediaton between different services based on a socalled "connector service". The composition of complex learning applications based on distributed services is not within the scope of the PRF.

The work described in [22], [23] utilize semantic web as well as web service technologies to enable adaptation to different learning contexts by introducing a matching mechanism to map between a context and available learning data. However, neither it considers approaches for automatic service discovery nor it is based on common standards. Hence, the reuse and automatic allocation of a variety of services or the mediation between different metadata standards is not supported. These issues apply to the idea of "Smart Spaces" for learning as well (cf. [24]).

Whereas the majority of the described approaches enable context-adaptation based on runtime allocation of learning data, all of them do not enable the automatic allocation of learning functionalities neither it does enable the integration of new functionalities based on open standards. Nevertheless, all approaches do not envisage mappings between different learning metadata standards to enable interoperability not only between learning contexts but also between platforms and metadata standards.

3 Semantic Web Service Based E-Learning Applications: Vision and Approach

This section describes our vision as well as the approach to support e-learning based on semantic web services.

3.1 Vision: Context-Adaptation Through Automatic Service Selection and Invocation

To overcome the limitations described above, we consider the automatic allocation and invocation of functionalities at runtime. A typical learning related service functionality provides the learner for instance with appropriate learning content or topic-specific discussion facilities. Learning processes are described semantically in terms of a composition of user objectives (goals) and abstract from specific data and metadata standards. When a specific learning goal has to be achieved, the most adequate functionality is selected and invoked dynamically regarding the demands and requirements of the actual specific context. This enables a highly dynamic adaptation to different learning contexts and learner needs.

This vision is radically distinctive to the current state of the art in this area, since it shifts from a data- and metadata-centric paradigm to a context-adaptive service-oriented approach. Moreover, using adequate mappings, our standard-independent process models can be translated into existing metadata standards in order to enable a reuse within existing standard-compliant runtime environments.

Addressing the limitations L1 and L2 identified in Section 0, we consequently reduce the efforts of creating learning process models (L3): one unique learning process model can adapt dynamically to different process contexts and can be translated into different process metadata standards.

3.2 Approach: Semantic Abstraction from Process Metadata, Functionalities and Data

Our approach is fundamentally based on utilizing SWS technologies to realize the following principles. To support these principles, we introduce several layers as well as a mapping between them in order to achieve a gradual abstraction (Figure 1).

1. **Abstraction from Learning Data and Functionalities.** To abstract from existing learning data and content we consider a Web Service Layer. It operates on top of the data and exposes the functionalities appropriate to fulfill specific learning objectives. This first step enables a dynamic supply of appropriate learning data to suit a specific context and objective. Web services at this layer may make use of semantic descriptions of available learning data. In order to abstract from these functionalities (Web services), we introduce an additional layer – the Semantic Web Service Layer. This layer enables the automatic selection, composition and invocation of appropriate Web services for a specific learning context. This is achieved on the basis of formal semantic, declarative descriptions of the capabilities of available services which enable the dynamic matching of service capabilities to specific user goals.

2. **Abstraction from Learning Process Metadata.** A first layer concerned with the abstraction from current learning process metadata standards is the Semantic Learning Process Model Layer. It allows the description of processes within the domain of E-Learning in terms of higher level domain concepts - e. g. learning goals, learners or learning contexts. This layer is mapped to semantic representations of current learning metadata standards in order to enable the interoperability between different standards. To achieve a further abstraction from domain specific process models – whether it is e. g. a learning process, a business process or a communication process – we consider an upper level process model layer – Semantic Process Model Layer. For instance, this layer supports the mapping between learning objectives and business objectives to support all kind of organizational processes.

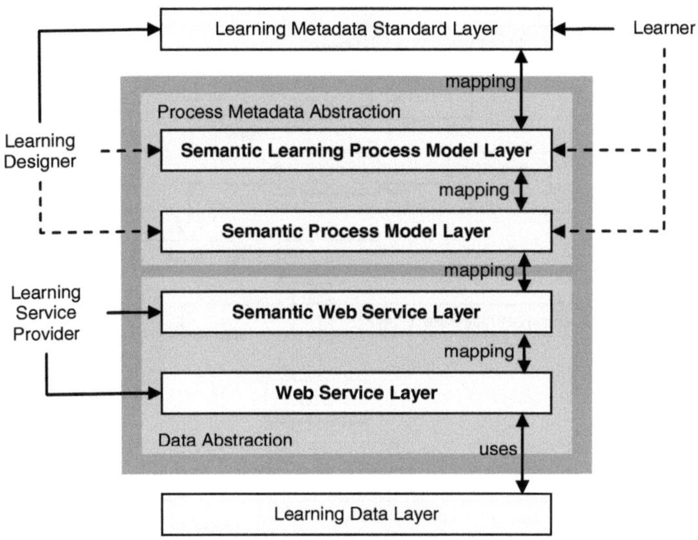

Fig. 1. Semantic layer architecture for supporting learning processes through SWS

Based on mappings between the described layers, upper level layers can utilize information at lower level layers. In particular, we consider mappings between a learning objective and a WSMO goal to enable the automatic discovery and invocation of a Web service (Web Service Layer) from, for instance, a standard-compliant learning application (Learning Application Standard Layer). As a result, a dynamic adaptation to individual demands of a learner within a specific learning context is achieved by using existing standard-compliant learning applications. It is important to note, that we explicitly consider mappings not only between multiple semantic layers but also within a specific semantic layer.

4 The Ontological Framework

This section describes an ontological framework aimed at implementing the introduced semantic layers.

4.1 Staged Ontological Mapping

To implement the described semantic layer architecture, we follow an approach of a staged ontological mapping between semantic models of a process at different levels of abstraction. Therefore, our approach considers different ontologies aimed at providing abstract semantic descriptions of data as well as processes.

The following figure gives an overview of the main ontological representations considered in our approach as well as their relationships:

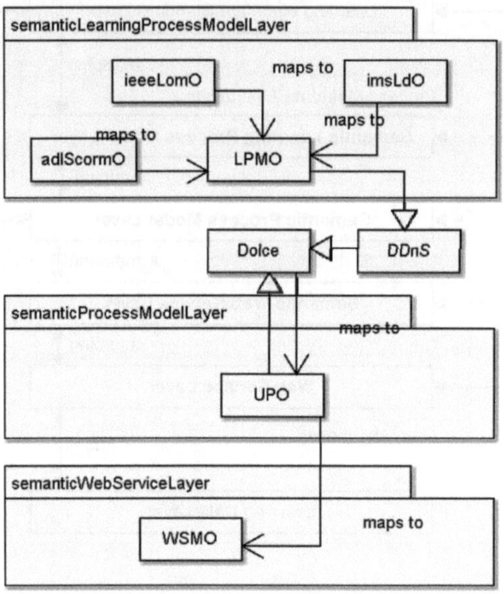

Fig. 2. Conceptual overview of proposed ontological framework

To enable mappings between different learning metadata standards, a higher level ontology is introduced, to model the learning process from a general point of view - independent from any supported platform or learning technology standard. This Learning Process Modelling Ontology (LPMO) implements the Semantic Learning Process Model Layer and is mapped to ontological representations of learning process models based on current e-learning metadata standards. Currently, representations of the following metadata standards are foreseen: adlScormO (The ADL SCORM 2004 Ontology); imsLdO (The IMS Learning Design Ontology); ieeeLomO (The IEEE LOM Ontology).

The next level in our staged mapping approach abstracts from the specialised process – e.g. learning process or business process – to a general process ontology. This is the Upper Process Ontology (UPO) which implements our Semantic Process Model Layer and is currently being developed as part of the SUPER project [25]. The UPO enables the description of a process independent from its specific purpose and can be mapped to domain specific process ontologies, e. g. the LPMO. In order to enable a high level of interoperability of our ontologies, we intend to align the LPMO as well as the UPO to the DOLCE foundational ontology [12] as well as to the DOLCE Descriptions and Situations ontology (DDnS) [11].

Furthermore, the UPO is mapped to the WSMO standard. Therefore, a gradual mapping between a standard learning application and WSMO entities is achieved based on these ontologies. It has to be highlighted, that our ontological architecture explicitly considers mappings not only between several semantic layers but also within a specific semantic layer. This enables for example the mapping of our LPMO concepts to other existing semantic descriptions of learning related concepts. Furthermore, it has to be taken into account that the proposed ontologies are currently implemented only partially, since this work is ongoing research at the moment.

4.2 Semantic Learning Process Model Layer

From an e-learning perspective, the LPMO has to be perceived as the central ontology within our architecture, since it describes the semantics of a learning process from a

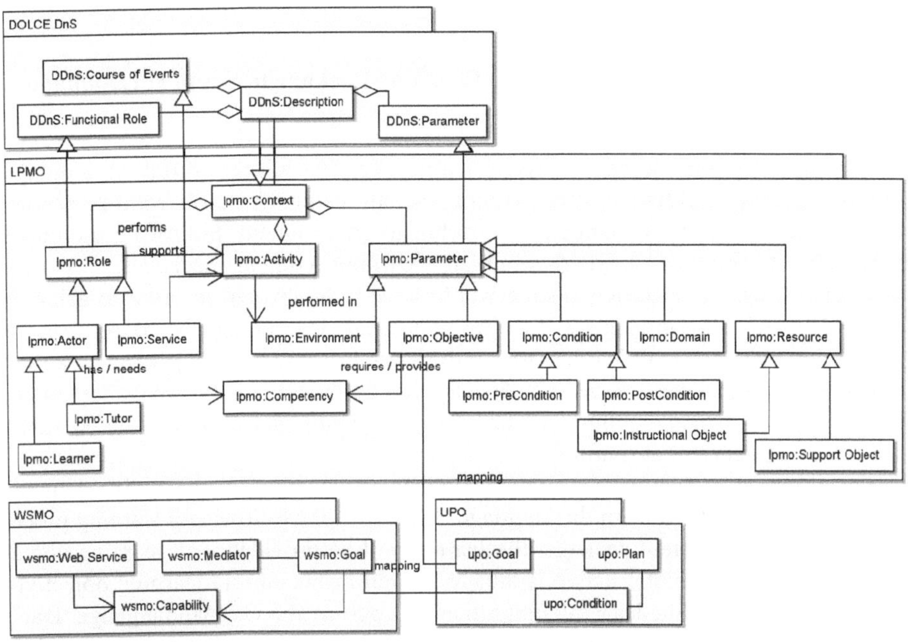

Fig. 3. Conceptual model of parts of the LPMO and key mappings to the UPO and the WSMO framework

general point of view and independent from any supported platform or learning technology standard. Figure 3 depicts an extract of the proposed LPMO containing some of its main concepts as well as some mappings to some key concepts within different semantic layers.

As shown below, a learning objective as defined in the LPMO is mapped to a upo:Goal – which represents a central concept within the Semantic Process Model Layer. This concept is furthermore mapped to the wsmo:Goal concept which represents one of the main concepts of the Semantic Web Service Layer and enables the mapping and matching of appropriate web services. Besides the proposed mappings between several semantic layers, mappings are also considered within a specific layer to enable a wide applicability of our approach. E. g. semantic concepts of our LPMO can be mapped to other existing semantic concepts representing learning-related entities within different approaches – e. g. learning process modules as defined in [19], [17].

5 A SWS Based Framework for E-Learning - Prototype Application Based on IMS Learning Design and WSMO

In order to validate the technical feasibility of the described approach, a first prototype was implemented. In this section, we describe an application based on IMS Learning Design as well as the WSMO framework. The application implements an initial use case by utilizing the semantic layers and fundamental concepts as introduced in 3.2.

5.1 Use Case: An Adaptive IMS LD Learning Package to Support Language Learning

Within our supported scenario, several learners request to learn different languages: English, German and Italian. It is assumed, that all learners have different preferences – e.g. their spoken native language or technical environment. Following the current approach of creating standard-compliant learning content packages, for every individual learner a specific package would have to be created in order to achieve a high level of appropriateness to the individual learner needs. Based on our application, we enable all learners to use the same learning content package – an IMS LD compliant content package. This is achieved by enabling a dynamic adaptation to the individual learner requirements based on a dynamic selection and invocation of semantic web services at runtime.

For example, a learner is authenticated as a person with the native language "English" and wants to learn the language "German". By following a learning process as defined in the content package, the learner will be provided with learning content appropriate to his specific native language as well as his current learning objective – an English-based online learning unit aimed at teaching the German language. Due to the dynamic adaptation at runtime, the standard-compliant learning process could suit all kind of different individual requirements. Since our approach is fundamentally based on the principles described in section 0, this scenario could be extended in the future to achieve a dynamic adaptation to all kind of different learning contexts or learner requirements.

5.2 A SWS Oriented Architecture

To implement a software architecture aimed at supporting our semantic web service based approach, several dedicated software layers have to be provided. The following figure illustrates the SWS oriented architecture for e-learning which is utilized in the prototype application:

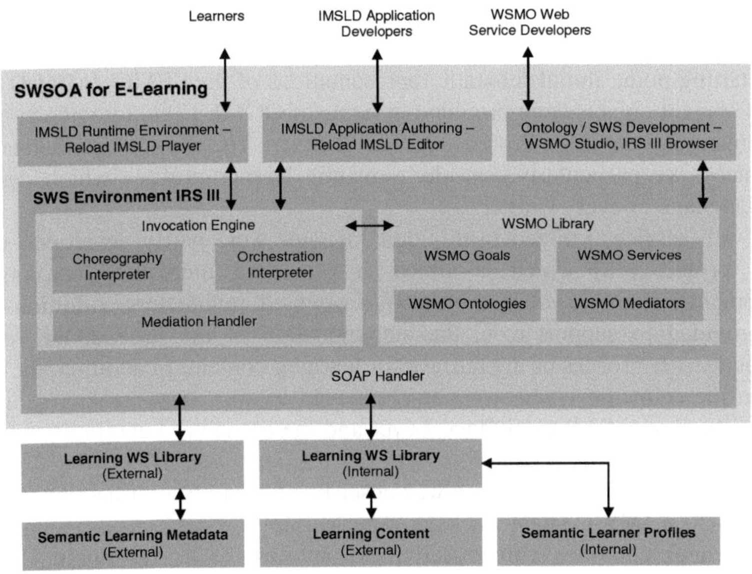

Fig. 4. SWS-based software architecture as utilized in the prototype application

The architecture depicted above is fundamentally based on the semantic layers described in section 3.2. In the figure, a SWS broker based on the WSMO framework serves as foundation for a dynamic selection, composition and invocation of web services. Services are distributed across different external repositories and provide functionalities based on existing learning data and metadata repositories. In addition, several user interfaces for developing and presenting learning applications as well as for developing formal semantic descriptions of web services are utilized. Our current implementation makes use of standard runtime environments and implements a SWS oriented architecture based on these infrastructural components. For WSMO runtime processing as well as development environment for WSMO, the SWS broker IRS III

[5] is used, whereas editing and runtime processing of IMS LD is supported by the Reload Learning Design Editor and Player [21].

5.3 Implementation Approach

To support the described scenario based on an SWS-based approach the following items had to be provided:

- An IMS LD-compliant content package describing the learning activities and objectives
- Web services able to achieve the objectives
- Semantic descriptions of available services based on WSMO
- Ontologies implementing the semantic layers as described above
- Mappings between the semantic layers as well as the IMS LD standard.

As starting point, initial semantic representations of the LPMO, IMS LD as well as utilized content objects were provided in terms of OCML [7] ontologies to implement the Semantic Learning Process Model Layer. To support individual learner preferences, we particularly consider semantic learner profiles which describe the native language of every learner.

The web services utilized in this demonstrator were partly developed within the LUISA project [18] which is aimed at providing innovative learning content management technologies based on a SWS oriented architecture. Additional services were provided to support e. g. the authentication of the learner, the retrieval of semantic learner profiles or the retrieval of learning content. In addition, the mappings between the semantic layers were implemented as web services – e. g. a mapping between the Semantic Web Service Layer and the Semantic Learning Process Model Layer.

Besides that, a learning process was described based on the IMS LD standard and included into a IMS Content Package. The learning process in our example defines some learning activities (imsld:Activities) as well as corresponding sequencing information. Instead of grounding these activities to static learning data, no static resources were associated with this learning process. In contrast, only references to the described WSMO-Goals were associated with every learning activity within the IMS LD metadata. At runtime, a WSMO-Goal then dynamically invokes a WSMO-web service which shows the appropriate capabilities to achieve the specific goal. The mapping between the IMS LD metadata and appropriate WSMO-Goals was achieved by associating IMS LD learning activities with HTTP-references to a web applet enabling to request the achievement of a specific WSMO-goal from the SWS broker.

5.4 Ontological Mappings

As described above, we created mappings between the initial implementations of semantic representations of the IMS LD standard, the LPMO and WSMO. This includes e. g. a mapping between the lpmo:Objective and the objective description used within the IMS LD metadata (imsld:Objective). Furthermore, semantic learning object descriptions based on the LPMO are mapped to learning content provided by the Open Learn Project [20] based on an initial ontology representing this specific learning content objects. Besides that, a web service implements the mapping between the language of a content object (ol:Language) and the native language of the learner (lpmo:Language).

It has to be highlighted, that our current prototype does not implement the mapping to the UPO. Instead of that, our learning process was mapped directly to WSMO, since the UPO currently is not supported by any software RTE. The following figure depicts the main ontological mappings as implemented in our prototype:

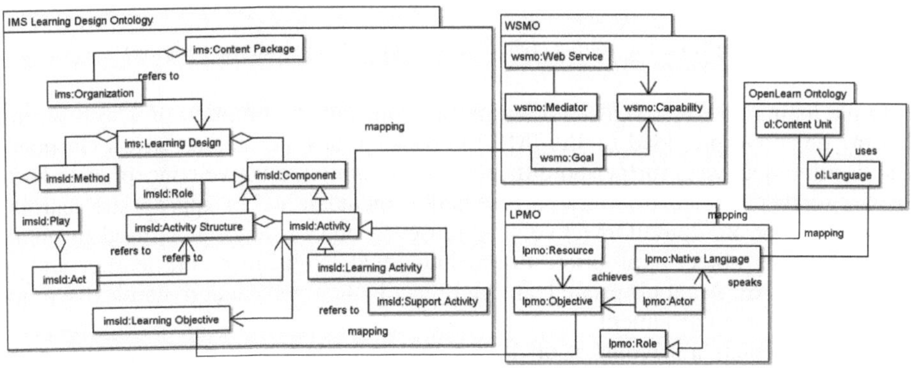

Fig. 5. Ontological mappings implemented and utilized in the prototype

5.5 Dynamic Adaptation at Runtime

In our example scenario, several web services are invoked to retrieve semantic learning metadata, learner profile descriptions and e-learning content as well as to map between different semantic concepts. An initial service first authenticates the learner and retrieves the semantic learner profile description. After providing an individual objective, our application dynamically selects and invokes semantic web services appropriate to the individual learner preferences and his specific objectives – as defined in the IMS LD metadata.

For example, a learner is authenticated as an English-speaking person (lpmo:Language=English) and uses an IMSLD package to learn the language German. Therefore, an imsld:Activity with the imsld:Objective “Learn German” is mapped to a WSMO-goal to achieve this learning activity. This triggers the selection, orchestration and invocation of different web services to achieve the described mappings and to retrieve appropriate learning content. The following OCML code listing shows the partial capability description of a web service able to provide learning content to teach the language German:

```
(DEF-CLASS ACHIEVE-IMSLD-OBJECTIVE-GERMAN-WS-CAPABILITY
   (CAPABILITY)
   ?CAPABILITY
   ((USED-MEDIATOR :VALUE ACHIEVE IMSLD-OBJECTIVE-GERMAN-MED)
   (HAS-ASSUMPTION
          :VALUE
          (KAPPA
              (?WEB-SERVICE) (= (WSMO-ROLE-VALUE ?WEB-
              SERVICE 'HAS-IMSLD-OBJECTIVE)"Learn
              German")))
    (HAS-NON-FUNCTIONAL-PROPERTIES :VALUE ACHIEVE-IMSLD-
    OBJECTIVE-GERMAN-WS-CAPABILITY-NON-FUNCTIONAL-
    PROPERTIES)))
```

Listing 1. Partial source code of a web service capability description

In the listing above, a WSMO description defines the assumption of a web service that the objective provided by the IMS LD package has the Value "Learn German". The imsld-Objective is furthermore mapped to the lpmo:Objective concept in order to invoke another service for retrieving semantic metadata of an appropriate learning object based on the lpmo:Objective. The retrieved object identifier is used to obtain an Open Learn object appropriate to the individual language of the learner and its current objective. An appropriate learning object is then presented dynamically in the IMS LD runtime environment.

Figure 6 depicts a screenshot of the Reload IMS LD Player while presenting the developed standard-compliant IMS Content Package and dynamically invoking SWS appropriate to fulfill the given learning objective "Learn German".

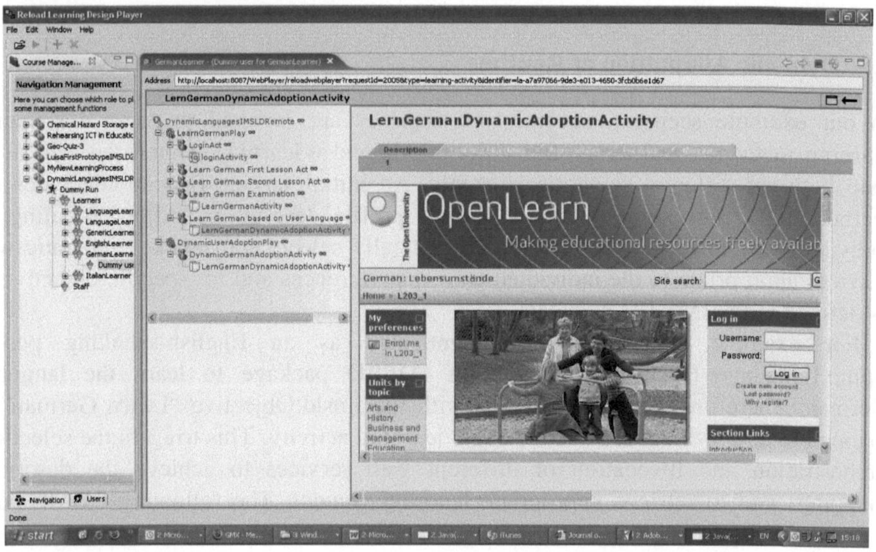

Fig. 6. Reload IMS Learning Design Player while dynamically invoking SWS for e-learning

The current prototype implements the basic approach of a standard-compliant SWSOA for e-learning as described here, and will be extended in the future in order to address existing limitations. Furthermore, the described approach was already applied to another e-learning standard - ADL SCORM 2004.

6 Conclusions

Our approach - the support of learning objectives based on a dynamic invocation of SWS at runtime of a learning process model - follows an innovative approach and is distinctive to the current state of the art in this area. By using semantic web as well as SWS technology this approach overcomes the limitations described in Section 0 and provides a high level of openness to reuse existing service and data repositories. Based on existing standards – SWS as well as E-Learning standards - new learning data as well as application functionalities can be integrated by our SWS oriented architecture. Furthermore, a high level of standard-compliancy and re-usability within existing runtime environments is supported. In particular, the following contributions should be taken into account:

- Dynamic adaptation to specific learning contexts at runtime
- Reuse and integration of available learning resources – services and data
- Automatic allocation of learning resources based on comprehensive semantics
- High reusability across learning contexts
- Platform- and standard-independence
- Decrease of development costs

Furthermore, our approach can lead to contributions for developing domain-specific SWS applications in general, since we consider mappings between the WSMO standard and higher-level process modeling as well as learning process modeling standards. This enables the development of complex SWS based applications and therefore several benefits are envisaged:

- Re-usability of SWS based applications based on semantic mappings with existing process metadata standards
- Utilization of established standard-compliant software environments to implement complex SWS based architectures

Since our framework is developed only in parts currently, next steps have to be concerned with the implementation of complete ontological representations of the introduced semantic layers as well as of current E-Learning metadata standards and their mappings. For example, currently the Semantic Process Model Layer is not used and semantic mappings between the Learning Process Model Ontology and available process metadata standards are only developed in extracts. Nevertheless, the availability of appropriate Web services aimed at supporting specific process objectives has to be perceived as an important prerequisite for developing SWS based applications. To provide more valid quantifications of the expected benefits, further case studies are needed to illustrate the formalized measurements introduced in the sections above. Besides that, future work could also be concerned with the mapping of semantic process models across different process dimensions – e. g. business processes or learning processes to enable a complete integration of a SWSOA in an organizational process environment.

References

1. Advanced Distributed Learning (ADL) SCORM 2004 Specification (http://www.adlnet.org).
2. Amorim, R. R., Lama, M., Sánchez, E., Riera, A., & Vila, X. A. (2006). A Learning Design Ontology based on the IMS Specification. Journal of Educational Technology & Society, 9 (1), 38-57.
3. M. Baldoni, C. Baroglio, V. Patti, and L. Torasso. Using a rational agent in an adaptive web-based tutoring system. In Proc. of the Workshop on Adaptive Systems for Web-Based Education, 2nd Int. Conf. on Adaptive Hypermedia and Adaptive Web-based Systems, pages 43-55, Malaga, Spain, 2002.
4. M. Baldoni, C. Baroglio, I. Brunkhorst, N. Henze, E. Marengo and V. Patti: A Personalization Service for Curriculum Planning. ABIS 2006 - 14th Workshop on Adaptivity and User Modeling in Interactive Systems, Hildesheim, October 9-11 2006.
5. Cabral, L., Domingue, J., Galizia, S., Gugliotta, A., Norton, B., Tanasescu, V., Pedrinaci, C.: IRS-III: A Broker for Semantic Web Services based Applications. In proceedings of the 5th International Semantic Web Conference (ISWC 2006), Athens, USA (2006).
6. Collis, B. and Strijker, A. (2004). Technology and Human Issues in Reusing Learning Objects. Journal of Interactive Media in Education, 2004 (4). Special Issue on the Educational Semantic Web [www- jime.open.ac.uk/2004/4].
7. J. Domingue, E. Motta and O. Corcho Garcia (1999). Knowledge Modelling in WebOnto and OCML: A User Guide, available from: http://kmi.open.ac.uk/projects/webonto/user_guide.2.4.pdf.
8. Dolog P., Henze, N., Nejdl, W., Sintek, M., Personalization in Distributed elearning Environments, In Proc. Of WWW2004 – The 13th international World Wide Web Conference, 2004.
9. Duval, E. (2002). 1484.12.1 IEEE Standard for learning Object Metadata, IEEE Learning Technology Standards Committee, http://ltsc.ieee.org/wg12/.
10. Fischer, G. and Ostwald, J. (2001). Knowledge Management: Problems, Promises, Realities, and Challenges, IEEE Intelligent Systems, 16-1(60-72). http://citeseer.ist.psu.edu/489331.html.
11. Gangemi, A., and Mika, P. Understanding the Semantic Web through Descriptions and Situations. In Meersman, R.; Tari, Z.; and et al., D. S., eds., Proceedings of the On The Move Federated Conferences (OTM'03), LNCS. Springer Verlag.
12. Gangemi, A., Guarino, N., Masolo, C., Oltramari, A., Schneider, L.: Sweetening Ontologies with DOLCE. In: A. Gómez-Pérez , V. Richard Benjamins (Eds.) Knowledge Engineering and Knowledge Management. Ontologies and the Semantic Web: 13th International Conference, EKAW 2002, Siguenza, Spain, October 1-4, 2002.
13. Henze, N., Personalized e-Learning in the Semantic Web. Extended version of 4. International Journal of Emerging Technologies in Learning (iJET), Vol. 1, No. 1 (2006).
14. Nicola Henze, Peter Dolog, and Wolfgang Nejdl: Reasoning and Ontologies for Personalized E-Learning. Educational Technology & Society, 2004, Vol. 7, Issue 4.
15. IMS Learning Design Specification (http://www.imsglobal.org).

16. Knight, C., Gašević, D., & Richards, G. (2006). An Ontology-Based Framework for Bridging Learning Design and Learning Content. Journal of Educational Technology & Society, 9 (1), 23-37.
17. Koper, R. (2004). Use of the Semantic Web to Solve Some Basic Problems in Education: Increase Flexible, Distributed Lifelong Learning, Decrease Teacher's Workload. Journal of Interactive Media in Education, 2004 (6). Special Issue on the Educational Semantic Web. ISSN:1365-893X [www-jime.open.ac.uk/2004/6].
18. LUISA Project - Learning Content Management System Using Innovative Semantic Web Services Architecture (http://www.luisa-project.eu/www/).
19. Naeve, A., Sicilia, M. A. (2006), Learning Processes and processing learning: from organizational needs to learning designs, ADALE workshop at the Adaptive Hypermedia conference, Dublin, June 20-23, 2006.
20. Open Learn Project: Online Educational Resources (http://openlearn.open.ac.uk/).
21. Reload Project (http://www.reload.ac.uk/).
22. Schmidt, A., Winterhalter, C. User Context Aware Delivery of E-Learning Material: Approach and Architecture, Journal of Universal Computer Science (JUCS) vol.10, no.1, January 2004.
23. Schmidt, A., Bridging the Gap Between E-Learning and Knowledge Management with Context-Aware Corporate Learning (Extended Version), In: Professional Knowledge Management (WM 2005), Post Proceedings, Springer, 2005.
24. Simon, B., Dolog., P., Miklós, Z., Olmedilla, D. and Sintek, M. (2004). Conceptualising Smart Spaces for Learning. Journal of Interactive Media in Education, 2004 (9). Special Issue on the Educational Semantic Web. ISSN:1365-893X [http://www-jime.open.ac.uk/2004/9].
25. SUPER – Semantics Utilized for Process Management within and between Enterprises (http://www.ip-super.org/).
26. WSMO Working Group, D2v1.0: Web Service Modeling Ontology (WSMO). WSMO Working Draft, (2004). (http://www.wsmo.org/2004/d2/v1.0/).

Semantic Composition of Lecture Subparts for a Personalized e-Learning

Naouel Karam, Serge Linckels, and Christoph Meinel

HPI, University of Potsdam, Germany
{naouel.karam,serge.linckels,christoph.meinel}@hpi.uni-potsdam.de

Abstract. In this paper we propose an algorithm for personalized learning based on a user's query and a repository of lecture subparts —i.e., learning objects— both are described in a subset of OWL-DL. It works in two steps. First, it retrieves lecture subparts that cover as much as possible the user's query. The solution is based on the concept covering problem for which we present a modified algorithm. Second, an appropriate sequence of lecture subparts is generated. Indeed, the different lecture subparts are only reachable when a given prerequisite is fulfilled, i.e., the learner must have a minimal background knowledge to be able to assimilate the requested learning object. Therefore, our algorithm takes into account the user's knowledge to generate a personalized lecture composition and suggests a flow of learning objects to the user.

1 Introduction

The Semantic Web aims to provide computer-processable information on the Internet. It extends the current Web through the use of standards, markup languages and related processing tools. The recent development of Semantic Web technologies has a great impact on e-Learning. Indeed, the semantic annotation of learning resources for the aim of automated retrieval and sequencing is fully in the stream of the Semantic Web. Recent efforts on standardization led to the definition of *Learning Objects* (LOs) as reusable units of educational content that can be sequenced into larger units to allow personalized learning [17,11].

The availability of LOs in electronic form increases dramatically. Hence, the task of finding the appropriate information becomes more and more awkward and time consuming.

In this paper we propose the algorithm *LectureComposer* for personalized learning that takes as input a user's query and a repository of LOs, both described in a subset of OWL-DL. The algorithm works in two steps. First, it retrieves LOs that cover as much as possible the user's query. Note that these LOs can belong to different original lectures. Second, it composes a sequence of LOs according to required prerequisites. The composition takes into account the user's knowledge about the subject in order to propose the best sequence for the retrieved LOs.

E. Franconi, M. Kifer, and W. May (Eds.): ESWC 2007, LNCS 4519, pp. 716–728, 2007.

For the first issue, we propose to use the concept covering problem, recently introduced for a subset of OWL-DL [9]. It is stated as follows:

> *Given an ontology $\mathcal{T}$ and a query description Q, find a combination of concepts from $\mathcal{T}$ that contains as much as possible of common information with Q and as less as possible of extra information w.r.t. Q.*

In our solution the concept covering problem allows to select a subset of LOs from the repository that covers as much as possible the user's query.

The work presented here has been developed in the context of the Web University project [2], which aims at exploring novel internet- and IT-technologies in order to enhance university teaching and research. Our solution can be used by learners to dynamically compose a personalized lecture according to their needs and according to their current knowledge about the subject. It is able to identify the missing information in the yielded results—i.e., the data that is not available in the repository—and the required parts that are needed to fulfill the user's background knowledge. Our solution is particularly interesting for education in a self-directed learning environment, where it fosters autonomous and exploratory learning. It will be integrated to the *e-Librarian Service* "CHESt" [14] which allows students to enter questions in natural language, and yields only semantically relevant multimedia resources to the students needs.

The remainder of the paper is organized as follows. Section 2 presents an overview of Description Logics and recalls the definition of the concept covering problem. Section 3 describes the formalization of the LO composition problem. Our proposed algorithm is detailed in section 4. An illustration of the algorithm is given in section 5. Section 6 concludes the paper and outlines future work.

2 Description Logics

Description Logics (DLs) [5] are a family of knowledge representation formalisms that allow to represent the knowledge of an application domain in a structured way and to reason about this knowledge. DLs play an important role in the definition of languages for the Semantic Web [3]. Indeed, the OWL sub-language OWL-DL is based on DLs.

2.1 Preliminaries

In DLs, the conceptual knowledge of an application domain is represented in terms of *concepts* (unary predicates) such as Network and Protocol, and *roles* (binary predicates) such as hasProtocol and hasTopology. Concepts denote sets of individuals and roles denote binary relations between individuals.

Based on basic concept and role names, complex concept descriptions are built inductively using concept constructors. The different DLs languages distinguish themselves by the kind of constructs they allow. Examples of concept constructs are the following:

- top-concept $\top$ and bottom-concept $\bot$ denote all the individuals in the domain and the empty set respectively,
- conjunction $\sqcap$, e.g., $\mathsf{StarTopology} \sqcap \mathsf{RingTopology}$ denotes topologies that are both, star and ring topology (mixed topology),
- value restriction $\forall r.C$, e.g., $\mathsf{Network} \sqcap \forall \mathsf{hasTopology.StarTopology}$ denotes networks that have only a star topology,
- existential restriction $\exists r.C$, e.g., $\mathsf{Network} \sqcap \exists \mathsf{hasProtocol.TCPIP}$ denotes networks that support the TCP/IP protocol.

Concept descriptions are used to specify terminologies that define the intentional knowledge of an application domain. Terminologies are composed of *inclusion assertions* and *definitions.* The first impose necessary conditions for an individual to belong to a concept, e.g., to impose that Internet is, among other things, a network that supports only the protocol TCP/IP, one can use the inclusion assertion: $\mathsf{Internet} \sqsubseteq \mathsf{Network} \sqcap \forall \mathsf{hasProtocol.TCPIP}$. Definitions allow to give a meaningful name to concept descriptions, e.g., to define that a home network is a LAN based only on a star topology one can write: $\mathsf{HomeNetwork} \doteq \mathsf{LAN} \sqcap \forall \mathsf{hasTopology.StarTopology}$.

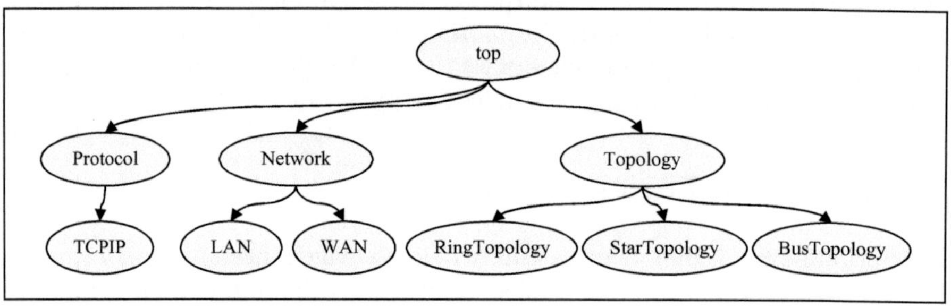

Fig. 1. Sample of a concept hierarchy about networking

DL systems provide various reasoning services. One of the most important is *subsumption*, which is the basis of the concept hierarchy (see figure 1). Formally, a concept description D subsumes a concept description C (noted $C \sqsubseteq D$) if every interpretation assigns to C a set of individuals included in the one assigned to D.

By opposition to early standard inferences in DLs, newly introduced inferences are called non standard inferences. Among these inferences we use the *least common subsumer* [4], which stands for the least concept description (w.r.t. subsumption) that subsumes a given set of concept descriptions.

Another non standard inference is the difference operation. Introduced in [16], it allows to remove from a given description all the information contained in another description. In some DLs, the difference may contain descriptions which are not semantically equivalent. Teege defines necessary conditions for a DL to have a semantically unique difference. Those DLs are said with structural subsumption. Among others, the language $\mathcal{EL}$—which allows for conjunction ($\sqcap$), existential restriction ($\exists r.C$) and the top concept ($\top$)—satisfies this property. We use this language in the context of our project. In DLs with structural subsumption, the subsumption test can be reduced to the test of inclusion between clause[1] sets. See [16] for further details.

This definition of difference requires that the second argument subsumes the first one. However, the difference $C - D$ between two incomparable descriptions C and D can be given by constructing the least common subsumer of C and D: $C - D = C - lcs(C, D)$.

2.2 The Concept Covering Problem

The concept covering problem [9] defines a cover of a concept C w.r.t. a terminology $\mathcal{T}$ as being the conjunction of some defined concepts in $\mathcal{T}$ that share some information with Q. Based on two non standard inferences in DLs—i.e., the least common subsumer (lcs), and the difference operation—a cover is formally defined as follows:

Definition 1. *Let $\mathcal{L}$ be a DL with structural subsumption, $\mathcal{T}$ be an $\mathcal{L}$-terminology and $S_\mathcal{T} = \{S_i, i \in [1, n]\}$ the set of concept definitions occurring in $\mathcal{T}$. A cover of a $\mathcal{L}$-concept description $Q \not\equiv \bot$ using the terminology $\mathcal{T}$ is a conjunction E of some names S_i from $\mathcal{T}$ such that $Q - lcs(Q, E) \not\equiv Q$.*

The best cover is defined based on the remaining information in the query (called *Rest*) and in the cover (called *Miss*).

Definition 2. *Let Q be an $\mathcal{L}$-concept description and E a cover of Q using $\mathcal{T}$. The rest of Q w.r.t. to E, written $Rest_E(Q)$ is defined as follows: $Rest_E(Q) \doteq Q - lcs_\mathcal{T}(Q, E)$.*

The missing information of Q w.r.t. E written $Miss_E(Q)$ is defined as follows: $Miss_E(Q) \doteq E - lcs_\mathcal{T}(Q, E)$.

The best cover is the one with the smallest Rest and Miss.

Definition 3. *A concept description E is called a best cover of Q using a terminology $\mathcal{T}$ iff:*

- *E is a cover of Q using $\mathcal{T}$, and*
- *there does not exists a cover E' of Q using $\mathcal{T}$ such that $(|Rest_{E'}(Q)|, |Miss_{E'}(Q)|) < (|Rest_E(Q)|, |Miss_E(Q)|)$, where $Rest_E(Q) = Q - lcs_\mathcal{T}(Q, E)$, $Miss_E(Q) = E - lcs_\mathcal{T}(Q, E)$, and $<$ stands for the lexicographic order.*

An algorithm to compute the best cover based on hypergraphs theory was introduced in [9]. It is defined for DLs with structural subsumption.

[1] A clause is a term of conjunction that cannot be decomposed into the conjunction of other terms.

3 The LO Composition Problem

3.1 The Notions of Learning Object and Composition Flow

We suppose that lectures (or other educational content) can be split into subparts. Each subpart is an educational entity that is about a precise subject in the lecture. We call such a subpart a Learning Object (LO). For example, we split a 90 minutes lecture about "Internetworking" that contains 80 slides into 27 LOs.

Related projects like [7] search for a set of LOs w.r.t. a user's query that is reachable with the user's knowledge. However, if the user's knowledge is not sufficient to reach the requested LO, no result is returned at all. Our solution returns a result even if the user's knowledge is not sufficient. In that case, the missing knowledge is identified and added to the proposed composition flow.

The problem is formalized as follows. Given a learning request Q and a repository of learning objects $\{LO_1, ..., LO_n\}$, find a composition of LOs that covers the user's query as much as possible.

Some LOs may require prerequisites. This means that the user's background knowledge must satisfy the prerequisites before he can reach the requested LOs. In our solution, the LOs are organized in a way that the user can acquire such required knowledge by reading other LOs. Those complementary LOs are detected by the system and added to the resulting composition flow.

3.2 Computing the Lecture Cover

The user's query and his background knowledge are denoted Q and $\mathcal{BK}$ respectively. The knowledge offered by a learning object LO_i and the prerequisites required to reach that LO are denoted $\mathcal{LO}_i$ and $\mathcal{PR}_i$ respectively. If the user has no knowledge about the subject, then $\mathcal{BK} = \top$. In the same way, when no prerequisites are needed to reach a LO_i, then $\mathcal{PR}_i = \top$.

For the sake of clarity, we use the following notations. Given a set of LOs $S = \{LO_1, ..., LO_n\}$:

- $\mathcal{LO}_S$ denotes the knowledge offered by all the LOs in S, i.e., the conjunction $\mathcal{LO}_1 \sqcap ... \sqcap \mathcal{LO}_n$,
- $\mathcal{PR}_S$ denotes the prerequisites needed to access all the LOs in S, i.e., the conjunction $\mathcal{PR}_1 \sqcap ... \sqcap \mathcal{PR}_n$.

Self-contained Set of LOs. A self-contained set of LOs w.r.t. a background knowledge $\mathcal{BK}$ must satisfy the following conditions:

- at least one LO is reachable with the user's background knowledge $\mathcal{BK}$, and
- every remaining LO must be reachable with the background knowledge $\mathcal{BK}$ augmented by the knowledge offered by some other LOs.

Formally, a self-contained set against a background knowledge is defined as follows.

Definition 4. *Given a set of LOs* $S = \{LO_1, ..., LO_n\}$, *S is a self-contained set against a background knowledge* $\mathcal{BK}$ *if:*

- *there exists at least one* $LO_i, 1 \le i \le n$ *such that* $\mathcal{BK} \sqsubseteq \mathcal{PR}_i$, *and*
- $\forall 1 \le i \le n$, *if* $\mathcal{BK} \not\sqsubseteq \mathcal{PR}_i$ *then there exists a set* $S_p \subseteq S$ *such that* $\mathcal{BK} \sqcap \mathcal{LO}_{S_p} \sqsubseteq \mathcal{PR}_i$.

Lecture Cover. We define a lecture cover according to a user's query as a self-contained set against the user's background knowledge that shares some information with the query.

Definition 5. *Let* $\mathcal{L}$ *be a DL with structural subsumption, Q a* $\mathcal{L}$*-concept description,* $S = \{LO_i, i \in [1,k]\}$ *a set of LOs and* $\mathcal{T} = \{\mathcal{LO}_i, i \in [1,k]\}$ *a* $\mathcal{L}$*-terminology describing the knowledge offered by S. A lecture cover* S_c *is a set of LOs* $S_c \subseteq S$ *such that:*

- *S is a self-contained set against* $\mathcal{BK}$, *and*
- $Q - lcs_{\mathcal{T}}(\mathcal{LO}_{S_c}, Q) \not\equiv Q$.

The best lecture cover is the one with the minimal non-covered part in the query.

Definition 6. *A best lecture cover of a query Q over a set of LOs S is a set* $S_c \subseteq S$ *such that:*

- S_c *is a lecture cover of Q, and*
- *there exists no lecture cover* S'_c *of Q using S such that* $|Q - lcs_{\mathcal{T}}(\mathcal{LO}_{S'_c}, Q)| < |Q - lcs_{\mathcal{T}}(\mathcal{LO}_{S_c}, Q)|$, *where* $<$ *is the lexicographic order from the definition 3.*

3.3 Computing the Flow

Once the lecture cover S_c is computed, the identified LOs are assembled into an appropriated sequence, called the *composition flow*. The method is the following:

- at each step:
  - compute the set of LOs reachable with the user's background knowledge ($\mathcal{BK}$), noted S_r, with $S_r \subseteq S_c$,
  - add the set S_r to the composition flow,
  - remove S_r from S_c,
  - update the user's knowledge ($\mathcal{BK}'$) with the knowledge of the LOs in S_r: $\mathcal{BK}' = \mathcal{BK} \sqcap \mathcal{LO}_{S_r}$,
- the process stops when no more LO remains in the lecture cover S_c.

Remark 1. We suppose that no cycle can occur in a self-contained set S. We define the notion of cycle in a set of LOs S as follows. Let LO_1 and LO_2 be LOs in S. We say that LO_1 directly *requires* LO_2 if $\mathcal{LO}_2 \sqcap C \sqsubseteq \mathcal{PR}_1$, where C is some concept description. We call "requires" the transitive closure of the relation "directly requires". Then, S contains a cycle iff there exists a LO in S that requires itself.

4 An Algorithm for Lecture Composition

Our algorithm to solve the LO composition problem is called *LectureComposer*. It is based on the algorithm *ComputeBCov* proposed in [9] for solving the best covering problem.

4.1 The Best Covering Algorithm

The problem of computing the best covers of a concept is reduced to searching for transversals with minimal cost of a hypergraph constructed from $\mathcal{T}$ and Q. Before describing the process we recall the definition of a hypergraph and a transversal.

Definition 7. *A hypergraph $\mathcal{H}$ is a pair (Σ, Γ) of a finite set $\Sigma = \{V_1, ..., V_n\}$ and a set Γ of subsets of Σ. The elements of Σ are called vertices, and the elements of Γ are called edges.*

A transversal of a hypergraph is a subset that hits all the edges of the hypergraph. Formally, it is defined as follows.

Definition 8. *A set $T \subset \Sigma$ is a transversal of $\mathcal{H}$ if for each $\epsilon \in \Gamma$, $T \cap \epsilon \neq \emptyset$. A transversal is minimal if no proper subset T' of T is a transversal.*

According to [9], we build the hypergraph $\mathcal{H}_{SQ}$ corresponding to a concept Q and a set of LOs $S = \{LO_i, i \in [1, k]\}$ as follows:

- each concept name $\mathcal{LO}_i$, $i \in [1, k]$ is associated with a vertex $V_{\mathcal{LO}_i}$, and
- each clause $A_i \in Q$ is associated to an edge e_{A_i} with $e_{A_i} = \{V_{\mathcal{LO}_i} \mid A_i \in_{\equiv} lcs(Q, \mathcal{LO}_i)\}$, where $\in_{\equiv}$ stands for membership modulo equivalence[2] of clauses.

The algorithm follows a classical approach for computing the minimal transversals with some improvements. It works in n steps, where n is the number of edges in the hypergraph. Starting from an empty set of transversals, it explores each edge of the hypergraph—one edge in each step—and generates a set of candidate transversals by computing all the possible unions between the candidates generated in the previous step and each vertex in the considered edge. At each step, the non-minimal candidate transversals are pruned. Heuristics and combinatorial optimization techniques [6] are used to discard non-minimal transversals at early steps.

As our definition of the best cover does not take into account a preference criteria between best covers, it is not necessary to compute all the transversals of the hypergraph. Thus, we adopt a deep first strategy for computing transversals as described in [13]. The idea is to compute a transversal of the partial hypergraph and then add the next hyperedge to it. Instead of computing all the transversals that follow, pick only one and add the next hyperedge. When no

[2] Two concept descriptions C and D are called *equivalent* ($C \equiv D$) if $C^{\mathcal{I}} = D^{\mathcal{I}}$ for all interpretations $\mathcal{I}$.

more hyperedge remains, the algorithm backtracks to the previous level and picks the next transversal, etc. In this way the first transversal is displayed quickly. As we need only one transversal, the process can stop after the computation of the first transversal. We call the modified algorithm *ComputeFirstBCov*.

In summary, our algorithm computes from all possible best concept covers only the first one that it finds. The advantage is that a solution—i.e., a best cover—is found quickly, because not all combinations in the hypergraph have to be explored.

4.2 The Composition Algorithm

Our algorithm (depicted in figure 2) takes as input a query Q and a set of LOs S. First, the best cover of Q w.r.t. S is computed (line 1). As explained in section 4.1, this corresponds to identifying the LOs that cover best the user's query. The result of the cover is the first transversal Tr of the corresponding hypergraph $\mathcal{H}_{SQ}$.

Require: a query Q, a set of LOs $S = \{LO_i, i \in [1, k]\}$.
Ensure: Tr
1: $Tr = ComputeFirstBCov(\mathcal{S}, Q)$
2: $P = \mathcal{PR}_{Tr} - lcs_{\mathcal{T}}(\mathcal{BK} \sqcap \mathcal{LO}_{Tr}, \mathcal{PR}_{Tr})$
3: **while** $P \not\equiv \top$ **do**
4:     **for** each edge $E \in \Gamma_P$ **do**
5:         $Tr \leftarrow$ the next transversal obtained by adding E to Γ.
6:     **end for**
7:     $P = \mathcal{PR}_{Tr} - lcs_{\mathcal{T}}(\mathcal{BK} \sqcap \mathcal{LO}_{Tr}, \mathcal{PR}_{Tr})$
8: **end while**

Fig. 2. The algorithm LectureComposer

Second, the prerequisites—to reach the LOs in the lecture cover—are updated (line 2). All the prerequisites that are not covered by the user's background knowledge or by the knowledge given by other LOs in Tr are identified.

Third, the best cover for the required prerequisites is computed. Technically, the hyperedges of each clause in the prerequisites' definitions are added incrementally following the deep-first strategy, and the transversal of the obtained hypergraph is computed (lines 4 – 6). The notation Γ_P is used to denote the set of hyperedges corresponding to the clauses of the concept description P.

Again the prerequisites are updated as described above (line 7).

The process is repeated until no more uncovered prerequisites remain (line 3).

We ensure that the algorithm terminates because all the prerequisites are in the repository and because no cycle can occur (see remark 1).

The next theorem proves that the algorithm LectureComposer returns a transversal Tr that corresponds to the best composed lecture cover of the query Q over a set of LOs S.

Theorem 1. *Let $\mathcal{L}$ be a DL with structural subsumption, Q a $\mathcal{L}$-concept description, $S = \{LO_i, i \in [1,k]\}$ a set of LOs. Then, $Tr =$ LectureComposer(Q, S) is the best composed lecture cover of Q over S.*

Proof. According to definitions 4, 5 and 6 about the best lecture cover of Q over S, Tr must satisfy the following properties:

(1) there exists at least one $LO_i, 1 \leq i \leq n$ such that $\mathcal{BK} \sqsubseteq \mathcal{PR}_i$,
(2) $\forall 1 \leq i \leq n$, if $\mathcal{BK} \not\sqsubseteq \mathcal{PR}_i$ then there exists a set $Tr \subseteq S$ such that $\mathcal{BK} \sqcap \mathcal{LO}_{Tr} \sqsubseteq \mathcal{PR}_i$,
(3) $Q - lcs_{\mathcal{T}}(\mathcal{LO}_{Tr}, Q) \not\equiv Q$, where $\mathcal{T} = \{\mathcal{LO}_i, i \in [1,k]\}$ is the $\mathcal{L}$-terminology describing the knowledge offered by S,
(4) there exists no lecture cover Tr' of Q using S such that $|Q - lcs_{\mathcal{T}}(\mathcal{LO}_{Tr'}, Q)| < |Q - lcs_{\mathcal{T}}(\mathcal{LO}_{Tr}, Q)|$, where $<$ is the lexicographic order of definition 3.

Points (3) and (4) follow directly from the definition of the best cover of a concept w.r.t. a terminology.

Let us prove point (2). The algorithm returns a set Tr verifying:

$$\mathcal{PR}_{Tr} - lcs_{\mathcal{T}}(\mathcal{BK} \sqcap \mathcal{LO}_{Tr}, \mathcal{PR}_{Tr}) = \top$$

This is equivalent to: $\bigsqcap_{1 \leq i \leq n} Pr_i - lcs_{\mathcal{T}}(\mathcal{BK} \sqcap \mathcal{LO}_{Tr}, \mathcal{PR}_{Tr}) = \top$, which again is equivalent to: $Pr_i - lcs_{\mathcal{T}}(\mathcal{BK} \sqcap \mathcal{LO}_{Tr}, \mathcal{PR}_{Tr}) = \top, \forall 1 \leq i \leq n$. From the definition of the difference operator we now that, if $C - D = E$ then $C \equiv E \sqcap D$, so we can write:

$$lcs_{\mathcal{T}}(\mathcal{BK} \sqcap \mathcal{LO}_{Tr}, \mathcal{PR}_{Tr}) \equiv \mathcal{PR}_i, \forall 1 \leq i \leq n$$

It follows that:

$$\mathcal{BK} \sqcap \mathcal{LO}_{Tr} \sqsubseteq \mathcal{PR}_i, \forall 1 \leq i \leq n$$

Thus, we have proven point (2). Let us now turn to point (1). We have proven in point (2) that $\mathcal{LO}_{Tr} \sqcap \mathcal{BK} \sqsubseteq \mathcal{PR}_i, \forall 1 \leq i \leq n$. Now we must prove that for at least one $i \in [0,1]$ we have $\mathcal{BK} \sqsubseteq \mathcal{PR}_i$. In other terms, this means that for at least one $i \in [0,1]$ we must prove that $\mathcal{LO}_{Tr} \not\sqsubseteq \mathcal{PR}_i$.

Let $j \in [0,1]$, we have: $\mathcal{BK} \sqcap \mathcal{LO}_{Tr} \sqsubseteq \mathcal{PR}_j$. For all $LO_i \in \mathcal{LO}_{Tr}$, $\mathcal{LO}_i \not\sqsubseteq \mathcal{PR}_j$, otherwise a cycle occurs in Tr. But as stated in remark 1, we suppose that S is acyclic. Thus, there exists at least one j such that $\mathcal{BK} \sqsubseteq \mathcal{PR}_j$ and we have proven point (1).

5 Illustrating Example

The example is based on a lecture about networking that can be found in the online tele-TASK archive [1]. A sample of the different LOs in the lecture with their corresponding prerequisites is shown in figure 3.

The example shows that the LO introducing computer networks ($\mathcal{LO}_1$) can be accessed by beginners because it requires no initial knowledge. The LO about

$\mathcal{LO}_1 \doteq \exists$communicationSystem.Network	$\mathcal{PR}_1 \doteq \top$
$\mathcal{LO}_2 \doteq \exists$network.WAN	$\mathcal{PR}_2 \doteq \exists$communicationSystem.Network
$\mathcal{LO}_3 \doteq \exists$network.LAN	$\mathcal{PR}_3 \doteq \exists$communicationSystem.Network
$\mathcal{LO}_4 \doteq \exists$communication.Protocol	$\mathcal{PR}_4 \doteq \exists$communicationSystem.Network
$\mathcal{LO}_5 \doteq \exists$protocol.TCPIP	$\mathcal{PR}_5 \doteq \exists$communication.Protocol
$\mathcal{LO}_6 \doteq \exists$structure.Topology	$\mathcal{PR}_6 \doteq \exists$communicationSystem.Network
$\mathcal{LO}_7 \doteq \exists$topology.StarTopology	$\mathcal{PR}_7 \doteq \exists$structure.Topology $\sqcap\exists$network.LAN
$\mathcal{LO}_8 \doteq \exists$topology.RingTopology	$\mathcal{PR}_8 \doteq \exists$structure.Topology
$\mathcal{LO}_9 \doteq \exists$topology.BusTopology	$\mathcal{PR}_9 \doteq \exists$structure.Topology

Fig. 3. Examples of LO descriptions and their corresponding prerequisites

star topology ($\mathcal{LO}_7$) requires some knowledge about topologies in general ($\mathcal{LO}_6$) and about LAN ($\mathcal{LO}_3$). Therefore, this LO can only be accessed if the user's background knowledge $\mathcal{BK}$ fulfills this requirements.

Let us suppose that the user wants to learn something about topologies in general and star topology in particular, and that he has a basic knowledge about computer networks. Formally, this is written as follows:

$$Q \doteq \exists\text{structure.Topology} \sqcap \exists\text{topology.StarTopology}$$
$$\mathcal{BK} \doteq \exists\text{communicationSystem.Network}$$

The associated hypergraph $\mathcal{H}_{SQ} = (\Sigma, \Gamma)$ consists on the vertices $\Sigma = \{V_{\mathcal{LO}_1}, ..., V_{\mathcal{LO}_9}\}$ and the edges $\Gamma = \{e_{\exists\text{structure.Topology}}, e_{\exists\text{topology.StarTopology}}\}$. The only minimal transversal is $Tr = \{V_{\mathcal{LO}_6}, V_{\mathcal{LO}_7}\}$ (see figure 4). It covers completely the query.

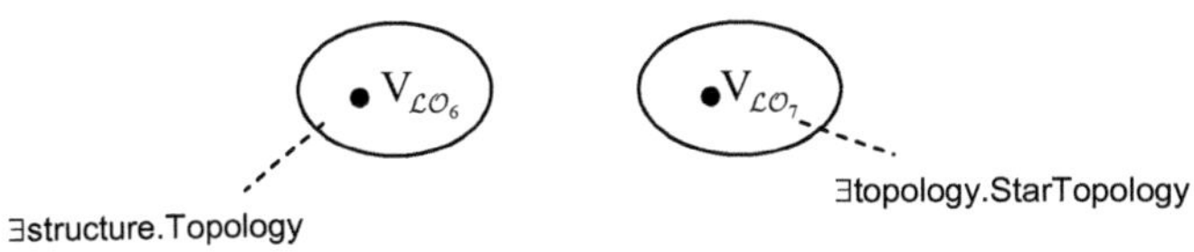

Fig. 4. The hypergraph $\mathcal{H}_{SQ}$

The prerequisites required for the set Tr are:

$$\mathcal{PR}_{Tr} = \exists\text{communicationSystem.Network} \sqcap \exists\text{structure.Topology} \sqcap \exists\text{Network.LAN}$$

The part of prerequisites that are not covered by the user's background knowledge or by other LOs is:

$$P = \exists\text{network.LAN}$$

The new hypergraph $\mathcal{H}'_{SQ}$ that is obtained by adding the hyperedges in Γ_P is shown in figure 5.

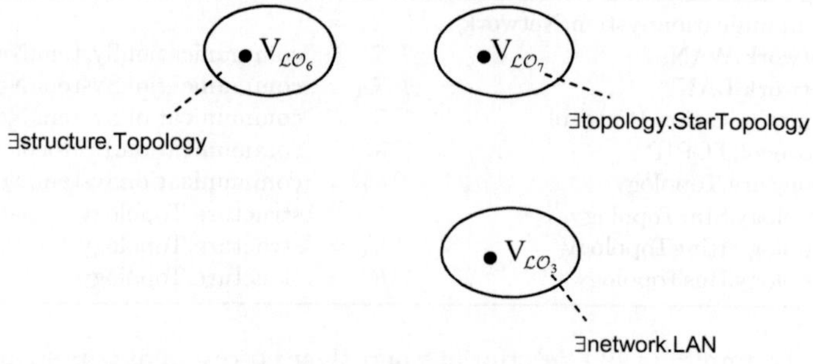

Fig. 5. The hypergraph $\mathcal{H}'_{SQ}$

The new minimal transversal is then $Tr = \{V_{\mathcal{LO}_6}, V_{\mathcal{LO}_7}, V_{\mathcal{LO}_3}\}$. The prerequisites needed for it are:

$$\mathcal{PR}_{Tr} = \exists \text{communicationSystem.Network} \sqcap \exists \text{structure.Topology} \sqcap \exists \text{network.LAN}$$

The part of prerequisites not covered by the user background knowledge or by other LOs is:

$$P = \top$$

Thus, the algorithm stops and returns LO_6, LO_7 and LO_3. Although, LO_3 is not directly requested by the user—i.e., it is not a cover of the user's query—it is needed to fulfill the missing prerequisite since the user must have some knowledge about LAN before being able to learn something about star topology.

6 Discussion

In this paper we have proposed a novel algorithm for personalized lecture composition based on lecture subparts. We used two non-standard inferences in DLs—i.e., the least common subsumer (lcs), and the difference operation—to compute the best cover of the user's query w.r.t. a repository of LOs. Then, the LOs are assembled into a composition flow. The algorithm takes into account the user's knowledge as well as the sequencing constraints between the LOs in order to retrieve a comprehensive and self-contained composition flow.

A lot of recent work has focused on personalized e-learning and composition of LOs [12,7,8,10]. The work presented in [7] is closely related to ours. The authors introduced a new definition of the concept covering problem that eliminates the limitation of the DLs to have structural subsumption. It is based on the concept abduction problem (CAP) which was introduced in [15] to provide an explanation when subsumption does not hold. But, the proposed algorithm returns a cover which is not necessarily the best one.

As DLs with structural subsumption are expressive enough for our application, we chose to use the algorithm of Hacid & al. [9] because it always returns the best cover. Also compared to [7], the novelty of our approach is that it always proposes a solution to the user. When the user's knowledge is not sufficient, our algorithm looks for complementary LOs and adds them to the composition flow in order to fill the missing knowledge.

Currently, we are improving our prototype that will be integrated in the Web interface of the online tele-TASK archive [1]. Experiments on real scenarios will be conducted on a set of lectures about "Internetworking". For this purpose, an appropriated domain ontology is under development.

In future work, we will improve our algorithm by taking into account further criteria for the best composition. We will consider minimizing a number of parameters like the number of retrieved LOs, optimizing the length of the composition flow, and the number of original lectures involved in the composition flow.

References

1. Tele-TASK – Teleteaching Anywhere Solution Kit. http://www.tele-task.de.
2. Web University project. http://www.hpi.uni-potsdam.de/~meinel/research/web_university.html.
3. F. Baader, I. Horrocks, and U. Sattler. Description logics as ontology languages for the semantic web. In *F. Baader, I. Horrocks, and U. Sattler. Description logics as ontology languages for the semantic web. In D. Hutter and W. Stephan, editors, Festschrift in honor of J org Siekmann, Lecture Notes in Artificial Intelligence.*, 2003.
4. F. Baader, R. Küsters, and R. Molitor. Computing Least Common Subsumers in Description Logics with Existential Restrictions. In T.Dean, editor, *Proceedings of the 16th International Joint Conference on Artificial Intelligence (IJCAI'99)*, pages 96–101. Morgan Kaufmann, 1999.
5. Franz Baader, Diego Calvanese, Deborah McGuinness, Daniele Nardi, and Peter F. Patel-Schneider, editors. *The Description Logic Handbook: Theory, Implementation and Applications.* Cambridge: University Press, 2003.
6. Boualem Benatallah, Mohand-Said Hacid, Alain Leger, Christophe Rey, and Farouk Toumani. On automating web services discovery. *The VLDB Journal*, 14(1):84–96, 2005.
7. Simona Colucci, Tommaso Di Noia, Eugenio Di Sciascio, Francesco M. Donini, and Azzurra Ragone. Semantic-based automated composition of distributed learning objects for personalized e-learning. In *The Semantic Web: Research and Applications 2nd European Semantic Web Conference ESWC 05*, volume 3532, pages 633–648. Springer-Verlag, 2005.
8. Robert G. Farrell, Soyini D. Liburd, and John C. Thomas. Dynamic assembly of learning objects. In *WWW Alt. '04: Proceedings of the 13th international World Wide Web conference on Alternate track papers & posters*, pages 162–169, New York, NY, USA, 2004. ACM Press.
9. M. Hacid, A. Leger, C. Rey, and F. Toumani. Computing concept covers: A preliminary report. In *Workshop on Description Logics*, 2002.
10. N. Henze. Personal readers: Personalized learning object readers for the semantic web, 2005.
11. A. Ip, A. Young, and I. Morrison. Learning objects - whose are they? In *Proceedings of the 15th Annual Conference of the National Advisory Committee on Computing Qualifications ISBN 0-473-08747-2*, pages 315–320, 2002.

12. Jelena Jovanovic, Dragan Gasevic, and Vladan Devedzic. Dynamic assembly of personalized learning content on the semantic web. In York Sure and John Domingue, editors, *ESWC*, volume 4011 of *Lecture Notes in Computer Science*, pages 545–559. Springer, 2006.
13. Dimitris J. Kavvadias and Elias C. Stavropoulos. An efficient algorithm for the transversal hypergraph generation. *J. Graph Algorithms Appl.*, 9(2):239–264, 2005.
14. Serge Linckels, Stephan Repp, Naouel Karam, and Christoph Meinel. The virtual tele-task professor—semantic search in recorded lectures. In Ingrid Russell, Susan Haller, J.D. Dougherty, Susan Rodger, and Gary Lewandowski, editors, *ACM SIGCSE'07 Technical Symposium on Computer Science Education*, pages 50–54, New York, NY, USA, 2007. ACM Press.
15. Tommaso Di Noia, Eugenio Di Sciascio, Francesco M. Donini, and Marina Mongiello. Abductive matchmaking using description logics. In Georg Gottlob and Toby Walsh, editors, *IJCAI*, pages 337–342. Morgan Kaufmann, 2003.
16. G. Teege. Making the Difference: A Subtraction Operation for Description Logics. In Jon Doyle, Erik Sandewall, and Pietro Torasso, editors, *KR'94: Principles of Knowledge Representation and Reasoning*, pages 540–550. Morgan Kaufmann, San Francisco, California, 1994.
17. Gottfried Vossen and Peter Jaeschke. Learning objects as a uniform foundation for e- learning platforms. *IDEAS*, 2003.

Caravela: Semantic Content Management with Automatic Information Integration and Categorization
(System Description)

David Aumüller and Erhard Rahm

University of Leipzig
{david,rahm}@informatik.uni-leipzig.de

Abstract. Semantic web content management poses much manual work onto the community. To reduce this labour we have devised Caravela[1], a generic approach to dynamic content integration and automatic categorization. Content and documents of different types can be integrated from diverse semi-structured sources and categorized along multiple dimensions. Automatic linking provides dynamic categorizations at no user cost. We illustrate our approach by an online bibliography categorizing scientific research publications.

1 Introduction

With the emergence of semantic technologies, methods for semantically annotating information are also used in web content management systems and collaborative environments such as wiki systems. A recent survey of systems for semantic annotation [17] lists several requirements of such systems, including support for user collaboration, ontologies, heterogeneous document formats, document evolution, persistent annotation storage, and automation. Current implementations have problems to fully meet these requirements. In particular, they incur *too much manual work* to add content, categorize content, and deal with the evolution of the content schema and ontologies.

To better meet the requirement of automation we have devised a new approach to semantic content management and created *Caravela*, a generic system for dynamic content integration and automatic categorization. In addition to functions of a web content management system it supports multiple taxonomies to semantically categorize different types of content (e.g. documents, movies, products). It provides functions to automatically integrate, transform, structure, and categorize content. Powerful automatic linking creates dynamic categorization without any user effort. Content can be enriched and periodically refreshed automatically from external web data sources. Furthermore, content type schema and taxonomy evolution are supported to meet changed user requirements. Caravela is fully operational and has e.g. already been used in several instances of a collaborative publication categorizer and a movie navigator.

[1] A *caravela* is a small, highly manoeuvrable ship, used for exploration.

E. Franconi, M. Kifer, and W. May (Eds.): ESWC 2007, LNCS 4519, pp. 729–738, 2007.

The remainder of this paper is structured as follows. We first illustrate the use of the generic platform by briefly introducing the publication categorizer application. We illustrate the underlying architecture of our system in section 3, describing the content repository supporting evolution. In section 4 we present the main contributions towards semi-automatic content management, in particular integration, transformation, and categorization. After presenting related work in section 5 we conclude with an outlook on further work.

2 Sample Application

Caravela is being used for the development and maintenance of web-based bibliographies to collect, structure, and classify scientific publications in a particular domain. While there exist many bibliographic utilities (comprehensive list e.g. on dmoz.org) most of them focus on the generation of references to include in own publications. However, there is little tool support for maintaining open, web-accessible bibliographies to collect relevant publications in dynamic areas, e.g. semantic web technologies. Such bibliographies should ideally categorize publications in semantically rich ways and require little manual work to add and update bibliographic entries. While wiki technology can help to spread the manual work among many users, automatic content integration and enrichment is very important to improve the utility of a web-based bibliography. The Caravela platform can be used to support such requirements.

Fig. 1 shows part of a screenshot of the publication categorizer for papers on schema evolution [13]. Publications are classified along multiple taxonomies, shown on the left. In the example domain we use separate taxonomies for research areas, publication venue, year, citation counts, etc. The mappings between publications and taxonomies may be many-to-many, e.g. for research areas. Categories exhibit an occurrence count indicating the number of corresponding publications.

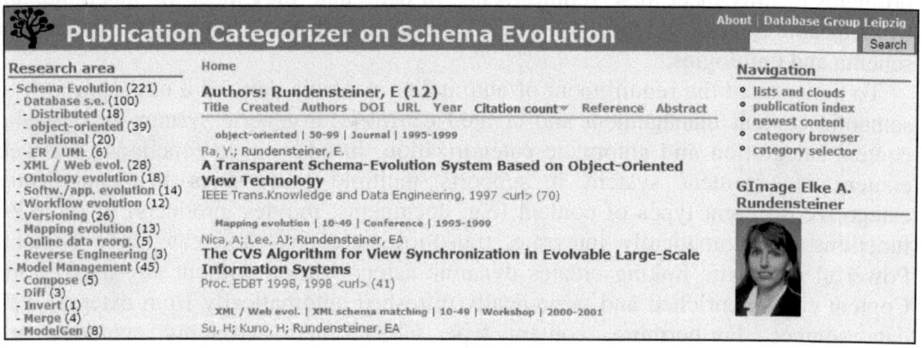

Fig. 1. Publication categorizer, showing publications sorted by citation count

All information to a single publication is presented on one web page, e.g. authors, title, venue, abstract, fulltext or reviews. To reduce manual work we can automatically integrate bibliographic data from external sources, e.g. from data sources like Google

Scholar, publisher web sites or bibliographic reference files. To enrich existing entries relevant data sources can dynamically be queried, e.g. to retrieve current citation counts. We also allow wiki-like manual insertion and editing of instances by interested users. Some taxonomies and some mappings are automatically generated from existing attribute values, e.g. for publication year, publication venue or citation count. Key terms like author names or conference names are automatically extracted and categorized ("automatic linking") and offered for navigational access. Standard features like navigation along the taxonomies, fulltext search, etc. are also supported. Lists of publications can be sorted by their attributes, e.g. their citation count or year.

3 Generic Content Representation

Fig. 2 shows the architecture of our generic approach to semi-automatic semantic web content management, Caravela. It consists of three layers providing data storage (repository), data handling, and data presentation. To limit the implementation effort and to focus on the new aspects for reducing manual work and semantic categorization, Caravela uses some functionality of an existing web content management system (Drupal, drupal.org). In this section we describe the model for generic content representation and its suitability to evolving content. The methods for content integration and categorization are described in section 4.

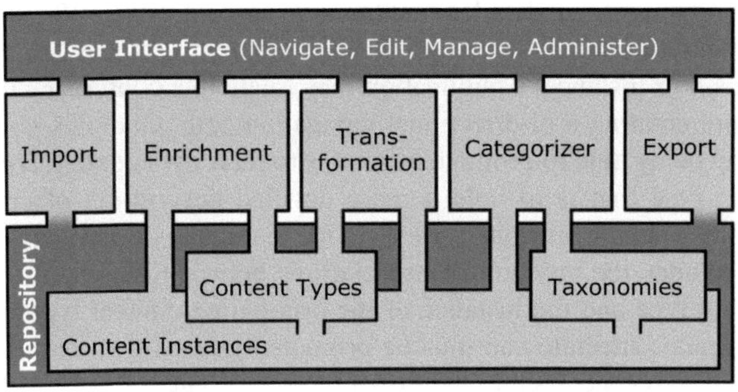

Fig. 2. Architecture of the Caravela platform

We use a relational database to persistently and generically store content and metadata, such as taxonomies. Fig. 3 below illustrates the generic repository structure. Each content item, e.g. publication or movie, is associated to a particular content type and described by several attributes. For example, the publication content type typically has attributes like title, authors, year, and venue. Attributes either store data in simple data types or comprise whole lists of constituents, e.g. a list of authors. We intentionally do not store the constituents separately but extract them dynamically when needed (see 4.3).

As user requirements change in managing content, the underlying content type needs to be adapted. Content types can be altered (or new ones added) by changing, extending or reducing their attribute lists. Operators for content transformation provide the means to establish the changes on the content instances. Thus, attribute values can be atomized by extracting parts of the values into another attribute, e.g. to extract special bits of data (consider e.g. dates, locations) from verbose text into its own attribute.

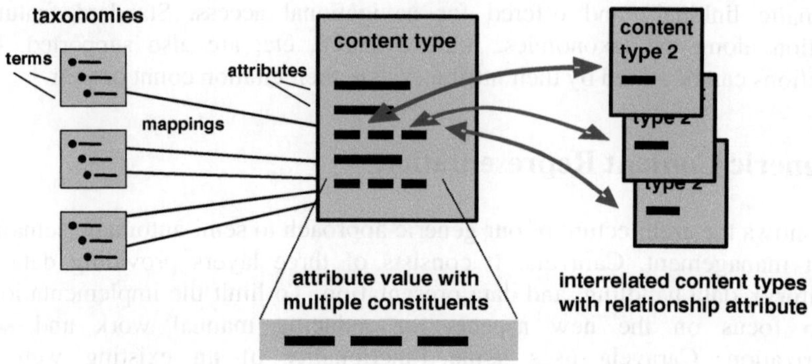

Fig. 3. Content types with various types of attributes and mappings to taxonomies

There may be multiple content types with interrelationships, e.g. a publication content type relating to a detailed conference content type. We use bidirectional relationships between content types so that the content instances are accessible from both directions. For instance, a publication may relate to a conference via a 'published in' relationship, creating a bi-directional navigation path. Complex content evolution is attributed by being able to promote attributes to their own content type. To establish a new content type that is to hold a more detailed description of content formerly residing within a single attribute, we provide data transformations across content types. This includes the creation of associations between the emerging instances of the new content type and the instance of the originating content type. For example, a 'publication venue' attribute can thus be promoted into its own 'conference' content type, or a list of movie actors into their own 'person' content type.

Each content type may have mappings to multiple taxonomies and each content instance may have multiple correspondences to one or more category terms per taxonomy. These mappings are stored in the database with links to the taxonomy terms (Fig. 3).

This simple and generic content representation model eases acquisition and storage of most types of content. Furthermore, content types and taxonomies can easily be added or changed thereby supporting schema evolution. The functionalities of Caravela can be generically extended by providing user-supplied scripts, e.g. to provide further integration (web scraping) and transformation capabilities. The current implementation offers content export to other semantic applications. Furthermore, multiple RSS-feeds provide dynamic content access for other applications.

4 Content Integration and Categorization

Semantic wiki applications (e.g. [2], [4], [11], [18]) can be used to collect any kind of content along with attribute value pairs and interrelating typed links. Such systems though impose the need to master specific syntax and/or lack automatisms to work with the data. With the approaches presented e.g. in [8] and [16], at least the initial population of information can be automated. The full potential of the Caravela platform though lies in automating tasks to save effort and quickly add and transform content. This involves similar data import and transformation tasks as for ETL processing (extract, transform, load) in data warehousing, i.e. information extraction (e.g. [1], [10]), data cleaning (e.g. [15]), and information integration (e.g. [5], [14], [16]), however for documents and web content, where often screen scraping has to take place (e.g. [7], [9]).

Caravela provides functionalities for automating content integration from the Web, content transformation, and automatic categorization. To help maintain high quality content and categorizations all information-editing tasks can also be performed manually. In the next section we outline the methods to integrate and adjust content from files and web sources. We then discuss how content instances are categorized along the taxonomies and how automatic linking is implemented and used, with special regards on the dynamic extraction of attribute constituents.

4.1 Generic Integration of Semi-structured Information

File import. Caravela provides a comprehensive import facility to integrate sets of items available in external data sources. To import data from external files, e.g. XML, RDF, RSS, or CSV files, we adopt a declarative mapping between these document variants to the attributes of a content type. These mappings are specified via a list of XPath expressions, denoting for each target content type attributes its source XML elements. A typical mapping to generate publication instances follows. Here, for each `RDF/item` construct in the source a publication content instance will be created and its attributes filled accordingly:

```
   rdf:RDF/item  -->  publication
./dc:title     -->  publication: title
./dc:creator   -->  publication: authors
```

Web data integration. On a more dynamic level Caravela supports the integration of external data by querying web services, search engines, and applying web/screen scraping. Prior to integration each content instance gathered by such services will be structured internally as list of attribute value pairs. User provided keywords are used to query a service and retrieve values for all attributes of a content type.

An ad-hoc (light-weight) schema matching takes place to map the available attributes of the external data source (web service) to the internal attributes of the according content type. As the schemas of content types usually consist only of few attributes, we use simple name-based matching using string matchers such as edit distance, stemming, n-grams, and phonetic matchers such as soundex. Especially the edit distance measure seems to provide good enough results in our context, as the threshold can be set to match plural and singular forms, e.g. ‘authors’ and ‘author’, or whether to accept 'actors' and 'authors' as match or not. To supply mappings that cannot be determined automatically the user can provide synonyms manually that get merged with the auto-generated mapping.

Querying a search engine or a web service with user supplied keywords usually yields multiple results. Thus, before the intended content instances are added into the system the user chooses the instances of interest from the list of returned result set. The user can also adjust and add details much like in a wiki page by editing and adding attribute value pairs directly in the result view (Fig. 4).

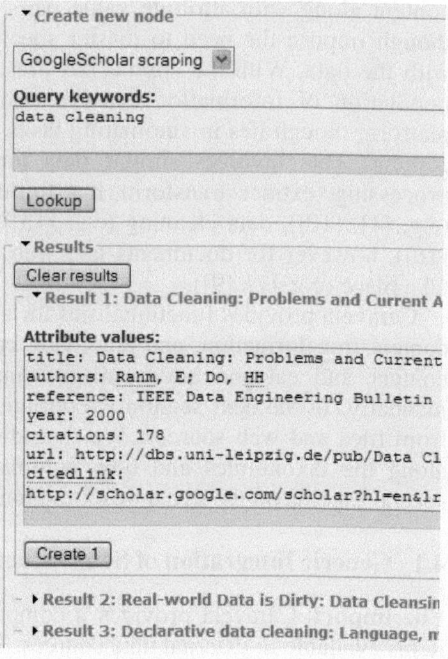

Fig. 4. Editable web data integration

Generally, before content instances get actually integrated into the system the user may decide whether to overwrite or skip already existing target instances, or merely append missing attribute values. Generic web/screen scraping is supported by providing scripts that take care of the web data extraction and transform the web data into attribute value pairs, e.g. via regular expressions. Any web service can be attached that offers querying the provided instances by keyword.

Content enrichment. Instead of creating new content instances by manually supplying keywords, attribute values from existing content instances can be used to automatically query the services to enrich and complete or update the content by incorporating related pieces of information from external data sources. To enrich existing content instances it is necessary to map an existing content instance to an external instance representing the same real world entity. Using multiple available attributes of a content instance as query keywords ensures a better object matching. In Caravela this content enrichment is available as bulk operation for a selection of existing content instances. As the parameters of such a query operation can be stored, it can also be made accessible as single-click action that triggers an update of attribute values of the currently displayed content instance only.

Content mashup. The external content object matching can be used to create a mashup, i.e. including external content live into a view instead of integrating the content into the repository. For instance, we present the results of a Google Scholar author search and author photographs or other images related to the author name in its own blocks on the right hand side of an author page. Other embeddable content of interest include maps and calendars depicting e.g. relevant conference locations and dates.

Data transformations. Integrating information from external data sources may yield inconsistencies such as differently coded values, legacy values, or free form values that may need to be transformed or aligned and mapped into consistent terms. Caravela provides operations to transform attribute values of content instances to achieve such data cleaning. Operations take the specified attribute values of a selection of content instances and replace them by the transformed values and/or fill other attributes with it. Operators for transformation are assembled by regular expressions for search and replace within or across attributes, e.g. replacing instances of unwanted values by defined ones (e.g. '1' and 'F' into female, or 'Very Large Data Bases' into 'VLDB'). As such search and replace rules may sometimes not suffice, conversion tables can be incorporated for look-up. More expressive content manipulation operators can be made available by user-supplied, pluggable scripts that supply functions to derive calculated values, e.g. to normalize author names into a consistent representation.

4.2 Categorization Along Multiple Taxonomies

Content instances in Caravela may be categorized along multiple taxonomies. Offering multiple hierarchies instead of merely one, each taxonomy may be more clearly defined and thus smaller in size, i.e. more easily to understand and maintain. Categorizing content results in a faceted classification available for navigation. This kind of navigation is getting more and more adopted in web applications, e.g. to narrow down product categories or to browse for images and other media (e.g. see [3], [11], [12], [19]). Often, the content in these applications is purely read-only from an end-user perspective or the navigation scheme or categorization is fixed, whereas in Caravela we offer category adjustments that instantly update navigation paths.

The various taxonomies are displayed each in a block on the left hand side. Along each category term the occurrence count indicates the number of correspondences belonging to the term and its descendants, thus adding up document instances assigned to more specific category terms (see Fig. 1). To avoid manual categorization work we provide several automatisms for the categorization along taxonomies. These include the categorization of content instances along given taxonomies, the creation of taxonomies from given content attribute values, and the extension of taxonomies by generating more general terms. We use regular expression and query patterns or incorporate user-supplied scripts to match and create terms.

One approach for automatic categorization is achieved by **deriving taxonomy correspondences** from given attribute values or parts thereof as specified via a regular expression pattern. Consider finding the corresponding decade for a given year. A substring comparison or numerical range containment fulfils this task, as e.g. a substring of the attribute value '1968' matches with the given category term '1960s' or the exact number matches the interval 1960—1969 using a 'between'-query. Existing attribute values can be used to **create a new taxonomy** from scratch, establishing according correspondences to the content instances. This is useful when there are many terms in attributes that are to form a category and/or no other sources available. Consider a taxonomy of all publication venues/conferences mentioned in the system. This approach first yields a flat taxonomy of terms, i.e. simply a 'controlled' vocabulary list, which may be extended by **generating more general**

terms to increase the expressiveness of the taxonomy. To derive the hypernym terms syntactic approaches as aforementioned can be used. From a flat taxonomy of single years e.g., we can derive a first level of more general terms by constructing the decades. Another level on top of that would be presented by the according centuries. This could e.g. be devised by taking the first two digits from the year, incrementing it by 1, and appending '[st|nd|rd|th] century' as suffix to name it appropriately. These functionalities can be extended in Caravela by providing scripts that contain functions to return a more general term to a given term. By querying thesauruses like WordNet such a script may come up with hypernyms not derivable by syntactical patterns. Another strategy for creating a rich taxonomy including more general terms would be to take the number of occurrences belonging to one term into account. Less frequent terms, e.g. years that only carry few correspondences to content instances, could be grouped with others under one more general term. Summarizing, Caravela offers the means to create whole category trees from available attribute values.

As categorization often underlies subtle semantic decisions, e.g. the assignations of publications to research areas, categorization of content cannot be fully automated. To ease manual categorization we devised the drag'n'drop category browser (using AJAX). It offers two modes: In the view displaying all available categories at once, documents can be moved freely around to adjust categorization. A second view presents each taxonomy individually along the list of documents not yet categorized into the current taxonomy. This further entices complete categorization of all content instances.

4.3 Dynamic Categorization by Automatic Linking and Weighting

Apart from taxonomical categorization Caravela offers a powerful dynamic categorization based on attributes. The key idea is that values from certain attributes are automatically and dynamically extracted and cross-linked to all content instances with the corresponding value. These attribute values can be offered for navigation, e.g. at the very spot where they appear in the content instances, or separately as (weighted) lists of grouped/aggregated attribute values (Fig. 5 and 6).

Fig. 5. Occurrence counts in attributes/constituents

As attributes may contain lists of values, the distinct values (constituents) are available via dynamic extraction as specified in the attribute definition. The default separator for constituents is a semicolon, but any other pattern may be defined. It can be chosen to define a split pattern as separator or a match pattern to identify the constituents or interesting parts of an attribute value. Any regular expression is allowed; this can be simply a comma or slash for a split pattern or more complex expressions for a match pattern. To display the according occurrence count behind each term (i.e. the number of content instances that contain the same term), the count is gathered using an SQL-query as in `select count(*) from content_type where attribute like '%term%'`. Each term carries an automatically created dynamic link to browse for the content instances containing that term, e.g. for publications by the same author or movies with the same actor. The creation of these dynamic cross links poses no effort to the user who after adding new content immediately benefits from the additional navigation path and the updated occurrence counts.

Aggregated (or grouped) lists of all distinct values or constituents of an attribute within the collection form another categorization scheme to distinguish more prominent attribute values or constituents. Instead of merely presenting the occurrence count or aggregated group count as number the constituents can be visually weighted to produce so called tag clouds. Here the occurrence counts get represented by font size or shades of gray. Thus, more frequent terms get represented larger or in a darker/deeper colour. Instead of representing the frequency (occurrence count) of the terms in the document collection, the weights can also be determined taking other attribute values into account. Regarding the publication categorizer a useful representation consists of author names weighted by their average or maximum citation count of their aggregated publications (Fig. 6). This highlights the more influential authors in the document collection. Such tag clouds are great means to start browsing a collection, as each attribute value or constituent links to appropriate overview pages. Again, there is no user effort in creating them.

Fig. 6. Author cloud as weighted by occurrences (shade) and citations (size)

5 Conclusion and Outlook

With the presented approach towards automatic semantic content management we keep the amount of manual work low. The proposed content repository model is applicable to a large variety of content, easy to maintain and to extend. By being able to integrate information from disparate and unstructured sources Caravela can be used to turn unstructured data into structured data of multiple formats. We provide automatisms for integrating, transforming, and categorizing content of varying type along multiple taxonomies, offering further automatically created dynamic links. By releasing the user from tedious manual work the community can collaboratively lay their strength on maintaining a high quality of the content. Periodically updating content by integrating information from external data sources helps to keep the managed data up to date, e.g. citation counts. Changed user requirements are attributed by schema and taxonomy evolution techniques. Caravela has been successfully applied for different applications, in particular for a powerful publication categorizer, which is well accepted by a growing user base. In further work we plan to provide workflow capabilities to support the repetitive execution of more complex information acquisition and content transformation tasks. The generic approach will be applied to more domains.

References

[1] A. Arasu, H. Garcia-Molina. Extracting structured data from Web pages. In *SIGMOD*, 2003

[2] D. Aumueller. Semantic Authoring and Retrieval within a Wiki. In *ESWC*, 2005

[3] V. Broughton. Faceted classification as a basis for knowledge organization in a digital environment: the bliss bibliographic classification as a model for vocabulary management and the creation of multidimensional knowledge structures. In *New Rev. Hypermedia Multimedia 7(1)*, 2002

[4] M. Buffa, F. Gandon. SweetWiki: semantic web enabled technologies in Wiki. In *Symposium on Wikis*, 2006

[5] W. Cohen. Some practical observations on integration of Web information. In *WebDB*, 1999

[6] A. Doan et al. Community Information Management. In *IEEE Bull. on Data Engineering*.

[7] G. Gottlob et al. The Lixto data extraction project: back and forth between theory and practice. In *PODS*, 2004

[8] A. Di Iorio et al. Automatic Deployment of Semantic Wikis: a Prototype. In *1st Workshop on Semantic Wikis*, 2006

[9] U. Irmak, T. Suel. Interactive wrapper generation with minimal user effort. In *WWW*, 2006

[10] A. Laender et al. A brief survey of Web data extraction tools. In *SIGMOD Record*, 31(2), 2002

[11] E. Oren etal. Annotation and Navigation in Semantic Wikis. In *SemWiki WS at ESWC*, 2006

[12] Schraefel, M.C. et al. The evolving mSpace platform: leveraging the Semantic Web on the Trail of the Memex. In *Hypertext*, 2005

[13] E. Rahm, P.A. Bernstein. An Online Bibliography on Schema Evolution. *ACM SIGMOD Record*, Dec. 2006

[14] E. Rahm et al. iFuice – Information Fusion utilizing Instance Correspondences and Peer Mappings. In *WebDB*, 2005

[15] E. Rahm, H.H. Do. Data Cleaning: Problems and Current Approaches. In *IEEE Data Eng. Bull. 23(4)*, 2000

[16] A. Sheth et al. Managing Semantic Content for the Web. In *IEEE Internet Computing 6(4)*, 2002

[17] V. Uren et al. Semantic annotation for knowledge management: Requirements and a survey of the state of the art. In *Journal of Web Semantics 4(1)*, 2005

[18] M. Völkel et al. Semantic Wikipedia. In *WWW*, 2006

[19] P. Yee et al. Faceted Metadata for Image Search and Browsing. In *ACM CHI*, 2003

The NExT System: Towards True Dynamic Adaptations of Semantic Web Service Compositions (System Description)

Abraham Bernstein and Michael Dänzer

University of Zurich, Department of Informatics, 8050 Zurich, Switzerland
{bernstein,daenzer}@ifi.unizh.ch

Abstract. Traditional process support systems typically offer a static composition of atomic tasks to more powerful services. In the real world, however, processes change over time: business needs are rapidly evolving thus changing the work itself and relevant information may be unknown until workflow execution run-time. Hence, the static approach does not sufficiently address the need for dynamism. Based on applications in the life science domain this paper puts forward *five requirements for dynamic process support systems*. These demand a focus on a tight user interaction in the whole process life cycle. The system and the user establish a continuous feedback loop resulting in a mixed-initiative approach requiring a *partial execution and resumption feature* to adapt a running process to changing needs. Here we present our *prototype implementation* NExT and discuss a preliminary validation based on a real-world scenario.

1 An Illustrating Scenario - As Is

Peter, the chemist in our scenario, needs to determine the 3D structure of a bio-molecule using NMR spectroscopy. Without having IT support, he uses his paper lab book to construct a rough experimental plan. He then starts the experiment. Only, he forgets to calibrate the spectrometer, a fact he quickly realises as the spectrometer returns first data, which shows a systematic and continuous shift over all values. At some later point, Peter stumbles on a problem with his experiment, which he does not know how to solve. He is unable to interpret a spectrum correctly and therefore to choose which measurement to perform as the next step. He reads a publication about a similar problem and makes a lengthy (formal descriptions are hard to explain in prose form) telephone call with his advisor, which is visiting a conference overseas. Anyhow, his advisor can help him and following his lead, he studies some intermediate results returned from the spectrometer. Peter then realises that he forgot to repeat a proceeding measurement with adapted parameter values rendering the current measurement totally useless. But even worse, he did not store the intermediate results during the execution, so he has to restart the execution from scratch.

E. Franconi, M. Kifer, and W. May (Eds.): ESWC 2007, LNCS 4519, pp. 739–748, 2007.

2 Introduction

The scenario shows that conducting meaningful experiments or exploratory activities requires a user to construct a complex and long-running sequence of atomic tasks which may have to be changed during all phases in their life cycle (*process choreography [1])*. Their size can be very large leading to long and complex interrelations. Furthermore, processes and their elements may change their degree of specificity (see Figure 1) over time: Underspecified processes can become well specified when more information becomes available and well-specified processes can become less specified (e.g., due to exceptions) – thus, the process moves along the Specificity Frontier [2]. A system acting in domains whose processes show varying degrees of specificity and dynamically move along the frontier must conform to such behaviour.

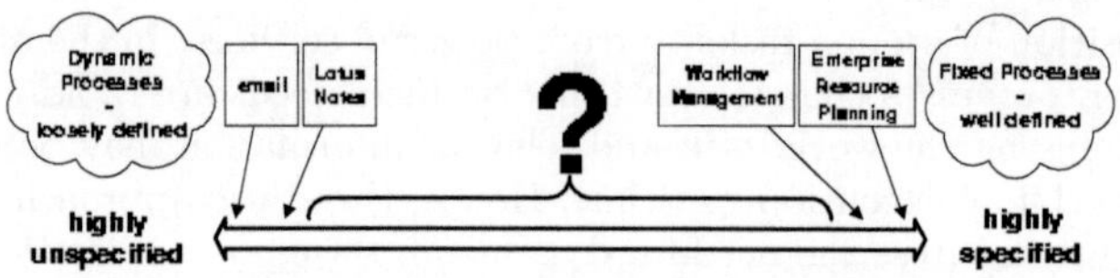

Fig. 1. The Specificity Frontier [2]

Usually, several potential realizations exist for an atomic task (such as Web or Grid Services or local procedure calls), so choosing the appropriate one (*process orchestration* [1]) turns out to be non-trivial. At run-time, exceptions can be thrown (e. g., hardwre malfunctions, software crashes), unforeseeable events may take place or the user wants to *intervene* when he observes something unusual, forcing the execution to halt and the system to react accordingly. The process must then be adapted in some way preserving its correctness and consistency. Finally, the execution must be resumed at the correct and optimal resumption point. Once the experiment has finished, all its related data must be documented (e.g., for publication in academia or to record that the process was maintained in legal environments).

In our opinion, a process support system acting in highly dynamic domains must focus on the user and keep him/her engaged in a tight interaction. Based on the system's domain knowledge and the explicitly given information, the system should provide contextual guidance to the user in all situations, especially in the complex creative phases of his work. More specifically, based on the preliminary work [2], we proclaim that such a system has to fulfill the following requirements:

$\mathcal{R}1$: Support users throughout the process choreography and orchestration steps.
$\mathcal{R}2$: Support partial executions and dynamic adaptations at run-time.
$\mathcal{R}3$: Integrate reasoners and planners to provide useful alternatives for the user.
$\mathcal{R}4$: Incorporate a *Case Base*. Then a *Case Based Reasoner [3]* can infer useful information from past cases (from both best and worst practices).

$\mathcal{R}5$: Support (semi-)automated *data mediation* to connect processes with different data formats which are transformable into each other.

In this paper we will present an overall approach for a process support system addressing these requirements. We focus on the second one and will show in more detail how partial executions, run-time adaptations and changes to parameter values can be supported. The remainder of this paper is structured as follows: In Section 3 we operationalize the requirements into concrete foundational challenges for our prototype NExT (*N*ext-generation *Ex*periment *T*oolbox). Section 4 then introduces the most important architectural and implementation aspects of NExT. A preliminary validation of the prototype in the context of the introductory scenario is discussed in Section 5, followed by a comparison with related work in section 6. We conclude with a summary and an outlook on future work.

3 Overall Operationalization of NExT

In order to assure and a clear separation of concerns, we divided NExT into two parts: the underlying knowledge bases (KBs) containing all the domain knowledge and a generic execution support system. The content of these KBs is provided in a formal, machine readable language, which is a pre-requisite for planning and reasoning ($\mathcal{R}3$, $\mathcal{R}4$). We identified three types of entities to store in separate online KBs: First a *Process Library* with models for all atomic tasks and templates for composite processes with a loose coupling to their concrete realizations allowing for their dynamic reassignment ($\mathcal{R}1$). Second, a *Data Entity Library* containing models for all data/object types to enable (semi-)automated data mediation in fulfillment for $\mathcal{R}5$. Last, but not least, a *Case Base* containing a collection of completed process executions enables both automated as well as human case based reasoning ($\mathcal{R}4$). Note that due to the KBs NExT exhibits significant network effects in the micro-economic sense: the more people use it the more attractive it becomes. If a sufficiently large group of people in a given domain publish their processes into the repositories then the possibility for knowledge exchange increases, collaborating in designing/executing processes is simplified, and their use as case bases and domain KBs increases the quality and diversity of planner/reasoner results.

We follow an approach known as Mixed-Initiative planning and execution [3]; the user and the NExT system work hand-in-hand informing each other with newly discovered facts. The more information and constraints the system receives from the user, the more (implicit) knowledge it can infer and present to him. He then can use this additional information to either retrieve even more information or make decisions, both of which become new input for the system. User and system are, thus engaged in a continuous feedback loop. In addition, the system continuously monitors newly arriving information (such as detected exceptions) and initiates an interaction with the user whenever necessary.

NExT guides the user by providing suggestions whenever she has to make decisions or she explicitly requests help. During the *process choreography* the system's degree of assistance ranges between suggestions, which processes are suitable for the next step, and the generation of whole process plans at once ($\mathcal{R}1$). During the *process orchestration* the system will (1) guide the users to concrete realizations and (2) help them to decide which one is suitable under the given constraints and user preferences ($\mathcal{R}1$). When two processes are chained together by a data flow and the types of their parameters are not "castable", then the system tries to resolve the mismatch or suggests solutions to the user ($\mathcal{R}5$). The execution history is recorded and contains the execution sequence of atomic tasks, the links to used realizations, and all intermediate results. This is an apparent prerequisite to build up cases ($\mathcal{R}4$) for CBR.

The NExT user interface (UI) attempts to integrate all the necessary tools in one common interface (the workbench metaphor). NExT's target audience are not computer scientists (but domain experts), so we tried to provide as simple as possible interaction approaches. A graphical data-flow style editor allows the user to easily create, start, pause, and adapt workflows and shows also the current state of the process during execution. Interactive browsers allow querying and browsing the KBs at any point in time and reasoners/planers/mediators act as wizard-like pop-ups to impart advise whenever asked.

3.1 Supporting Partial Executions and Adaptations

In contrast to pure static workflows, dynamically evolving processes in most cases cannot be fully specified before the start of the execution (e.g., some relevant information becomes only available at run-time). Therefore, we allow the user to start executing such processes, at least as long as the first steps in the sequence are well specified. Over time the amount of information rises and the process specification can be improved iteratively. Nevertheless, every problem that leads to a failure at runtime must be resolved. Hence, our concept of partial executions consists of four elements: (1) errors in the process specifications and/or exceptions and events must be detected before they affect the execution, (2) the process execution must be interruptible, (3) the user must be able to adapt the process to solve the problem, and (4) the execution must be re-continuable at a correct and optimal point to ensure the overall process consistency.

The process specification is validated each time before the execution starts (or re-continues) and at run-time, exceptions and events are caught by an exception handler. For further handling, we developed an ontology of possible incidents combined with adequate (semi-)automated resolution strategies (see Table 1).

Table 1. Excerpt of the problem ontology including the problem resolution strategies

Exception	Resolution Strategy
Hole in the sequence	1. Call planner to provide alternatives to fill the gap 2. Ask user to define a realization manually
Missing parameter makes a condition unsatisfiable	1. Query KB for processes, that produce the missing variables 2. Relax the condition 3. Remove processes whose effects make the condition unsatisfiable 4. At run-time, instantiate an input and let the user enter its value.

After catching a problem, NExT exploits this ontology to map each problem to an incident and determines then the priority for the problem resolution. Whenever a severe incident is detected that endangers the immediate continuation of the process the respective resolution strategy is applied instantly. On the other hand, minor, not time-critical problems are simply reported to the user which then can trigger the resolution manually (or the incident's severity rises over time above a threshold and then needs to be resolved immediately).

In most cases atomic tasks will not be interruptible when already under execution (except they explicitly support this behaviour). In most cases this issue can be addressed by interrupting the execution of the overall process when the execution of the current atomic task finishes. If the cause for the interruption is related to the outcome of the atomic task's execution then its outcome will have to either ignored, undone, or taken into account when it finished (in fact, this is an instance of the Specificity Frontier). Consider this logistics scenario: The plane transporting a piece of cargo for us is already airborne and we hear that the cargo staff at the destination airport is on strike. Thus, it will not be delivered at the demanded time, which is a hard constraint from our costumer. Since, it is not in our power to reroute the plane we have to adapt our process to the new circumstances.

Whenever possible the strategies attempt an automated, systems-led resolution of the exception. If this fails or the user intervenes, she is integrated in the loop (usually when too little information is known for the incident's resolution). We found that the majority of the resolution strategies include changes in the parameter values or adaptations of the process's control and/or data flow. Thus NExT must support such change operations and guide the user by the same mechanisms as during process creation phase. In addition it must be ensured that the process's new execution plan is consistent and of its execution trail/history remains correct and consistent. Before re-continuation of the process, the correct and optimal resumption point must be found. Whenever processes or parameter values that already were executed respectively computed are changed, it must be computed wether the execution path is still correct. If not, some processes must be rolled back to start over at a previous stage of the execution.

4 The NExT Prototype Implementation

In order to ensure domain independence our system's process meta-model defines the system's view on both processes and data entities ($\mathcal{R}3$, $\mathcal{R}5$ - Planner, Mediation). Code was written in terms of meta-model concepts whereas applications may inherit from or extend the meta-model for their own purposes. We describe processes by their IOPE, meaning the (semantic) notion of inputs, outputs, preconditions and effects (or post-conditions) and encode them in a declarative, formal and machine readable language. These are the minimal properties to use AI planners [4] ($\mathcal{R}3$). We furthermore differentiate between an *AtomicTask* and a *CompositeProcess*, whereas only the former can be related with one or several mappings to concrete realizations ($\mathcal{R}1$). The mapping contains the specific how (and where) to invoke a realization. A *CompositeProcess* on the other hand consists of a sequence of processes (potentially both atomic and composite). We have chosen to use OWL-S [5], because it supports most of the concepts we need out-of-the-box. When the execution of a process starts, the *HistoryTrail* is attached. All atomic tasks in their execution sequence and all intermediate values of all parameters are stored and define hereby a *Case* ($\mathcal{R}4$). *DataItems* can be nested to compose complex types ($\mathcal{R}5$ - mediation).

As our process execution engine we extended the Mindswap OWL-S API[1] with two features: First, we added a new type of grounding that an atomic task directly maps to a Java method. Second, we augmented the API with a facility to interrupt and resume a process execution. NExT, furthermore, provides a component to retrieve content for the user assistance ($\mathcal{R}1$, $\mathcal{R}3$-$\mathcal{R}5$) and a second component controlling the partial execution and dynamic adaptation aspect ($\mathcal{R}2$), which we present in more detail in the next section. The guidance component integrates several types of inferencing mechanisms:

- Deductive reasoners acting directly on the semantic model items. Specifically we used the Pellet reasoner [6] that came with the Mindswap API.
- A Case Based Reasoner can find past processes similar to the one in use. The current implementation relies on SimPack[2] [7] to retrieve similar entities.
- A plug-in interface to integrate several AI planners suitable for web service composition [8,9,10,11,12] into the system. Herby we can exploit their specialization on a certain planning aspect (e.g. to use planners addressing the changing information issue [13,14].

NExT is based on the Eclipse[3] framework. It is built as a workbench integrating graphical tools for all important purposes. A process editor allows the graphical creation and editing of workflows, their initiation and interruption. as well as monitoring all process-related information such as partial results during execution.

4.1 Supporting Partial Executions and Adaptations

We implemented a hierarchy with specific handlers for each type of incident in our ontology. These handlers encapsulate the incident itself, its severity, and implement its resolution strategy. To ease the development of these strategies general facilities for common steps are provided by the NExT system (such as UI widgets for user interaction or encapsulations for standard interactions with planners). We then extended the Mindswap OWL-S API to perform consistency checks on OWL-S process descriptions for design-time detection of problems and improved the exception handling within the execution engine for run-time detection. Both methods return an instance of incident stubs (or a list thereof). Depending on its severity, the incident is either added to a warning list for

[1] See http://www.mindswap.org/2004/owl-s/api
[2] See http://www.ifi.unizh.ch/ddis/research/semweb/simpack/
[3] See http://www.eclipse.org

detached resolution or the resolution strategy is immediately applied. As long as the problems are not resolved, The execution is postponed (when not yet started) or stays interrupted as long as not all severe problems are resolved.

Once the execution (re-) starts, the correct and optimal resumption point must be computed. If all changes took place after the current execution point, then we can simply continue the process. Else the algorithm attempts to roll back all the effects of the computation by applying the following strategy to each process step backwards until the first change:

1. When an inverse process is specified, invoke it. Proceed with next step.
2. When the process triggered no changes in the world state besides IO transformations, the corresponding values are set back. Proceed with next step.
3. The user is asked to perform the roll-back manually. To suggest potential solutions, the process library is queried to find a process with reversed input/output and pre-/post-condition.
4. Abort the execution.

Note, that we must consider that massive amounts of data can be generated during the execution. Hence, storing all intermediate results on all atomic tasks is unpractical. We therefore let the user define storage points in the process sequence at which the intermediate results are written on disk. Second, note, that finding the correct termination point for the strategy above shows some complexity too. Parallel execution of steps or loop construct may introduce dependencies between steps, which must be taken into account. The actually implemented algorithm takes these two points into consideration. With all these considerations, we propagate to fulfill $\mathcal{R}2$.

5 Preliminary Validation – The Introductory Scenario Revisited

Let us have a second look at our scenario. Peter conducts the same experiment, this time with a copy of NExT. First, he finds a similar project in the past, adopts its process sequence (see Figure 2 for a screenshot of NExT) and adapts it slightly to his needs. The "calibrate spectrometer" process is part of a) the pre-condition of the "run measurement" process and b) the standard "setup spectrometer" process template, so this time Peter does not forget this step. All the steps that can be automated such as spectrometer calibration or some simple analysis steps are executed automatically, but still some tasks need to be performed manually. Peter though encounters the same problem as before. But this time, the system provides him several potential solutions and he chooses the correct alternative amongst them. NExT re-sets the execution pointer to the correct position and continues the experiment avoiding its restart from scratch. In the end, Peter completes his experiment with success and much faster than earlier. In addition to his prose report, he uploads the whole case including all intermediate results, the history trail, and all additional information into a shared knowledge base of the journal. Furthermore, Peter is able to generalize a

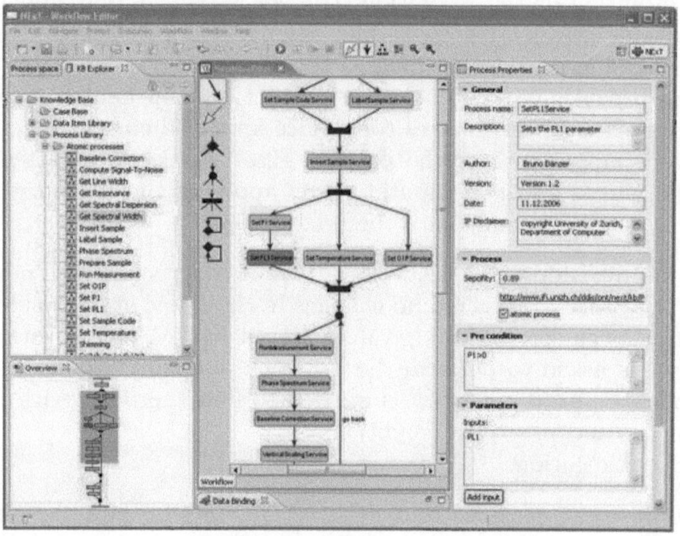

Fig. 2. Screenshot of NExT with the experimental sequence for a NMR case

part of the process sequence for a certain type of bio molecules into a template and publishes it in a NMR-community maintained Knowledge Base.

6 Related Work

Most of the Process Support Systems that have been developed in the past 30 years support either fixed, pre-defined, standard processes (e.g., workflow management systems) or informal ad-hoc dynamic processes (such as e-mail or groupware). The former use formal process definitions and can thus assist users during the workflow creation whereas the latter are not bound to strict rules to ensure flexible process adaptations at run-time. Only a few systems provide the base for both. The FAR [15] system implements an exception handler based on Event-Condition-Action (ECA) rules defined in a specific exception specification language. ADEPT$_{flex}$ [16] is based upon a graph-based workflow model and includes a complete and minimal set of (dynamic) change operations such as task insertion or deletion. Consistency and correctness are preserved hereby.

Modern systems from the life science community are oriented towards service orientation and grid computing. Prominent representatives thereof are Kepler [17], Pegasus [18], and Taverna [19]. They all provide the basic functionality to help users in the process life cycle, but none of them is focused on highly dynamic processes and tight user integration. Both Taverna and Kepler allow the user to manually pause an execution, Taverna can re-assign intermediate results during an interruption and Kepler allows in addition adaptations to the control and data flow. Pegasus on the other hand differentiates between the process and its realization, uses a partial-order planning [4] algorithm for guidance in the process composition and in combination with Virtual Data System [20] some interfaces support for data mediation are provided.

7 Future Work/Conclusion

In future, we want to deploy NExT in a life science environment to observe its practical usage for complex experiments. We plan to extend OWL-S by integrating the concepts of exceptions and events into the language. This would enable reasoning upon these concepts and thus improve the user guidance facilities. Furthermore, we will incrementally extend and refine our incident ontology and NExT's facilities for applying the resolution strategy. Also, we hope to exploit the ongoing research on both AI and non-AI composition algorithms to offer further guidance to users on the process composition and adaptation steps.

In this paper, we presented an approach for a process support system that assists its users throughout the whole process life cycle from creation to enactment, adaptation and publication in the end. The system aims at domains confronted with complex, long-running and highly dynamic processes. The process support system maintains a tight interaction with its human users: they want to be assisted in the creative work parts and they need to have the full control, but simple and monotonic tasks should be executed automatically to hold off the user from these time-consuming tasks.

As our main contribution we developed *five requirements for process support systems in complex experimental domains.* We have, furthermore, shown *a basic architecture and key implementation elements* of our NExT process support system based on Semantic Web technologies and AI planning and reasoning methodologies (planners, Case-Based Reasoning) that implements our vision. We especially focused on the *partial execution feature* ($\mathcal{R}2$) and showed how we detect problems, exceptions, and events at run-time (as well in design-time), allow for appropriate adaptations in the process, and resume the execution at the correct and optimal resumption point. We hope that such systems will enable the practical use of Semantic Web Services system in practice.

Acknowledgments. We are deeply indebted to Professor Konstantin Pervushin for his expertise in the NMR spectroscopy domain and to Professor Josef Joller and his students Markus Krähenbühl, Georg Kunz and Franco Sebregondi for their contribution in the runtime adaptation component.

References

1. Peltz, C.: Web services orchestration and choreography. Computer, Innovative Technology for Computing Professionals (2003)
2. Bernstein, A.: How can cooperative work tools support dynamic group processes? bridging the specificity frontier. In: Proceedings Computer Supported Cooperative Work (CSCW 2000), ACM Press (2000)
3. Veloso, M., Mulvehill, A., Cox, M.: Rationale supported mixed-initiative case-based planning. In: IAAI-97, Innovative Applications of Artificial Intelligence. (1997)

4. Ghallab, M., Nau, D., Traverso, P.: Automated Planning, theory and practice. Elsevier (2004)
5. Martin, D., Burstein, M., Hobbs, J., Lassila, O., McDermott, D., McIllraith, S., Narayanan, S., Paolucci, M., Parsia, B., Payne, T., Sirin, E., Srinivasan, N., Sycara, K.: Owl-s: Semantic markup for web services. (2004)
6. Sirin, E., Parsia, B., Grau, B.C., Kalyanpur, A., Katz, Y.: Pellet: a practical owl-dl reasoner. (Journal of Web Semantics)
7. Bernstein, A., Kaufmann, E., Kiefer, C., Bürki, C.: Simpack: A generic java library for similiarity measures in ontologies. Technical report, Department of Informatics, University of Zurich (2005)
8. Sirin, E., Parsia, B., Wu, D., Hendler, J., Nau, D.: Htn planning for web service composition using shop2. Journal of Web Semantics **1**(4) (2004) 377–396
9. Klusch, M., Gerber, A., Schmidt, M.: Semantic web service composition planning with owls-xplan. In: 1st International AAAI Fall Sympsoium on Agents and the Semantic Web. (2005)
10. Sheshagiri, M., desJardins, M., Finin, T.: A planner for composing service described in daml-s. In: International Conference on Automated Planning and Scheduling. (2003)
11. McIlraith, S., Son, T.: Adapting golog for composition of semantic web services. In: Proceedings of the 8th Intl. Conference on Knowledge Representation and Reasoning. (2002)
12. Ponnekanti, S.R., Fox, A.: Sword: A developer toolkit for web service composition. In: Proceedings Intl. WWW Conference. (2002)
13. Kuter, U., Sirin, E., Parsia, B., Nau, D., Hendler, J.: Information gathering during planning for web service composition. Journal of Web Semantics **3**(2) (2005)
14. Au, T.C., Kuter, U., Nau, D.: Web service composition with volatile information. In: Proceedings of the International Semantic Web Conference (ISWC). (2005)
15. Casati, F., Ceri, S., Paraboschi, S., Pozzi, G.: Specification and implementation of exceptions in workflow mangament systems. ACM Transactions on Database Systems **24**(3) (1999) 405–451
16. Reichert, M., Dadam, P.: Adeptflex - supporting dynamic changes of workflows without losing control. Journal of Intelligent Information Systems - Special Issue on Workflow Managment **10**(2) (1998) 93–129
17. Ludscher, B., Altintas, I., Berkley, C., Higgins, D., Jaeger-Frank, E., Jones, M., Lee, E., Tao, J., Zhao, Y.: Scientific workflow management and the kepler system. Journal for Concurrency and Computation: Practice and Experience, Special Issue on Scientific Workflows (2005)
18. Gil, Y., Ratnakar, V., Deelman, E., Spraragen, M., Kim, J.: Wings for pegasus: A semantic approach to creating very large scientific workflows. In: Proceedings OWL: Experiences and Directions. (2006)
19. Oinn, T., Addis, M., Ferris, J., Marvin, D., Senger, M., Greenwood, M., Carver, T., Glover, K., Pocock, M.R., Wipat, A., Li, P.: Taverna: A tool for the composition and enactment of bioinformatics workflows bioinformatics journal 20(17) pp 3045-3054, 2004. Bioinformatics Journal **20**(17) (2004) 3045–3054
20. Zhao, Y., Wilde, M., Foster, I., Voeckler, J., Dobson, J., Glibert, E., Jordan, T., Quigg, E.: Virtual data grid middleware services for data-intensive science. In: Middleware 2004, Concurrency, Practice and Experience. (2004)

WSMO Studio – A Semantic Web Services Modelling Environment for WSMO
(System Description)

Marin Dimitrov, Alex Simov, Vassil Momtchev, and Mihail Konstantinov

Ontotext Lab. / Sirma Group
135 Tsarigradsko Shose Blvd., Sofia 1784, Bulgaria
firstname.lastname@ontotext.com

Abstract. The Web Service Modelling Ontology (WSMO) provides a unique, highly innovative perspective onto the Semantic Web Services domain. Robust and easy-to-use tools play crucial role for the adoption of any technological innovation and indeed the overall value of the innovation can be severely undermined by the lack of proper tools supporting it. In this paper we present a prototype of an integrated modelling environment that supports and elaborates the innovative WSMO perspective.

1 Introduction

Robust and easy-to-use tools play a crucial role for the adoption of any new technology. Indeed, the overall value of a technological innovation can be severely undermined by the lack of proper tools to support it. The Web Service Modelling Ontology (WSMO, [1][2]) provides a unique, highly innovative perspective onto the Semantic Web Services domain.

Unfortunately, current tool support in the area is still lagging behind the theoretical advancements but real progress can be achieved only when the Semantic Web Services technology is easily usable. Another major problem with the tool landscape at present is that almost all of the available tools focus on only one of the relevant Semantic Web Services aspects (for example only ontology management or only service composition), while end users rarely focus on a single task and an integrated modelling environment will be more appropriate and productive.

In this paper we present a prototype that supports and elaborates the innovative WSMO perspective, making the technology accessible and easy to use for early adopters. In particular, we present *WSMO Studio*[1] – an open source integrated modelling environment for the Semantic Web Services domain, which is being developed within several EU-funded research projects[2].

[1] http://www.wsmostudio.org

[2] The work described in this document is partly funded by the European Commission under the IST projects DIP (FP6-507483), InfraWebs (FP6-511723), SemanticGov (FP6-027517) and SUPER (FP6-026850).

E. Franconi, M. Kifer, and W. May (Eds.): ESWC 2007, LNCS 4519, pp. 749–758, 2007.

This paper is organised as follows: section 2 presents the goals of *WSMO Studio.* section 3 provides details on present *WSMO Studio* functionality. section 4 attempts to evaluate the extent to which *WSMO Studio* has achieved the defined goals and contains a brief overview of several prominent Semantic Web Services tools and their relation to *WSMO Studio.* Finally, section 5 provides details on work in progress for domain specific extensions of *WSMO Studio.*

2 WSMO Studio Goals

The three main goals of *WSMO Studio* are:

1. Providing a prototype that supports and elaborates the WSMO approach to Semantic Web Services, making the technology accessible and easy to use for early adopters. As we have already pointed out, adequate tool support is crucial for the adoption of a technological innovation.
2. Providing an integrated environment that maximises the productivity of the user. Current SWS tools usually focus on only one aspect (ontology editing, service composition, etc.) but users most often need functionality that covers various tasks in an integrated manner.
3. Providing an extensible environment where 3rd party tool providers can easily add new functionality or modify and customise existing functionality.

With respect to the first goal, we have already presented in [3] a summary of the tasks that a Semantic Web Services modelling environment should cover, such as:

- ontology modelling,
- semantic annotation of existing web services,
- working with semantic repositories for publishing, browsing and querying of WSMO ontologies, services, goals and mediators,
- goal-based service discovery,
- specification of service compositions (i.e. orchestration and choreography interfaces),
- interaction with Semantic Web Service runtime environments (such as WSMX[3] and IRS-III[4]).

The second goal we have set is providing functionality that covers various related Semantic Web Services tasks in an integrated modelling environment, so that users will not need to use several different tools to accomplish their goals. As noted in [4] users rarely focus on only one aspect or task when they work – indeed, users usually play different roles and perform different related tasks, so the environment should provide functionality covering various perspectives and tasks.

[3] `http://www.wsmx.org`

[4] `http://kmi.open.ac.uk/projects/irs/`

Finally, extensibility is a requirement that is mostly ignored by the current Semantic Web Services tools. But as noted in [4]:

> *...open-ended extensibility is essential in the commercial IDE arena because no IDE vendor could possibly provide a sufficient set of useful tools to satisfy all customer needs. Which third party tool will be bundled as an add-in for a particular IDE is determined by market forces.*

This statement, based on commercial IDE experience, can equally well be applied for SWS modelling environments. Indeed, tool extensibility is even more important for an emerging domain such as the Semantic Web Services one, so that when the domain evolves, tools will be able to follow this evolution and provide the relevant functionality.

3 WSMO Studio Functionality

This section presents a summary of the main features of *WSMO Studio*, with respect to the requirements identified in [3]. A detailed overview of *WSMO Studio* is available in [5].

WSMO Studio is built on top of the Eclipse platform[5]. The functionality of *WSMO Studio* is available both as a standalone application and as a set of individual components (called *plug-ins*), that can be incorporated into 3rd party applications.

An important feature of Eclipse is that its component model presents a declarative specification of ways to extend an application, called *extension points* [6]. New plug-ins may extend existing plug-ins and may be easily incorporated into an application.

3.1 Core Components

The core runtime layer provides functionality common across all components of the *WSMO Studio*.

The most important component of this layer is the *wsmo4j* plug-in, based on the *wsmo4j* framework[6]. The functionality includes: creating WSMO models, validating models, export and import from various WSML formats (as defined in [7]) and languages such as RDF and a subset of OWL-DL.

The *wsmo4j* plug-in publishes several extension points, so that 3rd party extensions (for example a new parser) can be easily integrated.

In addition, the runtime layer contains several utility components for workspace management and a shared entity cache.

3.2 WSMO Editor

This plug-in provides the main User Interface for modelling of WSMO ontologies, goals and services. Import and export from WSML formats, RDF and a subset of

[5] `http://www.eclipse.org`

[6] `http://wsmo4j.sourceforge.net`

OWL-DL is provided via the core layer. Most of the User Interface elements of the WSMO editor are also available for extension and customisation via published extension points.

3.3 Choreography Editor

The Choreography editor (Figure 1) provides the User Interface for describing WSMO choreography interfaces, as defined in [8].

A WSMO choreography description is comprised of:

- a *state signature* that specifies the concepts and relations of an ontology that will be used to represent the choreography states, together with their respective roles.
- a set of *transition rules* that express the state changes.

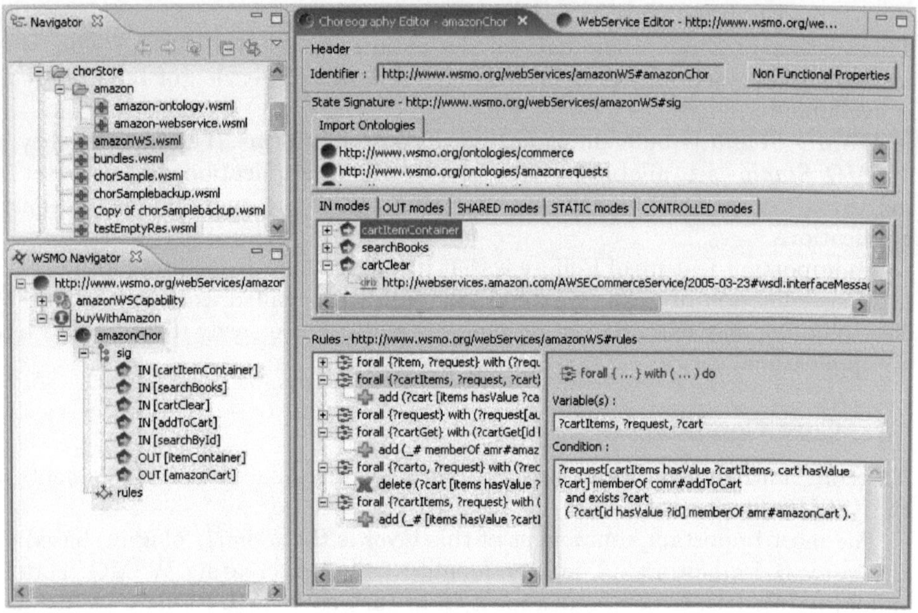

Fig. 1. WSMO Studio – choreography editor

The choreography descriptions, created by *WSMO Studio*, are used by execution environments (such as WSMX) during the execution of Semantic Web Services.

3.4 SAWSDL Editor

The SAWSDL editor (Figure 2) provides functionality for attaching semantic annotations to existing WSDL descriptions according to the SAWSDL recommendation [9]. SAWSDL is used at present as the grounding mechanism for

WSMO, i.e. mapping the semantic interface description of a service (in terms of imported ontologies, capability and choreography interface) to its WSDL interface.

The User Interface allows that semantic annotations are attached to the following WSDL elements: simple / complex XML types, messages, operations and interfaces.

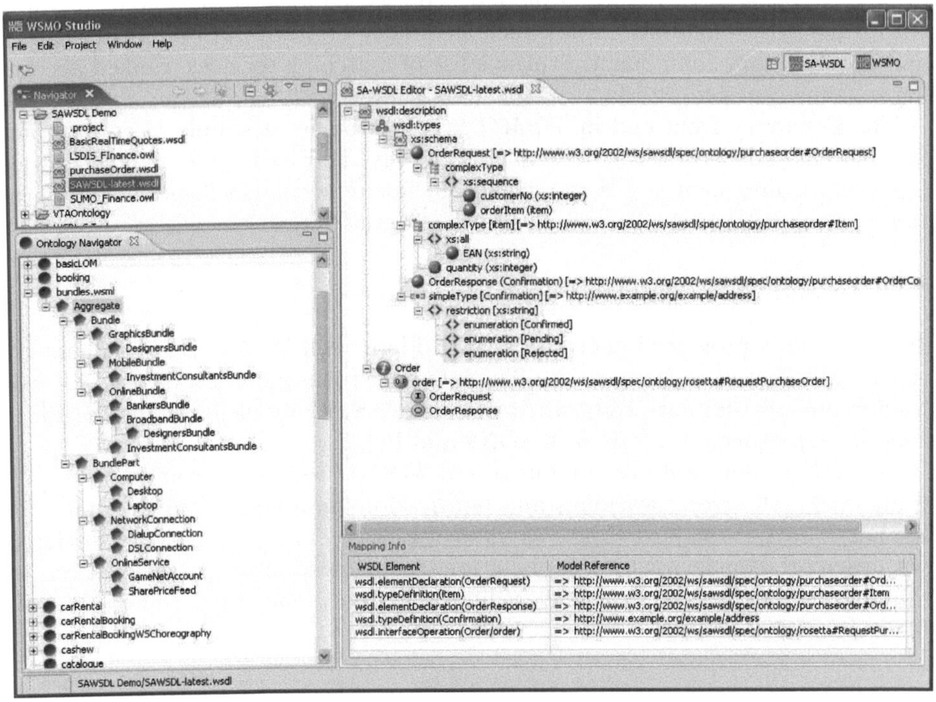

Fig. 2. WSMO Studio – SAWSDL editor

At present the SAWSDL editor does not provide support for the lifting/lowering schema, i.e. the XSLT expressions that define the exact mappings between complex XML types defined in WSDL documents and concepts in an ontology, but such functionality will be provided in future versions.

3.5 Repository Front-End

The Repository front-end provides an abstraction of an Semantic Web Service repository for storing and querying WSMO descriptions.

WSMO Studio will be able to interact with a particular repository as long as it provides a special adaptor based on the WSMO API repository interfaces[7].

[7] `http://wsmo4j.sourceforge.net`

At present *WSMO Studio* provides an integrated ORDI repository[8] as well as adaptors for remote IRS-III and WSMX repositories. Adaptors for other repositories will be developed in the future.

3.6 Service Discovery Front-End

Discovery of WSMO services at present is based on the idea of goal-based discovery [10], i.e. the user provides a request represented as a WSMO goal (possibly with some additional information such as quality-of-service or cost restrictions), and the discovery component returns a list of matching services, ranked according to some ranking criteria.

The Discovery front-end in *WSMO Studio* provides a simple User Interface for goal-based discovery engines. At present only the EPFL QoS-enabled Service Discovery Component [11] is supported but future versions will provide integration options with other discovery engines and matchmakers.

3.7 Integrated Validator and Reasoners

WSMO Studio provides integrated WSML-Flight and WSML-DL reasoners via the WSML2Reasoner framework[9]. The integrated reasoners can be used for validation and satisfiability tests when building WSMO ontologies. The currently supported reasoners are MINS, KAON2 and Pellet.

WSMO Studio contains an integrated WSML validator from the *wsmo4j* framework. All errors, warnings and notifications produced from the validator are listed in the standard *Problems* view of Eclipse. The information associated with each problem is: severity, explanation message and problematic location. To minimise the processing overhead, validation is performed only when a WSML file is opened or saved.

4 Evaluation and Related Tools

This section provides an evaluation of the extent to which *WSMO Studio* has achieved the defined goals and contains a brief overview of several prominent Semantic Web Services tools and their relation to *WSMO Studio*

4.1 Evaluation of WSMO Studio

With respect to the outlined goals in section 2, the following assessments can be made:

- *Functionality* – the *WSMO Studio* functionality, outlined in section 3, covers most of the basic tasks for Semantic Web Services modelling, namely support for modelling WSMO ontologies, services, goals; support for choreography

[8] `http://www.ontotext.com/ordi/`

[9] `http://dev1.deri.at/wsml2reasoner/`

descriptions; means for attaching semantic annotations to WSDL documents; front end to Semantic Web Service repositories and matchmakers; integrated reasoners and validators.

Other functionality, such as interaction with Semantic Web Services run-time environments (e.g. WSMX and IRS-III) or support for ontology mediation is still not available in *WSMO Studio* but is already provided by other tools (e.g. WSMT) and may be easily integrated into *WSMO Studio.*

- *Integration* – *WSMO Studio* provides functionality covering various Semantic Web Services modelling aspects in a single integrated modelling environment. Since *WSMO Studio* is based on the Eclipse platform, the functionality is provided by means of plug-ins, which can be also integrated into other products and applications.
- *Extensibility* – the Eclipse extensibility mechanism allows that functionality is added, replaced or customised by means of published extension points. *WSMO Studio* defines several such extension points, which allows customisation and replacement of specific User Interface elements, as well as adding new functionality (e.g. adaptors for new Semantic Web Services repositories or matchmakers, reasoners, etc.).

  WSMO Studio extensions have already been developed by 3rd parties within the DIP and InfraWebs research projects.

An additional aspect that influences all three goals is the *flexible licensing* – *WSMO Studio* is licensed under LGPL [12], which allows that individual components (plug-ins) are incorporated within and distributed with 3rd party applications without any restrictions on the licensing terms of the latter. Our expectation is that the Open Source licence increases the visibility of a tool and the chances for 3rd party contributions and adoption, though it is not possible to provide a reliable assessment on that.

4.2 Related Tools

Current tool support in the area of Semantic Web Services is still lagging behind the theoretical advancements. Integrated toolsets, that provide functionality covering many aspects of Semantic Web Services are still not readily available and tools generally lack means of extensibility, that could improve their reusability and adoption.

This section contains brief overviews of several prominent Semantic Web Services tools related to *WSMO Studio.*

The Web Services Modelling Toolkit[10] (WSMT) is an Eclipse based application which provides an advanced ontology viewer, an ontology mediation component and a WSMX management component as well as support for modelling WSMO ontologies and an integrated reasoner. Since WSMT is Eclipse based, its plug-ins can be used from within *WSMO Studio* and and vice versa.

[10] `http://sourceforge.net/projects/wsmt`

METEOR-S[11] is a Semantic Web Services toolset that provides functionality for adding semantics to web services standards like WSDL, UDDI and WS-BPEL. *Radiant* is an Eclipse based tool for adding semantic annotations to WSDL documents according to the SAWSDL and WSDL-S recommendations. *Lumina* provides an Eclipse based user interface for discovery of web services. Additionally, the METEOR-S toolset also includes a Service Discovery Engine and a semantically enhanced service registry based on UDDI.

OWL-S Editor[12] is a Protégé[13] plug-in for modelling OWL-S services. The editor provides means for describing atomic and composite processes (i.e. the OWL-S Process definition). The OWL-S Editor also provides some limited grounding support, by generating 'skeleton' OWL-S services from WSDL files. The options for integrating new functionality and extensions of existing functionality are quite limited.

Semantic Tools for Web Services[14] is a set of Eclipse plug-ins which offer two main functionalities: semantic annotation of existing WSDL documents into WSDL-S [13], and semantic matching and composition of web services (based on inferencing via the ABLE rule engine), where the queries are represented by WSDL files and the results are references to the matched WSDL service descriptions or a composition of services. The plug-ins are not extensible. STWS is not directly comparable to *WSMO Studio* since its focus is on WSDL-S based semantic annotations for WSDL.

5 Future Work

WSMO Studio has already been used as the Semantic Web Services modelling environment within two EU-funded research projects, DIP[15] and InfraWebs[16].

Within the scope of two active research projects, SUPER[17] and SemanticGov[18], *WSMO Studio* will be enhanced with domain specific extensions for the Business Process Modelling and eGovernment domains.

Within SUPER, *WSMO Studio* will provide functionality for modelling and querying of Semantic Business Processes using ontologies and Semantic Web Services, according to the Business Process Modelling Ontology (BPMO) [14][15]. In summary, BPMO provides an abstraction over business process modelling languages such as EPC [16] and BPMN [17], support for workflow patterns [18], and added semantics in terms of using ontology concepts for explicit modelling of the dataflow and attaching abstract WSMO goals to process tasks. The BPMO descriptions will be semi-automatically translated to sBPEL, which is the semantic version of WS-BPEL.

---

[11] `http://lsdis.cs.uga.edu/projects/meteor-s/`
[12] `http://owlseditor.semwebcentral.org/`
[13] `http://protege.stanford.edu/`
[14] `http://www.alphaworks.ibm.com/tech/wssem`
[15] `http://dip.semanticweb.org`
[16] `http://www.infrawebs.org`
[17] `http://www.ip-super.org/`
[18] `http://www.semantic-gov.org/`

Within SemanticGov, *WSMO Studio* will provide means for modelling of Public Administration services based on the Governance Enterprise Architecture (GEA) and the WSMO-PA [19][20] extension of WSMO for the Public Administration domain.

6 Conclusion

In this paper we presented a prototype of an integrated modelling environment for the Semantic Web Services domain, called *WSMO Studio*. The prototype has already been used for modelling of Semantic Web Services in several research projects.

Our future work will be focused on providing domain specific extensions for the application of WSMO in the eGovernment and Business Process Modelling domains.

References

1. Feier, C., Roman, D., Polleres, A., Domingue, J., Stollberg, M., Fensel, D.: Towards intelligent web services: Web Service Modeling Ontology (WSMO). In: In Proc. of the International Conference on Intelligent Computing (ICIC 2005), Hefei, China (2005)
2. Roman, D., Keller, U., Lausen, H., de Bruijn, J., Lara, R., Stollberg, M., Polleres, A., Feier, C., Bussler, C., Fensel, D.: Web Service Modeling Ontology. Applied Ontology **1** (2005) 77–106
3. Dimitrov, M., Simov, A., Momtchev, V., Ognyanov, D.: WSMO Studio - an integrated service environment for WSMO. In Bussler, C., Fensel, D., Keller, U., Sapkota, B., eds.: In Proceedings of the 2nd Workshop on WSMO Implementations (WIW 2005). Volume 134., Innsbruck, Austria (2005)
4. Rivieres, J.D., Wiegand, J.: Eclipse: A platform for integrating development tools. IBM Systems Journal **43** (2004)
5. Dimitrov, M., Simov, A., Momtchev, V., Konstantinov, M.: WSMO Studio Users Guide. (2006) Available online at http://www.wsmostudio.org.
6. Gruber, O., Hargrave, B.J., McAffer, J., Rapicault, P., Watson, T.: The Eclipse 3.0 platform: Adopting OSGi technology. IBM Systems Journal **44** (2005) 289–299
7. de Bruijn, J., Lausen, H., Krummenacher, R., Polleres, A., Predoiu, L., Kifer, M., Fensel, D.: D16.1: The Web Service Modeling Language WSML. WSML working draft, DERI (2005) http://www.wsmo.org/TR/d16/d16.1/v0.3/.
8. Roman, D., Scicluna, J., Fensel, D., Polleres, A., de Bruijn, J.: D14: Ontology-based choreography of WSMO services. WSMO working draft, DERI (2006) Available at online at http://www.wsmo.org/TR/d14/v0.4/.
9. Farrell, J., Lausen, H.: Semantic annotations for WSDL and XML Schema. W3C working draft, W3C (2006) Available online at http://www.w3.org/TR/sawsdl/.
10. Keller, U., Lara, R., Polleres, A., Toma, I., Kifer, M., Fensel., D.: D5.1: WSMO web service discovery. WSMO working draft, DERI (2004) Available online at http://www.wsmo.org/TR/d5/d5.1/v0.1/.
11. Vu, L., Hauswirth, M., Porto, F., Aberer, K.: A search engine for QoS-enabled discovery of Semantic Web Services. International Journal of Business Process Integration and Management (IJBPIM) (2006)

12. Free Software Foundation: GNU Lesser General Public License, version 2.1 (1999) Available at http://www.opensource.org/licenses/lgpl-license.php.
13. Akkiraju, R., Farrell, J., Miller, J., Nagarajan, M., Schmidt, M.T., Sheth, A., Verma, K.: Web service semantics – WSDL-S. W3C member submission, W3C (2005) Available at http://www.w3.org/Submission/WSDL-S/.
14. Hepp, M., Leymann, F., Domingue, J., Wahler, A., Fensel, D.: Semantic business process management: A vision towards using semantic web services for business process management. In: Proceedings of the IEEE ICEBE 2005, Beijing, China (2005) 535–540
15. Hepp, M., Belecheanu, R., Domingue, J., Filipowska, A., Kaczmarek, M., Kaczmarek, T., Nitzsche, J., Norton, B., Pedrinaci, C., Roman, D., Stein, S.: Business process modelling ontology and mapping to WSMO. SUPER technical report, Project IST-026850 SUPER (2006)
16. Mendling, J., Neumann, G., Nüttgens, M.: Towards workflow pattern support of Event-Driven Process Chains (EPC). In: 2nd GI Workshop XML4BPM - XML for Business Process Management, BTW 2005, Karlsruhe, Germany (2005) 23–38
17. Object Management Group: Business process modeling notation specification (BPMN). Technical report, Object Management Group (2006) Available at http://www.bpmn.org.
18. Aalst, W.M.P.V.D., Hofstede, A.H.M.T., Kiepuszewski, B., Barros, A.P.: Workflow patterns. Distributed and Parallel Databases **14** (2003) 5–51
19. Wang, X., Vitvar, T., Peristeras, V., Mocan, A., Goudos, S., Tarabanis, K.: WSMO-PA: Formal specification of public administration service model on semantic web service ontology. In: Proceedings of the Hawaii International Conference on System Sciences (HICSS), Waikoloa, Big Island, Hawaii (2007)
20. Peristeras, V., Mocan, A., Vitvar, T., Nazir, S., Goudos, S., Tarabanis, K.: Towards semantic web services for public administration based on the web service modeling ontology (WSMO) and the governance enterprise architecture (GEA),. In: Proceedings of the DEXA, Krakow, Poland (2006)

An Annotation Tool for Semantic Documents
(System Description)

Henrik Eriksson

Dept. of Computer and Information Science
Linköping University
SE-581 83 Linköping, Sweden
her@ida.liu.se

Abstract. Document annotation is a common technique for relating text and knowledge representation. Although the semantic web emphasizes the annotation of web pages, there are other types of documents that can benefit from ontology-based annotations. PDF documents combined with OWL ontologies form semantic documents that support professional printing, on-line viewing, and ontological models. PDFTab is an extension to the Protégé ontology editor that allows developers to annotate PDF documents with OWL-based ontologies. It is possible to add OWL ontologies to preexisting PDF documents and to relate document parts to concepts in the ontology. PDFTab integrates Adobe Acrobat with Protégé and allows users to switch seamlessly between the document and ontology views. PDFTab illustrates how it is possible to extend semantic-web techniques to the widely-used PDF format while maintaining strong support for ontology development and editing.

1 Introduction

The *semantic web* is a well-known approach to adding metadata in terms of ontologies to documents and to facilitating machine-to-machine communication [1]. Although the semantic-web languages and tools are inspired by the web and its underlying formats, such as HTML and XML, there are other possible application frameworks for the semantic-web approach. For example, there is a growing interest in multimedia annotation [2]. Furthermore, there are many types of textual documents that require formats other than HTML. For example, Adobe's Portable Document Format (PDF) is a widely-used electronic-document format, which supports on-line viewing and both desktop and professional printing [3].

The prospect of adding semantic annotations to these document formats is interesting because it makes it possible to extend the scope of the semantic web to include this type of electronic documents to form *semantic documents*. We have previously explored the benefits of semantic documents based on annotated PDF documents and their potential applications [4, 5]. One conclusion from this work is that, while semantic documents integrate well with regular electronic documents and allow viewing and printing using standard tools, they require new solutions for document annotation.

E. Franconi, M. Kifer, and W. May (Eds.): ESWC 2007, LNCS 4519, pp. 759–768, 2007.

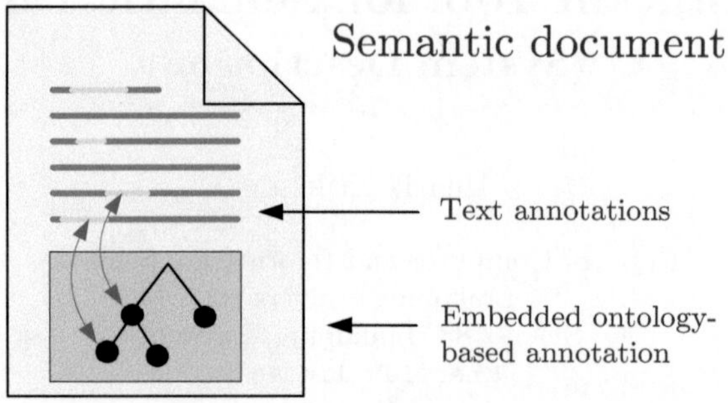

Fig. 1. Semantic documents. The ontology annotations are included in the internal document format.

Adequate tool support is a prerequisite for document annotation. Document formats such as PDF are difficult to create and modify manually (e.g., using a text editor). Thus, we need tools that can both read and write the document format and that can support markup in this format, for instance by selecting document parts and relating them to the ontology. Our approach is to extend the ontology editor Protégé [6] with a plug-in that supports seamless ontology development and annotation of PDF documents. The PDFTab extension allows developers to load PDF documents into Protégé and annotate them in a graphical user interface using Adobe Acrobat. Furthermore, PDFTab allows developers to relate document markup to individuals in the ontology. The advantage of the PDFTab tool architecture is that it retains the rich ontology-editing functionality of the Protégé environment while adding PDF support.

2 Semantic Documents

The semantic-document approach combines electronic documents with ontologies [4]. The principal idea is to store the ontologies in the internal document representation. Just as RDF and OWL statements can be invisible parts of web pages, semantic documents include ontologies not shown when users view and print the documents. Figure 1 illustrates the semantic-document approach. The documents contain the ontologies and the relationships between concepts in the ontologies and document parts. The rest of the document, including text, graphics, and formatting information, can remain unchanged. This approach allows PDF tools such as Adobe Acrobat to recognize and open the documents in the normal way.

While our approach in general does not assume a particular document format, there are several advantages of using PDF for prototyping and evaluating semantic documents.

1. PDF is a common format for printing and on-line publication. Since the early 1990s, authors and publishers have used PDF for document exchange and publishing.

2. PDF is an open and documented format [3]. There are many commercial and open source applications for generating, modifying, and viewing PDF documents.
3. The PDF specification allows for basic document annotations, such as text highlighting (by color) and the addition of textual notes (similar to MS Word).
4. PDF specifies document compressions and security (e.g, encryption and signing). These methods work for ontologies stored in PDF documents.

In addition, Adobe has added the extensible metadata protocol (XMP) to PDF [7]. XMP supports RDF-based metadata in the file formats of several Adobe products. Unfortunately, the way XMP uses the RDF structures in Acrobat makes it incompatible with OWL. Thus, we must store OWL-based metadata for semantic documents separate from XMP in PDF documents.

Compared to HTML, PDF is a much more complex format. One of the major advantages of the HTML format is that it allows document creation and editing with basic tools, such as text editors. Indeed, it is possible to add OWL annotations to HTML documents manually using a text editor (although this requires a lot of work). However, for a complex document format like PDF specialized annotation tools are essential. As mentioned previously, other widely-used formats, such as MS Word, may also be used as the basis for semantic-document markup. Tallis [8] discussed an interesting MS Word extension that supports semantic annotations. Authors can use this tool to add OWL statements to MS Word files interactively.

3 Annotation Model

Before presenting the tool architecture, let us first discuss the annotation model. There are several models for adding annotations to documents. In this section, we will discuss three of them. The first possibility is to add metadata to documents without relating the metadata to the document content or parts. XMP, for example, uses this method. The second modeling alternative is to relate the metadata to sections of the document text and other document parts. The advantage of the latter model is that it enables tight integration between documents and ontologies. For example, the model enables users and application programs to use the document text to look up parts of the ontology and vice versa.

Finally, it is possible to store metadata outside the documents, for instance in a separate metalevel database. The advantage of this approach is that no changes are required to the documents. However, the metadata do not follow the documents if they are copied, moved, or communicated to others electronically. Moreover, it is not possible to collect metadata from documents published on the web.

In our work on semantic documents and the PDFTab implementation, we choose the second approach; that is, to store the metadata in the document and to relate document annotations to the ontologies. Our motivation for using this model is that we want to integrate documents and ontologies and keep the metadata within the documents.

4 Tool Architecture

The goal for the tool architecture is to combine advanced ontology development with full support for PDF, such as fully compatible rendering. The strategy is to integrate two state-of-the-art tools to form an environment that supports both perspectives; documents and ontologies. The ontology part of the semantic-document annotation tool builds on the Protégé environment [6]. The PDF handling and rendering part uses Adobe Acrobat.

Protégé is a suitable basis for the ontology part because it has an advanced graphical user interface and it supports different ontology formats, including OWL [9]. Furthermore, Protégé is a widely-used tool with an active user community. The Java-based Protégé implementation features a core application programming interface (API) for manipulation of ontologies and knowledge bases. The basic system consists of this core API combined with standard plug-ins for graphical editing and storage of ontologies and knowledge bases. This extendability enables us to add new functionality for adding document views to the Protégé user interface and to interconnect Protégé with Adobe Acrobat.

There are many advantages of using Acrobat as the PDF-handling part. Acrobat is the standard tool for creating, manipulating, and viewing PDF documents. There are alternative PDF viewers, but many of these tools may have rendering incompatibilities for certain documents. Specifically, we could not find any Java-based PDF viewers with full PDF support. The main disadvantage of Acrobat, for our purpose, is that there is no support for communication and interoperability with Java programs. It is a major challenge to integrate Protégé with Acrobat because of the different programming languages and communication platforms used. Nevertheless, we believe that the best semantic-document tool solution can be accomplished by integrating these systems.

Figure 2 shows the overall tool architecture. Both Protégé and Acrobat use collections of standard extensions. The PDFTab extension interconnects Protégé and Acrobat using such plug-ins. PDFTab act both as a graphical plug-in that adds a tab to the main Protégé user interface and as a module for managing the communication with Acrobat. The Acrobat plug-in supports graphical markup of PDF documents and manages the communication with the Protégé plug-in. These plug-ins work in concert to enable the users to move seamlessly between the ontology and document views.

An important aspect of the architecture is support for multiple documents. The Adobe and Protégé extensions handle several documents, each with their own ontology. Protégé uses a *main ontology,* which can import subontologies. In Protégé, users can select the *active ontology* for editing. Likewise, Acrobat supports several open documents. PDFTab allows developers to associate a list of documents with the main ontology, much like importing the subontologies contained in the documents. The support for multiple documents enables developers to create an ontology structure consisting of a group of semantic documents, which refers to a common main ontology.

The current PDFTab implementation is a research prototype that allows users to experiment with semantic documents and to evaluate the architecture in

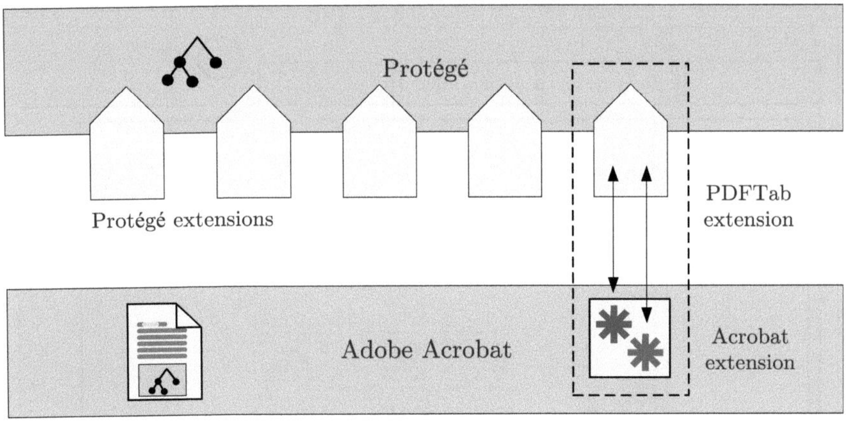

Fig. 2. The PDFTab tool architecture. Extensions of Protégé and Acrobat allow bidirectional communication.

practice. As such, the implementation has a number of limitations. Although Protégé is available on all platforms that support Java and the graphical user interface Swing and Acrobat is available on almost all major platforms (e.g., MS Windows, Apple Macintosh, Sun Solaris, and Linux), the current PDFTab implementation supports the MS Windows platform only. The Adobe plug-in API and inter-application communication protocol is platform dependent, and we choose to focus on the MS Windows platform. The implementation uses the MS Component Object Model (COM) for the communication between the applications, because Acrobat supports this technique.

5 User Interface

The overall design goal for the tool user interface is to provide an integrated environment for documents and ontologies. The user interface should follow the general interaction and visual style of both Protégé and Acrobat while making it easy for users to move between the documents and ontologies. Furthermore, the user interface should support handling of multiple documents. Finally, the other Protégé tabs for ontology editing should work as before, including custom and third-part tab extensions.

The user interface of PDFTab consists of two major parts. The Protégé part is a user-interface tab that adds a document view of the Protégé ontology. The Acrobat part consists of additional toolbar buttons for creating three types of semantic annotations and the graphical highlighting of document text and regions. Let us begin by discussing the Protégé document tab. Figure 3 illustrates the layout of the tool user interface with the document tab selected (Fig. 3a). The list of documents provides an overview of the documents associated with the main ontology in Protégé (Fig. 3b). The Acrobat view shows the selected document and allows users to browse it and add annotations (Fig. 3c). Protégé

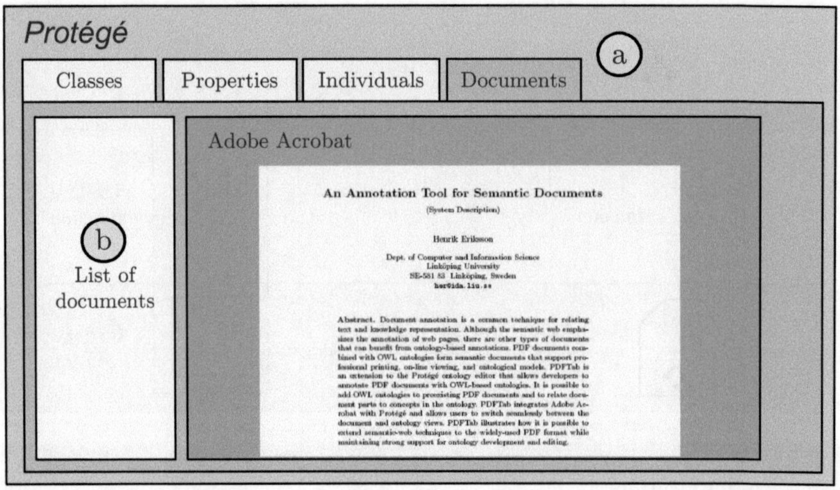

Fig. 3. The organization of the PDFTab user interface. (a) The tab for viewing PDF documents. (b) The list of the documents available. (c) The Acrobat view of the selected document.

uses Acrobat as a user-interface extension just like web browsers, such as MS Internet Explorer and Mozilla Firefox, use Acrobat to support viewing of on-line PDF documents. The main advantage of this user-interface layout is that users can easily move between the document view and one of the ontology views (e.g., classes, properties, and individuals).

5.1 Document Management

The support for multiple documents requires user-interface functionality for adding, removing, and inspecting documents. The list of documents (Fig. 3b) contains all documents associated with the project. Users can add existing documents to this list by selecting the document add button or by dragging documents to the list. In PDFTab, the documents must be plain PDF files created by an external application, such as Acrobat Distiller, or PDF-based semantic documents produced by PDFTab. Users can select items from the list of documents for inspecting metadata and for viewing and annotating in the Acrobat view (Fig. 3c). Alternatively, users can access a list of all annotations for the selected document. This annotation browser provides an overview of the document annotations, which is helpful for long documents with many annotations. Selecting an item in the annotation browser brings up the corresponding annotation individual, which is part of an annotation ontology.

5.2 Annotation Editing

Users edit annotations in the Acrobat view. PDFTab makes easy to add new annotations to the selected document. To create a new annotation, the user

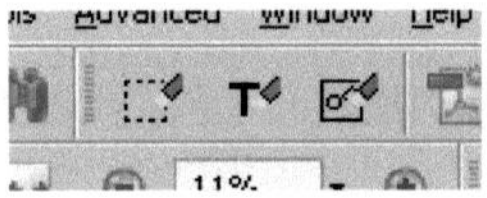

Fig. 4. Annotation buttons in the Acrobat view

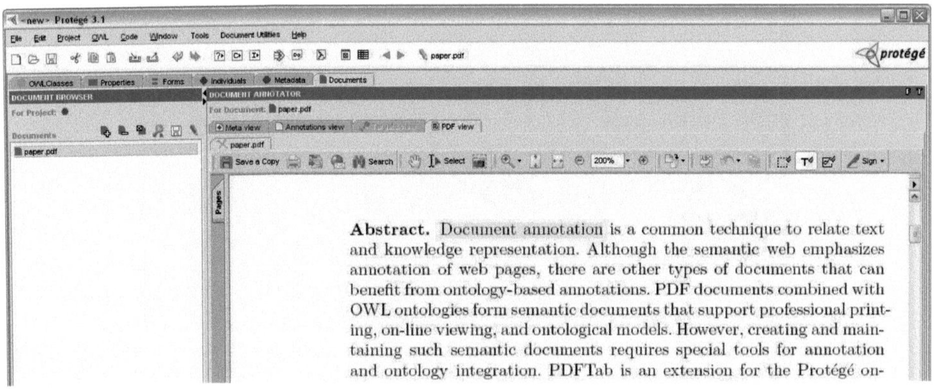

Fig. 5. Document annotation in PDFTab

selects an annotation button, such as the text annotation button, and highlights the text or area to annotate. Figure 4 shows the buttons for the rectangle, text, and graphics annotation tools added to Acrobat. By selecting one of these tool buttons, the user can select a rectangular area, a piece of text, or a figure in the document for semantic annotation. The Acrobat extension highlights the item selected and sends a request to Protégé to prepare the annotation on the Protégé side.

Figure 5 shows the Protégé user interface with the document view, which enables users to browse the document and navigate to the text to annotate. (This screen dump corresponds to the user-interface overview in Fig. 3.) Here, the user has selected the text annotation button and highlighted the text "Document annotation". After the user has marked the text, PDFTab creates automatically an OWL individual that represents the annotation, the *annotation individual,* and opens a form-based editor for this individual (see Fig. 6). In Protégé, users can custom-tailor forms by modifying the layout and selecting widgets for the fields [6]. This functionality enables users to adjust the form for the annotation to suit the subject domain. Furthermore, it is possible to navigate from the document to the ontology by double clicking on the highlighted text, which opens the editor for the annotation individual. Likewise, users can look up annotations in the document by clicking on buttons in the annotation-individual forms.

The annotation individual can refer to classes and individuals in the document's ontology or in another related ontology, such as the main ontology. By default, the annotation class has a *target property,* which is a pointer to a

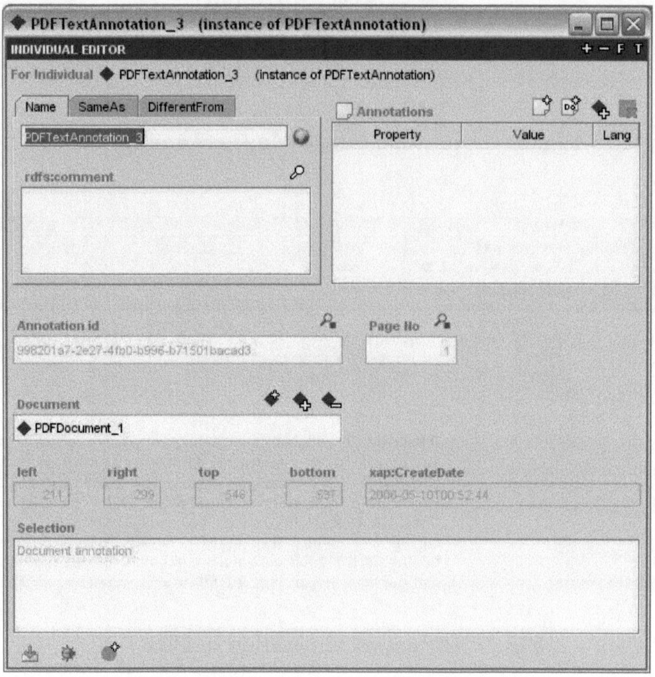

Fig. 6. Form for the annotation individual. This individual corresponds to the annotation in the document. The individual contains information about the annotation id, page number, position, size, creation date, selected text, and an optional reference to a corresponding domain individual.

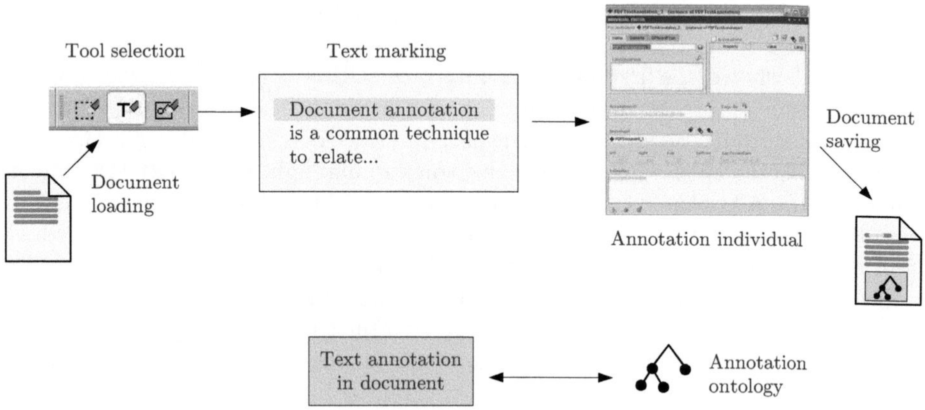

Fig. 7. The annotation process from document loading over annotation creation to document saving

target individual for the annotation. For example, the target individual can be a class modeling a domain concept or a domain-specific individual. In this case, the annotation individual acts as an intermediate between the document annotation and the domain ontology. This approach helps structuring the ontologies by separating annotation-specific information from domain-specific information.

Figure 7 summarizes the process of creating an annotation. The Protégé and Acrobat extensions load the PDF file (and extracts any previous metadata). Then the user selects the text annotation tool and marks the text to annotate. Next, the Acrobat extension initiates the creation process by highlighting the selected text and sending a creation message to the Protégé extension. Protégé then instantiates the text annotation class to create the annotation individual and opens a form-based editor for the annotation. When the user saves the document, Protégé will serialize all classes and individuals associated with the document and send this structure to Acrobat for inclusion in the document. Acrobat then saves the document as a file.

6 Discussion

Integrating documents and ontologies has many advantages. Documents are an important communication and storage format for human knowledge. Ontologies are a structured way of organizing and representing knowledge, especially terminologies and conceptualizations. However, to produce and use semantic documents in practice and extend the vision of the semantic web to documents beyond web pages, it is necessary to have the appropriate infrastructure for handling these documents.

PDFTab is a unique tool in the sense that it supports annotation of PDF documents and combines PDF with advanced ontology-editing support. The user interface combines ontology editing and document viewing in a seamless way and allows users to quickly move between the view, just like using a single application. Furthermore, we believe that the tool architecture is scalable. Both Protégé and Acrobat scale well for large ontologies and documents, respectively. PDFTab already supports multiple documents. Therefore, it is possible to extend the architecture to accommodate for massive document storage in repositories and databases by redesigning the mechanism for handling multiple documents.

The PDFTab approach of combining a standard Windows application with Protégé can serve as an architectural template for other development tools. For example, an interesting prospect is to combine Protégé with other office-oriented applications, such as MS Word and PowerPoint. Such hybrid tools can support new ways of developing ontologies and integrating them with other types of documents and application objects. We believe that the semantic web can benefit from new types of tools that expand the interoperability with common, everyday applications.

7 Conclusion

Extending the semantic web beyond web pages and HTML to complex document and multimedia formats requires new types of tools that can support both ontologies and the target format in an adequate manner. The PDFTab extension of Protégé illustrates the feasibility of efficient support for semantic documents. The combination of Adobe Acrobat and Protégé forms a solid foundation for state-of-the-art handling of documents and ontologies. Although the integration of such different tools requires advanced technical solutions, which adds to the complexity of the system, it is possible to overcome these difficulties and develop hybrid tools that combine the document and ontology views.

Acknowledgments

This work was supported in part by VINNOVA under grant no. 24478-1 and in part by Statistics Sweden.

References

[1] Berners-Lee, T., Hendler, J., Lassila, O.: The semantic web. Scientific American **284**(5) (2001) 34–43
[2] Stamou, G., van Ossenbruggen, J., Pan, J.Z., Schreiber, G.: Multimedia annotations on the semantic web. IEEE MultiMedia **13**(1) (2006) 86–90
[3] Adobe: PDF Reference Version 1.6. 5th edn. Adobe Press, Berkeley, CA (2004)
[4] Eriksson, H.: The semantic document approach to combining documents and ontologies. (in press)
[5] Eriksson, H., Tu, S.W., Musen, M.: Semantic clinical guideline documents. In: Proceedings of the AMIA 2005 Annual Symposium, Washington, DC (October 22–26 2005)
[6] Gennari, J.H., Musen, M.A., Fergerson, R.W., Grosso, W.E., Crubézy, M., Eriksson, H., Noy, N.F., Tu, S.W.: The evolution of Protégé: An environment for knowledge-based systems development. **58**(1) (2003) 89–123
[7] Adobe: XMP Specification. Adobe Systems Incorporated (2004)
[8] Tallis, M.: Semantic word processing for content authors. In: Proceedings of the Workshop on Knowledge Markup and Semantic Annotation at the Second International Conference on Knowledge Capture, Sanibel, FL (October 25–23 2003)
[9] Knublauch, H., Fergerson, R.W., Noy, N.F., Musen, M.A.: The Protégé OWL plugin: An open development environment for semantic web applications. In: Proceedings of the Third International Semantic Web Conference, ISWC 2004, Hiroshima, Japan (2004) 229–243

SWHi System Description: A Case Study in Information Retrieval, Inference, and Visualization in the Semantic Web

Ismail Fahmi, Junte Zhang, Henk Ellermann, and Gosse Bouma

Information Science Department and University Library,
University of Groningen
Broerstraat 4, 9712 CP Groningen, The Netherlands
{i.fahmi,junte.zhang,h.h.ellermann,g.bouma}@rug.nl
http://www.rug.nl

Abstract. Search engines have become the most popular tools for finding information on the Internet. A real-world Semantic Web application can benefit from this by combining its features with some features from search engines. In this paper, we describe methods for indexing and searching a populated ontology by using an information retrieval tool; its results are enriched with inference. For visualization purposes, all of the retrieved ontology instances are clustered based on their classes; and the clusters are linked using instance properties. The approach is illustrated using our SWHi (Semantic Web for History) prototype as a case study.

Keywords: semantic web, ontology, information retrieval, inference, visualization.

1 Introduction

The Semantic Web can be seen as a general web which describes units of information. Once this information is available in some web documents (which then become Semantic Web documents), people can gather and manipulate the exchanged information using Semantic Web technologies in various useful ways. Tim Berners-Lee in his Scientific American article "the Semantic Web" [2] illustrated some examples of how the semantic information can help everyday tasks, such as finding a health care provider, prescription treatments, making an appointment, and planning a trip.

In order for Semantic Web technologies to have an impact, they should be able to work together with existing general information retrieval technologies [6], and even provide new services which cannot be delivered by general Web search engines such as Google. The most popular Web search engine, Google[1] combines instant responses, huge repositories, advanced search methods, and a sophisticated relevance-ranking algorithm.

[1] Google www.google.com

E. Franconi, M. Kifer, and W. May (Eds.): ESWC 2007, LNCS 4519, pp. 769–778, 2007.

We begin the development of our Semantic Web application from a library point of view. In this case, Eric Miller [9] believes that "Libraries–digital libraries in particular–are important memory organizations that form a keystone for the development of the Semantic Web." The digital library stands on collections of annotated data, in the form of metadata. This should naturally make the digital library a successful primary adopter of the Semantic Web, and, on the other hand, challenge the Semantic Web to improve and widen services provided by digital libraries, which nowadays still heavily rely on information retrieval technology.

In our Semantic Web for History (SWHi) project, we combine some features from the Web search engine and the Semantic Web technology. From the Semantic Web technology point of view, the adoption of the Web search engine technology is expected to improve its search performance. In this paper we describe how the search engine tool Lucene[2] can be used to index ontology instances, parse user input queries, and retrieve matched instances. Given a plain list of ontology instances from the search results, Semantic Web technology will enrich the retrieved information. Inference will be applied during the indexing and enrichment steps. For visualization purposes, we organize the results in clusters based on classes of the retrieved instances (e.g. person, organization, document, subject, and year) and relations between instances in the classes. We present the clusters and their relationships using a two dimensional cluster map. Through this map, users can browse search results interactively, and explore interesting relationships in an ontology.

2 Motivation

2.1 Data Sources

The concept of a Semantic Web is promising but difficult to implement on most current Web documents. The Semantic Web requires semantic information which is typically not encoded in the documents. Considering the importance of such information, metadata is added into the document. On the other hand, many digital libraries do have metadata in place. Metadata is a key piece which describes every resource managed by the digital libraries.

Our SWHi application is developed from the digital library point of view, where our main data sources are repositories which provide metadata (subsection 3.2). This metadata is mapped and stored into an ontology based on an ontology schema. Furthermore, literal values in the metadata, for example describing title and description properties, are analyzed, from which we extract named entities, events and terminology. To enrich the ontology, we also extract new related information from selected Web documents.

2.2 Information Retrieval in the Semantic Web

Current popular web search engines (e.g. Google and Yahoo[3]) provide both simple and advanced search interfaces. While the advanced search interfaces

[2] Lucene lucene.apache.org

[3] Yahoo www.yahoo.com

provide more functionality, the simple search interfaces are preferred by most users. A Semantic Web application will typically also provide these kinds of search interfaces.

Using a simple search interface, a user can enter query terms regardless of the question in which fields the terms would exist. For example, a user may want to find any information (any document, year, person, or relation between persons) in our SWHi ontology related to a topic such as *French settling colonies involving Mr. Samuel Kirkland and General Washington in 1777.* Using a bag-of-words searching technique, she might simply type *kirkland washington 1777 french settle* into a search form. We face some issues while processing this query using an RDF query language such as SeRQL[3]. A query processor (which generates SeRQL queries) will face at least two problems. First, it does not know in which class or property a word can be found. To avoid this problem, a Semantic Web application such as OpenAcademia[4] requires users to type a keyword into the appropriate field (*author, title* or *year*) in its advanced search interface. Second, there are some limitations in the substring matching of SeRQL using a wildcard character '*'. Searching for *general** will match *general* and *generally* only at the beginning of a text. And searching for * *general* * (with a space between the wildcards and the word) will only match *general* in the middle of a text, but not at the beginning or end of the text. These problems can be solved using an information retrieval application such as Lucene which provides powerful, accurate, and efficient search algorithms. Besides the fielded searching feature, Lucene also supports phrase queries, wildcard queries, proximity queries, range queries and more[5].

Prior uses of information retrieval technology in a Semantic Web application can be found in QuizRDF[4], the Knowledge and Information Management (KIM) platform[1], OWLIR and Swoogle[6]. QuizRDF creates RDF resource indexes based on RDF Schema and retains this structure in its indexes. KIM uses Lucene engine to index and retrieve semantically annotated documents while OWLIR and Swoogle use the Haircut information retrieval engine to index and retrieve RDF documents based on character n-grams as indexing terms.

2.3 Visualization

In the Semantic Web, visualization is becoming more important. In our case, since the retrieved instances can be of any type (documents, persons, years, etc.), a common plain list presentation is not suitable. There are complex relationships among the resource instances which cannot be presented using a plain list. Moreover, this presentation typically only displays a small number of search results (in the range of 10-20 results per page). Documents obscured in the tail of a search result will likely never be accessed.

Various solutions to this problem have been proposed in the IR as well as the SW area. Currently, popular result presentations in the information re-

[4] OpenAcademia www.openacademia.org

[5] Lucene's Features lucene.apache.org/java/docs/features.html

trieval technology use topical clustering and mapping techniques. Vivisimo[6] and Grokker[7], for instance, both use clustering techniques to analyze and organize search results according to topics found in the retrieved document descriptions.

In the Semantic Web, the complex nature of relationships between concepts in an ontology has driven many efforts toward graphical visualization of ontology browsing and navigation [10,12,11,7]. For example, Cluster Map [7] is used to visualize instances of selected classes, organized by their classifications. This map is designed to aid users when navigating their search results and ontologies. The Spring embedding model [8,5] has been widely used to visualize collections of instances and ontologies[10,7]. It draws highly related entities close to each other with a directed edge and gives the effect of separation in a two-dimensional plane.

3 System Architecture and Data Source

3.1 Architecture

The general architecture of the SWHi system is shown in Fig. 1. It consists of three layers: Knowledge Management System (KMS), Semantic Web Application, and User Interface layers. The bottom layer, is responsible for processing (information extraction and semantic annotation), indexing, and storing historical resources (metadata and documents), also providing API for its upper layer to query the knowledge base. This layer highly depends on several third party tools, such as GATE[8], Sesame[9], and Lucene.

In the middle layer, several application modules which enable the Semantic Web were developed to carry out the following functions: processing data sources (text and metadata), processing user queries, and delivering results in several ways (network graph, cluster map, and time line). And the top layer provides interface to users which are being designed as simple and easy to use as possible.

3.2 Data Source and Ontology

It is obvious from the Fig. 1 that the ontology plays as a central role in the SWHi system. For the development of the SWHi ontology, we reuse existing ontology resources for structuring and storing historical information, namely: PROTON[10] base ontology, the types of American history imprints identified (automatically) in the metadata, the taxonomical subject classification by NewsBank/Readex[11], Dublin Core[12], and Friend of a Friend[13]. This ontology is stored using the Sesame 2, an RDF storage and querying framework.

---

[6] Vivisimo www.vivisimo.com

[7] Grokker www.grokker.com

[8] GATE gate.ac.uk

[9] Sesame www.openrdf.org

[10] PROTo ONtology proton.semanticweb.org

[11] Newsbank InfoWeb infoweb.newsbank.com/?db=EVAN

[12] Dublin Core dublincore.org

[13] FOAF xmlns.com/foaf/0.1/

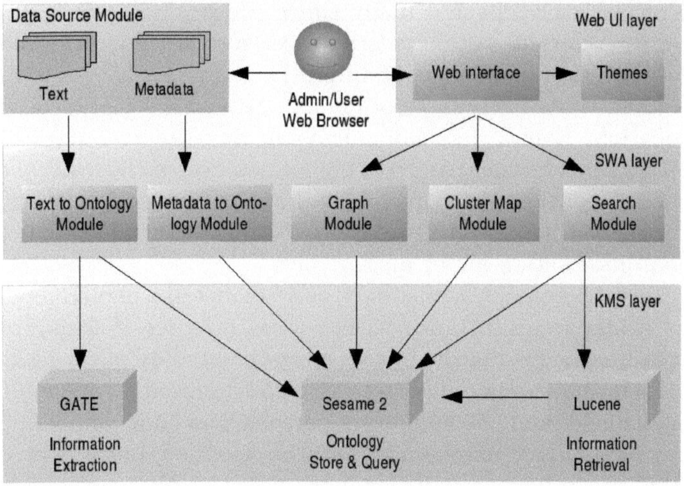

Fig. 1. The general architecture of the SWHi system

Our initial instances for this ontology are extracted from the Early American Imprints, Series I: Evans, 1639-1800[14]. This knowledge source gives insight in all published works of the 17th- and 18th-century America. Its metadata consist of 36,305 records, which are elaborately described (title, author, publication date, etc) with numerous values, and have been compiled by librarians in the format MARC21. In the future we will extend the data sources with other historical electronic journal metadata as well as full texts, like digitized version of Early American Imprints, Wikipedia, and biographical profiles that libraries have about historical US figures.

4 Indexing, Searching, and Inference

4.1 Indexing and Inference

Performance is very important in a real-word application. No matter how sophisticated and accurate the logic behind a search engine, if it cannot provide a fast search response, users will likely less appreciate it. To ensure that our system will have an acceptable response time and process query terms efficiently, we use the Lucene text search engine API to index our ontology instances. We make use of Lucene fielded data feature to store and index instance properties. When indexing, each instance in the SWHi ontology is added to a Lucene index as if it is a new document, and instance properties are added to the document as document fields.

In our experiments, getting information directly from an ontology through inference could cost an unacceptable processing time, especially if the inference

[14] Early American Imprints Series I www.readex.com

queries are complex or repeated many times. However inference is required to return the most relevant instances given user queries. For example, searching for instances with literal values containing the term "washington" in the SWHi ontology would give us 7 instances of Person class or 586 instances of all classes (Person, Location, Event, and Document) in any order. Using inference, instances with higher relevance can have higher position in the order. For example, a person who is known by many people and created many documents would get a higher score. For this purpose, we apply inference during an indexing step to get additional information about an instance.

Table 1 illustrates how an instance of Person class should be indexed by Lucene. For example, an instance with a label "George Washington" is being indexed. Each Lucene field of the instance is populated by querying the ontology repository. Then, the number of persons that Washington knows will be stored in the `foaf_knows` field, and the number of persons who know him will be stored in the `known_by` field. These numbers will not be searched, but will be used during query analysis which boost particular fields based on their values.

Table 1. The Lucene fields of Person class instance. The `container` field will used by search "All"

Field	Lucene type	Description
uri	Field.UnStored	a URI of an instance
rdfs_label	Field.Text	a short label of an instance
container	Field.Text	contains all data from other fields (not stored)
foaf_topic_interest	Field.Text	a textual list of topics
foaf_knows	Field.Keyword	the number of persons as String
known_by	Field.Keyword	the number of persons as String
dc_creator	Field.Text	a textual list of document titles
protont_involvedIn	Field.Text	a textual list of events
protont_startTime	Field.Keyword	a date as a String (YYYYMMDD)
protont_endTime	Field.Keyword	a date as a String (YYYYMMDD)

4.2 Searching and Enrichment

Retrieving and processing information from the SWHi ontology will be done through these processes: Query formulation, Query analysis and parsing, Search and retrieval, Enrichment, and Clustering.

Query Formulation. When performing a query, users often encounter difficulties whether they have to use "AND" or "OR" [13]. To overcome this problem, we present a *free-text search* interface to users. Using a form in this interface, users can type any query terms, such as a combination of terminology, person name, year, and location. For advanced users, we provide an advanced search interface where they can use multiple fields for their query terms.

Query Analysis and Parsing. All free-text search queries will be processed by Lucene. Since its searches are case-sensitive, a general best practice is to lowercase query terms during query analysis. The StandardAnalyzer which is used during the indexing will be used again to perform this task. In this step, we set different boosting factors to the index fields based on their importance to the class being searched. For example, the field `known_by` is a good indicator of how well-known the person was. This can be implemented using `FunctionQuery` feature of Lucene which can return a score based on fields' values. For advanced search queries, we generate SPARQL queries and send them directly to the Sesame repository.

Search and Retrieval. Given a free-text search query, Lucene searches its index to find all matched resources, and given an advanced search query, Sesame searches for instances from its ontology repository. Each time a search is performed, the Search Module retrieves URIs of instances in the search results and stores them into a cache memory. This will speed up the retrieval process when a user clicks on other pages of the same search results, which could happen if the number of instances exceed an allowed number of instances per page.

Enrichment. The goal of this step is to deduce new information given a list of URIs in the search results retrieved by the previous search process. The SWHi ontology will be used to enrich presented instances with important information and relationships. For this purpose, we define inference algorithms for each instance class. Since the search phase typically produces many search results, we have to optimize this enrichment phase to achieve an acceptable performance. For example, we limit only to the first 200 instances returned by Lucene that will be enriched for clustering, and 10 to 20 instances for a plain list presentation.

Clustering and Visualization. Clustering is performed to preprocess the results before they are presented in a cluster map to users. Since the results consist of various resource types, a visual representation technique would be a promising option. The underlying concept of our visualization is based on this Visual Information Seeking Mantra [13]: "*Overview first, then zoom & filter, then details on demand*".

We implement this principle by clustering the results. Our cluster data will be organized to support the following levels of presentation details:

Global view of results. Users can see a global view of all of the results in a two-dimensional graphical visualization. In this view, the results are organized into clusters based on the results' classes or topics. Grouping by topics seems to be more interesting than by classes, because users can see interactions between instances of different classes in a topic. Between the clusters, we draw relationships based on properties carried by the results in each cluster. Every instance in a cluster will be presented using a symbol.

Zoom view of a cluster. Users can zoom into a cluster to see interactions between instances in the selected cluster. For example, in a cluster of persons, users can see how persons in that cluster know each other.

Detailed description view of a cluster or a node. Users can read a detailed description of a cluster or a node (member of a cluster). For example, they can see the name of a cluster, the number of its nodes (cluster's members), a complete list of the nodes' titles and links, or a short description of a node.

5 Results

In this section, we describe the results of our system development and illustrate them with some examples. Assume users want to find information about "George Washington and wars" and type the keywords "+washington +wars" in a free-text search interface. After receiving search results, the first task of the user interface is to display a general view of the results. For this task we use the Cluster Map [7] as shown in Fig. 2.

The user interface in the figure is divided into three main panes: left, right, and bottom. The left pane shows the classification tree of the results, containing the names of the clusters, their children and the numbers of objects within each cluster. In this pane, the users perform cluster selection that will change the visualization of the clusters in the right pane. For example, *Document*, *Person*, and *Topic* clusters are selected together with some of their sub-clusters as shown in the right pane. Objects in the results are now classified according to their

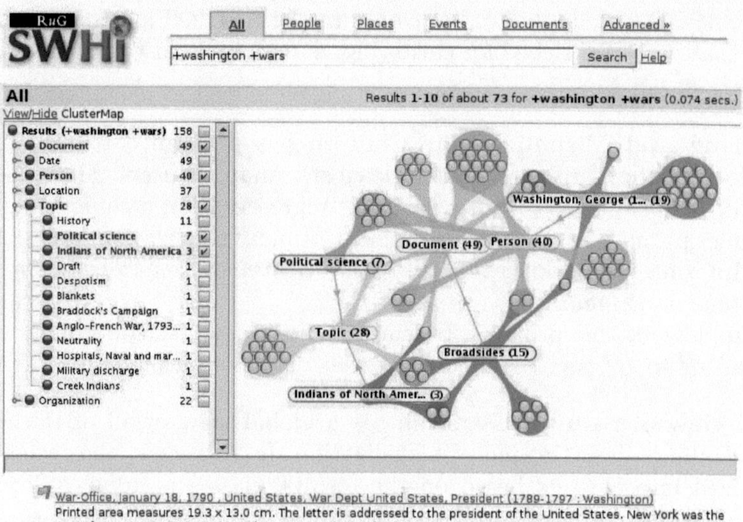

Fig. 2. Search results for the query "+washington +wars" visualized using ClusterMap

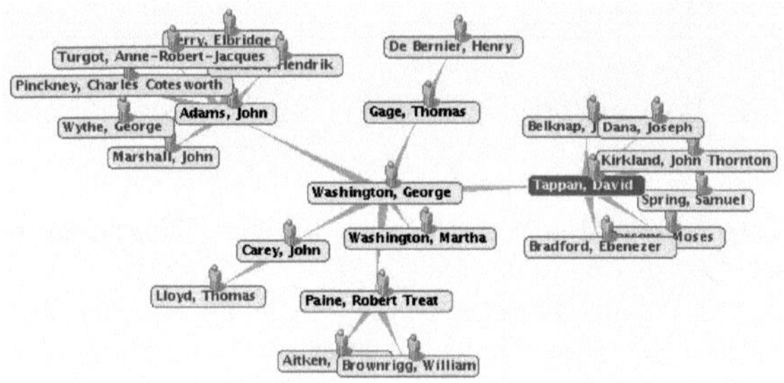

Fig. 3. Browsing objects and their relationships

clusters and intersections, which help users select objects satisfying their needs. In the figure they can click on a cluster of 7 objects (documents) related to a *Political science* topic. A list of objects (of any type, e.g., document, person, date, or location) in the cluster are then displayed in the bottom pane. Detailed information of an object will be presented when users click on a title in the list.

A different visualization strategy is shown in Fig. 3. While ClusterMap shows objects in the result set as clusters, the last strategy shows relationships between objects in a cluster or in the whole result set. For example, the figure shows how individuals in the *Person* class relate each other. Objects and their relations are not limited by the result set, but can be retrieved directly from the repository when requested information is not available in the set. This can be seen as another way of browsing the result set and repository. We implement this strategy using the TouchGraph[15] tool.

6 Summary and Future Work

This paper has described a case study in implementing information retrieval, inference, and visualization in the Semantic Web. The use of information retrieval technology is mainly motivated by pragmatic reasons which are to provide rich search functionalities and to return semantically related search results with high performance. However, in this application, the ontology keeps playing an important role, especially in shaping information structure in the indexes, in inferencing, and in clustering. Subject and terminology classes in the ontology help generating clusters based on resource topics. Visualization based on this grouping technique is interesting because it combines information from the ontology and information retrieval into a single graph.

The SWHi application is an on going project. We are also in a progress of enriching the ontology by implementing named entity, event, and terminology extraction.

[15] TouchGraph www.touchgraph.com

References

1. Kiryakov Atanas, Popov Borislav, Terziev Ivan, Manov Dimitar, and Ognyanoff Damyan. Semantic annotation, indexing, and retrieval. *Journal of Web Semantics*, 2, 2005.
2. Tim Berners-Lee, James Hendler, and Ora Lassila. The semantic web. *Scientific American*, 2001.
3. Jeen Broekstra and Arjohn Kampman. *An RDF Query and Transformation Language*, pages 23–39. Semantic Web and Peer-to-Peer. Springer Berlin Heidelberg, 2006.
4. John Davies, Richard Weeks, and Uwe Krohn. QuizRDF: Search technology for the Semantic Web. In *Proceedings of the 37th Annual Hawaii International Conference on System Sciences (HICSS'04)*, pages 112–119. BTexact Technologies, IEEE Computer Society, 2004.
5. P. Eades. A heuristic for graph drawing. *Congressus Numerantium*, 42:149–160, 2000.
6. Tim Finin, James Mayfield, Clay Fink, Anupam Joshi, and Scott R. Cost. Information Retrieval and the Semantic Web. In *Proceedings of the 38th International Conference on System Sciences*, 2005. Received Best mini-track paper award.
7. Christiaan Fluit, Marta Sabou, and Frank Harmelen van. Ontology-based information visualisation: Towards semantic web applications. Springer Verlag, 2005.
8. Thomas M.J. Fruchterman and Edward M. Reingold. Graph drawing by force-directed placement. *Softw. Pract. Exper.*, 21(11):1129–1164, 1991.
9. E Miller. Digital libraries and the semantic web. http://www.w3.org/2001/09/06-ecdl/slide1-0.html, 2001. Accessed 11 December 2006.
10. Paul Mutton and Jennifer Golbeck. Visualization of semantic metadata and ontologies. In *Seventh International Conference on Information Visualization (IV03)*, pages 300–305. IEEE, 2003.
11. Bijan Parsia, Taowei Wang, and Jennifer Golbeck. Visualizing web ontologies with cropcircles. In *End User Semantic Web Interaction WS*. ISWC 2005, 2005.
12. D.A. Quan and David R. Karger. How to make a semantic web browser. In *WWW '04: Proceedings of the 13th international conference on World Wide Web*, pages 255–265, New York, NY, USA, 2004. ACM Press.
13. Ben Shneiderman. The eyes have it: A task by data type taxonomy for information visualizations. In *IEEE Visual Languages*, pages 336–343, College Park, Maryland 20742, U.S.A., 1996.

Semantic Turkey: A Semantic Bookmarking Tool
(System Description)

Donato Griesi, Maria Teresa Pazienza, and Armando Stellato

AI Research Group, Dept. of Computer Science, Systems and Production
University of Rome, Tor Vergata
Via del Politecnico 1, 00133 Rome, Italy
{griesi,pazienza,stellato}@info.uniroma2.it

Abstract. In this work we introduce *Semantic Turkey*, a Semantic Extension for the popular web browser Mozilla Firefox. Semantic Turkey can be used to keep track of relevant information from visited web sites and organize collected content according to a personally defined ontology. Clear separation between knowledge data (the WHAT) and web links (the WHERE) is established into the knowledge model of the system, which allows for innovative navigation of both the acquired information and of the pages where it has been collected. This paper describes the architecture of the Semantic Turkey extension for Firefox, analyzes its development, shows its most interesting features and presents our plans for future improvements of the tool.

1 Introduction

In this work we introduce Semantic Turkey, a Semantic Extension for the popular web browser Mozilla Firefox [3], which can be used to annotate information from visited web sites and organize this information according to a personally defined ontology. Semantic Turkey should not be addressed as a "Semantic Web Browser" (whatever the nature of this term, which will probably take shape in the next future); it is intended as a personal desktop solution for organizing and managing the relevant information which is observed during web navigation, an advanced replacement for the traditional "Favorites" menu, offering clear separation between knowledge data (the WHAT) and web links (the WHERE), thus allowing for innovative navigation of the acquired information as well as of the pages where it has been observed.

2 Motivations and Approach Followed

Our research work, funded by the FILAS (Finanziaria Laziale di Sviluppo) agency under contract C5748-2005, has been centred on providing innovative methodologies and instruments for browsing the web and for organizing information of interest gathered during navigation. A specific point which emerged in our interviews inside FILAS is the emerging great need for efficient recovery of already visited pages (and, more in general, of already accessed knowledge): people are often exposed to large

E. Franconi, M. Kifer, and W. May (Eds.): ESWC 2007, LNCS 4519, pp. 779–788, 2007.

quantities of information, which are not always useful when seen for the first time, though difficult to recover when needed. The result is that people often become frustrated by the classical "I've seen it somewhere, but I don't remember where!" problem. We thus focused on finding interesting solutions for collecting, managing and retrieving data observed during web navigation. Our key goal was to overcome the limited usability of bookmarks lists, which:

- see weblinks as first class citizens. They can be categorized by implicitly adding them to a bookmarks folder, but they are no way separated from the knowledge they represent. More links could be related to the same subject, but there is no way to represent this aspect, except from considering the subject as a folder itself, thus betraying the intended equation: folder = category. Also, in some cases, it could be important to identify the portion of a page which contains the relevant information which caused it to be bookmarked. (e.g., "John Doe" is cited in a long web document which is very generic and not directly related to John Doe; we would like to take note of the page, still maintaining the focus on the real subject of our interest and immediately recognize where it has been identified).
- do not foresee any kind of multiple categorization. Any folder cannot belong to two or more different folders (a kind of multiple inheritance between categories), nor can any single weblink belong (with the possible exception of new systems adopting virtual foldering) to more than one folder (multiple instantiation).
- single knowledge resources cannot assume any kind of structure. It is not possible to further characterize a weblink, or to relate it with other ones (except putting them in the same folder/category).

Our project headed towards the development of a sort of "semantic notepad"[1] offering basic functionalities for:

1. capturing information from web pages, both by considering the page as a whole, as well as by annotating portions of their text
2. editing a personal ontology for categorizing the annotated information and, possibly, to exchange information with other people and exporting to other tools.
3. navigating the structured information as an underlying semantic net which, populated with the many relationships which bind the annotated objects between them, eases the process of retrieving the knowledge which was buried by the past of time For example, a user could discover that two persons which he has kept track of in separate sessions (by annotating their presence and some aspects of their profiles appearing in different web pages), work in the same place, or have any kind of connection he would not recall with any kind of traditional bookmarking/annotation service.
4. clearly separating the business model from the user interface, by adopting a "knowledge service" architecture. This way, the same architecture could be exploited for an enhanced personal web browser as well as for a shared environment for collaborative semantic tagging of web pages.

[1] "Taccuino" is the italian word for the term "Notebook". In our lab, we hate so much the silly Italian expression "Taccuino Semantico" (Semantic Notebook) that we started to use any kind of misspelling of its name, the funniest (and most used) of which was "Tacchino Semantico" (Semantic Turkey). The rest is history.

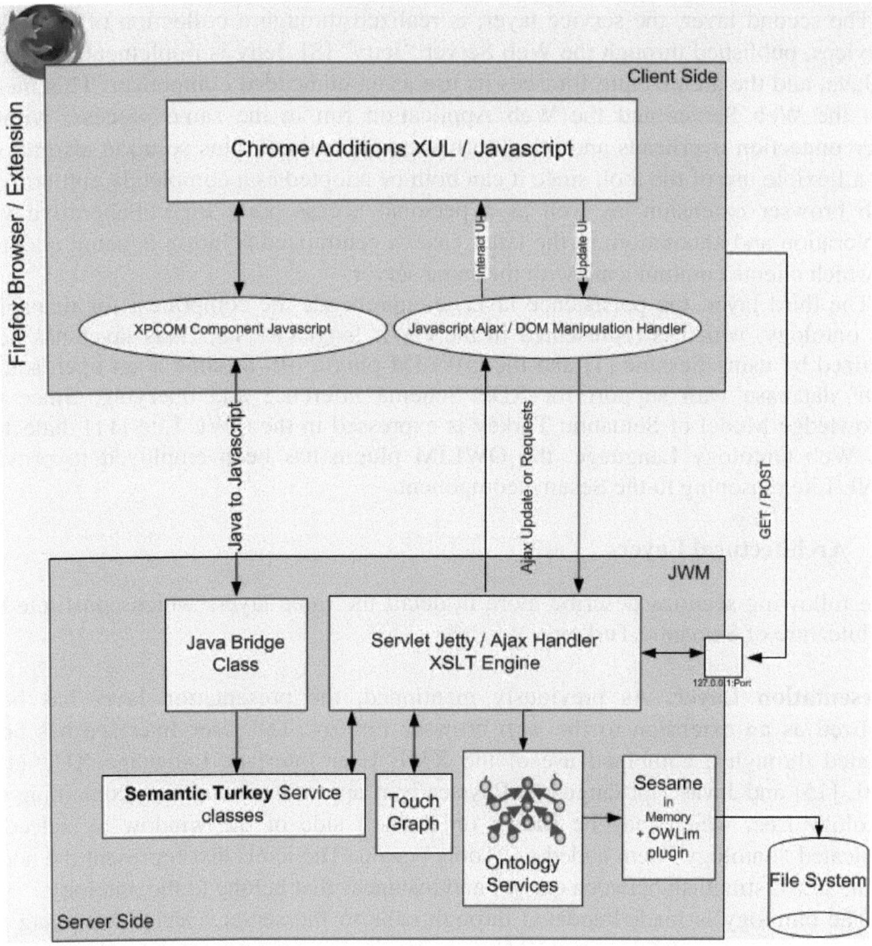

Fig. 1. Architecture of Semantic Turkey

3 Architecture

The architecture (Fig. 1) of Semantic Turkey consists in a web application, designed using a three layered approach.

The first layer, the presentation layer, has been developed as an extension for the web browser Firefox. Everything relating to user interaction is directly managed by the Firefox extension, thanks to a solution directly integrated in the browser. This approach has two main advantages: total reuse of the functionalities of a well assessed, stable and complete software for web browsing, and a non invasive offer for the user, who can still use the web browser he has been acquainted with.

The second layer, the service layer, is realized through a collection of Java Web Services, published through the Web Server "Jetty" [8]. Jetty is implemented entirely in Java, and the architecture foresees its use as an embedded component. This means that the Web Server and the Web Application run in the same process, without interconnection overheads and other sort of complications. This solution also allows for a flexible use of the tool, since it can both be adopted as a completely autonomous web browser extension, as well as a personal access point for collaborative web exploration and annotation: in the latter case, a centralized solution is being adopted, in which clients communicate with the same server.

The third layer, the persistence layer, comprehends the component for managing the ontology, which is represented in the OWL language [10]. This layer has been realized by using Sesame [1] and the OWLIM plugin [9]. Sesame is an open source RDF database with support for RDF Schema inference and querying. Since the Knowledge Model of Semantic Turkey is expressed in the OWL Lite [11] dialect of the Web Ontology Language, the OWLIM plugin has been employed to provide OWL Lite reasoning to the Sesame component.

3.1 Architectural Layers

The following sections describe more in detail the three layers which constitute the architecture of Semantic Turkey.

Presentation Layer. As previously mentioned, the presentation layer has been realized as an extension to the web browser Firefox. The User Interface has been created through a combined use of the XML User Interface Language XUL [17], XBL [15] and Javascript language. Physically it appears as a sidebar, containing the ontology tree, which may be shown on the left side of the window by selecting dedicated "ontology" item added in "Tools" menu. The icons that represent the nodes of the tree distinguish between classes and instances that belong to the ontology.

The ontology is loaded/updated through calls to the server, carried out using the Ajax [5] technique: the data – in XML format – is thus mainly exchanged between the two layers in an asynchronous way, to preserve good performance and to not penalize the activity of the browser.

The extension has also another prerogative, which is not an ordinary feature of the presentation layer: it has to assure that the web server is being loaded as an embedded component, at the start of the browser process. To do that XPCOM [16] components, written in JavaScript, have been developed for linking the chrome part and the Java part.

In order to load the Java component, the Simile Java Firefox Extension [12] has been used. This component allows to load java classes or jar packages, instantiate objects and to invoke static methods or methods of the object previously instantiated.

At the start of the browser process, after loading the java components (the java server code and the required libraries), a static method is being invoked with the role of instantiating the web server. This solution makes it possible to install all the application simply as a Firefox extension, without configuring other software.

Service Layer. This layer offers services which may be invoked through http requests submitted according to the Ajax paradigm, thus enabling communication between the client (Firefox extension) and the server. The server receives the requests coming from the client by GET or POST http calls, carries out the operations associated to these calls, and in case replies with an XML response. If a call implies the return of a XHTML page, a XSLT transformation is being performed, in order to decouple the data model with its manifestation in the presentation layer.

The majority of invocations to the server are being completed in an asynchronous way, so that, independently from the workload that is subjected the server, the browser can continue to respond to the user. This is a crucial issue for the usability of the application: expensive computations blocking normal behavior of the browser would otherwise not be tolerated by the user.

Besides supporting the communication with the client, the service layer provides the functionalities for definition, management and treatment of the data. Several objects are described through an ontological model (see next section), to represent both pure conceptual knowledge as well as application required information.

Finally, the service layer also provides another important functionality linked with the presentation layer. It allows for the capability of visiting the ontology through a graph view, using the TouchGraph library [14]. TouchGraph is an open source tool for visualizing networks of interrelated information. It renders networks of information concepts as interactive graphs that lend themselves to a variety of transformations. By engaging with the visual image, a user is able to navigate through large networks of information and to explore different ways of arranging the network's components on the screen. This functionality has been positively judged by the technophores, as it allows unexpected correlations to emerge from the network of information.

In order to access TouchGraph from presentation layer, a dedicated java applet and related servlet have been realized. The servlet works like a proxy, redirecting the applet loaded, with the correct parameters, to the client side.

Persistence Layer. Sesame provides the abstraction layer over ontological data. The foundation of the component is the Storage And Inference Layer (SAIL). This SAIL is an API that abstracts from the storage device used (in-memory storage, disk-based storage, RDBMS) and takes care of inference.

From the architecture perspective the Access APIs are the most important component. These APIs provide high-level access functionality to client applications, either locally or remotely (over HTTP or RMI).

Sesame can thus be deployed as an RDF database, with persistence in an RDBMS, or as a Java library for embedded use in applications. This last modality has been employed for the definition of the architecture. In our case, the ontology data is, by default, handled in memory and stored in the (local) File System, but it is possible to easily switch to the database storage backend for managing very large ontologies. Also, the ontology repository may be located in a different, remote, site, thus offering different possibilities for decentralizing the application.

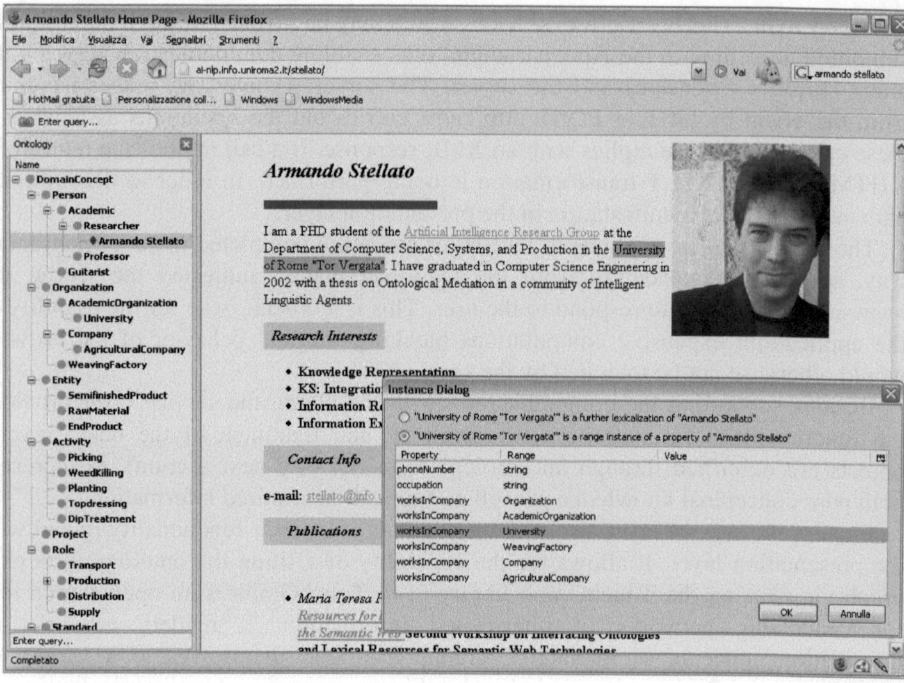

Fig. 2. Annotating concepts from a web page and establishing relationships between them

3.2 The Knowledge Model

The knowledge model of Semantic Turkey has been structured into four different layers of ontological knowledge:

1. *The Application ontology*: This ontology contains resources needed by Semantic Turkey to organize, retrieve and present information to the user.
2. *The Top Ontology* (which owl:import the Application Ontology): this ontology has originally been conceived inside our project for FILAS, and is thought for representing a minimal knowledge which should be shared across the different technophores. This ontology can simply be seen as a guideline for driving the personal annotations of each of the technophores, and could be used as well as a shared ontology for exchanging information between them.
3. *The Personal/Domain Ontology* (which owl:import the Top Ontology): The third ontological layer allows for a personalized organization of the knowledge which is extracted and collected from the web.
4. *The Knowledge Base* (which owl:import the Top Ontology), i.e. the set of instances which populate the personal ontology of the user.

The Application ontology is composed of resources useful for managing the annotation functionalities. These, among the others, include the classes:

- `Annotable` identifying the part of the ontology which can be annotated by the user
- `URL` which stores links to the visited pages

- `SemanticAnnotation` containing the annotations performed by the user, described by their URL, related concept etc…

and the properties:

- `has_location` linking URLs with *Annotable* concepts
- `observed_lexicalization` describing the form with which a given object appeared in a specific annotation. this property has been preferred to a more precise information, like reporting the byte offset of the annotation inside the page, to make retrieval of the annotated object more robust with respect to minor changes that occurred to the page over time.

The Application ontology is invisible to the user and is only exploited by the application to get the proper logic for administering the upper ontological layers. Key elements for the annotation process are expressed in terms of concepts from this ontology.

Resources originated from the Top ontology are read-only, and cannot be deleted as a consequence of any edit operation by the user. In a really general perspective, the Top Ontology could even be left empty (i.e. if there is no supposed shared conceptualization which must be adopted by users working on a common annotation framework; in this case, each user can build from scratch its own conceptualization, which will be thus constituted by the sole *Personal Ontology*), or external resources could be imported, possibly exchanging their content with other applications, like a mail browser (e.g. by adopting the FOAF ontology [4] for managing contacts) or a client for instant messaging. The *Personal Ontology* is the last conceptual layer which can be modeled according to personal preferences, perspectives and needs.

4 User Interaction

Semantic Turkey offers some basic editing operations for populating the *personal ontology* with annotations from visited web sites, as well as search and navigation functionalities which facilitate the recovery of already acquired knowledge.

4.1 Main Functionalities

The user may interact with the ontology panel to modify its personal ontology, through a series of operations, which we describe here, organized into categories.

Interaction with the browser. These mainly include drag&drop operations which allow to annotate information from the visited sites:

1. Drag and drop of a selection of a text from an html document displayed in the browser, on the icon that represents a class, in order to create an individual of that class. The selection will become the ID of the new individual and a new icon will be shown below the selected class
2. Drag and drop of a selection of text from an html document, on the icon that represents an individual, in order to characterize a property which that individual owns. A specific window will open, prompting the user to choose the fitting property. The selection will become the ID of a new individual that represents the

instance of the range of the property chosen. If the selected property is an object property, a new icon will be created relatively to the range class.

3. Drag and drop of a selection of text from an html document, on the icon that represents an individual, in order to define a further lexicalization for that individual. The user can choose, from the same panel described before, if the selection characterizes a range of a property or a new *observed lexicalization* (see section 3.2).

Direct Ontology Editing. These functionalities operate exclusively on the ontologies, as it should be important for the user to integrate its knowledge with information he would acquire through other media (communication with other people, radio, tv etc…). These include:

1. *Semantic Editing*. It is possible to create, modify and/or delete new classes/individuals/properties. All the operations are being carried out through specific panels that are activated by a context menu associated to the nodes of the tree, in a way much similar to traditional ontology editing tools, like Protégé [6] or TopBraid Composer [13]. By offering complete interaction with the ontology via the XUL interface (instead of an HTML interface, like in Piggy-Bank), the user is not diverted from his current navigation (i.e. the main browser panel is still focused on the visited web page, which would otherwise be replaced by the HTML UI) and may maintain its attention over the observed web page.

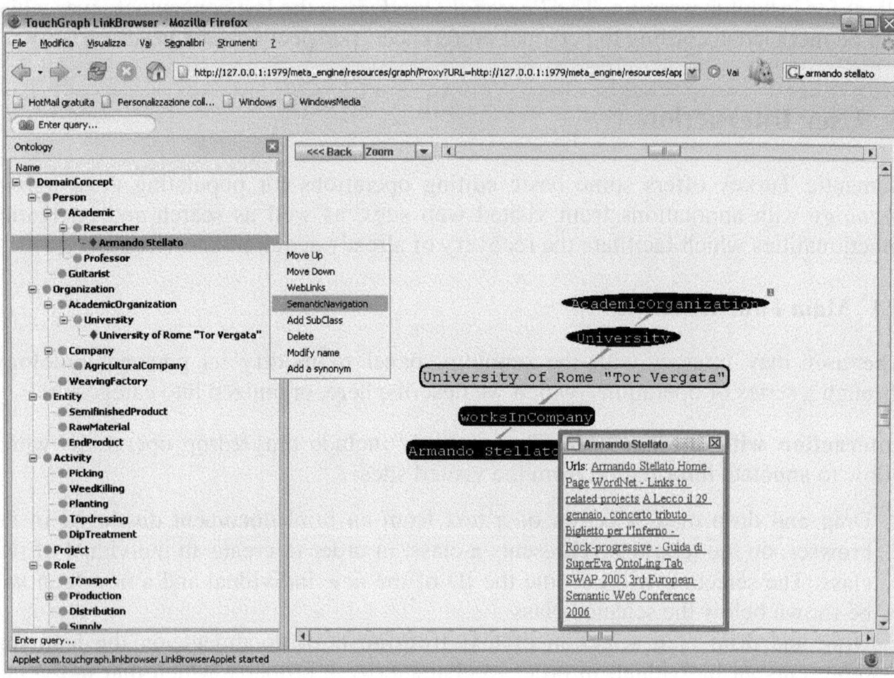

Fig. 3. Semantic Navigation: recalling ontology and web links for "Armando Stellato"

2. *Lexical Editing*. Add synonyms and documentation for the concepts. These alternative lexicalization provide several anchors for referring the same ontological entries. This solution facilitates retrieval of knowledge objects when the ontology reaches a considerable growth, or simply when its knowledge is transferred to other users. Advanced search functionalities over the ontology objects and their lexicalizations in different languages, have been made available thanks to an embedded indexing engine [7] and the adoption of a library implementing different string matching algorithms [2].

Semantic Navigation. As an additional feature, the user may graphically explore the ontology (Fig. 3), thanks to the *SemanticNavigation* component. A Java applet will be loaded on a new tab of the browser displaying the graph view of the ontology, allowing the user to navigate its content and get back to the pages related to the annotated knowledge. Conversely, Semantic Turkey reports to the user, through a dedicated status bar, the pages which have been previously annotated. When the user visits an already annotated page, an icon with the shape of a pencil is being shown in the lower part of the browser. If the icon is being clicked, the html text entries that represent the past annotations will be emphasized (providing the page still contains those entries) with a light background color.

5 Conclusions

In this paper Semantic Turkey, a special environment for supporting end users in annotating information caught from visited web sites, has been described.

Main objective of our first experience in developing Semantic Turkey has been to extend "usual" web browsing modalities, with a particular focus on efficient and intuitive retrieval of information already observed during past navigation. A key characteristic of this approach has been to separate the role of site bookmarking from the more complex aspect of knowledge management and, at the same time, to interweave both of them in a homogeneous perspective over the two dimensions of the Web: traditionally exposed documents and the new web of data fostered by the Semantic Web. We are now in the direction of refining the overall architecture to meet more general requirements which would make Semantic Turkey an open and reusable platform. In particular, the multilayered approach in the knowledge model must be flexible enough to allow the user to import and reuse any number of available ontologies, while an extension mechanism should make it easy to produce specific add-ons for adding new functionalities to the browser. The flexibility offered by the client-server paradigm in the overall architecture should also be exploited to offer the possibility of performing and handling concurrent accesses to remote ontology repositories, effectively transforming the system in a client front-end for collaborative ontology management.

Acknowledgements

Our research work on Semantic Turkey has been funded by the FILAS (Finanziaria Laziale di Sviluppo) agency under contract C5748-2005.

References

1. J. Broekstra, A. Kampman & F.v. Harmelen I. Horrocks & J. Hendler (ed.) Sesame: "A Generic Architecture for Storing and Querying RDF and RDF Schema". Springer Verlag, Proceedings of the First International Semantic Web Conference, Sardinia, Italy, pages 54-68, July 2002
2. W. W. Cohen, P. Ravikumar, and S. E. Fienberg. A comparison of string distance metrics for name-matching tasks. In Proceedings of the IJCAI-2003.
3. Firefox home page: http://www.mozilla.com/en-US/firefox/
4. Friend Of A Friend Ontology (FOAF): http://xmlns.com/foaf/0.1/
5. J.J. Garrett. "Ajax: A New Approach to Web Applications". Feb. 18, 2005 http://www.adaptivepath.com/publications/essays/archives/000385.php
6. J. Gennari, M. Musen, R. Fergerson, W. Grosso, M. Crubézy, H. Eriksson, N. Noy, and S. Tu. The evolution of Protégé-2000: An environment for knowledge-based systems development. International Journal of Human-Computer Studies, 58(1):89–123, 2003
7. Erik Hatcher and Otis Gospodnetić. Lucene in Action. Manning ed. 456 pages. 2004. ISBN: 1932394281
8. Jetty Java HTTP Servlet Server. http://jetty.mortbay.org/jetty/.
9. Kiryakov, D. Ognyanov & D. Manov OWLIM – a Pragmatic Semantic Repository for OWL. In Proc. of Int. Workshop on Scalable Semantic Web Knowledge Base Systems (SSWS 2005), WISE 2005, New York City, USA, 20 November 2005
10. Web Ontology Language: http://www.w3.org/TR/owl-features/
11. OWL Lite Description: http://www.w3.org/TR/2004/REC-owl-features-20040210/#s3
12. Simile Java Firefox Extension: http://simile.mit.edu/java-firefox-extension/
13. TopBraid Composer: http://topbraidcomposer.info/
14. Touchgraph Development Page: http://touchgraph.sourceforge.net/
15. Extensible Binding Language: http://www.mozilla.org/projects/xbl/xbl.html
16. XPCOM. http://www.mozilla.org/projects/xpcom/
17. XML User Interface Language (XUL) Project. http://www.mozilla.org/projects/xul/

The Web Service Modeling Toolkit - An Integrated Development Environment for Semantic Web Services
(System Description)

Mick Kerrigan, Adrian Mocan, Martin Tanler, and Dieter Fensel

Digital Enterprise Research Institute (DERI),
Leopold-Franzens Universität Innsbruck, Austria
firstname.lastname@deri.org

Abstract. The time of engineers is a precious commodity. This is especially true for engineers of semantic descriptions, who need to be highly skilled in conceptual modeling, a skill which will be in high demand as Semantic Web technologies are adopted by industry. Within the software engineering community Integrated Development Environments (IDEs) like the Eclipse Java Development Toolkit and NetBeans have proved to increase the productivity of engineers by bringing together tools to help engineers with their everyday tasks. This paper motivates the need for such an IDE for the Semantic Web and in particular describes the Web Service Modeling Toolkit (WSMT), an Integrated Development Environment for Semantic Web Services through the WSMO paradigm.

1 Introduction

The combination of Semantic Web and Web service technologies, to create Semantic Web Services (SWS), with the aim of automating the Web service usage process has been the aim of the WSMO[4], WSML[5] and WSMX[7] working groups over the last number of years. The research in this area has produced a conceptual model, a formal langauge and many back-end services for finding and using Web services that meet the requirements of end-users. However the process of creating the required semantic descriptions is a difficult task and the time of Ontology and Semantic Web Service engineers is a precious commodity.

Within the Semantic Web domain there is much ongoing research into how to present ontologies and semantic data to users. Each of the tools resulting from such research tend to address a given problem within the domain, with these individual research efforts rarely being aligned or integrated together. The lack of integration results in developers switching back and forward between different tools for different tasks. In this paper we introduce the Web Service Modeling Toolkit (WSMT)[1], an Integrated Development Environment for Semantic Web Services, aimed at increasing the productivity of Ontology and Semantic Web Service engineers.

[1] Available for download from http://wsmt.sourceforge.net

E. Franconi, M. Kifer, and W. May (Eds.): ESWC 2007, LNCS 4519, pp. 789–798, 2007.

An Integrated Development Environment (IDE) is defined as *a type of computer software that assists computer programmers to develop software*[2]. The main aim of an IDE is to improve the productivity of the developer by seamlessly integrating tools for tasks like editing, file management, compilation, debugging and execution. Before the creation of the WSMT, developers of semantic descriptions using the WSMO paradigm were forced to create their ontologies, web services, goals and mediators by hand in a text editor. This has many inherent problems, as due to the lack of validation and testing support it is very easy for errors to creep into these semantic descriptions, which go unnoticed by the developer until run-time. Many other tasks that are very easy in an IDE, can be hugely time consuming without one, for example registering a semantic description with an execution environment. Providing a fully integrated suite of tools for Semantic Web Services, that meets the needs and requirements of developers should aid in the adoption of the WSMO technology and reduce the overhead of creating SWS applications.

In the following section we provide the background for our work, in section 3 we describe the problems currently addressed by the WSMT, in section 4 we describe related work and finally in section 5 we provide some conclusions and future work.

2 Background

The Semantic Web aims to make the vast quantities of information available on the Web machine-understandable, by the use of ontologies to annotate Web content. Web service technologies have emerged as a contender for the next generation of Web applications, essentially lifting the Web from a static collection of information to a dynamic computational entity. Web services have machine-processable annotations that are well structured (using XML) and describe how to interface with these services. However these annotations are purely syntactic and not machine-understandable, thus large amounts of human effort is required to build Service Oriented Architectures. Semantic Web Services are the extension of ontologies to describe Web services such that a machine can reason about the functionality they provide, the mechanism to invoke them, and the data they expect as input and return as output. Once Web services are described semantically it allows for large parts of the Web service usage process to be automated. Services can be *discovered* based upon their functionality, can be *selected* based upon the quality of the service, heterogeneity issues with respect to the data they exchange or the process to invoke them can be *mediated.* It is important to note that the semantic descriptions to enable this functionality layer on top of the existing syntactic descriptions for existing Web services. Thus Semantic Web Services are not a reinvention of Web services but an enhancement to them. There are two main contenders in the field of Semantic Web Services, namely WSMO[4] and OWL-S[16][15]. Within the scope of this paper we focus on WSMO as the conceptual model for Semantic Web Services.

[2] http://en.wikipedia.org/wiki/Integrated_development_environment

The Web Service Modeling Ontology (WSMO)[4] is a conceptual model for creating semantic descriptions for Web services that can be used to resolve interoperability issues and automate the Web service usage process. WSMO is based on the Web Service Modeling Framework (WSMF)[5] and as such, is based on the four main elements of the WSMF: ontologies, web services, goals and mediators. The aim of WSMO is to solve the integration problem by describing Web services semantically and by removing ambiguity about the capabilities of Web services, the problems they solve and the process of interacting with them.

The Web Service Modeling Language (WSML)[10] is a formalization of the WSMO ontology, providing a language within which the properties of Semantic Web Services can be described. There are five language variants, based on Description Logic and Logic Programming. Each language variant provides different levels of logical expressiveness[10]. These variants are: WSML-Core, WSML-DL, WSML-Flight, WSML-Rule and WSML-Full. WSML-Core, which corresponds to the intersection of Description Logic and Horn Logic, provides the basis for all the variants, while WSML-Full unites the functionality of all variants. WSML Core is extended in the direction of more expressive Description Logic by WSML-DL and towards Logic Programming by WSML-Flight and WSML-Rule.

A Semantic Execution Environment (SEE) for Semantic Web Services can be used to bind service requestors and providers together at runtime using semantic descriptions of the user's goal and the provider's Web service. SEEs provide functionality for *discovering*, *composing*, *selecting*, *mediating*, and *invoking* Web services that match the end users requirements. There are currently two Semantic Execution Environments for WSMO, namely the Web Service Execution Environment (WSMX)[7] and IRSIII[2]. The SEE functionality is currently being standardized through the OASIS standardization.

3 The Web Service Modeling Toolkit

The Web Services Modeling Toolkit (WSMT) is an Integrated Development Environment (IDE) for Semantic Web Services implemented in the Eclipse framework. The WSMT aims to aid developers of Semantic Web Services through the WSMO paradigm, by providing a seamless set of tools to improve their productivity. As already mentioned a Integrated Development Environment (IDE) is defined as *a type of computer software that assists computer programmers to develop software*. IDE's like the Eclipse Java Development Toolkit (JDT)[3] and NetBeans[4] for developing java software have proven that good tool support can improve the productivity of engineers. It could be said that the time of ontology and Semantic Web Service engineers is a more precious commodity, as the number of people who are currently skilled in conceptual modeling is much less than those that can code in Java. This underscores the need for adequate tool support for working with semantic technologies and an IDE can tie together these tools in such a way that the whole application is more than the sum of the parts

[3] http://www.eclipse.org/jdt/
[4] http://www.netbeans.org

that make it up. The main advantage of using the IBM Eclipse framework is the existence of other useful plugin projects, for example the Eclipse Web Tools Platform (WTP)[5], which provides tools for Web service technologies like WSDL and XML. This enables developers to put the WSMT and the WTP together to build their Web services and semantically describe them with WSMO. The process of building ontologies or describing Web services semantically involves the creation of different types of documents. The main tasks in this process revolve around four main themes:

- **Editing:** Firstly the actual descriptions must be created. It is important that users of different skill levels are supported within the IDE, thus editing support at different levels of abstraction should be provided. Considering ontologies, it may be more convenient for the engineer to create an ontology using a textual representation and then to use a graph based representation to learn more about the ontology.
- **Validating:** The most common problem that occurs when creating semantic descriptions is incorrect modeling. It can be very easy for an engineer to make a mistake without any tool support. Validation of semantic descriptions is a non trivial task and validation at both the syntactic and semantic levels can vastly reduce the time an engineer spends debugging an ontology.
- **Testing:** Once valid semantic descriptions exist the engineer needs to ensure that they behave in the expected manner in their intended environment prior to deploying them. Having testing integrated into the development environment reduces the overhead of the user performing a lengthy, iterative, deploy-test scenario.
- **Deploying:** Ultimately the descriptions created within the development environment must be used in some run-time system. Deploying descriptions can also be a huge overhead on the engineer and having tool support in an IDE can prevent mistakes occurring at this crucial stage of the process.

The WSMT focuses on three main areas of functionality, namely the engineering of WSMO descriptions, creation of mediation mappings and interfacing with execution environments and external systems. Towards these aims, the Eclipse editors and views available within the WSMT are broken up into three Eclipse perspectives[6]. These perspectives are the WSML, Mapping and SEE perspectives respectively and are described in the next sections.

3.1 The WSML Perspective

Creating ontology descriptions is a non trivial task that requires the skills of a trained ontology engineer, and many tools exist for building ontologies in OWL[16] and RDF(S)[1]. In the case of WSMO the engineer must also be knowledgable about both Ontologies and Web services in order to creation semantic

[5] http://www.eclipse.org/webtools/

[6] Eclipse perspectives are used to group together editors and views that a given developer will use while performing a given set of tasks.

descriptions. As already described it is important that tool support exists for efficiently using the time of this skilled engineer. Within the WSML perspective a number of tools are provided that allow users of different skill levels to create, manage and interact with semantic descriptions in the WSMO paradigm. These tools include:

- **WSML Validation:** WSMO4J[7] is an object model for manipulating WSMO descriptions and is capable of parsing and serializing WSML documents to and from this object model. WSMO4J also provides a validator for each of the 5 WSML variants, which is exploited within the WSMT to validate the files that are located within the WSMT workspace as the user edits these documents. This validation ensures that the engineer of the semantic descriptions gets immediate feedback of errors they create, both in the syntax and the semantics of the semantic descriptions, as they create them.
- **WSML Text Editor:** Until recently ontology engineers using the WSMO paradigm would create there WSMO descriptions by hand in a text editor. It is very tempting when creating a toolkit to abstract away from a text editor and to provide more advanced editing support; However in many cases the engineer is more comfortable with editing the raw text of the semantic description. This is especially true with respect to WSML, as the WSML human readable syntax is a very lightweight syntax. Within the WSMT we cater for such users and provide them with additional features including syntax highlighting, syntax completion, in line error notification, content folding and bracket highlighting.
- **WSML Form based Editor:** Moving up from the WSML text editor, we have abstracted to a form based editor that provides the user with an intuitive interface for building semantic descriptions by completing forms. In this editor the user can create new elements related to their semantic description and specify their properties. Within the editor the user is always manipulating a WSMO4J object model, which can be serialized to the human readable syntax by saving the editor.
- **WSML Visualizer:** The WSML Visualizer[8] is a fully integrated graph based visualization and editing tool for WSML. The WSML Visualizer allows the engineer to learn more information about their semantic descriptions as they create them. Normally visualization solutions are bolted on top of existing ontology engineering solutions after the fact, by providing an integrated solution the user need not switch back and forth between an editing environment and a visualizer to understand the effects of changes to the ontology. The visualizer tries to resolve the graph scalability problem that is often found with visualization solutions by breaking down the visualization into multiple levels and allowing the user to browse these levels like in a web browser. The tool includes manipulation features like zoom and rotate, along with the ability to filter nodes of a certain type.
- **WSML Reasoner View:** Semantic descriptions are only useful if they can eventually be reasoned over and much work is ongoing within the WSML

[7] http://wsmo4j.sourceforge.net/

community to create reasoners for all 5 variants of the WSML language. This view exposes the functionality of the WSML2Reasoner framework, which currently provides access to the Pellet[14] description logic reasoner and MINS logic programming reasoner[8]. The purpose of this view is to allow the engineer of semantic descriptions to be sure that the descriptions they create yield the expected results when put into a reasoner.
- **Discovery View:** When engineers are creating Semantic Web Service descriptions, it must always be at the forefront of their mind that these Web Service descriptions will ultimately be discovered by a users Goal description. The discovery view can be used in a similar way to the reasoner view to allow the engineer to check that a specified Goal matches the expected set of the Web Services in the users workspace, i.e. that the Goal will behave as expected in a given discovery engine.

3.2 The Mapping Perspective

When enabling interoperability between two business partners it is important that the data exchanged is correct. However it is unlikely that both partners will use the same ontology to represent their data. Ontology to ontology mediation within the WSMX environment is considered as a semi automatic process, where mappings between two ontologies are created at design time and then applied automatically at runtime in order to perform instance transformation. Automatic approaches for creating mappings do exist but their accuracy is relatively low and we believe that for business to business integration an engineer must be involved in creating and validating the mappings. This is a non trivial task and the user should be guided through the process of creating the mappings and ensuring that they are correct. Within the WSMT all mappings are in the Abstract Mapping Langauge (AML)[13] syntax, which is formalism neutral, and later grounded to WSML within WSMX.

- **AML Validation:** The importance of validating semantic descriptions as the user creates them was already described in section 3.1, and this is equally true for the Abstract Mapping Language. Within the WSMT we currently provide validation for the syntax of AML documents and will extend this to semantic validation when a validator for the AML semantics is available.
- **AML Text Editor:** The Abstract Mapping Language Text Editor provides a text editor for the human readable syntax of the AML. It features similar features to that of the WSML human readable text editor including syntax highlighting, in line error notification, content folding and bracket highlighting. This editor enables the engineer to create or modify mappings through textual descriptions, this is especially useful for tweaking mappings that have been produced by other editors.
- **AML View Based Editor:** The AML View Based Editor provides a graphical means to create mappings between ontologies. It is often the case that the expert that understands the problem domain and is capable of aligning

[8] http://tools.deri.org/mins/

the two ontologies is not also a specialist in logics. The suggestion of possible mappings is done by using a set of algorithms for both lexical and structural analysis of the concepts. Additionally, the guidance is offered by decomposition and context updates. As described in [11], the graphical point of view adopted to visualize the source and target ontologies makes it easier to identify certain types of mismatches. This viewpoints are called *perspectives* and it is argued that by using combinations of these perspectives on the source and target ontologies, certain types of mappings can be created using only one simple operation, *map*, combined with mechanisms for ontology traversal and contextualized visualization strategies.

- **AML Mapping Views:** The AML Mapping Views have the role of providing a light overview on the mappings created either by using the AML Text Editor or the AML View Based Editor. Instead of seeing the full description of mappings it is also useful to see a more condensed version of this information with the entities in the source and in the target that are mapped and the conditions associated with them. For this purpose there been four types of Eclipse views defined (*Concept2Concept*, *Attribute2Attribute*, *Concept2Attribute* and *Attribute2Concept*), each corresponding to the combinations of the entities that can participate in a mapping.
- **MUnit Testing View:** Mappings must be updated as ontologies evolve, it can be hard for the engineer to be aware of the effects that these constant changes have. The MUnit unit testing view for the Abstract Mapping Language gives the engineer support to ensure that instances are correctly transformed. The user can define pairs of sources and targets, specifying that the result of transforming the sources, using the existing mappings, should be the targets. These tests can then be incrementally run by the engineer when validation of the mappings is required.

3.3 The SEE Perspective

Ultimately the purpose of creating semantic descriptions in WSMO and defining mappings between ontologies is to allow automation of the Web service usage process and to bind requester and provider at runtime with minimal human intervention. As described in section 2, Semantic Execution Environments, like WSMX and IRSIII provide automatic *discovery*, *composition*, *selection*, *mediation*, and *invocation* of Semantic Web Services. Many of the tasks performed by the Semantic Web Service engineer involve interfacing with these SEEs. Functionality is provided within the WSMT to reduce the effort normally spent interacting with these environments.

- **Integration with WSMX:** WSMX exposes a number of Web service interfaces that give access to its functions. These Web services can be split into two categories, namely services related to executing WSMX and services related to storing and retrieving descriptions. The SEE Perspective provides access to both of these types of services, users can right click on WSML and AML documents in the workspace and use the descriptions contained within these documents to invoke the entry points of WSMX i.e. AchieveGoal, InvokeWebService or these descriptions can be stored to the internal repositories of WSMX. The SEE perspective also shows all the descriptions currently stored within the WSMX repositories and allows the user to download these descriptions into their workspace.

– **Integration with IRSIII:** Similar functionality as for the WSMX system is available for the IRSIII system. One interesting difference is that all data within IRSIII is described in terms of the Open Universities OCML[12] format, and the WSMT must translate back and forth from this format when sending data to or retrieving data from IRSIII.

4 Related Work

This section outlines other efforts in the direction of creating integrated environments aiding users of semantic technologies. For this we have selected the most popular tool for creating ontologies in OWL, namely the Protégé OWL Plugin, along with WSMO Studio, which is the closest alternative to the WSMT for creating WSMO descriptions.

Protégé[6] is a free open source ontology engineering environment that can be extended to support different ontology formats. Through a collection of these extensions Protégé currently supports Frames, XML Schema, RDF(S) and OWL. The Protégé OWL Plugin[9] was developed at Stanford Medical Informatics[9] as a tool for editing the Web Ontology Language (OWL)[16]. The plugin attempts to abstract the user away from the underlying OWL RDF syntax and provides a number of graphical widgets that can be accessed by the user for building ontologies and creating instance data. The Protégé OWL Plugin can also be extended with plugins, existing plugins provide functionality for visualizing OWL ontologies, integrating with OWL reasoners etc. The main limitation of Protégé is that, while it is the standard toolkit used within the Semantic Web community for building ontologies, other tools are not integrated with this environment. Semantics will soon be present in many different parts of computer science and as already described integrating Semantic Web tools into the Eclipse framework allows for them to be used side by side with other toolkits like the Java Development Toolkit (JDT) or Web Tools Platform (WTP).

WSMO Studio[3] is a collection of tools for editing WSMO descriptions developed for the Eclipse Framework by OntoText Labs[10], Sirma. WSMO Studio offers the engineer functionality for editing WSMO descriptions through the WSML paradigm. The main difference between the WSMT and WSMO Studio is the level of abstraction. WSMO Studio focuses squarely on editing descriptions at the form based level and provides only basic functionality in a text editor for WSML, i.e. Syntax highlighting and no ontology visualization functionality.

---

[9] http://smi.stanford.edu/

[10] http://www.ontotext.com

With respect to Semantic Execution Environments, WSMO Studio has functionality for invoking the IRSIII system through its repositories view but does not provide any functionality for integrating with WSMX. The choice of Eclipse by both environments allows for the tools of each, which in many ways complement each other, to be placed in the same application.

5 Conclusions and Future Work

This paper has described the need for an Integrated Development Environment for the Semantic Web and for Semantic Web Services. It has also introduced the Web Service Modeling Toolkit (WSMT), which aims to provide such an IDE to improve the productivity of engineers of semantic descriptions. Section 3 introduced the concepts of **editing**, **validating**, **testing** and **deploying** in an IDE. The following table shows how the Web Service Modeling Toolkit meets these four concepts for each of the descriptions that it supports:

Feature	Ontology, Web Service & Goal	Mappings
Editing	Text, Form & Graph	Text & Dual Tree
Validating	Syntactic & Semantic	Syntactic
Testing	Reasoning View & Discovery View	MUnit View
Deploying	SEE Perspective	SEE Perspective

Future work in the WSMT includes automation of the process of ensuring that semantic description behave as expected, by the addition of unit testing functionality for WSML. This unit testing functionality will allow the user to ensure that their ontologies behave in the expected way as they evolve. Further integration with the WSMX and IRSIII systems will also be added to the SEE perspective, allowing the user to visualize the processes that are executing within the Semantic Execution Environments. Thus allowing engineers to debug their descriptions with respect to their behavior within a SEE.

Acknowledgments

The work is funded by the European Commission under the projects ASG, DIP, enIRaF, InfraWebs, Knowledge Web, Musing, Salero, SEKT, SEEMP, SemanticGOV, Super, SWING and TripCom; by Science Foundation Ireland under the DERI-Lion Grant No.SFI/02/CE1/I13 ; by the FFG (Österreichische Forschungsförderungsgesellschaft mbH) under the projects Grisino, RW^2, SemNetMan, SeNSE, TSC and OnTourism.

References

1. D. Brickley and R.V. Guha. Resource Description Framework (RDF) Schema Specification 1.0, W3C Recommentdation, 2000.
2. L. Cabral, J. Domingue, S. Galizia, A. Gugliotta, B. Norton, V. Tanasescu, and C. Pedrinaci. IRS-III: A Broker for Semantic Web Services Based Applications. In *Proc. of the 5th Int'l Semantic Web Conf (ISWC)*, 2006.
3. M. Dimitrov, A. Simov, V. Momtchev, and D. Ognyanov. WSMO Studio - an Integrated Service Environment for WSMO. In *WSMO Implementation Workshop 2005*, volume 134 of *CEUR Workshop*, 2005.
4. C. Feier, A. Polleres, D. Roman, J. Domingue, M. Stollberg, and D. Fensel. Towards Intelligent Web Services: The Web Service Modeling Ontology (WSMO). In *Proc. of the Int'l Conf on Intelligent Computing (ICIC)*, 2005.
5. D. Fensel and C. Bussler. The Web Service Modeling Framework (WSMF). *Electronic Commerce Research and Applications*, 1(2):113–137, 2002.
6. J. H. Gennari, M. A. Musen, R. W. Fergerson, W. E. Grosso, M. Crubzy, H. Eriksson, N. F. Noy, and S. W. Tu. The Evolution of Protégé: An Environment for Knowledge-Based Systems Development. Technical report, 2002.
7. A. Haller, E. Cimpian, A. Mocan, E. Oren, and C. Bussler. WSMX - A Semantic Service-Oriented Architecture. In *Proc. of the Int'l Conf on Web Services (ICWS)*, 2005.
8. M. Kerrigan. WSMOViz: An Ontology Visualization Approach for WSMO. In *Proc. of the 10th Int'l Conf on Information Visualization*, 2006.
9. H. Knublauch, R. W. Fergerson, N. F. Noy, and M. A. Musen. The Protégé OWL Plugin: An Open Development Environment for Semantic Web Applications. In *Proc. of the 3rd Int'l Semantic Web Conf (ISWC)*, 2004.
10. H. Lausen, J. de Bruijn, A. Polleres, and D. Fensel. WSML - A Language Framework for Semantic Web Services. In *Proc. of the W3C Workshop on Rule Languages for Interoperability*, 2005.
11. A. Mocan, E. Cimpian, and M. Kerrigan. Formal Model for Ontology Mapping Creation. In *Proc of the 5th Int'l Semantic Web Conf (ISWC 2006)*, 2006.
12. E. Motta. *Reusable Components for Knowledge Modelling. Case Studies in Parametric Design Problem Solving*, volume 53 of *Frontiers in Artificial Intelligence and Applications*. IOS Press, 1999.
13. F. Scharffe and J. de Bruijn. A language to specify mappings between ontologies. In *IEEE Conference on Internet-Based Systems SITIS6*, 2005.
14. E. Sirin, B. Parsia, B. Cuenca Grau, A. Kalyanpur, and Y. Katz. Pellet: A Practical OWL-DL Reasoner. *Submitted for publication at "Journal of Web Semantics"*, 2006. Available from: http://www.mindswap.org/papers/PelletJWS.pdf.
15. N. Srinivasan, M. Paolucci, and K. Sycara. Adding OWL-S to UDDI, Implementation and Throughput. In *Proc. of 1st Int'l Workshop on Semantic Web Services and Web Process Composition (SWSWPC)*, 2004.
16. W3C. OWL Web Ontology Language Reference. Technical report, 2004.

Understanding Large Volumes of Interconnected Individuals by Visual Exploration
(System Description)

Olaf Noppens and Thorsten Liebig

Institute of Artifical Intelligence
Ulm University
Ulm, Germany
{olaf.noppens|thorsten.liebig}@uni-ulm.de

Abstract. Ontologies are now used within an increasing number of real-world applications. So far, significant effort has been spend in building tools to support users in creating, maintaining, and browsing the terminological part of an ontology. On the other hand, only little work has been done in supporting the user to explore the manifold interconnected assertional knowledge in order to analyze, visualize, and understand this network of individuals. In this paper, we present a new efficient visualization and editing approach which allows to investigate relationships within large volumes of interlinked individuals in order to grasp the structure of the assertional knowledge more easily.

1 Motivation

An OWL ontology typically consists of two parts [2]: The terminological or schema part introduces concepts and properties and gives structure to them in terms of axioms using the available OWL Lite or OWL DL language constructs (also called TBox in Description Logics). The assertional or data part defines concrete individuals and relationships between those individuals (also called ABox) utilizing the concepts and properties of the terminology.

As an example, consider the ontology given by the widely known Lehigh University Benchmark (LUBM) [6]. This test suite has become a de-facto standard for measuring reasoning performance and consists of an ontology covering basic elements in the domain of tertiary education. For instance, its TBox defines terms like `Department`, `Student`, and `Course` as well as properties like `subOrganisationOf`, `takesCourse`, and `headOf`. A corresponding ABox covers virtual universities, faculties, research groups, etc. which are randomly populated with individuals such as `FullProfessor3`, `GraduateStudent27`, `GraduateCourse6`. These individuals are related via property instantiations, e. g. `GraduateStudent27 takesCourse GraduateCourse6`. A huge ontology of this kind obviously is characterized by a large number of individuals (i. e. students, departments, etc.), property instantiations (course attendees, employees, etc.) and a constant

E. Franconi, M. Kifer, and W. May (Eds.): ESWC 2007, LNCS 4519, pp. 799–808, 2007.

number of terms within the terminological schema. In fact, within a typical real-world application, the TBox is assumed to be much smaller in comparison to the amount of assertions in the ABox.

New reasoning systems as well as recent system optimizations have shown significant increase in speed for answering conjunctive queries with respect to LUBM test cases containing several hundred thousands of individuals [13,14,18]. Even if these results rely on synthetically generated data produced by a relatively simple benchmark generator such as the LUBM, they clearly demonstrate promising progress in the development of scalable reasoning systems.

On the other hand, little has yet been done to support ontology users or developers to visually edit or explore such large volumes of interrelated individuals. There is currently no representation approach or even a tool to gradually inspect an individual with respect to its direct and indirect fillers regarding a transitive property. For instance, within the university domain one could be interested in stepping through the `subOrganisationOf` relationship up to a specific institute in order to expand all the persons which are advised by the corresponding chair. Or within the Gene Ontology (GO) [4], which contains millions of individuals, users are presumably interested in investigating which specific DNA binding product interacts with which kind of receptors for example.

Our approach utilizes a user-driven visualization strategy, making use of animated expansion steps, clustering techniques, and different levels of detail views. It allows for operations on a single individual or on sets of individuals and takes property features such as transitivity, symmetry, etc. into account.

1.1 Visualizing Entailed Assertions

Note that our visualization approach inherently requires reasoning. Since OWL allows for property hierarchies as well as symmetrical, transitive, or functional properties, a reliable reasoning system such as RacerPro [7] is needed to make implicit property instantiations explicitly available. Without reasoning feedback an ABox visualizer would degrade to a syntax imaging tool, probably hiding the most important correlations. For instance, in case of `has-child` being a subproperty of a transitive property `has-descendant`, there implicitly exists a `has-descendant` link between an individual and all its children. Another example deals with the fact that in OWL individuals are not disjoint by default. As a consequence two fillers of a functional property (a property with at most one filler per source) will be merged to one individual by the inference engine. Thus, an ABox visualization should also render them as one object bearing two identifiers.

Therefore, the underlying presentation principle of our ABox visualizer is to always render the semantical implied relationships between individuals in order to make effectively visible what is entailed in the ontology. We argue that any serious ontology authoring or browsing tool should allow users to explore implicitly modeled as well as explicitly given information, while being able to distinguish between both. In this sense this component is a direct extension of our OWL TBox authoring tool OntoTrack [12].

1.2 Related Work

As mentioned before, there is poor tool support in browsing and editing ABox data. There are some OWL editors which allow for inspection of single individuals with help of selection lists and standard form elements such as Protégé [10] or SWOOP [8]. However, they only provide one level of detail and have no interface for exploring the fillers of more than one property in parallel. Moreover, list-style and table-based approaches are not adequate techniques to analyze large amounts of data and are less practical in showing a chain of property fillers in a clear and concise manner (e. g. to follow the transitivity of properties).

Tools such as GrOWL [11] try to offer a graph based interface which uses graphical icons to depict different types of nodes (class, property, or individual) as well as language constructs (negation, union, etc.). GrOWL provides a spring layout which dynamically adds nodes to the graph on user demand. However, this functionality is not sufficiently fine-graded enough for the task of gradually inspecting property fillers. In addition, the resulting graph is somewhat verbose and may distract the user due to frequent layout changes or node agglutination.

On the triple-based representation level of RDF, individuals are just as any other resources. Furthermore, since OWL is layered on top of RDF Schema, there is a lack of expressivity in RDF needed to capture the semantics of an OWL ABox appropriately. As said before, the graph representation of the RDF data model is conceptually different from the graph structure of an ABox even without considering entailed statements.

Nevertheless, RDF Gravity [5] effectively supports a user in filtering a RDF graph by means of selecting resources or specifying RDQL queries but is not ideal with respect to a user-friendly layout.

2 Interactive Exploring of Individuals

Exploring Large Volumes of Interrelated Individuals. When visualizing large data volumes there always is a trade-of between detailedness and overview. One lesson learnt from the visual analysis of large data sets in general is that it is not advisable to arbitrarily visualize all dependencies, particulars etc. at any time [9]. Hence, our approach tries to provide detail information on user demand while offering an overview at the same time. Here we adopt the single-view visualizing paradigm enabling selective detailed views which has turned out to be adequate for visualizing concept and property hierarchies as invented by our ontology authoring framework OntoTrack.

Our visualization paradigm is based on a user-directed exploration of interrelated individuals which does consider an individual as a first-class element but also allows to group them within clusters as following. Starting with a user selected individual, one can interactively exploit the property fillers (respectively datatype values in the case of datatype properties) of each property the individual is related to in a step-wise fashion. For instance, the LUBM ontology defines a property named `takesCourse` which relates students with courses, e. g. `GraduateStudent27 takesCourse Course10` as well as `Course21`.

Exploring on User-Demanded Expansion. Figure 1 shows that the root is related via the property `subOrganizationOf` with exactly one other individual. When hovering over an individual with the mouse pointer and clicking the middle mouse button, the tool offers a preview depicting all the properties together with their number of fillers (as long as they have at least one). From our experience, it is often the case that a "natural" exploration of an ontology will typically not only include a directed exploration with respect to a property but also its inverse direction even if there is no inverse property explicitly defined (to provide a bidirectional exploration). For instance, when having all courses a student takes one would like to explore other students of a specific course. In Figure 1 there are filler of four other properties to which the actual root is related to in an inverse fashion. In order to denote the different directions a small arrow indicates whether the actual root is the source of the property (right arrow) or a filler (left arrow).

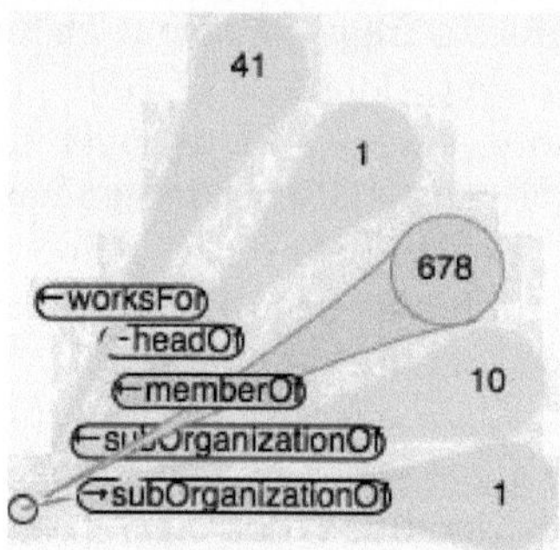

Fig. 1. Preview context menu for expanding property fillers

After expansion, all fillers with respect to the selected property are grouped within a so-called *property filler cluster* which will be drawn as a a club originating from the individual which is considered as the source of the expansion (e. g. `takesCourse` in Figure 2). Each individual within the cluster are rendered as circles whose labels are accessible via mouse-over tooltips. Due to a standardized package algorithm the clusters diameter approximates the number of individuals and allows to easily compare different filler sets by their rendering size.

At any time the user can guide his exploration by choosing a follow up root from the individuals within the currently visible clusters or may branch by selecting other properties for further expansion. To distinguish different properties different colors are used. In addition each club carries its corresponding property label. One can optionally switch of in-view label rendering. Then there will be a list of colored property names in order to denote which property is represented by which color. However, in both cases, the color of the property club and the property label is always the same for clubs which are based on the same property. In a similar manner, different colors are uses for the different types of fillers (i. e. individuals versus data values).

Figure 2 shows an example snapshot of a partially expanded LUBM ontology containing five universities (approx. 60 thousand overall individuals). Following the first expansion level at the very bottom of Figure 2 it is easy to perceive that the root individual (an undergraduate student) takes two courses. Furthermore, one of this courses has more than 30 other participants from which one takes four courses. The root individual also is member of an university with over 40 employees (see → `memberOf` ← `worksFor` expansion). Any changes in the layout such as (de-)expanding clusters are animated to easily grasp the differences between two exploration states. Beyond that the whole layout can continuously be zoomed or paned simply by mouse-down movements.

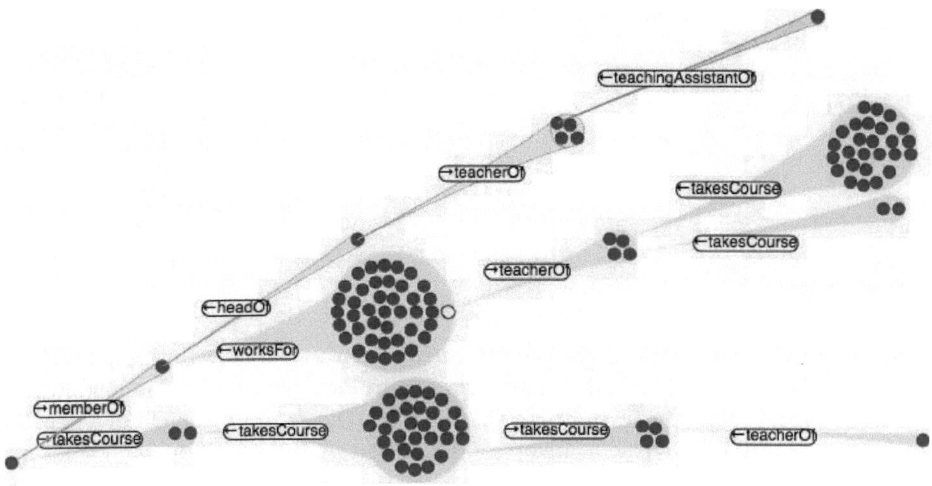

Fig. 2. Partial expanded property fillers

Close-By Detail Information. In our first prototype [15] we followed exactly the *detailed view* mode paradigm of OntoTrack to provide further information for individuals, such as name, via mouse-wheel usage. However, displaying individuals names within clusters results in diluting the correlation between the property filler cluster's diameter and its number of fillers. We therefore decided to provide an additional area for detailed information instead. When hovering over a property filler cluster with the mouse pointer, all contained individuals are displayed in the detailed view area at the right hand side of the visualization as shown in Figure 3. Here, further detailed levels can be activated or de-activated using the mouse-wheel, e. g. to display (in a non-graph-based way) all direct classes the selected individual is instance of, or even to display told information.

Identifiers, Labels, and Names. When showing the name of an individual the question arises what does the name mean and where it comes from. Our experience with our first prototype attested the feeling that for exploring an ontology more naturally, names as typically provided with a RDF label annotation is better suited. However, as these labels are just labels and does not carry any semantics or constraints there are not always used. For instance, to express that each individual which is instance of `Person` should have a name, this is typically modeled as restrictions with respect to a property `hasName`. Therefore, our visualization offers the possibility not only to use the local name of the URI or the label (if any) but also to specify a (datatype) property whose filler is used for labeling the graphical representation. Note that there must exist such a filler.

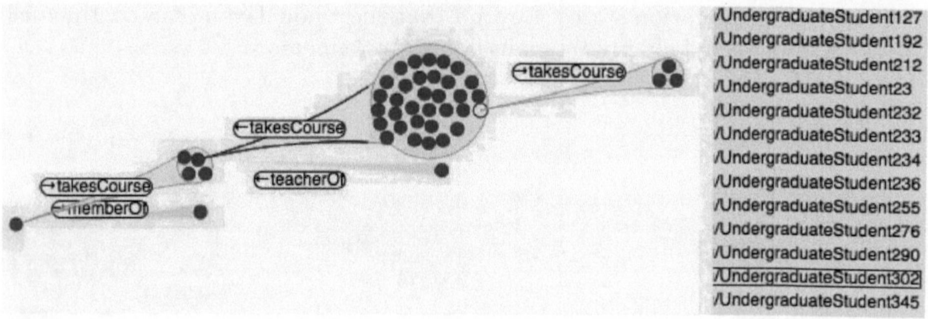

Fig. 3. Sampled detailed view for the individuals of the second property filler cluster `takesCourse`

Cloning in Favor of a Concise Visualization. From a logical perspective, one and the same individual can be related to another individual via different properties and may appear multiple times within the expansion path of a specific property, e. g. due to cycles or inverse exploration. A visualization can either use one single graphical representation for each individual or has to use a cloned graphical representation for different occurrences of the same individual. We decided to allow for clones representations in clusters containing one an the same individual because otherwise a path-directed expansion would no longer be possible. However, if a user hovers with the mouse pointer over an individual, all its visible representations are highlighted simultaneously.

Exploring at Different Levels of Granularity. In addition to expand a single individual it is also possible to expand the fillers of a whole cluster with respect to a property as shown in Figure 4. This can be done by choosing the club itself (border or background) rather than a specific individual as expansion source. The emerging filler cluster then contains the union of all fillers of the predecessor cluster. In case of already expanded clubs (with respect to that specific property), all their individuals move into the new cluster in an animated manner which may serve as a simple visual explanation of the underlying action semantics.

Distinguishing Between Inferred and Told Information. According to the semantics of a *transitive property* an individual is related to all directly or indirectly related fillers reachable via this property. Therefore, when expanding the fillers of a transitive property with respect to an individual (or cluster of

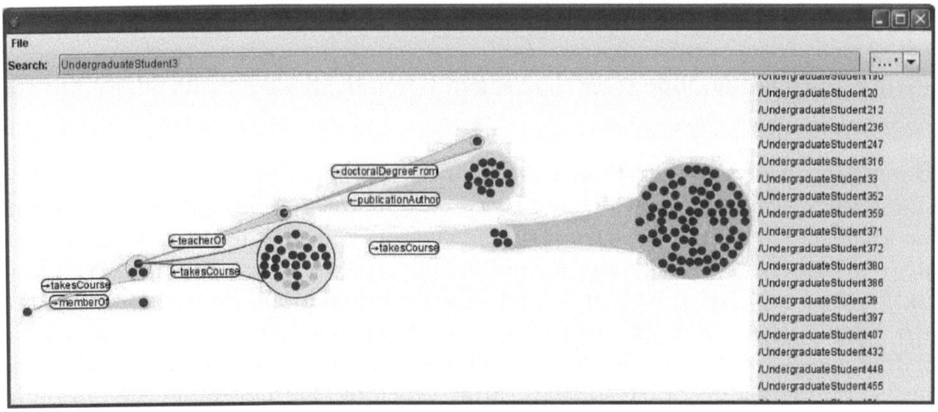

Fig. 4. Cluster expansion and instant highlighting of search matches

individuals) the transitive closure of related fillers are shown. One can, however, distinguish between directly and indirectly related fillers by expanding the next level. The first expansion will then only contains the *direct* property fillers whereas all other fillers are transferred into the subsequent property cluster in an animated fashion. For instance, the upper part of Figure 5 shows the first expansion level of the transitive property `subOrganizationOf`, which contains

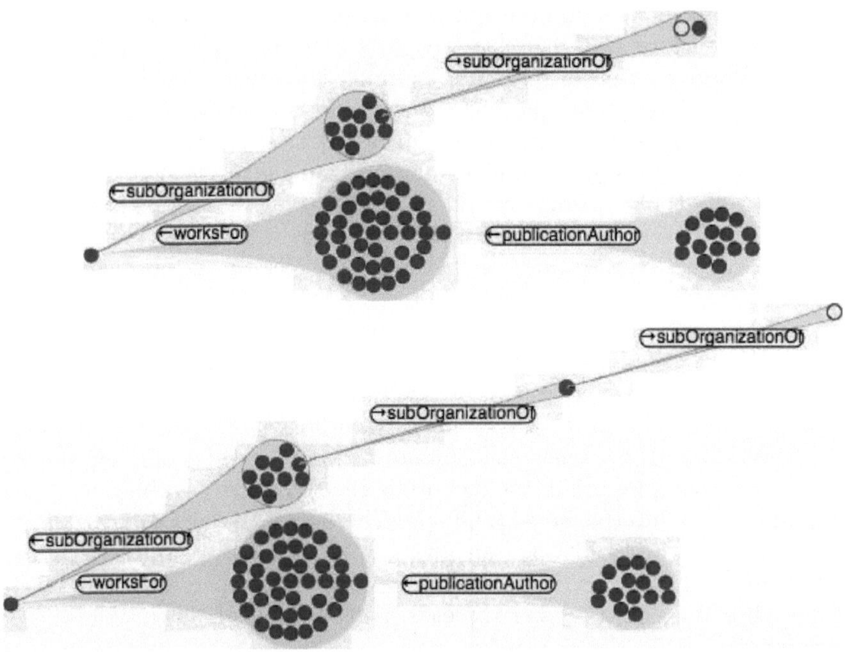

Fig. 5. Follow up expansion of a transitive property

two individuals which are reachable from the selected individual in the inverse expanded property filler cluster. After expanding the next level (lower part of Figure 5) it becomes apparent that one of the individuals now has moved to the next expansion level.

Searching and Editing Features. Our ABox visualizer also supports ONTOTRACK's search features which are based on dynamic queries [17] which also can be found in tools like SpaceTree [16]. For instance, during a string based search process the user starts typing an individual or property name and all matching entities are instantly highlighted. Each additional character or deletion in the search string directly results in an updated highlighting of the matching individuals as shown in Figure 4. In addition to browsing, the ABox visualizer also supports rudimentary editing features such as the removal and addition of property fillers. One can add an individual to a filler set by using the drag 'n drop paradigm. For instance, a user can drag an individual out of a cluster into another one or onto the work space. The result of the former is a changed filler set membership, the latter will remove the instance from the original filler set. However, removing an individual from all filler sets does not remove the individual from the underlying ontology. For those kind of editing actions additional authoring features still have to be developed.

3 Implementation and Current Work

During our implementation of a first prototype [15] we discovered several problems with respect to the visualization paradigm and the communication performance with our external reasoning system. These experiences were carefully considered when designing the current revision of our ABox visualizer. As a recent analysis has shown, ontologies typically consist of several hundreds of individuals [19]. Obviously, scalability and performance are very important issues for user-friendly tools. From a graphical perspective, we address scalability and performance with our decision for the Piccolo framework [3] because of our positive experience with large numbers of graphical objects in the ontology authoring framework ONTOTRACK.

Following from our description above, our ABox tool is linked with an OWL inference engine in order to be able to visualize entailed knowledge. Querying for all entailed knowledge at start up is obviously not a feasible solution when dealing with large volumes of data. As we are currently not able to use the standard DIG 1.1 interface [1] for reasoner communication because it only supports very basic ABox queries, we use in the meantime the RacerPro reasoner via its native syntax and query interface over TCP [7]. The use of the nRQL query language gives us also the possibility for an query optimization: we query for the first property fillers in advance which are successors of the current visible clusters[1]. The querying of the next answer tuples is then organized following the *last come,*

[1] The first ones are those which are "easily" to compute by the reasoner. For more information to that subject please refer to Racer's nRQL language.

first serve principle. If the user exploits other branches of the property fillers those will be computed first. The graphical representation also gives feedback about the process of querying, i. e. the question mark symbol is shown in the middle of a property filler cluster when some individuals are still missing until the reasoner answered the query for the remaining fillers. Note that the upcoming new version of DIG, namely DIG 2.0, will support more fine-graded and more expressive queries.

Current work deals with focus and thumbnail techniques in order to preserve an overview of the data during exploration operations. Whenever a cluster would occupy to much screen space, i. e. when the size will exceed a certain value, it will automatically switch to a more compact visualization using a thumbnail representation. We also plan to optimize the layout algorithm with respect to individual or club placing and better visualization of the starting points of property filler clusters. Our experience with the first prototype showed us some improvements we have implemented and therefore plan to conduct a user study in order to gain real-user feedback.

4 Summary and Outlook

In this paper we presented a novel approach for exploring large volumes of individuals within ontology languages such as OWL. In contrast to other visualization tools we focus on a visualization of semantically interrelated individuals. The system therefore incorporates a reasoning system to discover, for instance, direct and indirect fillers, symmetric, functional and transitive properties to provide a semantically correct visualization and not only told information. Our implementation allows for incremental inspection of filler sets, selective browsing and applies several abstractive visualization techniques not found in current tools. First experiences with volumes containing several thousands of individuals such as in the synthetical LUBM benchmark ontologies are encouraging.

References

1. Sean Bechhofer. The DIG Description Logics Interface: DIG/1.1. Technical report, University of Manchester, 2003.
2. Sean Bechhofer, Frank van Harmelen, Jim Hendler, Ian Horrocks, Deborah L. McGuinness, Peter Patel-Schneider, and Lynn Andrea Stein. OWL Web Ontology Language Reference. W3C Recommendation, February 2004.
3. Ben Bederson, Jesse Grosjean, and Jon Meyer. Toolkit Design for Interactive Structured Graphics. Technical Report CS-TR-4432, University of Maryland, January 2002.
4. The Gene Ontology Consortium. Gene Ontology: tool for the unification of biology. *Nature Genetics*, 25:25–29, May 2000.
5. Sunil Goyal and Rupert Westenthaler. RDF Gravity. Salzburg Research, 2004. http://semweb.salzburgresearch.at/apps/rdf-gravity/.
6. Yuanbo Guo, Zhengxiang Pan, and Jeff Heflin. LUBM: a benchmark for OWL knowledge base systems. *Journal of Web Semantics*, 3(2):158–128, July 2005.

7. Volker Haarslev and Ralf Möller. Racer: A Core Inference Engine for the Semantic Web. In *Proc. of the 2nd Int. Workshop on Evaluation of Ontology-based Tools (EON 2003)*, pages 27–36, 2003.
8. Aditya Kalyanpur, Bijan Parsia, Evren Sirin, Bernardo Cuenca Grau, and James Hendler. Swoop: A Web Ontology Editing Browser. *Journal of Web Semantics*, 4(2), 2006.
9. Daniel A. Keim. Visual exploration of large data sets. *Communications of the ACM*, 44(8):38–44, 2001.
10. Holger Knublauch, Ray W. Fergerson, Natalya F. Noy, and Mark A. Musen. The Protégé OWL Plugin: An Open Development Environment for Semantic Web Applications. In *Proc. of the 3rd Int. Semantic Web Conference (ISWC 2004)*, pages 229 – 243, 2004.
11. Sergey Krivov, Ferdinando Villa, and Rich Williams. GrOWL, visual browser and editor for OWL ontologies. *Journal of Web Semantics*, 2006.
12. Thorsten Liebig and Olaf Noppens. OntoTrack: A semantic approach for ontology authoring. *Journal of Web Semantics*, 3(2):116 – 131, 2005.
13. Ralf Möller, Volker Haarslev, and Michael Wessel. On the scalability of Description Logic instance retrieval. In *Proc. of the 29th German Conf. on Artificial Intelligence*, LNAI, pages 171–184, Bremen, Germany, June 2006. Springer.
14. Boris Motik and Ulrike Sattler. A comparison of reasoning techniques for querying large Description Logic ABoxes. In *Proc. of the 13th Int. Conf. on Logic Programming Artificial Intelligence and Reasoning (LPAR'06)*, volume 4246 of *LNCS*, pages 227–241, Phnom Penh, Cambodia, November 2006. Springer.
15. Olaf Noppens and Thorsten Liebig. Interactive Visualization of Large OWL Instance Sets. In *Proc. of the 3rd Semantic Web User Interaction Workshop (SWUI 06) at the ISWC'06*, 2006.
16. Catherine Plaisant, Jesse Grosjean, and Benjamin B. Bederson. SpaceTree: Supporting Exploration in Large Node Link Tree, Design Evolution and Empirical Evaluation. In *Proc. of the IEEE Symposium on Information Visualization (INFOVIS 2002)*, pages 57 – 64, Boston, USA, October 2002.
17. Ben Shneiderman. Dynamic queries for visual information seeking. *IEEE Software*, 11(6):70–77, 1994.
18. Evrin Sirin, Bijan Parsia, Bernardo Cuenca Grau, Aditya Kalyanpur, and Yarden Katz. Pellet: A practical OWL DL reasoner. *Journal of Web Semantics*, 2006. To appear.
19. Taowei David Wang, Bijan Parsia, and James Hendler. A Survey of the Web Ontology Landscape. In *Proc. of the 5th Int. Semantic Web Conference (ISWC 2006)*, 2006.

System Description: An Orienteering Strategy to Browse Semantically-Enhanced Educational Wiki Pages

Luciano T.E. Pansanato[1] and Renata P.M. Fortes[2]

[1] Universidade Tecnológica Federal do Paraná – Campus de Cornélio Procópio, Av. Alberto Carazzai, 1640, 86300-000 Cornélio Procópio, PR, Brazil
[2] Departamento de Ciências de Computação, Instituto de Ciências Matemáticas e de Computação, Universidade de São Paulo – Campus de São Carlos
Caixa Postal 668, 13560-970 São Carlos, SP, Brazil
luciano@utfpr.edu.br, renata@icmc.usp.br

Abstract. Wikis have been adopted along the years, aiming to provide an easy and simple support to people keep the information systems on the Internet up-to-date, and making possible efficient collaborative authoring. As the number of pages and corresponding contents increases, wiki users face difficulties when browsing for wiki pages. This paper presents a system prototype based on orienteering to browse semantically-enhanced educational wiki pages. The results of a user-based evaluation of the system prototype are also presented.

Keywords: Semantic Web, Metadata, Wiki, Interactive Retrieval.

1 Introduction

Information seeking activities play an important role in a wide range of tasks based on information systems. In Wikis, the support for information seeking activities is important to enable users to find the information they need. However, browsing for information in Wikis may be difficult when users are supported only by the traditional keyword search paradigm.

The keyword search paradigm is popular because of its ability to identify quickly the pages containing specific information. However, keyword search is not an efficient means of finding information in several situations. For example: (a) a user may be uncertain of what he is looking for until the available options are presented; (b) the target cannot be expressed by keywords or (c) the exact terminology used in the pages is unknown, and (d) in cases where a great deal of information and context should be obtained along with the pages (and not only the final page). In such situations, other paradigms such as traditional browsing may be more useful than keyword search [1].

We have observed this problem with the CoTeia[1] [2], a Computer Supported Collaborative Learning (CSCL) tool used to complement face-to-face lectures with

[1] http://incubadora.fapesp.br/projects/coteia/

E. Franconi, M. Kifer, and W. May (Eds.): ESWC 2007, LNCS 4519, pp. 809–818, 2007.

collaborative learning activities at Institute of Mathematical Sciences and Computing / University of São Paulo (ICMC/USP). CoTeia is a wiki-based asynchronous collaborative tool analogous to CoWeb [3]. CoTeia users, in particular faculty, frequently look for wiki pages to reuse their material. CoTeia has been used since 2001 and contains a significant amount of teaching material. Although CoTeia provides a keyword search tool, we have observed many cases where users engage in an exhaustive browsing to find what they are seeking.

We agree that typical wiki-based environments, such as CoTeia, can be enhanced by metadata, which may strongly influence information seeking techniques and tools. Semantic MediaWiki is also a well-known wiki behind Semantic Wikipedia [4]. However, the ease in using metadata to find wiki pages (or other kind of information objects) is related to the existence of appropriate support to information seeking strategies. The main objective of our research is to investigate and develop an interactive support inspired by an orienteering strategy [1, 5] to browsing for wiki pages, which uses an infrastructure of Semantic Web. Orienteering denotes a strategy in which people satisfy a particular information need through a sequence of small steps (or actions) to narrow the focus of the goal. At each step, prior information and local context are used to decide the next step.

We have developed a prototype for integrating a variety of tools to support an orienteering strategy. We argue that an information seeking environment should make available several categories of tools (including keyword search), enabling users to choose the appropriate tool or the best combination of tools (that is, the best strategy) in agreement with different levels of users' ability, background, preferences, and kind of information that they are looking for at moment. It enables users to prioritize different ways their choices of tools to each step in an orienteering strategy. We expected that users take advantage of this kind of environment to carry out information seeking tasks as an alternative to the one-size-fits-all approach of the keyword search paradigm.

Section 2 describes the orienteering strategy and the techniques we have explored to support information seeking activities. Sections 3 and 4 present the prototype and a preliminary user-based evaluation which points to its effectiveness. Sections 5 and 6 are dedicated to related work and conclusions respectively.

2 Strategies and Tools

In our work, we exploit the levels of information seeking activities as described by Bates [6]. Based on empirical studies of the information seeking behavior of experienced library users, Bates distinguishes four levels of activity:

1. A *move* is an identifiable thought or action that is a part of information seeking. For example, locating some portion of text in a wiki page.
2. A *tactic* is one or a handful of moves made to further an information seeking activity. For example, broadening or narrowing a query to retrieve a larger or smaller number of wiki pages.
3. A *stratagem* is a complex set of moves and/or tactics, and generally involves both a particular information domain anticipated to be productive by the user, and a

mode of tackling the particular organization of that domain. For example, finding all wiki pages which are related to a particular course.
4. A *strategy* is a plan for satisfying an information need, and may include combinations of all the previously mentioned types of information seeking activities. For example, performing a keyword search to retrieve wiki pages related to specific terms describing a subject, browsing through those considered relevant and then finding references in the text to books and articles.

We are interested in providing an interactive support to help users to find wiki pages at CoTeia. The support provided for moves and tactics is implemented as a set of tools and these are integrated in an environment. Combining the tools users may compose stratagems and strategies. Thus, in this work a *tool* is a software program that implements one or more methods or techniques to support an information seeking activity; and an *environment* is a software program that integrates diverse tools.

Subsection 2.1 describes the underlying orienteering strategy used to integrate the tools which has influenced the choice of the tools as well as some extensions. The tools are briefly discussed in Subsection 2.2.

2.1 Orienteering Strategy

Orienteering is a sport of finding one's way across country on foot using a map and a compass. The main strategy consists in using the information on the current position to make better choices about the way to reach the next checkpoint or the final target. Furthermore, the orienteer must constantly concentrate, make decisions, and keep track of the path covered.

A similar strategy has been described in studies of information seeking behavior in the literature [1, 5]. Orienteering involves using both prior and contextual information to narrow in on the actual information target, often in a series of steps (moves/tactics/stratagems, according to Bates, or simply actions), without specifying the entire information need up front. For example, first one can submit a query to a search engine to get into the proximity of the information that satisfy the target information need, and then explore the links retrieved to find the desired information. Other works also described this strategy [7, 8], though not under that name.

We have elected the following categories of techniques that could help the user to engage in an orienteering strategy: combining browsing and searching; showing context; previewing content; keeping interaction history; and narrowing toward the goal. The latter is a key feature of an orienteering strategy. It consists in starting with a general query and using small steps to narrow down the information space until the user finds what they are looking for.

2.2 Orienteering Tools

We have implemented common tools and extended others to incorporate metadata in their mechanism, i.e. a Resource Description Framework (RDF) model about a collection of wiki pages used in educational activities. Methods for automatic metadata generation were explored to populate the RDF model as described in previous work [9, 10]. The following tools were implemented:

- *Keyword search*. This tool is a search engine which searches previously indexed wiki pages for specified keywords and returns a list of those where the keywords were found. In spite of this approach being straightforward, it works for its purpose: to obtain a list of wiki pages relating to the keywords the user entered. Some search engines can use relevance ranking that many users find disconcerting: some pages can be ranked high even if they do not contain all the keywords.
- *Facet browse*. Faceted browsing is one way to use *faceted metadata* to allow users to find information. Metadata can have several facets: attributes in various orthogonal sets of categories. For example, in the domain of educational wiki pages, possible facets might be authors (professors or students). This tool (Facet browse) allows users to filter a set of items (e.g. wiki pages) by progressively selecting from only valid values (instances) of facets. The list of valid values is filtered to show only those that have results available. Thus, it is impossible to get an empty result. The combination of facets and hierarchy can help the user to decide how to start and to explore the collection.
- *Highlight*. This tool displays the occurrences of query terms within the context of the document retrieved. It is useful to support local exploration because most users do not read pages carefully when they scan text for what they are looking for. We extend this technique to highlight also names, locations, email addresses, and phone numbers. The knowledge about people's names and locations is harvested from ICMC/USP website and added inside the RDF model.
- *Flag*. This tool implements a preview technique that consists in automatically flagging a result which contains certain content so it can be found easily among other results. For example, results of documents that contain names, locations, email addresses, dates, or times. It is useful for users to be able to see at a glance whether the results they get in response to an action have a particular content.
- *Work memory*. The underlying technique of this tool consists in providing a special memory resource which can store results and some operations to handle them. Basically, we implemented the same memory functions of a typical calculator: M+ (sum), M- (difference), MR (recall), and MC (clear). This support is important in cases where the goal consists of a collection of results that should be obtained from a variety of different strategies, not just at the end of a strategy. For example, collecting links to wiki pages which contains educational material (for reuse) may involve the storage of intermediary results to compose the final list of those considered relevant.
- *Sort*. This tool sorts the results by relevance, as determined by the number and location of matched words in the wiki page, and by any metadata of wiki pages in the RDF model. While all search engines sort results by relevance as default, a few of them can sort by other option (e.g. Ask, http://www.ask.com/, allows to sort saved results by date or title).
- *Group*. Some search engines attempt to classify results automatically into concepts (or domains), such as Vivísimo, http://vivisimo.com/. The drawback of this

technique is that the categories automatically generated are not always well organized. This tool (Group) implements a straightforward improvement by exploiting the available metadata in the RDF model to organize the results according to the same category layout that is used by the tool for faceted browsing.
- *Undo and Restart.* The reverse of actions is supported by means of undoing and starting over the task. The Undo tool allows user to cancel any performed action and consequently, the results will reflect on the new sequence. The Restart tool allows user to begin again from scratch. These tools help to keep the interaction process under user control.

3 Prototype

We have developed a prototype which integrates the orienteering tools described in previous section. The prototype has been used to browse approximately 4,230 wiki pages stored in two *CoTeia repositories*[2]. Fig. 1 shows an example of interaction with the prototype. The interface is divided into three parts: the set of tools on the left, the history of actions on the top right, and the results and content area on the bottom right.

Using a tool often corresponds to a move, tactic, or stratagem (discussed in Section 2), and this interaction is called action in the interface. The sequence of performed actions during a strategy is showed in the history in the top right. The history serves as important aid to reduce working memory load. Any particular information seeking activity may include other behaviors and cognitions that cannot be captured by the prototype.

When the user performs an action, the results (i.e. a list of links to wiki pages in this case) to that action are presented. The next action has effect on the current set of results. This is an important aspect in the interaction process of the user with the prototype: the tools are integrated so that the results obtained in one are used as the input to the next.

Fig. 1 shows the result of a hypothetical interaction in which the user has chosen different tools to look for wiki pages with material (slides) used in a particular course. First, the user selected the term "SCE0225 Hipermídia" (A) in the facet Course to refine the entire collection of all wiki pages to a set of those related with the course. Second, the user performed a keyword search (B) to narrow the current set of wiki pages (from previous tool) to those which contain the word "aula" (lecture). Next, the user sorts the results alphabetically by Title (C) and follows a link to the wiki page entitled "Aula a Aula" (D). The wiki page is showed in the content area (E). Next, the highlight tool is used to highlight the word "Aula" (F) in the content of the wiki page (the highlight tool does not appear in the figure). Finally, the flag tool is used to puts a mark (a red lozenge) (G) on the results which contain one or more people's names.

The prototype was built using PHP and MySQL. We have used AJAX (Asynchronous JavaScript and XML) programming techniques to allow better user interactions in the interface. The RDF model is manipulated and searched using RAP (RDF API for PHP, http://sourceforge.net/projects/rdfapi-php/).

[2] (1) http://coteia.icmc.usp.br/coteia/ and (2) http://safedevel.icmc.usp.br/coweb/

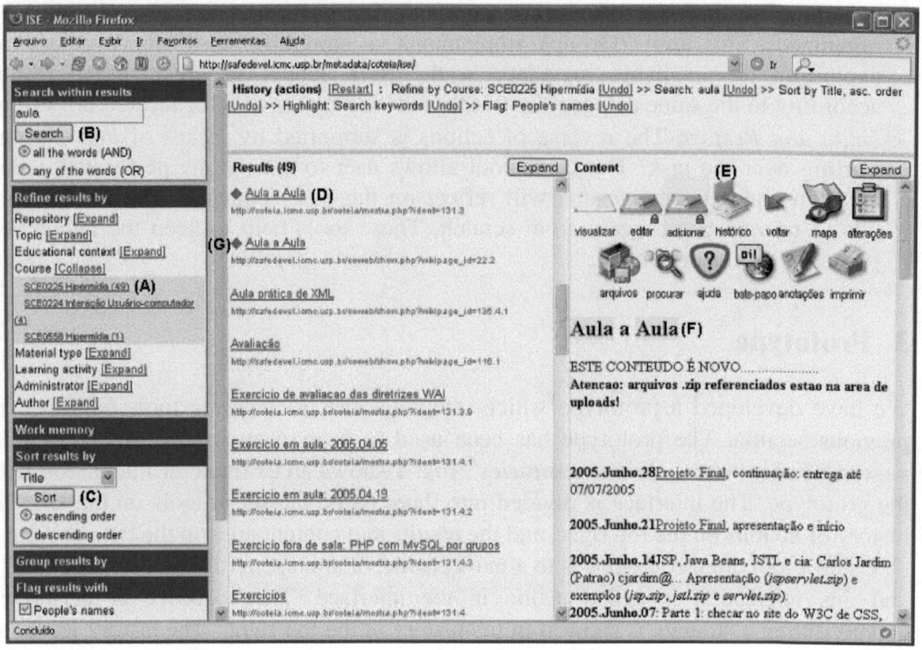

Fig. 1. Screenshot of the prototype

4 User-Based Evaluation

This section presents the evaluation we conduct to test and evaluate the prototype, and, through the users' interaction with the prototype, to gain knowledge about it.

4.1 Aims

We had three goals for the evaluation. First, we were interested in validating if the users could get relevant information using the prototype. Second, we wanted to confirm if the users use different strategies. Finally, we were interested in studying step-by-step the strategies that users follow to browsing for wiki pages using the prototype and their preferences concerning the available tools.

4.2 Methodology

We have combined a variety of methods to gather data about the user experience with the prototype. Our methodology combined: (a) Questionnaire, (b) Think Aloud, (c) Interview, and (d) User Log Recording. The participants consisted of 15 professors (11 women, 4 men) at ICMC/USP. They were all experienced users of search engines, searching for information daily. The participants were regular users of the CoTeia with 2-6 years of experience and weekly they access wiki pages for reading and/or editing. Thus, the participants are potential end-users of the prototype.

The participants completed a set of tasks that involved browsing for educational material for reuse. We defined five tasks in four different task scenarios (types), based on Shneiderman's definition [11]: specific fact-finding, extended fact-finding, open-ended browsing and exploration of availability (see Table 1). We were interested in covering the largest number of scenarios using the smallest number of tasks.

The tasks were presented to the participants in agreement with the concept of simulated work task situation [12]. A simulated work task situation is an open description of the context/scenario of a given situation. It then works as the trigger of the participant's information need and the base for relevance judgment. The objective is to ensure the largest possible realism by the involvement of potential users who, based on the simulated work task situation, develop individual and subjective information need interpretations. Individually, the participants use the prototype and assess relevance of the obtained results in relation to their perceptions of the information need and the underlying simulated work task situation.

Table 1. Five tasks in four scenarios

Tasks	Scenarios
T1: Look for a particular exercise.	specific fact-finding
T2: Look for the e-mail of a course monitor.	
T3: Look for evaluation criteria recently published by other professors.	extended fact-finding
T4: Look for a new approach used to teaching a particular topic.	open-ended browsing
T5: Look for the learning material available to a particular course.	exploration of availability

The evaluation was comprised of four sessions as follows: (a) a survey session, (b) a training session, (c) a task session, and (d) an interview session. The survey was designed to collect demographic information and participant's experience with search tools and experience with CoTeia. A training session was given, including an overview of the prototype so that the participant could become familiar with the tools and the environment. After the training session, participants completed tasks while thinking aloud about their strategies. The participants were instructed to articulate what they are thinking and what they feel while working with the prototype. The utterances were registered using audio recording. In connection with each task, the interactions of the participants with the prototype were registered using log-file recording. All participants were requested to complete the same tasks. No time limit was set for any of the tasks. After each task participants answered questions about relevance of the results obtained and the usefulness of the prototype concerning that task. After performing the tasks, participants were interviewed for their perceptions of the prototype that they had experienced.

4.3 Results and Observations

After each task, participants completed a short questionnaire about the results obtained and the usefulness of the prototype for the task. The judgments were made on 7-point Likert scales (1 = none, 7 = extreme). The participants assigned an average rating of 5.92 (SD = 0.90) for relevance and 6.07 (SD = 0.81) for usefulness. These results support our first goal: the users could get relevant information using the prototype.

By analyzing the user log we found out that participants used different tools to complete the tasks. The average number of different tools per task was 3.24 (SD = 1.48). We have also observed different number of times tools were used, different choices of tools, and different sequences chosen to perform each task. Additionally, these differences were detected among participants for a same task. In general, the results confirmed our hypothesis for the second goal: users use different strategies. We intend to investigate the users' strategies in detail to improve the integration of each tool within the environment.

Fig. 2 shows how often the tools were actually used. Facet Browse and Keyword Search were the most used tools. The tools provided ways of search/retrieval results, and participants chosen to begin with Facet Browse (62.5%) more frequently than with Keyword Search (37.5%). Unexpectedly, Work Memory has a somewhat high percentage of usage; however, few participants (3) made extensive use of Work Memory while others (9) did not consider this tool in their tasks.

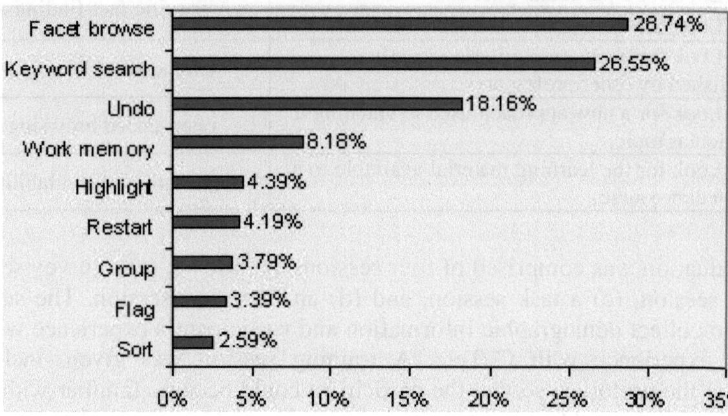

Fig. 2. Percentage of time tools were used

In general, participants considered available facets as representative of the wiki pages. *Course* (31.55%), *Learning Material Type* (16.67%), and *Learning Activity Type* (11.90%) were the most used metadata in faceted browsing. One of the participants requested that the term (year/semester) concerning Course should be included in the interface. These results have motivated the need for specific vocabularies.

Participants commented favorably about the approach of the prototype and the available tools. Some (6) followed a strategy and after having completed the task they said that could improve the strategy. Participants were specifically asked to comment about the control of the navigation process. All participants stated that they felt in control by using the history feature. Furthermore, when asked participants did not stated getting lost while using the prototype. Initially, some participants (3) did not understand that, after cancel (undo) an action, the results reflect the resultant sequence of actions. However, as the participants continued to use the prototype, they perceived this feature and tried to exploit it. At the end of the interview, most of participants indicated they intend to use the prototype in the future.

5 Related Work

Many Semantic Wikis have been described in the literature, e.g. Semantic MediaWiki [4], Platypus Wiki [13], WikSAR [14], SemperWiki [15]. These Wikis use semantic information (metadata) to offer improved navigation, browsing, and searching. In general, they are primarily focused on semantic navigation (e.g. providing additional information on the relation a link describes) and semantic search (e.g. allowing a semantic query on the underlying knowledge base). SemperWiki has also provided faceted browsing [16]; however, it does not integrate this technique with others.

Our proposal integrates several tools in an environment to help browsing and searching. The prototype provides an interactive support to information seeking strategies based on orienteering. Users may choose an appropriate tool or a combination of tools to take steps along the way to satisfy their information needs.

6 Conclusions

Browsing for information in Wikis can become difficult as the number of pages and corresponding contents increases. We present a prototype developed to demonstrate our initial ideas combining searching tools to compose a strategy of orienteering. An evaluation showed that the prototype is useful to find relevant information in CoTeia repositories and provided insight regarding the tools and the underlying orienteering strategy.

Concerning future work, our main efforts address experiments with other types of user, in particular students, to investigate their interactions and performance using the prototype. We also plan to recognize some repeating strategies that can yield to patterns.

Acknowledgments. Our thanks to FINEP, FAPESP and UTFPR for funding parts of this research.

References

1. Teevan, J., Alvarado, C., Ackerman, M. S., Karger, D. R.: The perfect search engine is not enough: A study of orienteering behavior in directed search. In: Proc. Conference on Human Factors in Computing Systems (2004) 415-422
2. Arruda Jr, C. R. E., Izeki, C. A., Pimentel, M. G. C.: CoTeia: Uma ferramenta colaborativa de edição baseada na Web. In: Proc. 8th Brazilian Symposium on Multimedia and Hypermedia Systems (2002) 371-374 (in Portuguese)

3. Guzdial, M.: Supporting learners as users. The Journal of Computer Documentation 23(2) (1999) 3-13
4. Völkel, M., Krötzsch, M., Vrandecic, D., Haller, H.: Semantic Wikipedia. In: Proc. 15th Int. Conf. on WWW (2006) 585-594
5. O'Day, V. L., Jeffries, R.: Orienteering in an Information Landscape: How Information Seekers Get From Here to There. In: Proc. ACM CHI (1993) 438-445
6. Bates, M. J.: Where should the person stop and the information search interface start? Information Processing and Management 26(5) (1990) 575-591
7. Navarro-Prieto, R., Scaife, M., Rogers, Y.: Cognitive strategies in Web searching. In: Proc. 5th Conference on Human Factors & the Web (1999)
8. Hölscher, C., Strube, G.: Web search behavior of internet experts and newbies. In: Proc. 9th Conf. on WWW (2000)
9. Pansanato, L. T. E., Fortes, R. P. M.: Strategies for Filling Out LOM Metadata Fields in a Web-Based CSCL Tool. In: Proc. Third Latin American Web Congress (2005) 187-190
10. Pansanato, L. T. E., Fortes, R. P. M.: Strategies for automatic LOM metadata generating in a web-based CSCL tool. In: Proc. 11th Brazilian Symposium on Multimedia and the Web (2005) 1-8
11. Shneiderman, B.: Designing information-abundant web sites: issues and recommendations. International Journal of Human-Computer Studies 47(1) (1997) 5-29
12. Borlund, P.: The IIR evaluation model: a framework for evaluation of interactive information retrieval systems. Information Research 8(3) (2003)
13. Tazzoli, R., Castagna, P., Campanini, S. E.: Towards a Semantic Wiki Wiki Web. In: Poster Track of the 3rd International Semantic Web Conference (2004)
14. Aumueller, D.: Semantic authoring and retrieval within a Wiki. In: Demos and Posters of the 2nd European Semantic Web Conference (2005)
15. Oren, E.: SemperWiki: a semantic personal Wiki. In: Proc. of the 1st Workshop on The Semantic Desktop, 4th International Semantic Web Conference (2005)
16. Oren, E., Delbru, R., Möller, K., Völkel, M., Handschuh, S.: Annotation and Navigation in Semantic Wikis. In: Proc. of the 1st Workshop on Semantic Wikis, 3rd European Semantic Web Conference (2006)

Efficient Content Creation on the Semantic Web Using Metadata Schemas with Domain Ontology Services
(System Description)

Onni Valkeapää, Olli Alm, and Eero Hyvönen

Helsinki University of Technology (TKK), Laboratory of Media Technology
University of Helsinki, Department of Computer Science
Semantic Computing Research Group (SeCo)
P.O. Box 5500, FI–02015 TKK, Finland
{onni.valkeapaa,olli.alm,eero.hyvonen}@tkk.fi
http://www.seco.tkk.fi/

Metadata creation is one of the major challenges in developing the Semantic Web. This paper discusses how to make provision of metadata easier and cost-effective by an annotation editor combined with shared ontology services. We have developed an annotation system supporting distributed collaboration in creating annotations, and hiding the complexity of the annotation schema and the domain ontologies from the annotators. Our system adapts flexibly to different metadata schemas, which makes it suitable for different applications. Support for using ontologies is based on ontology services, such as concept searching and browsing, concept URI fetching, semantic autocompletion and linguistic concept extraction. The system is being tested in various practical semantic portal projects.

1 Introduction

Currently, much of the information on the Web is described using only natural language, which can be seen as a major obstacle in developing the Semantic Web [1]. Since the annotations describing different resources are one of the key components of the Semantic Web, easy to use and cost-effective ways to create them are needed, and various systems for creating annotations have been developed [14,18]. However, there seems to be a lack of systems that 1) can be easily used by annotators unfamiliar with the technical side of the Semantic Web, and that 2) are able to support distributed creation of semantic metadata based on complex metadata annotation schemas and domain ontologies [19].

Metadata descriptions are usually based on ontologies of two kinds. First, an annotation ontology, i.e. a metadata schema, tells what kind of properties and value types should be used in describing a resource. For example, the Dublin Core schema uses 15 elements, such as *dc:title*, *dc.creator*, *dc:subject*, etc. Second, a set of domain ontologies are used to define vocabularies by which the values for metadata properties are given. This suggests that three kinds of tools are needed to address the problems

E. Franconi, M. Kifer, and W. May (Eds.): ESWC 2007, LNCS 4519, pp. 819–828, 2007.

of metadata creation. First, an annotation editor supporting the usage of different metadata schemas is needed. Second, we need services for supporting the usage of the domain ontologies (vocabularies) that are employed for the annotations. Third, tools for automating the creation of actual metadata descriptions in various ways, e.g., for finding suitable values for the elements, must be developed.

To test this idea, we have developed a system of three integrated tools that can be used to efficiently create semantic annotations based on metadata schemas, domain ontology services, and linguistic information extraction. These tools include, at the moment, an annotation editor system Saha[1] [19], an ontology service framework Onki[2] [9] and an information extraction tool Poka[3] for (semi)automatic annotation. The annotation editor Saha supports collaborative creation of annotations and it can be connected to Onki servers for importing concepts defined in various external domain ontologies. Saha has a browser-based user interface that hides complexity of ontologies from the annotator, and adapts automatically to different metadata schemas. The tool is targeted especially for creating metadata about web resources. It is being used in different applications within the National Semantic Web Ontology Project in Finland (FinnONTO)[4] [4].

In order to support the kind of annotation that is required in our project, we identified the following basic needs for an annotation system. These were also features that we felt were not supported well enough in many of the current annotation platforms:

- **Simplicity.** The system should, as a rule, hide technical concepts related to markup languages and ontologies from its user.
- **Adaptivity.** The system should be adaptable to different annotation cases with different kinds of contents to be described.
- **Quality.** When annotation is done by hand, the annotator should be guided to produce annotations in qualified and pre-defined form, if needed.
- **Collaboration.** The system should support collaborative annotation, where the annotation process can be shared among different annotators at different locations.
- **Portability.** The annotator should be able to use the system at any location without installing any special software.

2 Saha Annotation System

2.1 Utilizing Annotation Schemas

Ontologies may be used in two different ways in annotation: they can either serve as a description template for annotation construction (annotation schemas/ontologies) or provide an annotator with a vocabulary which can be used in describing resources

[1] http://www.seco.tkk.fi/applications/saha/
[2] http://www.seco.tkk.fi/applications/onki/
[3] http://www.seco.tkk.fi/applications/poka/
[4] http://www.seco.tkk.fi/projects/finnonto/

(reference/domain ontologies) [15]. An annotation schema has an important role in expressing *how* the ontological concepts used in annotations are related to the web resources being described. Without annotation schemas, the role of these concepts would remain ambiguous. In addition to explicitly expressing the relation between a resource and an annotation, the schema helps the annotator to describe resources in a consistent way and it can be effectively used to construct a generic user-interface for the annotation application.

Saha uses an approach similar to the one introduced in [8] to form its user interface according to an annotation schema loaded on it. Saha does not use any proprietary schemas, but instead will accept any RDF/OWL-based ontology as a schema. By schemas we mean a collection of classes with a set of properties. An annotation in Saha is an instance of a schema's class that describes some web resource and is being linked to it using the resource's URL (in same cases, URI). We make the distinction between the annotation of a document (e.g. a web page) and the description of some other resource (e.g. a person) that is somehow related to the document being annotated. In addition to containing classes used to annotate documents (*annotation classes*), an annotation schema used with Saha can also contain *reference classes* for describing resources other than documents. In other words, an annotation schema forms a basis for the local knowledge base (KB) that contains descriptions of different kinds of resources that may or may not exist on the web. Instances of the reference classes are used as values of properties in annotations.

Each annotation schema loaded to Saha forms an *annotation project*, which can have multiple users as annotators. In practice, an annotation project is Jena's[5] ontology model stored in a database. A model is comprised of the annotation schema and the instances of the schema's classes. It can be serialized to RDF/XML in order to use the annotations in external applications.

2.2 Architecture and User Interface

The main difference between Saha and ontology editors such as Protégé [12] is that Saha offers the end-user a highly simplified view of the underlying ontologies (annotation schemas). It does not provide tools to modify the structure (classes and properties) of ontologies, but rather focuses on using them as a basis for the annotations.

Saha is a web application implemented using the Apache Cocoon[6] and Jena frameworks. It uses extensively techniques such as JavaScript and Ajax[7]. The basic architecture of Saha is depicted in figure 1. It consists of the following functional parts: 1) annotators using web browsers to interact with the system, 2) Saha application running on a web server, 3) applications using the annotations created with Saha, 4) the Onki ontology service, 5) PostgreSQL database used store the annotations, and 6) the Poka information extraction tool.

The user interface of Saha, depicted in figure 2, provides an annotator with a view of the classes and properties of an annotation schema. The annotator can choose a

[5] http://www.hpl.hp.com/semweb/tools.htm
[6] http://cocoon.apache.org/
[7] http://en.wikipedia.org/wiki/Ajax_%28programming%29

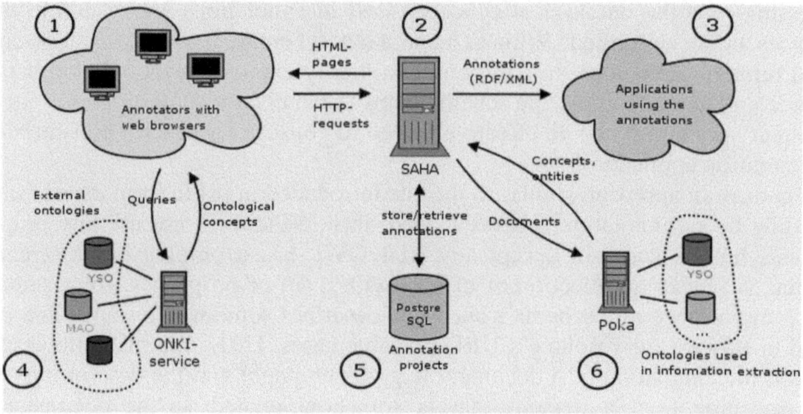

Fig. 1. Architecture of Saha

class from the class hierarchy (left side of the screen), view the annotations/KB-instances and create new ones. The lower part of the screen views the resource being annotated. In figure 2, an annotation belonging to class "Document" is being edited. The properties of the annotation, such as "Title", as well as fields to supply values for them are shown on the right side of the class hierarchy.

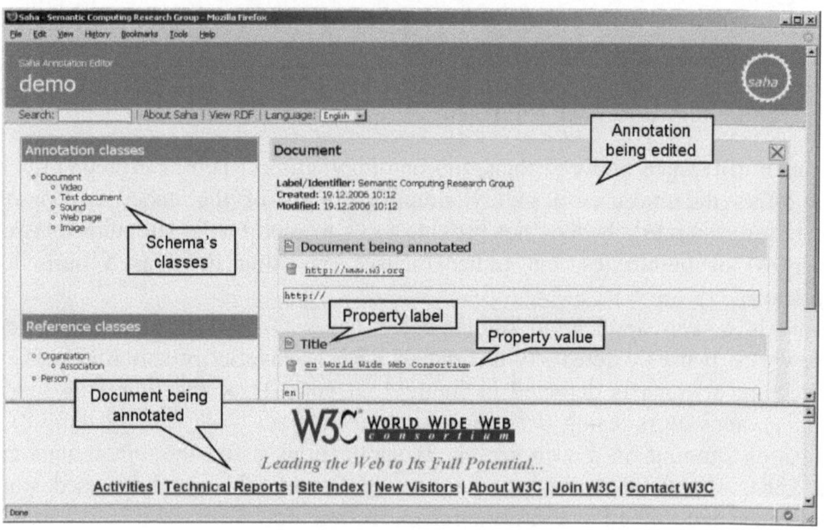

Fig. 2. The user interface of Saha

Properties of an annotation schema accept either literal or object values. In the latter case, values are KB-instances or concepts of some external domain ontology. KB-instances can be chosen using semantic autocompletion [5]. Here, the user types in a search word and selects a proper instance from the list populated by the system. If the proper KB-instance does not exist, user may also create a new one. *rdfs:range* or *owl:Restriction* is used to define the types of things that are allowed as values.

2.3 Setting Up an Annotation Project

Saha's annotation cycle starts by defining settings for an annotation schema. These settings will define 1) the way how the schema is visualized for the annotator, 2) how human readable labels (*rdfs:label*) are automatically created for new annotations and KB-instances, and 3) how different property fields are filled in the annotations. By visualization, we refer to e.g. defining a subset of schema's classes that are shown in the editor's class-hierarchy, or defining an order of the properties of a class in which they are shown to the annotator. Human readable labels, by turn, are needed when annotations or instances are represented in the user-interface. These labels can be, in many cases, formed automatically using property-values supplied by the annotator for the annotation/KB-instance. In Saha, properties can be filled manually or using integrated ontology services, which include the ontology server system Onki and the information extraction tool Poka to be presented in section 3. When using these services, we map a property of an annotation schema to the desired service. In the case of Onki, the values of the property will be concepts defined in some external domain ontologies, selected by an annotator using a dedicated Onki-browser. When Poka is used, values are ontological concepts or literals provided by the extraction tool. For example, an extraction component recognizing people's names could be coupled with the property *dc:creator*.

Settings for an annotation project are defined in a schema-specific RDF-file, which we call *meta-schema*. Although the use of a meta-schema is not compulsory, it is highly practical in most cases. At the moment, meta-schemas are done by hand, but we are developing an easy-to-use editor for the task.

3 Utilizing Ontology Services

3.1 Onki Ontology Services

One of the key features of Saha is its ability to connect to the Onki ontology service [9]. The Onki system has an important role in sharing ontological resources between different organizations and actors. In annotation, Onki enables the use of concepts of external domain ontologies as values of an annotation schema's properties. These ontologies are made available to the annotators through the Onki ontology server, which offers two interfaces to ontological information: searching and browsing. The first one is similar to the instance KB search described above. When using it, the annotator types a search word which is sent to the Onki ontology server character by character and matched with the concepts in the underlying ontology. Concepts matching to the query will be sent back to Saha and shown below the search field from which they can be selected by the user. The other option is to use a browser view of the Onki system. It is practical when the annotator does not get agreeable results using the semantic autocompletion, or wants to see the resources within the context of the class hierarchy. The Onki ontology browser can be opened in a new window by clicking a property field in Saha (see figure 3). After that, the annotator is able to browse the class hierarchy, and when a suitable concept is found, fetch it to the input form of Saha by clicking on the button "Fetch concept" on the Onki browser page. Both modes of using ontology services provided by Onki can be conveniently integrated to different web applications on the client side using Ajax.

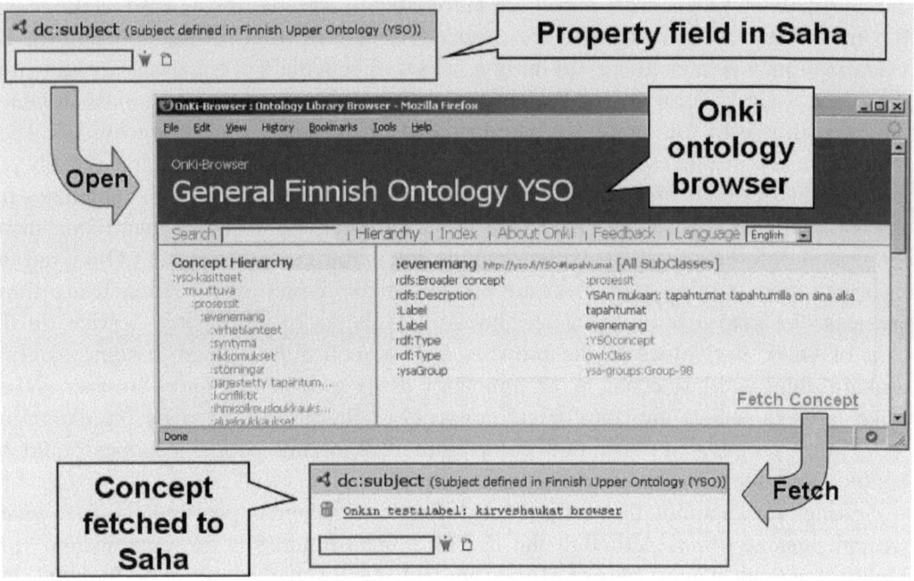

Fig. 3. Using the Onki ontology browser

3.2 Automatic Recognition of Concepts and Entities

Saha uses the ontology-based information extraction tool Poka to suggest concepts based on the documents being annotated. Poka can process a document to 1) recognize concepts of external ontologies and to 2) extract named entities using non-ontological tools.

In schema-based annotation, things to be extracted are defined by the properties of the annotation schema's classes. Accordingly, the function of an extraction component is to provide suitable concepts or entities to be used as values of those properties. Because Saha supports arbitrary annotation schemas, extraction tools must be adaptable in order to support different extraction tasks. In the Poka environment we have solved the problem of adaptivity in two ways. First, we have implemented generic non-ontological extraction components such as person name identifier and regular expression extractor. Second, user-defined external ontologies can be integrated with the system and used in concept recognition.

Extraction of Non-ontological Entities. In Poka, two extraction components for non-ontological entities have been implemented: person name extractor for Finnish language and regular expression extractor. The main idea in the rule-based name recognition tool is to first search for full names within the text at hand. After that, occurrences of the first and last names are mapped to full names. Simple coreference resolution within a document is implemented by mapping the individual name occurrences to the corresponding unambiguous full name if there exist one. Single first names and surnames without corresponding full names are discarded. Search for potential names is started from the uppercase words of the document. Predefined list of first names is utilized for recognizing potential full names. With morphosyntactic clues some hits can be discarded. For example, first names in Finnish rarely have certain morphological affixation such as "-ssa" (similar to the English preposition "in") or "-lla" (preposition "on") when they occur before the surname in the sentence. Poka makes use of the morphosyntactic analyzer and parser FDG[8] [17]. The FDG-parser's surface-syntactic analysis is also used for revealing proper names.

The names that are automatically recognized are suggested as potential new instances in Saha. The type of a new instance is a reference class of the annotation schema used in Saha, say *myAnnotation:Person.* If there exists an instance with the same name, the user can tell whether the newfound name refers to an existing instance or to a new one.

The regular expression extractor acts in a similar way as the name extractor. The difference is that the thing to be extracted is defined by the expression, not the component itself. Expressions can be utilized to find literal values or potential new instances from the document.

Extraction of Ontological Concepts. For ontological extraction, an ontology has to be integrated to the extraction system. By ontological extraction we mean 1) deduction of string representations of concepts *from* the ontology and 2) finding the occurences of the representations.

In Poka, the integration starts by defining a set of concepts in an ontology that are to be extracted from the documents. The ontology can be used in its entirety, or it can be only partly used by selecting e.g. instances or some sub-part of the ontology's hierarchy tree. After this, the human readable properties representing concept names, e.g. the values of the literal properties *rdfs:label* or *skos:prefLabel*, are chosen as targets for the recognition.

For the string matching in the extraction process, the string representations of ontological resources are indexed in the prefix trie. Since two or more concepts may share the same label, a trie entity can refer to multiple URIs. In some languages (e.g. Finnish), it is useful to lemmatize the concept representations for efficient extraction. This is because syntactical forms of words may vary greatly in languages with heavy morphological affixation [11]. Lemmatization of *both* the text and the concept names helps to achieve better recall in the extraction process.

Currently the adaptation of new extraction ontologies is done by system experts. Our future work involves developing a user interface for integrating ontological resources for extraction.

---

[8] http://www.connexor.com, Machinese Syntax

4 Discussion

4.1 Contributions and Applications

Ontology-based semantic annotations are needed when building the Semantic Web. Although various annotation systems and methods have been developed, the question of how to easily and cost-effectively produce quality metadata still remains largely unanswered. We tackled the problem by first identifying the major requirements for an annotation system. As a practical solution, an annotation system was designed and implemented which supports the distributed creation of metadata and which can utilize ontology services as well as automatic information extraction. It is designed to be easily used by non-experts in the field of the Semantic Web.

Saha is currently a working prototype. It is in trial use for the distributed content creation of the semantic health promotion portal TerveSuomi.fi [3,16]. Much of the content and metadata for the portal will be provided by health experts working at various health organizations in Finland. Saha has also been tested, among others, in metadata creation for the Opintie portal, a follow-up version of the educational semantic portal Orava [10], using Learning Object Metadata (LOM).

Full usability testing of Saha has not yet been conducted. Initial feedback from end users indicates that some intricate ontological structures, such as deep relation paths between resources, are difficult to comprehend. These difficulties, however, can be facilitated by proper design of annotation schemas.

4.2 Related Work

A number of semantic annotation systems and tools exist today [14,18]. These systems are primarily used to create and maintain semantic metadata descriptions of web pages.

Annotea [6] supports collaborative, RDF-based markup of web pages and distribution of annotations using annotation servers. Annotations created with Annotea can be regarded as semi-formal, since the system does not support the use of ontological concepts in annotations. Instead, annotations are textual notes which are associated with certain sections of the documents they describe.

The Semantic Markup Tool [8] has a user interface that is generated according to an annotation schema in a similar way as is done in Saha. It uses Information Extraction techniques to find different kinds of entities in documents and proposes them for values of the annotation's properties. The schemas it supports are relatively simple, and it cannot be thus used to describe more complex semantic relations. Moreover, the expressivity and adaptation of templates is not explicitly stated in [8]. The Ont-O-Mat system [2], in turn, can be used to describe diverse semantic structures as well as to edit ontologies. It also has a support for automated annotation. The user interface of the Ont-O-Mat is not, however, very well suited for the annotators unfamiliar with concepts related to ontologies and semantic annotation in general. Another example of the user interface of an annotation tool requiring understanding of the Semantic Web concepts can be found in SMORE [7].

Most of the current annotation systems, like the ones mentioned here, are applications that run locally on the annotator's computer. Because of this, the systems may not necessarily be platform independent and must always be installed on the user's system, before the annotation can begin. In Saha, these problems are addressed by implementing the system as a web application. By doing so, the system can be installed and maintained centrally and the requirements for the annotator's computational environment are minimal. The way Saha is designed and implemented also strongly supports the collaboration in annotation, making the sharing of annotations and new individuals (free indexing concepts) easy.

4.3 Future Work

Our future plans include using Saha to provide metadata for additional semantic portals as well as further develop the automation of the annotation. Currently, the coupling of the annotation schema's properties and information extraction components provided by the Poka are not fully utilizing the ontological characteristics. In other words, instead of using restrictions and constraints such as *rdfs:range* to define which of the schema's properties an automatically recognized resource matches to, we are currently using a meta-schema to do the mapping. However, our plans include using the property restrictions to do the matching in the future. We are also aiming to map the automatically extracted entities to ontologies in order to support property restriction with them as well. For example, *date* regular expressions would be mapped to a corresponding class of the reference ontology, say *myOnto:Date*. This way, the proper values for an object property are defined by the range (ontological restriction), not by the component itself.

Acknowledgements

This research is a part of the National Ontology Project in Finland (FinnONTO) 2003-2007, funded mainly by the Finnish Funding Agency for Technology and Innovation (Tekes) and a consortium of 37 companies and public organizations.

References

1. Dill, S., Tomlin, J., Zien, J., Eiron, N., Gibson, D., Gruhl, D., Guha, R., Jhingran, A., Kanungo, T., Rajagopalan, S. and Tomkins, A. (2003) SemTag and Seeker: Bootstrapping the Semantic Web via Automated Semantic Annotation. Proceedings of the 12th International World Wide Web Conference, WWW2003.
2. Handschuh, S. and Staab, S. (2002) Authoring and Annotation of Web Pages in CREAM. Proceedings of the 11th International Conference on World Wide Web, WWW2002.
3. Holi, M., Lindgren, P., Suominen, O., Viljanen, K. and Hyvönen, E. (2006) TerveSuomi.fi – A Semantic Health Portal for Citizens. Proceedings of the 1st Asian Semantic Web Conference, ASWC2006, poster papers.
4. Hyvönen E. (2006) FinnONTO—Building the Basis for a National Semantic Web Infrastructure in Finland. Proceedings of the 12th Finnish AI Conference STeP 2006.
5. Hyvönen, E. and Makelä, E. (2006) Semantic Autocompletion. Proceedings of the 1st Asian Semantic Web Conference, ASWC2006.

6. Kahan, J., Koivunen, M.R., Prud'Hommeaux, E. and Swick R.R. (2001) Annotea: An Open RDF Infrastructure for Shared Web Annotations, Proceedings of the 10th International World Wide Web Conference, WWW2001.
7. Kalyanpur, A., Hendler, J., Parsia, B. and Golbeck, J. (2005) SMORE – Semantic Markup, Ontology, and RDF Editor. Available at: http://www.mindswap.org/papers/SMORE.pdf
8. Kettler, B., Starz, J., Miller, W. and Haglich, P. (2005) A Template–based Markup Tool for Semantic Web Content. Proceedings of the 4th International Semantic Web Conference, ISWC2005.
9. Komulainen, V., Valo, A. and Hyvönen, E. (2005) A Tool for Collaborative Ontology Development for the Semantic Web. Proceedings of the International Conference on Dublin Core and Metadata Applications, DC 2005.
10. Känsälä, T. and Hyvönen, E. (2006) A Semantic View–Based Portal Utilizing Learning Object Metadata. Proceedings of the Workshop on Semantic Web Applications and Tools, the 1st Asian Semantic Web Conference, ASWC2006.
11. Löfberg, L., Archer, D., Piao, S., Rayson, P., McEnery, T., Varantola, K. and Juntunen, J.–P. (2003) Porting an English Semantic Tagger to the Finnish Language. In Proceedings of the Corpus Linguistics 2003 conference, pp. 457–464. UCREL, Lancaster University.
12. Noy, N., Sintek, M., Decker, S., Crubézy and M., Fergerson, R. (2001) Creating Semantic Web Contents with Protégé–2000. IEEE Intelligent Systems 2(16):60–71.
13. Popov, B., Kitchukov, I., Angelov, K. and Kiryakov, A. (2006) Co-occurrence and ranking of entities. Available at: http://www.ontotext.com/publications/CORE_otwp.pdf
14. Reeve, L. and Han, H. (2005) Survey of Semantic Annotation Platforms. Proceedings of the 2005 ACM Symposium on Applied Computing.
15. Schreiber, G., Dubbeldam, B., Wielemaker and J.,Wielinga, B. (2001) Ontology–Based Photo Annotation. IEEE Intelligent Systems, 16(3):66–74.
16. Suominen O., Viljanen K. and Hyvönen E. (2007) User-centric Faceted Search for Semantic Portals. Proceeedings of the 4th European Semantic Web Conference ESWC2007, forth-coming.
17. Tapanainen, P. and Järvinen, T. (1997) A Non–projective Dependency Parser. Proceedings of the 5th Conference on Applied Natural Language Processing, pp. 64–71.
18. Uren, V., Cimiano, P., Iria, J., Handschuh, S., Vargas-Vera, M., Motta, E. and Ciravegna, F. (2006) Semantic Annotation for Knowledge Management: Requirements and a Survey of the State of the Art. Journal of Web Semantics, 4(1):14–28.
19. Valkeapää, O. and Hyvönen, E. (2006) A Browser-based Tool for Collaborative Distributed Annotation for the Semantic Web. Proceedings of the Workshop on Semantic Authoring and Annotation, the 5th International Semantic Web Conference, ISWC2006.

Author Index

Printing: Mercedes-Druck, Berlin
Binding: Stein + Lehmann, Berlin

Lecture Notes in Computer Science

For information about Vols. 1–4401

please contact your bookseller or Springer

Vol. 4534: I. Tomkos, F. Neri, J. Solé Pareta, X. Masip Bruin, S. Sánchez Lopez (Eds.), Optical Network Design and Modeling. XI, 460 pages. 2007.

Vol. 4526: M. Malek, M. Reitenspieß, A.P.A. van Moorsel (Eds.), Service Availability. X, 155 pages. 2007.

Vol. 4523: Y.-H. Lee, H.-N. Kim, J. Kim, Y. Park, L.T. Yang, S.W. Kim (Eds.), Embedded Software and Systems. XIX, 829 pages. 2007.

Vol. 4519: E. Franconi, M. Kifer, W. May (Eds.), The Semantic Web: Research and Applications. XVIII, 830 pages. 2007.

Vol. 4517: F. Boavida, E. Monteiro, S. Mascolo, Y. Koucheryavy (Eds.), Wired/Wireless Internet Communications. XIV, 382 pages. 2007.

Vol. 4515: M. Naor (Ed.), Advances in Cryptology - EUROCRYPT 2007. XIII, 591 pages. 2007.

Vol. 4514: S.N. Artemov, A. Nerode (Eds.), Logical Foundations of Computer Science. XI, 513 pages. 2007.

Vol. 4510: P. Van Hentenryck, L. Wolsey (Eds.), Integration of AI and OR Techniques in Constraint Programming for Combinatorial Optimization Problems. X, 391 pages. 2007.

Vol. 4509: Z. Kobti, D. Wu (Eds.), Advances in Artificial Intelligence. XII, 552 pages. 2007. (Sublibrary LNAI).

Vol. 4506: D. Zeng, I. Gotham, K. Komatsu, C. Lynch, M. Thurmond, D. Madigan, B. Lober, J. Kvach, H. Chen (Eds.), Intelligence and Security Informatics: Biosurveillance. XI, 234 pages. 2007.

Vol. 4504: J. Huang, R. Kowalczyk, Z. Maamar, D. Martin, I. Müller, S. Stoutenburg, K.P. Sycara (Eds.), Service-Oriented Computing: Agents, Semantics, and Engineering. X, 175 pages. 2007.

Vol. 4501: J. Marques-Silva, K.A. Sakallah (Eds.), Theory and Applications of Satisfiability Testing – SAT 2007. XI, 384 pages. 2007.

Vol. 4500: N. Streitz, A. Kameas, I. Mavrommati (Eds.), The Disappearing Computer. XVIII, 307 pages. 2007.

Vol. 4493: D. Liu, S. Fei, Z. Hou, H. Zhang, C. Sun (Eds.), Advances in Neural Networks – ISNN 2007, Part III. XXVI, 1215 pages. 2007.

Vol. 4492: D. Liu, S. Fei, Z. Hou, H. Zhang, C. Sun (Eds.), Advances in Neural Networks – ISNN 2007, Part II. XXVII, 1321 pages. 2007.

Vol. 4491: D. Liu, S. Fei, Z.-G. Hou, H. Zhang, C. Sun (Eds.), Advances in Neural Networks – ISNN 2007, Part I. LIV, 1365 pages. 2007.

Vol. 4490: Y. Shi, G.D. van Albada, J. Dongarra, P.M.A. Sloot (Eds.), Computational Science – ICCS 2007, Part IV. XXXVII, 1211 pages. 2007.

Vol. 4489: Y. Shi, G.D. van Albada, J. Dongarra, P.M.A. Sloot (Eds.), Computational Science – ICCS 2007, Part III. XXXVII, 1257 pages. 2007.

Vol. 4488: Y. Shi, G.D. van Albada, J. Dongarra, P.M.A. Sloot (Eds.), Computational Science – ICCS 2007, Part II. XXXV, 1251 pages. 2007.

Vol. 4487: Y. Shi, G.D. van Albada, J. Dongarra, P.M.A. Sloot (Eds.), Computational Science – ICCS 2007, Part I. LXXXI, 1275 pages. 2007.

Vol. 4486: M. Bernardo, J. Hillston (Eds.), Formal Methods for Performance Evaluation. VII, 469 pages. 2007.

Vol. 4485: F. Sgallari, A. Murli, N. Paragios (Eds.), Scale Space Methods and Variational Methods in Computer Vision. XV, 931 pages. 2007.

Vol. 4484: J.-Y. Cai, S.B. Cooper, H. Zhu (Eds.), Theory and Applications of Models of Computation. XIII, 772 pages. 2007.

Vol. 4483: C. Baral, G. Brewka, J. Schlipf (Eds.), Logic Programming and Nonmonotonic Reasoning. IX, 327 pages. 2007. (Sublibrary LNAI).

Vol. 4482: A. An, J. Stefanowski, S. Ramanna, C.J. Butz, W. Pedrycz, G. Wang (Eds.), Rough Sets, Fuzzy Sets, Data Mining and Granular Computing. XIV, 585 pages. 2007. (Sublibrary LNAI).

Vol. 4481: J. Yao, P. Lingras, W.-Z. Wu, M. Szczuka, N.J. Cercone, D. Ślęzak (Eds.), Rough Sets and Knowledge Technology. XIV, 576 pages. 2007. (Sublibrary LNAI).

Vol. 4480: A. LaMarca, M. Langheinrich, K.N. Truong (Eds.), Pervasive Computing. XIII, 369 pages. 2007.

Vol. 4479: I.F. Akyildiz, R. Sivakumar, E. Ekici, J.C.d. Oliveira, J. McNair (Eds.), NETWORKING 2007. Ad Hoc and Sensor Networks, Wireless Networks, Next Generation Internet. XXVII, 1252 pages. 2007.

Vol. 4472: M. Haindl, J. Kittler, F. Roli (Eds.), Multiple Classifier Systems. XI, 524 pages. 2007.

Vol. 4471: P. Cesar, K. Chorianopoulos, J.F. Jensen (Eds.), Interactive TV: a Shared Experience. XIII, 236 pages. 2007.

Vol. 4470: Q. Wang, D. Pfahl, D.M. Raffo (Eds.), Software Process Dynamics and Agility. XI, 346 pages. 2007.

Vol. 4465: T. Chahed, B. Tuffin (Eds.), Network Control and Optimization. XIII, 305 pages. 2007.

Vol. 4464: E. Dawson, D.S. Wong (Eds.), Information Security Practice and Experience. XIII, 361 pages. 2007.

Vol. 4463: I. Măndoiu, A. Zelikovsky (Eds.), Bioinformatics Research and Applications. XV, 653 pages. 2007. (Sublibrary LNBI).

Vol. 4462: D. Sauveron, K. Markantonakis, A. Bilas, J.-J. Quisquater (Eds.), Information Security Theory and Practices. XII, 255 pages. 2007.

Vol. 4459: C. Cérin, K.-C. Li (Eds.), Advances in Grid and Pervasive Computing. XVI, 759 pages. 2007.

Vol. 4453: T. Speed, H. Huang (Eds.), Research in Computational Molecular Biology. XVI, 550 pages. 2007. (Sublibrary LNBI).

Vol. 4452: M. Fasli, O. Shehory (Eds.), Agent-Mediated Electronic Commerce. VIII, 249 pages. 2007. (Sublibrary LNAI).

Vol. 4451: T.S. Huang, A. Nijholt, M. Pantic, A. Pentland (Eds.), Artifical Intelligence for Human Computing. XVI, 359 pages. 2007. (Sublibrary LNAI).

Vol. 4450: T. Okamoto, X. Wang (Eds.), Public Key Cryptography – PKC 2007. XIII, 491 pages. 2007.

Vol. 4448: M. Giacobini et al. (Ed.), Applications of Evolutionary Computing. XXIII, 755 pages. 2007.

Vol. 4447: E. Marchiori, J.H. Moore, J.C. Rajapakse (Eds.), Evolutionary Computation,Machine Learning and Data Mining in Bioinformatics. XI, 302 pages. 2007.

Vol. 4446: C. Cotta, J. van Hemert (Eds.), Evolutionary Computation in Combinatorial Optimization. XII, 241 pages. 2007.

Vol. 4445: M. Ebner, M. O'Neill, A. Ekárt, L. Vanneschi, A.I. Esparcia-Alcázar (Eds.), Genetic Programming. XI, 382 pages. 2007.

Vol. 4444: T. Reps, M. Sagiv, J. Bauer (Eds.), Program Analysis and Compilation, Theory and Practice. X, 361 pages. 2007.

Vol. 4443: R. Kotagiri, P.R. Krishna, M. Mohania, E. Nantajeewarawat (Eds.), Advances in Databases: Concepts, Systems and Applications. XXI, 1126 pages. 2007.

Vol. 4440: B. Liblit, Cooperative Bug Isolation. XV, 101 pages. 2007.

Vol. 4439: W. Abramowicz (Ed.), Business Information Systems. XV, 654 pages. 2007.

Vol. 4438: L. Maicher, A. Sigel, L.M. Garshol (Eds.), Leveraging the Semantics of Topic Maps. X, 257 pages. 2007. (Sublibrary LNAI).

Vol. 4433: E. Şahin, W.M. Spears, A.F.T. Winfield (Eds.), Swarm Robotics. XII, 221 pages. 2007.

Vol. 4432: B. Beliczynski, A. Dzielinski, M. Iwanowski, B. Ribeiro (Eds.), Adaptive and Natural Computing Algorithms, Part II. XXVI, 761 pages. 2007.

Vol. 4431: B. Beliczynski, A. Dzielinski, M. Iwanowski, B. Ribeiro (Eds.), Adaptive and Natural Computing Algorithms, Part I. XXV, 851 pages. 2007.

Vol. 4430: C.C. Yang, D. Zeng, M. Chau, K. Chang, Q. Yang, X. Cheng, J. Wang, F.-Y. Wang, H. Chen (Eds.), Intelligence and Security Informatics. XII, 330 pages. 2007.

Vol. 4429: R. Lu, J.H. Siekmann, C. Ullrich (Eds.), Cognitive Systems. X, 161 pages. 2007. (Sublibrary LNAI).

Vol. 4427: S. Uhlig, K. Papagiannaki, O. Bonaventure (Eds.), Passive and Active Network Measurement. XI, 274 pages. 2007.

Vol. 4426: Z.-H. Zhou, H. Li, Q. Yang (Eds.), Advances in Knowledge Discovery and Data Mining. XXV, 1161 pages. 2007. (Sublibrary LNAI).

Vol. 4425: G. Amati, C. Carpineto, G. Romano (Eds.), Advances in Information Retrieval. XIX, 759 pages. 2007.

Vol. 4424: O. Grumberg, M. Huth (Eds.), Tools and Algorithms for the Construction and Analysis of Systems. XX, 738 pages. 2007.

Vol. 4423: H. Seidl (Ed.), Foundations of Software Science and Computational Structures. XVI, 379 pages. 2007.

Vol. 4422: M.B. Dwyer, A. Lopes (Eds.), Fundamental Approaches to Software Engineering. XV, 440 pages. 2007.

Vol. 4421: R. De Nicola (Ed.), Programming Languages and Systems. XVII, 538 pages. 2007.

Vol. 4420: S. Krishnamurthi, M. Odersky (Eds.), Compiler Construction. XIV, 233 pages. 2007.

Vol. 4419: P.C. Diniz, E. Marques, K. Bertels, M.M. Fernandes, J.M.P. Cardoso (Eds.), Reconfigurable Computing: Architectures, Tools and Applications. XIV, 391 pages. 2007.

Vol. 4418: A. Gagalowicz, W. Philips (Eds.), Computer Vision/Computer Graphics Collaboration Techniques. XV, 620 pages. 2007.

Vol. 4416: A. Bemporad, A. Bicchi, G. Buttazzo (Eds.), Hybrid Systems: Computation and Control. XVII, 797 pages. 2007.

Vol. 4415: P. Lukowicz, L. Thiele, G. Tröster (Eds.), Architecture of Computing Systems - ARCS 2007. X, 297 pages. 2007.

Vol. 4414: S. Hochreiter, R. Wagner (Eds.), Bioinformatics Research and Development. XVI, 482 pages. 2007. (Sublibrary LNBI).

Vol. 4412: F. Stajano, H.J. Kim, J.-S. Chae, S.-D. Kim (Eds.), Ubiquitous Convergence Technology. XI, 302 pages. 2007.

Vol. 4411: R.H. Bordini, M. Dastani, J. Dix, A.E.F. Seghrouchni (Eds.), Programming Multi-Agent Systems. XIV, 249 pages. 2007. (Sublibrary LNAI).

Vol. 4410: A. Branco (Ed.), Anaphora: Analysis, Algorithms and Applications. X, 191 pages. 2007. (Sublibrary LNAI).

Vol. 4409: J.L. Fiadeiro, P.-Y. Schobbens (Eds.), Recent Trends in Algebraic Development Techniques. VII, 171 pages. 2007.

Vol. 4407: G. Puebla (Ed.), Logic-Based Program Synthesis and Transformation. VIII, 237 pages. 2007.

Vol. 4406: W. De Meuter (Ed.), Advances in Smalltalk. VII, 157 pages. 2007.

Vol. 4405: L. Padgham, F. Zambonelli (Eds.), Agent-Oriented Software Engineering VII. XII, 225 pages. 2007.

Vol. 4403: S. Obayashi, K. Deb, C. Poloni, T. Hiroyasu, T. Murata (Eds.), Evolutionary Multi-Criterion Optimization. XIX, 954 pages. 2007.

MIX
Papier aus verantwortungsvollen Quellen
Paper from responsible sources
FSC® C105338

If you have any concerns about our products,
you can contact us on
ProductSafety@springernature.com

In case Publisher is established outside the EU,
the EU authorized representative is:
Springer Nature Customer Service Center GmbH
Europaplatz 3, 69115 Heidelberg, Germany

Printed by Libri Plureos GmbH
in Hamburg, Germany